Outdoor Power Equipment

Jay Webster

DELMAR
THOMSON LEARNING™

Australia Canada Mexico Singapore Spain United Kingdom United States

DELMAR
THOMSON LEARNING™

Outdoor Power Equipment
by Jay Webster

Delmar Staff:

Business Unit Director:
Alar Elken
Executive Editor:
Sandy Clark
Acquisitions Editor:
Vern Anthony
Developmental Editor:
Catherine Wein
Executive Marketing Manager:
Maura Theriault
Executive Production Manager:
Mary Ellen Black
Production Manager:
Larry Main
Production Editor:
Dianne Jensis
Production Editor
Betsy Hough
Channel Manager:
Mona Caron
Marketing Coordinator:
Kasey Young
Cover Design:
Michael Egan
Photo Sequences:
Photography by David Getty

Printed in the United States of America
4 5 6 7 8 9 XXX 05 04

For more information contact Delmar,
3 Columbia Circle, PO Box 15015,
Albany, NY 12212-5015.

Or find us on the World Wide Web at http://www.delmar.com
or http://www.autoed.delmar.com

Library of Congress Cataloging-in-Publication Data
Webster, Jay.
Outdoor power equipment / Jay Webster.
p. cm.
Includes index.
ISBN 0-7668-1391-6
1. Small gasoline engines. 2. Power tools. I. Title.
TJ790.W389 2000
631.3'7—dc21 00-029478

NOTICE TO THE READER

Publisher does not warrant or guarantee any of the products described herein or perform any independent analysis in connection with any of the product information contained herein. Publisher does not assume, and expressly disclaims, any obligation to obtain and include information other than that provided to it by the manufacturer.

The reader is expressly warned to consider and adopt all safety precautions that might be indicated by the activities herein and to avoid all potential hazards. By following the instructions contained herein, the reader willingly assumes all risks in connection with such instructions.

The Publisher makes no representation or warranties of any kind, including but not limited to, the warranties of fitness for particular purpose or merchantability, nor are any such representations implied with respect to the material set forth herein, and the publisher takes no responsibility with respect to such material. The publisher shall not be liable for any special, consequential, or exemplary damages resulting, in whole or part, from the readers' use of, or reliance upon, this material.

Contents

Photo Sequences

Preface

Outdoor Power Equipment is designed to be used for training outdoor power equipment technicians. These technicians work in the many shops that sell and repair outdoor power equipment such as lawn mowers, tillers, string trimmers, snow throwers, chain saws, and garden tractors. The textbook and supporting materials provide a comprehensive small-engine and outdoor power equipment training program for the high school, community college, or trade school levels.

The textbook is the result of 20 years of author's experience as a small-engine technology teacher and small-engine teacher trainer. The content is the result of years of small-engine curriculum development with input from present and former students; instructors from the high school, community college, and trade school level; and owners and employees of outdoor power equipment shops.

FEATURES

Several key features of this book differentiate it from other small-engine service textbooks:

- The operation and service of the most commonly used outdoor power equipment are discussed.
- There is strong emphasis on the shop skills needed by the outdoor power equipment technician.
- Troubleshooting and failure analysis are emphasized in every service chapter.
- Both two- and four-stroke engines are included.
- Photo sequences throughout the book show step-by-step service techniques.

Every effort was made during the development of this book to make it understandable and functional. It is divided into many short chapters so instructors can pick and choose the content that best suits their programs. It was written with a reading level suitable to the student's needs. Topics have been developed in a logical order from the simple to the complex. There is a strong emphasis on basic principles and basic shop procedures that will serve the student well as engines and equipment change over time.

A great deal of attention has been given to helping the student understand the text materials. Each chapter begins with a set of **learning objectives** to guide the student's learning. **New terms are highlighted** throughout the text and defined at their first use. **Chapter review questions** require the student to reexamine the chapter's most important topics. **Discussion topics and activities** at the end of each chapter direct the students to relate chapter information to their personal experiences and surroundings. These topics can be used for group or individualized instruction. **Warning messages** throughout the textbook help reinforce the instructor's shop safety program. **Cautions** throughout the textbook help prevent damage to tools or equipment. **Service tips** help the student learn some of the "tricks of the trade."

TEXT ORGANIZATION

This textbook is divided into six parts. Part 1 presents the shop information and skills needed by the technician with chapters on safety and certification, tools and measuring tools, and service information. Part 2 covers the operation of the small engine used in outdoor power equipment with chapters about four-stroke engine parts and operation, two-stroke engine parts and operation, and engine specification and selection. Part 3 presents the parts and operation of each of the engine's systems. There are chapters on the ignition system; fuel system; governor and throttle control system; lubrication system; cooling and exhaust system; manual starting systems; and electrical starting and charging systems. The chapters in Part

4 explain troubleshooting, preventive maintenance, ignition tune-up and fuel system tune-up. Part 5 presents the engine service techniques used to do failure analysis and rebuild a failed engine. There are chapters on engine disassembly and failure analysis; valve service; crankshaft, connecting rod and bearing service; cylinder and piston service; and engine assembly. Part 6 is unique to the small-engine textbook market in that it covers the operation and service of the most commonly used outdoor power equipment. There are chapters on rotary and reel walk behind mowers; lawn and garden tractors; edgers, tillers, and snow throwers; chain saws, string trimmers and leaf blowers; and portable pumps and blowers.

LAB MANUAL

The textbook is supported by the *Lab Manual for Outdoor Power Equipment*, which has an activity sheet for each chapter on how things work. The activity sheets provide tool and part identification exercises to reinforce student learning. The manual also includes a set of worksheets for the common job competencies required for trade proficiency as an outdoor power equipment technician. These worksheets list objectives, directions, warnings, and special tools required for each job, and provide spaces for the students to write observations and measurements as each of the jobs is accomplished.

ACKNOWLEDGMENTS

We greatfully acknowledge the individuals at Ohio Technical College for their valuable assistance in preparing the step-by-step photos in this text.

The author wishes to acknowledge the following people for their help in the development of this textbook:

Paul M. Bechwar, Manager, Engine Transmission Training, Tecumseh Products Company

Mike Barnett, Service Training Director, MTD Products, Incorporated

Andy Randle, Service Engineer, Ace Pump Corporation

Robert S. Coats, Technical Editor, Power Equipment Division, American Honda Motor Co., Inc.

Jan Peiffer, Public Relations Specialist, Simplicity Manufacturing, Inc.

Jenifer Crane, Cortaini & Morrison Advertising, STIHL

Rich Smith, Commercial Service Education Manager, The Toro Co.

Bill Mayberry, Clinton Engines Corporation

Paul Scholten, Engine Division, Kohler Co.

Rich Thuleen, Onan Corporation

Jay Webster, Jr., Chief Technical Officer, iwear.com

Joe Webster, Resource Specialist, Downey Unified School District

REVIEWERS

The author and Delmar would like to thank the following reviewers for comments and suggestions they offered during the development of the project. Our gratitude is extended to:

Jeffery Chapell
Cabool High School

Ken Gibbons
R & A Service

Arthur Green
Todd County Central High School

Carl Hawkins
Buchanan High School

Ray Hoover
Pulaski Technical College

Dave Lulich
Mauston High School

Chris Schaner
El Dorado High School

Dale Smith
Lewis & Clark Career College

Ronald Weaner
Dover High School

Tim Wyss
Clear Lake High School

Shop Information and Skills

Safety and Certification

OBJECTIVES

Upon completion and review of this chapter, you should be able to:

- List the basic systems of a small engine.
- Describe how to keep a safe work area.
- Explain how to prevent fires in the outdoor power equipment shop.
- List and explain the purpose of each type of personal protective equipment.
- Explain how to work safely with hand tools.
- Explain how to work safely with power tools.
- Describe the hazards of using cleaning chemicals.
- Explain how to operate an engine safely in the shop.
- Describe the hazards of working with storage batteries.
- List and describe four outdoor power equipment careers.
- Explain the purpose of container labeling and material safety data sheets.

TERMS TO KNOW

Aerosol spray cleaner
Air blowgun
Back brace
Bench grinder
Business owner
Compressed air
Cooling system
Decibel
Drill motor
Face shield
Fuel system
Goggles
Governor system
Hydraulic press
Ignition system
Lubricating system
Parts person
Personal protective equipment
Respirator
Safety container
Safety glasses
Service technician
Shop manager
Small engine
Solvent tank
Starting system

INTRODUCTION

Outdoor power equipment is equipment powered by small gasoline engines used to perform tasks such as lawn maintenance, garden preparation, snow removal, electrical generation, water removal, and tree service. Using good safety procedures will help you prevent accidents when working with outdoor power equipment.

SMALL ENGINES

A **small engine** (Figure 1-1) is a machine that uses the combustion of a fuel to develop power. All small engines have six systems. The **ignition system** produces a high-voltage spark inside the engine at the required time to ignite an air and fuel mixture. The **fuel system** stores fuel, mixes the fuel with the correct amount of air, and delivers the air and fuel mixture inside the engine. The **governor system** regulates the amount of air and fuel mixture to control engine speed. The **lubricating system** provides oil between moving parts to reduce friction, cool engine parts, and flush dirt away from engine parts. The **cooling system** dissipates heat from engine parts created by the combustion of the

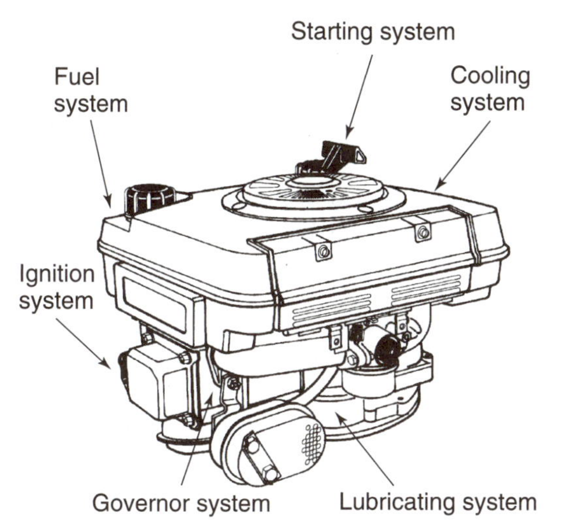

FIGURE 1-1 Small engines operate using ignition, fuel, governor, lubricating, cooling, and starting systems. (Provided courtesy of Tecumseh Products Company.)

air and fuel mixture. The **starting system** rotates the engine crankshaft to start the engine.

SAFE WORK AREA

A clean and organized work area helps prevent accidents. Your work area should be cleaned after each repair job. Small-engine shop customers often judge service quality by the appearance of the work area. In addition, safety codes may require certain shop procedures.

FIRE SAFETY

Outdoor power equipment shops often have gasoline, cleaning chemicals, and other materials that can catch fire easily. These materials are called *combustible*. They are fire hazards. Fuel, heat, and oxygen must be present to start a fire. A fire will go out if any one of these is removed.

Fire extinguishers are designed to extinguish specific fires quickly and safely. The work area must have the correct number and type of fire extinguishers. The local fire department usually sets these regulations.

Fire Extinguishers

You must know the location and proper use of fire extinguishers. Fires are classified as Class A, B, C, or D based on the combustible material involved. Fire extinguishers may be rated for more than one class of fire. See Figure 1-2. Class A fires involve ordinary combustibles such as paper, wood, cloth, rubber, plastics, refuse, and upholstery. Class A fire extinguishers have a letter A in a triangle and are color coded green on the label. Class A fire extinguishers put out a fire by coating and lowering the temperature of combustibles.

Class B fires involve combustible liquids such as gasoline, oil, grease, and paint. Class B fire extinguishers have a letter B in a square and are color coded red on the label. Class B fire extinguishers extinguish a fire by smothering or preventing oxygen from getting to the combustibles. This fire

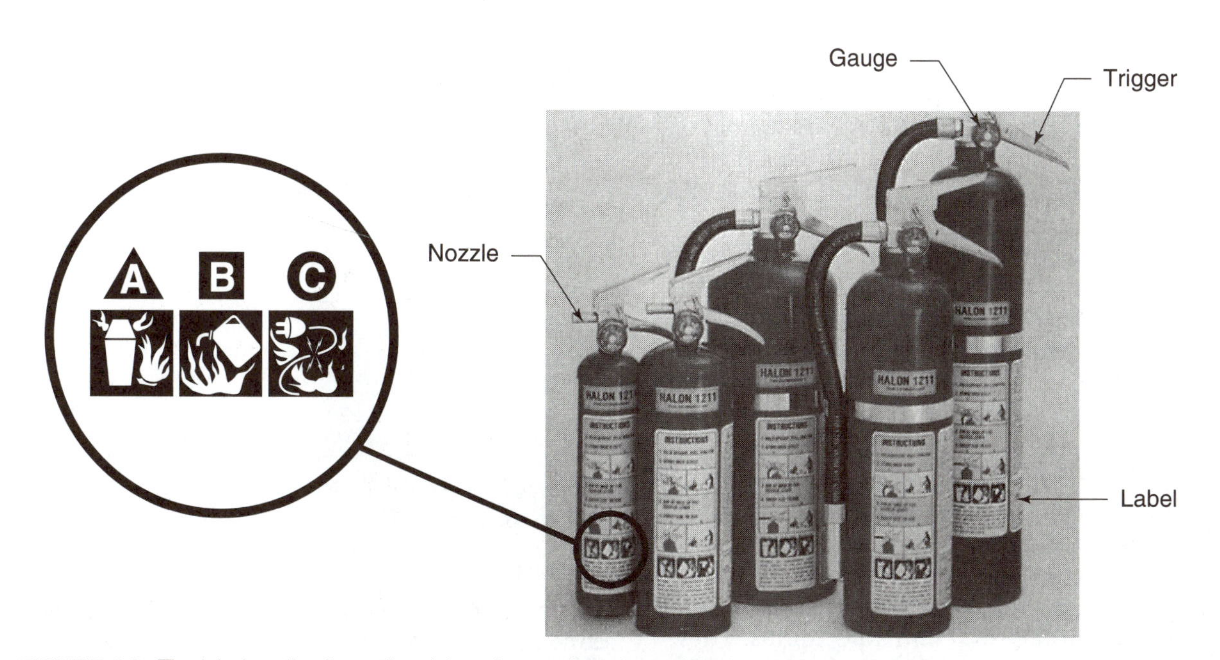

FIGURE 1-2 The label on the fire extinguisher shows what class of fire it can be used to fight.

extinguisher works best when discharged to cover the entire burning liquid surface.

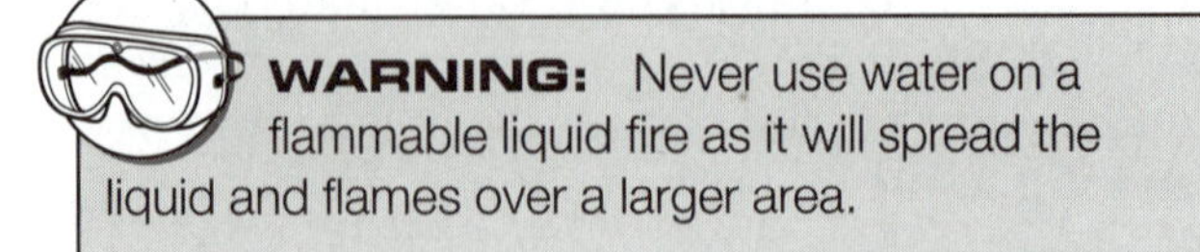

WARNING: Never use water on a flammable liquid fire as it will spread the liquid and flames over a larger area.

Class C fires involve electrical equipment such as motors, appliances, wiring, fuse boxes, breaker panels, and transformers. Class C fire extinguishers have a letter C in a circle and are color coded blue on the label. When fighting a Class C electrical fire, the power should be shut off as quickly as possible before using a fire extinguisher. Most outdoor power equipment shops have fire extinguishers rated for A, B, and C fires (Figure 1-2).

WARNING: Do not use water or other conducting fire extinguisher material on an electrical fire. If you do, you could get a severe electrical shock.

Class D fires involve combustible metals such as magnesium, potassium, sodium, titanium, and zirconium. Class D fire extinguishers have a letter D in a star and are color coded yellow on the label.

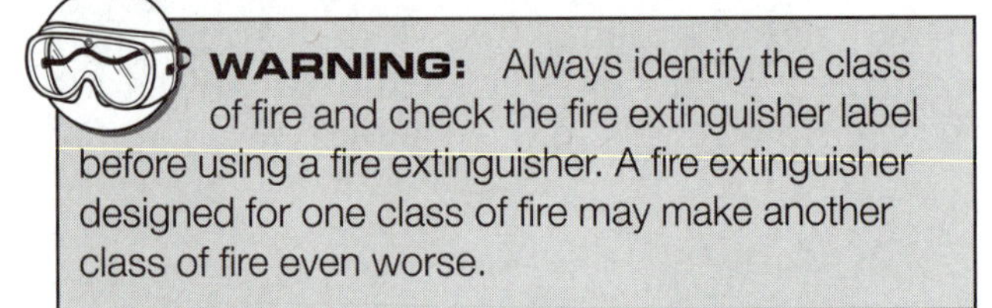

WARNING: Always identify the class of fire and check the fire extinguisher label before using a fire extinguisher. A fire extinguisher designed for one class of fire may make another class of fire even worse.

Fire extinguishers are pressurized to quickly discharge extinguishing material agents when you push the trigger. Leaks in a fire extinguisher can cause a loss of pressure over time. You should check fire extinguishers periodically to make sure they have the recommended pressure. The pressure gauge on top of the fire extinguisher shows the pressure of the fire extinguisher. State and local codes may require periodic inspection of fire extinguishers.

Flammable Liquids

Flammable liquids are commonly used in small-engine work areas and must be stored in a safety container. A **safety container** (Figure 1-3) is a metal container approved by the Underwriters' Laboratory (UL) for the storage of flammable liquids. These containers have a spring-loaded cap to prevent the escape of explosive vapors. There is a spark arrestor in the pour spout to prevent combustion inside the container.

Flammable liquids should be stored in an approved fireproof metal cabinet. The amount stored should be limited to the amount needed for immediate use. Local fire codes often limit the amount of flammable liquid that may be stored in a work area at a given time.

Proper labeling is required for flammable liquids. Labels for flammable liquids must comply with National Fire Protection Association (NFPA) standards. When pouring a flammable liquid from one metal container to another, keep the two containers in contact with each other, or use a bond wire to connect the two containers. This procedure prevents possible sparks caused by static electricity.

Fuel system service requires working with gasoline. Gasoline is combustible as a liquid, but is most dangerous in vapor form. Fires and explosions are caused when gasoline vapors are ignited. Vapors are heavier than air and settle in low spots along the floor. Any source of ignition can ignite gasoline va-

Metal Storage Cabinet

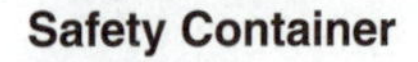

Safety Container

FIGURE 1-3 Flammable materials are stored in metal safety cabinets and in a safety container. (Courtesy of Gutman Advertising Agency.)

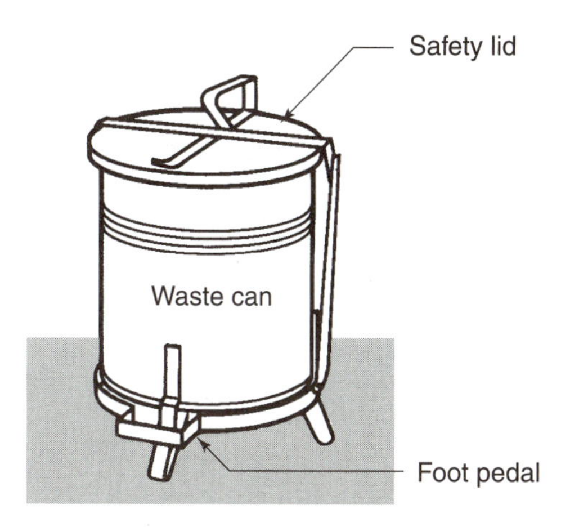

FIGURE 1-4 Dirty rags are stored in a fire safety container to prevent a fire.

pors. Common ignition sources are a lighted match or cigarette; an arc from an electric switch or motor; and the pilot light on a shop heater or a water heater. Clean up all gasoline spills immediately. Open doors and windows quickly to remove all the vapors.

Always do fuel system service outdoors or in a well-ventilated area. When you drain a carburetor or fuel line, collect the gasoline in an approved container. Dispose of the gasoline properly. Sometimes you may get gasoline on your clothing. Remove the clothing. Allow it to dry in a well-ventilated area away from any source of ignition.

Shop rags are used to remove grease, oil, gasoline, or solvent from parts and equipment. Used rags must be collected right after use and stored in a metal safety container (Figure 1-4). A foot pedal opens the lid. Gravity causes the lid to close automatically. The closed lid provides a seal to keep out oxygen and prevent combustion of the rags.

PERSONAL PROTECTIVE EQUIPMENT

Personal protective equipment (PPE) is safety equipment worn by an outdoor power equipment technician for protection against hazards in the work area. Personal protective equipment commonly used includes eye protection, ear protection, respiratory protection, and hand protection (Figure 1-5).

Eye Protection

Eye protection is used to prevent injury to the eyes or face. These injuries may be caused by flying particles, molten metal, chemical liquids or gases, or light rays. Most eye injuries are preventable by wearing eye protection. Cutting or grinding metal or using a hammer and chisel can cause flying objects. The best protection against flying object hazards is safety glasses. **Safety glasses** are glasses with impact-resistant lenses, special frames, and side shields. The lenses are made of special glass or plastic to provide maximum protection against impact. The frames are constructed to prevent the lenses from being pushed out of the frame during impact. Side shields prevent objects from entering the eyes from the side. Safety glasses are available in prescription form for people needing corrective lenses.

WARNING: For maximum protection, always wear safety glasses when working. Ordinary prescription glasses do not provide the required impact protection and do not have side shields.

A **face shield** is an eye protection device that covers the entire face with a plastic shield. The face shield is used for both eye and face protection from flying objects or splashing liquids. **Goggles** are an eye protection device secured on the face with a head band. Goggles cover the entire eye area and can be used over prescription glasses. The advantage of goggles over safety glasses is that they fit firmly against the face. This allows better distribution of force from the impact of the object. Some goggles have vents and baffles on top to provide protection from harmful vapors or fumes.

Goggles with clear lenses provide protection against flying objects or splashing liquids. Goggles with colored lenses also provide protection from harmful ultraviolet (UV) light rays from heating and welding equipment. UV rays can cause injury to the welder, observers, and other workers. Different welding tasks require certain colored lenses to provide adequate protection. The American Welding Society provides standards for the recommended eye protection.

Hearing Protection

Air-operated tools and running engines can produce high noise levels. A small-engine technician subjected to high noise levels can develop a hearing loss over a period of time. The severity of the hearing loss depends on level and duration of exposure. Ear protection devices are worn to prevent hearing loss.

Earplugs and earmuffs are common types of ear protection devices. Earplugs are made of moldable rubber, foam, or plastic and are inserted into ear

Guide to Personal Protective Equipment Selection

Hazard	Protective Equipment	Hazard	Protective Equipment
A. Protection from flying objects, wear whenever working in the shop	Occupational Safety Glasses	E. Hearing protection from loud noise levels	Ear Muffs
B. Full face and eye protection for flying objects and splashing liquid	Face Shield	F. Hearing protection from loud noise levels	Ear Plugs
C. Protection against flying objects, models available for welding and heating UV protection as well as eye protection against chemicals in the air	Goggles	G. Hand protection from cleaning chemicals, other types available for protection against metalworking hazards and electrical shock	Nitrile Gloves
D. Breathing protection from dangerous dust or chemicals	Respirator	H. Protection against loose clothing getting caught in rotating machinery	Shopcoat

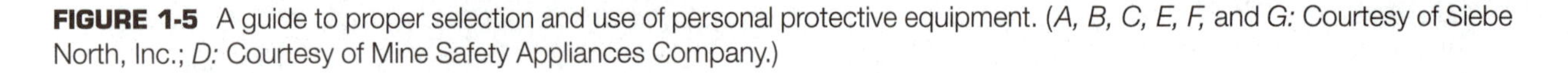

FIGURE 1-5 A guide to proper selection and use of personal protective equipment. (*A, B, C, E, F,* and *G:* Courtesy of Siebe North, Inc.; *D:* Courtesy of Mine Safety Appliances Company.)

canals. Earmuffs are worn over the ears. A tight seal around the earmuff is required for proper protection. Ear protection devices are assigned a noise reduction rating (NRR) number based on the noise level reduced. For example, an NRR of 27 means that the noise level is reduced by 27 decibels when tested at the factory. A **decibel** (dB) is a unit of noise intensity named after Alexander Graham Bell.

Respiratory Protection

Many hazardous chemicals and materials are used in making outdoor power equipment repairs. These chemicals can get into the air inside the shop. For example, spraying cleaning chemicals causes the chemicals to get into the air. Disassembling riding lawnmower brakes or some types of centrifugal clutches can sometimes cause dangerous asbestos dust to get into the air. The most common way for dangerous chemicals to get into the body is through breathing. Damage to the lungs or to other parts of the body can be caused by breathing dangerous chemicals.

A **respirator** is a safety device worn over the nose and mouth to protect against chemical and dust breathing hazards. There is a correct type of respirator for every breathing hazard. Some respirators work to clean or filter the air going into the lungs while more complex types supply clean air.

The filter type respirator is common to outdoor power equipment work. This respirator is used to filter out particles from the air before they can get into the nose and down to the lungs. A strap on the respirator is used to tighten it on the face over the nose and mouth. Breathing air is pulled through the two cartridges on the side of the respirator. Air is exhaled through the outlet at the bottom of the filter. Air going into the lungs is cleaned by filter elements located in the cartridges.

WARNING: Always use the correct type of respirator for the type of breathing danger. The respirator must fit and seal properly against the face to give the proper degree of protection. Facial hair can prevent proper sealing and make the respirator less effective.

The filter element must be inspected often. It can become plugged with dirt, paint, and other contaminants. Remove the cartridge cap to inspect the filter element. Replace any filter that shows signs of contamination. The respirator must have the correct type of filter element for the hazard. The respirator manufacturer provides a chart describing what number filter element must be used for protection from various hazards.

The filter element has a number printed on its surface to be used with a reference chart to determine what hazards the filter can protect against. Reference charts are usually provided in the package with the new respirator and in the package with replacement filter elements. Check the number against the chart. Some filters protect against paint hazards and others against particle hazards like asbestos.

SERVICE TIP: The fit and seal of a respirator on the face can be checked by installing the respirator then placing some food such as french fries under the nose. If the food odor can be detected, the respirator is not sealing properly.

Working around asbestos requires some special precautions because it has been found to cause lung cancer. Older types of brake shoes and clutches used on motor scooters or riding lawn equipment can contain asbestos fibers. Avoid blowing the dust off brake or clutch friction parts because the fibers can get into the air. Use only approved types of brake-washing solvents or special brake dust vacuum cleaners. Wear the correct type of respirator when working around a potential asbestos hazard.

Hand Protection

Hands are one of the most frequently injured parts of the body because they are used during every engine repair. There are two important aspects to hand protection. First, keep hands out of danger areas such as rotating parts. Identify the hand danger areas in the workplace and make an effort to keep hands out of that area.

The second part of hand safety is to wear hand protection when necessary. There are special protective gloves for many jobs that require hand protection. There are heavy work gloves for metalworking. Special gloves are used for handling chemicals such as those used to clean parts. Always use the correct type of glove for the hand hazards in the work area.

Clothing and Jewelry

The type of clothing worn in the shop is important to safety. Loose fitting clothing can get caught in rotating machinery. Wear short-sleeved shirts or wear shirts with the sleeves rolled up to prevent this type of accident. A necktie is very dangerous in the shop because it can get caught in rotating machinery. If a necktie is worn, it should be tucked into the shirt. Wearing a shop coat or overalls looks more professional and provides for safety better than casual clothing.

Jewelry is another safety problem. Wrist watches and rings can easily get caught in rotating machinery. A ring on the hand using a wrench can contact an electrical part and provide a path for electrical flow. A painful burn can result when electricity passes through the ring.

Always wear safety shoes in the shop. They have metal protection over the toe to prevent an injury if a heavy object falls on the foot. Safety shoes also grip slippery floors better than causal dress shoes. Long hair is another common hazard because it can get caught in rotating machinery. Always tie up long hair or wear a hat.

Back Protection

The back is one of the most frequently injured parts of the body and many of these injuries occur at work. A protective **back brace** that fits around the back and provides support, especially during lifting, is often required workplace wear.

Improper lifting is a common cause of back injury in the shop. Back injuries are not always caused by lifting too much weight. Many injuries are caused by lifting relatively small, light objects. Injury often occurs when lifting an object and twisting the body at the same time, which sometimes happens when the load being lifted is unbalanced.

Most back injuries can be prevented if the correct lifting procedure is followed (Figure 1-6). Avoid lifting any heavy engine or other heavy object without help. Always study the load to be lifted before doing any lifting. Look for a good place

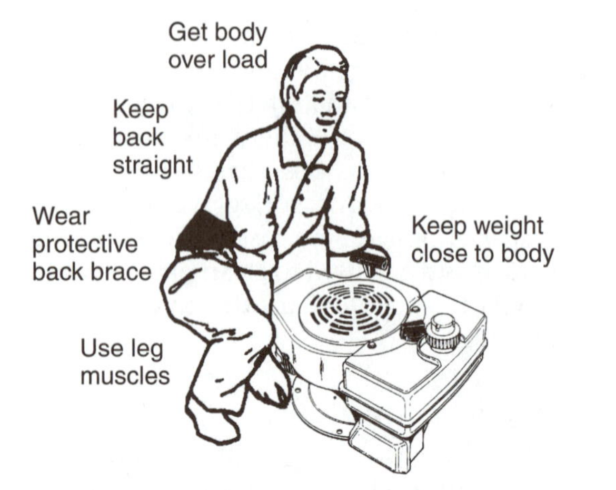

FIGURE 1-6 Wear a back brace and lift properly to protect the back from injury.

to get a hand hold. Determine where to lift and keep the load balanced. Place the body and legs as close to the object as possible. Get a good solid grip with the hands. Remember to bend the legs while keeping the back as straight as you can. Keep the object close to the body during the lift. Do not try to change the hand grip. Most important, do not twist the body during the lifting. When the object has to change direction, the feet should be moved in the new direction. When the load is set down, bend the legs while keeping the back straight.

HAND TOOL SAFETY

Most injuries occur to outdoor power equipment technicians while using tools and equipment. Following a few safe work procedures can eliminate these injuries. Hand tools seem safe but many accidents have been caused by poor use of hand tools. The most important thing to remember about hand tool safety is to use the correct tool for the job. Using a pair of pliers instead of a wrench or using a wrench handle as a hammer is dangerous.

Clean tools are an important part of hand tool safety. Keep your work area and your tools clean. Greasy tools can slip out of your hand and cause injury. Keep your hands as clean as possible when handling tools. When using tools with sharp edges like screwdrivers or chisels, point sharp edges of tools away from the body and fellow workers. Clamp small parts in the bench vise when using a chisel or screwdriver to protect the hands. Be sure that the screwdrivers are sharp and in good condition so they do not slip out of the screw heads and cause injury.

POWER TOOL SAFETY

Power tools are powered by electricity, hydraulic fluid, or compressed air. They have much more power than tools operated by hand. If not handled properly, power tools can cause very serious injuries.

Several general safety practices apply to all power tools. Make sure that all others are clear of the power tool before turning on the power. Always wear eye protection when operating a power tool. Be sure all machine safety guards are in correct position before starting the machine. Start the power tool and remain with it until it has been turned off and has come to a dead stop. Stay clear of power tools being operated by others. Avoid talking to or distracting someone using a power tool. To avoid being electrocuted do not use any electrical tool when standing on a wet or damp floor. Do not allow loose clothing to get caught in the rotating parts of a power tool.

Bench Grinder Safety

A **bench grinder** (Figure 1-7) is an electrically driven power tool that has an abrasive or wire wheel that is used to clean parts, remove metal, and sharpen tools. The abrasive grinding wheel can be hazardous because it is made from abrasive material that is bonded or glued together and can fracture and explode. The grinder has a guard or hood around the abrasive wheel to prevent injury if the wheel should explode. Never remove the hood from a grinder. Always throw away any grinding wheel that gets dropped on the floor because it could have an internal fracture.

A safety shield is installed in front of the abrasive wheel. The shield allows you to see the grinding while you are protected from flying chips or abrasives. Never remove this shield because flying

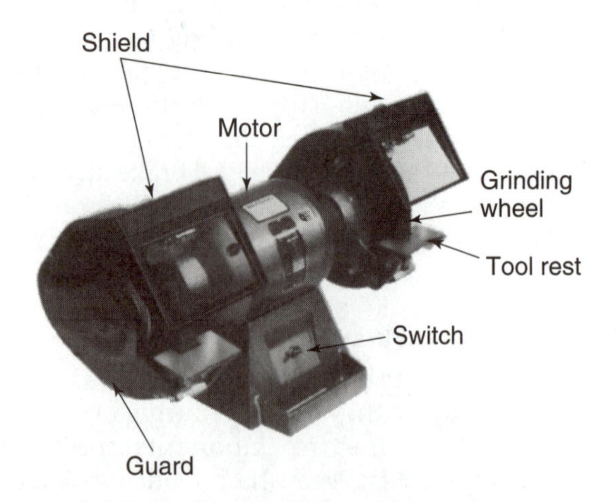

FIGURE 1-7 The bench grinder has a tool rest, guard, and shield to protect the user from flying objects.

chips or abrasives can cause injury. The shield does not provide complete protection from flying objects. Always wear eye protection when using the grinder, even if the safety shield is in place.

Support the part being ground against the tool rest in front of the grinding wheel during grinding. The tool rest prevents the object being ground from being pulled in between the wheel and guard. If this happens, a hand or glove could be pulled into the wheel. As the grinding wheel wears, the space between the wheel and the tool rest increases and the danger of pull-in increases. The tool rest must be inspected frequently and adjusted so that it is close to the abrasive wheel.

The tool rest adjustment is done with an adjusting bolt. Turn the grinder off and unplug the grinder from its electrical outlet before attempting to make the adjustment. Loosen the tool rest adjusting bolt. Move the tool rest toward the wheel until the space between them is a minimum of one-sixteenth of an inch and retighten the adjusting bolt.

WARNING: Never attempt to adjust any power machine that is not turned off and unplugged. Unintentional start-up could cause serious injury.

A. Battery Powered Drill

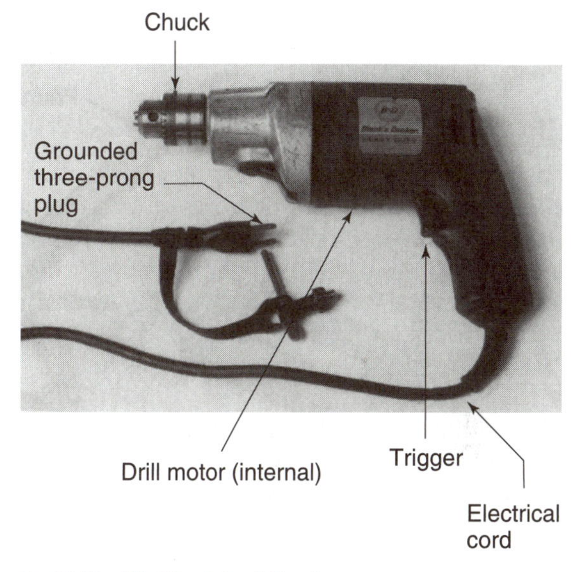

B. Drill with Electrical Cord

FIGURE 1-8 A drill motor may be powered by a battery pack or an electrical cord. (*A:* Courtesy of Black & Decker.)

Electric Drill Motor Safety

A **drill motor** is an electrically powered, hand-held motor that drives a drill bit used to drill holes. There are two general types of drill motors. Some are powered by a battery pack that fits in the handle (Figure 1-8A). The battery pack is removed and recharged periodically. The other type is powered by an electrical cord plugged into a wall outlet (Figure 1-8B). The power cord plug for this type of drill should be the three-prong grounded type and must be connected into a three-prong electrical wall outlet. The grounded power cord provides protection against shocks. Avoid the use of any three to two prong adapters. Do not cut any prongs off a power cord to fit them into a nongrounded outlet. When unplugging a power cord, always pull on the plug and not the power cord. Always stand on a dry surface when using any electrical equipment to avoid being electrocuted.

A drill motor in use can create a spark that could ignite flammable liquid vapors. Avoid the use of a drill motor if flammable vapors are present. A spark from a drill motor can also cause a storage battery explosion so avoid using a drill near any storage battery. Always leave the drill motor disconnected when not in use.

Like all other electrical tools, the drill motor should be disconnected from the wall outlet when installing or removing a drill bit. Be sure the chuck key is removed before starting the drill. Always wait for the drill motor to come to a complete stop before releasing the handle and placing it on the workbench. Drilling causes chips to be thrown off the drill bit. Always wear eye protection during drilling.

Hydraulic Press Safety

A **hydraulic press** (Figure 1-9) is a power tool that uses hydraulic power or a hydraulic ram to develop the high forces required to remove bearings or

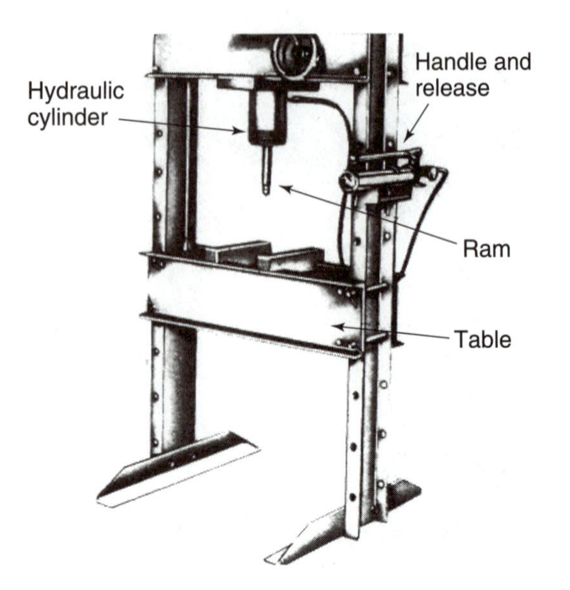

FIGURE 1-9 A hydraulic press uses hydraulic ram to apply hydraulic force on a part placed on the press table. (Courtesy of Snap-on Incorporated.)

straighten parts. The hydraulic press has an adjustable table that is raised or lowered by a hand wheel so that different sized parts can fit on the table. Pumping a handle attached to a hydraulic cylinder causes the ram to lower with several tons of force. A lever or knob releases the hydraulic pressure and retracts the ram to remove the component.

A hydraulic press is able to develop several tons of force. Bearings and other parts can explode under these pressures. Always wear eye and face protection when using a press. The part being pressed must be firmly supported on the table or it could slip during pressing and cause injury.

Compressed Air Safety

Compressed air is air that is pressurized by an air compressor and is available in the outdoor power equipment shop to operate air tools. Compressed air is commonly used through an air blowgun. An **air blowgun** is the gun-shaped attachment connected to the end of a compressed air line used to direct a stream of air. Any material, including air, under pressure can cause injury. The discharges of high-pressure air can cause injury to the face or body. Avoid pulling the air blowgun trigger when the air nozzle is pointed directly toward anyone. Do not look into the discharge nozzle while trying to find out if the nozzle is clogged. A worn air hose can burst under pressure, so always replace a worn hose to prevent accidents. Always check hose connections before turning on the air. Be sure to hold the

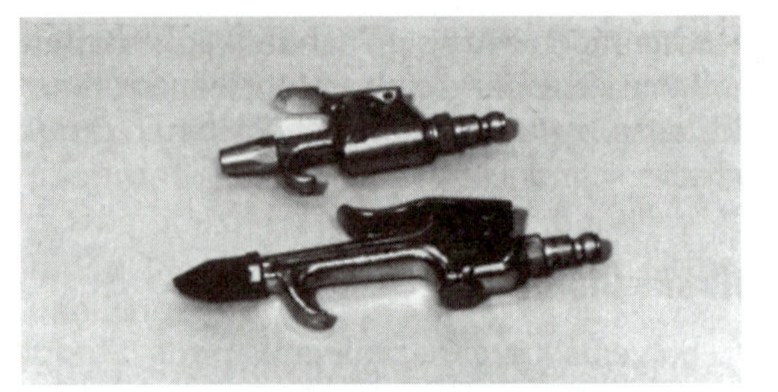

FIGURE 1-10 The safety air blowgun (top) has safety relief passages to prevent high air pressure from causing an air bubble to be pushed through the skin.

air hose to prevent it from whipping when turning air on.

In years past, it was common practice to use compressed air to blow solvent off parts that had been cleaned. This is dangerous because dirt and solvent blown by compressed air can cause injury. Solvent and dirt can be forced into the skin or the eyes or they may be inhaled. Avoid the use of compressed air to blow off parts whenever possible. Place parts washed in solvent on a clean rag to dry. There may be an occasion to use compressed air in fuel system work. It may be necessary to force air through fuel system parts to clean small passages. If compressed air must be used, wear eye and respirator protection to prevent injuries. Never use an air blowgun to dust off clothing or hair.

Another common air blowgun hazard occurs if the air blowgun is directed at the skin. An air bubble can be injected into the bloodstream and cause serious medical problems. This problem is more likely to occur if the skin has a cut, but it can also occur to uncut skin. Use approved air blowguns that have a relief passage that prevents too high a pressure (Figure 1-10).

Some outdoor power equipment uses air-filled or pneumatic tires. Compressed air in the shop is often used to inflate a tire. Tire inflation is hazardous because excessive pressure can cause the tire or rim to explode. Always use a reliable pressure gauge when inflating tires. Be careful not to exceed the recommended pressure. Always turn away from a tire when it is being inflated so that any explosion will not hit the face.

CLEANING CHEMICAL SAFETY

Parts cleaning is an important part of most engine repairs. Parts must be cleaned so that you can inspect the parts for wear and damage. Reassembled parts must be free of dirt that can contaminate the engine oil and cause wear. Most outdoor power equipment shops have several types of cleaning equipment.

Most small cleaning jobs are done with aerosol can cleaners. An **aerosol spray cleaner** is a spray can containing chemicals that break down dirt and grease and allow them to be removed. The dissolved grease can be then be air dried. Some types of cleaners are flushed off with water. The chemicals in these cleaners can be hazardous to eyes, skin, and lungs. Always read the warnings on the can and follow them (Figure 1-11). Wear eye protection, proper gloves, and a shop coat to prevent exposure to skin or eyes. Always do the cleaning in a well-ventilated area. Wear proper breathing protection.

Many shops use a solvent tank to clean engine parts. A **solvent tank** (Figure 1-12) is a metal tank containing cleaning solvent to flush grease and dirt off of parts. The solvent in the tank thins and washes away grease, oil, and sludge. Most solvent tanks have an electric motor that pumps solvent through a hose that may be used to flush off parts. On the

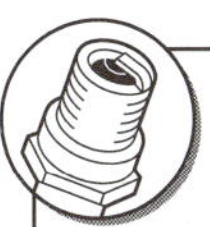

CASE STUDY

A few years ago, in the city of Los Angeles, paramedics responded to a 911 call from a service shop. When they got to the shop, they found a technician down on the floor. The technician was a white male whose skin had turned a bright pink color. From experience, the paramedics knew that the pink skin color was probably due to carbon monoxide poisoning. All the revival techniques failed and the technician died at the scene.

A death on the job normally results in an investigation by the police and investigators from California Occupational Health and Safety Administration. The investigation revealed that the technician did, indeed, die from carbon monoxide poisoning. The investigators set out to find out the source of this deadly gas.

The technician had been running the engine in the shop as he made tune-up adjustments to the carburetor and ignition system. Because it was a hot summer day, he had the large shop door fully open. Obviously, a running engine in an open shop could not be totally responsible for a lethal dose of carbon monoxide.

Investigators searched for other sources. A large dip tank for carburetors in the shop was found to have an open lid. Safety officials monitored the air surrounding the tank and found that the solution gave off significant levels of carbon monoxide.

Coworkers added the last piece to the puzzle when they reported that the technician was smoking just before he passed out. Carbon monoxide is a by-product of cigarette smoke. The combination of the engine, the carburetor tank, and the cigarette smoke provided enough carbon monoxide to kill. The moral to this story, as well as many other safety issues, is that attention to little things can make a difference.

FIGURE 1-11 Cleaning chemicals are available in aerosol cans that have a label explaining safety precautions. (Courtesy of CRC Industries Inc.)

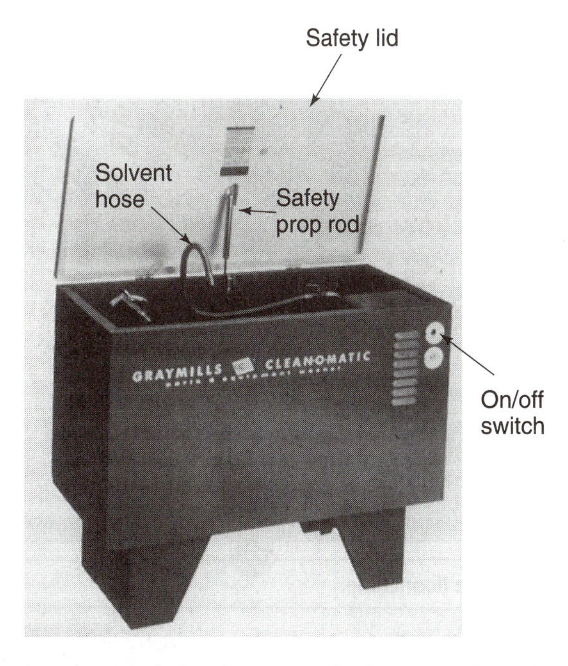

FIGURE 1-12 A solvent tank circulates cleaning solvent through a hose and nozzle to flush off parts. (Courtesy of Graymills.)

way to the nozzle, the solvent is circulated through a filter to keep it clean. Dirty solvent flows to the bottom of the tank. The lid of the solvent tank usually has a safety prop rod that holds the tank lid up. The prop rod is made in two parts that are held together with low temperature solder. If a fire should occur in the tank, the solder will melt, causing the prop rod to fall apart and allowing gravity to shut the top and smother the fire.

Many solvents used in these solvent tanks are flammable. Care must be taken to prevent an open flame around the solvent tank. Avoid smoking near the tank and have a Class B fire extinguisher in the area. Wear proper gloves when washing parts because the solvent can be absorbed through the skin and into the body, especially if there is a cut on the hand. Solvents that pass through the skin enter the bloodstream and can damage internal organs. Do not blow compressed air on hands wet with solvent. The solvent can be forced through the skin.

WARNING: Avoid the use of gasoline or paint thinners to clean parts. Gasoline and paint thinners are too flammable for parts cleaning. Gasoline is also extremely dangerous because the chemicals in gasoline can easily be absorbed through the skin and into the bloodstream.

ENGINE OPERATION SAFETY

Running an engine inside a shop can be hazardous. Engine exhaust gases contain carbon monoxide, which is hard to detect but is very poisonous. If an engine is operated without proper ventilation, carbon monoxide may build up to dangerous levels. Carbon monoxide poisoning begins with headaches and drowsiness and high exposures can lead to coma and death. Always connect the engine exhaust to the shop ventilation system (Figure 1-13) and make sure the ventilation system is working.

There are other sources of carbon monoxide besides engine exhaust. Shop heating systems with improper ventilation can create high levels of carbon monoxide. Some types of chemical parts cleaners give off carbon monoxide as they evaporate. Smoking is another common source of carbon monoxide. Technicians have been overcome by high exposures that have resulted when several small sources of the gas have built up to high levels. Many shops protect workers by installing carbon monoxide detectors that sense and warn of excessive levels.

Excessive engine speed is another engine operation hazard. An engine operated at too high a speed for too long a time can explode. In most cases, the broken engine parts are contained inside the engine, but parts sometimes break through and cause injury. Parts being rotated by the engine, such as mower and edger blades, can also come off during engine operation. Avoid excessive engine speeds. Wear eye and face protection.

Most outdoor power equipment shops repair riding type lawn equipment. Moving vehicles like these in and out of the shop must be done carefully to avoid accidents. Be sure to check for tools and other equipment that might be on the floor before driving a vehicle into the shop. Walk around the area to be sure it is cleared. Before starting the engine, test the brakes on any vehicle to be driven to make sure they work. Watch carefully to avoid hitting coworkers or customers when driving a vehicle in the shop. Drive slowly, carefully, and have a coworker provide guidance if there are hazards to be avoided.

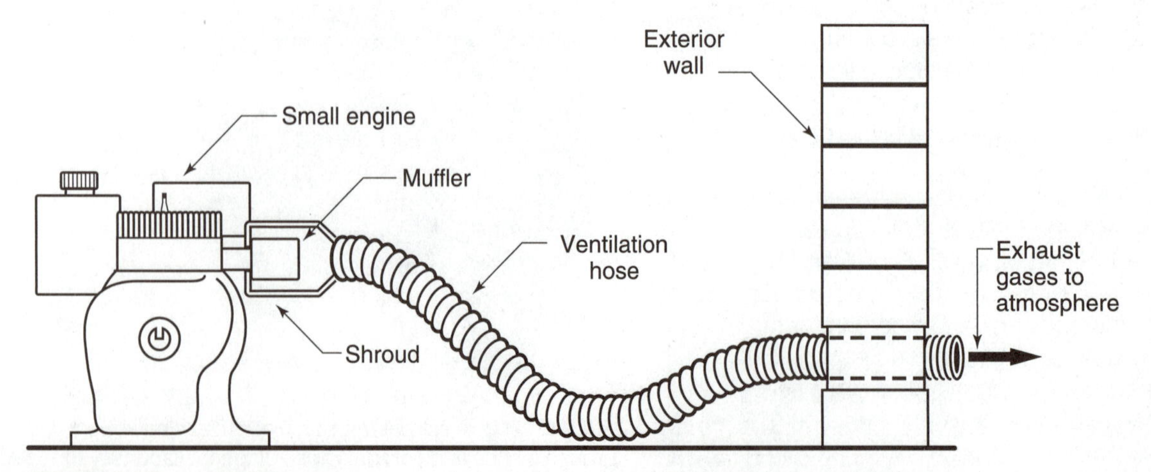

FIGURE 1-13 An engine operated in the shop must be connected to the shop ventilation.

There are several safety precautions to follow when running an engine on any riding type lawn equipment. Be sure to set any parking brake to prevent the equipment from rolling. Use wheel blocks to block the front and back of one of the wheels to prevent the vehicle from rolling if there is a problem with the parking brake.

WARNING: Accidents on riding type lawn equipment are consistently high on the list of published accident statistics. Each year people fall off the equipment and are injured in the rotating blades. Technicians testing this equipment after repairs should be extremely careful to avoid these types of accidents.

Never get under any riding equipment when someone else is working on it or when the engine is running. Avoid being in front of or behind this equipment when the engine is running to avoid being run over. Be cautious of hot exhaust system manifolds and moving engine parts when working on this equipment.

Always make sure an engine cannot start when you are working on it. Make sure the stop (kill) switch is in the off position. A stop switch may not be operating correctly. Always remove the wire from the spark plug and ground it. If the equipment has an ignition key, remove it so someone else cannot turn the key on. Allow rotating parts to stop turning before you work on an engine. Allow hot engine parts time to cool down.

STORAGE BATTERY SAFETY

A storage battery is used on many types of equipment to power the starting system. The battery is dangerous because it creates an explosive hydrogen gas especially when being charged. When handling batteries, avoid smoking or causing any source for ignition, such as a spark or open flame. Always follow manufacturer's recommended procedures when connecting or disconnecting a battery to prevent any sparking. A spark near a battery can cause an explosion. Always wear eye protection when servicing any battery.

The battery is full of an acid solution called electrolyte. Electrolyte is a mixture of sulfuric acid and water in a storage battery that is extremely corrosive and dangerous. A corrosive acid can quickly cause damage to eyes and skin. Chemical goggles and proper gloves should always be worn when handling batteries. If acid or battery corrosion gets on the skin, the area should be washed thoroughly. If electrolyte is splashed in the eyes, they should be washed with water in an eye wash or with a hose while a doctor is contacted immediately.

HAZARD COMMUNICATION

Many chemicals and other materials used in the outdoor power equipment shop can be hazardous. This is true in many other industries as well. The federal government recognizes this fact and has developed procedures to help workers protect themselves. One of the most important parts of this program is hazard communication, which is sometimes called the worker's "right to know."

The Occupational Safety and Health Administration (OSHA) is the federal agency in charge of workers' safety. Many states also have a similar agency. These organizations develop standards and conduct inspections to ensure that the workplace is safe and healthful.

One important part of OSHA's job is to enforce a hazard communication standard, which is called the right-to-know law. This standard makes sure all workers understand the hazards of any material they are using. There are two very important parts of the hazard communication program: container labeling and material safety data sheets. During your workday you may use any number of possibly hazardous materials. For example, you might use solvents, air conditioner refrigerants, paints, or battery electrolyte. The containers for these and all other hazardous materials must have a label. You should read this label before using the materials.

A typical label must identify the hazardous chemicals in the product. It must also tell you specific hazards. For example, the material might be poisonous or flammable. Necessary precautions should also be listed. For example, you might be warned to wear eye protection or to use the material in a well-ventilated area. First aid information is also provided on the label.

Unlabeled materials can be very dangerous. Many people have been injured because they did not know what was in a container. There may be times when you put material from a labeled container into another container. Always make a label for the new container describing the contents because you may not be the only person to use the material.

Much more detailed information about a material is found on a material safety data sheet (MSDS) (Figure 1-14). This information sheet provides detailed information about hazardous materials, hazardous ingredients, fire and explosion

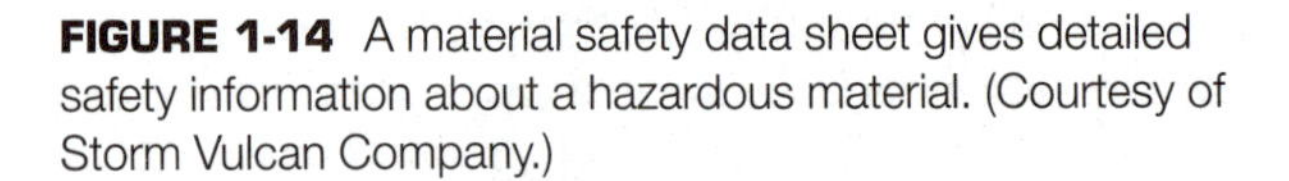

CRC MATERIAL SAFETY DATA SHEET SIL00

CRC Industries, Inc. • 885 Louis Drive • Warminster, PA 18974 • (215) 674-4300

PRODUCT NAME CLEAN-R-CARB (AEROSOL) #-MSDS05079
PRODUCT- 5079,5079T,5081,5081T
(Page 1 of 2)

1. INGREDIENTS	CAS #	ACGIH TLV	OSHA PEL	OTHER LIMITS	%
Acetone	67-64-1	750 ppm	750 ppm		2-5
Xylene	1330-20-7	100 ppm	100 ppm		68-75
2-Butoxy Ethanol	111-76-2	25 ppm	25 ppm	(skin)	3-5
Methanol	67-56-1	200 ppm	200 ppm		3-5
Detergent	-	NA	NA		0-1
Propane	74-98-6	NA	1000 ppm		10-20
Isobutane	75-28-5	NA	NA	1000ppm	10-20

2. PHYSICAL DATA : (without propellent)
Specific Gravity : 0.865 Vapor Pressure : ND
% Volatile : > 99
Boiling Point : 176 F initial Evaporation Rate : Moderately fast
Freezing Point : ND Vapor Density : ND
Appearance and Odor: pH: NA
A clear colorless liquid, aromatic odor

Solubility : Partially soluble in water.

3. FIRE AND EXPLOSION DATA
Flashpoint : -40 F Method : TCC
Flammable Limits : propellent LEL:1.8 UEL:9.5
Extinguishing Media : CO2, dry chemical, foam
Unusual Hazards : Aerosol cans may explode when heated above 120 F.

4. REACTIVITY AND STABILITY
Stability : Stable
Hazardous decomposition products
: CO2, carbon monoxide (thermal)

Materials to avoid : Strong oxidizing agents and sources of ignition.

5. PROTECTION INFORMATION
Ventilation : Use mechanical means to insure vapor conc. is below TLV.

Respiratory : Use self-contained breathing apparatus above TLV.

Gloves : Solvent resistant Eye & Face : Safety glasses
Other Protective Equipment: Not normally required for aerosol product usage.

FIGURE 1-14 A material safety data sheet gives detailed safety information about a hazardous material. (Courtesy of Storm Vulcan Company.)

data, health hazards, spill and leak procedures, and special precautions. Federal law requires that an MSDS be available for each hazardous material in your workplace. Sometimes they are posted in the shop or available in the shop office.

OUTDOOR POWER EQUIPMENT CAREERS AND CERTIFICATION

If you enjoy working on small engines and outdoor power equipment, you may wish to explore a career in this field. There are many opportunities for a career in the outdoor power equipment field, including service technician, shop manager, parts person, and business owner.

Career Opportunities

A **service technician** performs repairs on small engines and outdoor power equipment. An outdoor power equipment service technician works in shops that sell and repair outdoor power equipment (Figure 1-15). This job involves measuring, disassembling, machining, reassembling, and adjusting all types of engines and equipment. The service technician must have knowledge and experience repairing all types of engines and outdoor power equipment. This job requires you to have a high level of mechanical skill. It also requires strong people skills to work successfully with customers.

Working conditions for outdoor power equipment service technicians have improved over the years. Most small-engine shops are clean and well organized. Power equipment is available in most shops to reduce the amount of physical labor required on the job. The job can still be physically challenging. There are also numerous hazards in the workplace and you will be exposed to them throughout your working career.

Experienced service technicians with up-to-date training are generally well paid. The salary paid to a technician depends on the type of work; the employer; business conditions; and the availability of technicians. Technicians may be paid an hourly salary. Hourly employees are usually paid a straight wage for working a 40-hour work week. Overtime pay is generally paid for work beyond the basic 40-hour week. Sometimes technicians get a weekly salary and a commission or percentage of the shop's profit on parts sold to the customer. In smaller shops the technician may also sell equipment. Often a commission is paid for the equipment sale. Some shops pay the technician a percentage of the labor charges paid by the customer. Under this system, the harder and faster the technician works, the more money he can make.

Many shops provide paid benefits. Employers usually pay or share in the payment of employee health benefits, life insurance, and retirement plans. Some shops pay all or part of the technicians' uniform rental. Technicians usually buy their own general tools. The shop typically provides the specialized equipment.

The **shop manager** is the person who manages the sales and service at an outdoor power equipment shop. Usually a technician who has experience and training in business operations, the shop manager is often responsible for maintaining sales and service records, shop budgeting, sales, and inventories. The shop manager often meets with customers and fills out job estimates and invoices. The shop manager is responsible for the quality of the work done by the shop technicians. Managers are often paid a percentage of the shop sales and service profit.

FIGURE 1-15 A shop specializing in the sale and service of outdoor power equipment.

The **parts person** is the person responsible for ordering, cataloging, storing, and selling small-engine and equipment parts. These workers need a thorough knowledge of parts. They also need special training in reading catalogs, parts software, and inventory control. Parts specialists often work for larger companies that distribute small-engine parts and outdoor power equipment.

A **business owner** is the person who owns a shop that sells and services outdoor power equipment. Many technicians eventually open their own businesses. The business owner has total control of the business. If the business is successful, the owner can do well financially. There is also considerable risk to starting a business. A great deal of money has to be raised to start a business. Many small businesses fail because they do not make enough profit. Successful business owners do a great deal of research and get business training before they start a business.

Outdoor Power Equipment Certification

Once you decide to become an outdoor power equipment technician, the next step is to get the necessary training and experience to do the job. You need either a formal power equipment training program, on-the-job apprentice training, or a combination of the two. There are training programs in high schools, community colleges, and trade schools. Apprenticeships may be available in your local outdoor power equipment shop.

One very important thing you can do is to become an outdoor power equipment (OPE) certified technician. A technician can be certified by training and taking OPE certification tests. The OPE tests are promoted and coordinated by the Engine Service Association, Inc. (ESA). These tests measure the technician's knowledge in areas such as engine fundamentals, theory, servicing, failure analysis, troubleshooting, and repair. The tests provide certification in the areas of basic four-stroke engines, basic two-stroke engines, compact diesel engines, electrical; generators; and drivelines. The certified technician is allowed to wear the official OPE emblem (Figure 1-16) on his or her uniform.

The prestige of certification is beneficial to both the technician and the shop. The certified technician is recognized as a well-trained professional. Certified technicians are often paid at a higher rate than those without the certification. The shop is allowed to display the OPE sign (Figure 1-17), which helps develop customer confidence and promote sales and service.

FIGURE 1-16 The outdoor power equipment certification patch worn by certified technician. (Courtesy of Engine Service Association.)

FIGURE 1-17 The sign displayed where outdoor power equipment certified technicians work. (Courtesy of Engine Service Association.)

REVIEW QUESTIONS

1. List and describe the purpose of the systems common to all outdoor power equipment.
2. Describe the combustibles and types of fire extinguishers used on a type A fire.
3. Describe the combustibles and types of fire extinguishers used on a type B fire.
4. Describe the combustibles and types of fire extinguishers used on a type C fire.
5. Explain how to determine if a fire extinguisher is the correct type for a particular fire.
6. Explain why water should not be used on a flammable liquid fire.
7. Explain why nonconducting extinguishing agents must be used on an electrical fire.
8. Describe the steps to follow in safely storing and using gasoline in the shop.
9. Explain the difference between dress glasses and occupational safety glasses.
10. List and describe the common eye hazards found in the shop.
11. List and describe the eye hazards that can be protected against by goggles.
12. List and describe the eye hazards that can be protected against by a face shield.
13. List and describe the two types of personal protective devices used for hearing protection.
14. Describe how the filter type respirator is used to protect from breathing hazards.
15. Explain how to determine if the fit and seal of a filter respirator is correct.
16. Explain why the proper gloves must be used when handling chemicals.
17. Explain the proper procedure and personal protective equipment to use when lifting a heavy object.
18. List and explain the safety precautions to follow when using a bench grinder.
19. List and explain the safety precautions to follow when using compressed air.
20. Explain how to determine what safety procedures should be used when using spray can type cleaning chemicals.

DISCUSSION TOPICS AND ACTIVITIES

1. Develop a checklist for inspecting an outdoor power equipment shop for safety hazards. Use the checklist to inspect a shop.
2. Make a list of the kind and type of personal protective equipment that should be available in an outdoor power equipment shop.
3. Visit a local outdoor power equipment shop. Discuss employment opportunities in the outdoor power equipment service industry with the shop owner.

CHAPTER 2

Tools

OBJECTIVES

Upon completion and review of this chapter, you should be able to:

- Identify and explain the purpose of common wrenches.
- Explain the purpose and use of torque indicating wrenches.
- Describe the use and types of power wrenches.
- Identify and explain the purpose of the common types of pliers and screwdrivers.
- Identify and explain the purpose of the common types of punches and chisels.
- Describe the parts and use of the common types of drills and reamers.
- Explain the types and use of common types of hacksaws and files.
- Explain how to make and repair internal and external threads.
- Describe the types and uses of the common types of pullers.

TERMS TO KNOW

Adjustable wrench
Air ratchet wrench
Allen wrench
Ball peen hammer
Battery terminal pliers
Beam and pointer torque wrench
Box-end wrench
Brass hammer
Breaker bar
Breakover hinge torque wrench
Cape chisel
Center punch
Channel lock pliers
Chisel
Clutch screwdriver
Combination pliers
Combination wrench
Diagonal cutting pliers
Dial-reading torque wrench
Diamond point chisel
Die
Drill gauge
Drill index
File
Flat chisel
Hacksaw
Hammer
Impact wrench
Needle nose pliers
Nut driver
Offset screwdriver
Open-end wrench
Phillips screwdriver
Pin punch
Plastic tip hammer
Pliers
Puller
Punch
Ratchet handle
Reamer
Round nose chisel
Screwdriver
Screw extractor
Sliding T-handle
Snap ring pliers
Socket driver
Socket extension
Socket wrench
Speed handle
Standard screwdriver
Starter punch
Stripping and crimping pliers
Tap
Torque indicating wrench
Torx screwdriver
Twist drill
Vise grip pliers
Wrench

INTRODUCTION

The proper use and care of tools is one of the most important parts of being a technician. A number of tools are used on every repair job. Most of these tools are called "hand" because they are operated by the technician's own muscle power. A technician also uses power tools that get their power from electricity, compressed air, or hydraulics. Using the correct tool and using it correctly is an important part of a successful repair job.

WRENCHES

Many engine components are held together or fastened with threaded fasteners such as hexagonal or hex-shaped (six-sided) screws, bolts, and nuts. A **wrench** is a tool used to tighten or loosen hex head screws, bolts, and nuts. Since there are many different sizes of bolts and nuts, wrenches must also be made in different sizes to fit them.

The size of a wrench is determined by the size of the hex screw, nut, or bolt head on which it fits. The hex head has six corners and six flat areas called *flats*. The distance across the head from one flat to another determines its size (Figure 2-1). For example, if the number 14 is stamped on a wrench, it means that the opening of the wrench measures 14 millimeters across. It will fit on a hex head or nut that measures 14 millimeters across the flats.

Wrench sizes are given either in metric system or inch system units (Figure 2-2). Metric wrench sizes are given in millimeters such as 10 millimeters, 13 millimeters, 14 millimeters, 15 millimeters, 16 millimeters, 17 millimeters, and 18 millimeters. Sizes for inch system wrenches are given in fractions of an inch such as 5/16, 3/8, 7/16, 1/2, 9/16. Both inch and metric system fasteners are used in small engine work. You will need both metric and inch wrench sets.

WARNING: Be careful to use the right size wrench with any given size nut or bolt. If a wrench opening is larger than the hex, the corners of the nut or bolt head can be rounded off. There is also the greater danger of injury like bruised knuckles or even broken bones.

Open-End Wrenches

An **open-end wrench** (Figure 2-3A) is a type of wrench with an opening at the end that can slip onto the flats of a screw, bolt, or nut. Open-end wrenches are made in many different sizes and shapes. Most open-end wrenches have two open ends of different sizes so they can be used on two different sized fasteners. The wrench opening is usually at an angle of 15 degrees to the handle. One of the gripping surfaces or jaws is thicker than the other. This makes turning a fastener in a tight space much easier. A hexagonal nut can be turned continuously just by turning the wrench over between each swing (Figure 2-3B).

Open-end wrenches fit on only two flats of the fastener. Care must be taken to ensure that the

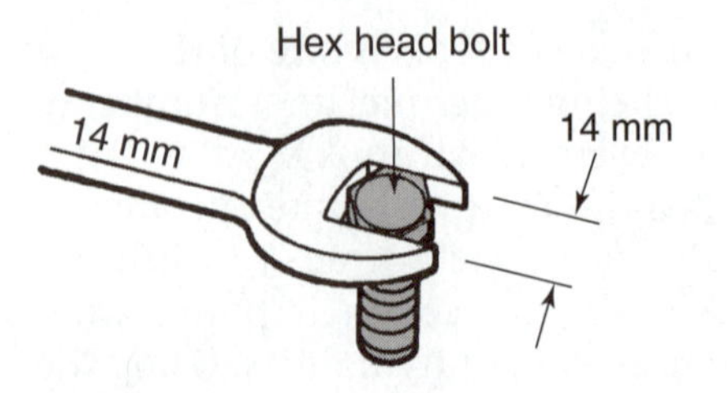

FIGURE 2-1 The size stamped on a wrench shows what bolt head or nut it fits.

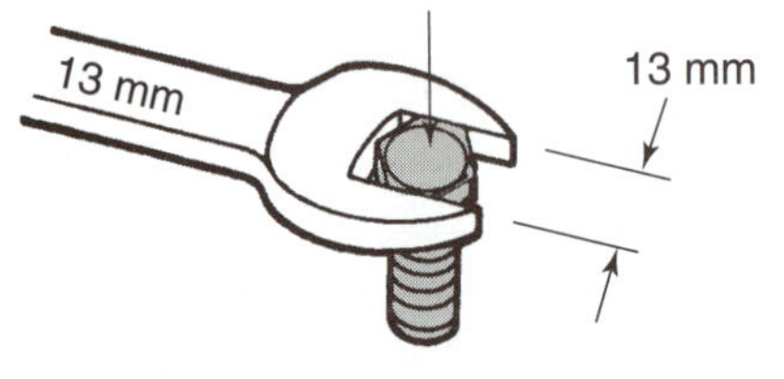

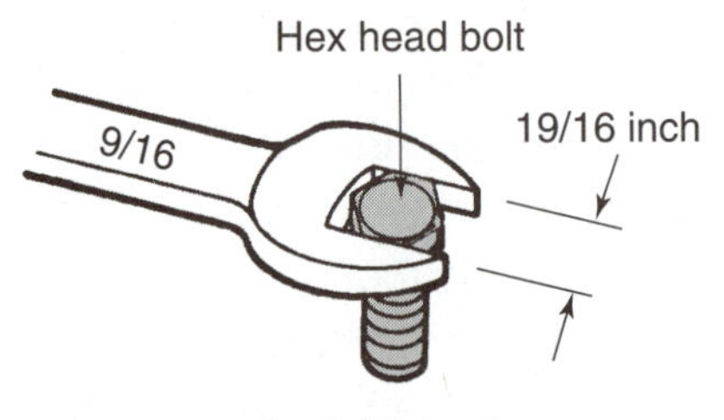

FIGURE 2-2 Metric and inch wrenches have size markings on the handle.

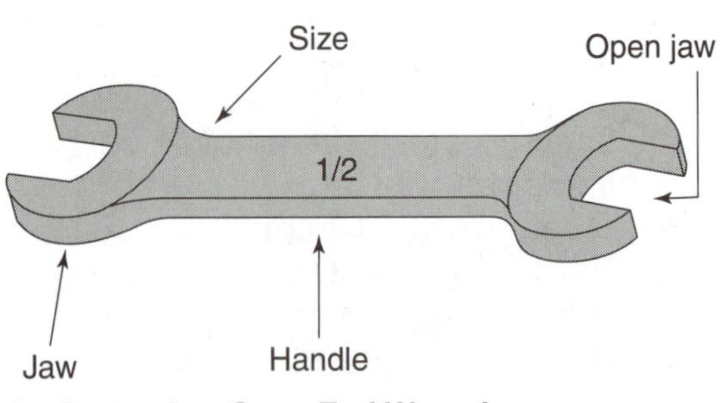

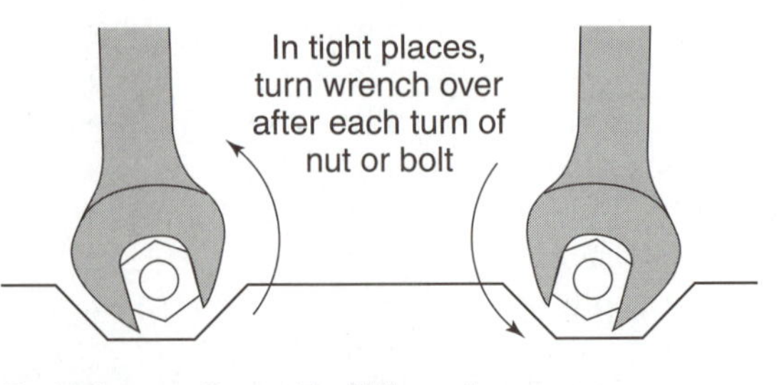

FIGURE 2-3 An open-end wrench has an opening in each jaw.

wrench does not slip off and cause injury. The best way is to pull on the wrench handle instead of pushing. When a wrench is pulled, it is usually away from some obstruction. When a wrench is pushed, it is usually toward an obstruction. A hand injury can result if the wrench slips off the fastener head. If an open-end wrench must be pushed toward an obstruction, do not wrap your fingers around the handle. Use the palm of your hand to push on the handle (Figure 2-4).

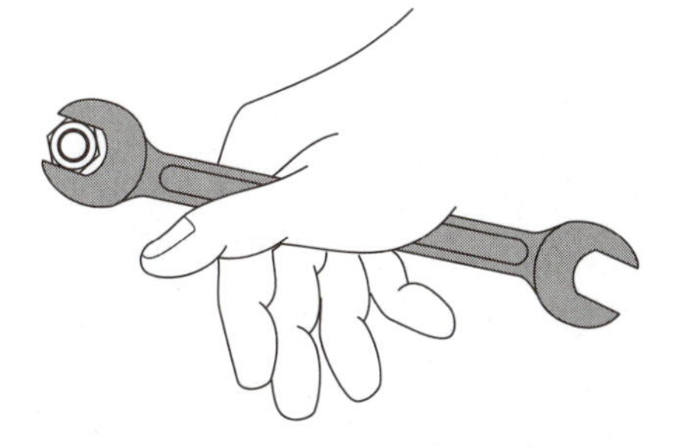

FIGURE 2-4 Correct way to push on an open-end wrench.

Box-End Wrenches

A **box-end wrench** is a wrench with jaws designed to fit on the corners of a hex-shaped fastener. Box-end wrenches do not fit on the flats like the open-end. They have corners called *points* that grip each of the six corners of the hex-shaped fastener (Figure 2-5A). Box-end wrenches allow you to use more force with less chance of the wrench slipping off the nut or bolt. Like other wrenches, they come in many inch and metric sizes. They also come with different numbers of points. The most common numbers of points are 6 and 12. The 6-point jaw is the strongest. The 12 points make it easier to engage the wrench when space for the handle is small. The angle of the wrench head is commonly offset at 15 degrees to the handle (Figure 2-5B). This helps when using the wrench in a tight space.

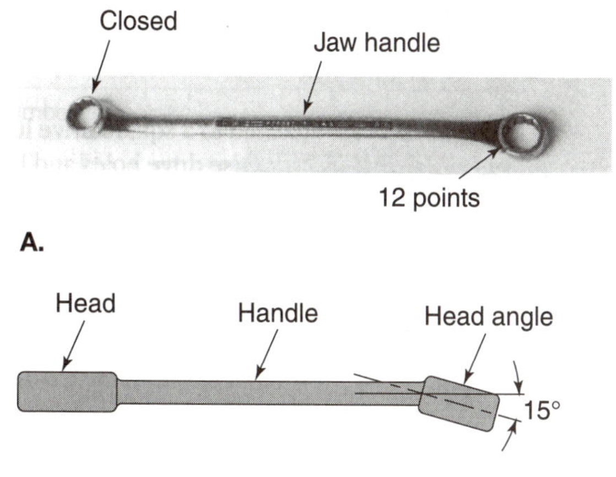

FIGURE 2-5 A box-end wrench fits all the way around the fastener head.

Combination Wrenches

A **combination wrench** (Figure 2-6) is a wrench with a box-end jaw at one end of the handle and an open end at the other. Combination wrenches are one of the most useful types of wrenches. The open-end side is used where space is limited. The box-end side is used for final tightening or to begin loosening. Combination wrenches are usually the same size at both ends.

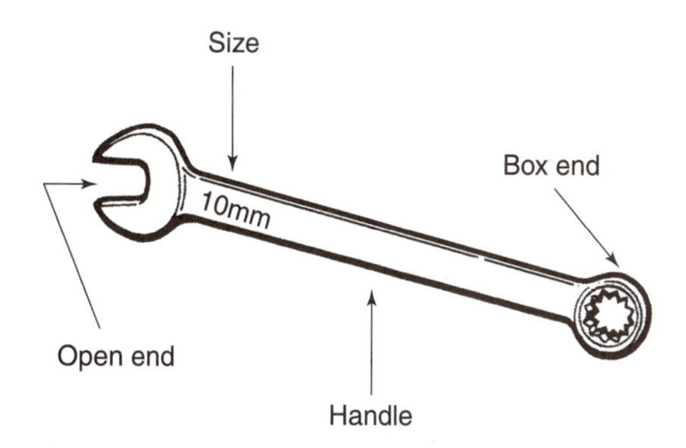

FIGURE 2-6 The combination wrench has an open end and a box end. (Courtesy of U.S. Navy.)

Socket Wrenches

A **socket wrench**, or socket, is a wrench that fits completely around a hex head screw, bolt, or nut (Figure 2-7) and can be detached from a handle. The socket is made detachable so that it can be removed from the handle. Many different sized sockets can be used with one handle. Sockets usually come in sets. Sets are made in all the inch and metric sizes.

Socket wrenches have a square hole at the handle end so they can be attached to a square drive lug

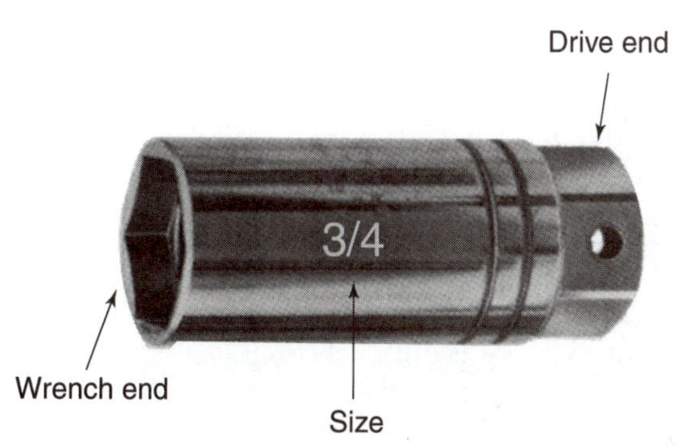

FIGURE 2-7 The socket wrench fits all the way around the fastener head. (Courtesy of Proto Tools.)

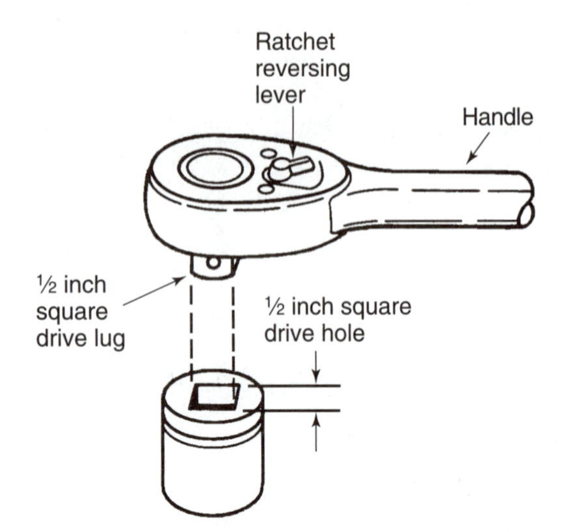

FIGURE 2-8 The square drive lug on the ratchet fits into the square hole on the socket.

on the handle (Figure 2-8). These square drive holes and lugs are made in different sizes (measured across the flats) for different amounts of turning force. The smallest size is ¼ inch. These are used for very small fasteners. Most outdoor power equipment work is done with a ⅜-inch drive socket wrench set. Large fasteners, such as flywheel nuts, require a ½-inch drive socket set.

Socket wrenches come in two basic lengths, standard and deep (Figure 2-9A). The standard length is used for most engine service jobs. The long or deep type is made longer to fit over a long bolt or stud. Deep sockets come in all the same common inch and metric wrench sizes as the standard size sockets.

Sockets are also available with 6, 8, and 12 points inside the socket to grip the corners of the hex fastener (Figure 2-9B). The 12-point socket is easiest to slip over a bolt or nut because of its many corners. The 6-point socket is the hardest to slip over the bolt or nut. The fewer the points or corners the stronger the socket. The 6-point socket is stronger than the 12 point.

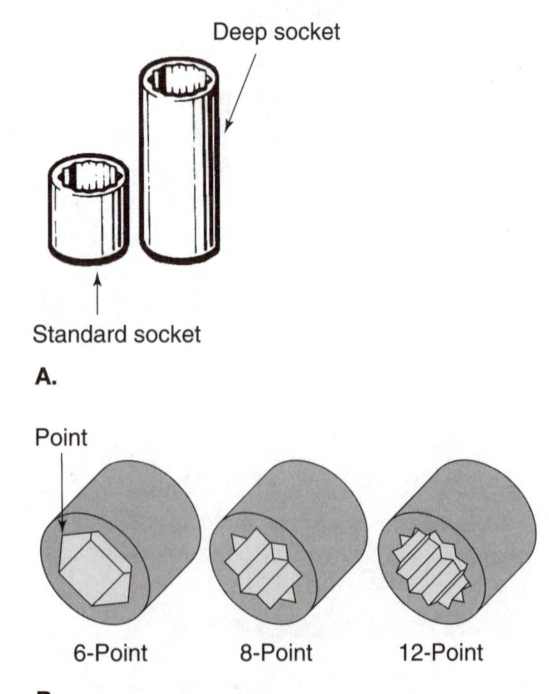

FIGURE 2-9 Sockets are made with different depths and different numbers of points.

> **SERVICE TIP:** In common shop talk, socket wrenches and attachments are often grouped together and just called "sockets."

A **socket driver** is the handle with a square drive lug that fits into the square hole in the socket wrench to drive a socket wrench. Many different types of drivers and other attachments are available to drive socket wrenches. The most commonly used type of driver is called a **ratchet handle** (Figure 2-10A). A ratchet handle, or ratchet, is a

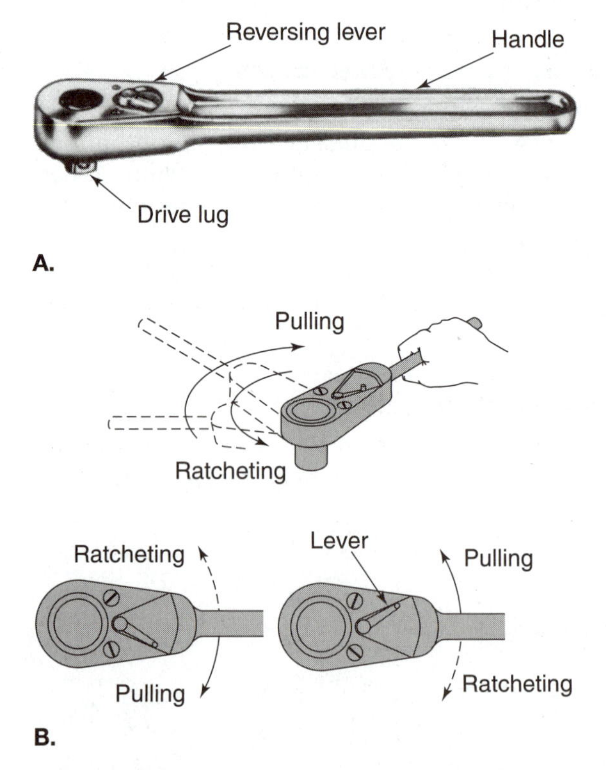

FIGURE 2-10 The ratchet driver has a reversing switch that allows it to freewheel in one direction. (*A:* Courtesy of Snap-on Incorporated.)

socket wrench drive handle that has a freewheeling or ratchet mechanism that allows it to drive a fastener in one direction and to move freely in the other direction (Figure 2-10B). A reversing lever on the ratchet allows switching the direction of drive and freewheeling. The bolt or nut is tightened or loosened by rotating the socket handle. This permits fast work in a small space because the socket does not have to be removed from the nut each time it is turned.

The **speed handle** (Figure 2-11A) is a socket driver shaped like a crank with a swivel handle at one end and a square drive lug at the other end. The speed handle is used for rapid driving of a socket wrench. The socket wrench is installed on the square drive end. When using the speed handle, you push on the end of the handle to hold the socket wrench firmly on the fastener. At the same time, you turn the handle rapidly in a direction to tighten or loosen. The speed handle is used when many bolts or nuts must be removed or replaced.

A **breaker bar** (Figure 2-11B) is another common socket driving tool. A breaker bar is a socket driver with a long handle, hinged at the drive lug end, used to break loose tight fasteners. The long handle provides a lot of turning force. The hinge on the drive lug end allows driving at different angles.

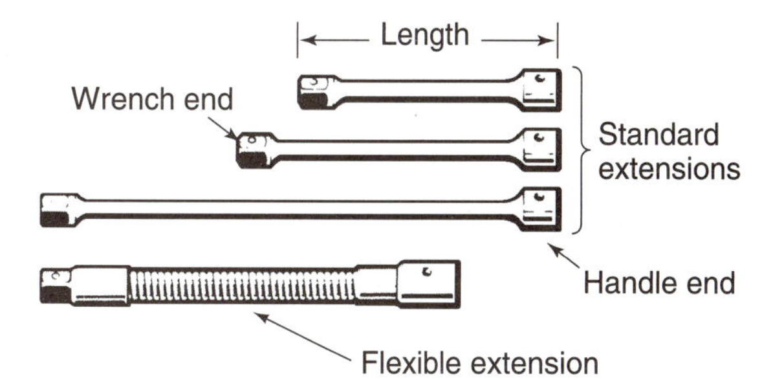

A. Socket Extensions

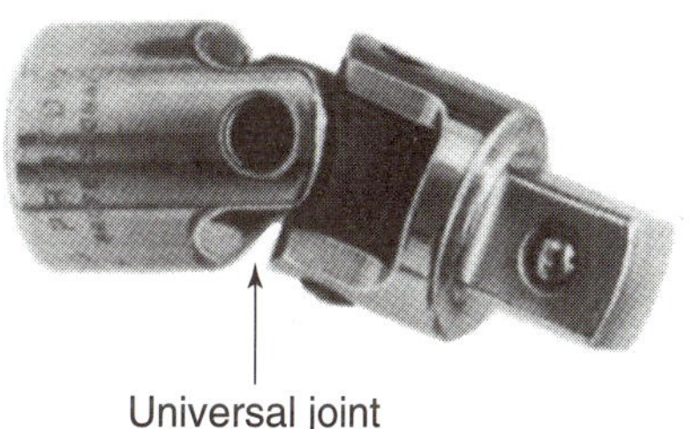

B. Universal Extension

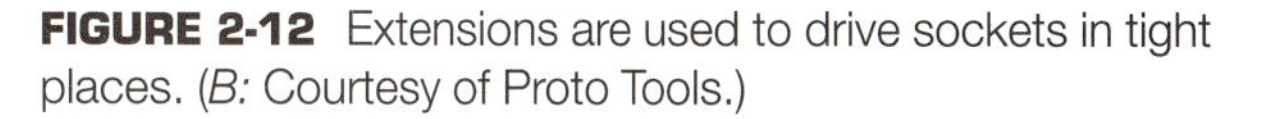

FIGURE 2-12 Extensions are used to drive sockets in tight places. (*B:* Courtesy of Proto Tools.)

A **sliding T-handle** (Figure 2-11C) is a socket wrench driver that has a handle shaped like a T. The T handle can be used on center or it can be moved outward to provide a longer lever for more turning power.

Socket wrenches are often attached to the drivers through extensions, Figure 2-12A. A **socket extension** is a shaft that can be connected from a socket driver to a socket wrench. The extension allows the socket to be used in an area where an ordinary handle would not have enough room to turn. Extensions come in different lengths for the different jobs. Flexible extensions that allow them to bend around obstructions are also available. A universal extension is used when the extension must operate at a sharp angle (Figure 2-12B). The universal extension is made in two parts that swivel to allow it to bend almost 90 degrees.

A **nut driver** (Figure 2-13) is a tool that combines a socket, extension, and screwdriver type

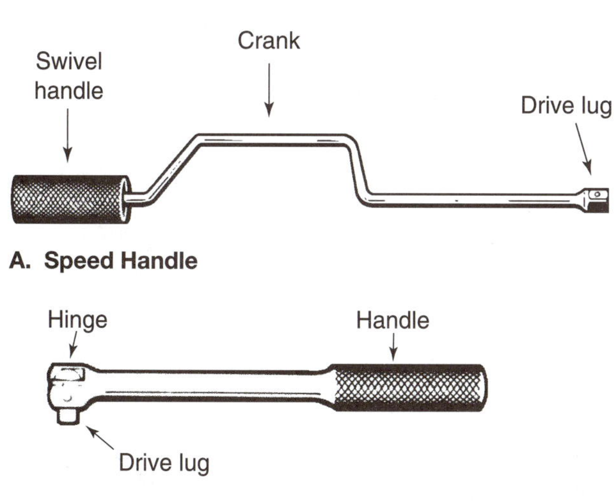

A. Speed Handle

B. Breaker Bar

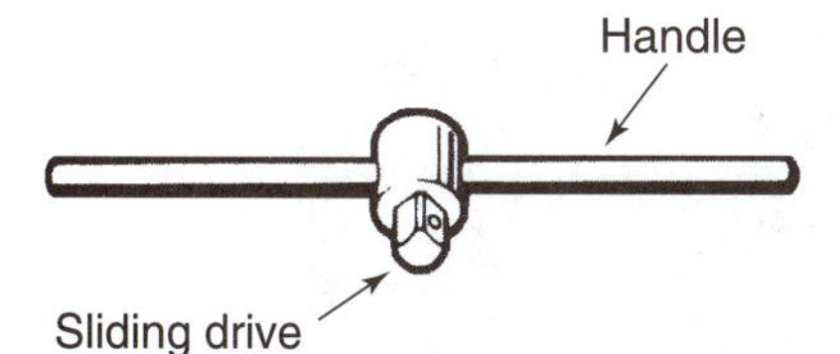

C. Sliding "T" Handle

FIGURE 2-11 Speed handle, breaker bar, and sliding T handle are common socket drivers. (Courtesy of U.S. Navy.)

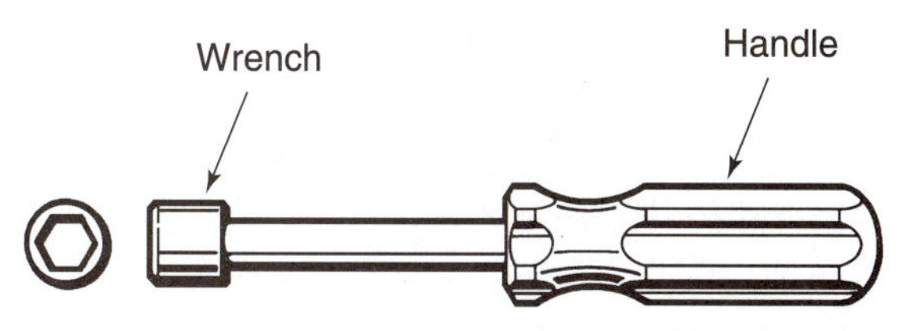

FIGURE 2-13 A nut driver is used to drive small sockets. (Courtesy of U.S. Navy.)

handle. Nut drivers are made in inch or metric size sets. They are made in small sizes for very small fasteners that can be removed or replaced with little turning effort.

Adjustable Wrenches

An **adjustable wrench** is a wrench with one stationary jaw and one jaw that may be adjusted using a threaded adjuster for different-sized fasteners (Figure 2-14A). Adjustable wrenches are made in different lengths from about 4 inches to about 20 inches. The longer the wrench, the larger the adjustable opening. For example, a 6-inch adjustable wrench opens $\frac{3}{4}$ inch wide while the 12-inch wrench opens $1\frac{5}{16}$ inches.

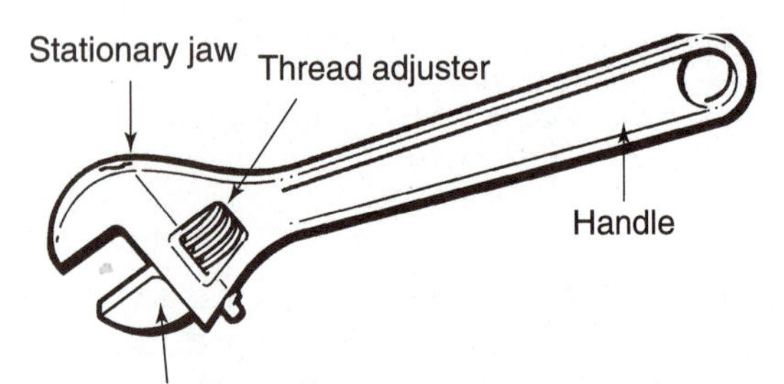

A. Parts of an Adjustable Wrench

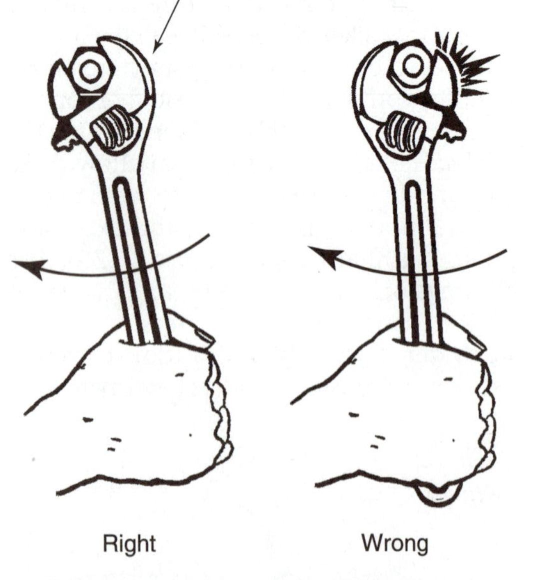

B. Right Way to Use an Adjustable Wrench

FIGURE 2-14 An adjustable wrench adjusts to different sizes but must be used correctly to avoid breakage. (*A:* Courtesy of U.S. Navy. *B:* Courtesy of General Motors Corporation, Service Technology Group.)

WARNING: The adjustable wrench is not as strong as a box-end wrench or an open-end wrench. The adjustable jaw can break if overloaded. Be careful to adjust the jaws to fit snugly against the flats of the nut or bolt head. When tightening, the wrench must be placed on the bolt or nut so that the force is exerted on the stationary jaw. The adjustable jaw can break and cause injury if the wrench is used incorrectly (Figure 2-14B).

SERVICE TIP: Technicians often incorrectly call the adjustable wrench a Crescent wrench, because the Crescent Tool Company made this type of wrench for many years. The name "Crescent" was on the handles of these wrenches. Now many different tool companies make this wrench. The correct name for the tool is adjustable wrench.

Allen Wrenches

An **Allen wrench** is a hex-shaped wrench that fits into the hollow head of Allen type screws (Figure 2-15). Allen head screws are often used to hold drive pulleys on engine crankshafts. They are available in sets and are sized according to the distance across the internal hex-shaped flats in which they fit. They are made in inch system sizes such as $\frac{3}{32}$ and $\frac{1}{8}$ and in metric sizes such as 6 millimeters, 8

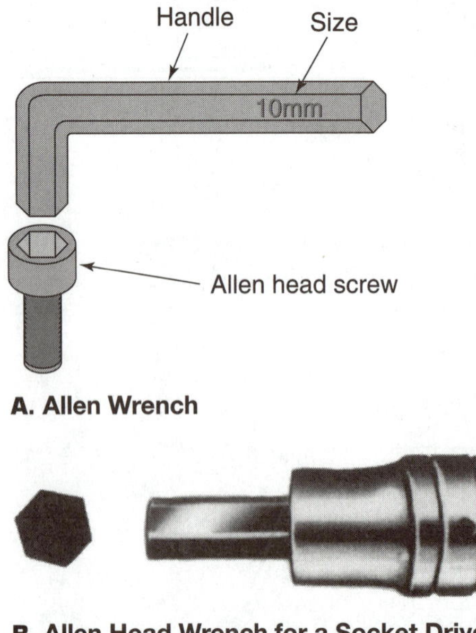

A. Allen Wrench

B. Allen Head Wrench for a Socket Driver

FIGURE 2-15 Allen wrenches may have their own handle or can fit on a socket driver. (*B:* Courtesy of Snap-on Incorporated.)

millimeters, and 10 millimeters. Allen wrenches are also made with square drive holes on one end to fit on socket drivers (Figure 2-15B).

TORQUE-INDICATING WRENCHES

Torque is a turning or twisting rotary unit of force that produces or tries to produce rotation. A **torque-indicating wrench,** or torque wrench, is a socket driver made to tighten a fastener to a specified amount of torque. Many engine fasteners must be tightened to exactly the correct torque with a torque wrench. Wrenches are made that measure torque in inch or metric units.

Torque measurements in the inch system are based on pounds-feet. This means a force, measured in pounds, is applied through a distance of one foot. For example, you use a wrench with a handle that is one foot long. You push on the end of the handle with 50 pounds of force. The torque you have applied is 50 foot-pounds (50 pounds × 1 foot). In the shop, these terms are reversed and torque is specified and measured in foot-pounds.

Sometimes smaller torque measurements are required. The foot can be divided into inches and measurements specified as inch-pounds. There are 12 inch-pounds in one foot-pound. A foot-pound specification can be changed to inch pounds by multiplying it by 12. For example, 20 foot-pounds is equal to 240 inch-pounds (20 × 12 = 240).

In the metric system the force is measured in Newtons or kilograms. The distance is measured in meters or centimeters. The three common measuring units are Newton-meters (abbreviated N•m); meter-kilograms (abbreviated m kg); and centimeter-kilograms (abbreviated cm kg).

The **beam and pointer torque wrench** is a type of torque wrench that has a shaft called a *beam,* which bends to show torque with a pointer and scale near the handle (Figures 2-16A and B). The handle of this wrench has a pivot inside that ensures that torque is applied through the center of the handle. You apply the force to the fastener by the wrench handle while looking at the torque scale (Figure 2-16C). Stop turning the wrench when the pointer lines up with the desired torque reading on the scale.

The **breakover hinge torque wrench** (Figure 2-17A) is a type of torque wrench that has a hinge system near the head and that swivels at a selected torque. You set the desired torque by turning the handle assembly (Figure 2-17B). The handle lines up with a torque scale on the bar of the wrench. For example, to set 100 foot-pounds, turn the handle

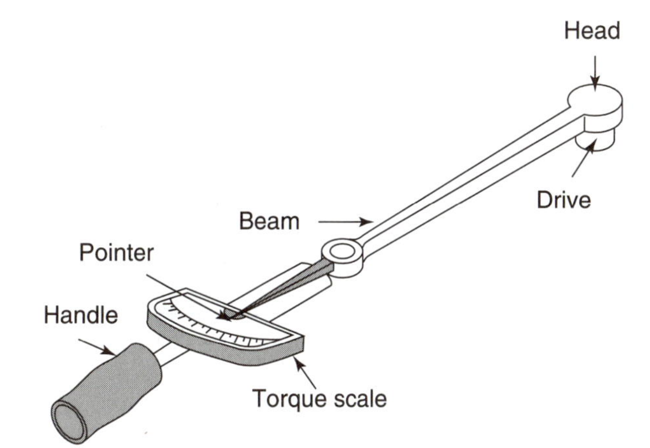

A. Parts of the Beam and Pointer Torque Wrench

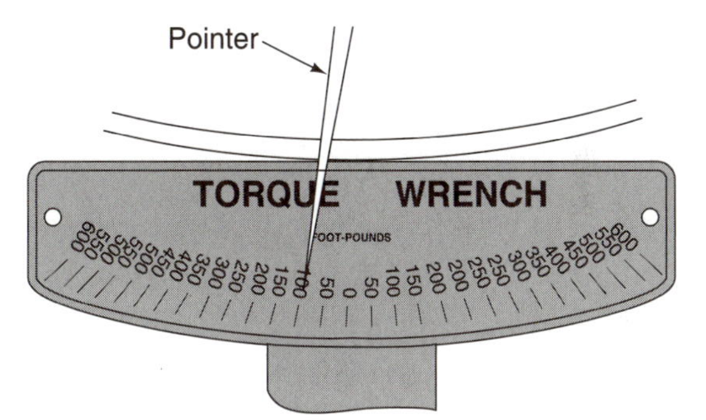

B. A Pointer and Scale Reading of 100-ft.-lb.

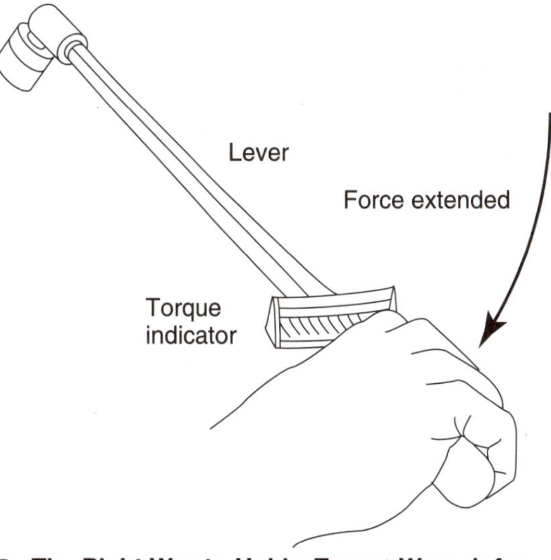

C. The Right Way to Hold a Torque Wrench for Accurate readings

FIGURE 2-16 A beam and pointer torque wrench uses the movement of a beam to measure tightening torque. (*C:* Courtesy of National Safety Council.)

until the 100 foot-pounds mark on the bar lines up with the zero mark on the handle. Begin to tighten the fastener with the wrench. When the torque on

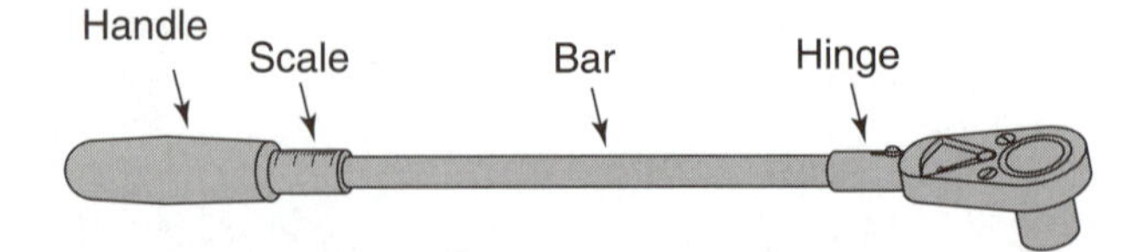

A. Parts of a Breakover Hinge Torque Wrench

B. Handle Setting at 100 ft.-lb.

FIGURE 2-17 A breakover hinge torque wrench measures torque on an adjustable scale at the handle.

the fastener reaches 100 foot-pounds, the hinge will swivel and make a clicking noise. The swivel motion and clicking noise is a signal to stop turning the wrench.

The **dial-reading torque wrench** (Figure 2-18A) is a torque wrench that has a dial on the handle with a needle that points to the amount of torque. There are many types of dial-reading torque wrenches. They are all operated just like the beam type. You read the torque by watching the needle inside the dial as it goes by the torque scale (Figure 2-18B). Stop turning the wrench when the needle lines up with the specified torque. Some torque wrenches have a digital rather than a dial reading.

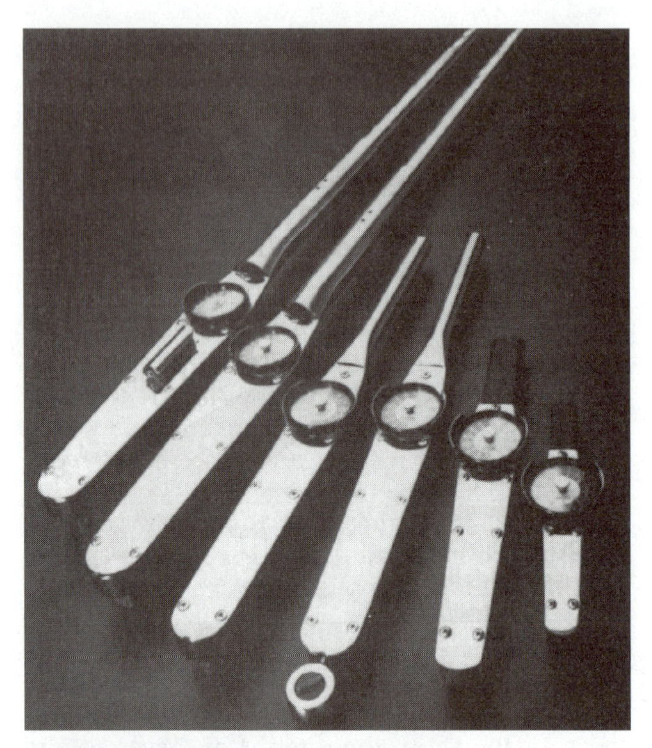

A. Dial Readout-type Torque Wrenches

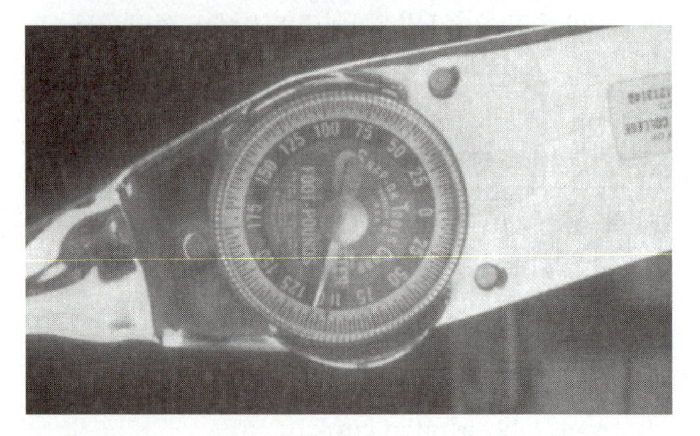

B. A Dial Reading of 100-ft.-lb.

FIGURE 2-18 A dial-reading torque wrench shows the torque measurement on a dial face. (*A:* Courtesy of Central Tools Inc.)

POWER WRENCHES

Hand-operated wrenches are slow to use when there are many fasteners to remove. Fasteners can be removed much faster using a power wrench. Power wrenches are powered by compressed air. Many other shop tools such as air hammers and drills can be powered by air (Figure 2-19).

The **air ratchet wrench** (Figure 2-20) is a ratchet handle with a square socket drive lug powered by compressed air for fast removal or installation of fasteners. The air wrench is connected by a hose to a shop compressed air source. Pulling the trigger causes an air-driven motor inside the handle to rotate a socket attached to the drive on the wrench. A reversing switch allows loosening as well as tightening.

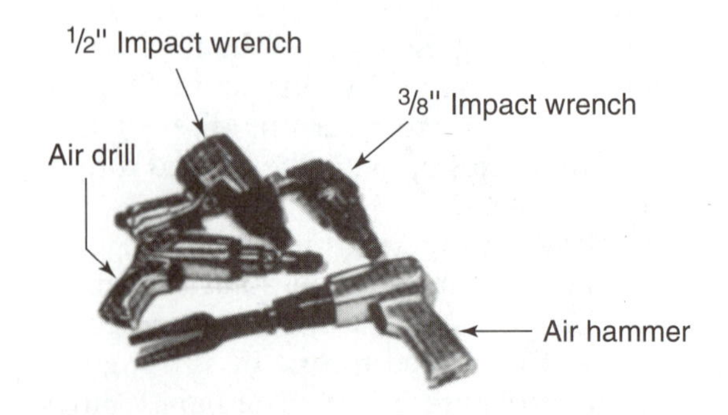

FIGURE 2-19 Air is used to power many shop tools. (Courtesy of Chicago Pneumatic Tool Co.)

FIGURE 2-20 An air-operated ratchet wrench uses air pressure to remove fasteners.

An **impact wrench** (Figure 2-21) is an air-operated wrench that provides an impact force while it drives a socket wrench to remove a fastener. The impact wrench has a larger air-operated motor than the air-operated ratchet wrench. This powerful air motor not only drives the socket but also vibrates or impacts it in and out. The force of the impact helps to loosen fasteners that are difficult to remove.

Many impact wrenches have no way of selecting tightening torque. The longer this type of wrench is held on the fastener, the more torque it applies. It is easy to damage fasteners or distort parts by overtightening with an impact wrench.

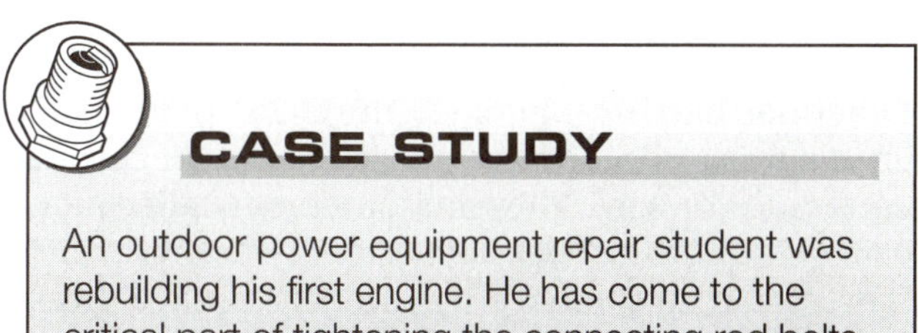

CASE STUDY

An outdoor power equipment repair student was rebuilding his first engine. He has come to the critical part of tightening the connecting rod bolts. He followed his instructor's instructions to look up the torque specification for these bolts. He knew that if the bolts were not tightened enough they might work loose and cause the engine to fail. He looked up the specifications in the shop manual and copied down the torque as 50. He got the breakover hinge torque wrench off the shop tool panel and adjusted it to 50 foot-pounds. He installed a 7/16 socket wrench on the torque wrench and began tightening the connecting rod bolts. He has just begun to tighten the bolt when he heard a loud snap. The snap was the sound of the bolt and part of the connecting rod breaking. The unhappy student learned the difference between inch-pound and foot-pound specifications. The 50 in the service manual was for inch-pounds. The student had just tried to tighten the small bolts to 600 inch-pounds!

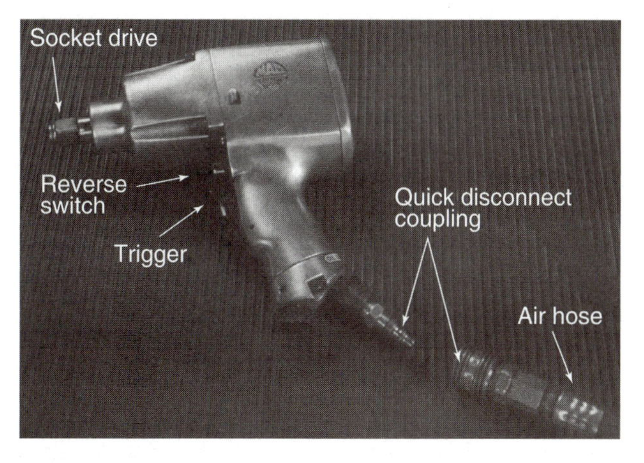

FIGURE 2-21 An impact wrench is used to remove hard to loosen fasteners.

You should use impact wrenches carefully. They are mostly used for disassembly, not for assembly.

The impact wrench is usually connected to a shop air hose with a part called a *quick disconnect coupling*. The quick disconnect coupling allows air lines to be attached to tools quickly without shutting off any air source. The part attached to the air hose is called the *female coupling*. The part attached to the wrench is called the *male coupling*. To attach the hose to the wrench, pull back the coupling slide on the female part. Insert it over the male coupling. Release the coupling slide to lock the two parts together. Disconnect the coupling by pulling the coupling slide and then pulling the male and female parts apart.

WARNING: Special, heavy-duty impact sockets must be used with impact wrenches (Figure 2-22). High-impact forces are applied at the drive end of an impact wrench. Impact sockets are thicker than standard hand wrench sockets. A standard socket can not withstand these impact forces and could break and cause serious injury. Always wear eye protection when using any power wrench.

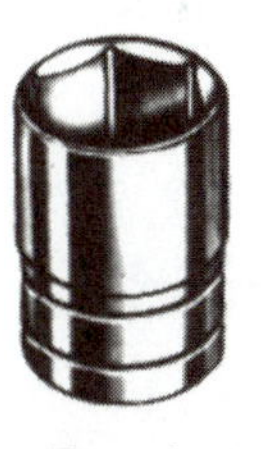

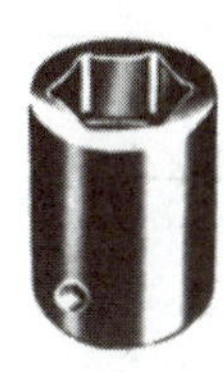

Regular **Impact**

FIGURE 2-22 Heavy duty sockets must always be used with impact wrenches. (Courtesy of Snap-on Incorporated.)

PLIERS

Pliers are a tool with two jaws operated with handles used to grip irregularly shaped parts and fasteners. The handles provide leverage that allows you to grip parts with great force. There are many special purpose pliers for many different repair jobs. Some have jaws designed for cutting parts like wire or cotter pins. Others are designed for specific kinds of gripping and holding tasks.

Combination Pliers

Combination pliers (Figure 2-23) are general purpose piers with a slip joint for two different-sized jaw openings. These pliers are used for gripping irregularly shaped objects. The slip joint where the two jaws are attached allows the tool to be set to two jaw openings. One is used for holding small objects. The other is used for holding larger objects. These pliers are used for pulling out pins, bending wire, and removing cotter pins. Some combination pliers have a side cutter to allow cutting of wire and cotter pins. They come in many different sizes. The size is determined by overall length. The 6-inch combination pliers are the most commonly used in outdoor power equipment repair.

Channel Lock Pliers

Channel lock pliers (Figure 2-24) are pliers with channels cut where the two jaws are attached so that the jaws may be adjusted to wide or narrow openings. They are sometimes called *water pump pliers*. The long channels allow the channel lock pliers to be used to grip large objects. The channels allow quick, nonslip adjustments. These pliers have long handles and jaws to reach and grip large objects. The jaw teeth are sharp and deep to grip pipes and hoses.

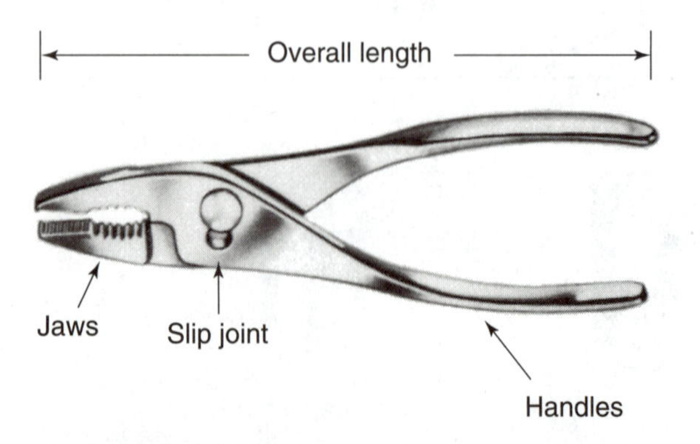

FIGURE 2-23 Combination pliers are used for gripping small objects. (Courtesy of Snap-on Incorporated.)

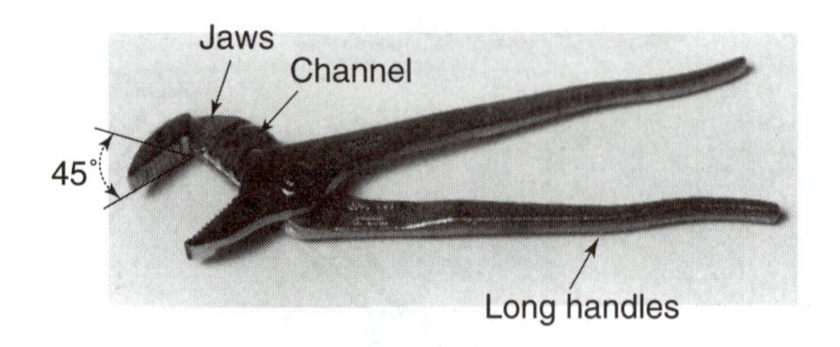

FIGURE 2-24 Channel lock pliers have long handles and large jaws for gripping large objects.

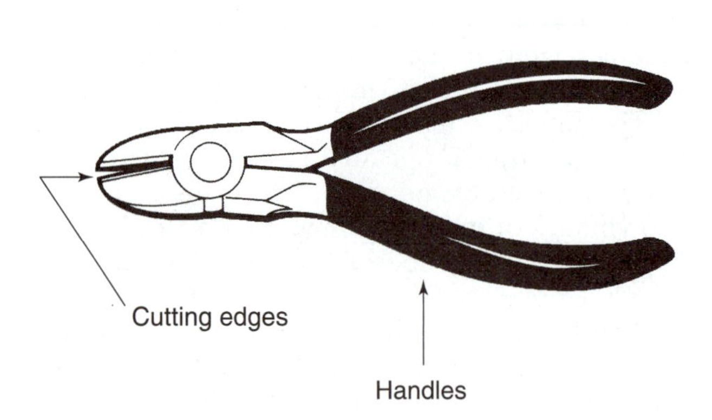

FIGURE 2-25 Diagonal cutting pliers are used to cut wire and cotter pins. (Courtesy of U.S. Navy.)

Diagonal Cutting Pliers

Diagonal cutting pliers (Figure 2-25) have hardened cutting edges on the two jaws used for cutting cotter pins or wire. They are sometimes called *diagonals*. The cutting edges are often used to grip and pull out cotter pins after the ends have been straightened out or cut off. They are made in many different sizes, based on overall length.

Needle Nose Pliers

Needle nose pliers (Figure 2-26A), or long-nose pliers, are pliers with long, slim jaws for gripping small objects or gripping objects in small or restricted spaces. Some needle nose pliers are equipped with side cutters. Others have the jaws bent at right angles for hard to get at places (Figure 2-26B).

Snap Ring Pliers

Snap ring pliers are pliers that are made to fit internal (Figure 2-27A) or external (Figure 2-27B) type snap rings and allow them to be removed or installed safely. Some engine parts are held together with rings that snap into place under tension. Internal snap ring pliers are used for internal type snap rings. They have jaws that close and grip

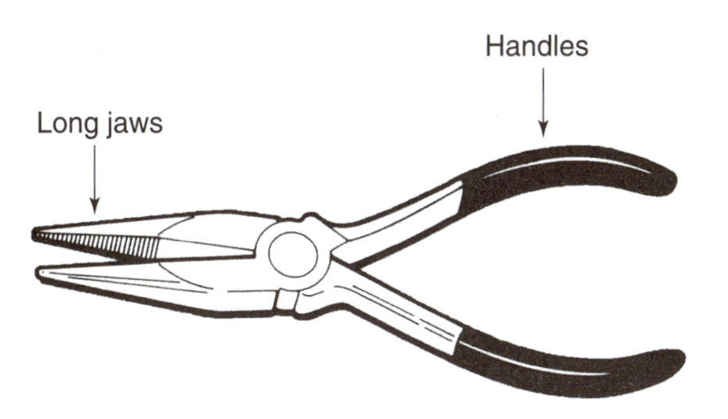

A. Needle Nose Pliers

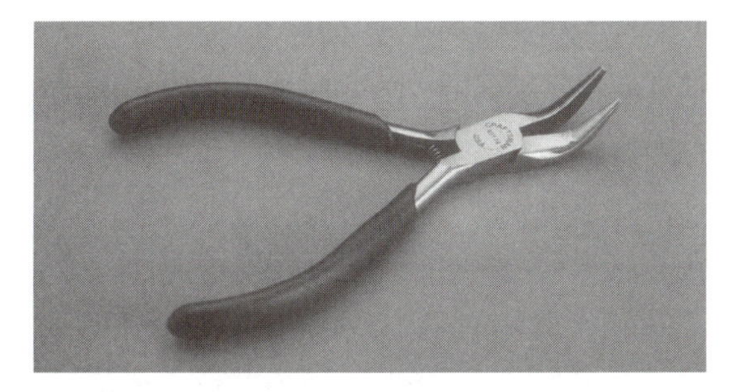

B. Bent Needle Nose Pliers

FIGURE 2-26 Needle nose pliers have straight or bent jaws and are used for gripping small objects. (*A:* Courtesy of U.S. Navy.)

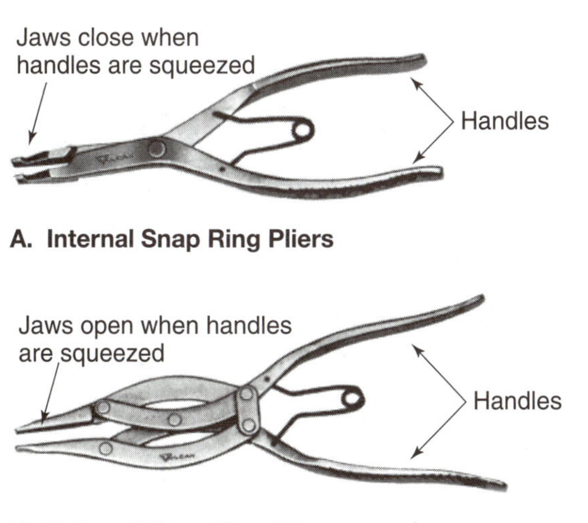

A. Internal Snap Ring Pliers

B. External Snap Ring Pliers

FIGURE 2-27 Internal and external snap ring pliers are used to safely remove snap rings.

when the handles are closed like an ordinary pair of pliers. External snap ring pliers have jaws that open when the handles are drawn together. They work to remove or install external type snap rings. The jaws of both types are made to safely fit the powerful snap rings. This allows them to be removed or installed without slipping off and creating a safety hazard.

WARNING: Snap rings are powerful springs. They must be handled with the proper tools and with great care. If not, they can fly off their retaining grooves and cause injury. Always use the right tool to remove or install snap rings. Always wear eye protection when working with snap rings.

Vise Grip Pliers

Vise grip pliers (Figure 2-28) are pliers with a lever-operated lock on the lower jaw that allow the jaws to be locked tightly on a part. The adjustable lower jaw is pulled back by a spring when the locking handle is released. You adjust the jaw opening size by turning a screw at the end of the primary handle. Vise grips do not take the place of wrenches. They are often used to hold nuts that have had their corners rounded off by the misuse of wrenches. They are also used to hold irregularly shaped parts and fasteners.

Battery Terminal Pliers

Battery terminal pliers (Figure 2-29) have jaws specially designed to grip terminal nuts on battery connections. These fasteners corrode rapidly from battery electrolyte. When a wrench no longer fits properly, battery terminal pliers can be used to grip them.

Stripping and Crimping Pliers

Stripping and crimping pliers (Figure 2-30) are special purpose pliers used to strip electrical wire insulation and crimp on solderless connectors. There are special cutting edges between the handles for stripping the insulation off different sizes of electrical wire. The jaws of the tool are made to squeeze or crimp different sizes of solderless terminals and

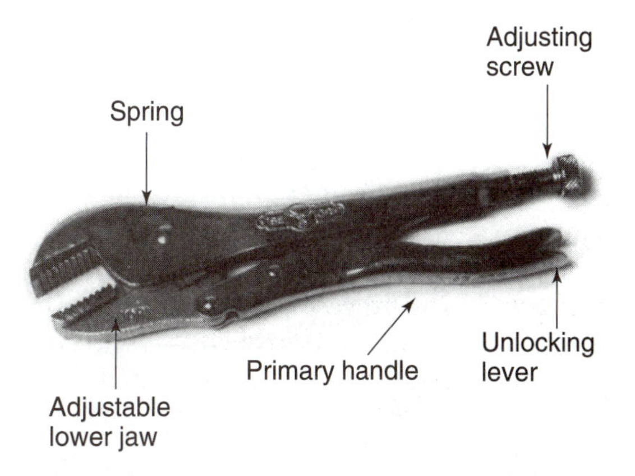

FIGURE 2-28 Vise grip pliers can be locked on to small parts to grip them tightly.

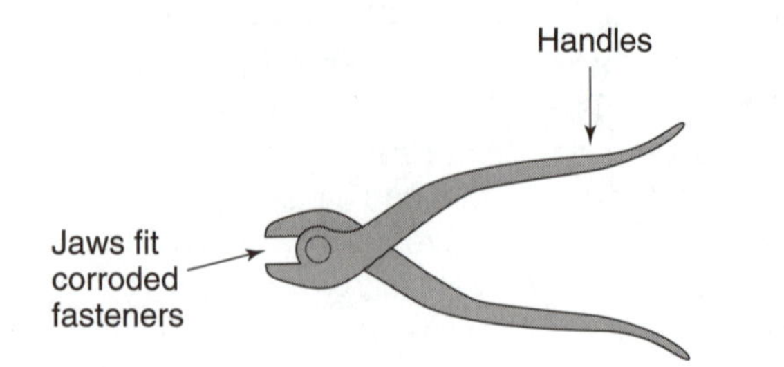

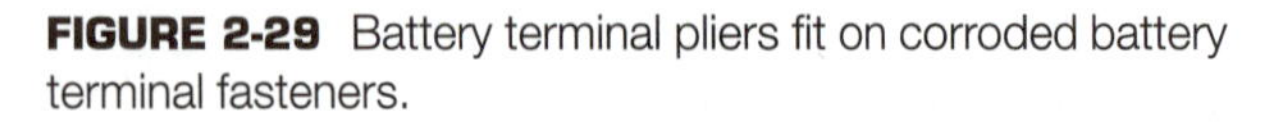
FIGURE 2-29 Battery terminal pliers fit on corroded battery terminal fasteners.

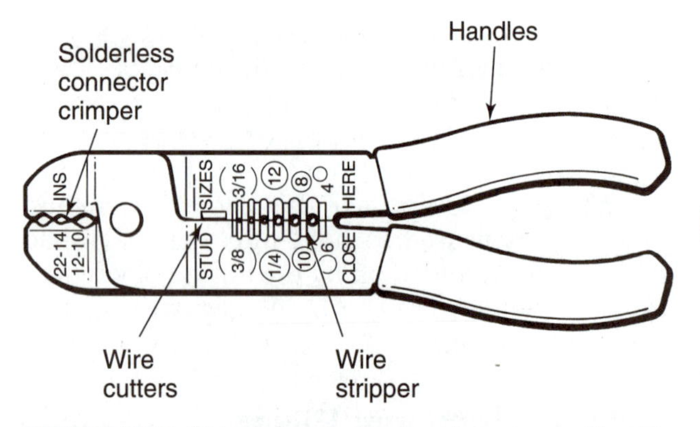

FIGURE 2-30 Stripping and crimping pliers are used to strip electrical wire and crimp on connectors. (Courtesy of U.S. Navy.)

connectors on the ends of wire. They also have a cutting surface for cutting electrical wire to length.

SCREWDRIVERS

A **screwdriver** is a tool with a handle at one end and a blade at the other used to turn or drive a screw. Because there are many different types of screws and screw heads, there are many different types of screwdrivers.

Standard Screwdrivers

The **standard screwdriver** (Figure 2-31A) is a screwdriver with a blade and tip designed to drive standard slotted screws. A standard screwdriver usually has a wood or plastic handle attached to a steel shank. The part that fits in the screw head is the blade or bit (Figure 2-31B).

The size of a screwdriver is determined by its overall length. The larger the screwdriver, the larger its blade. Standard screwdrivers are available in sets of different lengths. Some are made with interchangeable bits. The bits come in different sizes and fit into a single handle. Bits are also made that can be attached to common socket drivers.

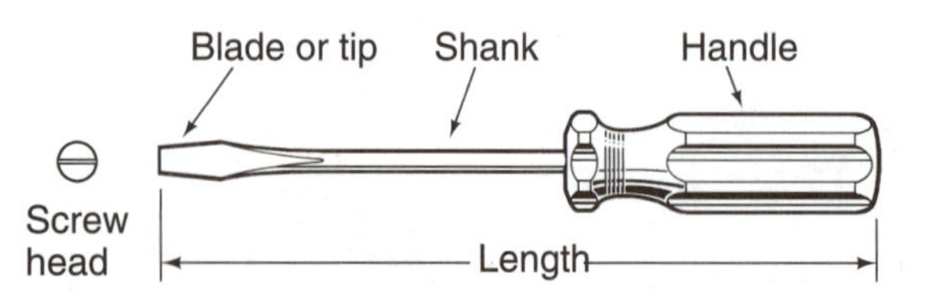

A. Standard Screwdriver

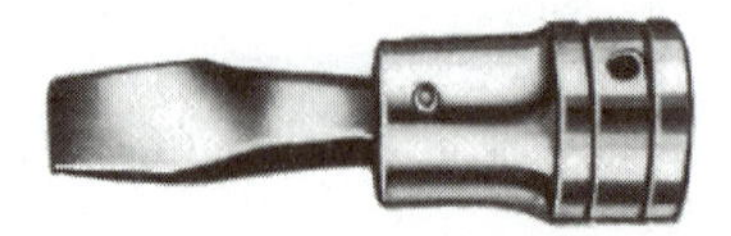

B. Screwdriver Bit for Socket Driver

FIGURE 2-31 A standard screwdriver has a blade that fits a slotted screw head. (*A:* Courtesy of U.S. Navy. *B:* Courtesy of Snap-on Incorporated.)

WARNING: Care must be taken to use a screwdriver with a blade that fits the slot in the screw snugly. If the fit is loose, the head of the screw may be damaged. More importantly, the screwdriver could slip out of the screw and stab the hand. Do not use a screwdriver to pry off parts. The tip of the blade is hardened to keep it from wearing. The hardness makes it brittle and may cause it to break. The broken tip could fly into the face and cause injury. Never use pliers to turn the blade of a screwdriver. They can slip off and cause injury. Heavy-duty screwdrivers are available with square shanks that can safely be used with a wrench.

Phillips Screwdrivers

The **Phillips screwdriver** (Figures 2-32A and B) is a screwdriver with a blade designed to fit into the slots of Phillips head screws. Phillips head screws have two slots that cross at the center. They do not extend to the edges of the head. They are made so that the screwdriver head will not slip out of the slots. The blades are sized on a numbering system from 0 to 6, 0 being the smallest and 6 the largest.

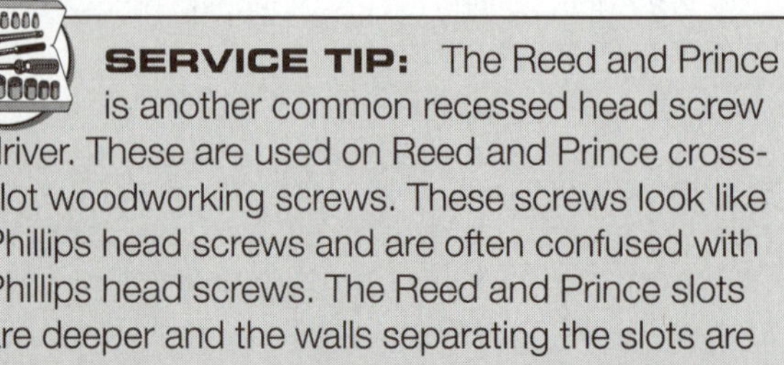

SERVICE TIP: The Reed and Prince is another common recessed head screw driver. These are used on Reed and Prince cross-slot woodworking screws. These screws look like Phillips head screws and are often confused with Phillips head screws. The Reed and Prince slots are deeper and the walls separating the slots are tapered. Do not mix up the two types of screwdrivers. Using the wrong screwdriver can result in damaged screw heads.

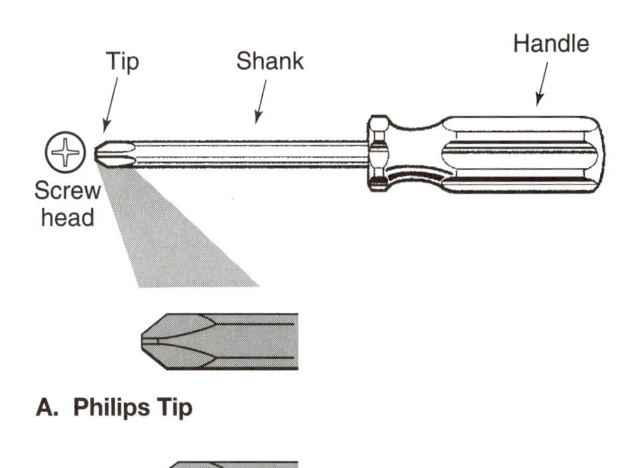

A. Philips Tip

B. Reed and Prince Tip

FIGURE 2-32 The Phillips screwdriver fits the recessed head of a Phillips head screw. (*A:* Courtesy of U.S. Navy.)

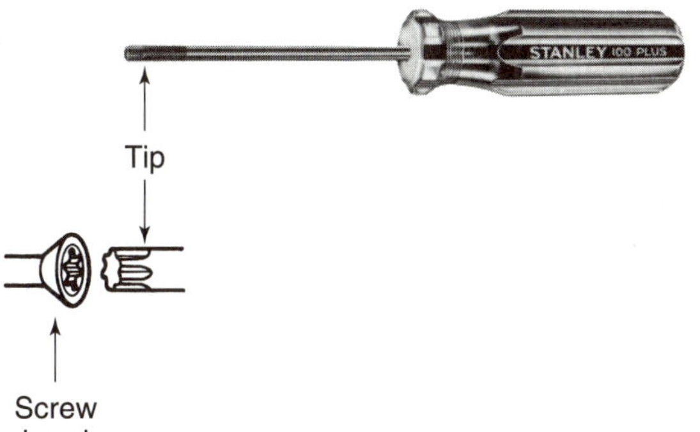

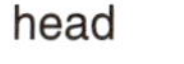

FIGURE 2-34 The Torx screwdriver fits the recessed head of a Torx screw head. (Courtesy of Stanley Tools.)

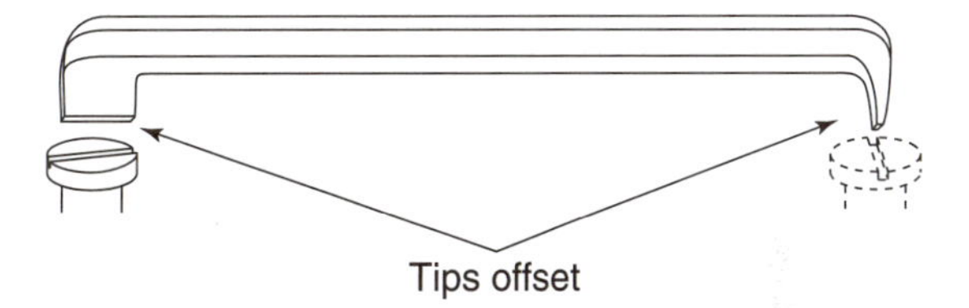

FIGURE 2-35 The ends of the offset screwdriver are reversed to allow turning in tight spaces. (Courtesy of General Motors Corporation, Service Technology Group.)

Clutch Screwdrivers

The **clutch screwdriver** (Figure 2-33) is a screwdriver with a figure eight-shaped blade made for turning recessed clutch head screws. Clutch head screws are also known as figure eight screws because the screw head looks like the number eight. Clutch type screw heads are made to keep the screwdriver blade from slipping off the screw head.

Torx Screwdrivers

The **Torx screwdriver** (Figure 2-34) is a screwdriver with a blade made to fit the deep, six-sided recessed slot in a Torx-type screw head, which allows good turning power while holding the screwdriver tight in the screw head. These screwdrivers are available in sets that fit the common size Torx screw heads. They are often made with interchangeable bits on the end of the shanks.

Offset Screwdrivers

The **offset screwdriver** (Figure 2-35) is a screwdriver with blades made at an offset angle to allow driving screws in tight spaces. Offset screwdrivers have blades at opposite ends that are at right angles to each other. The ends of the offset screwdriver are changed after each swing to continue to turn a screw in a tight area. These screwdrivers come in common blade or bit sizes.

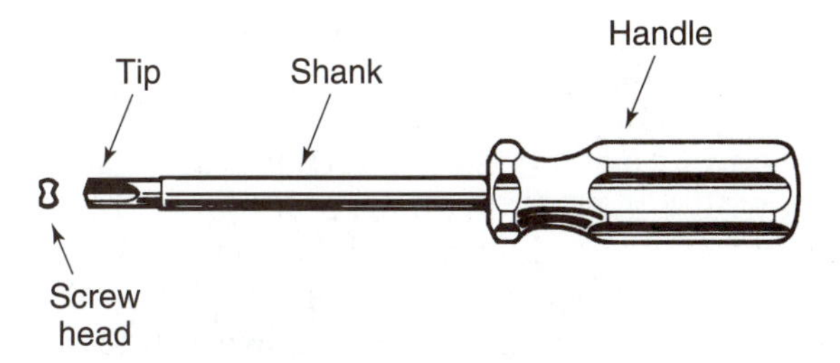

FIGURE 2-33 The clutch head screwdriver fits the figure eight-shaped clutch head screw. (Courtesy of U.S. Navy.)

HAMMERS

A **hammer** is a tool with a handle and a head used to drive parts or tools. The handles are often made from wood or plastic. The heads are made from different materials for different service procedures. You will need to know how to select and use several different types of hammers.

WARNING: Be sure that the hammer's head is securely attached to the handle before using any hammer. A loose head can fly off and cause serious injury. Avoid striking a hardened steel surface with a steel-headed hammer because particles of steel may break loose from the hammer or steel surface and fly off and cause injury.

A **ball peen hammer** (Figures 2-36A and B) is a type of hammer with a steel head that has one round face and one square face. The rounded face

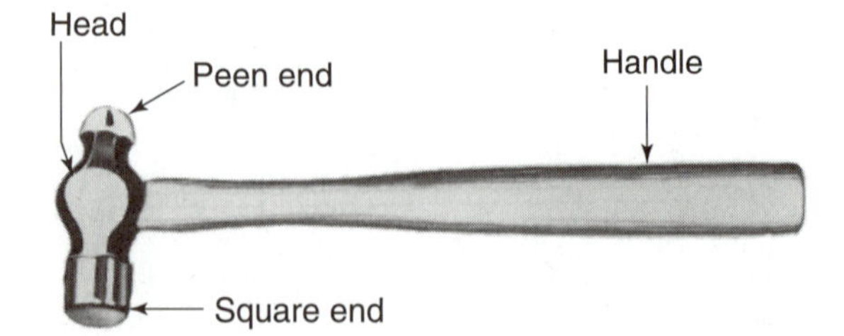

A. Ball Peen Hammer

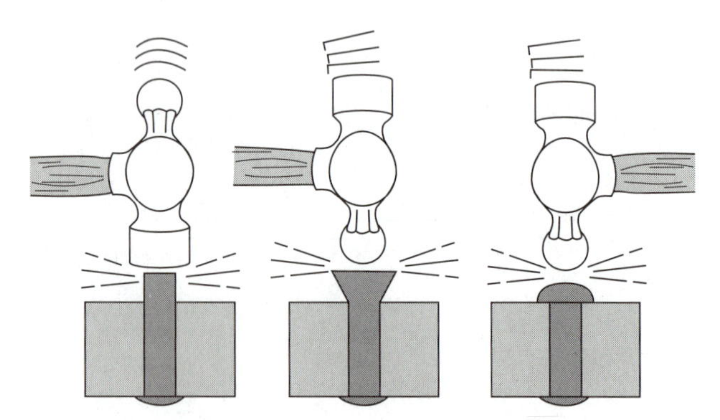

B. Using a Ball Peen Hammer to Set a Rivet Head

FIGURE 2-36 A ball peen hammer has a square end for general hammering and a peen end for rivet setting.

on the head is called the ball peen and is used to form or peen over the end of a rivet. The square end is used for general hammering such as driving a punch or chisel. Ball peen hammers are classified in different sizes according to the weight of their head. Small ones weigh as little as 4 ounces and large ones weigh more than 2 pounds.

CAUTION: Do not hammer on engine parts with the hardened head of a ball peen hammer. The head is harder than the part and can damage the part. To protect part from damage, always use a hammer that is softer than the material being hammered on.

There are two common types of soft face hammers. The **plastic tip hammer** (Figure 2-37A) is a type of hammer with plastic tips on the head for protecting the surface of parts being hammered. These hammers have heads that are much softer than the ball peen steel head hammers. They can be used to align or adjust parts that might be damaged by a steel-headed hammer. The **brass hammer** (Figure 2-37B) is a type of hammer with a head made from soft, heavy brass. It is used to hammer on parts without damaging them. Hammers with brass heads are much heavier than plastic. They can be used when more force is required. Brass hammers are often used for driving pins and other parts that would be damaged by steel hammers.

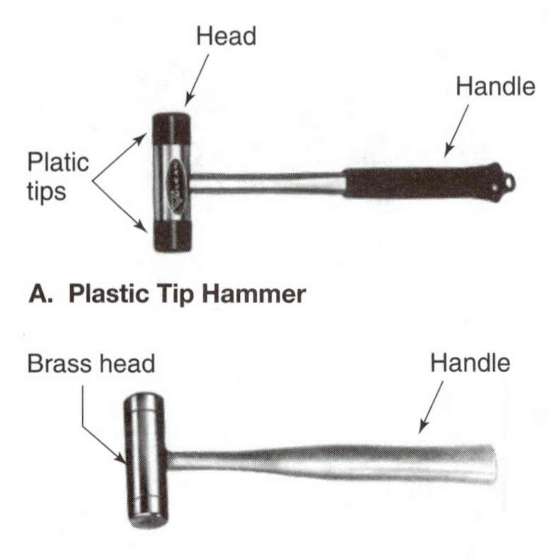

B. Brass Head Hammer

FIGURE 2-37 Plastic and brass head hammers are used to hammer on parts without damaging them.

PUNCHES

Punches are tools used with ball peen hammers to drive pins or to mark the center of a part to be drilled, Figure 2-38. The most common types of punches are the pin punch, starter punch, and center punch. A **starter punch** (Figures 2-38A and B) is a punch with a taper on the end that allows the starting of pin removal. A **pin punch** (Figures 2-38B and C) is a punch used with a hammer to drive out pins from engine parts. These two different kinds of punches may be needed to remove a pin or rivet. A starter punch is used to break the pin loose. A pin punch that is smaller than the hole is used to drive the pin out. Always use the largest starter and pin punch that will fit the hole.

A starter punch may also be called a *drift punch* or simply a *drift*. Drifts are usually made of brass. They are used to drive out highly finished parts without damage. Long, tapered steel drifts are often used to align two parts so that a bolt or screw can be installed.

A **center punch** (Figure 2-38D) is a short, steel punch with a hardened conical point ground to a 90-degree angle used to mark the centers of holes to be drilled. If the center punch is used correctly, the depression it makes in the metal will guide the point of the drill bit. This makes sure the hole is drilled in the correct spot.

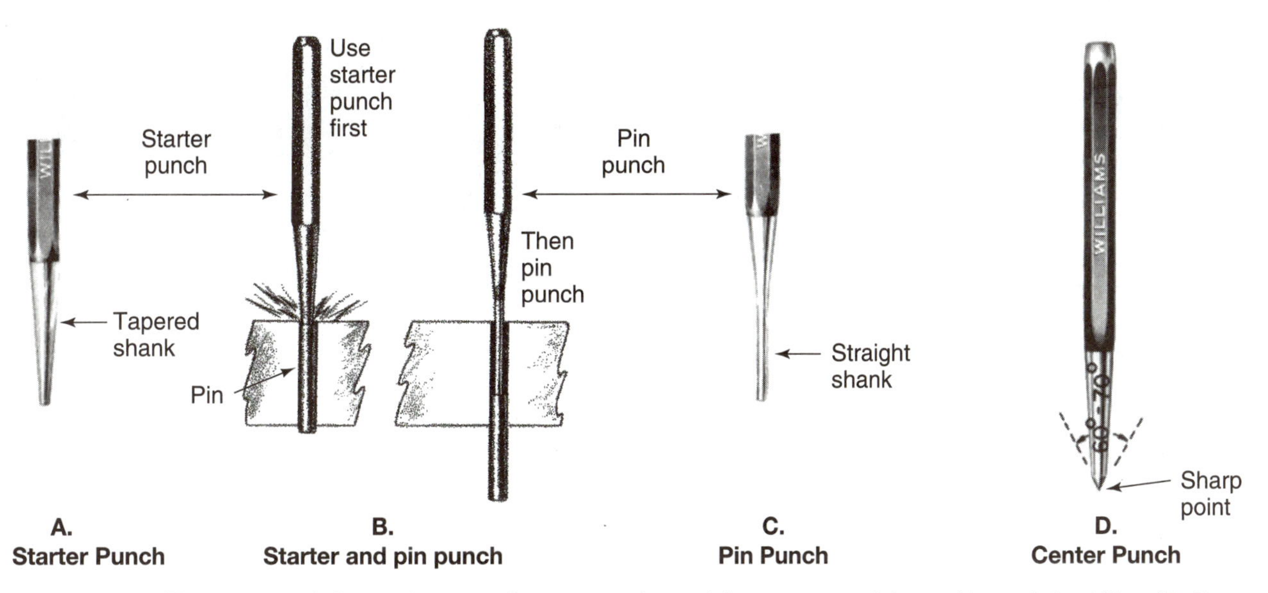

FIGURE 2-38 The starter and pin punch are used to remove pins and the center punch is used to mark for drilling. (*B:* Courtesy of General Motors Corporation, Service Technology Group.)

CHISELS

Chisels are bars of hardened steel with cutting edges ground on one end, driven with a hammer to cut metal, Figure 2-39. Chisels are used for splitting nuts that cannot be removed with a wrench. The **flat chisel**, or cold chisel (Figure 2-39A), is a type of chisel used for general cutting of steel. It is driven with a ball peen hammer. The heavier the chisel, the heavier the hammer required to drive the chisel. The **cape chisel** (Figure 2-39B) has a narrow cutting edge for cutting narrow grooves such as keyways. The cape chisel is made with the cutting

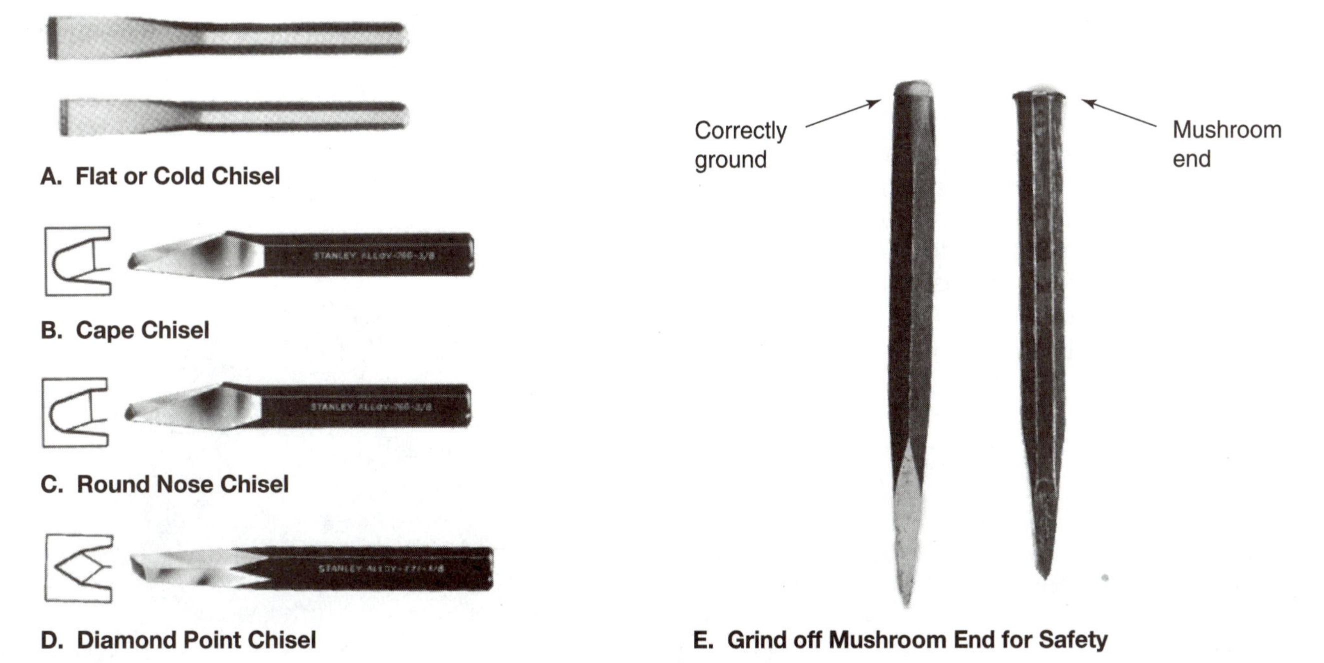

FIGURE 2-39 Different types of chisels are used to cut different shapes in metal. (*A:* Courtesy of K.D. Tool Company.)

edge slightly wider than the shank to make sure that it will not bind when used to cut a narrow groove. The **round nose chisel** (Figure 2-39C) has a single bevel cutting edge used mostly for cutting semicircular grooves and inside rounded off corners. The **diamond point chisel** (Figure 2-39D) has a diamond-shaped cutting edge that is used for cutting V-shaped grooves and for squaring up the corners of slots.

WARNING: Wear eye protection when using a chisel. Make sure the part to be cut is held securely in a vise. Hammering sometimes curls over or mushrooms the upper end of a chisel. The head must be ground smooth because the chips may fly off a mushroomed head and cause injury (Figure 2-39E).

TWIST DRILLS

The **twist drill** (Figure 2-40), or drill, is a cutting tool mounted or chucked in an electric drill motor to cut holes. The end of the drill is called the *point.* The spiral part the drill is called the *body and flute.* The part that fits in the electric drill motor is the *shank.*

Twist drills are made in four different size groups:

- Fractional sizes from 1/64 inch to 1/2 inch and larger in steps of 1/64 inch
- Letter sizes from A to Z
- Number sizes 1 to 80
- Millimeter sizes

A **drill index** is a set of drills in either fractional, metric, letter, or number sizes. Drill size charts list each drill and give decimal and metric equivalent sizes. The size of a twist drill is stamped on the shank. After much use, the size may be difficult to read. A **drill gauge** (Figure 2-41) is a metal plate with holes identified by size that is used to determine the size of a drill. Place the drill to be measured in the holes until you find the hole that best matches the drill.

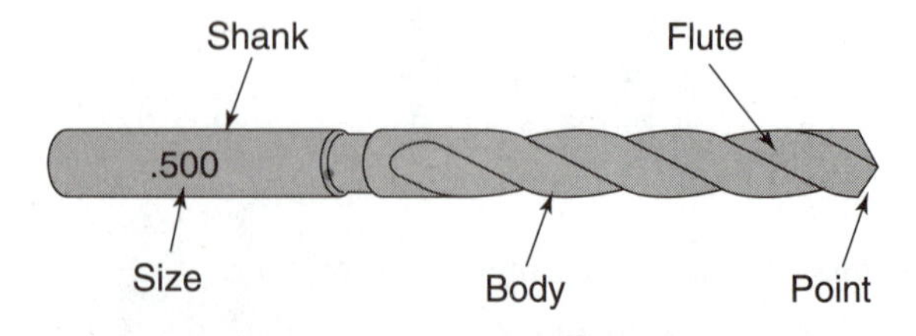

FIGURE 2-40 A twist drill is installed and driven in a drill motor to machine a hole to a precise metric or inch size.

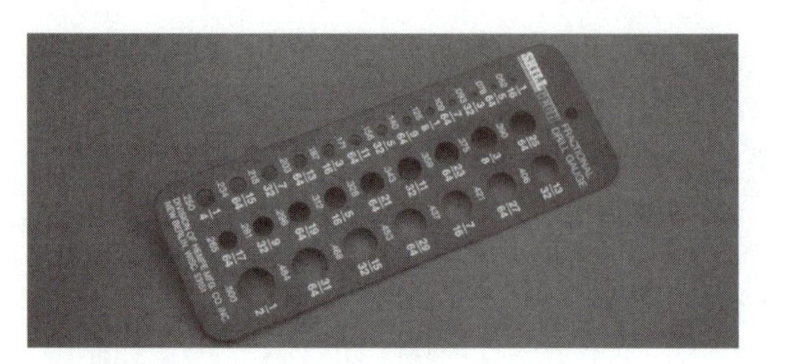
FIGURE 2-41 The size of a twist drill can be found by inserting it into a drill gauge.

A twist drill must be sharp to do a good job of cutting. The point of the drill is sharpened with a bench grinding wheel. Cutting different materials requires different cutting angles. Getting the correct angle on a drill point usually requires a special holding device that fits on the grinder tool rest.

Twist drills are most often used in portable electric drill motors (Figure 2-42) sometimes just called drills. The drill is installed in the drive end of the drill motor called a chuck. The three jaws of the chuck are tightened over the twist drill shank using a chuck key. There are three common sizes of portable drill motors: 1/4, 3/8, and 1/2 inch. The size of the chuck determines the largest drill size that will fit inside. For example, a 1/4-inch drill is approximately the largest drill that will fit in a 1/4-portable drill motor.

WARNING: Wear eye protection when operating a drill to protect against flying metal chips.

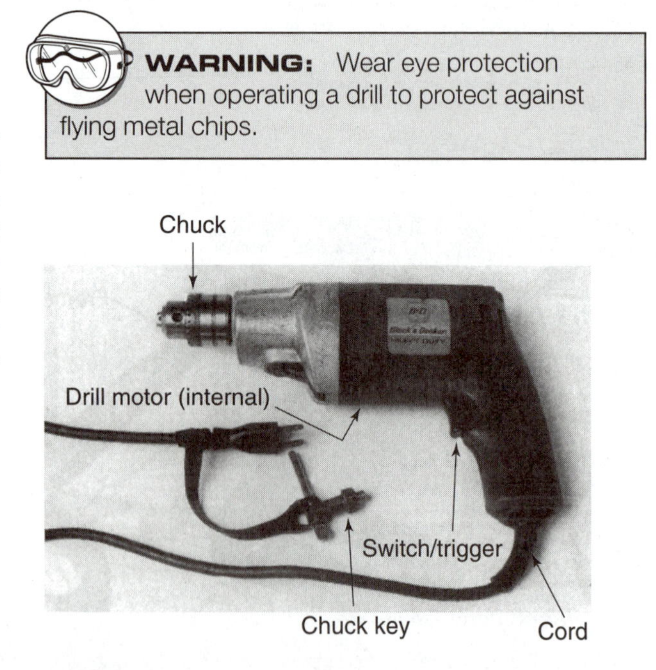

FIGURE 2-42 A portable electric drill motor is used to drive twist drills.

REAMERS

A **reamer** (Figure 2-43) is a tool with cutting edges used to remove a small amount of metal from a drilled hole. A reamer is used when a very precise hole is necessary for a precision fit. Some engine parts have bushings that must be finished to size with a reamer. Reamers are made in many different sizes. The size is stamped on the shank. Adjustable reamers are available that adjust to different sizes.

> **CAUTION:** A reamer should be turned with a wrench or with a tap wrench in a clockwise direction. Turning a reamer backward will quickly dull its cutting edges.

HACKSAWS

A **hacksaw** (Figure 2-44A) is a type of saw that uses a blade made to cut metal. A hacksaw may be used to cut exhaust pipes and other metal parts during a repair job. The two main parts of a hacksaw are the frame and the blade. Hacksaw frames are made so that they can be adjusted to fit 8-, 10-, or 12-inch long blades.

The hacksaw blade installed in the frame does the cutting. Blades are made in different lengths and also with different numbers of teeth per inch (TPI). The TPI is marked on the hacksaw blade. The common TPI are 14, 18, 24, and 32. The thicker the material to be cut, the fewer the TPI required on a hacksaw blade. For example, a blade with 18 TPI is used for most general sawing. Thin metal tubing would be sawed with a blade with more teeth such as 32 TPI.

To properly install a hacksaw blade in a frame (Figure 2-44B), have the teeth pointing toward the front of the frame and away from the handle. A blade that is installed backward will not cut properly and the blade will dull rapidly. Hold the saw with one hand on the handle and the other on the frame. Push the saw forward and down to cut. Release the pressure to back the saw up for the next stroke.

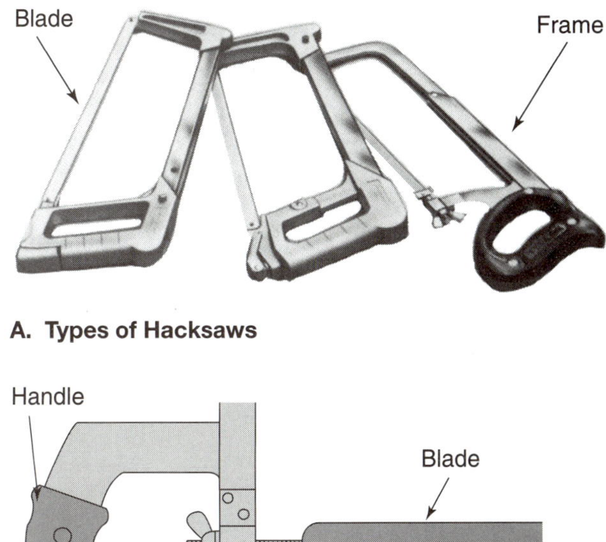

FIGURE 2-44 A hacksaw may be equipped with different length and teeth per inch blades and is used to cut metal for repair jobs. (*A*: Courtesy of Snap-on Incorporated.)

FILES

A **file** (Figure 2-45A) is a hardened steel tool with rows of cutting edges used to remove metal for polishing, smoothing, or shaping. The face is the part of the file with the cutting teeth. The edge is the side of the face. The part that goes into the handle is the tang. The tapering part between the tang and the face is called the *heel*. Files are made in different lengths, measured from the tip to the heel.

The file tang is shaped to fit into the handle. Always install a handle before filing to protect yourself from the sharp tang. The handle is installed by placing the tang into the handle then striking the handle on a workbench (Figure 2-45B).

Files with cutting edges that run in only one direction are called *single cut files*. Files made with cutting edges that cross at an angle are *double cut*

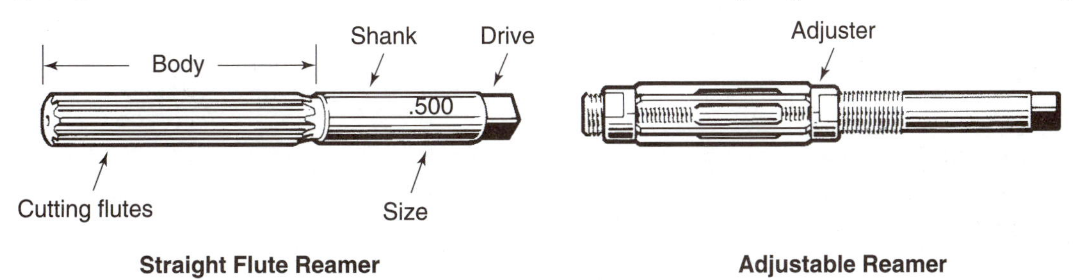

FIGURE 2-43 A reamer is a precision tool used to finish a drilled hole to an exact size. (Courtesy of U.S. Navy.)

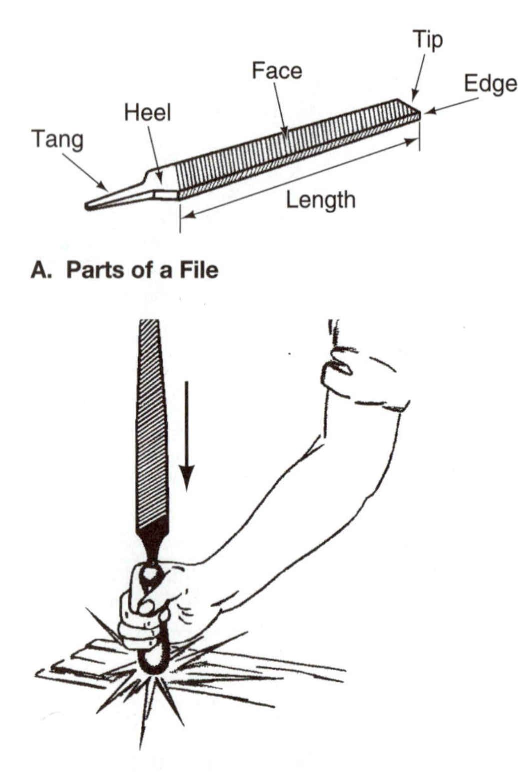

A. Parts of a File

B. Setting a File Handle

FIGURE 2-45 Files are used to remove small amounts of metal for polishing or shaping. (*B:* Courtesy of General Motors Corporation, Service Technology Group.)

files. The cutting edges may be spaced close together or wide apart. The wider they are spaced, the faster the file will remove metal. A file with cutting edges that are closer together will remove less metal. They are used to smooth or polish a metal surface. Files are made in different shapes such as flat, half-round, round, triangular, and square to file different shapes.

When you use a file, grip the handle with one hand while pushing down on the file face with the other hand (Figure 2-46). A file is designed to cut in only one direction. Raise the file on the return stroke. Dragging it backward dulls the cutting edges. Always put small parts in a vise for filing. The teeth on the file will get clogged with metal filings. Remove them by tapping the file handle on the workbench. You can also brush the teeth with a special file cleaning brush called a *file card*.

> **WARNING:** Wear eye protection when using a file. Do not use a file without a handle. Be sure to mount small parts in a vise when filing. Do not use a file as a pry bar because it is very brittle and could break and cause injury.

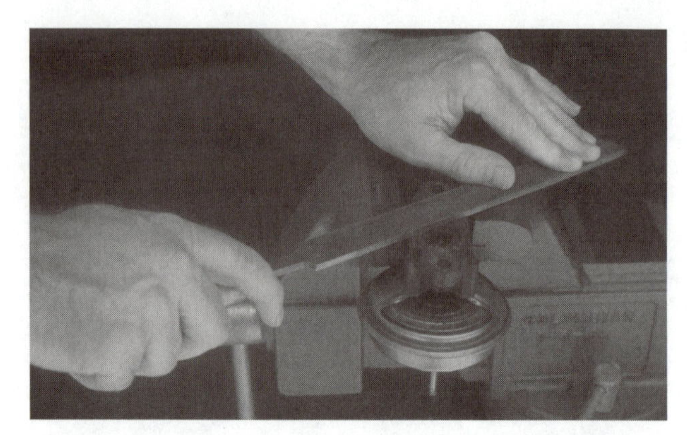

FIGURE 2-46 The correct way to hold a file when filing.

THREADING TOOLS

Many outdoor power equipment repair jobs require repairing or replacing the threads on parts. A **tap** (Figure 2-47A) is a tool used to make or repair inside threads. The tap has a square drive so that it can be driven by a tap wrench. The part that cuts the threads is called a *cutting flute*. The thread size the tap cuts is shown on the shank. A tap of the correct size is installed in a holding tool called a *tap wrench* (Figure 2-47B). The tap is then turned in the hole allowing the cutting flutes to make a new thread or to repair damaged ones. Taps are available for common inch and metric thread sizes.

Taper taps are slightly smaller on the end to help get them started in the hole. They are used to tap threads all the way through a hole. A bottom-

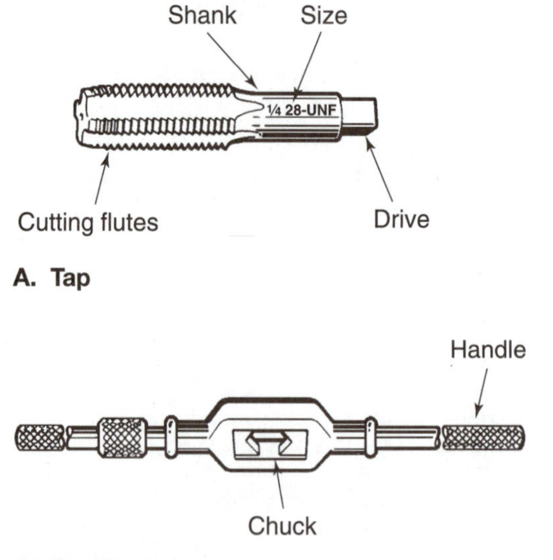

A. Tap

B. Tap Wrench

FIGURE 2-47 Taps held in a tap wrench are used to make or repair damaged internal threads in engine parts. (Courtesy of U.S. Navy.)

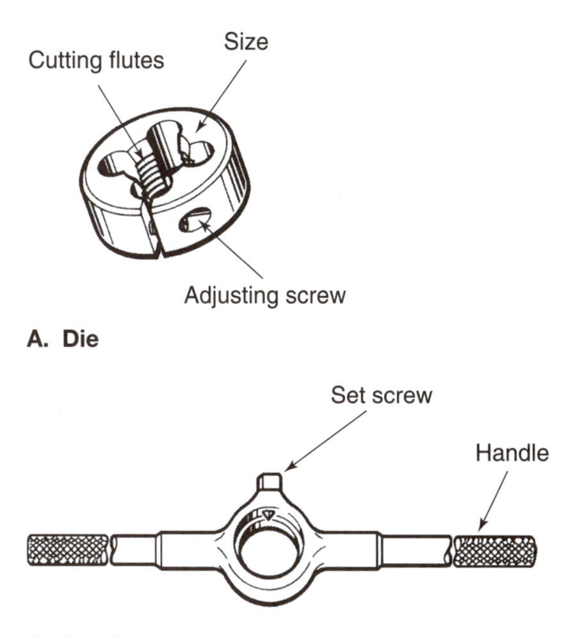

FIGURE 2-48 Dies held in a die stock are used to make or repair damaged threads in engine parts. (Courtesy of U.S. Navy.)

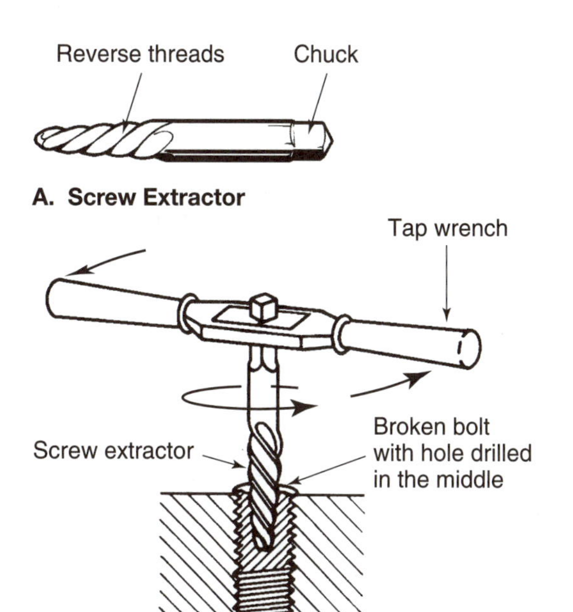

FIGURE 2-49 A screw extractor is used with a tap wrench to remove a broken fastener. (*A:* Courtesy of U.S. Navy.)

ing tap has a squared off end and is used to cut threads down to the bottom of a blind hole.

When a tap is used to repair threads no drilling is necessary. When threads are to be tapped in a new hole, the hole has to be the exact size to provide enough clearance for the tap. This drill size is called the *tap drill size*. For example, to tap a new ¼-inch UNC thread in a part, it must first be drilled with a number 7 drill. Charts are used that show the tap drill size for each size and type of thread to be tapped.

A **die** is a tool used to repair or make outside threads (Figure 2-48A). The part that cuts the threads are called cutting flutes. The size of the threads a die cuts is marked on top of the die. Most dies have an adjusting screw for changing the depth of cut. The die is installed and driven in a tool with two long handles called a *die stock* (Figure 2-48B). The die is held in place in the die stock with a set screw. The die is rotated down over the part to machine new threads or repair damaged ones. Dies come in inch and metric thread sizes.

A **screw extractor** (Figure 2-49A) is a tool used to remove a bolt or screw that has broken off in a threaded hole. The correct procedure for extracting the broken bolt or screw is as follows (Figure 2-49B). First, drill a hole of the correct diameter into the center of the broken fastener. Install the correct size screw extractor into the drilled hole. Use a tap wrench to turn the screw extractor counterclockwise. The reverse threads on the extractor will dig into the broken screw and pull it out.

PULLERS

A **puller** is used to remove gears, bearings, shafts, and other parts off shafts or out of holes. These tools are used for many common engine service procedures. A puller is used to pull a gear, bearing, or pulley off an engine crankshaft. Rust and corrosion on the shaft can cause these parts to be very difficult to remove. The forcing screw type of puller has jaws that fit on the part to be removed (Figure 2-50). A forcing screw is rotated against the shaft the part is mounted on. The forcing screw has a centering point on the end. It centers the screw in the center of the shaft. The forcing screw threads provide the leverage to separate the parts.

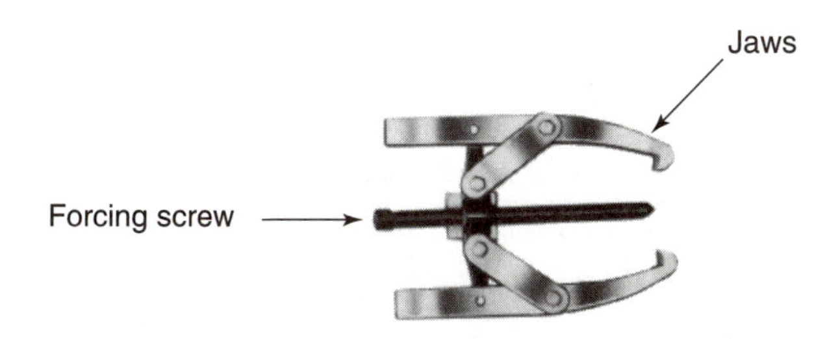

FIGURE 2-50 The forcing screw puller is used to remove a pulley from a shaft. (Courtesy of Owatonna Tool Company.)

Forcing screw type pullers are made with either two or three jaws to grip the part. A three-jaw puller is usually better than a two-jaw puller because it grips the part in three places. Many parts, however, have shapes that do not allow gripping in three places. These must be pulled with a two-jaw puller.

To use this puller, adjust the puller jaws and cross arms until the jaws fit properly on the pulley. Choose the correct size wrench that fits the forcing screw. Thread the screw down in contact with the pulley shaft. Make sure the point of the forcing screw is centered on the shaft. If not, readjust the puller cross arms or jaws. Slowly and cautiously tighten the forcing screw. Watch carefully for any evidence that the puller is trying to slip off the part or the forcing screw is trying to slip off center on the shaft. Make any necessary readjustments. If the puller is secure, tighten the puller forcing screw and pull the pulley from its shaft.

> **WARNING:** If the puller jaws do not fit the part or the cross arm will not allow proper placement of the jaws, select another size puller. Using an incorrect size puller or not adjusting the puller correctly can allow the jaws to slip off during pulling and cause injury. Always wear eye protection when using a puller because parts can break under pulling forces and cause eye injury.

Parts such as bearings and bushings may be installed inside blind holes with a press fit. The removal of these parts requires a slide hammer type puller (Figure 2-51). This puller has a jaw that grips the part to be removed from the hole. A heavy slide is then moved rapidly up the puller shaft until it hits a stop at the end. The inertia of the heavy slide provides the hammering force to pull the part out of the hole.

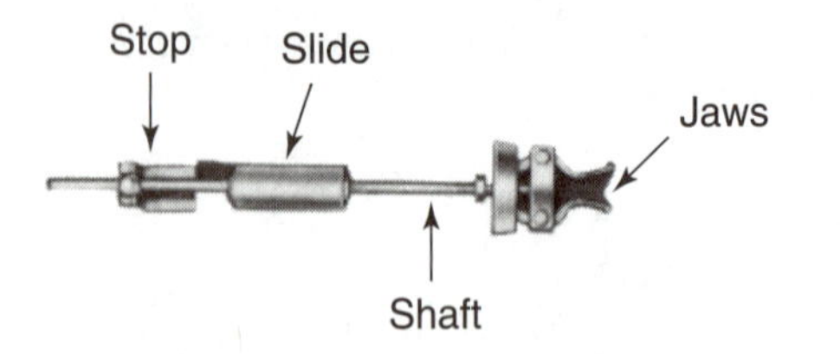

FIGURE 2-51 The slide hammer puller is used to pull parts from a blind hole. (Courtesy of Owatonna Tool Company.)

SPECIAL TOOL SETS

Most engine makers supply special tool kits or individual tools (Figure 2-52). These tools are used for specific service procedures on specific models of engines. They make many service jobs easier to perform. Some service procedures are impossible without the special tools. There is often a basic tool kit with tools for common service procedures like flywheel removal or engine rebuilding. There may be a more specialized tool kit available for a specific service procedure such as main bearing replacement. Most shops have the basic and specialized tool kits available for the engines they service regularly. Engine makers often provide a manual showing the specialty tools.

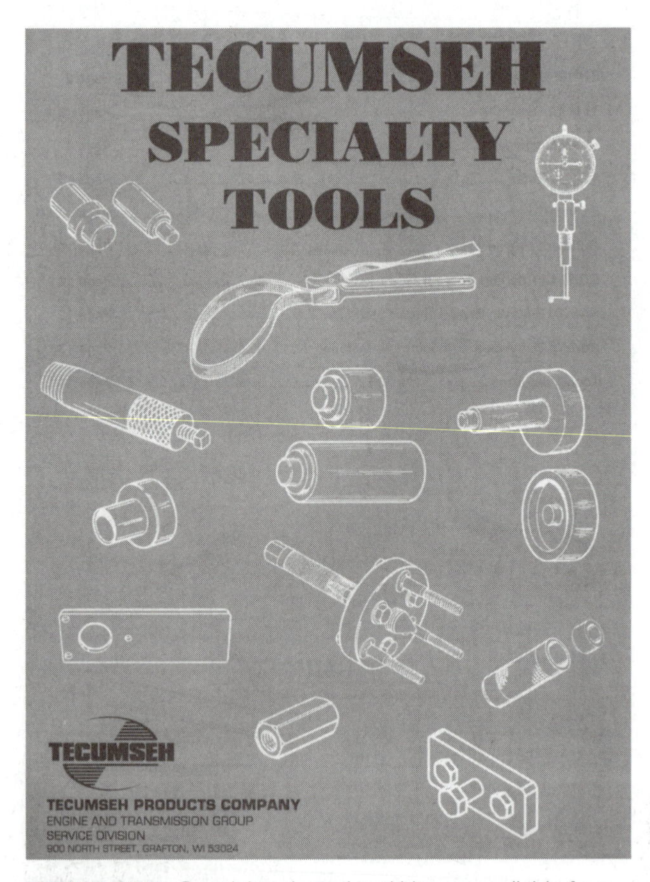

FIGURE 2-52 Special tools and tool kits are available from each engine manufacturer for specific repair operations.

REVIEW QUESTIONS

1. Explain the relationship between the size of a wrench and the size of a fastener head.
2. Describe when it is best to use an open-end wrench and when it is best to use a box-end wrench.
3. List and describe the three common sizes of square socket drives.
4. List and describe three types of socket drivers.
5. Describe how to use an adjustable wrench correctly during loosening and tightening.
6. List and describe the three types of torque wrenches.
7. List and describe the two types of power wrenches.
8. Explain why impact sockets must be used with an impact wrench.
9. List and describe three uses for combination pliers.
10. List and describe two uses for diagonal cutting pliers.
11. List and describe the two basic types of snap ring pliers.
12. List and describe the three functions of stripping and crimping pliers.
13. Explain why it is dangerous to use a screwdriver as a prybar.
14. List and describe three types of screwdrivers used for recessed head screws.
15. Explain why it is dangerous to strike a hardened surface with a ball peen hammer.
16. Explain why it is necessary to use two types of punches to remove a pin.
17. List and describe three types of chisels.
18. List and describe the four size groups used for twist drills.
19. List and describe the purpose of the two types of taps.
20. Explain how to use a three-jaw puller to remove a pulley from a shaft.

DISCUSSION TOPICS AND ACTIVITIES

1. Use a tool catalog to develop a wish list of tools you need.
2. Convert the following foot-pound torque specifications into inch-pounds: 5 foot-pounds, 12 foot-pounds, 50 foot-pounds.
3. You are using a 1-foot-long wrench and you are exerting a force of 20 feet on the wrench handle. How many foot-pounds of torque are you applying?

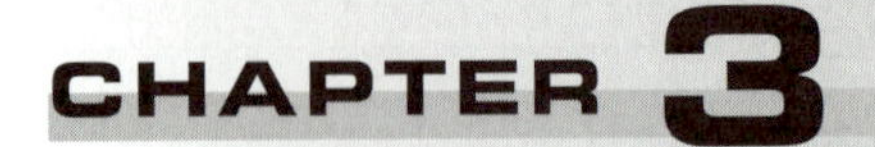

Measuring Tools

OBJECTIVES

Upon completion and review of this chapter, you should be able to:

- Use the inch measuring system to measure engine parts.
- Use the metric measuring system to measure engine parts.
- Describe how to convert between measuring systems.
- Measure with a rule.
- Measure with a feeler gauge.
- Measure with an outside micrometer.
- Measure with an inside micrometer.
- Measure with a depth gauge.
- Measure with a telescoping gauge.
- Measure with a small hole gauge.
- Measure with a dial indicator.

TERMS TO KNOW

Conversion factor
Depth micrometer
Dial indicator
Feeler gauge
Inch measuring system
Inside micrometer
Metric measuring system
Outside micrometer
Rule
Small-hole gauge
Stepped feeler gauge
Tape rule
Telescoping gauge

INTRODUCTION

The small engine and each of its parts is made to close tolerances. Keeping these tolerances is an important part of service. Each engine maker gives the service technician specifications. These tolerances and clearances must be correct or the engine will not work properly. You must be able to measure precisely to use these specifications.

INCH-MEASURING SYSTEM

The most common type of measurement made in small-engine work is length or distance. The **inch measuring system** is a length- or distance-measuring system based on the inch. Measurements based on the inch are the most commonly used in the United States. The inch can be abbreviated as "in." or by the symbol ". The inch can be divided into fractions or decimals. For example, a quarter of an inch can be specified as ¼ inch or 0.250 inch.

In the fractional method, an inch ruler can be divided into the following units (Figure 3-1):

1 in.	(one inch)
½ in.	(one-half inch)
¼ in.	(one-quarter inch)
⅛ in.	(one-eighth inch)
1⁄16 in.	(one-sixteenth inch)
1⁄32 in.	(one thirty-second inch)
1⁄64 in.	(one sixty-fourth inch)

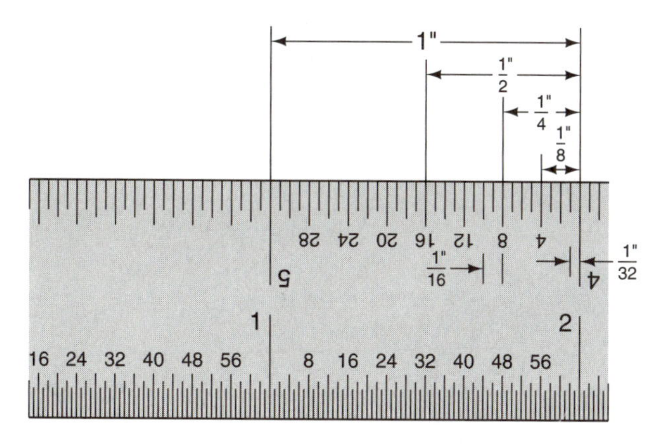

FIGURE 3-1 Dividing an inch into fractions.

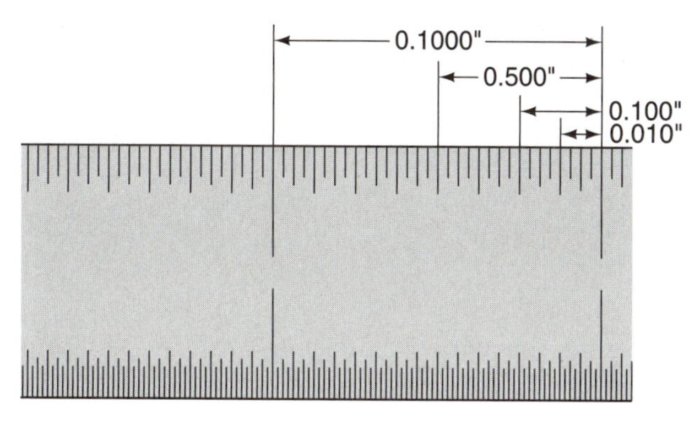

FIGURE 3-2 Dividing an inch into decimals.

The smallest division of a typical ruler is 1⁄64 of an inch, which is about equal to the thickness of a fingernail. This is the smallest measurement that most people can see.

Often you will need to make measurements much smaller than 1⁄64 of an inch. This is possible by dividing an inch using the decimal system. With this system, the inch is divided by ten, this in turn by ten, and so on (see Figure 3-2):

1.0 in.	(one inch)
0.1 in.	(one-tenth inch)
0.01 in.	(one-hundredth inch)
0.001 in.	(one-thousandth inch)
0.0001 in.	(one ten-thousandth inch)

Many small engine parts must be measured to within one-thousandth of an inch (0.001). A human hair is about three-thousandths of an inch (0.003) thick.

METRIC MEASURING SYSTEM

Used by most of the countries in the world, the **metric measuring system** is a length- or distance-measuring system based on the meter. The United States is one of the last countries to use the metric system. Many engines made in the United States use the inch system of measurement, but engines made in Europe and Japan use the metric system. Eventually, the metric system will replace the inch system in the United States. You must be able to measure in both systems.

Number of Meters	Prefix	Symbol
1 000	kilometer	km
1	meter	m
0.01	centimeter	cm
0.001	millimeter	mm

FIGURE 3-3 Common metric system units, prefixes, and symbols.

The metric system uses different names for the units of measurement. See Figure 3-3. For example, the measurement of a human hair may be made in millimeters and a soccer field in meters. The names are formed by adding a prefix to the word "meter."

Another part of the metric system is that each prefix can be abbreviated with a symbol. The symbol for the meter is m. The prefix can be added to this symbol. For example, 1,000 meters may be written 1,000 m. Since a kilometer is equal to 1,000 meters, it may be written 1 kilometer or 1 km. One-thousandth of a meter may be written 1 millimeter or 1 mm.

CONVERTING BETWEEN MEASURING SYSTEMS

Many engines made in the United States have specifications in the inch system; imported engines have specifications in the metric system; some engines have specifications in both systems. You may have a metric measuring tool but a specification you need to measure in the inch system. Or you could have an inch measuring tool and need to measure a specification in the metric system. You may find it necessary to convert units from one system to another.

A specification from one system can be converted to the other by multiplying by a number called a **conversion factor.** Conversion factors are given in Figure 3-4. Suppose you need to convert the specification 16 mm to units in the inch system:

- First look up the conversion factor in a chart and find that it is 0.03937

To Find		Multiply	×	Conversion Factors
millimeters	=	inches	×	25.40
centimeters	=	inches	×	2.540
centimeters	=	feet	×	32.81
meters	=	feet	×	0.3281
kilometer	=	feet	×	0.0003281
kilometer	=	miles	×	1.609
inches	=	millimeters	×	0.03937
inches	=	centimeters	×	0.3937
feet	=	centimeters	×	30.48
feet	=	meters	×	0.3048
feet	=	kilometers	×	3048.
yards	=	meters	×	1.094
miles	=	kilometers	×	0.6214

FIGURE 3-4 Common inch to metric conversion factors.

- Multiply the number of millimeters by the conversion factor or 0.03937
- 16 mm × 0.03937 = 0.63
- 16 mm is equal to 0.63 of an inch

RULE

A **rule** is a measuring tool made from a flat length of wood, plastic, or metal graduated in inch or metric units. The rule gets its name from the word ruler. It is a simple but important measuring tool.

Inch rules are usually 6 or 12 inches long. Most rules divide the inch into $\frac{1}{8}$, $\frac{1}{16}$, and $\frac{1}{32}$ divisions. Precision machinist rules have $\frac{1}{32}$ and $\frac{1}{64}$ inch divisions (Figure 3-5). When you use a rule, place one end of the rule at one end of the length to be measured. Then, find which of the rule marks most nearly lines up with the other end of the length to be measured.

Inch rules may also be divided into decimal units. A decimal rule divides each inch into ten divisions. Each of these divisions is 0.10 of an inch. If a part being measured lines up with six of these divisions, it measures 0.60 of an inch.

Metric system rules (Figure 3-6) are often subdivided into centimeters and millimeters. Some metric rules are further divided into .5 millimeter spaces. A typical metric rule is 100 millimeters long and every 10th mark is equal to 1 centimeter.

FIGURE 3-5 An inch rule divided into $\frac{1}{32}$ and $\frac{1}{64}$ inch. (Courtesy of L.S. Starrett Company.)

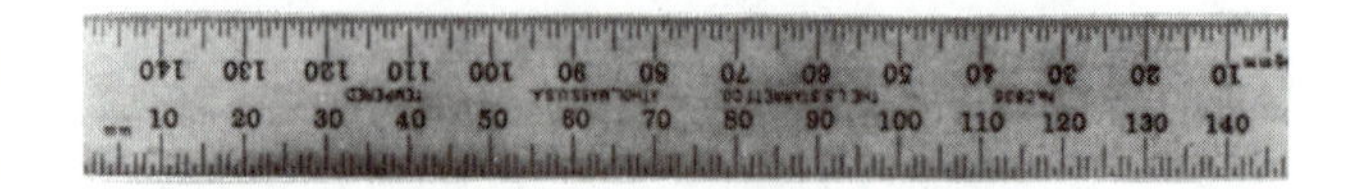

FIGURE 3-6 A metric rule divided into millimeters. (Courtesy of L.S. Starrett Company.)

FIGURE 3-7 A 10-foot-long tape rule. (Courtesy of L.S. Starrett Company.)

A **tape rule** (Figure 3-7) is a metal tape divided into metric or inch units that can be pulled out of a housing. Tape rules come in a variety of lengths and are often used to make long measurements. When not in use, the tape rolls back into the holder.

FEELER GAUGE

A **feeler gauge** (Figures 3-8A and B) is a measuring tool using precise thickness blades or round wires to measure the space between two surfaces. A feeler gauge set has a holder that has many common size flat blades or round wires. This tool is used to measure spaces such as valve clearances and spark plug gaps.

The thickness is written on the gauge blades or wires in thousandths of an inch or hundredths of a millimeter (Figures 3-9A and B). You can measure the space between two parts by finding the blade that fits in the space. The blade should slide in and out of the space. It should touch both sides at the same time without being forced. If the gauge and the space are the same size, the gauge will feel tight as it is moved in and out. This is where it gets the name feeler. The blades should be wiped after every use with a clean oily cloth to remove dirt and prevent rust.

A **stepped feeler gauge** (Figure 3-10) is a type of feeler gauge that has blades with two thickness-

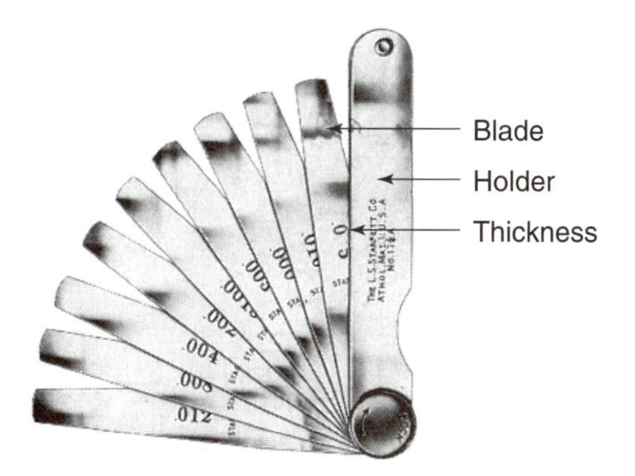

A. Flat Blade Feeler Gauge Set

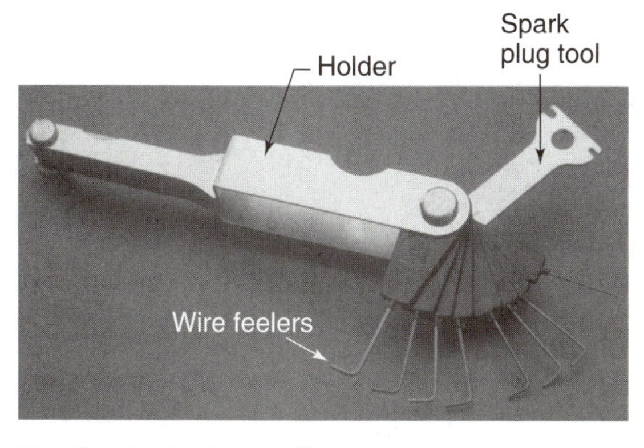

B. Wire Feeler Gauge Set

FIGURE 3-8 Feeler gauges are used to measure the space between two parts. (*A:* Courtesy of L.S. Starrett Company.)

es. The tip of each blade is ground 0.002 inch thinner than the rest of the blade. These are sometimes called *go-no-go gauges*. For example, if the tip of the blade is 0.010 inch thick, the rest of the blade is made 0.012 inch thick. Suppose you use this blade to measure a space and the tip slips in but the rest of the blade does not. You have found that the space is between 0.010 inch (go) and 0.012 inches (no-go) wide.

OUTSIDE MICROMETER

The **outside micrometer**, or mike, is a measuring tool designed to make measurements of the outside of a part. Micrometers are made in different sizes and shapes and divided into either metric or inch units.

The main parts of an outside micrometer (Figure 3-11) are handle, frame, anvil, spindle, sleeve, and thimble. The measuring surfaces are at the ends of the stationary anvil and the movable spindle. The spindle is the end of a screw that threads into the sleeve. The other end of the screw is

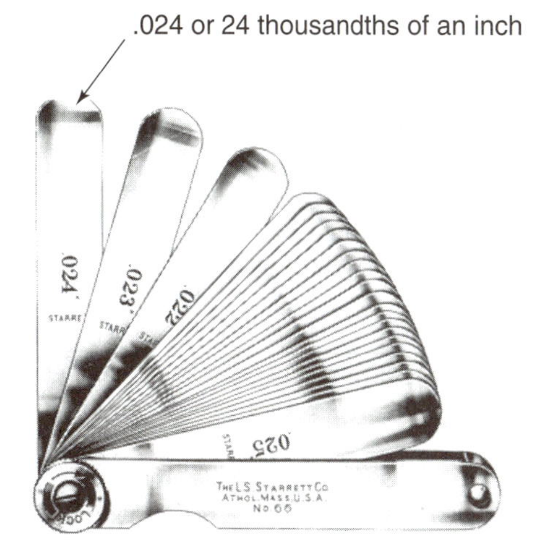

A. Feeler Gauges Marked in Thousandths of an Inch

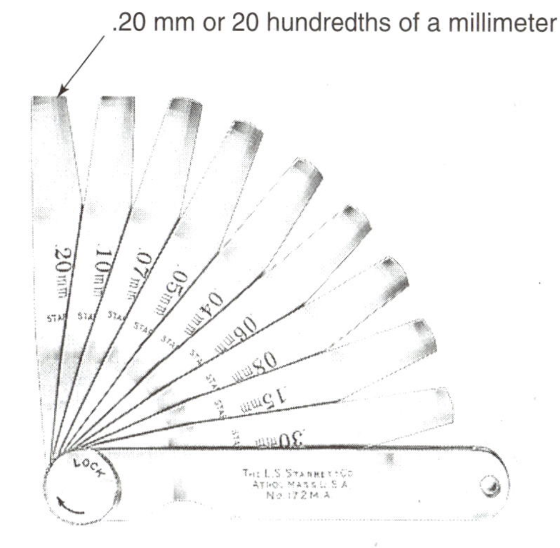

B. Feeler Gauge Marked in Hundredths of a Millimeter

FIGURE 3-9 Feeler gauges are made in inch and metric measurements. (Courtesy of L.S. Starrett Company.)

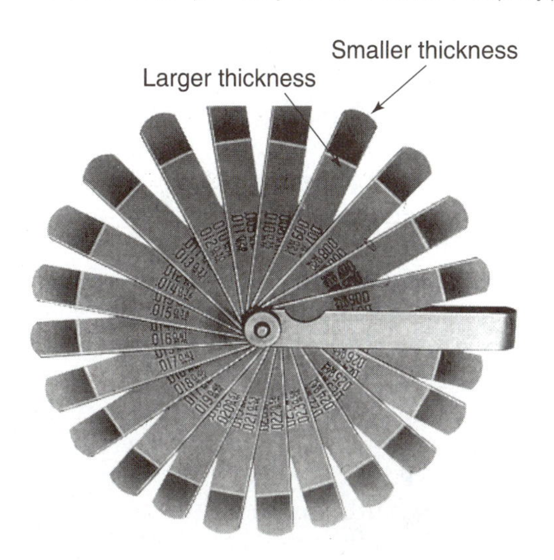

FIGURE 3-10 A feeler gauge can have stepped or different thickness blades. (Courtesy of L.S. Starrett Company.)

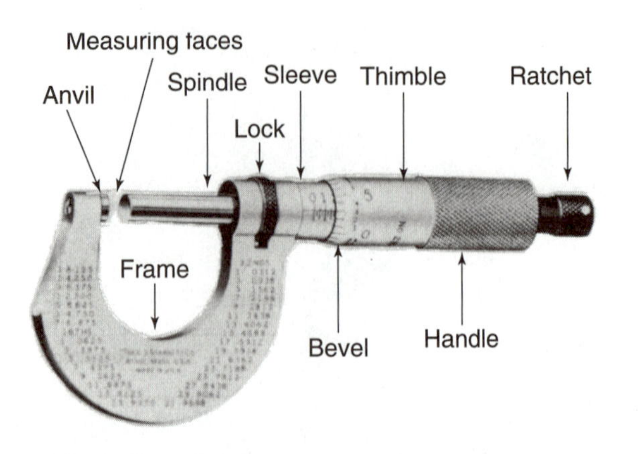

FIGURE 3-11 The parts of an outside micrometer. (Courtesy of L.S. Starrett Company.)

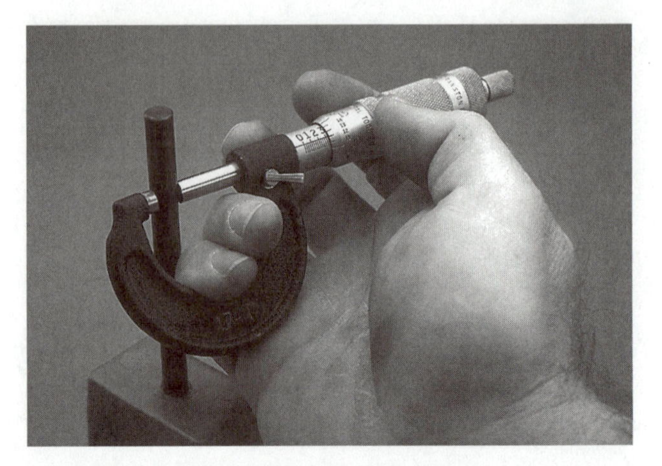

A. Holding a Micrometer with One Hand

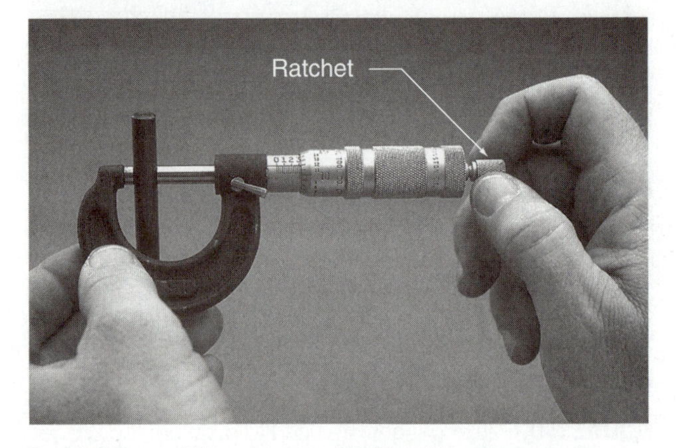

B. Holding a Micrometer with Two Hands

FIGURE 3-12 An outside micrometer can be held with one or two hands.

attached to the thimble. Turning the thimble moves the spindle toward or away from the anvil.

The micrometer must be held properly when measuring (Figure 3-12A). You can hold it with one hand by putting one finger through the frame. Then use your thumb and forefinger to turn the spindle. This position provides good control over the position of the anvil and spindle. Another way to hold a micrometer is to grip the frame with both hands (Figure 3-12B). Remove one hand from the frame to make adjustments with the thimble, then return the hand to the frame. Some technicians prefer to hold the micrometer this way when measuring small objects.

Use a clean rag to wipe any dirt from the measuring surfaces of the spindle and anvil before measuring. Any dirt here will make a large error in the reading. Then wipe off the area of the part to be measured. Place the micrometer over the part in the area to be measured. Gently rotate the thimble to bring the spindle and anvil in contact with the part. The anvil and spindle must be at right angles to the part being measured. If they are not, the reading will not be accurate. Center the spindle and anvil exactly across the diameter to be measured. If you do not, the reading will be undersize. Rock the micrometer slightly as the spindle is turned down the last few thousandths. This will help to get correct positioning.

The amount of force used to tighten the spindle down onto the part is very important. If it is too tight or too loose, you will not get an accurate reading. The spindle and anvil should just contact the part lightly. There should be a slight drag when the micrometer is moved back and forth. If the spindle is forced down too hard, the reading will be incorrect. Even worse, the micrometer can be damaged.

Many micrometers have a ratchet knob on the end of the thimble handle. The ratchet has a friction clutch that allows the spindle to turn until it comes in contact with the part being measured. The clutch part then spins instead of turning the thimble, which prevents overtightening the spindle against the part.

Most micrometers have a lock to prevent the thimble from turning after a measurement. Engaging the lock prevents the reading from changing. When locked, it can be removed from the part.

Reading an Inch Micrometer

You read an inch system micrometer by looking at the divisions on the sleeve and thimble (Figure 3-13). The spindle screw is made so that one revolution moves the spindle 0.025 inch toward or away from the anvil. This means that 40 turns of the screw will move the spindle exactly 1 inch ($40 \times 0.025 = 1.000$).

A scale on the sleeve is divided into 40 divisions. Each division is equal to 0.025 inch. Start

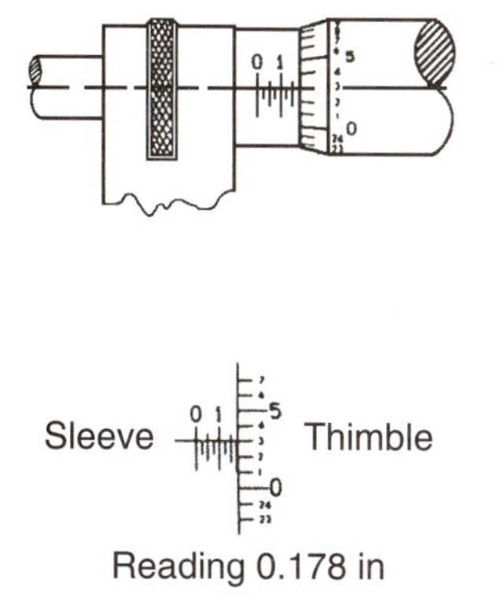

Sleeve Thimble

Reading 0.178 in

FIGURE 3-13 An inch micrometer measurement of 0.178 in. (Courtesy of L.S. Starrett Company.)

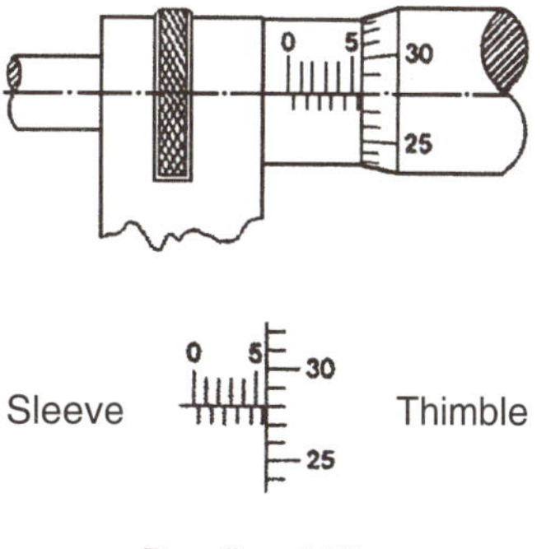

Sleeve Thimble

Reading 5.78 mm

FIGURE 3-14 An metric micrometer measurement of 5.78 mm. (Courtesy of L.S. Starrett Company.)

with the spindle against the anvil and turn the screw out. Every revolution of the thimble uncovers one of the divisions on the sleeve. Every fourth division on the sleeve is numbered. The numbers start with the zero mark when the spindle is against the anvil. The next numbered division is at 0.100 inch (1/10 of an inch) from the closed position. The three unnumbered divisions between zero and one are at 0.025, 0.050, and 0.075 inch.

The bevel on the front of the thimble is also divided into equal parts. Since the thimble and spindle travel 0.025 inch per revolution, there are 25 divisions on the bevel. These divisions allow you to read the amount of spindle travel for a part of a revolution. A revolution from one thimble mark to the next is 1/25 of a revolution. This moves the spindle 0.001 inch.

You read a micrometer measurement by addition. Add together:

1. The last visible numbered division on the sleeve
2. The unnumbered sleeve divisions
3. The divisions on the bevel of the thimble
4. The total is your measurement.

An easy way to remember is to think of the units as making change from a $10 bill. Count the figures on the sleeve as dollars, the vertical lines on the sleeve as quarters, and the divisions on the thimble as cents. Add up the change and put a decimal point instead of a dollar sign in front of the numbers.

The most common size micrometer is 0–1 inch. This size can be used to measure a part that is less than 1 inch. Micrometers are made in other sizes such as 1 to 2 inches, 2 to 3 inches, and larger to measure large components. Larger micrometers are read in the same way as the 0–1 inch micrometer.

Reading a Metric Micrometer

A metric micrometer has the same parts and works just like an inch system micrometer (Figure 3-14). One complete revolution of the thimble moves the spindle exactly 0.500 millimeters. Two complete revolutions of the thimble moves the spindle exactly 1 millimeter. The line on the sleeve is divided into millimeters from 1 to 25 millimeters. Each millimeter is subdivided in half. Two revolutions of the thimble moves the spindle a distance of 1 millimeter.

CASE STUDY

Making an error in reading a precision measuring tool reading can lead to a very expensive mistake. A student in a small-engine class was preparing to machine a cylinder oversize with a boring bar. The new oversize piston had to be measured and the boring bar set up to machine this size. The student measured the piston and set up the boring bar. The cylinder was machined oversize and removed from the boring bar.

When the student tried the piston in the cylinder, there was an obvious problem. The piston was way too loose in the cylinder. The instructor remeasured the piston and checked the setting on the boring bar. The piston measured 3.050 inch. The student had incorrectly measured the piston as 3.500. The cylinder was machined so large that no oversize piston was available to fix the problem. The engine had to be scrapped but the student learned a lesson. Careful measuring saves time and expense.

The beveled edge of the thimble is graduated in 50 divisions. Every fifth line is numbered from 0 to 50. A complete revolution of the thimble moves the spindle 0.5 mm. Each division on the thimble is equal to 1/50 of 0.5 mm or 0.01 mm.

Reading a metric micrometer is similar to the inch type. Add the reading in millimeters you see on the sleeve to the reading in hundredths of a millimeter on the thimble.

INSIDE MICROMETER

The **inside micrometer** (Figures 3-15A and B) is a measuring tool used to measure the inside of holes or bores. You might use an inside micrometer to measure the wear in an engine cylinder. Measuring rods of different lengths come with the micrometer. You can do different ranges of measurement by assembling different rods in the micrometer head.

Inside micrometers are made with both inch and metric system scales. The inside micrometer is read like that of the outside micrometer. More practice is required to get an accurate measurement with an inside micrometer. You must hold it perfectly straight at right angles to the centerline of the hole being measured (Figure 3-16). Then move one end back and forth slightly to get the maximum reading. Always take two or three additional readings as an accuracy check.

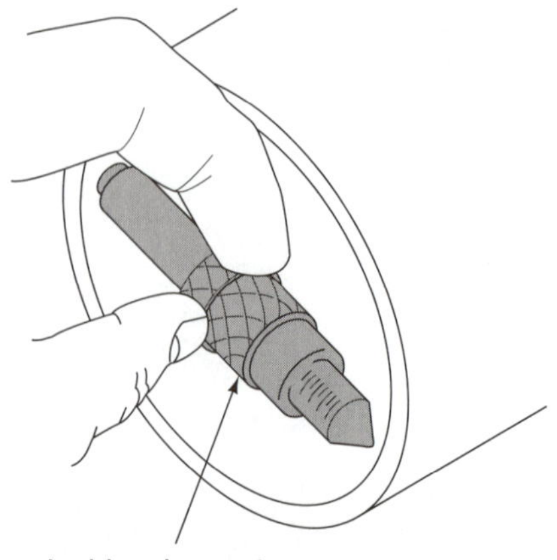

FIGURE 3-16 Using an inside micrometer to measure a bore.

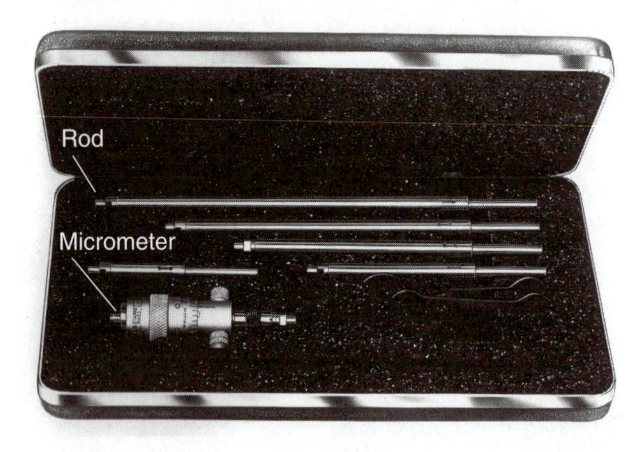

A. Inside Micrometer Set

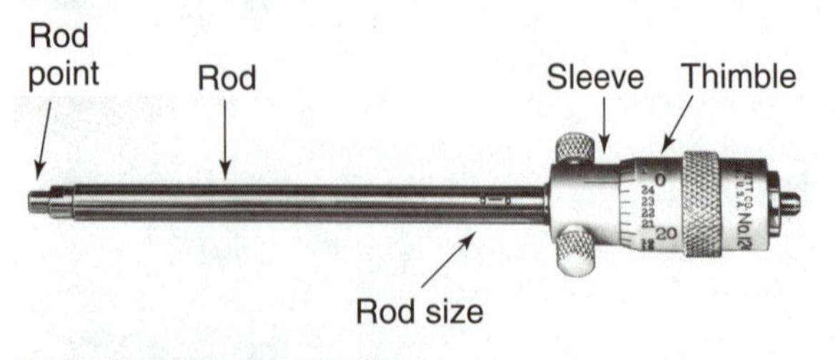

B. Inside Micrometer Parts

FIGURE 3-15 An inside micrometer is used to measure the inside of a bore. (Courtesy of L.S. Starrett Company.)

DEPTH MICROMETER

The **depth micrometer** (Figures 3-17A and B) is a tool used to measure the distance between two parallel surfaces. Depth micrometers have the same thimble and sleeve as other micrometers and are made to take readings in either inch or metric units. They come in sets with different length depth rods for different ranges of measurement.

To measure with a depth micrometer, place the base on one surface to be measured. Then rotate the thimble to move a depth rod down to the lower surface. The distance the rod moves down is the depth. The depth is read just like any micrometer measurement on the thimble and sleeve.

TELESCOPING GAUGE

A **telescoping gauge** (Figure 3-18) is a measuring tool with spring-loaded plungers used together with a micrometer to measure the inside of holes or bores. Telescoping gauges are made in sets to measure from small to very large bores.

The telescoping gauge has a handle that is attached to two spring-loaded plungers. Using the handle, place the telescoping gauge into the bore (Figures 3-19A and B). Release the lock screw on the handle. The spring-loaded plungers will come out and touch the side of the bore. Rock the gauge back and forth to be sure it is square in the bore. Turn the lock screw to lock the plungers in position. Remove the telescoping gauge from the bore. The exact size of the hole is found by measuring across the two ends of the plungers with an outside micrometer.

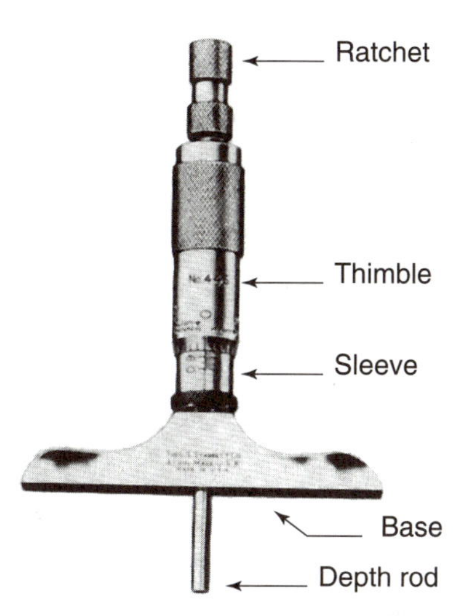

A. Parts of a Depth Micrometer

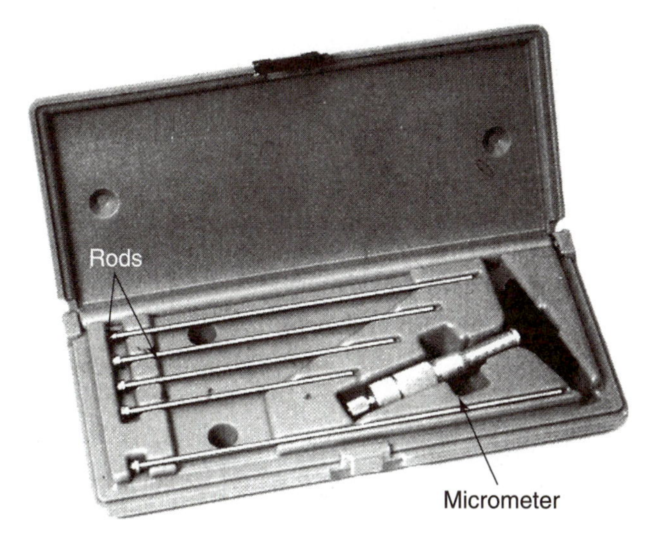

B. Depth Micrometer Set

FIGURE 3-17 A depth micrometer is used to measure a height difference between two surfaces. (*A:* Courtesy of L.S. Starrett Company. *B:* Courtesy of Central Tools Inc.)

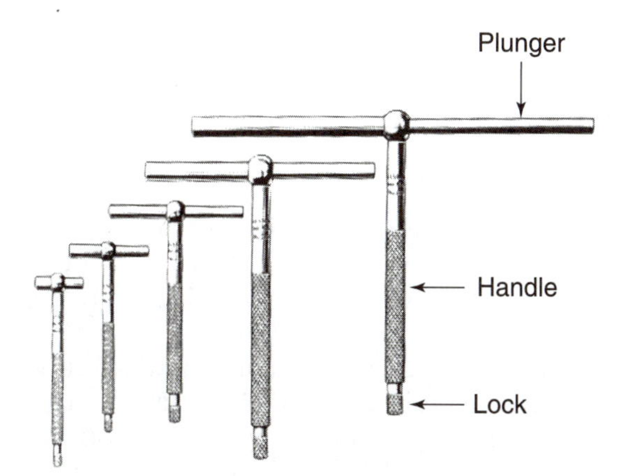

FIGURE 3-18 Telescoping gauges are used to measure the inside of a bore. (Courtesy of L.S. Starrett Company.)

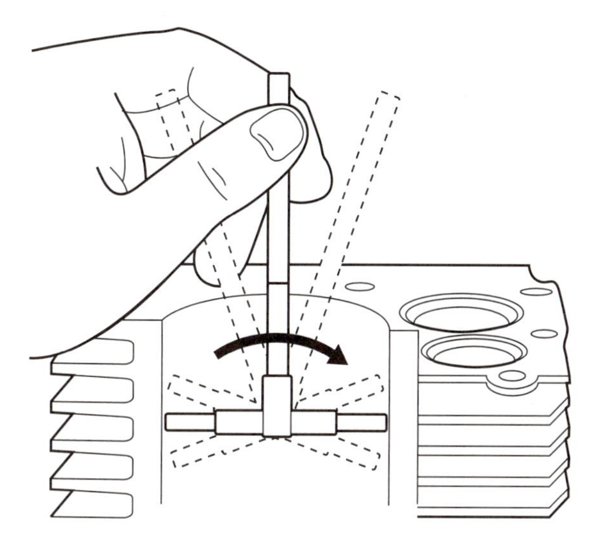

A. Adjusting the Gauge to Bore Size

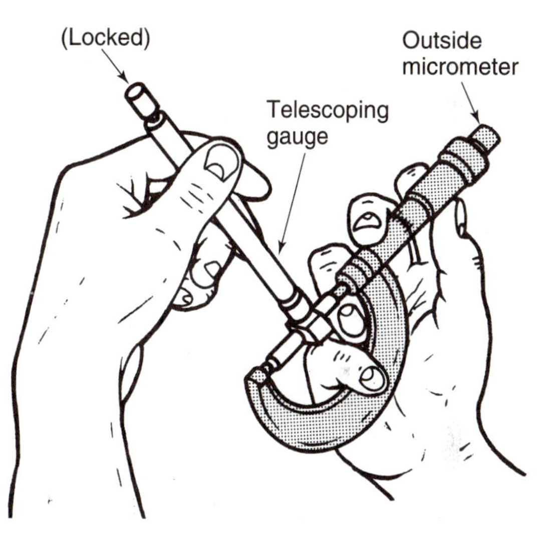

B. Measuring the Telescoping Gauge

FIGURE 3-19 The telescoping gauge is inserted into the bore then measured with an outside micrometer.

SMALL-HOLE GAUGE

A **small-hole gauge** (Figures 3-20A and B) is a measuring tool with a round expandable head that is used together with an outside micrometer to measure the inside of small holes. Parts, such as valve guides, have very small holes. These holes are too small to use an inside micrometer or telescoping gauge. Small-hole gauges are made in sets to cover different diameters of holes.

The small-hole gauge has a round head. The head diameter can be changed with a small wedge. The wedge slides up or down inside the head by turning the handle adjuster. Put the round head of the gauge into the hole to be measured. Turn the handle until the round head touches the sides of the

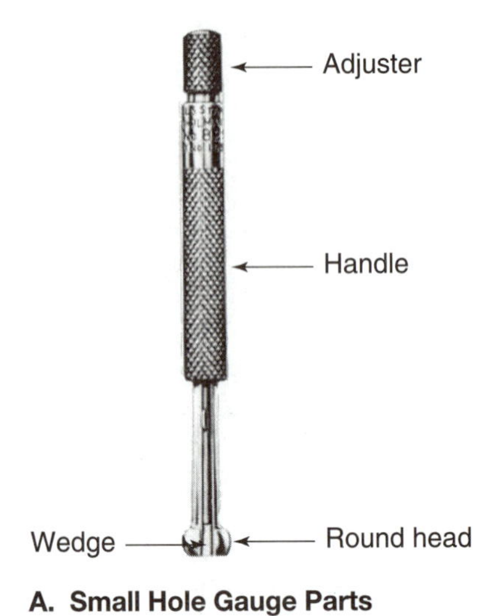

A. Small Hole Gauge Parts

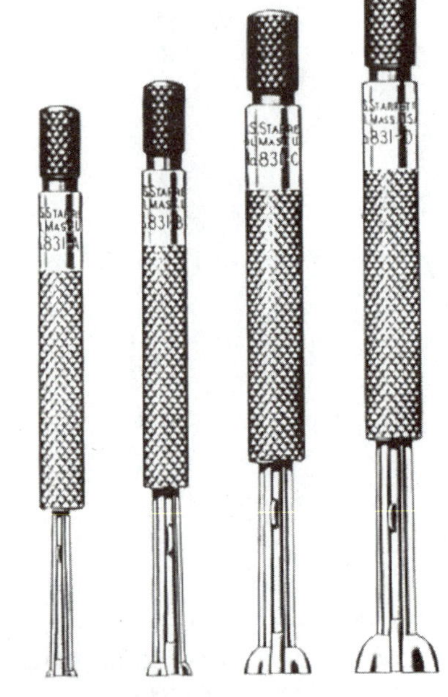

B. Small Hole Gauge Set

FIGURE 3-20 Small-hole gauges are used to measure small holes. (Courtesy of L.S. Starrett Company.)

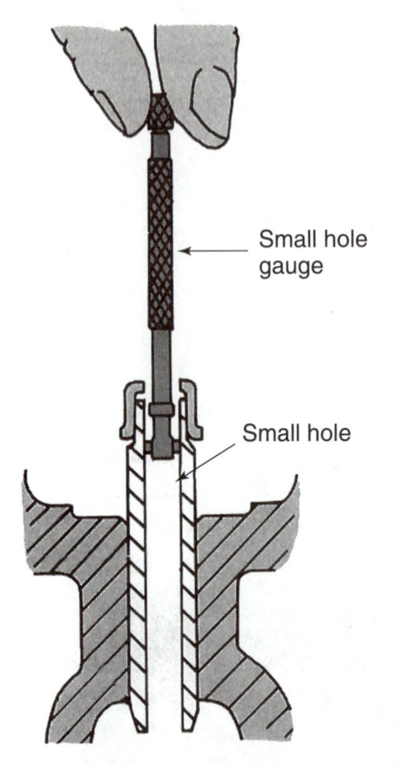

A. Expand the Small Hole Gauge to the Hole Size

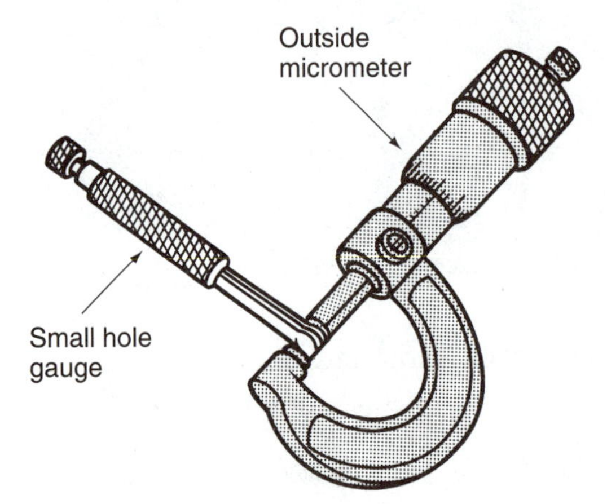

B. Measuring the Small Hole Gauge

FIGURE 3-21 The small-hole gauge is adjusted to the size of the hole then measured with an outside micrometer. (*A:* Courtesy Ford Global Technologies, Inc.)

hole. Remove the gauge from the hole. Use an outside micrometer to measure the diameter of the expanded round head (Figures 3-21A and B). The size of the head is the same size as the hole.

DIAL INDICATOR

The **dial indicator** is a tool that measures very small movements on the face of a dial. The movement may be called *play, end play, free play,* or *run out.* Dial indicators are made that measure in inch or metric units. The dial indicator is used with attachments and support arms that allow it to be mounted on the part to be measured. They may be attached to the part with small special C clamps. They may be attached with magnetic mounting bases.

The dial indicator uses a small rod called a *plunger* connected to a pointer by a gear. Plunger movement is shown by pointer movement on the dial face (Figure 3-22). The parts inside the dial increase the movement of the pointer, which allows you to see small measurements. The measurement

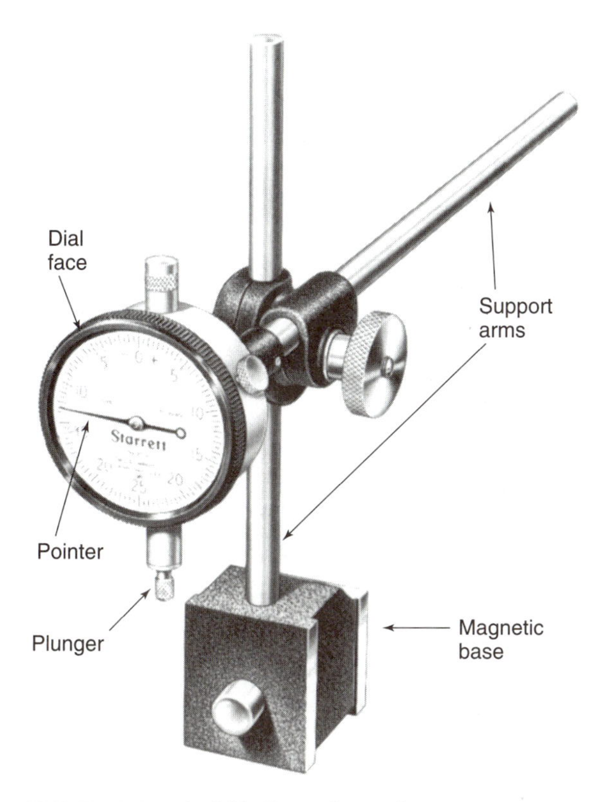

FIGURE 3-22 A dial indicator is used to measure movement with a plunger connected to a pointer. (Courtesy of L.S. Starrett Company.)

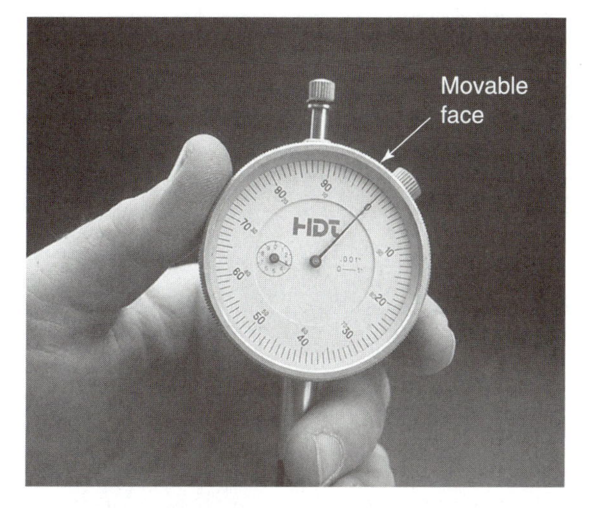

A. Setting the Dial Indicator Face to Zero

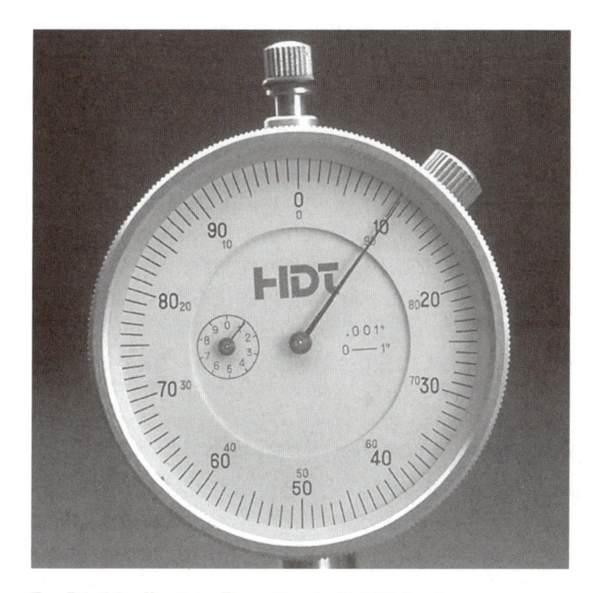

B. Dial Indicator Reading is 0.010 Inch

FIGURE 3-23 The movable face is rotated to align the pointer with zero then the play can be measured by the needle.

shown by the pointer may be in either thousandths of an inch or hundredths of a millimeter. The dial face on some indicators is set up so you read both to the right or to the left of 0. Others are set up to read from the 0 only in a clockwise direction.

When you mount a dial indicator, adjust the support arms as short as possible. The dial indicator mounting must be rigid. You can get an inaccurate reading if the arms are too long or the clamping is too loose. Mount the indicator plunger so that it is at 90 degrees against the part. Always read the dial indicator looking straight on the face. Looking at it from the side can cause a visual error. Remember that a dial indicator is a precision instrument like a watch. It must be handled with great care.

Reading an Inch Dial Indicator

Mount the dial indicator on the part to be measured. Push the spring-loaded plunger against the part to be measured. Push until the needle moves a small distance. This is called preloading the indicator. After preloading, rotate the movable dial until the zero of the dial face lines up with the pointer (Figure 3-23A).

The typical inch-reading dial indicator face has graduations that start at zero at the top. There are 50 divisions to the bottom of the dial. The divisions continue back up to the top. Each division is equal to 0.001 inch (one thousandth of an inch). Each ten divisions on one side of the face is marked as 10, 20, 30, 40, and 50. The same numbers are shown on the other side so that movement on both sides of zero can be measured.

Move the part to be measured back and forth against the plunger while looking at the dial face. The measurement is shown by pointer movement to the left or right of zero. For example, the pointer moves to the left to 0.010 inch (Figure 3-23B). Then it moves past zero to the right until it lines up

A. Metric Dial Indicator Reading Zero **B. Dial Indicator Reading 0.10 mm**

FIGURE 3-24 The metric dial indicator is preloaded and read similar to the inch type.

with the 0.010 inch. The reading is found by adding both sides of the reading together. The left movement is 0.010 inch. The right movement is 0.010 inch. The total reading is 0.020 thousandths of an inch. Movement in one direction is measured by looking at the single movement of the pointer.

Reading a Metric Dial Indicator

A metric dial indicator is mounted, preloaded, and read just the same as the inch dial indicator. The only difference is that the units on the face are divided in metric units (Figures 3-24A and B). Usually, there are 50 divisions on one side of the face and 50 on the other. Each ten divisions are marked as 10, 20, 30, and 40. There is a zero at the top and a 50 at the bottom. Each of these divisions is 0.01 mm (one hundredth of a millimeter).

To measure with the metric dial indicator, preload the plunger. Rotate the face until the pointer is lined up with the zero on the dial face. Move the part to be measured while looking at the reading on the dial face. For example, the needle swings to the left during part movement until the pointer points to 0.10 millimeter (ten hundredths of a millimeter). Then it swings to the right of the zero until it lines up with the 0.10 millimeter (ten hundredths of a millimeter) line. The measurement is the total of the two readings or 0.20 millimeter (20 hundredths of a millimeter). Movement in one direction is measured by looking at the single movement of the pointer.

REVIEW QUESTIONS

1. List and describe the common fractional divisions of the inch in the inch system of measurement.
2. List and describe the common decimal divisions of the inch in the inch system of measurement.
3. Describe how to convert from units in the inch to units in the metric measuring system.
4. Explain how to use a feeler gauge to measure the space between two surfaces.
5. Explain how to use the thimble and sleeve readings to read an inch micrometer.
6. Explain how to use the thimble and sleeve readings to read a metric micrometer.
7. Describe the two ways to hold an outside micrometer while making a measurement.
8. Explain how an inside micrometer is used to measure inside a bore.
9. Explain how a depth micrometer is used to measure the difference in height between two surfaces.
10. Explain how a telescoping gauge is used to measure inside a bore.

11. Explain how a small-hole gauge is used to measure inside a small hole.
12. Describe how to preload a dial indicator plunger.
13. Describe how to set up and read an inch dial indicator.
14. Describe how to read a dial indicator measurement when the needle swings from one side of zero to the other side.
15. Describe how to set up and read a metric dial indicator.

DISCUSSION TOPICS AND ACTIVITIES

1. Use an inch and metric rule to measure:

   Thickness of a quarter
   Diameter of a quarter
   Width of your thumb
   Length of your shoe

2. Use a metric and inch outside micrometer to measure:

   The thickness of a hair
   The diameter of a paper clip
   The thickness of a pencil lead

3. Use a conversion chart to convert the following inch measurements to metric units:

   1 inch
   2 feet
   55 miles

Measuring Electricity, Pressure, and Vacuum

OBJECTIVES

Upon completion and review of this chapter, you should be able to:

- Define and explain the terms *amperage, voltage,* and *resistance.*
- Use a digital multimeter to measure amperage, voltage, resistance, and continuity.
- Use a pressure gauge to measure pressure.
- Use a vacuum gauge to measure vacuum.

TERMS TO KNOW

Amperage	Multimeter
Analog meter	Pressure gauge
Continuity	Resistance
Digital meter	Vacuum gauge
Electrical circuit	Voltage

INTRODUCTION

The measuring tools that have been described previously are all used to measure length or distance. Outdoor power equipment technicians often have to make other types of measurements. The electrical equipment on snow throwers, tillers, and garden tractors require that the technician understand and be able to measure electricity. Technicians also often have to measure air pressure on tires and engine intake manifold vacuum.

MEASURING ELECTRICITY

Servicing the engine's ignition system and electrical accessories requires that you be able to measure electricity. Three terms used to describe electricity are amperage, voltage, and resistance.

Electricity can be difficult to understand because it cannot be seen. It may be helpful to think of electricity like water in a water system (Figure 4-1). The water in your house starts in a water tank or reservoir. This is the water source. The water flows through pipes to get to your home. The water reservoir creates a pressure to move water to your house. Water pressure is measured in pounds per square inch (psi). In the same way, there must be an electrical pressure to move current through a wire. The pressure to move electrical current is called voltage. **Voltage**, abbreviated V, is the amount of pressure pushing the current through an electrical circuit. Voltage is measured in volts.

The rate of water flowing through a pipe is measured in gallons per minute. The rate of electrical flow though an electrical system is called *current flow*. **Amperage,** abbreviated A, is the rate of current flowing in a wire. Amperage is measured in amps. Large water pipes are used when a large flow of water is needed. Large wires are used when a large amount of current is needed.

An open water faucet offers very little resistance to the flow of water. As the faucet is closed, it offers more and more resistance. A closed faucet stops water flow. An open faucet allows the most flow. The same thing happens in an electrical circuit. **Resistance** is the electrical resistance to current flow in a circuit. Resistance is measured in ohms.

Digital Multimeter

Analog or digital electrical test meters are used to measure amps, volts, and ohms in a circuit. The **analog meter** (Figure 4-2A) is an older style test instrument that is read by observing a pointer in re-

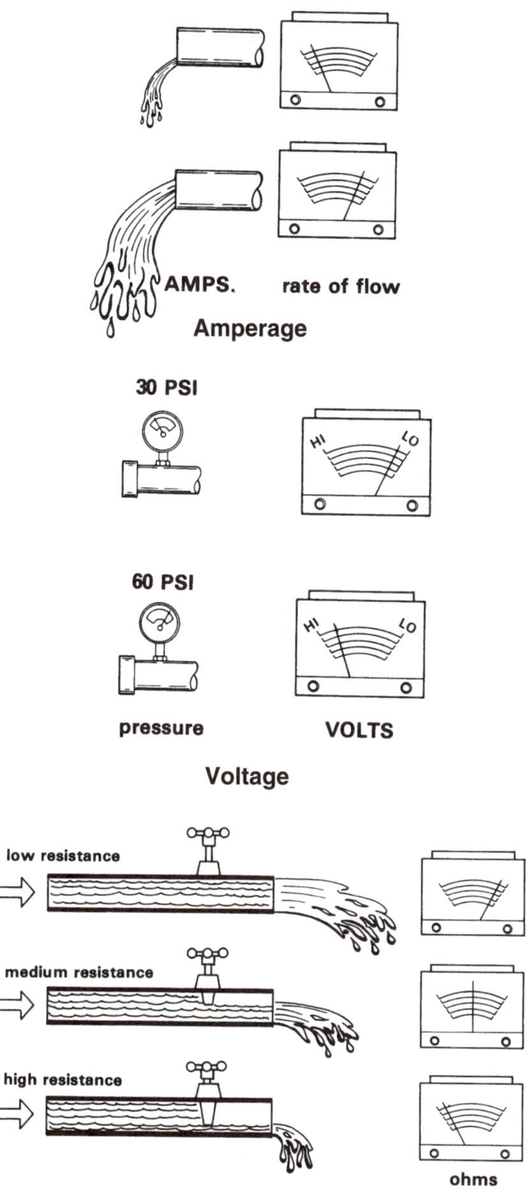

FIGURE 4-1 Water flow through pipes can be compared to voltage, amperage, and resistance. (Provided courtesy of Tecumseh Products Company.)

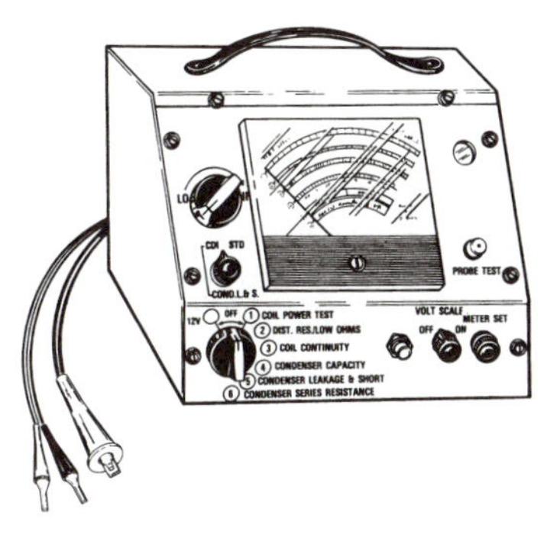

A. Analog Meter

B. Digital Meter

FIGURE 4-2 Analog or digital meters can be used to test electrical parts. (*A:* Provided courtesy of Tecumseh Products Company. *B:* Courtesy of TIF Instruments, Inc.)

lation to a scale. For example, a wrist watch with moving hands is called an analog watch. Most meters, like many wrist watches, have a digital readout. The **digital meter** (Figure 4-2B) is a test instrument that shows test results as numbers on a small screen. These numbers show the exact reading in volts, amps, or ohms. Digital and analog meters work the same way and are connected the same way.

The most common digital meter used for electrical service is called a **multimeter** (Figure 4-3). A multimeter, or digital volt-ohmmeter (DVOM), is an electrical test instrument used to measure voltage, amperage, resistance, and continuity. A selector switch on the front allows you to switch from one type of measurement to another.

The multimeter has several scales and a set of test leads or wires. The leads have either probes with points or alligator clips on the end that are used to connect the multimeter into electrical circuits. One wire is red and is called the positive (+) lead. The other wire is black and is called the negative (–) lead.

CAUTION: Before connecting test leads from a meter to a circuit, be sure to select the correct scale and function on the meter. If not, the meter or the electrical system being serviced can be damaged.

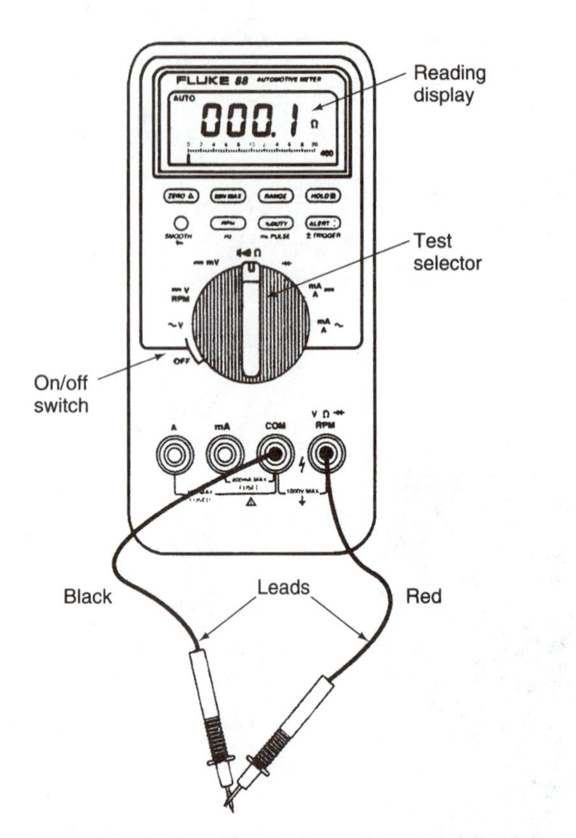

FIGURE 4-3 A multimeter is used to measure voltage, amperage, continuity, and resistance in a circuit by selecting different functions on the meter.

Before you use a multimeter, study its instruction manual. There are many types of meters. They may have different scales and switch positions. Know what each switch, scale, and button on the meter does before you use the meter. Using a multimeter is a five step process:

1. Determine the required function and set the selector switch in position to measure that function.
2. Set the meter function switch to the position to measure the highest quantity expected.
3. Connect the meter test leads to the circuit or component.
4. Read the meter by observing the numbers on the screen display.
5. Disconnect the meter test leads from the circuit or component.

Meter switches, scales, and displays often use abbreviations and symbols (Figure 4-4). An abbreviation is a letter or combination of letters that stand for a word. For example, V is often used for voltage. A symbol is a picture that shows an idea. Symbols are often used on meters because they do not depend on any language a person speaks. For example, the symbol for a positive connection is the + sign.

Measuring Voltage

Voltage in a circuit is measured with the voltage function of the multimeter. Leave the meter in the off position until it is connected into the circuit. When measuring voltage, always connect a meter across the circuit in parallel. Connect the leads with the correct polarity. Connect the positive red tester lead (+) to the positive (+) side of the circuit. Connect the negative black tester lead (–) to the negative (–) side of the circuit.

Move the meter selector switch to the voltage measuring position (Figure 4-5). Many meters have the ability to measure both alternating current voltage (AC) and direct current voltage (DC). Alternating current voltage is the type in home outlets and in the output of portable generators. Direct current is the type in engine ignition and starting systems.

Common Meter Abbreviations

AC	Alternating current	ma	Milliamperes
DC	Direct current	µA	Microamperes
V	Volts	kΩ	Kilohms
mV	Millivolts	mΩ	Megohms
kV	Kilovolts	µF	Microfarads
A	Amperes	nF	Nanofarads

Common Meter Symbols

ϟ	Dangerous or high voltage		Diode
~	AC	)))))	Diode test
⎓	DC		See service manual
	AC or DC	⚠	See manual for explanation
+	Positive	H	Hold
–	Negative	)))))	Audio beeper
⏚	Ground		Capacitor
±	Plus or minus	%	Percent
Ω	Ohms-resistance		

FIGURE 4-4 Common abbreviations and symbols used on multimeters.

FIGURE 4-5 Setting the selector switch on the voltage measuring position.

The next step is to select the correct voltage range. Range is the amount of voltage to be measured. The multimeter will have several voltage ranges such as 200 millivolts (a millivolt is one thousandth of a volt), 2 volts, 20 volts, 200 volts, and 1,000 volts. Always select a range that is higher than the voltage you expect to measure. This will prevent damage to the meter. For example, you expect a test reading of 12 volts. You should set the test selector switch at the 20 volts scale.

After the correct range is selected, switch the multimeter to the on position. The digital reading on the meter now displays the difference in voltage between the points where the voltmeter leads are attached. When you are finished, turn the meter off. Remove the test leads from the circuit.

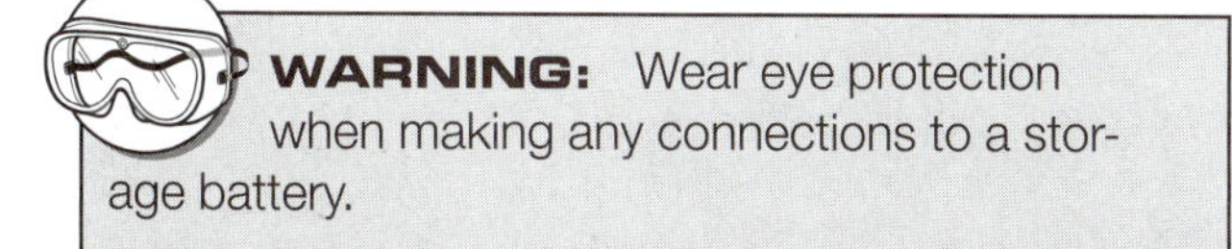

WARNING: Wear eye protection when making any connections to a storage battery.

Another common use of the multimeter voltage function is measuring the voltage in a 12-volt battery (Figure 4-6). First, locate the positive and negative terminals on the battery. Negative terminals are marked "Neg." or may have a – sign. The positive terminal is often larger. It may be marked "Pos." or have a + sign. Move the selector switch to the voltage function position to measure 20 volts. Touch the end of the red, positive lead from the multimeter to the positive terminal of the battery. Touch the end of the black, negative lead from the multimeter to the negative terminal of the battery. Turn the meter to the on position and look at the display. The digital display shows the amount of voltage in the battery.

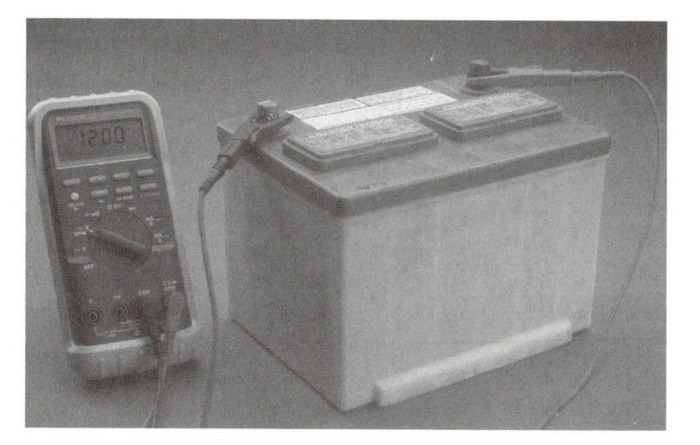

FIGURE 4-6 Measuring battery voltage with a multimeter.

Measuring Amperage

The amperage function of a multimeter is used to measure amperage in an electrical circuit (Figure 4-7). The circuit must be disconnected, or broken, to connect the ammeter in series. A common test is to measure current flow into a battery.

Make sure the power is off in the circuit before connecting the test meter. Turn the meter off before connecting it to the circuit. Disconnect the negative terminal of the battery. Connect the positive lead of the multimeter to the battery cable. Connect the negative lead of the multimeter to the battery terminal.

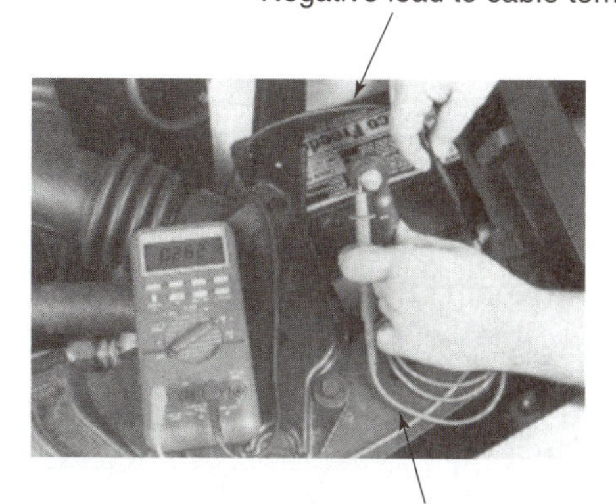

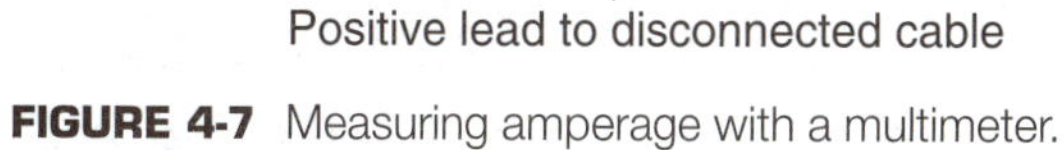

FIGURE 4-7 Measuring amperage with a multimeter.

> **CASE STUDY**
>
> The low voltage created by a battery will not shock but it can cause a painful burn. A student was removing a battery for testing from a lawn and garden tractor. The instructor heard a yelp of pain and went to investigate. The student was holding her hand and appeared to be in considerable pain. The instructor administered first aid and got the story. She was using a wrench to loosen a battery cable connection. The wrench completed a circuit between the battery and a ring on her finger. The current flow through the gold ring caused it to get red hot. The red hot ring left a very painful burn on her finger. The instructor used this incident to remind students that jewelry can be very unsafe in the shop.

Set the tester selector switch to the correct DC amperage function. There may be three or more ranges. Select the highest amperage range. Turn the meter power on. Turn the circuit you are testing on and read the amperage. Read the amperage on the display. Turn off the circuit power. Turn the meter to off. Disconnect the meter from the circuit.

> **CAUTION:** Ammeters must always be connected in series with a circuit and with correct polarity. The ammeter has a very low resistance and if connected incorrectly, it may be damaged. Always start measuring using the highest range on the ammeter.

Measuring Resistance

The resistance of a wire or electrical part is another common multimeter measurement. The component or wire must be disconnected from the circuit voltage before testing resistance.

The multimeter has several resistance scales. The scales can be selected by the test selector switch. They are commonly marked R × 1, R × 100, and R × 10,000. The R × 1 scale is used for very low resistance measurements. The R × 100 scale is used for resistance measurements in hundreds of ohms. The R × 10,000 scale is used for high resistance in the thousands of ohms. Always start on the highest resistance scale and then select lower scales as required.

Many multimeters require that the test leads be removed from the part of the meter used to check voltage or current and be plugged into a special jack to measure ohms. Check the tester manual for correct tester lead hook up.

> **CAUTION:** Always disconnect the wire or component from the circuit before testing for resistance. Circuit voltage can damage the multimeter ohmmeter circuit.

To measure the resistance of a wire, remove the wire from the circuit (Figure 4-8). Adjust the function switch to the correct function and highest range. Connect one tester lead to one end of the wire. Turn the multimeter on and connect the other tester lead to the other end of the wire. Either lead can be used because the polarity is not important when testing the resistance of a wire. The meter shows the resistance of the wire in ohms. Turn off the meter and disconnect the meter test leads.

Measuring Continuity

An **electrical circuit** is a complete path for current flow in an electrical system. The circuit must be unbroken or continuous to have a current flow. **Continuity** is a condition in an electrical circuit in which there is a complete path for current flow. Often you must find out if a circuit is complete, or has continuity. You may need to find out if a circuit component such as a fuse, wire, or other electrical component has an open circuit or "open." If a part or circuit has continuity it will allow electrical cur-

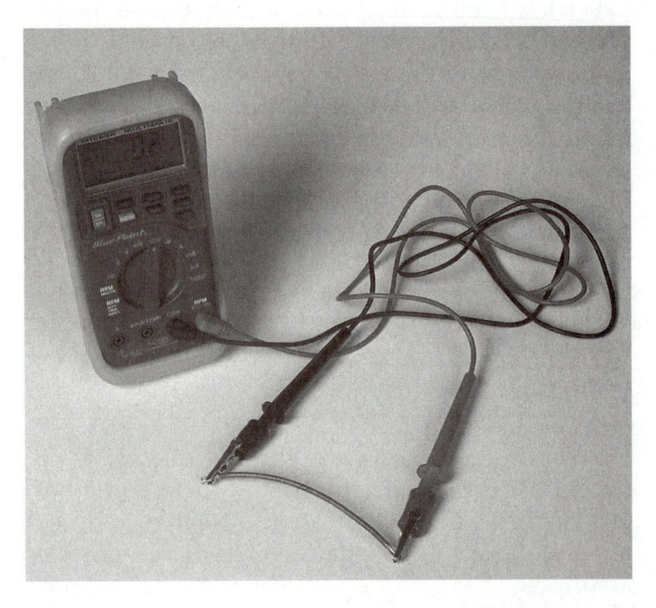

FIGURE 4-8 Measuring resistance with a multimeter.

rent to flow. If an electrical part or wire does not have continuity, the circuit will not work.

Continuity can be measured with a multimeter. The first step in measuring continuity is to remove the part to be tested from the circuit. The continuity part of the multimeter can be damaged if a part has circuit voltage. If a fuse or wire is to be checked for continuity, both ends of the wire or fuse should be disconnected from the circuit. The meter sends a small current through the part or circuit from its internal battery. A buzzer in the meter beeps anytime there is continuity.

To check continuity of a wire, turn the selector switch to the continuity test position (Figure 4-9). Turn the meter to the on position. Test the meter operation by touching the two test leads together. The buzzer will beep showing that there is continuity through the meter leads.

Touch one of the test leads to one end of the fuse and hold it in place. Either lead can be used because the polarity is not important when testing the continuity of a fuse or wire. Touch the other lead to the other end of the fuse. The buzzer will beep if the fuse has continuity. If the buzzer does not beep, try moving the leads around for better contact. If there is no beep, the wire or fuse is defective and is open.

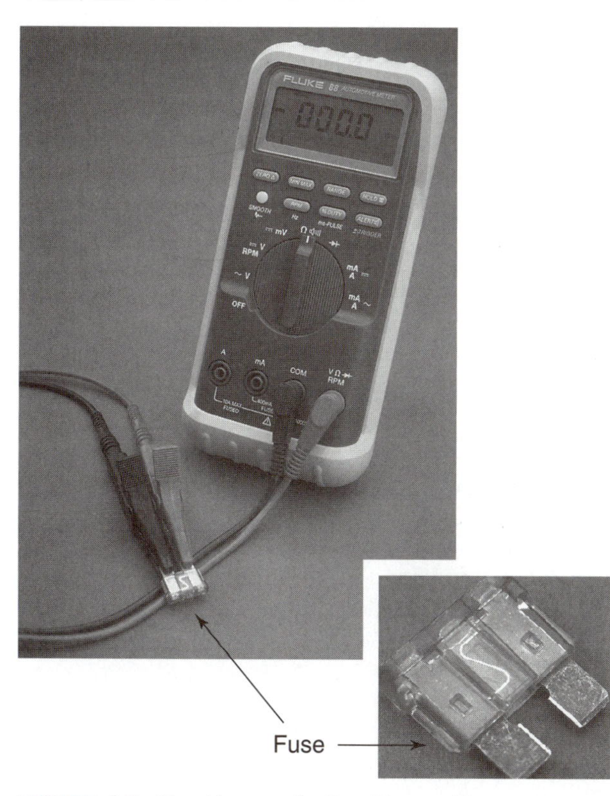

FIGURE 4-9 Checking continuity with a multimeter.

CAUTION: The continuity function of a multimeter has its own battery. The electrical component to be tested must always be removed from the circuit. Do not connect a continuity tester to an external voltage source or it will be damaged.

MEASURING PRESSURE

Pressure measurements such as tire air pressure or cylinder compression pressure are made with a **pressure gauge.** An inch system pressure gauge measures in psi or pounds per square inch. A metric pressure gauge measures in kPa or kilopascals.

A dial type tire air pressure gauge (Figures 4-10A and B) is an example of a pressure measuring tool. A hose on the tool is connected to the pressure source. This might be a tire or an engine cylinder. Pressure comes into the gauge from the hose. The pressure gauge has a face divided into inch or metric units. The

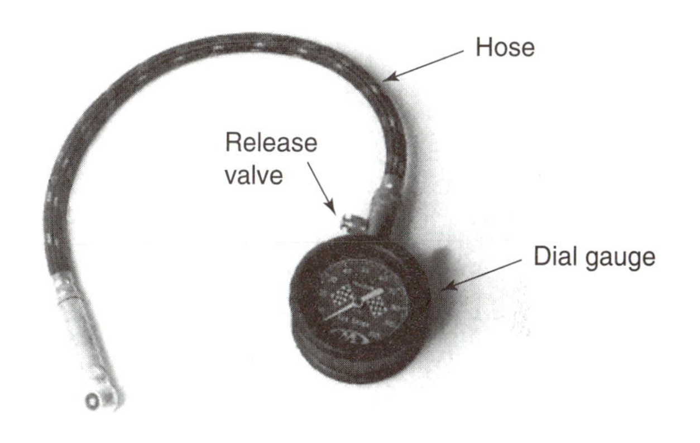

A. Dial Type Pressure Gauge

B. A Pressure Reading of 32 psi

FIGURE 4-10 A dial type tire gauge is an example of a pressure gauge.

pressure causes the gauge needle to rotate and show the pressure. A release valve on the hose connection allows you to release the pressure after testing.

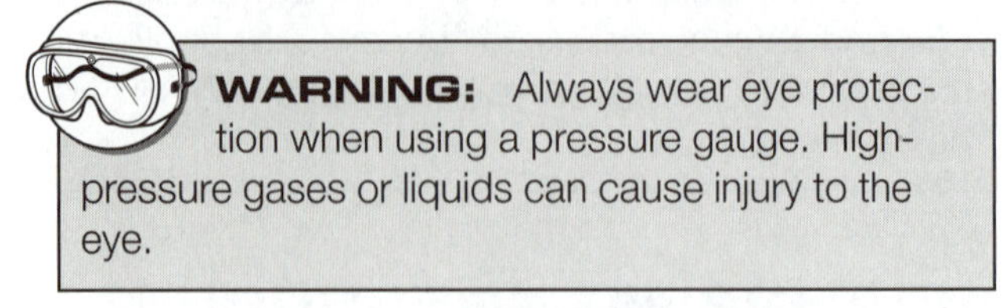

WARNING: Always wear eye protection when using a pressure gauge. High-pressure gases or liquids can cause injury to the eye.

MEASURING VACUUM

When an engine is running, the intake stroke in the cylinders creates a low pressure. This low pressure pulls in the mixture of air and fuel. In engine work this low pressure is often called a "vacuum." A true vacuum is a complete absence of air and is found only in the laboratory or out in space. Engine intake manifold vacuum is really just a very low pressure. A **vacuum gauge** is a tool used to measure low pressure such as intake manifold vacuum (Figure 4-11). Gauges for measuring this pressure are typically calibrated in inches H_2O (inches of water) and inches Hg (inches of mercury).

FIGURE 4-11 A low pressure gauge is used to measure intake manifold vacuum. (Courtesy of Mac Tools Distributors.)

REVIEW QUESTIONS

1. Define and explain the amperage in an electrical circuit.
2. Define and explain voltage in an electrical circuit.
3. Define and explain resistance in an electrical circuit.
4. Explain why you should select a higher range when setting up a multimeter.
5. List the steps in connecting a multimeter to measure voltage.
6. List the steps in connecting a multimeter to measure amperage.
7. List the steps in connecting a multimeter to measure resistance.
8. Explain why continuity is needed in an electrical circuit.
9. List the steps used to measure continuity in an electrical wire.
10. Explain the purpose of a pressure gauge.

DISCUSSION TOPICS AND ACTIVITIES

1. Locate some scrap electrical parts. Use a multimeter to check the resistance of the parts.
2. Locate a selection of electrical wires. Use a multimeter to check each wire for continuity.
3. Locate a selection of electrical fuses. Use a multimeter to check each fuse for continuity.

CHAPTER 5

Threaded Fasteners

OBJECTIVES

Upon completion and review of this chapter, you should be able to:

- Identify and explain the use of the common types of threaded fasteners.
- Explain how inch and metric threads are sized and designated.
- Explain the purpose and use of inch and metric fastener grade markings.
- Describe the steps to follow to determine a thread size.
- Repair damaged internal and external threads.
- Describe the types and uses of thread compounds.

TERMS TO KNOW

Allen head screw
Antiseize compound
Belleville lock washer
Binding head screw
Bolt
Castellated nut
Clutch head screw
Conical flat washer
Drive nut
Flanged cap screw
Flat head screw
Flat washer
Grade markings
Helical spring lock washer
Hex flange nut
Hex-head cap screw
Hex nut
Jam nut
Lock washer
Nut
Penetrating fluid
Phillips head screw
Pitch gauge
Screw
Self-tapping screw
Set screw
Single thread lock nut
Single thread nut
Slotted screw
Square nut
Stud
Stud remover
Tensile strength
Thread diameter
Threaded fastener
Thread-locking compound
Thread pitch
Thread series
Toothed washer lock nut
Toothed lock washer
Truss head screw
Washer
Wing nut

INTRODUCTION

Outdoor power equipment is assembled with bolts, nuts, screws, and other small parts called fasteners. Fasteners may be threaded or nonthreaded. Nearly every engine repair requires you to remove and replace these fasteners. Fastener use is very important to each repair. Problems like a damaged screw can cause complete engine failure.

THREADED FASTENERS

A **threaded fastener** is a type of fastener that uses the wedging action of threads to hold two parts together. Common threaded fasteners are bolts, screws, studs, and nuts. Bolts, screws, and studs have outside or external threads; nuts have inside or internal threads (Figure 5-1).

Screws

A **screw** (Figures 5-2A and B) is a fastener that fits through a hole in one part and into a threaded hole in a second part to hold the two parts together. A screw has a head at one end and a threaded shank at the other end. The threads on the screw fit into

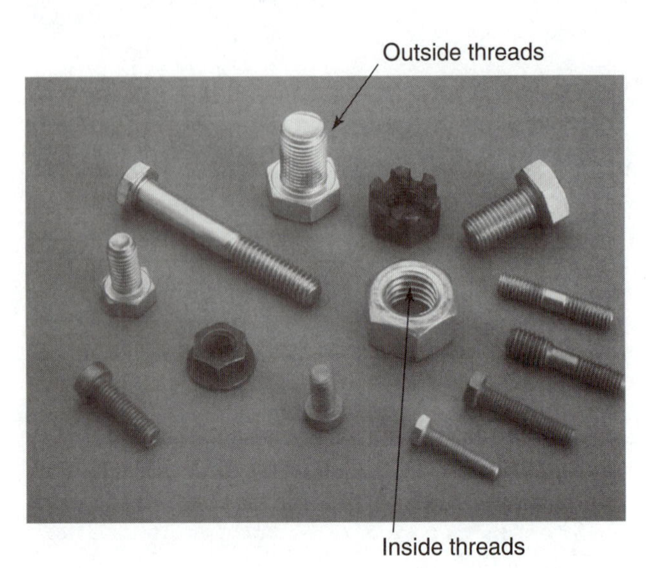

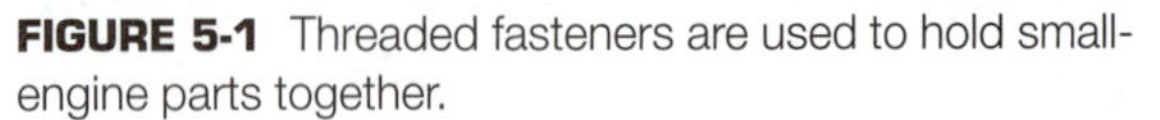

FIGURE 5-1 Threaded fasteners are used to hold small-engine parts together.

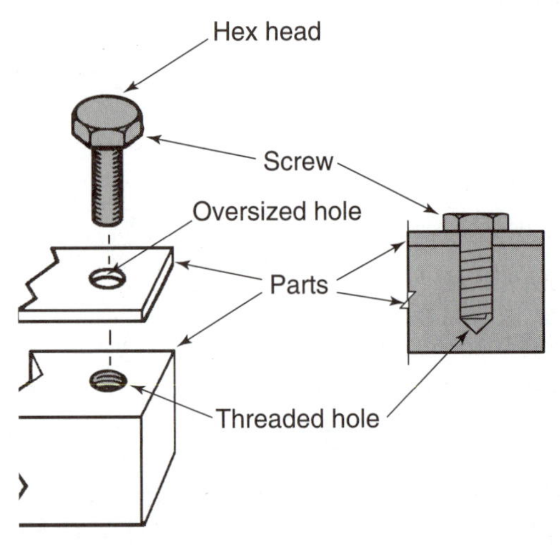

A. Screw Fits in Threaded Hole

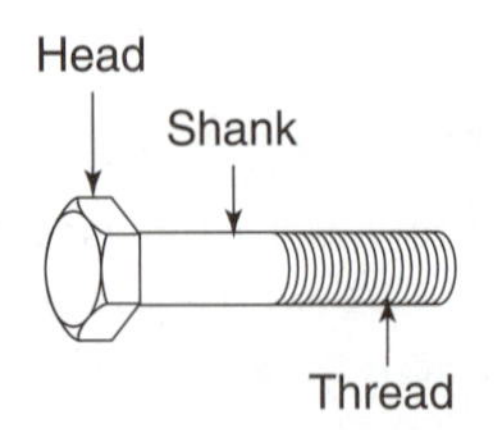

B. Parts of a Screw

FIGURE 5-2 A screw fits through a hole in one part and threads into the second part.

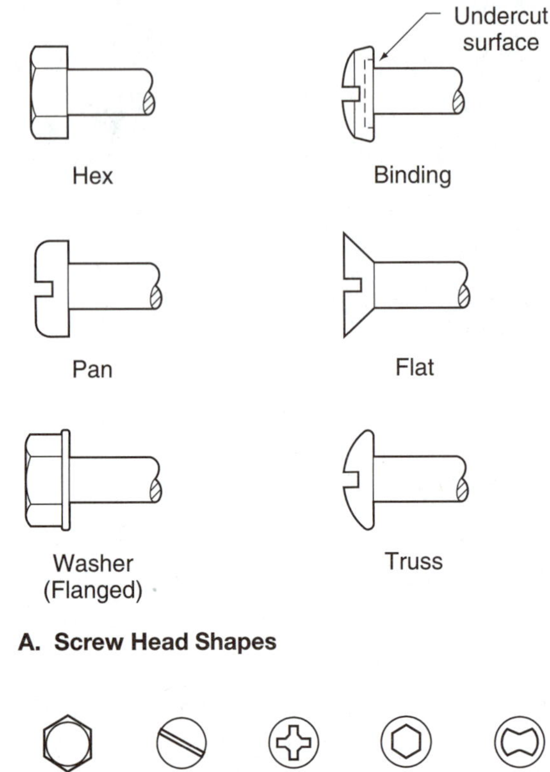

FIGURE 5-3 Screws are identified by the shapes of the heads and types of slots in the head.

the threads in the hole. The wedging action of the threads holds the parts together.

Screws are identified by the shapes of their heads (Figure 5-3A). The **hex-head cap screw**, or cap screw, is a screw with a hexagonal head. The cap screw is the most common screw used in engine work. It is used to hold larger parts like the cylinder head on the engine. The box, open-end, combination, and socket wrenches are made to drive the cap screw. The **flanged cap screw** is a cap screw that has a flange formed under the head. The flange eliminates the need for a washer.

There are many other types of screws used in outdoor power equipment work. The **flat head screw** is a screw with a head made so it can fit flush or flat with the surface. The flat head screw fits into a tapered hole in the part. The hole is called a *countersink*. The **truss head screw** is a screw with a large head for extra holding power. Truss head screws are often used to hold parts made from aluminum. The **binding head screw** is a screw with an undercut under the head. Binding head screws are used to connect wires to electrical parts. The undercut protects the end of the wire from fraying.

Screws are also identified by the types slots in their heads (Figure 5-3B). A **slotted screw**, or machine screw, is a screw with a head made to fit the common screwdriver. Slotted screws are used to attach small parts on outdoor power equipment.

Screws are made with slots for different types of wrenches and screwdrivers. These are used when the part must be tightened more than is possible with a slotted screwdriver. A **Phillips head screw** is a screw with a cross-shaped slot in the head for use with a Phillips screwdriver. They are used to keep the screwdriver from slipping out of the slot. An **Allen head screw** is a screw with a head with an inside, hex-shaped hole. The hex-shaped hole fits an Allen head wrench. Allen screws are often used to hold drive pulleys on engine crankshafts. The **clutch head screw** is a screw with a figure eight slot. It is used with a clutch head screwdriver.

A special type screw is often used to hold pulleys on engine crankshafts (Figure 5-4). The **set screw** is a screw with a sharp point at the end used to lock a part to a rotating shaft. The set screw is threaded through a pulley. The sharp end contacts the crankshaft. As it is tightened, it digs into the crankshaft and holds the pulley in place.

The **self-tapping screw** (Figure 5-5) has a thread that makes an internal thread as it is screwed into a nonthreaded part. Self-tapping screws are used when there is no place behind the part to install a nut. These screws can only be used on thin or soft metals.

Bolts

A **bolt** is a threaded fastener that fits through a hole in two parts and into a threaded nut to hold the parts together (Figure 5-6). Two wrenches are used when bolts are tightened or loosened. One wrench is used to drive the bolt. The other wrench is used to keep the nut from turning. The only difference between most bolts and screws is their use. A screw is used in a threaded hole. A bolt is used with a nut. The hex-head cap screw becomes a bolt when it is used with a nut.

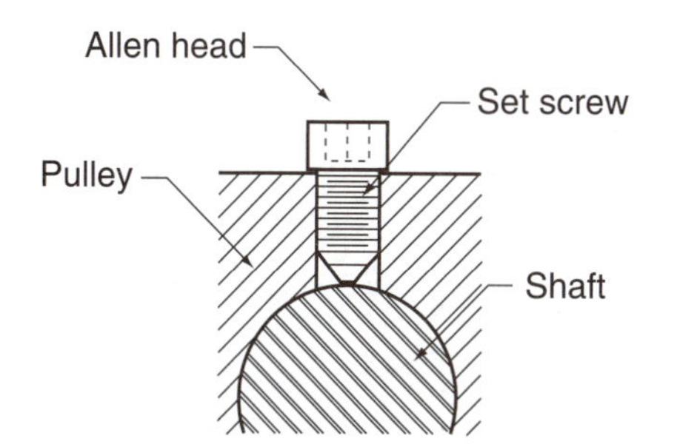

FIGURE 5-4 A set screw is used to hold a pulley on a shaft.

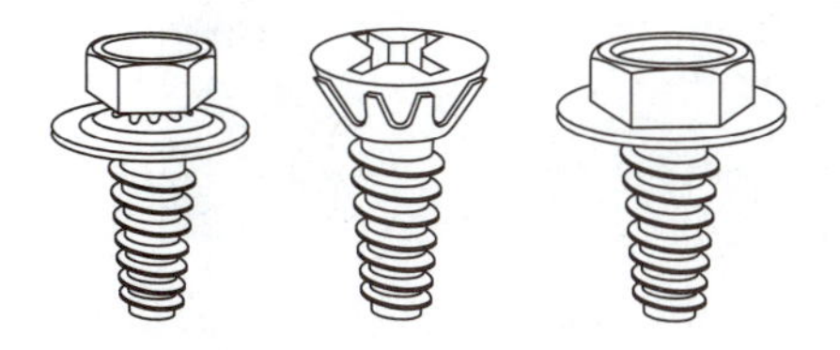

FIGURE 5-5 Self-tapping screws make their own threads in soft metal.

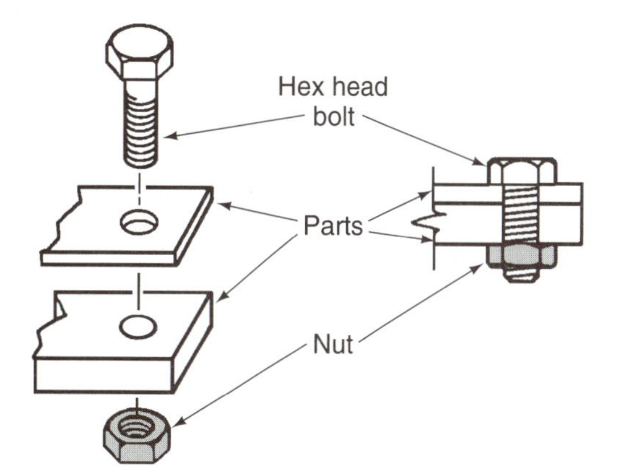

FIGURE 5-6 A bolt fits through two parts and threads into a nut.

SERVICE TIP: In common use many hex-head engine screws are called bolts. For example, cylinder head bolts are called bolts. Because they are threaded into the cylinder block, they are not really bolts but screws.

Some types of bolts are used only with nuts (Figure 5-7). These bolts have round heads. Some have a shoulder under the head. The shoulder seats in the part. The bolt is prevented from turning while the nut is being installed. This is needed

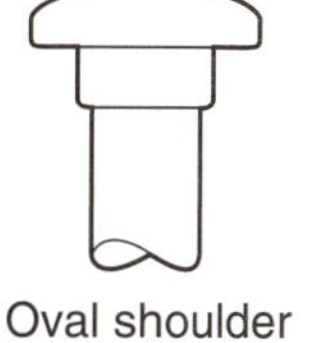
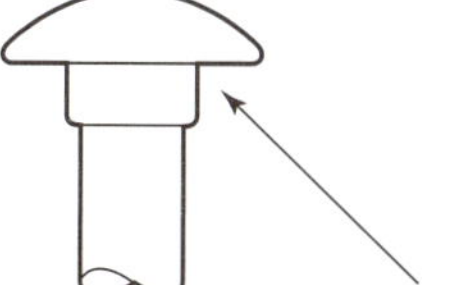

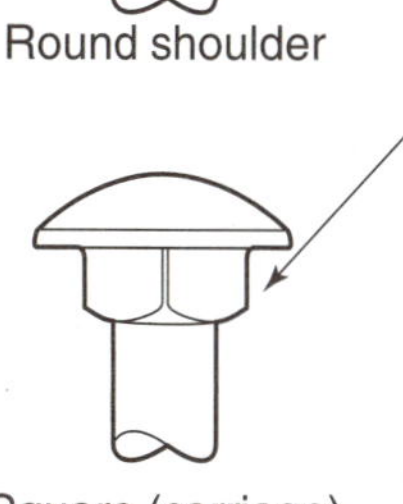
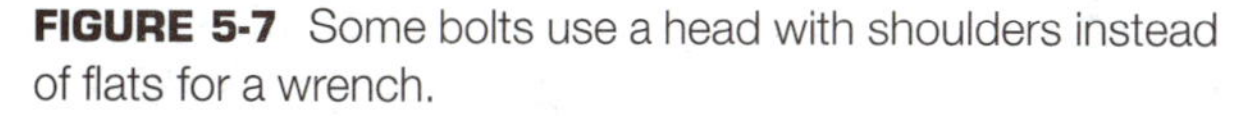

FIGURE 5-7 Some bolts use a head with shoulders instead of flats for a wrench.

because the round head does not have a place for a wrench or screwdriver.

Studs

A **stud** is a threaded fastener with threads on one end for a threaded hole and on the opposite end for a nut (Figures 5-8A and B). Studs are used where the positioning of a part is important. A continuous thread stud is threaded all along its length. Most engine studs are the double end type. They have threads only on the ends.

Studs can get damaged from overtightening or from cross threading a nut on the threads. A stud is replaced by unscrewing it from the part. A new one the same size is installed. Studs can be difficult to replace because they are often in place for a long period of time. They are installed in areas where there is constant heating and cooling. Corrosion and rust often builds up between the threads on the stud and the part.

The first step in stud removal is to use penetrating fluid. **Penetrating fluid** is thin oil used to remove rust and corrosion between two threaded parts. Soak the area of the stud threads with penetrating fluid. Sometimes the stud will need to be soaked overnight before you can remove the stud.

Before you remove the old stud you must measure its height. Use a scale to measure the distance the old stud extends up from the part surface. You will need this measurement later when installing the new stud.

A tool called a **stud remover** is used for removing studs. It grips a stud and is used with a wrench to remove a stud. To remove the stud, install the correct size stud remover over the stud (Figures 5-9A and B). The jaws on the tool will grip the outside of the stud. Install a socket driver on the stud remover drive. Remove the stud by turning the tool in a counterclockwise direction.

A stud can also be removed using two nuts (Figures 5-10A, B, and C). Locate two nuts with the cor-

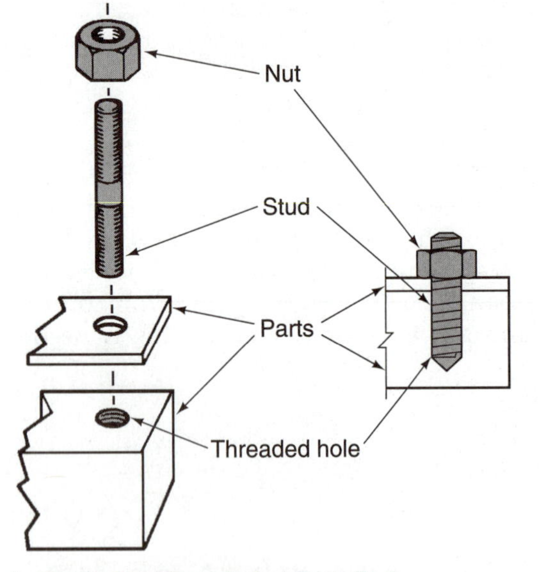

A. Studs Are Threaded at Both Ends

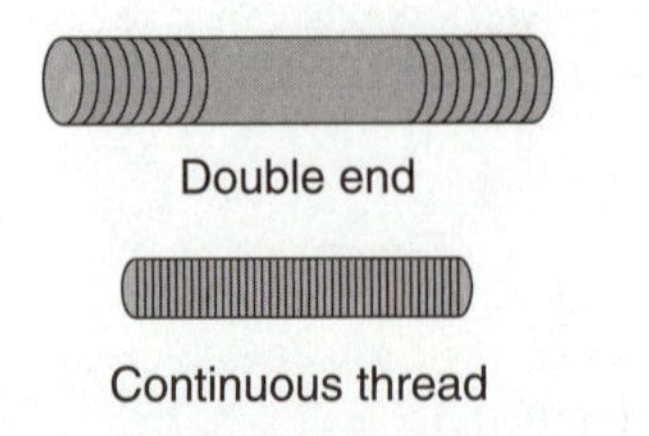

B. Stud Types

FIGURE 5-8 A stud has threads at both ends and is removed with a stud remover or with two nuts.

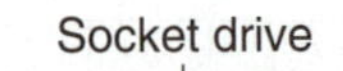

A. Stud Removing Tool

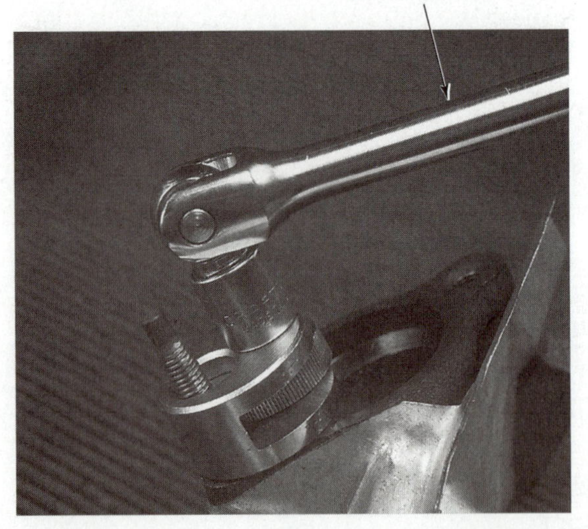

B. Using a Stud Remover to Remove a Stud

FIGURE 5-9 A stud driver and a socket handle are used to remove a stud.

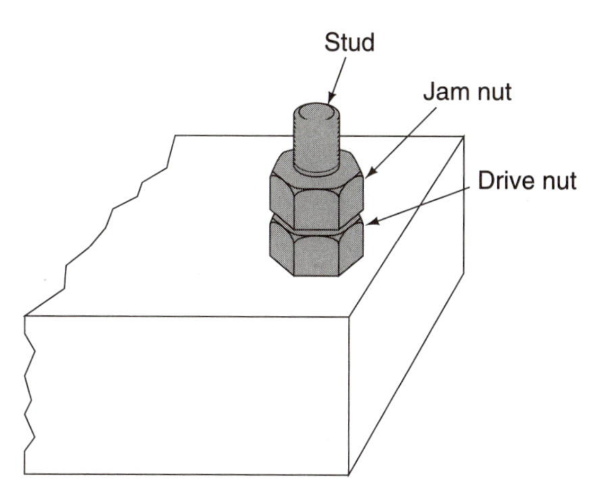

A. Install Two Nuts

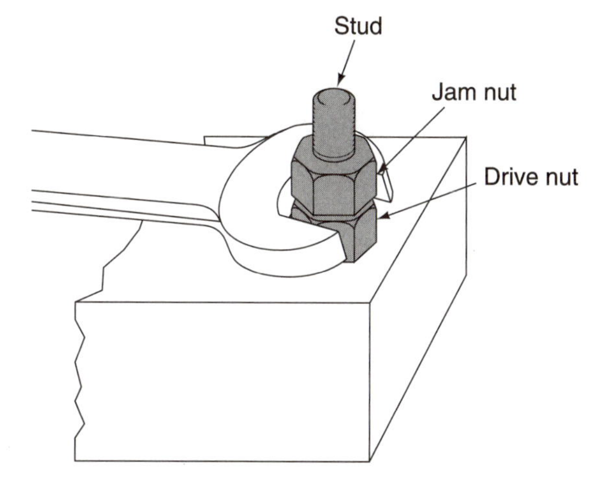

B. Drive Stud with Bottom Nut

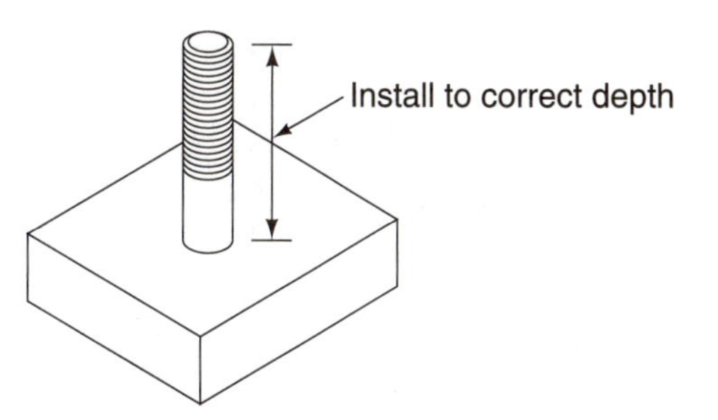

C. Install to Correct Depth

FIGURE 5-10 Two nuts can be used to remove a stud.

rect thread to fit on to the stud. Start one nut and thread it all the way down to the bottom of the stud. This is called the **drive nut.** Start another nut and thread it down until it contacts the first nut. This is called the **jam nut.** Use a wrench on the bottom drive nut to prevent it from turning. Use another wrench on the jam nut and tighten or jam it against the drive nut. The jam nut will now lock the drive nut in place on the stud.

Use an open end wrench to rotate the bottom drive nut in a counterclockwise direction. This will cause the stud to rotate and come out of the threaded hole.

Inspect the inside threads when the old stud has been removed. They may be rusty or damaged. Clean up the threads by rotating the correct size tap through the threaded hole. Compare the new stud with the old one to be sure they are exactly the same thread size and length.

Start the new stud by hand, making sure it enters the threads squarely. Turn the stud in as far as possible by hand before using any tools. Then, use two nuts to drive the stud into the part. Use the depth measurement made on the old stud as a gauge to get the new one in the same depth.

Nuts

A **nut** is a type of threaded fastener with inside thread and is used with bolts and studs. The many types of nuts are identified by their shapes (Figure 5-11). The **square nut** has four sides or flats for a wrench. Most nuts used on engine bolts or studs are hex-shaped. A **hex nut** has six sides or flats so that it can be driven with box-end, open-end, or socket wrenches. A **hex flange nut** is a nut with a flanged surface below the hex used to take the place of a washer.

Sometimes special nuts are required where vibration can cause them to loosen (Figure 5-12). The nut that holds the rotating blade on the end of a crankshaft is an example. A **castellated nut** (Figure 5-12A), or slotted nut, has a hole drilled through its side that, when tightened, will match up with a hole in a bolt or stud. A cotter pin or safety wire is then pushed through both parts. The pin or wire is then bent to hold it in place and prevent the nut from loosening. The **toothed washer lock nut** (Figure 5-12B) has a permanently installed toothed lock washer. When tightened, the teeth on the washer dig in to the part surface. The **single thread lock nut** (Figure 5-12C) has arched prongs that grip the bolt or stud threads to prevent loosening. A single thread lock nut is used when thin sheet metal parts are held together with screws.

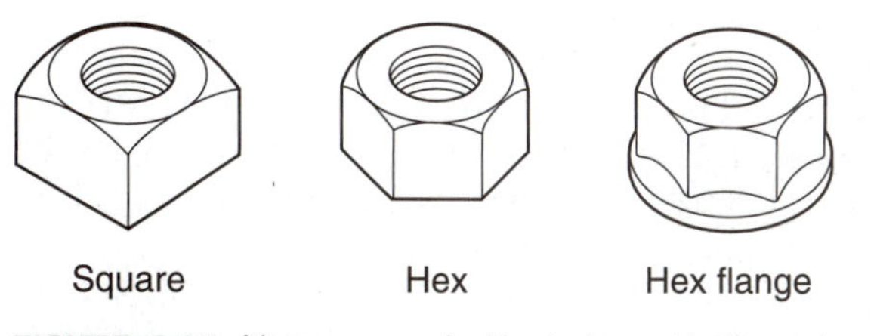
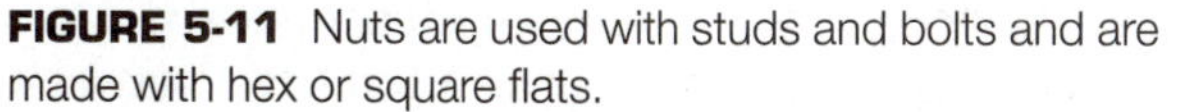

FIGURE 5-11 Nuts are used with studs and bolts and are made with hex or square flats.

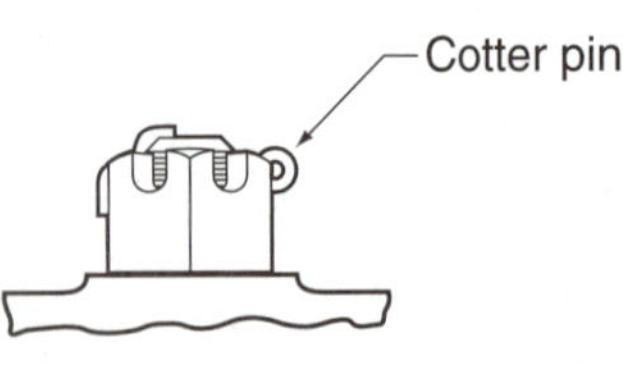

A. Castellated Nut

B. Toothed Washer Nut

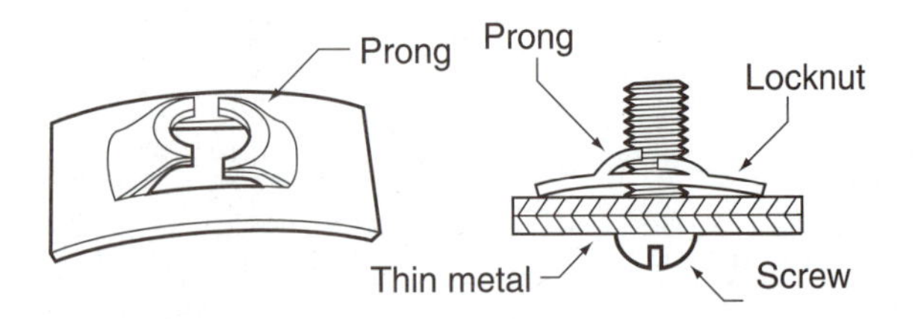

C. Single Thread Locknut

FIGURE 5-12 Lock nuts are used to prevent parts from working loose.

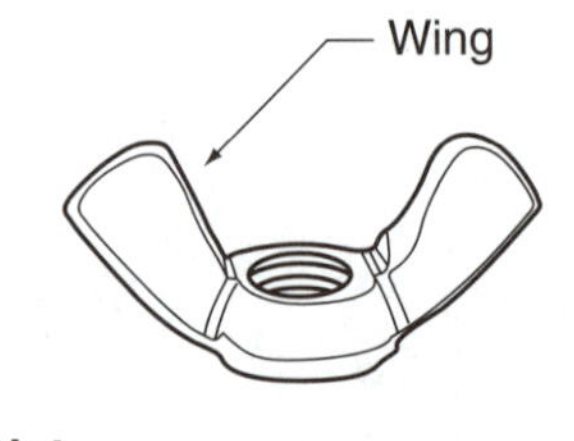

A. Wing Nut

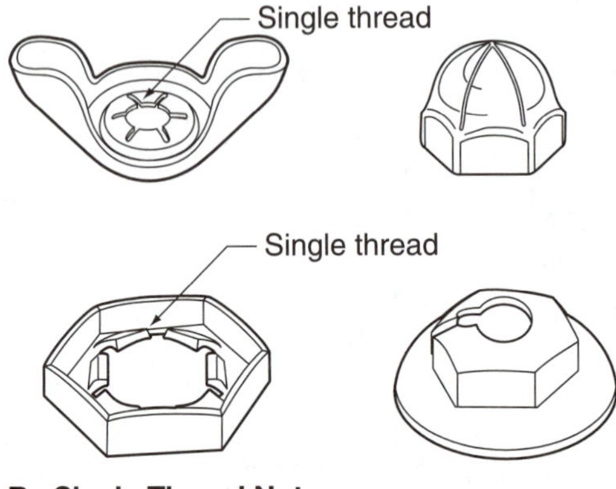

B. Single Thread Nuts

FIGURE 5-13 Speed nuts allow fast assembly and disassembly.

Some nuts, called *speed nuts,* are made for fast installation or removal (Figure 5-13). They are used on parts that need to be taken apart often. The **wing nut** (Figure 5-13A) has small gripping handles called *wings* that allow the nut to be loosened or tightened by hand. The **single thread nut** (Figure 5-13B), or pal nut, has one single thread for fast installation or removal. The single thread nut can be wing- or hex-shaped.

CAUTION: Once a lock nut has been used and removed it loses its locking ability. Lock nuts should always be replaced when parts are reassembled.

Washers

A **washer** (Figure 5-14) is a fastener that is used with nuts or screw heads to protect surfaces and prevent loosening. Washers made from nonmetallic materials can be used to insulate parts of an electrical system. Washers with conical surfaces can be used like springs. They can take up space and remove clearance in parts.

A **flat washer** (Figure 5-14A) is a washer with a flat surface used to spread tightening forces over a wider area. The flat washer is often used between a screw head and an engine part. It may also be used under a stud nut. The flat washer also prevents a machined surface from being scratched. Flat washers are made with many different outside and inside dimensions and are identified by their inside diameter. The **conical flat washer** (Figure 5-14B) has a conical shape used to take up space and remove play from an assembly.

A **lock washer** (Figure 5-15) is designed to prevent fasteners from vibrating loose. There are many different types of lock washers. The **helical spring lock washer** (Figure 5-15A) has two offset ends that dig into the fastener head and the part to

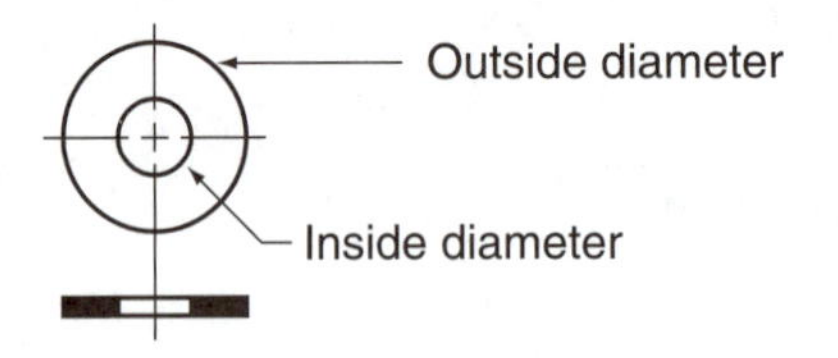

A. Plain Flat Washer

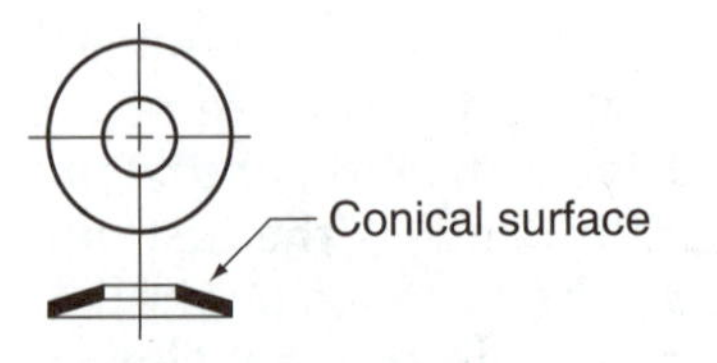

B. Conical Flat Washer

FIGURE 5-14 Flat washers are used between fasteners to spread out tightening forces.

A. Helical Spring Lock Washer

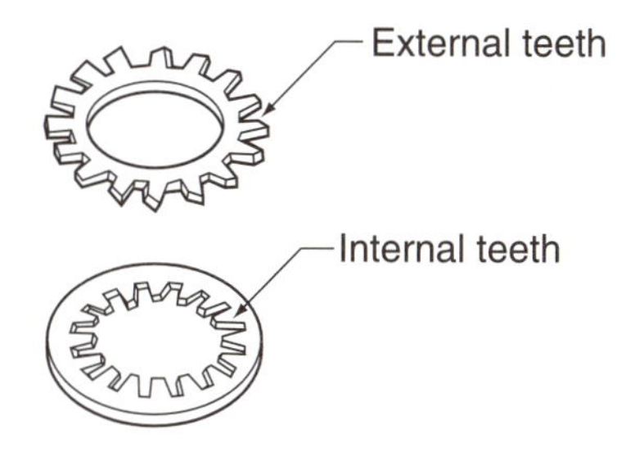

B. Toothed Lock Washer

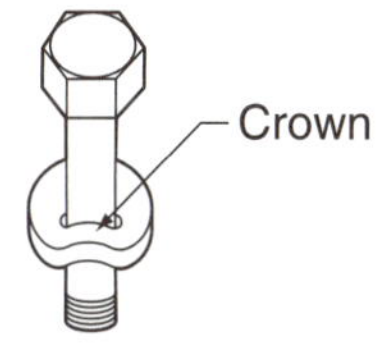

C. Belleville Lock Washer

FIGURE 5-15 Lock washers are used to prevent fasteners from getting loose.

lock them together. As the nut is tightened, the washer ends are compressed like a spring and dig into the surfaces. The **toothed lock washer** (Figure 5-15B) is a lock washer with small internal or external teeth that are made to lock against the fastener and part.

A **Belleville lock washer** (Figure 5-15C) uses a bent or crowned surface to lock a fastener to a part surface. The crowned surface is compressed when the fastener is tightened, providing a pressure to prevent loosening. Belleville washers are used in parts that have different temperatures and different expansion rates. For example, some cylinder head bolts are Belleville washers.

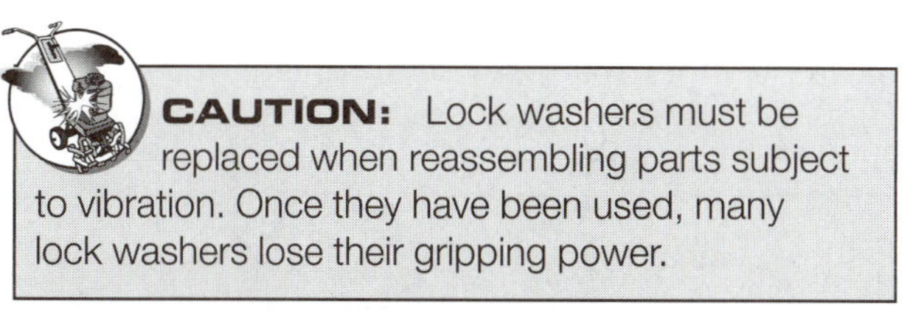

CAUTION: Lock washers must be replaced when reassembling parts subject to vibration. Once they have been used, many lock washers lose their gripping power.

THREAD SIZE AND DESIGNATION

The threads on threaded fasteners are identified by a size and designation system. There is a different size and designation system used with inch and metric threads. Thread sizes are important to the outdoor power equipment technician. Damaged or missing fasteners must be correctly identified before you can get the right replacement.

There are several basic parts of a thread size and designation (Figures 5-16A and B). The **thread diameter** is the largest diameter on an internal or external thread. The **thread pitch** is the distance between the peaks or crests of the internal or external threads. The **thread series** is the number of threads found in a specified length of threads.

Inch System Thread Sizes

An example (Figure 5-17) of an inch system designation is ¼-28 UNF × 2. The ¼ represents a bolt, screw, or stud thread diameter of ¼ inch. When used with a nut, the fraction is the inside diameter of the thread. Small-engine fastener diameters are made in inch fractions such as ¼, 5/16, ⅜ inch.

The 28 in the designation is the pitch. Pitch is the number of threads per inch. The next set of letters is the thread series designation. The two common ones are UNC and UNF. UNC means the threads are Unified National Coarse. UNF means the threads are Unified National Fine. All ¼ inch UNF fasteners have 28 threads per inch. A coarse thread fastener that is ¼ inch in diameter is designated ¼-20 UNC and has 20 threads per inch.

Most technicians know the types only as coarse and fine. Coarse has fewer threads per inch than fine (Figure 5-18A and B). The difference is easily seen and they must not be mixed. Trying to tighten a coarse thread screw into a hole with fine

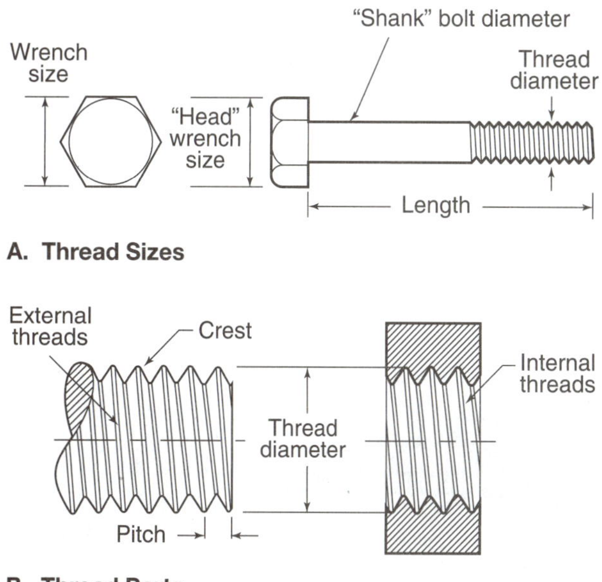

A. Thread Sizes

B. Thread Parts

FIGURE 5-16 Parts of a threaded fastener.

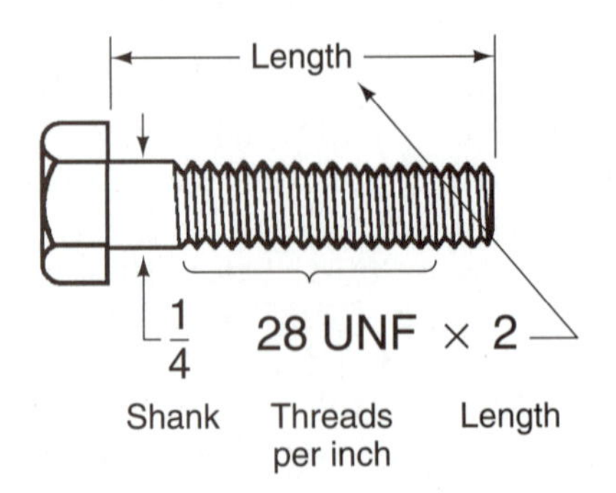

FIGURE 5-17 A complete thread designation for an inch-sized fastener.

threads will damage the threads. Coarse threads are used in aluminum parts. They provide greater holding strength in soft materials. Fine threads are used in many harder materials such as cast iron and steel. They provide excellent holding strength in hard material.

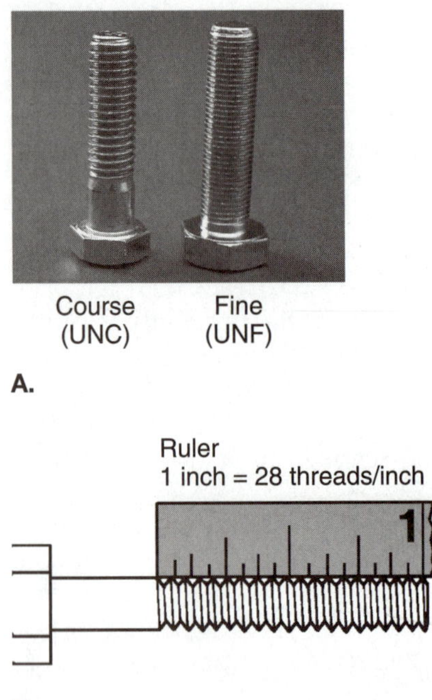

FIGURE 5-18 Fine and coarse threads have different numbers of threads per inch.

The 2 in the designation indicates that the length of a bolt, screw, or stud is 2 inches. The length of a screw is measured from the end of the shank to the bottom of the head. The length of a stud is measured from one end to the other. Bolts, screws, and studs are made in many different lengths. The lengths are in inches and fractions of an inch. The × in the designation is used for the word "by" to introduce its length.

The size of the head on a hex-head screw or bolt is often confused with the thread diameter. The head size is related to the thread size but is not part of the fastener designation. The head size is the distance across the flats of the hex. For example, a cap screw with a ¼-inch thread diameter has a head size of 7⁄16 inches. A 7⁄16-inch wrench size is used to remove or replace the ¼-inch diameter cap screw.

Metric System Thread Sizes

Metric system fasteners are designated similar to the inch system but in metric units. Metric thread specifications were developed by the International Organization for Standardization (ISO). Metric threads are sometimes called ISO threads.

An example (Figure 5-19) of a metric thread designation is M 12 × 1.75 × 40. The M means that the fastener has metric threads. The number 12 is the thread diameter in millimeters of a bolt, screw, or stud. It is also the inside diameter of a nut. The 1.75 number is the pitch in millimeters. The pitch is the space between the crests to two threads that are next to each other. The number 40 is the length of the bolt or screw in millimeters. Length is measured from the end of the fastener to the bottom of the head.

Metric system fasteners use the pitch number to show fine and coarse threads. They do not use a coarse or fine designation. For example, a fine thread metric bolt may be indicated by M 8 × 10. A bolt of the same diameter with a coarse thread is designated M 8 × 1.25. The larger pitch number indicates a wider space between threads.

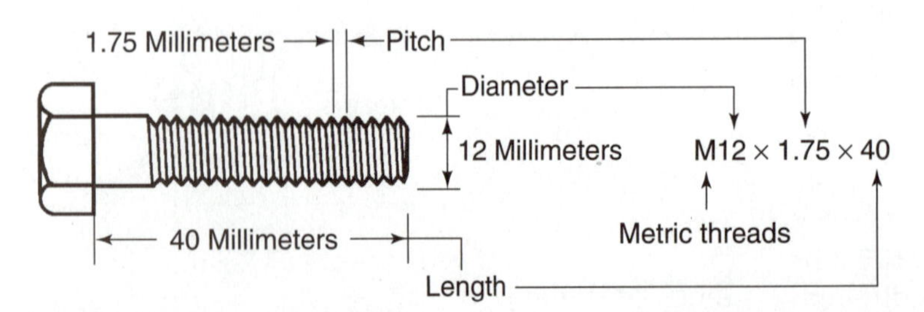

FIGURE 5-19 A complete thread designation for a metric-sized fastener.

The head size for a metric hex-head bolt or screw is the distance across the flats just as in the inch system. The size of the head determines what metric wrench size must be used on the fastener. Common metric head sizes range from 9 millimeters up to 32 millimeters. The most commonly used small-engine metric head sizes are 10, 12, 13, and 14 millimeters.

CAUTION: Start a fastener by hand and make sure it threads on properly before using a wrench. Threading a fine thread fastener into a coarse thread hole will result in part damage. Mixing inch and metric system fasteners also will result in part damage.

SERVICE TIP: Take care during disassembly to keep fasteners organized and identified. Place disassembled fasteners in plastic bags and identify their location to save time on assembly.

FASTENER GRADE MARKINGS

Fasteners are made in different strengths for different parts. For example, the cap screw used to attach the connecting rod cap to the connecting rod is under great stress. The cap screws that hold the blower housing have a much easier job. The connecting rod fasteners are made with more strength than those used on the blower housing.

Fastener strength is shown by grade markings. **Grade markings** are marks on the fastener used to show the grade, material, and tensile strength. **Tensile strength** is the maximum load in tension (pulling force) that a fastener can hold before it breaks. A low-quality fastener will break under a low pulling force. A high-quality fastener can stand a high pulling force.

The grade marking system is different for inch and metric fasteners. Inch fasteners use Society of Automotive Engineers (SAE) standards. The SAE system uses lines marked on the head of the hex-head screw or bolt. The more lines on the head of the fastener, the higher its tensile strength.

The SAE grade marking system (Figure 5-20) uses four basic categories. A fastener with no lines on the head has the lowest tensile strength. These are graded SAE 1, 2, and 3. The next category has three lines on the head and is graded SAE 5. Those with four lines are called SAE grade 6. Those designated SAE grade 7 have 5 lines. The highest grade is SAE 8 with six lines.

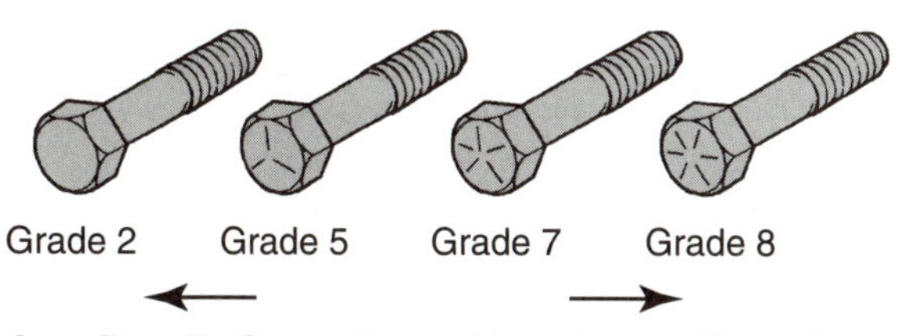

FIGURE 5-20 Inch-sized fastener strength is determined by grade marking lines on the head. (Courtesy of General Motors Corporation, Service Technology Group.)

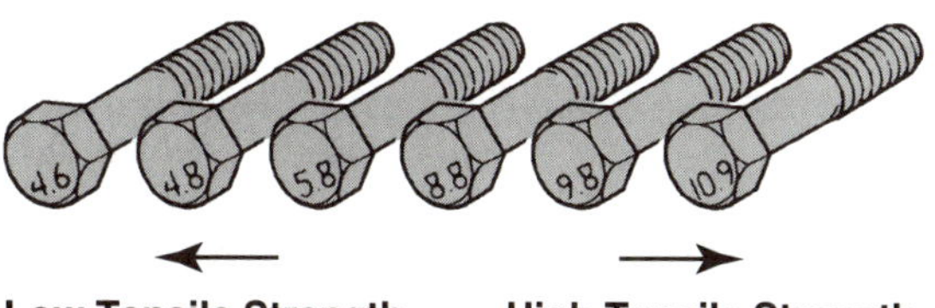

FIGURE 5-21 Metric-sized fasteners use numbers for grade markings. (Courtesy of General Motors Corporation, Service Technology Group.)

Metric system fasteners have a similar grading system, but they use numbers on the head (Figure 5-21). The higher the number, the higher the tensile strength. A fastener with a 4.8 on the head is not nearly as strong as a 10.9. A metric fastener without any number is similar in strength to an SAE grade 0.

Nuts are also graded according to their tensile strength (Figure 5-22). A different marking system is used for inch and metric-sized nuts. Inch system nuts are graded with dot marks stamped on the top of the nut. A low quality nut has no dots. A high quality nut has up to six dots. The metric system grading system uses a number stamped on the top of the nut. A nut without a number has low tensile strength. One marked with a number 10 has high tensile strength.

WARNING: Use the proper strength fastener required for the job. Low-quality fasteners can fail, causing possible injury or damage to the equipment.

THREAD IDENTIFICATION

When a fastener is damaged or missing, you must be able to measure and identify the thread on a fastener. A new fastener can then be purchased that is the correct size. The easiest method of determining

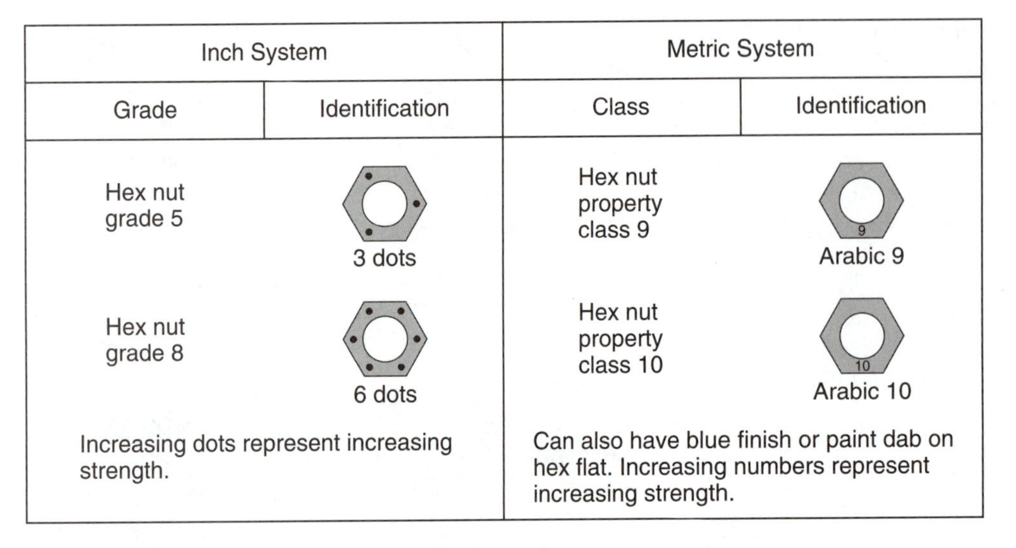

Inch System		Metric System	
Grade	Identification	Class	Identification
Hex nut grade 5	3 dots	Hex nut property class 9	Arabic 9
Hex nut grade 8	6 dots	Hex nut property class 10	Arabic 10
Increasing dots represent increasing strength.		Can also have blue finish or paint dab on hex flat. Increasing numbers represent increasing strength.	

FIGURE 5-22 Inch and metric grade markings used on nuts.

fastener size is to use another fastener with a size you know. For example, one of three identical blower housing screws is missing. You can use the other screws to find its size. Locate a nut that threads on to the screw properly. Then take the nut when you go to buy the new screw. Be sure that the new one is the right size and grade.

Often you will not have an identical fastener. You will have to measure the fastener shank diameter and thread. The most accurate way to measure is to use an outside micrometer. Measure and record the diameter of the fastener shank using the correct size inch or metric micrometer.

The next step is to find the thread pitch. Pitch is measured with a tool called a **pitch gauge** (Figure 5-23). A pitch gauge is a tool with multiple blades that have teeth that are used to match up with threads in order to identify their pitch. There is a pitch gauge made for metric threads and one for inch threads.

The first step in using a pitch gauge is to select either the inch or metric gauge. This can be determined by the shank diameter measurement. If the

CASE STUDY

One Monday morning, a customer in an angry mood returned a rotary lawnmower to a shop. He had picked the mower up the previous Friday after having the blade replaced. As the customer pushed the mower into the shop, it was not too difficult to see the cause of his anger. The mower deck on his almost new mower had a large gaping hole in one corner. The blade had come off the end of the crankshaft and ripped through the mower deck. The mower was damaged beyond repair. The end of the engine crankshaft was bent so even the engine was not usable.

The shop owner turned the mower on its side and saw the cause immediately. The hex-head cap screw that threads into the crankshaft to hold the blade had broken. The owner checked with the technician who had done the job. The technician had lost the original cap screw and replaced it with one he found in the shop. The original bolt was a high tensile strength SAE grade 6 bolt. The replacement was a SAE grade 0 bolt. The very unhappy shop owner had to replace the owner's $700 lawnmower because of a 50¢ cap screw.

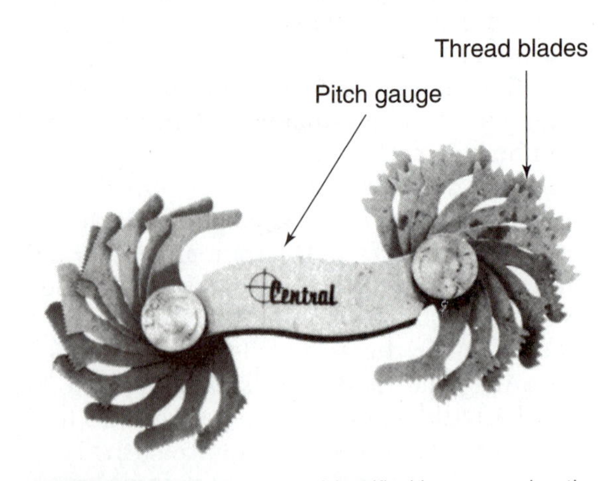

FIGURE 5-23 Fasteners are identified by measuring the shank and matching the threads to a pitch gauge. (Courtesy of Central Tools Inc.)

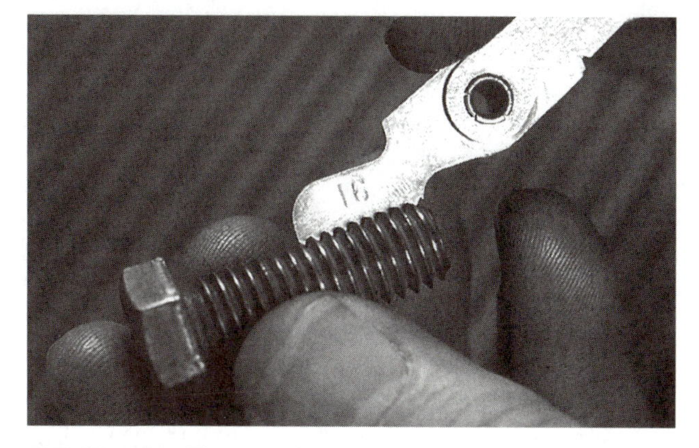

FIGURE 5-24 The 16-inch pitch gauge fits these threads and shows the pitch to be 16 threads per inch.

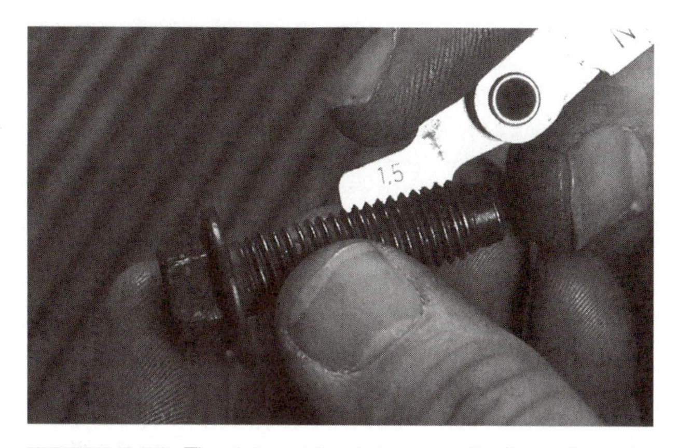

FIGURE 5-25 The 1.5 metric pitch gauge fits these threads and shows the pitch to be 1.5 millimeters.

shank is a standard inch measurement such as ¼ inch (0.250 inch), use an inch system pitch gauge. If the measurement is 10 millimeters, use a metric gauge.

Place each of the pitch gauge blades on the fastener threads until you find a match. The teeth on the blade must fit perfectly into the threads of the fastener. When the correct blade is found, read the number on the blade. A number 16 on an inch-sized blade means 16 threads per inch (Figure 5-24). A number 1.5 on a metric-sized blade means a pitch of 1.5 millimeters (Figure 5-25).

THREAD REPAIR

Threads are often damaged by corrosion, overtightening or cross threading. Corrosion can eat away the thread surface. Overtightening can stretch the threads out of shape. Cross threading, often called *stripping,* is a common type of thread damage (Figure 5-26). It may be caused by threading fasteners together with threads that do not

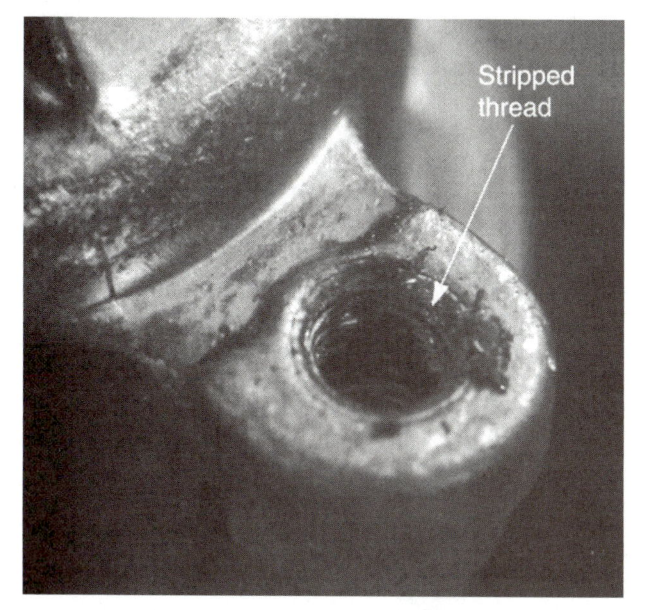

FIGURE 5-26 A threaded hole with damage from cross threading.

match. Another common cause is starting a fastener with a wrench instead of by hand.

Damaged fasteners are not repaired. They are simply replaced. Damaged internal threads in major parts need to be repaired. Sometimes damaged threads in parts such as the side cover or cylinder assembly cannot be repaired. This means the engine will have to be scrapped.

Damaged internal threads can often be repaired with the correct size tap. The tap is driven down the damaged hole to cut away the damaged threads. Taps are available in sets for cutting either inch- or metric-sized threads. The thread size marking on the shank of the tap must be matched to the thread to be repaired (Figure 5-27). You can find

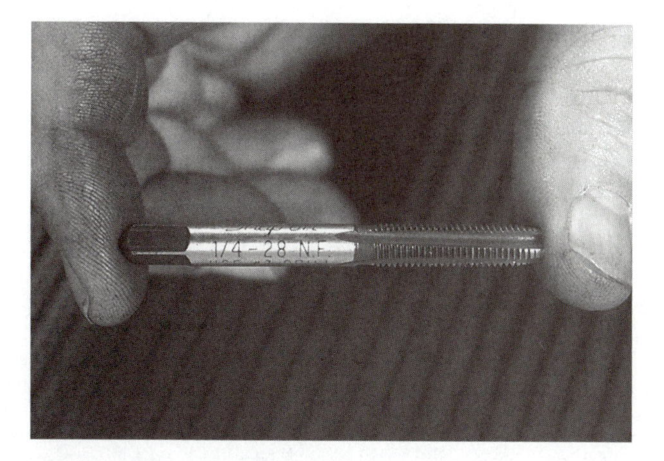

FIGURE 5-27 A tap with the size ¼ 28 UNF on the shank.

the thread size by using the fastener that fits in the hole and a pitch gauge.

There are two different types of threaded holes (Figure 5-28A): A through hole goes all the way through a part; a blind hole (Figure 5-28B) goes in part way and stops. A tap with a tapered end, called a *taper tap,* is used for through holes. A bottoming tap, with a square end, is used for blind holes. A typical internal thread repair is shown in the accompanying sequence of photographs.

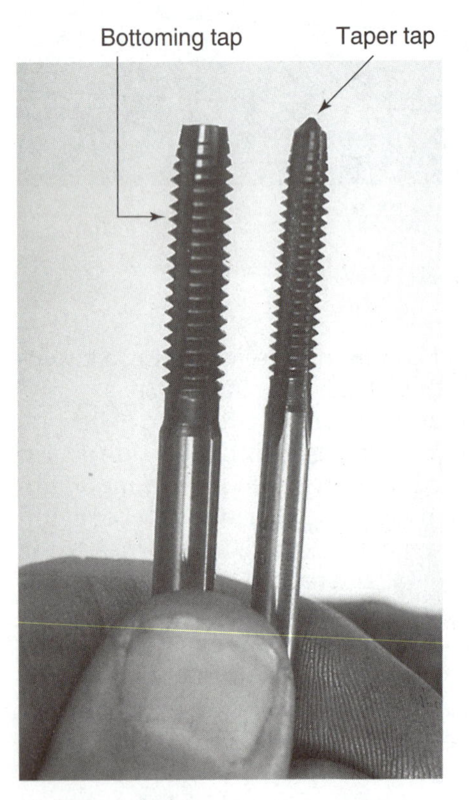

A. Types of Taps

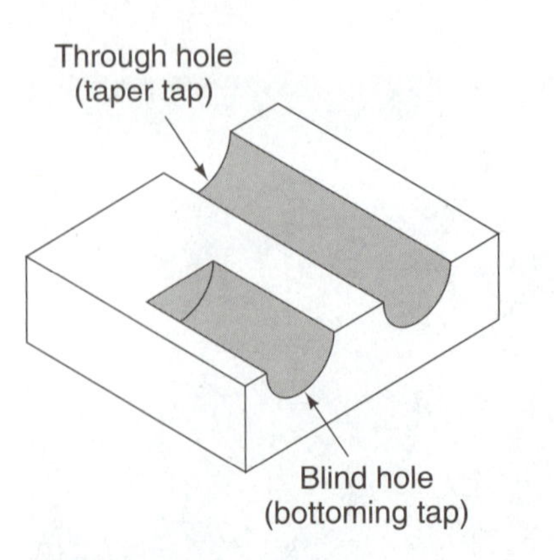

B. Types of Holes

FIGURE 5-28 A bottoming tap is used for blind holes and a taper tap for through holes.

CAUTION: Incorrect use of a tap can damage a part beyond repair. If possible, locate a scrap engine and practice cutting threads on a similar hole before attempting to repair a good engine.

Find the correct tap and install it in a tap wrench. Carefully start the tap squarely in the hole. Slowly turn it clockwise down the threads to the bottom. Then, rotate it counterclockwise back out of the hole. Clean the metal chips left by the tap cutting out of the hole. Metal chips can damage the engine. Screw in the correct size fastener and make sure it threads in properly.

WARNING: When using a tap to cut threads, wear eye protection to prevent an eye injury from flying metal chips.

THREAD COMPOUNDS

Two types of thread-coating compounds (Figure 5-29) are used in engine work. Fasteners that are used around rust and corrosion can stick together. They are very difficult to remove. When fasteners stick in their threads, it is called *seizing.* A seized fastener may have to be removed by cutting it apart. Sometimes engines are scrapped because of seized fas-

A. Anti-Seize Compound B. Thread-Locking Compound

FIGURE 5-29 Antiseize compounds prevent threads from seizing and thread-locking compounds prevent them from loosening. (Courtesy of Saverio Bono, *Auto Technology: Theory & Service.*)

Repairing Damaged Internal Threads with a Tap

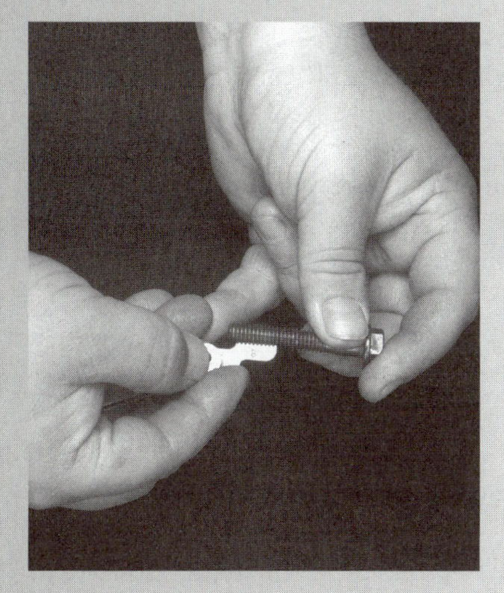

1. Determine the thread size of fastener with a pitch gauge that fits into the damaged thread.

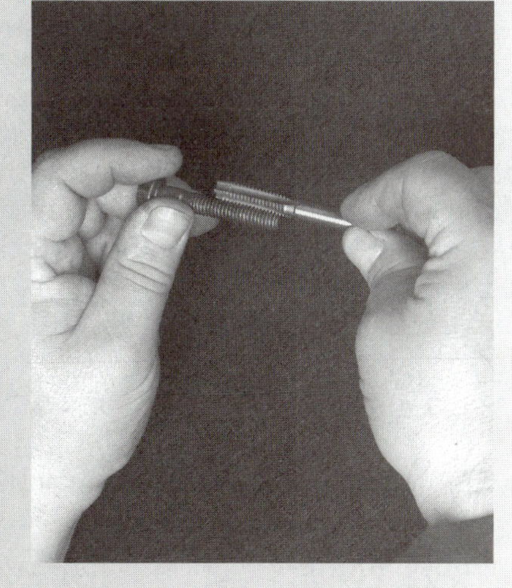

2. Select the correct size and type of tap for the threads to be repaired.

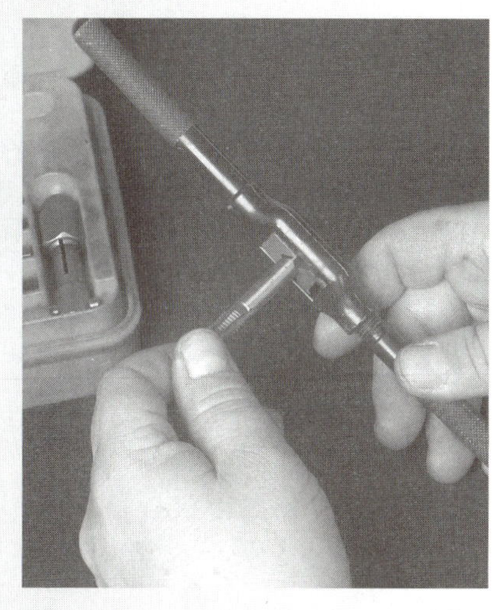

3. Install the tap into the tap wrench.

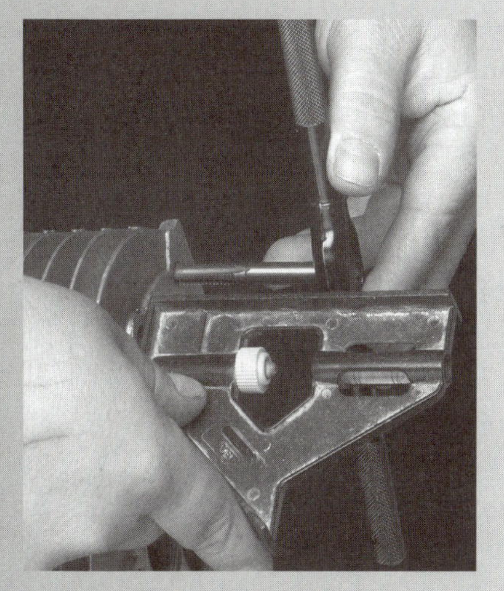

4. Start the tap squarely in the threaded hole using a machinist square as a guide.

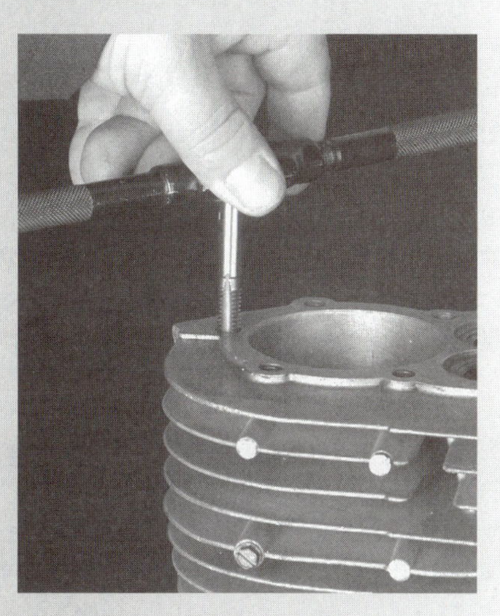

5. Rotate the tap clockwise down the threaded hole the complete length of the threads.

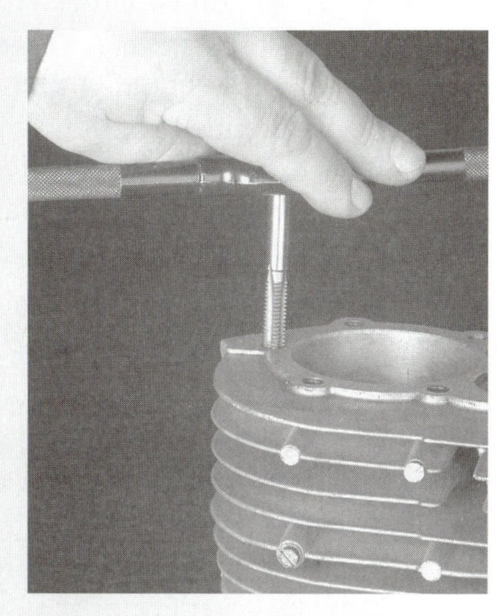

6. Rotate the tap back out of the hole by turning it counterclockwise.

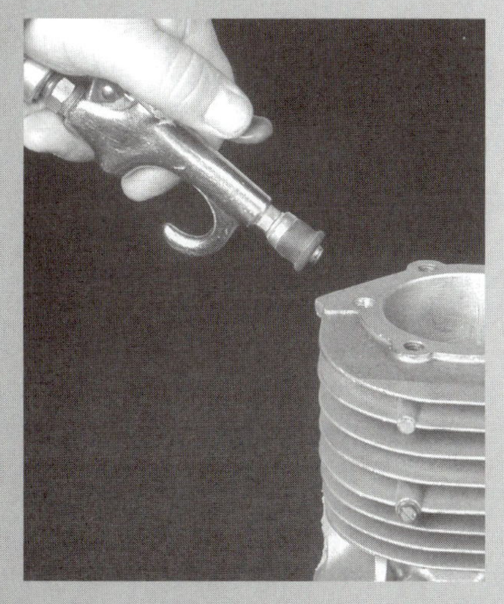

7. Clean the metal chips left by the tap out of the hole.

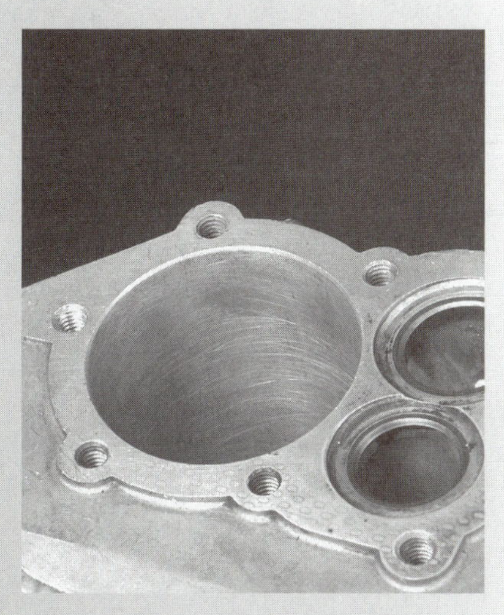

8. Inspect the threads left by the tap to be sure they are acceptable.

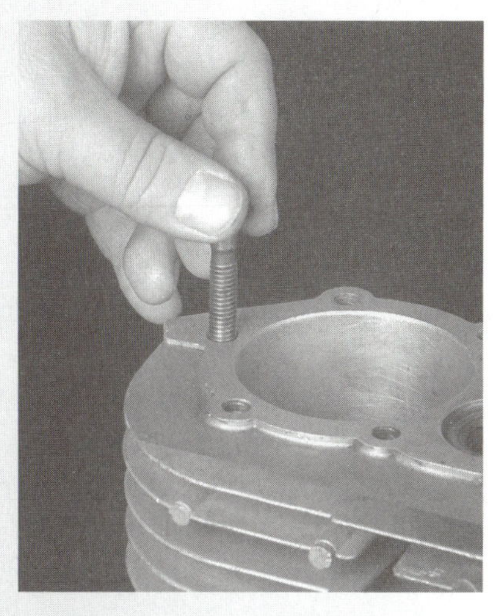

9. Test the threads by threading the correctly sized fastener into the threaded hole.

teners. An **antiseize compound** (Figure 5-29A) is a material used on the fastener threads during assembly that prevents seizing.

Sometimes a fastener gets so much vibration that lock washers cannot prevent it from loosening. A **thread-locking compound** (Figure 5-29B) is a material used on studs and other fasteners that locks the threads together to prevent them from vibrating loose. Thread-locking compounds are available in different grades with different locking power. For example, one grade is used for parts that may be taken apart in the future. Another grade is available for fasteners to lock them together permanently. Always follow the engine maker's recommendations when using thread-locking compounds.

REVIEW QUESTIONS

1. List and describe the main types of screws used in outdoor power equipment repair.
2. Explain the procedure used to remove and replace a stud with a stud-removing tool.
3. Explain the procedure used to remove and replace a stud with two nuts.
4. List and describe the main types of nuts used in outdoor power equipment repair.
5. List and describe the main types of lock washers used in outdoor power equipment repair.
6. Identify and describe the main types of lock nuts used on outdoor power equipment fasteners.
7. Explain the purpose of a flat washer.
8. List and explain the parts of an inch thread designation.
9. List and explain the parts of a metric thread designation.
10. Explain how thread grade markings are used when replacing fasteners.
11. Describe how to use a known fastener to identify a lost fastener.
12. Explain the purpose and use of a pitch gauge.
13. Describe how to identify the size fastener.
14. Explain how a damaged internal thread can be repaired with a tap.
15. List and describe the purpose of two types of thread compounds.

DISCUSSION TOPIC AND ACTIVITIES

1. Collect a sample of the fasteners you can find in the shop. Identify the type and purpose of each of the fasteners.
2. Use a micrometer and metric pitch gauge and identify the designation for a selection of metric fasteners.
3. Use a micrometer and inch pitch gauge and identify the designation for a selection of inch fasteners.

CHAPTER 6

Nonthreaded Fasteners

OBJECTIVES

Upon completion and review of this chapter, you should be able to:

- Identify and explain the purpose of keys used on outdoor power equipment parts.
- Identify and explain the purpose of pins used on outdoor power equipment parts.
- Describe the use and safe removal and installation of internal and external snap rings.
- Explain how to select the correct size and type of replacement electrical wire for outdoor power equipment.
- Describe how to select and install terminals and connectors on electrical wires.

TERMS TO KNOW

Clevis pin
Cotter pin
Dowel pin
Electrical connector
Electrical terminal
E-ring
External snap ring
Internal snap ring
Key
Pin
Roll pin
Snap ring
Tapered pin

INTRODUCTION

Many outdoor power equipment parts are held together with fasteners that do not use threads. The main types of nonthreaded fasteners are keys, pins, and snap rings. Many outdoor power products have electrical systems. You may need to use electrical wire and connectors to make repairs.

KEYS

A **key** (Figure 6-1) is a small metal fastener used with a gear or pulley to lock it to a shaft. One half of the key fits into a key seat on the shaft. The other half fits into a keyway on the gear or pulley.

Keys are identified by their shapes (Figure 6-2). The most common shapes of keys are square, flat, Pratt and Whitney, and Woodruff. Many small engines use Pratt and Whitney or Woodruff type keys to hold the flywheel to the magneto end of the crankshaft.

The flywheel key on some engines is made from a special soft material (Figure 6-3). This key has an important job on lawnmowers. If the lawnmower blade hits a solid object, the blade will stop suddenly. The engine crankshaft will also begin to stop. The heavy flywheel will resist stopping. This force will distort the flywheel key. The distorted key prevents the magneto parts in the flywheel from working, which stops the engine and prevents damage to the equipment.

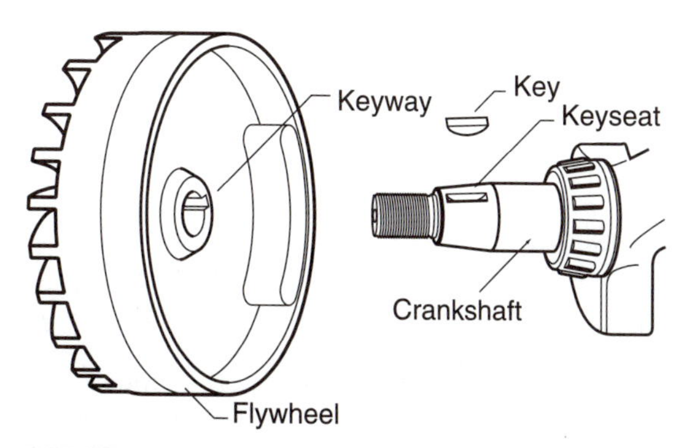

FIGURE 6-1 Keys are often used to hold the flywheel and drive pulleys to the crankshaft.

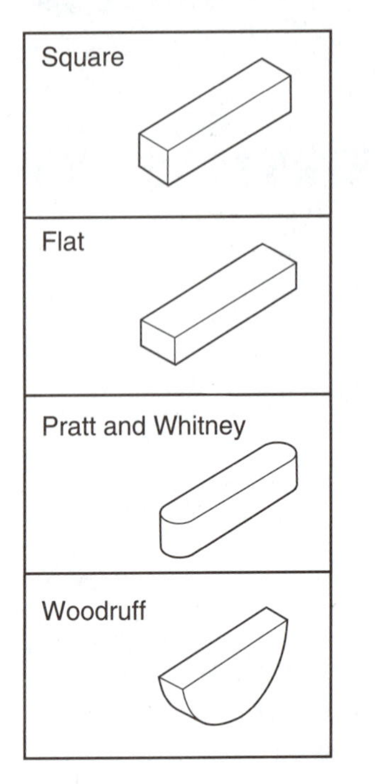

FIGURE 6-2 Keys are identified by their different shapes.

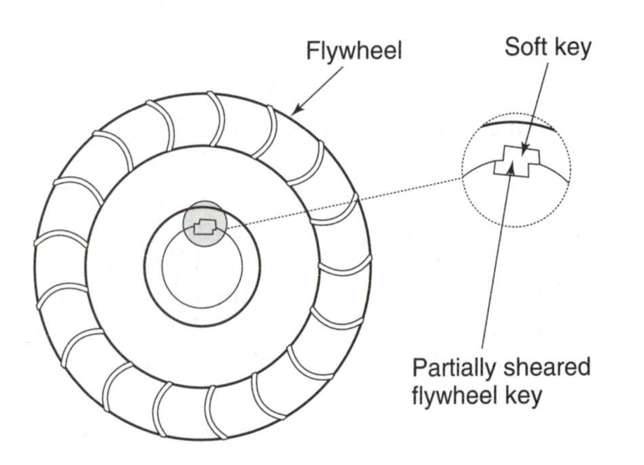

FIGURE 6-3 A soft key can distort when a lawnmower blade hits an object and shuts down the engine.

A square key is often used to hold a drive pulley to the end of the crankshaft (Figure 6-4). The crankshaft often has a long, square key seat. The drive pulley is mounted on the crankshaft. It can then be moved anywhere along the shaft. The square key is used because it can be moved back and forth with the pulley. A set screw is used to hold the pulley in position.

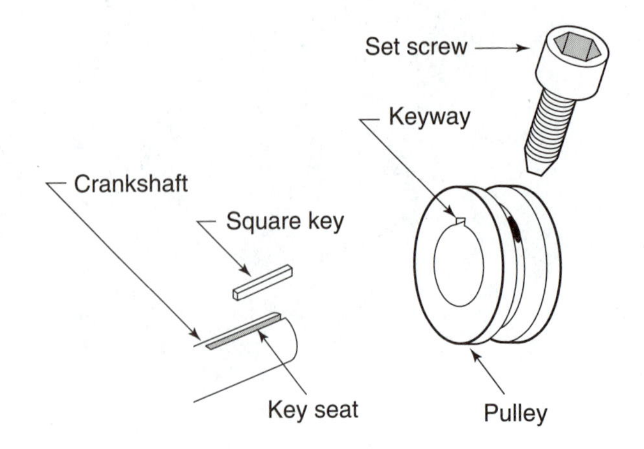

FIGURE 6-4 A square key and a set screw are used to hold a pulley to a crankshaft.

PINS

A **pin** (Figure 6-5) is a small, round metal fastener that fits into a drilled hole to hold two parts together. Pins are often small. They require special pliers or a punch and hammer for installation and removal. A **dowel pin** is straight for most of its length with slightly tapered ends. A **tapered pin** has a large end that tapers down to a small end. A **Clevis pin** has a flange at one end and a small hole for a cotter pin or safety wire at the other end. A **cotter pin** has a split shaft that is installed through holes in fasteners and the split ends are bent over to prevent part loosening.

A **roll pin** is a hollow pin with a split down its length. The roll pin is installed in a hole that is smaller than the diameter of the pin. The split al-

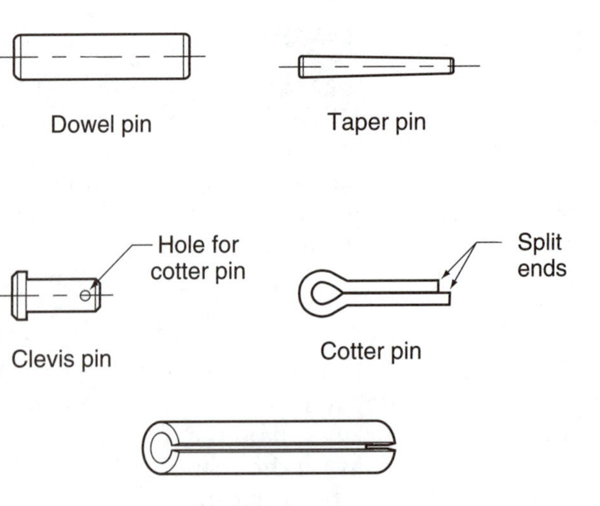

FIGURE 6-5 Different pins used to hold two parts together.

lows it to compress when installed. This compression locks it in place. Roll pins come in different lengths and different diameters. They are removed and replaced using a pin punch and hammer.

> **CAUTION:** Roll pins can lose their gripping power when they are removed. When possible, they should be replaced with new ones during assembly.

SNAP RINGS

A **snap ring**, or retaining ring, is an internal or external expanding ring that fits in a groove (Figure 6-6). The force of expansion or contraction is used to hold parts in place. Retaining rings are made from high-quality metal. This allows them to be expanded or contracted by a tool to put them on or take them off. Once in place, the tension on the ring causes it to "snap" back to its original shape and stay in place.

The **internal snap ring** is a retaining ring that is compressed to fit into a groove machined inside a hole. The **external snap ring** is a retaining ring that is expanded to fit over and into a groove machined on a shaft. The internal ring is compressed for installation. The external ring is expanded for installation.

Most snap rings (Figure 6-7A) are removed and replaced with special pliers. Snap ring pliers (Figure 6-7B) expand or contract snap rings to get them on or off. Most snap rings are gripped on the ends by snap ring pliers. Many snap rings have small holes

Groove

Internal Snap Ring

Shaft

Groove

External Snap Ring

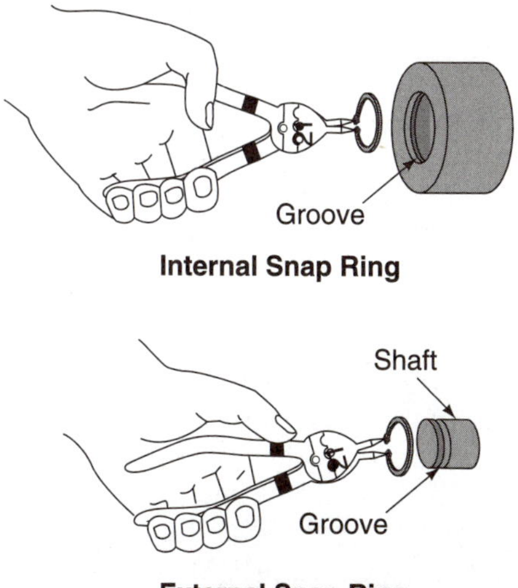

FIGURE 6-6 Snap rings may be internal to fit an inside groove or external to fit a groove in a shaft.

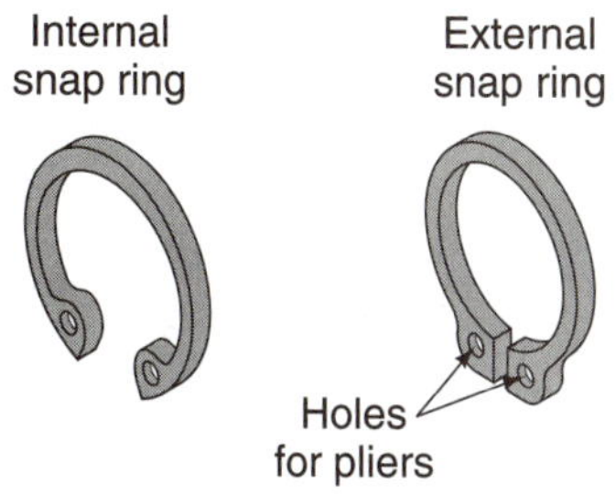

Internal snap ring

External snap ring

Holes for pliers

A. Types of Snap Rings

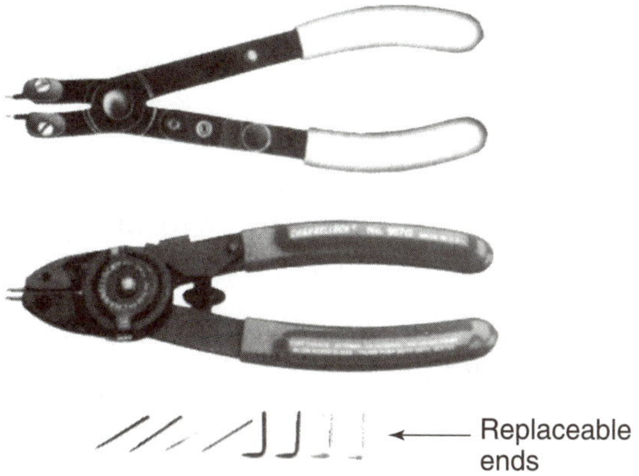

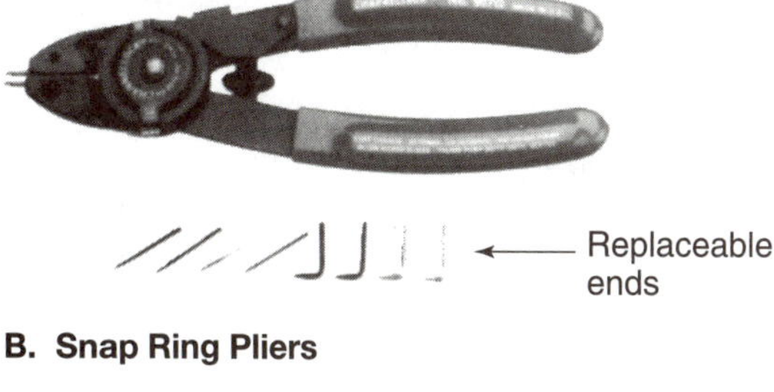

Replaceable ends

B. Snap Ring Pliers

FIGURE 6-7 Many snap rings have holes for the ends of snap ring pliers. (*B:* Courtesy of MAC Tools Distributors and Lisle Corp.)

on the ends of the ring. Pliers for these rings have jaws that fit these holes. Snap ring pliers often have replaceable ends to fit different sized snap rings.

The **E-ring** (Figure 6-8), or E-clip, is a very small external snap ring. It is too small for snap ring pliers and it has several slots. A small screwdriver is used to pry it off the part. E-rings can be installed

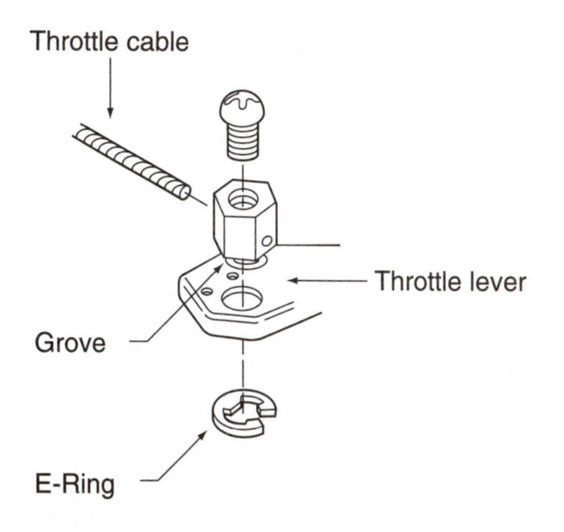

FIGURE 6-8 E-rings or E-clips are often used on throttle linkage.

into their groove by hand. They are often used to connect a throttle cable to a throttle lever.

CAUTION: All internal and external snap rings lose some holding ability when they are repeatedly expanded or contracted. Snap rings also may be easily distorted during removal. Replace snap rings whenever possible.

WARNING: Wear eye protection when removing or installing snap rings. Snap rings may fly off the tool and cause injury.

CASE STUDY

A customer returned a recently rebuilt chain saw to an outdoor power equipment shop. The shop had done an upper end (new piston rings) rebuilding job on the engine. The customer reported that the engine ran well and that he had cut down several trees. All of a sudden the saw made a strange noise and stopped suddenly. The shop owner and the technician who did the overhaul began to inspect the saw.

The first thing they noticed was that the rope starter would not crank the engine. The engine was totally locked up. The technician disassembled the engine and removed the cylinder. There was a deep gouge in one side of the cylinder. Further inspection revealed the source of the problem. One of the snap rings used to retain the piston pin had come out of the groove. The piston pin worked its way out of the piston pin bore and contacted the cylinder wall. The pin had welded itself into the cylinder and locked the piston up. It was obvious that the retaining snap ring was the problem. The question was whether it was defective or improperly installed. The other snap ring was still in the piston. The shop owner used some pliers on this ring to remove it and found the problem immediately. The snap ring has lost its tension, probably from being expanded and contracted too many times. New snap rings should have been used during the engine service. The engine damage was so extensive that the engine was beyond rebuilding. The shop had to buy the customer a new chain saw.

ELECTRICAL WIRE AND ELECTRICAL FASTENERS

You may need to do repairs to outdoor power equipment electrical parts. Riding lawnmowers and garden tractors have extensive wiring systems. You will need to know about wires and electrical system fasteners to make these repairs.

Electrical Wire

Electrical wires are sometimes called *leads* because they lead the current where it has to go. Electrical wire has an inside core covered with insulation (Figure 6-9). The inside core directs or conducts the electricity where it has to go. This is where it gets the name "conductor." The core is made from copper or aluminum, both of which allow good electrical flow.

An insulation material is made to stop electricity from passing through. Plastic and rubber are examples of insulators. The insulation on an electrical wire stops electrical current from getting out of the core. This allows the electricity to go where it has to go. The insulation material is colored and may have stripes to identify different wires.

The amount of electrical current that can flow through a wire is determined by the size of the inside core. The larger the area of the core, the more current can flow through the wire. The smaller the area, the higher the resistance to electrical flow.

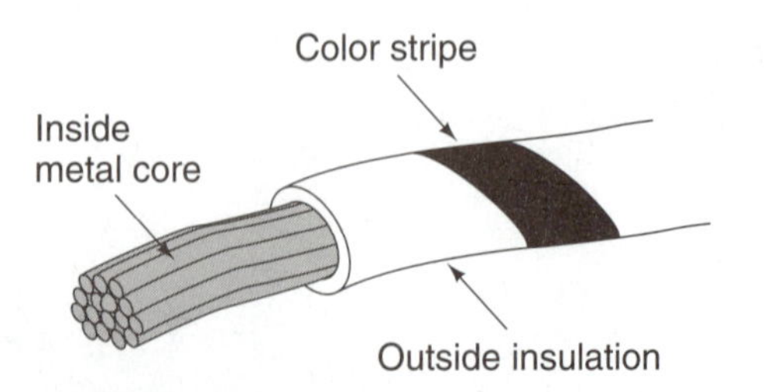

FIGURE 6-9 Electrical wire has an outside insulation and an inner core.

The amount of current the wires in an electrical system can carry is very important. Outdoor power equipment makers choose wiring to match the needs of the electrical system. You must carefully select the correct size replacement wires. Wires that are too big or too small can cause problems. Wire length is also important. The longer the wire, the more resistance it offers to current flow.

Wire size is set by one of two wire size systems. The American wire gauge (AWG) is the most common wire size system (Figure 6-10A). This system gives a number from 0 to 20 to the wire. These numbers specify the size of the wire's inner core. The higher the number, the **smaller** the size of the

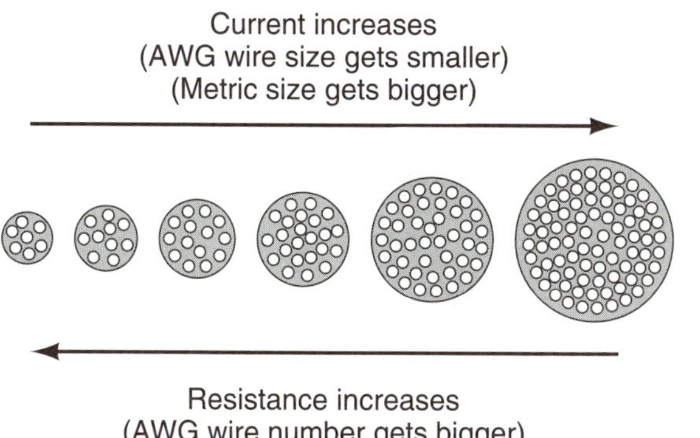

A. American Wire Gauge System

Wire Gauge Sizes	
Metric Size (cubic millimeters)	Wire Size
0.5	20
0.8	18
1.0	16
2.0	14
3.0	12
5.0	10
8.0	8
13.0	6
19.0	5

B. Metric System Wire Sizes

FIGURE 6-10 Current flow increases and resistance decreases as the wire core size gets bigger.

inner core. The lower the number, the **larger** the size of the inner core. The smallest AWG wire inner core is 20 gauge. The largest AWG wire is 0 gauge.

Wire sizes can also be set by metric system measurements (Figure 6-10B). The cross-sectional inner core of the wire is measured and specified in cubic millimeters. A typical wire size in this system is 0.8. This means the wire has a cross-sectional area of 0.8 cubic millimeters. The larger the number, the larger the size of the inner core. A wire with a small core (similar to 20 AWG size) would be 0.5 cubic millimeters. A large wire (similar to 5 AWG size) would be 19.0 cubic millimeters.

Wires are sold in packages or on spools. The packages give the wire size. Wires are made with different colored insulation and different colored stripes. The colors are used to help you trace the wires in the system when troubleshooting. When possible, use the same colors for the replacement wires.

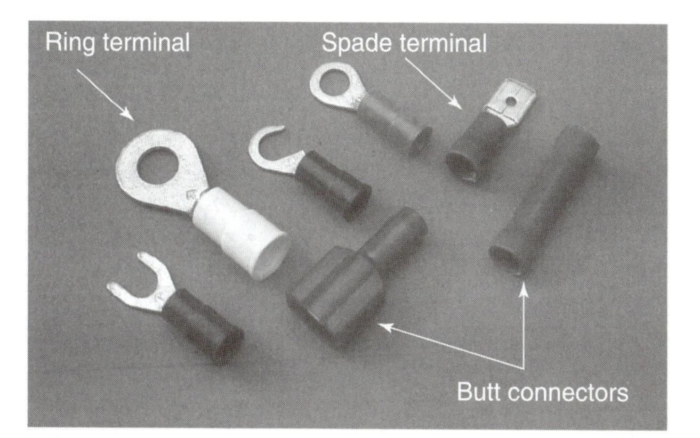

FIGURE 6-11 Solderless connectors are used to repair wiring on electrical systems.

Terminal and Connector Fasteners

Terminals and connectors are fasteners used to connect wires into an electrical system. An **electrical terminal** is a metal fastener that is attached to the end of a wire so that it can be connected into an electrical system. An **electrical connector** is a fastener that allows two or more wires to be connected together. Some connectors are used to permanently connect the wires. Others are made to snap together and apart.

There are many different types of terminals and connectors (Figure 6-11). Some are installed using a soldering gun. Most outdoor power equipment electrical wire repairs are done with solderless terminals and connectors. Terminals are made with different types of connecting ends. They are also made in sizes for different gauges of wire. Solderless butt connectors are made to connect the ends of wires together quickly.

Stripping and crimping pliers (Figure 6-12) are used to strip the insulation off the end of the wire. There is a set of different sized stripping cutters behind the jaws. They are also used to crimp on the solderless connectors. The pliers have jaws that crimp the tabs of terminals and butt connectors.

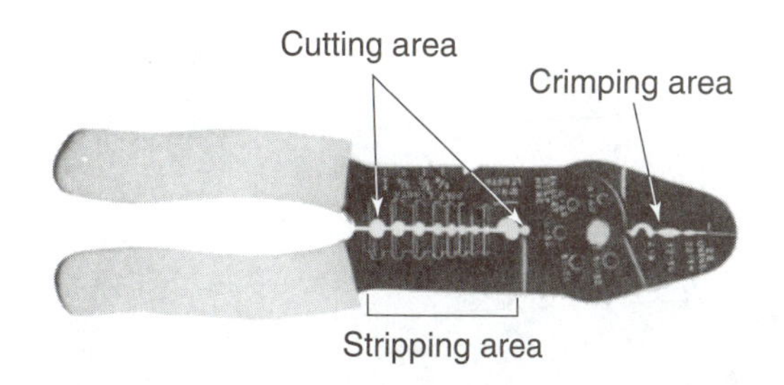

FIGURE 6-12 Wire stripping and crimping pliers are used to strip insulation from wire and install solderless connectors. (Courtesy of Vaco Products Company.)

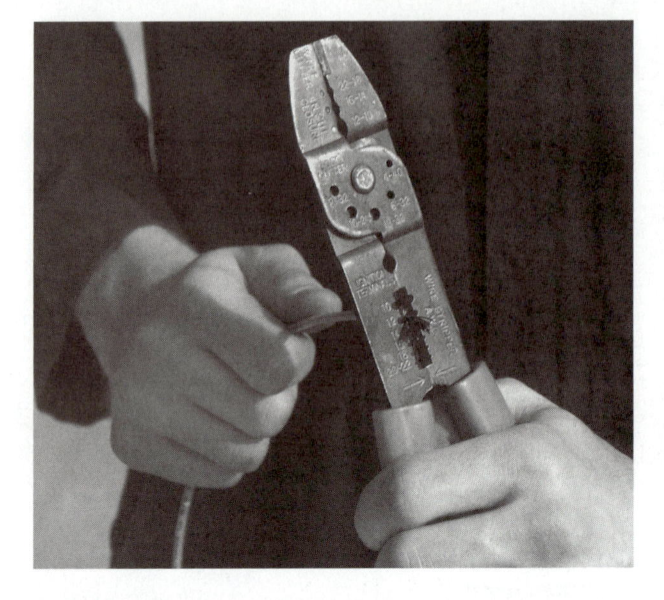

FIGURE 6-13 Use the correct size stripping cutter to strip the end of the wire.

To install a solderless terminal, strip the insulation off the end of the wire (Figure 6-13). Place the wire in the correct size stripping cutter. Squeeze the handles to cut the insulation. Pull on the pliers while holding the wire to remove the cut insulation.

The solderless terminal has several parts (Figure 6-14). The metal part has the terminal end and a set of crimping tabs. The tabs form the hole for the end of the wire. Terminals are sized according to the size wire that will fit into the hole. The terminal must be the correct size for the wire. Too small a terminal will not allow the wire to fit into the crimping area. Too large a hole will not give enough holding power when the tabs are crimped. The terminal has a plastic insulator that fits over the crimping area. The insulator prevents electricity from getting to any other noninsulated wire or part.

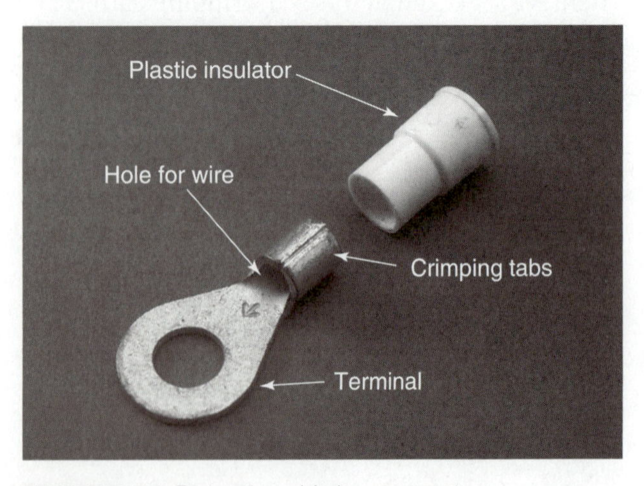

FIGURE 6-14 Parts of a solderless connector.

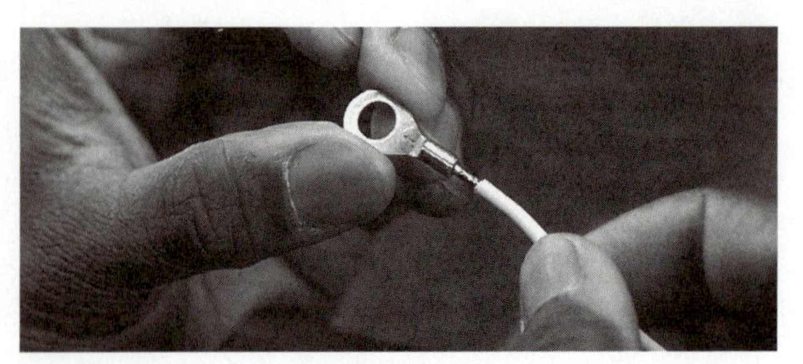

FIGURE 6-15 Insert the stripped wire end into the solderless connector crimping tabs.

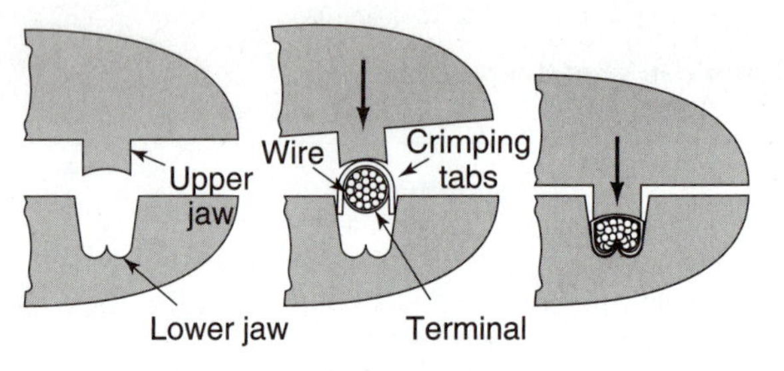

A. Crimping the Connector Tabs

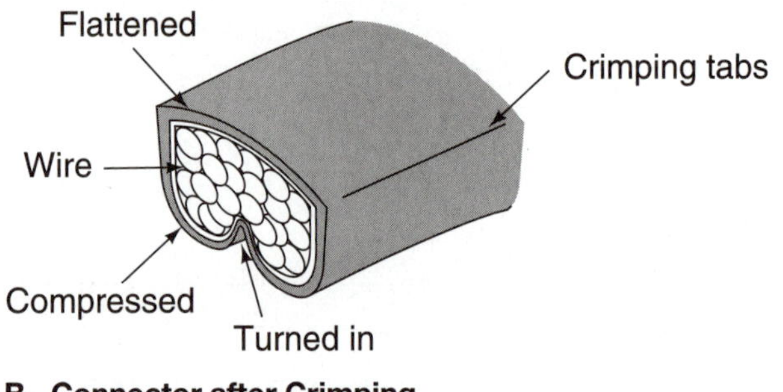

B. Connector after Crimping

FIGURE 6-16 Crimping the crimping tabs into the wire core.

Place the stripped end of the wire into the correct size solderless terminal (Figure 6-15). Use the crimping jaws to squeeze the crimping tabs down on the wire (Figures 6-16A and B). The crimping will lock the wire in place. It also makes a good electrical connection. Some pliers have different size crimping jaws for different sized terminals.

Butt connectors are used to join two wires (Figure 6-17). They are made in different sizes, matched to the sizes of wire to be joined. The butt connector has a metal sleeve covered with a plastic insulator. The ends of the two wires are placed into the metal sleeve. The center of the metal sleeve works like crimping tabs. Use the crimping jaws of the pliers to crimp the center of the sleeve down on the two ends of the wire to lock them in place.

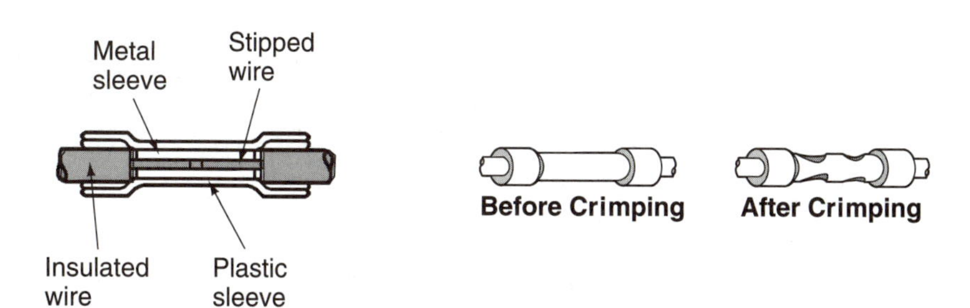

FIGURE 6-17 Butt connectors are crimped in the middle to hold both wires in place.

REVIEW QUESTIONS

1. List and describe the main types of keys.
2. Describe two uses of keys on an engine crankshaft.
3. Explain the purpose of using a soft metal key on some crankshafts.
4. Explain why a set screw is used with a key on some crankshaft pulleys.
5. List and describe the main types of pins used to assemble engine parts.
6. List and describe the two main types of retaining rings.
7. Explain how retaining ring pliers are used to remove and replace retaining rings.
8. List and describe the two basic parts of electrical wire.
9. Explain how wire core diameter affects wire electrical resistance.
10. List and explain how the AWG system is used to describe wire size.
11. Describe how the metric system is used to describe wire size.
12. List and describe the three uses for electrical stripping and crimping pliers.
13. Explain how electrical stripping and crimping pliers are used to strip insulation from electrical wires.
14. Explain how electrical stripping and crimping pliers are used to crimp a solderless terminal to a wire end.
15. Describe how electrical stripping and crimping pliers are used to crimp a solderless butt connector to a wire end.

DISCUSSION TOPICS AND ACTIVITIES

1. Locate and identify a selection of nonthreaded fasteners in the shop.
2. Use some scrap wires to practice stripping and installing solderless connectors.
3. A lighting circuit for a lawn and garden tractor uses 16 gauge wire. What do you think the effect on the lights would be if you used 20 gauge wire to rewire the circuit?

Service Information

OBJECTIVES

Upon completion and review of this chapter, you should be able to:

- Identify an engine's model designation.
- List and explain the purpose of the three main types of service information used by technicians.
- Use an equipment owner's manual to look up service information.
- Use a shop service manual to look up specifications and service procedures.
- Use a troubleshooting guide to troubleshoot an engine problem.
- Use a parts guide to look up a part number for a replacement part.
- Use an engine evaluation record to record engine failure information.
- Describe how to properly care for a customer in a small-engine shop.

TERMS TO KNOW

Engine evaluation record	Serial number
Engine model number	Service bulletin
Equipment owner's manual	Shop manual
Estimate	Specification
Flat rate schedule	Troubleshooting
Graphic symbols	Troubleshooting guide
Invoice	Warranty
Part number	Warranty claim
Parts manual	

INTRODUCTION

The small engine has become more and more complicated over the years. In years past, a service technician might have worked on one or two types of engines that did not change much from year to year. In those days you could keep all the repair information you needed in your head. Today there are many more types of engines and equipment. Each new model uses more complex systems. One of your most important tools is up-to-date service information.

ENGINE IDENTIFICATION

The first step in finding service information is to identify the engine you are working on. Each engine has an identification decal with a name and logo, which is often located on the engine blower housing (Figures 7-1A and B).

Most engine makers have several types and sizes of engines, each identified with a different model number (Figures 7-2A and B). An **engine model number** is a set of numbers and letters used to identify the size and type of engine. You need to locate the model number for the engine being serviced. Service information may be different for different models of engines. Model numbers are often stamped on the blower housing.

The model numbers are often used like a code. The code is used with a decoding chart (Figure 7-3). Usually found in the shop service manual, the decoding chart shows information about the engine. You can find out engine displacement, crankshaft direction, and starter type using this chart.

Engines also have a serial number (Figure 7-4). A **serial number** is a set of numbers and letters used to identify engine production information.

A. Engine Maker's Name Decal

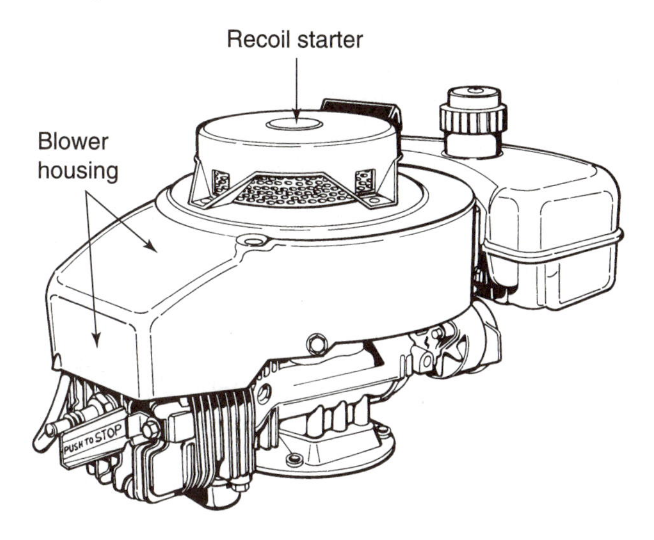

B. Engine Name Locations

FIGURE 7-1 The engine maker's name decal is usually located on the blower housing. (Provided courtesy of Tecumseh Products Company.)

The serial number is often stamped on the blower housing or on the engine crankcase. The shop service manual may have a decoding chart to explain this information.

SERVICE INFORMATION

The technician uses three general types of service information. These are:

1. Step-by-step service information
2. Specifications or measurement information
3. Troubleshooting information

Step-by-step information gives the steps to follow in doing a repair job (Figure 7-5). The steps are often numbered in order. Two very important rules apply when following service steps. First, make sure you have the steps for the exact engine model you are working on. Second, be sure you follow the steps exactly and in the order they are given.

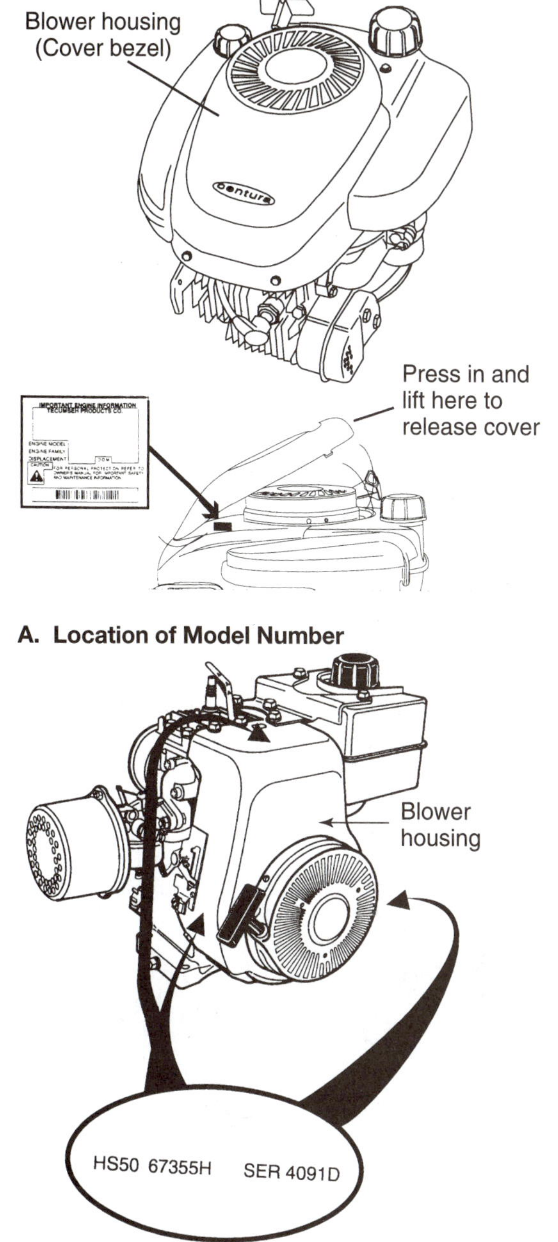

B. Location of Model Numbers

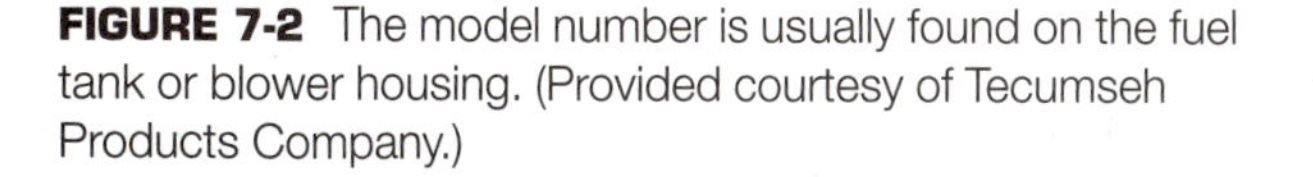

FIGURE 7-2 The model number is usually found on the fuel tank or blower housing. (Provided courtesy of Tecumseh Products Company.)

A **specification** is the information and measurements used to troubleshoot, adjust, and repair an engine. The word "specifications" is often shortened to just "specs." Measuring tools are used to measure these specifications. Examples of specifications are crankcase oil capacity, spark plug electrode gap, and ignition timing.

Specifications may be found in the step-by-step information. They are often given in tables. All

Interpretation of Model Number

The letter designations in a model number indicate the basic type of engine.

OHH — Overhead Valve Horizontal
OHM — Overhead Valve Horizontal Medium Frame
OHSK — Overhead Valve Horizontal Snow King
OVM — Overhead Valve Vertical Medium Frame
OVRM — Overhead Valve Vertical Rotary Mower
OVXL — Overhead Valve Vertical Medium Frame Extra Life
OHV — Overhead Valve Vertical

The number designations following the letters indicate the basic engine model.

The number following the model number is the specification number. The last three numbers of the specification number indicate a variation to the basic engine specification.

The serial number or D.O.M. indicates the production date of the engine.

Using model OHV16-204207A, serial 5215C as an example, interpretation is as follows:

OHV16-204207A is the model and specification number.

OHV	Overhead Valve Vertical
16	indicates the basic engine model.
204207A	is the specification number used for properly identifying the parts of the engine.
5215C	is the serial number or D.O.M. (Date of Manufacture)
5	is the last digit in the year of manufacture (1995).
215	indicates the calendar day of that year (215th day or August 3, 1995).
C	represents the line and shift on which the engine was built at the factory.

FIGURE 7-3 A decoding chart is used to get information from the model number. (Provided courtesy of Tecumseh Products Company.)

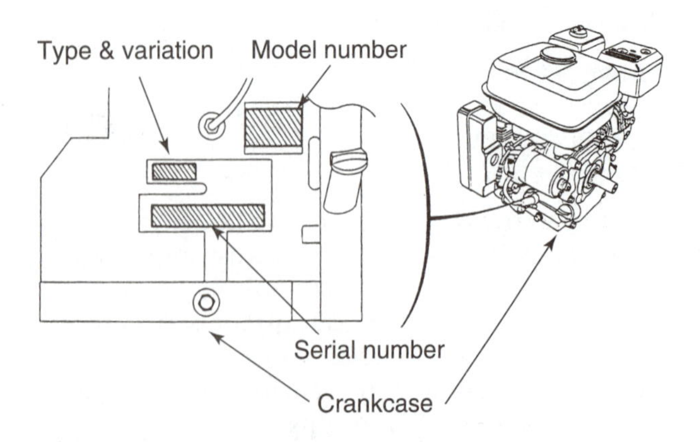

FIGURE 7-4 Common locations for engine serial numbers. (Courtesy of American Honda Motor Co., Inc.)

REWIND STARTERS

Disassembly Procedure

1. Untie knot in rope and slowly release spring tension.
2. Remove retainer screw, retainer cup (cam dog on snow-proof type), starter dog and spring, and brake spring.
3. Lift out pulley—turn spring and keeper assembly to remove. Replace all worn or damaged parts.

Assembly Procedure

1. Place rewind spring and keeper assembly into pulley—turn to lock into position. The spring should have a light coating of grease on it.
2. Place pulley into the starter housing.
3. Install brake spring, starter dogs, and dog return springs.
4. Replace retainer cup (cam dog on snow-proof type); and retainer screw. Tighten to 115–135 inch pounds.

FIGURE 7-5 An example of a step-by-step procedure. (Provided courtesy of Tecumseh Products Company.)

specification tables have several parts. The heading at the top shows the engine service area covered by the specifications (Figure 7-6). The different engine models are listed across the chart. The column below each model shows the specifications for that model. Read down the chart to find the specification you need. Then read across to the specification for that engine.

SERVICE TIP: Use a rule or piece of paper as a guide to read across a specification table. Write the specification down. Do not try to memorize specifications, because this could result in a service error.

Troubleshooting is a step-by-step procedure followed to locate and correct an engine problem (Figure 7-7). Troubleshooting information, which is very important to the technician, may be given in step-by-step procedures or in charts. Troubleshooting information must be for the same problem as the engine you are working on. It also must be for the same engine model you are working on. Follow the troubleshooting steps in order until the problem is located and corrected.

SERVICE MANUALS

Several types of manuals are used when servicing an engine. The most common types of manuals are

Specifications	OVRM40		OVRM50-6.75 OVRM105		OVRM120		OHH50 OHSK50	
	Standard English	Metric mm	Standard English	Metric mm	Standard English	Metric mm	Standard English	Metric mm
Displacement (in^2) (CC)	9.06	148.50	10.49 Note (A)	171.93 Note (A)	11.9 Note (A)	195.04 Note (A)	10.49	171.93
Stroke	1.844	46.838	1.938	49.225	1.938	49.225	1.938	49.225
Bore	2.500 2.501	63.500 63.525	2.625 2.626 Note (B)	66.675 66.700 Note (B)	2.795 2.796 Note (B)	70.993 71.018 Note (B)	2.625 2.626	66.675 66.700
Ignition Module Air Gap	.0125	.3175	.0125	.3175	.0125	.3175	.0125	.3175
Spark Plug Gap	.030	.762	.030	.762	.030	.762	.030	.762
Valve Clearance In./Ex.	.004 .004	.1016 .1016	.004 .004	.1016 .1016	.004 .004	.1016 .1016	.004 .004	.1016 .1016
Valve Seat Angle	46°		46°		46°		46°	
Valve Seat Width	.035 .045	.889 1.143	.035 .045	.889 1.143	.035 .045	.889 1.143	.035 .045	.889 1.143
Valve Guide Oversize Dimension	INT. EX. .2807 .2787 .2817 .2797	INT. EXT. 7.130 7.079 7.155 7.104	INT. EX. .2807 .2787 .2817 .2797	INT. EX. 7.130 7.079 7.155 7.104	INT. EX. .2807 .2787 2817 .2797	INT. EX. 7.130 7.079 7.155 7.104	.2807 .2817	7.135 7.155
Crankshaft End Play	.006 .027 Note (C)	.1524 .6858 Note (C)	.006 .027 Note (C)	.1524 .6858 Note (C)	.006 .027 Note (C)	.1524 .6858 Note (C)	.006 .027	.1524 .6858
Crankpin Journal Dia.	.8610 .8615	31.869 21.882	.9995 1.000	25.362 25.400	.9995 1.000	25.362 25.400	.9995 1.000	25.362 25.400
Crankshaft Dia. Flywheel End Main Brg.	.9985 .9990	25.362 25.375	.9985 .9990	25.362 25.375	.9985 .9990	25.362 25-375	.9985 .9990	25.362 25.375
Crankshaft Dia. P.T.O. Main Brg.	.9985 .9990	25.362 25.375	1.0005 1.0010	25.413 25.425	1.0005 1.0010	25.413 25.425	.9985 .9990	25.362 25.375
Conn. Rod Dia. Crank Brg.	1.0005 1.0010	25.413 25.425	1.0005 1.0010	25.413 25.425	1.0005 1.0010	25.413 25.425	1.0005 1.0010	25.413 25.425
Camshaft Bearing Diameter	.4975 .4980	12.637 12.649	.4975 .4980	12.637 12.649	.4975 .4980	12-637 12.649	.4975 .4980	12.637 12.649
Piston Dia. Bottom of Skirt	2.4950 2.4520	63.373 63.378	2.6204 2.6220	66.558 66.599	2.6204 2.6220	66-558 66-599	2.6204 2.6220	66.558 66.599
Ring Groove Side Clearance 1 st & 2nd Comp.	.0020 .0050	.051 .127	.0020 .0050	.051 .127	.0020 .0050	.051 .127	.0020 .0050	.051 .127
Ring Groove Side Clearance Bottom Oil	.0050 .0035	.013 .089	.0050 .0035	.013 .089	.0050 .0035	.013 .089	.0050 .0035	.013 .089
Piston Skirt to Cylinder Clearance	.0040 .0058	.102 .147	.0030 .0056	.076 .142	.0030 .0056	.076 .142	.0030 .0056	.076 .142
Ring End Gap	.010 .020	.254 .508	.010 .020	.254 .508	.010 .020	.254 .508	.007 .017	.178 .432
Cylinder Main Bearing Diameter	1.0005 1.0010	25.413 25.425	1.0005 1.0010	25.413 25.425	1.0005 1.0010	25.413 25.425	1.0005 1.0010	25.413 25.425
Cylinder Cover/Flange Main Brg. Diameter	1.0050 1.0010	25.413 25.425	1.0050 1.0010	25.413 25.425	1.0050 1.0010	25.413 25.425	1.0050 1.0010	25.413 25.425

FIGURE 7-6 A chart for engine specification. (Provided courtesy of Tecumseh Products Company.)

Problem: ENGINE OVERSPEEDING

Steps to follow to correct the problem:

1. If the engine runs wide open (faster than normal), shut the engine off or slow it down **immediately**.
2. Check the condition of the external governor shaft, linkage, governor spring, and speed control assembly for breakage or binding. Correct or replace binding or damaged parts.
3. Follow the governor adjustment procedure and reset the governor—see "Service" in this chapter.
4. Run the engine. Be ready to shut the engine off if an overspeed problem still exists. If the problem persists, the engine will require disassembly to inspect the governor gear assembly for damage, binding, or wear.
5. See Chapter 9 under "Disassembly Procedure" to disassemble the engine.
6. Remove the governor gear assembly. Repair or replace as necessary.

FIGURE 7-7 A troubleshooting procedure is used to locate the cause of a problem. (Provided courtesy of Tecumseh Products Company.)

the equipment owner's manual, shop manual, troubleshooting guide, parts manual, service bulletin, and engine evaluation record.

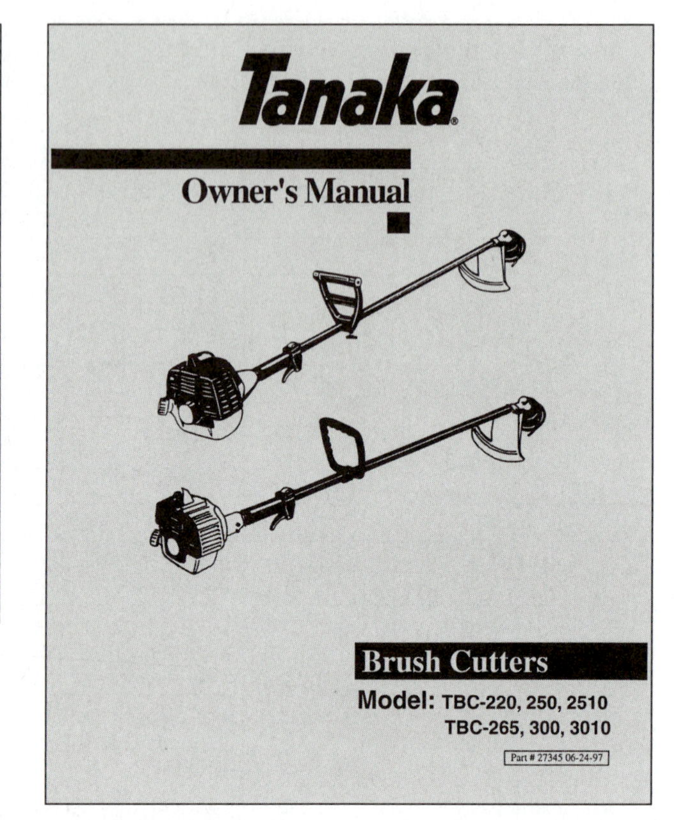

FIGURE 7-8 An equipment owner's manual. (Courtesy of Tanaka/ISM.)

Equipment Owner's Manual

An **equipment owner's manual** (Figure 7-8) is a booklet given with new outdoor power equipment that explains the engine and equipment operation. The purpose of the manual is to explain to the new owner how to safely operate the equipment. These manuals often have some maintenance information and specifications.

Shop Manual

The **shop manual** (Figure 7-9) is a booklet with detailed service information supplied to technicians by the equipment or engine maker. Shop manuals may be called *manufacturer's repair manuals, mechanics' handbooks,* or *shop service manuals*. A shop manual may cover one engine model or several similar models. Equipment service manuals often cover one model of equipment.

The first step in using an engine shop manual is to find the engine model. Then, find the correct shop manual for that engine. The typical shop manual has a table of contents (Figure 7-10) located at the front. An engine is divided into major service areas such as ignition, carburetion, governors, and lubrication. A section in the shop service manual is given to each of these areas. For example, suppose you need to repair a carburetor. Turn to the pages in the carburetor section. You will find step-by-step repair procedures, specification charts, and troubleshooting information on carburetors.

Troubleshooting Guide

A **troubleshooting guide** is a chart or diagram showing the steps to follow to locate and correct a service problem. The troubleshooting guide can be a separate booklet or it may be a section of the shop manual. These charts or diagrams are there to help you save time in identifying and solving problems. They list common problems, possible causes, and corrective actions to fix the problems.

The road map troubleshooting chart (Figure 7-11) is one common type of troubleshooting guide. It has the troubleshooting steps arranged like roads on a road map. You follow the steps just the way you would follow a highway road map. You use this guide by locating the troubleshooting chart for the engine's problem. For example, the problem is that the engine will not start. Find the troubleshooting chart that begins with "Engine will not start." The first step on the roadmap is to "Check for spark" then "Check if spark plug is wet or dry." If the spark plug is wet, one set of steps is followed. If it is dry,

FIGURE 7-9 Shop manuals are available for different engine models.

CONTENTS

SPECIFICATIONS	1
SERVICE INFORMATION	2
MAINTENANCE	3
AIR CLEANER, MUFFLER	4
RECOIL STARTER, FAN COVER	5
CARBURETOR	6
FUEL TANK, GOVERNOR ARM, CONTROL BASE	7
FLYWHEEL, IGNITION COIL, STARTER MOTOR	8
CYLINDER HEAD, VALVES	9
CYLINDER COVER, GOVERNER	10
CRANKSHAFT, PISTON	11
REDUCTION UNIT	12
OPTIONAL PARTS	13

FIGURE 7-10 Example of a shop manual table of contents. (Courtesy of American Honda Motor Co., Inc.)

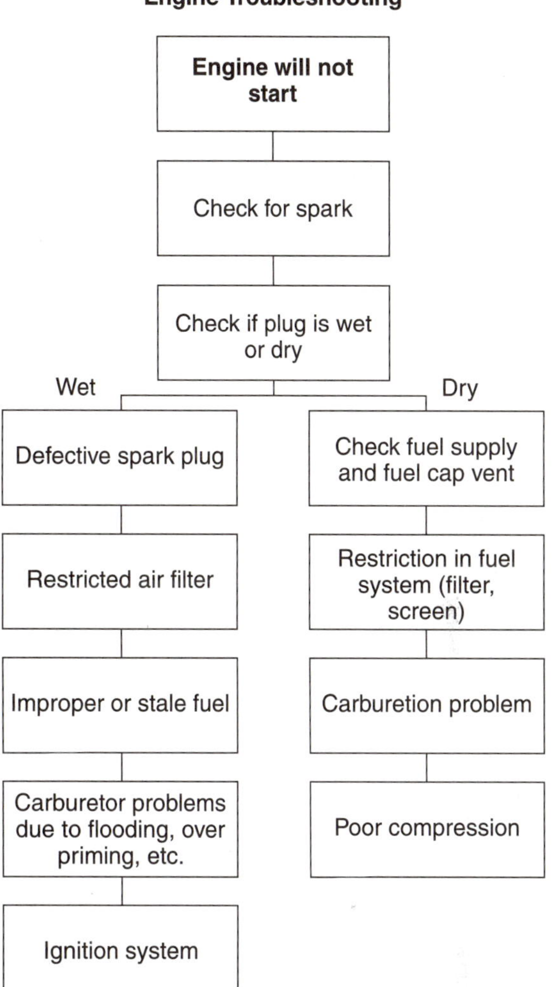

FIGURE 7-11 A road map type troubleshooting guide. (Provided courtesy of Tecumseh Products Company.)

a different set of steps is followed until the problem is corrected.

The troubleshooting chart (Figure 7-12) is another type of guide. The top of the chart shows the area to be covered. For example, one area is carburetion. Problems, such as plugged air filter, are shown on the left side of the chart. The symptoms are listed after each of the problems. The operating conditions in which these problems occur are shown at the top of the chart. These are start, idle, accelerate, and high speed. To use the chart, first identify the problem. Then, identify the operating conditions when it happens. For example, the engine hesitates. This happens during acceleration. Look under the accelerate category and find the problem called "hesitates." Then, look down the chart to find possible causes. The causes listed are problems such as leaky carburetor gasket, air bleed restricted, stuck or dirty ball check, plugged tank filter or vent, fuel pick-up restricted; or main nozzle restricted. Check each of these problems until you find the cause of the problem.

Some troubleshooting charts and repair procedures use graphic symbols (Figure 7-13, page 85). **Graphic symbols** are small road sign type symbols used to show steps on troubleshooting guides or re-

CARBURETION TROUBLESHOOTING

	START			IDLE					ACCELERATE			HIGH SPEED				
AIR SYSTEM PROBLEMS	Hard Starting	Fuel Leak at Carburetor	Engine Floods	Will Not Idle	Rich Idle	Idles with Needle Closed	Hunts - Erratic Idle	Idles Fast - Lean	Will Not Acceler- ate	Over Rich Accelera- tion	Hesitates	Will Not Run at High Speed	Low Power	Hunts at High Speed	Runs with Needle Closed	Engine Over- speeds
Plugged Air Filter	●			●	●				●	●		●	●			
Leaky Carburetor Gasket				●			●	●			●					●
Throttle or Choke Shafts Worn	●			●			●	●					●	●		●
Choke Not Functioning Properly	●															
Plugged Atmospheric Vent		●	●													
Air Bleed Restricted	●			●	●		●		●		●			●		
Damaged or Leaky "O" Rings		●					●	●						●		●
DIAPHRAGM SYSTEM PROBLEM																
Damaged Diaphragm	●	●		●				●					●		●	
Stuck or Dirty Ball Check				●		●			●		●			●		
Diaphragm Upside Down	●															
FUEL SYSTEM PROBLEM																
Plugged Tank Filter or Vent	●								●		●	●	●			
Fuel Pick-up Restricted	●			●			●		●		●	●		●		
Idle Port Restricted				●			●									
Damaged Adjustment Needles	●			●	●	●	●		●			●	●	●	●	●
Incorrect Float Height			●				●		●	●			●	●		
Main Nozzle Restricted	●								●		●	●	●	●		
Dirty, Stuck Needle and Seat	●	●	●										●			
Fuel Inlet Plugged	●			●			●						●	●		

FIGURE 7-12 A chart type troubleshooting guide. (Provided courtesy of Tecumseh Products Company.)

pair procedures. These symbols speed up troubleshooting because you can see at a glance exactly what to do. You do not have to read directions.

Parts Manual

A **parts manual** is a book or list that shows illustrations and lists the identification number of each engine or equipment part (Figure 7-14). Most engine or equipment repairs require the replacement of worn parts with new ones. Getting the exact replacement part is an important part of each repair job. Parts are ordered and purchased using a part number. A **part number** is a number given to each of the engine or equipment parts by each engine or equipment maker. Each engine and equipment maker has its own parts numbering system.

To use a parts manual, first find the engine model. Then, turn to the pages of the parts manual that cover that engine model. Each engine part will be shown in exploded view. Each part will be labeled with a part number. Sometimes several related parts are replaced during a repair job. These

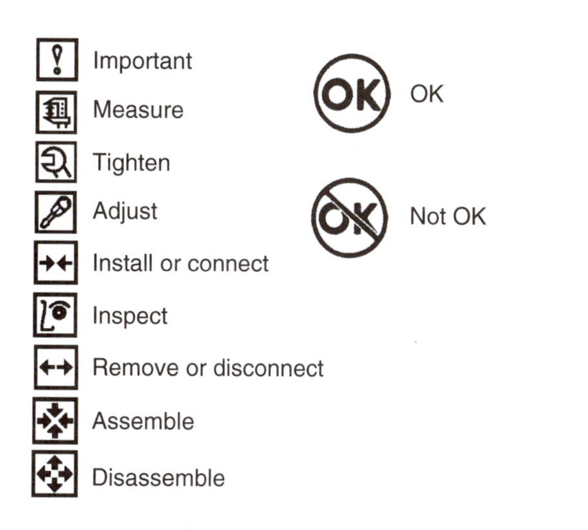

FIGURE 7-13 Graphic symbols make troubleshooting steps quicker to use. (Courtesy of General Motors Corporation, Chevrolet Motor Division.)

parts can be grouped as a parts assembly. They may be given a parts assembly number. You order a specific part or parts assembly using the part number for identification.

Some engine manufacturers publish their parts manuals on microfiche. Microfiche is a card with film that is viewed in a reader. The film has pages from the manual reduced to a very small size. The reader enlarges the reduced image so that it can be easily seen. Microfiche parts manuals take up much less space than book type manuals. The pages are read just like the pages in a book type manual.

Computers are also used for parts information. Some engine makers have parts manuals on computer software. Others have parts manuals on the internet. The exploded view of the parts and the parts numbers are shown just like a manual.

Service Bulletins

Engine and equipment repair procedures and specifications sometimes change after a shop manual has been published. Technicians in the field learn about service changes and service problems through a service bulletin (Figure 7-15). A **service bulletin** is a revised service procedure or specification given by engine or equipment makers to technicians in the field. The bulletin is often a one- or two-page description of a service procedure. The bulletin is read and then placed in the shop manual for reference.

Engine Evaluation Record

An **engine evaluation record** (Figure 7-16, page 88) is a form with questions used to guide the technician through engine inspection, disassembly, and failure analysis. These forms often have several pages. They are divided into sections such as owner and equipment information, usage and maintenance information, preliminary examination, and tear down analysis. You follow the steps and write in the necessary information in each section.

Following an engine inspection data record helps you do a complete job. It makes sure you do not forget any failure cause. Some makers require this form be filled out for a warranty claim. A **warranty claim** is a request by an engine or equipment owner for a repair to be covered (paid for) by the engine or equipment maker.

A **warranty** is a legal document provided by the engine or equipment maker that says that certain parts of the engine or equipment will be repaired at no cost to the owner. The failure has to happen within a certain amount of time or hours of service. Authorized service shops do the warranty work after proper verification. Shops are paid by the engine or equipment maker. Often a technician must inspect the engine or equipment to find out if it should be covered by a warranty. A technician would look to see if it has been maintained properly, abused, or has too many hours.

CUSTOMER CARE

Customer care is a very important part of service. The customer has many repair shops from which to choose. The shop and technician that treat customers with promptness, courtesy, and professionalism will have repeat customers and a growing business. You must always remember that the customer is the reason for the job.

The first step in the customer care process is to greet the customer as soon as they enter shop. The first impression you make with the customer is often a lasting one. Customers sense if the shop has concern for them and their reasons for being there. A good technician attitude can calm customers who may be worried about a visit to the shop.

You should show a good strong smile when greeting the customer. When possible, address the customer by name. Be sincere and thank the customer for coming to the shop. Always be well groomed and wear clean shop clothing. Begin the meeting with an introduction. Be a good listener and use good eye contact. Make a good first impression by looking and sounding confident. Be careful with promises, because customers expect them to be kept.

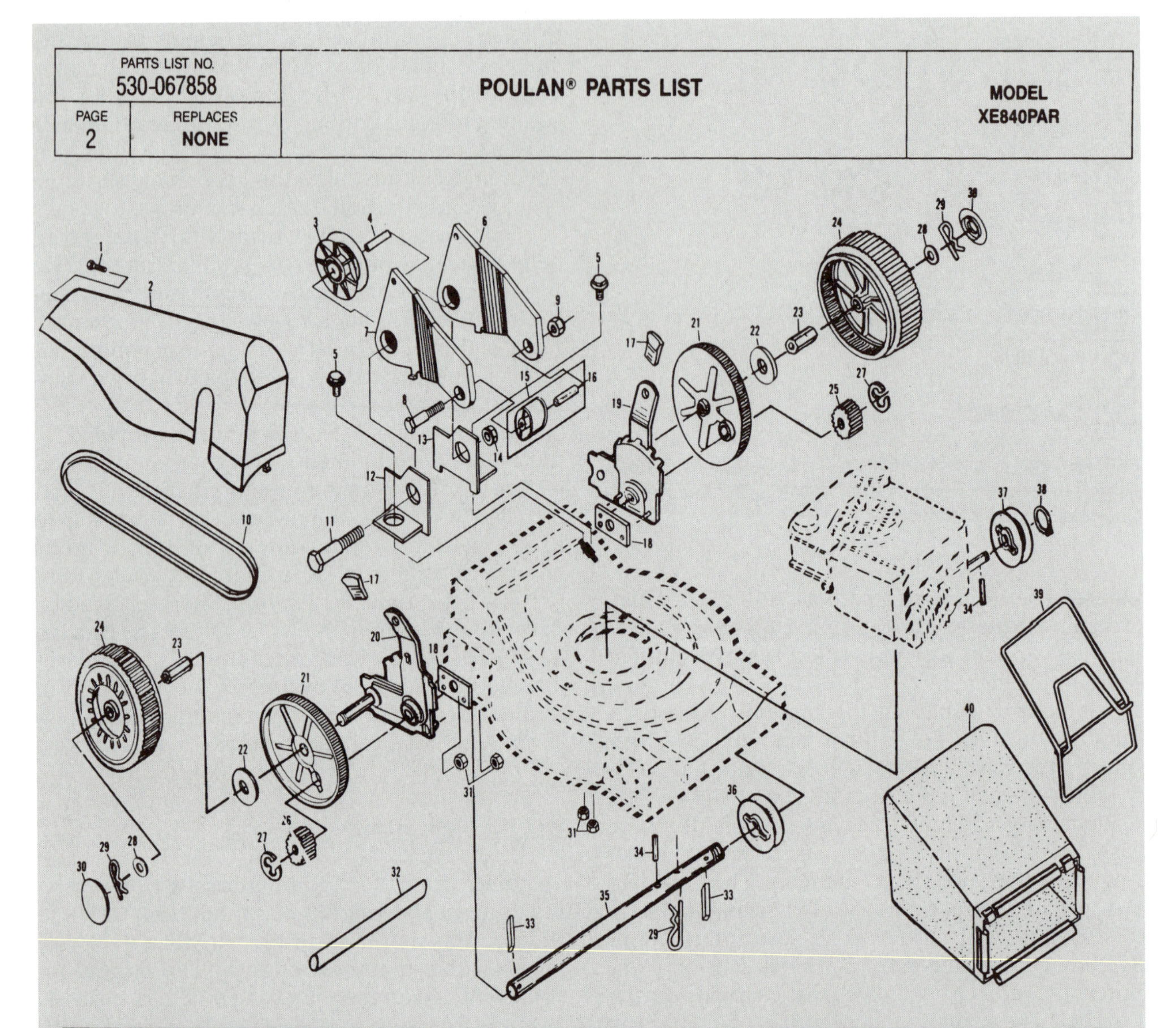

Key No.	Part No.	Description
1	532-750097	Hex Washer Head Screw
2	532-850043	Drive Cover
3	532-088516	Idler Pulley
4	532-850037	Pulley Spacer
5	532-065231	Hex Head Machine Screw
6	532-850033	Idler Bracket (Left)
7	532-850035	Idler Bracket (Right)
8	532-053109	Hex Head Bolt
9	532-050962	Hex Locknut
10	532-850028	Drive Belt
11	532-050884	Hex Head Bolt
12	532-850031	Pivot Bracket (Right)
13	532-850029	Pivot Bracket (Left)
14	532-063124	Locknut
15	532-048160	Idler Assembly Kit (Incl. Ref. #16)
16	532-088576	Idler Spacer
17	532-077865	Selector Knob
18	532-850049	Spacer Plate
19	532-850982	Selector Assembly (Left)
20	532-850983	Selector Assembly (Right)
21	532-088080	Dust Cover
22	532-067725	Washer
23	532-088446	Wheel Bushing
24	532-850707	Wheel & Tire Assembly
25	532-850016	Drive Pinion (Left)
26	532-850017	Drive Pinion (Right)
27	532-053753	E-Ring
28	532-052160	Flat Washer
29	532-085179	Retainer Clip
30	532-800544	Hubcap
31	532-063601	Locknut
32	532-850373	Drive Shaft Cover
33	532-850018	Drive Pawl
34	532-088028	Roll Pin
35	532-850025	Drive Shaft
36	532-088082	Driven Pulley
37	532-850027	Drive Pulley
38	532-058334	Retaining Ring
39	532-088614	Catcher Frame
40	532-800563	Grass' Bag Assembly

FIGURE 7-14 Parts and part numbers can be found in a parts manual or parts list. (Courtesy of Poulan/Weed Eater.)

ISSUED: January, 1999

SUBJECT: Rich Running Engines Above 7,000 Feet

MODELS OR TYPES AFFECTED: HM, HMSK, OHSK

Tecumseh Products Company has been working with the California Air Resource Board (CARB) and the Environmental Protection Agency (EPA) to improve engine performance at elevations above 7,000 feet.

The result of these cooperative efforts is the availability of leaner service jets for the specific models listed on the attached page. These changes to the engines carburetor jetting will require some questioning and education of the customer by the sales person prior to the sale or the technician if the sale was already made. Part of the sale of new emissionized product **must** be questioning the customer regarding the elevation at which the equipment will be used.

If the unit was already sold and the customer is commenting about a rich running condition, poor run quality or excessive smoke, check the main jet number compared to the chart. By following this chart you can install the correct jet for the customers operational elevation while maintaining the engines emissions compliance and performance.

Please note under the regulation guidelines there are specific altitudes at which these changes can be made. The first range is 6,000-11,000 feet above sea level, the second is over the 10,000 feet. The change to carburetor calibration for these conditions requires changing the main metering jet which is part of the bowl nut.

The main jets are identified with a stamped in number on the bottom (see illustration). Using the chart below select the correct service jet kit. The kit will include instructions, jet(s) and a bright colored decal advising the customer that the unit has been re-jetted to compensate for high altitude. The decal will further advise the operator to have the unit re-jetted if taken out of that range.

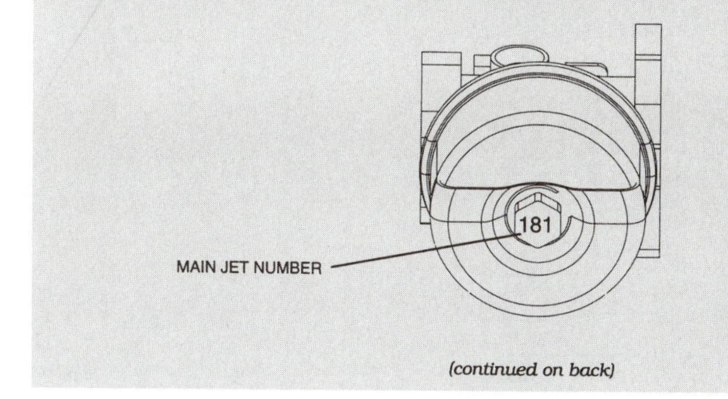

(continued on back)

FIGURE 7-15 Service bulletins have the most up-to-date information. (Provided courtesy of Tecumseh Products Company.)

Some angry and upset customers are going to come into the shop. This anger may be as a result of a past experience in a shop. It could have nothing at all to do with you or getting an engine or equipment serviced. Customers have spent a lot of money on their outdoor power equipment. Getting it serviced is usually an inconvenience and is likely to cause them an unplanned expense. It is common for some customers to be unhappy when they enter the shop. You should remain calm and not to react defensively. Try not to take the customer's conduct personally. Focus on the customer's problem, not on the customer's behavior.

Once the customer is greeted, the next step is to conduct an interview and fill out an estimate (Figure 7-17). An **estimate**, or repair order, is a form that details the customer's complaint and lists the probable labor, part, and material costs required for a repair. The estimate is often combined with another form called an invoice. An **invoice**, or bill, is a form that lists the labor, parts, and material and other costs owed by the customer.

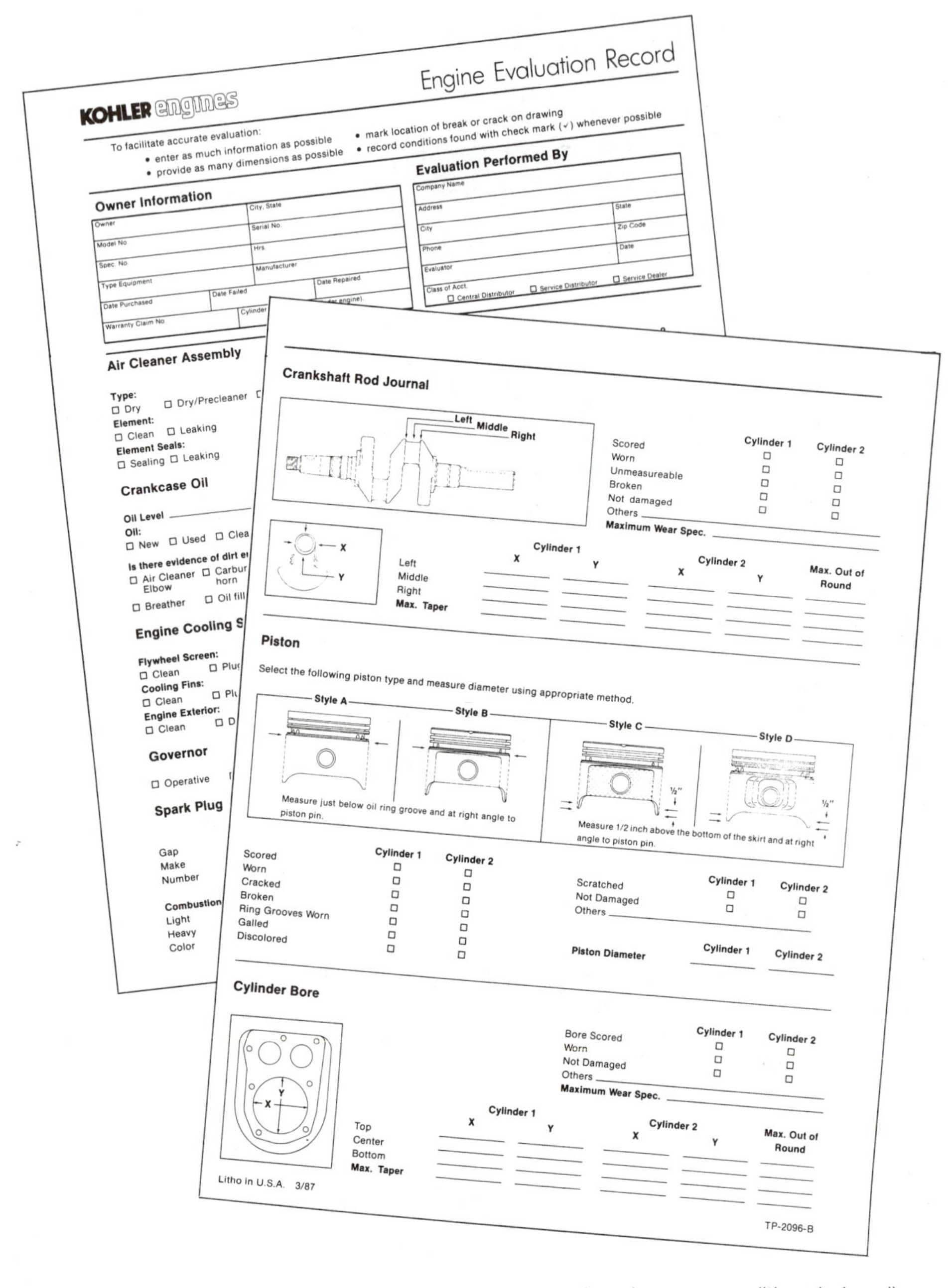

KOHLER engines

Engine Evaluation Record

To facilitate accurate evaluation:
- enter as much information as possible
- provide as many dimensions as possible
- mark location of break or crack on drawing
- record conditions found with check mark (✓) whenever possible

Owner Information

Owner | City, State
Model No. | Serial No.
Spec. No. | Hrs.
Type Equipment | Manufacturer
Date Purchased | Date Failed | Date Repaired
Warranty Claim No. | Cylinder

Evaluation Performed By

Company Name
Address | State
City | Zip Code
Phone | Date
Evaluator
Class of Acct. ☐ Central Distributor ☐ Service Distributor ☐ Service Dealer

Air Cleaner Assembly

Type: ☐ Dry ☐ Dry/Precleaner
Element: ☐ Clean ☐ Leaking
Element Seals: ☐ Sealing ☐ Leaking

Crankcase Oil

Oil Level
Oil: ☐ New ☐ Used ☐ Clea
Is there evidence of dirt e
☐ Air Cleaner Elbow ☐ Carbur horn
☐ Breather ☐ Oil fill

Engine Cooling S

Flywheel Screen: ☐ Clean ☐ Plu
Cooling Fins: ☐ Clean ☐ Pl
Engine Exterior: ☐ Clean ☐ D

Governor

☐ Operative

Spark Plug

Gap
Make
Number

Combustion
Light
Heavy
Color

Crankshaft Rod Journal

Left Middle Right

	Cylinder 1	Cylinder 2
Scored	☐	☐
Worn	☐	☐
Unmeasureable	☐	☐
Broken	☐	☐
Not damaged	☐	☐
Others	☐	☐

Maximum Wear Spec.

	Cylinder 1 X	Cylinder 1 Y	Cylinder 2 X	Cylinder 2 Y	Max. Out of Round
Left					
Middle					
Right					
Max. Taper					

Piston

Select the following piston type and measure diameter using appropriate method.

Style A — Style B — Measure just below oil ring groove and at right angle to piston pin.

Style C — Style D — Measure 1/2 inch above the bottom of the skirt and at right angle to piston pin.

	Cylinder 1	Cylinder 2
Scored	☐	☐
Worn	☐	☐
Cracked	☐	☐
Broken	☐	☐
Ring Grooves Worn	☐	☐
Galled	☐	☐
Discolored	☐	☐

	Cylinder 1	Cylinder 2
Scratched	☐	☐
Not Damaged	☐	☐
Others		

Piston Diameter — Cylinder 1 — Cylinder 2

Cylinder Bore

	Cylinder 1	Cylinder 2
Bore Scored	☐	☐
Worn	☐	☐
Not Damaged	☐	☐
Others		

Maximum Wear Spec.

	Cylinder 1 X	Cylinder 1 Y	Cylinder 2 X	Cylinder 2 Y	Max. Out of Round
Top					
Center					
Bottom					
Max. Taper					

Litho in U.S.A. 3/87

TP-2096-B

FIGURE 7-16 An engine evaluation record used to record engine part condition during disassembly. (Courtesy of Kohler Co.)

In some shops, the owner may fill out these forms. In other shops, technicians do this task. The estimate is what the technician uses to guide the repair job. The estimate and invoice are also legal documents. They say that the shop understands the engine or equipment problem. They say what the approximate cost to the customer will be to make the needed repairs. The customer signs and agrees to pay for the shop's services.

CAUTION: An estimate or repair order that is incorrect, incomplete, misspelled, or illegible can cause a bad repair. It can result in possible legal problems for the shop, and it can give the customer a poor impression. There is a good reason for all the information asked for on repair order forms, especially when writing up a repair covered by a warranty.

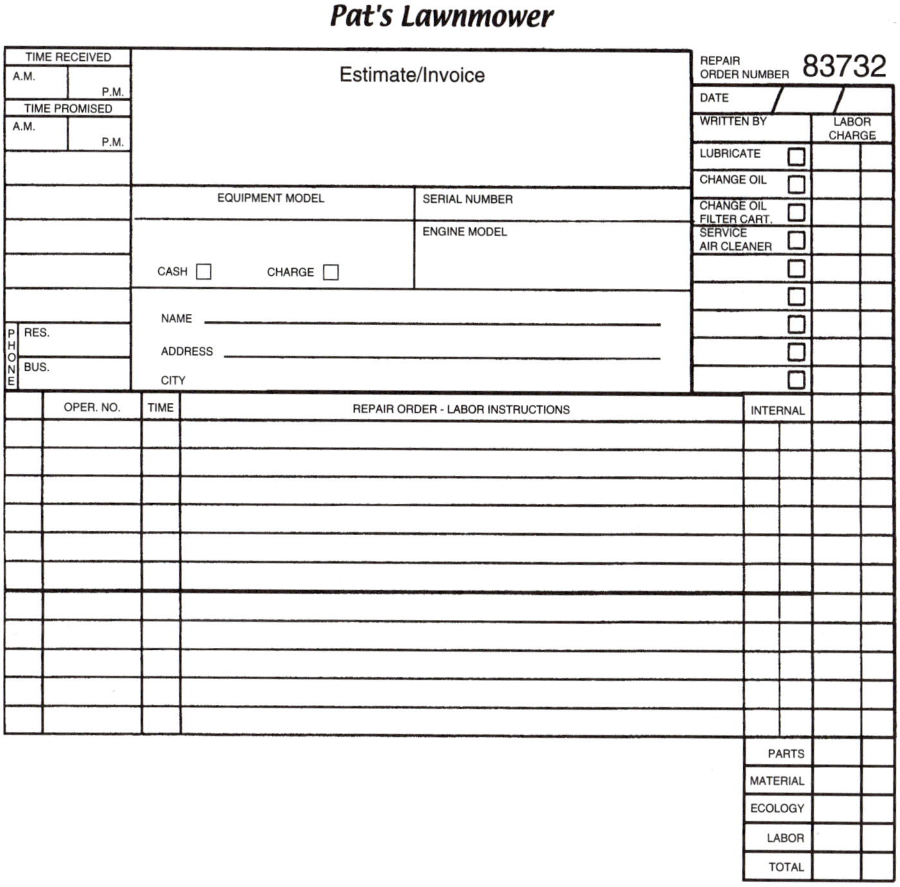

Pat's Lawnmower

Estimate/Invoice

TIME RECEIVED
A.M. P.M.
TIME PROMISED
A.M. P.M.

REPAIR ORDER NUMBER 83732
DATE / /
WRITTEN BY
LABOR CHARGE
LUBRICATE
CHANGE OIL
CHANGE OIL FILTER CART.
SERVICE AIR CLEANER

EQUIPMENT MODEL
SERIAL NUMBER
ENGINE MODEL
CASH CHARGE

NAME
ADDRESS
CITY
PHONE RES.
BUS.

OPER. NO. | TIME | REPAIR ORDER - LABOR INSTRUCTIONS | INTERNAL

PARTS
MATERIAL
ECOLOGY
LABOR
TOTAL

FIGURE 7-17 An estimate and invoice form used to estimate and bill a repair job.

An estimate and invoice are often a printed form or a software form on a computer screen. There are many different styles, but they all ask for the same basic information. The form must fully identify the customer and the engine or equipment. It must also fully show the customer complaint, the cause, and the correction. The form has a printed repair order or invoice number. That number is used on only that one form. It shows up on all the carbon copies of the form. The number is used as a reference when discussing the form.

There are blank spaces at the top of the form for identifying the engine or equipment. There are spaces for information about the customer. Below the area for customer information is a set of lined spaces. These spaces may be called "Instructions" or "Labor Instructions." These spaces are filled in with the customer's complaint, possible cause, and corrections. The complaint is the reason the customer has for bringing the engine or equipment to the shop. You listen carefully to what the customer says. Then, you question the customer about the problems. Write the complaint legibly and clearly on the form.

You will need to apply your technical skills to help find the cause of the problem. Questions to the customer must focus in on the exact problem. For example, if the customer complains that the engine stalls, ask:

- Does it stall all the time?
- Does it stall when it is first started?
- Does it stall only after it has been running a while?
- Does it stall when accelerating from idle?
- Is it intermittent or all the time?
- When was the last time the engine was worked on?
- What type of work was done?
- Did the problem seem to be solved?

When you know the cause of the problem, fill out the estimate part of the form. An estimate must be given to the customer showing the cost of the repair before a repair is done. Customers expect accurate estimates of what repairs will cost when they leave equipment in a repair shop.

A repair job consists of two basic parts. One is the cost of the replacement parts. The other is the

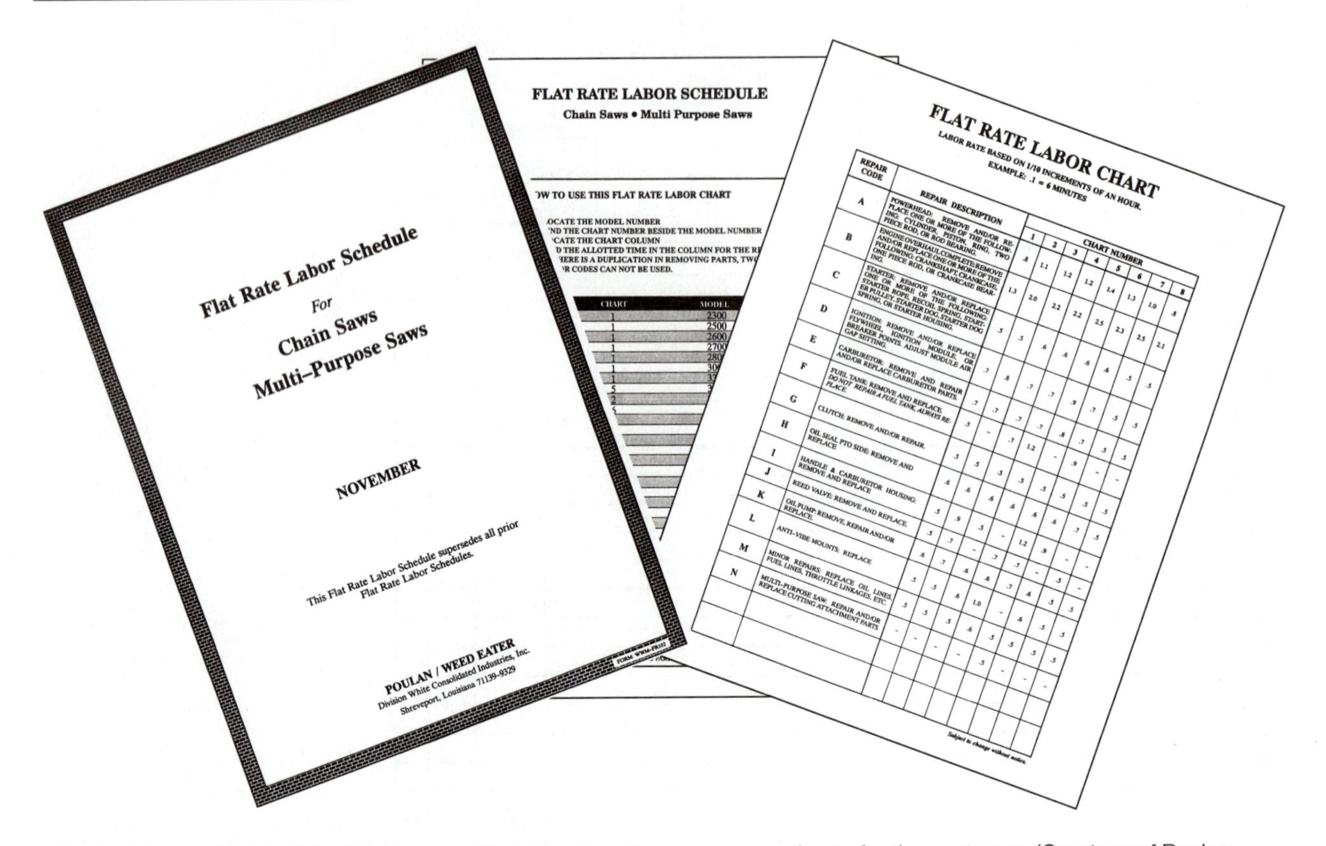

FIGURE 7-18 A flat rate schedule is used with parts prices to prepare an estimate for the customer. (Courtesy of Poulan Weed Eater.)

cost of the technician's time to make the repair. This is called the *labor cost.* A customer is charged for replacement parts and for the time it takes to do a repair job.

Sometimes a repair job is estimated using a flat rate schedule (Figure 7-18). A **flat rate schedule** is a guide that lists the labor time to be billed for equipment repairs. Flat rate schedules do not show labor prices. They show the amount of time that the job should take. Labor time is usually shown in hours and tenths of an hour. This means that .1 or 1⁄10 of an hour is equal to 6 minutes. Multiply the shop hourly labor rate by the time shown in the labor guide to determine the total labor charge. For example, a flat rate schedule for a chain saw lists the time to remove a clutch at .5 of an hour. The labor rate charged by the shop is $15 per hour. The labor estimate to remove the clutch is .5 hour multiplied by $15.00 or $7.50.

An estimate also includes the cost of the parts that require replacement. The cost of parts changes rapidly. The best way to check parts prices is to call the parts supplier. Materials such as shop rags, solvent, or cleaners are sometimes charged to the customer. They are part of the estimate. Some shops charge the customer an ecology fee for the removal or disposal of hazardous materials such as old gasoline, engine oil, and batteries.

To do an estimate, first determine what parts must be replaced. Then, look up or call for the cost of the parts. Add this to the labor charge. Then, add sales taxes to get the total estimate for the customer. In some areas, sales taxes may be charged only for the parts. In other areas sales taxes are charged for the labor. This depends on the laws where the repair shop is located. Any material or ecology fees are added to the total.

Many areas have laws about giving customers cost estimates. You should find what that law requires. Usually, these laws require that an estimate of repair costs be given to customers in advance of the work being done. If the actual charges are over the estimated charges, you may have to check with the customer before doing the work. This is why estimate and or invoice should be legible and accurate. When the law requires a written estimate, be sure the customer gets a copy of the form. Be sure that his or her signature is on the form.

Promising a time and date to perform the work is often a part of the estimate. Parts availability can determine whether work can be accepted and promised on any given day. Most shops have commonly used parts in stock. Large and unusual parts are sometimes hard to get. Customers should be told that parts problems could cause a delay in finishing the job.

CASE STUDY

A customer rolled a lawnmower into the shop one Saturday for repair. The technician greeted her and began to fill out a job estimate. The lawnmower, she told him, probably needed a tune up. It had started to misfire under load and finally stopped running. The technician gave the engine a quick look and wrote up an estimate for a new spark plug, fuel filter, air filter, and carburetor adjustment. The technician promised the job would be completed by Friday afternoon of the next week. The estimate was signed and the lawnmower moved into the to be repaired line.

The technician was busy with other projects all week and got around to the tune up on Friday afternoon. He had left this job until late in the day because he knew he could do this one fast. He started the job by removing the spark plug. The minute he saw the tip of the old spark plug, he knew he was going to have problems. The spark plug had heavy mechanical damage. Something had hit the spark plug. He grabbed the rewind starter pull rope and gave it a pull. The engine was locked up. He now knew the engine had stopped because of major parts failure. He quickly removed the cylinder head and found a broken piston caused by a broken connecting rod. He used the dip stick to check the oil level and found no oil.

The tune-up estimate had just mushroomed into a major and expensive engine overhaul. He now had the embarrassing job of calling the customer and revising the cost and explaining why she would not be able to mow her lawn that weekend.

REVIEW QUESTIONS

1. Explain how to locate an engine model designation on an engine.
2. Explain how to use an engine model number decoding chart to get technical information from an engine model number.
3. List four examples of the type of information provided by a model designation.
4. List and describe three general types of service information used by the technician.
5. Explain how step-by-step service procedures are used during a repair.
6. Explain how to use a specification table to locate information.
7. Explain the purpose of troubleshooting information.
8. Explain the purpose and types of information found in an equipment owner's manual.
9. Explain the purpose and types of information found in a shop service manual.
10. Explain how to use a road map type troubleshooting guide to locate the cause of a problem.
11. Explain how to use a diagnosis chart troubleshooting guide to locate the cause of a problem.
12. Explain the purpose and provide two examples of graphic symbols.
13. Explain the purpose of a warranty claim.
14. Explain how to use a parts manual to look up a part number.
15. Explain the purpose of a service bulletin.

DISCUSSION TOPICS AND ACTIVITIES

1. Pick an engine in the shop and locate the model and serial numbers. Use the shop manual for the engine and make a list of all the information you can get from the model number.
2. Using the same engine, see if you can locate the correct owner's manual, shop manual, troubleshooting guide, and parts manual.
3. Locate an engine parts manual, then see if you can find the part number for a piston and connecting rod.

Engine Operation

Four-Stroke Engine Operation and Cylinder Block Parts

OBJECTIVES

Upon completion and review of this chapter, you should be able to:

- Explain the parts and operation of a basic engine.
- Identify and explain the operation of each of the four strokes in a four-stroke cycle.
- Explain the parts and operation of a multiple cylinder four-stroke engine.
- Identify and explain the operation of each of the four-stroke engine cylinder parts.

TERMS TO KNOW

Ball bearing
Bearing clearance
Bottom dead center (BDC)
Bushing
Combustion
Combustion chamber
Compression
Compression ring
Compression stroke
Connecting rod
Connecting rod bearing
Connecting rod big end
Connecting rod journal
Connecting rod small end
Counterbalance system
Crankcase
Crankcase cover
Crankshaft
Cycle
Cylinder
Cylinder block
Cylinder head
Cylinder head bolts
Cylinder head gaskets
Cylinder liner
Cylinder wall
Exhaust port
Exhaust stroke
Flywheel
Four-stroke cycle engine
Horizontal crankshaft
Inline cylinder engine
Intake port
Intake stroke
Internal combustion engine
Main bearing
Main bearing journals
Multiple-cylinder engine
Oil control ring
Opposed cylinder engine
Piston
Piston clearance
Piston head
Piston pin
Piston pin bore
Piston ring
Piston skirt
Plain main bearing
Power stroke
Precision insert bearing
Ring belt
Ring expander
Ring groove
Ring land
Rod cap
Saddle bore
Stroke
Tapered roller bearing
Timing gear
Top dead center (TDC)
V-cylinder engine
Vertical crankshaft
Wiper ring

INTRODUCTION

An engine uses fuel and air to develop power. A fuel, such as gasoline, is mixed with air, and this mixture is burned, causing an expanding gas. The expanding gas is used to develop power. **Combustion** is the burning of a mixture of air and fuel. An **internal combustion engine** is a type of engine in which combustion takes place inside the engine.

BASIC ENGINE PRINCIPLES

The internal combustion engine is often called a *piston engine*. These engines use a piston to develop power. All piston engines have the same basic parts: cylinder, piston, combustion chamber, connecting rod, crankshaft, and flywheel. A **cylinder** is a hollow metal tube, closed at one end, used for burning the air and fuel. The **piston** is a round metal part that fits inside the cylinder. It gets the force from the burning air and fuel mixture. The

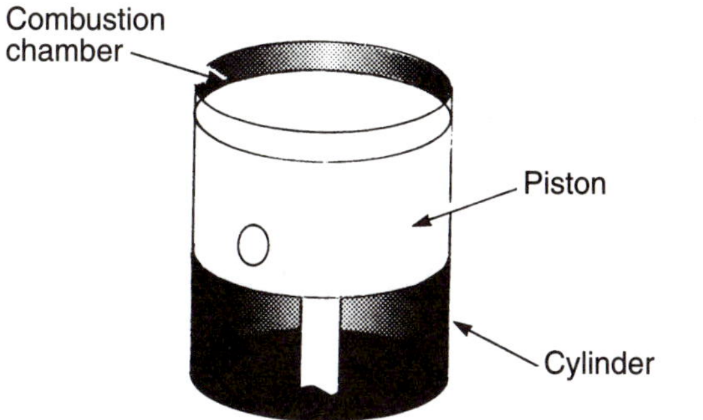
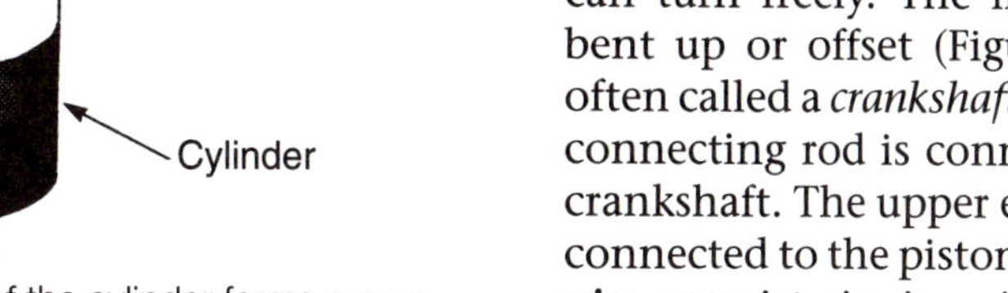

FIGURE 8-1 The closed end of the cylinder forms a combustion chamber.

combustion chamber (Figure 8-1) is a small space between the piston and the top of the cylinder where the burning of the air and fuel mixture takes place. The expanding gas forces the piston down the cylinder. A **connecting rod** is a rod connected to the center of the piston that transfers movement to the crankshaft. The **crankshaft** is a shaft that changes the up-and-down movement of the piston to rotary motion.

The connecting rod and crankshaft work like the spokes on a wheel (Figure 8-2). Think of the connecting rod as connected to a piston at one end. The other end is connected to a pin on the spoke of a wheel. The piston is forced down. The connecting rod attached to the pin on the spoke of the wheel also moves down, causing the wheel to turn. The up-and-down movement of the piston is called *reciprocating motion*. The crankshaft changes reciprocating motion to rotary motion.

The crankshaft has its ends mounted so that it can turn freely. The middle of the crankshaft is bent up or offset (Figure 8-3). The offset part is often called a *crankshaft throw*. The lower end of the connecting rod is connected to the middle of the crankshaft. The upper end of the connecting rod is connected to the piston with a piston pin. A **piston pin,** or wrist pin, is a pin that goes through the top of the connecting rod and through the piston. The piston pin allows motion between the piston and connecting rod and allows the connecting rod to follow the crankshaft's motion (Figure 8-4).

The combined action of the piston, connecting rod, and crankshaft is like pedaling a bike. You move your legs up and down to pedal a bike. The crank arms on the bicycle change up-and-down motion to rotary motion. During pedaling, your legs have to move up so the pedals can be pushed down again. The engine's piston has to be returned to the top of the cylinder for the same reason. The part that does this is called a **flywheel** (Figure 8-5). A flywheel is a heavy wheel on the end of the crankshaft that uses its rotating weight (inertia) to return the piston to the top of the cylinder. When the piston is forced down, the flywheel goes around with the crankshaft. Since it is heavy, it does not slow down easily. Flywheel inertia keeps the crankshaft turning. This movement causes the piston to go back up to the top of the cylinder.

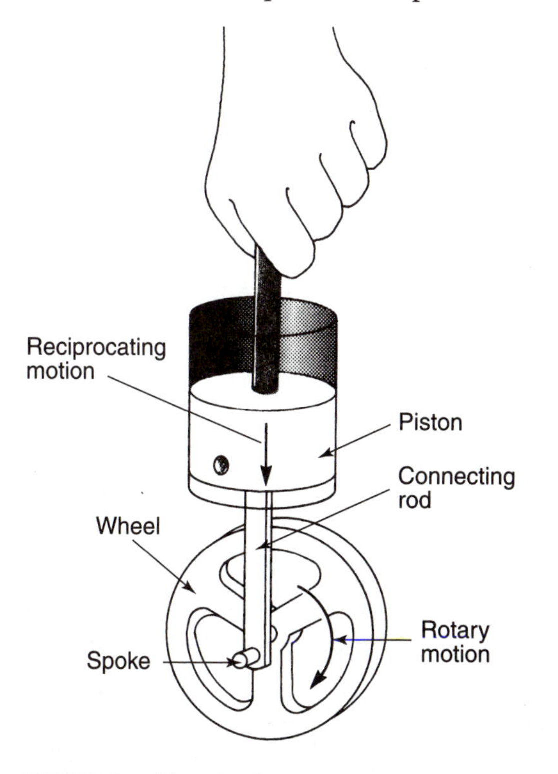

FIGURE 8-2 The wheel turns when the piston is pushed down.

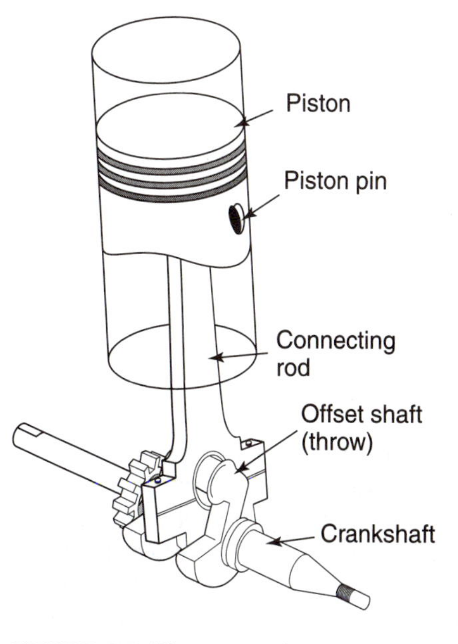

FIGURE 8-3 The connecting rod is connected to an offset shaft on the crankshaft.

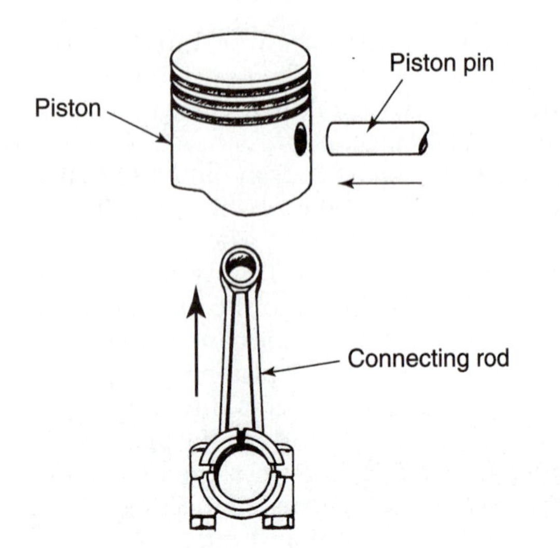

FIGURE 8-4 The piston pin allows the connecting rod to follow piston movement.

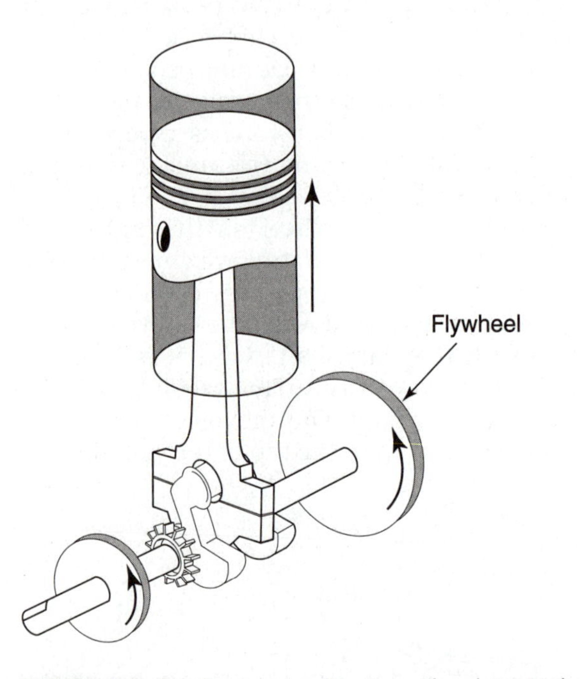

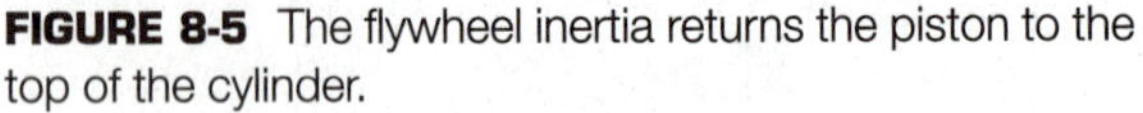

FIGURE 8-5 The flywheel inertia returns the piston to the top of the cylinder.

The piston gets pushed down in the cylinder by combustion. Flywheel inertia pushes the piston back up to the top of the cylinder. The movement of the piston in the cylinder is called a stroke. A **stroke** is piston movement from the top of the cylinder to the bottom, or from the bottom of the cylinder to the top. The upper and lower limits of piston movement are called top and bottom dead center (Figure 8-6). **Top dead center (TDC)** is the highest position a piston can go in a cylinder. **Bottom dead center (BDC)** is the lowest position a piston can go in a cylinder.

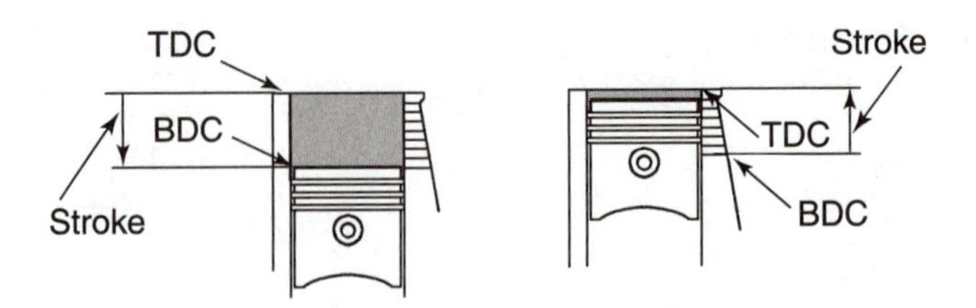

FIGURE 8-6 A stroke is piston movement from top to bottom or bottom to top of the engine.

FOUR-STROKE CYCLE

In many engines, the power is developed using four piston strokes. This kind of engine is called a four-stroke cycle engine. A **cycle** is a sequence of events that is repeated over and over. A **four-stroke cycle engine** is an engine that develops power in a sequence of events using four strokes of the piston. They are commonly called *four-stroke engines*.

The four-stroke cycle engine needs a way of getting an air and fuel mixture into the combustion chamber. It also has to get burned exhaust gases out of the combustion chamber. This is done with two passages or ports in the combustion chamber. The ports can be opened or closed as needed. The **intake port** is a passage in the combustion chamber used to let an air and fuel mixture into the cylinder. The **exhaust port** is a passage in the combustion chamber used to get the burned exhaust gasses out of the cylinder.

Intake Stroke

The intake stroke is the first stroke in the four-stroke cycle engine. The **intake stroke** (Figure 8-7) occurs when the piston moves down, pulling the air and fuel mixture into the combustion chamber. The intake stroke begins as the piston moves rapidly down the cylinder. The rapid piston movement causes a low pressure (less than atmospheric pressure) in the combustion chamber. At the same time, the intake port is opened (by an intake valve). There is atmospheric pressure on the air and fuel mixture in the intake port. The higher pressure in the intake port forces the air and fuel mixture into the lower pressure area of the combustion chamber. The piston goes down to BDC. The crankshaft has turned one-half revolution (180 degrees). The combustion chamber is filled with a mixture of air and fuel. It is ready for the compression stroke.

Compression Stroke

The compression begins as the piston starts back up the cylinder. The **compression stroke** (Figure 8-8) is a stroke that occurs when the piston moves to

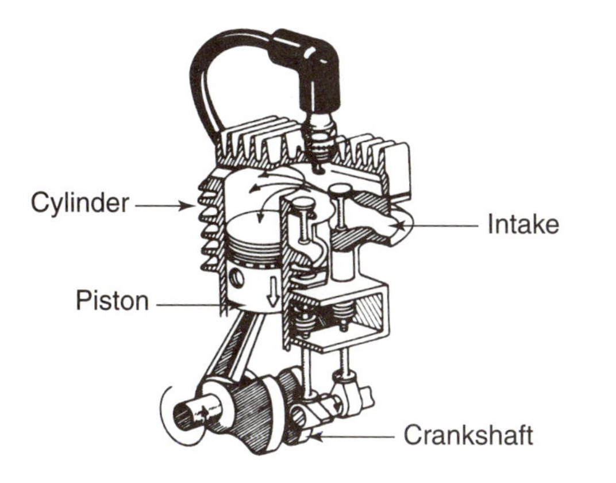

A. The Intake Stroke

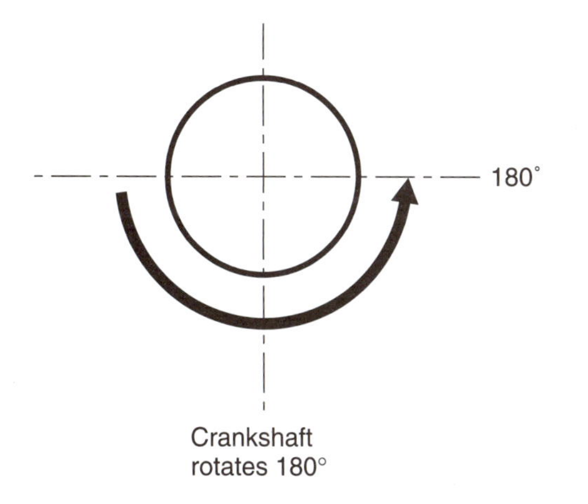

B. The Crankshaft Rotates One-Half Revolution

FIGURE 8-7 The piston moves down, pulling in an air and fuel mixture on the intake stroke. (*A:* Provided courtesy of Tecumseh Products Company.)

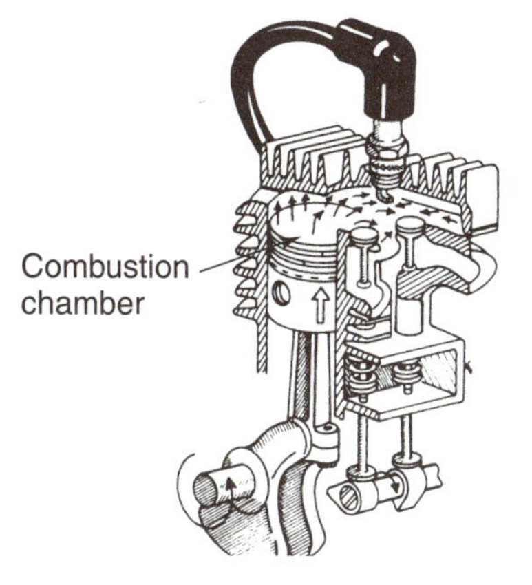

A. The Compression Stroke

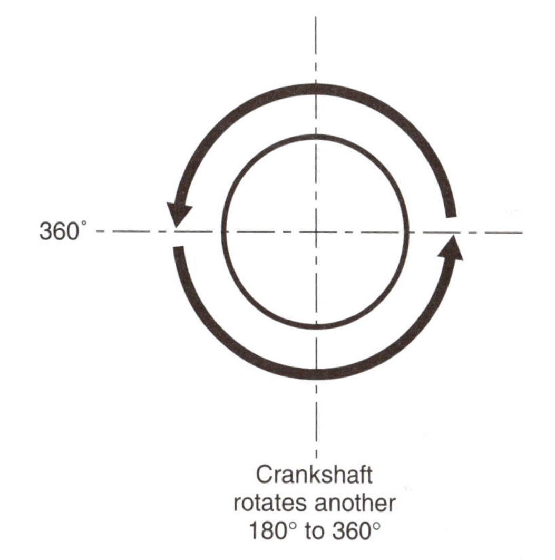

B. The Crankshaft Rotates Another One-Half Revolution

FIGURE 8-8 The piston moves up, compressing the air-fuel mixture on the compression stroke. (*A:* Provided courtesy of Tecumseh Products Company.)

the top of the combustion chamber with both ports closed. As the piston goes up to TDC, it squeezes the air-fuel mixture into a smaller space. **Compression** is the squeezing or compressing of the air-fuel mixture in the combustion chamber. The more a gas is compressed, the more heat (power) that results during combustion. During the compression stroke, the crankshaft has turned another half a revolution (180 degrees). During the two (intake and compression) strokes, it has turned one complete revolution (360 degrees).

Power Stroke

The power stroke starts when the piston reaches the top of the compression stroke. The exhaust and intake ports remain closed. The **power stroke** (Figure 8-9) occurs when the compressed air-fuel mixture in the combustion chamber is ignited to force the piston down the cylinder. When the piston reaches BDC, the power stroke is over. The crankshaft has turned another half revolution (180 degrees). It has now rotated a total of 540 degrees.

Exhaust Stroke

The burned air-fuel mixture must be forced out to allow room for a new air-fuel mixture. The exhaust stroke begins as the piston starts back up the cylinder. The **exhaust stroke** (Figure 8-10) occurs when the exhaust port is opened (by an exhaust valve) and the piston moves up to push exhaust gases out of the cylinder. The exhaust stroke is over when the piston reaches TDC and the exhaust port is closed. The crankshaft has rotated another half revolution (180 degrees).

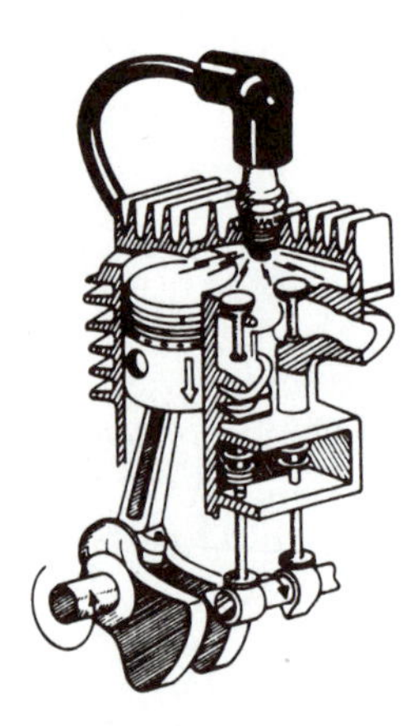

A. The Power Stroke

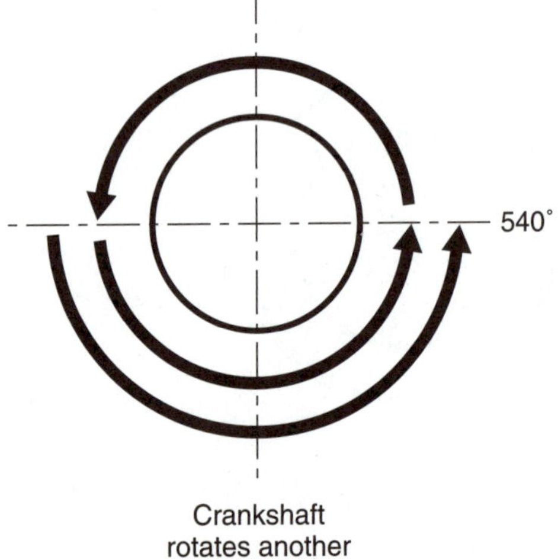

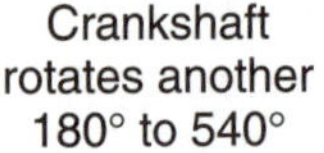

B. The Crankshaft Rotates Another One-Half Revolution

FIGURE 8-9 The burning air and fuel mixture pushes the piston down on the power stroke. (*A:* Provided courtesy of Tecumseh Products Company.)

Exhaust

A. The Exhaust Stroke

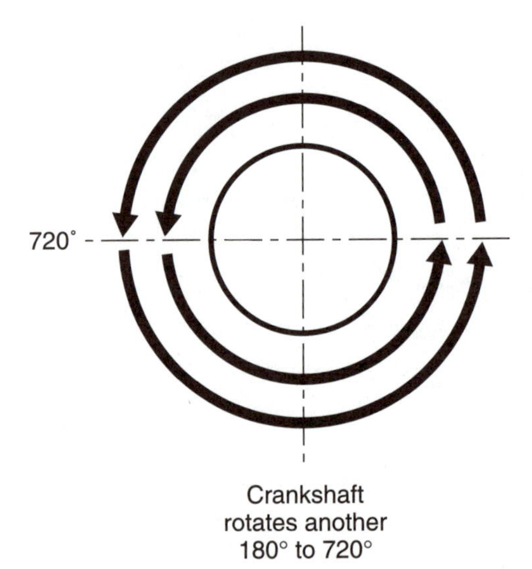

Crankshaft
rotates another
180° to 720°

B. The Crankshaft Rotates Another One-Half Revolution

FIGURE 8-10 The piston moves back up and pushes the burned gases out of the cylinder on the exhaust stroke. (*A:* Provided courtesy of Tecumseh Products Company.)

The four strokes have occurred in two crankshaft revolutions (720 degrees). Power is actually developed during only one of the four strokes. During the other three strokes, the engine is pumping and compressing air and fuel or exhaust gasses. When the piston starts down for another intake stroke, the four-stroke cycle begins all over again.

CASE STUDY

Outdoor power equipment instructors often teach students the four-stroke cycle by having them remove the cylinder head from an engine and identify the strokes. This is a good idea because students can see how the engine actually works. The students can identify the intake stroke because the intake valve opens as the piston moves down the cylinder. Both valves are closed on the compression stroke as the piston moves back up the cylinder. Both valves stay closed as the piston moves down the cylinder on the power stroke. The exhaust valve opens as the piston begins to move back up the cylinder on the exhaust stroke.

During one class many of the students were having a problem identifying the strokes. There seemed to be mass confusion. After helping several groups of students get it right, the instructor figured out the problem. He had forgotten to explain which direction to rotate the crankshaft. An engine crankshaft rotates clockwise when viewed from the flywheel end. The rotation direction is counterclockwise when looking at the PTO (power takeoff) end of the crankshaft. The students were rotating the engines by turning the flywheel. Those that were rotating them the wrong direction could not get the intake, compression, power, and exhaust stroke in the correct order.

MULTIPLE CYLINDER FOUR-STROKE ENGINES

Four-stroke cycle engines can be made with any number of cylinders. Engines for small outdoor power equipment have a single cylinder. When increased power is required there are two options: A single cylinder engine with a larger cylinder can be used, or an engine with more than one cylinder, called a **multiple-cylinder engine** is also available.

Multiple-cylinder engines have different cylinder arrangements. The **inline cylinder engine** (Figure 8-11) is a multiple-cylinder engine with cylinders arranged in a row. The crankshaft is longer than that of a single cylinder engine. It has two offsets (throws), one for each piston and connecting rod. The crankshaft offsets are 180 degrees to each other. When the piston in one cylinder is at TDC, the piston in the other cylinder is at BDC. When the engine is running, one piston will be on the intake or the power stroke. The other cylinder will be on the compression or the exhaust stroke.

The **V-cylinder engine** is a multiple cylinder engine with cylinders arranged in two rows in the shape of a V. The crankshaft is located between and below the cylinders. The V-cylinder engine can have any number of cylinders. Outdoor power product engines usually have just two. A two-cylinder V-engine is often called a V-twin. The twin V-cylinder crankshaft often has a single offset (Figure 8-12). Both connecting rods are attached to one offset. There is a 90-degree angle between the centerline of the two cylinders. This allows each cylinder to be on a different stroke in the four-stroke cycle. One piston will be on the intake or the power stroke. The other will be on the compression or the exhaust stroke. Some twin V-cylinder engine crankshafts have an offset for each cylinder.

The **opposed cylinder engine** (Figure 8-13) is a multiple-cylinder engine with cylinders opposite each other. The crankshaft fits in between the cylinders. Two opposed cylinders are used on small engines. The crankshaft for this engine has two offsets that are at 180 degrees to each other, which allows each piston to be on a different stroke in the four-stroke cycle. One piston will be on the intake or the power stroke. The other will be on the compression or the exhaust stroke.

CRANKCASE AND MAIN BEARINGS

Up to now, the engine that has been described has been simplified, with only a few basic parts. An actual engine has many more parts. All four-stroke engines have the same general parts.

The **crankcase** (Figure 8-14) is the main housing that supports each end of the crankshaft and allows it to rotate. The crankcase is also the reservoir for the engine's lubricating oil. It is made from cast aluminum on most engines. Older engines have a cast-iron crankcase.

A **crankcase cover** is a removable cover on the crankcase that allows access to the crankshaft.

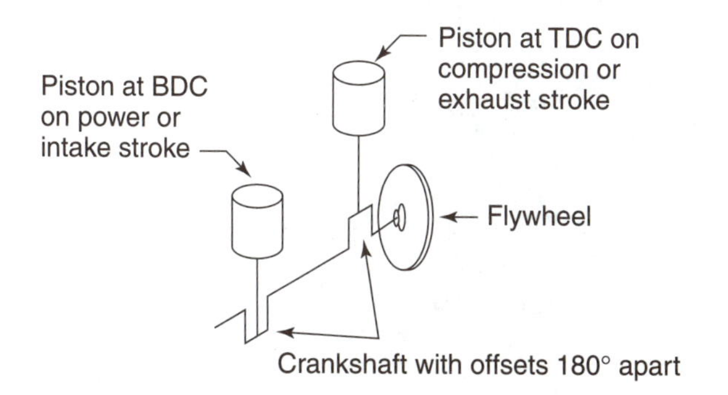

FIGURE 8-11 A multiple cylinder engine with inline cylinders.

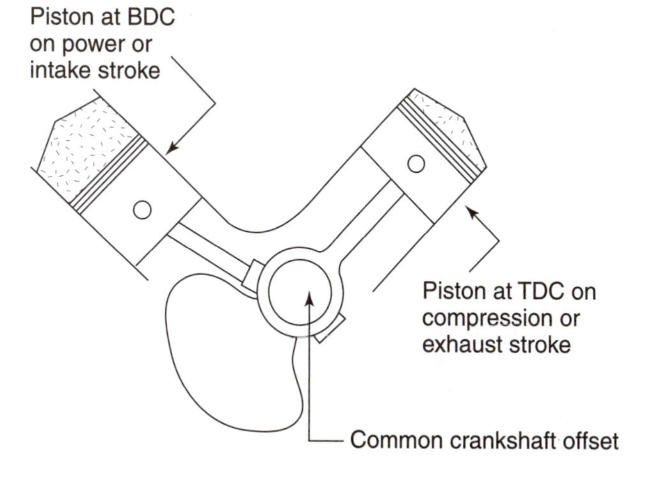

FIGURE 8-12 A multiple cylinder engine with a V-twin cylinder arrangement.

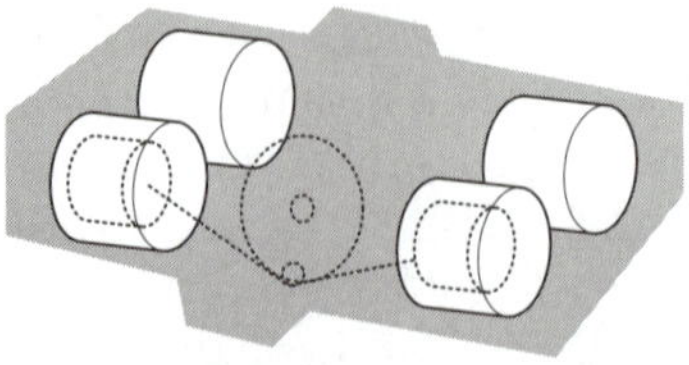

Horizontal or opposed

FIGURE 8-13 A multiple cylinder engine with opposed cylinders.

FIGURE 8-14 The crankcase cover is removed to access inside parts.

Engines that have a horizontal crankshaft have a crankcase with a side cover. Engines with a vertical crankshaft have a crankcase with a bottom cover. This type is called a *base plate, sump,* or *bottom cover.* A gasket is used between cover and crankcase housing to seal in lubricating oil.

A hole is machined in the crankcase and in the side cover for main bearings. A **main bearing** is a bearing used to support each end of the crankshaft, allowing it to rotate. There are two main bearings in a single cylinder engine. One is located on the PTO side of the engine. The PTO is the side of the engine where the crankshaft is used to drive the equipment. This end is under the most stress. The flywheel (or magneto) is located on the other side. The flywheel end (FWE) bearing is the side where the magneto and flywheel are mounted. Multiple-cylinder engines may have two offsets (throws) for two connecting rods on the crankshaft. These engines may have a third center main bearing between the two crankshaft offsets.

There are several common types of main bearings. The crankcase material can determine what type of main bearing is used. A bearing must be made from a softer material than the part that it supports. Soft materials are used for bearing surfaces so that wear occurs to the bearing and not to the more expensive crankshaft.

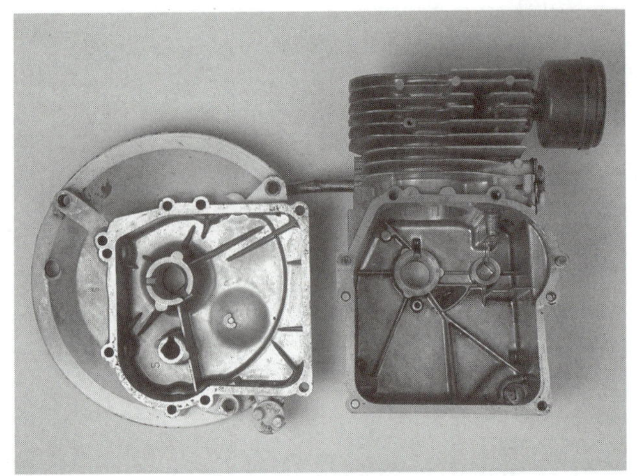

FIGURE 8-15 Plain main bearings in the crankcase and cover.

The simplest type of main bearing for an aluminum crankcase is a plain bearing (Figure 8-15). A **plain main bearing** is machined directly in the aluminum of the crankcase or cover.

Often used for the main bearings, a **bushing** (Figure 8-16) is a thin sleeve made from a soft bearing material pressed into a hole to provide a bearing surface. Bushings are made from soft metals such as bronze. Some are made from bronze formed on a steel backing. The outside diameter of the bushing is slightly larger than the machined hole in the crankcase or cover. It is driven or pressed into place. This tight fit, called a *press fit,* locks it in place. Bushings are more expensive than plain bearings machined directly in aluminum.

FIGURE 8-16 A bushing type main bearing in a crankcase cover.

They have the advantage of being replaceable during engine service.

The plain or bushing main bearing inside diameter is slightly larger than the outside diameter of the crankshaft. **Bearing clearance** is the small difference in size between a shaft and a bearing. This difference makes a space for lubricating oil, which is used to reduce friction and wear. Oil grooves or holes in the bearing allow oil to flow into the bearing clearance.

SERVICE TIP: The main bearings in the crankcase and side cover or base plate are machined together with the cover installed during manufacturing. This procedure insures perfect alignment between the two bearings and allows the crankshaft to turn with very little friction. However, this means that the cover from one engine cannot be used on any other engine.

Larger and more expensive engines made for heavy-duty work may have ball or tapered roller main bearings. A **ball bearing** is a bearing that uses a set of caged balls between an inner and outer bearing surface called a race (Figure 8-17). The ball bearing may be held in place on the crankshaft or in the crankcase cover (Figure 8-18). The inner race of the ball bearing fits on the crankshaft. The outer race fits into the crankcase or cover. Sometimes ball bearings are used on just the PTO end of smaller engines.

A **tapered roller bearing** (Figure 8-19) is a bearing that uses a set of caged tapered roller bearings between an outer and inner race. The roller bearing inner race, cage, and roller fit on the crankshaft. The outer race fits in the crankcase or cover. Tapered roller bearings are often used on both the PTO and FWE sides of larger engines.

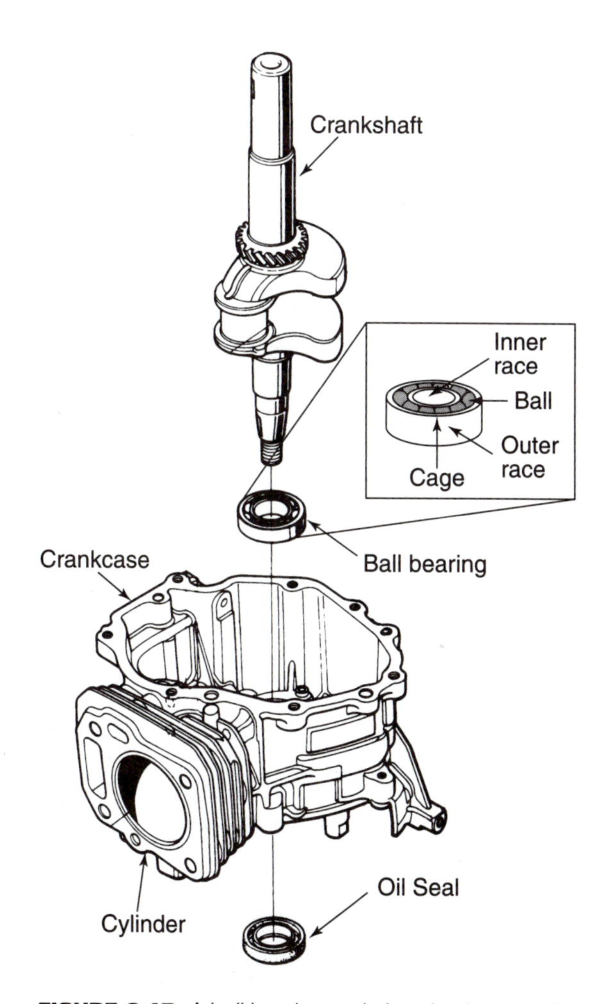

FIGURE 8-17 A ball bearing main bearing in a crankcase. (Courtesy of American Honda Motor Co., Inc.)

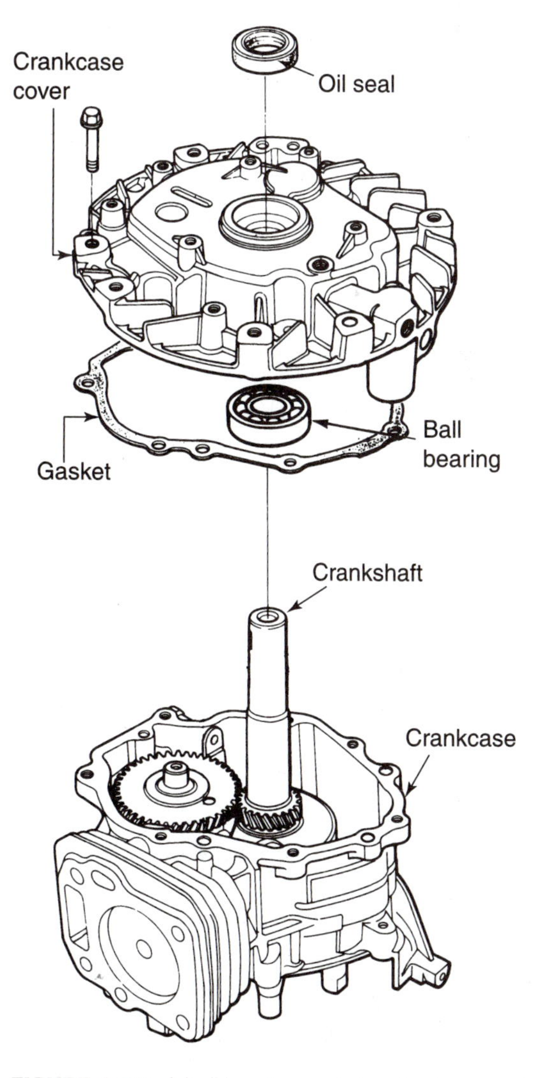

FIGURE 8-18 A ball bearing main bearing in a crankcase cover. (Courtesy of American Honda Motor Co., Inc.)

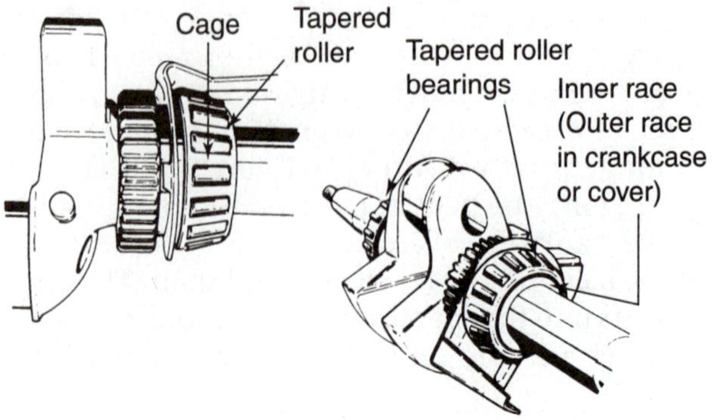

FIGURE 8-19 A tapered roller main bearing on a crankshaft. (Provided courtesy of Tecumseh Products Company.)

Ball and roller bearings both work in the same way. The inner race is held tightly in place on the crankshaft. The outer race is held tightly in place in the crankcase or cover. The balls or rollers rotate on their two races. There are no parts turning on either the crankshaft or the crankcase (or crankcase cover). The cage allows the balls or rollers to rotate but prevents them from touching each other. The advantage of ball or roller bearings is that there is no wear on the crankshaft or side cover. All wear occurs to the bearing. The bearing is easily replaced during service.

CYLINDER AND BLOCK

Most small engines have a cylinder and crankcase that are made as one part. A **cylinder block** (Figure 8-20) is a crankcase and cylinder assembly combined into one casting. Some engines have a separate cylinder, which is attached to the crankcase by long studs. This type is more expensive to make. The cylinder can be replaced easily when it is worn out.

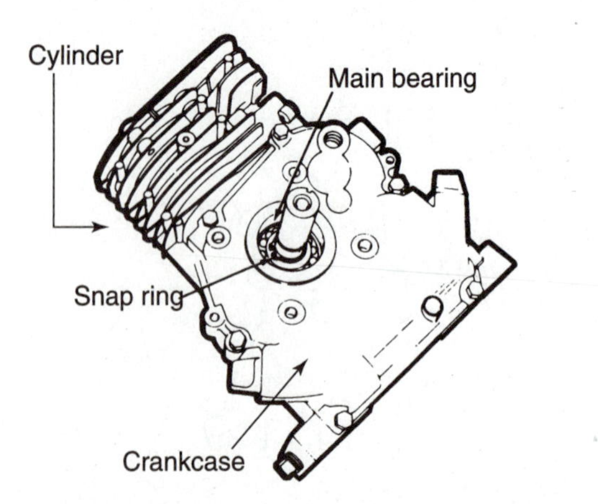

FIGURE 8-20 The crankcase and cylinder are combined into a cylinder block. (Provided courtesy of Tecumseh Products Company.)

FIGURE 8-21 The cylinder walls support the piston as it moves up and down.

The **cylinder wall** (Figure 8-21) is the inside surface of the cylinder where the piston slides up and down. The space or clearance between the cylinder wall and the piston must be to exact specifications. The clearance allows free up-and-down movement of the piston.

Cylinder walls have very severe operating conditions. The heat from combustion in the combustion chamber passes in to the cylinder. Very high temperatures are created that cause distortion. Distortion can change the space between the piston and the cylinder wall. The sliding contact of the piston and piston rings wears away at the cylinder wall surface. The top of the cylinder is where it gets the hottest. This is the hardest part of the engine to lubricate. The top part of the cylinder wears faster than any other part of the engine. Cylinder design and material are very important to long engine service.

In many older engines, the cylinder block assembly is made from cast iron (Figure 8-22). Cast iron is an excellent material for the cylinder walls, because it has materials in it that provide a very slippery surface for the piston and piston rings. The problem with cast iron is its weight. Engines with cast iron blocks are extremely heavy.

Newer engines use aluminum cylinder blocks (Figure 8-23) because aluminum is much lighter. Aluminum allows heat to pass though it better than cast iron, making the engine easier to cool. Less expensive aluminum block engines have the cylinder walls machined directly in the aluminum

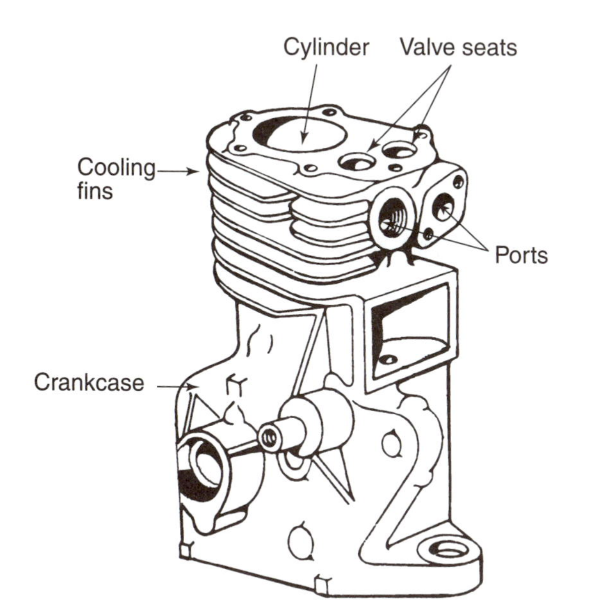

FIGURE 8-22 An older cylinder block made from cast iron.

casting, providing an aluminum surface for the piston and piston rings. Aluminum cylinder walls wear more rapidly than cast-iron ones. These engines wear out more rapidly than those made from cast iron.

More expensive aluminum engines take care of cylinder wear with a **cylinder liner** (Figure 8-24), a cast-iron sleeve installed in an aluminum cylinder block when it is made. The cylinder liner provides an excellent cylinder wall surface. It does not add much weight to the aluminum cylinder block. The cylinder casting has cooling fins around the cylinder work with the cooling system to remove heat from the cylinder.

FIGURE 8-23 An aluminum cylinder block with aluminum cylinder walls.

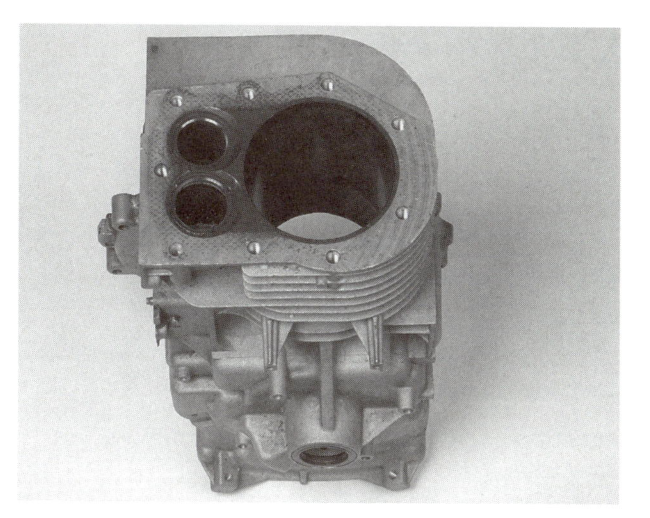

A.

B.

FIGURE 8-24 An aluminum cylinder block *(A)* with a cast-iron cylinder liner or sleeve *(B)*.

Both cast-iron and aluminum cylinder walls are made in the same way. The cylinder hole is made with a cutting tool called a *boring bar*. The cylinders are then finished to exact size with an abrasive tool called a *hone*. The purpose of honing is to get a crosshatch pattern on the cylinders. Gravity causes lubricating oil to run down the cylinder walls. Cylinder walls, piston, and piston rings will wear fast without oil. The crosshatch grooves from honing holds lubricating oil on the cylinder wall (Figure 8-25).

CYLINDER HEAD AND CYLINDER HEAD GASKET

The **cylinder head** (Figure 8-26) is the top of the cylinder that is used to form the combustion chamber. The cylinder head is a separate part on four-stroke engines so it can be removed to service the valve mechanism. Cylinder heads for engines that have the valve mechanism in the cylinder block have no moving parts. Engines that have the

FIGURE 8-25 A crosshatch pattern on the cylinder walls improves cylinder lubrication. (Courtesy of Sealed Power.)

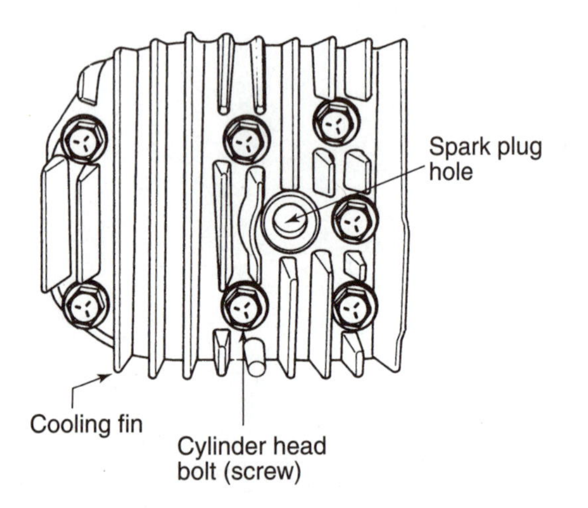

FIGURE 8-26 An aluminum cylinder head with cooling fins and spark plug hole. (Provided courtesy of Tecumseh Products Company.)

valve mechanism located in the cylinder head have more complex cylinder heads. Cast-iron cylinder heads are used on engines with a cast-iron block. Aluminum cylinder heads are used on engines with an aluminum block. They all have cooling fins that remove heat from the combustion chamber. There is a threaded hole in the cylinder head for a spark plug.

Cylinder head bolts are the hex head fasteners used to attach the cylinder head to the cylinder block. These fasteners are often called *head bolts*. They are actually screws because they thread into holes in the cylinder block and not into a nut (Figure 8-27). It is common to find several different lengths of head bolts in a cylinder head. This may be done because some parts of the cylinder head are thicker than other parts. Sometimes the cylinder head is the same thickness but the threaded holes are different depths.

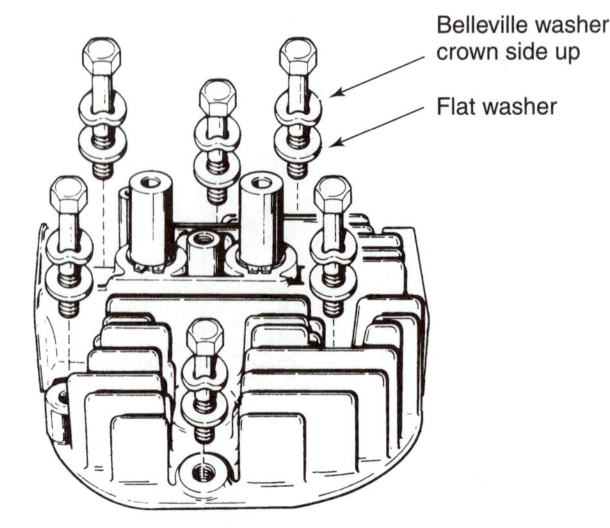

FIGURE 8-27 Cylinder head bolts fasten the cylinder head to the cylinder block. (Provided courtesy of Tecumseh Products Company.)

The engine will loose power if the high pressure (compression) developed in the combustion chamber gets out. There must be a good seal between the top of the cylinder block and the mating surface of the cylinder head. A **cylinder head gasket** (Figure 8-28) is a gasket used between the cylinder head and cylinder block to prevent the loss of pressure from the combustion chamber. These gaskets are commonly made like a sandwich. They have a soft heat-resistant material in the middle with a layer of thin metal such as aluminum, steel, or copper on each side. The head bolts squeeze the gasket between the block and cylinder head surfaces. The gasket material is pushed into any irregularities in the metal surface, forming a pressure-tight seal.

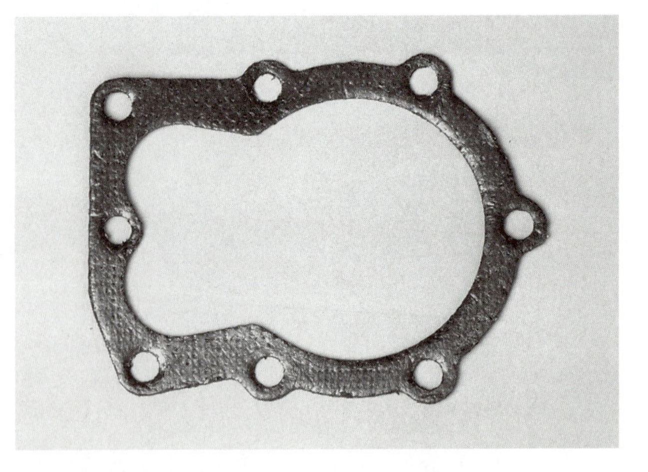

FIGURE 8-28 The head gasket seals between the cylinder head and cylinder.

CRANKSHAFT

The crankshaft (Figure 8-29) is a cast-iron part that is ground and machined to very close tolerances. **Main-bearing journals** are the round surfaces on either end of a crankshaft that fit in the main bearings. One of the main-bearing journals is called the PTO journal; the other is called the FWE journal.

The **connecting rod journal** is the offset part of the crankshaft where the connecting rod is attached. This part is also called the *throw* or *crank pin*. Crankshafts for multiple-cylinder engines can have more than one crankshaft journal (Figure 8-30). A twin cylinder crankshaft can have one wide connecting rod journal for two connecting rods. Some have two separate connecting rod journals.

The connecting rod journal, connecting rod, and piston assembly is heavy. When the crankshaft is rotating, this weight can cause a high centrifugal force. The force is on one side of the crankshaft. The crankshaft would vibrate without a balancing force. Crankshaft counterweights are weights opposite to the connecting rod journal that balance the rotating weight of the connecting rod journal, connecting rod, and piston assembly. Counterweights prevent vibration and allow the engine to run smoothly.

A **timing gear** is a gear on one end of the crankshaft used to drive the engine's camshaft. This gear is sometimes called a *crank gear*. The timing gear meshes with a gear on the camshaft. Oil holes and passages called *galleries* are sometimes drilled throughout the crankshaft. They are used to route oil into the main and connecting rod bearings. The FWE of the crankshaft is often threaded for a flywheel nut. Key seats are commonly machined in both the PTO and FWE. These are used for keys that drive the flywheel and drive pulley.

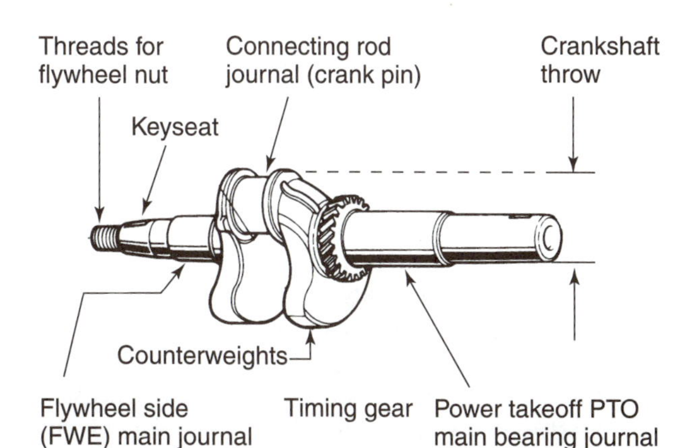

FIGURE 8-29 The parts of a single throw crankshaft. (Courtesy of American Honda Motor Co., Inc.)

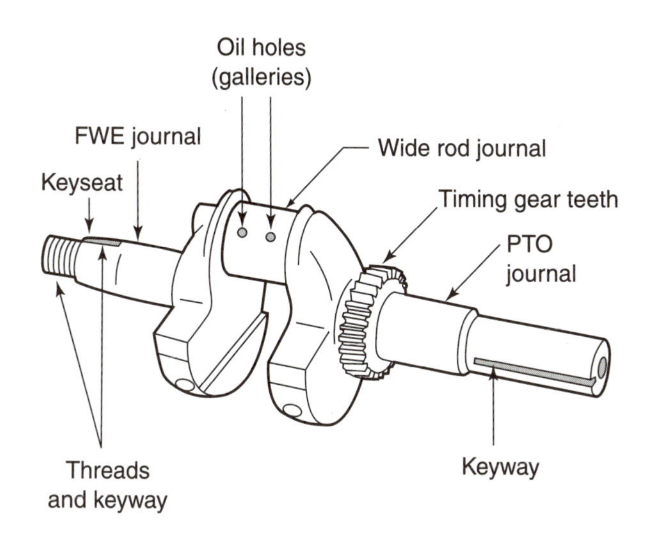

A. V-Twin Crankshaft

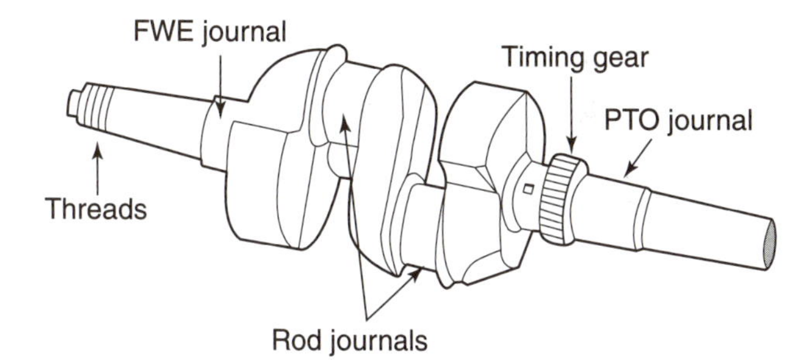

B. Opposed Twin Crankshaft

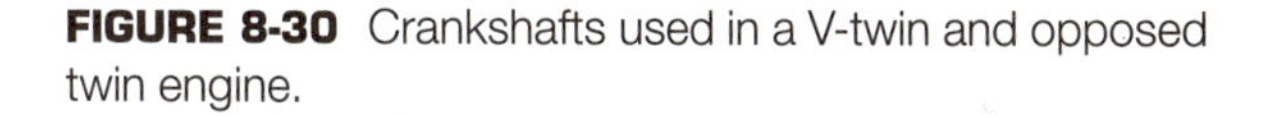

FIGURE 8-30 Crankshafts used in a V-twin and opposed twin engine.

A **vertical crankshaft** is a crankshaft that fits in an engine in a vertical direction (Figures 8-31A, B). These engines have a bottom-mounted crankcase cover or base plate. Rotary type lawnmowers use a vertical crankshaft engine. A **horizontal crankshaft** fits in an engine in a horizontal direction (Figures 8-31C, D). This type of engine has a side-mounted crankcase cover. Engines with horizontal crankshafts are used to power equipment such as edgers or reel type mowers.

Some engines have counterbalance systems (Figure 8-32). A **counterbalance system** is a weighted shaft driven off the crankshaft used to reduce engine vibration. The counterweight shaft has a gear on the end. This gear is driven by the crankshaft timing gear. The gear causes the shaft to turn in a direction opposite to the crankshaft. Counterweights on the shaft turn in opposite direction to the crankshaft counterweights. The opposing weights create opposing forces that stop vibrations, which results in a smooth running engine.

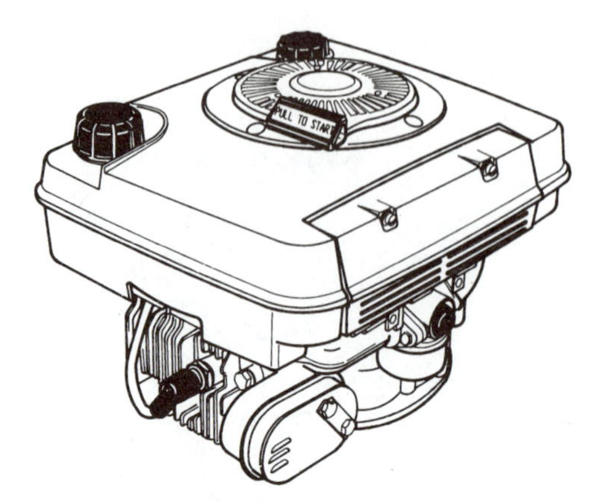

A. Vertical Engine

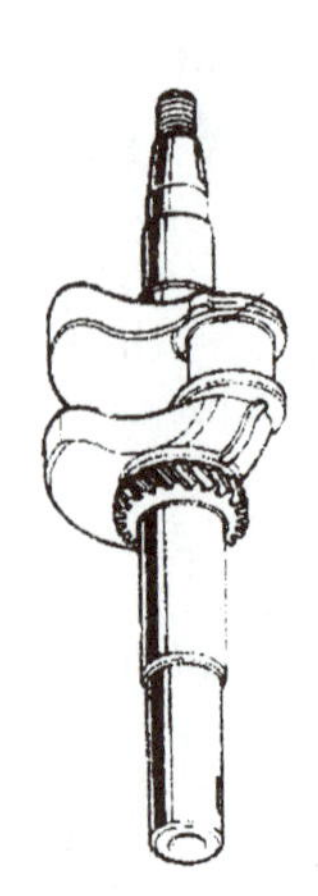

B. Vertical Crankshaft

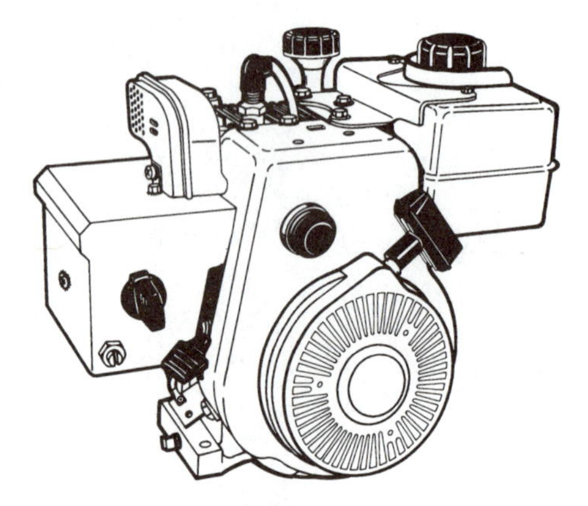

C. Horizontal Engine

D. Horizontal Crankshaft

FIGURE 8-31 Engines may have a horizontal or vertical crankshaft direction. (*A* and *C:* Provided courtesy of Tecumseh Products Company. *B* and *D:* Courtesy of American Honda Motor Co., Inc.)

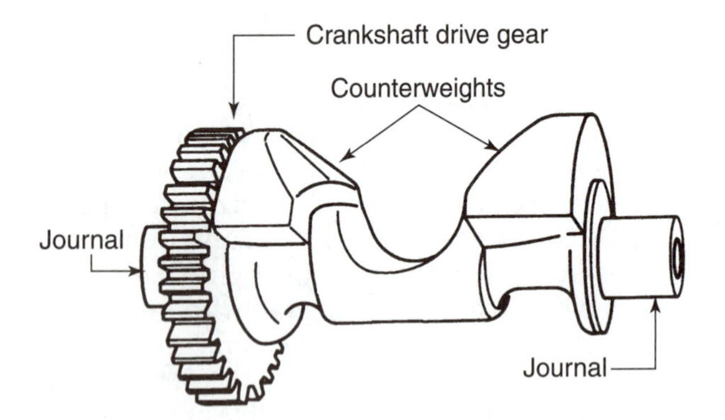

FIGURE 8-32 The Ultra Balance® system reduces engine vibration. (Provided courtesy of Tecumseh Products Company.)

PISTON

The piston must be strong. It gets the full force of the combustion pressures on the power stroke. The piston must also be as light as possible. The piston and connecting rod are parts that reciprocate or go up and down. The lower the weight of reciprocating parts, the higher the engine's power.

Most outdoor power equipment engines use cast aluminum pistons. All pistons have the same basic parts (Figure 8-33). The **piston head** (Figure 8-34) is the top of the piston where it gets its push combustion pressures. The head of most small-engine pistons is flat. Some pistons are contoured on top. A domed head is common on many high-performance engines. This head takes up more room in the combustion chamber, resulting in higher combustion pressures. Notches (indentations) are often made in the domed section of the piston head to provide clearance for overhead type intake and exhaust valves. Irregular and dished-shaped heads are used to move the air-fuel mixture past the spark plug. This movement can improve combustion. A **ring groove** is a groove in the piston to hold the piston rings. Most four-stroke pistons have three ring grooves. A **ring land** is the raised space between the ring grooves. The **ring**

belt is the ring and land area of the piston. The **piston skirt** is the piston area below the ring belt. The skirt contacts the cylinder wall to guide the piston as it moves up and down. Some pistons have cutout areas called *slippers*. They work to reduce the cylinder contact area and lower friction. The **piston pin bore** is a hole in the piston for the pin used to connect the piston to the connecting rod. There are high forces on the piston pin area during the power stroke. The underside of the piston pin bore area has supports called *bosses*.

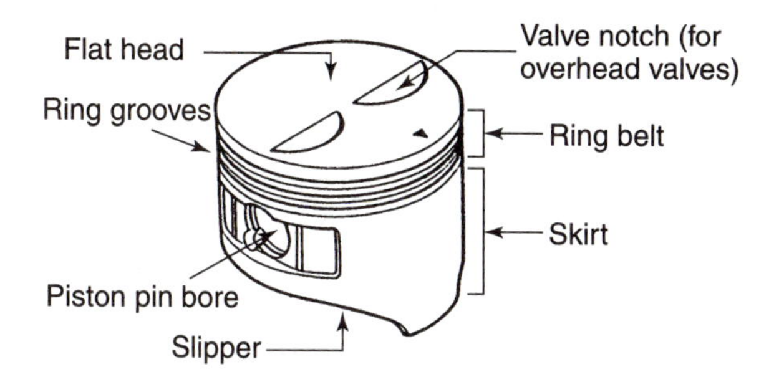

FIGURE 8-33 Parts of a piston. (Courtesy of American Honda Motor Co., Inc.)

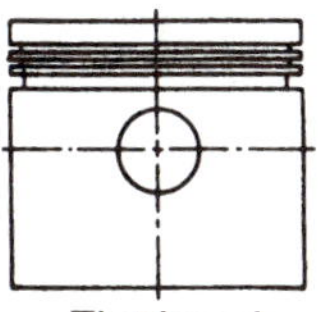

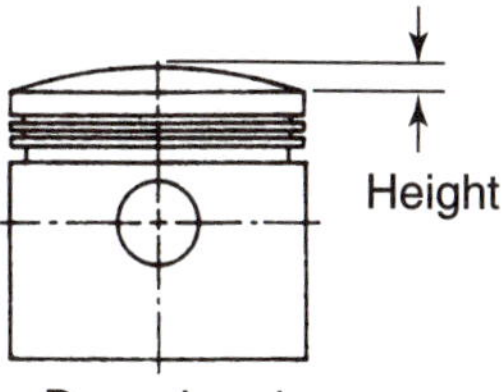

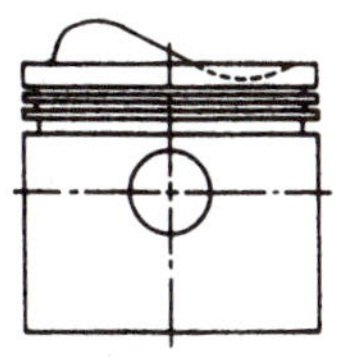

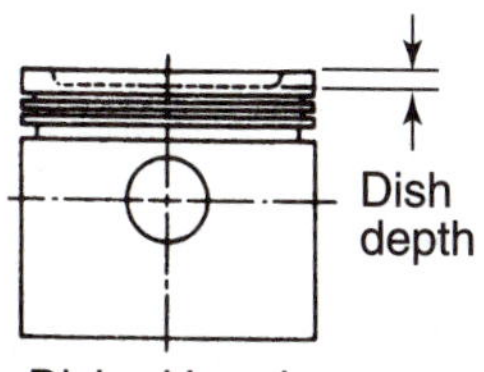

FIGURE 8-34 Pistons have different-shaped heads. (Courtesy of CHAMPION-Federal Mogul.)

The piston is made slightly smaller than the cylinder so it can move up and down the cylinder freely. The **piston clearance** is the small space between the piston skirt and the cylinder wall that allows the piston to move freely in the cylinder.

The piston will not be properly supported if there is too much piston clearance. The skirt can rock against the cylinder wall and the contact between the skirt and cylinder wall will lead to the breaking of the piston skirts. Too little clearance is also a problem. High combustion temperatures heat the piston. As it heats, it expands at a much higher rate than the cylinder walls. The piston can become wedged or seize in the cylinder.

Some engines prevent piston seizing with an oval piston skirt (Figure 8-35). When the piston is

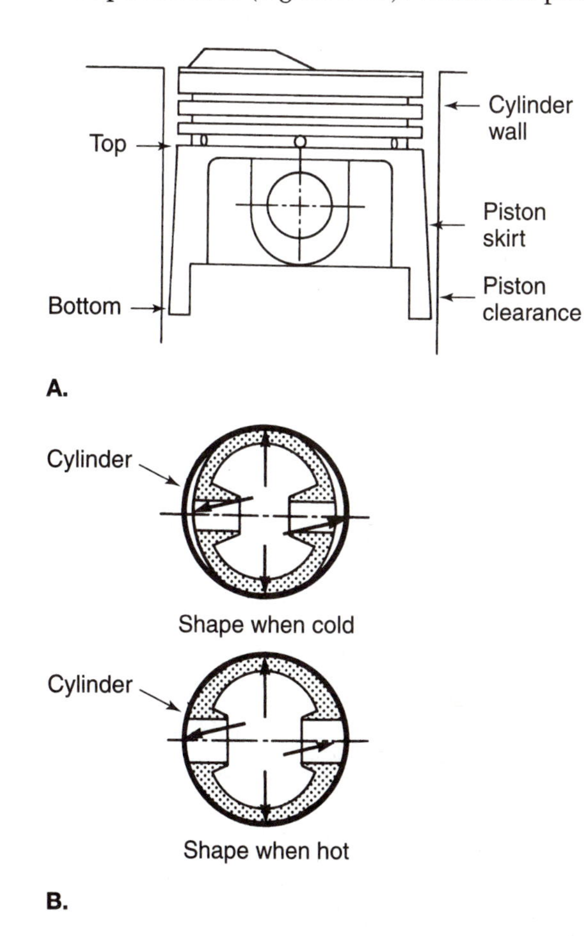

FIGURE 8-35 Piston clearance is used to keep the piston from seizing or slapping. (*A:* Courtesy of CHAMPION-Federal Mogul. *B:* Courtesy of Sunnen Products Company, St. Louis, Missouri.)

cold, the oval parts of the skirt push against the cylinder walls. The oval skirt stops expansion across the larger diameter as the piston heats up. The piston gets round and has a bigger area of skirt contact as it warms up. The skirt on these pistons is often tapered outward at the bottom. This is often called a *barrel-shaped design*.

PISTON RINGS

A **piston ring** (Figure 8-36) is an expandable metal ring in the ring groove of a piston that makes a sliding seal between the combustion chamber and crankcase. Four-stroke engines usually have three piston rings. The **compression ring,** located in the groove or grooves near the piston head, is used to seal in the compression pressure. The **wiper ring,** located in the center ring groove, is used to control oil and compression pressure. The **oil control ring,** located in the bottom ring groove, is used to prevent excessive oil consumption.

The surface of the piston ring that contacts the cylinder wall is called the *face*. The two ends of the piston ring are separated by a gap, called an *end-gap*. The end gap is wide when the rings are not in the engine. The gap closes up when the rings are installed in the piston and cylinder.

Most four-stroke engines use a compression ring made of cast iron or steel. The compression ring fits in the top piston ring groove. Some compression rings are rectangular. They form a simple mechanical seal against the cylinder wall. They are larger than the cylinder diameter when outside the engine. When installed, they push out against the cylinder wall and make a seal.

Compression rings often use ring twisting action and compression pressures to improve the seal. The compression ring is often made with a notch machined on the upper inside edge. The notch allows compression pressures from the combustion chamber to get between the ring land and the rear of the piston ring (Figure 8-37) and push the ring face out against the cylinder wall for a better seal.

The wiper ring (second compression ring) is located in the second ring groove. This ring often has a notch cut on the lower part of the face. The top part of the face is used to seal compression pressure that gets by the compression ring. The top of the ring face tips away from the cylinder wall (Figure 8-38). This happens on all strokes except the power stroke. During the power stroke, compression pressures square it up. During the other three strokes, the lower outside corner of the ring has a strong contact with the cylinder wall. The lower inside corner of the ring makes a good seal in the ring groove. The corner of the flat face works like a wiper. It pushes oil on the cylinder walls back down toward the crankcase.

> **CAUTION:** When installing piston rings with notches or grooves, be sure to install rings in the correct ring groove and in the correct direction in order for them to work properly.

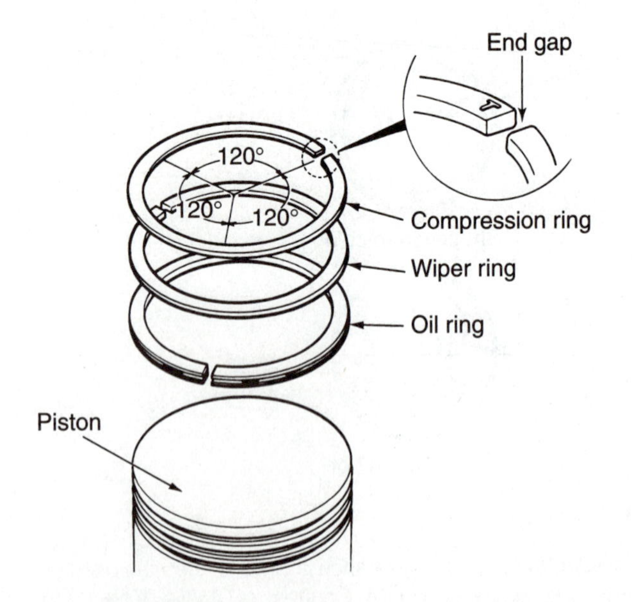

FIGURE 8-36 Most engines use three piston rings. (Courtesy of American Honda Motor Co., Inc.)

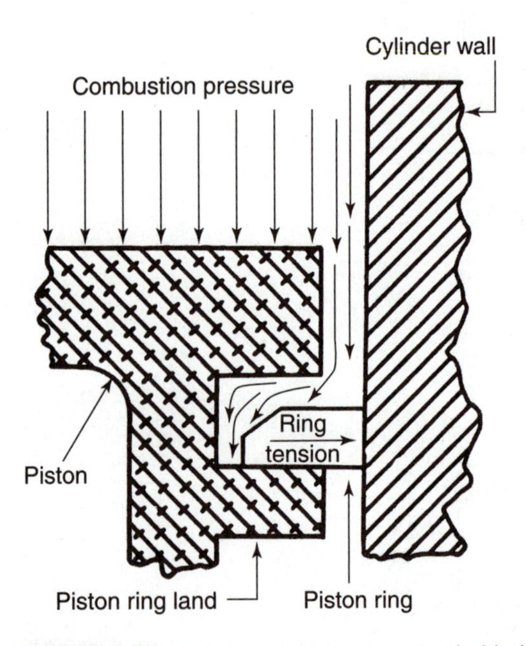

FIGURE 8-37 Compression pressures get behind the compression ring to help it seal. (Courtesy of CHAMPION-Federal Mogul.)

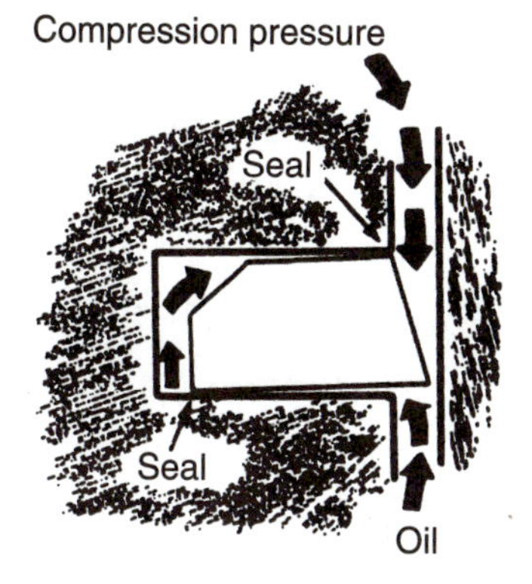

FIGURE 8-38 The wiper ring tips in its groove to help it seal and wipe oil. (Courtesy of Sealed Power.)

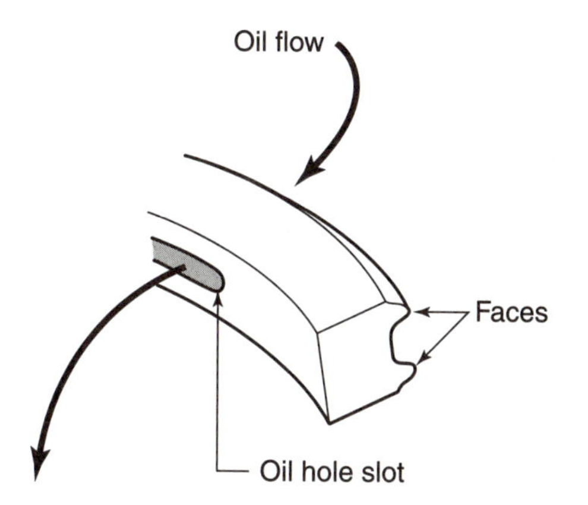

A. One-Piece Oil Control Ring

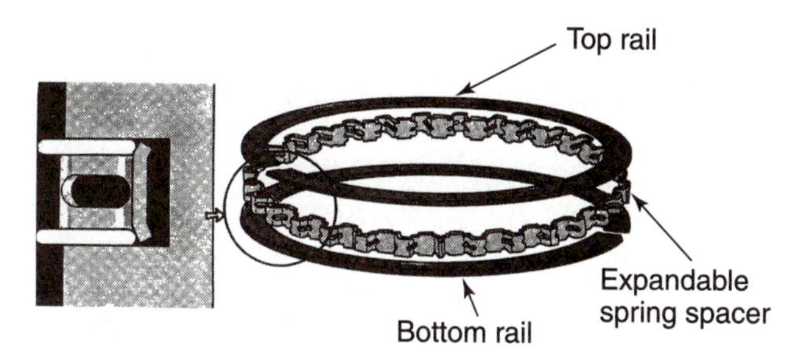

B. Multi-Piece Oil Control Ring

FIGURE 8-39 Oil rings may be one piece or multipiece. (*B:* Courtesy of Sealed Power.)

The oil control ring prevents excessive oil from getting from the crankcase into the combustion chamber. Oil in the combustion chamber causes oil consumption and spark plug fouling. The one-piece oil control ring (Figure 8-39A) has two narrow faces separated by oil passages. The multipiece oil control ring (Figure 8-39B) has an expandable spring spacer and two rails. The spring spacer is slightly larger around than the cylinder. When it is

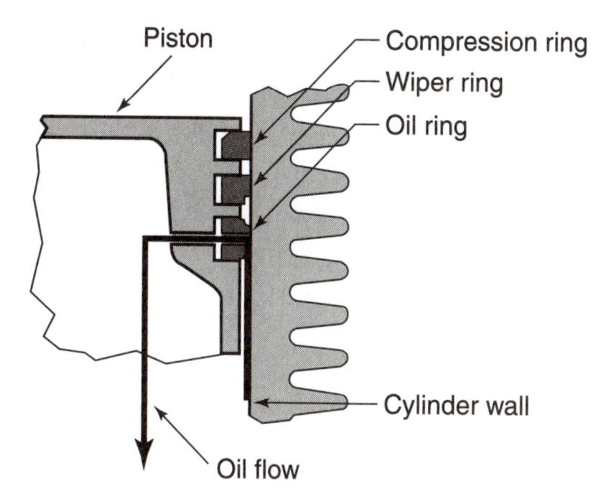

FIGURE 8-40 Oil is wiped off the cylinder wall and flows through the oil ring and piston ring groove.

on the piston and in the cylinder, it pushes the rails against the cylinder wall. The rails seal against the cylinder like the faces of the one-piece oil control ring. The faces scrape oil off the cylinder wall and direct it through the ring passages. The oil flows through holes in the piston oil control ring groove. Then it falls back down into the crankcase (Figure 8-40).

As the face cast-iron oil ring wears, its spring tension against the cylinder wall gets smaller. Also the width of the faces gets bigger, causing reduced oil control. Some engines use a ring expander to solve this problem. A **ring expander** (Figure 8-41) is an expandable spring used behind oil control rings to increase ring tension. The expander is slightly larger around than the cylinder. When assembled behind the piston, it pushes out on the ring, forcing the ring against the cylinder wall for improved oil control.

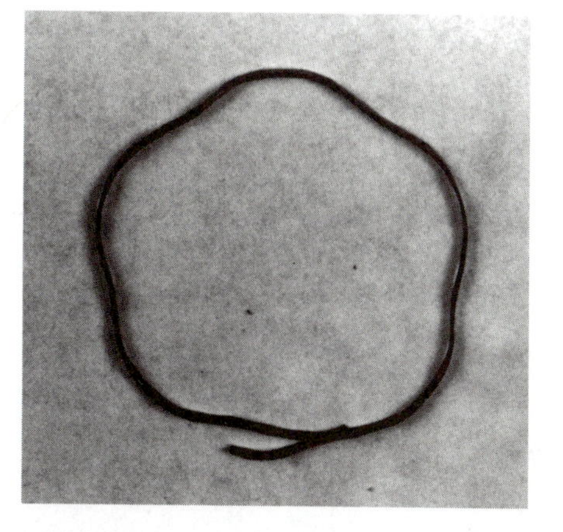

FIGURE 8-41 An expander is used to push some rings out against the cylinder wall.

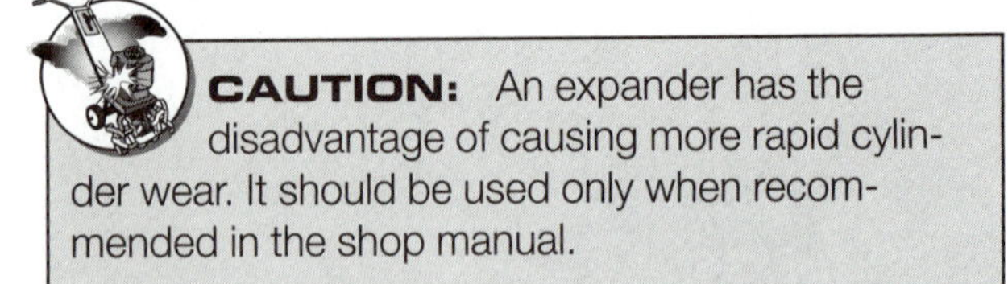

CAUTION: An expander has the disadvantage of causing more rapid cylinder wear. It should be used only when recommended in the shop manual.

CONNECTING ROD

The connecting rod (Figure 8-42) must be strong. All the forces of combustion are transferred to it on the way to the crankshaft. It must also be light because it is reciprocating weight. Connecting rods are generally made from aluminum. The center of the rod usually has an I beam shape to improve strength without adding weight.

The **connecting rod big end** is the end of the connecting rod attached to the crankshaft connecting rod journal. The **saddle bore** is the large diameter hole in the big end of the connecting rod that fits around the crankshaft journal. The big end of a four-stroke connecting rod is made in two parts, which allows it to be installed on the crankshaft.

A **rod cap** is the removable part of the connecting rod that allows it to be assembled around the crankshaft journal. Sometimes studs are used in the connecting rod. Sometimes bolts that fit through the rod are used. Nuts are used to retain the rod cap. Many engines use screws (often called *bolts*) that pass through the cap and fit into threads in the rod.

High-quality fasteners are used on the rod cap. High stresses can break the fasteners. Vibration can cause fasteners to get loose. Either problem can cause engine failure. Special locking fasteners are used on some engines. Other engines have sheet metal locking tabs. These are bent over the connecting rod screw heads after installation. The locking tabs prevent the rod cap fasteners from vibrating loose (Figure 8-43).

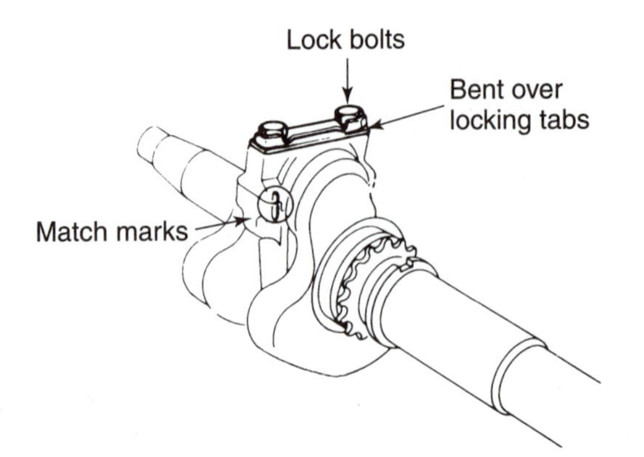

FIGURE 8-43 Retaining tabs bent over bolt heads to prevent fasteners from getting loose. (Courtesy of American Honda Motor Co., Inc.)

The rod cap and connecting rod big end must form a perfect circle around the crankshaft journal. The saddle bore is machined to very close tolerances with the cap in place. The connecting rod and cap are then marked with match marks. A match mark is a special notch cut in two parts that are lined up during assembly to be sure they fit in the correct direction. The cap must always be installed with the match marks together. The match marks are usually made so that they face the technician during engine assembly. Reversing the cap would cause the saddle bore to be out-of-round. The parting line or split between the connecting rod cap and connecting rod may be at any angle. Horizontal and vertical splits (Figure 8-44) are the most common.

The **connecting rod small end** is the end of the connecting rod attached to the piston. This end has the machine holes for the piston pin. Inexpensive engines do not have a separate bearing or bushing. The pin fits directly in the aluminum rod small end. Larger and more expensive engines have

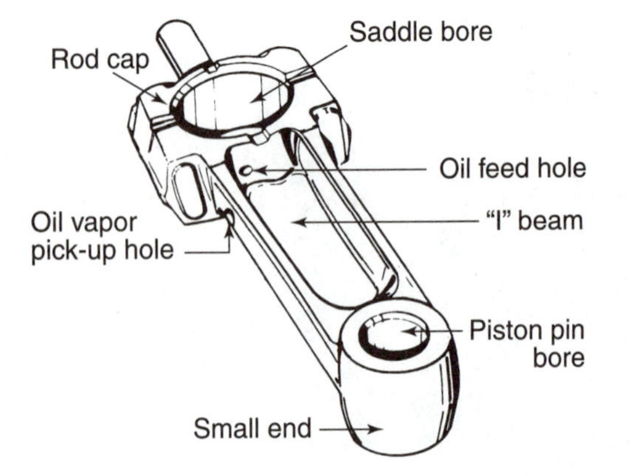

FIGURE 8-42 Parts of a connecting rod. (Provided courtesy of Tecumseh Products Company.)

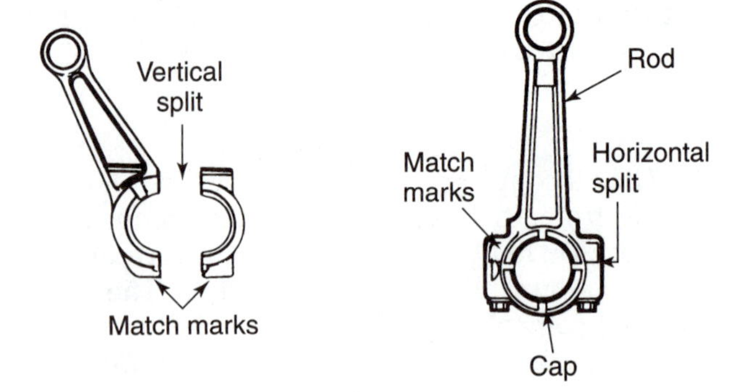

FIGURE 8-44 Horizontal and vertical split rod caps. (Provided courtesy of Tecumseh Products Company.)

a separate pin bushing installed in the rod. These are replaced during engine service.

CONNECTING ROD BEARING

A bearing surface is required between the saddle bore and the crankshaft journal. A **connecting rod bearing** allows the connecting rod to rotate freely on the crankshaft journal. Engines that use aluminum connecting rods use the surface machined in the saddle bore as a bearing. This makes the rod inexpensive. The connecting rod has to be replaced when there is wear in the rod-bearing area.

Larger and more expensive engines use a precision insert rod bearing (Figure 8-45). A **precision insert bearing**, or insert, is a thin sleeve bearing made in two halves and inserted in the connecting rod big end and connecting rod cap. These are commonly called *rod bearings*. Tabs on the bearings fit into matching notches in the saddle bore to lock them in place. Inserts are made with a steel backing. A bearing material is applied over the backing. Common bearing alloy materials are tin, copper, and cadmium. Oil holes may be drilled in the insert for lubrication.

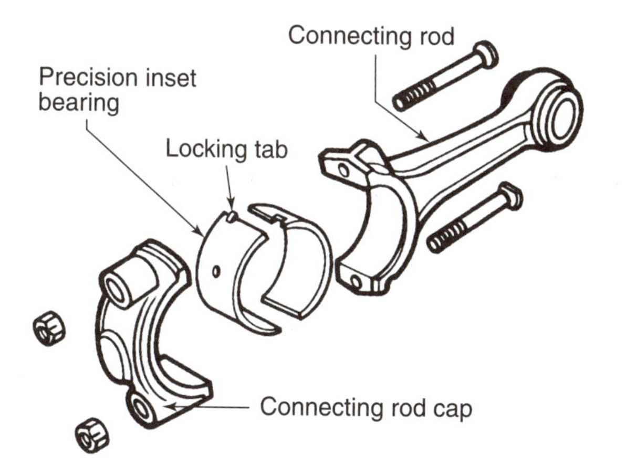

FIGURE 8-45 Some connecting rods use a precision insert bearing between the connecting rod and crankshaft journal.

PISTON PIN

The connecting rod is connected to the piston with a piston pin (Figure 8-46). The piston pin is often called a *wrist pin*. The motion of the connecting rod and piston is like the human wrist and hand. The piston pin is made of high-quality steel. The full force of the combustion pressure goes through it from the piston to the connecting rod. It is usually hollow to reduce reciprocating weight.

The piston pin often "free floats" in the piston and connecting rod. The piston pin is made slightly smaller than the holes in the piston and the connecting rod. It can "float" or rotate in both parts. It is held in place by two retaining rings, usually one on each side of the piston. The retaining rings fit into grooves in the piston. There is an advantage to the floating fit. The pin can seize up in either the piston or rod. The parts will not lock up. The fit or clearance between these parts is one of the closest in the engine.

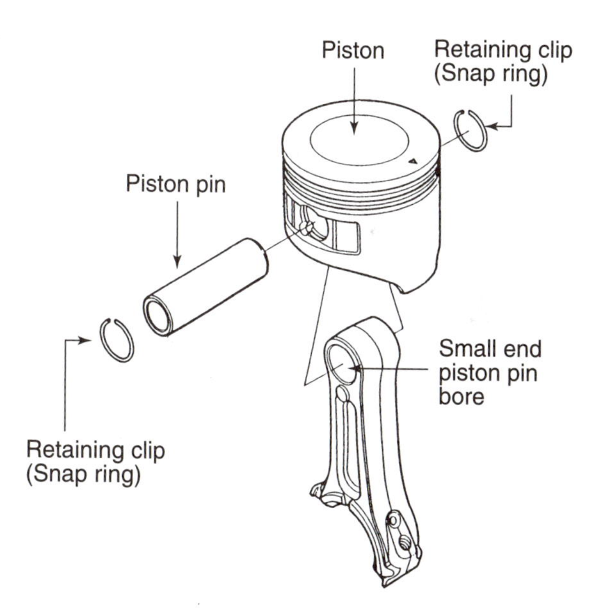

FIGURE 8-46 Retaining rings are used to hold the piston pin in the piston and rod. (Courtesy of American Honda Motor Co., Inc.)

REVIEW QUESTIONS

1. List and explain the purpose of the basic parts that make up a piston engine.
2. Explain how up-and-down piston movement is changed to rotary motion at the crankshaft.
3. Define and explain the action in the cylinder during the intake stroke.
4. Define and explain the action in the cylinder during the compression stroke.
5. Define and explain the action in the cylinder during the power stroke.
6. Define and explain the action in the cylinder during the exhaust stroke.

7. Define and explain the term *stroke* in relation to TDC and BDC.
8. Explain the purpose and construction of the crankcase.
9. Explain the difference between the PTO and the magneto side of an engine.
10. List and describe four types of crankshaft main bearings.
11. Explain the advantage of a ball or roller main bearing over a plain bearing.
12. List and describe the two different materials used for cylinder walls.
13. Explain why a cast-iron liner is used in some aluminum cylinders.
14. Explain the parts and purpose of a head gasket.
15. List and describe the parts of a crankshaft.
16. List and describe the purpose of the parts of a piston.
17. Explain why notches are used on a wiper ring.
18. Explain the purpose and operation of the oil control piston ring.
19. Identify and explain the parts on the connecting rod.
20. Explain how the piston pin is retained in the piston and connecting rod.

DISCUSSION TOPICS AND ACTIVITIES

1. Remove the cylinder head from an engine. Rotate the engine and identify the four strokes of the four-stroke cycle.
2. Rotate the engine until the piston is ready for an intake stroke. Scribe a line on the crankshaft. Rotate the engine through each of the four strokes and observe the number of degrees of crankshaft rotation. How many degrees of rotation occur during each stroke?
3. Disassemble a scrap engine. Spread the parts out on the bench and try to identify each part.

CHAPTER 9

Four-Stroke Engine Valve System Parts

OBJECTIVES

Upon completion and review of this chapter, you should be able to:

- Describe the parts and operation of an L-head valve system.
- Describe the parts and operation of an overhead valve system.
- Identify the parts and explain the operation of valves, valve springs, and retainers.
- Identify the parts and explain the operation of valves, valve guides, and valve seats.
- Explain the parts and operation of the camshaft and how it works to time valve opening.
- Identify the parts and explain the operation of the valve lifter, push rod, and rocker arm.
- Explain the purpose of valve clearance and describe how it is adjusted.

TERMS TO KNOW

Cam gear
Camshaft
Exhaust valve
Intake valve
L-head valve
Overhead valve (OVH)
Push rod
Rocker arm assembly
Timing marks
Valve
Valve clearance
Valve cover
Valve face
Valve guide
Valve head
Valve lifter
Valve locking groove
Valve margin
Valve rotator
Valve seat
Valve spring
Valve spring retainer
Valve stem
Valve timing
Valve tip

INTRODUCTION

The valve system is an assembly of parts responsible for getting the fresh air and fuel mixture into the cylinder on the intake stroke. It is also responsible for getting the burned exhaust gasses out of the cylinder on the exhaust stroke. The valve system is sometimes called the valve train.

VALVE SYSTEMS

A **valve** is a part that is opened or closed to control the flow of air and fuel or burned exhaust gas through the engine's ports. The valves work like a plug in the bottom of a bathtub (Figure 9-1). When the plug is pushed into the hole, water cannot get out. If the plug is pulled up out of its hole, water can get past.

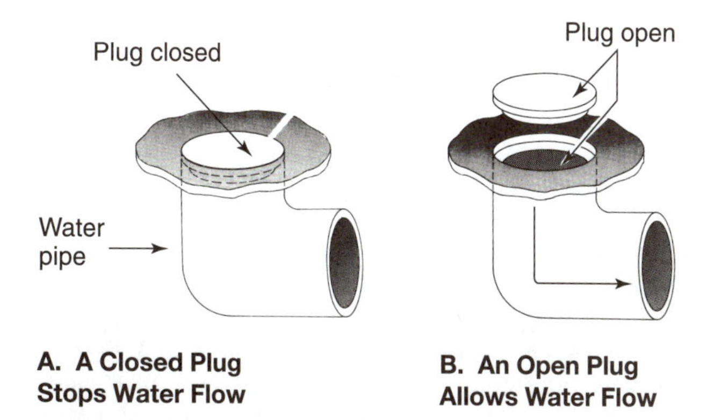

FIGURE 9-1 When it is down, a bathtub plug stops water and allows water flow when it is up.

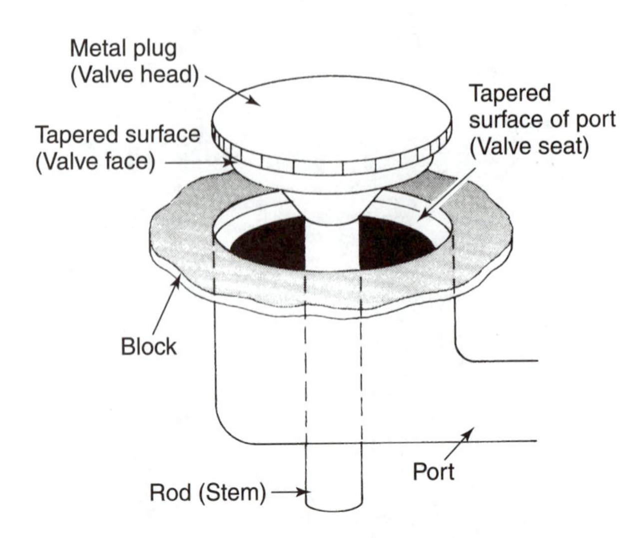

FIGURE 9-2 A valve is like a plug with a rod attached to the bottom.

An engine valve is basically a round metal plug connected to a rod (Figure 9-2). The head of the valve is tapered. The port area has a matching taper. When closed, the tapered part of the valve seals tightly against the port.

The valves are opened and closed by cam lobes (eccentrics) on a shaft (Figure 9-3). The lobes contact the ends of the valves (through valve lifters). The valve is closed when the lowest part of the cam lobe is under the valve stem. As the lobe turns, the high part of the lobe pushes up on the valve stem. The valve opens. As the cam turns further, the lobe passes under the stem. The valve closes.

There are two basic kinds of valve mechanisms, L-head and overhead. The **L-head valve system** (Figure 9-4) is a valve system where the valves are located in the cylinder block. This system is commonly called a *flathead*. Many small engines use the L-head valve system, because it requires fewer parts and is less expensive.

The **overhead valve (OHV) system** (Figure 9-5) is a valve system that has the valves located in the cylinder head above the piston. Use of the overhead valve system in small engines is increasing. It is more efficient at getting the gases in and out of the combustion chamber. The result is a more powerful engine and one that has fewer emissions.

VALVES

There are two valves in every cylinder. The **intake valve** is opened on the intake stroke to allow the air-fuel mixture into the cylinder. The **exhaust valve** is opened on the exhaust stroke to allow gases to escape from the cylinder.

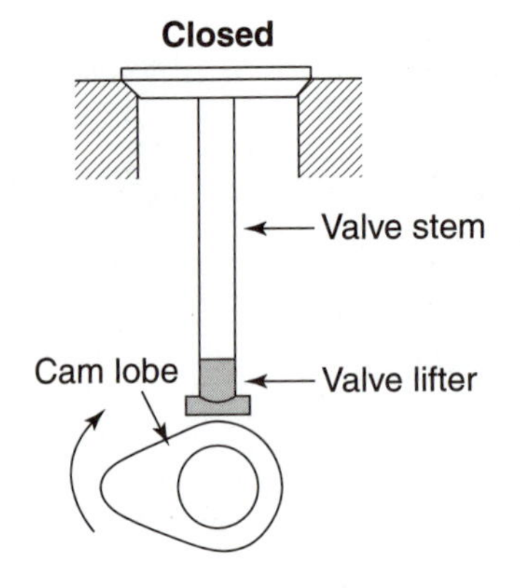

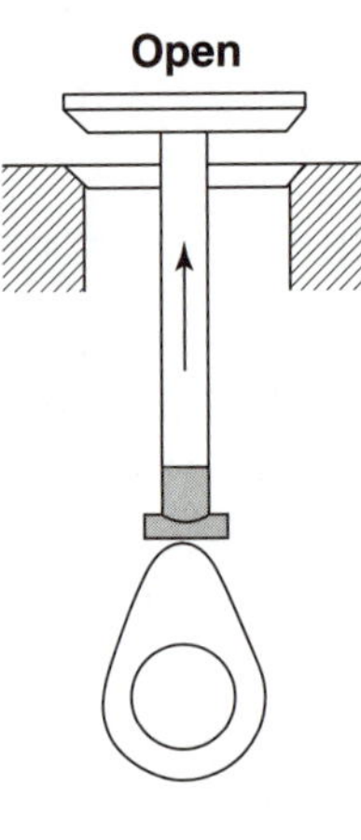

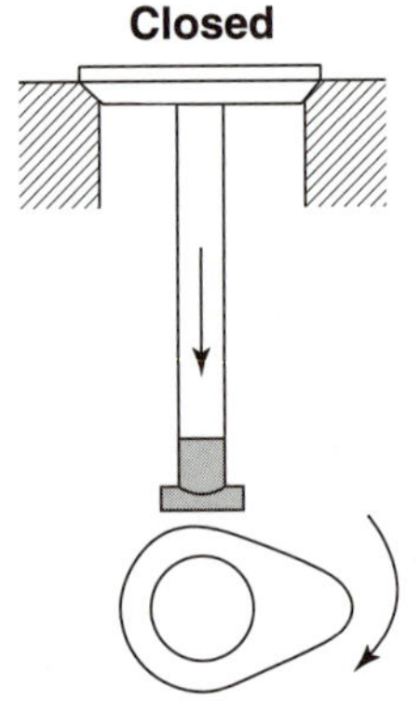

FIGURE 9-3 A cam lobe rotating below the valve stem can open and close the valve.

The intake and exhaust valves both have the same basic parts (Figure 9-6). The **valve head** is the part of the valve that when closed, seals the intake or exhaust port. The **valve face** is the precision ground area of the valve head that seals against the valve seat. The **valve margin** is the round area of the valve head above the valve face. The **valve stem** is the shaft connected to the valve head. The **valve tip** is the end of the valve stem. The **valve-locking groove** is a groove on the end of the valve stem used to retain a valve spring.

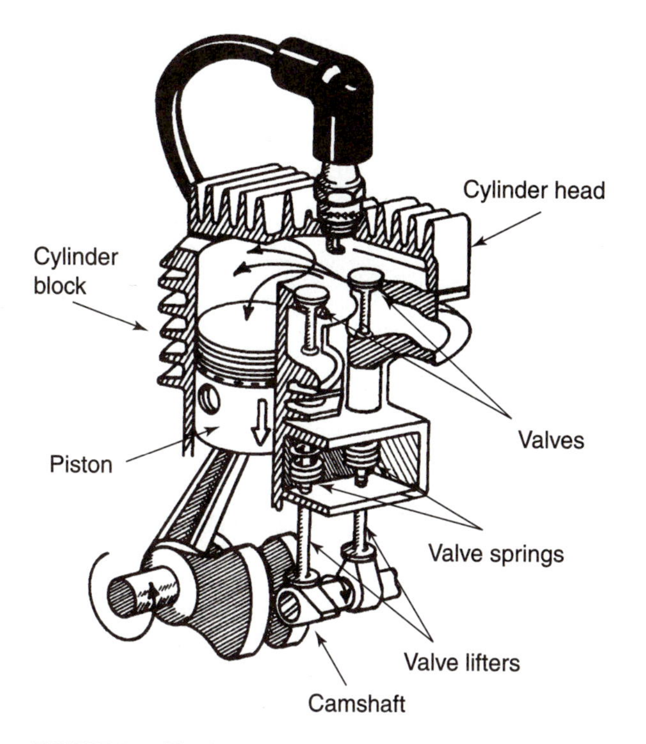

FIGURE 9-4 The L-head valve system has the valves located in the cylinder beside the piston. (Provided courtesy of Tecumseh Products Company.)

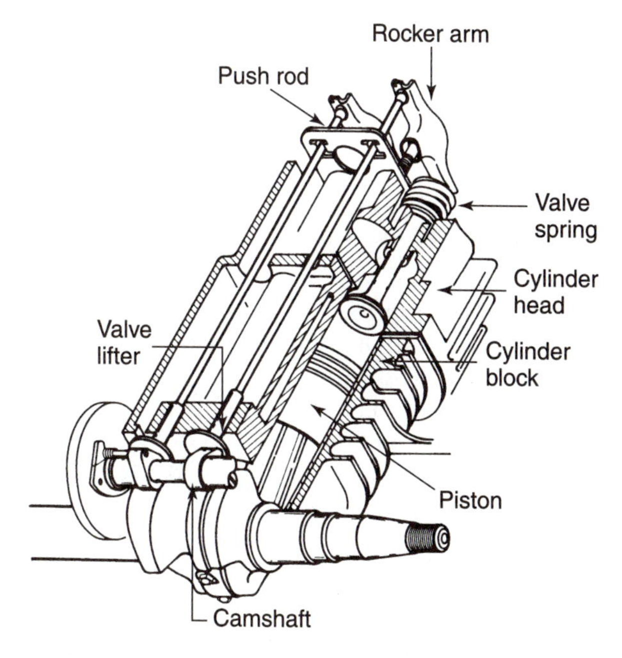

FIGURE 9-5 The overhead valve system has the valves located in the cylinder head above the piston. (Provided courtesy of Tecumseh Products Company.)

The intake valve head is usually larger in diameter than the exhaust valve (Figure 9-7). It controls the slow-moving, low-pressure intake mixture. Exhaust valves are often smaller because exhaust gases leave the cylinder under higher pressures.

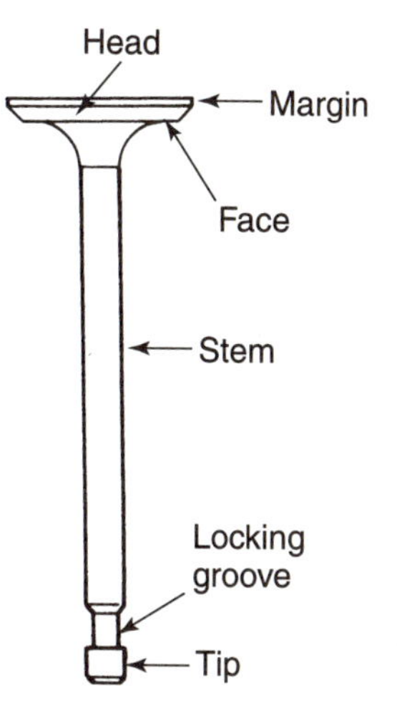

FIGURE 9-6 Parts of a valve. (Provided courtesy of Tecumseh Products Company.)

FIGURE 9-7 Intake valve heads are often larger than those on exhaust valves.

Valves are made of very high quality steel because they get very hot during combustion. The cool intake mixture cools the intake valve that operates cooler than the exhaust valve. The materials and surface coatings that are used on valves depend on expected temperatures. The exhaust valve is usually made of austenitic stainless steel alloy. The valve face may be hardened with a hard alloy. This is called *hardfacing*.

CAUTION: The hardfacing on valve faces is often just a few thousandths of an inch (hundredths of a millimeter) thick. Sometimes these valves cannot be reconditioned by grinding. Always check the shop manual.

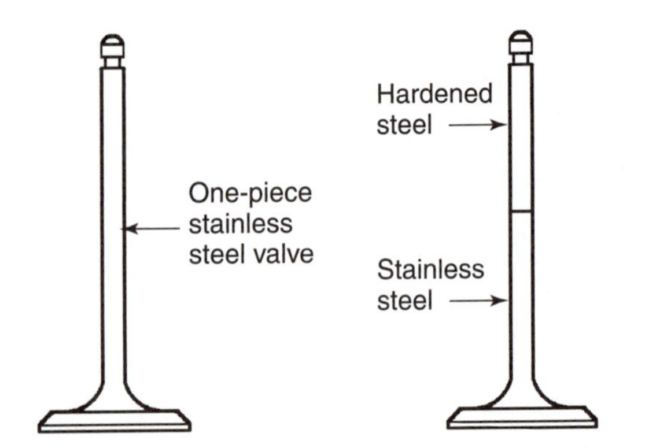

FIGURE 9-8 Valves are made in one piece or welded together from different materials.

Valve stems may be made from the same material as the valve head. The stem fits in the valve guide. Up and down movement in the guide can wear out the stem. Stem and guide wear can cause excessive clearance, which can lead to oil consumption and excessive hydrocarbon emissions. Sometimes a hardened steel stem is welded to a stainless steel head (Figure 9-8). Some valves use stems that are hardfaced. Both these methods prevent wear to the valve stem. This allows smaller valve guide clearances.

VALVE SPRING AND RETAINER

A cam lobe opens the valve and a spring closes the valve. The **valve spring** (Figure 9-9) is a coil spring used to hold the valve in a closed position. Valve springs are made from high-quality steel. When the engine is running, they expand and compress many times per second. Eventually they will lose their tension and must be replaced.

SERVICE TIP: Valve springs often have progressive coils that are tighter (closer together) at one end than the other. Progressive valve springs eliminate the need for a second dampening spring inside the main spring. When assembling these springs, the tight coils are positioned toward the valve head. This means they go down on an OHV engine and up on an L-head engine.

A **valve spring retainer** is a part that holds the compressed valve spring in position on the valve stem (Figure 9-10). The bottom of the spring sits on the cylinder block of an L-head and on the cylinder head of an OHV engine. The valve stem fits through

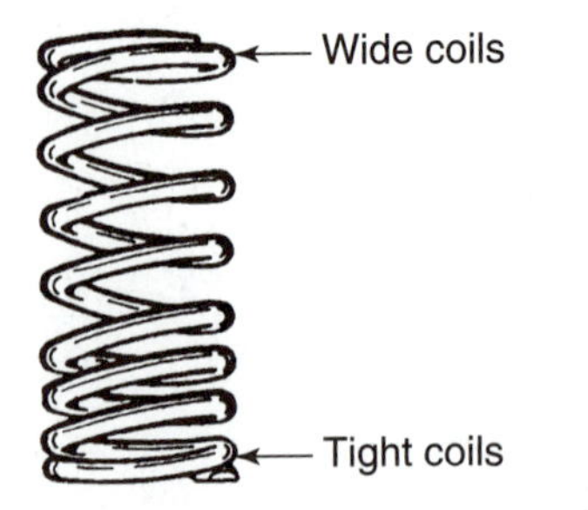

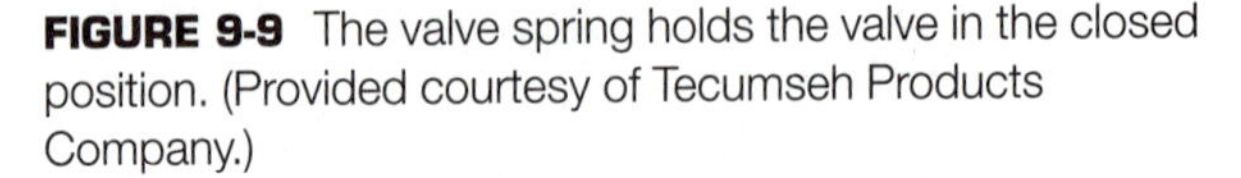

FIGURE 9-9 The valve spring holds the valve in the closed position. (Provided courtesy of Tecumseh Products Company.)

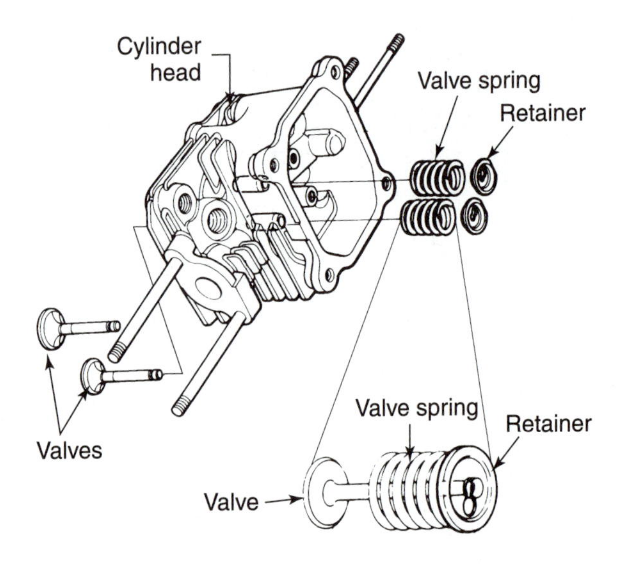

FIGURE 9-10 The valve spring sits on the cylinder head or block and is held in place with a retainer. (Courtesy of American Honda Motor Co., Inc.)

the center of the spring. The spring is compressed, and the retainer (along with a retainer washer in some designs) sits on the bottom of the spring. It fits in to the valve stem locking groove.

There are three common types of valve spring retainers (Figure 9-11). One is a washer-shaped part

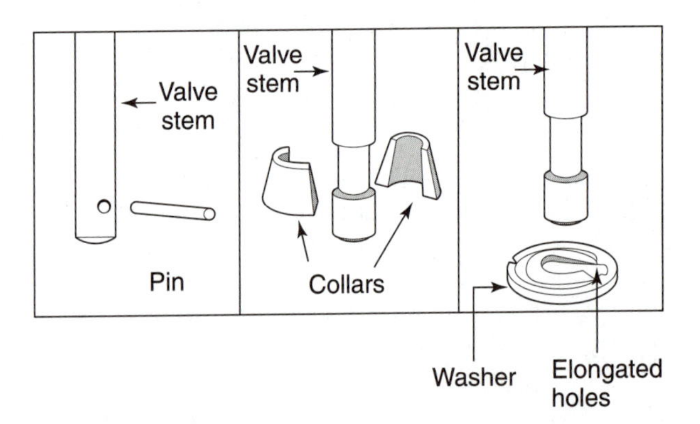

FIGURE 9-11 Three types of valve spring retainers.

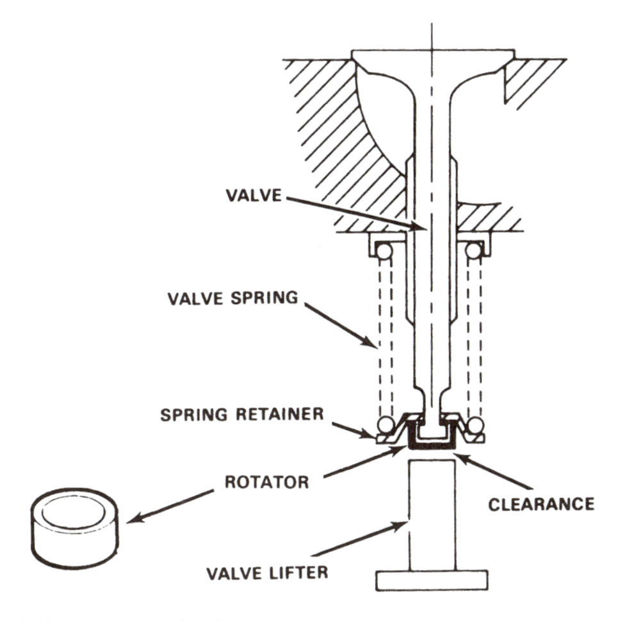

FIGURE 9-12 A valve rotator rotates the valve to even out wear. (Courtesy of American Honda Motor Co., Inc.)

with a hole roughly in the shape of a number eight (8). The larger diameter of the hole is used to slip the retainer over the valve stem. The retainer is then centered. The small part of the hole fits in the valve stem locking groove. Another type uses two split valve locks or collars. These are inserted into the valve stem locking grooves. They fit into a separate retainer washer, locking the spring to the valve. Older engines have a small hole in the valve stem. A pin is driven into the hole to hold the retainer washer in place.

VALVE ROTATOR

Some engines use a **valve rotator** (Figure 9-12), a part connected between the valve stem and valve spring retainer that causes valve rotation. A valve rotator has a spring system that causes the valve to rotate when it is lifted off its seat. Rotation helps to even out valve-to-seat wear and dissipates heat more evenly, which helps the engine last longer.

VALVE SEAT

The **valve seat** (Figure 9-13) is a precision ground opening in the port that makes a pressure tight seal with the valve face. The seats are in the cylinder block on L-head engines and in the cylinder head on OHV engines. The valve seat reaches high temperatures and takes constant pounding from the valve opening and closing.

Valve seats are made from cast iron or steel in order to handle these forces. Seats in cast-iron cylinder blocks or heads can be made from cast iron. They can be machined directly in the cast iron. Some engines have seats made as separate parts. On some engines they can be replaced during service. On other engines they are cast in place during manufacturing and are not replaceable. Aluminum cylinder heads or blocks must have separate valve seats. These are often made of cast-iron

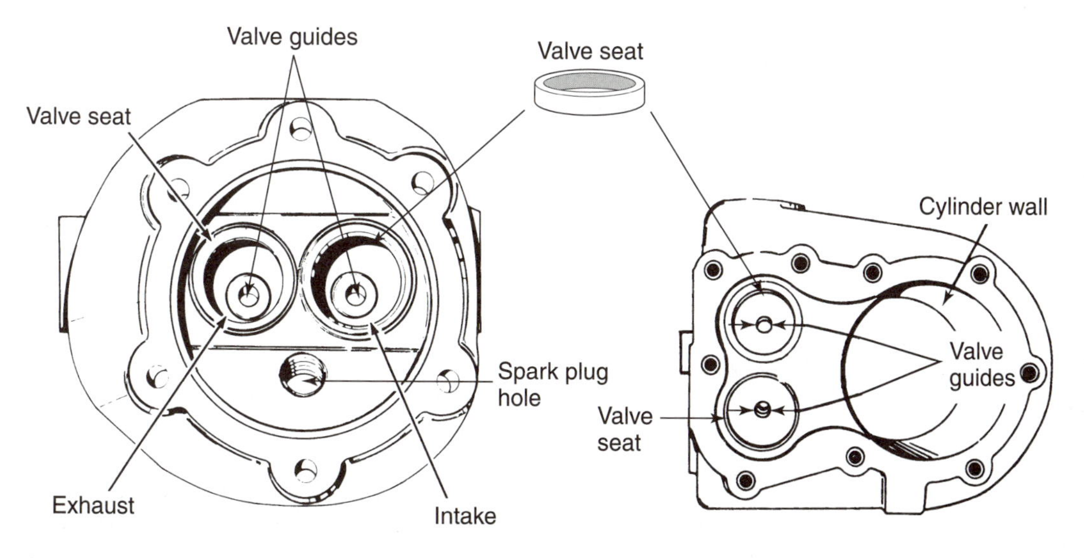

A. Valve Seats in Cylinder Head

B. L-Head Valve Seats in Cylinder Block

FIGURE 9-13 Valve seats are in the L-head block and the overhead valve cylinder head. (Provided courtesy of Tecumseh Products Company.)

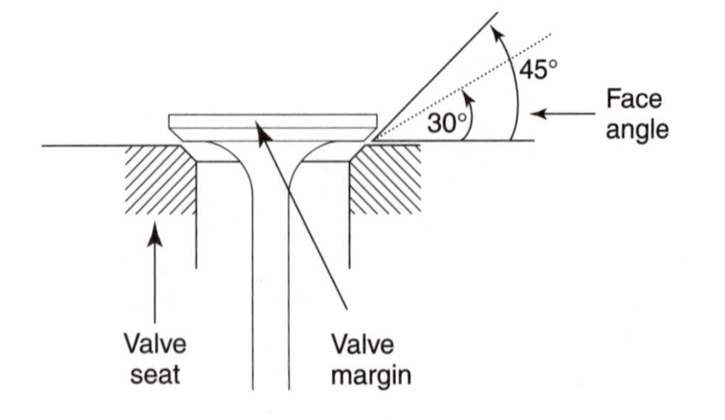

FIGURE 9-14 The valve face seals against the valve seat.

or steel seats and installed with a press fit. Sometimes the seats are hardened by hardfacing.

The valve sealing surface is machined on both the valve face and valve seat by grinding. The angle ground on the valve seat matches the angle ground on the valve face, usually 45 degrees. Some engines use a 30-degree angle (Figure 9-14).

VALVE GUIDE

The **valve guide** (Figure 9-15) is a part that supports and guides the valve stem in the cylinder block or cylinder head. The guide keeps the valve centered over the valve seat. Many of the less expensive aluminum engines have the guide machined directly in the cylinder block. More expensive engines use a valve guide that may be removed and replaced during engine service. Made of cast iron or a softer material such as bronze alloy, replaceable guides are pressed or driven into the cylinder head or cylinder block (Figure 9-16).

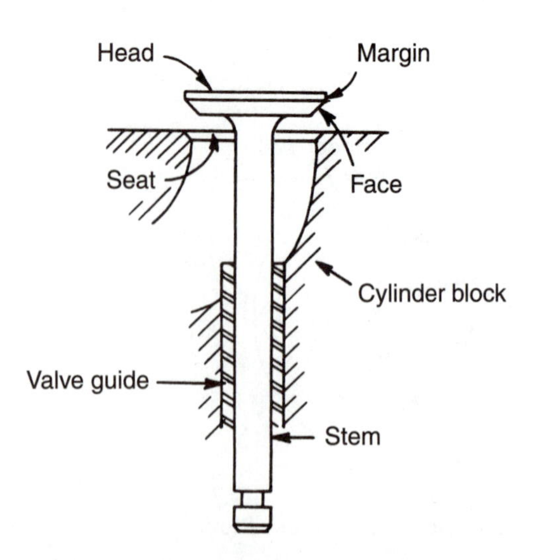

FIGURE 9-15 The valve guide supports the valve stem. (Courtesy of Clinton Engines Corporation.)

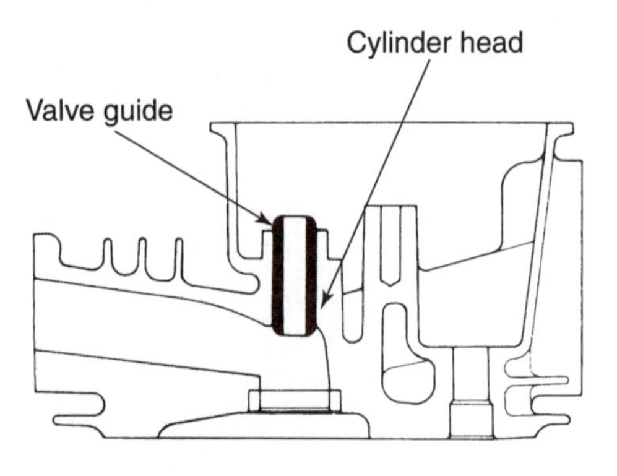

FIGURE 9-16 Replaceable valve guides may be used in the cylinder block or cylinder head. (Courtesy of American Honda Motor Co., Inc.)

The valve guide and valve seat both help cool the valve. The valve head gets hot during the power and exhaust stroke. Heat passes through the valve head and through the seat when the valve is closed. Heat that travels through the valve stem passes through the valve guide.

The clearance between the valve stem and the guide must allow free movement of the valve. It must also allow a small amount of oil to get between the stem and guide for lubrication. If there is too much clearance, too much oil can get between the stem and guide and the oil will get into the combustion chamber. Oil in the combustion chamber fouls the spark plug. This is more of a problem on the intake valve stem. There is a vacuum in the cylinder when this valve is open. Too much clearance can also cause increased hydrocarbon emissions when fuel from the combustion chamber escapes past the valve stem.

CAMSHAFT

The **camshaft** is a shaft with cam lobes (Figure 9-17) used to open the intake and exhaust valves at the correct time for the four-stroke cycle. The camshaft for a single cylinder engine has one lobe for the intake valve and another for the exhaust valve. Each of the cam lobes is ground into a shape that provides the correct valve action. The shaft of the cam is concentric with a round part of the lobe called the *base circle*. The nose of the lobe is higher than the base circle and provides the valve opening distance or lift. The heel of the lobe is opposite the nose. When the heel is under the valve stem, it allows the valve to close.

The camshaft has a bearing journal at each end (Figure 9-18). The journals fit into camshaft bearings in the crankcase and cover. These bearings are

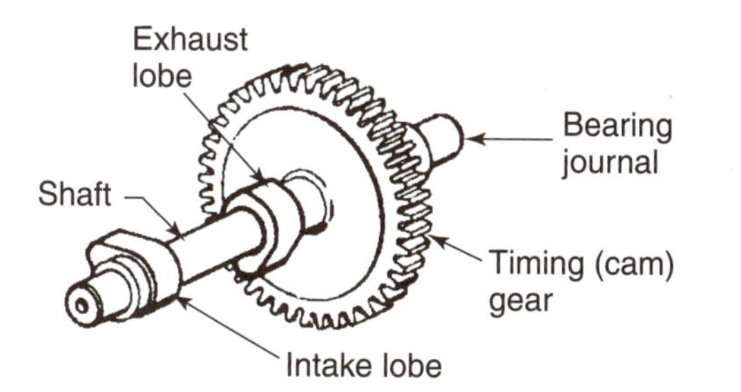

A. Cam Shaft

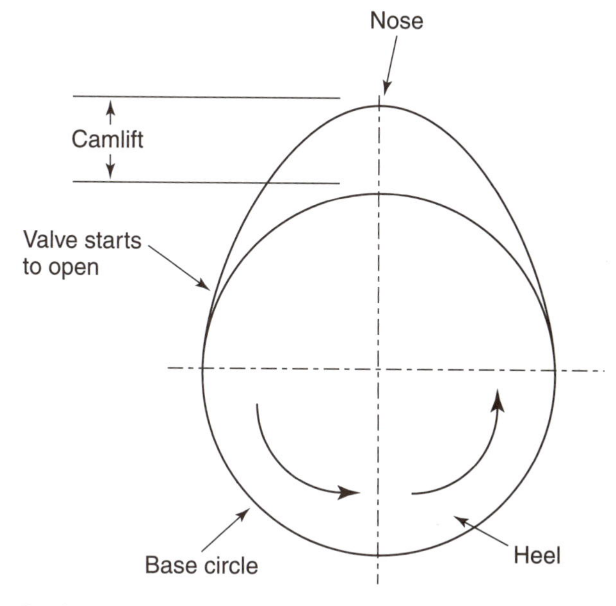

B. Cam Lobe

FIGURE 9-17 Parts of a camshaft and camshaft lobe. (*A:* Provided courtesy of Tecumseh Products Company. *B:* Courtesy of DaimlerChrysler Corporation.)

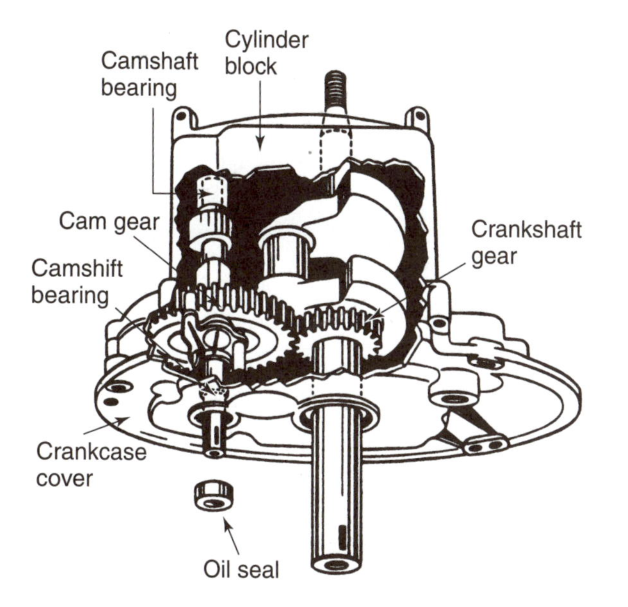

FIGURE 9-18 The camshaft is supported in bearings in the cylinder block and crankcase cover. (Courtesy of Clinton Engines Corporation.)

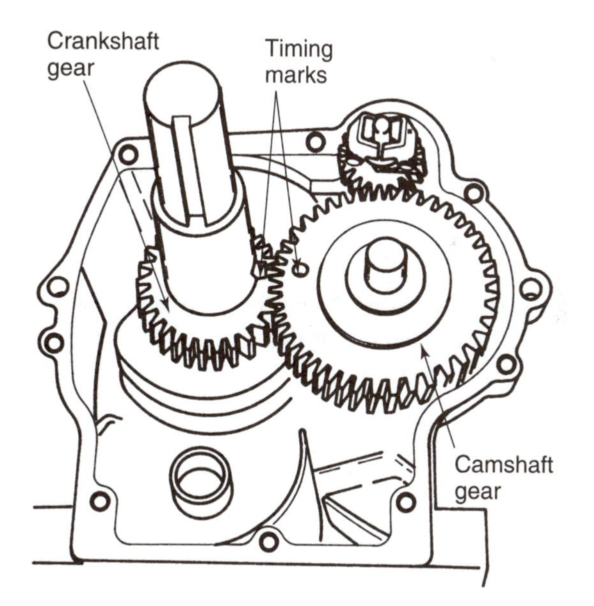

FIGURE 9-19 Timing marks on the camshaft and crankshaft gear. (Provided courtesy of Tecumseh Products Company.)

commonly machined in the aluminum of the cover and crankcase. They receive less wear than main bearings because the forces are lower. Also, the camshaft turns half as fast as the crankshaft.

The **camshaft gear**, or timing gear, is a gear on the end of the camshaft driven by a gear on the crankshaft. When the crankshaft turns, it turns the camshaft. The camshaft gear is twice as big as the crankshaft gear, so the camshaft turns only one-half as fast as the crankshaft. This is required because the valves open only on the intake and exhaust strokes. They are closed on the compression and power strokes. **Timing marks** (Figure 9-19) are marks on the camshaft and crankshaft gear that are aligned together so that the camshaft opens the valves at the correct time in relation to crankshaft position.

VALVE TIMING

Valve timing is the opening and closing of the intake and exhaust valves in relation to crankshaft position. Valve timing is controlled by the shape of the camshaft lobe. The meshing of the camshaft and crankshaft timing gears also affects valve timing. Valve timing is specified in degrees of crankshaft rotation. One revolution of the crankshaft is

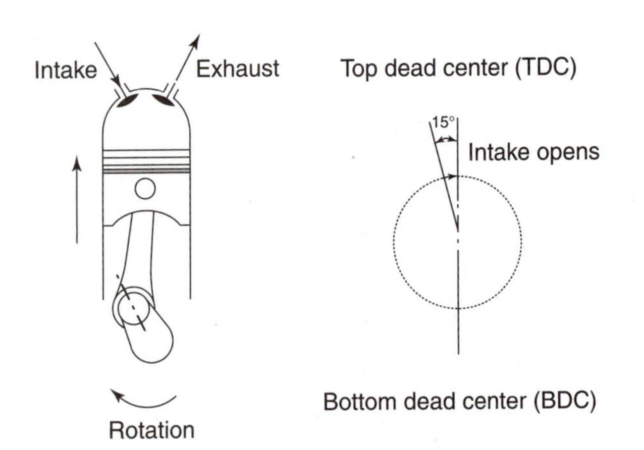

FIGURE 9-20 The intake valve begins to open.

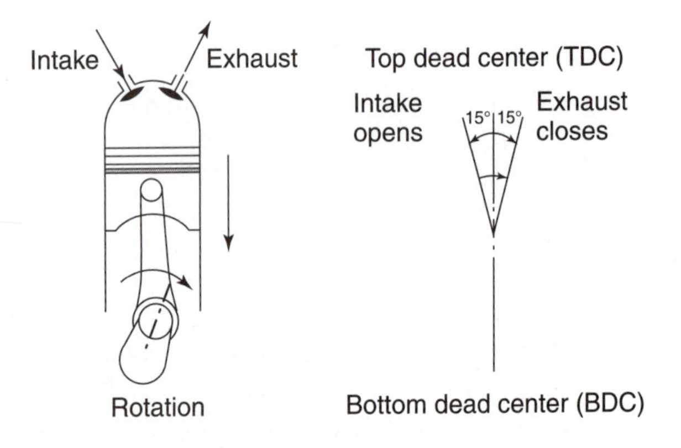

FIGURE 9-21 The exhaust valve closes.

360 crankshaft degrees. Exact valve timing specifications differ between engines. Timing is similar in all four-stroke engines.

The intake valve begins to open 15 crankshaft degrees before the piston is at TDC (Figure 9-20), giving the valve a head start. At TDC, the valve is well off its seat. The air-fuel mixture can then flow easily past the open valve.

As the piston reaches and passes TDC, the exhaust valve is still in the process of closing. At high engine speeds, the burned gases flow past the exhaust valve into the muffler. This flow helps draw in the air and fuel mixture. Valve overlap is the short time (measured in crankshaft degrees) when both the intake and exhaust valves are open. At 15 crankshaft degrees past TDC, the exhaust valve finally closes (Figure 9-21).

The intake stroke continues as the piston moves downward. The air and fuel mixture is drawn into the combustion chamber. The piston reaches the bottom of its stroke. Then it starts to rise in the cylinder for the compression stroke. The intake valve remains open, allowing the moving column of incoming air and fuel to continue filling the combustion chamber. This continues long after the piston changes direction. The intake valve does not close until 50 crankshaft degrees past BDC. The intake valve total opening period or duration is 15 degrees before TDC plus 180 degrees to BDC plus 50 degrees after BDC, a total of 245 crankshaft degrees (Figure 9-22).

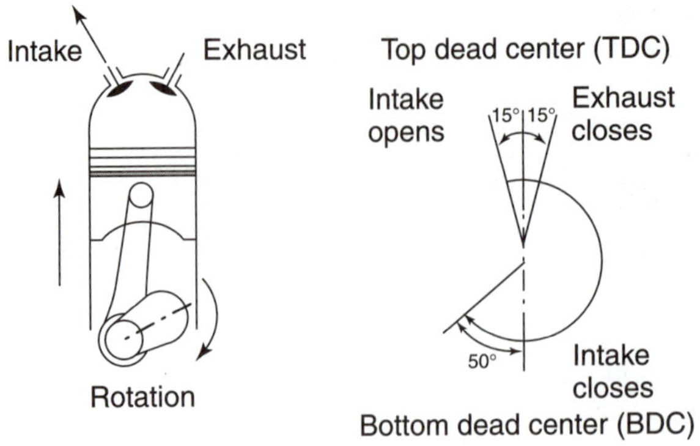

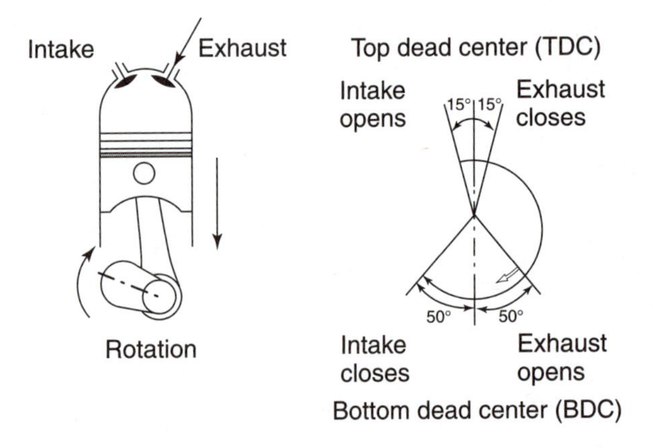

FIGURE 9-22 The intake valve closes.

Intake Exhaust Top dead center (TDC) Intake opens 15° 15° Exhaust closes Rotation 50° 50° Intake closes Exhaust opens Bottom dead center (BDC)

FIGURE 9-23 The exhaust valve opens.

The piston continues upward on the compression stroke, compressing the air-fuel mixture. Just before reaching TDC, the mixture is ignited. At TDC, the power stroke begins. The piston is forced downward once again. The exhaust valve begins to open at 50 crankshaft degrees before the piston reaches BDC (Figure 9-23), well before the power stroke has actually been completed. The exhaust gases leave the cylinder because of their own pressure. This reduces the engine's effort to push out the burned gases on the upward stroke of the piston.

The piston completes its downward travel and once again rises in the cylinder for the exhaust stroke. At 15 crankshaft degrees past TDC, the exhaust valve closes (Figure 9-24). The total duration or opening period of the exhaust valve is 50 degrees before BDC plus 180 degrees to TDC plus 15 degrees after TDC, a total of 245 degrees. These

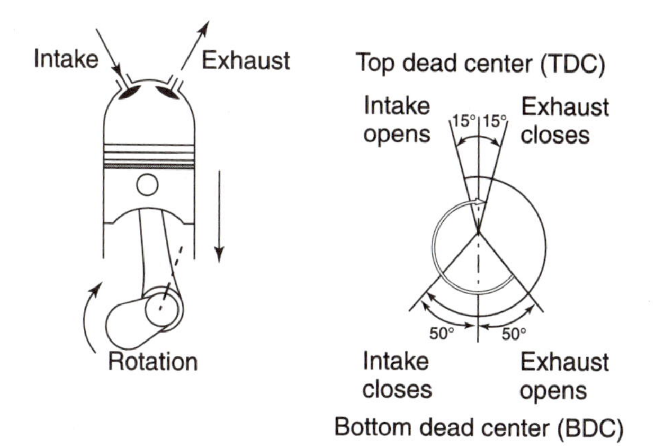

FIGURE 9-24 The exhaust valve closes.

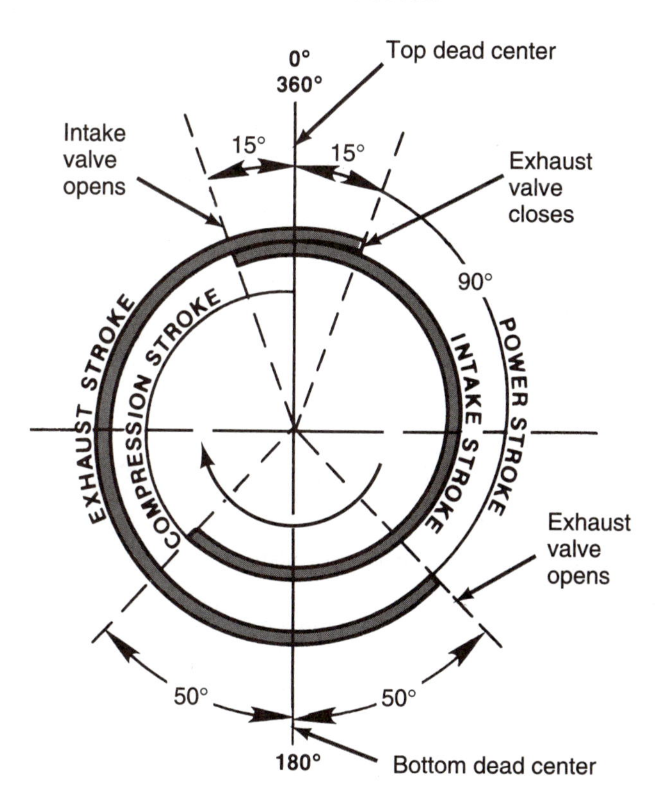

FIGURE 9-25 A valve timing diagram shows valve action in crankshaft degrees.

events are often shown in a valve timing diagram. A valve timing diagram (Figure 9-25) is a graph that shows the valve opening and closing through two revolutions or 720 degrees of crankshaft rotation.

CASE STUDY

An outdoor power equipment student was struggling to get the engine he had rebuilt started. It is unusual for a properly rebuilt engine not to start after several pulls on the starting rope. The instructor went over to the student to see if he could help. The instructor installed a spark tester to make sure the engine was getting an ignition spark. The ignition was providing a good strong spark. Maybe it's a fuel problem, he commented, and removed the spark plug. The insulator end of the spark plug was damp with fuel. Fuel was not the problem. The list of possible problems was getting shorter. The instructor grounded the spark plug wire so he would not get shocked and put his thumb over the spark plug hole. He had the student pull on the rope starter.

The instructor knew he should feel a strong compression push against his thumb when the piston comes up on each compression stroke. He also knew that he should feel a weaker push against his thumb when the piston comes up on the exhaust stroke. This engine had a definite problem. There was poor compression and there was no difference between the power and exhaust strokes.

The instructor had observed the assembly of this engine and knew the valves were adjusted properly. He knew that the valves were tested for sealing during assembly and they were not the problem. He had the student remove the cylinder head. With the cylinder head removed, he had the student rotate the engine and observe the piston and valves. The problem was obvious. The intake valve was open when it should be closed on the compression stroke. The instructor knew only one thing could cause this problem. The student had not lined up the timing marks between the camshaft and crankshaft timing gears properly. The unhappy student had to disassemble the engine to fix the problem.

VALVE LIFTER (TAPPET), PUSH ROD, AND ROCKER ARM

The camshaft does not work directly on the valves. A system of parts is used to transfer the movement of the camshaft to each of the valves. These parts are different on the L-head and overhead designs. Both systems use a part called a valve lifter. A **valve lifter** (Figure 9-26), or tappet, is a valve mechanism part that transfers cam lobe up-and-down movement to the valve stem.

The valve lifter is often made from cast iron. It usually has a small end at the top and a large, mushroom-shaped end (Figure 9-27), which rides

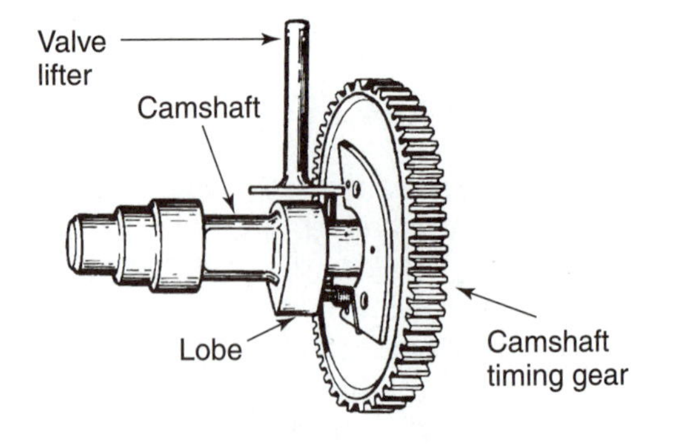

FIGURE 9-26 The valve lifter rides on the camshaft lobe. (Provided courtesy of Tecumseh Products Company.)

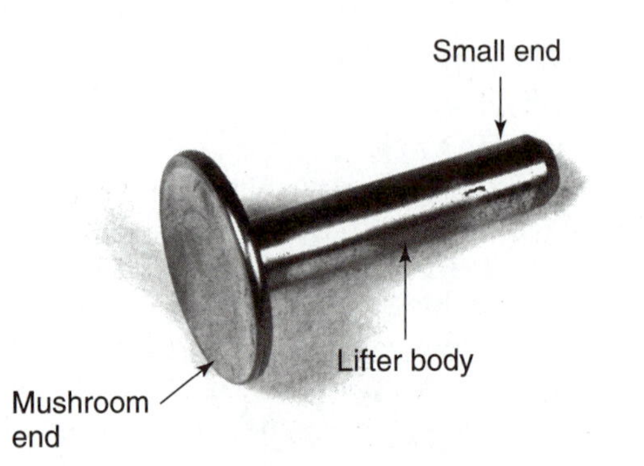

FIGURE 9-27 Parts of a valve lifter.

on the camshaft lobe. The lifter body is supported by a hole machined in the cylinder block.

The valve lifter is the only part between the cam lobe and the valve stem in an L-head valve mechanism. As the camshaft lobe moves upward, the valve lifter moves upward. The valve lifter pushes on the valve stem to open the valve.

Overhead valves are located above the piston in the cylinder head. The camshaft and valve lifters are located in the cylinder block. Cam lobe motion must be transferred up to the cylinder head to push on the valves. A push rod and rocker arm assembly are used to transfer the motion (Figure 9-28). A **push rod** is a part that transfers valve lifter motion from the cylinder block up to the rocker arm in the cylinder head. Push rods must be light and strong. They are commonly made from tubular steel.

The **rocker arm assembly** is an assembly on the cylinder head that changes the upward push from a push rod to a downward motion to open the valve. Some rocker arm assemblies have a rocker arm and a rocker shaft (Figure 9-29). The shaft is mounted on the cylinder head with studs and shaft supports. The rocker arm is a stamped steel or cast aluminum part that fits on the rocker shaft. The hole in the rocker arm is slightly larger than the diameter of the shaft to allow the rocker arm to rock up and down.

The pivot type rocker arm is more common (Figure 9-30). It has an individual stud-mounted pivot for each rocker arm. The rocker arm is attached to the pivot by an adjusting nut and lock

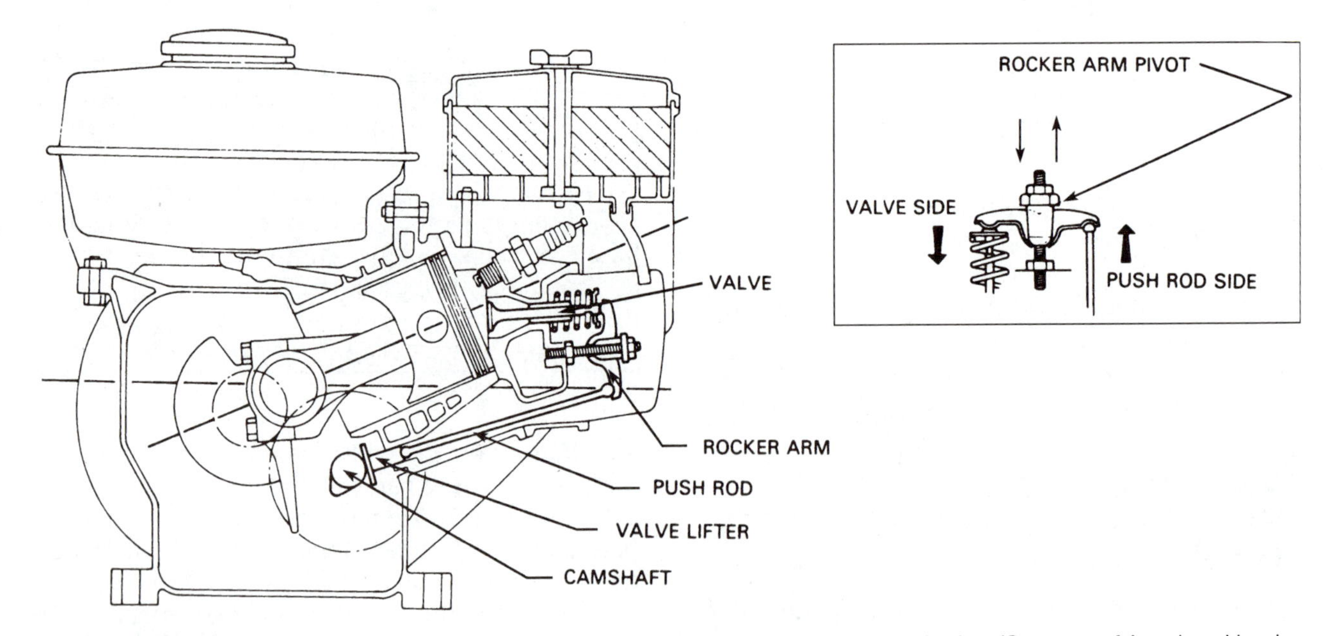

FIGURE 9-28 The push rod and rocker arm transfer camshaft motion up to the overhead valve. (Courtesy of American Honda Motor Co., Inc.)

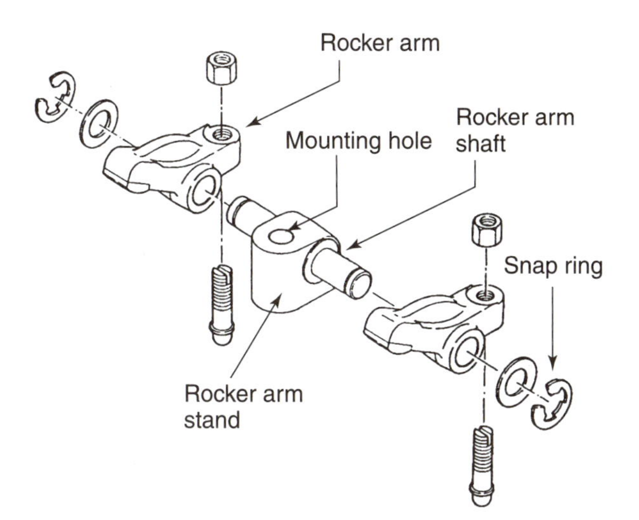

FIGURE 9-29 Rocker arms can be mounted on a rocker shaft. (Courtesy of Onan Corp.)

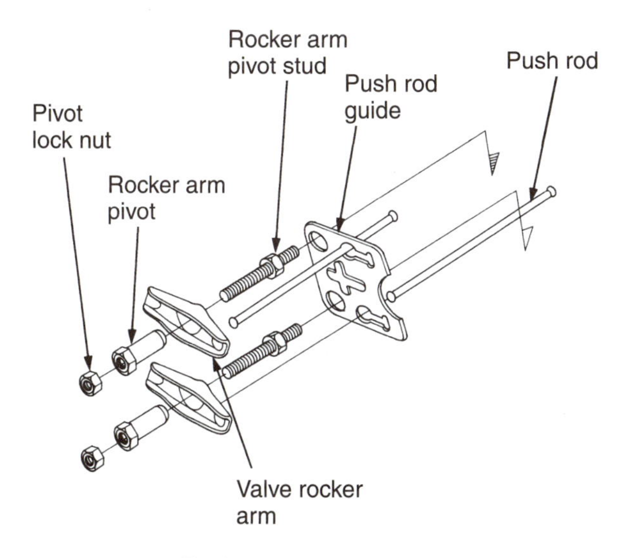

FIGURE 9-30 Rocker arms can be mounted on a stud and pivot. (Courtesy of American Honda Motor Co., Inc.)

nut that are threaded on the mounting stud. One end of the rocker arm contacts the push rod. The other end contacts the valve stem. When the push rod moves upward, one end of the rocker arm moves upward, causing the opposite end to move down and push the valve open.

The OHV parts require access for service. A **valve cover** (Figure 9-31) is a removable part used to get access to the rocker arm area. The valve cover is sometimes called a *cylinder cover*. Lubricating oil is routed up to the valve area to lubricate the rocker arms and valve stems. A gasket is used between the valve cover and the cylinder head to prevent oil loss. Seals are sometimes required around the screws that retain the valve cover.

Some OHV engines use an overhead cam. The overhead cam design has a camshaft mounted above the cylinder in the cylinder head, eliminating the need for push rods and rocker arms. One advantage of this design is lower reciprocating weight because of the elimination of the push rods. A disadvantage of this design is higher cost caused by a more complex cylinder head and the need for a long chain to drive the camshaft.

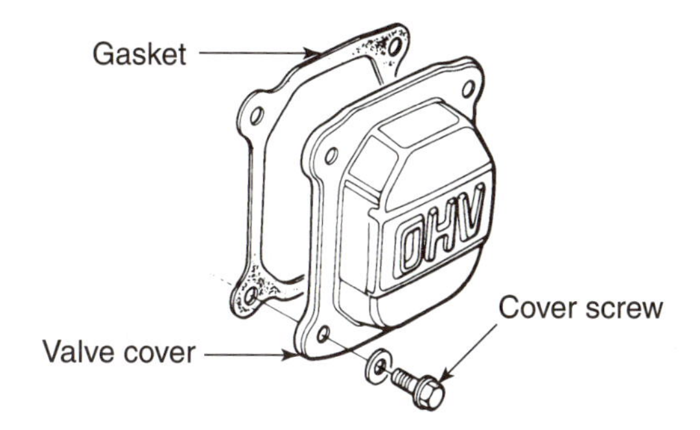

FIGURE 9-31 A valve cover is used for access to overhead valve rocker arms. (Courtesy of American Honda Motor Co., Inc.)

VALVE CLEARANCE

Both the L-head and OHV parts get hot when the engine is running. Heat causes the valves and other parts to expand. As the valve stem heats up, it gets longer. The longer stem prevents the valve face from coming into contact with the valve seat when the lifter is on the heel of the cam lobe. The valve will not seat properly. Heat will not be able to pass from the valve into the valve seat. The valve will be destroyed.

These problems are eliminated by **valve clearance,** a small space in the valve parts that allows for heat expansion. Valve clearance is sometimes called *valve lash*. L-head valve clearance is adjusted by changing the length of the valve lifter. Making the lifter longer reduces valve clearance. Making it shorter increases valve clearance. Less expensive engines use a nonadjustable valve lifter. The end of the valve stem is ground during engine assembly to adjust clearance. More expensive engines use an adjustable valve lifter. The lifter adjuster is turned with a wrench to make the lifter longer or shorter.

Overhead valves are adjusted by changing the space between the rocker arm and the valve stem. The end of the rocker shaft type rocker arms have an adjuster. The adjuster can be turned with a wrench to increase or decrease the clearance. Pivot-mounted rocker arms are adjusted with the center pivot, which can be adjusted up or down in relation to the rocker arm (Figure 9-32).

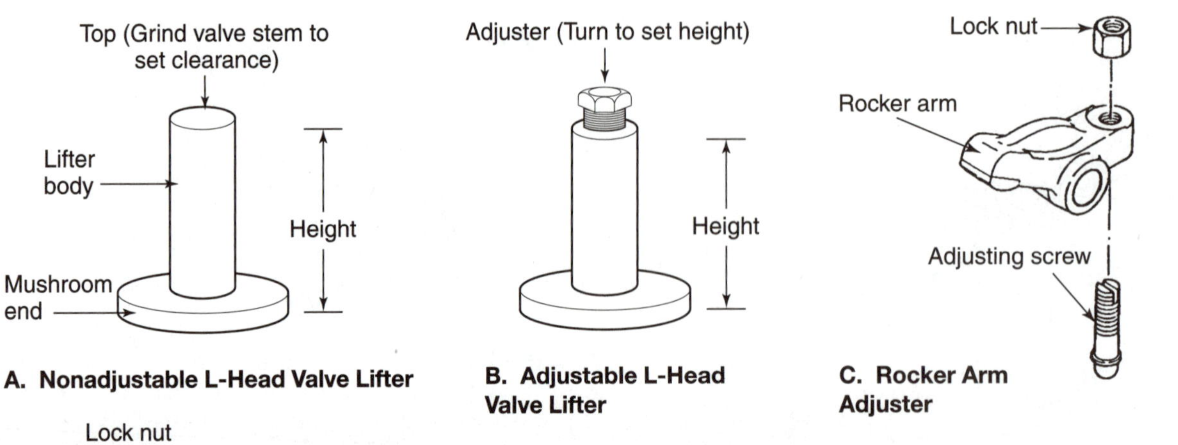

A. Nonadjustable L-Head Valve Lifter

B. Adjustable L-Head Valve Lifter

C. Rocker Arm Adjuster

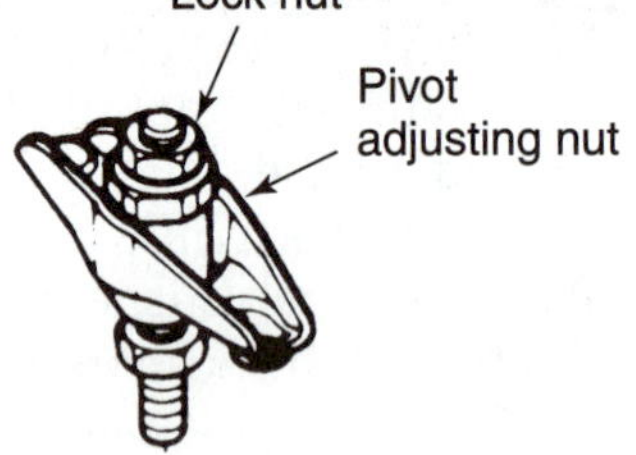

D. Rocker Arm Pivot Adjuster

FIGURE 9-32 Valve adjustment methods for L-head and overhead valve systems. (Courtesy of Onan Corp. *D:* Courtesy of American Honda Motor Co., Inc.)

REVIEW QUESTIONS

1. List and describe the operation of the parts of an L-head valve system.
2. List and describe the parts of an overhead valve system.
3. Describe the parts and materials used in an intake and exhaust valve.
4. Explain the purpose of the valve spring and valve spring retaining washer.
5. List the three common methods used to retain the valve spring on the valve.
6. Explain the purpose of a valve rotator.
7. Explain how the valve seat and valve face work to seal compression pressures in the combustion chamber.
8. Explain why replaceable valve seats are required in an aluminum cylinder block.
9. Explain the purpose of a valve guide.
10. Explain the problems caused by too much clearance between the valve guide and valve stem.
11. List and describe the parts of a camshaft lobe.
12. Explain how valve opening is related to degrees of crankshaft rotation.
13. Explain why valve clearance is needed in a valve system.
14. Explain how valve clearance is adjusted on an L-head valve system
15. Explain how valve clearance is adjusted on an overhead valve system.

DISCUSSION TOPICS AND ACTIVITIES

1. Disassemble the valve system parts on a scrap engine. Identify and explain the operation of each of the parts.
2. Remove a cylinder head from an L-head engine. Make an index mark on the crankshaft. Rotate the engine through the four strokes. Observe where the valves open in relation to crankshaft position.
3. Construct a cardboard disk graduated in 360 degrees. Tape the disk to the engine flywheel. Move the piston up to top dead center on the intake stroke. Rotate the engine through all four strokes and note the number of crankshaft degrees each valve is open. Make a valve timing diagram for the engine.

CHAPTER 10

Two-Stroke Engine

OBJECTIVES

Upon completion and review of this chapter, you should be able to:

- Describe the basic parts and operation of a two-cycle engine.
- Explain the operation of reed valve loop-scavenged port control two-stoke engine.
- Explain the operation of third port loop-scavenged port control two-stroke engine.
- Describe the operation of cross-scavenged port control two-stroke engine.
- Identify and describe the parts of a two-stroke engine.

TERMS TO KNOW

Bypass passage
Cross scavenged two-stroke
Half keystone piston ring
Knock pin
Multiple piece crankshaft
Needle bearing
Reed valve
Reed valve loop-scavenged engine
Rotary crankcase valve
Scavenge phase
Third port
Third port loop-scavenged engine
Two-stroke cycle engine
Two-stroke exhaust port
Two-stroke intake port
Two-stroke lower end
Two-stroke upper end

INTRODUCTION

A **two-stroke cycle engine**, or two-stroke, is an engine that uses two piston strokes and one revolution of the crankshaft to develop power. Many small engines use the two-stroke cycle. They require fewer parts than the four-stroke cycle engine. Two-stroke engines have had a major impact on outdoor power equipment because they are light and portable. They can be easily carried and used in any position. Two-stroke engines also run at much higher speeds than four-stroke engines. They are used on many types of hand-carried equipment such as string trimmers, blowers, and chainsaws.

BASIC PARTS AND OPERATION

The basic parts of a two-stroke engine (Figure 10-1) are the same as those in the four stroke. The main difference is that there is no valve system. There is a combustion chamber with a cylinder and piston. The piston is connected to a crankshaft by a connecting rod. A flywheel is mounted to the end of the crankshaft.

The two-stroke develops power using intake, compression, power, and exhaust strokes in the cylinder, the same as the four-stroke engine. The difference is that they occur in just two piston strokes. One stroke is power and exhaust. The other stroke is intake and compression. The crankshaft turns one complete turn or revolution (360 degrees) during these two strokes.

There are no valves in a two-stroke engine. There are two ports located in the cylinder. The **two-stroke exhaust port** is a passage in the cylinder that allows exhaust gases to flow out of the cylinder. The **two-stroke intake port** is a passage in the cylinder that allows the air and fuel mixture to enter the cylinder. The intake port opens into a

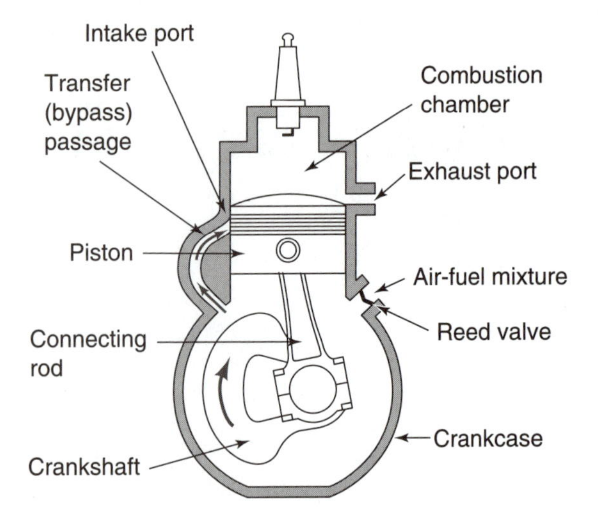

FIGURE 10-1 Parts of a basic two-stroke cycle engine.

passageway to the crankcase. This is called a *bypass* or *transfer passage*. The **bypass passage**, or transfer passage, is a passageway that connects the intake port to the crankcase. The two-stroke crankcase is a housing for the crankshaft. It may also be a route for fuel and air into the cylinder. A valve is often used to control the flow in and out of the crankcase. A **reed valve** is a spring-loaded flapper type valve that controls the flow of air and fuel into the crankcase.

The two cycles begin when the piston moves up in the cylinder. The space below the piston gets bigger, causing a low pressure (partial vacuum) in the crankcase. The air-fuel mixture is pulled into the crankcase through the reed valve. At the same time, the piston covers the intake and exhaust ports, trapping the air and fuel above the piston. The piston compresses the mixture in the combustion chamber. As the piston gets near the top of its upward stroke, there is a spark from the spark plug. The spark starts the air-fuel mixture burning. The combustion pressure forces the piston down the cylinder. The connecting rod rotates the crankshaft.

As the crankshaft turns, the piston moves down. The crankcase area below the piston becomes smaller. The smaller space puts the air-fuel mixture under pressure The pressure in the crankcase closes the reed valve, preventing the air-fuel mixture from escaping. The mixture gets compressed more and more tightly in the crankcase.

Finally, the piston goes far enough down the cylinder to uncover the exhaust port. This allows the burned exhaust gases to go out of the cylinder through the exhaust port. As the piston reaches the bottom of its stroke, it uncovers the intake port. The mixture in the crankcase flows through the bypass passage and intake port and into the cylinder.

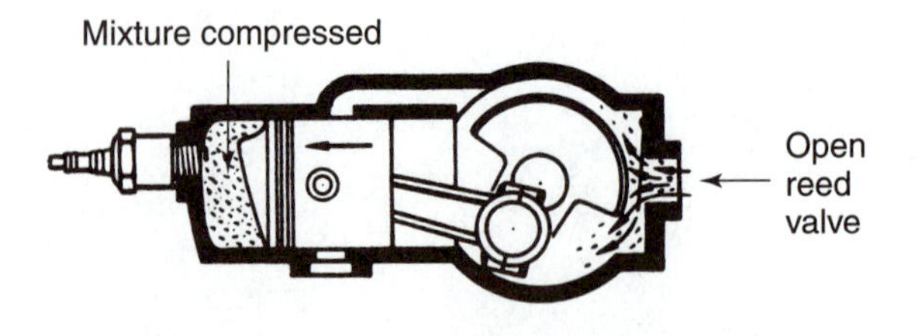

A. Compression—Piston Moves Up, Pulling the Mixture into the Crankcase

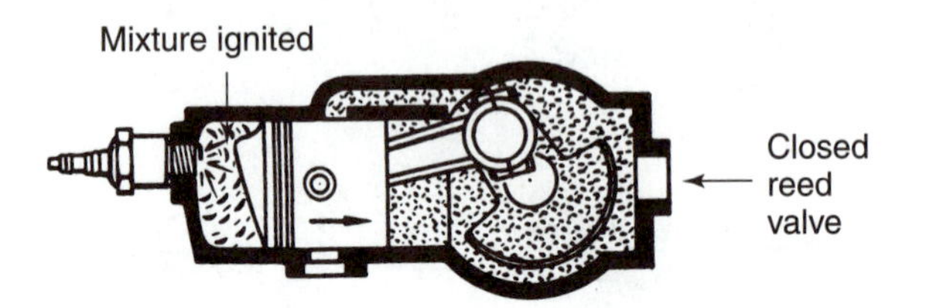

B. Power—The Burning Mixture Pushes the Piston Down the Cylinder

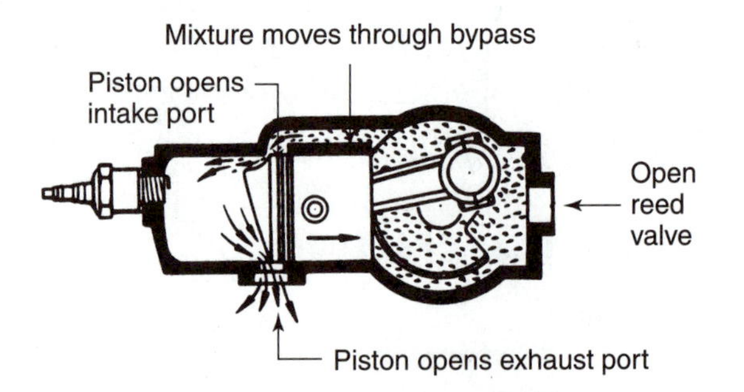

C. Exhaust—Piston Uncovers the Ports, Burned Gases Escape, and a New Mixture Gets In

FIGURE 10-2 Basic two-stroke cycle engine operation. (Courtesy of Clinton Engines Corporation.)

The piston begins to move back up the cylinder to start the cycle again. The cycle of events has occurred in two strokes of the piston. The crankshaft has completed one revolution (360 degrees) (Figure 10-2).

TWO-STROKE CYCLE ENGINE PORT CONTROL

The air-fuel mixture gets into the engine through ports. Different engines use different ways of controlling the mixture flow. This is called the scavenge phase of engine operation. **Scavenge phase** is the replacement of exhaust gas in the two-stroke engine combustion chamber with a new mixture of air and fuel. Engines are identified by the method used on the scavenge phase. The three common scavenge systems are reed valve loop-scavenged, third port loop-scavenged, and cross-scavenged.

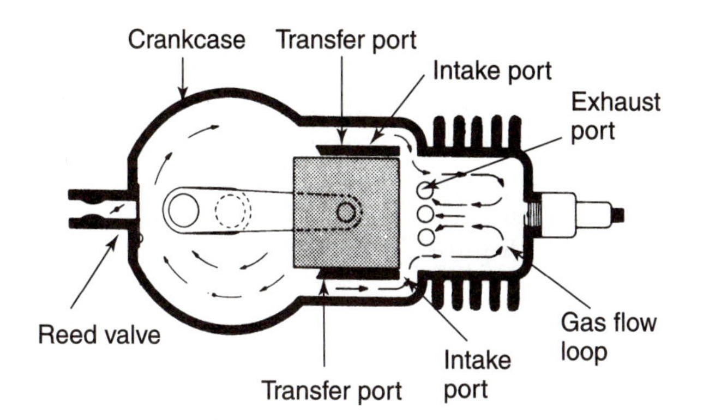

FIGURE 10-3 Parts of a loop-scavenged two stroke. (Courtesy of Tecumseh Products Company.)

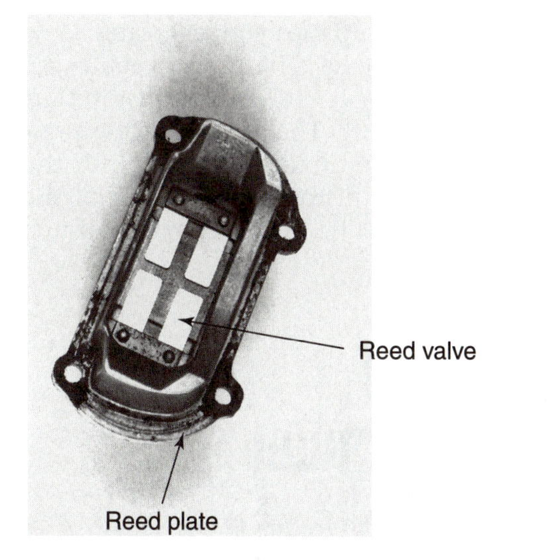

FIGURE 10-4 A reed valve is a thin metal strip seated on a reed plate.

Reed Valve Loop-Scavenged

A **reed valve loop scavenged engine** is an engine that uses intake ports and a reed valve connected to the crankcase for scavenging (Figure 10-3). There are several ports located around the cylinder. The ports on opposite sides are used to transfer two streams of air and fuel into the cylinder. The two streams meet at the cylinder wall opposite the exhaust ports. Here, they move (deflect) upward. The streams then move (deflect) downward. This forces the burned gases through the exhaust ports. The streams form a loop as they flow through the cylinder. The engine gets its name from this loop.

The reed valve controls the gas flow (Figure 10-4). The reed valve has a mounting plate called a *reed plate*. A number of thin spring steel plates fit over holes in the plate. Each of these are called *reeds* or *reed valves*. The reed valves may act as their own spring. Some have a separate spring.

The reed valve opens and closes with crankcase pressure (Figure 10-5). When the piston moves up, a low pressure is created under the piston in the crankcase. This low pressure gets stronger than the reed valve spring pressure. The reed valve is pulled open, allowing air and fuel to flow into the crankcase. As the piston moves down in the cylinder, the pressure in the crankcase increases. The reed valve spring pressure closes the reed valve.

Third Port Loop-Scavenged

The **third port loop-scavenged engine** uses a third port controlled by the piston skirt for scavenging (Figure 10-6). The exhaust ports are several round holes in the cylinder. They are controlled by

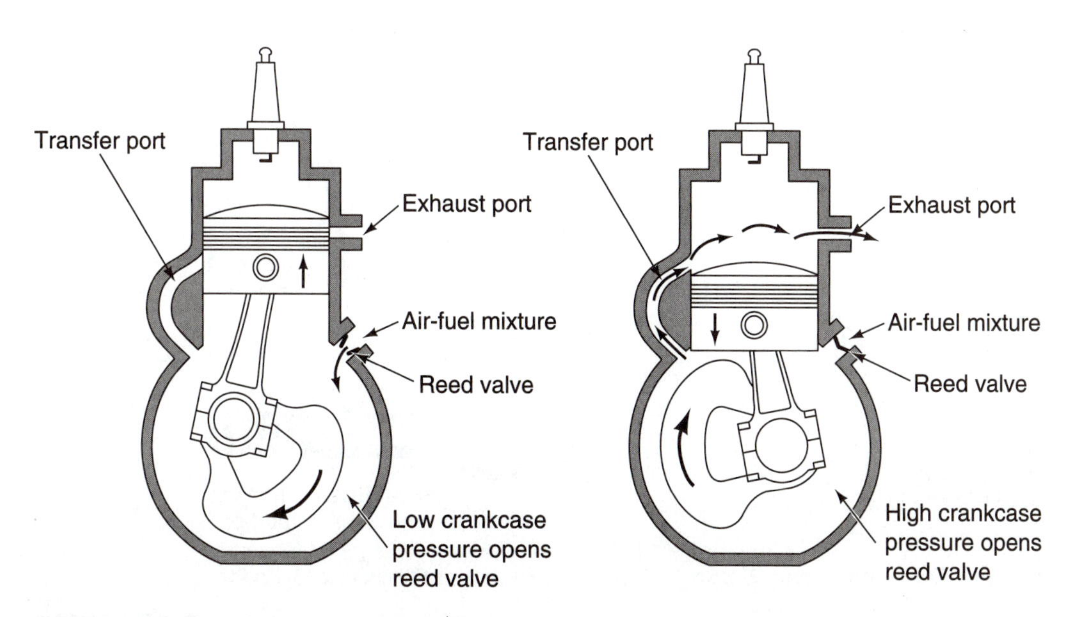

FIGURE 10-5 The reed valve is opened by low pressure and closed by a spring.

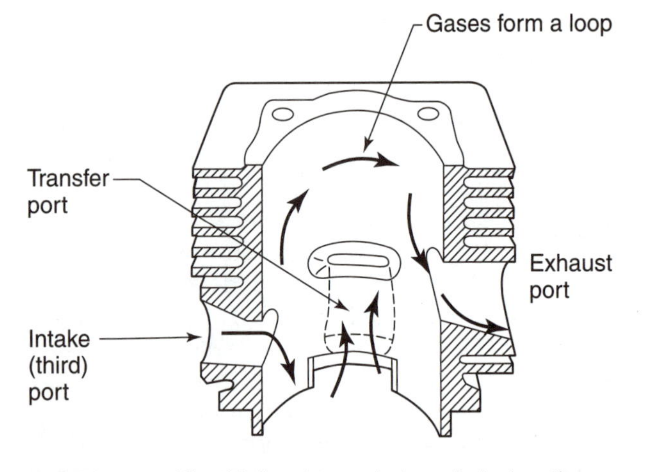

FIGURE 10-6 The third port two-stroke cylinder has three ports.

piston position to allow exhaust gases to flow out of the cylinder. These ports are the highest ones in the cylinder. They are uncovered first as the piston moves down the cylinder. The transfer ports are two large oval passageways connected to the crankcase located on either side of the cylinder.

When the crankcase pressure gets high enough, the air-fuel mixture in the crankcase can flow through the transfer port into the cylinder. The **third port**, or intake port, is a two-stroke engine port connecting the air-fuel source (carburetor) to the crankcase.

The skirt of the piston is used to control the flow in and out of the crankcase. The piston moves up in the cylinder. The skirt covers the transfer and third (intake) ports. As it goes higher, the piston head covers the higher exhaust ports. The air-fuel mixture is compressed above the head of the piston (Figure 10-7A).

Piston movement up the cylinder creates a low pressure in the crankcase. The piston gets to the top of the cylinder (Figure 10-7B). The piston skirt uncovers the third port. The low crankcase pressure causes the air-fuel mixture to flow into the crankcase. When the piston is near the top, the spark plug creates a spark. This ignites the air-fuel

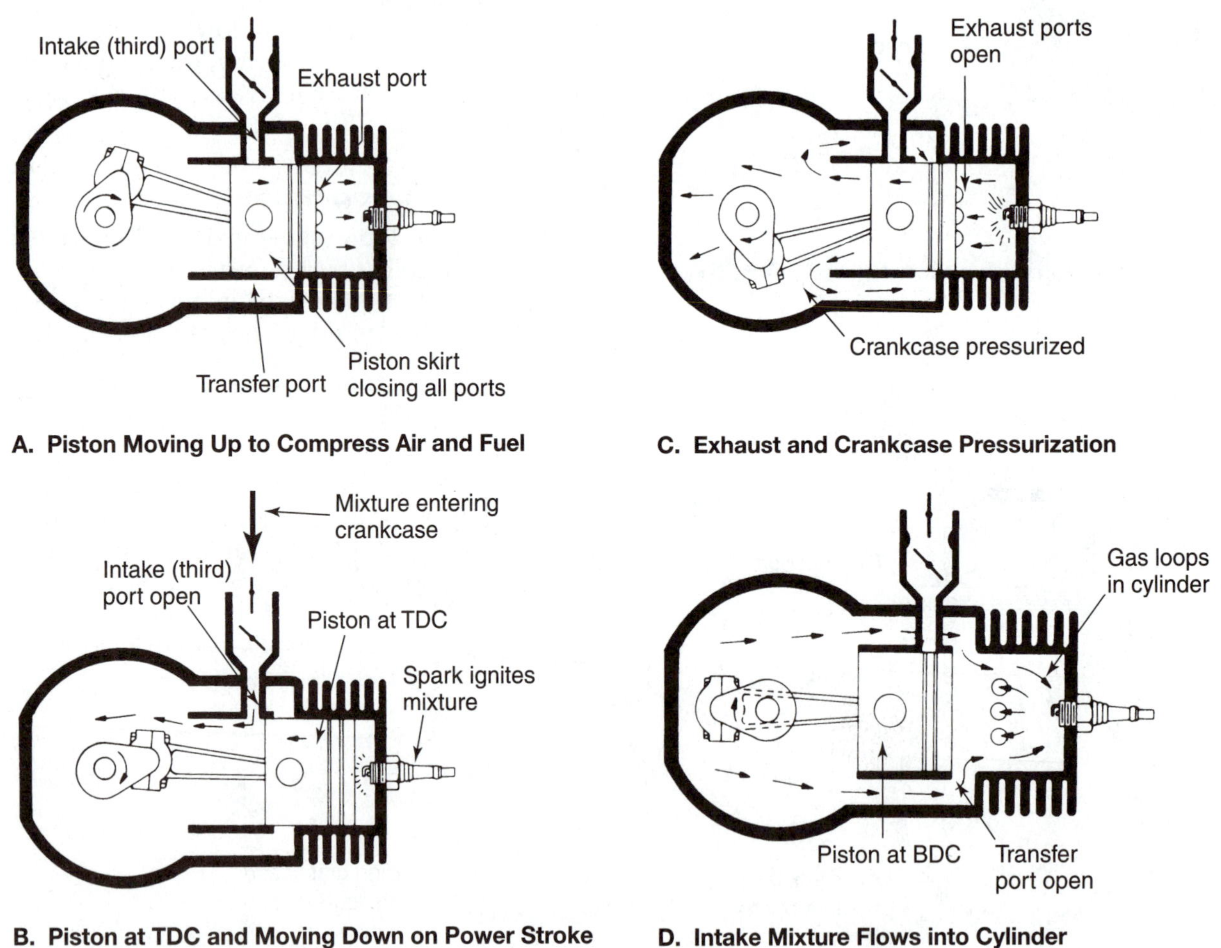

FIGURE 10-7 Third port loop-scavenged engine operation. (Provided courtesy of Tecumseh Products Company.)

mixture. The burning mixture expands and pushes the piston down. The connecting rod causes the crankshaft to rotate.

The piston is pushed by combustion pressure until it uncovers the exhaust ports (Figure 10-7C). The high pressure in the cylinder drops when exhaust gases begin to flow out the exhaust ports. The piston reaches the bottom of the cylinder (Figure 10-7D). Here, the exhaust ports are fully uncovered and the transfer port is open. Crankcase pressure forces any remaining exhaust gas out of the cylinder. It then replaces it with a fresh mixture of air and fuel. The flow of gases in the cylinder at this time makes a loop. The loop is the same as in the reed valve two-stroke engine.

The loop-scavenged engine produces more horsepower for its weight than many other two-stroke engines. This engine does a good job of completely removing the exhaust gases. Some engines have exhaust gases left at the end of the power stroke. When exhaust gases mix with new air-fuel mixtures, the engine loses power.

Some third port engines use pistons with holes or slots in the skirt area. These slots line up with passages in the cylinder when the piston is at the bottom of its stroke. The air-fuel mixture passes through the skirt slots to get into the transfer port.

Cross-Scavenged

A **cross-scavenged two-stroke** (Figure 10-8) is an engine that uses a deflector on the head of the piston for scavenging. This engine has exhaust ports and transfer ports in the cylinder. The transfer ports allow the air-fuel mixture to come up from the crankcase.

The piston moves down the cylinder on a power stroke. As it uncovers the exhaust ports, exhaust begins to flow out of the cylinder. As the piston moves down more, it uncovers the transfer ports. The air-fuel mixture flows into the cylinder. A deflector on the head of the piston causes an upward flow of the intake gases on one side of the cylinder. The gases are directed up until they hit the top of the combustion chamber. The other side of the piston deflector causes the gases to go down toward the exhaust ports.

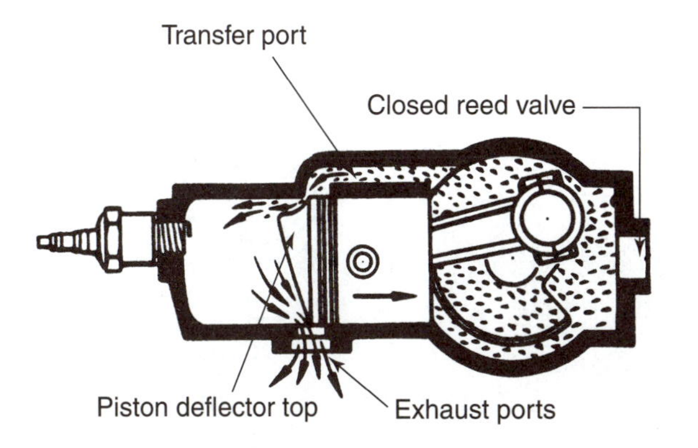

FIGURE 10-8 A cross-scavenged two-stroke uses a deflector piston to direct intake and exhaust flow. (Provided courtesy of Tecumseh Products Company.)

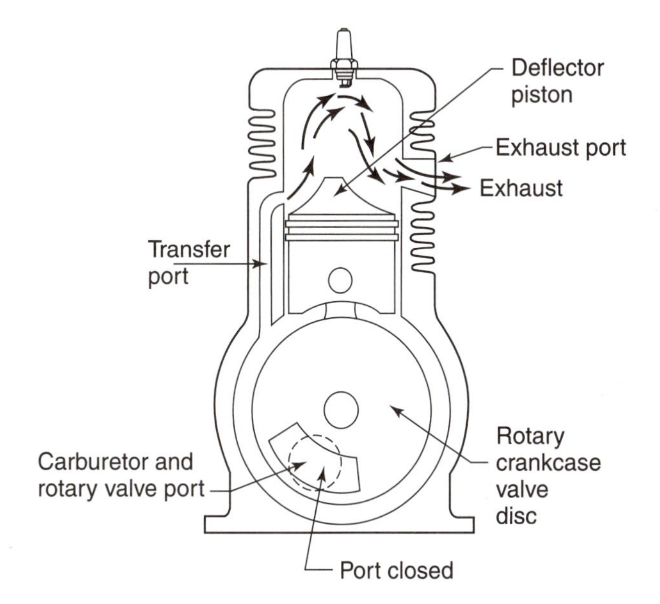

FIGURE 10-9 A cross-scavenged two-stroke with a rotary crankcase valve.

The piston deflector does a good job of directing the gases and separating the intake and exhaust flow. The flow of gases is across the cylinder, therefore it is called a *cross-scavenged system.*

Cross-scavenged engines may use a reed valve. Some use a rotary valve to control the flow of air and fuel into the crankcase. A **rotary crankcase valve** is a disc on the engine's crankshaft used to control the flow of air and fuel into the crankcase (Figure 10-9). The valve has a shape that allows it to open a port during part of the crankshaft revolution. When open, air and fuel are allowed to enter the low pressure of the crankcase. As the crankshaft rotates more, the larger part of the valve covers over the port. This seals the crankcase.

SERVICE TIP: Piston installation is very important on two-stroke engines. If a loop-scavenged piston is installed in the wrong direction, the skirt holes or slots in the piston may not line up with the transfer port. The scavenging will not work right if a deflector type piston is installed backward. The engine will not run properly.

CASE STUDY

A student in an outdoor power equipment class finished the assembly of a two-stroke engine for a small edger. The engine was mounted on the edger and ready to be started for the first time. The student mixed up a gallon of two-stroke oil and fuel and filled the fuel tank. After rechecking all the adjustments, he began pulling on the starter rope. The engine was very hard to start but finally began to run. It smoked out the exhaust for a while but soon the smoke cleared up.

When the engine warmed up the student increased the throttle. The engine did not have the characteristic two-stroke sound. It would not reach a high rpm. The instructor and student both thought that the carburetor was probably the problem. The instructor directed the student to adjust the carburetor main mixture to see if it had any effect on engine speed. The engine would not run any faster even with the mixture screw adjusted to maximum fuel delivery. The instructor disassembled the carburetor to see if there was any problem that would not allow enough mixture into the engine. There were no obvious problems with the carburetor.

There was another, similar engine in the shop that was not being used. This engine ran fine. The instructor and student switched ignition and fuel systems from the good engine to the problem engine. The edger engine continued to run poorly. This eliminated an ignition or fuel problem. The cause had to be deeper. The engine had to be disassembled to find the problem. When the cylinder was pulled up off the piston assembly, the problem was obvious. The deflector top piston was assembled in the wrong direction. The exhaust and intake streams coming into the cylinder were not aimed in the correct direction and the engine could not develop any power.

ENGINE PARTS

Two-stroke engine parts are smaller and lighter than those of a four-stroke engine. Two-stroke engines are sometimes identified as the upper or lower end. The **two-stroke upper end** parts are located at the top of the engine such as cylinder head, cylinder, piston, piston rings, and piston pin. The **two-stroke lower end** parts are located at the bottom of the engine such as crankcase, main bearings, crankshaft, and connecting rod.

Crankcase and Main Bearings

The crankcase houses and supports the crankshaft. The crankcase and cylinder assembly is cast into one part in many two-stroke engines (Figure 10-10). Other engines use a separate crankcase and cylinder assembly (Figure 10-11). A mount for the reed valve may be cast into the crankcase. Crankcase assemblies are made from aluminum. Aluminum is lightweight and allows good heat flow (dissipation).

The crankcase is made to allow the installation and removal of the crankshaft. Some crankcases have a removable cover on one side to permit access (Figure 10-12). Others are made in two halves that can be separated. Some engines have a crankcase that splits in the vertical direction (Figure 10-13). Other engines have a crankcase that splits in the horizontal direction (Figure 10-14).

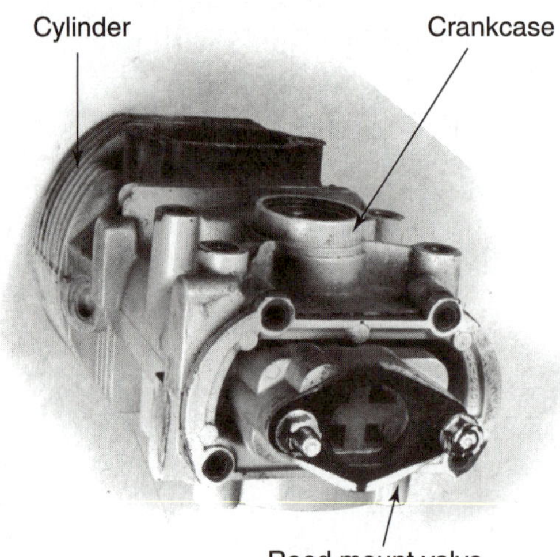

FIGURE 10-10 A two-stroke crankcase and cylinder combined into one part.

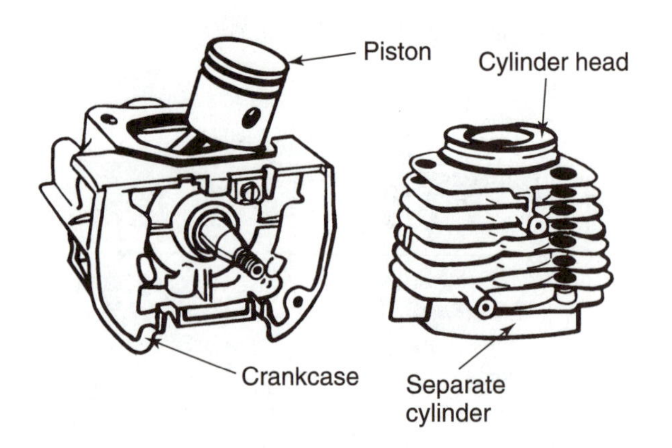

FIGURE 10-11 A two-stroke engine can have a separate cylinder. (Courtesy of Poulan/Weed Eater.)

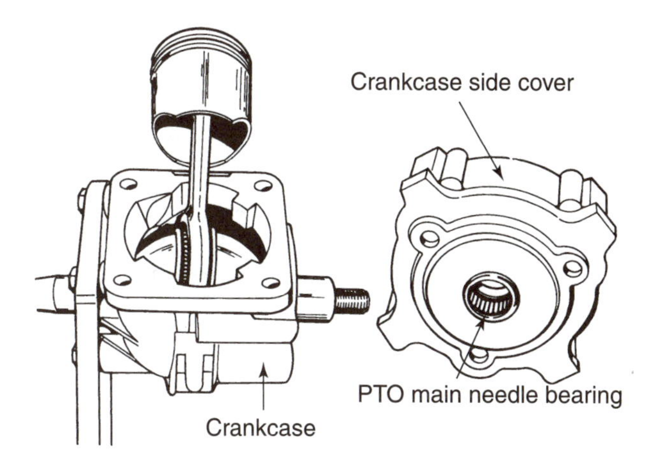

FIGURE 10-12 A crankcase with a removable side cover. (Provided courtesy of Tecumseh Products Company.)

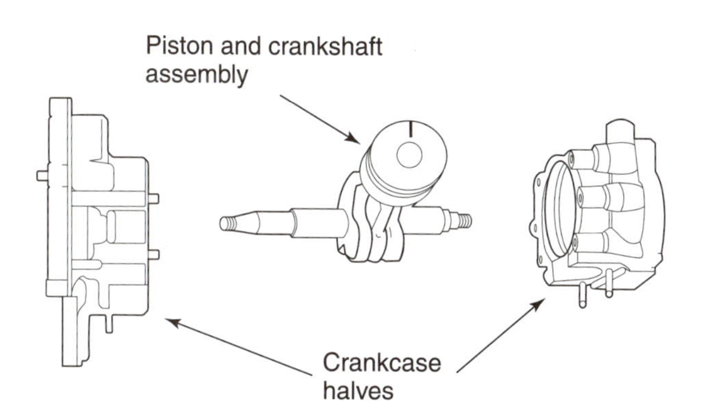

FIGURE 10-13 A crankcase that splits apart vertically. (Courtesy of Tanaka/ISM.)

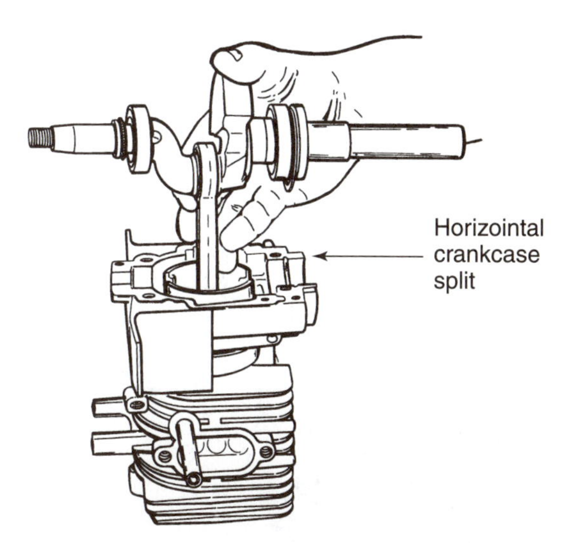

FIGURE 10-14 A crankcase that splits apart horizontally. (Provided courtesy of Tecumseh Products Company.)

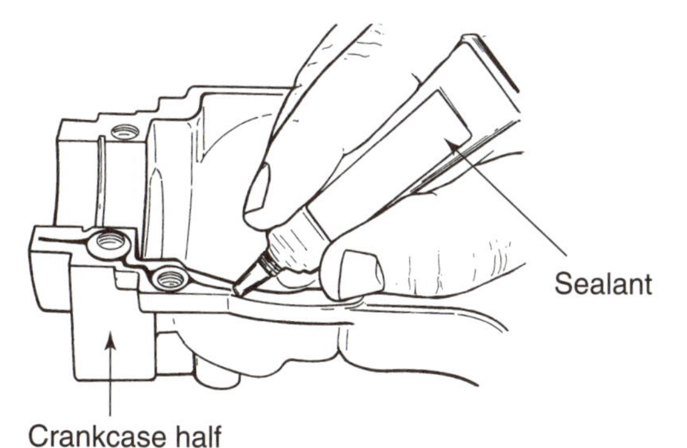

FIGURE 10-15 The crankcase halves are sealed to keep pressure in the crankcase. (Provided courtesy of Tecumseh Products Company.)

A pressure is developed inside the crankcase to force the air-fuel mixture up the transfer ports. This requires a pressure-tight seal between crankcase covers or halves. Sometimes the seal is made with a gluelike material called *sealant* (Figure 10-15). Other engines use gaskets. A separate part called a *crankshaft seal* is often used around the rotating crankshaft. The crankshaft seal prevents pressure loss around the crankshaft ends.

A main bearing is installed in the crankcase for each side of the crankshaft. Two-stroke engines operate at high speed. They have lubricating oil mixed with the air-fuel mixture. These conditions require ball or needle main bearings. Plain bearings cannot be lubricated well by a fuel and oil mixture. Some two-stroke engines have ball bearing main bearings for both ends of the crankshaft (Figure 10-16). The ball bearings have a set of balls supported between an inner and outer race.

Many two-stroke engines use needle bearings for both the main and connecting rod bearings. A **needle bearing** is a small diameter roller bearing often held in a case or cage that works as an outside race (Figure 10-17). Needle main bearings have the roller case installed in the crankcase or crankcase cover. The bearing case fits in the aluminum of the crankcase or cover, because steel bearings can not rotate directly on aluminum. When assembled, the needles operate directly on the crankshaft main journal.

Cylinder and Block

Two-stroke cylinders are made from aluminum with a cast iron liner. The cylinder has ports and transfer passages cast into the cylinder. The aluminum makes the engine light in weight. It allows heat to pass quickly out of the cylinder. The cast iron liner makes a good bearing surface for the piston rings and piston skirts. Two-stroke engines do

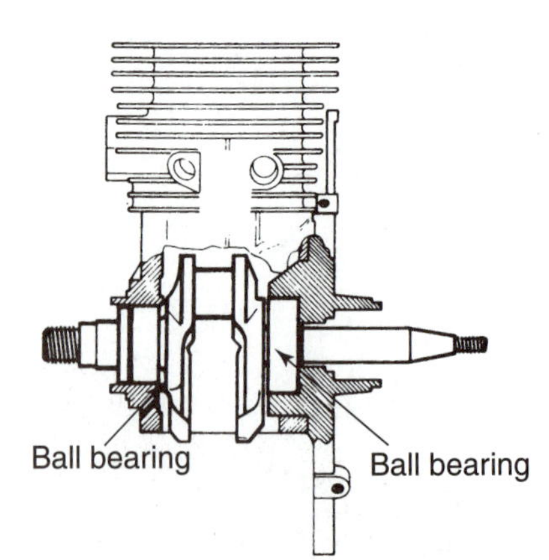

A. Ball Bearing Main Bearings

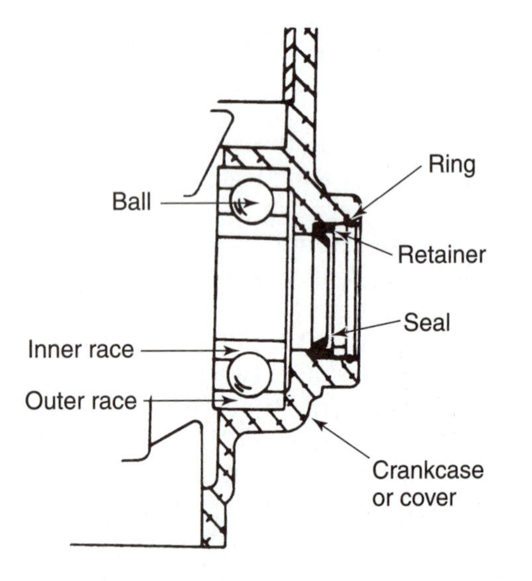

B. Ball Bearing Parts

FIGURE 10-16 Ball bearings used for two-stroke main bearings. (Provided courtesy of Tecumseh Products Company.)

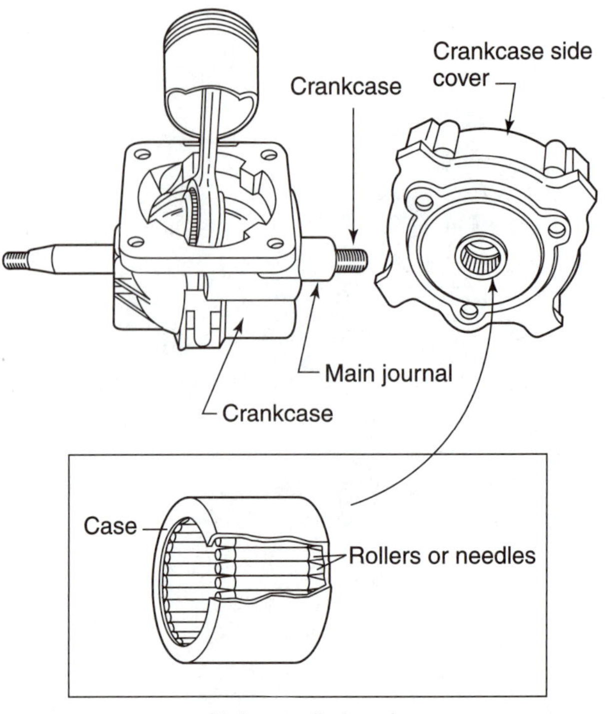

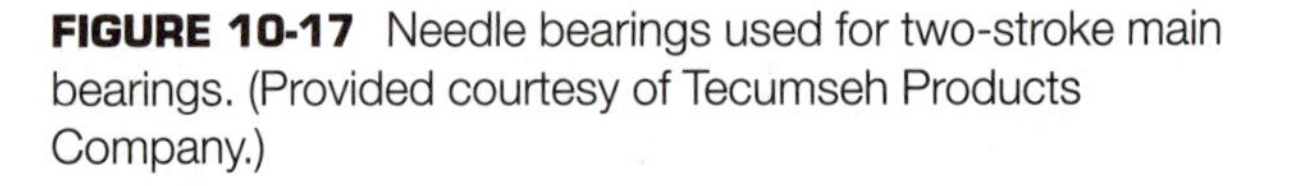

FIGURE 10-17 Needle bearings used for two-stroke main bearings. (Provided courtesy of Tecumseh Products Company.)

not use aluminum as a cylinder wall material. The high piston speeds and cylinder wall lubrication problems require cast iron.

Some engines have the cylinder cast as one piece with the crankcase (Figure 10-18). This is called a *cylinder block*. Other engines have a separate cylinder mounted to the crankcase with screws or studs. The separate cylinder has the advantage of being replaceable during engine service.

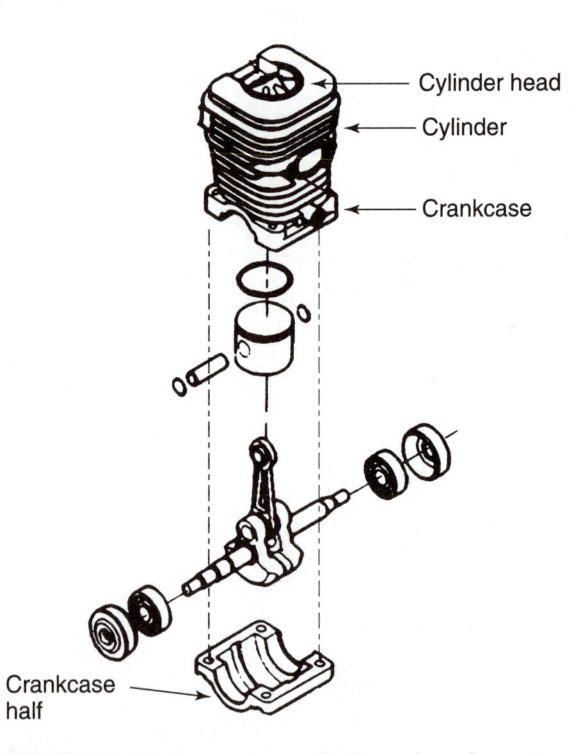

FIGURE 10-18 A one-piece cylinder and crankcase assembly. (Courtesy of Poulan/Weed Eater.)

Cylinder Head

The purpose of the cylinder head is to form a combustion chamber at the top of the cylinder. Cylinder heads for smaller two-stroke engines may be cast as one piece with the aluminum cylinder block, eliminating the need for a cylinder head gas-

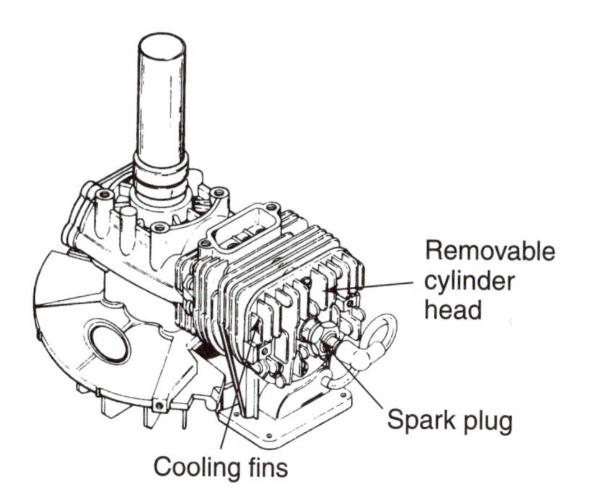

FIGURE 10-19 A two-stroke engine with a removable cylinder head. (Provided courtesy of Tecumseh Products Company.)

ket. The piston assembly is removed and installed from the bottom.

Other engines have a removable aluminum cylinder head (Figure 10-19). It can be mounted with bolts, screws, or studs. A cylinder head gasket fits between the cylinder and cylinder head to form a seal. The cylinder head has cooling fins to remove heat from the combustion chamber. Threads for a spark plug are located in the center.

Combustion chamber shape is very important in two-stroke engines. Many engines use a round or hemispherical shape combustion chamber (Figure 10-20). These have a small "squish area" near the edge of the cylinder. The squish area forces the mixture to flow past the spark plug at the end of the compression stroke, improving combustion on the power stroke.

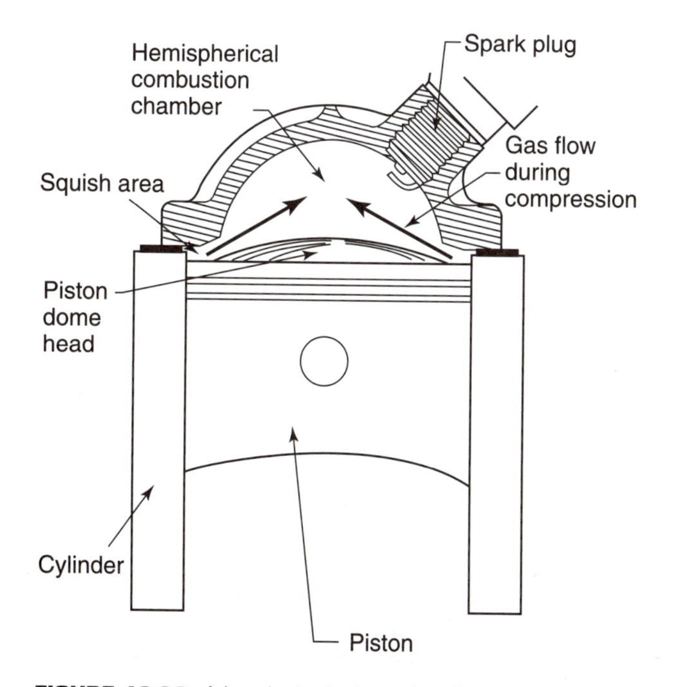

FIGURE 10-20 A hemispherical combustion chamber with a squish area for improved combustion.

Piston and Piston Rings

Smaller loop-scavenged engines use a piston with a flat top (Figure 10-21). Larger engines may use a piston with a domed top. Domed pistons increase compression pressures and improve gas flow in the combustion chamber. A cross-scavenged engine piston has a deflector on the head that controls intake and exhaust gas flow. The piston may have a slot in the skirt used to open and close the intake port (Figure 10-22).

The ring area of the piston often has two ring grooves. The ring grooves may have a small pin called a knock pin. A **knock pin**, is a stop in the piston ring groove that prevents the piston ring from rotating around the ring groove (Figure 10-23). The ends of the ring must not be allowed to rotate. They could catch in the ports and break the ring.

Two-stroke engines often use two piston rings (Figure 10-24). Both are compression sealing rings. The rings prevent compression pressure from leaking out of the combustion chamber area. Oil control rings are not used because of the type of lubrication used on two-stroke engines.

Many engines use a simple rectangular-shaped ring. Some engines use a piston ring called a *half keystone ring* (Figure 10-25). A **half keystone piston ring** is a piston ring with a top side that tapers downward and allows compression pressures to force the ring against the cylinder wall, improving its seal with the cylinder wall. The ring is made

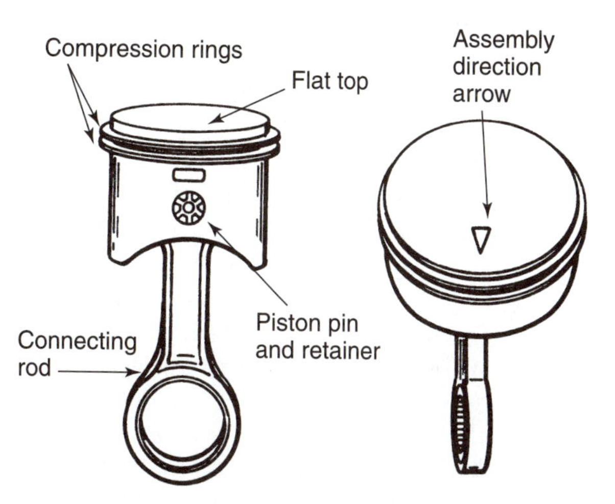

FIGURE 10-21 A loop-scavenged piston with a flat top. (Provided courtesy of Tecumseh Products Company.)

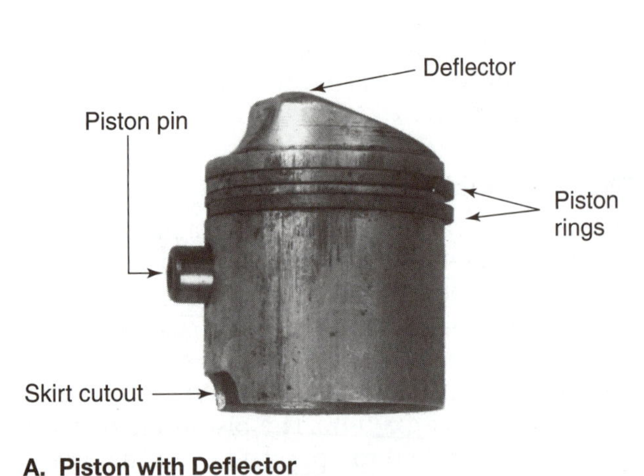

A. Piston with Deflector

Deflector top
Skirt slot

B. Piston with Deflector and Skirt Slot

FIGURE 10-22 Pistons can have a deflector head and slots for mixture control.

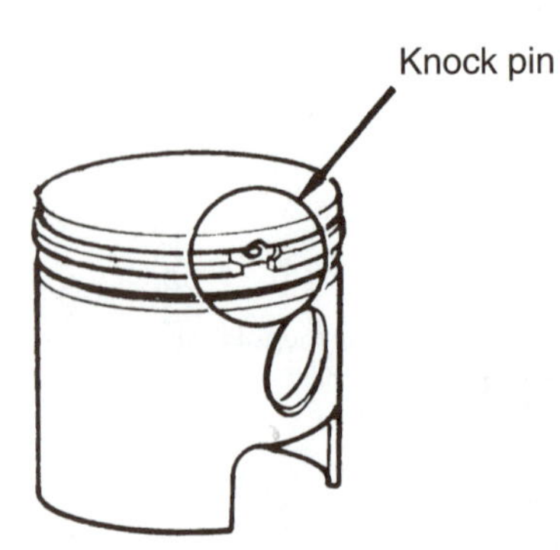

FIGURE 10-23 A knock pin prevents rings from rotating in their grooves. (Courtesy of Tanaka/ISM.)

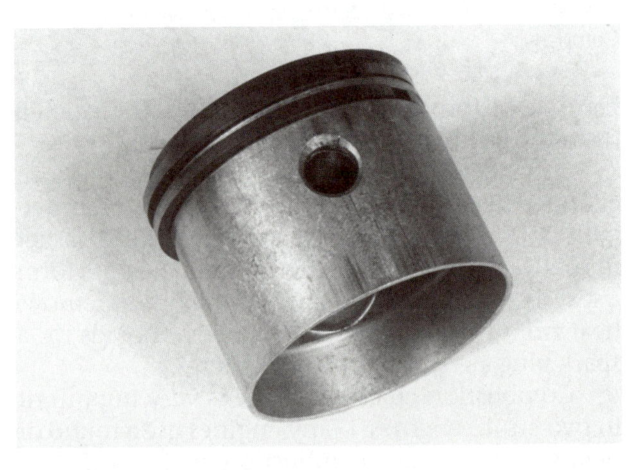

FIGURE 10-24 A piston with two rectangular-shaped compression rings.

smaller than the ring groove so it can move slightly up and down as the piston changes direction. The sliding movement prevents carbon (burned oil) from forming on the cylinder wall.

Connecting Rod and Piston Pin

The connecting rods are often made of cast iron to provide needed strength. This also allows the use of needle bearings. They have an I-beam shape for high strength and low weight. Needle bearings are often used in both the big and small ends. They allow high-speed operation. They also can be lubricated easily.

Many engines use a one-piece connecting rod (Figure 10-26). This connecting rod does not have a removable rod cap. It has a large big end saddle bore that can be slipped over the crankshaft and needle bearings. Some engines use a multiple piece crankshaft. The connecting rod journal is installed through the one-piece connecting rod.

The big end (crankshaft end) of the connecting rod may be made in two parts using a rod cap (Figure 10-27). The case for the needle bearing is made in two parts. One half is installed in the rod and the other in the cap. The cap can be taken apart to remove the connecting rod for service.

The piston pin is made from high-quality steel. It is usually hollow to lower its weight. The piston pin usually is held in the connecting rod and piston with two retaining rings.

Crankshaft

Many two-stroke engines use a one-piece cast-iron crankshaft similar to that used on four-stroke engines. The crankshaft has two main bearing journals and an offset connecting rod journal. One end of the crankshaft is the PTO side. The other is the FWE. Needle bearings are often used on the con-

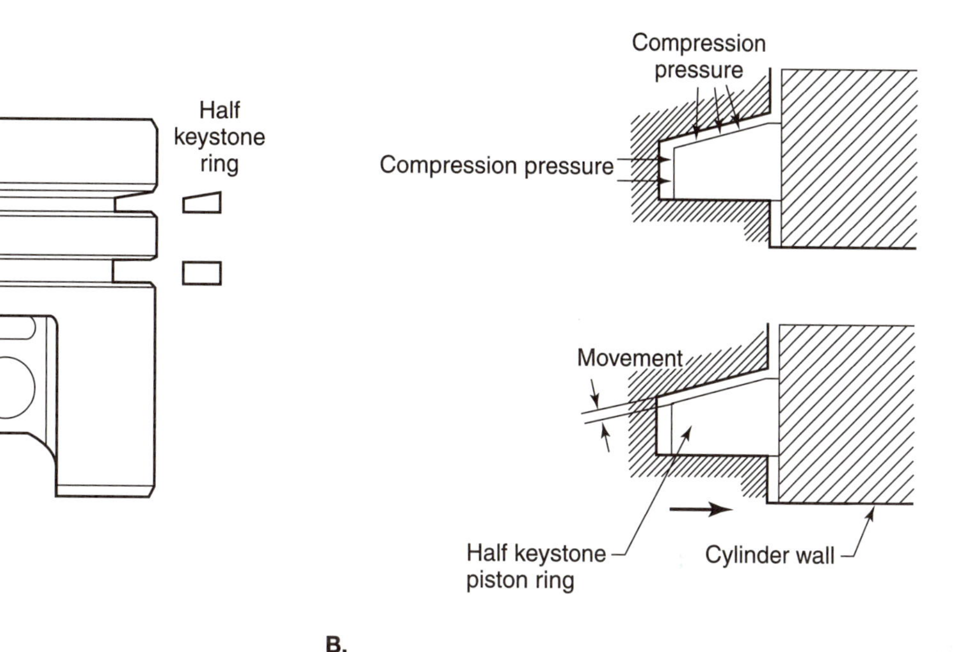

FIGURE 10-25 A half keystone ring moves in the ring groove to eliminate carbon buildup. (*A:* Provided courtesy of Tecumseh Products Company.)

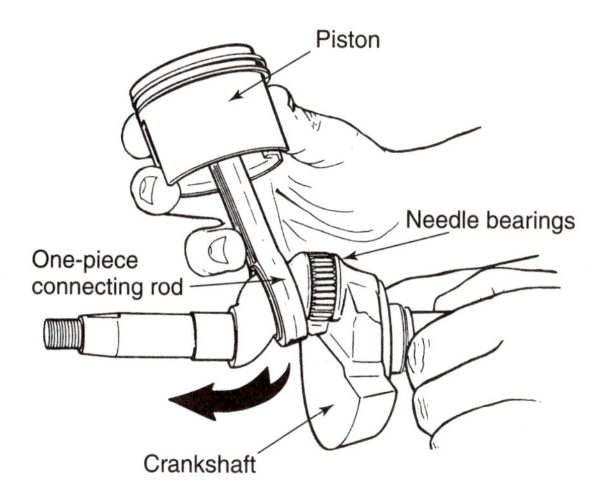

FIGURE 10-26 A one-piece connecting rod that slips over the crankshaft. (Provided courtesy of Tecumseh Products Company.)

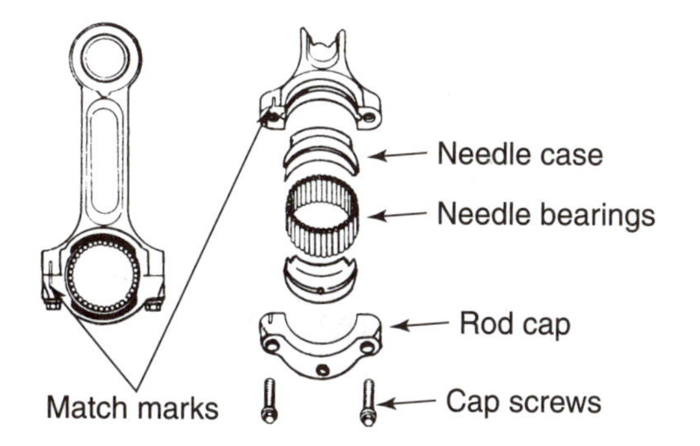

FIGURE 10-27 A connecting rod with a removable rod cap. (Provided courtesy of Tecumseh Products Company.)

necting rod journal. A single or double counterweight is cast opposite the crankshaft journal to balance the reciprocating weight of the piston and connecting rod.

Two-stroke engines may use a multiple piece or built-up type crankshaft (Figure 10-28). The **multiple piece crankshaft** is a crankshaft made in separate parts that is assembled by pressing together. The crank pin is a removable part of a multiple piece crankshaft that forms the connecting rod

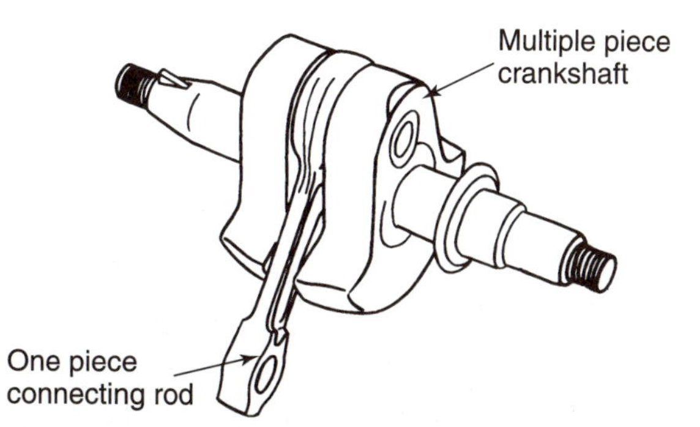

FIGURE 10-28 A multiple piece crankshaft is usually supplied with the connecting rod already installed. (Courtesy of Poulan/Weed Eater.)

journal. The crank pin is pressed through the two counterweights. It fits through the center of a connecting rod and bearing. This allows the use of a caged roller bearing for the connecting rod. These crankshafts are commonly supplied as an assembly and are not taken apart in the shop.

REVIEW QUESTIONS

1. List and describe the operation of the basic parts of a two-stroke engine.
2. Explain the purpose and operation of a reed valve.
3. Explain how a low pressure is formed in the crankcase.
4. Explain how a high pressure is formed in the crankcase.
5. Explain the purpose of the two-stroke intake ports.
6. Explain the purpose of the two-stroke exhaust ports.
7. Describe the parts and operation of the reed valve loop-scavenged engine.
8. List the parts and describe the operation of a third port loop-scavenged engine.
9. List the parts and describe the operation of the cross-scavenged port control engine.
10. Explain the purpose of the deflector on top of a cross-scavenged port control engine piston.
11. Explain why piston direction is important in a two-stroke engine.
12. Explain the purpose of a slot in a skirt of a piston used in a third port loop-scavenged engine.
13. List and describe the main types of crankcase designs used on two-stroke engines.
14. List and describe the parts of a needle-bearing assembly.
15. List and describe the main types of cylinder designs used on two-stroke engines.
16. Explain why a knock pin is used in a piston ring groove.
17. Explain why a two-stroke engine does not use an oil control piston ring.
18. List and describe the two methods of assembling a connecting rod without a cap on the crankshaft.
19. Describe the bearings used at the small and big end of the connecting rod.
20. List and describe the parts of a built-up crankshaft.

DISCUSSION TOPICS AND ACTIVITIES

1. Disassemble a scrap two-stroke engine. Identify each of the parts. Determine what method the engine uses for scavenging.
2. Locate a two-stroke engine with a removable cylinder head. Remove the cylinder head. Rotate the engine crankshaft and explain each event in the two-stroke cycle.
3. Place a two- and four-stroke engine side by side. Rotate the crankshaft on each engine and identify each cylinder event in both engines. How are they similar? How are they different?

CHAPTER 11

Engine Specifications and Selection

OBJECTIVES

Upon completion and review of this chapter, you should be able to:

- Describe and explain the engine internal size measurements of bore, stroke, displacement, and compression ratio.
- Define and explain torque.
- Define and explain horsepower.
- Explain how a dynamometer is used to measure torque and horsepower.
- Explain how to read and interpret a performance curve.
- Explain how engine brake horsepower and volumetric efficiency is determined.
- Use an engine specification sheet to repower an outdoor power product.

TERMS TO KNOW

American National Standards Institute (ANSI)
Bore
Brake horsepower (BHP)
Compression ratio
Consumer Product Safety Commission (CPSC)
Displacement
Dry weight
Efficiency
Engine dynamometer
Engine specification sheet
Horsepower
Horsepower curve
Mounting base
Original Equipment Manufacturer (OEM)
Outdoor Power Equipment Institute (OPEI)
Performance curve
Recommended maximum operating horsepower
Recommended speed range
Repowering
RPM
Stroke
Thermal efficiency
Torque
Torque curve
Volumetric efficiency

INTRODUCTION

Engines for outdoor power equipment are identified and compared in a number of ways. One engine may be compared against another by its internal size specifications. All engines used on outdoor power equipment have a specified horsepower and torque rating. Horsepower, torque, and efficiency ratings are commonly used when selecting equipment or selecting a new engine for present equipment.

INTERNAL ENGINE SIZE SPECIFICATIONS

Engines are made in different internal sizes to meet different requirements. A walk-behind lawnmower may have a small internal size engine. A garden tractor may have a large one. Engine internal size is measured in the area where power is developed.

Bore and Stroke

The **bore** is a measurement of the diameter of an engine's cylinder (Figure 11-1). The **stroke** is the distance the piston moves from TDC to the BDC of the

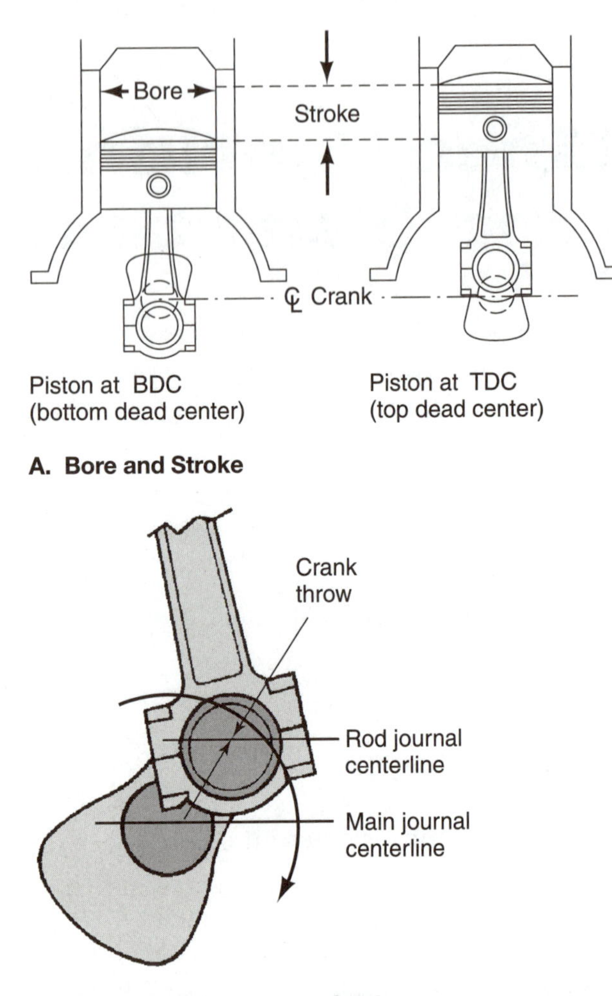

FIGURE 11-1 The bore is the cylinder diameter and stroke is determined by crankshaft dimensions.

cylinder. Stroke size is determined by the crankshaft. The distance between the centerlines of the connecting rod and main bearing journals controls the movement of the piston. The crankshaft moves the piston up and down. The stroke is two times (up and down) the distance between the centerlines of the connecting rod and main bearing journals. This distance is often called the *crankshaft throw*.

The larger the bore and longer the stroke, the more powerful the engine. The bore and stroke are given in inches or millimeters. An engine could have a bore of 3½ inches and a stroke of 4 inches. A metric engine could have a bore of 84 mm and a stroke of 88 mm. Sometimes just the numbers are given. For example: 3½ × 4. The bore is always the first number.

Displacement

The bore and stroke of an engine are used to find its displacement. **Displacement** is the size or volume of the cylinder measured when the piston is at the bottom of the cylinder (Figure 11-2). The bigger the bore and longer the stroke, the larger the cylinder volume or displacement. Some engines have more than one cylinder. The displacement of all the cylinders is added together to find the total displacement.

The displacement of an engine may be given in cubic inches or in cubic centimeters. Cubic centimeters are abbreviated cc's. Cubic inch displacement is abbreviated CID. The larger an engine's displacement, the more powerful the engine. The displacement of many small engines is 6 to 8 cubic inches. For example, A Honda G-300 engine has a bore of 3.0 × 2.4 inch (76 × 60 millimeters). The displacement of this engine is 16.6 cubic inches (272 cubic centimeters). In the metric system, cubic centimeters are converted to liters. An engine with a 1,000 cubic centimeter displacement is called a 1-liter engine. A 2,000 cubic centimeter engine is called a 2-liter engine.

SERVICE TIP: You can measure the bore and stroke of an engine and determine its displacement. The formula used to calculate displacement is:

$$\text{Displacement} = \frac{\text{Bore}^2 \times \text{pi} \times \text{Stroke}}{4}$$

The bore is squared, or multiplied by itself. The symbol pi stands for 3.14. For example, an engine has a bore of 3 inches and a stroke of 3 inches:

$$\text{Displacement} = \frac{3^2 \times 3.14 \times 3}{4}$$

$$= \frac{9 \times 3.14 \times 3}{4}$$

$$= 21.20 \text{ cubic inches}$$

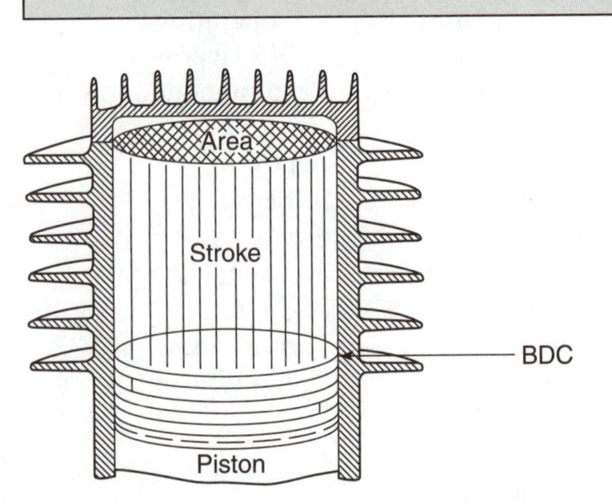

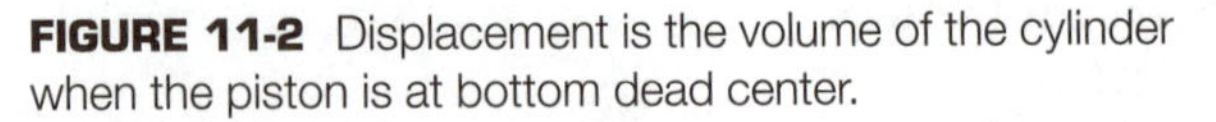

FIGURE 11-2 Displacement is the volume of the cylinder when the piston is at bottom dead center.

FIGURE 11-3 An engine with a compression ratio of 8 to 1.

Compression Ratio

An engine's compression ratio is another important internal engine measurement. One of the four strokes of the four-stroke-cycle engine is called a compression stroke. The piston moves up while both the intake and exhaust valves are closed. The air-fuel mixture is compressed in the compression chamber. The more the mixture is compressed, the higher the engine's compression. The higher the compression, the more power the engine can develop.

The amount of compression an engine has is specified by its compression ratio. **Compression ratio** is the ratio between the cylinder volume with the piston at bottom dead center compared to the volume when the piston is at top dead center.

Compression ratio is found by measuring the area of a cylinder. One measurement is made when the piston is at the bottom of its stroke. Another is made when the piston is moved to the top of its stroke. For example, a piston is at the bottom of its stroke. The cylinder volume is measured as 8 cubic inches. The piston is moved to the top of its stroke. The volume is now only 1 cubic inch. The 8 inches of-and fuel mixture has been compressed into a 1-cubic-inch area, a compression ratio of 8 to 1 (Figure 11-3). Small engines generally have compression ratios between 6 to 1 and 8 to 1.

HORSEPOWER AND TORQUE

The most common way of comparing engines is by the amount of power they develop. Engine power is measured in horsepower. The idea of horsepower comes from a time long ago when engines were compared to the power of a horse. Engine torque is an important part of horsepower.

Torque is a rotary unit of force that causes or tries to cause rotation. When you use a wrench to tighten a bolt, you are applying torque to the bolt. When the bolt is tight, you will not be able to turn it any more. You are still applying torque even though the bolt does not turn.

The torque is developed when the engine's piston turns the crankshaft. The piston can be compared to the strength in your arm. If you have a strong arm, you can apply a lot of force to a wrench handle (Figure 11-4). An engine with a large piston and a high compression ratio can apply a strong force to the crankshaft. If you use a wrench with a long handle, you can apply a lot of torque. If the engine crankshaft has a long stroke, it will develop a lot of torque. Torque is given in pounds-feet in the inch system. It is given in Newton-meters in the metric system.

Torque is an important part of horsepower. **Horsepower** is a unit of engine power measurement determined by engine speed and torque. Horsepower can be computed with the formula: Horsepower = rpm × torque ÷ 5252. Torque is the turning effort developed by the engine. The **rpm** is the rotational speed of an engine crankshaft in revolutions per minute. The faster an engine crankshaft rotates, the more power it can develop. The 5252 is a number called a constant. It is used to divide the answer to get the correct units.

There is another power measurement unit used with engines based on a metric system unit called the watt. In this system, 1,000 watts (or one kilowatt) equals 1.34 horsepower. This means that 746 watts equals 1 horsepower.

MEASURING HORSEPOWER

Engine horsepower is most often determined in the laboratory. Engine makers measure the horsepower of their engines. Horsepower can be found by measuring an engine's torque at a specific rpm. An **engine dynamometer** (Figure 11-5) is a device used to measure engine torque and calculate engine horsepower. Engine dynamometers are used to measure and rate horsepower. The dynamometer does not measure horsepower directly. It measures torque and rpm. These measurements are then used to determine horsepower mathematically. Many dynamometers do this calculation automatically. They give the operator a horsepower figure.

Most dynamometers measure torque of an engine by changing the rotating torque to a stationary torque. The stationary torque is then measured

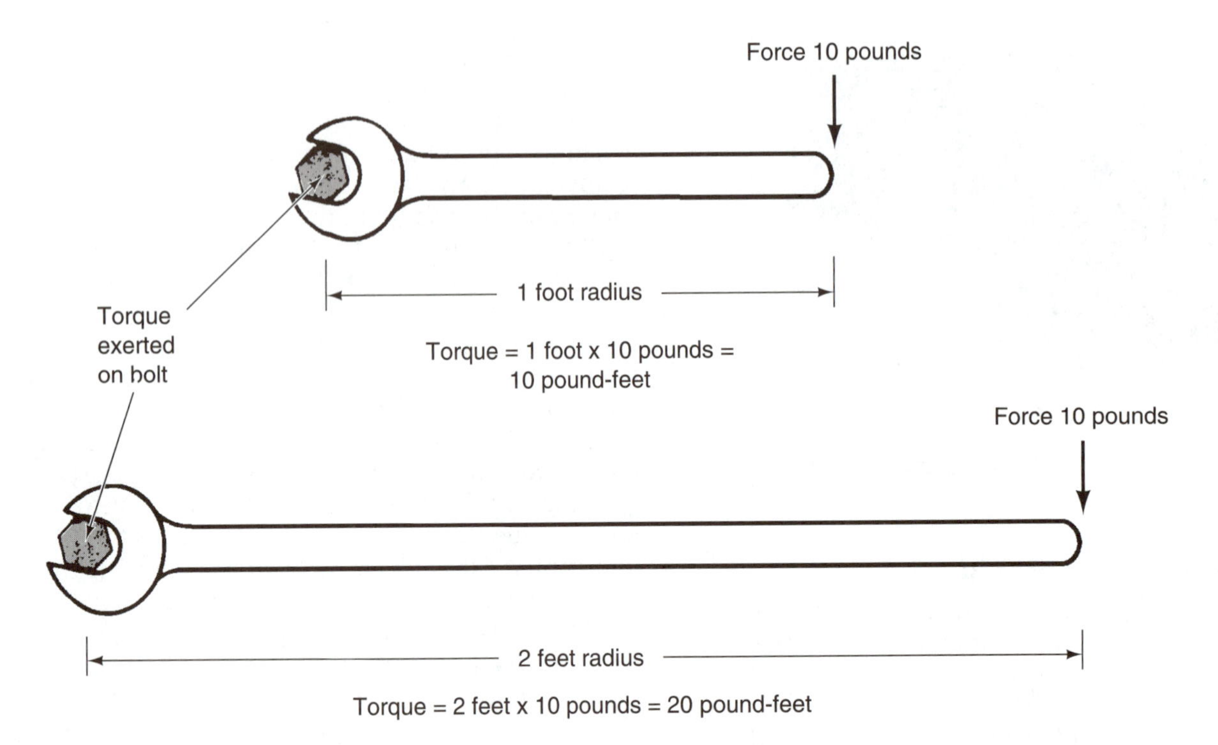

FIGURE 11-4 Torque can be increased with a longer wrench handle or more force on the end of the wrench.

with a hanging scale, load cell, or strain gauge. The measurement is taken at the end of a torque arm.

Most dynamometers use a hydraulic water brake to change rotating torque to stationary torque. Torque is controlled by the amount of water in the absorption unit. A water valve controls the flow of water through the absorption unit. As the water flow increases, the amount of water in the absorption unit also increases.

FIGURE 11-5 An engine dynamometer measures engine rpm and torque and determines horsepower.

The engine to be tested is mounted on the dynamometer mounting plate. The plate has rubber shock mounts to reduce vibrations. The PTO end of the engine crankshaft is attached to the water brake unit. A rubber isolator coupling is used to reduce vibrations into the test gauges. A special heavy flywheel may be mounted on the crankshaft to prevent fluctuations on the test gauges. The water brake housing is attached to a torque arm. The torque arm is connected at the other end to a force-measuring scale. A throttle control is attached to the engine carburetor and used to adjust engine test speeds. A water supply hose is connected from a water inlet to the water brake and an outlet hose connected to a drain.

WARNING: Always wear eye, face, and ear protection to prevent injury from excessive noise or parts damage when testing an engine on a dynamometer.

The engine is started and warmed up prior to testing. The engine speed is adjusted from idle to wide open throttle (WOT) using the throttle control. A tachometer on the dynamometer shows en-

FIGURE 11-6 A water brake and torque-measuring load cell on a dynamometer.

gine speed in rpm. The engine is loaded by turning the load control knob. This adjusts the amount of water in the water brake. It loads the engine from no load to maximum load. The water brake is rotated by the water force inside its housing. This rotation results in a torque on the torque arm. The torque arm pushes on a force-measuring load cell. The load cell is connected to a torque-reading force gauge. The higher the engine's force on the load cell, the higher the torque reading (Figure 11-6).

The operator works the throttle control and load knob at the same time. The tachometer and force gauge show engine torque at different engine speeds. The results of the torque and rpm measurements are then used to find horsepower. Most laboratory dynamometers do the calculation and provide a direct horsepower reading. Engine makers rate and advertise the horsepower of their engines. Most engine dynamometers measure a type of horsepower called brake horsepower. **Brake horsepower (BHP)** is horsepower measured at the flywheel or PTO end of the crankshaft. Brake horsepower is named for an old braking device that was wrapped around an engine flywheel to measure horsepower.

Brake horsepower is affected by several operating conditions. Air density is the amount of oxygen in the air. The engine runs best on dense air. The more oxygen that enters the combustion chamber, the better the combustion. The density of the air is affected by temperature. Hot air has less oxygen. Air density is affected by altitude. The higher the altitude, the lower the air density. Barometric pressure and humidity can also affect air density.

The brake horsepower engines can only be compared if they have the same operating conditions. This problem is eliminated in two ways. Engines must be tested and rated under the same conditions. Correction formulas have to be used for engines when testing engines under different conditions. The test standards are developed by the American National Standards Institute (ANSI) and Society of Automotive Engineers (SAE). They update and publish these standards regularly. The current standard is ANSI/SAE J1349-JAN90.

Performance Curves

A **performance curve** is a graph that shows the horsepower and torque of an engine compared to engine speed. These curves often show recommended maximum operating horsepower and rpm range. The performance curve is part of the technical specification sheet available for each engine.

A **torque curve** is a graph showing torque at different engine rpm. A **horsepower curve** is a graph showing horsepower at different rpm. Engine torque and horsepower are often shown on the same graph. Horsepower and torque measurements are shown along one side of the graph. Engine rpm is shown along the top or bottom of the graph. When the test is complete, all the marks are connected together to form a horsepower and torque curve. A torque and horsepower curve (Figure 11-7) allows both torque and horsepower to be compared to engine operating speed.

Most engines have similar horsepower and torque curves. The horsepower does not start at zero because an engine does not run at zero speed. The curve is cut off at the bottom. Horsepower increases as the engine speed and load increase. The curve shows the rpm where the engine reaches its maximum horsepower. An engine is capable of running faster than the speed at which it reaches maximum horsepower. The horsepower begins to decrease after reaching the maximum point. The reasons for the decrease are related to the torque curve.

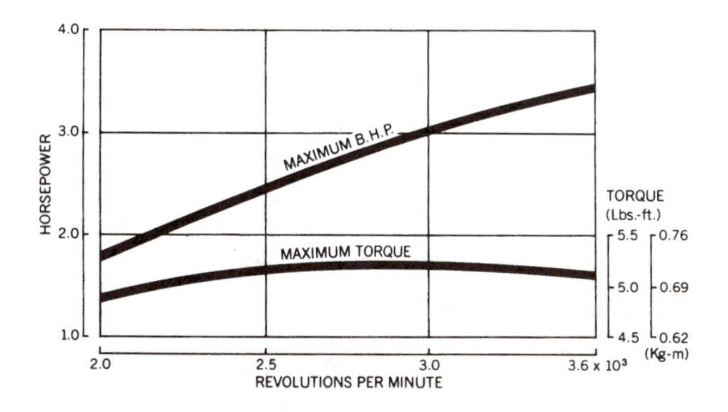

FIGURE 11-7 A torque and horsepower curve. (Courtesy of American Honda Motor Co., Inc.)

The torque curve shows the engine torque at different speeds in pounds-feet or Newton-meters. The difference between the torque curve and the horsepower curve shows how the engine works at different loads and speeds. The horsepower curve continues to climb as the engine speed increases. It continues until maximum horsepower is reached. This is also true with the torque curve. However, the torque curve reaches its maximum point earlier.

Engine torque changes widely over the normal range of crankshaft speeds. At very low speeds (200–300 rpm), an engine develops only enough torque to keep itself running. As engine speed and load increase, torque increases, reaching the speed where torque is the highest. This is where the engine maker rates the torque. It is the most efficient engine operating speed. At this point, the cylinders are taking in the biggest and most efficient air and fuel mixture. The exhaust gases in the cylinder are being forced out most effectively.

The torque curve drops off rapidly after the high point. At higher engine rpm, there is less time for air and fuel to enter the cylinder. There is also less time for the exhaust gasses to leave the cylinder. This results in a weaker push on the pistons and less torque. Other factors, such as internal engine friction and pumping losses, work to cause a drop in torque. Pumping losses are the energy used in pumping in air and fuel and exhausting it.

The horsepower curve is directly affected by the torque curve. Torque is one of the parts in the horsepower formula. The horsepower curve does not directly correspond with the torque curve. Horsepower is affected by engine speed. The horsepower increases past the peak of the torque curve. This is possible because the engine rpm increases beyond this point. At some point the torque drops off so much that more rpm cannot cause an increase in horsepower.

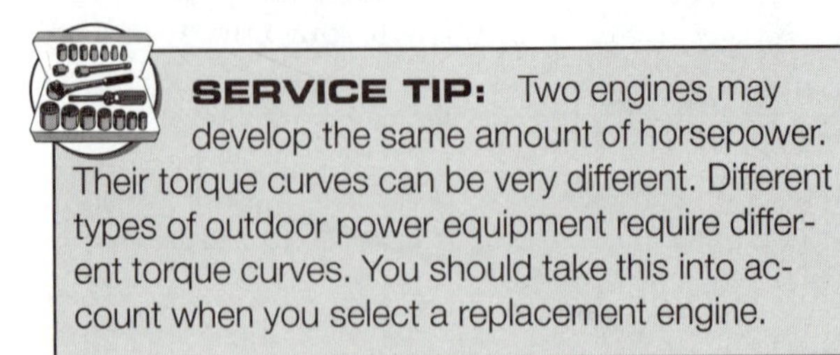

SERVICE TIP: Two engines may develop the same amount of horsepower. Their torque curves can be very different. Different types of outdoor power equipment require different torque curves. You should take this into account when you select a replacement engine.

ENGINE EFFICIENCY

The purpose of any engine is to change energy into useful work. **Efficiency** is the measure of how well an engine converts energy into work. The internal combustion engine changes chemical energy of

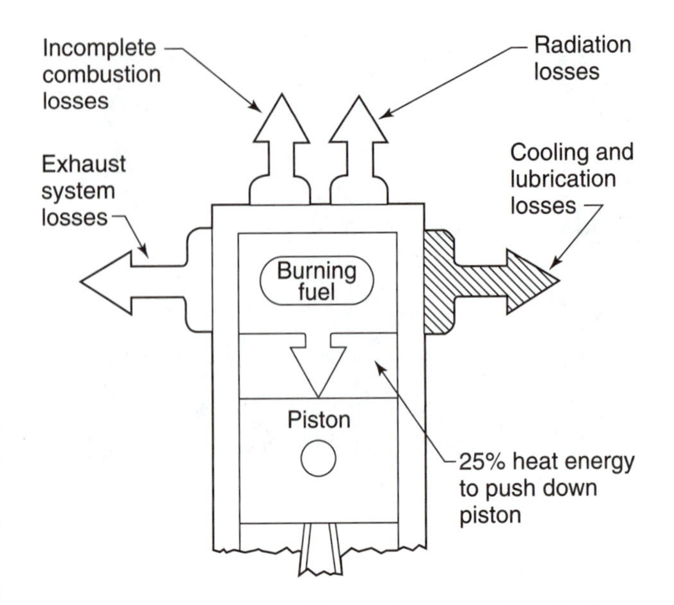

FIGURE 11-8 The thermal losses from an engine give it an efficiency rating of 25 percent.

gasoline into heat (thermal) energy. **Thermal efficiency** is the percentage of heat energy available in the fuel that is changed into power at the engine's crankshaft. No engine is 100 percent efficient. There is always a loss of energy through heat and friction. Engines lose heat out the exhaust. They also lose heat through the cooling and lubrication system. There are also losses due to heat radiation and incomplete combustion. Only about 25 percent of the total energy in gasoline is changed into useful work at the crankshaft PTO (Figure 11-8).

Volumetric efficiency is another type of engine efficiency rating. **Volumetric efficiency** is the actual volume of a cylinder compared to the volume that is filled during engine operation (Figure 11-9).

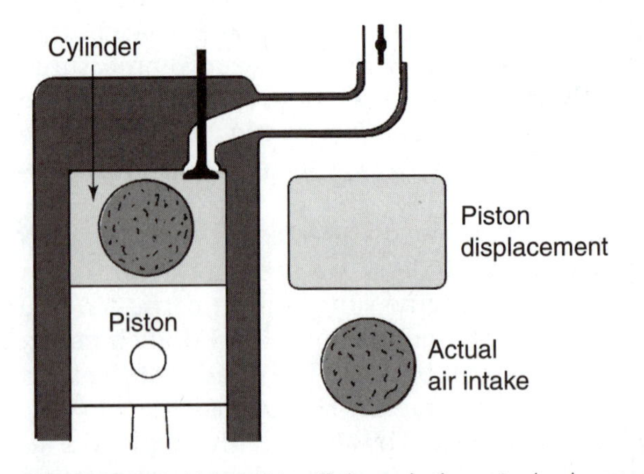

FIGURE 11-9 Volumetric efficiency is the actual volume of a cylinder compared to the volume that is filled during engine operation. (Reproduced by permission of Deere & Company, © 1991. Deere & Company. All Rights reserved.)

At high engine speeds, the valves are open for a very short time. The cylinders are not completely filled with the air-fuel mixture.

Volumetric efficiency is measured on an engine dynamometer with fuel and air flow equipment (Figure 11-10). The flow of air going into the engine is measured with a sensitive air flow drum and a pressure-measuring gauge called a *manometer*. Air flow is measured in cubic feet per minute (cfm). Fuel flow is measured in pounds per hour with a rotameter. Air flow, fuel flow, and engine horsepower are used to calculate volumetric efficiency.

ENGINE SELECTION

When an engine is worn out, you and your customer will have to make a decision. Should the engine be serviced or replaced? A total engine overhaul can cost the customer a lot of money. This cost can be compared with the price of replacing the old engine with a new one.

Repowering

Repowering is the selection and installation of a new replacement engine for an outdoor power product. Most repowering is done by installing the same type of engine on the equipment. You simply order the correct replacement engine from your parts supplier. These are called OEM engines. **Original equipment manufacturer (OEM)** is a description used with parts and equipment supplied by the original engine or equipment manufacturer. Sometimes there are minor changes in the replacement engine. These are often updated internal parts or systems. An installation kit often is supplied with the engine. These kits have the installation parts and procedures necessary to install the engine.

FIGURE 11-10 Volumetric efficiency is measured by measuring fuel and air flow into an engine.

There will be times when you need to repower outdoor power equipment with a different engine. Older equipment may not have an OEM engine available, you may need to increase power or torque, or you may want to update the type of engine used on older equipment. Repowering with a different engine is a more complicated process.

Outdoor power products are designed and built to strict safety standards to protect people and property from potential hazards. A standard is an accepted practice used to make sure that equipment is manufactured, installed, and operated safely. Standards for outdoor power equipment may be specified by several organizations. The **American National Standards Institute (ANSI)** is a national organization that helps identify industrial and public needs for national standards. The **Consumer Product Safety Commission (CPSC)** is a federal commission that evaluates the safety of all types of consumer products including outdoor power products. The **Outdoor Power Equipment Institute (OPEI)** is a trade group that represents the manufacturers of outdoor power equipment.

WARNING: When you repower, an outdoor power product must have the same safety standards as the OEM installation. You must be sure that things like guards and other safety devices work the same after your installation. Fuel line and exhaust gas routing must meet safety standards. Failure to do so could result in an injury to the operator. You and the shop could be held liable in a lawsuit.

Engine Specification Sheets

The first step in repowering is to get an engine specification sheet for the replacement engine. An **engine specification sheet** (Figure 11-11) is a list showing the performance data, the internal specifications, and the external dimensions of a replacement engine. These sheets are available from the engine makers and parts suppliers. You use them to compare information from the OEM engine with the possible replacement engine.

External Engine Dimensions

The external size of an engine is one of the first checks you should make. The replacement engine must fit into the same space as the old engine. The specification sheet will show the exact height,

FIGURE 11-11 Engine specification sheets show the performance data, the internal specifications, and the external dimensions of a replacement engine.

width, and length of the engine (Figure 11-12). You must measure the old engine to find out if it will fit.

The **mounting base** (Figure 11-13) is the part of the engine crankcase or cover that has holes for mounting to the equipment. The size of the mounting base and the spacing of the mounting holes are different for different engines. You must measure the old engine mounting base and compare it to the specification sheet. You may find that new mounting holes will have to be drilled in the equipment to mount the engine.

The specification sheet gives the weight of an engine, usually specified as dry weight. The **dry weight** of an engine is its weight without oil or gasoline. The weight of a replacement engine is important. A replacement engine that is heavier or lighter than the OEM can change how the equipment operates, which can lead to safety problems.

The PTO shaft dimensions are another important consideration (Figure 11-14). Most engine makers have many styles of PTO shafts. Some are made in different diameters and lengths, some are

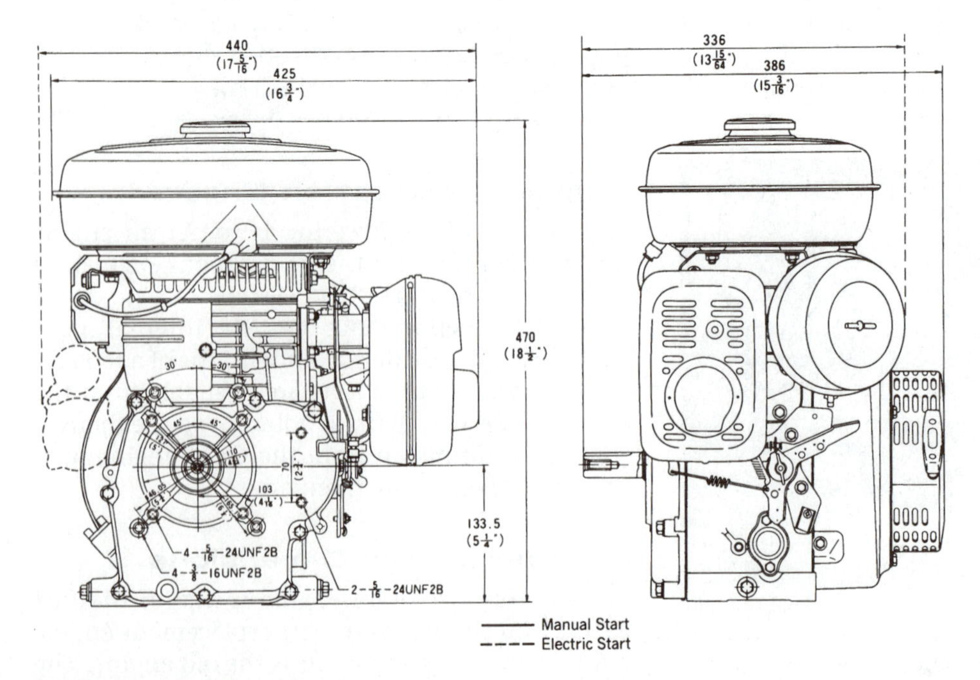

FIGURE 11-12 External engine size is shown on an engine specification sheet. (Courtesy of American Honda Motor Co., Inc.)

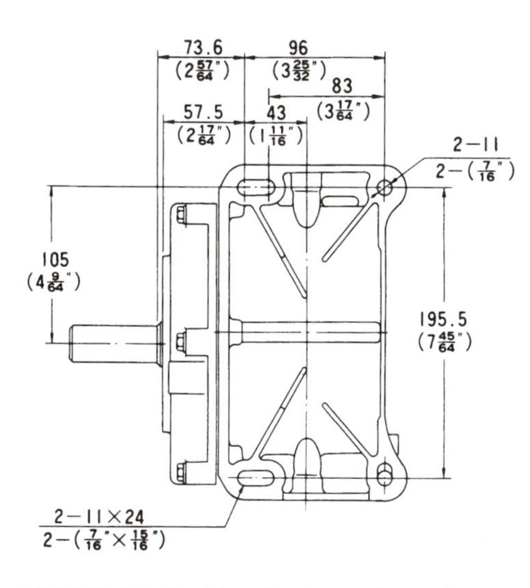

FIGURE 11-13 Mounting base dimensions are shown on an engine specification sheet. (Courtesy of American Honda Motor Co., Inc.)

tapered, others have a keyway. The PTO must match up to the drive parts such as pulleys or clutch on the equipment.

Engine Performance

The engine specification sheet shows the engine's internal engine specification and performance curve (Figure 11-15). Common specifications listed are:

- Engine type (number of cylinders, horizontal or vertical crankshaft, overhead or L-head valves)
- Displacement (cubic inches or cubic centimeters)
- Bore and stroke (bore × stroke in inches or millimeters)
- Maximum horsepower (brake horsepower at a specific rpm)
- Maximum torque (torque in pounds-foot or kilogram per cubic meter at a specific rpm)
- Ignition system type (solid state or breaker point)
- Starting system type (recoil or electric)
- Air cleaner type (dual or single element)
- Fuel tank capacity (in gallons or liters)
- Dry weight (in pounds or kilograms)
- External dimensions (length × width × height)
- PTO type and dimensions (tapered or keyway)

The specification sheet also has the engine performance curve. It is very important to match the engine performance to the equipment. Sometimes you will want to find a replacement engine with the same performance as the OEM engine. At other times you may want an engine with different performance.

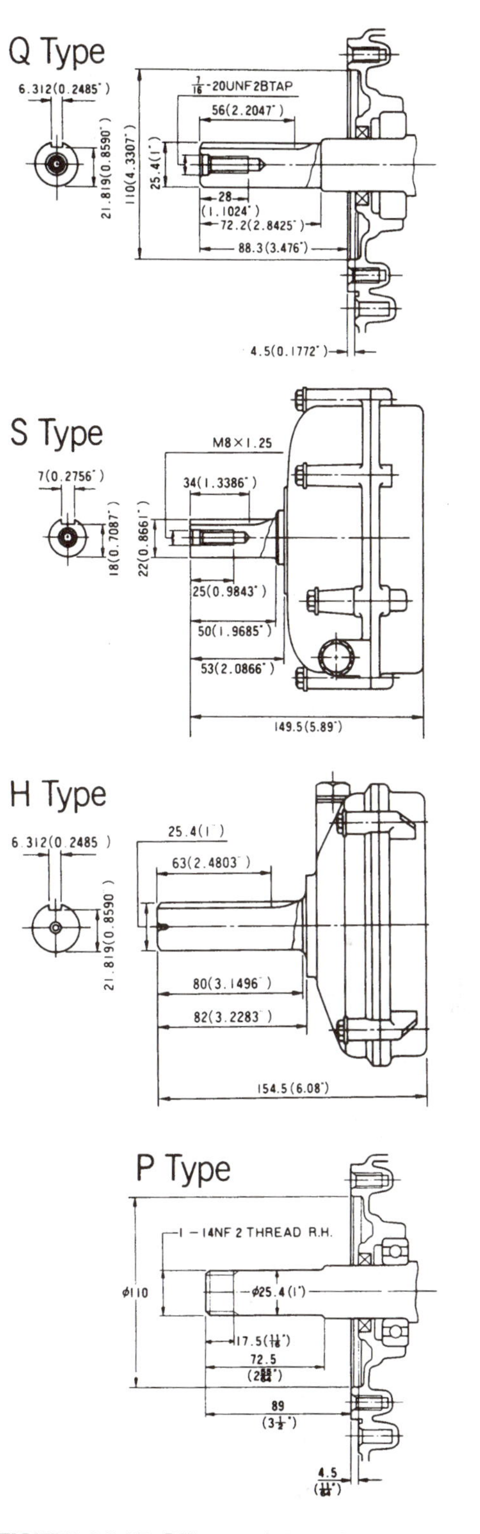

FIGURE 11-14 Different sizes and types of PTO shafts are shown on an engine specification sheet. (Courtesy of American Honda Motor Co., Inc.)

The performance curve shows the engine's horsepower and torque curve. It shows the rpm where maximum torque and maximum horsepower is

Engine Type	Single cylinder, side valve, four-stroke, forced-air-cooled, (horizontal shaft)
Piston Displacement	16.6 cu. in. (272cc)
Bore × Stroke	3.0 × 2,4 in. (76 × 60 mm)
Compression Ratio	6.5:1
Max. Horsepower	7.0 HP/3,600 rpm (SAE No. J607a)
Max. Torque	10.9 lbs.-ft./2,500 rpm (1.5 Kg-m/2,500 rpm)
Ignition System	Capacitor Discharge Ignition (CDI)
Ignition Timing	20° B.T.D.C.
Starting Systems	Recoil starter
Alr Cleaner	Dual element
Fuel Tank Capacity	1.59 gal. (6.0 liters)
Dry Weight (Q Type)	48.5 lbs. (22.0 kg)
Dimensions L × W × H (Q Type)	15³⁄₁₆ × 16¾ × 18½ in. (386 × 425 × 470 mm)

FIGURE 11-15 Engine specifications are listed on the engine specification sheet. (Courtesy of American Honda Motor Co., Inc.)

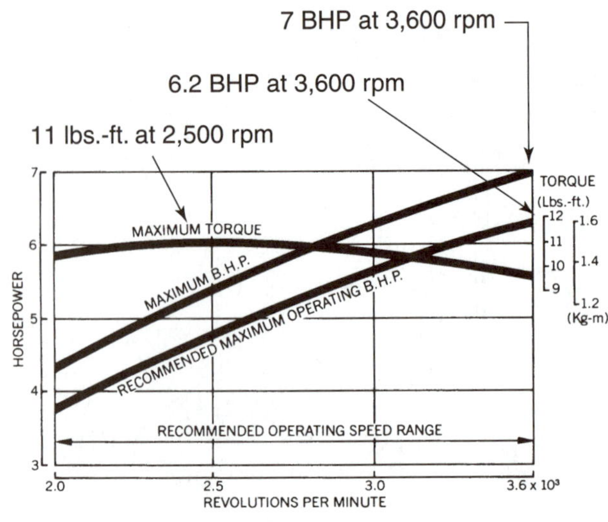

FIGURE 11-16 A performance curve is used to compare one engine with another. (Courtesy of American Honda Motor Co., Inc.

reached. You need to know how much horsepower and torque your repowering project requires. You also need to know the engine speed where horsepower and torque are at their maximum. Two engines with the same horsepower may have very

CASE STUDY

An outdoor power equipment repair shop got involved with repowering an old tiller. The tiller was 20 years old and powered by a 6-horsepower cast-iron engine. The tiller was used only during the springtime and was in good shape. The engine had suffered from improper off-season storage problems. It had a great deal of sludge in the oil and the fuel system had corrosion problems. The engine ignition had points and condenser ignition that was getting unreliable. The customer requested that the shop repower the tiller with a late model engine.

The technician assigned the job made some measurements on the mounting space, PTO size, and PTO height. The technician determined that a used 10-horsepower engine available in the shop would fit on the tiller. After negotiating a price for the engine, the technician went about adapting it into the tiller. New base plate holes were drilled and new throttle linkage had to be built. When the job was done, all the belt guards and other safety features on the old tiller were functional.

The happy customer took the tiller home but did not get around to using it until the spring planting season. The next news the shop heard about the tiller was in the form of a lawsuit filed by the customer against the shop. The lawsuit specified damages against the shop for injuries the customer suffered when he used the tiller. The customer complained that the higher horsepower and torque of the new engine caused him to lose control of the tiller and be thrown to the ground. The customer alleged that he had sustained a back injury as a result of the accident. The lawsuit further specified that the shop was negligent in installing too powerful an engine on the old tiller.

The case was settled before trial. The shop had to hire an attorney to defend the case. In the settlement, the shop agreed to pay the customer's medical bills and several thousand dollars in damages. The technician and shop owner both considered this a very expensive lesson.

different maximum horsepower engine speeds. Similarly, two engines with the same maximum torque may have very different maximum torque speeds. You have to carefully consider the torque and horsepower needs of the equipment. You can use the OEM engine performance curve as a starting point.

The performance curve also shows the recommended maximum operating horsepower and recommended maximum operating speed range (Figure 11-16). The **recommended maximum operating horsepower** is 85 to 90 percent of the maximum horsepower developed by an engine. This is the horsepower specification manufacturers recommend for normal use of their engine. The **recommended speed range** is the engine maker's recommended range of engine operating rpm.

WARNING: An engine should not be operated above the recommended speed range. Excessive speed can cause parts failure and personal injury.

Operating Conditions

There are several operating conditions that could have an effect on engine choice. Three common conditions are dirty air, vibration, and high temperature. You should talk with your customer and determine engine use conditions. A mower used in a dirt-filled meadow may kick up a great deal of dirt and dust. An engine for this application must have a good air cleaner system. Some engines may not be equipped with an adequate system. A riding mower may be used on a well-maintained lawn. Another mower may be used on a rut-filled meadow. Vibration problems may cause early parts failure. An engine should be selected that offers a vibration-free engine and accessory mounting.

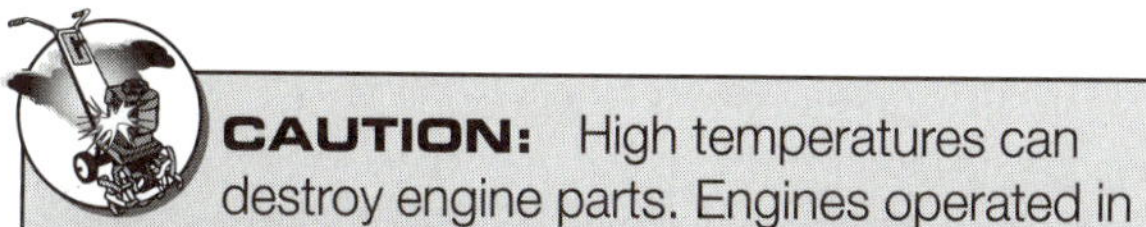

CAUTION: High temperatures can destroy engine parts. Engines operated in hot areas must have very good cooling and lubrication systems. Some engines have cooling and lubrication systems that are more efficient than others.

REVIEW QUESTIONS

1. Describe how the bore and stroke is used to determine engine displacement.
2. Explain how the compression ratio of an engine is calculated.
3. Explain the relationship between compression ratio and the power developed by an engine.
4. List and explain the formula for horsepower.
5. Define torque and explain how it relates to engine horsepower.
6. Describe how horsepower of an engine is measured with an engine dynamometer.
7. Explain the purpose of a performance curve.
8. Explain why maximum torque and maximum horsepower occur at different engine rpm.
9. Describe why the brake horsepower of an engine has to be corrected to standard conditions.
10. Define and explain the thermal efficiency of an engine.
11. Define and explain volumetric efficiency of an engine.
12. Explain how to use an engine's specification sheet performance curves to determine maximum horsepower.
13. Explain how to use an engine's specification sheet to determine maximum and recommended horsepower.
14. Explain how to use a specification sheet to determine engine external size.
15. List and describe the operating conditions that affect engine repowering choice.

DISCUSSION TOPICS AND ACTIVITIES.

1. Use thick oil and a graduated measuring device to determine the (displacement) volume of a cylinder with the piston at top dead center.
2. Use the thick oil to measure the combustion chamber volume on the same engine. How can you use this information to determine compression ratio?
3. Measure the bore and stroke of a scrap shop engine. Use the formula for displacement to calculate the engine displacement:

$$\text{Displacement} = \frac{\text{Bore}^2 \times \text{pi} \times \text{Stroke}}{4}$$

PART 3

Engine Systems

Ignition Systems—Magneto

OBJECTIVES

Upon completion and review of this chapter, you should be able to:

- Explain how electricity and magnetism are used to create induction in a magneto.
- Describe and explain the operation of a basic magneto armature, magnet, and coil.
- Describe the parts and operation of a magneto coil.
- Explain the operation of the armature and coil primary and secondary windings in induction.
- Identify and describe the parts of a breaker point-triggered magneto system.
- Explain the operation of a breaker point-triggered magneto system.
- Describe the parts and explain the operation of a transistor-triggered magneto.
- Describe the parts and explain the operation of a capacitor discharge magneto.

TERMS TO KNOW

Armature
Breaker points
Capacitor discharge ignition (CD)
Condenser
Conductor
Diode
Electricity
Ground circuit
Ignition module
Induction
Insulator
Magneto
Magneto coil
Primary winding
Retrofit part
Secondary winding
Semiconductor
Transistor

INTRODUCTION

The ignition system creates a high voltage electric spark at the spark plug. The spark causes the air-fuel mixture in the combustion chamber to burn. The combustion pushes the piston down for a power stroke.

The ignition has a low voltage and a high voltage system. Low voltage electrical current is changed to high voltage by a magneto. A **magneto** is an ignition part that uses magnetism to create a low voltage that is changed to a high voltage by induction. The high voltage goes through spark plug wires to the spark plug. The low voltage parts are often called the *low tension* or *primary circuit.* The high voltage parts are often called the *high tension* or *secondary circuit.*

ELECTRICITY

You need a basic understanding of electricity and magnetism to understand how a magneto works. Electricity and magnetism work because of atoms. Atoms are very small particles that make up everything in the universe. An atom is so small that it is not visible. It can only be seen under the most powerful electron microscope.

Atoms are made up of even smaller particles. They are constructed much like planets (Figure 12-1). They revolve like planets around the sun. An atom has a center called a *nucleus,* which is made up protons and neutrons. The sun is like the nucleus. Other small particles called *electrons* circle in orbits around the nucleus, in much the same way

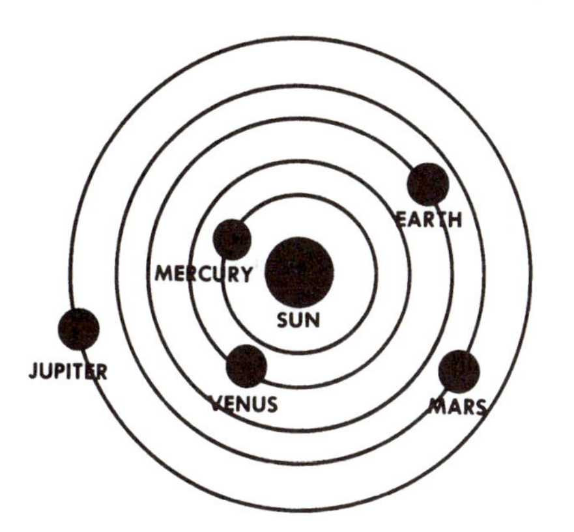

A. Solar system

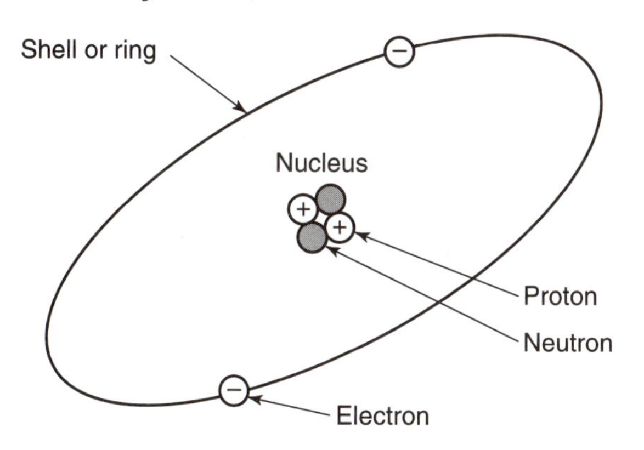

B. Atom

FIGURE 12-1 Atoms are constructed much like our solar system.

that the planets circle the sun. Electrons travel at a very fast speed.

The particles that make up the atom have positive and negative electrical charges. The symbol + is used to show a positively charged particle. The symbol – is used to show one with a negative charge. Two particles that are positively charged push away from each other. Two negatively charged particles push away from each other. A positively charged particle and a negatively charged particle attract each other. This positive to negative attraction is what holds the atoms.

The nucleus is made up of protons with a positive charge and neutrons with a neutral charge. The electrons have a negative charge. They orbit at a fixed distance away from the nucleus. An atom may have one, two, or three rings of electrons, depending on the number of electrons it contains. Each ring requires a specific number of electrons.

The electrons stay in their orbit around the nucleus because of the electrical attraction, much like the gravitational pull of the sun on the Earth. The electrons that orbit closest to the nucleus are

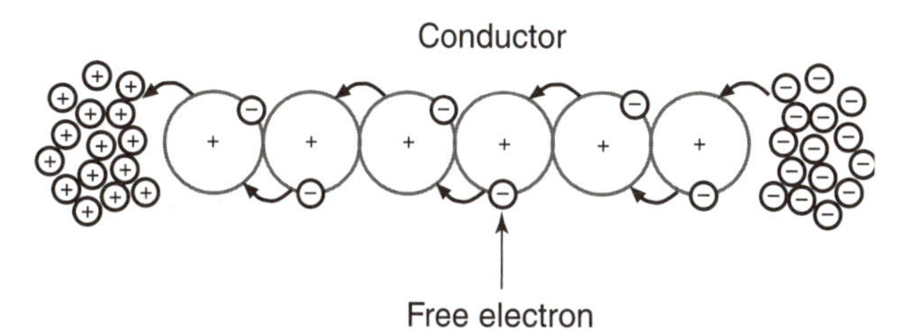

FIGURE 12-2 Electricity is the movement of free electrons from one atom to another.

strongly attracted. They are called *bound electrons*. The electrons that are farther away from the pull of the nucleus can be forced out of their orbits. These are called *free electrons*. Free electrons can move from one atom to another. This movement is known as electron flow.

Electricity is the movement or flow of electrons from one atom to another (Figure 12-2). A condition of imbalance is necessary to have a movement of electrons. In a normal atom, the positively charged nucleus balances the negatively charged electrons, holding them in orbit. If an atom loses electrons, it will become positive in charge. It will attract more electrons in order to get its balance.

A **conductor** is any material that allows a good electron flow and conducts electricity. A good conductor must be made of atoms that give off free electrons easily. Also, the atoms must be close enough to each other so that the free electron orbits overlap. Ignition systems use copper and aluminum wires to conduct electricity. They allow good electron flow.

An **insulator** is a material with atoms that will not part with any of its free electrons and will not conduct current. The copper wire in an ignition system is covered with a plastic insulation material. The insulation prevents the current from escaping out of the wire.

A circuit is a path that allows current to flow to operate an electrical device. Any circuit, no matter how complicated, is made up of several essential parts (Figure 12-3). There must always be a source of electrical pressure or voltage. For example, the voltage source may be a battery that causes current flow to light a light bulb. The light bulb will offer resistance to the current flow (electrical load). Wires or conductors connect the battery and light bulb. One wire (conductor) connects the positive battery terminal to the light. Another wire (return conductor) completes the circuit from the light back to the negative terminal of the battery. In order for current to flow in a circuit, the path must be unbroken. In fact, the term circuit means circle.

Electricity can also flow through metal parts on

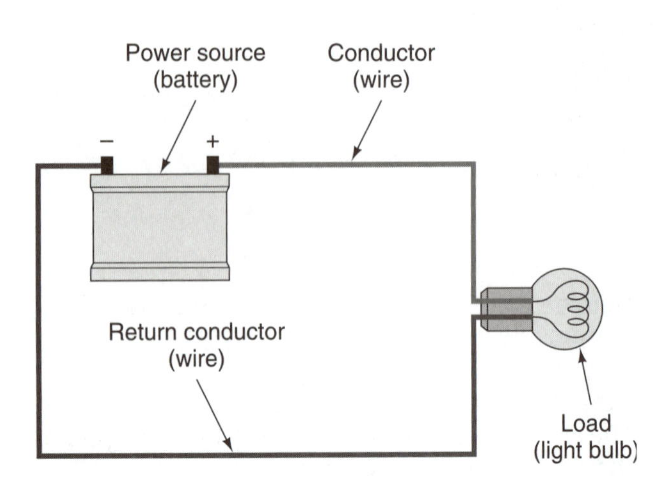

FIGURE 12-3 Parts of an electrical circuit. (Courtesy of Ford Global Technologies, Inc.)

the engine or equipment. Metal parts can be used in place of one of the wires in a circuit. Instead of a wire from the light bulb to the battery, a wire could be connected from the battery to a metal part on the engine (Figure 12-4). A wire from the light could be

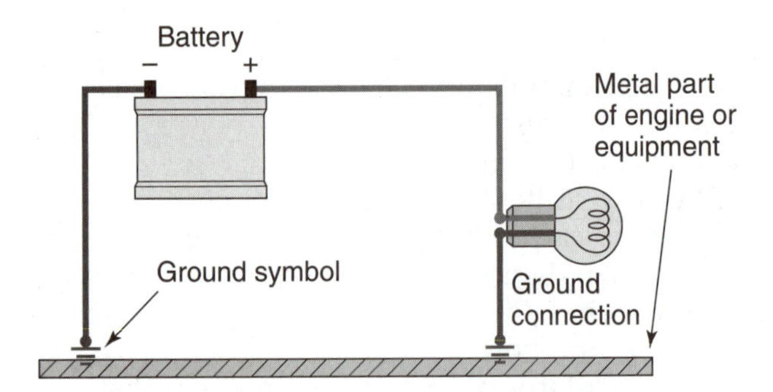

A. Ground Instead of a Return Wire

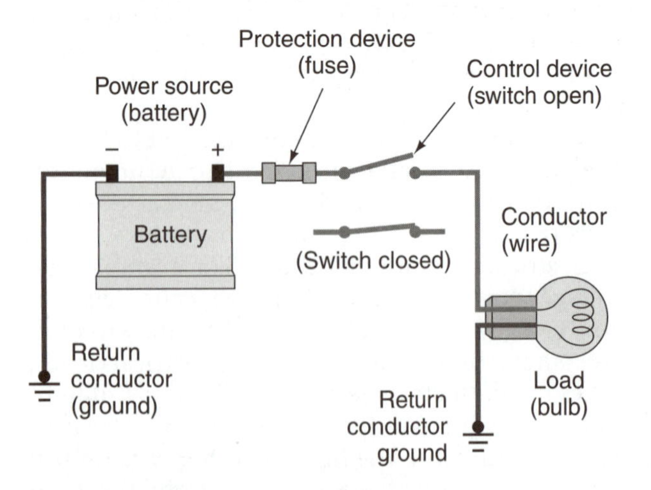

B. Circuit Controlled with a Switch

FIGURE 12-4 Electrical parts can be connected to metal instead of a wire to make a complete circuit. (Courtesy of Ford Global Technologies, Inc.)

connected to the same metal part. Electricity will flow through the wire and into the light. Current will flow through the light into the metal on the engine or equipment and back to the battery. This completes the circuit and the light lights up. The **ground circuit** is the electrical circuit connected to metal on the engine. The ground circuit allows one wire to be used to make a complete circuit. Most engine electrical parts have one grounded connection. A ground symbol on a wiring diagram is used to show how an electrical part is grounded.

Most circuits also have a switch so that the circuit can be turned on and off. A wiring diagram will show a symbol for the open or closed switch. Many circuits also contain a protection device called a *fuse*. If too much current starts to flow in a circuit, the fuse will melt and open the circuit. This protects the more expensive electrical components from damage.

MAGNETISM

Magnetism is used in the operation of the magneto. The word *magneto* comes from the word magnet. Each electron has a circle of magnetic force around it. In an unmagnetized piece of iron, the electron orbits are not arranged in any pattern (Figure 12-5). In a magnetized piece of iron, the electron orbits are lined up, causing the circles of magnetic force to be added together. All the magnetic forces work together. The piece of iron has a strong magnetic field.

Magnetos have two magnets in the engine flywheel (Figure 12-6). The magnetism from these magnets may be used to make electricity. Each of the two magnets in the flywheel is like one leg of a horseshoe magnet. One leg of the horseshoe magnet has a north and the other has a south pole. Magnetic lines of force are attracted from one leg to

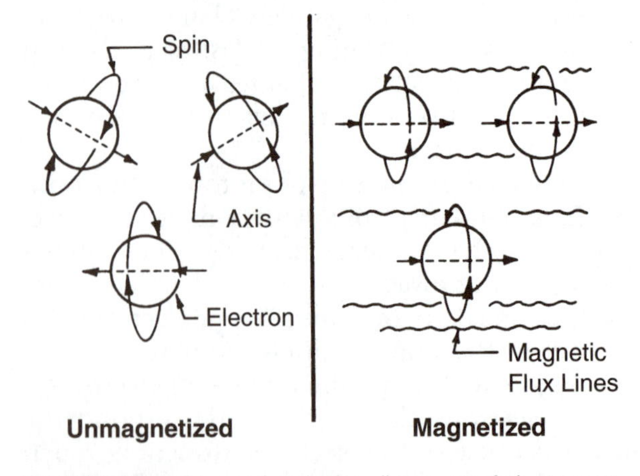

FIGURE 12-5 Magnetism is the alignment of electron orbits. (Courtesy of Kohler Co.)

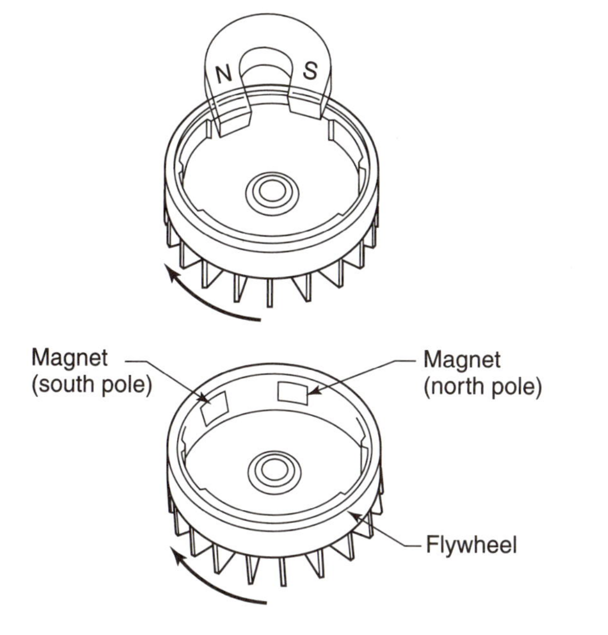

FIGURE 12-6 Flywheel magnets have a north and a south pole.

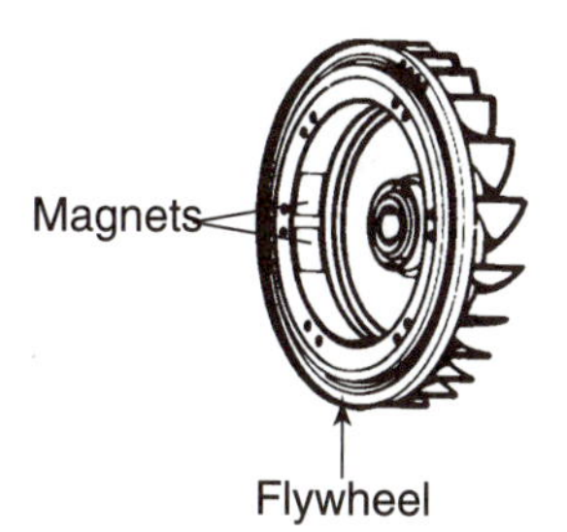

FIGURE 12-7 Magnets cast into the flywheel. (Provided courtesy of Tecumseh Products Company.)

the other. This magnetic attraction is used to develop electrical flow in the magneto.

The magneto magnets may be on the inside or the outside of the engine's flywheel. They are usually cast into the flywheel aluminum when it is made (Figure 12-7). On some engines, they are mounted to the flywheel with an adhesive or with screws.

INDUCTION

Induction is the transfer of electrical energy from one conductor coil to another without the coils touching each other. A coil, also called a *winding,* is a wire that is wrapped in circles. Induction is used in magneto coils. When current flows through a coil, a magnetic field is created in the coil. The coil with current flowing in it is placed near another coil without current. The two coils will influence each other.

Induction can be shown using two coils placed

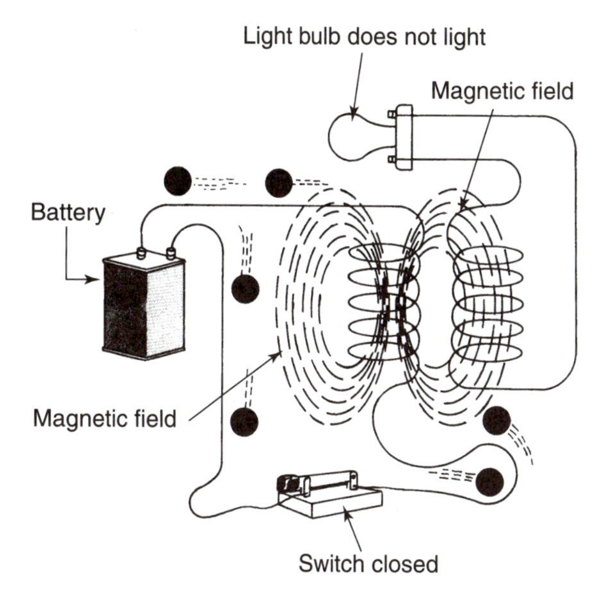

FIGURE 12-8 Current flow in one coil creates a magnetic field around both coils.

next to each other (Figure 12-8). A battery, switch, and wiring is connected to one coil. It has a complete circuit. A light bulb is connected to the second coil. It is not connected to the battery. The switch on the first coil is closed. Current flows in the coil and creates a magnetic field around both coils. The bulb connected to the second coil will not light.

There is no electrical current in the second coil. If the switch is opened, the magnetic field around the first coil will collapse. This collapsing magnetic field causes current to flow in the second coil. The light connected to the second coil will light (Figure 12-9). It will only go on for a fraction of a

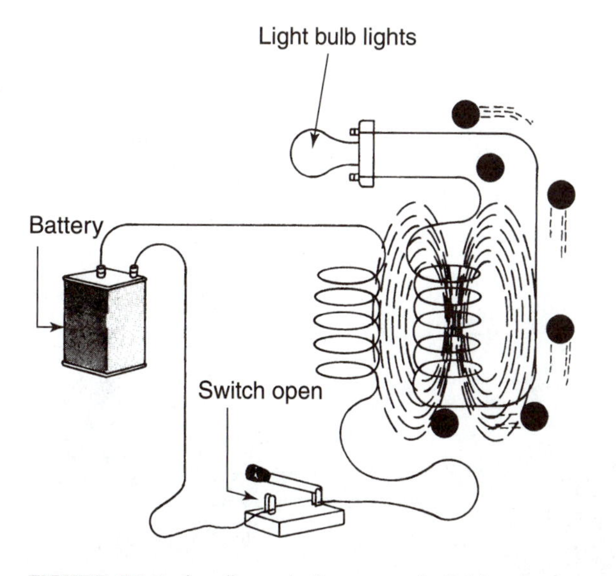

FIGURE 12-9 A collapse in the magnetic field around one coil causes current flow in the other coil.

second. If the first coil is charged, then collapsed again, the light will light again.

ARMATURE AND COIL

The magneto uses magnets, an armature, and coils of wire to create electricity through induction. An **armature** is a set of thin, soft iron strips used to make a path for magnetism. The thin iron strips, called *laminations,* are riveted together to form the armature.

There are two- and three-leg armatures. An armature leg is a separate laminated iron path for magnetism. A three-leg armature is a magneto armature with three paths for magnetism (Figure 12-10). A two-leg armature is a magneto armature with two paths for magnetism (Figure 12-11).

The armature works with the magneto coil to create high voltage electricity. The **magneto coil** is a set of two separate, fine copper wire coils in an insulated housing (Figure 12-12). The coils are wrapped around the middle of the armature. The coils and armature become one part.

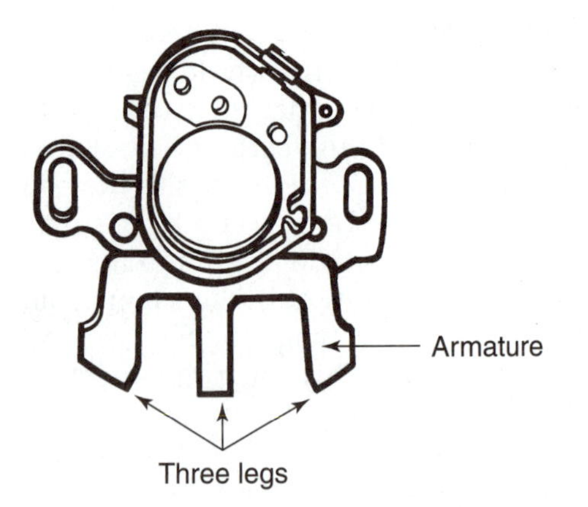

FIGURE 12-10 An armature with three legs. (Provided courtesy of Tecumseh Products Company.)

FIGURE 12-11 An armature with two legs.

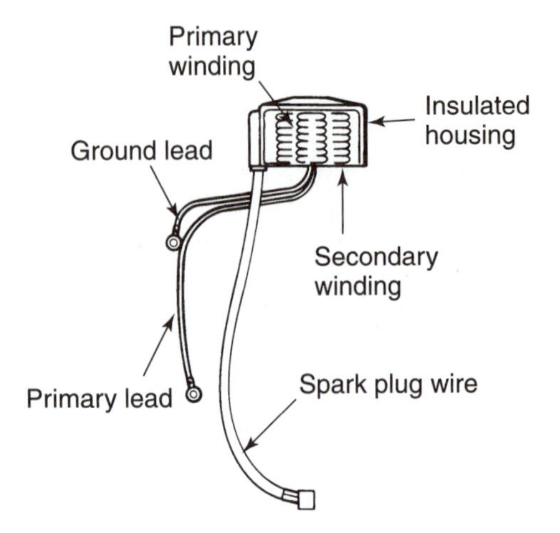

FIGURE 12-12 The magneto coil has a primary and secondary winding inside an insulated housing. (Provided courtesy of Tecumseh Products Company.)

One of the magneto coils is called the primary winding. The other is called the secondary winding. The **primary winding** is a magneto coil that is influenced by the magneto magnets to develop low voltage. The primary winding uses slightly heavier wire than the secondary. It is wound about 150 times (turns). It is usually wound on the inside of the secondary winding. The primary winding is made for low voltage (200–300 volts). One end of the primary winding, called the *ground lead,* comes out of the housing. The other end, called the *primary lead,* comes out of the housing.

The **secondary winding** is the magneto coil that develops high voltage through induction. The secondary winding is wrapped over the primary winding. It is made from very thin wire with about 10,000 turns. The secondary winding is made to create a high voltage (10,000–25,000 volts). This high voltage is needed to jump the spark plug gap in the cylinder. One end of the secondary winding is grounded inside the coil. The other end comes out of the coil housing. When outside the coil, it is called the *spark plug wire.* It may also be called the *secondary lead* or *high-tension lead.*

The armature and coil are located next to the flywheel (Figure 12-13) so that they can be influenced by the magnets. If the magnets are on the outside of the flywheel, the coil and armature assembly will be mounted above the flywheel. If the magnets are on the inside the flywheel, the coil and armature assembly will be mounted under the flywheel. These are called either *internal* or *external*

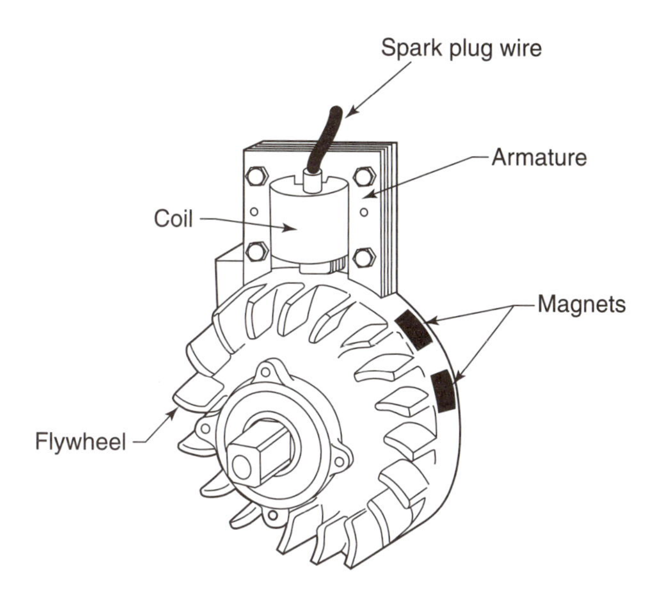

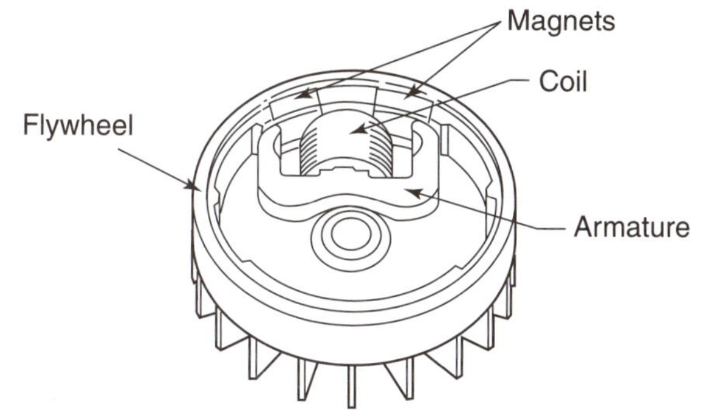

B. Armature Under the Flywheel

FIGURE 12-13 Magneto armatures are mounted on top or under the flywheel.

coils. The operation of the internal and external coils is the same.

MAGNET AND COIL OPERATION

The flywheel magnets work with the primary winding to develop low voltage electricity (Figure 12-14). The magnets go around as the flywheel turns. At some point in the rotation, the magnet's north pole lines up with the center leg of the armature. Magnetic lines of force from the magnet go through the armature. The magnetism goes up the armature center leg. Then it goes down the left leg. It goes into the magnet's south pole. The south pole is lined up under the left leg. A magnetic field builds up around the primary winding.

As the flywheel turns some more, the magnet's north pole lines up under the right leg of the ar-

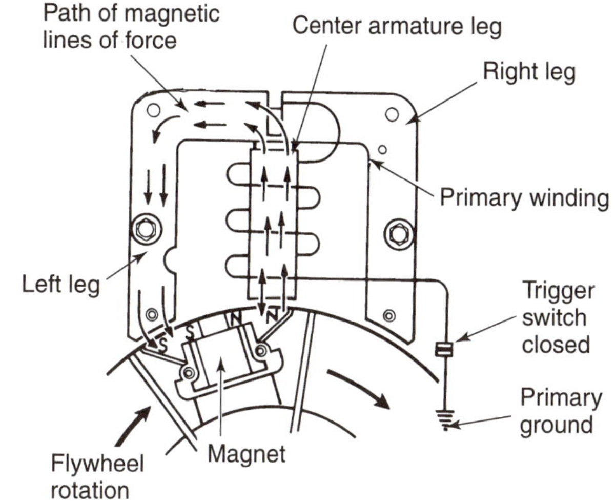

FIGURE 12-14 Magnetic flow when the magnets are under the left and center armature leg. (Provided courtesy of Tecumseh Products Company.)

mature (Figure 12-15). The magnetic lines of force move up the right leg, then move down the center leg to the magnet's south pole. The flow of magnetism through the center leg of the armature has reversed direction. The reversing magnetic field causes a small voltage in the coil primary winding.

A very small voltage is developed in the primary winding, but it is not enough to get the air-fuel mixture in the cylinder burning. Induction is used to make a high voltage in the secondary winding. The coil primary winding is connected to a mechanical or electronic trigger switch. The trigger switch is closed while the voltage is being developed in the

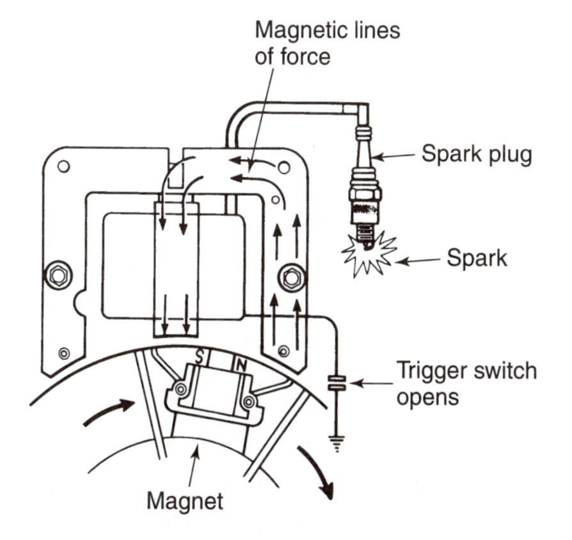

FIGURE 12-15 Magnetic flow reverses direction when the magnets line up under the center and right armature leg. (Provided courtesy of Tecumseh Products Company.)

coil primary. The closed switch makes a complete circuit for the coil primary winding.

The small primary winding voltage makes a strong magnetic field around the secondary winding. This magnetic field must be collapsed quickly to create a voltage in the secondary winding. The magnetic field is collapsed by opening the trigger switch. The open switch takes away the ground path for current flow in the primary winding. The collapsing lines of magnetic force cut across the many turns of wire in the secondary winding, making a high voltage current in the secondary winding. The high voltage current flows out of the coil into the spark plug wire. From here it goes to the spark plug to ignite the air-fuel mixture.

BREAKER POINT TRIGGERED MAGNETO SYSTEMS

The magneto needs a switch to open (trigger) the coil primary winding circuit. An electronic trigger switch is used on newer engines. Older engines (those usually made before 1985) use a mechanical trigger switch. **Breaker points** are mechanical trigger switches that open and close the primary winding circuit. Breaker points are sometimes called *contact points or points.*

Breaker Point Operation

The breaker points are often mounted under the flywheel, which protects them from dirt and water. The breaker points are two small, round, metal contacts (Figure 12-16). One of the contacts is on a moving arm, which is mounted on a pivot. This is called the *moving point.* The other contact, called the *stationary point,* does not move.

One style of breaker points uses a small rod called a *plunger* to move the arm. The breaker points are opened and closed by the crankshaft.

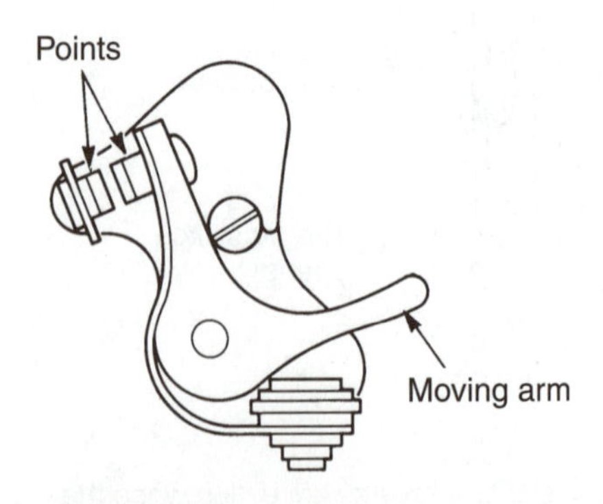

FIGURE 12-16 The breaker points are a mechanical switch that opens and closes the primary circuit.

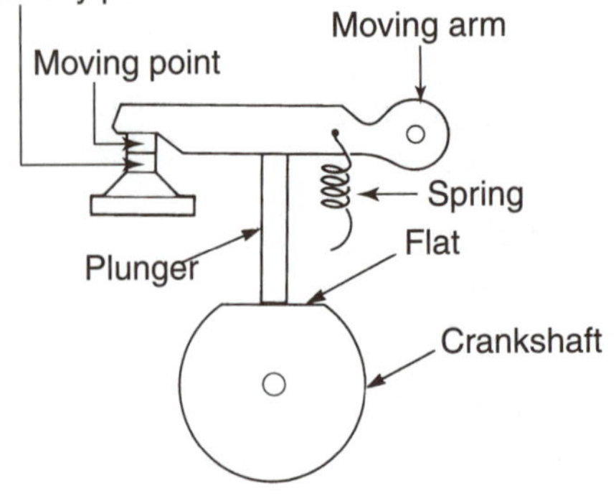

FIGURE 12-17 Plunger-operated breaker points in closed position.

The plunger rides on the crankshaft. The crankshaft has a small flat spot to open the points. The breaker points are closed when the plunger is on the flat spot (Figure 12-17). A small spring helps hold the points closed.

As the crankshaft turns, the flat spot rotates away from the plunger. The larger diameter of the crankshaft pushes up on the plunger (Figure 12-18). The plunger pushes up on the moving point arm. The breaker points open and break the primary circuit.

Many engines have breaker points that are opened by a cam lobe (Figure 12-19). This system also uses a stationary and movable breaker point. The cam lobe is driven by the crankshaft through a key. The movable breaker point is on a moving arm and pivot. The pivot is made from an insulation material. It is in contact with the cam.

The points are closed when the low part of the cam is in contact with the movable arm. The points are held closed by a flat metal spring. The cam turns with the crankshaft. The high point of the cam lobe pushes on the point moving arm. This opens the breaker points (Figure 12-20).

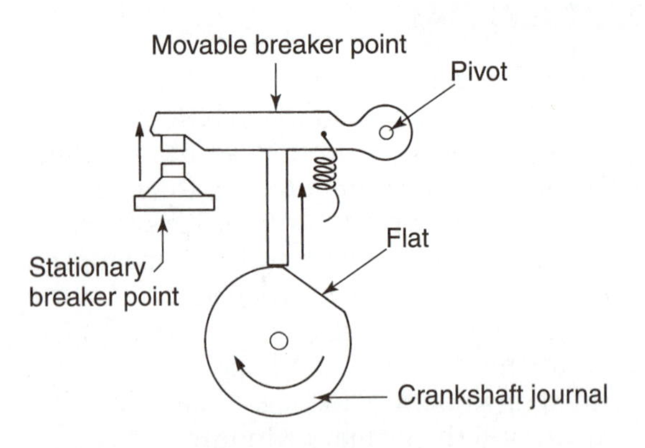

FIGURE 12-18 Rotation of the crankshaft opens the plunger-operated breaker points.

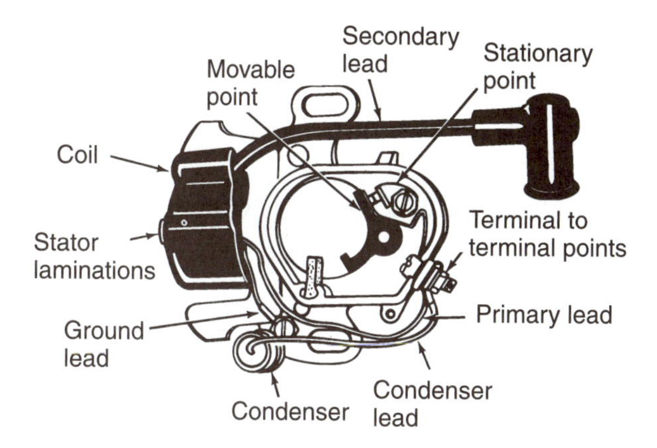

FIGURE 12-19 Parts of a pre-1985 cam-operated breaker point system. (Provided courtesy of Tecumseh Products Company.)

Breaker Point Triggered Coil Operation

The breaker points trigger the coil primary circuit. The flywheel magnets make a small voltage in the primary winding. One end of the primary winding goes to the armature. The other end is connected electrically to the moving point arm. Current can flow through the primary winding and into the moving arm. When the breaker points are closed, current flows across from the moving point to the stationary point (Figure 12-21). Current flows from the stationary point to ground, completing the circuit. A magnetic field builds up in the magneto coil.

When the flywheel rotates further, the breaker points open. Current cannot flow through the open breaker points (Figure 12-22). The electricity flowing in the primary winding stops quickly. A high voltage develops in the secondary winding. The voltage may be as high as 25,000 volts. The high voltage flows out of the secondary winding through the spark plug wire to the spark plug.

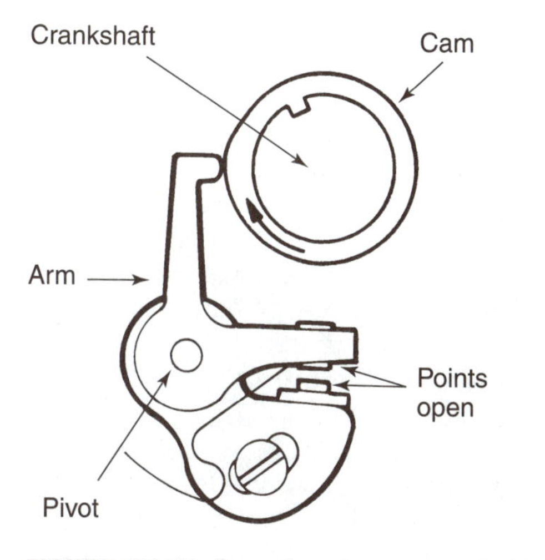

FIGURE 12-20 Operation of cam-operated breaker points.

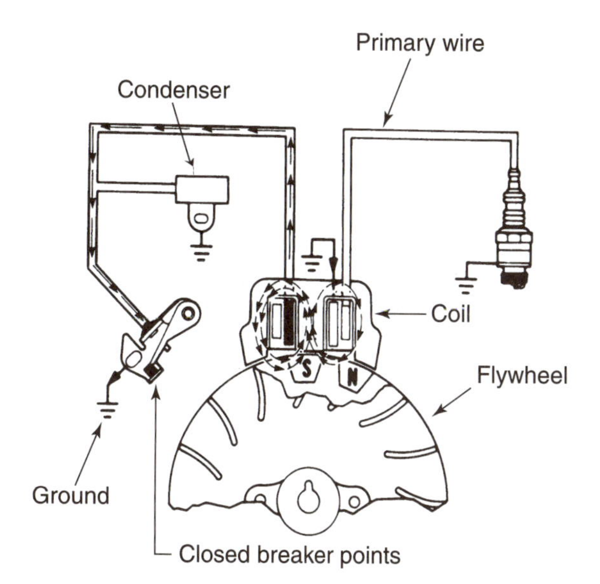

FIGURE 12-21 Current flow through the primary circuit when the breaker points are closed. (Courtesy of McCulloch Corporation.)

Condenser

Making a high voltage secondary winding depends on a rapid collapse of the primary winding magnetic field. Current must stop flowing the instant the breaker points open. If it does not, no high voltage will develop. As the breaker points open, electricity will try to jump (arc) across the open breaker points. If this happens, there will be no magnetic flow in the coil. No high voltage will develop in the secondary winding. Arcing across the

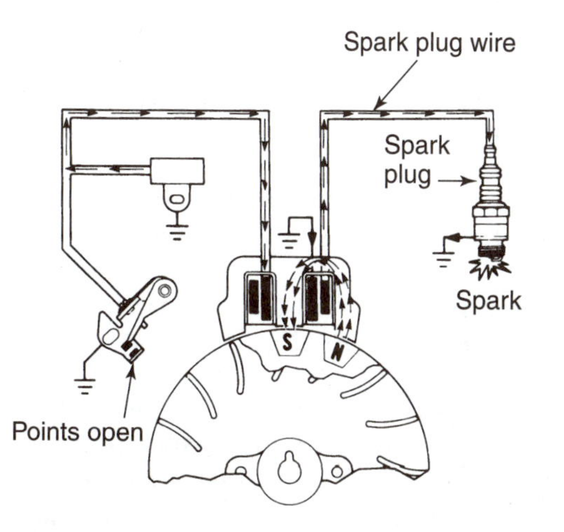

FIGURE 12-22 Current flow through the secondary circuit when the breaker points open. (Courtesy of McCulloch Corporation.)

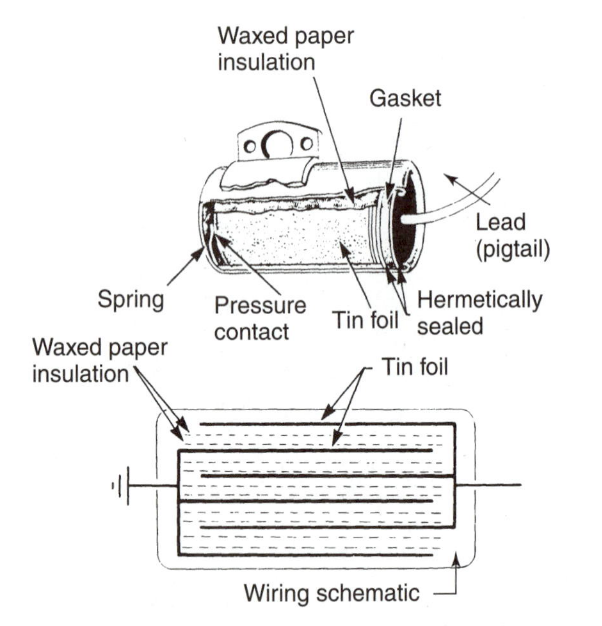

FIGURE 12-23 A condenser prevents arcing across the open breaker points.

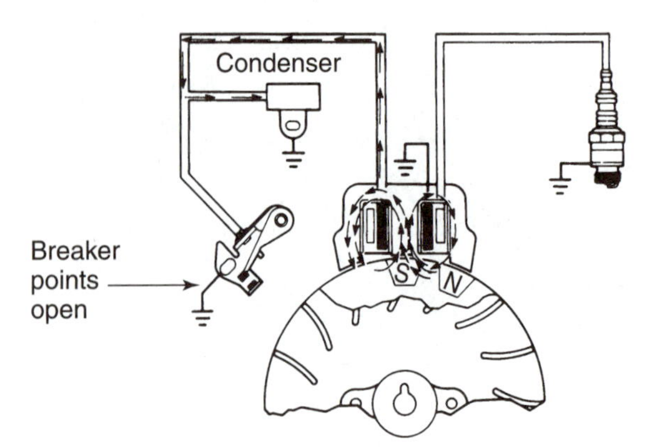

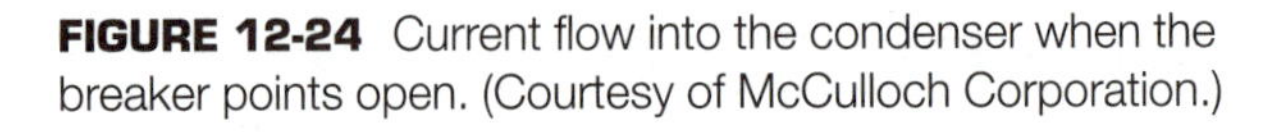

FIGURE 12-24 Current flow into the condenser when the breaker points open. (Courtesy of McCulloch Corporation.)

breaker points will cause the breaker points' contact surfaces to be burned and destroyed.

A **condenser** (Figure 12-23) is an electrical part that stores electrical current to prevent it from jumping across the open breaker points. It is also called a *capacitor*. The condenser is shaped like a small can with a small insulated wire (pig tail) that comes out of one end. The wire is connected to the moving point. The condenser is mounted on a bracket close to the breaker points.

The condenser works like a tiny storage tank. It has two long sheets of conductor foil inside, which are separated by several sheets of insulated paper. When the breaker points open, current wants to jump across the opening. Instead, it takes the path of least resistance. It goes through the condenser wire and flows into the conductor foil sheets (Figure 12-24). The condenser is in a charged condition. It keeps this charge until the breaker points close. Current then flows out of the condenser to ground. This flow has no effect on breaker point operation.

Some engines combine the condenser and the stationary breaker point into one part (Figure 12-25). The stationary breaker point is part of the condenser end cap. The primary circuit wire is connected through a small hole and is held in place with a small spring. The operation is the same as with the separate condenser.

CASE STUDY

Breaker point ignition systems are a common cause of engine performance problems. They are usually the source of a "no spark" problem on older engines. Most technicians try to convince customers with these older engines to retrofit the ignitions with solid state ignitions. Sometimes the customer or parts availability prevent this conversion.

A technician had just finished replacing the breaker points on an older edger engine. The engine had run when it came into the shop. It simply needed a new spark plug and breaker points as part of a periodic maintenance. With these jobs done, the technician pulled on the rope to start the engine. The engine spun but did not make even a pop like it wanted to start. The technician installed a spark checker on the engine and it showed no spark.

The technician removed the breaker points cover and inspected the breaker points. The wires seemed to be correctly installed. He rotated the engine and watched the breaker points. They opened and closed fine. He reached for his feeler gauge to recheck the points spacing again. As he held the feeler gauge he noticed some oil on the feeler gauge blade. He thought perhaps some of this oil may have been transferred to the new breaker points when he had adjusted them. He sprayed some electrical parts cleaner on a clean paper towel and wiped off the breaker points surface. He assembled the ignition. The engine started on the first rope pull.

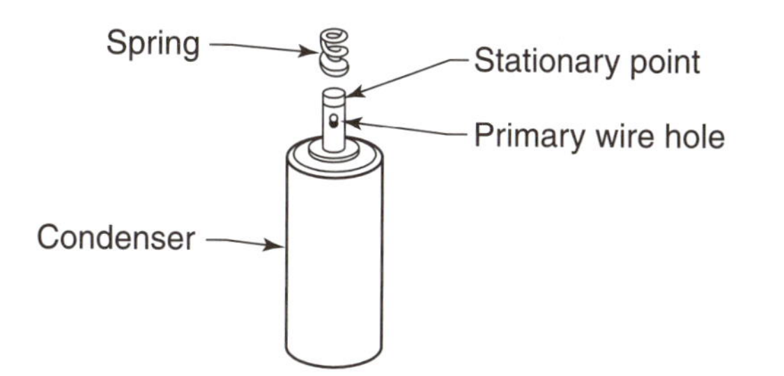

FIGURE 12-25 A condenser combines with a stationary breaker point.

ELECTRONICALLY TRIGGERED MAGNETO SYSTEMS

There are problems with a breaker point-triggered primary winding. The breaker point contact surfaces wear and burn from constant current flow. At some point, the burned breaker points resist current flow. The engine is hard to start and does not run well.

The plunger or movable breaker point cam surface also wears. This can change the breaker point spacing (gap). The engine will begin to run poorly. Water and dirt can get into the breaker point area and prevent proper operation. The breaker points must be changed regularly to prevent these problems.

Electronically triggered magneto systems use solid state electronics to trigger the coil primary winding. The reliability problems with mechanical breaker points are avoided because there are no moving parts. The electronic parts can be sealed to avoid problems from dirt or water. The electronic system is maintenance free because it does not re-

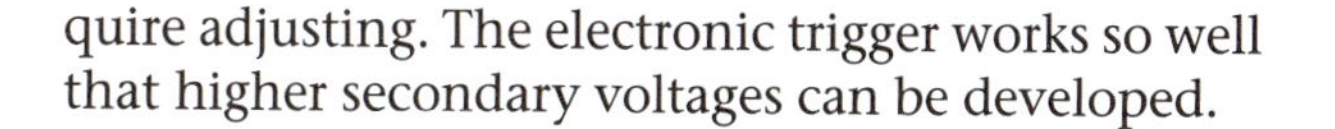

quire adjusting. The electronic trigger works so well that higher secondary voltages can be developed.

Transistor Operation

Many electronically triggered magneto systems use a transistor to trigger the coil primary winding. A **transistor** (Figure 12-26) is a solid state electrical device used to open and close an electrical circuit. A transistor can be turned on or off to open or close a circuit.

A transistor is a chemical device that has electrical functions. Transistors range in size from about the size of a dime to smaller than the eye can see. The typical ignition system transistor has a metal housing with two electrical connectors. Inside the transistor are semiconductor materials. A **semiconductor** is a material that allows current flow like a conductor under certain conditions but acts like an insulator and stops current flow under other conditions.

A symbol is used for a transistor when it is shown on electrical diagrams (schematic). The symbol shows its inside circuits and its current flow direction (Figure 12-27). The transistor has three electrical circuits inside called the emitter (E), collector (C), and base (B).

The main current goes through a transistor circuit from the emitter to the collector (Figure 12-28). Current cannot pass through this circuit unless a small amount of current is allowed to pass through the emitter to base circuit. A small amount of current is used to control a large amount of current.

A transistor can be controlled electrically. The voltage developed by the rotating magneto magnets can be used to control the transistor. The correct voltage polarity must be applied to the emitter to base circuit. When it is, current will flow in this circuit. At the same time, current will flow in the emitter to collector circuit. When the polarity of

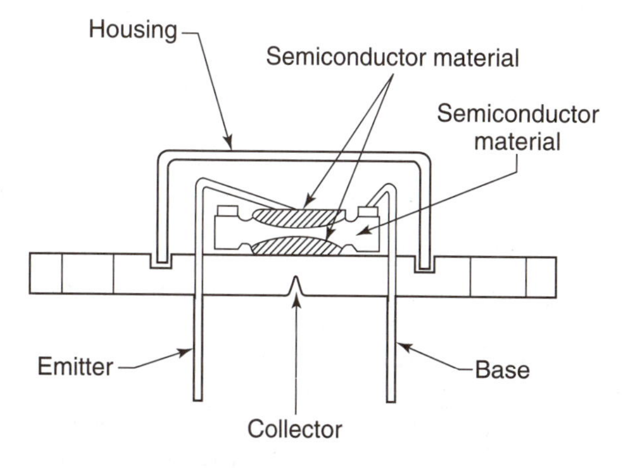

FIGURE 12-26 A transistor is a solid state electrical device used to open and close an electrical circuit.

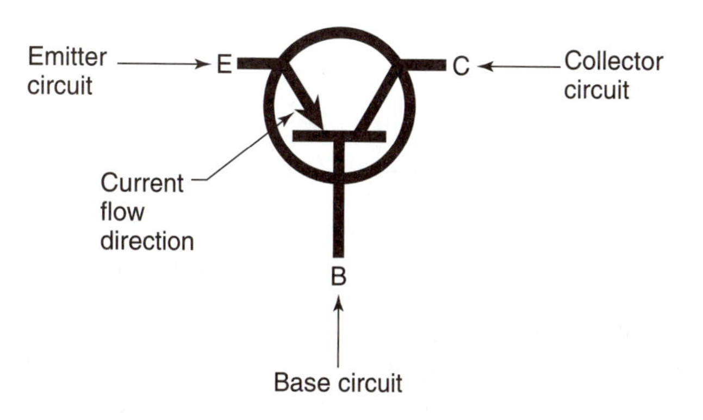

FIGURE 12-27 A symbol for a transistor shows current flow direction.

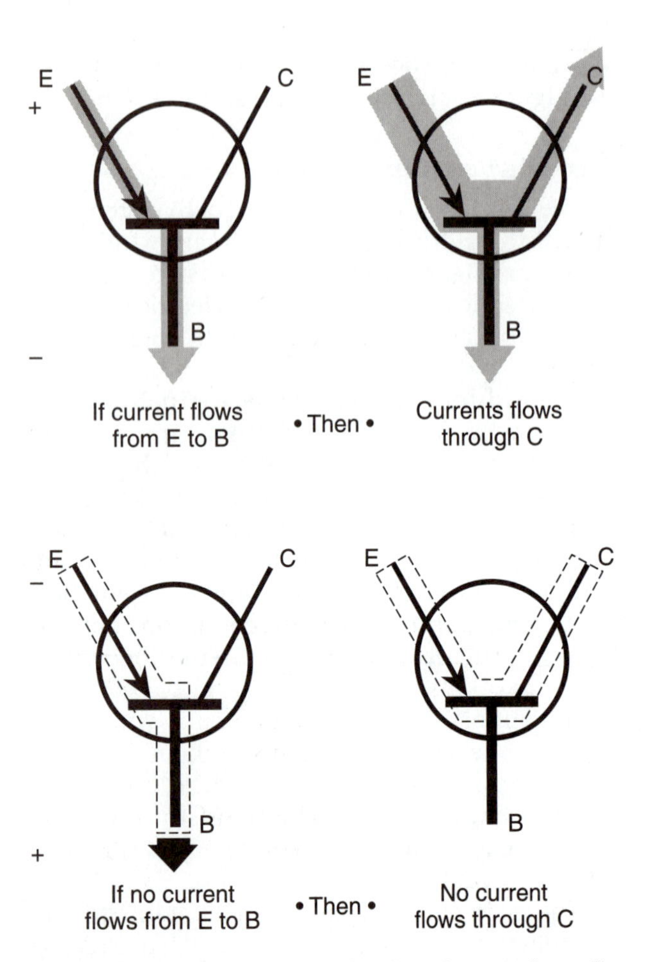

FIGURE 12-28 Reversing the direction of current flow will turn the transistor circuit on and off.

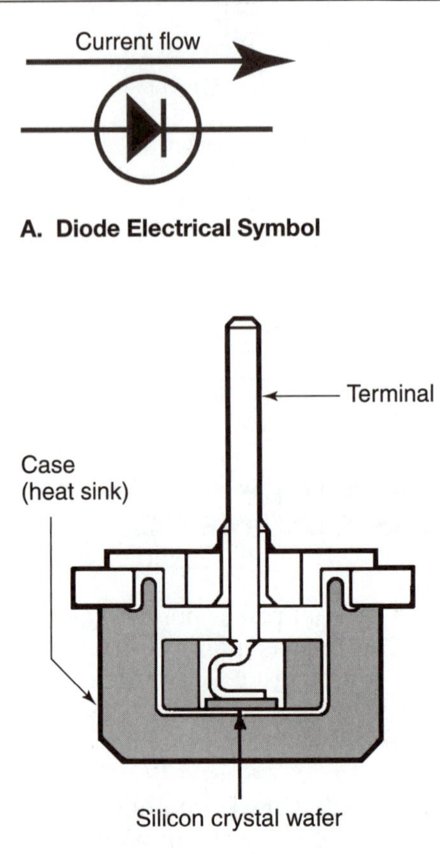

FIGURE 12-29 A diode allows current flow in one direction and stops current flow in the opposite direction. (*A:* Courtesy of General Motors Corporation, Service Technology Group. *B:* Courtesy of Ford Global Technologies, Inc.)

the voltage is reversed, the emitter to base circuit is turned off. No current can flow in the emitter to collector circuit.

Diode Operation

A diode is often used in electronically triggered magneto systems. A **diode** is a solid state electrical device that allows current flow in one direction and stops current flow in the opposite direction. A diode symbol has an arrow that shows current flow direction (Figure 12-29A).

The diode is a small quarter size or smaller part (Figure 12-29B). It has a metal case. The case is called a heat sink because it takes electrical heat away from the diode. The diode has one electrical connector (terminal). The terminal is connected to a silicon wafer (semiconductor material) inside the diode. The silicon wafer controls current flow. Current flows through the diode when positive (+) voltage is connected to the diode terminal and a negative (–) voltage is connected to the diode case. With these electrical connections, the wafer will not allow current to flow in the opposite direction. Current cannot flow from the case to the terminal.

TRANSISTOR-TRIGGERED MAGNETO

A transistor-triggered magneto system has some of the same parts as a breaker point system. There are magnets in the flywheel that create a voltage in a primary winding. The coil and armature are like that used on breaker point systems. The main difference is the addition of a part called the module (Figure 12-30). An **ignition module** is an electronic part that contains the transistor and circuits used to trigger the primary winding. The module is often housed and sealed with epoxy in the same housing as the magneto coil.

When the flywheel magnets rotate past the armature, a current is created in the coil primary winding. The current develops a magnetic field in the secondary winding. The module transistor cir-

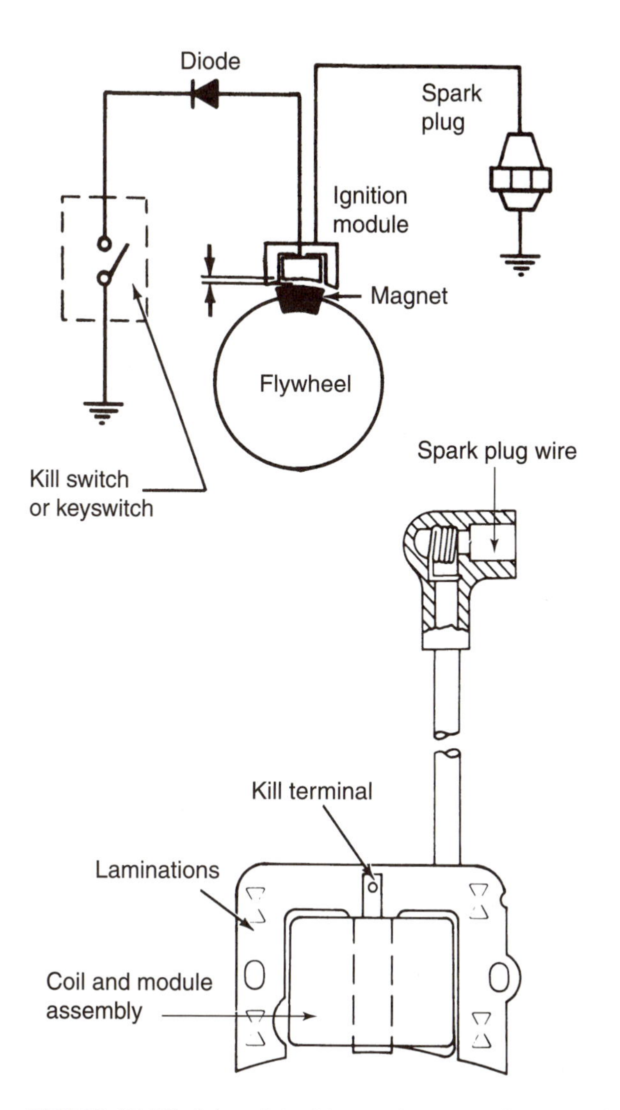

FIGURE 12-30 A transistor-triggered magneto has a module combined with the coil. (Courtesy of Kohler Co.)

cuit opens and closes the primary circuit. When open, a high voltage current develops in the coil secondary winding. The high voltage goes out the spark plug wire to the spark plug. A stop (kill) wire is typically connected to the module to stop the ignition for shut down.

CAPACITOR DISCHARGE IGNITION

The **capacitor discharge ignition (CDI)** is another type of electronically triggered ignition system It uses energy stored and discharged from a capacitor to create a high voltage. The capacitor discharge parts are housed in a sealed module inside the magneto coil assembly (Figure 12-31).

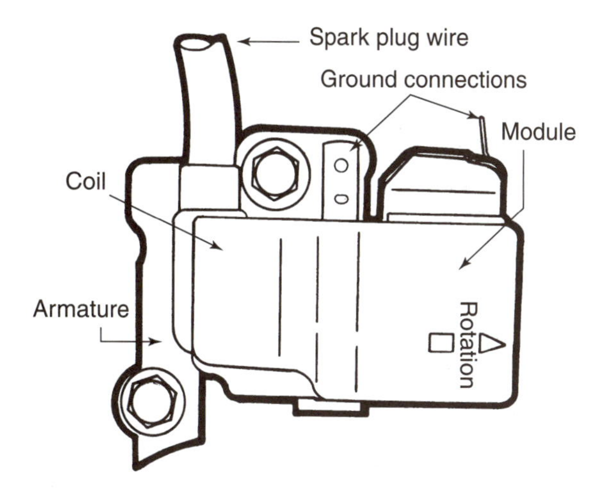

FIGURE 12-31 The capacitor discharge magneto has a module located in the magneto coil. (Provided courtesy of Tecumseh Products Company.)

A capacitor (condenser) has two foils (plates) separated by an insulator. When current enters the capacitor, electrons build up on one foil. Their negative charge repels a like number of electrons on the other foil. In this condition, the capacitor is charged. Energy is stored in the charged capacitor. When the current flow is stopped, the energy stays in the capacitor. When a conductor is connected across the two plates will it discharge, or regain electron balance. A small capacitor can store a large charge and provide a large discharge.

A capacitor discharge ignition system uses the storing and discharge to develop high voltage. The output from a charged capacitor goes into the magneto coil primary. This builds a strong magnetic field in the secondary winding. The magnetic field is collapsed and creates a high voltage in the secondary winding. The capacitor is then electrically disconnected from the coil and recharged. The discharge into the primary can occur again for the next firing cycle.

Flywheel magnets are used just as in other systems. As they move under the armature, a small current is developed in a charge coil (Figure 12-32). The pulse transformer changes the low voltage direct current into a higher voltage alternating current. The alternating current is changed back to direct current by a diode. The current is used to build up a charge in the energy storage capacitor. The output of the pulse transformer charges the capacitor to approximately 400 volts.

The 400 volts stored in the capacitor must be discharged at just the right time for ignition. As the magnets continue rotating, they go past a trigger coil. The flywheel magnets develop a small signal

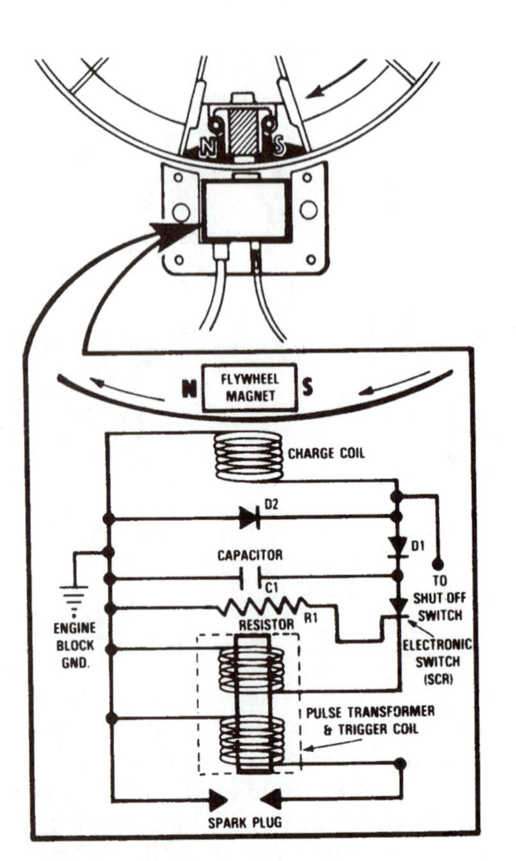

FIGURE 12-32 Wiring diagram for a capacitor discharge (CD) ignition system. (Provided courtesy of Tecumseh Products Company.)

current in a trigger coil. The signal from the trigger coil goes to a switching transistor, called a *silicon-controlled rectifier* (SCR). The signal current is used to close an electronic switch (SCR). It then switches the circuit to discharge the capacitor into the coil primary. The high voltage in the magneto coil goes to the spark plug. The SCR then switches the circuitry again. The capacitor is allowed to charge and get ready for the next discharge.

RETROFIT SOLID STATE IGNITIONS

A **retrofit part** is a part manufactured by an engine maker used to update a system manufactured earlier. The Magnetron®[1] system developed by Briggs & Stratton is an example. The Magnetron is an OEM system for newer engines and a retrofit system for older engines. The retrofit Magnetron comes as a kit for installation on older engines that had breaker point ignition. The primary wire leading to the breaker points is cut and spliced to the Magnetron. The kit contains parts and instructions for installation.

[1]Magnetron® is a registered trademark of Briggs & Stratton Co.

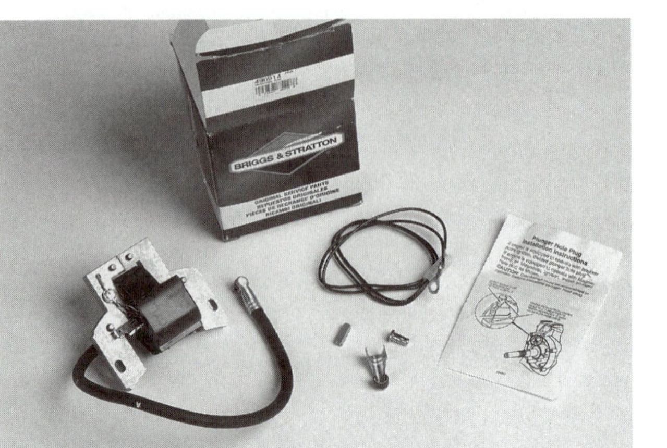

FIGURE 12-33 An installation kit to retrofit breaker points and condenser engines.

The Magnetron has a trigger coil and an integrated circuit chip. The integrated circuit chip contains the solid state circuit to trigger the coil primary winding. The solid state circuit has two transistors and a diode connected to the trigger coil that work together with the flywheel magnets to develop and collapse a magnetic field in the primary winding. A high voltage is developed in the coil secondary just as in other systems. These parts are sealed in a nylon case.

Solid state ignition retrofit kits are also available with the trigger coil built into the magneto coil and armature assembly. These systems are made to replace breaker points systems used on older engines. Systems are available for both two- and four-stroke engines. The old magneto coil, armature, and breaker points are removed. The retrofit coil and armature with a solid state trigger are installed in their place (Figure 12-33).

TWO-CYLINDER IGNITION SYSTEMS

A two-cylinder engine has an ignition system that can fire two spark plugs (Figure 12-34). All the systems use a separate spark plug wire and spark plug for each system. Most systems use the same magnets, armature, coil, and solid state module used on the single cylinder. The magneto makes a spark on each crankshaft rotation. The spark is sent through both spark plug wires. Both spark plugs get the spark at the same time. One of the cylinders is ready for a power stroke on each engine crankshaft revolution. The spark causes ignition in this cylinder.

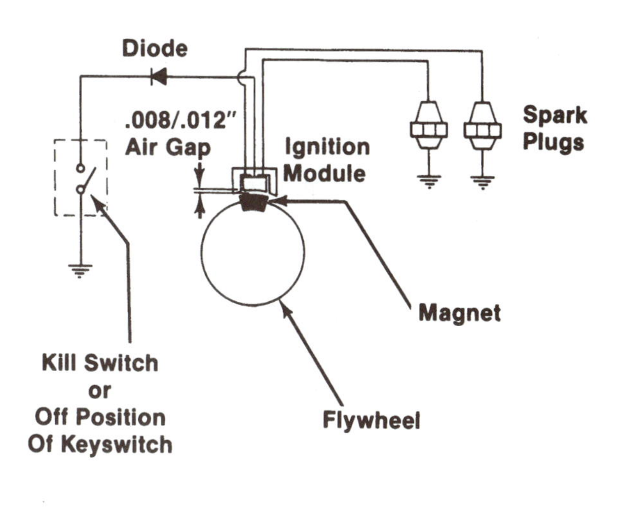

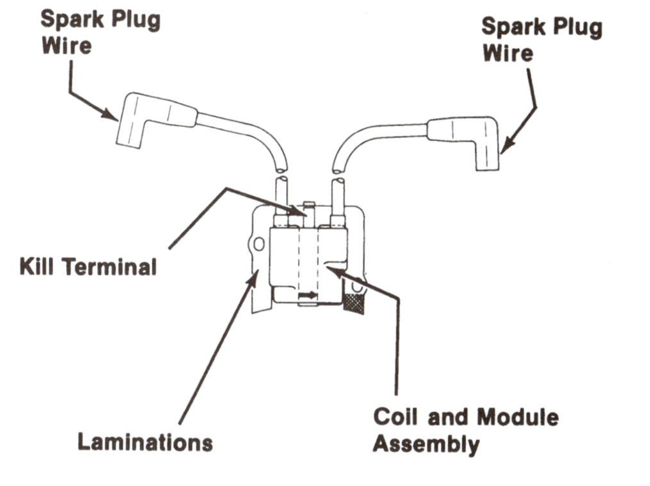

FIGURE 12-34 An ignition system for a twin cylinder engine. (Courtesy of Kohler Co.)

The other cylinder will be on exhaust stroke. The spark in this cylinder will not have any affect.

BATTERY IGNITION

The battery ignition is used on some larger engines with 10 or more horsepower. Higher voltages are possible with a battery ignition than a magneto system. The battery can develop a stronger voltage than that generated by magnets and armature.

The battery ignition uses a storage battery instead of a magnet and armature to develop primary current. Ignition primary current flows from the battery to an ignition coil (Figure 12-35). The battery ignition coil has a primary and secondary winding like a magneto coil. There is a lamination in the center for the magnetic field. The primary winding is connected at each end to one of two primary terminals. The secondary winding is connected to the large terminal in the center of the coil.

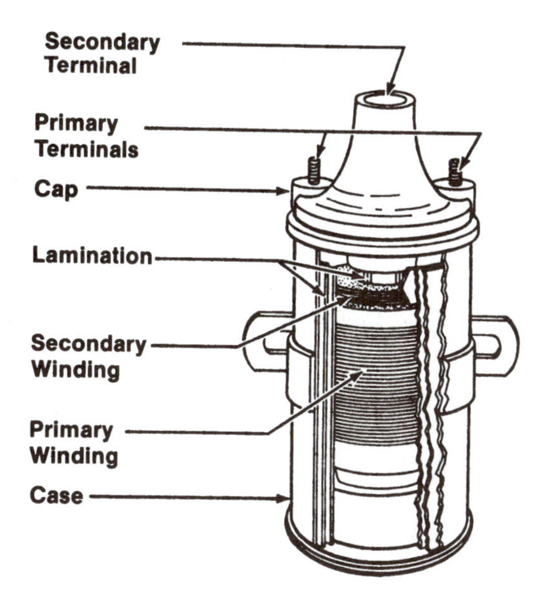

FIGURE 12-35 An ignition coil used on a battery ignition system. (Courtesy of Kohler Co.)

A magnetic field is developed when primary current flows through the primary coil. The breaker point or solid state trigger collapses the magnetic field. High voltage is created in the secondary winding. High voltage flows out of the coil center terminal to the spark plug (Figure 12-36).

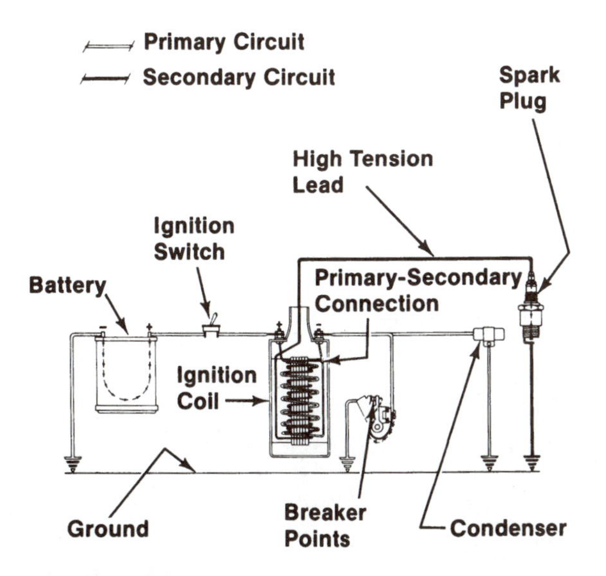

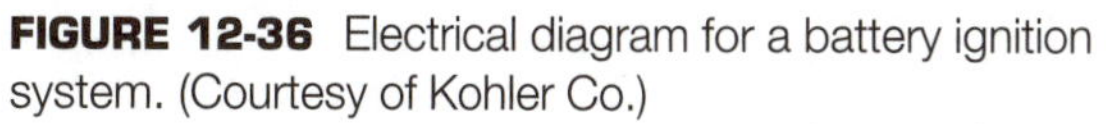

FIGURE 12-36 Electrical diagram for a battery ignition system. (Courtesy of Kohler Co.)

REVIEW QUESTIONS

1. Describe and explain how the electron movement in atoms creates electricity.
2. Explain the difference between an insulator and a conductor.
3. Describe the elements needed for a complete electrical circuit.
4. Explain how induction is used to transfer electrical energy from one coil to another.
5. Explain how the coil primary and secondary winding work to induce a high voltage.
6. Describe how the magnets on the flywheel create a current in the coil primary winding.
7. List and describe the parts of a plunger-operated set of breaker points.
8. List and describe the parts of a cam-operated set of breaker points.
9. Explain the operation of a breaker point-triggered magneto coil system.
10. List the problems with breaker point types of ignition systems.
11. Explain the purpose and operation of an ignition condenser.
12. Describe the purpose and symbol for a transistor.
13. Describe the purpose and symbol for a diode.
14. Explain the operation of a transistor-triggered magneto coil system.
15. Explain the operation of a capacitor discharge magneto coil system.

DISCUSSION TOPICS AND ACTIVITIES

1. Disassemble the parts of a breaker point-triggered magneto system. Rotate the engine and observe breaker point operation. Trace the wiring and explain the current flow from primary to secondary circuit.
2. Disassemble the parts of a transistor-triggered magneto system. Identify the parts. Trace the current flow from the primary to the secondary circuit.
3. An engine is operating at 3,400 rpm. The breaker points are operated by a cam on the crankshaft. How many sparks per second does the ignition system develop?

CHAPTER 13

Ignition Systems—High-Voltage Components

OBJECTIVES

Upon completion and review of this chapter, you should be able to:

- Identify the parts and explain the operation of spark plug wires.
- Identify the parts and explain the operation of a spark plug.
- Explain the size system used to identify spark plugs.
- Describe the different electrode designs used on spark plugs.
- Explain the purpose of the codes used to identify spark plugs.
- Describe the purpose and operation of an ignition kill switch.
- Describe the purpose and operation of safety interlocks.

TERMS TO KNOW

Kill switch
Safety interlock
Spark plug
Spark plug firing end
Spark plug gap
Spark plug heat range
Spark plug reach
Spark plug thread diameter
Spark plug wire

INTRODUCTION

The ignition system high-voltage system is often called the *secondary circuit* or *high tension* circuit. The high-voltage system routes the high voltage developed in the magneto coil to the engine's combustion chamber to create combustion. The main high-voltage components are the spark plug wire and the spark plug. This system may also include a stop (kill) switch and safety interlocks.

SPARK PLUG WIRES

The **spark plug wire** (Figure 13-1) is a highly insulated cable used to route the high voltage from the coil secondary winding to the spark plug. The spark plug wire may also be called a *secondary wire, high voltage wire, spark plug lead, high tension wire,* or *ignition cable*. Spark plug wires must be able to handle high voltage without leakage. They must also be able to withstand water, oil, vibration, and abrasion.

The spark plug wire is connected to the magneto coil secondary winding. This connection is often soldered and sealed with epoxy when it is made. The wire has a center conductor covered with an insulator material. The conductor may be metal or a nonmetal such as carbon impregnated fiberglass. The insulator may be made from heat-resistant hypalon or silicone rubber. Some wires have layers of glass braid material between the insulator and conductor to make the wire stronger.

The spark plug wire has a metal spark plug terminal connector soldered or crimped on its end. The terminal connector fits on to the spark plug terminal, allowing current flow into the spark plug. A silicone rubber boot is often used over the terminal connector. The boot makes a moisture free seal between the connector and spark plug terminal.

SPARK PLUG

The **spark plug** (Figure 13-2) is an ignition system part that gets the high-voltage electricity from the

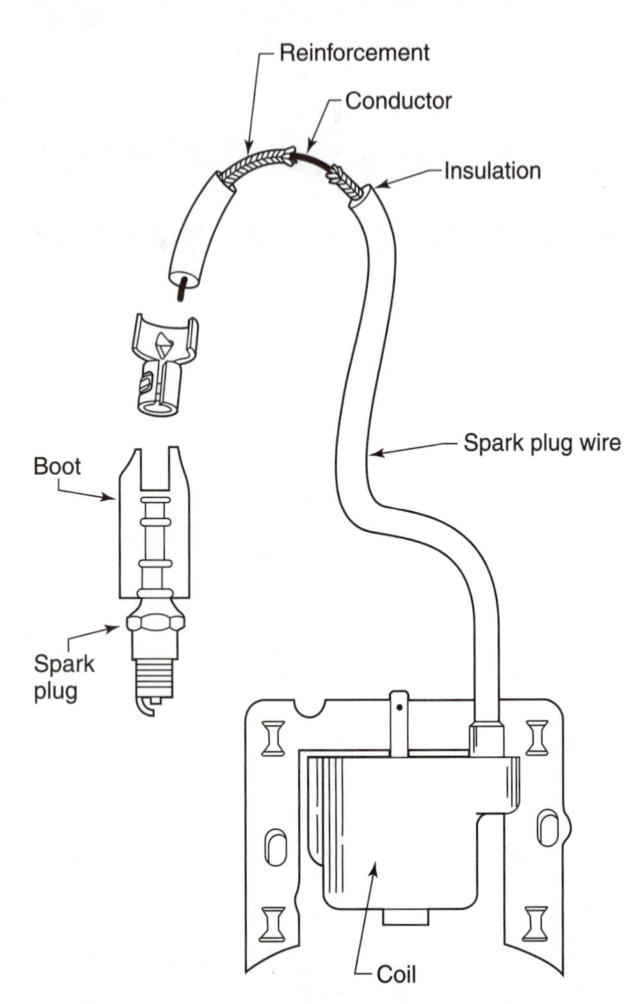

FIGURE 13-1 Parts of a spark plug wire.

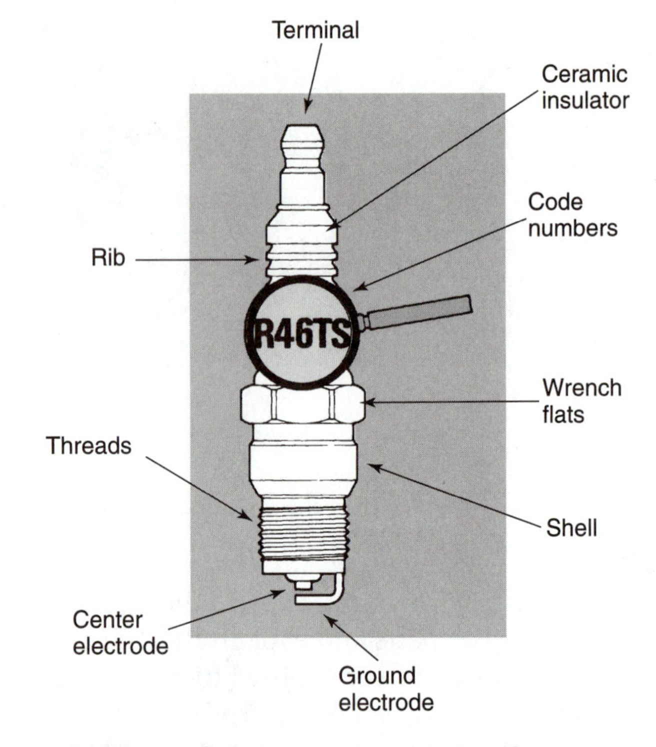

FIGURE 13-2 Outside parts of a spark plug. (Courtesy General Motors Corporation, Service Technology Group.)

magneto and creates a spark in the combustion chamber. The spark is used to ignite the air-fuel mixture. The spark plug must deliver this high-voltage spark under conditions of high temperature and high combustion chamber pressure.

Spark Plug Construction

The top part of the spark plug has a terminal. The spark plug cable is connected to the terminal. The terminal is connected to a thick wire that goes through the middle of the spark plug, called a *center electrode*. The center electrode makes a path for high-voltage current to the spark plug gap. The high-voltage current must not be allowed to escape or leak away. A ceramic insulator fits around the center electrode to prevent leakage. The ceramic insulator has ribs on its outside diameter. The ribs increase the distance between the terminal and the nearest ground. This helps eliminate current leakage (flashover). Flashover is a problem when the outside of the ceramic is dirty or wet.

The bottom of the spark plug has a metal housing called a *shell*. The shell has threads to thread the spark plug into the cylinder head. There are hex-shaped flats on the shell for spark plug sockets. The shell has a side (ground) electrode. When the spark plug is installed, the shell and side electrode have an electrical ground. The side electrode is a small distance away from the center electrode. The two electrodes form a gap. The **spark plug gap** is the space between the spark plug center and side (ground) electrode (Figure 13-3). The ignition voltage jumps the spark plug gap to create a spark for ignition.

Spark Plug Operation

High-voltage secondary current flows from the magneto coil through the spark plug wire. The current enters the spark plug at the terminal end. Current goes down the center electrode to the spark plug gap. The current overcomes the resistance of the air and fuel mixture. It jumps the gap to the side (ground) electrode. A spark is created as the current jumps the gap. The spark ignites the combustible mixture of air and fuel.

The spark jumps across the gap after it overcomes the resistance of the air-fuel mixture (Figure 13-4). The voltage required to jump the spark plug gap changes under different conditions. The wider

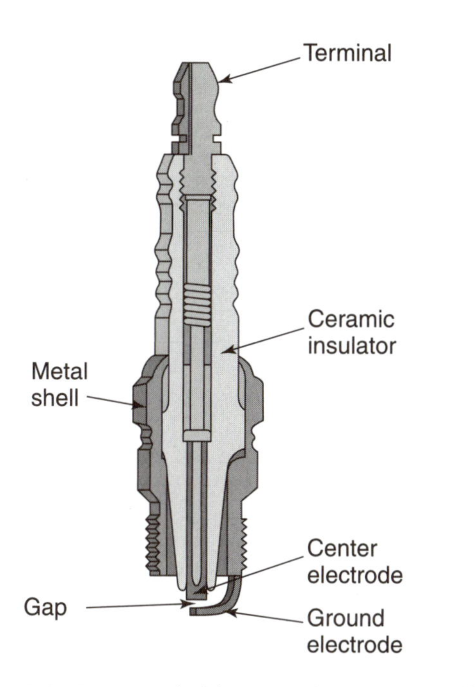

FIGURE 13-3 Inside parts of a spark plug. (Courtesy of Cooper Automotive/Champion Spark Plug.)

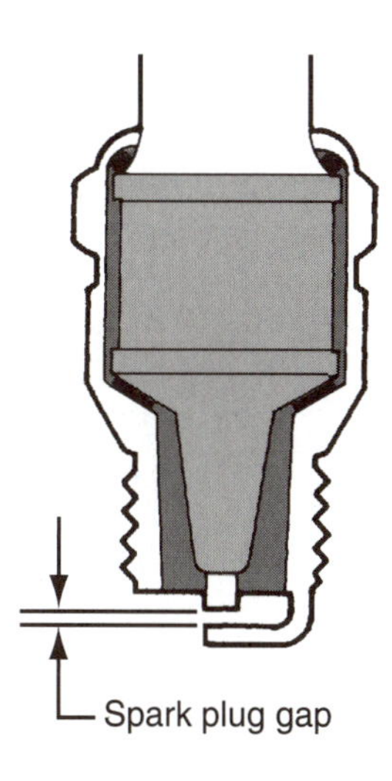

FIGURE 13-4 The high voltage jumps the gap to create a spark. (Courtesy General Motors Corporation, Service Technology Group.)

the spark plug gap, the higher the required voltage. Electrode condition is very important. Dirty or burned (eroded) electrodes require a high voltage. A lower voltage is required if the electrodes are clean and sharp. Also, higher voltages are required for higher compression pressures.

Spark Plug Sizes

To work properly, the spark plug must be the right size for the combustion chamber. Different sizes of

FIGURE 13-5 Spark plugs have different thread diameters.

spark plugs are required for different engine designs. **Spark plug thread diameter** is the diameter of the threaded shell measured in millimeters. Spark plugs are made with different thread diameters (Figure 13-5). Metric threads are used so the plugs can be used in both imported and U.S. made engines.

Spark plug reach is the length of the threaded section of a spark plug shell. There are several common reach dimensions manufactured for different engines (Figure 13-6). The thickness of the combustion chamber determines what reach is necessary. Overhead valve engines often use a longer reach than the L-head design.

SERVICE TIP: If an incorrect reach is used, there can be a number of problems. If the reach is too long, the firing end may be hit by the piston or valve. The firing end may overheat and break off if it is too long. The broken parts can damage the top of the piston or valve. If the reach is too short, the spark plug can misfire. Engine power may be decreased

Heat Range

The **spark plug firing end** is the part of the center electrode and insulator that fits in the combustion chamber. This end of the spark plug reaches temperatures that may exceed 2,000°F (776°C). Temperatures higher than 1,500°F (582°C) can overheat the firing end, causing the firing end to glow red hot. The glowing firing end in the combustion chamber will ignite the air fuel mixture at the wrong time and cause a loss of engine power.

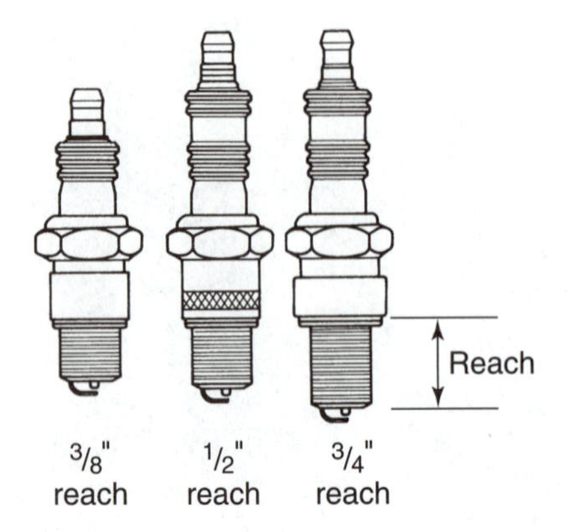

FIGURE 13-6 Spark plugs have different reaches. (Provided courtesy of Tecumseh Products Company.)

Firing end temperatures that are too low can also cause problems. The firing end must get hot enough to burn off oil and fuel deposits that try to form on the electrodes and insulator. A buildup of deposits fouls the spark plug. High voltage will not be able to jump the gap and the engine will misfire or stop running.

Spark plug firing end temperature is controlled by removing heat. The heat is removed through the engine's cylinder head. The path of heat flow is up the ceramic insulator to the metal shell. From here, it goes into the engine's cylinder head cooling fins and into the air.

Spark plugs are made to operate within a specific temperature range. Different spark plugs have different heat ranges. **Spark plug heat range** is a specification for the transfer of heat from the spark plug firing end. Heat range is determined by the distance the heat flows from the firing end up to the shell (Figure 13-7). If the path is a long one, the firing end will remain at a high temperature, and it is called a *hot spark plug*. If the path for heat flow is short, heat is removed more easily from the firing end. This type of spark plug is called a *cold spark plug*. Each engine maker specifies the heat range that will work best in their engines.

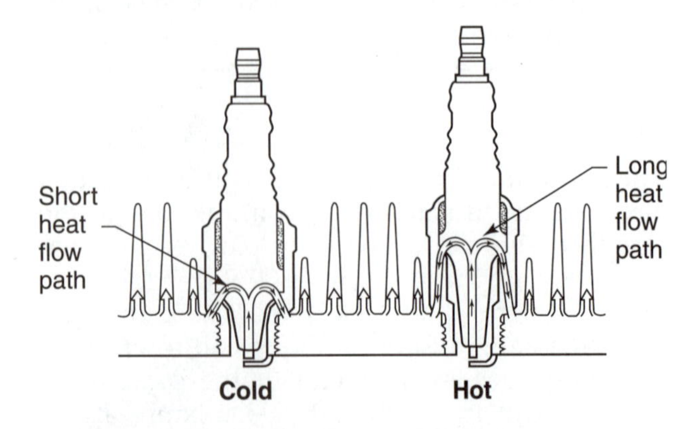

FIGURE 13-7 Cold and hot spark plug heat range.

SERVICE TIP: Sometimes a spark plug heat range that is hotter or cooler is installed for certain operating conditions. Engines that run at low speed for a long time may need a hotter heat range. A hotter plug is often used in a worn engine that is oil fouling the plugs. Cooler heat ranges are used for engines that run at high speed for long periods.

Electrode Type

Several different kinds of spark plug electrodes are used (Figure 13-8). The conventional electrode has a ground electrode that extends completely under the center electrode. This type is used in most four-stroke engines. Two-stroke engines have oil mixed with the gasoline. When the oil enters the combustion chamber, it can burn and stick on the spark

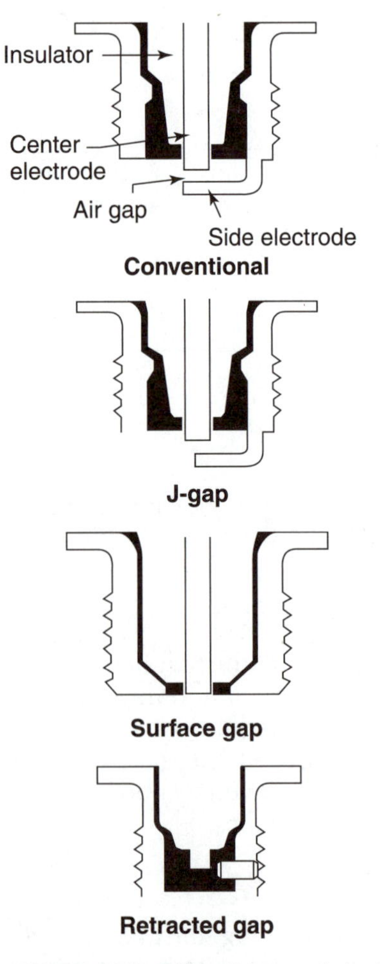

FIGURE 13-8 Different types of electrode gaps.

FIGURE 13-9 Spark plugs are matched to an engine by the code on the insulator.

plug electrodes. This oil buildup is called *fouling*. Fouling can prevent the spark from jumping the spark plug gap. Several different electrode designs are used on two-stroke spark plugs to prevent fouling. The J-gap, retracted gap, and surface gap electrode have less surface for oil buildup.

Spark Plug Codes

A spark plug must have the correct thread diameter, heat range, electrode type, and reach for an engine. Spark plug makers use a code system to identify these features on their spark plugs. The code is printed on the ceramic insulator of the spark plug. See Figure 13-9. A code chart for interpreting the code system is available from each spark plug maker.

CAUTION: Each spark plug manufacturer uses its own code system. A higher number may indicate a hotter heat range for one maker. It could be a colder heat range for another. Always use the correct code chart when selecting spark plugs for an engine.

CASE STUDY

A customer appeared at the local outdoor power equipment shop one Saturday morning with a 20-year-old reel type mower. The engine was an old cast-iron model that had clearly seen better days. The owner's complaint was that the engine was misfiring badly. This problem had been going on for some time. He said that the problem would go away when he installed a new spark plug but would come back soon after. The customer dug into his pocket and showed the shop owner a handful of spark plugs.

The shop owner checked out the old spark plugs. They were all fouled from oil burning in the combustion chamber. This was not unexpected given the age of the equipment. The shop owner removed the spark plug from the engine and checked it out. It was beginning to foul just like the others. The shop owner put his finger into the exhaust pipe on the engine and it came out black with oil.

The owner discussed the possibility of a new mower with the customer. The customer was very attached to the old mower and could not afford a new one. The shop owner had his technician look up the available heat ranges for spark plugs that would fit this engine. He selected a spark plug that was two steps hotter than the one normally installed in the engine. A hotter spark plug would do a better job of burning the oil off the plug so it would stay clean enough to spark. The shop owner saw the mower owner later that month and got the report that the problem had been solved.

KILL (STOP) SWITCH

A running engine is stopped by stopping current flow in the ignition system. A **kill switch** is a switch used to stop current flow in the ignition system and stop the engine. A magneto stop switch kills the ignition by providing a path to ground for either the primary or secondary current flow.

The simplest kill switches are those used on the ignition secondary circuit. One common type uses a metal lever next to the spark plug and spark plug wire (Figure 13-10). The lever is attached to the cylinder head or blower housing and is made so that the operator can push it in contact with the spark plug terminal and wire. When this contact is made, current going to the spark plug gets a new low-resistance path to ground. The current passes through the lever to ground instead of through the spark plug. The moment the spark stops at the spark plug, the engine stops running.

Many engines are used on equipment that requires a stop switch in a more convenient position. These engines use an electrical switch (Figure 13-11). Common switch types are two-position on/off

FIGURE 13-10 A tiller engine with spark plug terminal kill switch. (Courtesy of The Toro Company.)

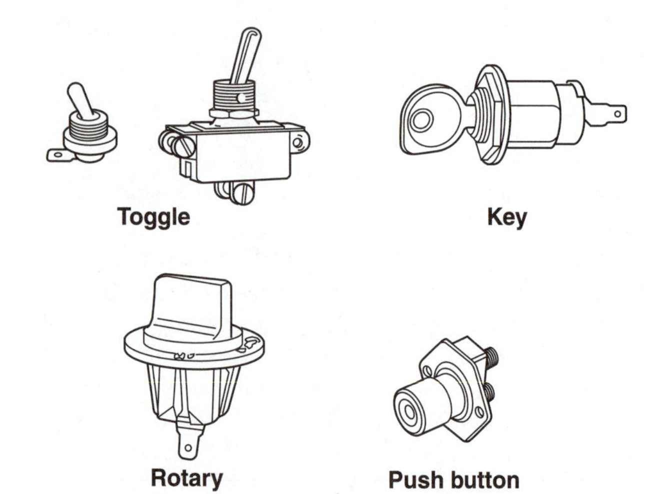

FIGURE 13-11 Different types of primary circuit kill switches.

toggle, multiple position rotary, push button, or key. Sometimes the kill switch is combined with an electric starter switch.

The kill switch controls a circuit that can ground the magneto coil primary circuit (Figure 13-12). The circuit works the same way for breaker points or electronic triggered ignitions. A ground wire is connected to the magneto coil primary winding. This wire is connected to a terminal. Another wire runs up to the stop switch from the terminal. When the switch is in the on or run position, the circuit is not grounded. The magneto works normally. When the switch is moved to the off or stop position, it provides a path to ground. The coil primary winding cannot build up

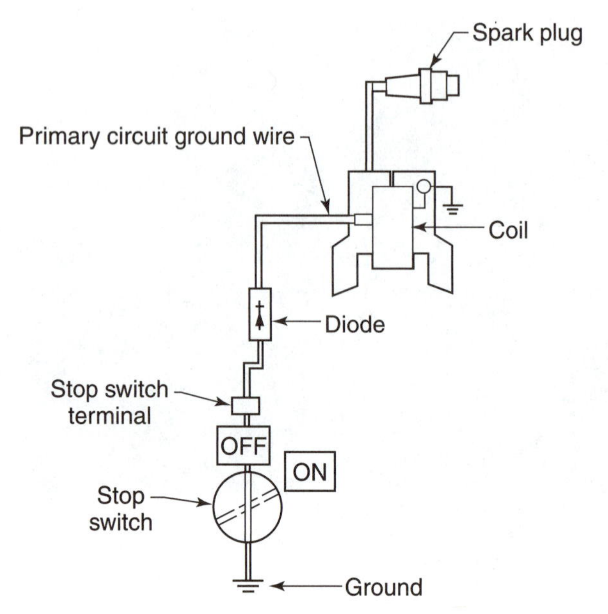

FIGURE 13-12 An electrical circuit for a primary ground circuit.

a magnetic field and this prevents ignition, stopping the engine. A diode is used on some circuits to protect the coil from any current coming into it in the wrong direction.

The stop switch is usually connected to a wiring harness (Figure 13-13). The wiring harness routes the stop switch wire from the switch to the coil and armature assembly on the engine. There is

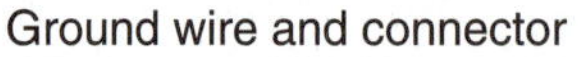
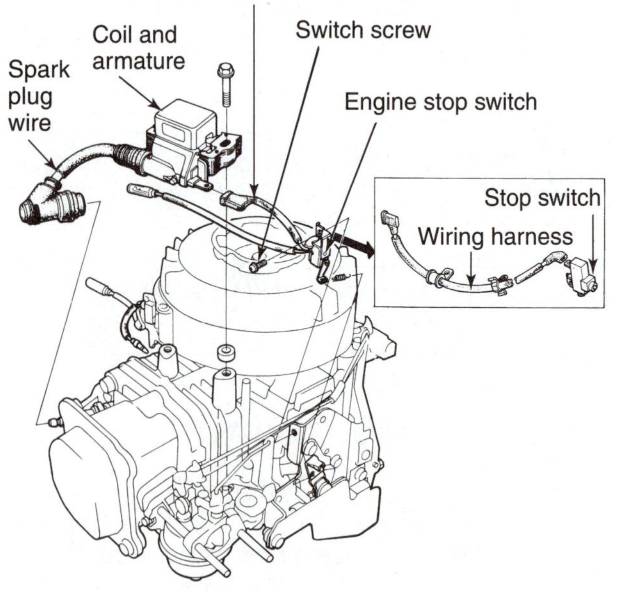

FIGURE 13-13 Parts of a primary ground kill switch. (Courtesy of American Honda Motor Co., Inc.)

often a push-on connector to connect the kill switch wire to the coil and armature assembly.

CAUTION: If the stop switch wire is disconnected on some systems, the engine cannot be stopped with the kill switch. Always make sure the kill switch is functional before using any power equipment.

WARNING: You should always allow the engine to come down to idle speed before killing the ignition. If you kill an engine at running speed, unburned fuel can pass through the engine into the muffler. The hot muffler can ignite the fuel and cause a backfire.

CAUTION: Unless you have an emergency, do not stop an engine by removing the spark plug wire from the spark plug. Removing the spark plug causes a high-voltage (spike) in the secondary system. The high voltage may get into the primary circuit where it can damage the coil or solid state parts in the module.

SAFETY INTERLOCKS

Many ignition systems have safety interlocks designed into the circuit. A **safety interlock** is a switch or circuit that prevents ignition if there is an unsafe condition on the power equipment. The American National Standards Institute (ANSI) has established standards for safety interlocks for outdoor power equipment.

Interlocks are required on many types of walk behind and hand-held power equipment. Mowers require a hand-activated interlock that prevents engine start up if the operator's hands are not on the equipment handle. This safety feature prevents the operator from getting hands or feet in contact with a rotating blade. Another common type prevents engine starting if a blade guard or grass bag is not in place.

Interlocks are also used to prevent engine damage. This kind of interlock has an oil pressure switch that senses low oil pressure caused by low oil level. When low oil pressure is sensed, it automatically kills ignition.

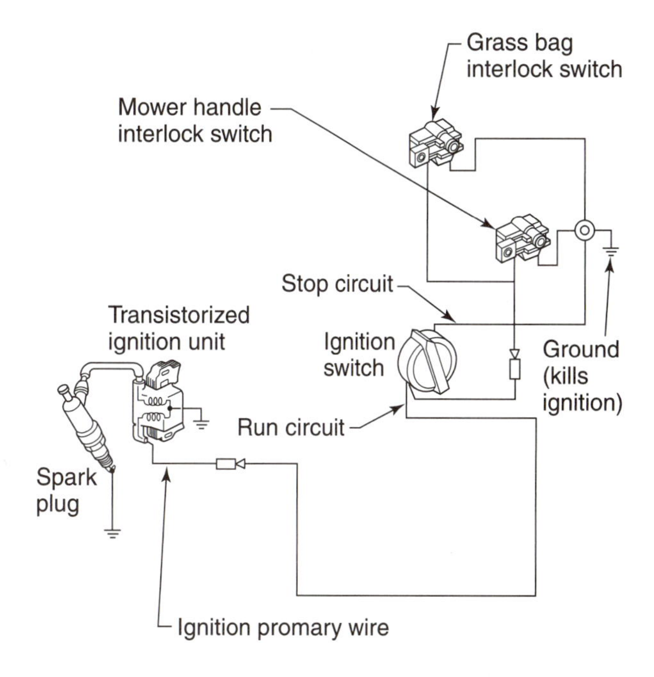

FIGURE 13-14 A safety interlock circuit with a mower handle and grass bag interlock switch.

Safety interlock circuits work like the ignition kill switch circuit (Figure 13-14). The kill circuit works when the operator turns the kill switch to the off position. This position closes a circuit for the ignition primary wire to a ground and stops ignition. Interlocks are connected to the run side of the stop switch. Interlocks have a wire that connects them to the ignition primary wire. When the mower grass bag is in position, the interlock switch is open. The open switch allows normal ignition. The interlock switch closes when the grass bag is not in position. The closed switch sends ignition primary current to ground. Ignition is prevented and the engine is stopped. The same type of switch is used for a mower handle position interlock.

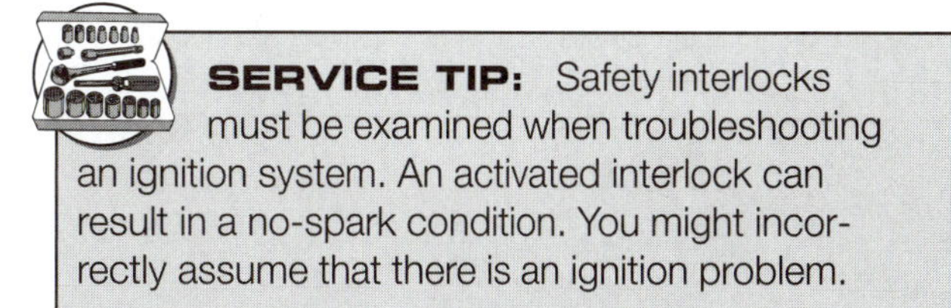

SERVICE TIP: Safety interlocks must be examined when troubleshooting an ignition system. An activated interlock can result in a no-spark condition. You might incorrectly assume that there is an ignition problem.

REVIEW QUESTIONS

1. Describe the parts and operation of a spark plug wire.
2. Describe the parts of a spark plug.
3. Explain the current flow from the spark plug terminal to the ground electrode.
4. Describe the path for heat flow from the spark plug into the cylinder head cooling fin.
5. Explain the difference between a hot and cold heat range spark plug.
6. Explain why different engines use different spark plug thread diameters.
7. Explain the difference between two spark plugs that have different reach dimensions.
8. Explain the purpose of the code printed on a spark plug.
9. List the common items found on a spark plug code.
10. List the common electrode types used on two-stroke engines.
11. Describe the operation of a kill switch on an ignition secondary circuit.
12. Explain the operation of a kill switch in an ignition primary circuit.
13. Explain the purpose of a safety interlock.
14. Describe the operation of a safety interlock for a mower grass bag.
15. Explain why a safety interlock is used on a walk behind mower handle.

DISCUSSION TOPICS AND ACTIVITIES

1. Locate several scrap ignition coils. Cut the insulation off the spark plug wire and identify the parts of the high voltage wire.
2. Collect a number of used spark plugs from around the shop. Compare and list the differences you can find between the spark plugs.
3. Examine the operational controls for a lawn mower. Make a list of all the interlocks you can find.

Fuel System—Fuels, Tanks, and Pumps

OBJECTIVES

Upon completion and review of this chapter, you should be able to:

- List and describe the different fuels used in small engines.
- Describe the combustion process and octane rating system.
- Describe and explain the sources and types of engine emissions covered by EPA regulations.
- Describe the parts and construction of a fuel tank.
- Identify and describe the different types of fuel lines and fittings.
- Explain the purpose and identify the types of fuel filters.
- Describe the parts and operation of a mechanical fuel pump.
- Describe the parts and operation of an impulse fuel pump.
- Explain the purpose and operation of a primer.

TERMS TO KNOW

Clean air standards
Detonation
Fuel filter
Fuel fittings
Fuel line
Fuel pump
Fuel shut-off valve
Fuel strainer
Fuel tank
Fuel tank filler cap
Gasohol
Gasoline
Gravity feed fuel system
Hydrocarbon
Impulse fuel pump
Intake manifold
Liquified petroleum gas (LPG)
Mechanical fuel pump
Normal combustion
Octane number
Outlet filter screen
Oxides of nitrogen
Preignition
Primer
Stale fuel
Volatility

INTRODUCTION

The purpose of the fuel system is to provide the air-fuel mixture for engine operation. The fuel system stores enough fuel for an extended period of engine operation. It delivers the fuel to the engine. It also mixes the fuel with the proper amounts of air for efficient burning in the cylinder.

FUELS

An engine needs a fuel that can be mixed with air and burned in the combustion chamber. The fuel used must have the correct combustion characteristics to burn properly. Most small engines use gasoline, but some engines use gasohol or liquefied petroleum gas.

Gasoline

Gasoline is the fuel used in most engines. **Gasoline** is a hydrocarbon-based engine fuel refined from crude petroleum oil. Gasoline is made (formulated) with many different chemicals that help to prevent fuel system problems. They also work to ensure normal combustion in the cylinder.

Combustion starts when the spark plug ignites the compressed air-fuel mixture (Figure 14-1). The combustion process should not be an explosion. The ignition should start a flame that burns across

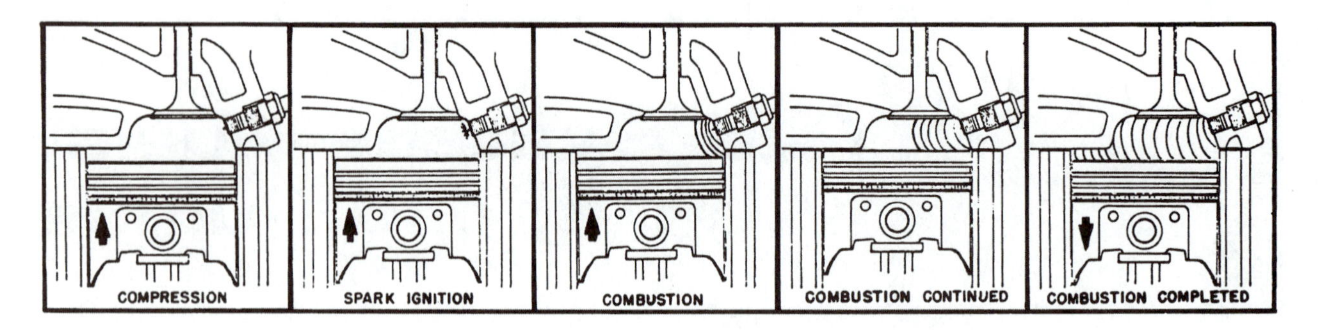

FIGURE 14-1 Normal combustion gives a smooth push on the piston. (Courtesy of CHAMPION-Federal Mogul.)

the combustion chamber. **Normal combustion** is when the spark plug ignites the air-fuel mixture and a wall of flame spreads smoothly across the combustion chamber. Normal combustion pushes the piston smoothly down the cylinder.

Preignition is combustion that is started by a hot part or deposit in the combustion chamber (Figure 14-2). Carbon deposits often glow red hot and cause hot spots on the spark plug firing end or piston head. This causes ignition at the wrong time, or preignition. The engine will run rough and make a knocking sound. Preignition can lead to valve and piston damage.

Sometimes detonation occurs instead of normal combustion. **Detonation** is when a part of the air-fuel mixture in the combustion chamber explodes rather than burns (Figure 14-3). Detonation is also called *spark knock* or *ping* and is caused when the burning flame spreads too fast, raising the heat and pressure on the unburned part of the air-fuel mixture. The unburned mixture bursts into a second flame. There is an explosion when the two flames meet. The explosion causes strong hammering pressures on the piston and other engine parts. It also results in a noise that sounds like a ping or knock. Detonation can damage spark plugs and eventually pistons.

High combustion chamber pressures and temperatures cause detonation. Engines with high compression ratios are more likely to detonate. Det-

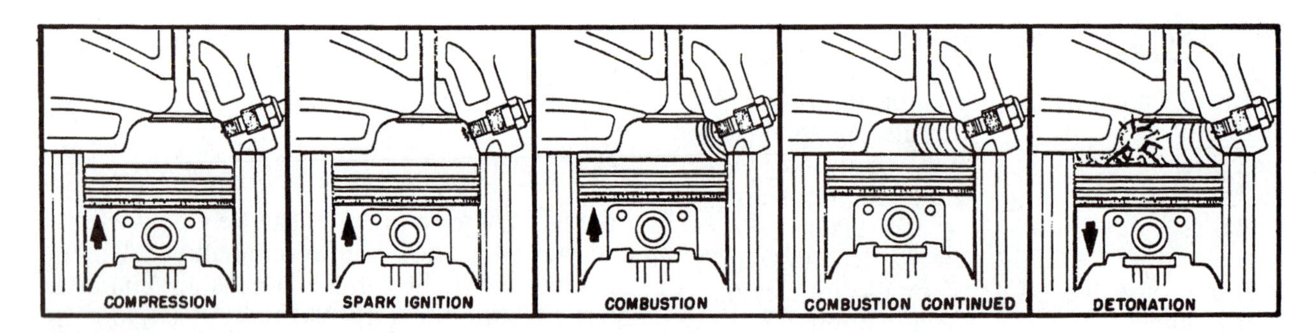

FIGURE 14-2 Preignition is caused by a hot spot in the combustion chamber. (Courtesy of CHAMPION-Federal Mogul.)

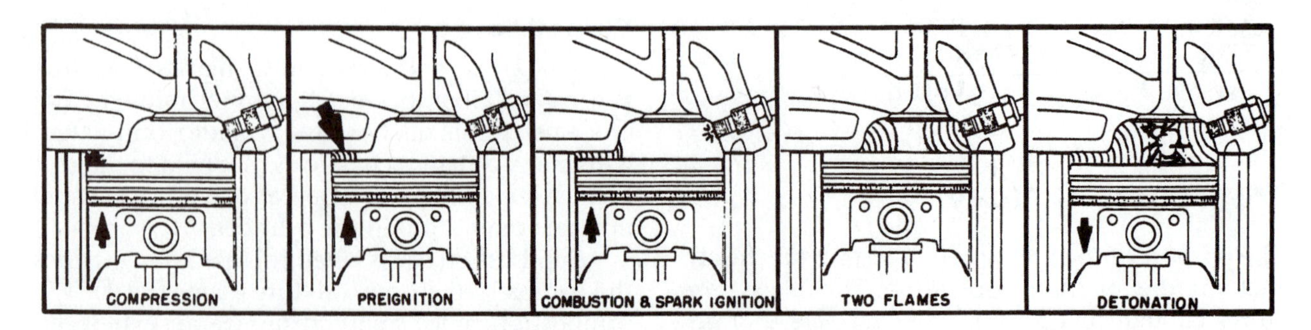

FIGURE 14-3 Detonation causes two flame fronts and a violent push on the piston. (Courtesy of CHAMPION-Federal Mogul.)

onation can occur from high engine heat, which can be caused by cooling system problems, lean mixtures, or too much spark advance.

Materials are added to gasoline when it is made to reduce detonation. The ability of a fuel to stop detonation is called its *antiknock quality*. Gasoline antiknock qualities are rated by an octane number. The **octane number** is a rating of how well a gasoline prevents detonation in the combustion chamber.

Federal law requires that the octane rating of a gasoline be displayed on the service station pump. Engine makers specify the octane rating required for their engines. This information can be found in the owner's manual and the shop service manual (Figure 14-4). A rating of 87 octane or higher is common for most engines.

An older method of improving a gasoline's antiknock value was addition of tetraethyl lead. The lead slows down the wall of flame in the combustion chamber and reduces detonation. Fuel with tetraethyl lead is called leaded fuel. The problem with leaded fuel is that lead is highly poisonous. Lead particles leave the combustion chamber with the exhaust gases. Leaded fuels have been phased out.

Unleaded fuels have been developed with high antiknock ratings. All newer engines are designed to use unleaded fuels. Some older engines require leaded fuel. The lead lubricated older types of materials used in valves and valve seats. Lead additives are available to mix with gasoline.

Tecumseh Products Company strongly recommends the use of fresh, clean, unleaded regular gasoline in all Tecumseh Engines. Unleaded gasoline burns cleaner extends engine life, and promotes good starting by reducing the build up of combustion chamber deposits,. Unleaded regular, unleaded premium or reformulated gasoline containing no more than 10% Ethanol, or 15% MTBE, or 15% ETBE may also be used.

Leaded fuel is generally not available in the United States and should not be used if any of the above options are available.

Never use gasoline, fuel conditioners, additives or stabilizers containing methanol, white gas, or fuel blends which exceed the limits specified above for Ethanol, MTBE, or ETBE because engine/fuel system damage could result.

Regardless of which of the approved fuels are used, fuel quality is critical to engine performance. Fuel should not be stored in an engine or container more than 30 days prior to use. This time may be extended with the use of a fuel stabilizer like TECUMSEH'S, part number 730245.

FIGURE 14-4 Fuel recommendations in a shop manual. (Provided courtesy of Tecumseh Products Company.)

Gasoline is made from a mixture of liquids with different volatility ratings. **Volatility** is how easily a fuel changes from a liquid to a vapor state. Some of the liquids in gasoline must evaporate at low temperature to allow the engine to be started easily when it is cold. As the engine begins to warm up, other less volatile liquids vaporize.

Gasoline manufacturers make summer and winter fuels with different volatility ratings. Owners that store gasoline during one season and use it during another may have poor engine performance.

Gasoline that is stored for a period of time picks up water. Oxygen degrades the fuel in a process called oxidation. **Stale fuel** is gasoline that has oxidized or has picked up water. Eventually thick (gum), varnishlike deposits will build up in the fuel. These deposits can cause clogged fuel system passages. Varnish can also cause the intake valve stem to stick in the valve guide.

Engine manufacturers often recommend that owners store only a 30-day supply of fuel. There are fuel additives available that will slow the formation of fuel deposits during storage. Gasoline should be stored in a cool ventilated area to slow oxidation. The fuel tank should be either completely empty or completely full when not in use. A full tank has less air and lowers the chance of oxidation.

Gasohol

Gasohol is an engine fuel made by blending gasoline with alcohol distilled from farm grains. Alcohol is a clean burning fuel. Small amounts (around 10 percent) of alcohol are blended with gasoline to make gasohol. This mixture burns with a combustion process similar to straight gasoline.

Burning fuels with high percentages of alcohol requires major changes in engine and fuel system parts. Alcohol is a solvent and can damage rubber, plastic, and brass fuel system parts. Some types of alcohol can cause corrosion to metals used in the casting of carburetors. Water is absorbed by alcohol. This can cause any water in a fuel tank to settle in the bottom and from there go out the fuel outlet. When the water enters the combustion chamber, it can cause misfire and overheating.

SERVICE TIP: Oxygenated or reformulated (RFG) gasoline is used in many parts of the country to reduce automotive exhaust emissions. Gasoline may be oxygenated with alcohol. This gasoline can cause some of the same problems as gasohol.

CAUTION: Too much fuel alcohol can be especially harmful to two-stroke engines. It can prevent lubricating oil used in these engines from staying in suspension with the fuel. Always check the owner's manual or shop service manual for fuel recommendations before using gasohol in either a four- or two-stroke engine. Some engine makers have a tester to find the percentage of alcohol in the fuel.

Liquefied Petroleum Gas

Some engines are operated on a fuel called liquefied petroleum gas. **Liquefied petroleum gas (LPG or LP)** is a butane- or propane-based fuel that is stored under pressure as a liquid but released as a gas. Liquefied petroleum gas is very clean burning. It burns with lower amounts of harmful exhaust emissions. Its disadvantages are lower heat content, bulky storage, and fewer places to fill the fuel storage tank.

CLEAN AIR STANDARDS

Outdoor power product engines develop emissions that contribute to air pollution (smog). They develop the same emissions as the automobile. Unlike the automobile, small engines in the past have not been subject to strict regulations. Air quality experts have determined that the small engine is a major source of pollution. Research conducted by the EPA shows that mowing a lawn for half an hour makes as much smog as driving a new car 172 miles.

Two-stroke engines are of special concern. The port control on two strokes allows a great deal of unburned fuel to go straight through the cylinder and out the exhaust pipe. The lubricating oil is used once and then expelled as part of the exhaust. Using a chainsaw for two hours can produce as much smog as driving a car coast to coast.

EPA Standards

Clean air standards have recently been developed and implemented by the federal Environmental Protection Agency (EPA). **Clean air standards** are EPA and state regulations that limit the pollutants emitted by small air-cooled engines. Some state air quality agencies have also developed regulations. The current regulations became effective in 1997. They cover most outdoor power equipment with engine sizes less than 26 horsepower (19 Kw).

Engines are classified as hand-held or nonhand-held under these regulations. Hand-held engines are engines mounted on equipment that is carried by the operator. A chainsaw is an example of the hand-held category. Most hand-held engines are the two-stroke type. Nonhand-held type engines are those that are on equipment not carried by the operator. Most of these engines are the four-stroke type.

CASE STUDY

A customer came in to the outdoor power equipment shop pushing an almost new lawnmower. It was the same expensive mower the shop had sold the customer last year. The customer's complaint was that it would not start. He said that he had pulled on the starter rope for an hour and the engine had not even tried to start.

The shop owner was used to these situations. It was the beginning of spring and many people were getting their lawnmowers out of the long winter storage. They were starting their spring mowing as the grass was just starting to grow. The shop owner removed the filler cap from the gasoline tank and smelled the fuel. The fuel smelled like old paint or varnish. Even though he knew the answer, he asked the customer how he had prepared the mower for winter storage. The questioning look on the customer's face was answer enough.

The fuel had not been drained from the fuel tank and it had turned stale from oxidation. The entire fuel system had to be flushed and cleaned to remove the gum and varnish. The shop owner resolved to do a better job of educating his customers about winter storage and stale fuel problems.

Emission Sources

The sources of harmful engine emissions are:

- fuel vapors (fumes)
- crankcase vapors (fumes)
- exhaust emissions

Fuel vapors are gasoline fumes that escape from the engine. Fuel vapors can escape from vented fuel tanks and other fuel system parts. Nonvented fuel systems have been developed to reduce these emissions.

Crankcase vapors are fuel vapors that enter the air from the engine crankcase. Fuels vapors can leak past the piston rings and enter the engine crankcase.

Older engines allowed these harmful crankcase vapors to vent into the atmosphere. Crankcase breathers that route the vapors back into the engine air intake, where they are burned, are now used on most engines.

Harmful exhaust emissions, pollutants that come out the engine exhaust, are much more difficult to correct. Improving the combustion process is the best way to lower exhaust emissions. Engines with overhead valves and improved combustion chambers develop lower emissions. The best way of lowering emissions from older engines is to keep the ignition and fuel system in good operating condition.

Emissions

An engine develops three main air pollutants: hydrocarbons, carbon monoxide, and oxides of nitrogen (Figure 14-5). A **hydrocarbon (HC)** is an engine emission caused by incomplete combustion in the combustion chamber or evaporation from the fuel system and crankcase. When hydrocarbons are subjected to sunlight, they can combine with other pollutants and cause smog.

Hydrocarbon exhaust emissions are difficult to control. During heavy engine load (maximum power), the engine needs a mixture with large amounts of fuel. There is not enough oxygen to burn all the fuel. Unburned hydrocarbons go out the exhaust.

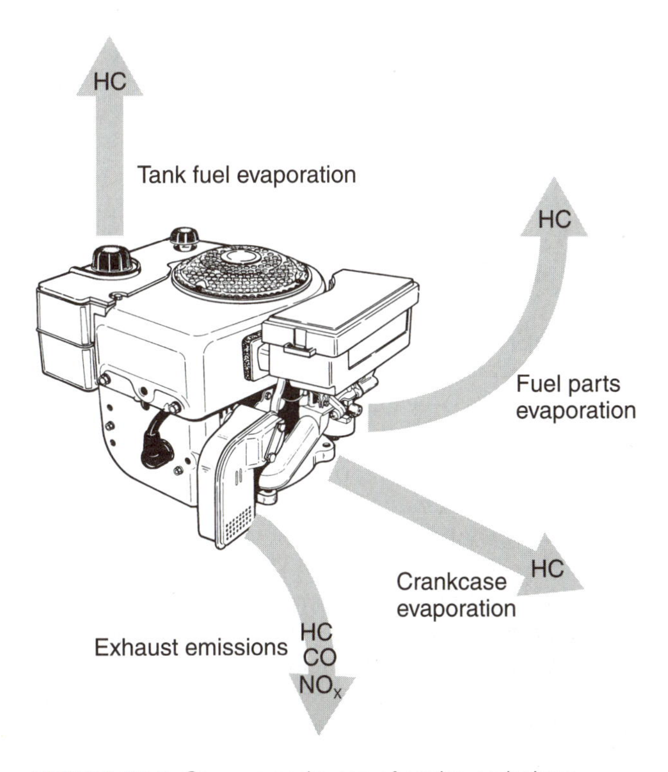

FIGURE 14-5 Sources and types of engine emissions. (Provided courtesy of Tecumseh Products Company.)

High hydrocarbon exhaust emissions may mean that the engine is not running properly. It may be misfiring because of a mechanical or ignition or fuel system problem. Also, if the combustion chamber is too cold, the fuel on the cylinder walls will not ignite and will leave the combustion chamber as unburned hydrocarbons.

> **CAUTION:** Spilling fuel during filling is a major source of small engine hydrocarbon emissions. Always use a funnel or nonspill nozzle to slowly fill the tank. Do not fill the tank all the way to the top. Leave room in the fuel tank for expansion. Be careful not to spill the fuel during filling. Make sure the gas cap is tightened properly. Store and transport gasoline cans out of direct sunlight.

Carbon monoxide (CO) is an exhaust emission formed when there is incomplete combustion. High carbon monoxide emission levels can be caused by incorrect carburetor adjustments. They are also caused by a restricted air flow to the combustion chamber. A dirty air filter will increase carbon monoxide emissions.

Oxides of nitrogen (NO_x) are engine emissions formed when combustion temperatures reach high levels. Oxides of nitrogen can combine with unburned hydrocarbons in the atmosphere to produce photochemical smog.

> **CAUTION:** Clean air standards prohibit a technician from removing or rendering inoperative any emission reduction device.

FUEL TANKS, LINES, AND FITTINGS

The **fuel tank** is a storage container attached to the engine or equipment that provides the source of fuel for engine operation. The fuel tank may be made from plastic or steel. Tanks are often attached to an accessible part of the engine's blower housing (Figure 14-6).

The **fuel tank filler cap** (Figure 14-7) is a cap on top of the fuel tank that can be removed to refuel the engine. Some fuel tank caps have small slots or holes called vents. The vents prevent a vacuum and allow atmospheric pressure into the fuel tank. Atmospheric pressure moves fuel from the tank into

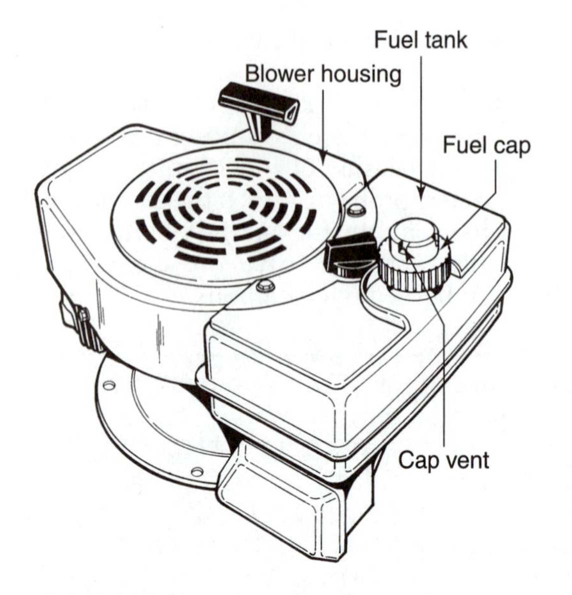

FIGURE 14-6 The fuel tank is often mounted to the blower housing. (Provided courtesy of Tecumseh Products Company.)

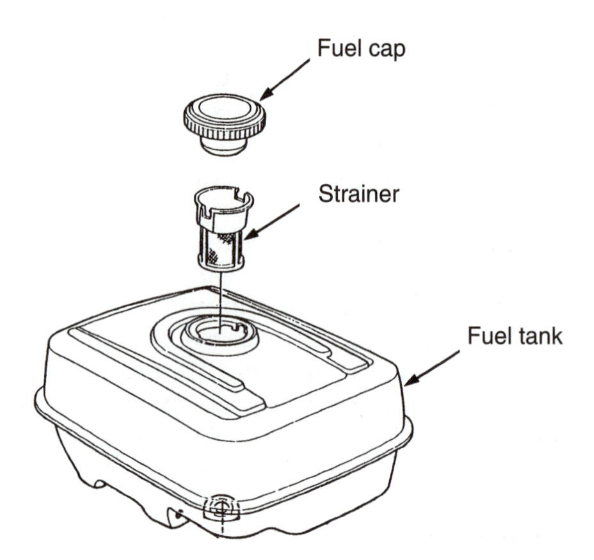

FIGURE 14-8 A fuel strainer in the opening of a fuel tank. (Courtesy of American Honda Motor Co., Inc.)

the rest of the fuel system components. Plugged vents will prevent the engine from running. The cap also has a set of baffles (thin plates). The baffles prevent any dirt or dust that comes in the vent from getting into the fuel.

Some fuel tanks use a fuel strainer under the fuel cap. A **fuel strainer** (Figure 14-8) is a wire mesh screen in the fuel tank opening that prevents fuel contamination during filling. The strainer is removable so that it can be cleaned periodically.

The fuel tank is connected to the other fuel system parts with lines and fittings (Figure 14-9). **Fuel fittings** are small connecting parts used to connect fuel lines to fuel system parts. The fuel tank outlet is located near the bottom of the tank. An **outlet filter screen** is a screen in the fuel tank outlet that strains the fuel as it leaves the tank. The outlet filter screen is attached to an outlet fitting on the bottom of the tank. The screen is often a long mesh tube. The tube has a lot of surface area for straining without becoming plugged. An outlet fuel line is connected to the outlet fitting.

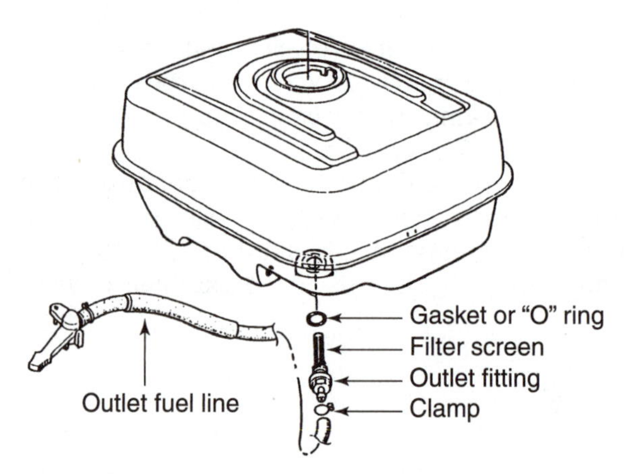

FIGURE 14-9 An outlet filter screen located on the bottom of the fuel tank. (Courtesy of American Honda Motor Co., Inc.)

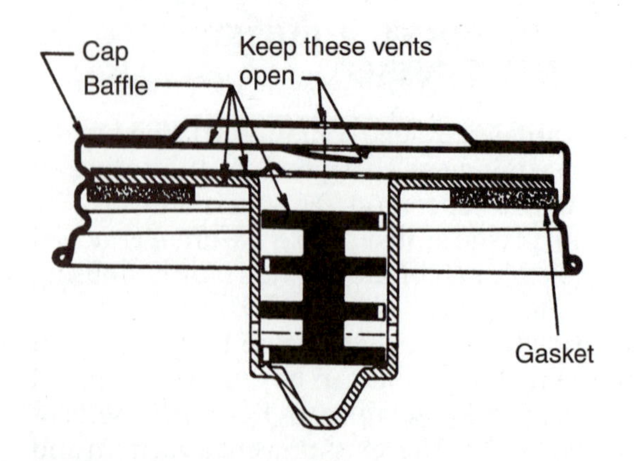

FIGURE 14-7 A fuel tank filler cap with vents and baffles. (Courtesy of Clinton Engines Corporation.)

Sometimes the fuel outlet assembly includes a **fuel shut-off valve**, which is a valve in the fuel tank outlet that allows the fuel flow from the tank to be shut off (Figure 14-10). The valve has a handle which, when turned, opens or closes the fuel tank outlet. A packing nut (nut with a gasket) fits around the valve. The packing nut prevents fuel from leaking out of the valve area. Fuel shut off is needed when:

- Transporting the engine
- Preparing the engine for long-term storage

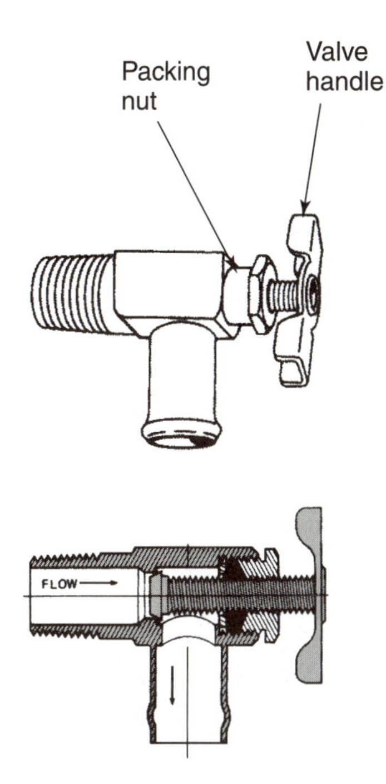

FIGURE 14-10 A fuel shut-off valve assembly. (Courtesy of Dana Corp.)

■ Preventing flooding during hot shut-down
■ When working on the fuel system

A **fuel line** is a metal, flexible nylon, or synthetic rubber line used to route fuel to fuel system parts. Nylon and synthetic rubber lines are usually called fuel hoses because they are flexible. Hoses are connected to hose fittings. Hose fittings have a fuel pipe with sealing rings formed on the end. A fuel hose is pushed on the fuel pipe and over the sealing rings. A clamp is used behind the sealing rings to keep the fuel line in position (Figure 14-11). Common types of clamps are spring, worm drive, and rolled edge. The spring types are removed and replaced with pliers. The other types are removed and replaced with screwdrivers.

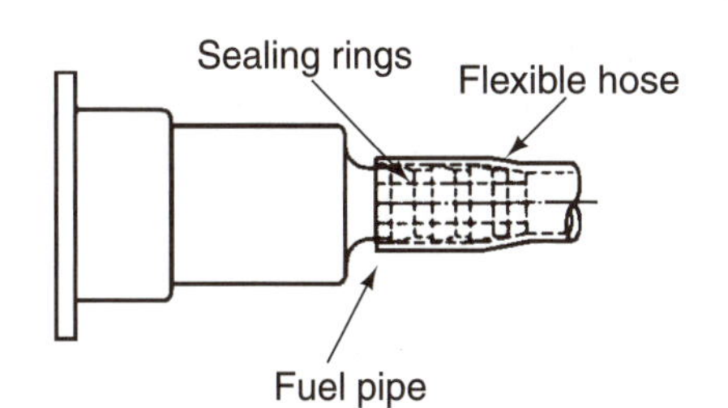

A. Flexible Fuel Hose Fitting

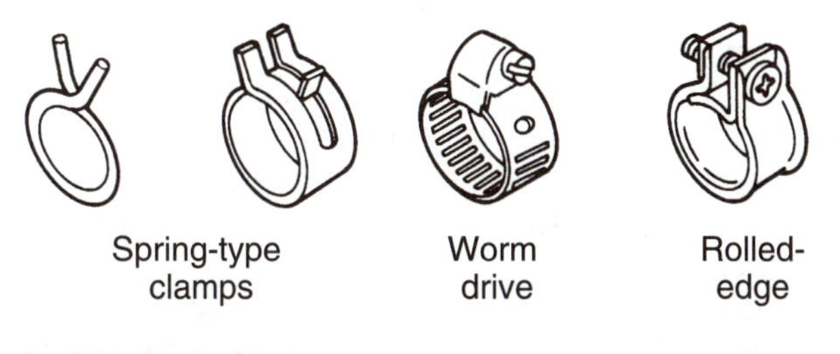

B. Fuel Hose Clamps

FIGURE 14-11 Flexible fuel hoses are sealed with sealing rings and held in place with clamps.

CAUTION: Copper fuel lines may be found on older engines. Copper tubing should be avoided when repairing or replacing fuel lines because it has poor fatigue strength. It breaks sooner than steel when subjected to vibration and movement. A fuel leak can cause a fire.

SERVICE TIP: The sealing rings on fuel line hoses prevent fuel from leaking. The hose clamps prevent the hose from working loose. A leaking hose can usually not be corrected by tightening the hose clamp. Overtightening the clamp distorts the hose and creates more leakage. Leaking hoses should always be replaced.

Metal fuel lines are usually made from steel. Older engines sometimes have copper lines. Metal fuel lines are joined with a connector fitting and flare nut (Figure 14-12). The connector fitting is threaded on both ends. It has flats for a wrench. A flare nut is used to join the line to the connector fitting. A sleeve or a flare is used between the connector and fuel line for a gasoline tight seal. A sleeve is a small, sliding, soft metal part that fits

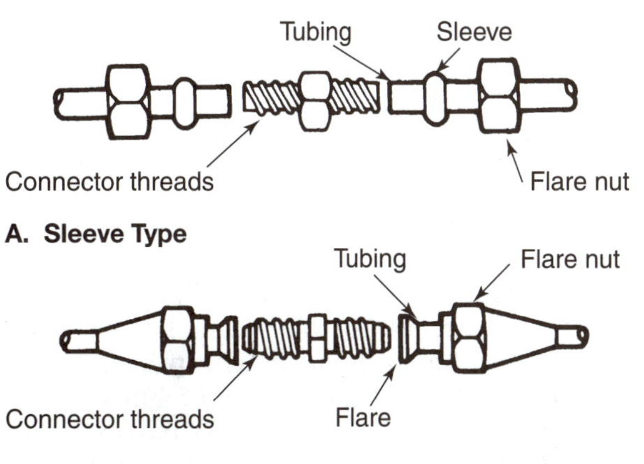

A. Sleeve Type

B. Flare Type

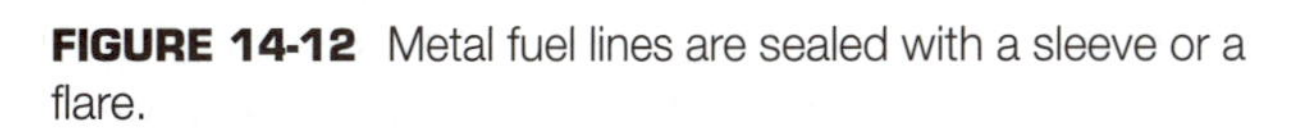

FIGURE 14-12 Metal fuel lines are sealed with a sleeve or a flare.

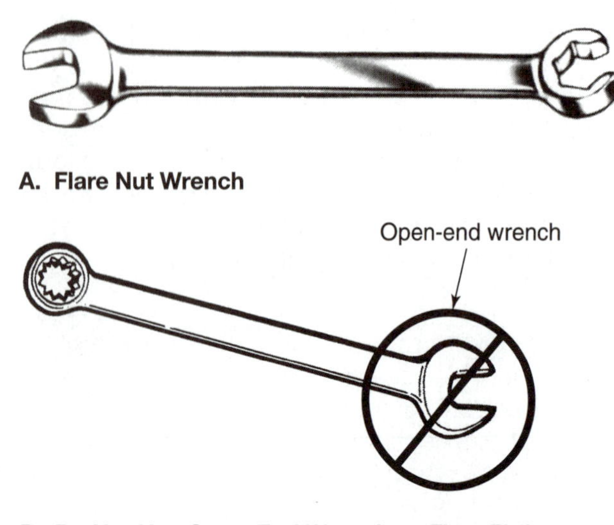

FIGURE 14-13 Use a flare nut wrench, not an open-end on fuel line fittings. (*A:* Courtesy of Snap-on Incorporated. *B:* Courtesy of U.S. Navy.)

around the fuel line. It is squeezed between the nut and the connector for the seal. A flare is a funnel-shaped end formed on the fuel line. When tightened, the flare is forced against the connector fitting to seal the connection.

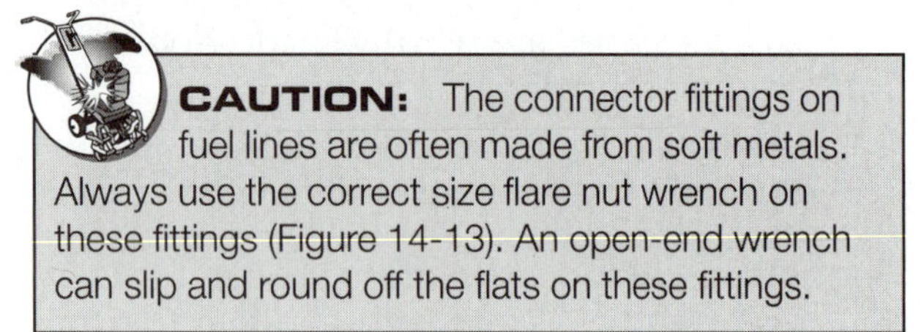

CAUTION: The connector fittings on fuel lines are often made from soft metals. Always use the correct size flare nut wrench on these fittings (Figure 14-13). An open-end wrench can slip and round off the flats on these fittings.

FUEL FILTERS

Clean fuel is very important in any fuel system. Dirt can plug small fuel and air passages in the carburetor. Water in the fuel can cause engine performance problems. Filtering the fuel eliminates these problems. Most engines have a replaceable fuel filter. The **fuel filter** is a filter element connected into the fuel line used to clean fuel before it enters the carburetor.

Most newer engines use an in-line disposable fuel filter (Figure 14-14). The filter has a plastic housing with a paper filter element inside. The filter is connected to the fuel line by fuel hoses. Fuel moving through the fuel line must pass through the paper filter element. Dirt and water are trapped on the outside of the paper filter material. There is often a fuel direction arrow on the housing that shows in which direction they should be installed.

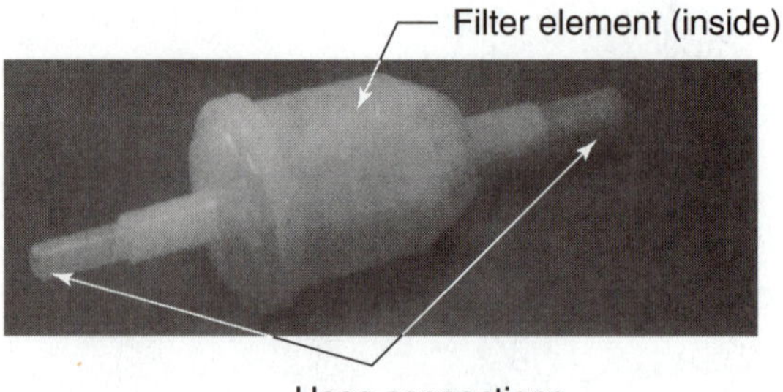

FIGURE 14-14 A disposable fuel filter is often installed with fuel hoses.

FUEL PUMPS

Many engines are mounted and used in a stationary or upright position. These engines often use a gravity feed fuel system. A **gravity feed fuel system** has the fuel tank mounted above the carburetor so that gravity forces fuel from the tank to the carburetor. The gravity feed system does not require any fuel pump.

Other engines use a remote fuel tank (lawn or garden tractor) or are operated in different positions (leaf blower). These engines cannot depend on gravity to move the fuel, so a fuel pump is used on these engines. A **fuel pump** is a pump that creates a low pressure to pull fuel out of the fuel tank then creates a pressure to move the fuel to the carburetor. Some fuel systems use a mechanical or impulse fuel pump.

Mechanical Fuel Pump

The **mechanical fuel pump** (Figure 14-15) is a pump that uses a diaphragm operated by an eccentric on the engine camshaft to pump fuel from the tank to the carburetor. Mechanical pumps are usually mounted on the side of the engine block. The pump has a lever (rocker arm) that enters the engine and rides on a camshaft eccentric (lobe). As the camshaft eccentric rotates, it causes the rocker arm to move up and down. The lever is connected to a diaphragm through a pull rod. A diaphragm is a flexible pumping element in the fuel pump pumping chamber that, when moved, changes the volume of the chamber. The pumping chamber may be located above or below the diaphragm. There is an inlet and outlet check valve located in the pumping chamber.

Pumping (intake) starts when the cam eccentric pushes up on the pump rocker arm. The rocker arm pulls the pull rod and diaphragm down. The rapid movement of the diaphragm causes a low pressure in the pumping chamber. The low

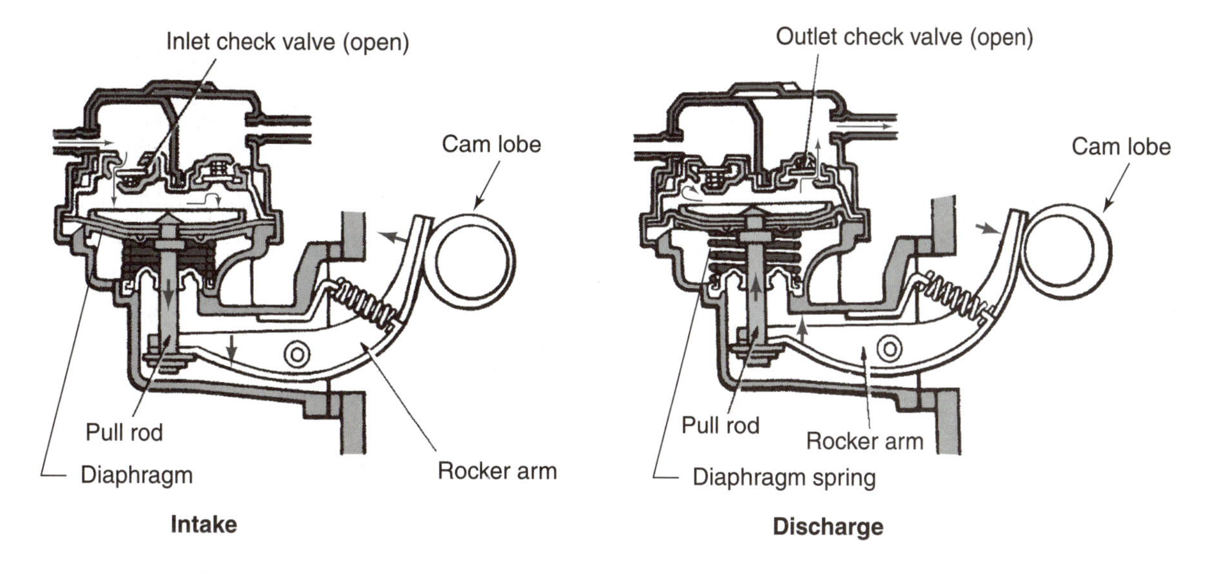

FIGURE 14-15 Parts and operation of a mechanical fuel pump.

pressure overcomes the inlet check valve spring pressure. The inlet check valve opens. Atmospheric pressure in the fuel tank forces fuel in past the open inlet check valve. During this time, the outlet valve is held closed by a spring and by the low pressure in the chamber.

Fuel moves out of the pump (discharge) when the camshaft eccentric allows the diaphragm spring to move the diaphragm upward. Upward movement of the diaphragm builds pressure in the pumping chamber. The pressure closes the inlet check valve and opens the outlet check valve. The fuel then flows from the pumping chamber. It goes out an outlet passage into the fuel line connected to the carburetor.

Impulse Fuel Pump

The **impulse fuel pump** (Figure 14-16) is a pump operated by pressure impulses in the intake manifold or the crankcase. The **intake manifold** is a part that delivers the air-fuel mixture from the carburetor into the intake port. Two-stroke engines usually use an impulse from the crankcase. Four-stroke engines use an intake manifold impulse.

The impulse fuel pump has a center housing and two end covers (Figure 14-17). A thin plastic diaphragm fits between each of the end covers and the center housing. There are gaskets between the diaphragms, end covers, and center housing. The gaskets make a pressure-tight seal inside the pump. One of the end covers has a fitting for a flexible line. The line is called the *pulse hose*. The pulse hose

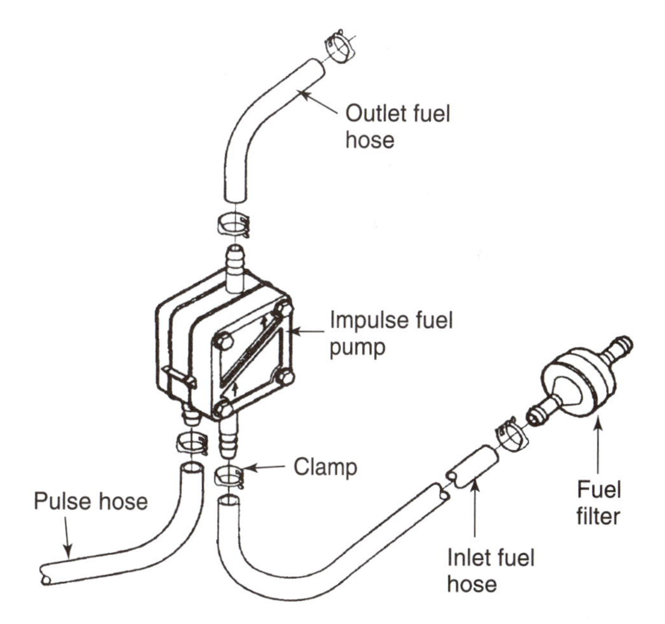

FIGURE 14-16 A pulse fuel pump is operated by crankcase or intake manifold pressure pulses. (Courtesy of Onan Corp.)

is connected to the crankcase or the intake manifold.

When the engine is running, the pressure in the crankcase or intake manifold goes up and down (pulses). This pulse occurs as the piston goes up and down. The pulse hose sends this pulse to the fuel pump diaphragms. The rising and falling pressure pulse causes the diaphragms to move up and down. The moving diaphragms create a

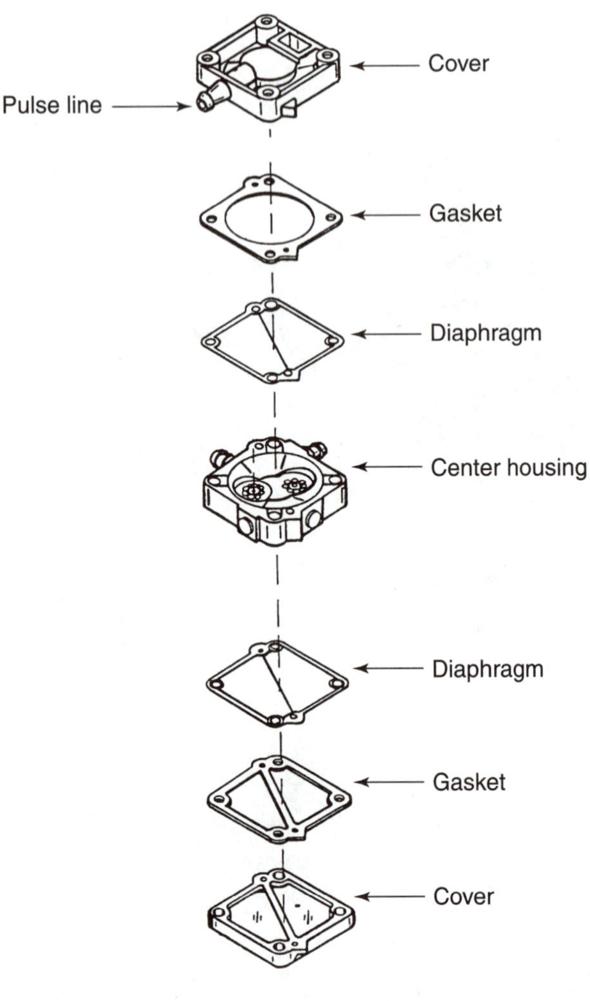

FIGURE 14-17 The inside parts of a pulse type fuel pump. (Provided courtesy of Tecumseh Products Company.)

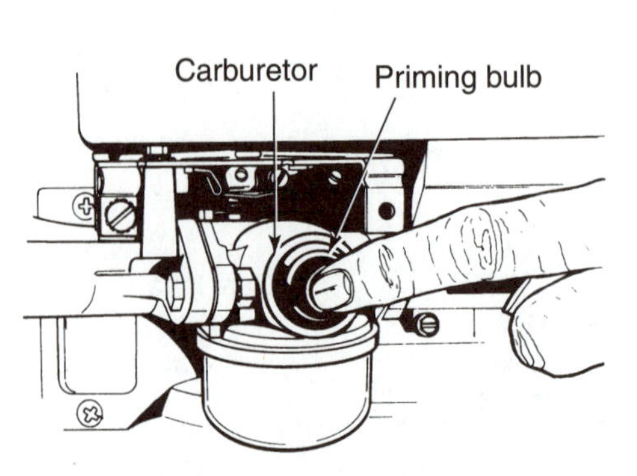

FIGURE 14-18 A carburetor-mounted primer. (Provided courtesy of Tecumseh Products Company.)

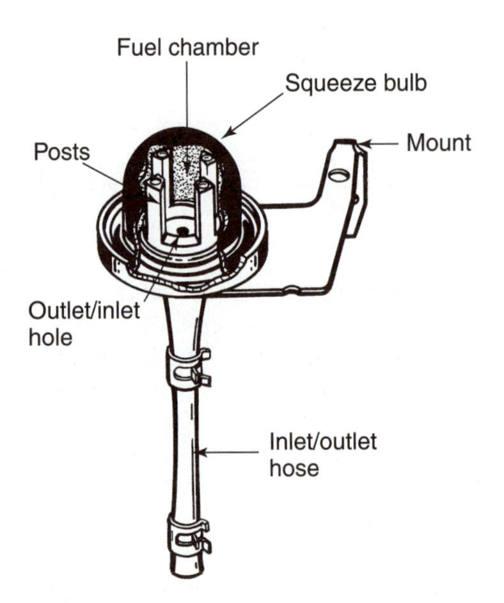

FIGURE 14-19 A primer assembly that is a separate part. (Provided courtesy of Tecumseh Products Company.)

vacuum to pull fuel into the pump inlet. Then a pressure is created to move the fuel out the outlet to the carburetor.

Primer

A **primer** is a rubber squeeze bulb used to force fuel into the air entering the combustion chamber to start the engine (Figure 14-18). The squeeze bulb is mounted to the side of some carburetors. It can be a separate assembly mounted on the engine. The bulb is connected directly or through a hose to a fuel source. The fuel source can be either in the carburetor or fuel tank.

To start a cold engine, the operator squeezes the bulb (Figure 14-19). This forces fuel out of the bulb fuel chamber past a ball check valve and directly into the engine air intake. When the bulb is released, a low pressure is created inside the bulb. The low pressure pulls fuel from the fuel source back into the bulb. A ball type check valve is used in the system. The check valve prevents fuel from being forced back into the fuel source (tank or carburetor) during priming.

REVIEW QUESTIONS

1. Describe the process of normal combustion in a combustion chamber.
2. Describe the process of detonation and pre-ignition in a combustion chamber.

3. Explain the octane system for rating gasoline.
4. Describe the problems caused by too much alcohol in gasoline.
5. Explain why small-engine emissions are regulated.
6. List the emission sources in a small engine.
7. Explain how hydrocarbon emissions are developed in an engine.
8. Describe how carbon monoxide emissions are developed in an engine.
9. List and explain how vapor and exhaust emissions are controlled.
10. Explain how to properly fill a gas tank to prevent hydrocarbon emissions.
11. Describe the parts of a fuel tank cap and explain why the cap is vented.
12. Describe the purpose and operation of a fuel filter.
13. Describe the parts and operation of a mechanical fuel pump.
14. Describe the parts and operation of an impulse fuel pump.
15. Describe the parts and operation of a primer.

DISCUSSION TOPICS AND ACTIVITIES

1. Locate an owner's manual for a Briggs & Stratton, Honda, Kohler, and Onan engine. List the recommended gasoline octane requirements for each engine.
2. Use the same manuals to make a list of the fuel storage and handing requirements for each engine.
3. Fuel mileage on a vehicle such as a garden tractor can be computed if you know how much fuel was consumed and how many miles were driven. A tractor goes 100 miles on 8 gallons of fuel. What is the fuel mileage in miles per gallon?

Fuel System—Carburetors and Air Cleaners

OBJECTIVES

Upon completion and review of this chapter, you should be able to:

- Describe the parts and explain the operation of the basic carburetor.
- Identify the parts and explain the operation of a vacuum carburetor.
- Identify the parts and explain the operation of a float carburetor.
- Identify the parts and explain the operation of the diaphragm carburetor.
- Identify the parts and explain the operation of a suction feed diaphragm carburetor.
- Explain the purpose and operation of an automatic choke.
- Describe the parts and explain the operation of the common types of air cleaners.

TERMS TO KNOW

Air bleed
Air cleaner
Atomization
Automatic choke screw
Carburetor
Carburetor throat
Choke
Choke lever
Diaphragm carburetor
Downdraft carburetor
Emulsion tube
Float
Float bowl
Float carburetor
Foam air filter element
High-speed adjustment screw
Inlet needle valve
Inlet seat
Jet
Low-speed adjustment screw
Manual choke
Metering
Oil bath air cleaner
Paper air filter element
Sidedraft carburetor
Suction feed diaphragm carburetor
Throttle valve
Updraft carburetor
Vacuum carburetor
Vaporization
Vent
Venturi

INTRODUCTION

The **carburetor** is a fuel system part that mixes fuel with air in the proper proportion to burn inside the engine. Although there are many different types of carburetors, they all have similar parts and operate on the same basic principles. Air that enters the carburetor mixes with the fuel and enters the engine combustion chamber. The air must be filtered by an air cleaner to prevent abrasives from damaging engine components.

CARBURETOR FUNDAMENTALS

The carburetor meters, atomizes, and distributes the fuel into the engine intake air. The carburetor must do this during many different operating conditions. These conditions include cold or hot starting, idle speed, part throttle, acceleration, and full power.

Gasoline must be atomized before it can be ignited inside an engine. **Atomization** is the breaking up of the liquid fuel into fine particles so that it can be mixed with air to form a combustible mixture. In a carburetor, gasoline is drawn into the incoming air stream as a spray. The spray is then atomized or broken into fine droplets to form a mist. The air-fuel mist goes into the intake manifold where the fuel is vaporized.

Liquid gasoline will not burn inside an engine

combustion chamber. Liquid gasoline must be changed into vapor before it will burn. **Vaporization** is the changing of a liquid such as gasoline into a vapor or gas. Vaporization can occur only when the liquid is heated enough to boil. For example, if you heat water in a pan on the stove, the heat is transferred to the water, raising its temperature. When the water reaches its boiling point, the liquid (water) changes to a vapor (steam). Different liquids have different boiling points. Gasoline has a much lower boiling point than water.

The intake stroke of the piston creates a low pressure in the intake manifold. This pressure is far less than atmospheric pressure. A liquid's boiling point drops as its pressure gets lower. The gasoline mist boiling point is so low that it boils at room temperature. Heat from the air around each fuel drop causes the gasoline to vaporize.

Metering is the mixing of fuel and air needed for efficient burning inside the engine's combustion chamber. Metering is based on the amount of oxygen that must be mixed with gasoline to get it to burn. About 15 pounds of air are needed to completely burn 1 pound of gasoline. This is a 15:1 ratio of air and fuel. More fuel is called a *rich mixture*. Less fuel is called a *lean mixture*.

Air-fuel ratios are based on weight (Figure 15-1), not on volume (size). The volume of air and fuel changes with pressure or temperature. Weight is not affected by such changes. For example, 9,000 gallons of air are required to burn 1 gallon of gasoline. The 9,000 gallons of air weigh 90 pounds. A gallon of gasoline weighs about 6 pounds. The air-fuel ratio is 15:1 by weight (90 divided by 6 equals 15).

Most engines operate on air-fuel ratios from 11:1 (very rich) to 17:1 (slightly lean). A rich mixture is needed for cold starting, acceleration, and full power. A leaner mixture can be used for idle and part throttle. The carburetor mixes these different air and fuel ratios for different operating conditions.

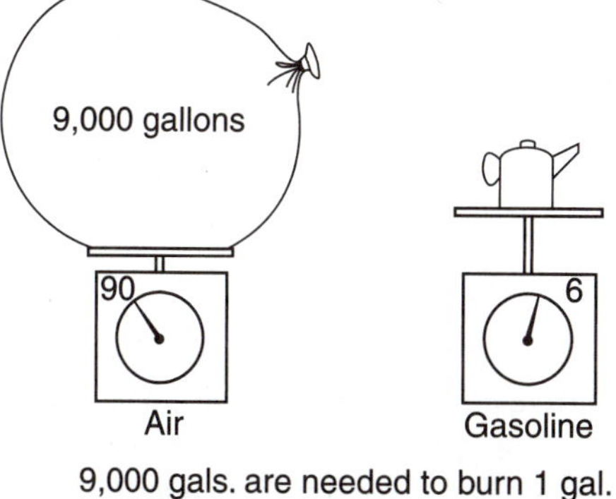

FIGURE 15-1 Air-fuel ratio of a mixture is determined by weight.

Basic Carburetor

A carburetor is shaped like a tube that is open at both ends (Figure 15-2). Air is pulled in the open end of this tube when the piston is on the intake stroke. The moving piston creates a low pressure in the cylinder. Air goes through the tube because the pressure is higher in the tube than in the cylinder. Air goes into the engine through the open intake valve.

The middle of the carburetor tube has a restricted area called a **venturi** (Figure 15-3), which is an area in a carburetor air passage used to create a low pressure for fuel delivery. Air has a difficult time getting through the venturi. As the air does get through, it speeds up. The speed up causes a low pressure in the venturi. The low pressure is used to pull in fuel.

A small amount of fuel is stored under the venturi (Figure 15-4). A hollow pickup tube runs from the fuel to the venturi. As air goes through the

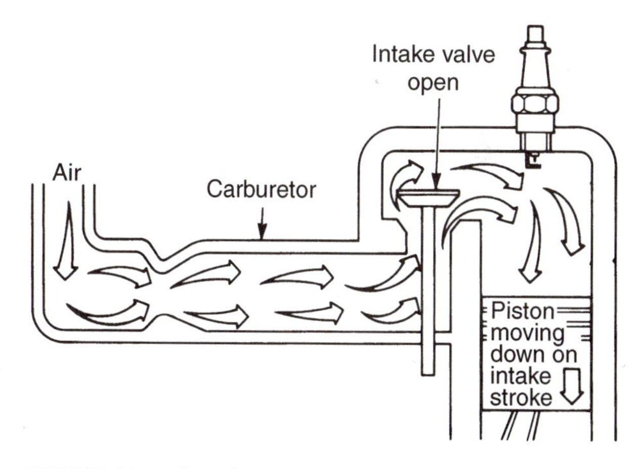

FIGURE 15-2 A carburetor is a tube open at both ends and attached to the intake port.

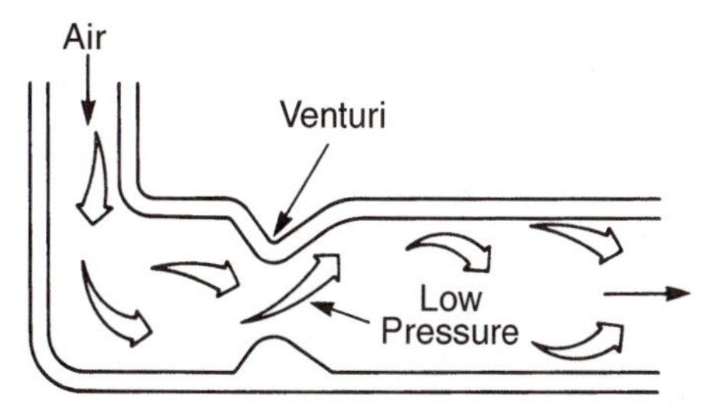

FIGURE 15-3 The venturi is a restricted part of the carburetor that causes a low pressure.

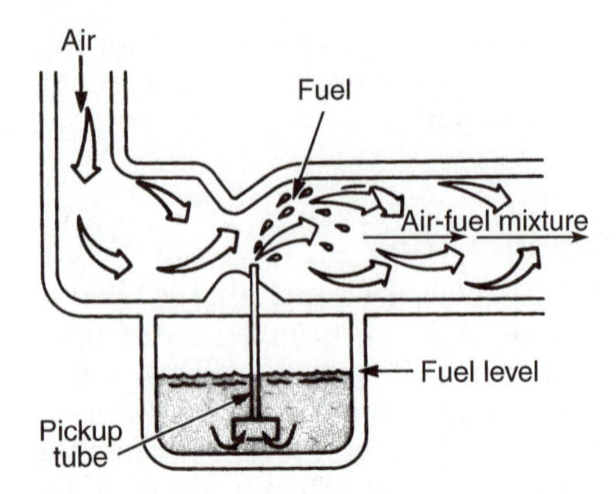

FIGURE 15-4 The pickup tube allows fuel to mix with the air.

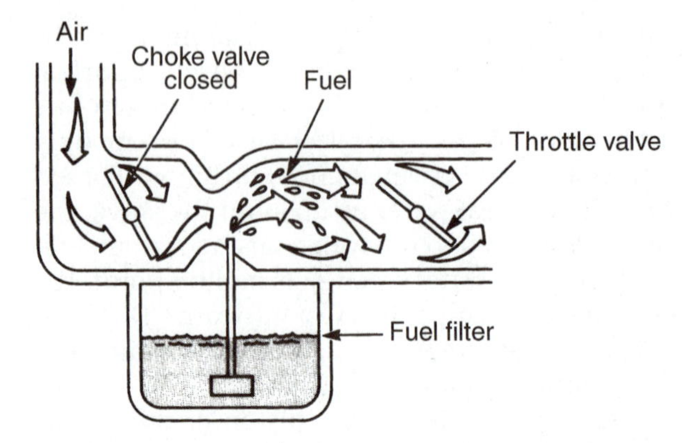

FIGURE 15-6 The closed choke valve provides a rich mixture to start a cold engine.

venturi, the low pressure pulls fuel up the pickup tube. The fuel mixes with the air in the venturi. The mixture of air and fuel goes into the engine's combustion chamber.

The amount of air-fuel mixture that gets into the engine determines engine speed. If a large amount of air-fuel mixture enters, the engine speed increases. If just a small amount of air-fuel mixture gets in, the engine speed decreases. A **throttle valve** (Figure 15-5) is a valve that is opened or closed by the operator to regulate the amount of air-fuel mixture that enters the engine. The throttle valve is a round plate that fits in the end of the carburetor tube. It is sometimes called a *throttle plate* or *butterfly valve*. A closed throttle valve allows little air and fuel to get into the engine. The engine runs slowly with a closed throttle valve. As the throttle valve is opened, more air-fuel mixture gets into the engine and the engine runs faster.

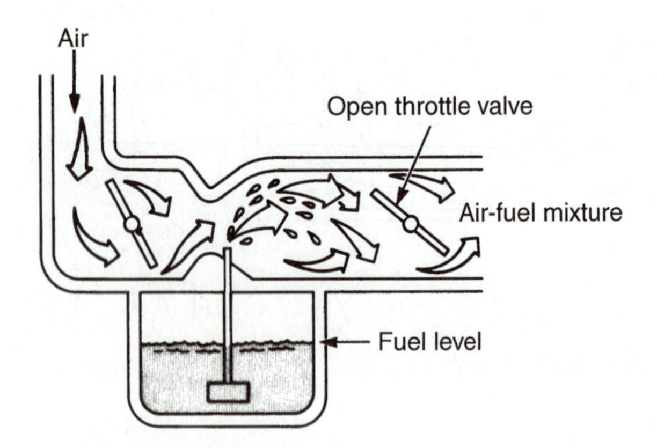

A. Open Throttle Valve

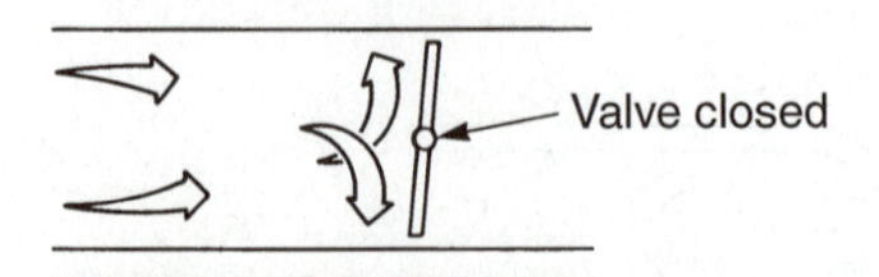

B. Closed Throttle Valve

FIGURE 15-5 The throttle valve controls the air-fuel mixture going into the engine.

When an engine is cold, the intake and combustion chamber parts are cold. These cold parts make the vaporization of gasoline difficult and the engine hard to start. Many carburetors use a choke to help start a cold engine. The **choke** (Figure 15-6) is a carburetor valve that restricts air flow and creates a rich mixture for cold starting. The choke valve is a small, round valve that fits in the carburetor air inlet (throat). When closed, the choke valve limits the air that can get into the carburetor throat. A strong vacuum is created in the venturi. The vacuum pulls a large amount of fuel out of the pickup tube. The rich air-fuel mixture goes into the cylinder to start the cold engine.

The choke valve is connected to a lever on the outside of the carburetor. The **choke lever** is a linkage used to manually open and close the choke valve. Some carburetors have a choke that opens and closes automatically.

Adjustment Screws

The engine needs a large volume of air and fuel when it runs at high speed. Fuel for high-speed operation comes from the venturi pickup tube. The amount of fuel that goes up the pickup tube is regulated and adjusted with a screw. A **high-speed adjustment screw** is a carburetor screw used to regulate the amount of fuel that goes up the main pickup tube during high-speed operation. The screw has a tapered end that fits in the side of the carburetor and into the pickup tube. The screw is used to increase or decrease the size of the passageway for fuel to move up the pickup tube. If the screw is turned all the way in, the smallest amount

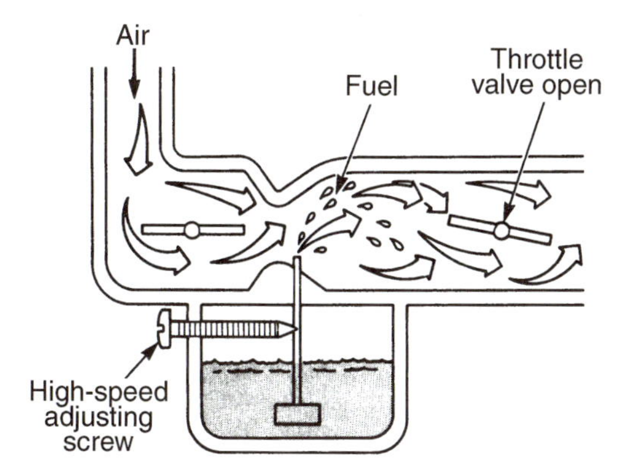

FIGURE 15-7 The high-speed adjustment screw controls the fuel coming up the pickup tube.

of fuel can go up the tube. If the screw is turned all the way out, the largest amount of fuel can go up the tube. A spring is used on the adjustment screw to prevent it from vibrating loose. Adjusting the high-speed screw is a service adjustment done on a carburetor.

The throttle valve is in a closed position when the engine is running slowly or at idle speed. A closed throttle valve allows very little air to flow through the venturi. Very little fuel is pulled up the pickup tube. An additional amount of fuel must go into the engine so it will run. This is done with a small hole behind the throttle valve. During idle, fuel is pulled through a passage and out the hole. The amount of fuel that comes out the hole is controlled and regulated by an adjustment screw.

The **low-speed adjustment screw** is a carburetor screw used to regulate the amount of fuel that gets behind the closed throttle plate during low-speed or idle operation (Figure 15-8). The screw has a pointed end that goes through the side of the carburetor and into a low-speed fuel passage. If the screw is turned all the way in, very little fuel (or air-fuel

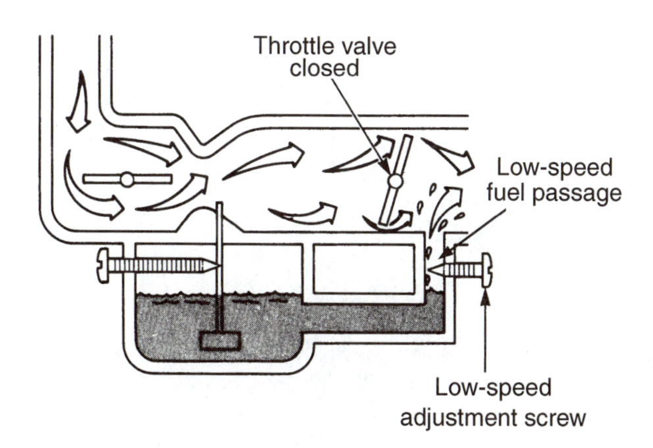

FIGURE 15-8 The low-speed adjustment screw controls the fuel coming out behind the throttle valve.

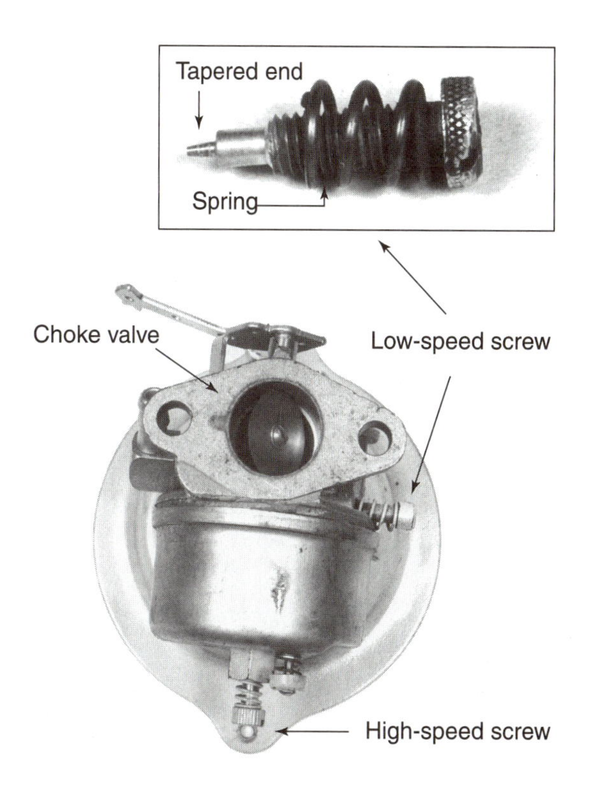

FIGURE 15-9 A carburetor with a high- and low-speed adjustment screw.

mixture in some designs) can come out the hole. If the screw is turned out, more fuel can get out the hole (Figure 15-9). Adjusting the low-speed screw is another service adjustment done on a carburetor.

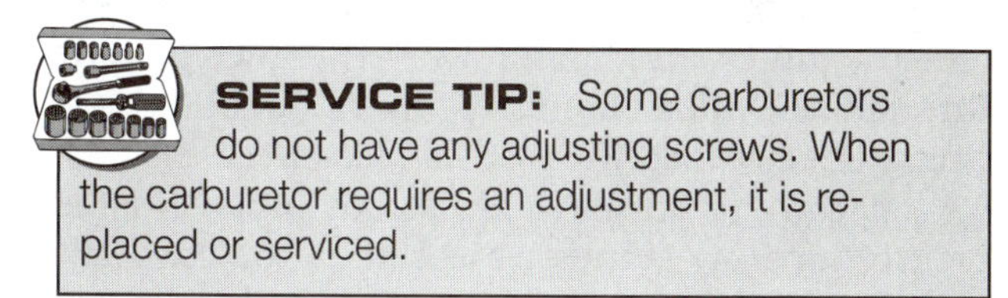

SERVICE TIP: Some carburetors do not have any adjusting screws. When the carburetor requires an adjustment, it is replaced or serviced.

Jets, Emulsion Tubes, Bleeds, and Vents

Fuel and air are routed through calibrated passages in the carburetor body. These passages have to be the correct size to get the proper air-fuel mixture. The common passages in carburetors are jets, emulsion tubes, and air bleeds (Figure 15-10). A **jet** is a carburetor part with a small, calibrated hole for regulating the flow of fuel, air, or a mixture of air and fuel. A fixed orifice jet, sometimes called a *pilot jet,* is a jet with a hole that is not adjustable. An adjustable orifice jet is a jet with hole that can be made larger or smaller with a screw.

An **emulsion tube**, a long tube installed in a carburetor passageway, is another common carburetor part. It works like a jet to allow a controlled

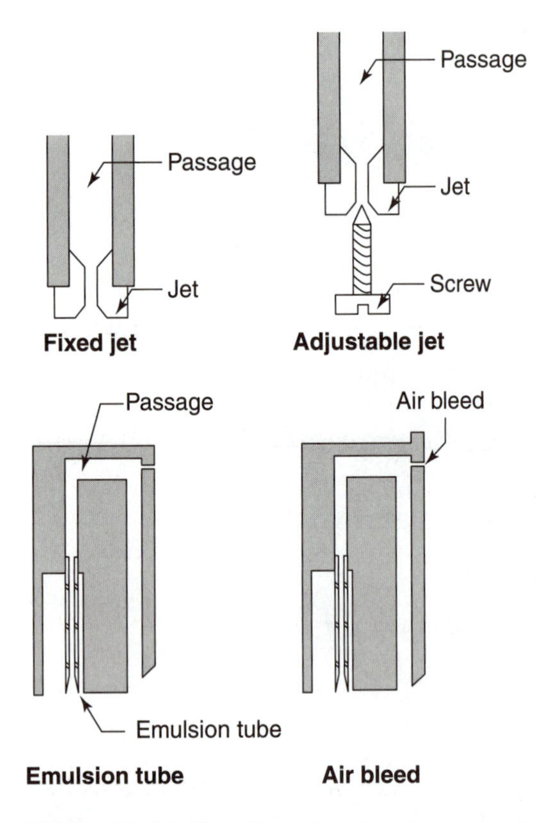

FIGURE 15-10 Flow-through carburetor passages are controlled by jets, emulsion tubes, and air bleeds.

amount of fuel or air through a passageway. The tube has a series of holes along its length. Air enters the holes and mixes with the fuel passing through the tube. Emulsion tubes get their fuel from a space called the *emulsion tube well.*

All carburetors have air bleeds and vents to control air flow through the carburetor passages. An **air bleed** is a small air passage that allows atmospheric pressure to push on the fuel in the carburetor.

A **vent** is an air passage that allows atmospheric pressure to act on the fuel in a float bowl (Figure 15-11). Many carburetors have an internal float bowl vent located between the air cleaner and the venturi. The atmospheric pressure in this area changes as the air cleaner element gets dirty. The internal vent compensates for this pressure change. Without this compensation, a dirty air cleaner would cause the engine to run too rich. With this compensation, less pressure pushes on the fuel when the air cleaner is dirty. The mixture becomes leaner because of the lowering of atmospheric pressure in the float bowl. Having the vent below the air cleaner also ensures that the vent does not get plugged up by dirt.

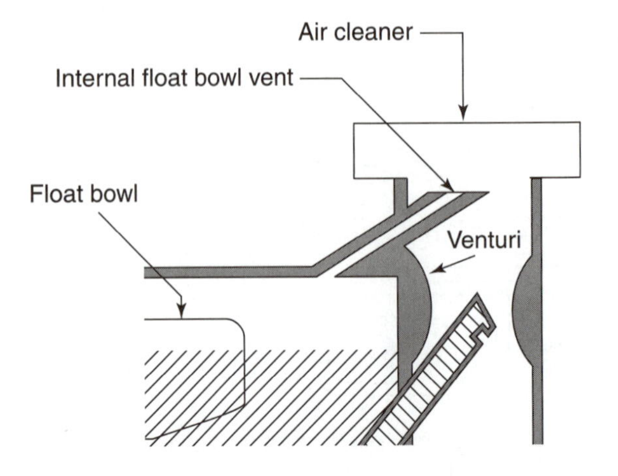

FIGURE 15-11 An internal float bowl vent regulates atmospheric pressure on the fuel in the float bowl.

VACUUM CARBURETOR

The vacuum carburetor is one of the most common carburetors. These carburetors are often installed on the smaller, least expensive engines. They are often on walk behind lawn mowers and edgers. A **vacuum carburetor** uses vacuum to pull fuel out of the fuel tank and mixes it with the air entering the engine (Figure 15-12). Vacuum carburetors are also called *suction carburetors*. These carburetors are always mounted on top of the fuel tank. All vacuum carburetors work the same basic way.

A vacuum carburetor has a very simple one-piece housing (Figure 15-13). The housing is basically a tube with an opening at one end for intake air to enter. A choke valve in the air opening can be opened or closed to regulate air flow. The other end of the housing is an outlet for the air-fuel mixture. The mixture goes out the outlet and enters the en-

FIGURE 15-12 Engine with a vacuum carburetor.

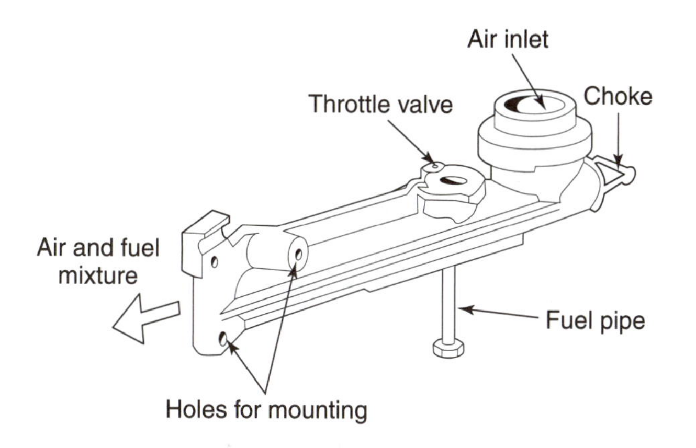

FIGURE 15-13 Parts of a vacuum carburetor.

gine intake port. This end has screw holes to mount the carburetor to the engine.

There is a throttle valve inside the carburetor just past the air entrance. Below the throttle valve is a tube called the *fuel pipe* (Figure 15-14). The fuel pipe brings fuel up into the carburetor from the fuel tank. The bottom of the fuel pipe has a small screen that stops dirt from going up the pipe and into the carburetor. A small ball check valve fits in the bottom of the pipe. The ball check valve allows fuel to go up the pipe. It will not let fuel run back out of the pipe.

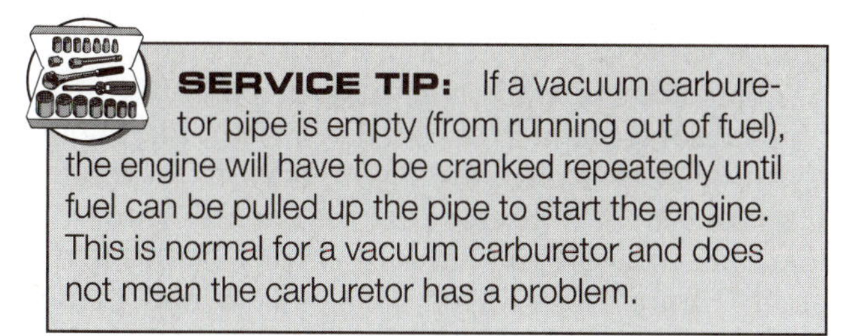

SERVICE TIP: If a vacuum carburetor pipe is empty (from running out of fuel), the engine will have to be cranked repeatedly until fuel can be pulled up the pipe to start the engine. This is normal for a vacuum carburetor and does not mean the carburetor has a problem.

The fuel tank fits on the bottom of the carburetor. The carburetor fuel pipe goes down into the bottom of the fuel tank. The cap on the fuel tank is vented to allow atmospheric pressure into the tank. Without a vent, a vacuum could form in the

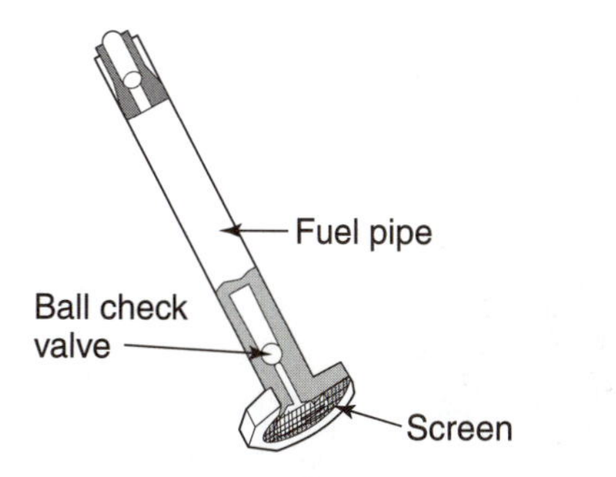

FIGURE 15-14 The fuel pipe has a screen and ball check valve.

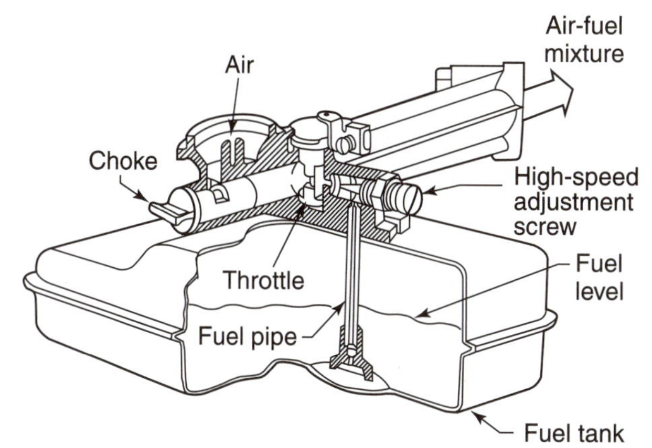

FIGURE 15-15 Air coming through the carburetor pulls fuel up the fuel pipe and into the engine.

tank, which would prevent fuel from going up the fuel pipe.

A low pressure (vacuum) is created inside the carburetor as the piston moves down on the intake stroke. This low pressure pulls fuel up the fuel pipe (Figure 15-15). The throttle valve is in the flow of air entering the carburetor. It slows the air down and creates a low pressure like a venturi. The low pressure helps pull fuel up the fuel pipe. The fuel is mixed with the intake air passing through the carburetor housing. The amount of fuel that comes out the fuel pipe is regulated by a high-speed adjustment screw.

There are two discharge (metering) holes in the carburetor housing (Figure 15-16). Fuel can enter through either or both of these metering holes. When the throttle is open, there is a high vacuum in the carburetor. This vacuum causes maximum flow

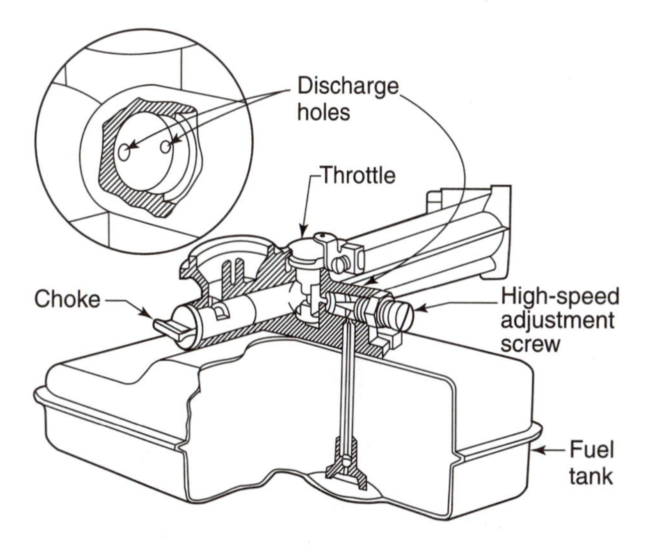

FIGURE 15-16 Fuel enters through one or both fuel discharge holes.

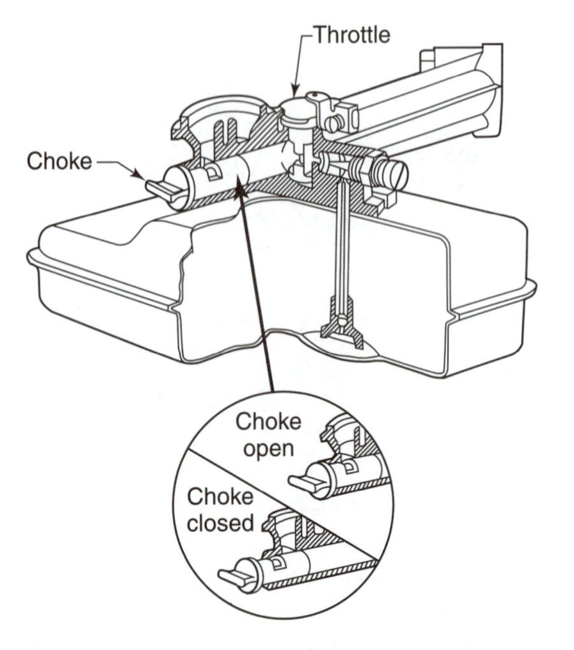

FIGURE 15-17 The choke is opened to let air in and closed to stop the air.

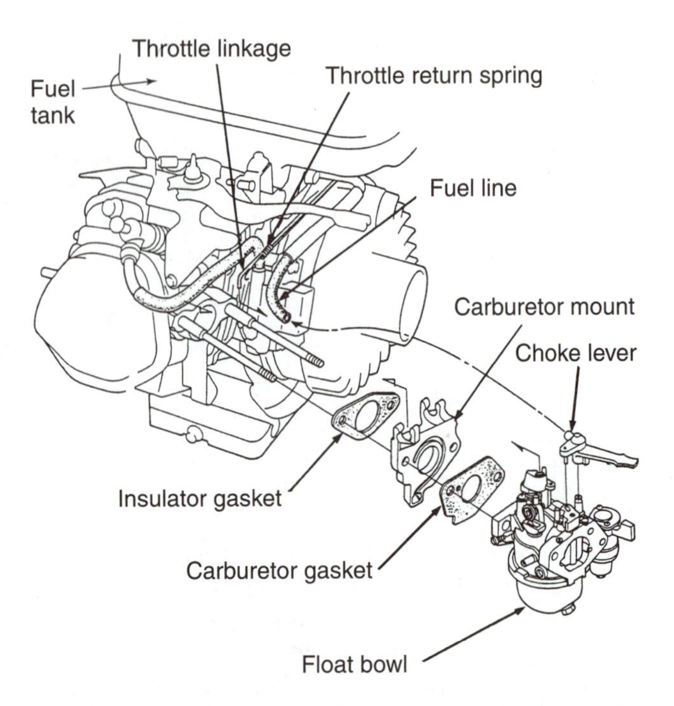

FIGURE 15-18 Engine with a float carburetor. (Courtesy of American Honda Motor Co., Inc.)

out of both holes. When the throttle is fully closed, only one of the metering holes has a vacuum. This hole allows a small amount of fuel flow for idle. This allows the engine to run with a closed throttle.

The choke is a simple sliding plate (Figure 15-17). It is pulled out by the operator (or linkage on the throttle control) to close off the air. The closed choke restricts air flow and creates a high vacuum in the carburetor. The high vacuum causes a maximum amount of fuel to be delivered from the fuel pipe. This rich mixture helps in starting a cold engine. The operator pushes the choke in when the engine starts.

FLOAT CARBURETOR

Many larger, more expensive engines use a **float carburetor** (Figure 15-18), which is a carburetor that has an internal fuel storage supply controlled by a float assembly. The fuel tank on the float carburetor system is attached to another part of the engine. It is often mounted higher than the carburetor. Gravity causes fuel to flow from the tank through a fuel line to the carburetor. Some engines use a fuel pump to transfer the fuel from the tank to the carburetor. A float assembly in the carburetor controls the flow of fuel from the tank.

Float Operation

The fuel line from the fuel tank provides fuel to the carburetor. The fuel line is connected to the carburetor fuel inlet fitting. The inlet has a part called the *fuel inlet seat*. The **inlet seat** is a carburetor part that houses and provides a matching seat for the tapered end of the float inlet needle valve. The fuel goes through the inlet seat into the carburetor float bowl.

The **float bowl** is a carburetor part that provides a storage area for fuel in the carburetor. There is a small vent in the top of the bowl to let in air. The **float** is a carburetor part that floats on top of fuel and controls the amount of fuel in the float bowl (Figure 15-19). Floats are hollow and made from copper or plastic. A hinge is attached to the float. The hinge is connected to the side of the float bowl by a pin and pivot. These parts allow it to go

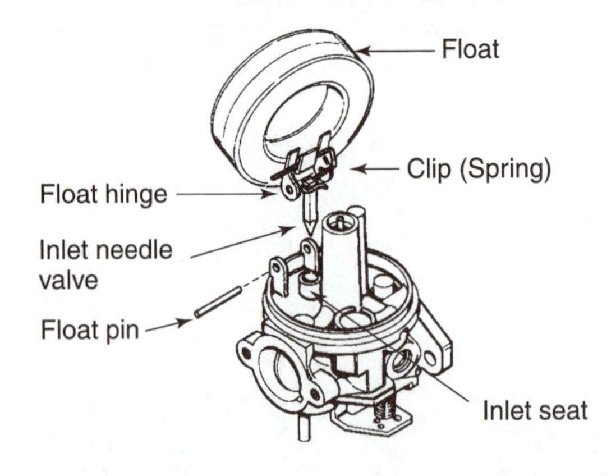

FIGURE 15-19 Parts of a float and inlet valve assembly. (Provided courtesy of Tecumseh Products Company.)

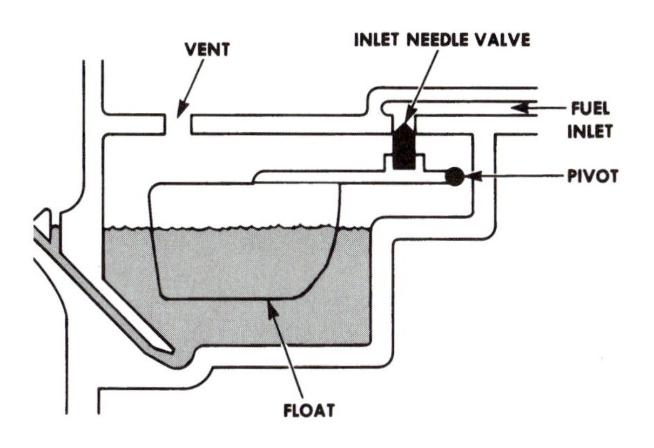

A. Float Parts

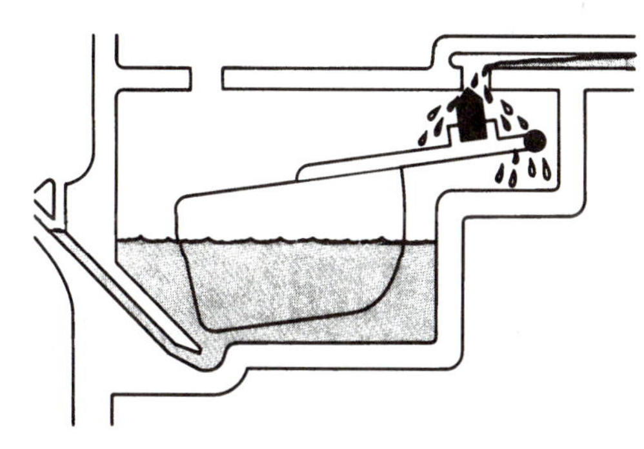

B. Inlet Valve Open

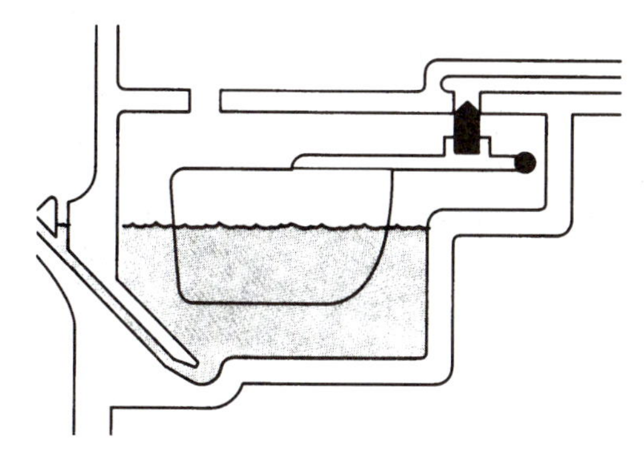

C. Inlet Valve Closed

FIGURE 15-20 Float movement up and down opens and closes the needle inlet valve.

up or down on top of the fuel level. The **inlet needle valve** is a carburetor part that fits on the float and controls fuel flow through the fuel inlet. The needle valve has a sharp end.

As the engine runs, fuel in the float bowl is used up. The fuel level in the bowl drops. As the fuel level drops, the float drops. When the float moves down, the inlet needle valve connected to it moves out of the inlet seat. Fuel is allowed to come in through the inlet passage from the fuel tank. The fuel level rises as more fuel comes into the bowl. The float also rises, pushing the inlet needle valve into the inlet seat. When the inlet needle valve rises far enough, it "seats." This prevents fuel from getting into the bowl. Flow from the fuel tank is stopped until the fuel level in the float bowl drops again. This action keeps repeating to maintain the required level of fuel in the float bowl (Figure 15-20).

Float Carburetor Types

All float carburetors use a float to control the fuel level. There are, however, many different styles of float carburetors. These carburetors are commonly identified by the direction of air flow into the carburetor throat. A **carburetor throat** is the part of the carburetor that directs air flow in toward the venturi.

Updraft, downdraft, and sidedraft are common carburetor designs (Figure 15-21). An **updraft**

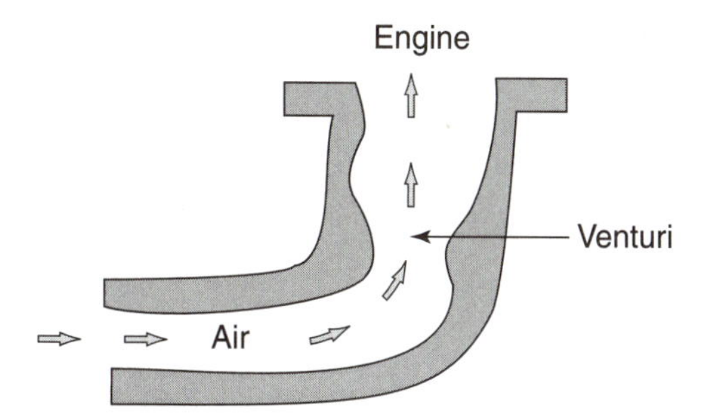

A. Updraft

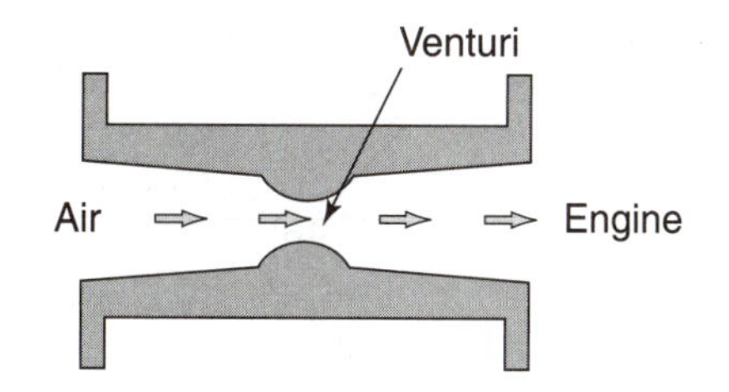

B. Sidedraft

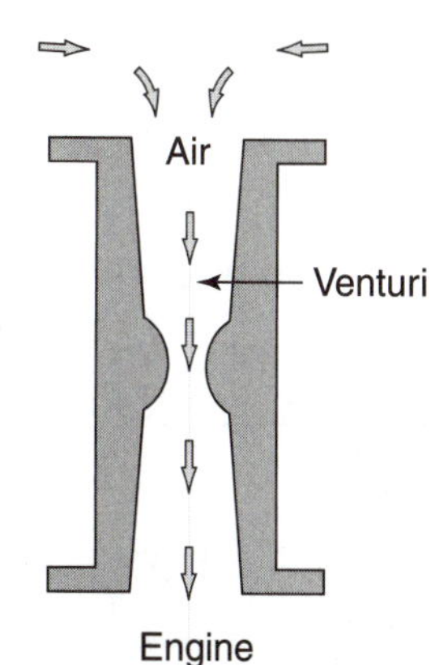

C. Downdraft

FIGURE 15-21 Float carburetors may be updraft, sidedraft, or downdraft.

carburetor is a carburetor in which the air flows into the venturi in an upward direction. Updraft carburetors are commonly installed on older, larger engines. A **downdraft carburetor** is a carburetor in which the air flows into the venturi in a downward direction. Downdraft carburetors are used on some multiple cylinder engines. An intake manifold is used to connect a downdraft carburetor to the intake ports of each cylinder. A **sidedraft carburetor** is a carburetor in which the air flows into the venturi from the side. The sidedraft carburetor is very common and is used on many different sizes and styles of engines.

Float Carburetor Operation

Most float carburetors operate in the same basic way. The operation of the carburetor can be divided into different modes (circuits):

- Float system operation
- Idle system operation
- Part throttle operation
- High-speed operation
- Choke system operation

The float system operates during all engine speeds (Figure 15-22). It provides fuel for all the other carburetor systems. Fuel flows from the fuel tank to the carburetor by gravity or a fuel pump. Fuel enters the carburetor through the inlet fitting. It goes past the inlet needle valve and begins filling the carburetor bowl. As the bowl fills, the float rises, raising the inlet needle valve toward the inlet seat (Figure 15-23). When the inlet needle closes, fuel flow into the bowl stops. Fuel remains at this level until engine operation begins to draw fuel from the bowl. When the fuel level drops again, the float

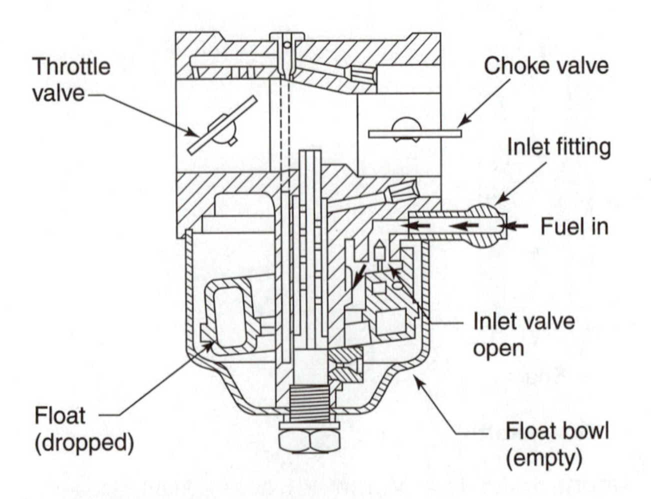

FIGURE 15-22 Fuel inlet to the float bowl.

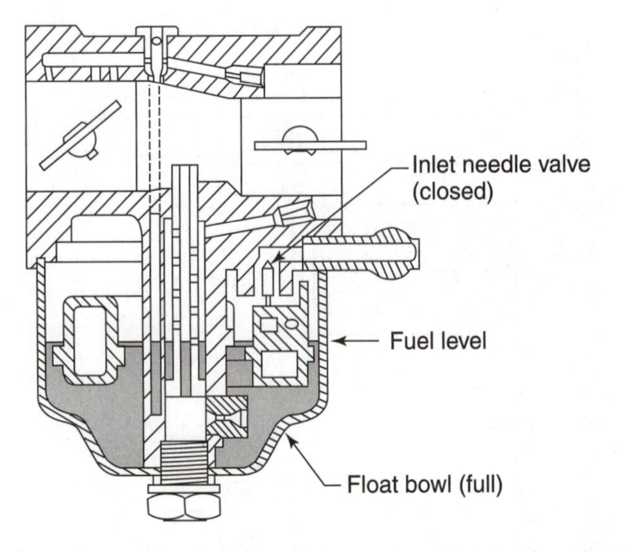

FIGURE 15-23 Needle valve operation with a full float bowl

moves down causing the inlet needle valve to move away from the inlet seat. Fuel again flows into the float bowl. This happens over and over to provide a constant fuel supply.

When the engine is idling, the throttle valve is in the closed (or nearly closed) position. The idle system delivers air-fuel mixture to the intake port side of the throttle valve (Figure 15-24). Without this system, the engine could not run at idle speed.

When the cylinder is on an intake stroke, low pressure is created in the intake port. The carburetor throttle plate area also has this low pressure. Atmospheric pressure in the carburetor bowl pushes fuel through the fixed high-speed jet. Fuel continues up a small passage called the *idle passage*.

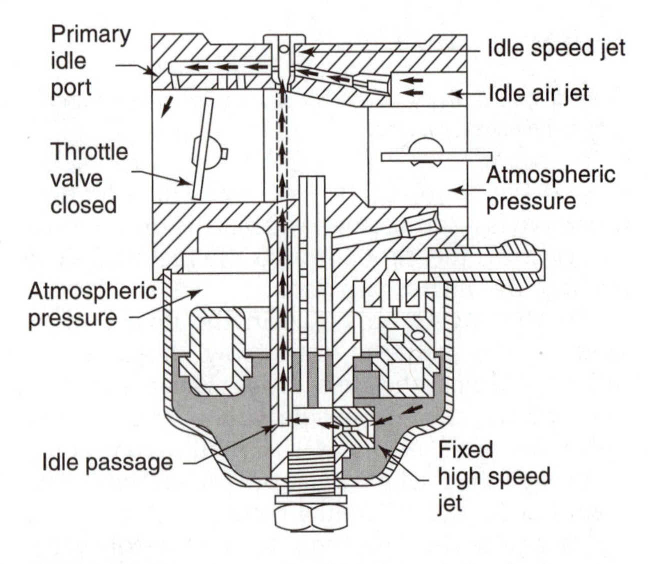

FIGURE 15-24 Float carburetor idle system operation.

The atmospheric pressure from the intake stroke also pulls air into the throat of the carburetor. Some of this air goes through a passage called the *idle air jet*. As the fuel comes up the idle passage, it enters the center of the idle speed jet. Here it mixes with air from the idle air jet. The air and fuel mixture is then pulled through a passage called the *primary idle port*. It then goes out into the carburetor throat. Here, it mixes with air flowing through the carburetor throat and goes into the engine's cylinder.

When an operator (or governor) wants a speed increase, the throttle linkage is used to open the throttle valve. The carburetor uses the part throttle system (Figure 15-25) when the throttle valve is open part of the way. The part throttle system has the same air-fuel flow as the idle system with one exception. There are several secondary idle ports in the carburetor throat. These are uncovered as the throttle plate opens. The secondary ports give additional routes for the air and fuel mixture for part throttle engine speeds. The part throttle system can operate momentarily as the throttle passes from idle to high speed. It can also operate continuously if the throttle stays in the part throttle position.

When the operator (or governor) moves the throttle linkage past the part throttle position to more fully open, the carburetor uses the high-speed system (Figure 15-26). The intake stroke causes a low pressure in the carburetor throat. Atmospheric pressure pulls air through the venturi in the middle of the carburetor throat. There is a drop in pressure at the venturi. Atmospheric pressure pushes fuel through the fixed high-speed jet. From there it goes through a large passage called the *main pickup tube*.

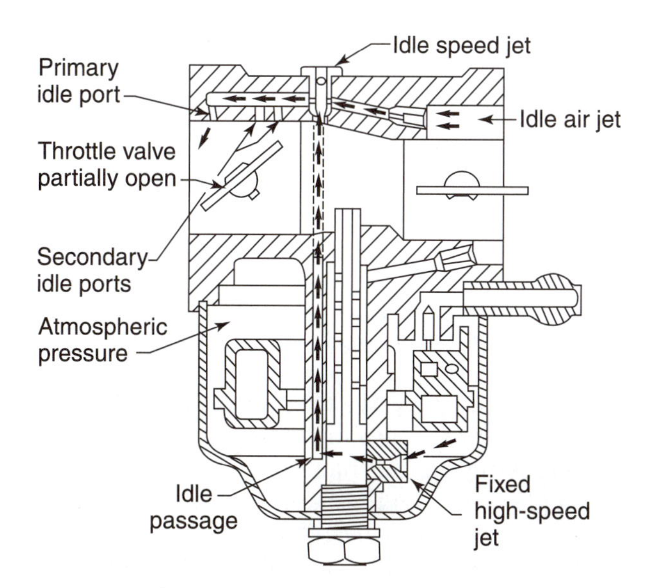

FIGURE 15-25 Float carburetor part throttle system operation.

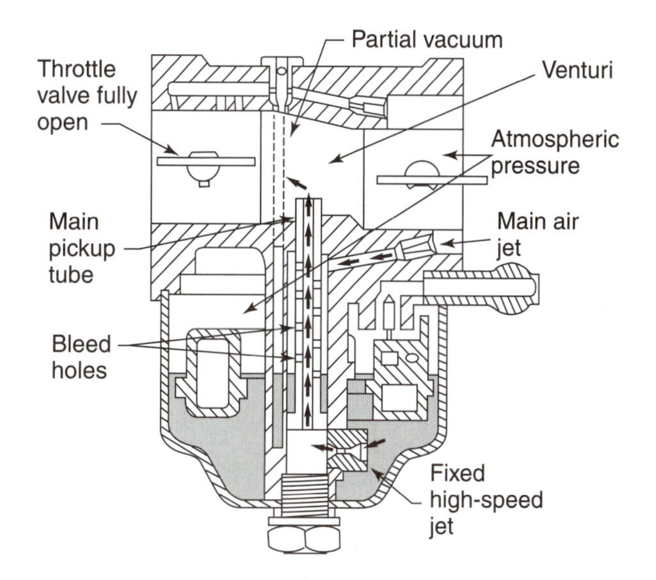

FIGURE 15-26 Float carburetor high-speed system operation.

Atmospheric pressure also pushes air through a large air passage called the *main air jet*. From the main jet air flows to the outside of the main pickup tube. This air enters through the main pickup tube bleed holes. There, it mixes with the fuel coming up the inside of the main pickup tube. The air-fuel mixture is pushed up and out the main pickup tube into the incoming air at the venturi.

A choke system (Figure 15-27) is used when starting a cold engine. The choke valve fits in the carburetor throat air entrance. Some choke valves have small air bleed holes. Others are slightly smaller than the carburetor throat. This allows a small amount of air to get past the closed valve. When the choke valve is closed (engine is on an

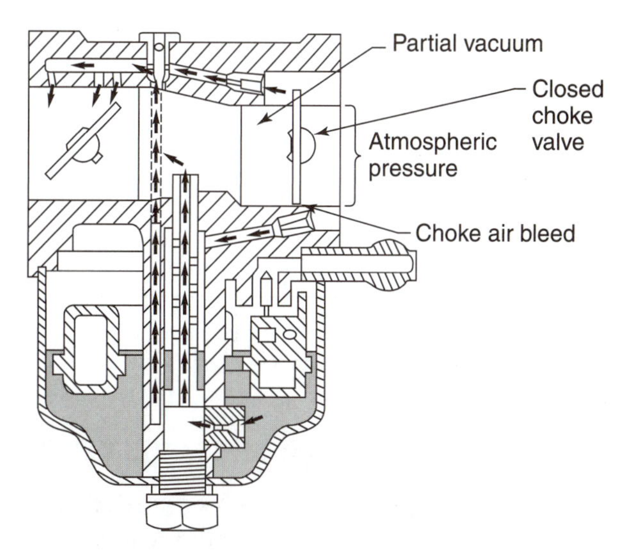

FIGURE 15-27 Float carburetor choke system operation.

CASE STUDY

A technician was preparing a brand new lawnmower for delivery to a customer. She had checked the engine controls and the oil level and had filled the fuel tank with fresh fuel. She set the choke for a cold start and pulled the starter rope. After a couple of pulls, the engine started and sounded good. She moved the choke linkage to off position and allowed the engine to run and warm up. She intended to make any necessary carburetor adjustments after the engine was warm. She went into the office to do some paperwork while the mower was warming up.

While in the office she heard the engine start to change idle speed. It began to speed up then started to misfire. By the time she got to the mower, the engine had quit running. The first thing she noticed was that there was fuel dripping off the bottom of the carburetor. At first she thought a fuel line might be leaking. A quick inspection ruled this problem out. She removed the air cleaner and quickly found the source of the fuel. It was coming out of the carburetor throat. This could mean only one thing. The float bowl was overfilling.

She considered this a very unusual problem for a new engine. The problem had to be in the float or inlet needle. A sticking inlet fuel valve is a common problem on older carburetors with stale fuel, but this is a new carburetor with fresh fuel. She removed the carburetor top and located the source of the problem. The float was sunk in the fuel when it should have been floating on top. She removed the hollow float and inspected it closely. There was a small hole caused by a manufacturing defect. The hole allowed fuel to enter the float and caused it to sink. The sunken float held the fuel inlet valve open and allowed fuel to overflow and flood the engine. A new float fixed the problem.

intake stroke), there is a low pressure in the entire carburetor throat. Fuel is pulled out of the idle and high-speed system passages. This provides a very rich mixture to help the engine start.

DIAPHRAGM CARBURETOR

A **diaphragm carburetor** (Figure 15-28) is a carburetor that has a flexible diaphragm to regulate the amount of fuel available inside the carburetor.

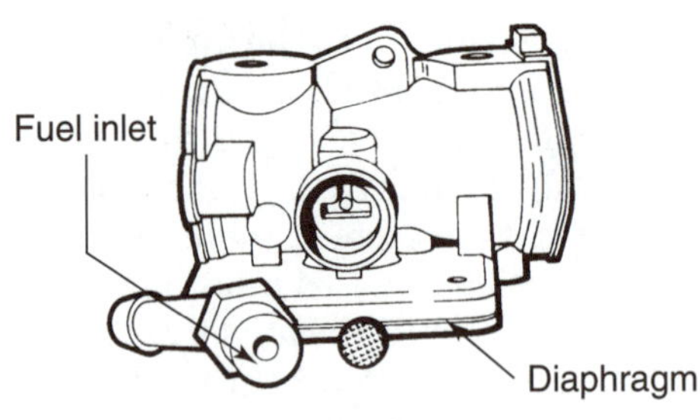

FIGURE 15-28 A diaphragm carburetor uses a diaphragm to regulate the amount of fuel in the carburetor. (Provided courtesy of Tecumseh Products Company.)

Float and vacuum carburetors work only on engines that are used upright. A float- or vacuum carburetor-equipped engine cannot be turned on its side. The float or fuel tube would not be able to regulate the fuel level. The engine would run out of fuel and stop. Hand-held outdoor power equipment such as a chainsaws, leaf blowers, and string trimmers must work in any position. Two-stroke engines used on this equipment use a diaphragm carburetor, because it works in any position.

Diaphragm Operation

The top part of the diaphragm carburetor is the same as a float carburetor. It has a throat, choke valve, throttle valve, and venturi. The diaphragm carburetor does not have a float bowl. Instead, it uses a diaphragm similar to that used in a fuel pump (Figure 15-29). The diaphragm controls a small amount of fuel in a fuel chamber. A fuel inlet needle valve similar to that in a float carburetor is used to control fuel flow into the carburetor.

The diaphragm is made from a flexible material. It is stretched across a small space called a *chamber*. The center of the diaphragm has a metal tab (or lever on some designs) that contacts the inlet needle valve. The inlet needle valve works the same as the needle valve in a float carburetor. The space below the diaphragm is called an *air chamber,* which has an air vent that allows atmospheric pressure below the diaphragm. The air chamber provides the space for diaphragm up-and-down movement.

Atmospheric pressure on the fuel in the tank causes fuel to flow from the fuel tank to the fuel inlet. The spring pushes down on the control lever, causing the needle valve to drop down and allowing fuel to come in around the inlet needle valve. As the fuel fills up the chamber, it pushes down on the diaphragm (Figure 15-30). Downward movement of the diaphragm causes the control lever to pivot upward. This movement pushes up on the inlet needle valve, closing the fuel inlet. When fuel is used up, the diaphragm comes back up allowing

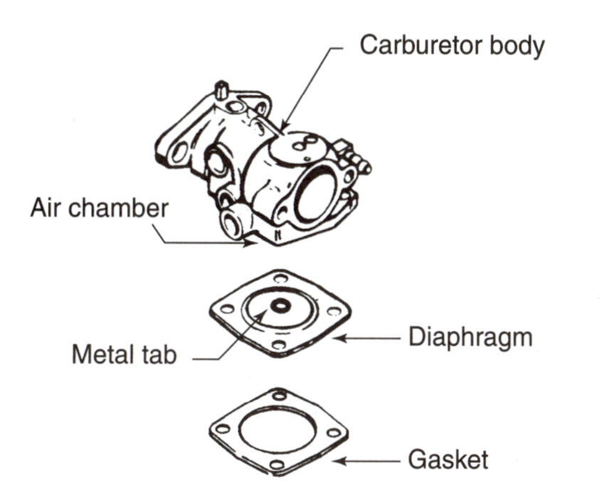

A. Diaphragm Parts

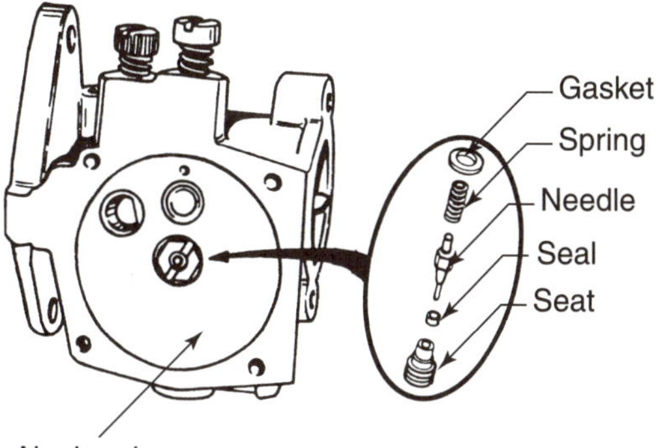

B. Needle Valve Location

FIGURE 15-29 Parts of the diaphragm and needle valve assembly. (Provided courtesy of Tecumseh Products Company.)

the inlet needle valve to open to let fuel in again. The diaphragm does not depend upon gravity. It will work the same way no matter what the position of the engine. The engine can run even if it is upside down.

Some diaphragm carburetors have two diaphragms: One is used to control the fuel flow into the carburetor, the other is an impulse fuel pump. The pump section pumps fuel from the fuel tank to the carburetor.

Diaphragm Carburetor Operation

The diaphragm carburetor provides the correct air-fuel mixtures for several modes (circuits) of operation (Figure 15-31). These include:

- Cold starting
- Idle
- Intermediate speed
- High speed

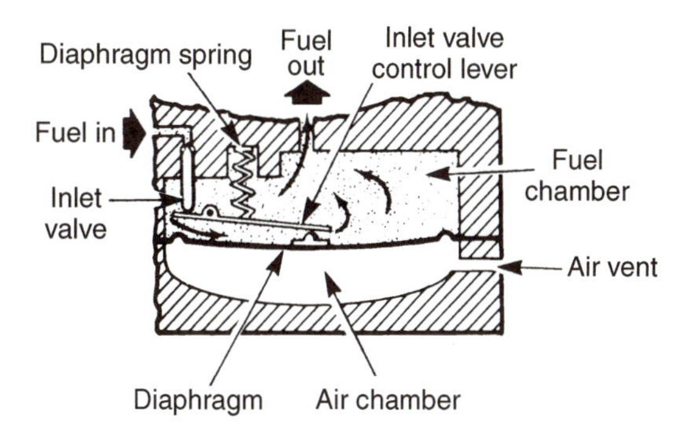

A. Diaphragm Parts

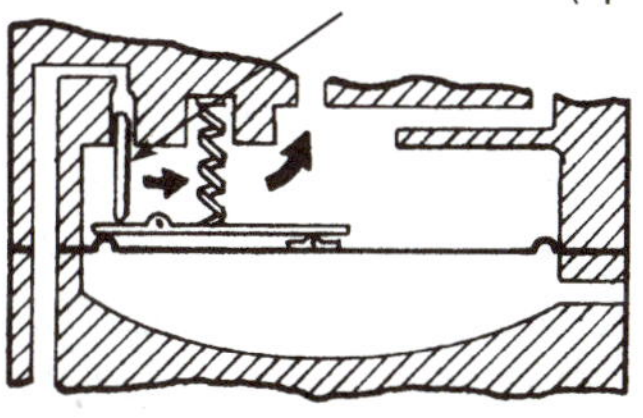

B. Fuel Entering the Fuel Chamber

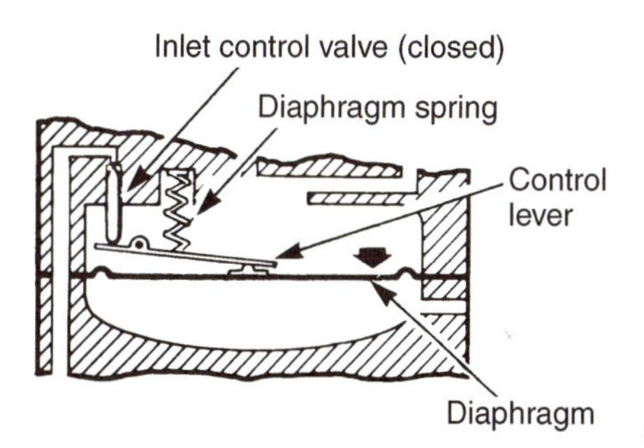

C. Diaphragm Stopping Fuel Flow

FIGURE 15-30 Operation of the diaphragm as it regulates fuel.

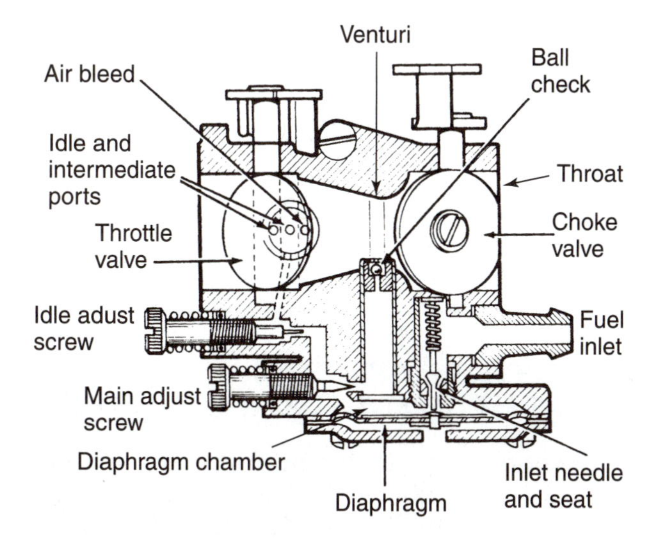

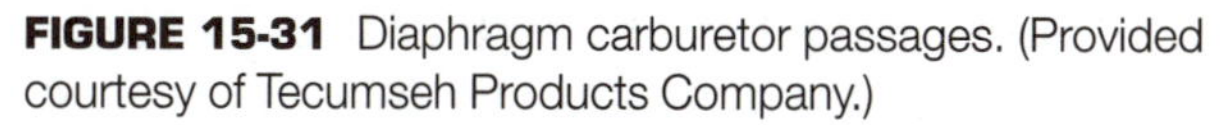

FIGURE 15-31 Diaphragm carburetor passages. (Provided courtesy of Tecumseh Products Company.)

When the engine is cold, a rich mixture is required for starting. The choke system has a valve (sometimes called a *shutter*) in the carburetor throat. In the choke mode, the choke valve is closed. The only air that can get into the engine enters through openings around the choke valve. When the engine is cranked during starting, the intake stroke creates a low pressure in the venturi. The low pressure pulls fuel from the diaphragm chamber up the main nozzle. Fuel is also pulled from idle fuel discharge ports. The fuel mixes with the air that passes around the choke valve. A very rich air and fuel mixture is used to start the cold engine.

A small amount of fuel is needed to keep the engine running during idle speeds. The throttle valve is almost closed during idle. A small idle discharge port is located on the engine side of the closed throttle valve. The low pressure in this area pulls fuel from the diaphragm chamber. Fuel goes past an idle adjusting screw and is delivered behind the throttle valve. The fuel is mixed with air that gets through the almost-closed throttle valve. Additional air comes through an idle air bleed passage. The idle adjusting screw adjusts the amount of fuel that is delivered out the idle discharge port.

When the throttle valve is moved past the idle position, it uncovers two more discharge ports, called the *intermediate ports*. They provide more fuel to mix with the air flowing into the engine. The fuel flows from the diaphragm chamber past the idle mixture adjusting screw. Fuel and air flow is the same as in the idle mode. The additional fuel from the intermediate ports allows the engine to operate at faster speeds.

The high-speed mode is used when the throttle valve is opened further. Air flows through the carburetor throat at high speed. The venturi further accelerates the air flow and creates a low pressure in the venturi area. This low pressure pulls fuel into the air stream through a delivery tube called the *main nozzle*. Fuel gets into the main nozzle through a passageway from the diaphragm chamber. Fuel going up the main nozzle must pass the main adjusting screw, which is used to adjust the amount of fuel for high-speed operation.

SUCTION FEED DIAPHRAGM CARBURETOR

The **suction feed diaphragm carburetor** is a carburetor that combines the features of a vacuum carburetor and the impulse fuel pump (Figure 15-32). This carburetor is used on four-stroke engines. These engines are not usually used in a variety of positions. The carburetor is mounted on the top of the fuel tank. It meters fuel into the air in the same way as the vacuum carburetor. Some carburetors have the diaphragm mounted in a side chamber. Others have the diaphragm located between the carburetor body and the fuel tank (Figure 15-33).

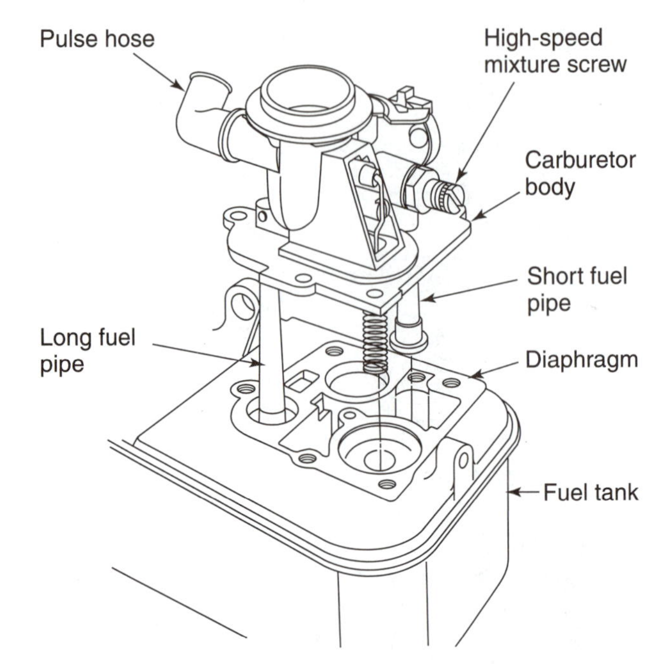

FIGURE 15-32 Parts of a suction feed diaphragm carburetor.

This carburetor is different from the vacuum carburetor. It has two different length fuel pipes (Figure 15-34). The longer fuel pipe goes into the fuel tank and is used to pull fuel out of the tank. The shorter fuel pipe goes into a small chamber of the fuel tank. The chamber is called the *fuel cup* or *fuel well*. A diaphragm fits between the carburetor and the fuel cup. The diaphragm works like an impulse fuel pump, transferring fuel between the tank and the fuel cup. This system gives a constant level of fuel regardless of fuel tank level.

A pulse hose connects the pumping chamber to the intake manifold (or crankcase in some designs). When the engine is running, the pulse hose transmits a pulse to the diaphragm chamber. The diaphragm moves up and down with the pressure pulses, pumping fuel up the long fuel pipe into the fuel tank cup. Fuel goes out of the fuel cup into the venturi through the short fuel pipe.

AUTOMATIC CHOKE

Many carburetors use a manual choke. The **manual choke** is a choke valve that must be operated by

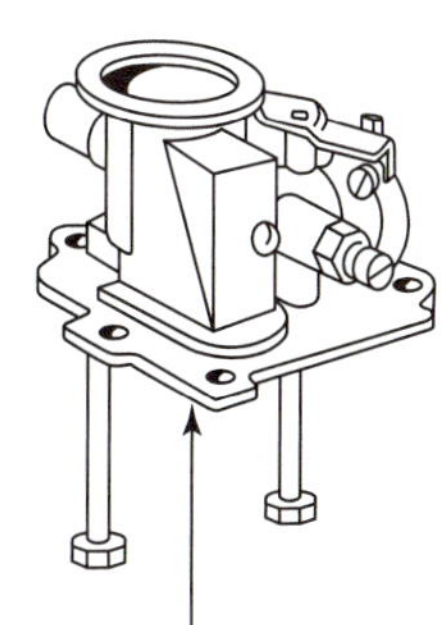

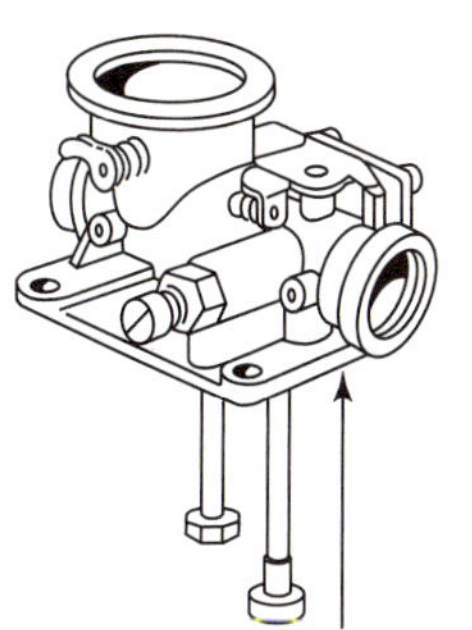

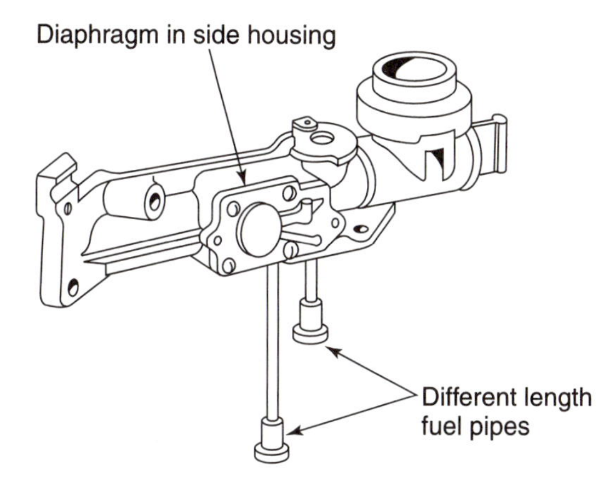

FIGURE 15-33 Suction feed diaphragm carburetors have two fuel pipes.

hand by moving a choke lever or through linkage from an engine control lever or switch. Some carburetors use an automatic choke. An **automatic choke** is a choke valve connected to a diaphragm or bimetal spring that automatically opens or closes the choke valve.

The diaphragm type automatic choke uses a diaphragm mounted under the carburetor (Figure 15-35). It is connected to the choke valve shaft by a link. A spring under the diaphragm holds the choke valve closed when the engine is not running. When the engine is started, low pressure is created in the cylinder during the intake stroke. The low pressure acts on the bottom of the diaphragm through a small passage. The low pressure pulls the diaphragm down. The link attached to the diaphragm also travels down. The link pulls the choke valve into the open position.

The vacuum under the diaphragm leaks away when the engine is stopped. The spring moves the diaphragm in a direction to close the choke valve. The choke is ready for another starting cycle.

The bimetal type automatic choke uses a coiled spring (Figure 15-36) made from two different metals. The two metals have different amounts of heat

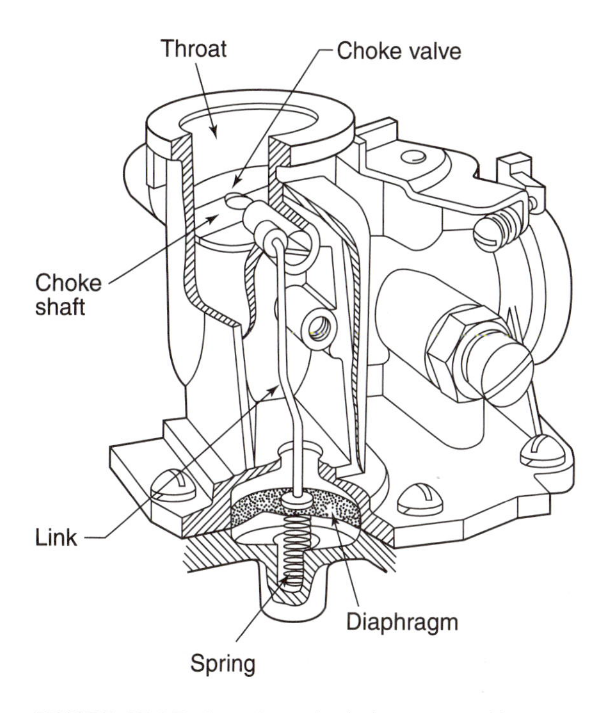

FIGURE 15-35 An automatic choke operated by a diaphragm.

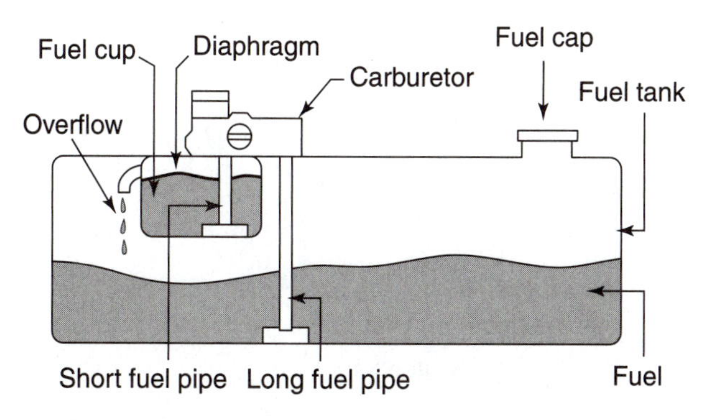

FIGURE 15-34 Operation of the suction feed diaphragm carburetor.

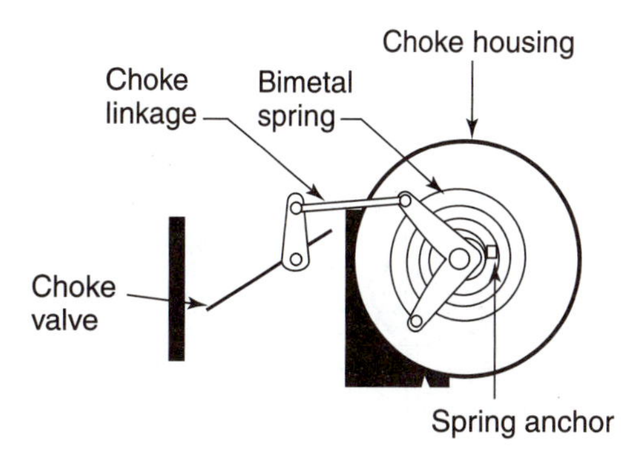

FIGURE 15-36 An automatic choke operated by a bimetal spring.

expansion. The two metals cause the spring to coil or uncoil as it gets hot and cold. The choke spring is mounted in a small carburetor housing next to the choke valve. One end of the choke spring is connected (directly or by linkage) to the choke valve. The other end is anchored to the housing.

When the engine is cold, the bimetal spring contracts. The end of the spring moves the choke valve to a closed position. As the engine warms up, the spring expands. The expanding spring moves the choke valve into an open position.

AIR CLEANER

The engine pulls in a large amount of air even during low-speed operation. Engines on outdoor power equipment often operate in dusty or dirty conditions. The intake air is full of abrasive dirt particles. These abrasives must be prevented from entering the engine. An **air cleaner** is a filter element that traps dirt before it can enter the engine air intake system. The three types of air cleaners are oil bath, paper filter element, and foam filter (oil-wetted polyurethane).

Oil Bath Air Cleaner

The oil bath air cleaner is a common type of air filter used on older engines. An **oil bath air cleaner** is an air cleaner that directs incoming engine air over an oil sump that catches dirt particles (Figure 15-37). Like other air cleaners, the oil bath type is mounted to the carburetor. All the air going into the carburetor must go through the air cleaner first. The housing is the main part of the oil bath air cleaner. The housing has air passages that direct the incoming air. Incoming dirty air comes in the top of the housing. It is then directed down the inside of the housing. The housing passages cause the air to change direction and start back up the housing.

There is a small oil sump in the bottom of the housing. The air entering the housing is made to change direction over the oil. Dirt in the air is too heavy to make the quick turn. It keeps going straight into the oil. The oil traps the dirt and the cleaned air goes through an oiled screen. Any dirt left in the air sticks to the oiled screen. Cleaned air then goes down the center of the housing and into the carburetor.

The oil bath air cleaners are not used on newer engines. The oil in the housing requires frequent service. The complicated air routing inside the housing limits the air flow into the engine, which limits engine power.

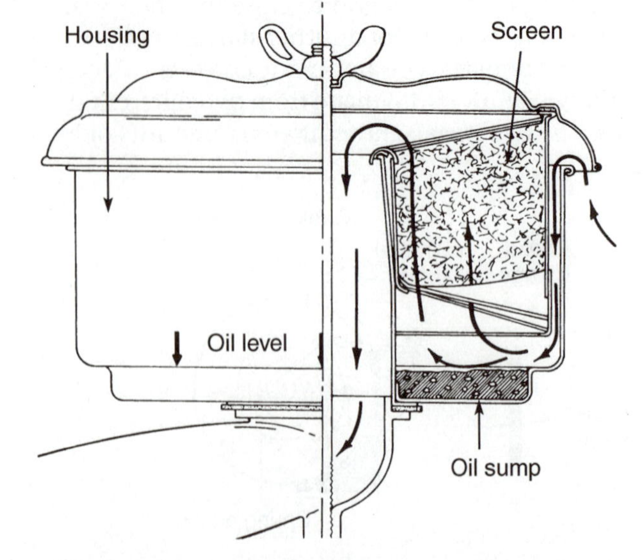

FIGURE 15-37 An oil bath air cleaner has sump with oil to clean the air. (Courtesy of Kohler Co.)

Paper Filter Air Cleaner

Many engines use an air cleaner housing that has a paper filter element. A **paper air filter element** is a paper air cleaner filter that allows air to pass through while trapping dirt particles (Figure 15-38). The paper filter is called a *filter element.* It is made from a filter paper that is folded into pleats. The pleats provide a large surface area for air to pass through. A fine mesh metal screen is often used to support the paper element. The screen also protects against a fire hazard if an engine backfires. A backfire occurs when a flame escapes from the combustion chamber up through the carburetor throat. The filter element has a rubber seal on the top and bottom. These provide an air tight seal for the filter in the housing. Sealing is very important. Any air

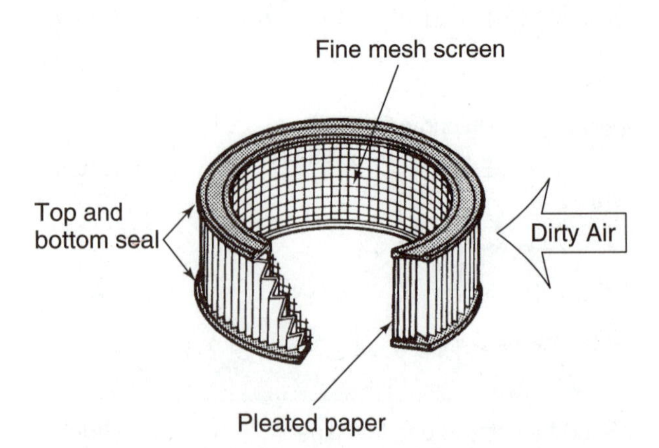

FIGURE 15-38 A paper filter element traps dirt and allows clean air to pass.

that does not go through the filter on the way into the engine could contain dirt.

The paper filter element fits inside an air cleaner body or housing (Figure 15-39). A metal cover usually covers the top of the filter. One or more screws or studs hold the cover and paper filter in the air cleaner body. Sometimes a gasket is used between the air cleaner body and the carburetor mounting area.

Foam Filter Air Cleaner

A **foam air filter element** (Figure 15-40) is a flexible polyurethane foam air cleaner filter material that filters intake air. The foam material has tiny pores (microscopic holes) that trap dirt in the intake air. Foam air filter elements are used on many engines. They may be used either as the main filter element or along with a paper filter as a precleaner (prefilter). A precleaner is a thin foam wrap around a paper air filter element. The precleaner filters the larger dirt particles before they get to the paper element. These dual action filters are often used on engines that are operated in dusty conditions.

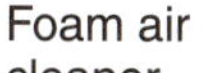

FIGURE 15-40 A foam type air filter element.

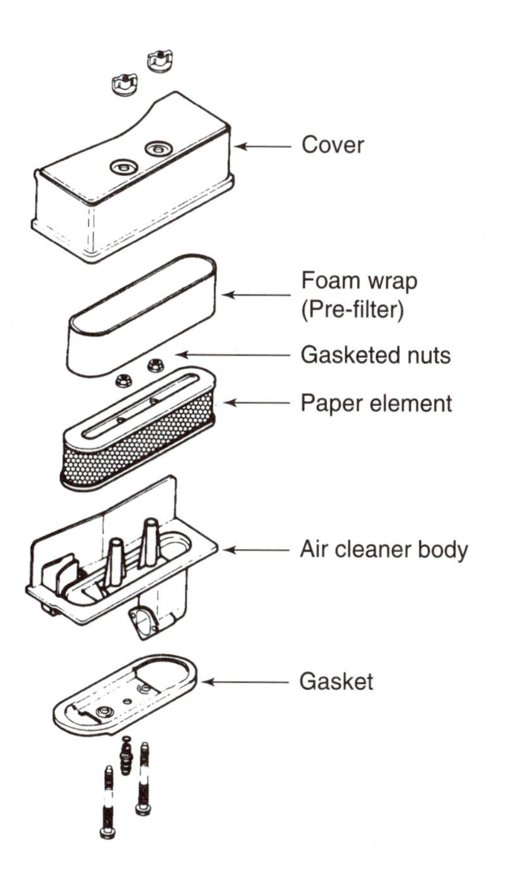

FIGURE 15-39 A paper filter element with a foam prefilter. (Provided courtesy of Tecumseh Products Company.)

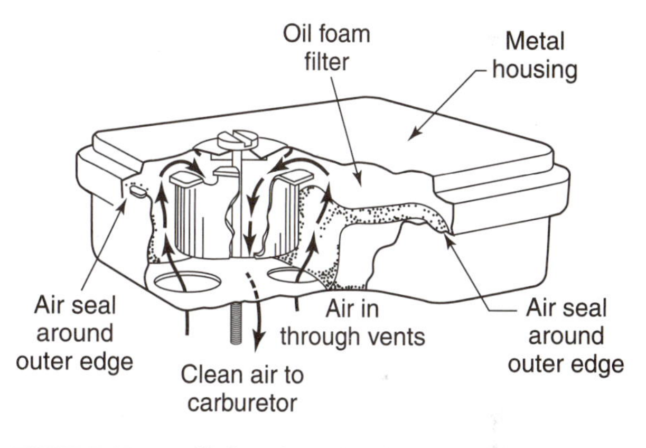

FIGURE 15-41 Air flow through a foam filter element.

Foam air filter elements are common on vacuum carburetors (Figure 15-41). Air enters the foam air cleaner through large holes in the bottom of the metal housing. Air must go through the foam filter pores to get into the engine. The pores filter out dirt before it can get into the engine. The foam filter or precleaner wrap is usually coated (wetted) with a light coat of engine oil. The oil coating improves the foam filter's dirt trapping ability.

REVIEW QUESTIONS

1. Explain why a carburetor must atomize and vaporize fuel before it can be delivered to the cylinder.
2. List and describe the operation of the parts that make up a basic carburetor.
3. Explain the purpose of the high- and low-speed adjustment screw.
4. Describe the parts and operation of a vacuum carburetor.
5. Describe the parts and explain the operation of a float system.
6. List and describe the three common types of float carburetors based on air flow direction.
7. List and describe the purpose of the five different operating modes for a float carburetor.
8. Explain the fuel-air flow through a float carburetor when the engine is at idle speed.
9. Explain the fuel-air flow through a float carburetor when the engine is at part throttle.
10. Explain the fuel-air flow through a float carburetor when the engine is at high speed.
11. Explain the parts and operation of the choke valve on a float type carburetor.
12. Describe the parts and operation of a diaphragm used in a diaphragm carburetor.
13. List and describe the purpose of the four different operating modes for a diaphragm carburetor.
14. Describe the parts and operation of the suction feed diaphragm carburetor.
15. Describe the parts and explain the operation of a diaphragm-operated automatic choke.
16. Describe the parts and explain the operation of a bimetal spring type choke.
17. Describe the parts and explain the air flow through an oil bath air cleaner.
18. Describe the parts and explain the air flow through a paper element air cleaner.
19. Describe the parts and explain the air flow through a foam element air cleaner.
20. Explain the purpose of the foam precleaner used on a paper air filter element.

DISCUSSION TOPICS AND ACTIVITIES

1. Disassemble a float carburetor. Identify each of the parts and trace the flow of air and fuel in each of its operating modes.
2. Disassemble a diaphragm carburetor. Identify each of the parts and trace the air and fuel flow in each of its operating modes.
3. A 15 to 1 air fuel ratio is based on 90 pounds of air needed to burn 6 pounds of fuel. What is the weight of air and fuel in a 17 to 1 air fuel ratio?

Governor and Throttle Control Systems

OBJECTIVES

Upon completion and review of this chapter, you should be able to:

- Identify the parts and explain the operation of a manual throttle control.
- Explain the purpose of a governor control system.
- Describe the parts and operation of an air vane governor system.
- Explain the parts and operation of a centrifugal governor system.
- Describe the parts and operation of a flywheel brake.

TERMS TO KNOW

Air vane governor
Centrifugal governor
Dead man lever
Flywheel brake
Governor
Manual throttle control system
Tachometer

INTRODUCTION

The speed of an engine is regulated by the throttle valve opening. Some engines use a manual throttle control, which allows the operator to directly set the engine speed. This control system is often used on engines used on riding type outdoor power equipment. Many engines that are used on outdoor power equipment like walk behind mowers and edgers use a governor to control engine speed.

MANUAL THROTTLE CONTROL

A **manual throttle control system** is an engine speed control in which the operator sets the engine speed by positioning the carburetor throttle valve. Engines with manual throttle control often use a remote control linkage. The throttle may be operated at a distance from the carburetor or the controls may be installed directly on the engine. More commonly they are on the handle of the equipment (Figure 16-1). The operator sets the speed by moving throttle linkage.

The throttle linkage on walk behind or hand power equipment is operated with a hand lever. A foot pedal may be used on a tractor or riding mower. The movement of the lever or pedal usually does not go directly to the carburetor. It goes to a control lever assembly on the side of the engine (Figure 16-2).

A throttle cable is used to transfer movement from the lever or pedal. A throttle cable has a flexible metal housing. The housing is solidly mounted (commonly clamped). A cable inner wire is free to move back and forth inside the cable housing. It goes from the hand or foot control to the control lever assembly. The control lever assembly is a plate with levers and screw stops for the engine controls. One end of the cable inner wire is connected to the throttle control lever on the equipment. The other is connected to the control lever assembly. Linkage connects the carburetor throttle valve to the control lever assembly. As the operator moves the throttle control, the inner wire moves the carburetor throttle valve (through the control lever assembly).

Many carburetors are equipped with a throttle return spring (Figure 16-3). The return spring is

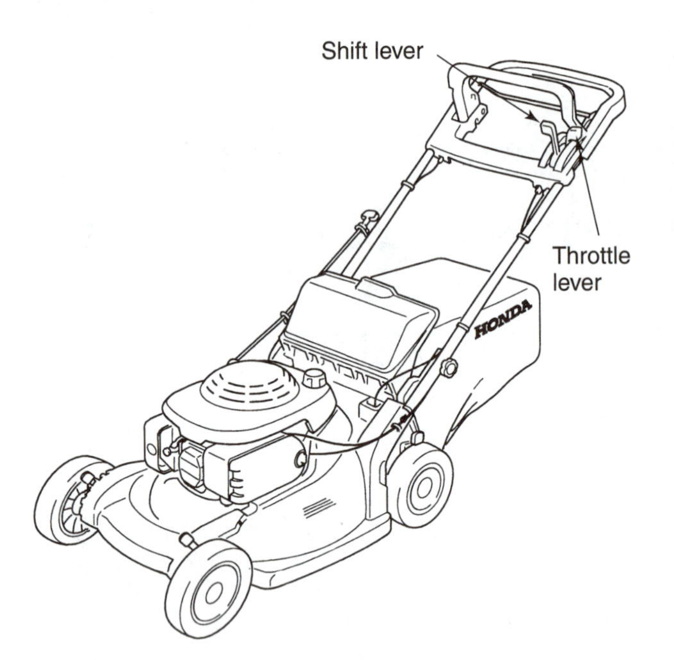

FIGURE 16-1 Handle-mounted remote engine control. (Courtesy of American Honda Motor Co., Inc.)

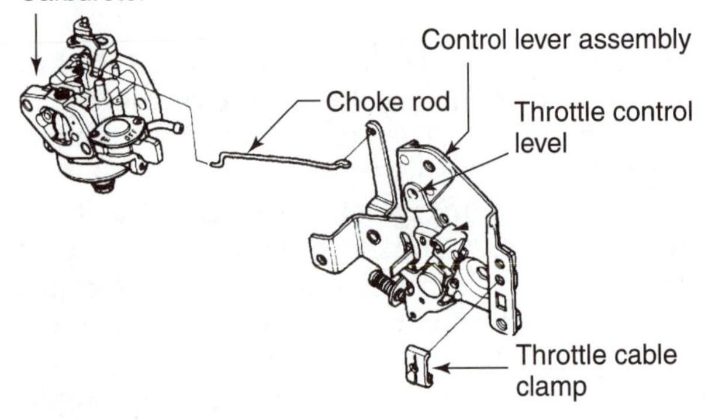

FIGURE 16-2 A control lever assembly on the side of the engine. (Courtesy of American Honda Motor Co., Inc.)

connected to the carburetor throttle valve to hold the throttle valve in the closed position. The spring pressure works against (opposes) throttle valve opening. The spring pressure makes sure the throttle valve closes when the operator releases the throttle control. Sometimes an additional return spring is located on the lever or pedal assembly.

Choke and Kill Switch Linkage

Manual throttle control systems often have linkage that operates the choke. The choke is activated when the operator selects start position. A linkage rod is connected to the manual control lever. The linkage rod is connected to the choke lever on the carburetor. The choke lever moves the choke valve in the carburetor throat into a closed position. After engine start up, the operator can set the manual control to the run position. The linkage rod moves the choke lever to open the choke valve.

Many engines have an ignition kill switch that is part of the throttle control lever assembly (Figure 16-4). The kill switch has a contact that is activated by the throttle control lever. In the stop position, the throttle control lever touches the kill switch contact. This provides a path to ground for the primary wire. The ignition primary ground kills the ignition. When the manual control is set on start or run, the throttle lever is moved out of contact with the kill switch. This removes the ground path and allows (enables) ignition.

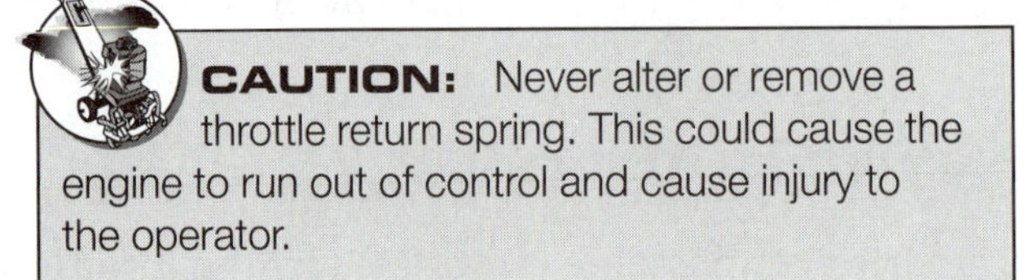

CAUTION: Never alter or remove a throttle return spring. This could cause the engine to run out of control and cause injury to the operator.

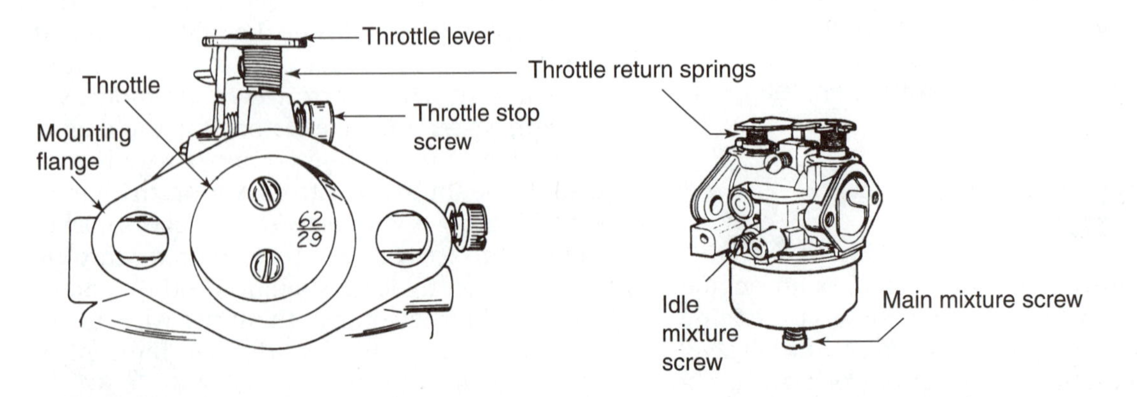

FIGURE 16-3 The throttle return spring returns the throttle valve to the closed position. (Provided courtesy of Tecumseh Products Company.)

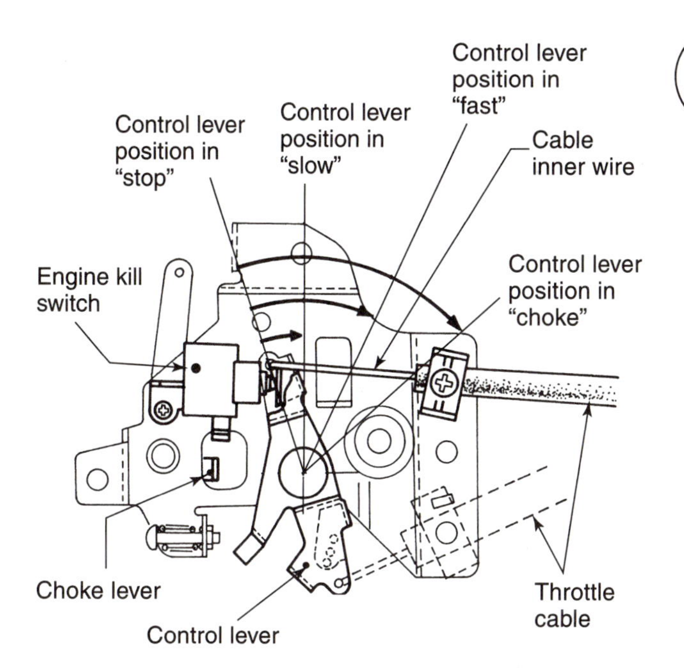

FIGURE 16-4 A control lever assembly for throttle, choke, and kill switch. (Courtesy of American Honda Motor Co., Inc.)

Dead Man Lever

Lawn and garden equipment often has a special type of safety control called a dead man lever. A **dead man lever** is an engine speed control that automatically returns engine speed to idle or stops the engine or blade when the operator's hand is removed (Figure 16-5). Equipment with a dead man lever can only be operated if the lever is depressed. When an operator loses control of the equipment handle, the hand often comes off the lever, causing an automatic shutdown. The operator does not have to make a decision to shut down the equipment. If the operator were to fall down, the equipment automatically shuts down. This is why it is called a "dead man" lever.

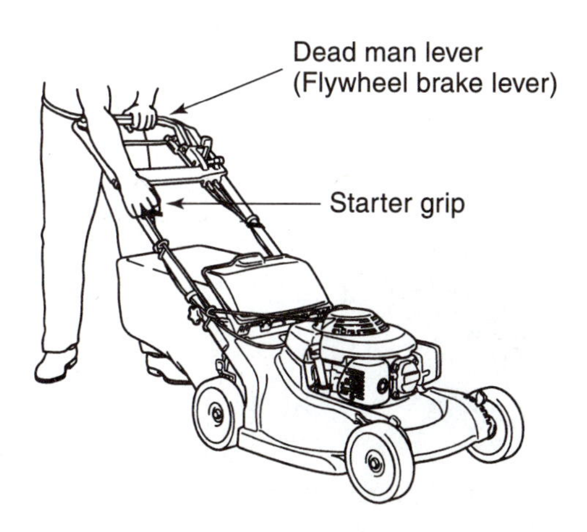

FIGURE 16-5 A dead man lever kills the engine if the hand is not on the handle. (Courtesy of American Honda Motor Co., Inc.)

CASE STUDY

The push for equipping lawn and garden equipment with dead man levers came from incidents like this one. Two gardeners were finishing up a lawn job. The house owner requested that they square off the top of her 3-foot-high hedge. The gardeners had a hedge trimmer in the truck so they agreed to do the job. After they finished mowing they stored the lawnmower and got out the hedge trimmer. They could not get the hedge trimmer to start.

They did not want to disappoint their customer so they came up with this brilliant idea. They could start their rotary lawnmower and lift it over the hedge, using the rotary blade to cut the hedge. They moved the mower over near the hedge, started the engine, and set the throttle to the fast position. One of the gardeners got on each side of the mower and gripped it under the deck and lifted. They raised the mower high enough to top the hedge. With one gardener on each side of the hedge, they started trimming.

Their problem started when the hedge got wider and they had to stretch their arms to support the mower. The mower was also getting heavier as the work continued. Finally one gardener lost his grip and the mower weight shifted. As they both struggled to gain control, fingers came in contact with the rotating blade. Both men lost fingers from both hands.

GOVERNOR THROTTLE CONTROL

A **governor** is a throttle control system that senses engine load and automatically adjusts engine speed up or down. The governor throttle control system has three main functions:

- Protect the engine from overspeeding
- Maintain a safe blade or equipment speed
- Match the engine speed to engine load demands

Any engine develops its maximum horsepower at a specific rpm. For example, an engine's maximum horsepower might be 4 horsepower at an

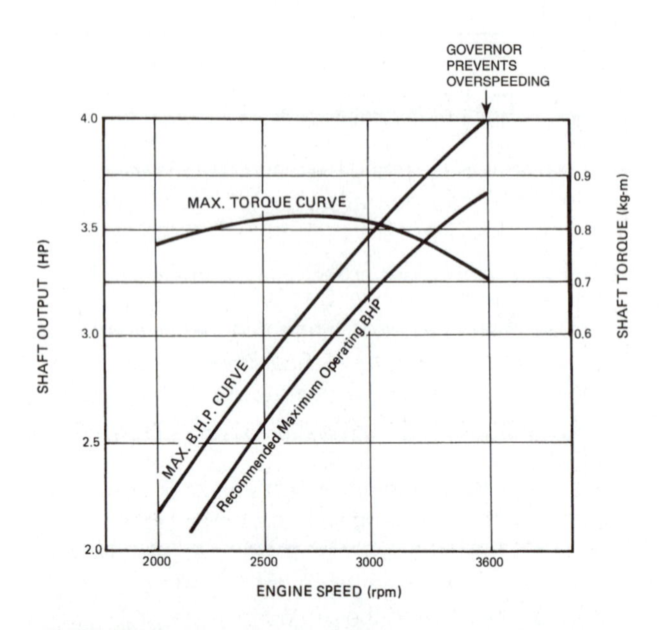

FIGURE 16-6 The governor protects the engine from excessive speed. (Courtesy of American Honda Motor Co., Inc.)

engine speed of 3,600 rpm (Figure 16-6). Operating the engine below 3,600 will result in less power but longer engine life. Operating the engine above 3,600 rpm will result in a much shorter engine life. High engine speeds create more heat, friction, and wear. An engine operated at too high a speed for too long can break a connecting rod or piston. The governor system prevents excessive engine speed.

Safe Blade Tip Speeds

Many engines are used to turn blades for lawn service. These blades are often mounted directly to the power takeoff end of the crankshaft. A rotary mower, for example, has a blade that rotates under the mower. Blade speeds are measured by the distance the blade tip rotates in a minute. The measurement is in feet per minute (fpm). Blade lengths range from 18 to 25 inches. The longer the blade, the faster the tip moves (fpm) at any given engine rpm.

The recommended (industry standard) safe maximum blade tip speed is 19,000 feet per minute. The governor is designed to maintain the correct engine rpm for a safe blade rotational speed. The recommended safe tip speed may be exceeded if the governor is not adjusted correctly (too high rpm). The safe tip speed can also be exceeded if a longer blade is installed. A blade that is rotated beyond its recommended safe tip speed can have metal failure. The blade can break up send broken pieces over a wide area.

Blade tip speed can be determined with a tachometer, measuring tape, and blade speed formula. Engine maximum speed is measured with a tachometer. A **tachometer** is an electronic meter used to measure engine rpm. The cutter blade length is measured with a measuring tape. The blade speed formula is engine speed × blade length × 0.262

For example, a rotary lawnmower engine has the cutter blade connected directly to the crankshaft. The length of the cutter blade is measured diagonally across the cutter edges and determined to be 18 inches (460 millimeters). A tachometer is connected to the engine. The rpm is measured as 3,600 rpm with the engine at no load and the throttle control at wide open (WOT):

Blade speed in feet per minute

$$= \text{engine speed} \times \text{blade length} \times 0.262$$
$$= 3{,}600 \times 18 \times 0.262$$
$$= 16{,}978 \text{ fpm}$$

The blade speed of 16,978 fpm is less than 19,000 fpm recommended maximum (Figure 16-7).

> **WARNING:** Do not alter or disconnect a governor. Excessive engine rpm could cause any engine driven blade or equipment to fail and cause personal injury to the operator. Always follow manufacturer's recommendations for blade tip speed.

Engine Speed and Load

The governor matches the engine speed to the engine load. For example, during lawn mowing you might set engine speed to fast and start cutting the grass. As you mow, you may get into some high grass. The engine has to work very hard to cut the high grass. The

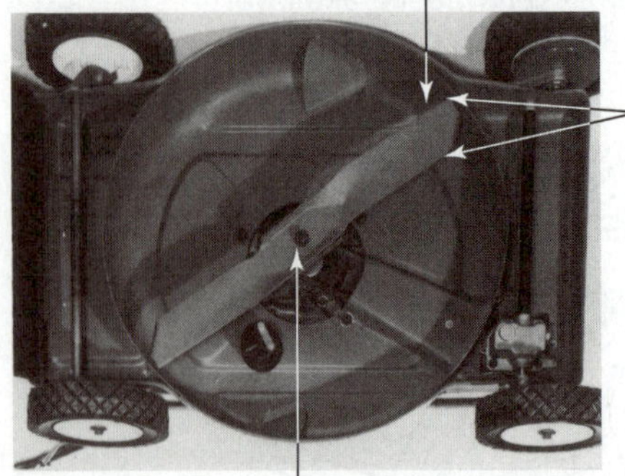

FIGURE 16-7 Blade length and crankshaft speed are used to determine safe blade speed. (Courtesy of The Toro Company.)

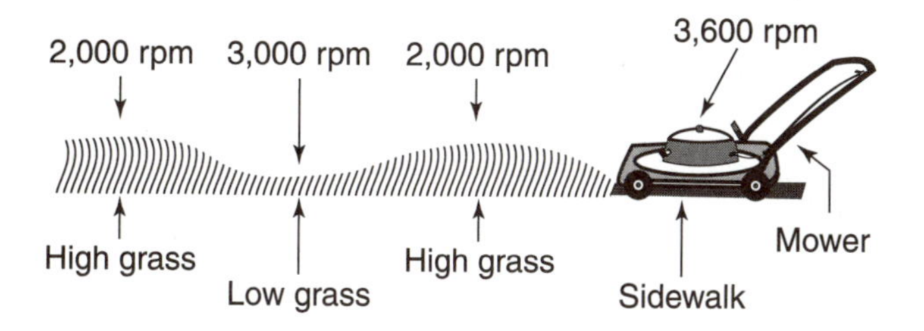

FIGURE 16-8 A nongoverned engine changes speed as engine load increases or decreases.

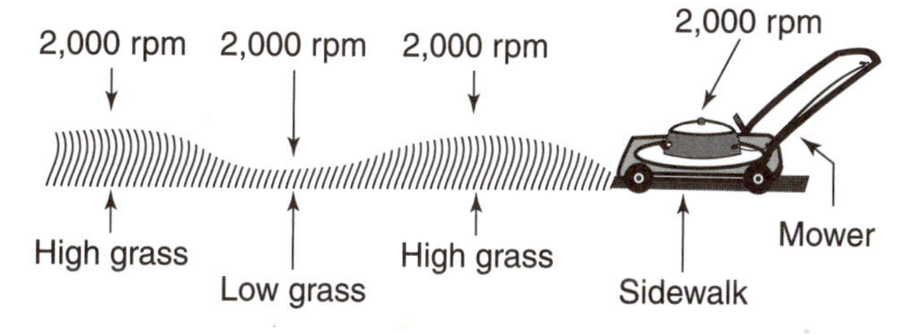

FIGURE 16-9 The governor matches the engine speed to engine load.

engine will start to slow down because of the high engine load. As you move the mower over a sidewalk the blade stops cutting any grass. The engine speeds up because of low engine load. Without a governor, you would have to increase and decrease engine speed all the time (Figure 16-8).

The governor keeps the engine running at a steady speed by opening or closing the throttle valve. The governor closes the throttle valve for low engine loads. It opens the throttle for more power for high engine loads. The governor senses these load conditions and makes the throttle adjustments automatically (Figure 16-9).

AIR VANE GOVERNOR

An **air vane governor** is a governor that uses air flow coming off the flywheel to regulate throttle opening to engine load. The air vane governor is also called a *pneumatic governor*. Air vane governors are used on both two- and four-stroke engines.

An air vane is a flat piece of plastic or steel mounted on a pivot above the flywheel (Figure 16-10). The air vane is connected to the carburetor throttle valve linkage by a small linkage rod. This rod is called a *governor link*. The throttle lever is mounted on a pivot. The lever is connected to one end of a spring called the *governor spring*. The other end of the governor spring is connected to the throttle valve linkage. This is the same connection as the governor link rod.

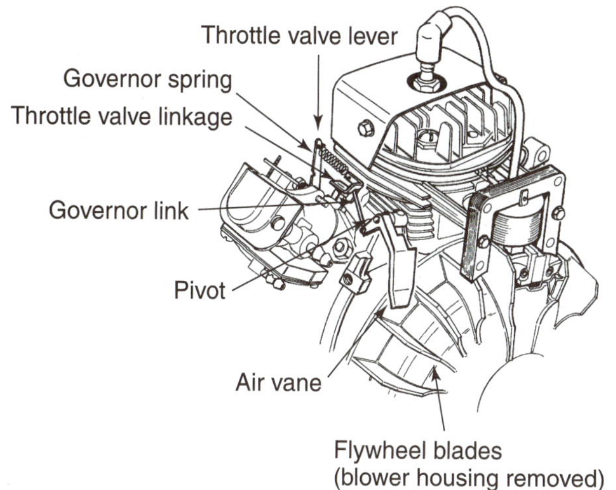

FIGURE 16-10 Parts of an air vane governor assembly. (Provided courtesy of Tecumseh Products Company.)

The air vane governor uses two opposing forces to match engine speed to engine load. The air flow off the flywheel pushes against the air vane. This force goes to the throttle valve through the governor link. The air vane movement senses high or low engine load. The faster the engine runs (lower the engine load), the higher the flywheel air flow. High air flow against the vane causes the governor link rod to move the throttle valve toward a closed position. Engine speed slows down.

The opposing force comes from the governor spring (Figure 16-11). The governor spring is connected to the throttle valve linkage. The spring tension is in a direction to hold the throttle valve open. The spring tension is set to hold a high enough engine speed for high engine load. As the engine load gets lower (low grass) its speed begins to increase. High air flow on the air valve will overcome the governor spring pressure and slow the engine (Figure 16-12). The engine slows down when the mower cuts high grass. The lower air flow on the vane allows the governor spring tension to increase engine speed. The governor blade constantly moves back and forth in response to engine load and maintains a constant engine speed.

A remote throttle control is often used on engines that have an air vane governor. The remote control lever is connected to the handle of the mower equipment. A throttle cable connects the control lever to the throttle lever and governor spring. The air vane is connected to the throttle valve through the governor link rod (Figure 16-13).

The air flow on the flywheel works through the air vane and governor link rod. These parts close the throttle valve in response to low engine load. The higher the air flow, the lower the engine load.

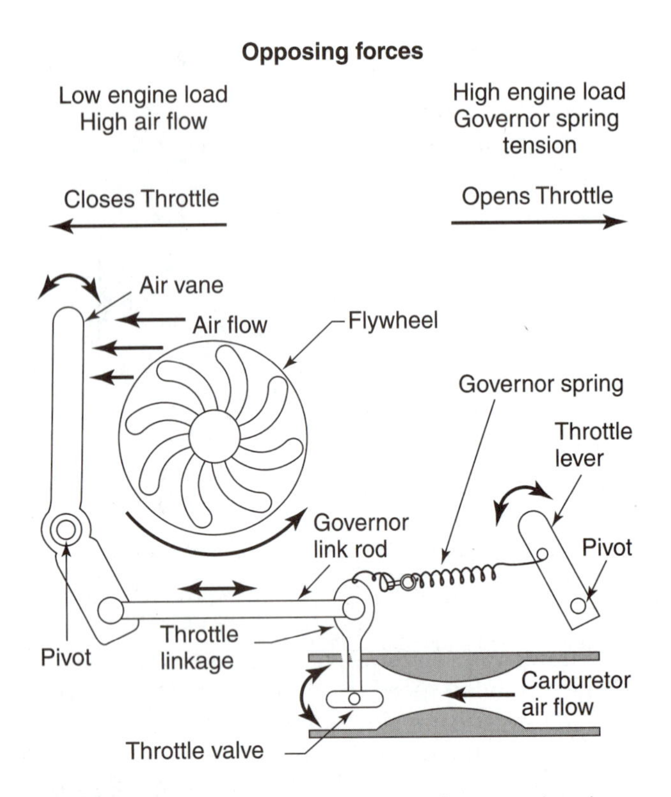

FIGURE 16-11 Opposing forces control governed engine speed.

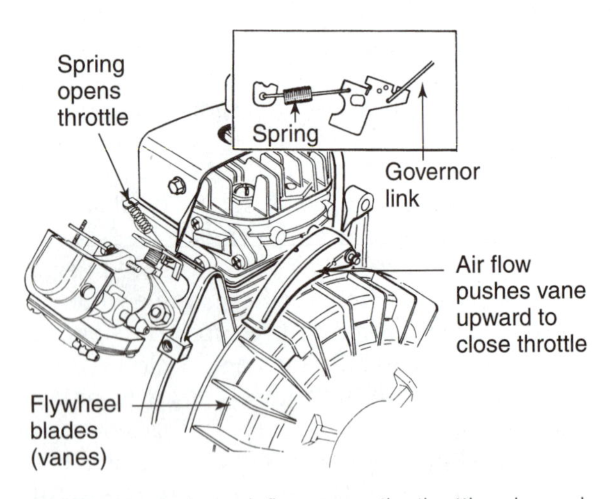

FIGURE 16-12 Low air flow opens the throttle valve and high air flow closes the throttle valve. (Provided courtesy of Tecumseh Products Company.)

Under these conditions, a stronger force is developed to close the throttle valve. The governor spring tension opposes this force and tries to hold the throttle valve open.

Governor spring tension is changed when the operator selects a speed setting with the control lever. Movement of the control lever from slow to

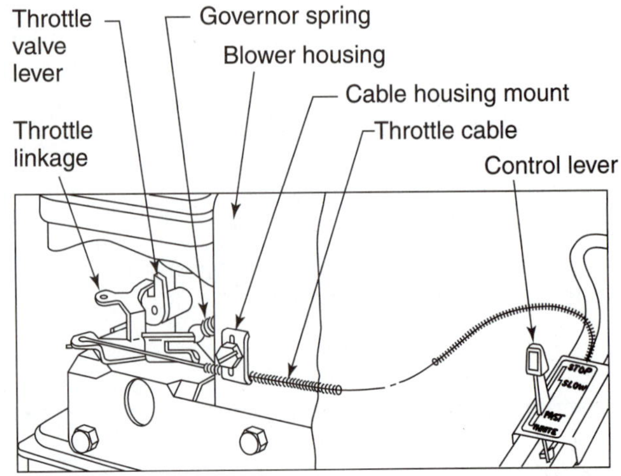

FIGURE 16-13 Parts of a remote throttle lever governor control.

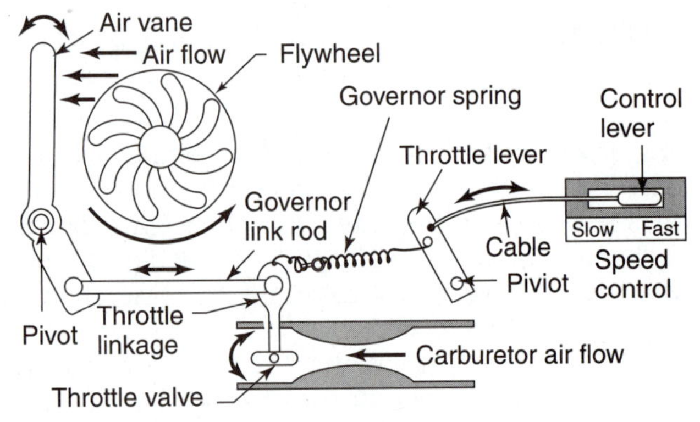

FIGURE 16-14 Operation of a remote throttle lever governor control.

fast increases governor spring tension (Figure 16-14). The higher governor spring tension holds the throttle valve open with more force. This causes a higher governed engine speed. Moving the lever back to slow lowers the governor spring tension. This results in a lower governed engine speed.

CENTRIFUGAL GOVERNOR

Many engines use a centrifugal governor. A **centrifugal governor** is a governor that uses centrifugal force to regulate throttle opening. These governors are also called *mechanical governors*. The centrifugal force is generated by rotating flyweights inside the engine crankcase The centrifugal governor works to maintain a constant engine rpm as engine load increases or decreases.

The centrifugal parts of the governor are located inside the engine crankcase (Figure 16-15). The

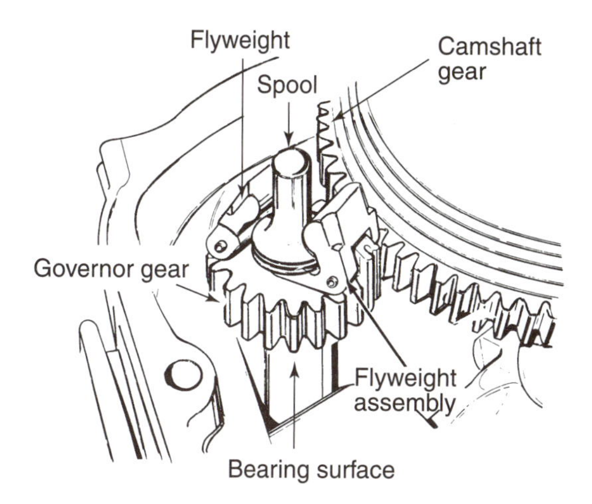

FIGURE 16-15 The centrifugal governor meshes with the camshaft gear. (Provided courtesy of Tecumseh Products Company.)

centrifugal governor is driven off the engine's camshaft gear. The governor has a governor gear that fits on a shaft. The shaft fits into a bearing surface in the crankcase. The teeth on the governor gear mesh with the teeth on the camshaft gear. When the engine is running, the rotating camshaft causes the governor gear to rotate. Two specially shaped weights, called *flyweights,* fit through a pivot on the governor gear. The flyweights have an arm formed on the end. The arms contact a spool that fits over the governor shaft.

As the engine runs, the camshaft gear rotates the governor gear at high speed. The speed of the governor gear creates a centrifugal force, which causes the flyweights to move outward (Figure 16-16). The outward movement is held back by the mounting pivots. The flyweight arms move in a direction to push the spool up the governor shaft. Linkage in contact with the spool transfers the movement outside the engine. As the engine slows down, spring pressure on the linkage causes the flyweights to retract. The spool moves back down the governor shaft.

The movement of the governor spool is transferred outside the engine by a governor rod (Figure 16-17). The rod goes through the engine crankcase cover. It is supported on bearings so that it can rotate freely in the cover. There is a seal on the end of the rod. The seal prevents engine oil from leaking out of the engine between the rod and the cover. A lever on the inside end of the rod contacts the governor spool.

Movement of the governor spool up and down in response to engine speed causes the governor rod to rotate back and forth (Figure 16-18). The governor rod is connected to a governor lever on the outside of the engine. A governor spring is connected to the lever. The governor spring opposes the force of the governor weights. The governor spring tension keeps the governor rod in contact with the governor spool.

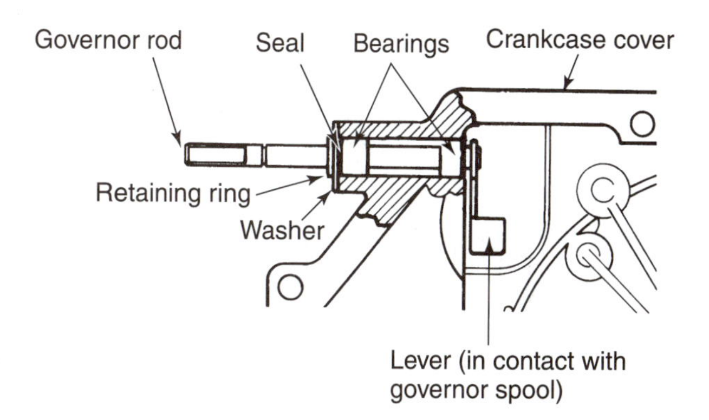

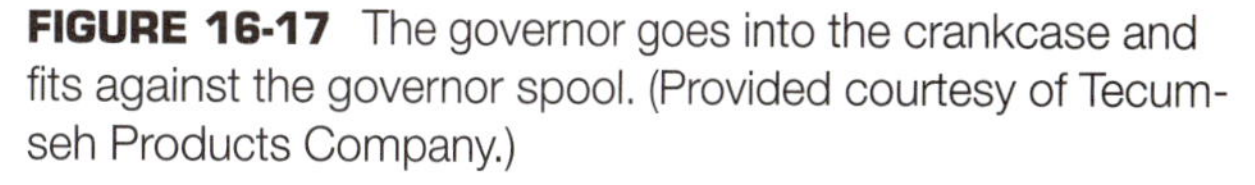
FIGURE 16-17 The governor goes into the crankcase and fits against the governor spool. (Provided courtesy of Tecumseh Products Company.)

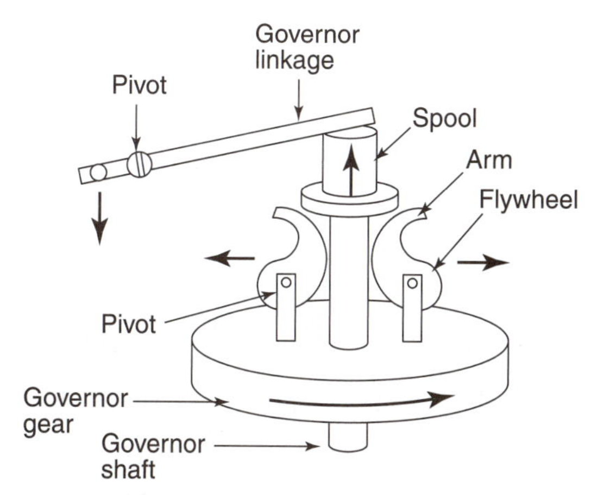

FIGURE 16-16 Operation of the centrifugal governor.

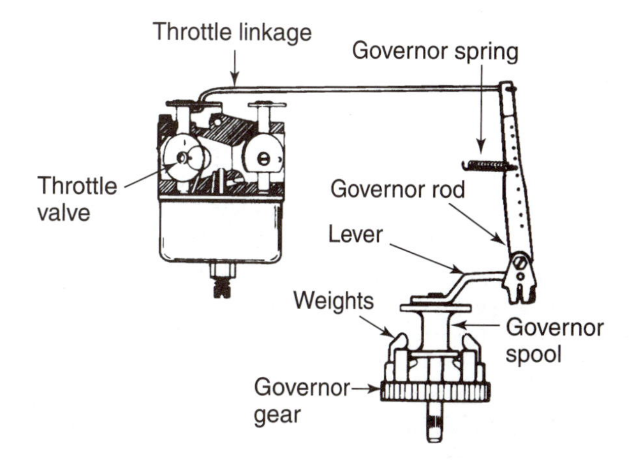

FIGURE 16-18 Governor linkage parts. (Provided courtesy of Tecumseh Products Company.)

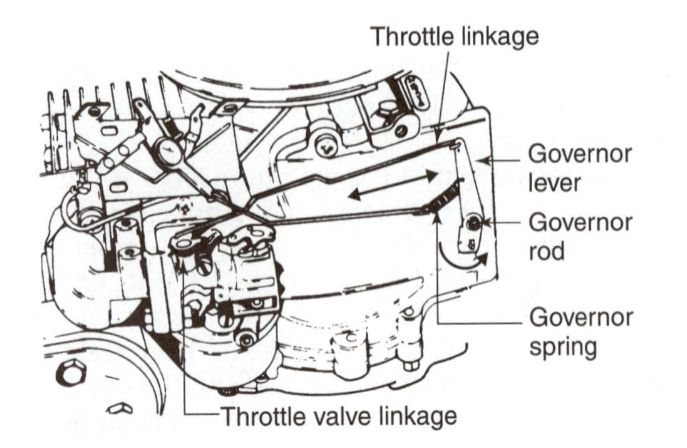

FIGURE 16-19 Movement of the governor linkage. (Provided courtesy of Tecumseh Products Company.)

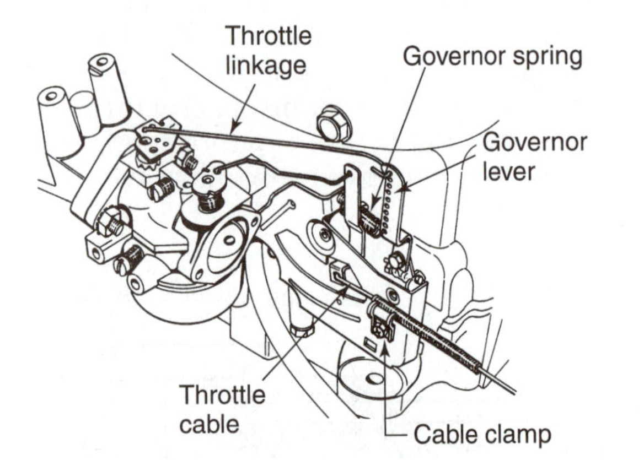

FIGURE 16-20 Remote throttle control for a governor system. (Provided courtesy of Tecumseh Products Company.)

The governor lever is connected through a throttle linkage rod to the carburetor throttle valve (Figure 16-19). Movement of the governor rod causes movement to the governor lever. The lever moves the throttle linkage and throttle valve. The governor spring tension opposes the linkage movement and works to hold the throttle open. The governor spring also retracts the flyweights when the engine speed and centrifugal force drops.

A throttle cable may be connected to the centrifugal governor linkage to provide remote throttle control (Figure 16-20). Movement of a control lever on the equipment handle moves the throttle cable. The cable movement changes the tension of the governor spring. Increasing governor spring tension causes a faster governed speed. Decreasing governor spring tension causes a slower governed speed.

FLYWHEEL BRAKE

The Consumer Product Safety Commission has developed standards for mower blade safety. The CPSC is concerned about operators whose hands, feet, or other body parts were injured from contact with the rotating mower blade. Blades under power often caused these injuries. Injuries were also caused by blades continuing to rotate when the engine was shut down. For example, a rotary mower blade continues to rotate for a period of time after engine shut down. Continued rotation is caused by momentum stored in the rotating blade.

The CPSC standards require mower blades to stop automatically. They must stop within three seconds after the operator's hand is removed from the dead man lever on the mower handle. Mower makers have developed a safety brake on the engine flywheel to satisfy this requirement. A **flywheel brake** is a brake that stops the rotating flywheel when the operator's hand is removed from the equipment handle dead man switch (Figure 16-21).

The flywheel brake has a metal brake shoe surfaced with a friction material. The brake shoe can be moved into contact with a braking surface on the rotating flywheel. The braking surface may be on the outside diameter or inside diameter of the flywheel. The brake shoe is attached to a movable linkage with a large spring. The linkage is connected to a cable. The cable runs up to the lever on the handle. A starter (rope or electric) is used with this system that can only be operated while holding the equipment handle.

During starting, the operator has to activate the dead man lever on the handle. If the dead man lever is not activated the engine will not start. As the lever is activated, it pulls on the brake control cable. The cable moves the linkage. The linkage pulls the brake shoe away from the flywheel. At the same time it compresses the spring. As long as the dead man lever is depressed, the control cable

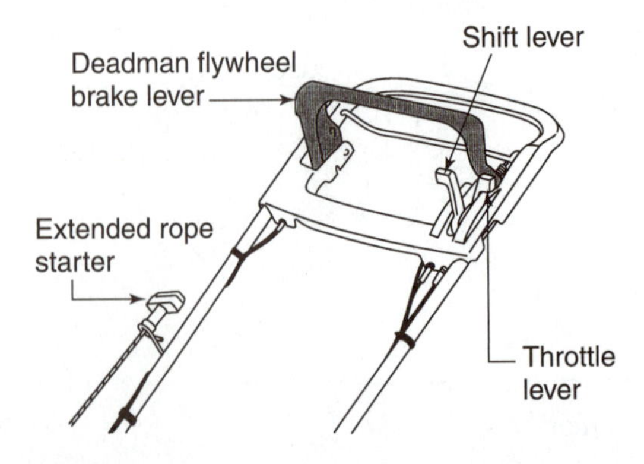

FIGURE 16-21 The dead man lever operates the flywheel brake. (Courtesy of American Honda Motor Co., Inc.)

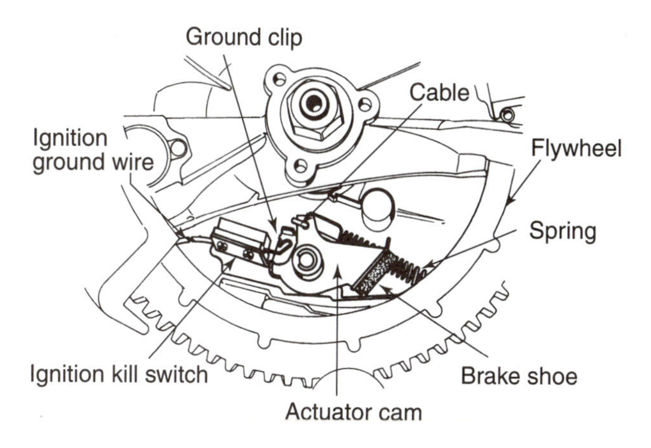

FIGURE 16-22 Flywheel brake operation in the off position. (Provided courtesy of Tecumseh Products Company.)

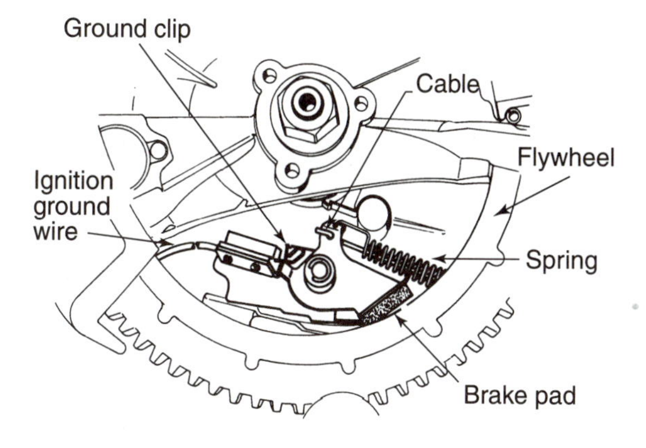

FIGURE 16-23 Flywheel brake operation in the on position. (Provided courtesy of Tecumseh Products Company.

and linkage keep the brake shoe away from the flywheel (Figure 16-22).

When the operator releases the dead man lever, the control cable is released. The brake spring forces the brake shoe into contact with the flywheel. The engine (and cutter blade) comes to a quick stop. Sometimes the brake also works a kill switch to shut down the engine (Figure 16-23).

REVIEW QUESTIONS

1. Describe the parts and operation of a remote throttle control.
2. List the three main functions of a governor throttle control system.
3. Explain why a governor is necessary to prevent engine damage.
4. Explain why a mower blade should not be rotated at excessive speed.
5. Explain how to calculate blade tip speed.
6. Explain why a governor is required to match engine speed to engine load.
7. Describe the parts of an air vane governor assembly.
8. Explain the operation of the air vane governor during high engine load.
9. Explain the operation of the air vane governor during low engine load.
10. Describe the parts of a centrifugal governor assembly.
11. Explain the operation of a centrifugal governor during high and low engine load.
12. Describe the parts and operation of a flywheel brake.
13. Explain how a dead man lever is used to activate a flywheel brake.
14. Describe the purpose of a tachometer.
15. Describe the parts and operation of a manual throttle control.

DISCUSSION TOPICS AND ACTIVITIES

1. Locate a shop engine with an air vane governor. Remove the blower housing. Move the air vane back and forth. Observe the operation effect on the carburetor throttle.
2. Locate a shop engine with a centrifugal governor. Remove the crankcase cover and identify the governor parts. Move the flyweights back and forth and observe the effect on the governor linkage.
3. A rotary mower has a 20-inch blade that is rotating at 3,600 engine rpm. What is the blade tip speed in feet per minute? Is the tip speed safe?

CHAPTER 17

Lubrication Systems

OBJECTIVES

Upon completion and review of this chapter, you should be able to:

- Explain how the lubrication system reduces friction in an engine.
- Describe and explain engine oil viscosity and service ratings.
- Identify the parts and explain the oil flow through a splash lubrication system.
- Identify the parts and explain the oil flow through a barrel pump lubrication system.
- Identify the parts and explain the oil flow through a full pressure lubrication system.
- Explain how a premix of gasoline and oil is used to lubricate two-stroke engines.
- Describe the operation of an oil pressure warning switch.
- Describe the parts and operation of a crankcase breather.
- Explain the parts and operation of a low oil level alert system.

TERMS TO KNOW

API service rating
Barrel oil pump
Crankcase breather
Engine oil
Friction
Full pressure lubrication system
Low oil level alert system
Oil clearance
Oil dipper
Oil drain plug
Oil fill plug
Oil filter
Oil pressure
Oil slinger
Oil sump
Premix
Premix lubrication system
Pressure lubrication
Relief valve
Rotor oil pump
SAE viscosity rating
Splash lubricating system
Viscosity

INTRODUCTION

The lubrication system circulates engine oil between the moving parts of an engine. The oil reduces friction and wear between parts that rotate or slide on each other. The circulating oil also helps cool the engine by carrying heat away from hot engine parts. The oil is also used to clean or flush dirt and other abrasives off engine parts. Oil on the cylinder walls helps piston rings seal compression pressures.

REDUCING FRICTION

Friction is the resistance to motion created when two surfaces rub against each other. Friction is caused when two metal parts rub on each other. If you push a book across a table top, you can feel a resistance. This is the friction between the book and table. The rougher the table and book surfaces, the more the friction. The two surfaces tend to lock together. If you put a weight on the book, it takes even more effort to move it across the table. As the amount of pressure between two things increases, their friction increases. The type of material the two things are made of also affects the friction. If the table is made of glass, the book slides across it easily. If it is made of rubber, it is very difficult to push the book across.

Friction has to be reduced between engine parts for several reasons. Power is needed to overcome

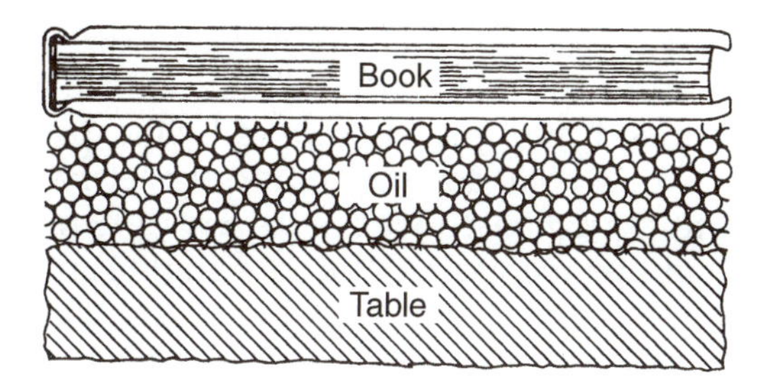

FIGURE 17-1 An oil film lifts the book up and reduces friction between the book and table.

friction. The lower the friction between engine parts, the more usable power an engine can develop. Friction between two things causes them to heat up and to wear. You can demonstrate this to yourself rubbing your hands together rapidly. You will feel them heat up. Friction between the skin of your two hands causes the heat.

Oil is used to reduce friction on engine parts. If you were to pour a layer of oil on the table top, you would be able to push the book more easily (Figure 17-1). The friction has been reduced between the book and table. The oil forms a thin layer called an *oil film*. The film of oil gets under the book and actually lifts it off the table surface. Oil does not completely eliminate friction. Friction can be reduced to a point where long engine life may be expected.

An oil film is used between engine parts to reduce friction and wear from metal-to-metal contact. When a rotating shaft fits in a bearing, there is a small space between the bearing and shaft. This space is made for the oil film. **Oil clearance** is a space between two parts where oil can flow to create an oil film between the parts. The oil clearance space is usually only 0.002–0.003 inch (0.05–0.08 millimeter).

Oil goes into the oil clearance area between the bearing and the shaft (Figure 17-2). The rotation of the shaft causes a film of oil to wedge between shaft and bearing. The shaft is lifted slightly so that it does not rest on the bearing but on an oil film. When the oil film wedge is formed, there is no metal-to-metal contact.

ENGINE OIL

Engine oil, or lubricating oil, is the most common fluid used to provide lubrication. **Engine oil** is a petroleum- or synthetic-based lubricant circulated between engine parts to prevent friction and wear. Lubricating oil can be made (refined) from crude petroleum pumped from oil wells. Synthetic oil is made from nonpetroleum sources. Oil used in an

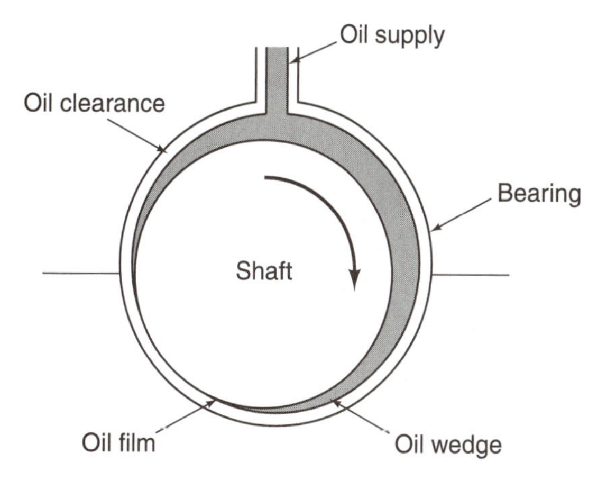

FIGURE 17-2 Oil is wedged between the bearing and the shaft to reduce friction.

engine must have the correct viscosity and service rating for that engine.

Oil SAE Viscosity Ratings

The **SAE viscosity rating** of oil is a rating of the thickness or thinness of the oil determined by a rating system developed by the Society of Automotive Engineers (SAE). **Viscosity** is the thickness or thinness of a fluid. High viscosity fluids are thick and flow sluggishly. Thick pancake syrup is an example of a fluid with a high viscosity. Low viscosity fluids flow very freely. Oils used in engines must flow freely in cold conditions but must have enough thickness during times of high temperature.

The SAE viscosity ratings give oil that is thin a low viscosity number, like SAE 10. Thicker oil gets a higher number, like SAE 50. Oil may have a single viscosity rating, like SAE 30 (Figure 17-3). It is called

FIGURE 17-3 Oil with a single SAE viscosity rating. (Courtesy of Quaker State Corporation.)

FIGURE 17-4 Oil with an SAE multiviscosity rating. (Courtesy of Quaker State Corporation.)

a *single viscosity oil*. Other common single viscosity oils are SAE 10, SAE 20, SAE 40, and SAE 50.

The higher the viscosity number, the better the protection when the engine is hot. High viscosity oils stay in between parts during high temperature. Low viscosity oils can get too thin and flow out of the oil clearance during high temperature. The lower the viscosity number the more freely the oil flows. The lower viscosity oil gives better protection when the engine is cold. Cold, thick oils may have trouble getting into oil clearance areas. Thinner oils can get into the oil clearance area faster.

Oils are available with multiple viscosity ratings (Figure 17-4). Multiviscosity oil is rated with more than one viscosity number such as SAE 10-40. This means it flows freely like SAE 10 when the weather is cold, but has a viscosity like SAE 40 when it is hot.

Multiviscosity oils have a larger percentage of additives and a smaller percentage of base oil. The additives are chemicals that improve the oil. Single viscosity oil like SAE 30 has about 95 percent of base oil and 5 percent additives. SAE 10W-30 has 90 percent base oil and 10 percent additives. SAE 10W-40 multiviscosity oil has only 80 percent base oil and 20 percent additives. This is how the oil gets its wide viscosity rating (from 10 to 40).

CAUTION: Viscosity improver is an additive that helps the oil keep its viscosity under different temperatures. This additive contains chemicals that can break down under the high temperatures found in air cooled engines. Check the owner's manual before using SAE 10W-40 in an air-cooled engine because some engine makers do not recommend using this oil.

Small-engine makers often recommend single viscosity oils for warm weather operation. They may recommend multiple viscosity oils like SAE 5W-30 or SAE 10W-30 for very cold service. Engines with thick high viscosity can be very difficult to start in cold temperatures. The cold engine and thick oil prevent the engine from being cranked fast enough to start. A viscosity rating with a W after it, like SAE 20W, means the oil is rated for starting in cold temperatures.

The viscosity number is printed on the side of the oil container. The owner's manual and shop service manual for the engine specifies what viscosity should be used. Sometimes the oil viscosity information is found on a decal located on the engine (Figure 17-5).

The viscosity rating of an oil always stays the same. The actual viscosity of an oil changes as engine temperature and ambient temperature change. Ambient temperature is the air temperature (weather) around the engine. Oil gets thicker in cold weather. Oil gets thinner during high engine loads and in warm weather. Oil viscosity ratings are used to select oil based on expected temperature range. Some owner's and shop service manuals have a chart showing the viscosity to use for different engine ambient temperatures (Figure 17-6).

Oil API Service Ratings

Engine oil is also rated on how well it holds up under severe service conditions in the engine. The **API service rating** is an oil rating system that specifies how well an oil performs under severe engine service. The American Petroleum Institute (API) service ratings are related to automobile (not small engine) warranty standards. The small engine maker determines which of these ratings is acceptable to meet their engine protection standards.

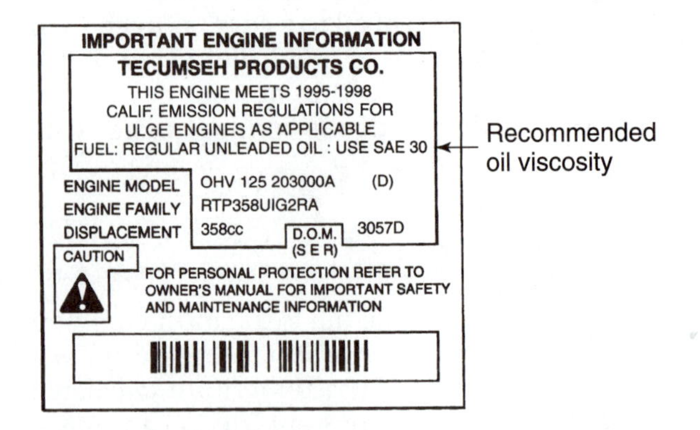

FIGURE 17-5 A viscosity recommendation may be found on an engine decal. (Provided courtesy of Tecumseh Products Company.)

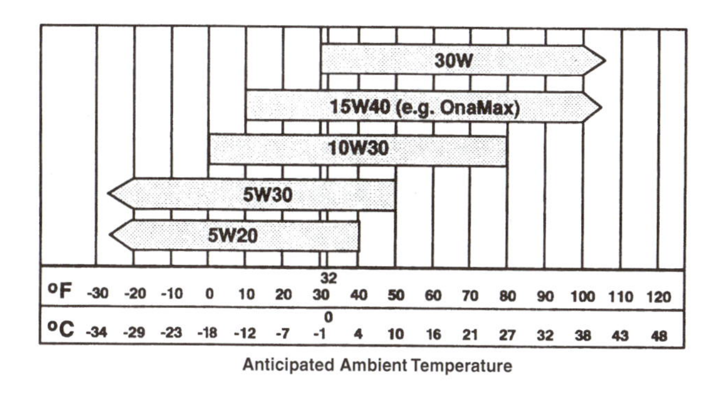

FIGURE 17-6 Oil viscosity recommendations for different ambient temperatures. (Courtesy of Onan Corp.)

The rating system uses two letters. The first letter is either an S or C. The letter S stands for spark-ignition gasoline engines. Oil rated for diesel engines uses a C in the rating. The second letter stands for the service rating. The higher the letter the higher the rating. Common rating are SA (the lowest rating) through SJ. As standards are increased, higher letters are used in the rating system. Oil often has both an S and a C rating.

The API rating for the oil is printed on the oil container. The engine owner's manual and shop service manual will specify which service classification should be used. A higher classification can always be substituted, but never a lower one.

The API label on an oil container also shows if oil has energy-conserving abilities. Newer oils are made to reduce friction enough to decrease fuel consumption. If the oil has this ability, it is shown on the label. The SAE viscosity rating, API service rating, and energy conservation designation are all typically found on a symbol located on the oil container (Figure 17-7).

CAUTION: Oil service (API) ratings are revised regularly. Always match the API service rating and viscosity rating on the oil container to the specifications for the engine. Using the wrong type of oil can cause engine damage. Synthetic oils have the same viscosity and service ratings as petroleum-based oils. They may not work properly with the seals and gaskets used in older engines. They should only be used if specified by the owner's manual or service manual.

The oil container may have a starburst certification mark (Figure 17-8). This certification mark shows that the oil meets current standards for passenger cars developed by the International Lubricant Standardization and Approval Committee (ILSAC). The certification was developed to help the consumer. Oil that is certified has passed certification requirements. The center of the symbol shows the specific application, "For Gasoline Engines."

A. Oil Container

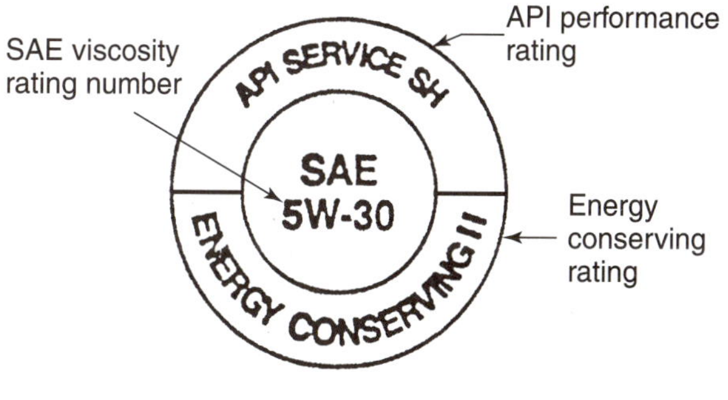

B. API Service Symbol

FIGURE 17-7 The API service rating is found on a symbol on the oil container. (*A:* Courtesy of Quaker State Corporation. *B:* Courtesy of The Valvoline Company.)

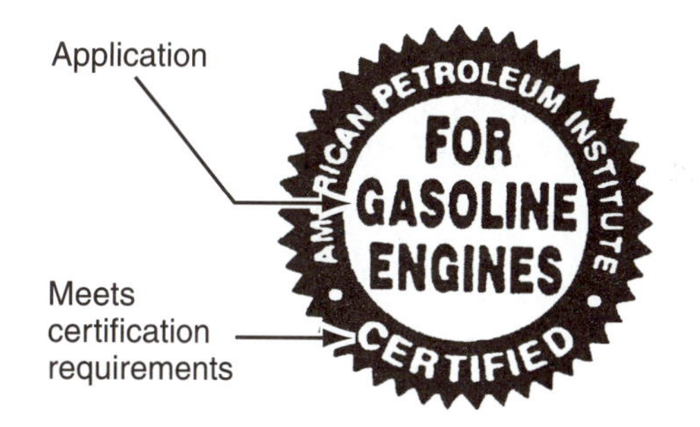

FIGURE 17-8 International Lubricant Standardization and Approval Committee (ILSAC) starburst symbol. (Courtesy of the Valvoline Company.)

FOUR-STROKE LUBRICATING SYSTEMS

Four- and two-stroke engines often use different types of lubricating systems. Lubrication systems differ in the way lubricating oil is circulated into the engine parts. The four-stroke engine may use a splash, barrel pump, or full pressure lubricating system.

Splash Lubrication

Many of the less expensive small four-stroke engines use a splash lubrication system. A **splash lubrication system** is a lubrication system that depends on the splashing of oil on engine parts for lubrication.

Oil for the splash lubrication system is stored in an oil sump. The **oil sump** is a lubricating oil storage area at the bottom of the engine crankcase that provides the necessary quantity of oil for splash lubrication. There are two removable plugs in the oil sump. An **oil drain plug** is a removable plug located in the bottom of the oil sump used to drain engine oil. An **oil fill plug** is a removable plug located in the oil sump used to check or change oil.

The lubricating oil in the sump has to be splashed on the parts requiring lubrication. Horizontal crankshaft engines commonly use an oil dipper (Figure 17-9). An **oil dipper** is a dipper attached to a connecting rod used to splash oil in the sump on engine parts. The dipper is fastened to the bottom of the connecting rod. It rotates as the connecting rod and crankshaft rotates.

Vertical crankshaft engines cannot use a dipper because the connecting rod is not submerged in the oil. Some vertical crankshaft engines use an **oil slinger** (Figure 17-10), a gear with paddles that is used to splash oil on engine parts. The oil slinger gear meshes with the camshaft gear and as the camshaft turns, the slinger turns. The small paddles on the slinger dip into the oil in the sump and splash it on the engine parts.

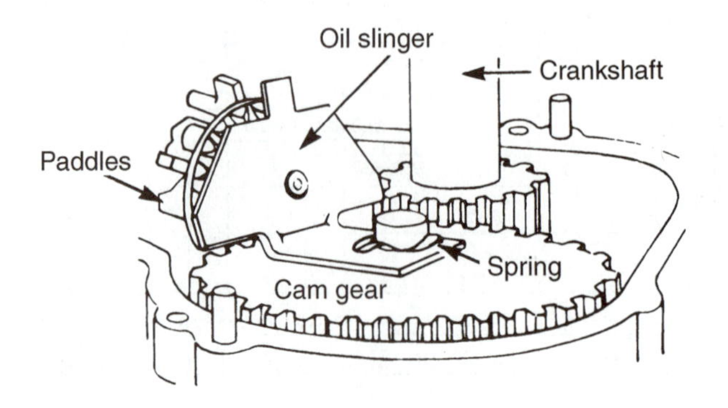

FIGURE 17-10 Splash lubrication systems for vertical crankshaft engines use an oil slinger.

When the engine is running, the dipper or slinger dips into the oil and splashes the oil upward. Some of the oil is splashed on the cylinder. The oil is carried upward by the piston skirt and oil control piston ring. This provides an oil film between the piston and cylinder wall. The piston pin gets lubricated by splash and gravity from oil running down the inside of the piston.

The oil splashed up by the dipper or paddle falls back down and into oil holes in some engine parts. An oil hole in the connecting rod sends oil into the crankshaft connecting rod journal (Figure 17-11). Oil flows by gravity between the connecting rod and crankshaft journal to lubricate this important area.

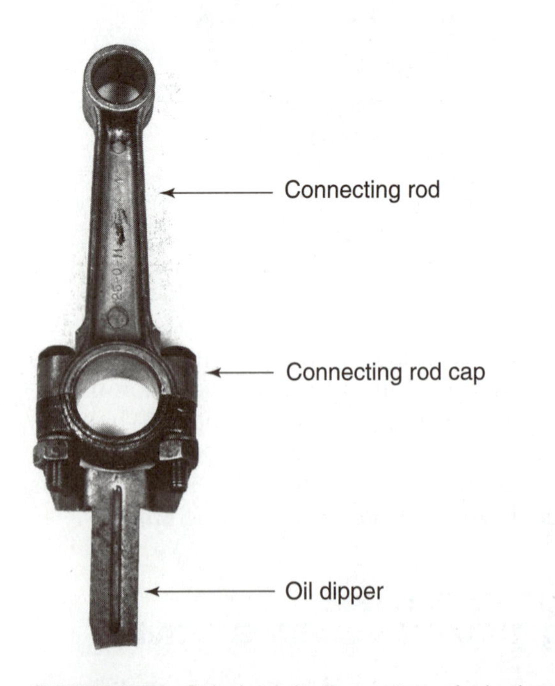

FIGURE 17-9 Splash lubrication systems for horizontal crankshaft engines use a dipper on the connecting rod.

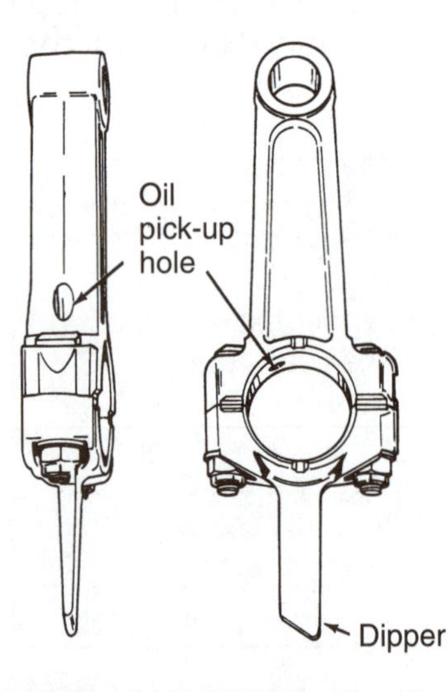

FIGURE 17-11 Oil flows through a hole in the side of a connecting rod to lubricate the crankshaft connecting rod journal. (Provided courtesy of Tecumseh Products Company.)

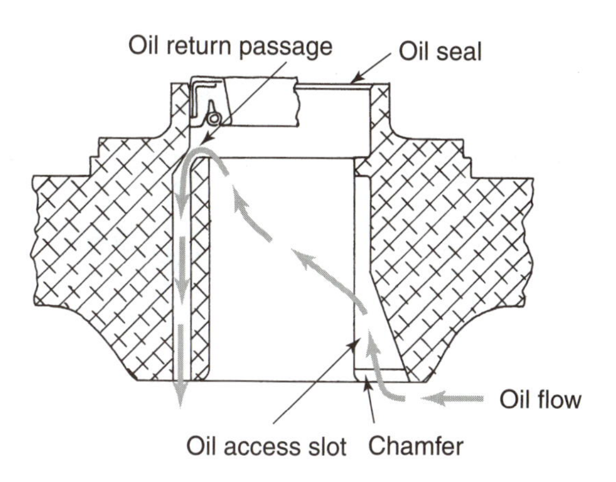

FIGURE 17-12 Passages and slots in a main bearing for lubricating oil. (Courtesy of Clinton Engines Corporation.)

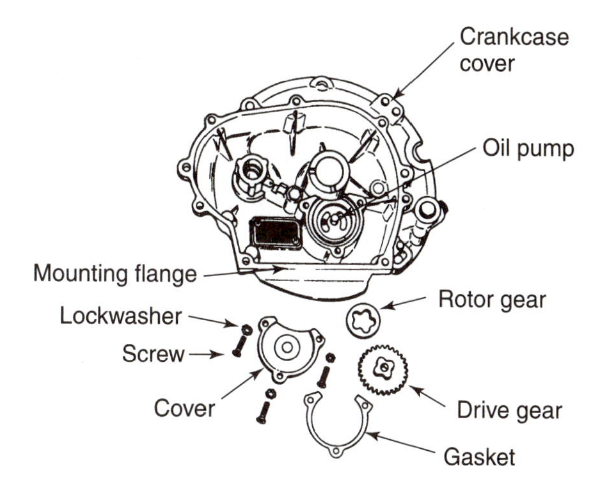

FIGURE 17-13 Some vertical crankshaft engines have a gear type pump in the crankcase cover. (Provided courtesy of Tecumseh Products Company.)

There are holes, slots, chamfered areas, or passages in each of the main bearings, (Figure 7-12). Gravity causes oil to run into the main bearings. An oil seal prevents oil from getting out of the crankcase. There are often return slots to send the oil back to the sump. An oil film is developed between the crankshaft journal and main bearing. The cam shaft bearings and valve lifters are usually lubricated in the same way. Valve guides and valve stems are usually lubricated directly by splashed oil.

Barrel Pump Pressure Lubrication

Pressure lubrication is a lubrication system in which a pump pressurizes the oil and forces it through passageways into engine parts. Pressure lubrication works better than splash lubrication. The oil under pressure provides a protective lubrication film between the parts. The constantly moving oil in a pressure system also does a better job of carrying away heat and flushing away dirt from the parts.

Vertical crankshaft engines are more difficult to lubricate with splash lubrication than horizontal crankshaft engines. The upper main crankshaft bearing is located a long way from the oil level in the sump. Getting oil up this high is difficult for a splash system. Some vertical engines use a gear type pump (Figure 17-13) similar to those used on full pressure lubrication systems. Some vertical crankshaft engines use a barrel oil pump. A **barrel oil pump** is an oil pump commonly used on vertical crankshaft engines to deliver oil under pressure to engine parts. A barrel oil pump has two main parts: a body and a plunger. The body (Figure 17-14) has a large hole at one end that fits over an eccentric shaft on the camshaft. The other end of the body is shaped like a cylinder. The cylinder is the pumping chamber. The plunger is basically a long piston that fits inside the body cylinder. The opposite end of the plunger has a ball shape. It is anchored in a recess in the crankcase cover.

The barrel pump works like a piston and cylinder. When the plunger moves out of the body, it creates a low pressure. This movement is called an *intake stroke.* When the plunger moves into the body, it is called a *compression stroke.* The pump pulls in oil on the intake stroke. It pushes oil into engine parts on the compression stroke.

The pump is submerged in the oil sump. There is a constant source of oil around the pump body. The plunger is mounted so that it is always stationary. The rotating camshaft eccentric causes the pump body to move in relation to the stationary

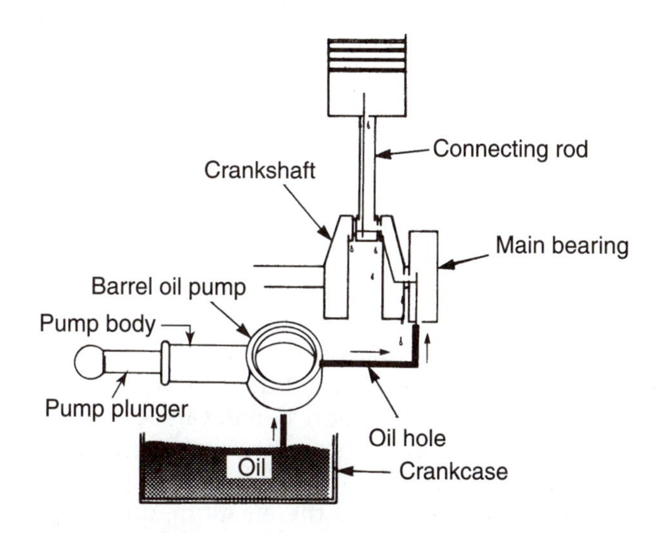

FIGURE 17-14 The barrel pump body is moved up and down an eccentric on the end of the camshaft. (Provided courtesy of Tecumseh Products Company.)

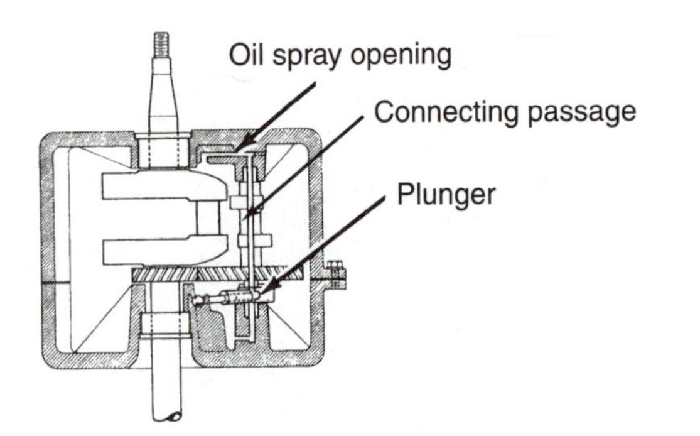

FIGURE 17-15 Oil circulation passages in the top main bearing area. (Provided courtesy of Tecumseh Products Company.)

plunger. Since the pump plunger cannot move, the body slides in and out over the plunger. Sump oil surrounding the pump is pulled into the body on the intake stroke. Oil leaves the pump through an oil hole and is forced into engine parts on the compression stroke.

Oil holes and oil passages are used to get the oil into the engine parts. Oil is pumped under pressure from the sump through a passageway drilled in the center of the camshaft. After it enters the camshaft, it goes to the top camshaft bearing journal. A connecting passageway allows oil to flow over to the top crankshaft main bearing. Oil is forced between the top main bearing journal and bearing. There is a spray mist hole in the connecting passage. It directs a spray of oil down at the piston, connecting rod, and other internal parts.

The lower main bearing and camshaft bearing are under the oil level in the sump. They do not require pressurized oil. A crankshaft oil passage gets oil under pressure from the top main bearing (Figure 17-15). Oil goes from the crankshaft oil passage through the crankshaft to the connecting rod journal. Oil flows between the crankshaft rod journal and connecting rod. An oil drain hole in the top main bearing allows any extra oil to drain out of the main bearing area and back into the sump.

Full Pressure Lubrication

Some of the more expensive engines use a **full pressure lubrication system** in which an oil pump provides oil under pressure to all the engine bearings. The full pressure lubrication system can be used on both horizontal and vertical crankshaft engines.

The main parts of the full pressure lubrication system (Figure 17-16) are the oil sump, oil pickup screen, oil pump, oil filter, and oil passageways. The oil sump provides a reservoir of oil at the bottom of the engine. There is a drain plug at the bottom of the sump for oil changing. The sump is filled through an oil filler cap located on the top of the engine.

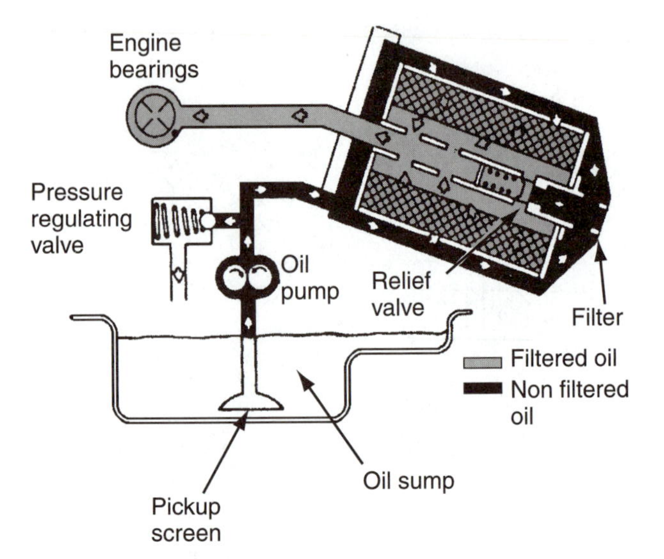

FIGURE 17-16 Parts of a full pressure lubrication system.

An oil pickup screen is located in the oil sump. The pickup screen is connected to the pump through a pickup tube. The pickup tube to pump connection is sealed with an O ring. The pickup screen prevents dirt from entering the oil pump.

The rotor type oil pump is common on full pressure lubrication systems (Figure 17-17). A **rotor oil pump** uses the movement of two rotors to move oil under pressure into the parts that require lubrication. A rotor pump is also called a *georotor pump*. A rotor is a round part with a set of circular (spherical) lobes. The rotor oil pump has an inner and outer rotor. The engine drives the inner rotor. The inner rotor can be driven by a shaft that engages the end of the crankshaft. Some are driven by the end of the camshaft. A drive gear in mesh with the crankshaft or camshaft is often used to drive a rotor. The lobes on the inner rotor drive the outer rotor.

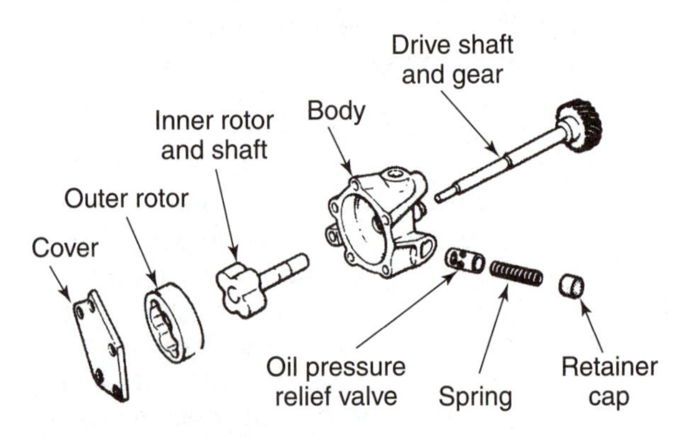

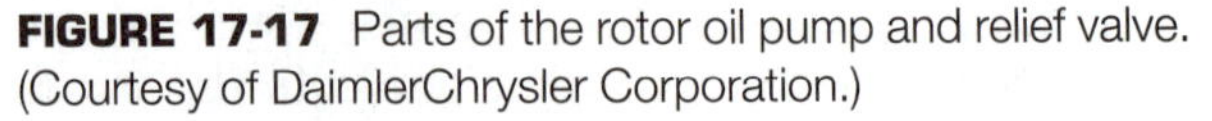
FIGURE 17-17 Parts of the rotor oil pump and relief valve. (Courtesy of DaimlerChrysler Corporation.)

The internal and external lobes fit closely together. As the rotors rotate, the internal and external lobes separate. The separation causes a low pressure in the pump housing. Oil is pulled into the housing by the low pressure. Further movement of the rotors causes the lobes to come back together. Oil between the lobes is squeezed or pressurized. Oil exits the pump body under pressure.

The oil pressure in the full pressure system has to be regulated. **Oil pressure** is the pressure created by an oil pump throughout the full pressure oil circulating system. Oil pressure is regulated with a relief valve. A **relief valve** is a valve assembly that controls the oil pressure in a full pressure lubrication system. The relief valve is located next to the oil pump outlet. The relief valve is usually built into the pump. Some engines have a relief valve installed in a cylinder block oil passage.

The relief valve is a piston (or ball)-shaped valve backed up by a spring. The piston and spring fit in a small cylinder. The piston controls a passageway back to the sump called the *dump orifice*. As engine speed increases, oil pressure builds up in the system. When the oil pressure gets to a preset amount, the piston moves against its spring, allowing the piston to uncover the dump orifice. Oil goes into the dump orifice and back to the sump. The oil pressure decreases. As pressure drops, the spring moves the piston in a position to close the orifice. This action repeats as often as necessary to regulate the oil pressure.

Oil goes out of the oil pump and into an oil filter assembly. An **oil filter** is a paper filter element used to clean engine oil before it enters the engine parts (Figure 17-18). The oil filter assembly has a paper filter element inside a metal housing (shell). The bottom of the housing has a threaded ring that fits on a mount outside of the cylinder block. A large O ring seals the filter against a machined surface on the cylinder block. Inlet and outlet passages direct oil in and out of the filter (Figure 17-19). When it is time to change the filter, the metal shell and filter element are changed as a unit.

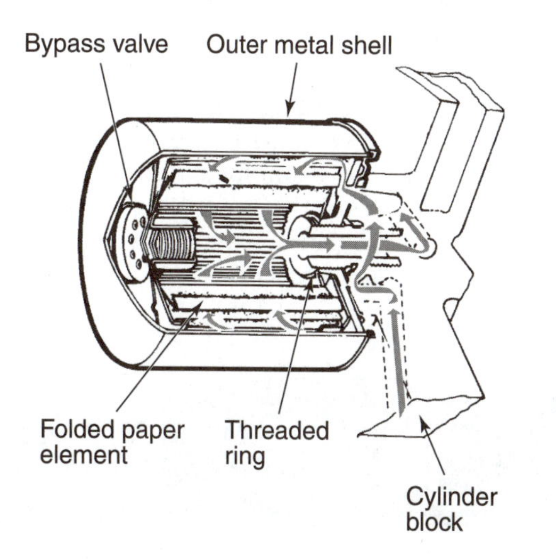

FIGURE 17-18 Oil flow through an oil filter assembly. (Courtesy of DaimlerChrysler Corporation.)

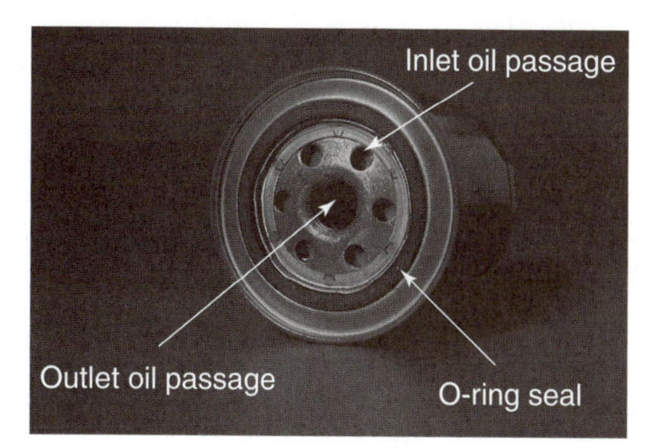

FIGURE 17-19 Inlet and outlet passages in an oil filter. (Courtesy of AlliedSignal AUTOMOTIVE, Filters and Spark Plugs.)

Filter elements are made from pleated paper (Figure 17-20). The folds in the paper provide a large surface area for filtering. As the oil goes through the paper, the dirt and acids stick to the outside. Only the cleaned oil gets through.

After a long period of use, the oil filter element may become clogged. A clogged filter will not allow oil to pass through the element. A bypass valve (Figure 17-21) inside the filter assembly prevents this problem. When the filter element becomes clogged, the pressure inside the filter housing increases. The high pressure pushes the bypass valve open. The oil then goes around the filter element, instead of through it. It goes directly into the engine. Oil gets to the engine parts, but it does not get cleaned by the filter.

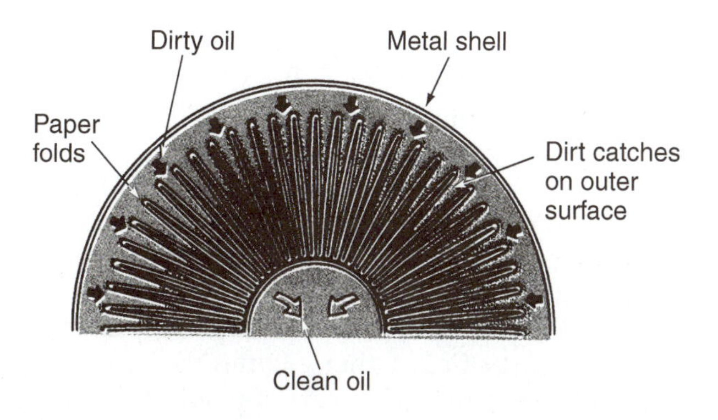

FIGURE 17-20 Oil flows through a paper filter element to clean the oil. (Courtesy of Dana Corp.)

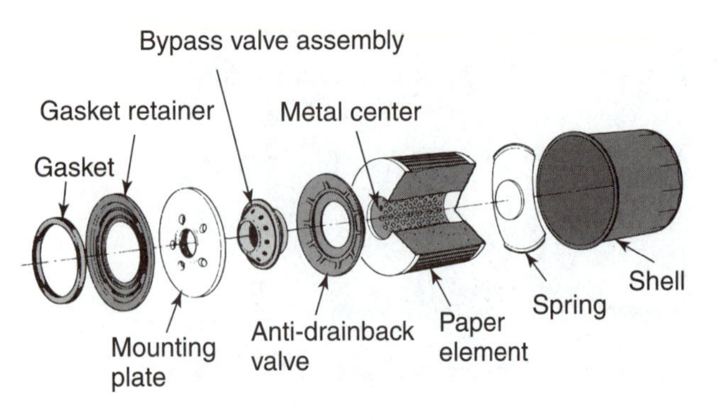

FIGURE 17-21 The bypass valve protects the system if the filter gets clogged. (Courtesy of General Motors Corporation, Service Technology Group.)

When the oil leaves the filter assembly, it enters the cylinder block oil passages and flows into several long passageways drilled in the cylinder block. Oil flows through passages drilled in the camshaft and crankshaft then into main bearings. The crankshaft has passages drilled from the main journals to the connecting rod journals for oil flow. Oil under pressure is allowed to escape between the connecting rod journal and bearing. This oil splashes on other parts such as cylinder walls and valve components. When the oil has passed through the part requiring lubrication, it runs down the inside of the engine. It collects back in the oil sump.

TWO-STROKE LUBRICATION SYSTEMS

Most two-stroke engines are lubricated by mixing oil with the incoming air-fuel mixture. Most two-stroke engines are lubricated by premix lubrication. A **premix lubrication system** is a two-stroke engine lubrication system in which oil and gasoline are mixed together in the fuel tank and enter the engine on the intake stroke. Both oil and gasoline go in the fuel tank. They are usually mixed together by the operator. Fuel is available in some areas with oil already mixed. **Premix** is a commercially available mixture of fuel and oil used in two-cycle engines.

Two-stroke lubricating oil (Figure 17-22) is a special type of oil that mixes with the gasoline and stays mixed for a long time. The mixture is different for different engines. Some engines use a 32 to 1 mix. This mixture has 32 parts of gasoline for each one part of oil. Other engines use a 50 to 1 mix. This mixture has 50 parts of gasoline for each one part of oil. There is always more gasoline than oil in the mixture.

FIGURE 17-22 Two-stroke oil is mixed with gasoline for lubrication.

The correct mixture can be found in the equipment owner's manual. The correct mixture is also often shown on a fuel tank decal (Figure 17-23). There are mixing containers available to mix oil and gasoline. These containers have space for the required amount of fuel and exactly one gallon of gasoline. Always shake the container after filling to be sure the oil gets into suspension in the gasoline.

The oil and fuel mixture goes into the engine's crankcase on the intake stroke through the reed valve. The oil in the mixture floats around in the crankcase as a mist. Gravity causes the tiny oil droplets to fall on the engine parts and provide the lubrication. Oil droplets fall on the crankshaft connecting rod, main bearings, piston pin, and cylinder walls. Some of the oil also goes with the fuel into the combustion chamber. This oil is burned along with the fuel. A two-stroke engine with this type of lubrication system often has a smoky exhaust.

The main advantage of the premix lubrication system is that it has no moving parts. While the premix system is simple, it has disadvantages. The

FIGURE 17-23 A gasoline and oil mixing ratio decal. (Courtesy of The Toro Company.)

operator can mix the oil and gasoline incorrectly. If there is not enough oil, the engine parts will not receive enough lubrication. The engine can quickly wear out. If too much oil is added, the engine will smoke and excessive carbon will build up on engine parts. Eventually, the spark plug will be fouled with carbon and the engine will run poorly.

> **CAUTION:** The mixture must be correct for the engine or engine damage and performance problems can result. Always consult the owner's or shop service manual regarding two-stroke oil and gasoline mixtures.

LOW OIL LEVEL ALERT SYSTEM

Some engines use a **low oil level alert system** that warns the operator and prevents engine operation if engine oil level is too low. This system prevents engine damage from an insufficient amount of oil in the crankcase. It automatically stops the engine (kills the ignition) before the oil level falls below a safe level. The system uses a light-emitting diode as a warning lamp located near the engine controls (Figure 17-24). The lamp lights up to warn the operator that the engine has stopped due to low oil level.

The low oil level alert system uses a float-operated switch located in the crankcase (Figure 17-25). The switch has a float inside that rides on the level of oil in the crankcase. An electrical wire called the *switch lead* is connected to the float. There is a magnet on the bottom of the float. An electrical switch is located below the float. As long as the oil level is high enough, the float and magnet stay out of contact with the switch.

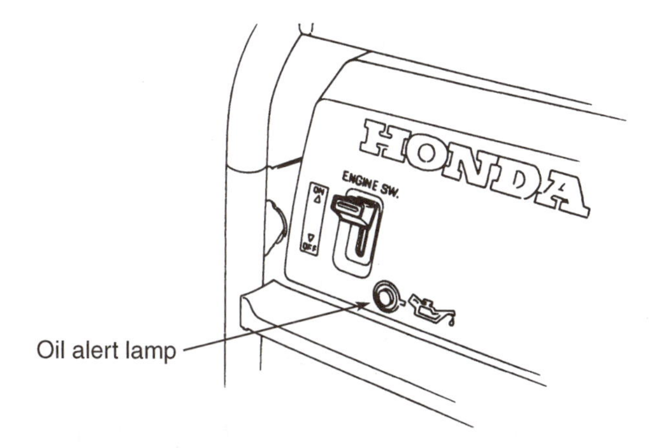

FIGURE 17-24 A warning lamp for a low oil level alert system. (Courtesy of American Honda Motor Co. Inc.)

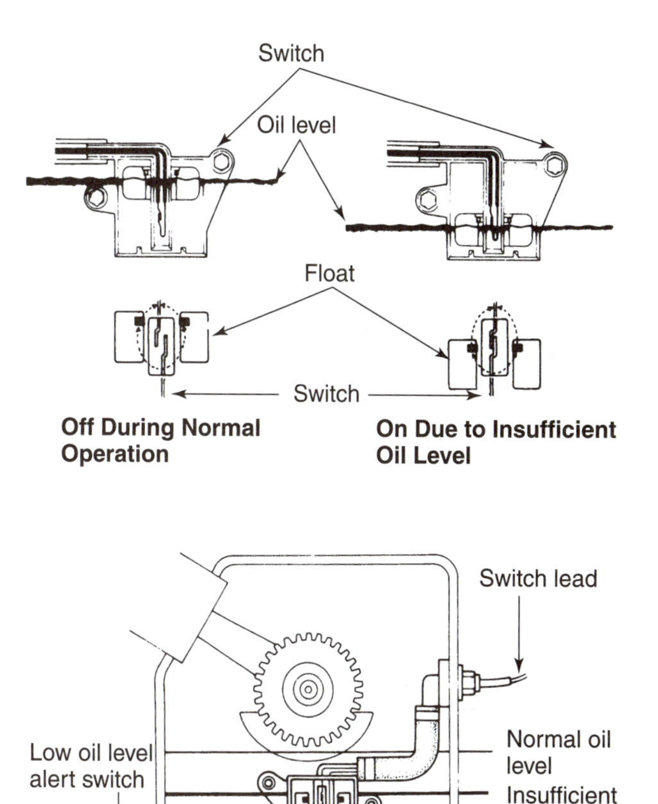

FIGURE 17-25 Operation of the low oil level switch in the crankcase. (Courtesy of American Honda Motor Co. Inc.)

If the oil level drops too low, the float and magnet drop far enough to approach the switch. The magnetism will cause the switch contact to close. When the switch is turned on, an electronic signal is sent to the transistorized ignition coil. The ignition coil is short-circuited and the engine is stopped.

A voltage is generated in the primary ignition coil winding as the engine comes to a stop. This voltage causes the warning light to flash until the flywheel comes to a stop, warning the operator of the low oil problem. If the starter rope is pulled after the engine has stopped, the engine will not start and the warning light will flash. If oil is added, the switch allows ignition and turns off the voltage source for the light.

CRANKCASE BREATHER

Four-stroke cycle engines often have a crankcase breather. A **crankcase breather** is a valve assembly that vents the pressure buildup in a four-stroke

CASE STUDY

A customer brought a 2-year-old lawnmower into the shop. The customer's complaints were that the engine exhaust created a cloud of white smoke and the engine used excessive amounts of oil. The owner was angry because she had just purchased the mower two seasons ago and has put very few hours on the equipment. She was in no mood to pay for an engine overhaul.

The technician assigned the job did not think the engine could be worn enough to cause these problems in such a short time. He inserted his finger in the exhaust pipe and it came out wet with oil. Clearly this engine was burning oil. His first thought was that a manufacturing defect in the cylinder or piston ring might be the cause. He started a systematic check to determine the source of the problem. He pulled the dip stick and found that the oil level and oil condition looked good. He removed the air cleaner in preparation for doing a compression check.

With the air cleaner removed, he noticed something he did not expect to see. The carburetor throat was wet with oil. He turned the air cleaner over and pulled off the hose from the crankcase breather to the air cleaner. The hose was dripping with oil. He removed the crankcase breather assembly and located the source of the problem. The reed valve in the breather was stuck in the open position. The fiber valve had a small burr that caused it to hang up. He changed the breather and started the engine. The exhaust was clean, proving that the breather was the problem.

engine crankcase. When the engine is running, the upward piston movement causes a low (negative) pressure in the crankcase. The negative pressure is desirable because it helps prevent oil consumption.

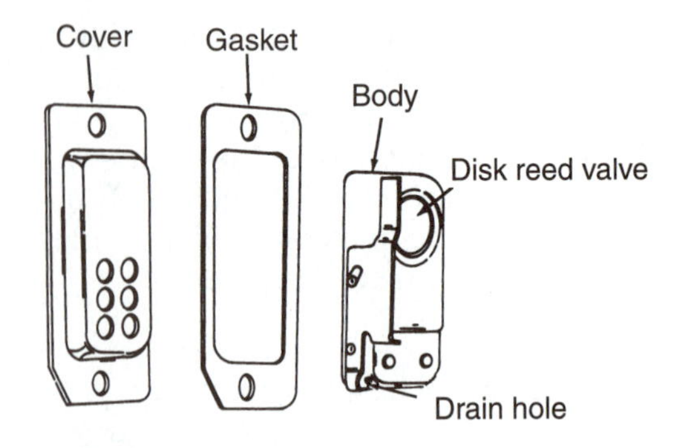

FIGURE 17-26 The crankcase breather prevents a pressure buildup in the crankcase. (Provided courtesy of Tecumseh Products Company.)

Piston downward movement develops a positive pressure in the crankcase. A positive pressure in the crankcase is not good because it can cause oil to be forced out of the crankcase through gaskets and seals.

The crankcase breather assembly is often mounted on the side of the L-head engine crankcase. The overhead valve engine breather is mounted on the top of the valve cover. The breather has a metal base that mounts on the engine. Inside the base is a small reed valve (Figure 17-26), usually made from a fiber material that may act as its own spring or be backed up with a separate spring. The reed valve is in the closed position when the pressure in the crankcase is low. When pressure in the crankcase rises above the setting of the reed valve, the valve opens. Pressure is vented past the open valve.

There is an oil and hydrocarbon mist in the crankcase. The oil comes from lubricating oil splash. The hydrocarbons come from gasoline vapors in the combustion chamber that leak past the piston rings. On older engines, the gas was vented outside the engine. This gas is a source of air pollution or smog. Newer engines collect the vented gas in a breather tube. The tube sends the gasses to the air cleaner, carburetor, or intake manifold. The oil and hydrocarbon mist is mixed with incoming engine air and burned in the cylinder.

REVIEW QUESTIONS

1. Describe the purposes of the lubrication system.
2. Explain why reducing friction in an engine is important.
3. Explain how an oil film can be used to reduce friction.
4. Define viscosity and describe the importance of viscosity to engine oil.
5. Describe how the viscosity of engine oil is rated.
6. Describe the purpose of the American Petroleum Institute oil service ratings.

7. Describe the parts and operation of a splash lubrication system for a horizontal crankshaft engine.
8. Describe the parts and operation of a splash lubrication system for a vertical crankshaft engine.
9. Describe the oil circulation in a barrel pump type pressure lubrication system.
10. Explain the operation of a barrel type oil pump.
11. Describe the parts and operation of a rotor type oil pump used on a full pressure lubrication system.
12. Describe the parts and operation of an oil filter used on a full pressure lubrication system.
13. Describe how premix is used to lubricate a two-stroke engine.
14. Describe disadvantages of the premix lubrication system.
15. Explain the purpose of a crankcase breather used on a four-stroke engine.

DISCUSSION TOPICS AND ACTIVITIES

1. Look up the oil viscosity and service ratings for several shop engines. Make a chart and identify the differences between engines.
2. Disassemble one engine with a splash lubrication system and another engine with a full pressure lubrication system. Trace the oil flow through each of the parts.
3. An oil viscosity chart shows that a 20-20W oil is not recommended for temperatures below –7°C. Use a conversion chart to convert this temperature to degrees F.

CHAPTER 18

Cooling and Exhaust Systems

OBJECTIVES

Upon completion and review of this chapter, you should be able to:

- Describe the parts and operation of the forced air cooling system.
- Explain the flow of air thorough an air cooling system.
- Describe the parts of a liquid cooling system.
- Explain the flow of coolant through a liquid cooling system.
- Describe the parts and operation of a muffler.
- Describe the parts and operation of a spark arrester.

TERMS TO KNOW

Air cooling system
Back pressure
Blower housing
Carbon
Coolant
Coolant pump
Coolant reservoir
Cooling fan
Cooling fins
Debris screen
Draft air cooling system
Ducting
Exhaust manifold
Forced air cooling system
Liquid cooling system
Muffler
Radiator
Radiator hose
Radiator pressure cap
Spark arrester
Thermostat

INTRODUCTION

The heat created in the combustion chamber of an engine can reach temperatures as high as 3,600°F (1,982°C). The engine's heat is used to produce useful power. Some heat flows out the exhaust system on the exhaust stroke. A significant amount of the heat enters the engine's parts and can quickly damage them. Too much heat causes the breakdown of lubricating oil. The expansion of parts from heat causes a loss of oil clearances that leads to part seizing. The heat can eventually cause parts to melt or warp. The heat must be carried away (dissipated) from the engine parts.

Engines have an air or liquid cooling system to remove the heat. Most small engines are air cooled. They use air circulated around the engine parts to take away the heat. Some larger, stationary engines use liquid cooling in which a liquid coolant circulates around hot engine parts to carry away the heat.

AIR COOLING SYSTEMS

An **air cooling system** is a cooling system in which air circulates over the hot engine parts to remove the heat. Air may be directed over the parts by a natural draft. More commonly, air is forced over the parts by a forced circulation system.

The engine parts that get the hottest are the cylinder and cylinder head. They have cooling fins formed around their outsides. **Cooling fins** (Figure 18-1) are thin fins formed on the outside of an engine part used to get the greatest amount of hot metal into contact with the greatest amount of air. When the engine is running, heat builds up in the cylinder head and cylinder. It moves (transfers) out into the cooling fins. Air flows around the cooling fins and carries away the heat.

Cooling fins are efficient at removing the excess heat from the cylinder and cylinder head. The 3,600°F (1,982°C) combustion temperature meets a

FIGURE 18-1 Air cooling fins are formed on the cylinder.

layer of cooled gas along the edges of the cylinder. The cooled gases act as an insulator, preventing the very hot combustion temperatures from reaching the cylinder wall. When the heat goes into the cylinder, it has lowered to about 1,200°F (649°C). Heat moves outward toward the end of the fin. Cooling air takes the heat away and lowers the temperature (Figure 18-2). The temperature on the end of the fins is often as low as 100°F (38°C).

Draft Air Cooling

A **draft air cooling system** is a cooling system in which air is circulated over hot engine parts by the movement of the engine through the air. Draft air cooling is used on highway vehicles like motorcycles. The movement of these vehicles through the air is used to force air over hot engine parts. As the motorcycle moves, it pushes through the air. The movement causes the air to flow over the motorcycle and through the cooling fins. There is a natural flow of air as long as the motorcycle is moving. The faster the motorcycle goes, the more air flows over the engine. When the motorcycle is stopped and the engine is running, there is no air flow over the engine. If the engine runs too long under these conditions, it will overheat.

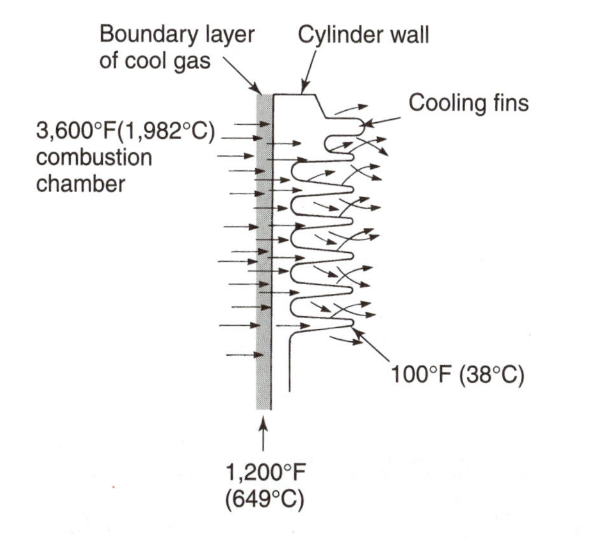

FIGURE 18-2 The cooling fins lower temperatures by passing the heat into the air.

Forced Air Cooling

Engines used on outdoor power equipment do not move or do not move fast enough for draft cooling. Air has to be forced around the parts with a **forced air cooling system,** a cooling system in which air is circulated over hot engine parts by an air pump to remove the heat.

The main parts of a forced air cooling system are cooling fins, flywheel (fan) blades, blower housing, and engine ducting (Figure 18-3). The cylinder and cylinder head have cooling fins to control heat and direct air flow. These parts are often made from aluminum. Heat passes (dissipates) through aluminum faster than other materials such as cast iron.

The flywheel (fan) (Figure 18-4) is used as a pump to move air around the engine parts. The flywheel has a set of curved blades around its diameter. The blades work like a centrifugal pump. As the flywheel rotates, the curved blades contact the air. Air flows over the curved shape of the blades. Air is moved rapidly out of the center of the flywheel. This movement causes a low pressure in the center of the flywheel. The low pressure pulls in air to the flywheel center. Centrifugal force causes the air to be thrown off the outside tip of the blade under pressure.

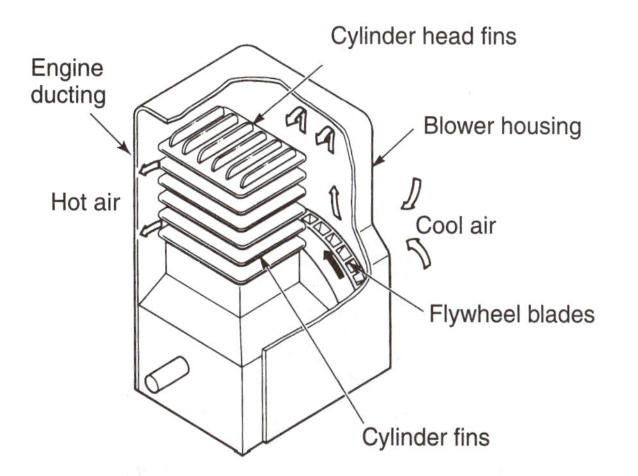

FIGURE 18-3 Cooling air is pumped into the blower housing and guided around cooling fins by ducting.

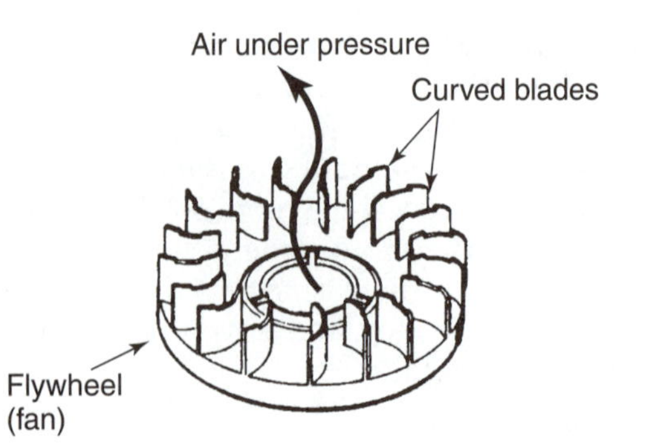

FIGURE 18-4 The curved flywheel (fan) blades use centrifugal force to pump air. (Provided courtesy of Tecumseh Products Company.)

The flywheel fits inside a blower housing. The **blower housing** (Figure 18-5) is a cooling system part that directs the cooling air flow into and out of the flywheel. The blower housing is also called a *fan cover*. There is a large air inlet hole in the middle of the blower housing. Cool air is pulled in the inlet through a debris (trash) screen. A **debris screen** is a screen in the center of the blower housing that filters incoming cooling air on the way to the flywheel. The debris screen keeps grass and other foreign objects out of the housing. Many engines have a debris screen that rotates with the flywheel. Others have a stationary debris screen on the blower housing.

The shape of the blower housing directs the cooling air. Air goes up to the top of the blower housing and is guided from the blower housing to the hot engine parts. Air flows through each of the cooling fins.

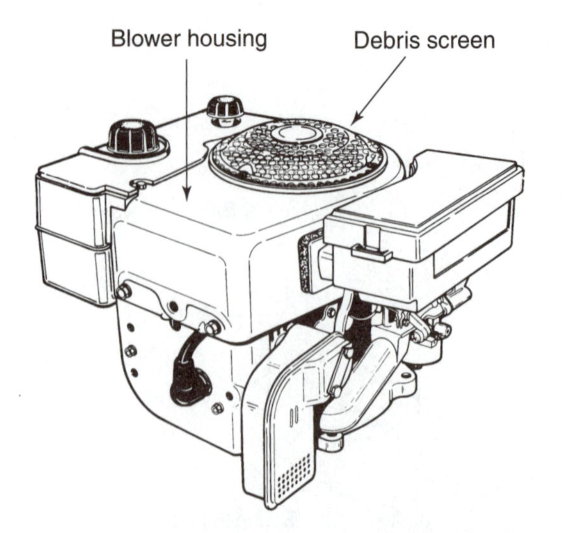

FIGURE 18-5 The blower housing guides air through a debris screen and into the flywheel. (Provided courtesy of Tecumseh Products Company.)

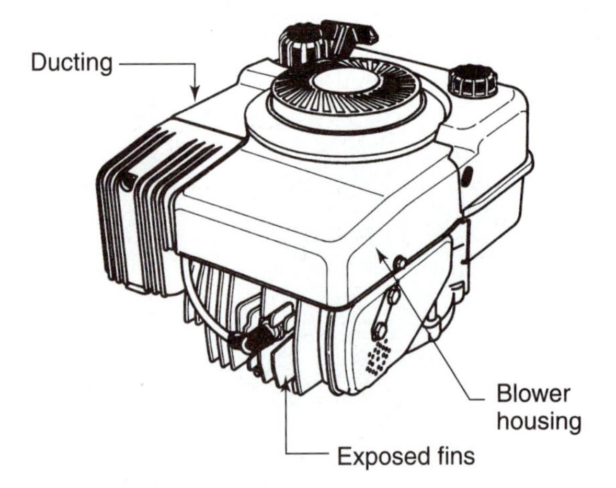

FIGURE 18-6 An engine with partial ducting over the cooling fins. (Provided courtesy of Tecumseh Products Company.)

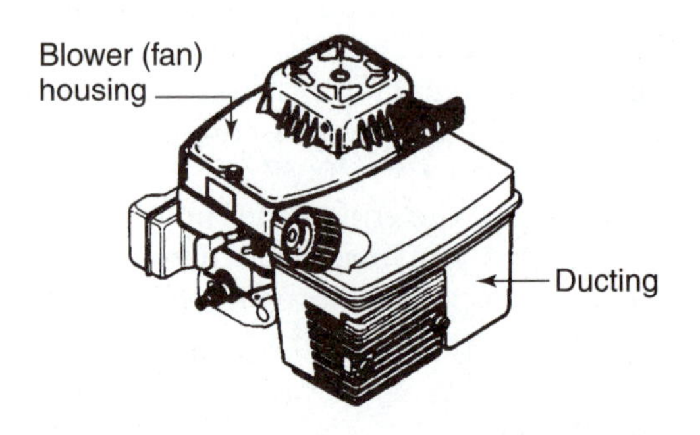

FIGURE 18-7 An engine with full ducting over the cooling fins. (Provided courtesy of Tecumseh Products Company.)

The cylinder and cylinder head cooling fins may be covered with sheet metal or plastic ducting. **Ducting** is sheet metal or plastic covers that guide the flow of cooling air over the cooling fins. Many engines have partial ducting (Figure 18-6), which is ducting that leaves some of the cooling fins exposed. Some engines have full ducting (Figure 18-7) in which all the cooling fins are covered.

SERVICE TIP: When engines are assembled, all the cooling ducting must be replaced. Missing ducting can cause the engine to overheat. Ducting must be able to dissipate heat.

LIQUID COOLING SYSTEMS

Liquid cooling systems are often used on large portable generators that have two-cylinder engines. The **liquid cooling system** is a cooling sys-

CASE STUDY

An outdoor power equipment student completed a semester-long project of adapting a chain saw engine to a small minibike. The student got the minibike frame and fenders chrome plated to improve their appearance. At the same time he got the engine cooling ducting chrome plated to match the frame and fenders. The completed project appeared very impressive.

The student completed the installation of the engine and was ready for a test ride. The engine started up and sounded very good. The test ride was done in an area behind the shop. The student returned shortly to the shop and described a high-speed misfire that started in the middle of the test ride. The instructor suggested some carburetor checks to make and the student took the minibike out for another test ride. The engine worked well for the first couple of laps and then began to misfire again.

When the minibike was back in the shop the instructor took another look. The engine had an unusual odor. The instructor touched the outside of the blower housing and could tell that the engine was overheating. The instructor focused on the chrome plated engine ducting. He wondered if the chrome plating was preventing proper heat dissipation. He had the student switch the ducting with one from a shop engine. The minibike went out on another test ride with the painted ducting. The engine ran fine this time. The chrome plating prevented engine heat from leaving the engine.

tem in which a liquid is circulated around hot engine parts to carry off the heat (Figure 18-8). There are coolant passages in the block around the cylinder and in the cylinder head near the valve. The passages are sometimes called *water jackets*. Heat from combustion goes through the metal of the cylinder head and cylinder wall into the coolant passages. The heat then goes into the liquid coolant circulating through the passages.

Liquid cooling systems can take care of more heat than can air cooling systems. Liquid holds heat better than air. A multiple cylinder liquid cooled engine is less expensive to make, because liquid cooled cylinders are cast together in a block. The finned cylinders have to be made separately. Liquid cooling passages also reduce engine noise so that the engine runs more quietly.

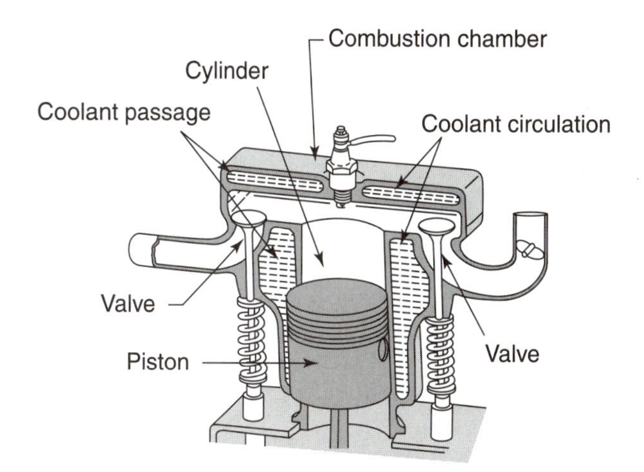

FIGURE 18-8 The liquid cooling system uses coolant to carry heat away from hot parts.

The main parts of a liquid cooling system are coolant, coolant pump, radiator, thermostatic valve, radiator hoses or pipes, radiator pressure cap, and reservoir tank. The system circulates liquid around hot engine parts to carry off the heat. The heat then passes into the air through the radiator.

Coolant

Coolant is a liquid circulated around hot engine parts in a liquid cooling system. In years past, water was used as the coolant in the liquid cooling system. Improved coolants have been developed over the years. Water causes rust in the water coolant passages. Rust prevents good heat transfer. Rust can scale off and clog the radiator core.

Cooling systems use a mixture of ethylene glycol and water as the coolant (Figure 18-9). Ethylene

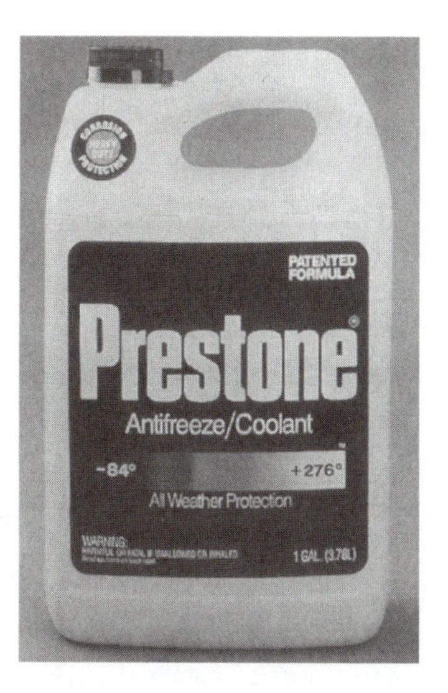

FIGURE 18-9 Coolant helps prevent freezing and boiling. (Courtesy of First Brands Corporation.)

glycol is also used as antifreeze, a chemical mixture that prevents the cooling system liquid from freezing. If the coolant freezes, expanding ice can cause severe parts damage. Ethylene glycol protects the system from freezing in temperatures well below zero.

When the liquid in a cooling system boils, the coolant turns to steam. The coolant pump is made to pump a liquid. The pump cannot circulate steam. Coolant circulation stops. The engine temperature gets dangerously high. Ethylene glycol has a boiling temperature far above that of water. The boiling temperature of water is 212°F (100°C) at sea level. A half-and-half mixture of ethylene glycol and water has a boiling temperature of approximately 225°F (109°C).

Coolant Pump

The **coolant pump** (Figure 18-10) is a centrifugal pump used to circulate the coolant through the coolant passages and into the radiator. The coolant pump is sometimes called the *water pump*. The pump has a housing with coolant passages. A shaft in the middle of the pump housing is supported on a bearing so that it is free to rotate. When the engine is running, a belt or gear drives the shaft. The shaft drives an impeller (small wheel) with curved blades. As the impeller spins, a low pressure is created in the center of the impeller. The low pressure draws in coolant. Centrifugal force causes the coolant to be thrown off the tip of the blades under pressure.

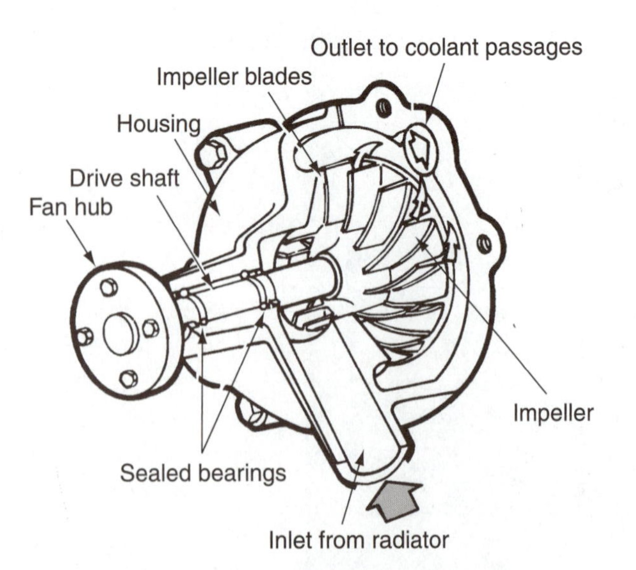

FIGURE 18-10 The coolant pump uses an impeller with blades to pump coolant through the system. (Courtesy of DaimlerChrysler Corporation.)

The pump circulates coolant throughout the cooling system. Coolant is pumped from the bottom tank of the radiator. It goes into the engine, the cylinder, and the cylinder head. After circulating around the coolant passages, coolant enters the radiator upper tank.

Radiator

The coolant removes heat from the hot engine parts. Heat must then be removed from the coolant. The hot coolant is pumped out of the engine and into a heat exchanger (radiator). The **radiator** (Figure 18-11) is a heat exchanger used to remove heat from the coolant in the liquid cooling system.

The radiator is mounted in front of the engine where air can flow through its center. The radiator has a top tank, a bottom tank, and a center core. Hot coolant is pumped out of the engine and into the top tank of the radiator. It then enters the radiator core. The coolant flows through small core pipes made from a good heat-conducting metal like copper or aluminum. The heat passes out of the liquid and into the wall of the core pipes. Each core pipe has a large number of air fins. Air circulates through the core. As the air goes past the fins, it takes the heat from the fins. The cooled liquid then flows

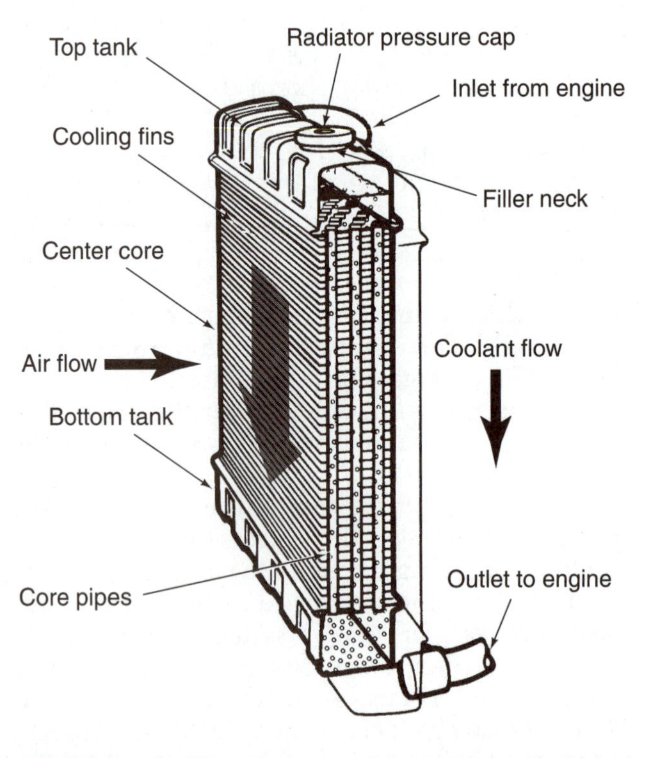

FIGURE 18-11 The radiator passes heat in the coolant into air passing through the radiator core. (Courtesy of DaimlerChrysler Corporation.)

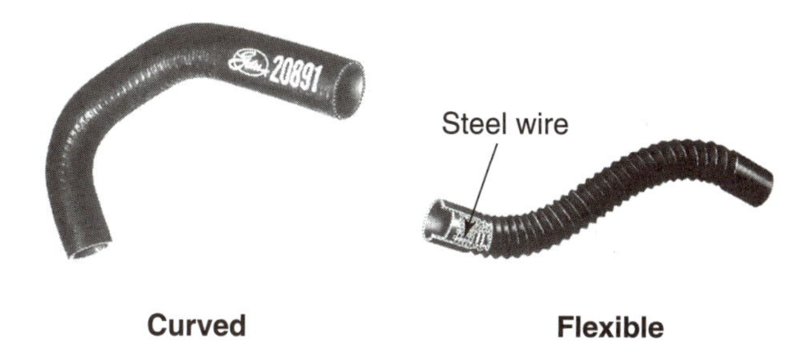

FIGURE 18-12 Radiator hoses connect the radiator to the coolant pump. (Courtesy of The Gates Rubber Company.)

into the bottom tank of the radiator. Coolant is pumped back into the engine to pick up more heat.

Coolant flows from the engine to the radiator and from the radiator to the engine through radiator hoses. A **radiator hose** (Figure 18-12) is a hose connecting the radiator to the engine cooling system. Engines are often mounted on rubber mounts to reduce vibration. The rubber mounts allow the engine a little movement as it is accelerated and decelerated. The flexible radiator hoses prevent engine movement from being transferred to and damaging the radiator. The hoses are often made from butyl rubber reinforced with steel wire. They may be flexible or permanently curved.

Thermostat

Engine temperature is regulated to make sure it does not run too hot or too cold. If the temperature gets too high (overheating), engine parts can be damaged. If the temperature is too low (overcooling), there can be high fuel use and sludge buildup in the oil. A thermostat is used to regulate cooling system temperature. A **thermostat** (Figure 18-13) is a valve that controls the flow of coolant into the radiator from the engine to regulate engine operating temperature. The thermostat is usually located in a housing where the upper radiator hose is connected.

Many thermostats have a wax pellet inside. The wax pellet is a part that gets longer when heated and shrinks when cooled. The pellet is connected through a piston to a valve. When engine coolant gets hot, the pellet pushes against the piston. The piston opens the valve. The open valve allows coolant to flow from the engine into the radiator (Figure 18-13B).

When engine coolant temperature is low, the pellet shrinks. This allows a spring to close the valve. The circulation of coolant through the radiator is stopped. Coolant then goes through a bypass passage back into the block. The opening and closing of the thermostat valve regulates coolant flow

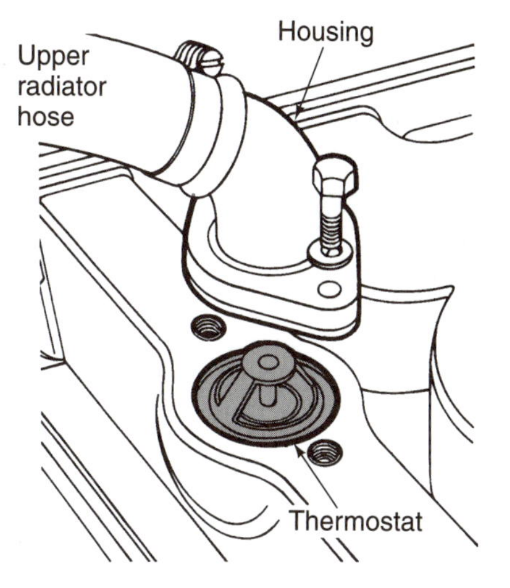

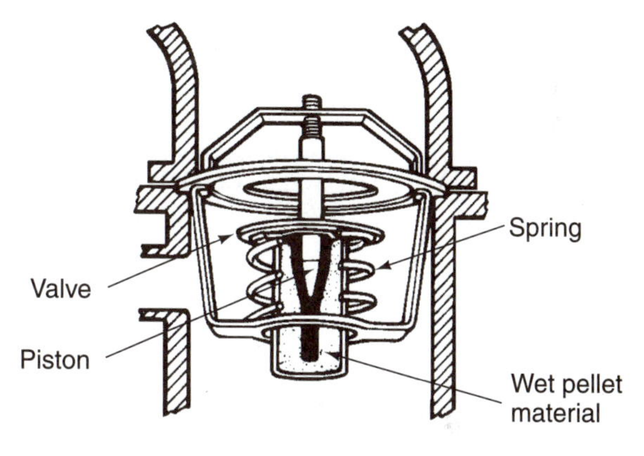

FIGURE 18-13 A thermostat regulates flow through the radiator to regulate engine operating temperature. (Courtesy of DaimlerChrysler Corporation.)

into the radiator (Figure 18-14). The engine temperature stays within operating temperature limits.

Cooling Fan

A cooling fan is used to circulate air through the radiator core. The **cooling fan** is a bladed air pump used to pull air through the radiator core to cool the coolant in the radiator (Figure 18-15). Most fans are driven by a fan belt on the engine. Some engines have a fan that may be driven by an electric motor. Electric fans often have a temperature switch. The switch turns the fan on and off as needed.

> **WARNING:** An electric fan can switch to on even when the engine is not running. Avoid having hands or tools near an electric fan.

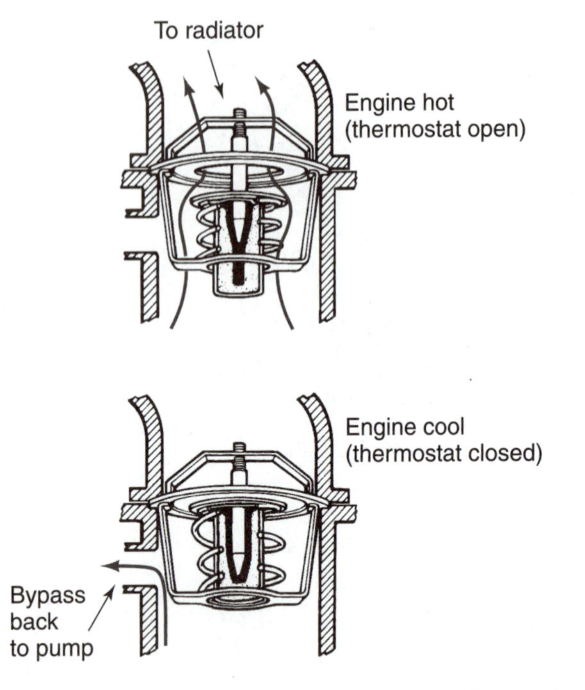

FIGURE 18-14 The thermostat valve opens when coolant is hot and closes when coolant is cold. (Courtesy of DaimlerChrysler Corporation.)

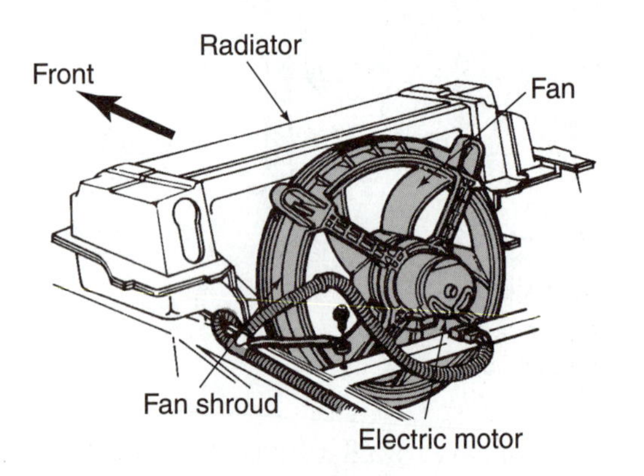

FIGURE 18-15 The cooling fan pulls air through the radiator core.

Radiator Pressure Cap and Coolant Reservoir

The radiator pressure cap (Figure 18-16) fits on top of the radiator in the radiator cap neck. The **radiator pressure cap** is used to access the coolant in the radiator and to regulate pressure and vacuum in the cooling system.

The radiator pressure cap increases the pressure in the cooling system, which raises the boiling temperature of the coolant. If allowed to boil, the coolant will turn into steam and expand violently. The coolant pump cannot circulate steam. If circulation stops, the engine temperature will rise to a dangerous level.

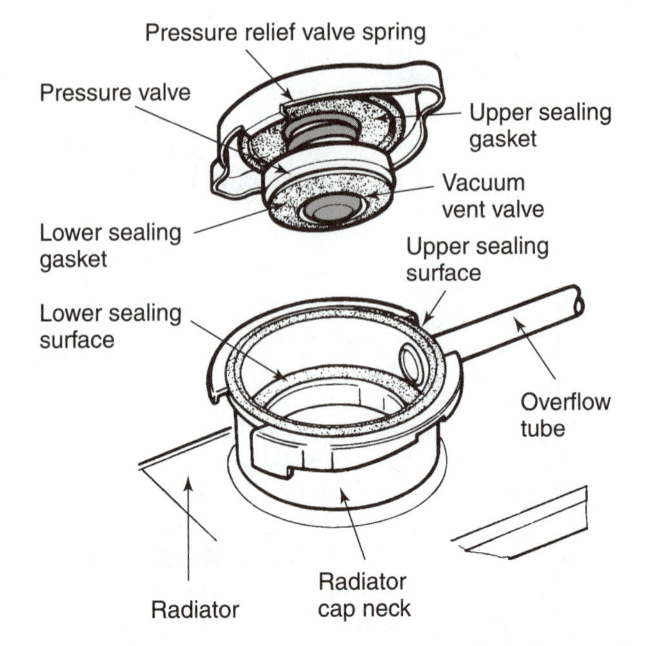

FIGURE 18-16 The radiator pressure cap fits in the top of the radiator. (Courtesy of DaimlerChrysler Corporation.)

At atmospheric pressure, water boils at 212°F (100°C). A half-and-half mixture of antifreeze and water boils at about 226°F (109°C). Each pound of pressure that the radiator cap holds raises the coolant boiling temperature about 3°F. For example, a radiator pressure cap increases coolant pressure to 12 pounds. The boiling point is raised from 226°F (109°C) to approximately 260°F (126°C).

The radiator pressure cap has a pressure valve and a vacuum valve (Figure 18-17). The pressure valve regulates cooling system pressure. The pressure valve presses against a seat in the radiator filler neck. There is a spring on the valve that holds it closed. The pressure required to open the valve is determined by spring tension. Valve opening regulates the pressure in the system. The pressure valve stays closed until the regulated pressure is reached. If the pressure goes higher, the valve opens and coolant or steam pressure escapes through the overflow tube. When the pressure goes down, the valve closes and the cycle begins again.

CAUTION: Each radiator cap is rated for a specific pressure. The pressure the radiator cap is set for is printed on the top of the cap. Each pound of pressure the coolant is raised causes the boiling point to be raised approximately 3°F. The engine shop manual specifies the correct pressure rating for a cooling system. Caps with lower or higher settings should not be used or the cooling system could be damaged.

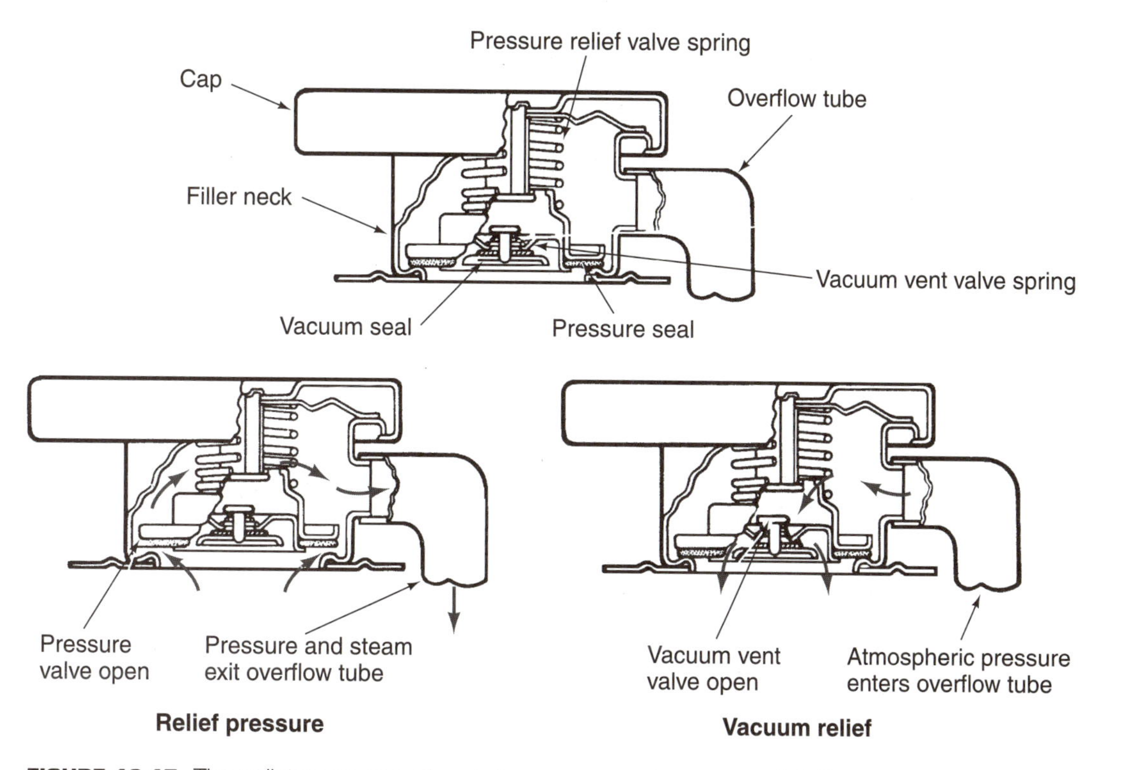

FIGURE 18-17 The radiator cap operation during vacuum and pressure relief. (Courtesy of DaimlerChrysler Corporation.)

The vacuum valve in the radiator cap prevents a vacuum in the cooling system. A vacuum causes thin radiator tubes or flexible radiator hoses to collapse. The vacuum valve normally remains closed to seal the cooling system from the outside air. When the engine is shut down, temperature and pressure decrease. This could cause a vacuum. A vacuum causes the vacuum valve to open. When it opens, air enters the system and prevents a vacuum.

Coolant Reservoir

The pressure relief valve in the radiator cap is connected to an overflow tube, which is connected to the reservoir tank. The **coolant reservoir** (Figure 18-18) is a container connected to the radiator and used to recover coolant that moves out of the radiator as it is heated. During pressure relief, the coolant goes through the tube and into the reservoir. When the system has cooled off, the coolant is drawn through the overflow by vacuum and goes back into the radiator.

The tank is often made of clear plastic so the level of coolant is visible. There are lines on the reservoir tank. There is a "full" or "max" line and a "low " or "min" line. The coolant level should be between them. A cap on top of the reservoir is used to add coolant, if necessary. The coolant level in the bottle normally goes up and down as the system heats and cools.

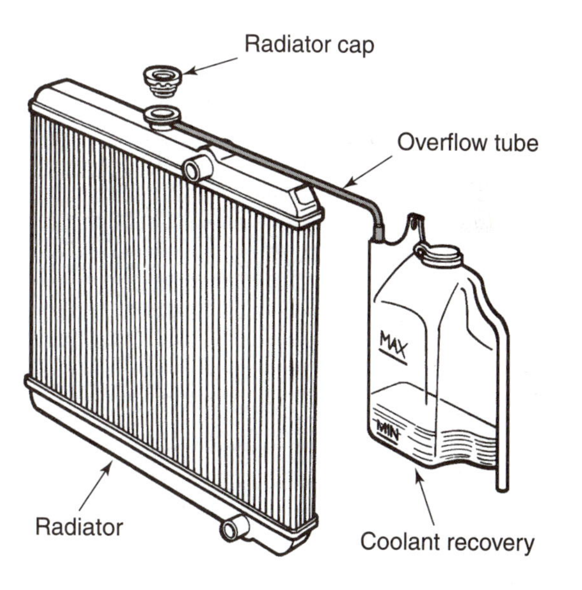

FIGURE 18-18 A reservoir is connected to the overflow tube to recover vented coolant. (Courtesy of DaimlerChrysler Corporation.)

EXHAUST SYSTEMS

Exhaust gases are routed out of the engine during the exhaust stroke. In the four-stroke engine this occurs when the exhaust valve opens. In the two-stroke engine this occurs when the piston uncovers the exhaust port. The exhaust system must provide for a smooth flow of exhaust gases. The exhaust system must also lower exhaust noise.

Any unnecessary bends or restrictions in the system slows the flow of exhaust out of the engine. A smooth flow of exhaust gases prevents a back pressure in the exhaust port. A **back pressure** is a positive pressure in the exhaust port that opposes the flow of gases out of the port. A back pressure stops all the exhaust gases from getting out of the cylinder during the exhaust stroke. Any leftover exhaust gases take up space needed for the incoming air and fuel mixture. Less air and fuel mixture to burn means lower engine power.

An exhaust system of the correct size and shape can increase engine power. Exhaust systems can be made to create a negative pressure in the exhaust port. These systems develop a flow that can pull (extract) exhaust gases out of the cylinder. This type of exhaust flow increases engine power.

Outdoor power equipment exhaust systems usually have a muffler and spark arrester (Figure 18-19). A **muffler** is a part that cools and quiets the exhaust gases coming from an engine. Single cylinder engines often mount the muffler directly to the exhaust port outlet. A flange and gasket assembly is used to fasten the muffler to the exhaust port.

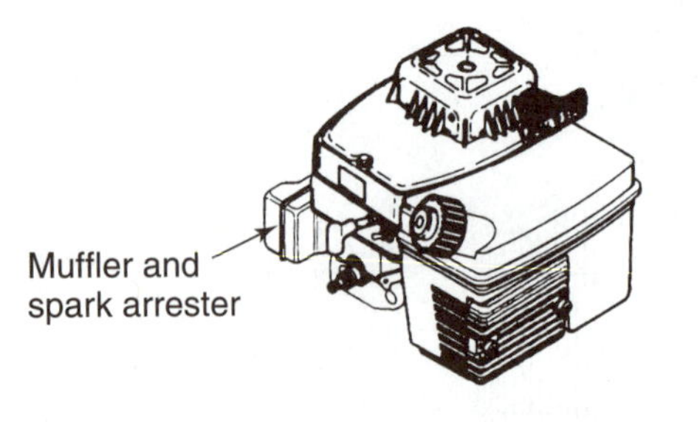

FIGURE 18-19 Muffler and spark arrester on a two-stroke engine. (Provided courtesy of Tecumseh Products Company.)

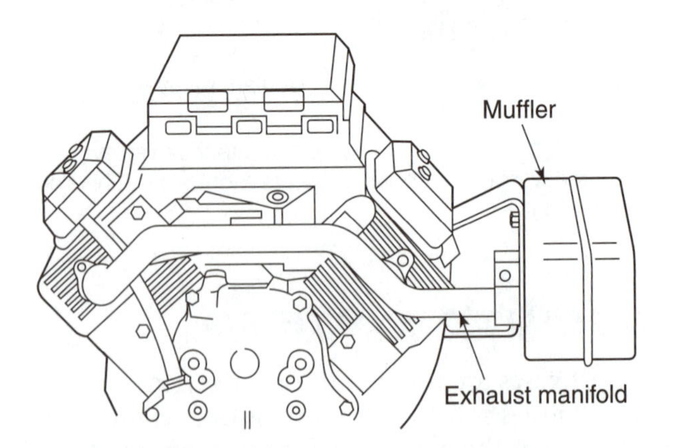

FIGURE 18-20 An exhaust manifold on a multiple cylinder engine connects both cylinders to the muffler.

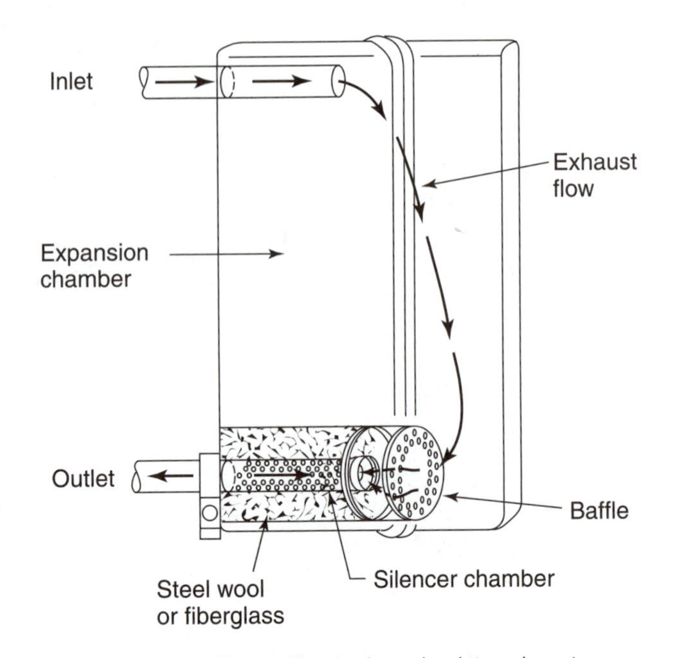

FIGURE 18-21 The muffler cools and quiets exhaust gases with an expansion and silencer chamber.

Multiple cylinder engines often use an exhaust manifold. An **exhaust manifold** is a pipe assembly that routes exhaust gas from the exhaust port to the muffler. The multiple cylinder exhaust manifold routes exhaust gas from each cylinder to one muffler (Figure 18-20).

A muffler is basically a metal can with an expansion chamber (Figure 18-21). An expansion chamber is an area of the muffler where high pressure exhaust gases can expand and cool. The cooled gases are routed through a set of baffles. Baffles are metal plates with small holes. Gases go from the baffles into a silencer chamber. The pipe that routes the exhaust gases through the silencer chamber has many holes that allow the gas to expand into an area filled with steel wool or fiberglass. The silencer chamber dampens out the noise and quiets the exhaust. The cooled and quiet exhaust gas is then routed out of the muffler through a hole or pipe. Some systems have a separate exhaust pipe attached to the muffler outlet.

The muffler gets very hot when the engine is running. Some engines have a metal cover over the muffler. This metal cover is called a *muffler protector*. It helps protect the muffler from damage and the operator from burns.

The exhaust leaving an engine often has carbon particles. **Carbon** is a solid particle formed when oil is burned in the combustion chamber. Carbon particles coming out the exhaust are red hot sparks. These sparks can easily start a fire if they fall on any combustible material. Sparks must be con-

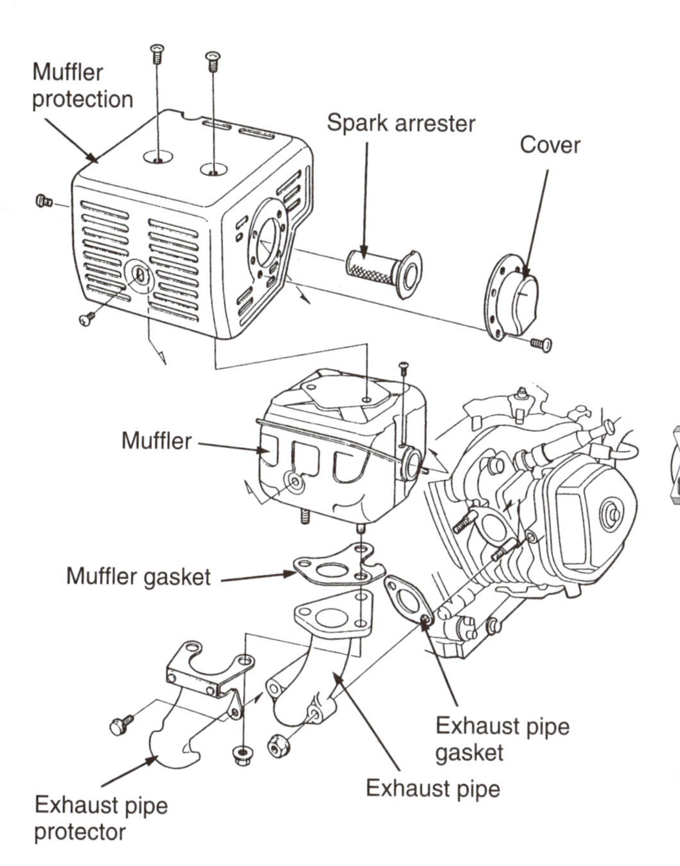

tained inside the muffler or in a separate part called a spark arrester. A **spark arrester** is an exhaust system part that traps sparks coming out an engine exhaust to prevent a fire (Figure 18-22). The spark arrester is a small screen installed in the exhaust flow. The screen filters out the carbon pieces as the exhaust gas passes through. The spark arrester may be removable so that it can be cleaned periodically to remove the carbon buildup. A plugged spark arrester prevents good exhaust flow and causes poor engine performance.

> **SERVICE TIP:** Do not run a four-stroke engine without the muffler attached. When the engine is stopped, the super-heated exhaust valves could be exposed directly to cool air entering the open exhaust port. This can cause some types of valve materials to warp.

FIGURE 18-22 (Left) A muffler assembly with a muffler protector and spark arrester. (Courtesy of American Honda Motor Co., Inc.)

REVIEW QUESTIONS

1. Describe the purpose and operation of cooling fins.
2. Explain the operation of a natural draft air cooling system.
3. List and describe the parts of a forced air cooling system.
4. Explain how the flywheel fins pump air through an air cooled engine.
5. Explain the purpose of coolant passages in a liquid cooling system.
6. Explain why engine coolant is better than water in cold conditions.
7. Explain why engine coolant is better than water in hot conditions.
8. Explain how a coolant pump works to circulate coolant in an engine.
9. Explain how a radiator works to exchange heat from the coolant to the air.
10. Explain how and why a radiator pressure cap maintains system pressure.
11. Explain how and why a radiator pressure cap prevents a system vacuum.
12. Explain how a thermostat works to regulate engine operating temperature.
13. Explain the purpose of the exhaust system.
14. Describe the parts and operation of a muffler.
15. Describe the parts and operation of a spark arrestor.

DISCUSSION TOPICS AND ACTIVITIES

1. Disassemble a shop engine and identify the cooling system parts. Trace the flow of air through the cooling system.
2. Locate a scrap muffler and spark arrestor. Use a hacksaw to cut the parts open. Trace the flow of exhaust through the parts.
3. Each pound of radiator cap pressure raises the coolant boiling temperature about 3°F. What temperature (above boiling 212°F) could the cooling system be increased to without boiling, if a 14-pound pressure cap were used?

CHAPTER 19

Manual Starting Systems

OBJECTIVES

Upon completion and review of this chapter, you should be able to:

- Explain the purpose of a manual rewind starter.
- Describe the parts and operation of a ball rewind starter.
- Describe the parts and operation of a dog rewind starter.
- Explain the parts and operation of a ratchet recoil starter.
- Explain the parts and operation of a gear starter.
- Describe the parts and operation of windup spring starter.
- Explain the purpose and operation of a compression release.
- Describe the steps to follow to troubleshoot a manual starter.
- Describe the steps to follow to replace a starter spring.

TERMS TO KNOW

Ball rewind starter clutch	Ratchet recoil starter
Compression release	Rewind starter
Dog rewind starter clutch	Rewind starter clutch
Gear starter	Starter housing assembly
Manual starter	Starter rewind spring
Pull rope	Windup spring starter

INTRODUCTION

The engine is started by rotating the crankshaft fast enough to draw a mixture of air and fuel into the cylinder. It has to be rotated fast enough for the magneto to create a spark to ignite the mixture. Older engines were started with a rope wound around a crankshaft pulley. Pulling on the rope rotated the crankshaft. Engines often did not start on the first pull. The operator had to wind the rope around the pulley and pull on the rope again. Newer engines use a much easier starting system. These starting systems may be operated manually or by electricity.

A **manual starter** is an engine starter that requires the operator to rotate the crankshaft manually for starting (Figure 19-1). Many manual starting systems use a rewind starter. A **rewind starter** is a manual starter with a spring a mechanism that automatically rewinds a rope back into the starting position. With a rewind (recoil) starter, the operator does not have to wind a rope around the pulley.

Rewind rope starting systems also use a clutch mechanism. The clutch disconnects the rope from the crankshaft when the engine starts. If the rope were not disconnected, it would pull the operator's hand into the engine. Manual starting systems are identified by starter clutch. Common types of rewind starter clutches are the ball and dog. Manual starters are also used that operate with gears or springs to rotate the engine.

BALL REWIND STARTER

The rewind starter has a starter housing (case) assembly and a rewind starter clutch (Figure 19-2). The **starter housing assembly** is a part of the manual starter that has the pull rope, starter pulley

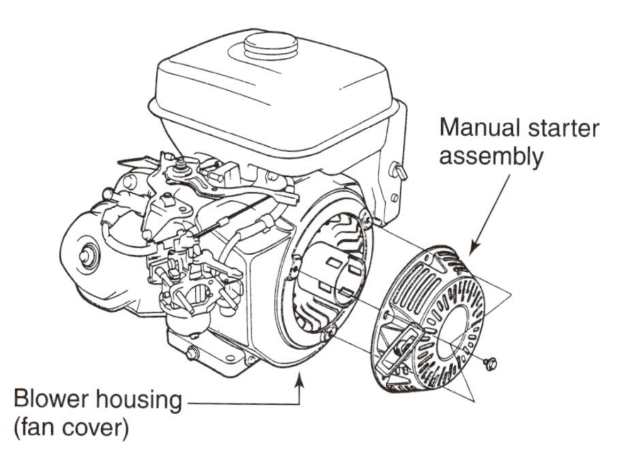

FIGURE 19-1 The manual starter housing is attached to the engine blower housing. (Courtesy of American Honda Motor Co., Inc.)

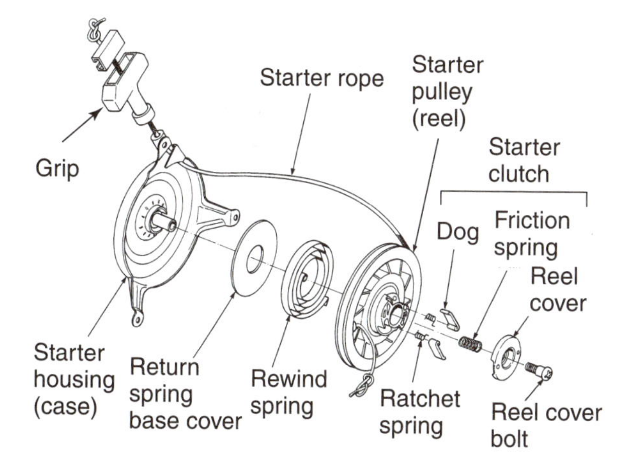

FIGURE 19-2 Parts of a rewind starter assembly. (Courtesy of American Honda Motor Co., Inc.)

(reel), and rewind (return) spring. The starter housing assembly fits on the blower housing. The starter pulley is the part that rotates to rotate the crankshaft. The starter pulley engages the engine's crankshaft through a rewind clutch. The **pull rope** is a rope used on the rewind starter pulley to rotate (crank) the crankshaft for starting. One end of the pull rope is attached to the pulley. The other end has a handle or grip for the operator. The handle or grip seats against the starter housing when not in use. The **starter rewind spring** rewinds the pull rope on the starter pulley after each engine starting. Long and narrow with a hook formed on each end, the spring is coiled in a circle with the outer end hook attached to the starter housing. The inner end is connected to the starter pulley.

When the operator pulls on the rope, the pulley rotates. The pulley is held in the center by its engagement to the rewind clutch. It is supported on the outside by its attachment to the blower housing. As the pulley is rotated, it compresses the rewind spring inside the starter housing. When the operator lets go of the handle, the spring causes the pulley to rotate in the opposite direction. This rapidly rewinds the rope on the pulley back into the starting position. The operator can crank the engine again if necessary.

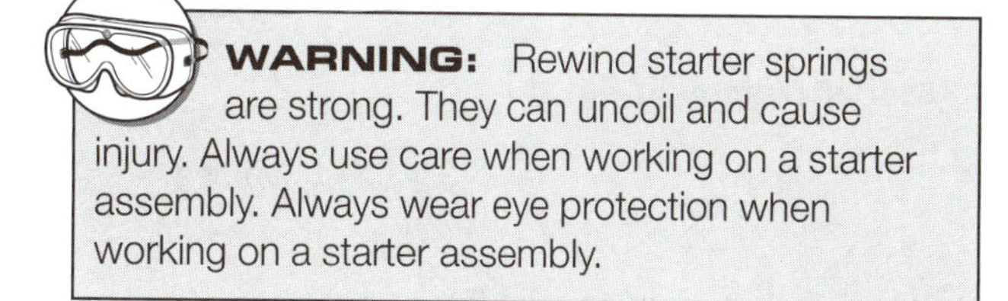

WARNING: Rewind starter springs are strong. They can uncoil and cause injury. Always use care when working on a starter assembly. Always wear eye protection when working on a starter assembly.

The pulley has to be rotated clockwise to crank the engine for starting. The pulley has to be rotated counterclockwise to rewind the rope. During rewind, the pulley must be disconnected from the engine's crankshaft. If the pulley is not disconnected, the spring will not return the rope. The pulley has to be disconnected when the engine starts. If it is not, the rope might pull the operator's hand into the engine. The **rewind starter clutch** is a part of the manual starting system that disconnects the starter pulley from the crankshaft when the engine starts.

The **ball rewind starter clutch** (Figure 19-3) is a clutch that uses a ball and pawl assembly to disconnect the starter pulley from the crankshaft when the engine starts. The clutch has a housing assembly with a threaded hole. The threaded hole is used to screw the housing on to the end of the crankshaft. The other end of the clutch has a ratchet that engages the pulley in the starter housing. A ratchet is a part that turns freely in one direction and locks up in another. The clutch ratchet locks the clutch housing to the pulley in one direction. When locked, the housing rotates the crankshaft. In the opposite direction, the ratchet unlocks the pulley from the housing.

The inside of the ball clutch (Figure 19-4) has several irregularly shaped channels around its outer edge. A steel ball or balls rides in one or more of the channels. The end of the ratchet has a hook-like shape called a *pawl*. During cranking, the rotating pawl wedges a ball between its end and a narrow part of the clutch housing. When the engine starts, centrifugal force causes the balls to fly out in the widest part of the housing. In this position, they cannot wedge between the pawl and the housing. The housing and the ratchet are disconnected from each other.

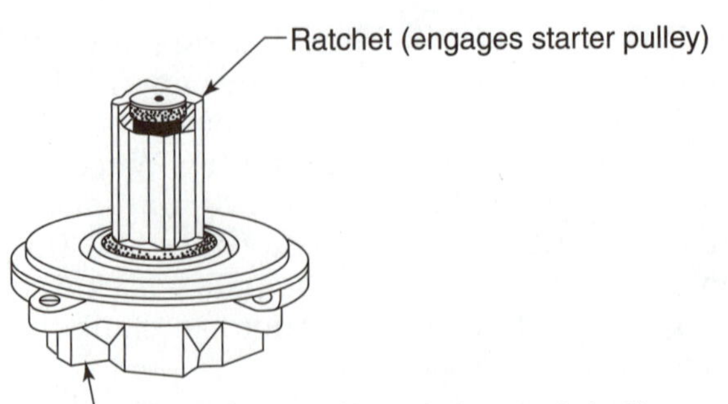

A. Starter Clutch

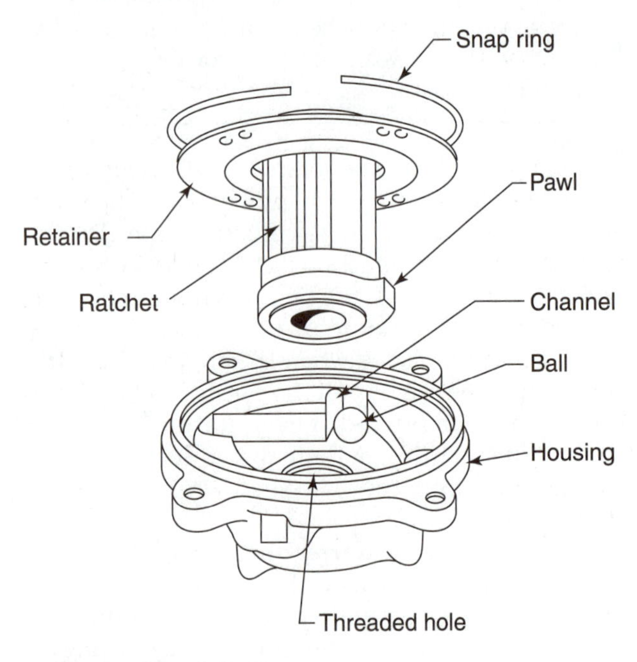

B. Starter Clutch Parts

FIGURE 19-3 Parts of a ball rewind starter clutch.

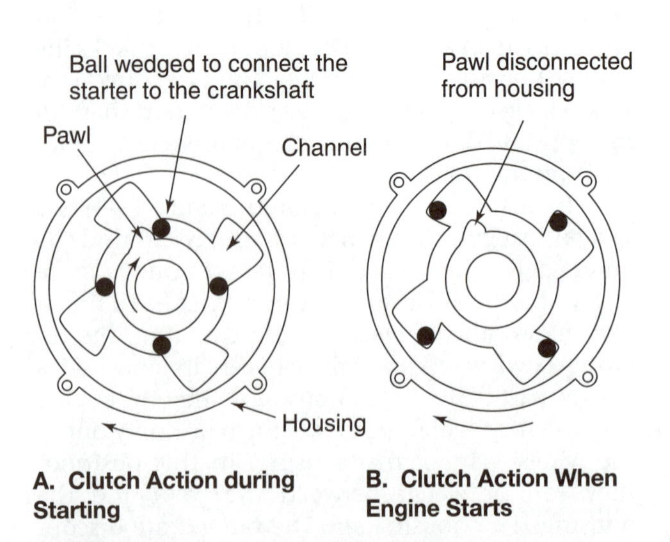

A. Clutch Action during Starting

B. Clutch Action When Engine Starts

FIGURE 19-4 Operation of the ball rewind starter clutch.

CASE STUDY

An outdoor power equipment student had just finished overhauling an engine for an edger. She added oil and poured fresh fuel into the fuel tank. The engine was ready for its first start up. She had to pull on the starter rope only two times before the engine started and ran. Initially the engine sounded fine. She allowed the engine to idle and warm up to make final carburetor adjustments.

After a short period of running, the engine began to make a high pitched scraping noise. The class gathered around to check out the unusual sound. With most of the class watching, the pull rope grip separated from the pull rope and the rope disappeared into the blower housing. The noise coming out of the blower housing got even louder. The student killed the engine to locate the source of the problem.

The instructor directed the student to remove the blower housing. He had her lock up the flywheel and remove the ball starter clutch. When the starter ball clutch was disassembled, the problem was obvious. The steel clutch balls had worn out the aluminum channels in the starter clutch housing. The balls had wedged in a position to lock up the clutch instead of allowing it to freewheel. This had caused the clutch to pull the rope in and had created the noise. With a new ball clutch and a new starter rope, the engine sounded fine again.

DOG REWIND STARTER

The dog clutch is another common type of rewind starter. The **dog rewind starter clutch** uses of a set of starter dogs to connect and disconnect the starter pulley from the engine when the engine starts. The dog type rewind clutch has a starter housing assembly with rope, pulley, and spring. These parts work the same way as in the ball type. The clutch mechanism has two basic parts. A hub is attached to the end of the crankshaft and is rotated to start the engine. The starter pulley is rotated with the pull rope. The pull rope is rewound with a rewind spring.

The pulley and hub are connected and disconnected by one or more small levers called dogs (Figure 19-5). The dogs are attached to the starter pulley. They are short, flat parts made from a high-quality steel. When the rope is pulled, a cam in the

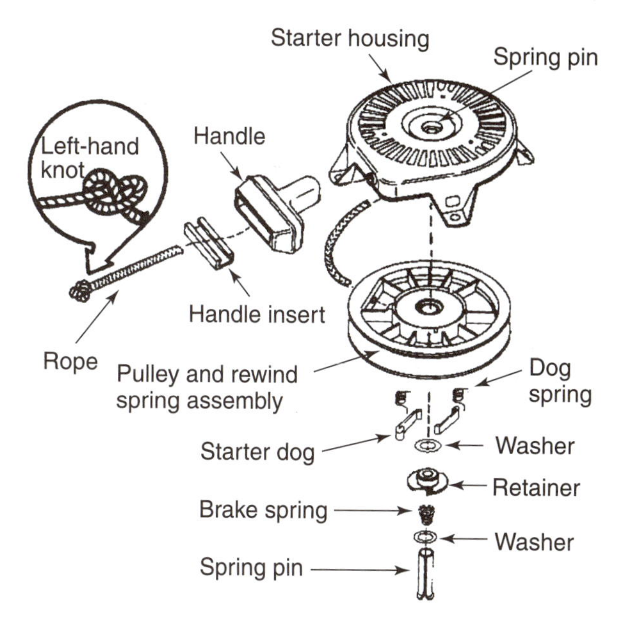

FIGURE 19-5 Parts of a dog rewind starter clutch. (Provided courtesy of Tecumseh Products Company.)

pulley pushes the dog or dogs outward. When a dog moves outward, it engages notches cut into the hub. The pulley is locked to the hub to crank the engine. When the engine starts or the rope rewinds, centrifugal force causes the dog to come out of contact with the hub (retract). The hub is disconnected from the pulley (Figure 19-6). The rewind spring works the same as on the ball type clutch.

Some clutches use one dog. Larger horsepower engines that require more cranking effort use two or more. Some clutches use small springs to position the dogs. Other dog type clutches use friction shoes instead of notches. The friction shoes connect the hub to the pulley by friction.

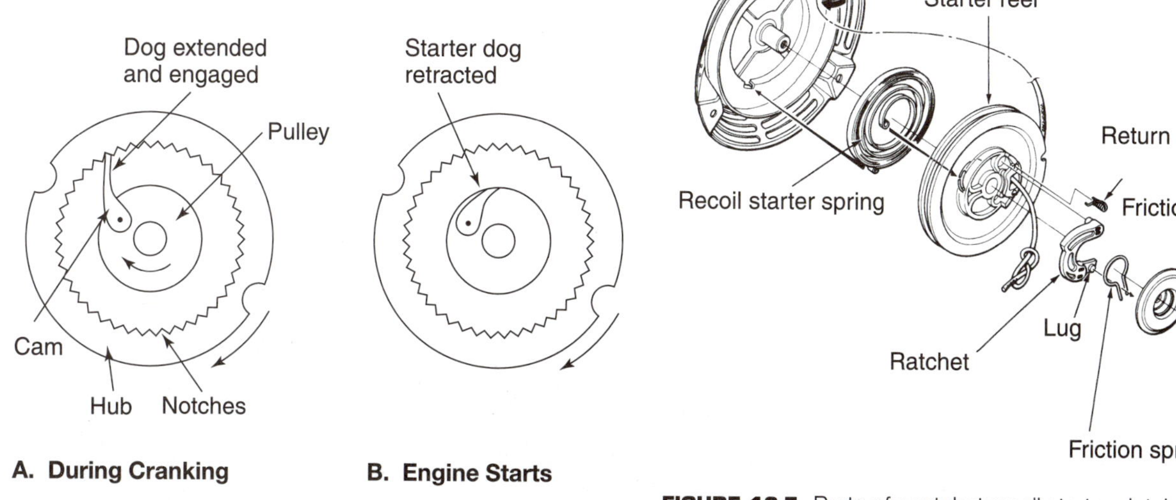

FIGURE 19-6 Operation of a dog rewind starter clutch.

RATCHET RECOIL STARTER

A **ratchet recoil starter** (Figure 19-7) uses a ratchet assembly to connect and disconnect the starter rope from the engine. The ratchet recoil starter works like the dog type. A ratchet clutch is used instead of dogs. The starter parts are housed in a recoil starter case. The case is mounted above the flywheel on the fan cover (blower housing). The flywheel is rotated for starting through a drive hub. The hub fits in the center of the flywheel. The starter rope enters the starter case and is wound around a starter reel. One end of a recoil starter spring is attached to the case. The other is attached to the end of the starter reel.

Pulling on the rope rotates the starter reel. A spring-loaded ratchet assembly fits on the starter reel. The ratchet assembly rotates with the starter reel. During cranking, the ratchet pivots on its mount. It wedges against and drives a friction plate. The friction plate, in turn, drives the flywheel hub and cranks the engine.

When the engine starts, the engine drives the friction plate. Flywheel rotation provides a force

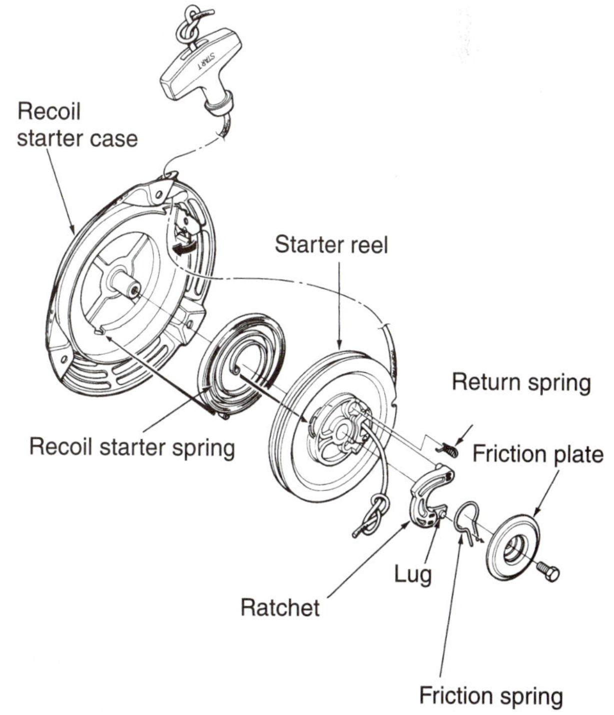

FIGURE 19-7 Parts of a ratchet recoil starter clutch. (Courtesy of American Honda Motor Co., Inc.)

back through the ratchet assembly. This force causes the ratchet assembly to move out of contact with the friction plate. The starter reel is prevented from rotating. The recoil starter spring rewinds the rope back on the starter reel. The system is ready for another starting cycle.

GEAR STARTER

Some vertical crankshaft engines use a gear type starter. A **gear starter** (Figure 19-8) is a starter that has a gear that engages another gear on the flywheel to crank the engine. The gear starter allows the operator to pull straight up on the rope to crank the engine. The flywheel has a ring of gear teeth formed on its outside diameter. The starter assembly has a pulley attached to a threaded shaft. A starter gear with internal threads rides on the shaft. When the rope is pulled, the pulley turns the shaft. The threads force the starter gear to go (walk) down the shaft and engage the flywheel gear. The starter gear causes the flywheel gear to turn to crank the engine. During engine start, the rotating flywheel gear makes the starter gear walk back out of contact. The spring on the shaft returns the starter gear to its released position. The pulley is rewound with a rewind spring as in other starters.

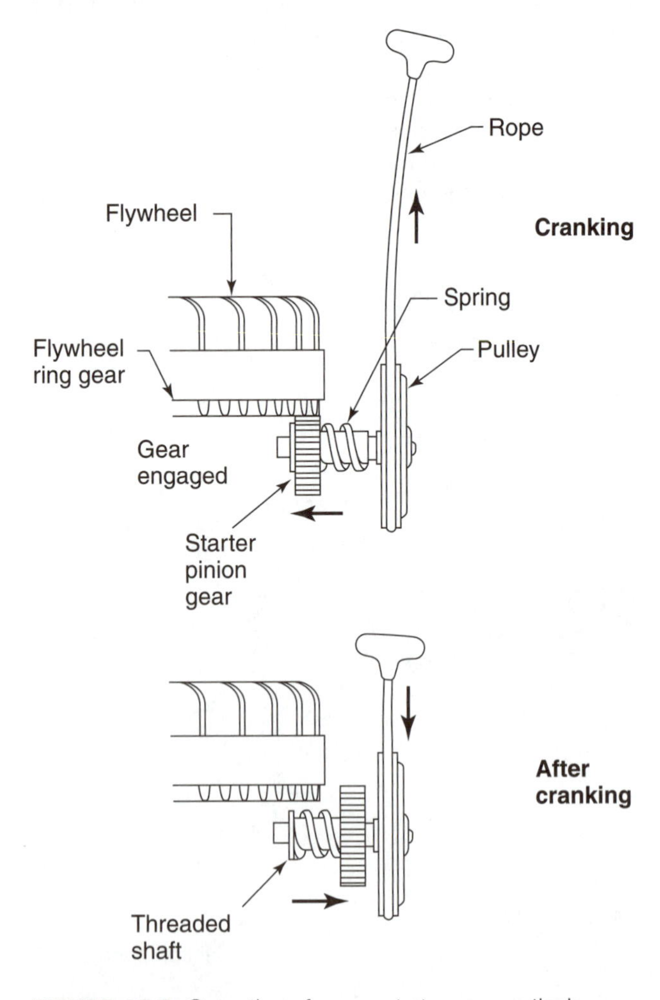

FIGURE 19-8 Operation of a gear starter on a vertical crankshaft engine.

WINDUP SPRING STARTER

Some starter systems use a windup spring instead of a rope to crank the engine. A **windup spring starter** (Figure 19-9) uses a handle to wind up a large spring and the spring is then released to crank the engine. These starters have a fold-up handle on the blower housing. A control knob or lever is used to control the spring.

The strong spring is called a *windup spring*. When the handle is rotated, the windup spring is wound up tightly (Figure 19-10). A spring-loaded ratchet catch prevents the handle and spring from unwinding. When the spring is wound up, the operator pushes on a knob or lever to unlock the spring. The spring is connected to a rewind pulley like those used in the ball clutch. The spring energy rotates the pulley. The pulley cranks the engine through the ball clutch. If the engine fails to start, the operator uses the handle to load the windup spring again.

COMPRESSION RELEASE

When an engine is being cranked, the starter is working against engine compression. The higher the compression pressure, the more effort required

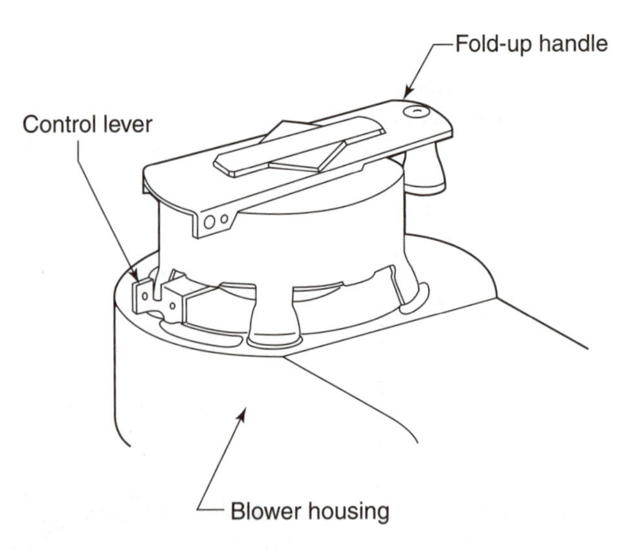

FIGURE 19-9 Outside parts of a windup spring starter.

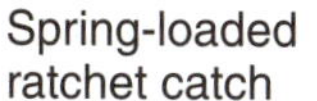
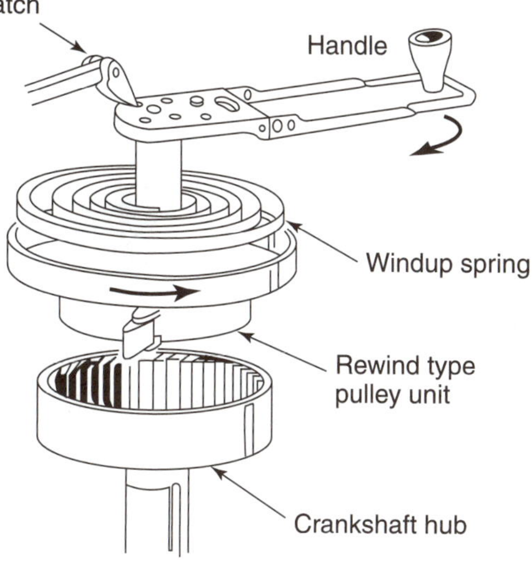

FIGURE 19-10 Operation of the windup starter.

to crank the engine. Some engines use a compression release to reduce compression for easier starting. A **compression release** (Figure 19-11) is a system that lifts the exhaust valve to release compression pressure during cranking. Compression release systems use a flyweight system on the engine camshaft. The flyweight is attached to the camshaft drive gear. During starting, the flyweights are held in an inward position by a spring. In this position, a tab sticks out of one of the flyweights above the camshaft exhaust lobe. It lifts the exhaust valve off its seat for the first part of the compression stroke. The open exhaust valve allows compression to leak out of the compression chamber. A much lower compression pressure is developed in the cylinder. The engine can be cranked easily.

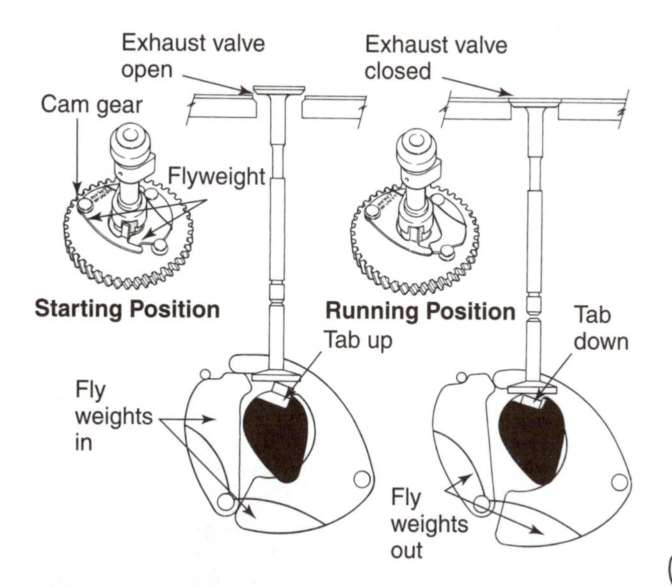

FIGURE 19-11 A compression release holds the exhaust valve off its seat to lower engine compression for starting. (Courtesy of Kohler Co.)

As the engine starts, centrifugal force acts on the flyweights. They move outward against spring pressure. This movement causes the tab to drop into a recessed area on the camshaft lobe. In this retracted position, the tab is below the surface of the lobe. It no longer has any effect on the valve. The engine can operate at full compression and full power. When the engine is stopped, the spring returns the flyweights to the inward position. The compression release is again in position to aid in starting.

MANUAL STARTER TROUBLESHOOTING AND SERVICE

Manual starting systems commonly work for a long time without a problem. When they do have a problem the most common causes are:

- Broken pull rope
- Broken rewind spring

A broken pull rope is usually obvious. The broken end of the rope comes out of the starter housing when pulled. A broken rewind spring is also easy to diagnose. The rope will fail to rewind back in to the pulley after it is pulled.

When you find either of these problems, you can either replace the entire starter assembly or replace the spring and rope. When the unit is disassembled for either the rope or spring, you should replace both parts. If there have been enough starting cycles to wear out a rope, the spring will probably be getting to the end of its service life. You may find that the cost of a replacement starter assembly is less than buying individual parts for the repair. If you decide to replace the individual parts, check the shop service manual for the procedure and any special tool that may be required. The procedure for replacing a common type of rewind starter is shown in the accompanying sequence of photographs.

WARNING: Wear eye protection when repairing a rewind starter to protect against the possibility of a flying spring.

PHOTO SEQUENCE 2

Rewind Starter Repair

1. Remove the fasteners that hold the rewind starter assembly to the blower housing and remove the rewind starter.

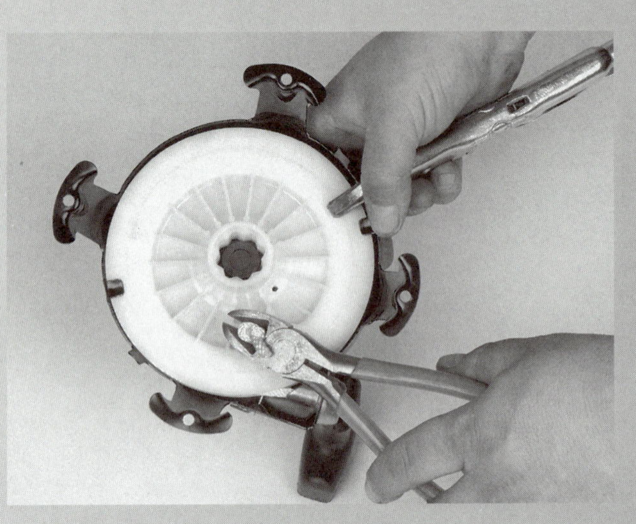

2. Pull the rope out of the housing and use vise grip pliers to lock the pulley in position. Cut the pull knot at the pulley end and pull the rope out of the housing.

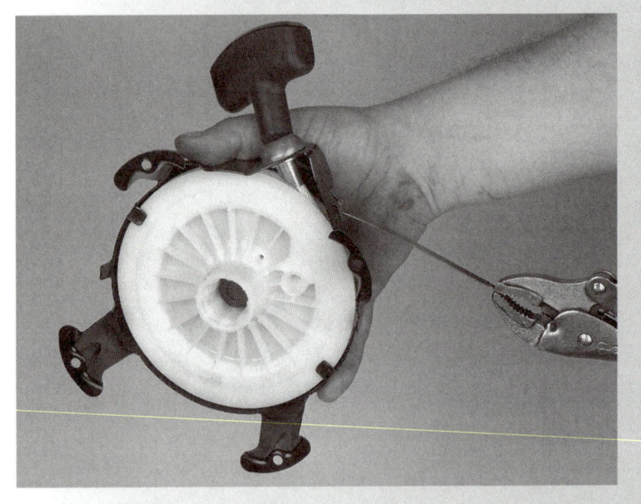

3. Remove the vise grip pliers and allow the spring tension to unwind the pulley. Use vise grip pliers on the pulley end of the rewind spring to pull it out of the housing.

4. Bend the tabs (tangs) on the starter housing and remove the starter pulley.

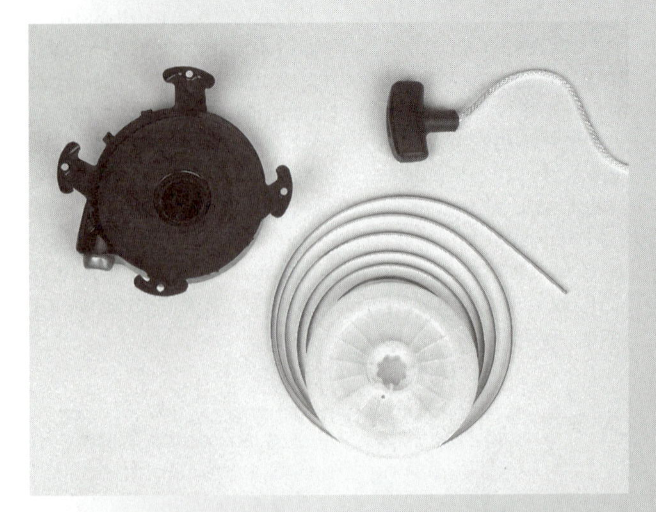

5. Clean and inspect all the starter parts. Lubricate the new spring with engine oil.

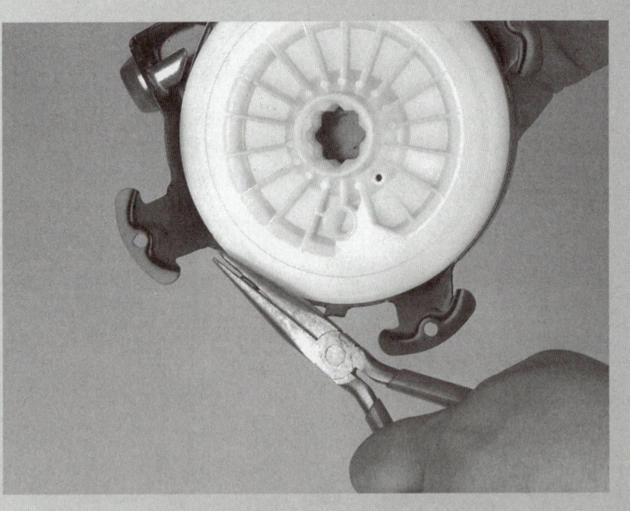

6. Install the pulley back into the housing and bend the tabs back into position to hold the pulley.

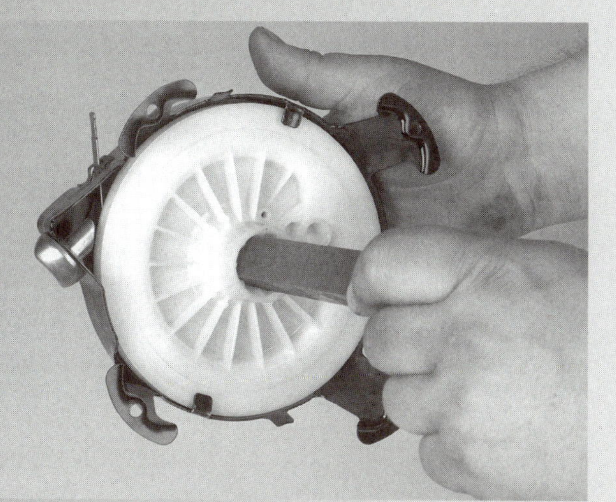

7. Insert the spring into the housing in the correct direction and engage it in position on the pulley. Turn the pulley in the direction it would turn during engine starting and feed the spring into the housing hole.

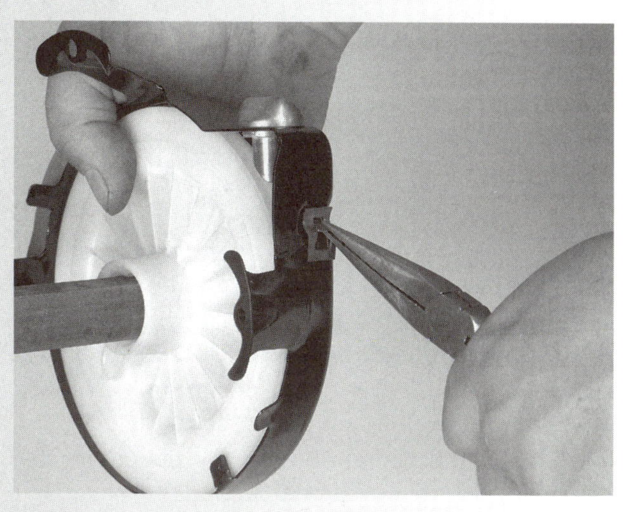

8. Use needle nose pliers to anchor the end of the spring in the housing notch.

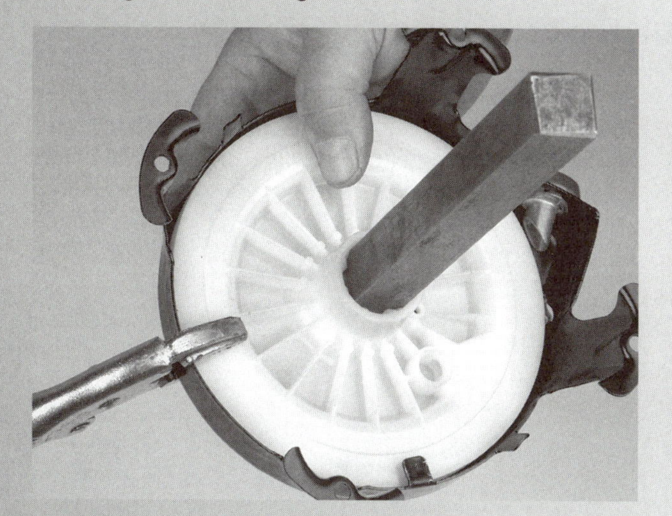

9. Wind the pulley and spring in the rope-pulling direction until they stop moving. Rotate the pulley in the opposite direction until the pulley rope hole lines up with the housing rope hole. Use vise grip pliers to prevent the pulley from rotating.

10. Melt one end of the nylon rope with a match to keep it from fraying.

11. Insert the end of the rope through the housing and into the pulley hole. Tie a knot to secure the end of the rope.

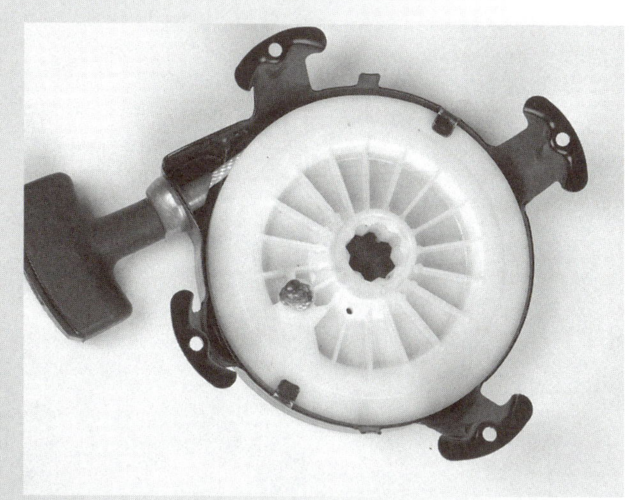

12. Remove the vise grip pliers and allow the pulley to rotate and pull the rope into the housing.

REVIEW QUESTIONS

1. Explain the purpose of a rewind starter.
2. Explain the purpose of a starter rewind clutch.
3. Describe the purpose of a pull rope in a manual starter.
4. Describe the parts and operation of a ball rewind starter clutch during engine cranking.
5. Explain the operation of a ball rewind clutch during engine starting.
6. Describe the parts and operation of a dog rewind starter clutch during cranking.
7. Explain the operation of a dog rewind starter clutch during engine starting.
8. Describe the parts and operation of the ratchet recoil starter during cranking.
9. Explain the operation of the ratchet recoil starter during engine starting.
10. Describe the parts and operation of the gear starter during engine cranking.
11. Explain the operation of the gear starter during engine starting.
12. Describe the parts and operation of a windup spring starter during cranking.
13. Explain the operation of the windup spring starter during engine starting.
14. Explain the purpose a compression release system.
15. Describe the parts and operation of a compression release system.

DISCUSSION TOPICS AND ACTIVITIES

1. Disassemble a ball rewind clutch. Move the balls in and out of their notches and note how the clutch engages and disengages.
2. Disassemble a ratchet recoil clutch. Identify the parts and note how the clutch engages and disengages.
3. A gear type starter has 120 teeth on the starter flywheel ring gear. There are 12 teeth on the starter pinion gear. What is the gear ratio?

Electrical Starting and Charging Systems

OBJECTIVES

Upon completion and review of this chapter, you should be able to:

- Describe the parts and operation of the storage battery.
- Identify the parts and explain the operation of a starter motor.
- Describe the parts and operation of a starter drive.
- Identify the parts and explain the operation of a solenoid.
- Trace the flow of electricity through a starter control circuit.
- Explain how a conductor in a magnetic field can generate electricity.
- Describe the parts and operation of an alternator.
- Describe the parts and operation of a charging system rectifier and regulator.
- Trace the flow of electricity through a charging system circuit.

TERMS TO KNOW

Alternating current
Alternator
Battery cell
Battery vent cap
Charging system circuit
Direct current
Electrical starting system
Electromagnetic induction
Flywheel ring gear
Inertia starter drive
Motor feed circuit
Rectifier
Solenoid
Starter control circuit
Starter drive
Starter motor
Starter overrunning clutch
Starter pinion gear
Starter switch
Storage battery
Voltage regulator

INTRODUCTION

Larger outdoor power equipment requires powerful engines. These engines often have large displacement and high compression ratios. Many of these engines have multiple cylinders. Large powerful engines are difficult for an operator to start manually. Hand cranking some of these engines would take a great deal of strength. These engines usually use electrically operated starting systems. An **electrical starting system** uses electrical power to crank the engine for starting. Electrical starting systems also make the equipment easier to use. The operator has to simply turn a switch to start the engine.

The electrical starting system requires:

- Battery
- Starter motor
- Solenoid
- Starter key switch
- Connecting cables and wiring

The system may be equipped with a charging system to recharge the battery when the engine is running.

STORAGE BATTERY

The electrical starting system needs a power source. A battery provides the electrical energy to crank the engine. Most small engines use a storage battery similar to that used in automobiles and motorcycles. A **storage battery** is a rechargeable battery that uses lead plates in electrolyte to develop electrical power.

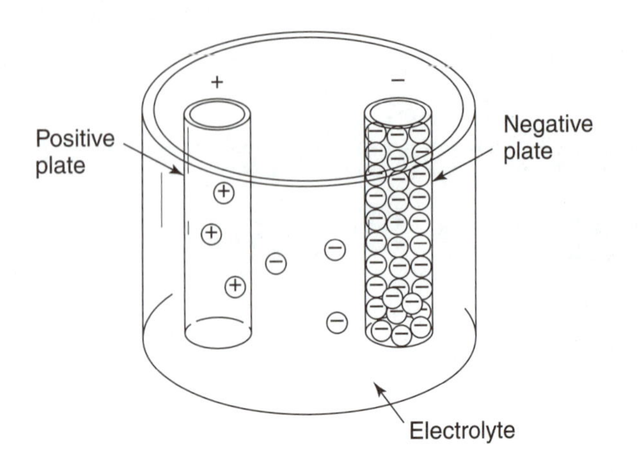

FIGURE 20-1 Two different kinds of lead in electrolyte make a simple battery.

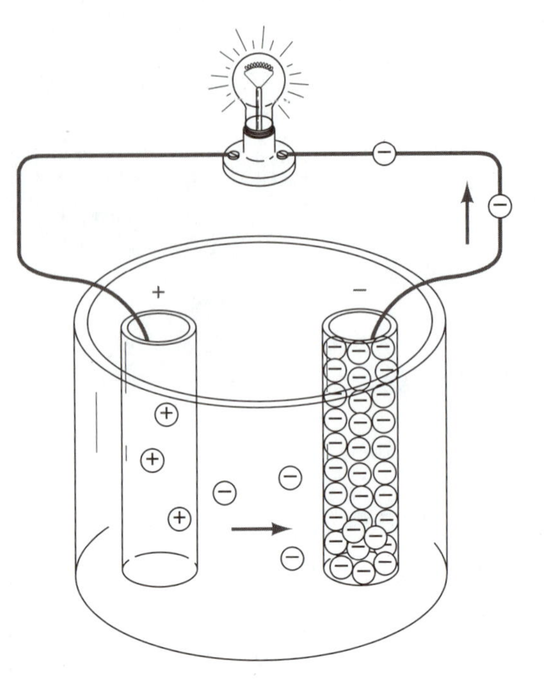

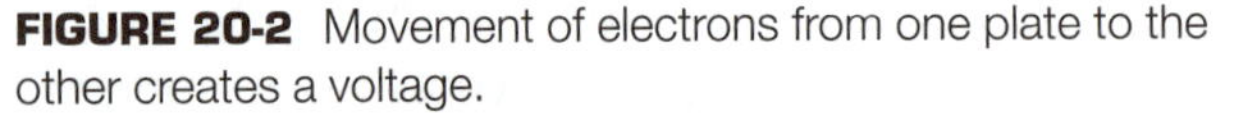

FIGURE 20-2 Movement of electrons from one plate to the other creates a voltage.

The storage battery gets energy from the engine's charging system. The battery stores energy until needed. It provides this energy to the electrical starting systems on demand. The storage battery changes electrical energy into chemical energy during charging. It changes chemical energy back into electricity during discharging.

Battery Cell

A simple battery cell has two plates (metal strips) made of different types of lead (Figure 20-1). One of the plates is positive. The other plate is negative. The plates are placed in a container of acid. The acid is called electrolyte. The electrolyte reacts chemically with the positive plate. The positive plate releases electrons with negative charges. They transfer to the other plate, giving it a negative charge. When the battery is fully charged, the negative plate has an excess of electrons.

In a simple battery connected into a circuit with a light bulb (Figure 20-2), the excess electrons are pulled away from the negative plate and flow to the light bulb. The light bulb lights up. As chemical action continues inside the cell, the electrolyte becomes weaker and weaker. Eventually the negative and positive plates become more and more alike chemically. When all the electrons are used up, the light bulb goes out. The battery cell is discharged.

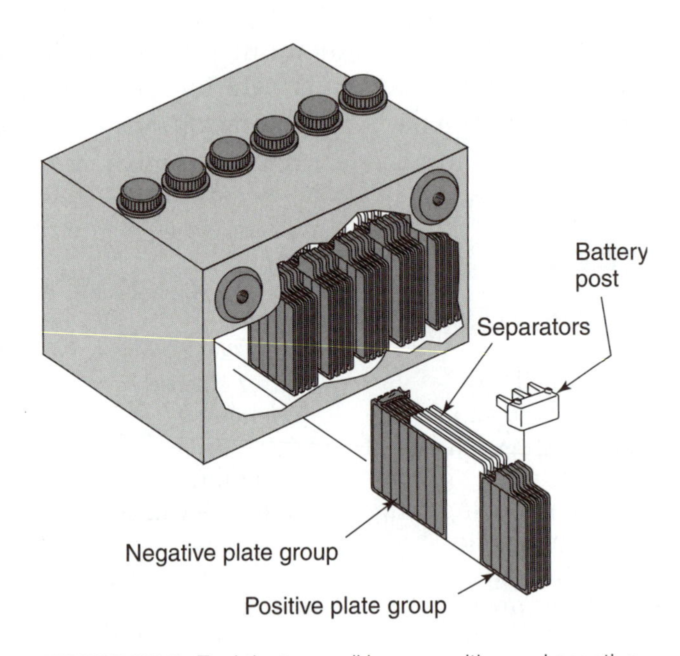

FIGURE 20-3 Each battery cell has a positive and negative plate group. (Courtesy of Delco Remy Division.)

Battery Parts

A **battery cell** (Figure 20-3) is a group of positive and negative lead plates placed into a container with electrolyte. Separators fit between each plate to prevent them from touching each other. The cell is the basic component of the battery. No matter what the cell size or the number of plates, the voltage of a fully charged cell is about 2 volts. More voltage is possible by connecting cells together. The negative terminal of one cell can be connected to the positive terminal of another. This is called a *series connection*. The voltage is the sum of all the cells connected together. The battery voltage is the sum of the voltage of its cells. Six cells connected together provides 12 volts.

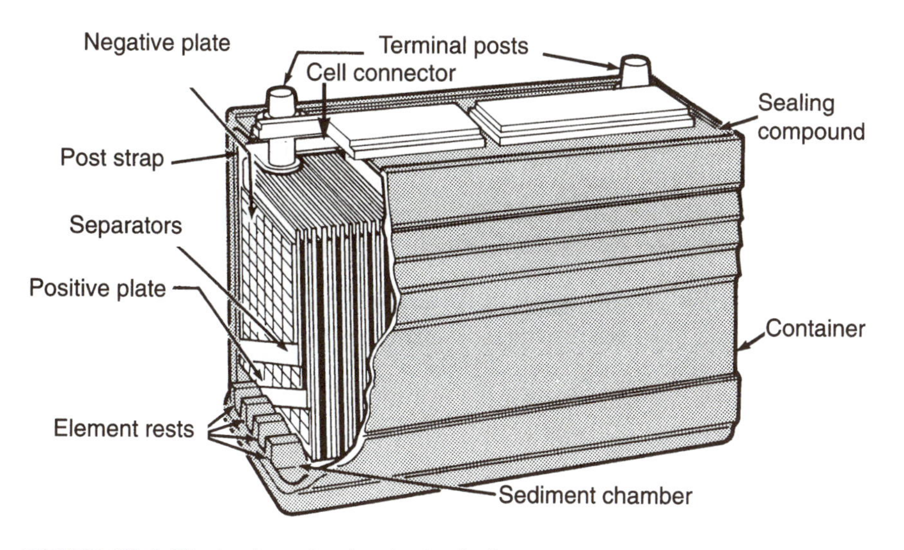

FIGURE 20-4 The basic parts of a storage battery.

Battery containers (Figure 20-4) are one molded piece, usually made of hard rubber or plastic. The container has a separate compartment for each cell. Battery containers must handle extremes of heat and cold, vibration, and acid. The bottom of each cell compartment has element rests below the plates that provide a place for sediment to collect below the plates. The element rests prevent short circuits from sediment that falls from the plates. The cells are arranged with the negative terminal of one cell next to the positive terminal of the next cell. Cell connectors connect the cells in series. There are five cell connectors in a 12-volt battery.

The battery has a plastic or rubber cover that makes an acid-tight seal. Some batteries use a vent cap over each cell (Figure 20-5). The **battery vent cap** is a cap that fits in a battery cell for inspection and for adding water. A hole in the vent cap allows hydrogen gas to escape as the battery is charged. Other batteries have a pry off strip on the top of the battery. There is no vent in the strip. To add water, the strip has to be removed.

Many batteries are maintenance free and do not have cell caps or vents (Figure 20-6). Instead, the battery is sealed. This battery does not require frequent inspection and water filling.

The battery has two terminals to connect it into a circuit (Figure 20-7). One terminal is positive (+) and the other is negative (–). The terminals may be marked with a + or – sign. They may be identified with the letters "Pos" or "Neg." The positive terminal is often slightly larger to prevent installing cables on the wrong terminal. There are three common types of terminals: Tapered terminals have a slight taper; stud terminals have threads; side terminals are located on the side of the battery.

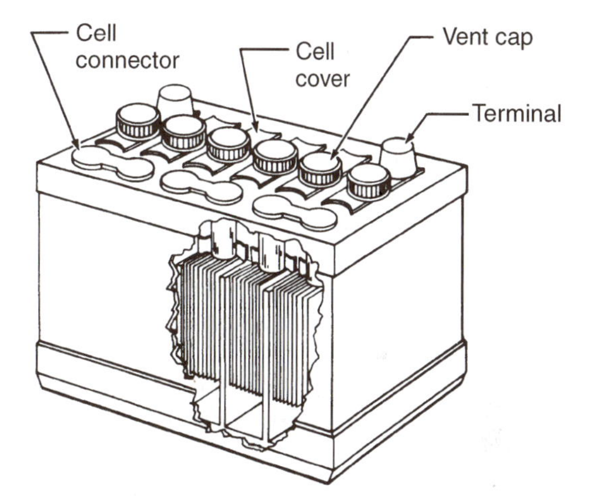

FIGURE 20-5 A battery with vent caps on the top of each cell. (Courtesy of Kohler Co.)

FIGURE 20-6 A maintenance-free battery is completely sealed. (Courtesy of Delco Remy Division.)

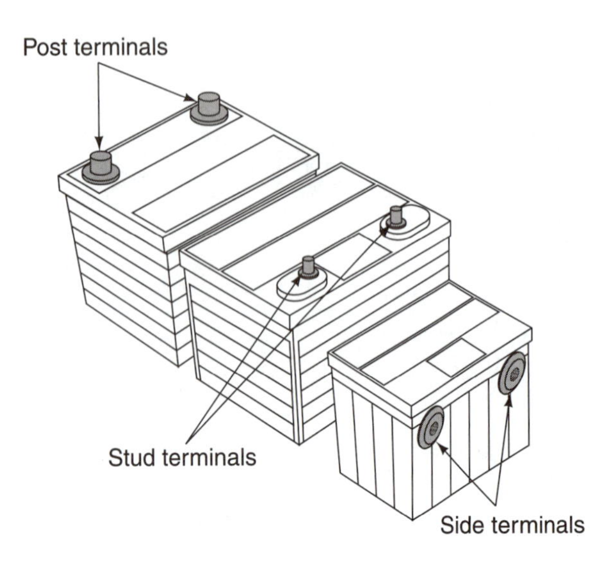

FIGURE 20-7 Three common types of battery terminals.

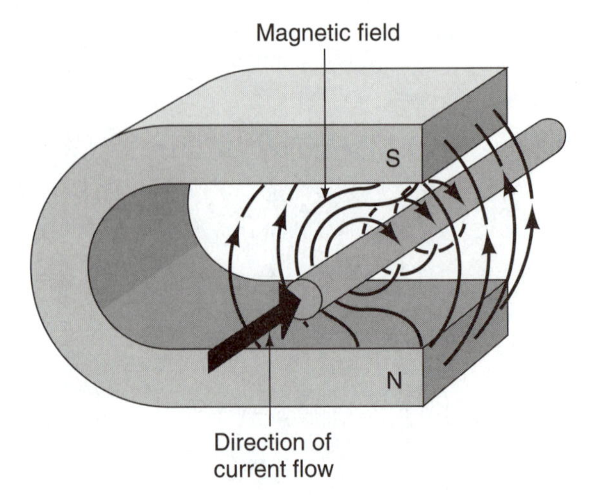

FIGURE 20-9 Two magnetic fields are created when current flows through a conductor. (Courtesy of General Motors Corporation, Service Technology Group.)

STARTER MOTOR

The **starter motor** is an electric motor used to crank the engine for starting (Figure 20-8). The starter motor drives a small gear called a *pinion gear.* A **starter pinion gear** is a small gear driven by the starter motor that rotates the engine crankshaft. The **flywheel ring gear** is a ring of gear teeth on the flywheel that are engaged by the starter pinion gear to crank the engine. The starter motor is mounted on the side of the engine with the pinion gear near the flywheel. The starter motor is mounted either horizontally or vertically. The starter motor direction depends on the engine's crankshaft direction.

Basic Motor Operation

The starter motor uses magnetism to change electrical energy into mechanical motion. The basic operation of a starter can be shown with a horseshoe magnet and a wire (conductor) with current flow. When current flows through the wire, there are two separate magnetic fields (Figure 20-9). One magnetic field comes from the horseshoe magnet. The other comes from the current flow through the wire.

The direction of the magnetic lines of force between the two magnet legs (poles) is upward. Magnetic lines of force always leave a north pole and enter a south pole. The wire's magnetic field creates circles around the wire. The result is more magnetic lines of force on the left-hand side of the wire than on the right. There is a strong magnetic field on one side of the wire. There is a weaker magnetic field on the other side. The wire will move from the strong field to the weak field. The movement will be from left to right (Figure 20-10).

A basic starter motor uses a wire formed into a loop (Figure 20-11). The loop is placed between two magnetic poles. The poles are called pole shoes.

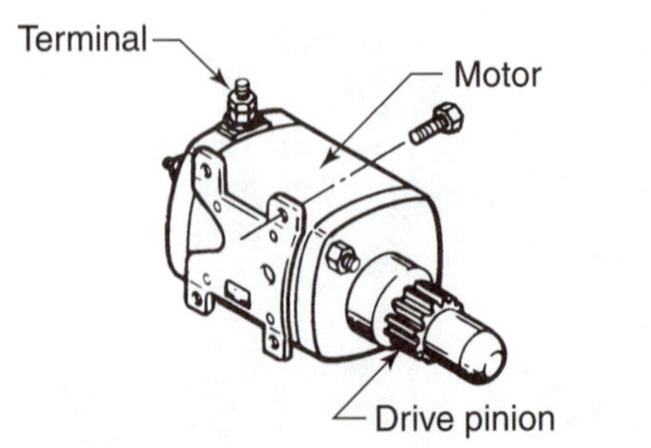

FIGURE 20-8 An electric starter motor used to crank the engine. (Courtesy of Simplicity Manufacturing.)

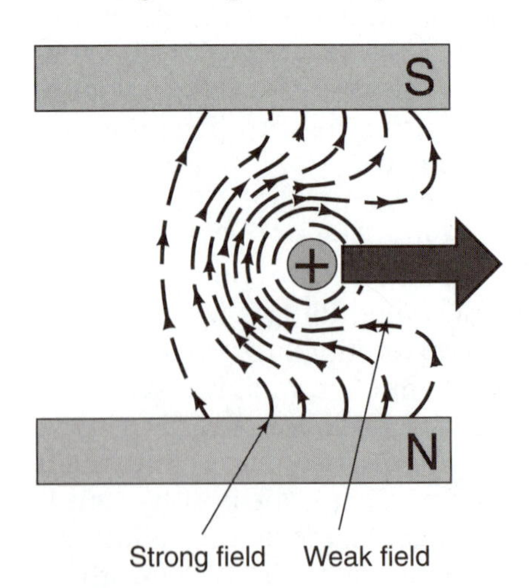

FIGURE 20-10 The conductor moves from the strong to the weak field. (Courtesy of General Motors Corporation, Service Technology Group.)

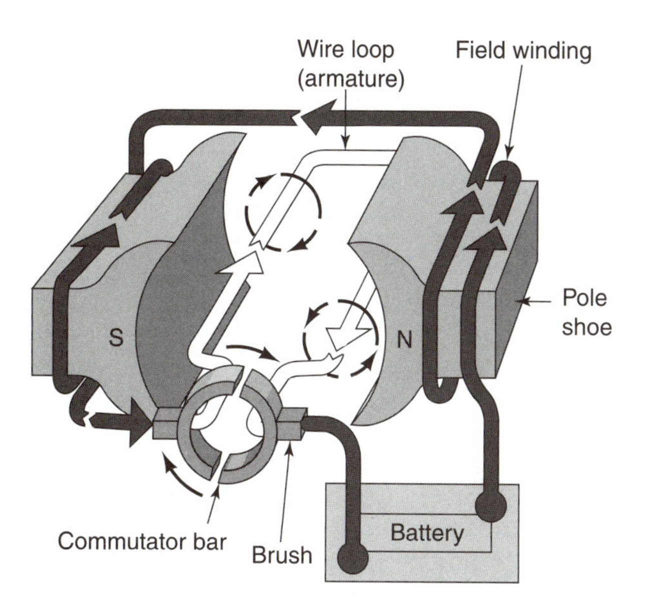

FIGURE 20-11 Parts and current flow through a basic motor. (Courtesy of General Motors Corporation, Service Technology Group.)

One is a north pole and the other a south pole. Circular-shaped contacts are connected to the wire loop. The contacts are called commutator bars. Sliding electrical contacts called *brushes* ride on the rotating commutator bars. The brushes and commutator bars allow a complete electric circuit into and out of the wire loop.

One brush is connected electrically to the battery. The other is connected to a wire that is wrapped around each pole shoe. This wire is called a *winding*. The winding makes the magnetic field of the magnets stronger.

Current flows from the battery through the winding to a brush and commutator bar. It flows through the wire loop to the other commutator bar and brush and then flows back to the battery. The magnetic fields created by this current flow cause a strong magnetic field on one side of the loop. The loop rotates in the magnetic field.

Starter Motor Parts

The starter motor needs more power than a single loop could develop. Starter motors use many loops on a part called an armature (Figure 20-12). The armature is an assembly of wire loops and commutator bars. Each of the loops is held in place by a laminated iron core. The laminated iron core is a set of thin iron plates that help concentrate and direct the magnetic lines of force. The loops are connected to commutator bars made from copper. The bars are insulated from each other with a material called *mica*.

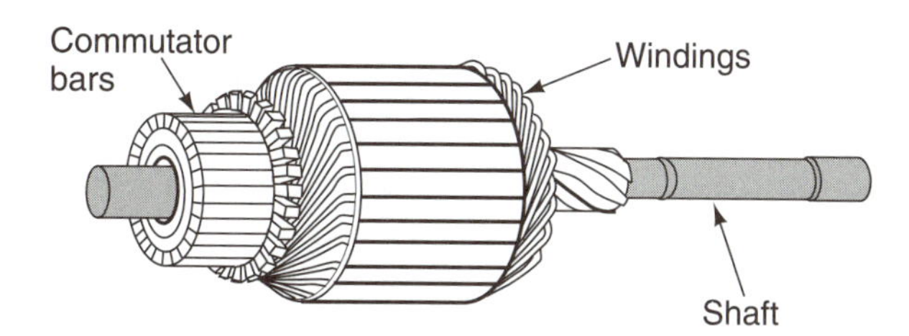

FIGURE 20-12 Parts of a starter motor armature. (Courtesy of Kohler Co.)

The armature fits on a shaft so that it can rotate and drive the starter pinion gear. A steel armature shaft goes through the iron core laminations. It supports the armature in the housing assembly. The loops and commutator bars are insulated from the shaft.

The starter motor is a three-piece assembly. It has a center housing (frame), drive end cap, and brush end cap. The center housing contains the pole shoes and field windings The field windings have many wraps (turns) that increase the magnetic field strength. The field windings are protected by an insulation wrapping. The field windings and the brushes are connected electrically to a terminal. The terminal is usually located at the top rear of the center housing. The drive end cap fits on one end of the center housing. The brush end cap fits on the other end. There are bearings or bushings in the end and brush cap. These support the armature on its shaft. Long bolts hold the three-piece assembly together.

Brushes ride on the commutator and direct full battery current into the armature (Figure 20-13). The brushes are often made from copper or carbon to get good electrical contact. The brushes fit in

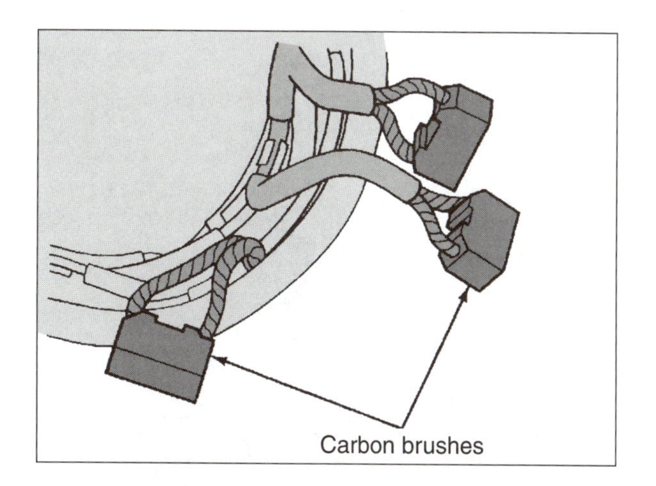

FIGURE 20-13 Brushes ride on the armature commutator to make a sliding electrical contact. (Courtesy of Ford Global Technologies, Inc.)

brush holders. The brush holders are attached to the brush end cap. A small spring in the brush holder pushes the brushes firmly against the commutator. Wires attached to the brushes direct current into and out of the armature.

STARTER DRIVES

The **starter drive** is the part used to connect and disconnect the rotating starter motor armature shaft to the engine for cranking. The starter drive has a small steel or nylon pinion gear that fits on the end of the starter armature shaft. The pinion gear meshes with the ring gear on the engine's flywheel. When the operator turns the key switch, the starter motor drives the pinion gear. The pinion gear drives the flywheel and cranks the engine.

There is a gear reduction between the pinion and flywheel ring gear (Figure 20-14). The gear reduction is necessary to increase starter motor torque. This allows a small electric motor to crank a larger gasoline engine. There are 15 to 20 teeth on the ring gear for each tooth on the drive pinion gear. This means that the armature will rotate about 15 to 20 times for every engine revolution. The armature must rotate at 1,500 to 2,000 rpm in order to crank the engine over at 100 rpm.

The pinion gear cannot stay meshed with the flywheel ring gear after the engine starts, because the gear reduction would cause the starter motor armature to spin at very high speeds. Engine speeds above 1,000 rpm would damage the armature and brushes. To avoid this, the starter drive disengages the pinion gear from the ring gear. This happens as soon as the engine begins to run. There are several types of starter drives. Many starter motors use an inertia starter drive. The **inertia starter drive** uses inertia to engage and disengage the starter motor pinion from the flywheel ring gear.

The inertia drive pinion has screw threads on its inner bore (Figure 20-15). These threads match screw threads on the outside of a hollow sleeve. The sleeve fits loosely over the starter motor armature shaft. The sleeve is connected through a roll pin to a clutch retainer. The retainer is keyed to the armature shaft.

When the operator turns the starter motor switch, the armature shaft accelerates rapidly. The rotational force goes from the armature to the clutch retainer to the sleeve. The pinion fits loosely on the threads of the hollow helix sleeve. The rotating threads between the pinion and the sleeve cause the pinion to travel forward on the sleeve. The pinion goes into mesh with the flywheel ring gear. When the pinion reaches the end of its travel, it contacts a stop on the sleeve. In this position, there are no longer any threads for the pinion to slide on. The pinion begins to rotate with the armature. In mesh with the flywheel ring gear, the pinion cranks the flywheel to start the engine.

When the engine starts, the pinion is driven at a higher speed than the armature and helix sleeve. The threads between the sleeve and pinion cause the pinion to travel back down the sleeve. It goes out of mesh with the flywheel ring gear.

Some engines use a solenoid and overrunning clutch to drive the pinion. These starter motors have a solenoid mounted on top of the starter housing (Figure 20-16). The solenoid (explained in the next section) shifts the pinion in and out of mesh with the flywheel. When the starter switch is moved to start, the solenoid moves the pinion into mesh. When the starter switch is moved to run, the solenoid moves the pinion out of mesh.

The armature has to be protected from rotating too fast (overspeeding). Pinion overspeeding can occur when the engine starts and the pinion is still

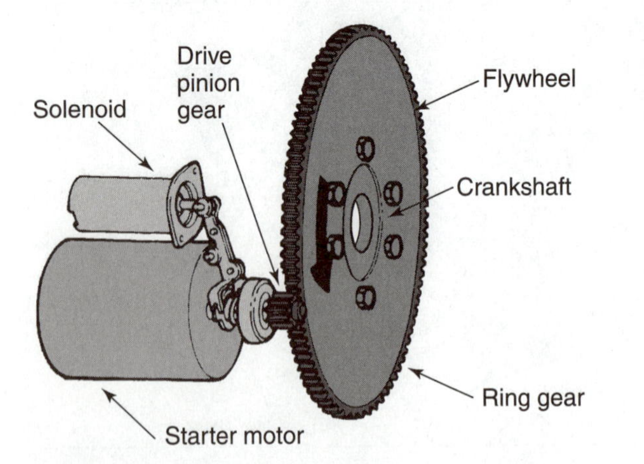

FIGURE 20-14 The small pinion and large flywheel ring gear provide a gear reduction. (Courtesy of Ford Global Technologies, Inc.)

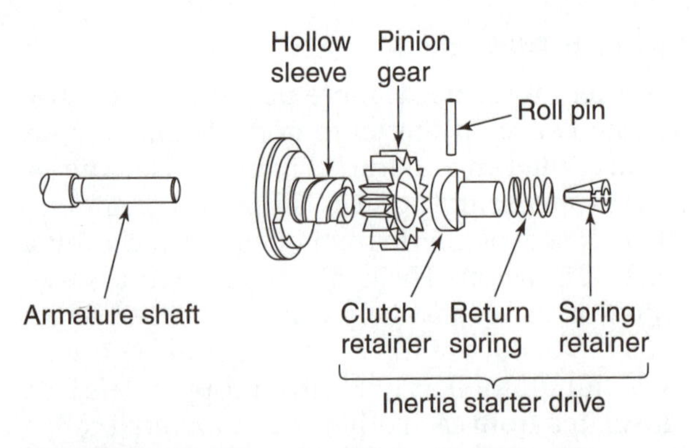

FIGURE 20-15 The parts of an inertia starter drive.

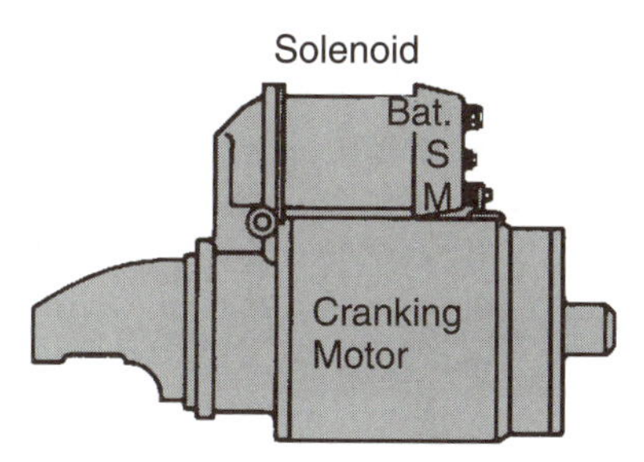

FIGURE 20-16 A starter with a solenoid mounted on top. (Courtesy of General Motors Corporation, Service Technology Group.)

in mesh with the flywheel. The **starter overrunning clutch** (Figure 20-17) is a starter drive part that uses rollers in notches to lock and unlock the pinion to the armature to protect the starter motor from overspeeding.

The overrunning clutch has an outside part called a shell that has tapered notches inside. Steel rollers fit in the notches. Small accordion type springs fit into the notches. The springs push against the rollers. The rollers are wedged against the pinion gear and the small part (taper) of the notches. The pinion gear has a collar on the end. The collar is an inner race for the clutch.

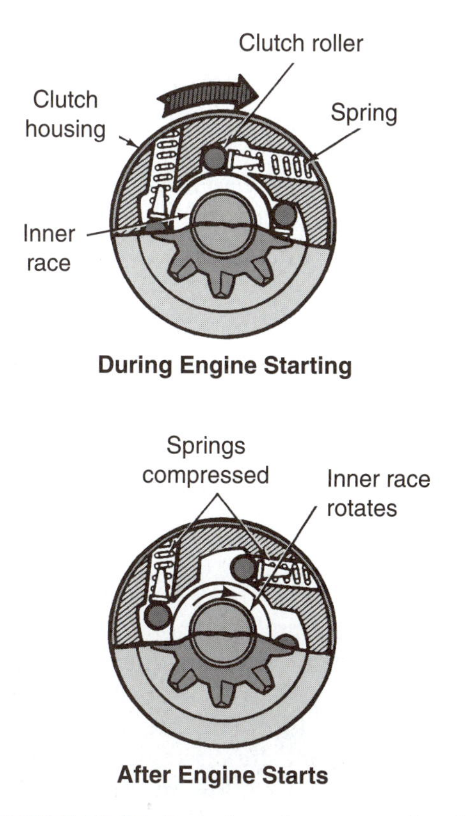

FIGURE 20-17 Operation of an overrunning clutch.

When the starter motor is cranking, motor torque goes from the shell to the pinion gear through the rollers. When the engine starts, the flywheel ring gear drives the pinion gear faster than the armature, causing the rollers to move away from the tapers. When the rollers get into the wider part of the notches, they are no longer able to drive the inner race and the pinion gear. The clutch disconnects the two parts, allowing the pinion gear to overrun or freewheel. The clutch protects the starter motor armature from overspeeding even if the operator continues to operate the starter motor after engine start-up.

SOLENOID

The starter motor operation is controlled by a solenoid. The **solenoid** is a magnetic switch used to control the electrical circuit between the battery and starter motor. The solenoid can be mounted on the starter motor. If so, it is connected to a shift lever to move the pinion gear. Solenoids that do not shift the pinion can be mounted in any location on the equipment or engine.

The solenoid switch has a housing that has two electrical coils. One is called a *hold-in winding* and the other a *pull-in winding*. Both coils are wound around a hollow cylinder that is a movable plunger. A round contact disc is connected to one end of the movable plunger. There are two electrical terminals located near the contact disc. One of the terminals is connected to the battery. The other terminal is connected to the starter motor. When the disc is out of contact with the terminals, the circuit between the battery and starter motor is open. When the disc makes contact with the two terminals, the circuit is closed. The closed circuit sends battery current across the disc to power the starter motor.

The plunger and disc are controlled by the starter switch (Figure 20-18). The switch controls the electrical circuit for the two solenoid windings. The hold-in winding contains many turns of fine wire. The pull-in winding contains the same number of turns of larger wire. When the operator closes the starter switch, current flows from the battery to the solenoid switch terminal. It flows through the hold-in winding and then to ground back to the battery. Current also flows through the solenoid pull-in winding to ground. The two windings together create a strong magnetism that pulls the plunger into the center of the windings. This moves the contact disc in a position to close the circuit between the solenoid and the battery terminals. With the circuit between the battery and starter motor closed, the engine is cranked.

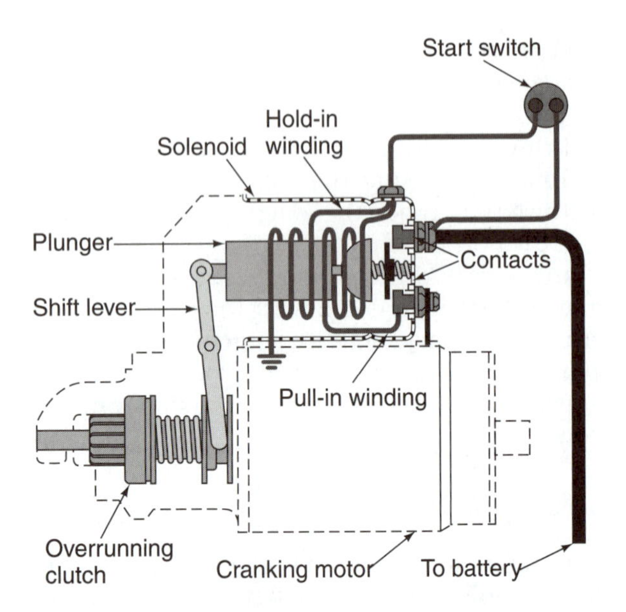

FIGURE 20-18 Parts and wiring for a solenoid mounted on a starter motor. (Courtesy of General Motors Corporation, Service Technology Group.)

The pull-in winding helps the hold-in winding pull the plunger into the core. Once the plunger stops, much less magnetism is needed to hold the plunger in the cranking position. When the contact disc is in contact, the pull-in winding is shorted. No current flows through the winding, reducing current draw on the battery. It also reduces the amount of heat created in the solenoid.

When the operator releases the starter switch, the magnetic field collapses. The return spring in the cylinder moves the plunger to the at rest position. The cranking cycle is complete.

STARTER SWITCH AND WIRING

The **starter switch** is a multipurpose switch that controls the starter motor through the solenoid and may also be used as the ignition on/off switch (Figure 20-19). The switch is often mounted on the equipment handle near the operator's hand. The switch often has three positions: start, run, and off. The start position directs battery current to the solenoid and powers the starter motor and the magneto. The run position stops the current flow to the starter motor but continues to power the magneto. The off position grounds current flow to kill the magneto ignition.

The starting system wiring has two circuits. One is the starter control circuit. The other is the motor feed circuit. The **starter control circuit** is the starter system wiring circuit that connects the starter switch to the solenoid and the magneto. The control circuit has small gauge wiring. It is often routed with other wires through a wiring harness. The **motor feed circuit** is the heavy gauge cable wiring circuit that connects the battery to the starter motor. The heavy cable carries the high current required to operate the starter motor.

CHARGING SYSTEM

Engines that use a storage battery for starting must have some way of charging the battery. The **charging system** develops and regulates a voltage used to recharge the storage battery. The main parts of the charging system are the alternator, rectifier, and voltage regulator. These parts are connected into a charging system circuit that includes the starter switch and battery.

Generating Electricity

The charging system generates electricity by electromagnetic induction. **Electromagnetic induction** is a process that generates electricity by moving a conductor through a magnetic field (Figure 20-20). There is a magnetic field around a wire when current passes through the wire. This idea is used in electromagnetic induction. If a magnetic field is moved past a wire, the magnetic lines of force are cut. A voltage develops in the wire. The same thing happens if the wire is moved in the magnetic field. When this wire is connected to a circuit, current can flow. The direction of current flow depends if the wire moves up or down in the magnetic field.

The amount of voltage generated by electromagnetic induction depends on:

- The number of magnetic lines of force that are cut
- The number of turns of the wire
- The speed at which the wire passes through the magnetic lines of force

In electromagnetic induction, a magnet is rotated inside a wire loop (conductor). Lines of magnetic force cut across the conductor, creating a voltage that causes a current to flow through the completed circuit. Current flow direction depends on the magnet's poles. When the north pole is at the top and the south pole is at the bottom, the flow of current is in one direction. If the magnet is rotated 180 degrees, the south pole will be at the top and the north pole at the bottom. The lines of force cut the conductor in the opposite direction. The voltage created causes current to flow in the opposite direction.

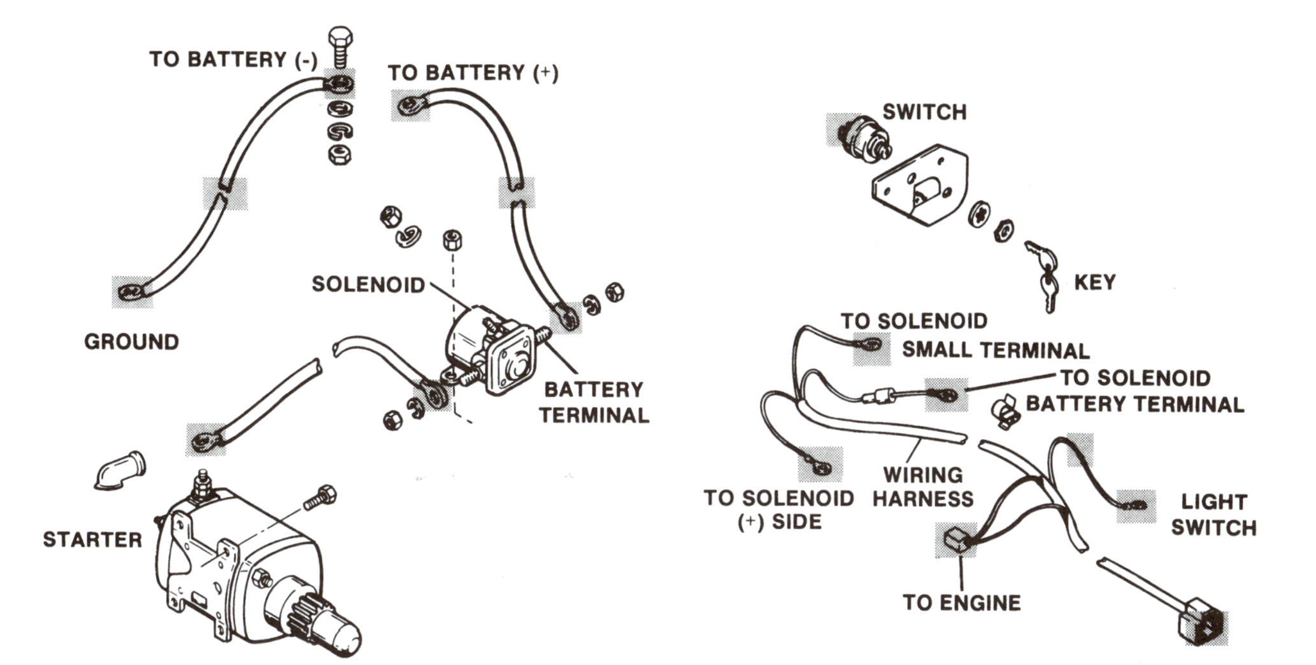

A. Starting System Electrical System

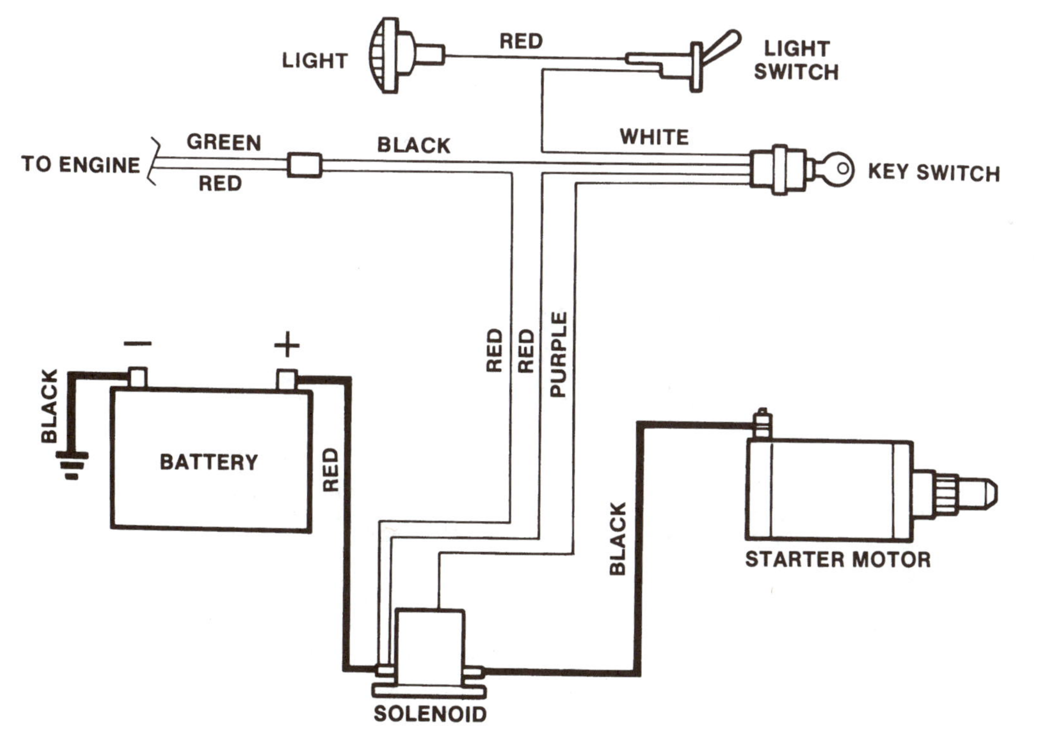

B. Starting System Electrical Diagram

FIGURE 20-19 A starting system electrical system and wiring diagram. (Courtesy of Simplicity Manufacturing.)

The direction of current flow reverses every half-revolution (Figure 20-21). Current that reverses like this is called *alternating current.* **Alternating current** is electrical current that changes from positive (+) to negative (–) at a regular cycle. Charging system alternating current has to be changed to direct current. The battery needs direct current for charging. **Direct current** is electrical current that does not change from positive (+) to negative (–).

CASE STUDY

A snow thrower with an electric starting system was the subject of an electrical system troubleshooting demonstration in an outdoor power equipment class. A student had brought his father's snow thrower to class because it had a problem with the starting system. When the key switch was moved to the start position, the system made a clicking noise but would not crank the engine. The student's father wanted to know if he should replace the battery, solenoid, or starter motor.

The small-engine instructor gathered his class around to show the student a quick way to troubleshoot this common starting system problem. The instructor explained to the students how to make a quick visual check of all the starting system wiring. He pointed out that the battery connections are often the problem and they must be checked for corrosion. The wiring all looked fine.

Before doing the testing, the instructor removed the spark plug wire and grounded it for safety. The snow thrower starting system had a separately mounted solenoid. The instructor showed the students the two large terminals on the solenoid. One terminal is for battery power in from the battery and the other for battery power to be directed to the starter motor. The instructor used a heavy gauge jumper cable and momentarily jumped across these two solenoid terminals. The starter motor cranked the engine at normal cranking speed.

The quick test eliminated the battery as the source of the problem because it cranked the engine fine. Since the starter motor cranked the engine, it was working fine. The test determined that the solenoid was most likely the problem. A new solenoid was installed and the starting system was functional again.

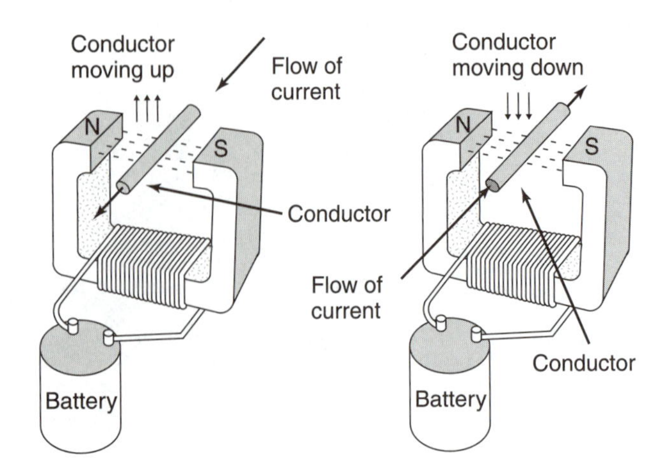

FIGURE 20-20 Developing electricity through electromagnetic induction.

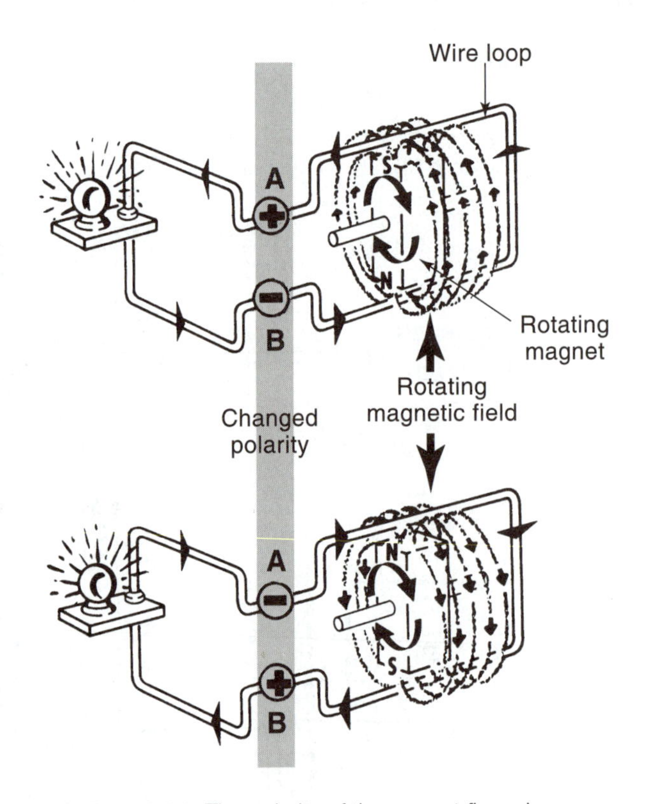

FIGURE 20-21 The polarity of the current flow changes as the magnet rotates through the wire loop. (Courtesy of Delco Remy Division.)

Alternator

The **alternator** is a generator that develops alternating current that is changed to direct current to charge the battery. An alternator has a stator and flywheel magnets. A stator is a stationary part of an alternator. It is made up of coils of wire wound around a metal lamination. The coils of wire provide the conductor for current generation. The metal laminations concentrate the magnetic lines of force.

The higher the number of stator coils, the higher alternator output. A single-coil stator can develop about .5 amps, which is enough to charge a small battery. A four-coil stator can develop 3–5 amps, enough to charge a larger battery. Riding tractors and mowers may have lights and other accessories. The alternator will need more output to keep the battery charged. Sometimes there are separate coils on the stator just for the tractor lights. Stators with

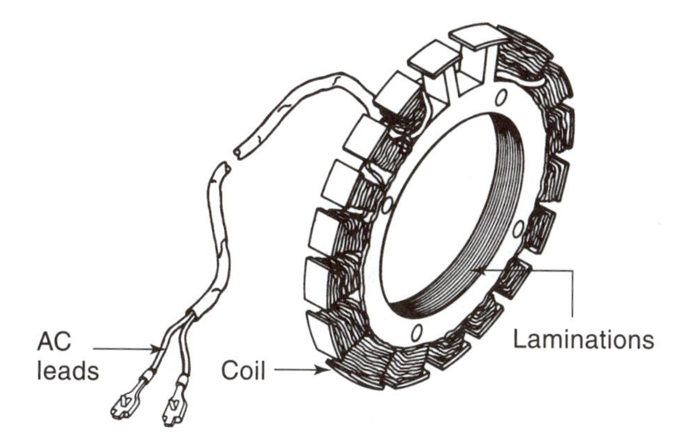

FIGURE 20-22 A stator with enough coils to develop 15 amps of electrical current. (Courtesy of Kohler Co.)

8–18 coils are common for this equipment (Figure 20-22).

The magnetic field for electrical generation is provided by flywheel magnets. The magnets are arranged in a circle, often called a magnet ring. The individual magnets are called *pole pieces*. The stator coils are mounted under the flywheel. They must be close to the rotating flywheel magnets. The flywheel magnets have a magnetic field. The lines of magnetic force run from the north pole to the south pole. As the flywheel rotates and the position of the magnets change, the direction of the magnetic field changes (alternates). The current that flows out the stator output (AC) wire is an alternating current (Figure 20-23).

Rectifier and Regulator

The alternating current developed by the alternator is changed to direct current for battery charging. A

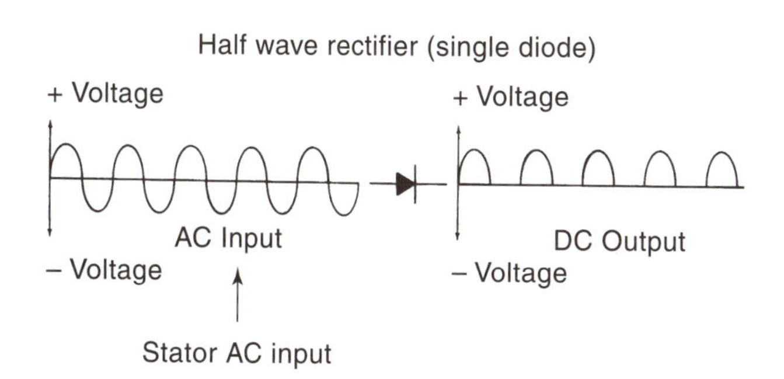

FIGURE 20-24 A single diode rectifier changes half of the output to direct current. (Provided courtesy of Tecumseh Products Company.)

rectifier is used to change alternating current to direct current. A **rectifier** is a charging system part that changes the alternating current output of the alternator to direct current.

Rectifiers are solid state electrical parts that use one or more diodes to change (rectify) alternating to direct current. A diode allows current flow in only one direction. It prevents electrical flow in the opposite direction. A single diode can be connected to rectify the stator coils output. The single diode allows only the positive half of the alternating current through (Figure 20-24). It does not allow the negative half of the alternating current through. The result is a direct current. This is called *half-wave rectification*. Single diode, half-wave rectification rectifiers are simple and inexpensive. They are commonly used in low-output systems.

Higher output systems use a system that converts the entire alternating current output to direct current. Four diodes are used in a circuit called a *bridge* (Figure 20-25). The four-diode bridge circuit output is called *full-wave rectification*. Full-wave rectification is used on high-output charging systems.

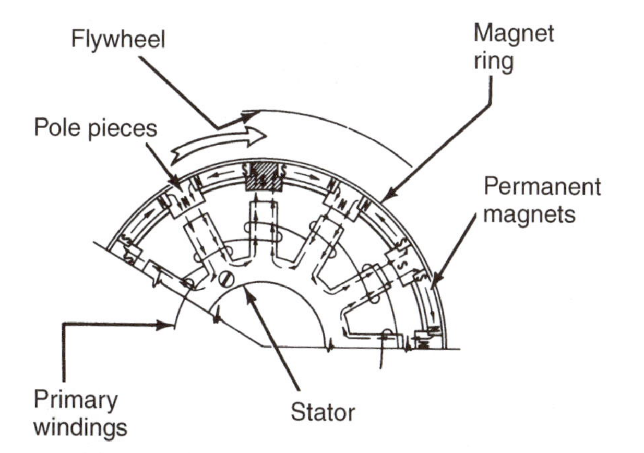

FIGURE 20-23 The magnet ring rotates over the stator coils and develops alternating current. (Courtesy of Kohler Co.)

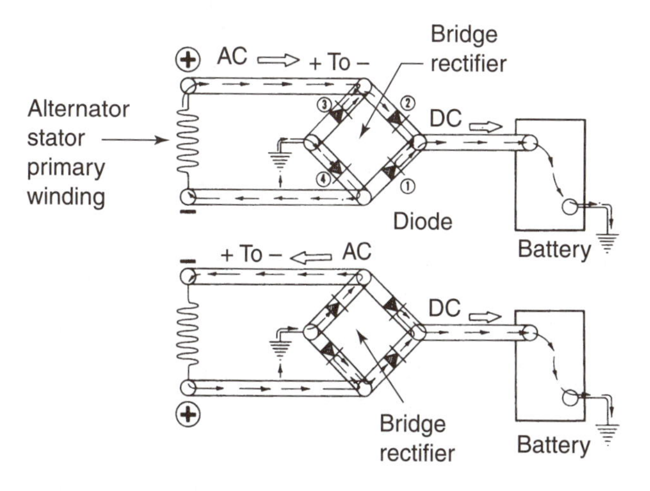

FIGURE 20-25 A four-diode bridge circuit changes all of the output to direct current. (Courtesy of Kohler Co.)

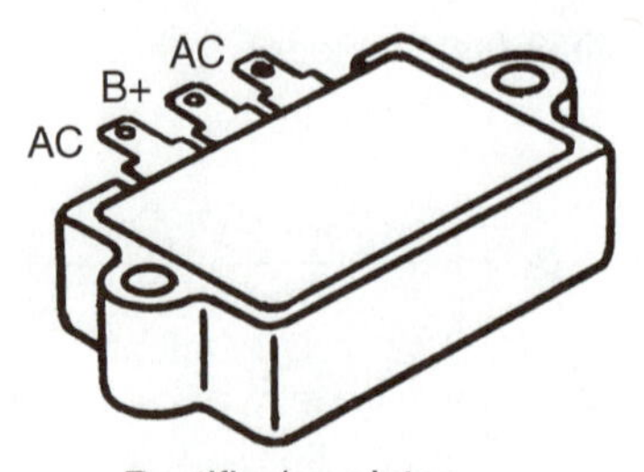

FIGURE 20-26 A rectifier-regulator rectifies current and protects against high voltage. (Courtesy of Kohler Co.)

Some charging systems also include a regulator. The charging system must be protected against too high a voltage. Excessive voltage can destroy the battery and other electrical parts. The **voltage regulator** is a charging system part that senses battery charge and limits alternator output to prevent battery overcharging.

The current developed by an alternator is self-limiting. Stator coil construction limits current to safe levels. There is no need for an external current-limiting device. Voltage regulation is not automatic. Voltage can reach excessive levels in some high-output systems.

Voltage is regulated with a zener diode. A zener diode can change its resistance when a specified voltage is reached. A zener diode operates the same as a rectifying diode to a certain point. It allows current flow in one direction and blocks current flow in the opposite direction. When a specified voltage is reached, it can stop current flow. The zener diode is connected to the stator output circuit. It stops current flow if the voltage reaches a predetermined setting. Voltage regulators are often combined in the same solid state part as the rectifier. The part is called a *rectifier/regulator* (Figure 20-26).

Charging System Circuit

The **charging system circuit** (Figure 20-27) is the wiring system that connects the charging system parts together. The output of the alternator is connected to the rectifier and regulator. The starter switch is connected to the charging circuit with small gauge wire. This connection makes a route for current flow into the battery. Headlight and other electrical system parts may be included in the electrical system. A small light bulb called the *charge indicator light* is often used in the output circuit. When the light is illuminated, it shows that the charging system is working. In some designs, the light comes on to show the system is not working.

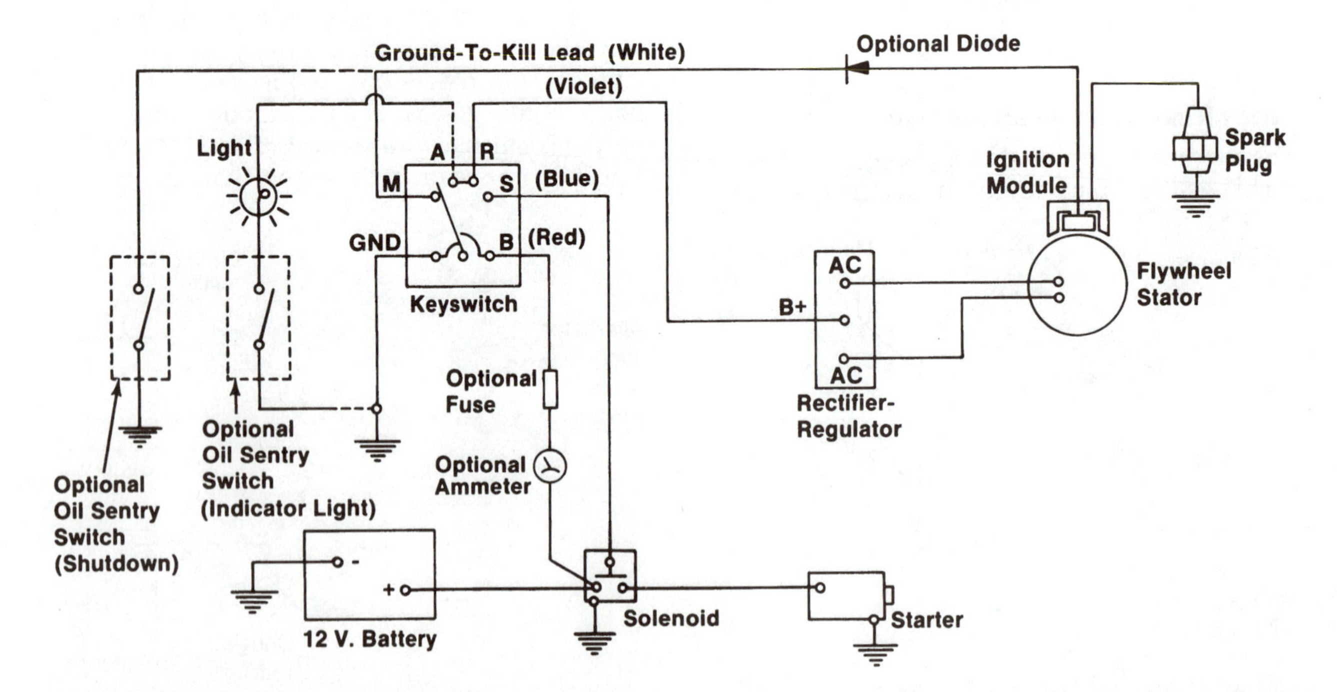

FIGURE 20-27 A charging system circuit. (Courtesy of Kohler Co.)

REVIEW QUESTIONS

1. Explain how a lead acid battery cell works to develop a voltage.
2. List and describe the components that make up the elements of a battery cell.
3. List the basic parts and explain the operation of the basic motor principle used in a starter motor.
4. Explain why the starter pinion gear is smaller than the ring gear used on the engine flywheel.
5. Describe the parts and operation of an inertia starter drive.
6. Describe the parts and operation of an overrunning clutch starter drive.
7. Explain how the overrunning clutch in the starter drive system protects the armature from overspeeding.
8. Explain the purpose of a starter solenoid in a starting system circuit.
9. Describe the operation of the starter solenoid when the starter key switch is in the start position.
10. Describe the operation of the starter solenoid when the starter key switch is in the run position.
11. List and describe the two wiring circuits used to connect the components of the starting system.
12. Explain how a conductor and a magnetic field can be used to generate electricity.
13. List and describe the main parts of an alternator.
14. Explain the operation of a single diode used in a half-wave rectifier system.
15. Explain the operation of a four-diode bridge used in a full-wave rectifier system.

DISCUSSION TOPICS AND ACTIVITIES

1. Disassemble a shop starter motor. Identify the parts of the motor and drive. Trace the current flow through the motor.
2. Disassemble a shop alternator. Identify each of the parts. Trace the current flow through the alternator.
3. A starter flywheel ring gear has 30 teeth for each tooth on a starter pinion. The engine is being cranked at 100 rpm. What is the rpm of the starter armature?

Troubleshooting, Maintenance, and Tune-up

CHAPTER 21

Troubleshooting

OBJECTIVES

Upon completion and review of this chapter, you should be able to:

- Troubleshoot an engine to locate the cause of a starting problem.
- Troubleshoot an engine to locate the source of a performance problem.
- Troubleshoot an engine to locate the source of a vibration problem.
- Troubleshoot an engine to locate the source of an engine noise.
- Troubleshoot an engine to determine the cause of high oil consumption.
- Troubleshoot an engine to determine the cause of overheating.

TERMS TO KNOW

Compression gauge
Compression test
Cylinder balance test
Cylinder leakage tester
Engine misfire
Fuel supply test
Oil consumption
Service tachometer
Spark tester
Surging

INTRODUCTION

Troubleshooting is a step-by-step procedure followed to locate and correct the cause of an engine problem. There is often a difference between the cause of the problem and the reason an engine ends up in a shop for repair. The customer may bring into the shop an engine that runs poorly. Your troubleshooting may show that the air-fuel mixture is too rich. Your conclusion may be that there is a fuel system problem. The cause of the problem could be an operator who does not use the choke properly. Remember, you have to find the real cause, not just the symptom of the problem. If you do not, the engine may be back in the shop with the same problem.

When you fill out the repair estimate form, try to find out the history of the problem. A broken piston ring may be the problem. The cause may be an operator who does not check or change the oil. Ask the customer how the engine is used and operated.

A troubleshooting procedure is a systematic check of the most likely problems. Unless you follow a logical procedure, time can be lost looking in the wrong places. Do not overlook the obvious. Make these checks (Figure 21-1) before you troubleshoot:

- Is there oil in the engine? Check engine oil level and condition.
- Is there fresh fuel in the fuel tank? An engine with no fuel or stale fuel will not start or run.
- Are the engine controls working? A bad kill switch or stuck choke cable will cause engine performance problems.

Engines are made to operate reliably for long periods of time. After a high number of service hours the engine may begin to experience one or more of the following problems:

- Fails to start or starts with difficulty
- Has low power, runs rough, or misfires
- Vibrates
- Makes internal noise
- Consumes large quantities of oil
- Overheats

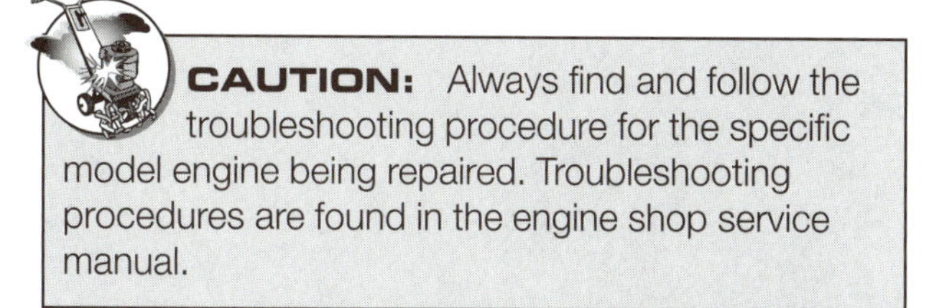

CAUTION: Always find and follow the troubleshooting procedure for the specific model engine being repaired. Troubleshooting procedures are found in the engine shop service manual.

TROUBLESHOOTING STARTING PROBLEMS

The engine must have ignition, fuel, and compression to start and run. The engine will not start if there is a loss of ignition. If the engine does not get enough fuel, it will fail to start or be hard to start. The same problem occurs if it gets too much fuel. There will not be any combustion if the air-fuel mixture is not compressed on the compression stroke. You should follow a step-by-step procedure as shown in the accompanying sequence of photographs to find out if the problem is ignition, fuel, or compression.

Spark Testing

An engine that will not start or run may have an ignition system problem. An engine with a magneto problem does not produce the high-voltage spark required to ignite the air-fuel mixture. The basic test to determine if the ignition system is operating properly is to check for spark at the end of the spark plug cable (Figure 21-2).

WARNING: The spark plug wire has high voltage. You can get shocked if you touch the spark plug cable terminal when turning the flywheel. Be sure that there is no fuel in the area that might be ignited by the spark and cause a fire or explosion.

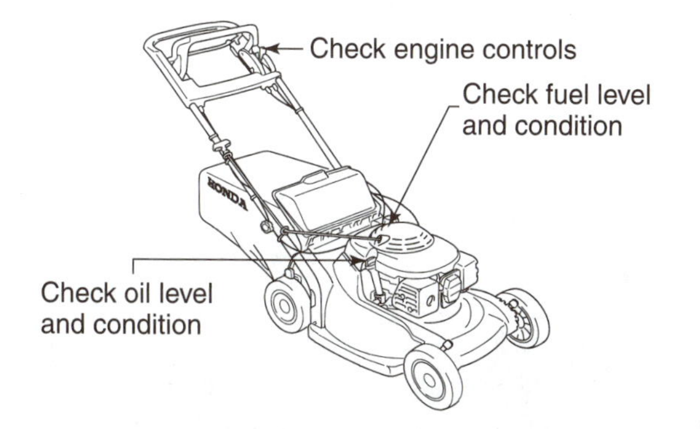

FIGURE 21-1 Check oil, fuel, and engine controls before you start troubleshooting. (Courtesy of American Honda Motor Company, Inc.)

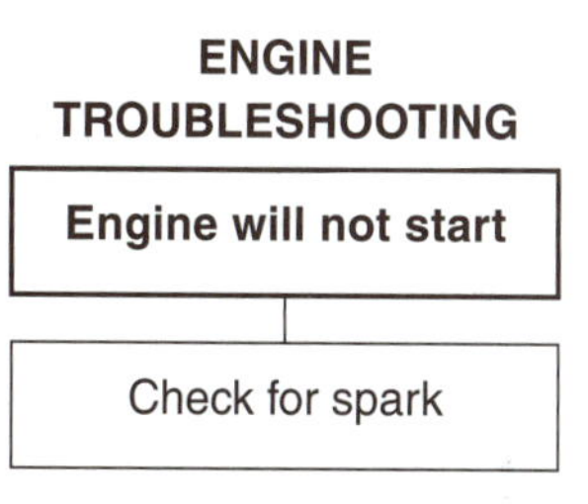

FIGURE 21-2 When an engine will not start, check for spark.

To check for spark, remove the spark plug wire from the spark plug terminal (Figure 21-3). Most spark plug wires have an insulated boot that fits over the spark plug terminal. Always remove the wire by pulling on the insulated boot and not the spark plug wire. Pulling on the spark plug wire can cause it to separate from the connector. This can result in a poor electrical connection.

CAUTION: Always use the shop manual recommended spark tester and procedure for checking spark. Incorrect spark checking procedure can damage a capacitive discharge or solid state ignition.

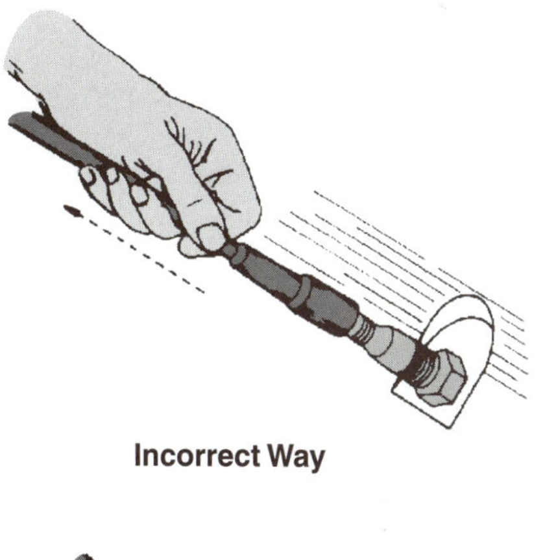

FIGURE 21-3 Remove the spark plug cable by twisting and pulling on the boot, not the wire. (Courtesy of Cooper Automotive/NAPA Belden.)

PHOTO SEQUENCE 3

Troubleshooting an Engine That Will Not Start

1. Remove the fuel tank cap and check fuel level and condition. Add fresh fuel if necessary.

2. Use a spark tester to determine if there is spark for ignition. Service the ignition if there is no spark.

3. Remove the spark plug and inspect the insulator tip for correct fuel mixture. Service the fuel system if the insulator tip shows a lean or rich mixture.

4. Ground the spark plug and turn the kill switch to the off position for safety.

5. Test for compression by placing your thumb over the spark plug hole.

6. Pull on the starter rope. Good compression will blow your thumb off the spark plug hole. Service the engine if the compression is low.

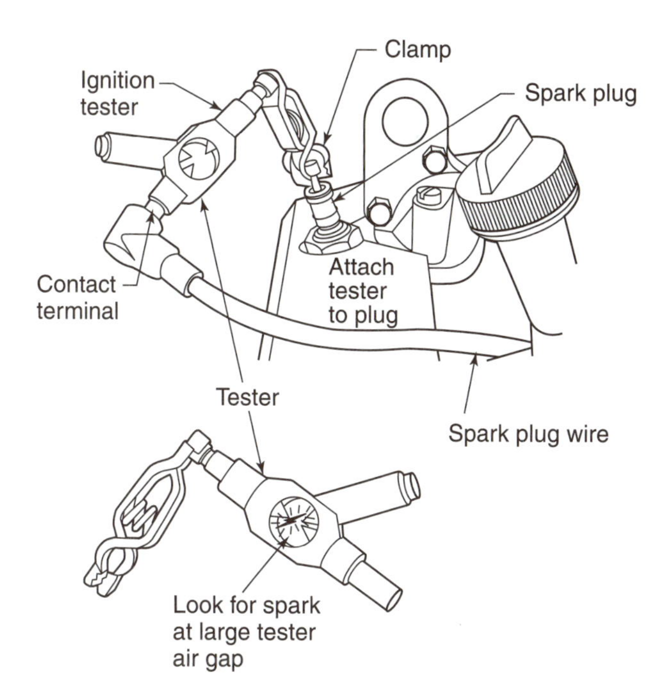

FIGURE 21-4 Testing for spark with an ignition tester.

A **spark tester** (Figure 21-4) is used to test for spark by viewing the spark across an air gap. Connect the tester clamp to the spark plug terminal. Connect the spark plug wire terminal to the tester contact terminal. When the tester is connected, crank the engine. You should see a spark jump the tester air gap. The large tester air gap acts like a normal spark plug gap under compression pressure. If a spark jumps the tester gap, the magneto is working. If there is no spark, there is a magneto problem. Look for a bright, blue spark. This shows a strong ignition system. A dull, yellow spark shows a weak ignition system. If you find a strong spark and the engine will not start, the problem is most likely in another system. You will need to do more troubleshooting.

SERVICE TIP: Engines that use a soft key on the flywheel can have a partially sheared key that will cause the ignition to occur at the wrong time. The engine can pass the spark test but still have an ignition problem. You should suspect this problem if you find the spark plug is wet with fuel.

If an engine does not have a good strong spark, the ignition system needs more testing and service. Each part of the ignition system must be tested and repaired as necessary. Servicing the ignition system is part of an ignition tune-up. Common causes of no spark are:

- Dirty or burned breaker points on a breaker point triggered magneto
- Sheared flywheel key on engines with soft flywheel keys
- Incorrect ignition timing
- Defective magneto coil

Spark Plug Inspection

An engine that has ignition spark but does not start may have a fuel system problem (Figure 21-5). The engine must have the correct air-fuel mixture to start and run smoothly. Too much or too little air or fuel in the mixture may make the engine hard to start. It can even prevent it from starting at all.

The first step in finding a fuel problem is to remove and inspect the spark plug. The spark plug firing end is in the middle of the air-fuel mixture that enters the combustion chamber. Air-fuel mixture condition will be visible on the spark plug firing end insulator tip. The insulator tip is the area around the center electrode.

When the correct air-fuel mixture burns, it leaves behind a very small amount of brown or tan fuel deposit. These deposits build up in the combustion chamber and on the firing end of the spark plug. A brown or tan color shows that the air-fuel mixture is correct (Figure 21-6), so you should look for the problem in another system.

An air-fuel mixture that has too much fuel is called *rich*. When a rich mixture is burned in the combustion chamber, it leaves behind black soot (powder). The black soot comes from unburned gasoline. A spark plug insulator tip that is black or wet with gasoline probably has a fuel system problem (Figure 21-7). A rich mixture is hard to ignite and burn. Gasoline and soot will soon cover the spark plug electrode surfaces and prevent the spark from jumping the electrode gap. A spark plug insulator tip that is wet with fuel will not spark and the engine will not start.

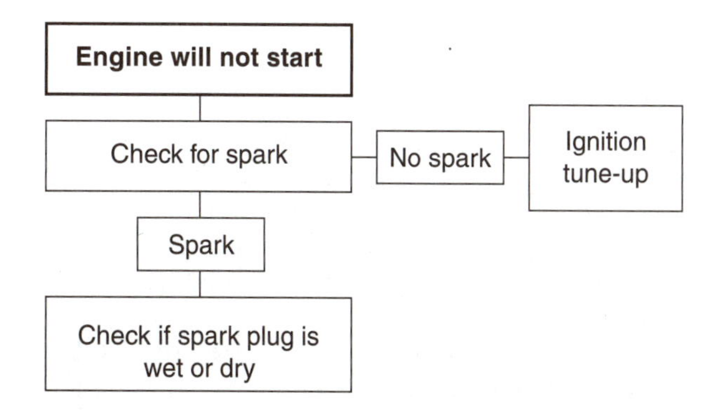

FIGURE 21-5 If the engine has spark, check the spark plug.

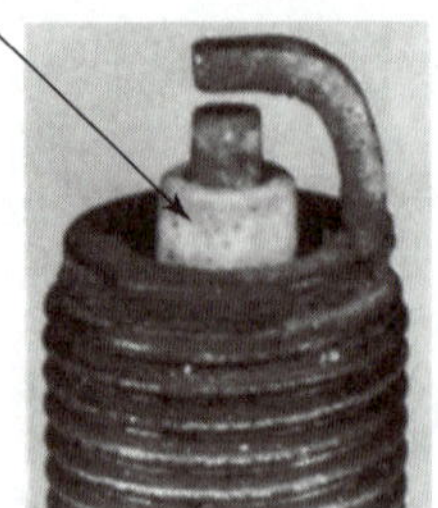

FIGURE 21-6 The correct air-fuel mixture leaves a brown or tan insulator tip. (Courtesy of CHAMPION-Federal Mogul.)

FIGURE 21-7 A black, wet spark plug insulator tip is caused by a rich mixture. (Courtesy of CHAMPION-Federal Mogul.)

SERVICE TIP: Another possible cause for a wet or soot-covered spark plug could be a defective spark plug. A spark plug that does not spark will not burn the fuel in the combustion chamber. Troubleshoot this problem by installing a new spark plug.

Common fuel system problems (Figure 21-8) that would create a wet spark plug electrode tip are:

- Stuck carburetor choke
- Restricted air filter
- Stale fuel
- Incorrect carburetor settings
- Plugged carburetor air passage

A lean air-fuel mixture does not have enough fuel. An engine that does not start or run may have a lean mixture. A lean mixture does not leave any gasoline deposit in the combustion chamber or on the spark plug insulator tip. A dry, white insulator tip (Figure 21-9) shows an air-fuel mixture that is too lean to burn properly.

Common fuel system problems (Figure 21-10) that result in a lean mixture are:

- Empty fuel tank
- Clogged fuel tank screen

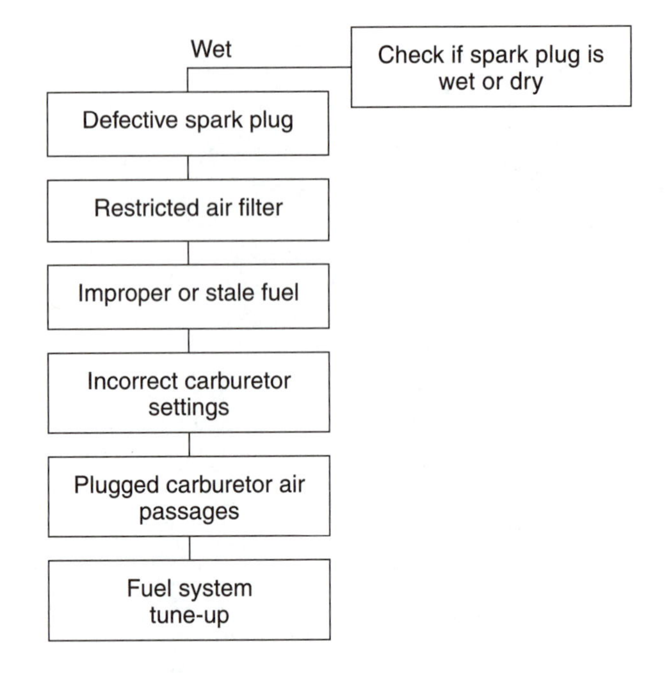

FIGURE 21-8 Common problems when the spark plug shows a rich mixture.

FIGURE 21-9 A dry, white spark plug insulator tip is caused by a lean mixture. (Courtesy of CHAMPION-Federal Mogul.)

- Clogged fuel filter
- Plugged carburetor fuel passage

Always check any fuel tank shut-off valve to make sure that it is open. Sometimes dirt, water, or grass gets into the fuel tank and clogs the fuel line at the bottom of the tank. Some fuel tanks have a screen in the outlet fuel line to keep dirt from going to the carburetor. This screen can be full of dirt so the fuel cannot get through. Older fuel tanks have a vented fuel cap. A plugged fuel cap vent can also stop fuel from going to the carburetor.

WARNING: A fuel supply test must be done in a well-ventilated area away from any source of ignition. Gasoline vapors can cause a fire. Do not do this test on a hot engine because spillage can cause a fire. Dispose of the gasoline properly.

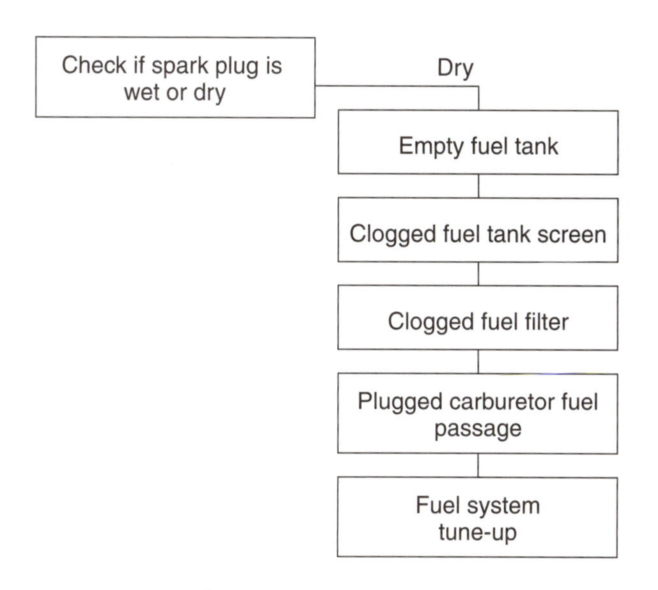

FIGURE 21-10 Common causes when the spark plug shows a lean mixture.

Fuel system restrictions are determined with a fuel supply test. A **fuel supply test** is a troubleshooting procedure used to determine if enough fuel is getting from the fuel tank to the carburetor.

Use a set of vise grip pliers to clamp the flexible fuel line between the fuel filter and carburetor (Figure 21-11). Disconnect the fuel line from the carburetor inlet. Hold the fuel line over a metal container. Release the vise grip pliers to allow fuel to flow out of the fuel line. The flow should be as wide as the inside diameter of the fuel line. Low or no flow means a restriction. Move the vise grip pliers to clamp the fuel line between the fuel tank and filter. Remove the fuel filter and repeat the test. If the flow improves, the problem is a clogged fuel filter. If the flow is still restricted, the problem is in the fuel tank screen. If the flow is not restricted, the problem is in the carburetor. Troubleshooting and servicing the carburetor is part of a fuel system tune-up.

FIGURE 21-11 A fuel supply test is done by disconnecting the fuel line and checking flow.

Compression Test

An engine may have ignition and fuel, but it will not start if there is not enough compression (Figure 21-12). The air-fuel mixture has to be compressed or there will be no power stroke. The amount the mixture is compressed determines the pressure developed on the power stroke. A **compression test** is a test of the pressure in an engine cylinder to determine engine condition. A compression test checks the condition of piston rings and valves. It also checks for a defective head gasket.

You can do a simple test for compression by checking flywheel rebound (Figure 21-13). Turn

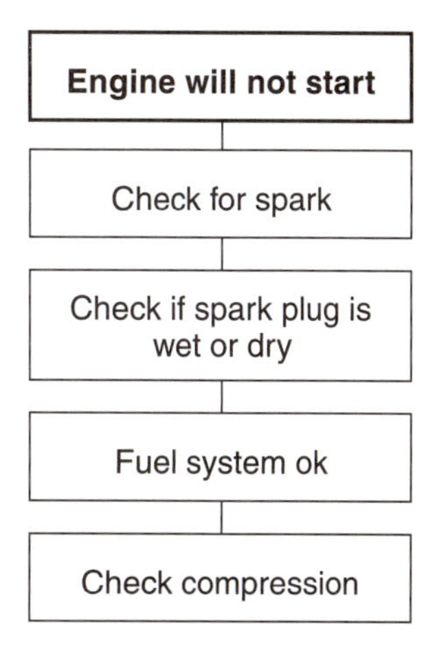

FIGURE 21-12 If the ignition and fuel system are working, check engine compression.

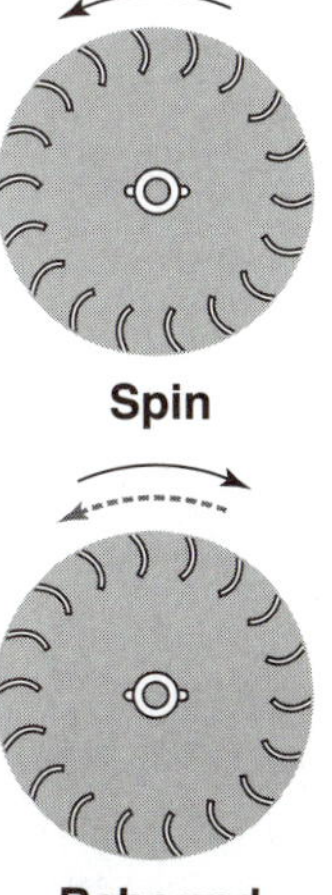

FIGURE 21-13 A rebound test is used to check engine compression.

the magneto kill switch to off before doing the rebound test. Rotate the flywheel by hand counterclockwise (when viewed from the flywheel side of the engine). You should feel resistance against the flywheel when the piston comes up on the compression stroke. Rotate the flywheel rapidly against this resistance. You should feel a sharp flywheel rebound (bounce back). Rebound shows there is high enough compression for engine starting. If there is slight or no rebound, there is not enough compression for engine starting.

The exact compression of the engine can be determined with a **compression gauge,** a tester with a pressure gauge used to measure the amount of compression pressure in an engine cylinder (Figure 21-14). The pressure gauge has a scale divided into pounds per square inch or kilograms per square centimeter. The tester is attached to the engine cylinder by a push-in or screw-in type fitting. The push-in type fitting has a tapered rubber tip. It is held in the spark plug hole by hand. The screw-in type has a flexible hose that is threaded into the spark plug hole.

Turn the magneto kill switch to off before doing the compression test. Remove the spark plug. Both spark plugs should be removed from a twin cylinder engine. This will prevent too much engine drag during cranking. Open the carburetor throttle valve fully to allow good air flow into the cylinder. Install the screw-in or push-in attachment to the compression gauge in the spark plug hole (Figure 21-15). Crank the engine several rotations while watching the compression gauge. Stop cranking when the gauge shows no further pressure increase. Note and record the pressure. If the engine has two cylinders, repeat these steps for each cylinder.

Take the reading with the engine at normal operating temperature (if possible). This will show piston ring and valve sealing under normal operating conditions. Twin cylinders must be tested the same number of strokes for a good comparison.

Compare the compression readings with compression specifications in the shop service manual. A good cylinder will test close to the specified compression. A bad cylinder will test below specifications. Most four-stroke engines have a normal compression of around 75 pounds per square inch. Many two-stroke engines have a normal compression of about 90 pounds per square inch.

A low compression reading may indicate an improperly seated valve or worn or broken piston

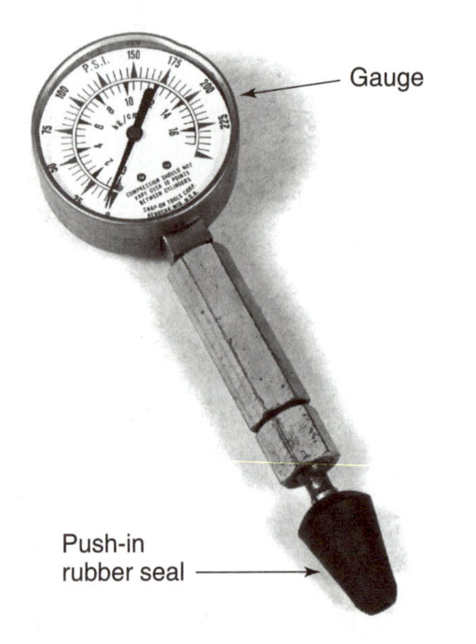

A. Push-In Compression Tester

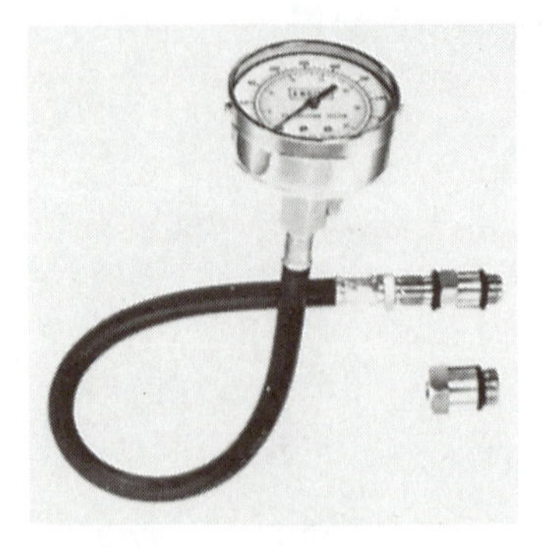

B. Screw-In Compression Tester

FIGURE 21-14 Push-in and screw-in compression gauges are used to check engine compression. (*B:* Courtesy of K-Line Industries, Inc.)

FIGURE 21-15 Testing engine compression with a compression gauge.

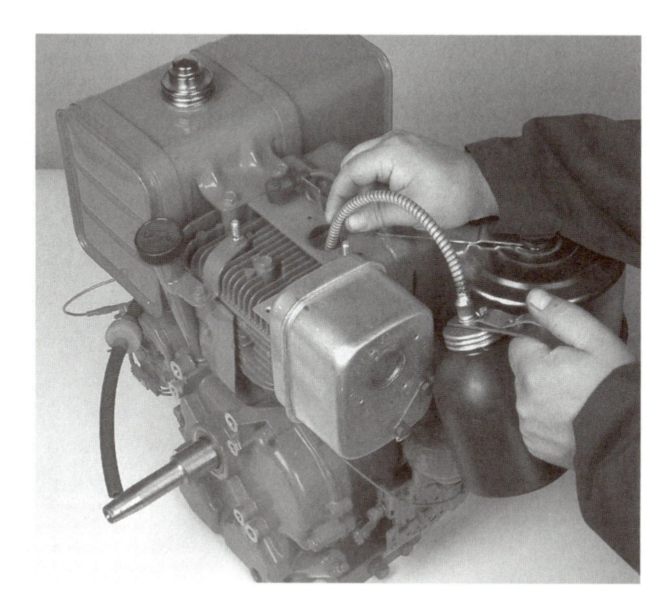

FIGURE 21-16 Squirt engine oil in the cylinder and retest to find the cause of low compression.

rings. Worn piston rings are shown by low compression on the first stroke that tends to build up on the following strokes. You can troubleshoot for ring problems by squirting a small amount of engine oil into the cylinder (Figure 21-16). Test the compression again. The engine has a ring sealing problem if the reading improves.

Valve problems are shown by a low compression reading on the first stroke. The compression will not build up on the following strokes. You will not get an improved reading with oil in the cylinder. Leaking head gaskets give the same test results as valve problems. A typical procedure for checking engine compression is shown in the accompanying sequence of photographs.

Cylinder Leakage Test (CLT)

The condition of piston rings and valves can be determined with a **cylinder leakage tester**, a tester that fills the cylinder with air and checks for leakage around piston rings and valve seats (Figure 21-17). The cylinder leakage tester has a pressure gauge and an air fitting. The air fitting is connected to shop air. The spark plug is removed from the cylinder to be tested. An outlet hose from the tester fits in the spark plug hole.

The piston in the cylinder to be tested is positioned at top dead center on the compression stroke. Follow the tester instructions to set the air pressure and zero the gauge.

Air from the tester fills the cylinder. A good cylinder has a very slow rate of leakage. A bad cylinder leaks down rapidly. You can determine the source of the leakage by listening to air escaping from the engine. Air leakage from the air cleaner shows a leaking intake valve. Air escaping through the muffler shows a leaking exhaust valve. Air leakage from the crankcase breather shows worn piston rings.

FIGURE 21-17 A cylinder leakage tester is used to determine the condition of internal engine components.

TROUBLESHOOTING ENGINE PERFORMANCE PROBLEMS

There are three common engine performance problems:

- Engine lacks power
- Engine misfires
- Engine surges or runs unevenly

An engine that will not run at high speed with an open throttle will not develop enough power (Figure 21-18). Check the air intake for an obstruction. Look for a dirty air filter, an oil-saturated air filter, or grass in the air filter entrance. Improper oil level or viscosity can reduce engine power. An exhaust system restriction prevents proper exhaust flow and reduces engine power. The air filter, engine oil, and exhaust system should be checked and serviced during the regular periodic maintenance.

An improperly adjusted carburetor can cause a too rich or too lean air-fuel mixture. A carburetor with internal problems such as a leaking fuel inlet needle and seat can also cause mixture problems. Air-fuel mixture problems result in low engine power. An obstructed muffler or spark arrestor will

PHOTO SEQUENCE 4

Measuring Engine Compression

1. Remove and ground spark plug wire.

2. Remove spark plug.

3. Install compression gauge into spark plug hole.

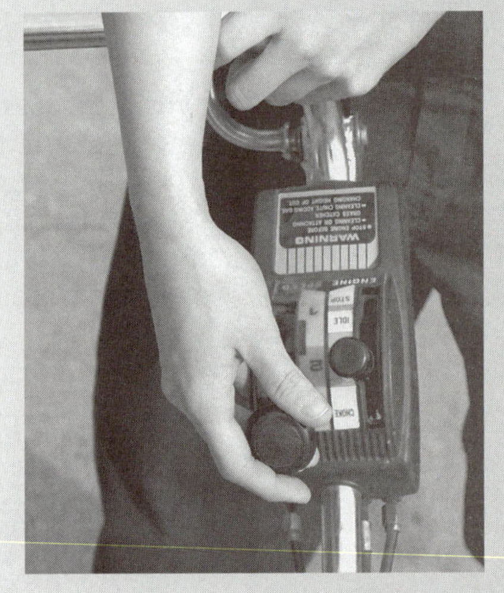

4. Move the throttle to the fully open (fast) position

5. Pull starter rope and observe engine compression.

6. Remove the compression gauge and squirt oil into cylinder.

7. Install compression gauge into cylinder and retest compression again.

8. Compare compression reading with specifications in the shop service manual.

9. Remove compression gauge and install spark plug and spark plug wire.

cause a loss of power. Troubleshooting and servicing the fuel system are part of a fuel system tune-up.

Internal engine problems can also cause low engine power. Improper valve clearance adjustment affects engine power. Valve clearance should be checked and adjusted as recommended by the engine maker. Low cylinder compression pressures reduce engine power. Compression problems such as worn piston rings, leaking valves, and blown head gaskets can be found with a compression test.

An **engine misfire** (Figure 21-19) is an engine performance problem in which the air fuel mixture is not ignited on each engine power stroke. Technicians commonly describe this problem as an engine "miss." The first check to make on a misfiring engine is the spark plug. An incorrect, worn, or fouled spark plug is a common cause of misfiring.

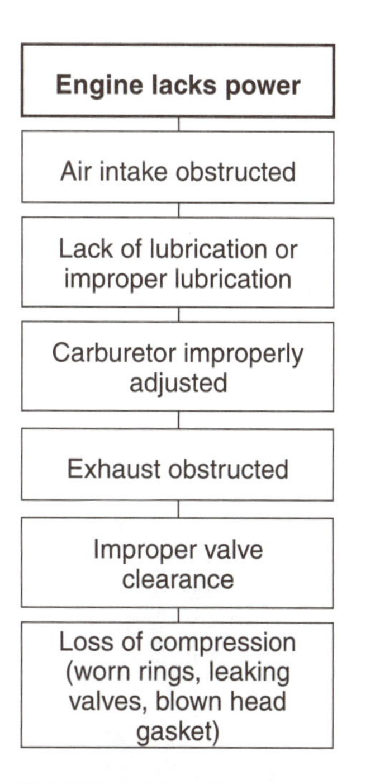

FIGURE 21-18 Common causes for an engine that lacks power.

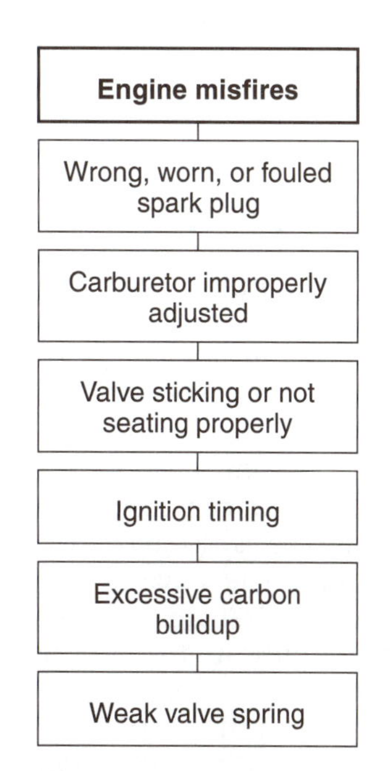

FIGURE 21-19 Common causes for an engine that misfires.

CASE STUDY

A customer brought a garden tiller into the outdoor power equipment shop. The engine had suddenly stopped running in the middle of a tilling job. The customer had checked the fuel and spark before he brought it into the shop. Neither of these areas seemed to be the problem.

Before giving the customer an estimate, the shop owner gave the engine a quick inspection. When he pulled on the starter rope, the engine cranked much too fast. The owner knew this was an indication of compression loss. The owner directed his technician to perform a compression test on the engine.

The technician removed and grounded the spark plug cable for safety and removed the spark plug. He inserted the compression gauge into the cylinder. He then pulled on the starter rope several times and observed the compression gauge. There was barely 20 pounds per square inch of compression in the cylinder. The technician removed the gauge and squirted a small amount of oil into the cylinder. The cylinder compression was tested again and the compression was not affected by the oil. The conclusion was that the engine has a valve sealing problem. The customer was given an estimate for a valve service job.

Later, when the technician removed the cylinder head the cause of the problem was obvious. The exhaust valve had seized in the valve guide. The valve was stuck in the open position.

Check the spark plug for the proper application or a fouled condition and replace as necessary. A carburetor adjustment or internal problem can cause a rich or lean air-fuel mixture. Rich or lean mixtures can cause a misfire. Adjust the carburetor or remove the carburetor for cleaning. Any of these problems indicate the need for an engine tune-up.

Several internal engine problems can cause a misfire. Valves that are not sealing can cause this problem. The problem can also be caused by valves that stick (seize) in their guides. The valve sealing or sticking problems mean that the engine needs major service. A compression pressure test can be used to locate these problems.

Excessive oil consumption past the piston rings can result in a carbon buildup in the combustion chamber that can prevent proper combustion and create a misfire. The cylinder head can be removed and the carbon cleaned out, but more likely an engine overhaul will be required for a permanent solution. Weak intake or exhaust valve springs are another source of a misfire and generally indicate a complete engine overhaul.

Surging (Figure 21-20) is an engine performance problem that occurs when the engine runs fast and slow at the same throttle setting. A common cause for this problem is a plugged fuel tank vent or a dirty air filter. The air filter may need replacement or cleaning. Check the carburetor adjustment or clean the carburetor as part of a fuel system tune-up.

The governor and throttle linkages are common sources of surging. Check all linkages to the governor and throttle. The governor shaft, throttle shaft, and all pivot points need to be checked for binding. Any binding problems require cleaning or parts replacement.

A problem with the ignition system that causes an intermittent spark to the spark plug will cause the engine to surge. The ignition system will require testing to locate this type of problem. An ignition system tune-up should be done when an intermittent spark is suspected.

Multiple cylinder engine problems can be more difficult to locate than those in a single cylinder engine. The problems may be in one or both cylinders. There are some multiple cylinder engine

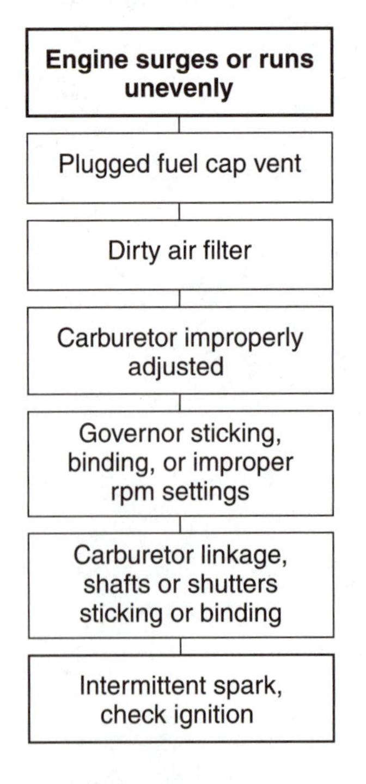

FIGURE 21-20 Common causes for an engine that surges or runs unevenly.

performance areas that affect both engine cylinders. Other problems affect only one of the cylinders. Carburetor and ignition timing problems commonly affect both cylinders.

Multiple cylinder engine problems that commonly affect only one cylinder are:

- Spark plug
- Spark plug wire
- Head gasket
- Leak in one end of an intake manifold
- Valve leakage
- Piston rings

Multiple cylinder performance problems can be isolated to one cylinder or the other using a cylinder balance test. A **cylinder balance test** is a test performed on a multiple cylinder engine to find out if each cylinder is creating the same amount of power. The cylinder balance test is performed with a service tachometer and a screwdriver with an insulated handle. A **service tachometer** (Figure 21-21) is a tester that is connected to the engine ignition system to display engine rpm.

WARNING: To avoid a high-voltage shock, use a screwdriver with an insulated handle when grounding out ignition systems.

CAUTION: Check the engine service manual before grounding out any ignition system. Some types of solid state ignitions can be damaged by this procedure. Do not operate the engine at top speed for long periods of time. An engine with an internal problem can be damaged by this procedure.

To do a cylinder balance test, operate the engine until it reaches normal operating temperature. Stop the engine. Connect a spark tester to both spark plugs. Connect the tachometer to the ignition system. Follow the instructions provided with the tachometer. Start the engine and open the throttle momentarily to top speed. Check the spark at both ignition testers. If the spark is equal at both testers, the problem is not in the ignition. An ignition problem is shown by intermittent spark at one or both of the spark testers.

The next step is to observe the engine rpm on the tachometer. Ground out one cylinder with an insulated handle screwdriver (Figure 21-22). Contact the alligator clip on the tester and at the same time a good ground on the engine. With the cylinder grounded, the engine rpm will drop. Note the amount of rpm loss you see on the tachometer display. Remove the screwdriver and allow the engine to get back to idle speed. Repeat the procedure on the other cylinder and note its rpm loss.

Compare the test results to cylinder balance specifications in the shop service manual. The two cylinders are in balance if they have the same rpm loss during the test. This means they are doing the same amount of work. The difference between the rpm loss of the two cylinders should less than 75 rpm.

An engine may have a performance problem and good cylinder balance. This means it has a problem common to both cylinders. You may find a difference between the two cylinders greater than 75 rpm. The cylinder with the smallest rpm loss is not doing as much work. Inspect the low cylinder for a low compression problem.

The cylinder balance test may also show that a cylinder is not working at all. You may ground out one cylinder and find that there is no rpm loss. This means there is no power being developed by this cylinder. If you ground out the good cylinder, the engine will stop running.

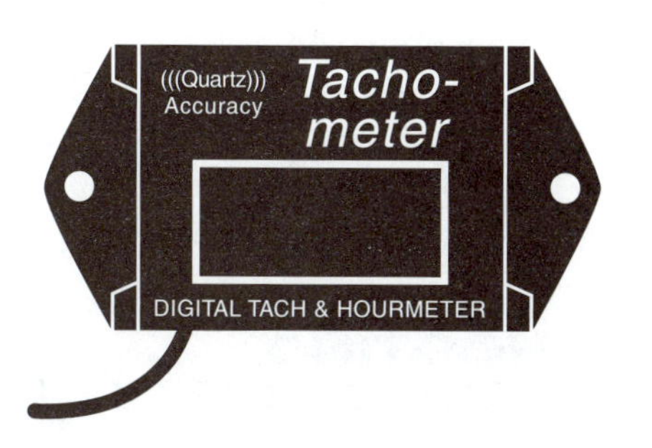

FIGURE 21-21 A service tachometer is used to measure engine rpm.

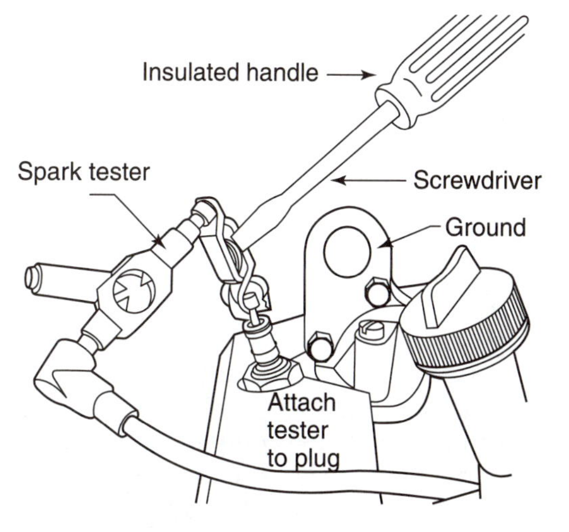

FIGURE 21-22 An insulated screwdriver is used to ground out a cylinder and check cylinder balance.

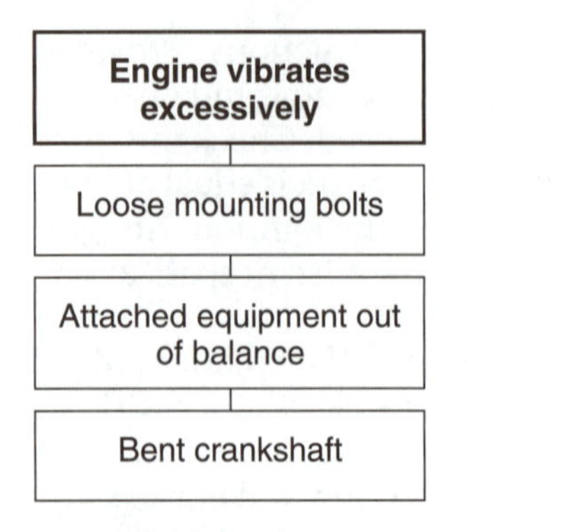

FIGURE 21-23 Common causes for an engine that vibrates excessively.

TROUBLESHOOTING VIBRATION PROBLEMS

Engine vibration is a condition in which the engine shakes back and forth as it runs (Figure 21-23). Vibration can be caused by loose engine mounting bolts. Check the bolts to make sure they are tight. Another common source of vibration is the equipment driven by the engine. Equipment like blade or tines may be out of balance or out of alignment. One way to check this problem is to disconnect the engine from the equipment. Remove the drive chain or belt and run the engine. If the vibration goes away the problem is in the equipment. Sometimes you can remove and replace the equipment with known good parts. For example a rotary lawnmower blade can be replaced with another blade. Start the engine and see if the vibration goes away.

A bent crankshaft power takeoff (PTO) shaft is another cause of vibration. You can check the crankshaft PTO end for bends with a dial indicator. Set up the dial indicator with the plunger on the crankshaft PTO surface. Rotate the crankshaft and note any runout on the dial indicator gauge. Excessive runout will cause a vibration problem.

TROUBLESHOOTING ENGINE NOISES

Abnormal knocking sounds are a sign of engine trouble (Figure 21-24). Knocking noises usually mean the engine needs major service. You can become an expert at finding the cause of a knock by learning its sound and location. Engine equipment should be disconnected during troubleshooting. Loose or improperly adjusted equipment will cause knocks. This way you will be sure that the noise is inside the engine and not from the equipment.

Carbon can build up in the combustion chamber. This is a common problem if the engine is burning oil. The piston top can contact the carbon and create a knock. If this problem is suspected, remove the cylinder head. Remove the carbon and replace the cylinder head using a new head gasket.

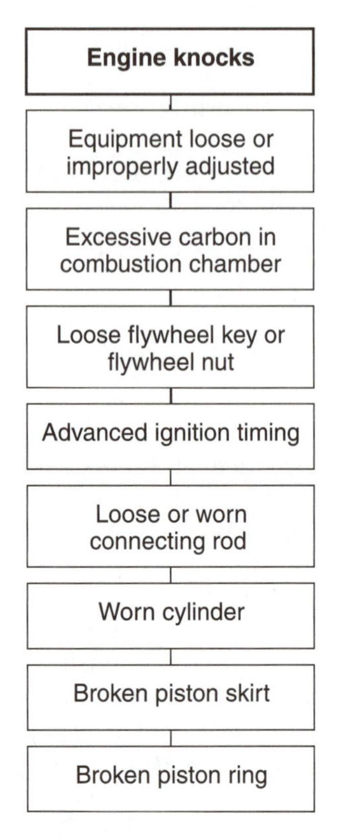

FIGURE 21-24 Common causes for an engine that knocks.

SERVICE TIP: Ignition timing that is too far advanced can cause abnormal combustion and a knocking noise. If the engine has adjustable timing, check and adjust the timing. Then recheck the engine for noise.

Knocking is also caused from too much clearance between moving parts. A loose flywheel key or loose flywheel nut can cause a knocking noise. Engine knocks are common from worn connecting rod bearings. Worn connecting rod bearings make a lighter knock when the engine is not under load. Knocks can be caused by piston rocking (slapping) in a worn cylinder. Broken piston skirts and piston rings also make noise.

TROUBLESHOOTING ENGINE OIL CONSUMPTION

Oil consumption is lubricating oil loss caused by oil leakage or oil burning inside the combustion

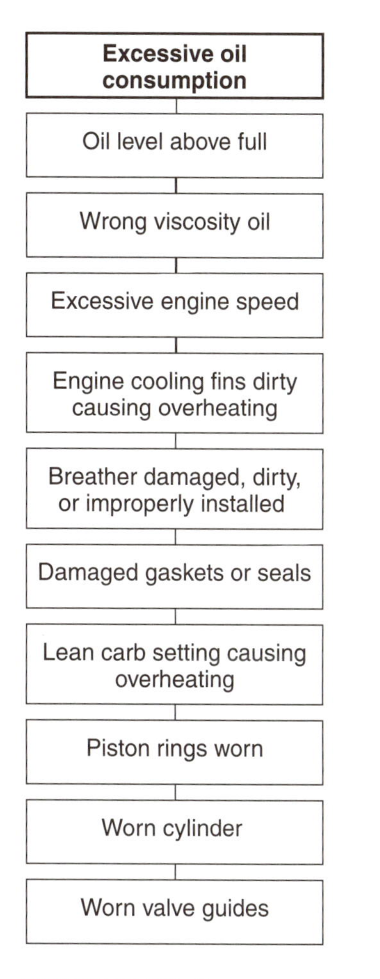

FIGURE 21-25 Common causes of excessive oil consumption.

chamber (Figure 21-25). Oil consumption is often noticed when:

- Engine oil level drops rapidly
- Blue or white oil smoke come out of the exhaust
- Oil on the outside surfaces of the engine

The first check to make is oil level, oil viscosity, and oil condition. An overfilled crankcase will create a pressure that can force the oil out of the engine. Using the wrong viscosity oil or not changing the oil can result in thin oil. Thin oil is more likely to leak into the combustion chamber or outside the engine.

Excessive engine speed can overheat the engine and result in a loss of oil viscosity. A lean carburetor adjustment or a cooling system problem can cause overheating. Any of these conditions can cause oil consumption.

Inspect the crankcase breather for dirt, damage, or improper installation. A plugged or damaged breather can cause oil to be pulled out of the crankcase. The oil will go into the air intake and into the combustion chamber. Damaged gaskets or seals can allow oil to leak out of the crankcase.

Worn internal parts can cause excessive oil to enter the combustion chamber. Common wear areas are piston rings, cylinder, and valve guides. Piston rings and cylinder condition can be determined by measuring engine compression. Compression readings that improve with oil in the cylinder show ring and cylinder wear.

TROUBLESHOOTING OVERHEATING PROBLEMS

Engine overheating can cause engine performance and oil consumption problems (Figure 21-26). Even more importantly, overheating can destroy internal engine parts. When an engine overheats, external engine parts will be very hot and there will be an overheated smell from the engine.

One common cause of overheating is an overloaded (overworked) engine. Operating the engine at too high a speed for long periods of time can cause overheating. A high load is often caused by dull cutting blades. Cutting excessive amounts of material will cause a high load. For example, the mower depth may be set too low on a lawn mower. Too high an engine idle rpm setting can also result in overheating.

An overheating can be caused by a lubrication or cooling system problem. Low oil level can result in overheating. Lubricating oil carries heat away

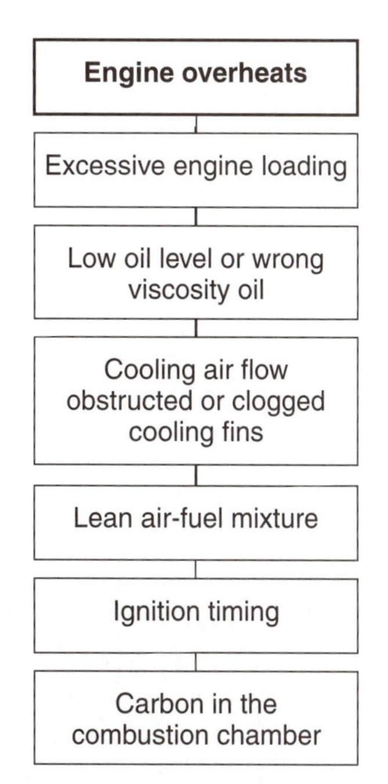

FIGURE 21-26 Common causes of engine overheating.

PHOTO SEQUENCE 5

Removing Carbon from a Cylinder Head (Decarbonizing)

1. Remove and ground the spark plug wire for safety. Remove the spark plug.

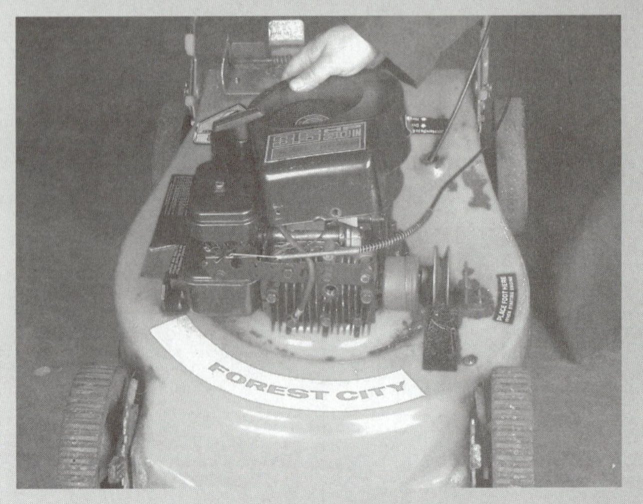

2. Remove the blower (fan) housing.

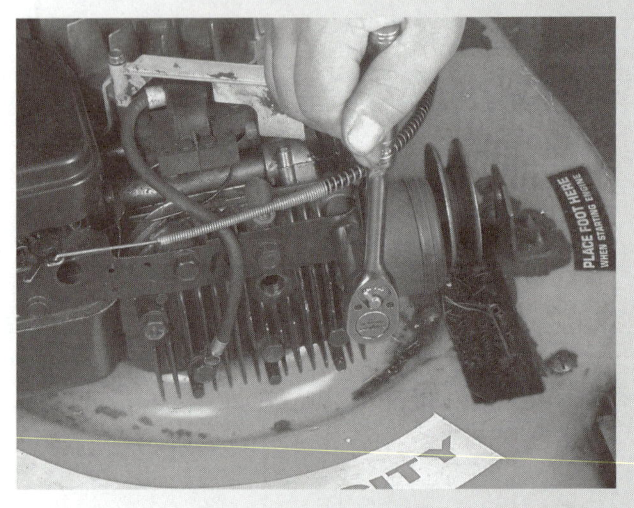

3. Remove the cylinder head bolts and lift off the cylinder head.

4. Remove the cylinder head gasket.

5. Use a scraper to remove carbon from the piston head and valves.

6. Use a scraper and wire brush to remove carbon from the cylinder head combustion chamber.

7. Turn the crankshaft and check each valve for damage.

8. Use a tap to clean up any damaged head bolt threads.

9. Compare the new head gasket to the old one to be sure they are the same.

10. Install the head gasket and cylinder head and start the cylinder head bolts.

11. Tighten the cylinder head bolts in the correct order to the specified torque.

12. Install the blower housing, spark plug, and spark plug wire.

from engine parts. If there is not enough oil, heat will build up on the parts. Cooling air flow is slowed by clogged cooling fins, ducting, or debris screen. These problems will result in overheating.

A lean air-fuel mixture burns much hotter than a correct mixture. The fuel in the mixture has a cooling effect on engine parts. A lean mixture will cause the engine to overheat.

Ignition timing affects engine temperature. A properly timed ignition allows combustion chamber heat to be dissipated through the engine parts. Incorrect ignition timing allows heat to build up in the combustion chamber causing the engine to overheat.

Carbon buildup in the combustion chamber can also cause overheating. The carbon coating stops heat flow through the cylinder head and into the cooling fins. The cylinder head will overheat. The high temperature in the cylinder head will transfer into the cylinder and other internal parts. Overheating from mixture, ignition, and carbon buildup are corrected by a fuel and ignition tune-up.

SERVICE TIP: A two-stroke engine with an incorrect amount of oil in the fuel mixture will overheat. Drain and refill the fuel tank with the correct mixture to diagnose this problem.

REVIEW QUESTIONS

1. Explain how to use a spark tester to troubleshoot an engine starting problem.
2. Explain why you should check the fuel in the tank before troubleshooting.
3. Describe the appearance of a spark plug insulator tip for an engine that has a too rich air-fuel mixture.
4. Describe the appearance of a spark plug insulator tip for an engine that is not getting enough air-fuel mixture.
5. Describe the appearance of a spark plug insulator tip for an engine that is getting the correct air-fuel mixture.
6. List and explain four common fuel system problems that can cause a too rich air-fuel mixture and prevent an engine from starting.
7. List and explain four common fuel system problems that can cause a too lean air-fuel mixture and prevent an engine from starting.
8. Explain how to use a compression gauge to test the compression of an engine.
9. Explain how to use a compression gauge to determine if low compression is due to worn piston rings or leaking valves.
10. List and explain three areas to check for an air obstruction when an engine lacks power.
11. List and explain two common fuel system problems that can cause an engine to lack power.
12. List and explain two ignition system problems that can cause an engine to misfire.
13. List and explain two internal engine problems that can cause an engine to misfire.
14. List and explain three common causes for engine surging.
15. List and explain five problems that could cause one cylinder of a multiple cylinder engine to malfunction.
16. Explain how to perform and interpret a cylinder balance test.
17. List and explain three common causes for an engine with a vibration problem.
18. List and explain four common causes for abnormal noises or knocks inside the engine.
19. Explain how to diagnose an external oil leak from a gasket or seal.
20. List and explain three common causes of engine overheating.

DISCUSSION TOPICS AND ACTIVITIES

1. Have a friend disable (bug) the ignition system on a shop engine. Troubleshoot the engine to find the problem.
2. Have a friend disable (bug) the fuel system on a shop engine. Troubleshoot the engine to find the problem.
3. An engine is burning oil and uses a pint every hour of operation. How many quarts of oil will the engine use in 20 hours of operation?

CHAPTER 22

Preventive Maintenance

OBJECTIVES

Upon completion and review of this chapter, you should be able to:

- Explain the purpose and importance of doing regular maintenance.
- Explain how to use a regular maintenance schedule to maintain an engine.
- Check and change engine oil.
- Service an air and liquid cooling system.
- Service a crankcase breather.
- Service a paper and oil bath air filter.
- Change a fuel filter.
- Adjust engine valve clearance.
- Prepare an engine for off-season storage.
- Determine storage battery condition.

TERMS TO KNOW

Battery carry tool
Battery charging
Battery terminal brush
Battery terminal puller
Dipstick
Engine oil level
Fogging
Maintenance schedule
Oil filter tool
Open cell voltage check
Preventive maintenance
Slow charger
Sludge

INTRODUCTION

An engine requires regular maintenance in order to work correctly for a long period of time. Properly cared for engines may last as long as 2,500 hours. **Preventive maintenance** is the maintenance service done at regular times to increase the life of an engine. Preventive maintenance is the most important factor in long engine service life.

MAINTENANCE SCHEDULES

Preventive maintenance is required on all the engine's systems. Maintenance is required after a number of months or hours of operation. The owner's manual shows what maintenance jobs should be done and when. Some engine makers also list this information in a maintenance guide or shop service manual. Most engine makers give specific service jobs to do at the beginning and end of a season. These are used when equipment is used just during the summer (mower) or winter (snow thrower) season.

The time between regular maintenance jobs is often listed in a **maintenance schedule** (Figure 22-1), a chart listing the maintenance jobs and the recommended times for doing the jobs. The common times used by engine makers are:

- Before each use
- 20 hours or every season (monthly)
- 50 hours or every season (every 3 months)

MAINTENANCE SCHEDULE

	ITEM / REGULAR SERVICE PERIOD (3) Perform at every indicated interval		Before Each Use	First 20 Hours	Every 50 Hours	Every 100 Hours	Every 300 Hours	Refer to page
	Blade condition and blade bolt tightness	Check	O					17 57
	Grass bag (HRB215)	Check	O					22
•	Engine oil	Check	O					18
		Change		O		O		45
•	Air Cleaner	Check	O					20
		Clean-Replace*			O(1)			47
•	Spark plug	Clean-Adjust				O		48
		Replace					O	
	Spark arrester (optional equipment)	Clean				O		79
•	Idle speed	Check-Adjust					O(2)	50
	Flywheel brake cable	Adjust		O		O		52
	Flywheel brake pad	Check-Adjust		O(2)		O(2)		–
	Drive clutch (SDA) cable	Adjust		O		O		53
	Shift cable (SDA)	Adjust		O		O		54
	Throttle cable	Adjust		O(2)		O(2)		–
•	Valve clearance	Adjust					O(2)	–
•	Fuel tank	Clean					O(2)	–
•	Fuel line	Replace	Every 2 years (2)					–

- • Emission related items.
- * Replace the paper element only.
- (1) Service more frequently when used in dusty areas.
- (2) These items should be serviced by an authorized Honda servicing dealer, unless you have the proper tools and are mechanically proficient. Refer to the Honda shop manual for service procedures.
- (3) For commercial use, log hours of operation to determine proper maintenance intervals.

FIGURE 22-1 A maintenance guide lists maintenance jobs and when they should be done. (Courtesy of American Honda Motor Co., Inc.)

- 100 hours or every season (every 6 months)
- 100–300 hours (every year)

ENGINE OIL MAINTENANCE

Checking and changing engine oil is one of the most important service jobs. Doing this job regularly increases engine service life. A low oil level prevents proper engine lubrication that results in internal parts damage. Dirty oil allows abrasive dirt and metal to circulate throughout the engine parts. This can cause fast parts wear.

Check Engine Oil Level and Condition

The **engine oil level** is the level of the oil in the engine crankcase available for the lubrication system. Engine oil level should be checked each time the engine is used. This is the best way to make sure the engine has enough oil for safe operation. The oil should also be checked for condition to determine if the oil should be changed.

The engine has to be turned off before checking the oil level. This will give the oil time to run off engine parts and go back into the crankcase. When the engine is running, oil is splashed around

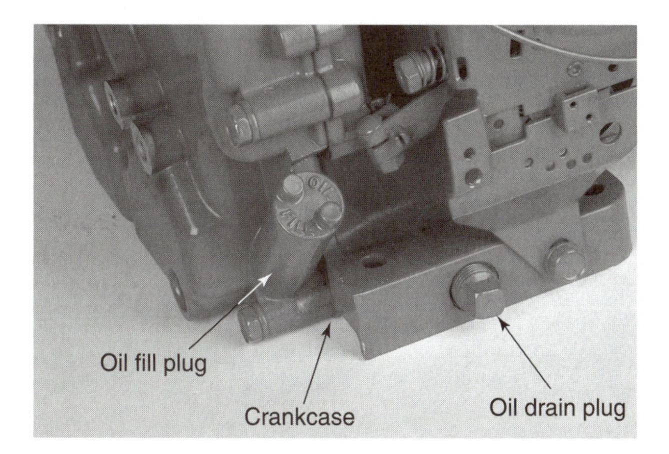

FIGURE 22-2 An engine with a fill plug hole for checking and adding engine oil.

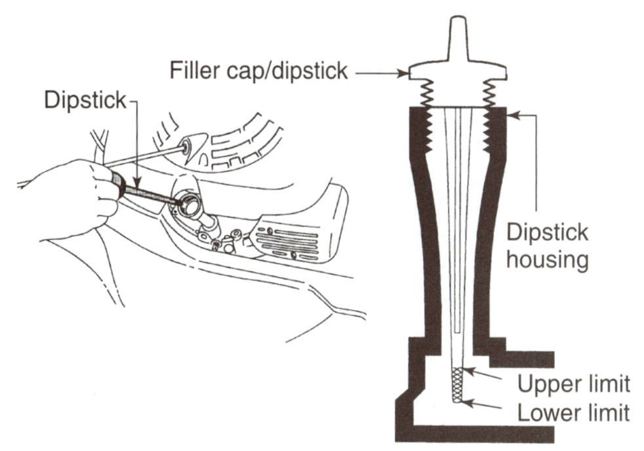

FIGURE 22-3 A dipstick is used to check oil level in the crankcase. (Courtesy of American Honda Motor Co., Inc.)

inside the crankcase. The oil level reading may be off if the oil level is checked just after shut down.

The oil level is checked on smaller engines by checking the oil level in the oil fill plug hole (Figure 22-2). There are usually two plugs on the outside of the engine. The oil fill plug is a removable plug on the crankcase used to check and add engine oil. The oil drain plug is a removable plug on the bottom of the crankcase used to drain engine oil. The higher of the two plugs is the oil fill plug. The fill plug is usually made of plastic and is often shaped like a wing nut so it can be removed by hand.

Make sure the power equipment is on a level surface before checking oil level. Carefully wipe the area around the fill plug with a shop rag. This will stop dirt from entering the crankcase when the plug is removed. Remove the oil filler plug. Check the oil level by looking into the oil fill hole. If oil is needed, add it until it comes to the top of the oil fill hole. Replace the oil fill plug after checking. Wipe off any oil spilled on the outside of the engine. Oil on the outside of the engine attracts dirt.

Many engines use a dipstick for checking oil level (Figure 22-3). A **dipstick** is a long, flat metal part with a handle and gauge marks used to measure engine oil level. The accompanying sequence of photographs shows how to check engine oil in a dipstick equipped engine. Locate the dipstick. It is usually on the side of the engine. Some engine makers paint the handle a bright color so it can be found easily. Pull the dipstick out of its housing. Wipe the measuring end clean with a clean shop rag.

There is no standardized system for the measuring marks on a dip stick. Some dipsticks have two lines. One line is marked "operating range" and another "add." Many just have cross hatch lines that show the operating range. The marks are identified in the equipment owner's manual. Push the dip stick firmly back into its housing until it seats. Pull it back out again. The oil level can be seen on the measuring marks of the dipstick.

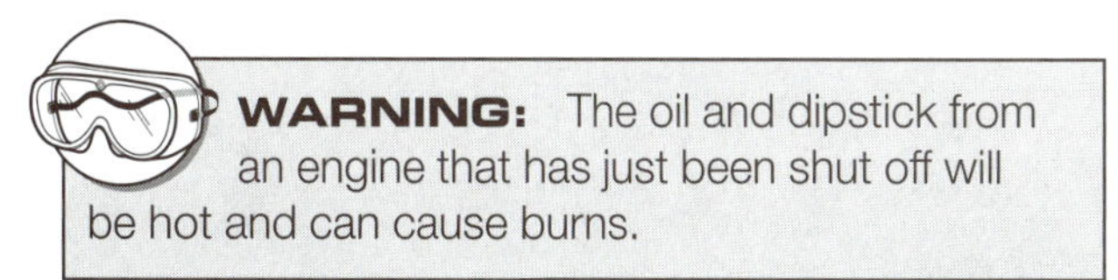

WARNING: The oil and dipstick from an engine that has just been shut off will be hot and can cause burns.

Oil on the dipstick can be used to determine oil condition (Figure 22-4). The longer the engine oil stays in the engine, the more dirt gets into the oil. Heat, moisture, and dirt eventually cause the oil to lose some of its protective qualities. The oil's condition can be determined by looking at its color on the dipstick. Oil that has not been contaminated has a clear, light brown color. Oil that has been in the engine a long time turns black. It will have a dirty appearance. Dirty, black oil should be changed.

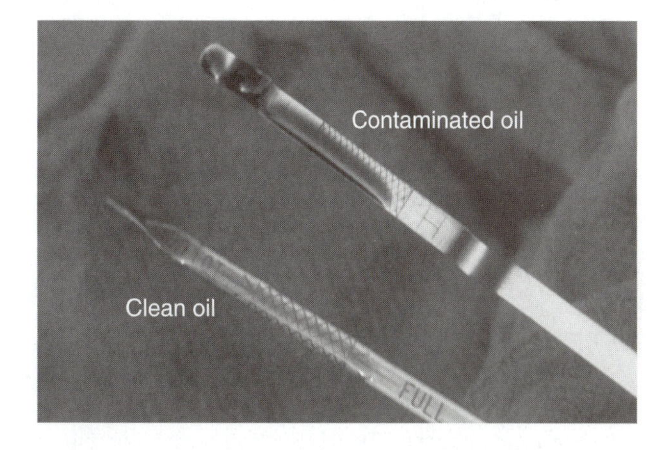

FIGURE 22-4 Clean and contaminated oil on a dipstick.

PHOTO SEQUENCE 6

Checking Engine Oil on Dipstick Equipped Engines

1. Make sure the power equipment is on a level surface before checking oil level. Wipe the area around the dipstick to prevent dirt from entering the engine.

2. Locate the dipstick on the side of the engine. Some engine makers paint the handle a bright color so it can be found easily.

3. Pull the dipstick out of its housing and wipe the measuring end clean with a clean shop rag.

4. Push the dipstick firmly back into its housing until it seats.

5. Pull it back out again. The oil level can be seen on the measuring marks of dipstick.

6. Push the dipstick back into place. Make sure it is firmly seated to keep water and dirt out of the engine.

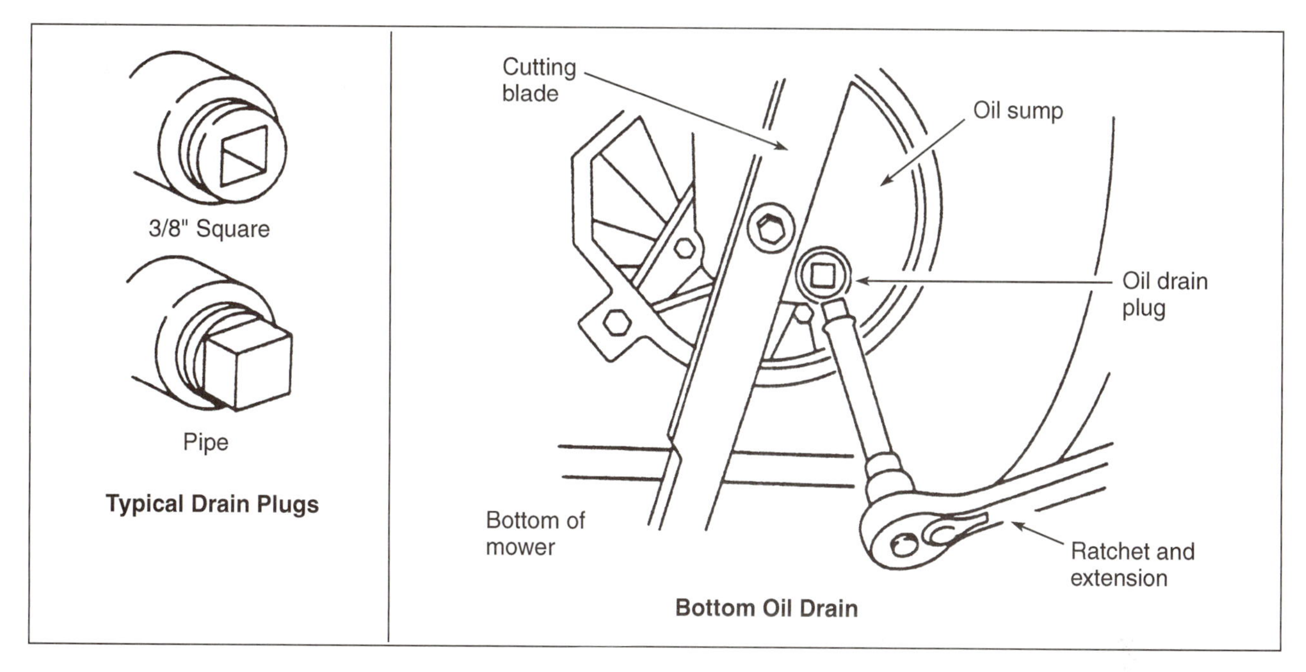

FIGURE 22-5 A drain plug used to drain engine oil out of the bottom of the crankcase. (Courtesy of MTD Products, Incorporated.)

Oil is often added to a dipstick engine through the dipstick hole or housing. The handle for the dipstick is used as an oil filler cap. Some engines use a separate oil filler cap. After checking the oil level and condition, push the dipstick back into place. Make sure it is firmly seated. This will keep water and dirt out of the engine.

Change Engine Oil and Oil Filter

Regular oil changing is the most important service job for long engine life. Hot oil inside the engine joins with oxygen. The oil can break down chemically. This is called oxidation. Water forms inside the cylinders during combustion. The water can enter the oil and cause sludge. **Sludge** is a mudlike mixture of oil and water that can plug oil passages inside the engine. There is a chemical in the oil called sulfur. If the sulfur and water join, sulfuric acid can form. The acid can damage engine parts such as bearings.

Engine oil also gets contaminated. Dirt, dust, and metal can get into the oil. Dirt and dust often contain abrasive particles that can damage engine parts. Metal particles come off engine parts as they wear. As the engine runs, oil flushes dirt and metal particles off engine parts. The abrasive dirt and metal particles get into the oil. Dirty oil circulated around the engine parts will damage the parts. The dirty oil must be drained regularly.

Engine oil should be changed after a period of engine running. This allows the oil to get hot. It also allows the dirt inside the engine to mix in with the oil. Dirt falls out of the oil when it cools.

Engine oil is drained on many engines by removing the drain plug (Figure 22-5). The drain plug is located on the bottom of the crankcase. Some larger engines have two drain plugs, one on each side of the crankcase. Either or both of these plugs may be removed to drain the oil. Other engines do not use a drain plug; these engines are tipped on their side to drain the oil out of the dipstick housing (Figure 22-6). Place an oil pan below the drain plug. Use the correct size wrench to remove the drain plug. As soon as the drain plug is removed, oil will pour out. Allow enough time for the oil to completely drain out.

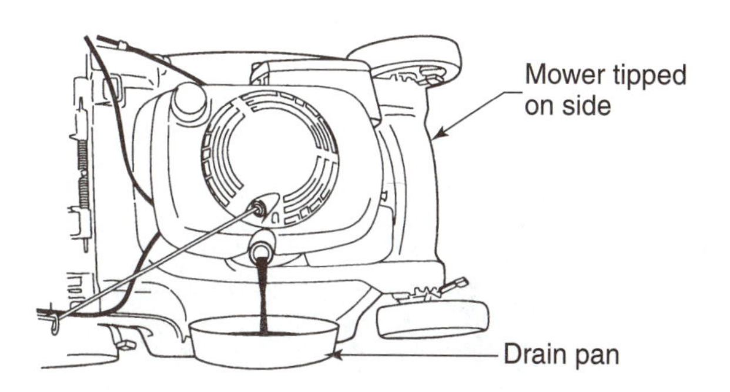

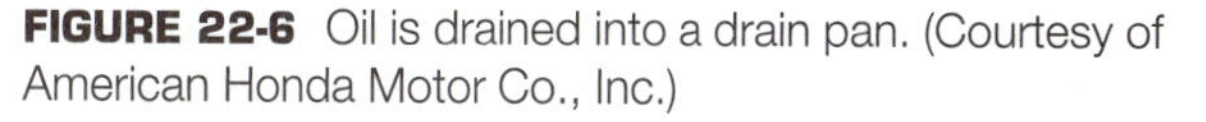
FIGURE 22-6 Oil is drained into a drain pan. (Courtesy of American Honda Motor Co., Inc.)

> **WARNING:** Do not get your hands in the stream of draining engine oil. Oil can be very hot. Always wear eye protection when changing oil to prevent hot oil from splashing into the eyes.

Clean the drain plug and check the thread area for wear (Figure 22-7). A drain plug with damaged threads does not tighten properly. A damaged plug will need to be replaced. Many smaller engines use drain plugs with self-sealing tapered pipe threads. Some larger engines have drain plugs with a sealing washer. The purpose of the washer is to prevent leaks around the drain plug. The sealing washer is often made of soft metal such as aluminum. It may have a sealant applied to its surface. A new sealing washer should always be installed on a drain plug.

Start the drain plug by hand. Then use the correct sized wrench for tightening. Some engine makers have a drain plug tightening torque specification in the shop service manual. If there is a torque specification, use a torque wrench to tighten the drain plug. Not tightening enough can cause leakage. Tightening too much can cause stripped threads.

> **CAUTION:** Do not use a large adjustable wrench to tighten a drain plug. The long wrench handle will apply too much torque to the plug. The drain plug threads may be stripped. The jaws of the adjustable wrench can round off the flats on the plug. Always use a wrench that fits the plug properly.

Some larger engines use an oil filter. The filter is usually changed each time the oil is changed. The constant heating and cooling of the filter can make it very tight and difficult to remove. The oil filter is removed by gripping the outside of the filter housing with an oil filter tool. An **oil filter tool** is a tool made to grip and turn the housing of an oil filter for removal. There are many different types of oil filter tools. They are made in different sizes and are adjustable to fit different sizes of oil filters (Figure 22-8).

FIGURE 22-7 Inspect the drain plug threads and replace the sealing washer.

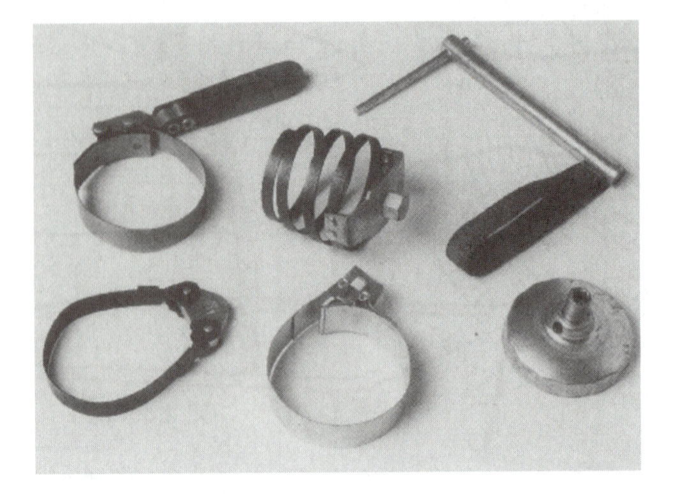

FIGURE 22-8 Different types of oil filter tools.

Place the oil filter tool around the oil filter housing and adjust it to fit the housing size (Figure 22-9). Turn the tool counterclockwise to loosen the filter until it can be turned by hand. Allow the oil to drain into the oil drain pan. Set the filter in the drain pan to drain and then dispose of it properly.

Wipe the filter mounting surface with a clean shop rag. Put a light coat of oil on the rubber gasket of the new filter (Figure 22-10) to help it seat properly. Install the filter by turning it clockwise by

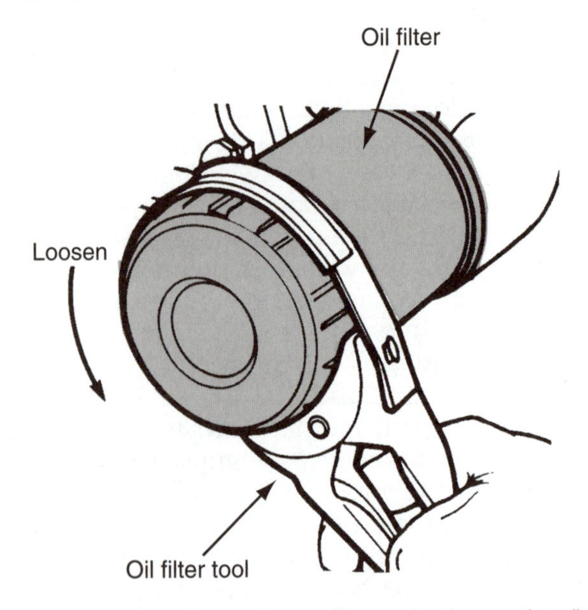

FIGURE 22-9 The oil filter tool is used to loosen the oil filter for removal. (Courtesy of DaimlerChrysler Corporation.)

FIGURE 22-10 Coat the oil filter gasket with oil and fill the filter with oil.

hand until it seats snugly against its mount. Then turn the oil filter approximately ½ to ¾ of a turn more for final tightening.

> **CAUTION:** Do not use a tool to tighten the filter, because this would cause it to be tightened too much.

> **SERVICE TIP:** If the filter mounting allows, fill the filter with engine oil before installing it on the engine. This will reduce the time required to fill the filter. Oil will get to the engine parts faster on startup. Engine wear will be prevented during the first startup after the filter change.

The crankcase is refilled with the correct type and amount of oil. On smaller engines, oil is filled through an oil fill plug. An oil filler cap or dipstick housing is used on larger engines. The correct amount of oil to refill can be found in the owner's manual or shop manual. This specification is often called *oil capacity*.

Use a clean funnel or a pouring container that will fit in the oil fill hole to prevent spilling oil on the engine (Figure 22-11). Open the container of oil and pour through the funnel into the engine. Fill with the recommended amount of oil. Most oil containers have graduation marks that allow you to measure parts of a quart. Check the oil level by looking through the oil fill plug hole or with marks on the dipstick. Install the filler cap, dipstick, or fill plug.

Start the engine and allow it to warm up to operating temperature. Check the area around the

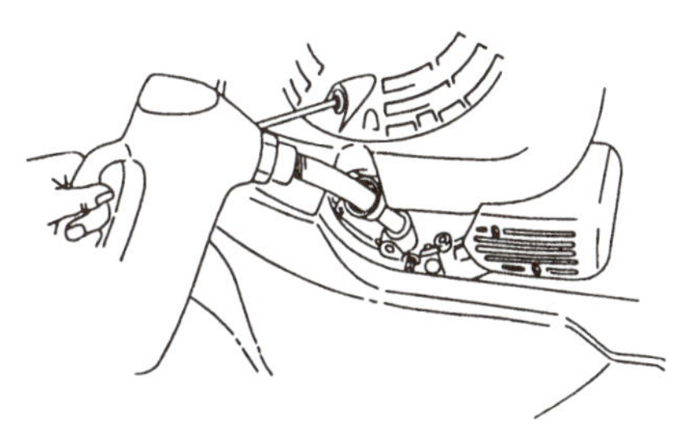

FIGURE 22-11 Fill the crankcase with the recommended type and amount of oil. (Courtesy of American Honda Motor Co., Inc.)

drain plug and oil filter for leaks. An oil filter leak often causes large amounts of oil to pour out. If the filter leaks, shut the engine off quickly. Either the filter is not tight enough or the filter gasket is defective. If the filter is not found to be loose, install a new filter. Add the correct amount of oil again.

If there are no leaks, allow the engine to warm up to operating temperature. Keep checking for leaks at the filter or drain plug. Make one final check of the oil level. When oil is pumped into the new filter, the oil level will drop. More oil may need to be added. Wipe off any oil spilled on the engine.

> **CAUTION:** Used engine oil and the used oil in an old filter must be disposed of properly. Oil thrown in the trash can end up in a landfill where it can seep into our groundwater. Dispose of the used oil and oil filter correctly so that they do not end up harming the environment. The oil and oil filter should be taken to a recycle center.

COOLING SYSTEM MAINTENANCE

The cooling system removes heat to prevent engine parts from overheating. A poor cooling system can cause engine parts to reach too high a temperature. Overheated parts can fail, causing total destruction of the engine. Engine overheating can also cause breakdown of engine oil and loss of lubrication. Engines need regular cooling system service to prevent these problems.

Many engines are installed on equipment like lawn mowers and chain saws that run where there is a lot of dirt or grass in the air. Some of this dirt and grass gets through the blower housing debris screen. It can end up in the cooling fins. The debris screen and cooling fins can get so plugged up that air cannot pass through properly. Heat is not taken away, and the engine parts can overheat.

> **WARNING:** Engine parts are hot when the engine has been running. Always allow the engine to cool down before working on the cooling system.

Smaller engines use a stationary screen on the blower housing. Larger engines use a screen that rotates with the flywheel. On some engines, air enters through vents in a finger guard assembly on the rewind starter. Check and clean the screen before each use of the engine (Figure 22-12).

The engine will overheat if the cooling fins are not cleaned regularly. The blower housing is removed to clean the cooling fins (Figure 22-13).

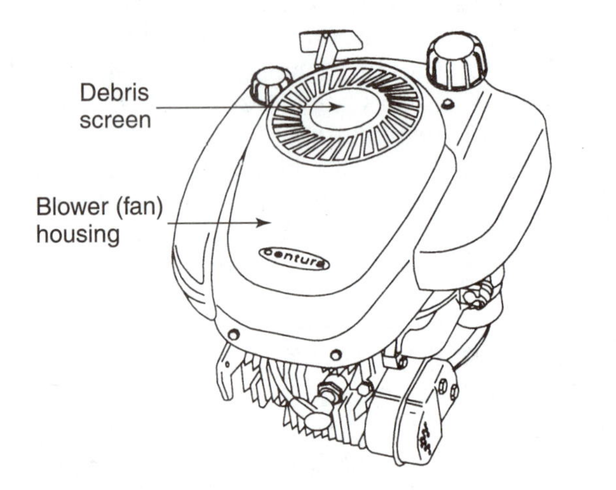

FIGURE 22-12 Check and clean the debris screen before each engine use. (Provided courtesy of Tecumseh Products Company.)

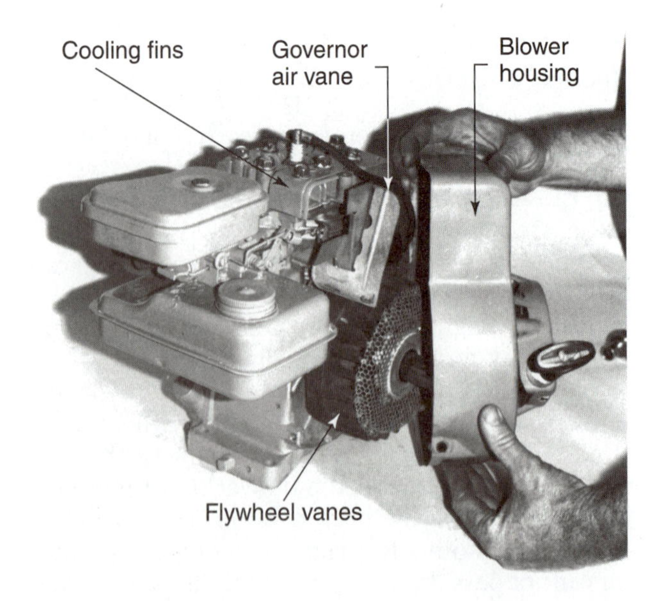

FIGURE 22-13 Cooling areas to check and remove debris.

Remove the blower housing by removing the screws that hold it to the engine. Then lift the blower housing off the engine. Wipe the inside of the blower housing with a shop rag to remove grass and dirt. Check the cooling fins on the cylinder head and cylinder. Use a wire brush or shop rag to remove all the debris. Sometimes a buildup of debris can prevent movement of the air vane governor. Clean this area. Move the air vane back and forth to check for free movement. Replace the blower housing and install the screws.

CRANKCASE BREATHER MAINTENANCE

A crankcase breather is a valve usually located on the side of the engine. It is used to allow crankcase pressure to escape into the engine air intake. Multiple cylinder engines may use more than one breather assembly. Overhead valve engines may use breathers in the valve covers.

After a period of time, the crankcase breather valve disc may get covered with sludge. The valve may stick open or closed. The valve assembly should be inspected at regular intervals. Some valves may be disassembled for service. Others are replaced instead of serviced.

The valve assembly is often mounted with one or more screws. Remove the screws and remove the breather. Inspect the breather disc valve (Figure 22-14). If it is covered with sludge, wash it in solvent. Replace the breather if the valve shows signs of wear or will not open and close easily. Check the breather tube for cracks or tears and replace as necessary. Use new gaskets and seals when replacing the breather assembly.

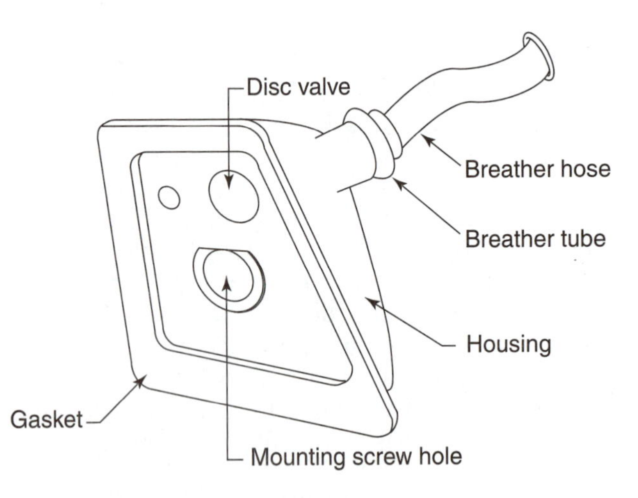

FIGURE 22-14 Check the operation of the crankcase breather disc.

SERVICE TIP: A problem with the crankcase breather can cause too much pressure to build up in the crankcase. Crankcase oil can be forced out of the crankcase seals and gaskets. Be sure to check the breather anytime you find oil on the outside of an engine. Always check the valve before replacing gaskets or seals.

AIR CLEANER MAINTENANCE

The air filter element in the air cleaner traps dirt and debris before it can enter the engine. As the air cleaner traps debris, it begins to fill up. The air filter element can become clogged if it is not changed regularly. The air cleaner can fill up very fast when the air around the engine is dirty. A plugged air cleaner will slow down the air flow into the engine and cause low engine power.

CAUTION: Always use a shop rag to clean the area around the air cleaner before removal to prevent dirt from falling into the carburetor intake when the air cleaner is lifted off.

Engine makers specify a regular air filter element change or cleaning time in the owner's manual or shop service manual. The air cleaner is serviced by removing the filter element from the air cleaner housing (Figure 22-15). Air cleaners are mounted to the engine in several ways. The foam and paper air cleaners often use a cover screw that goes through the air cleaner cover and into the air cleaner housing. Some filters are held in place with a nut. The nut fits on a stud on the air cleaner base. A wing nut is often used so that a wrench is not required. Larger air cleaners used on multiple cylinder engines may have a latch. A latch allows quick disassembly.

Oil Bath Air Cleaner Maintenance

The oil bath air cleaner has two main parts, a cover and a bowl (Figure 22-16). Remove the nut or screw attaching the cover. Lift the cover off the bowl. Lift the bowl off the carburetor. Properly dispose of the dirty oil in the bowl. The cover has a wire screen inside. Clean the screen in solvent and coat it with engine oil. Wash the bowl in solvent and wipe it clean with a shop rag. Place the bowl back on the carburetor. Pour clean engine oil into the bowl. There is a line on the bowl marked "oil level." Fill the bowl

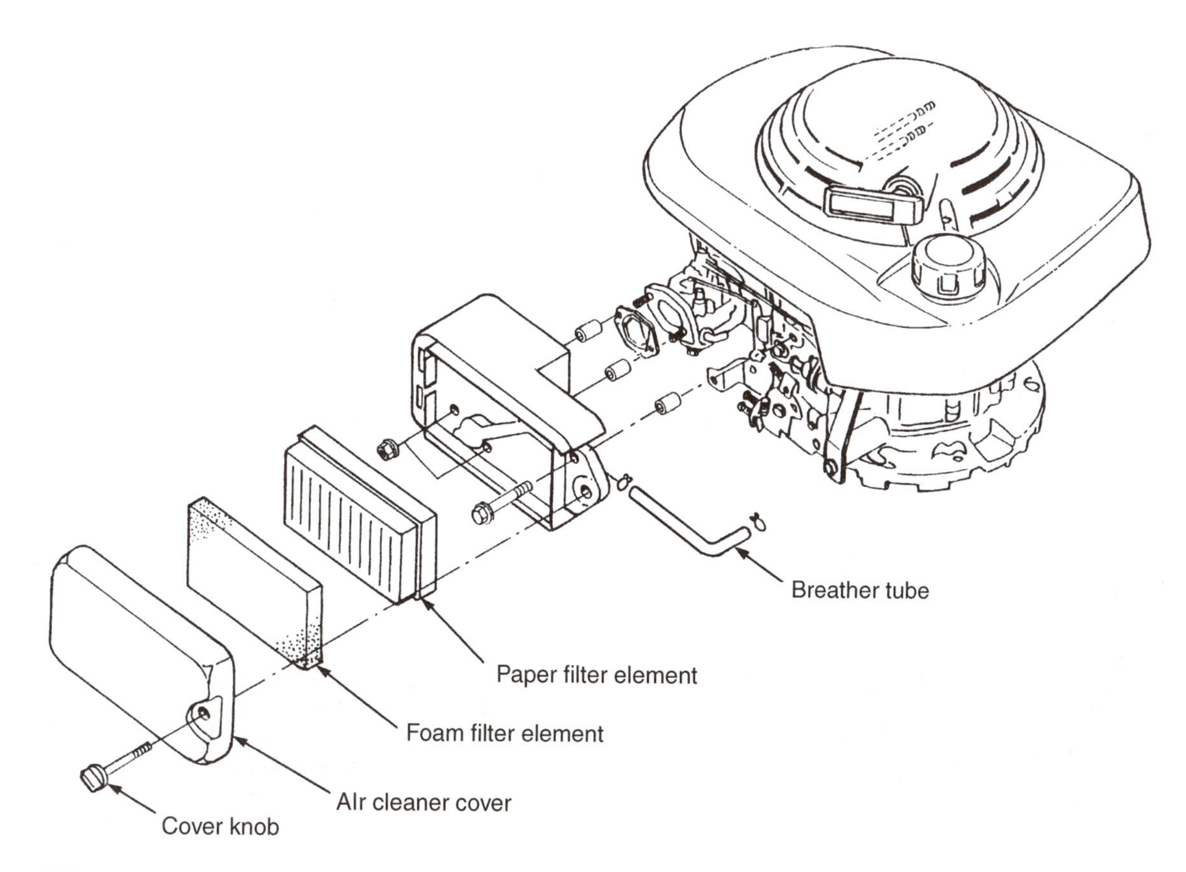

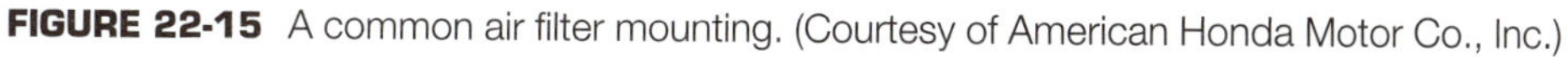

FIGURE 22-15 A common air filter mounting. (Courtesy of American Honda Motor Co., Inc.)

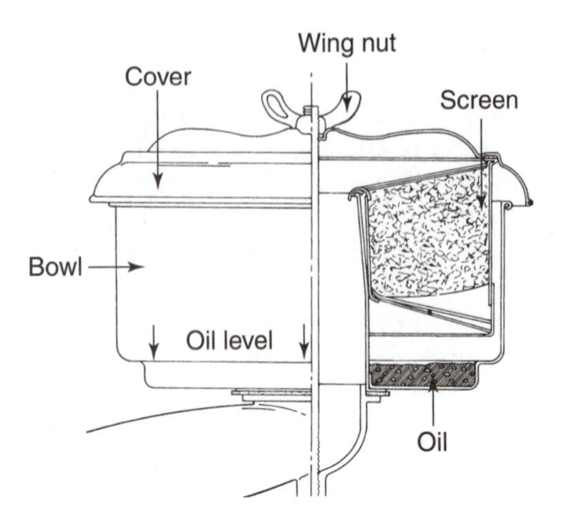

FIGURE 22-16 Parts of an oil bath air cleaner assembly. (Courtesy of Kohler Co.)

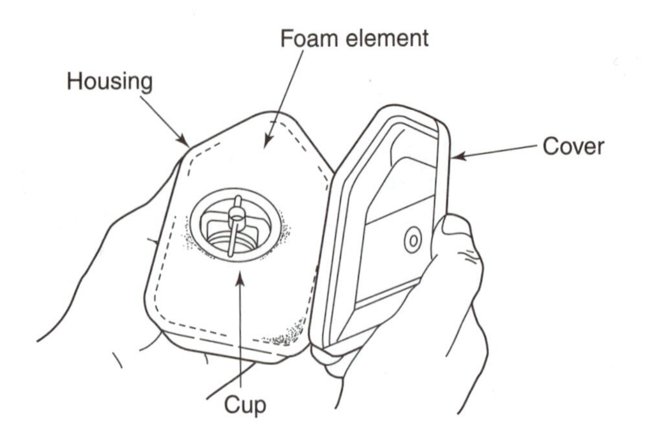

FIGURE 22-18 Parts of a foam air cleaner assembly.

to this level. The bowl should not be filled over the line. Intake air can cause oil to be pulled into the carburetor. Replace the cover and replace the screw or nut.

Foam Air Cleaner Maintenance

The foam filter element is held between the housing and cover. A metal cup supports the center part of the foam element. Remove the retaining screw. Lift off the cover, element, and housing as a unit (Figure 22-17). Remove the cover and pull the element and cup out of the housing (Figure 22-18). Remove the cup from the element.

The foam element can dry out and break apart (Figure 22-19). Carefully check it for any tears or broken areas. Also look for a lot of dirt buildup on the filter element. A dirt buildup makes the filter working area smaller. Filters that show this type of wear should not be cleaned. These should be replaced. Match up the replacement filter to the old one. It must be exactly the right size to fit into the air cleaner housing.

If the filter element is in good shape, wash the foam in the soapy water. Squeeze the foam in a dry rag to remove the water. Place it on a clean rag and allow it to air dry. When dry, pour a small amount of engine oil on the foam. Squeeze the foam in a clean shop rag to get out most of the oil. Wash the cover, body, and cup in solvent and dry it with a clean shop rag. Put the parts together and install them on the engine.

FIGURE 22-17 Removing a foam air cleaner assembly.

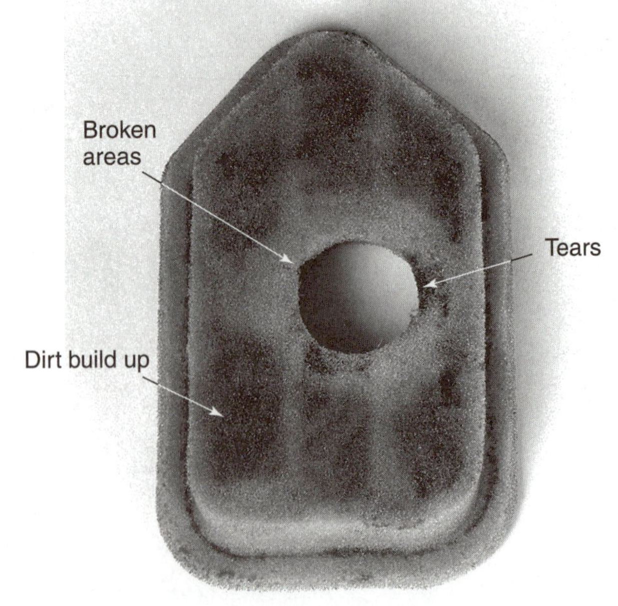

FIGURE 22-19 Check the foam air filter for tears, broken areas, and dirt.

Paper Air Cleaner Maintenance

The paper air cleaner has a cover (housing), a paper filter element (cartridge), and a base (Figure 22-20). Remove the screw, nut, or latch and remove the cover. Lift out the paper air filter element. It is normal to find dirt on the outside of the filter element. This is the material that has been trapped by the filter. Often, dirt on the filter can be removed by tapping the filter lightly on a bench surface. You can insert a light bulb on the inside of the filter element. Look through the element toward the light. You will see any tears or plugged areas of the filter element.

Look for a damaged seal or gasket on the sealing areas of the element (Figure 22-21). Tears on the element and damaged seals or gaskets are signs that the filter may have allowed debris into the engine. Look for areas on the element that are wet with oil. These can mean that the crankcase breather is allowing too much oil to get into the air cleaner. If you find these problems, the filter should be changed.

CAUTION: Paper filters cannot be cleaned in solvent. Solvent causes the paper to come apart. Do not use shop air to blow out a filter. The air pressure damages the filter paper.

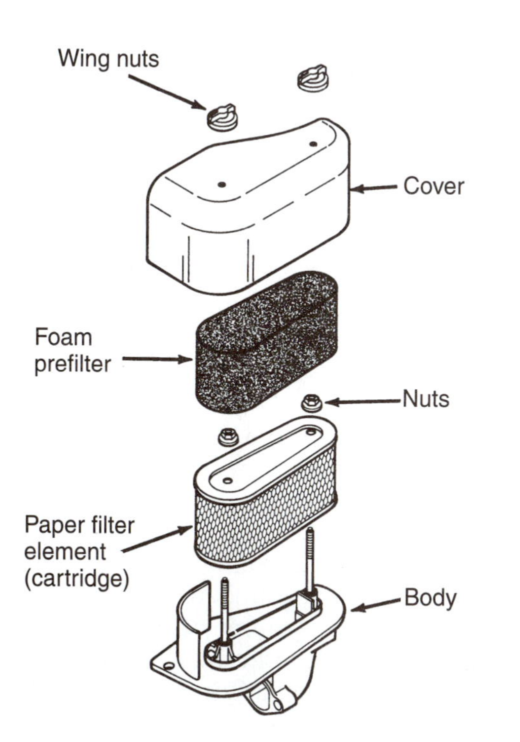

FIGURE 22-20 Parts of a paper air filter assembly. (Provided courtesy of Tecumseh Products Company.)

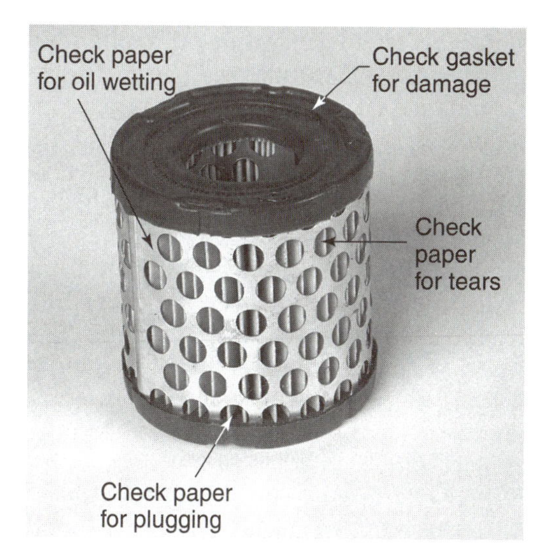

FIGURE 22-21 Check the paper air filter for oil wetting, plugging, tears, and gasket damage.

Place the new filter element next to the old one and compare the two filter elements. Both must be exactly the same size. Make sure the gaskets on the top and the bottom of both filter elements are exactly the same. Place the new air filter element on the base and install the cover. The gasket surface must be aligned on both the top and bottom. The shop service manual may say to oil or grease the gasket to improve its seal. Tighten the latch, screw, or nut until snug.

CAUTION: The air filter gasket must fit correctly and seal properly. A leak at the gasket means that air will go directly into the engine around the gasket without going through the filter element. Abrasives can get into the engine and shorten engine life.

Many engines use an outer foam precleaner and an inner paper air filter in combination. The foam precleaner is serviced just like the foam air cleaner. The paper filter is cleaned just like the standard paper filter.

FUEL FILTER MAINTENANCE

WARNING: Be careful not to spill gasoline on a hot engine part when changing a fuel filter. It could ignite and result in a fire. Collect gasoline in a metal container and dispose of it properly.

Many engines have an in-line fuel filter between the fuel tank and carburetor. The fuel filter can become plugged with dirt that gets in the gasoline. Clean fuel is very important in any fuel system. Dirt can plug small carburetor fuel passages. Regular change times for the fuel filter are usually listed in the owner's manual or shop service manual.

Fuel filters are changed by removing the filter fuel lines. Flexible nylon or synthetic rubber fuel lines are often used to connect fuel filters. The connector pipe on the end of the fuel filter has sealing rings. The fuel line fits over the sealing rings. A clamp is used behind the sealing rings to keep the fuel line in place. The clamp can be a spring, worm drive, or rolled edge type. The spring type is removed and replaced with pliers. The other types are removed and replaced with a screwdriver.

If the fuel tank has a shut off valve, turn the valve to off. Many tanks do not have a shut off valve. Use vise grip pliers to clamp the tank outlet fuel line closed. This will prevent fuel leakage from the tank when lines are disconnected. Place a metal container under the filter to catch gasoline that leaks out of the filter. Remove the two clamps and separate the fuel lines from the filter (Figure 22-22).

Compare the old filter to the new one to be sure it is the correct part (Figure 22-23). The filter must be installed in the correct direction in the fuel line. Some filters have an arrow that shows fuel flow direction. The arrow should point toward the carburetor. Some filters have the words "inlet" and "outlet" to show fuel flow direction. The outlet end goes toward the carburetor.

Inspect the fuel lines before you install the filter. Flexible fuel lines can get brittle as they age. When they get brittle, they can crack and leak. Replace any fuel lines that have cracks. Always use the recommended type of fuel line hose when replacing fuel lines.

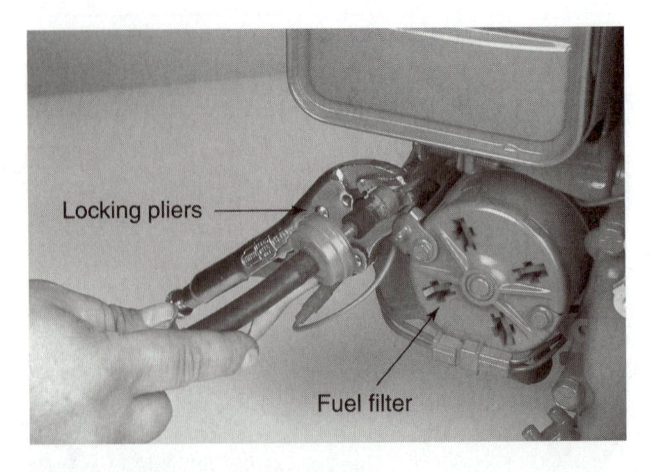

FIGURE 22-22 Removing the fuel filter fuel lines to remove the fuel filter.

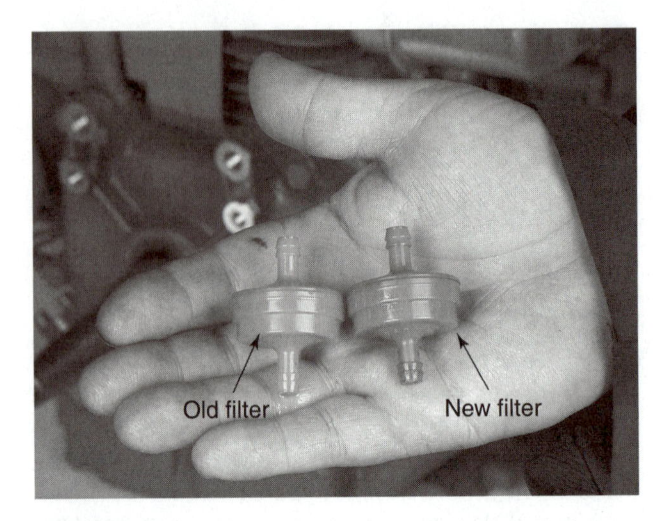

FIGURE 22-23 Compare the old and new filters to be sure they are the same.

Install the new filter by pushing the fuel lines on to the inlet and outlet. Install the lines far enough on the connector to get a good seal. Install the clamps. Be sure the clamps are in the correct place and tightened securely. Start the engine and check all the connections for leaks. Run the engine for several minutes and watch for leaks. If there are leaks, check the clamps for tightness.

SERVICE TIP: You may find a lot of sediment in the fuel line from the carburetor to the fuel filter. This sediment will quickly plug the new fuel filter. Remove and flush the fuel tank. Remove the outlet fitting and screen and flush the tank with clean gasoline. Reassemble the tank with a new or cleaned outlet screen.

VALVE CLEARANCE ADJUSTMENT

Valve clearance is a small space between the valve lifter and valve stem or rocker arm and valve stem. Valve clearance allows for heat expansion in the valve parts. Many less expensive engines have the valve clearance set by grinding the end of the valve stems. This is done during engine assembly and is not adjustable after engine assembly. More expensive engines have an adjustable valve clearance. This type must be regularly checked and adjusted as necessary. Some shop service manuals specify this adjustment as part of a tune-up. Incorrect valve clearance can affect how the engine runs.

If the valve clearance is too wide, the engine will be noisy. Excessive valve clearance can cause wear to camshaft and valve lifter contact areas. Push rods may also be bent. If the clearance becomes large enough, valve timing will be changed and the result will be poor engine power.

Valve clearance that is too narrow can cause even more serious problems. Too little valve clearance does not allow for heat expansion. The valve will not be able to close on its seat. This prevents the valve from sealing the combustion chamber. The result will be low compression and poor engine power. The valve will also become very hot. Heat will not be able to move away from the valve head into the valve seat. The high temperatures can melt the valve head.

Check the steps in the shop service manual before valve clearance is measured and adjusted. The shop service manual will give a step-by-step job for the valve adjustment. It also gives the specifications for clearance. Some valve mechanisms must be adjusted only when the engine is cold. Others must be adjusted only when the engine is hot. Be careful to follow the correct steps. Engine temperature changes the clearance.

Some engines have different clearance specifications for the intake and exhaust valves because of higher exhaust valve temperatures. Different metals with different expansion rates are often used in the two valves. If this is the case, the exhaust valve clearance will normally be larger.

The valve clearance is measured when the piston is at top dead center (TDC) on the compression stroke. At TDC both valves are fully closed. The valve lifter is resting on the heel, or low point, of the cam lobe. The shop service manual shows how to rotate the crankshaft to get the piston at TDC. Many engines have an alignment mark on the flywheel. The mark is lined up with another line on part of the engine when the piston is at TDC (Figure 22-24).

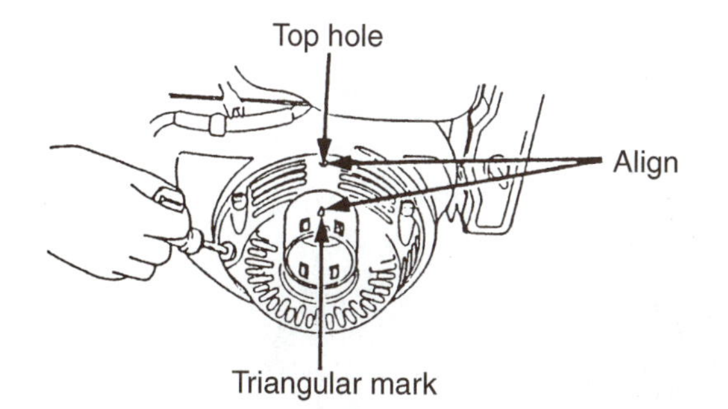

FIGURE 22-24 Alignment marks on the engine and crankshaft are used to align the piston at top dead center. (Courtesy of American Honda Motor Co., Inc.)

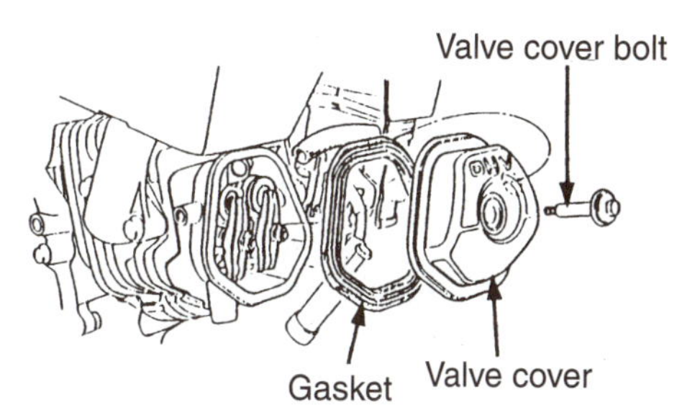

FIGURE 22-25 Removing the valve cover. (Courtesy of American Honda Motor Co., Inc.)

Overhead valve engine valve clearance is often adjusted with adjustable rocker arms. Remove the nuts or screws and remove the valve cover (Figure 22-25). The clearance is measured using a feeler gauge (Figure 22-26). Select a feeler gauge that is the same as the clearance specification. Try to slide the feeler gauge into the clearance space. If the feeler gauge blade will not slide into the space, the clearance is too small. If the feeler blade fits into the space too loosely, there is too much clearance. The feeler gauge should fit in and out with a slight drag.

The valve clearance is usually adjusted by turning a rocker arm pivot nut in the center of the rocker arm. Loosen the lock nut and turn the

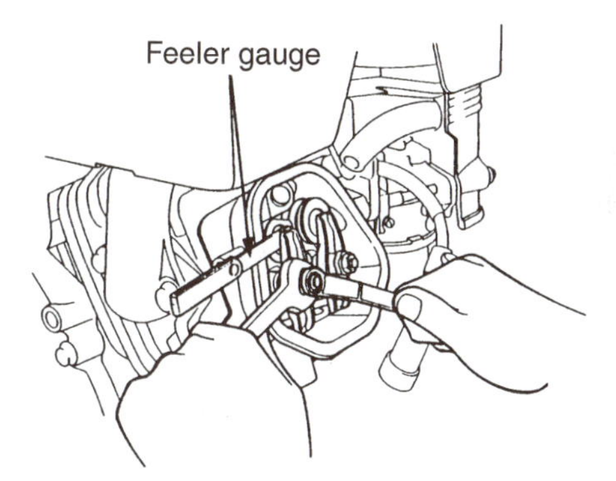

A. Adjusting Clearance

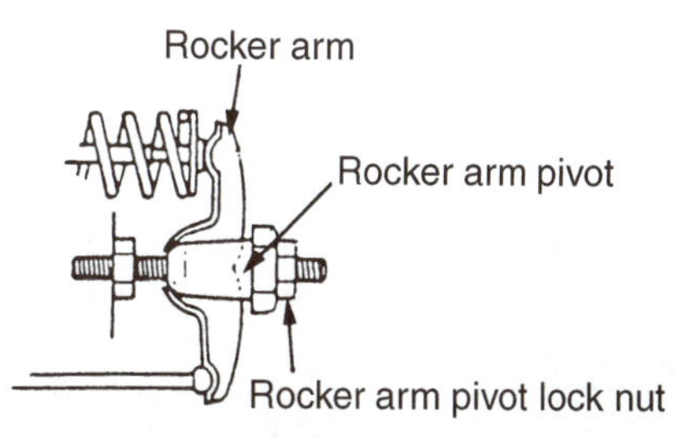

**B. To Increase Valve Clearance, Screw Out.
To Decrease Valve Clearance, Screw In.**

FIGURE 22-26 Valve clearance is measured with a feeler gauge and adjusted by turning the rocker arm pivot. (Courtesy of American Honda Motor Co., Inc.)

PHOTO SEQUENCE 7

Adjusting Overhead Valve Clearance

1. Allow engine to cool. Remove and ground the spark plug cable for safety. Use a screwdriver to loosen and remove the valve cover retaining screws.

2. Remove valve cover and old valve cover gasket.

3. Rotate engine and observe rocker arms. Position piston at top dead center, both valves fully closed. There should be clearance between the rocker arms and valve stems.

4. Look up the specifications for intake and exhaust valve clearance in the shop service manual.

5. Insert the correct size feeler gauge between the intake valve stem tip and the rocker arm.

6. If necessary, loosen the pivot lock nut and turn the pivot screw out to increase clearance or in to decrease clearance.

7. Insert the correct size feeler gauge between the exhaust valve stem tip and the rocker arm.

8. If necessary, loosen the pivot lock nut and turn the pivot screw out to increase clearance or in to decrease clearance.

9. Install a new valve cover gasket and install the valve cover. Install the spark plug cable.

adjustment nut in to make the clearance smaller. Turn the nut out to make the valve clearance larger. Use one wrench to turn the adjuster. Use another wrench to loosen or tighten the lock nut.

Use a new valve cover gasket when installing the valve cover. Reusing the old gasket may cause an oil leak. Replace the valve cover nuts or screws. Tighten the nuts or screws according to the steps in the shop service manual. Sometimes seals are installed around the valve cover fasteners. Always use new seals to prevent oil leaks. A typical procedure for adjusting the valves on an overhead valve engine is shown in the accompanying sequence of photographs.

OFF-SEASON STORAGE

Sometimes engines are used on equipment that is operated only during a certain season. A lawn mower may be used only during the spring and summer. A snow thrower may be used only during the winter months. When an engine is not going to be used for a period of months, it must be prepared for storage as shown in the accompanying sequence of photographs.

Clean the Equipment

The equipment should be cleaned before it is stored. Use a degreaser or soap and water to clean the equipment. Dirty engine and equipment parts may attract moisture. Moisture will cause part corrosion and rust.

CAUTION: Care must be taken when using a high-pressure washer or steam cleaner to wash equipment. Steam and washing chemicals can get into electrical parts like the magneto and kill switch. These parts can be damaged by moisture and cleaning chemicals. High-pressure cleaning can also remove engine decals that are required for engine identification and service.

Oil Internal Parts

Lubricating oil will run off the piston rings and cylinder walls during long periods of nonuse. Rust can form on the cylinder walls and piston rings. The rust may cause the piston rings to stick (seize) to the cylinder wall. The rust will prevent the engine crankshaft from rotating. Rust on the cylinder wall can score (scratch) the rings when the engine is restarted.

Turn the kill switch to the off position. Disconnect the spark plug wire and ground the wire for safety. Remove the spark plug. Rotate the engine crankshaft until the piston is at TDC on the compression stroke. Fill an oil squirt can with clean engine oil. Use the squirt can to put several squirts (one tablespoon) of oil in the combustion chamber through the spark plug hole (Figure 22-27). Replace the spark plug and the spark plug wire.

Internal engine parts can be protected from rusting by fogging. **Fogging** is a method of coating an engine with oil through the carburetor to prevent rust during storage. Fogging requires an old

PHOTO SEQUENCE 8

Preparing an Engine for Off-Season Storage

1. Use a degreaser or soap and water to clean the equipment.

2. Drain the fuel from the fuel system. Start the engine and allow it to run until the carburetor runs out of fuel.

3. Disconnect the spark plug wire and ground the wire for safety.

4. Remove the spark plug. Rotate the engine crankshaft until the piston is at TDC on the compression stroke.

5. Use a squirt can to put several squirts (one tablespoon) of oil in the combustion chamber through the spark plug hole.

6. Replace the spark plug and the spark plug wire.

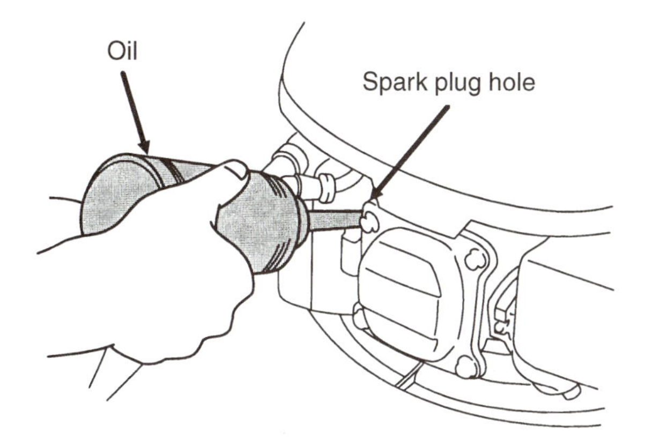

FIGURE 22-27 Squirt oil into the cylinder to prevent rusting during storage. (Courtesy of American Honda Motor Co., Inc.)

spark plug and a squirt can of new engine oil. Install the old spark plug. An old spark plug is used because fogging causes oil fouling Remove the air cleaner. Start the engine and allow it to run at part throttle. Squirt oil into the carburetor throat until the engine stalls. The oil entering the intake will coat the engine parts and prevent rust.

When the engine is restarted after storage, the oil will cause the exhaust to smoke. The smoke will clear up after a period of engine operation. If there is too much oil in the cylinder the spark plug may be fouled and will have to be replaced.

Drain the Fuel Tank

All the fuel from the engine's fuel system should be removed before storage. Gasoline should not be allowed to remain in a fuel system a long time. Oxygen in the air causes gasoline to oxidize. Oxidation can turn the gasoline into a thick gum or varnish. Gasoline that has oxidized smells like paint or varnish. The oxidized fuel will plug carburetor passages and prevent proper carburetor operation.

Some equipment makers recommend a fuel stabilizer be added to the fuel to prevent oxidation. Fuel with stabilizer does not have to be drained for short-term storage. Check the storage instructions in the owner's manual before using fuel stabilizer. Follow the directions on the fuel stabilizer (Figure 22-28).

> **WARNING:** Do not drain the fuel out of an engine that is hot from running. Gasoline can ignite from contact with hot engine parts. Do not store equipment with gasoline in the fuel tank in any heated garage or basement.

FIGURE 22-28 Fuel stabilizer is added to fuel to prevent oxidation.

Some engines have a drain plug on the bottom of the carburetor (Figure 22-29). Other engines require that you remove the fuel line between the fuel tank and carburetor. Allow the fuel to drain into a metal container. Engines with suction carburetors do not have a fuel line. These tanks are drained by removing the tank cap and turning the engine on its side. Allow the fuel to flow out the tank into a metal container. Start the engine and let it run until the engine runs out of fuel. This will remove all the fuel in the carburetor. Dispose of the fuel properly.

Remove the Battery

If the equipment has a battery, it should be removed. Store the battery in a warm, dry area on a

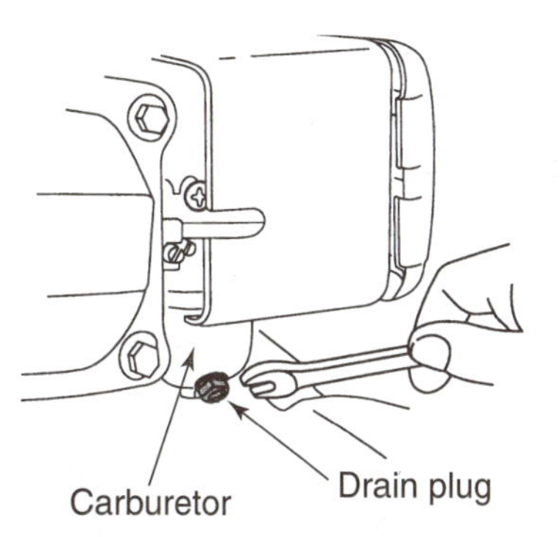

FIGURE 22-29 A carburetor with a drain plug used to remove fuel for storage. (Courtesy of American Honda Motor Co., Inc.)

wooden board. Do not store it on concrete. The dampness of concrete will cause it to discharge. If possible, store the battery connected to a battery charger. Do not allow the battery to freeze. Frozen battery electrolyte will break the battery case.

BATTERY MAINTENANCE

Storage batteries are used on engines with electrical starting systems. The battery needs regular checking and service. The battery is a common source of starting system problems. If it fails, the engine will not crank over for starting.

WARNING: Wear eye protection when working on a battery. Always wear personal protective equipment to prevent injury from electrolyte splash that can injure skin and eyes. Never smoke or create any spark around a battery or it might explode.

Battery Visual Checks

Anytime an engine fails to crank or cranks slowly, begin by checking for signs of damage. The battery should be checked for:

- A cracked or bulged case or cover
- Signs of electrolyte leakage from the case or cover
- Damaged insulation on battery cables
- Corrosion buildup on terminals and posts
- Loose or missing holding hardware
- Electrolyte level (if the battery has cell vent caps)

The battery must be replaced if it has a cracked case or electrolyte leakage (Figure 22-30). Damaged cables should be replaced. Corrosion can be cleaned off the post and terminals. Check the top of the battery for dirt or electrolyte. Electrolyte on the top of the battery may be the result of overfilling. Dirt on top of the battery can cause it to discharge when the equipment is not in use. Check the battery hold-down hardware to be sure it is tight and in place. Missing or loose hold-down hardware can allow the battery to vibrate when the engine is running. Vibration can damage the inside parts of the battery.

Electrolyte level checking is not possible or necessary on maintenance-free batteries. The electrolyte level should be checked monthly on batteries with cell vent caps. During warm weather equipment use, these batteries lose water out of the cells. If the water drops too low, the battery will fail.

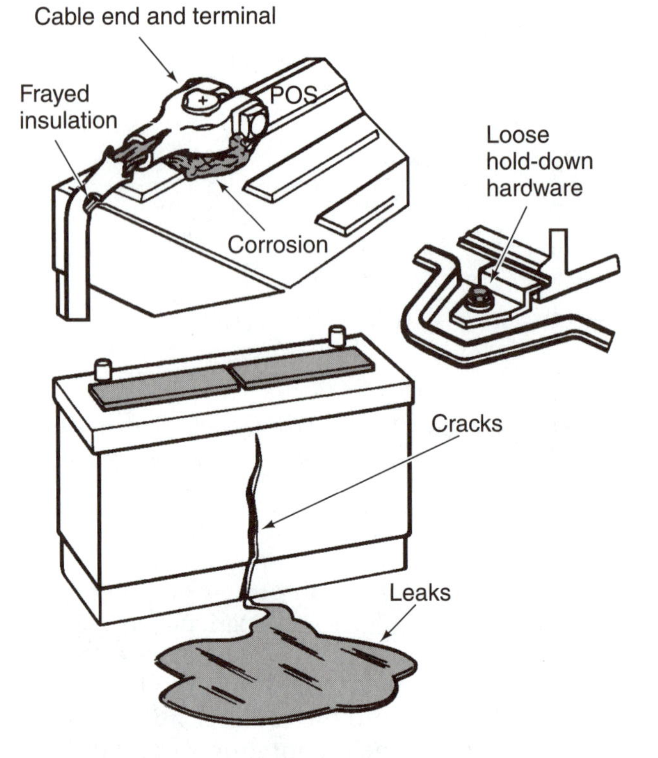

FIGURE 22-30 Inspect the battery for corrosion, damage, or loose hold-down hardware. (Courtesy of DaimlerChrysler Corporation.)

Older batteries have cell vent caps. Low maintenance batteries have a pry-off cell cover. Remove the cell vent caps by unscrewing them. Use a screwdriver to pry off the cell cover. The electrolyte level should be higher than the plates. Batteries have a guide ring built into the top of the case above the cell. The electrolyte level should be at the level of the guide ring.

Add water to the cell if the electrolyte level is low. Always use distilled water because regular tap water may have a high salt and mineral content. These materials can damage the battery. Add the water with a plastic or rubber container made for battery filling (Figure 22-31). Do not use a metal funnel because metal could cause a short between the plates. Be careful not to overfill. Overfilling will dilute the electrolyte. It can also cause an acid buildup on the outside of the battery.

Clean Battery Case and Cables

WARNING: Always wear eye protection when cleaning a battery or battery cable. Always wear personal protective equipment to prevent injury from electrolyte splash that can injure skin and eyes.

FIGURE 22-31 Filling a battery cell with distilled water.

Clean the outside of the battery when there is a buildup of dirt or corrosion. Dirt, moisture, and corrosion on the top of the battery provide a path for current flow between the battery terminals, which can cause the battery to self-discharge. If the battery has cell vent caps or a pry-off strip, they must be installed tightly before cleaning the battery.

> **WARNING:** Prior to disconnecting any hold-down hardware, the battery must be disconnected from the circuit. If this is not done, an accidental touch of a wrench between the positive terminal and ground could create a direct short. The short could cause a burn or a possible battery explosion. Always disconnect the ground cable first, because this cable is less likely to cause a spark when it is connected or disconnected.

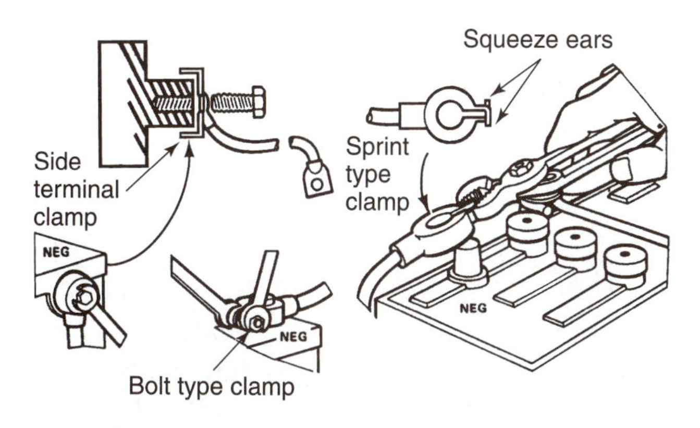

FIGURE 22-32 Removing a battery cable connection. (Courtesy of DaimlerChrysler Corporation.)

Remove the battery from the battery tray or mount before cleaning. Always use the correct size wrench to loosen battery posts. Adjustable wrenches or pliers can damage the fasteners. Remove screw type clamps by holding the screw end and driving the nut end with a wrench. Remove spring type clamps by squeezing the ears on the terminal. This will release the terminal from the post (Figure 22-32).

Sometimes a battery terminal puller is needed to pull the terminal off the battery post. A **battery terminal puller** is a special puller made to pull battery terminals off of battery posts (Figure 22-33).

> **WARNING:** Use a battery carry tool to carry the battery. Avoid holding the battery against the body when carrying because electrolyte can get on skin or clothing and cause injury.

Remove all the battery hold-down hardware. Use a properly adjusted battery carry tool to lift and remove the battery from the tray. A **battery carry tool** clamps around a battery and allows carrying the battery by the tool handle (Figure 22-34). Place the battery in an area where it can be safely cleaned.

Mix a battery cleaning mixture of one tablespoon of baking soda and one quart of water. This mixture will neutralize the battery electrolyte (acid). Brush the mixture over the battery case (Figure 22-35). Allow the mixture to sit for a few minutes. Flush the case off with water. Dry the battery with paper towels and dispose of the towels properly.

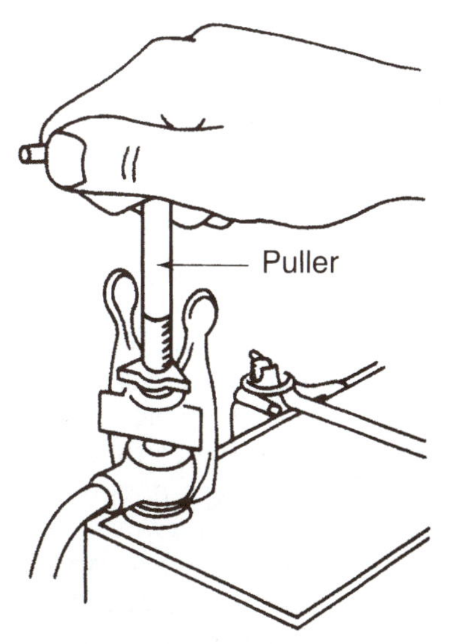

FIGURE 22-33 Using a puller to remove a battery cable. (Courtesy of DaimlerChrysler Corporation.)

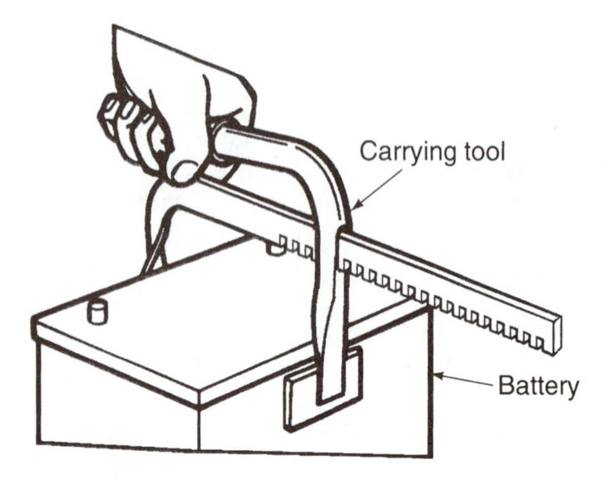

FIGURE 22-34 Carry the battery with a battery carry tool. (Courtesy of DaimlerChrysler Corporation.)

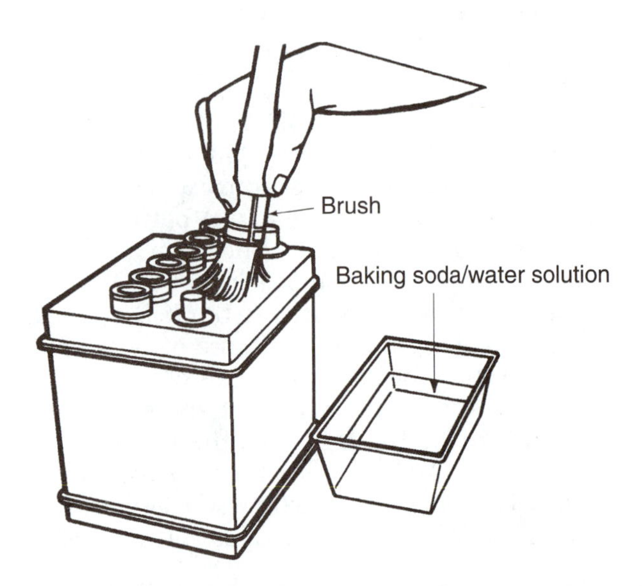

FIGURE 22-35 Clean the battery with a solution of baking soda and water. (Courtesy of DaimlerChrysler Corporation.)

Use a wire brush to remove the corrosion and flaking paint from the battery tray and hold-down hardware. Brush the baking soda mixture over the parts and allow them to dry. When the parts are dry, paint them with a rust-resistant paint.

There must be a good electrical contact between the battery posts and the battery cable terminals. Corrosion on the terminals and posts increases electrical resistance and can cause starting problems. A corrosion buildup destroys the cable connections. Clean the cable terminal and battery post with a battery terminal wire brush. A **battery terminal brush** (Figure 22-36) is a wire brush used to clean corrosion from the battery post and battery cable terminal. One side of the brush is made to clean inside the cable terminal. The other side is made to fit over and clean the battery post. Use the baking soda and water mixture to remove the corrosion. Be sure the post and terminal contact surfaces are clean and bright after cleaning.

FIGURE 22-36 Clean battery cable terminals with a battery terminal brush.

Place the battery back into the tray using the battery carry tool. Replace the hold-down hardware and tighten the fasteners properly. Hold-down hardware that is too loose can allow the battery to vibrate. Hardware that is too tight can break the battery case.

WARNING: When connecting a battery to a circuit, always connect the negative terminal last to prevent sparks and a possible battery explosion.

First, install the positive cable terminal on the positive battery post. Then install the negative cable terminal on the negative battery post. The terminals (after installation) and hardware may be coated with a protective spray to resist corrosion. Do not allow the protective spray to get between the terminal and post. This can cause excessive resistance. Test the battery connections by trying the starter.

Test Battery Open Circuit Voltage

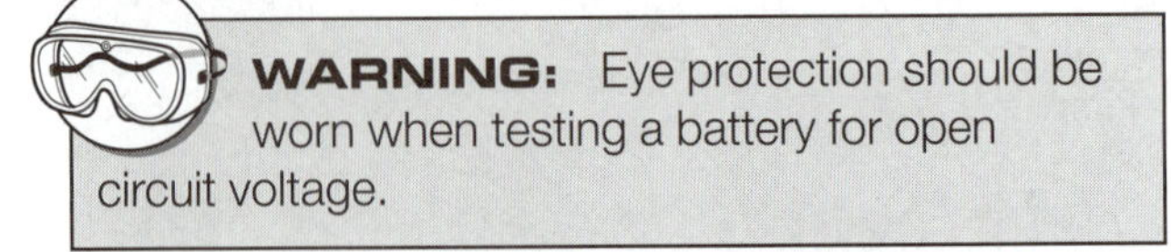

WARNING: Eye protection should be worn when testing a battery for open circuit voltage.

An **open cell voltage check** is a battery test used to determine if a battery is fully charged. This is often called *state of charge*. An open cell voltage test is done with a digital volt-ohmmeter. The battery temperature must be between 60 and 100 de-

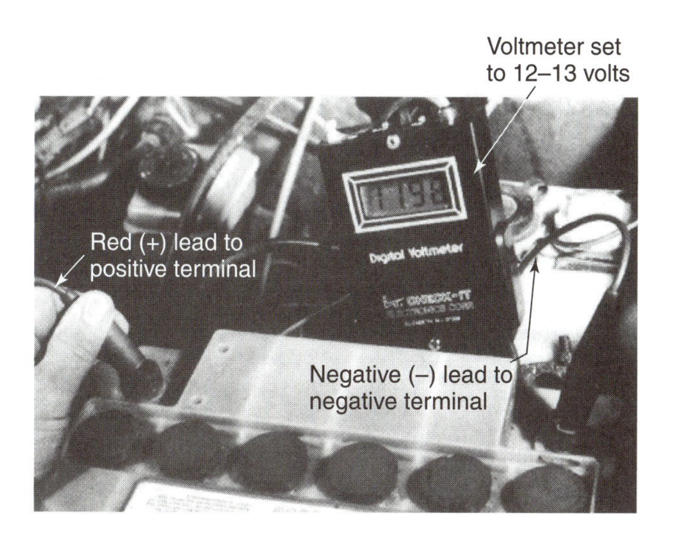

FIGURE 22-37 Testing battery open circuit voltage to determine state of charge.

grees before doing this test. A very hot or cold battery will give bad test results. If the battery has just been on a charger, wait at least 20 minutes after charging to do the test. This will allow the battery voltage to settle down.

Disconnect the ground battery cable from the negative battery post. Turn on the digital voltohmmeter. Select the correct voltage scale to measure 12 to 13 volts. Connect the red (positive) test lead from the voltmeter to the positive battery terminal. Connect the black (negative) test lead to the negative battery post (Figure 22-37). Check and note the voltage reading on the meter. A reading of 12.65 volts or higher shows the battery is fully charged (state of charge is 100 percent). A voltage of 11.70 volts or lower shows the battery is completely discharged (state of charge is 0 percent) (Figure 22-38).

Charge a Battery

WARNING: Eye protection should be worn when charging a battery. Do not charge a battery with frozen electrolyte as it could overheat and explode. Do not go higher than the engine maker's battery charging limits. Do not plug the charger into the wall outlet until the battery has been connected to the charger.

Recharge a battery when the open circuit voltage readings show it is at less than a full charge. Often a battery can be charged in or out of the equipment. If it is charged in the equipment, both cables must be disconnected before connecting a charger. Failure to do this may damage the engine's

BATTERY OPEN CIRCUIT VOLTAGE AS AN INDICATOR OF STATE OF CHARGE

Open Circuit Voltage	State of Charge
12.6 or greater	100%
12.4 to 12.6	75-100%
12.2 to 12.4	50-75%
12.0 to 12.2	25-50%
11.7 to 12.0	0-25%
11.7 or less	0%

FIGURE 22-38 Battery open circuit voltage shows state of charge.

charging system. **Battery charging** is using direct current from a battery charger to restore battery chemical potential. A battery must be charged with the correct amount of current (rate) and for the correct length of time.

CASE STUDY

Batteries can be very dangerous if proper precautions are not taken. Many experienced technicians have had bad experiences with a battery. A discharged battery was removed from a snow thrower and left on the shop slow charger for the entire day. As the technician was closing the shop for the day, he remembered the battery. He was in a hurry to close up so did not go to the trouble to turn the charger off. He just disconnected the positive charger connection from the battery. The removal of the charger connector with the charger on caused a spark. Hydrogen gas caused by the charging process had collected above the battery. The gas ignited and the flame traveled into the battery. The gas inside the battery ignited and caused the battery to explode (Figure 22-39). The exploding battery sent out a cloud battery acid and chunks of battery lead in a wide circle.

The technician who was closing up the shop had left his safety glasses in his tool box. He went to the hospital with acid burns to his face and eyes. He recovered without any permanent eye damage. He now has a great deal more respect for battery safety.

A discharged battery may often be brought back to satisfactory condition by slow charging. A **slow charger** is a battery charger used to charge a battery at a low amperage rate. A slow charger uses a 2–10 amperes rate to charge a battery. Connect the slow charger to the battery by connecting the charger negative (black) lead to the negative battery post. Connect the positive charger (red) lead to the positive battery post. Turn the charger on.

CAUTION: Using an automotive type fast charger on a small battery can cause it to explode. Always read and follow engine makers recommendations before battery charging. A battery that has been discharged for a long time should not be fast charged because it can overheat and explode.

As the battery charges, the open circuit voltage increases. Check the battery open circuit voltage at regular times throughout the charge. The average length of time necessary to slow charge a battery ranges from 12 to 16 hours. The battery is fully charged when the open circuit voltage stops increasing.

FIGURE 22-39 This battery exploded because of careless use of a battery charger.

REVIEW QUESTIONS

1. Explain the purpose and common service intervals of a preventive maintenance schedule.
2. Describe how to check oil level and condition on an engine with a dipstick.
3. List the steps to follow in draining and refilling engine oil and oil filter.
4. Describe the procedure used to clean the debris screen and cooling fins of an air-cooled engine.
5. Explain why a poor cooling system can cause engine failure.
6. Explain how to inspect and service a crankcase breather.
7. Explain why periodic air cleaner service is important to long engine life.
8. Explain the steps to follow in servicing a foam type air cleaner.
9. Explain the steps to follow in servicing a paper element type air cleaner.
10. Describe the steps to follow in changing an in-line fuel filter.
11. Explain the purpose of valve clearance in a valve mechanism.
12. Explain how a feeler gauge is used to measure valve clearance.
13. Describe the steps to follow in preparing an engine for off-season storage.
14. List and describe the areas to check when performing a battery visual inspection.
15. Describe the procedure used to test a battery cell for open voltage.

DISCUSSION TOPICS AND ACTIVITIES

1. Find two different engine makers' maintenance schedules. Make a list of the differences between the maintenance jobs and maintenance intervals for the two engines.
2. Use a digital voltmeter to measure open circuit voltage of several shop batteries. Decide which of the batteries can still be used on an electrical starting system.
3. Using graph paper and an engine maintenance guide, construct a graph showing what maintenance jobs should be done and when they should be performed.

CHAPTER 23

Ignition System Tune-up

OBJECTIVES

Upon completion and review of this chapter, you should be able to:

- Troubleshoot an ignition system.
- Troubleshoot, remove, and install a spark plug.
- Troubleshoot and service flywheel magnets, key and keyway.
- Troubleshoot and service breaker points and condenser.
- Measure and adjust ignition timing.
- Static test the magneto coil.
- Dynamic test the magneto coil, solid state trigger and insulation.
- Install the flywheel and set the armature gap.

TERMS TO KNOW

Advanced ignition timing
Armature air gap
Capacitance
Flywheel holder
Flywheel puller
Fouling
Ignition timing
Impact nut
Microfarad
Retarded ignition timing
Spark plug gapping tool
Spark plug wrench
Starter clutch wrench
Stator plate
Thread chaser
Thread insert

INTRODUCTION

The ignition and fuel systems must operate correctly for maximum engine performance. The engine needs a strong, high-voltage spark at just the right time. The fuel system must provide exactly the right air-fuel mixture. Both the ignition and fuel systems require periodic adjustment and service to operate correctly. These service jobs are often called a *tune-up*.

IGNITION SYSTEM TROUBLESHOOTING

Ignition system problems can cause hard starting or no start, low engine power, or engine misfire. Ignition system troubleshooting always starts with a spark check. Do an ignition tune-up when you find a no spark or weak spark condition. An ignition tune-up is also a preventive maintenance procedure. The owner's manual may recommend a tune-up at specific time intervals. The tune-up will prevent engine performance problems.

An ignition tune-up includes:

- Replacement of the spark plug
- Inspection of the spark plug cable
- Inspection of the flywheel hub and key
- Setting the armature or coil air gap
- Coil testing
- Adjusting ignition timing

Engines that have breaker point ignitions require breaker point and condenser replacement.

SPARK PLUG SERVICE

The minute a spark plug is installed in an engine, it begins to wear out. The center and ground electrodes start to burn and wear away. Oil and gasoline build up on the electrodes. A dirty or worn out spark plug may not allow the spark to jump across the electrodes. The engine may be hard to start, misfire, or stop running. During a tune-up, the spark plug is replaced with a new one.

Troubleshooting

The spark plug may be at fault if an engine has a performance problem not caused by fuel or compression. Always start by testing the ignition for spark. If the spark testing shows a strong spark at the correct time, troubleshoot the spark plug. Install a new spark plug and see if it cures the ignition problem.

You can learn a lot about the engine by inspecting the old spark plug. Carefully check the appearance and condition of the spark plug firing end. Check the color and condition of the insulator firing end and electrodes. This will tell how well the engine has been operating. It will also show if the spark plug is the right one for the type of operation.

Often the same type of spark plug used in two similar engines will differ greatly in appearance. The causes of these differences are:

- Piston ring condition
- Valve condition
- Lubrication and cooling system condition
- Carburetor adjustments
- Kind of fuel used
- Engine operating conditions

Fouling (Figure 23-1) a rapid build-up of oil or gasoline on the spark plug electrode is a common spark plug problem. Fouling can cause a weak spark or no spark across the electrodes. The result is hard starting or misfire especially under load. You should determine if the fouling is due to oil or fuel. Wipe some of the material off the electrode and observe it closely. Gasoline fouling looks like black or gray dry soot. Oil looks like dirty black liquid.

Gasoline fouling is due to a fuel system problem. Common problems are an improperly adjusted or malfunctioning carburetor, poor quality of gasoline, or clogged air filter. Oil fouling is often caused by burning excessive oil, too high an oil level, or a plugged crankcase breather. Another cause of fouling is the use of a spark plug with too cold a heat range.

Electrode wear (Figure 23-2) is another common source of problems. The spark across the electrodes causes the center electrode to wear away. Wear makes the gap between the center and ground electrode larger and larger. In addition, the sharp edges of both electrodes will get rounded. The ignition system must develop higher voltages to push a spark across the worn electrode gap. Spark plug electrode wear often allows the engine to run satisfactorily at low speeds. The engine will misfire under load. Higher loads cause higher combustion chamber pressures. The spark has a harder time jumping the electrode gap under pressure.

FIGURE 23-1 Common spark plug fouling problems. (Provided courtesy of Tecumseh Products Company.)

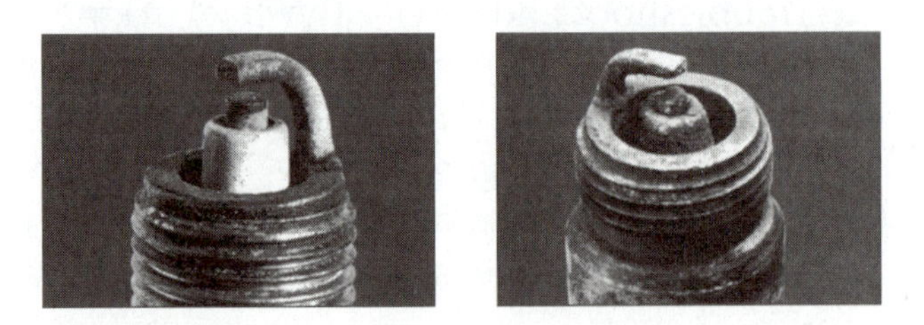

FIGURE 23-2 Normal and worn spark plugs. (Courtesy of Cooper Automotive/Champion Spark Plug.)

Removing the Spark Plug

> **WARNING:** Make sure the kill switch is off when the spark plug cable is removed from the spark plug and grounded. If the engine is rotated, the magneto can cause the loose spark plug cable to spark. The high voltage can cause a shock or ignite fuel vapors.

Spark plugs are removed with a spark plug wrench. A **spark plug wrench** is a deep socket wrench lined with a rubber insert to protect the porcelain spark plug insulator. Spark plug sockets

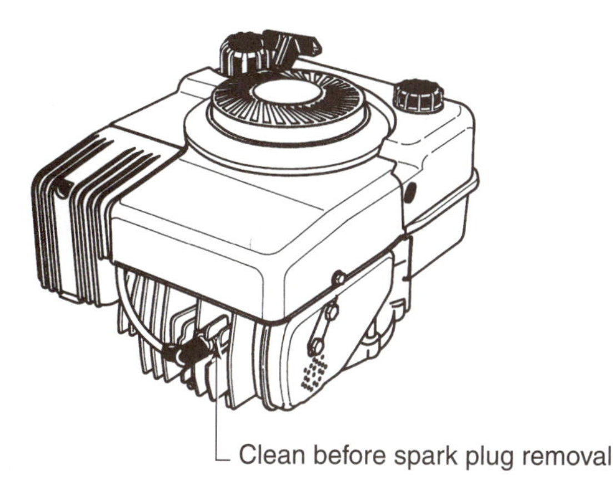

FIGURE 23-3 Clean the area around the spark plug before removing the plug. (Provided courtesy of Tecumseh Products Company.)

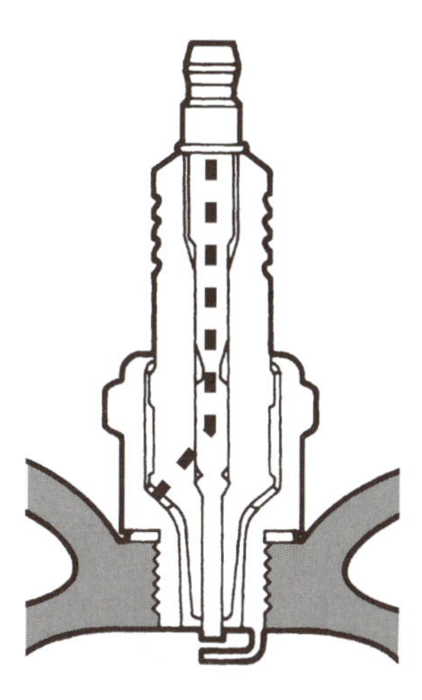

FIGURE 23-5 A broken insulator can cause current leakage from the center electrode to the shell. (Courtesy of General Motors Corporation, Service Technology Group.)

are available in many sizes to fit different size spark plugs. A standard socket driver handle is used to turn the wrench.

The first step in spark plug removal is to clean the cylinder head area around the spark plug (Figure 23-3). This prevents dirt from entering the engine combustion chamber when the spark plug is removed. Remove the spark plug wire from the spark plug terminal by pulling on the boot. Place the spark plug wrench over the spark plug. Make sure the wrench fits all the way over the spark plug shell (Figure 23-4). The spark plug insulator will be broken if the wrench is not down all the way. A broken insulator can be difficult to see. It can cause current leakage from the center electrode to the shell and keep the engine from getting ignition (Figure 23-5).

After removal, check the spark plug gasket. The gasket makes a gas tight seal between the spark plug and the cylinder head seat. It also conducts heat away from the spark plug into the cylinder head. If the gasket is not tightly seated, leaking combustion gases will cause spark plug overheating. If the gasket is flattened too much, the spark plug shell may be distorted or cracked.

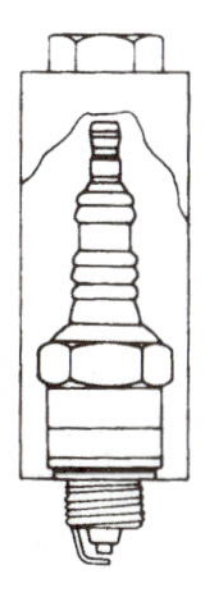

FIGURE 23-4 The spark plug socket must fit all the way over the spark plug to protect the plug insulator.

Spark Plug Gapping and Installation

A replacement spark plug must be selected with care. The replacement spark plug must be the correct thread size, reach, and heat range. Spark plug specifications are part of the code printed on the spark plug insulator (Figure 23-6). The code is different for different spark plug manufacturers. Always check the shop service manual or a spark plug application chart to help determine the correct replacement spark plug. Spark plugs used on overhead valve engines have a longer reach than those used on L-head engines. Be careful not to install a spark plug with too long a reach (Figure 23-7). Engine damage can occur if the piston hits the spark plug.

SERVICE TIP: Do not select a new spark plug using the number on the old spark plug. The last person that installed the spark plug may have used one that is incorrect.

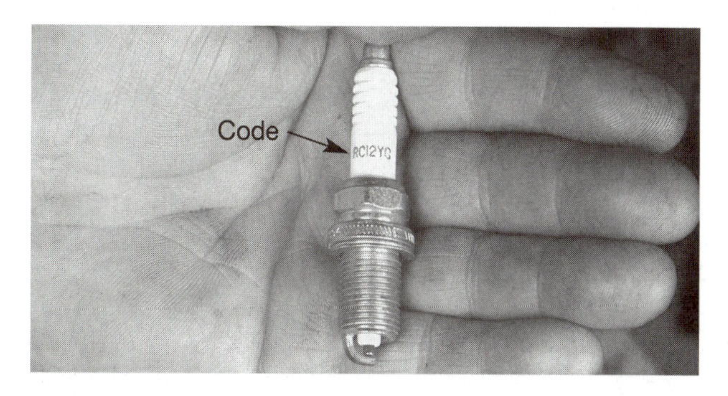

FIGURE 23-6 The spark plug code shows if the spark plug is correct for the engine.

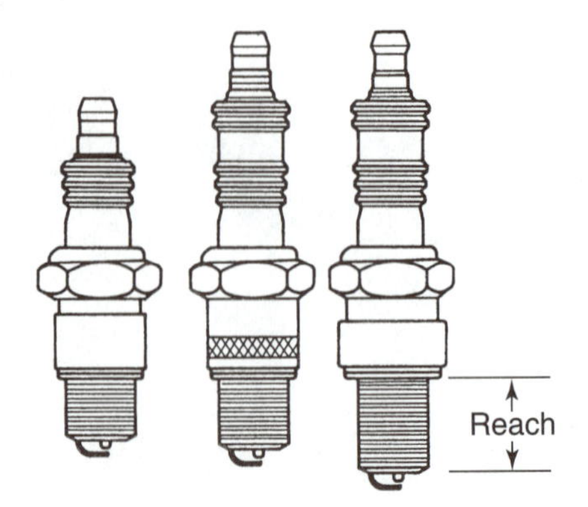

FIGURE 23-7 The spark plug reach must be correct for the thickness of the cylinder head. (Provided courtesy of Tecumseh Products Company.)

The shop manual will specify a recommended heat range for a replacement spark plug. The shop manual may recommend different heat range spark plugs for different operating conditions. Compare the old plug to a spark plug analysis chart. You may find that the spark plug firing end appears overheated. You should select a replacement spark plug one heat range cooler. If the spark plug firing end is fouled from oil burning, select a spark plug that is one heat range hotter.

The electrode gap must be measured and adjusted to specifications before the new spark plug is installed. The spark plug gap is the distance between the spark plug ground and center electrode. The gap is measured in thousandths of an inch or hundredths of a millimeter. An incorrect gap can decrease ignition system performance. Different combustion chambers and ignition systems require different sized gaps.

Spark plugs are gapped with a **spark plug gapping tool** (Figure 23-8), which is a tool with wire feeler gauges for gap measuring and a ground electrode bending tool for gap adjusting. To measure the gap, look up the specifications. For example, some engines use a gap of 0.030 inch (0.076 millimeter). Select the feeler wire on the gapping tool that is the same as the specified gap (Figure 23-9). Gently push the wire into the space between the electrodes. It should just fit with a gentle push. If it will not go in, the gap is too small. If it goes in and is loose, the gap is too large.

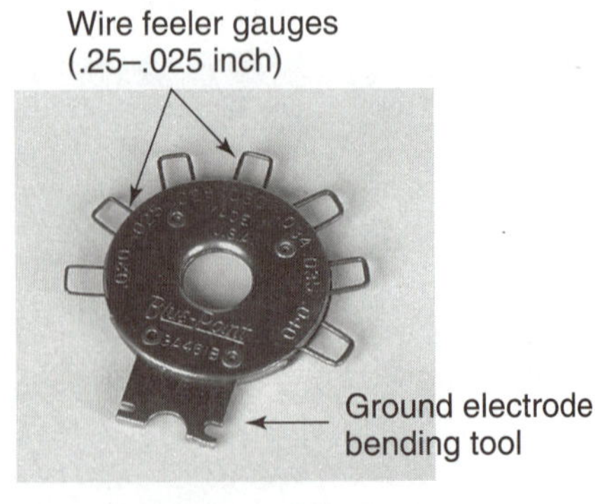

FIGURE 23-8 Spark plug gaps are measured and adjusted with a spark plug gapping tool. (Courtesy of Lisle Corp.)

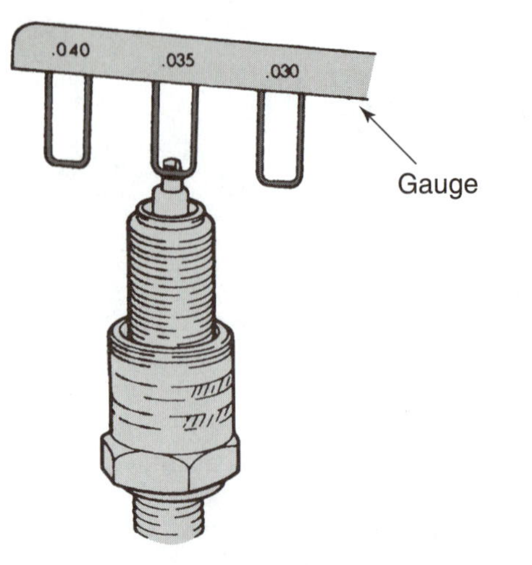

FIGURE 23-9 Measure the gap by sliding the wire gauge into the gap. (Courtesy of DaimlerChrysler Corporation.)

The hook on the gapping tool is used to adjust the gap (Figure 23-10). The gap is adjusted by bending the ground electrode. To make the gap smaller, bend the ground electrode closer to the center electrode. To make it larger, bend the ground electrode away from the center electrode. After changing the gap, recheck it with the wire gauge. The center electrode should never be bent. Sidewise pressure on the center electrode may crack or break the insulator tip.

SERVICE TIP: Gapping pliers are available for adjusting spark plug gaps. They should not be used because they can damage the spark plug's internal seals.

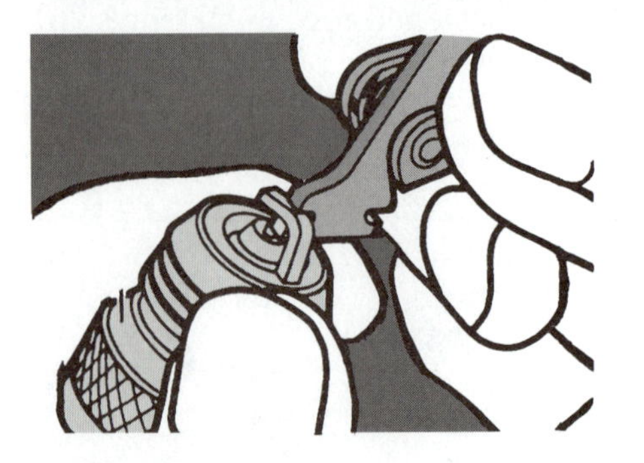

FIGURE 23-10 Adjust the gap by bending the ground electrode with the bending tool. (Courtesy of Sun Electric Corporation.)

Always start the spark plug back into the cylinder head threads by hand. Do not use a wrench until the spark plug cannot be turned any longer by hand. Starting the spark plug with a wrench can cause cross threads. Use the correct size spark plug socket wrench to tighten the spark plug. The correct spark plug tightening is important for proper engine performance. If the spark plug is not tight enough, the gasket can leak. There will be a loss of compression pressure. Overtightening the spark plug will damage the cylinder head spark plug threads. This can also cause compression pressure leakage.

Some engines have torque specifications for spark plugs. If you can find torque specifications, use a torque wrench to tighten the spark plug (Figure 23-11). Tightening procedures in the shop manual are often given without a torque specification. These often recommend that the spark plug be tightened until snug. Then, tighten the spark plug a specific number of additional turns. Connect the spark plug wire boot to the spark plug terminal. Removing and replacing a spark plug is shown in the accompanying sequence of photographs.

Spark Plug Thread Repair

Damage to the spark plug threads in the cylinder head is a common problem. Most engines have a cylinder head made from aluminum. Some engines use a steel insert for the spark plug threads. Most engines have the threads cut directly in the aluminum cylinder head. Aluminum is a much softer material than the steel spark plug threads. The aluminum cylinder head threads are often damaged by cross threading and seizing.

FIGURE 23-11 Using a torque wrench to tighten a spark plug to the correct torque.

Cross threading is common in aluminum threads. Cross threading is caused when the spark plug is not started straight into the threads This happens when the spark plug is not started into the threads by hand. Another problem is using too much force to tighten the spark plug. This causes the threads to be stretched out of shape or broken.

Seizing is a problem caused by the use of steel and aluminum threads together. Over time electrolysis can cause material from the aluminum threads to transfer to the steel threads. The two sets of threads can fuse or seize together. You may find this problem when attempting to remove the spark plug. A seized spark plug will appear to be locked in the threads. As you turn the seized spark plug, the aluminum threads are damaged.

Damaged spark plug threads can be repaired or a new cylinder head must be installed. Cylinder heads for L-head engines are relatively inexpensive. Major damage to the threads in these heads is usually not repaired. It is easier to replace the cylinder head. Overhead valve cylinder heads are more expensive and justify thread repair.

Minor thread damage can be repaired using the correct size tap or thread chaser. A **thread chaser** (Figure 23-12) is a cutting tool that is rotated through a thread to clean and straighten damaged threads. Thread chasers are available for all common spark plug thread sizes. The thread chaser has undersize tap threads on the end to help start the tool in the hole. There are regular size tap threads above the undersized threads. Some thread chasers have a spring-loaded cutter to clean up the spark plug gasket seat. Start the thread chaser in the threads by hand. Then, carefully rotate it down the hole using a wrench on the driver end of the tool. The larger tap threads will then cut and repair the damaged threads in the hole.

Threads that are too damaged for thread chasing can be repaired with a thread insert. A **thread insert** (Figure 23-13) is a new steel thread that is inserted in a specially tapped hole in a damaged aluminum thread. The new thread insert is the same size as the original thread. The damaged threads are first drilled oversize with the correct size drill.

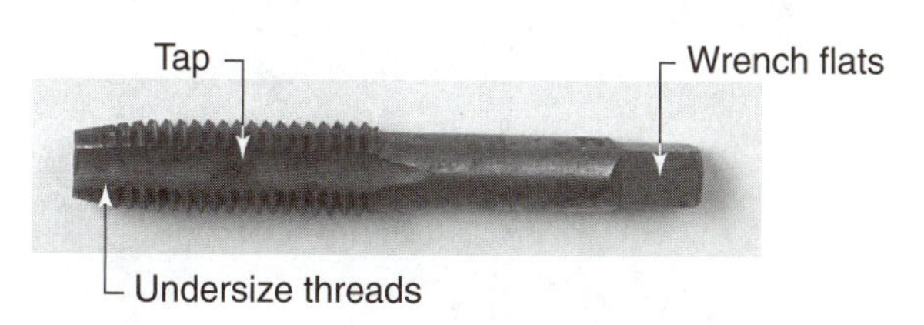

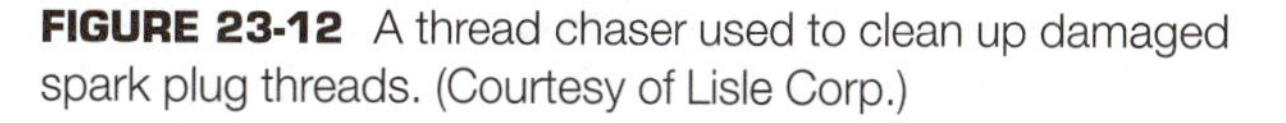

FIGURE 23-12 A thread chaser used to clean up damaged spark plug threads. (Courtesy of Lisle Corp.)

PHOTO SEQUENCE 9

Removing and Replacing a Spark Plug

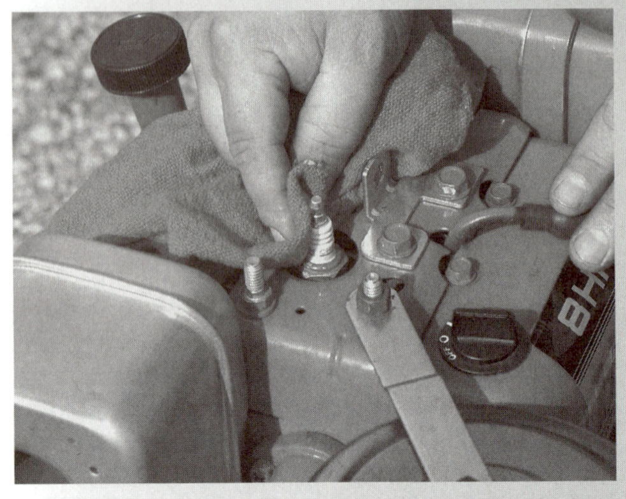

1. Allow engine to cool. Wipe the cylinder head area around the spark plug.

2. Remove and ground the spark plug wire.

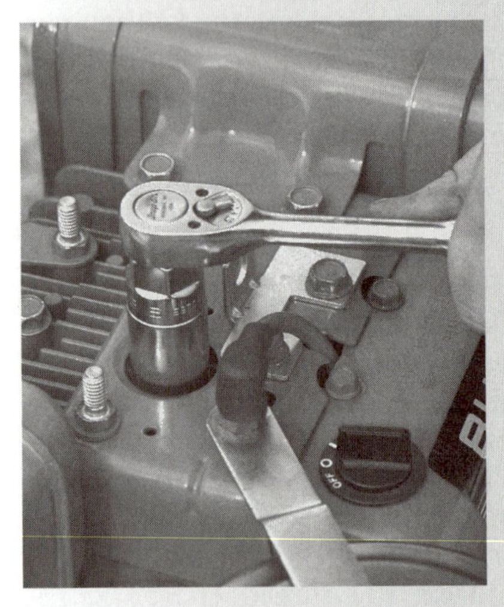

3. Remove the old spark plug with a spark plug socket.

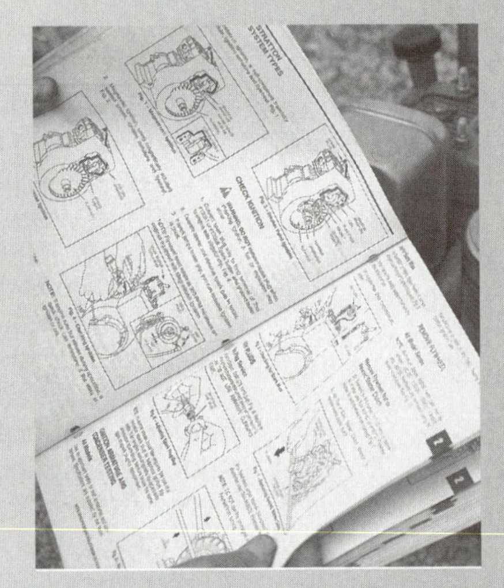

4. Check the shop service manual for the correct spark plug and the spark plug gap.

5. Check and, if necessary, adjust the gap on the new spark plug.

6. Inspect the cylinder head spark plug threads for damage. Repair threads if necessary.

7. Start the new spark plug into its threads by hand.

8. Tighten the spark plug to the recommended torque with a torque wrench.

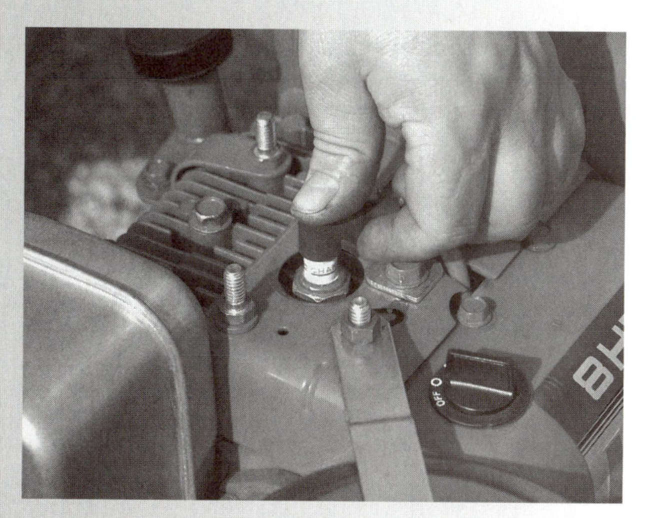

9. Install the spark plug wire on the spark plug terminal.

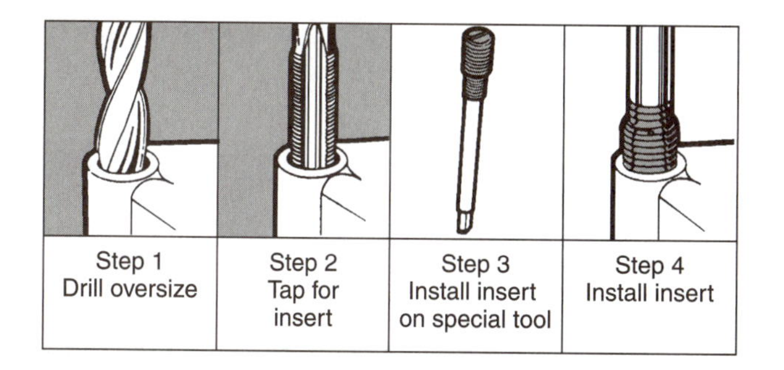

FIGURE 23-13 A damaged cylinder head spark plug hole can be repaired by installing a thread insert. (Courtesy of General Motors Corporation, Service Technology Group.)

FIGURE 23-14 The tools and parts to install a thread insert are available in a kit.

The drilled hole is tapped with a special tap. The tapped threads are made to fit a helical steel insert. The insert is rotated into the hole with a special driving tool. When the insert is in place, it forms the new threads. The threads are the same size as the original. Thread inserts come in a kit (Figure 23-14) that includes an insert, an insert tap, and a driver tool. Some kits include a special sized drill.

FLYWHEEL MAGNETS, KEY, AND KEYWAY SERVICE

Troubleshooting may show a no spark condition at the spark plug wire. This means the problem is not at the spark plug. The problem is in one of the magneto parts. Each magneto part must be inspected and tested to find the problem.

Troubleshooting

Start the magneto troubleshooting by checking out the flywheel. It must often be removed to check out the other parts of the magneto. After the flywheel is removed, the other magneto parts can be inspected and tested.

Several parts of the flywheel are very important to magneto operation. Weak flywheel magnet magnetism will prevent proper coil operation. The result will be no spark or weak spark to the spark plug. Weak magnetism is usually caused by dirt or rust on the flywheel magnets. Flywheel magnets can also loose their magnetism over time.

The position of the rotating flywheel magnets in relation to the magneto coil armature sets ignition timing. Anything that changes the relationship between the magnets and armature changes ignition timing. Common problems are distorted flywheel key, loose flywheel nut, damaged flywheel flange, or damaged keyway. Any of these problems can cause a no spark or weak spark condition.

Removing the Flywheel

To remove the flywheel, remove the blower housing that covers the flywheel. Remove the blower housing fasteners and lift the blower housing off the engine. The flywheel holding nut will be visible. The flywheel usually has a tapered hole in the middle. The tapered hole fits on a matching taper on the crankshaft flywheel end (FWE). The crankshaft taper has a keyway. The flywheel taper has a keyseat. A key fits in the keyway and keyseat to lock the flywheel on the crankshaft. The end of the crankshaft is threaded. A large nut or threaded rewind starter housing is used to hold the flywheel on the crankshaft.

Flywheels may be removed in several ways. First, the nut or rewind clutch housing must be removed. To remove the nut, the flywheel must be prevented from rotating. A **flywheel holder** is a tool used to prevent the flywheel from turning when the flywheel nut or other retainer is removed. Flywheel holders are special tools provided by each engine maker. Many two-stroke engines have small air vanes. These would be difficult to grip with a holding tool. A strap wrench (Figure 23-15) is often used on these flywheels. The strap fits around the flywheel rim. The strap wrench handle is used to prevent the flywheel from turning. Some holding tools are used to grip the flywheel vanes (Figure 23-16). Other holders lock in to fit the hub near the center of the flywheel. Others holders fit into the air vanes on the rim of the flywheel. Some have handles and others sit on the bench.

Some engines have a starter clutch that threads onto the crankshaft. The clutch housing takes the place of a flywheel nut. The clutch housing is removed to remove the flywheel. The starter clutch housing does not have flats for a wrench. A special wrench, called a **starter clutch wrench** (Figure 23-17), is available from the engine maker to remove the starter clutch. It has lugs that fit into the starter clutch housing to remove the starter clutch.

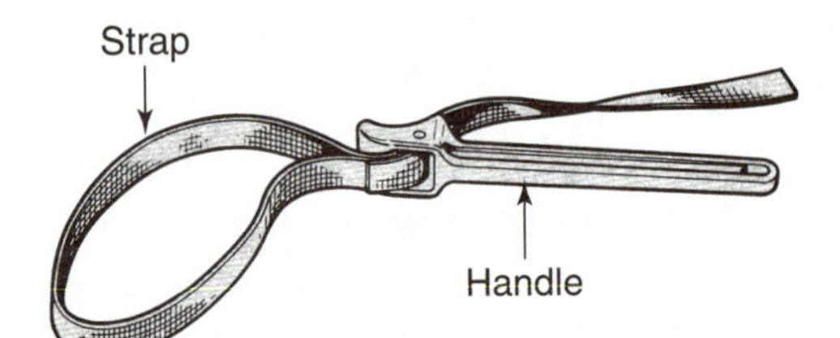

A. Strap Wrench

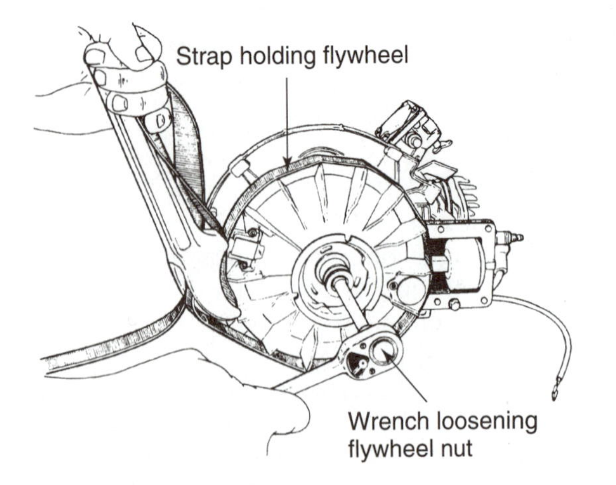

B. Using a Strap Wrench

FIGURE 23-15 A strap wrench is used to hold the rim of the flywheel. (Provided courtesy of Tecumseh Products Company.)

FIGURE 23-16 A flywheel vane holder used to prevent the flywheel from rotating.

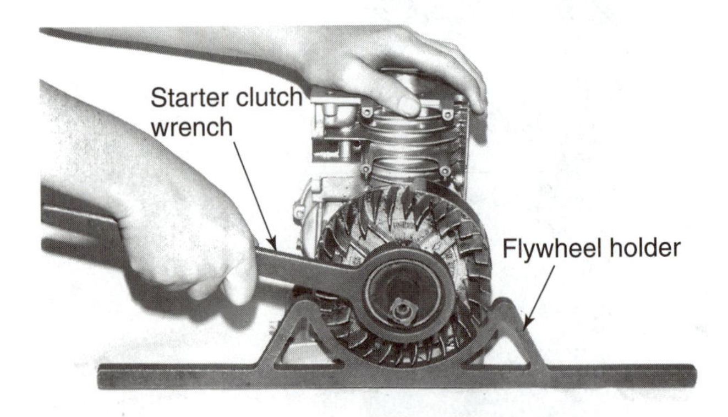

FIGURE 23-17 The starter clutch wrench is used to remove the starter clutch.

CAUTION: Most rewind starters have right-hand threads. Some flywheel nuts, however, have left-hand threads. Closely inspect the visible threads or check the shop service manual to make sure you turn the wrench in the correct direction.

Put the flywheel holder or strap wrench on the flywheel to prevent it from turning. The strap wrench holder is wrapped around the flywheel rim in a direction to oppose the wrench-turning force. The hub holder is placed in position on the center of the flywheel. Hold the flywheel from rotating with one hand. Use the correct size socket wrench or starter clutch wrench to loosen and remove the flywheel nut.

With the nut removed, the flywheel is ready to be removed from the crankshaft taper. Sometimes the flywheel will come off by hand. Often, the tapered fit between the two parts keeps them locked together. The tapered fit may be exceptionally tight. There may be corrosion between the aluminum flywheel and cast-iron crankshaft. Either of these conditions makes the flywheel difficult to remove.

WARNING: Wear eye protection when using pulling tools. Parts can break under load and injure your face or eyes.

CAUTION: Do not use an automotive style puller (Figure 23-18) to remove a flywheel. This kind of puller grips the outside rim of the flywheel. The flywheel is very thin in the center and very little pressure from this kind of puller will bend or even break the flywheel.

FIGURE 23-18 An automotive type puller can damage a flywheel.

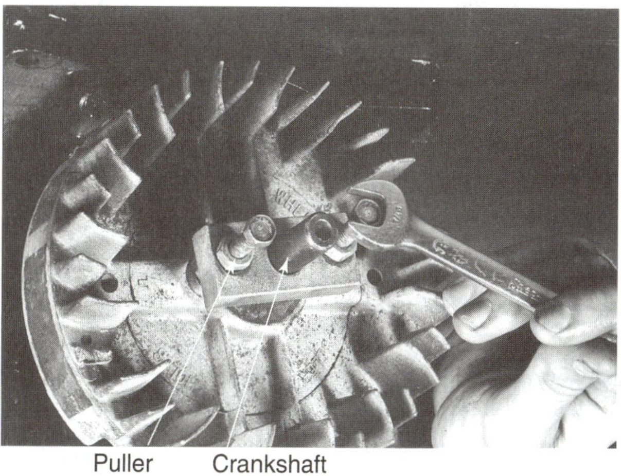

FIGURE 23-19 A flywheel puller used to remove the flywheel from the crankshaft taper.

Using a flywheel puller is the most usual way of removing a flywheel from the crankshaft taper. A **flywheel puller** (Figure 23-19) is a tool used to separate the flywheel from the crankshaft taper. Flywheel pullers are available from engine makers to fit their flywheels. Many flywheels have holes near the hub to allow mounting the puller. The flywheel puller often has a flange that is attached to the flywheel. It has a center forcing screw that is rotated against the end of the crankshaft to pull off the flywheel. Another common type works without the center forcing screw. The center of the puller rests on the crankshaft. Two mounting screws fit into the flywheel. The two mounting screws are installed then rotated evenly to pull off the flywheel.

Select the correct puller for the engine. Install the puller on the flywheel. Be sure to install the puller screws well into the flywheel threads. The high pulling force could cause the screws to be pulled out of the threads, which will damage the flywheel and make it difficult to remove. Carefully rotate the center or twin forcing screws to pull off the flywheel.

Some two-stroke flywheels are pulled using an impact nut and pry bar (Figure 23-20). An **impact nut** is a hardened nut that is threaded onto the crankshaft and used as a surface for hammering on the flywheel for removal. Remove the flywheel nut and thread the impact nut on in its place. This protects the threads on the crankshaft. Place a pry bar or screwdriver under the flywheel. Pry upward with the pry bar or screwdriver. At the same time, tap the impact nut squarely with a hammer. If

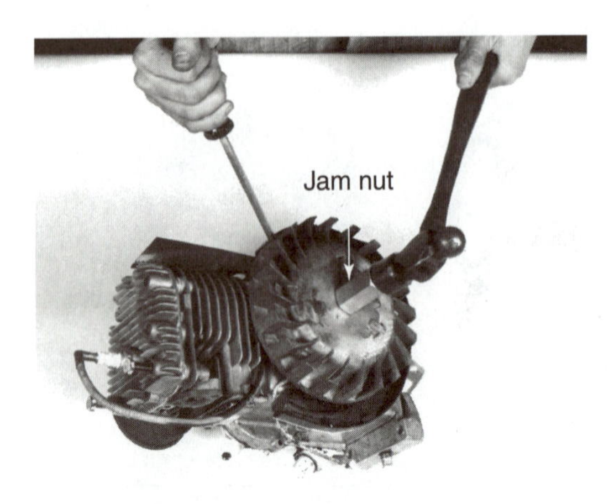

FIGURE 23-20 A screwdriver and jam nut are used to remove some small flywheels.

necessary, rotate the flywheel a half turn and repeat the steps. The prying and impact force work together to unlock the flywheel taper. You will see the flywheel loosen from the crankshaft. Remove the impact nut and flywheel.

CAUTION: Hammering directly on the crankshaft or crankshaft nut with a steel hammer can damage the crankshaft threads.

SERVICE TIP: Sometimes the time-consuming puller operation can be avoided by lightly tapping on the end of the crankshaft with a soft face hammer. A soft face hammer blow often shocks loose the grip of the taper. Be sure to use only a soft face hammer. If possible, install the flywheel nut on loosely to avoid damage to the crankshaft.

Inspecting the Flywheel

The flywheel magnets can lose their magnetism. To check for magnetism, place an unmagnetized screwdriver approximately ¾ inch (19 millimeters) away from the magnet (Figure 23-21). It should be strongly attracted to each magnet in the flywheel. Most magnets are cast into the flywheel. Some older flywheels have replaceable magnets. A magnet that has lost magnetism must be replaced. Either replace the individual magnet or replace the flywheel.

The key and keyway in the hub area is an important area for inspection (Figure 23-22). The key positions the flywheel in relation to the crankshaft. This, in turn, determines the position of the flywheel magnets in relation to the magneto armature. The position of the magnets in relation to the armature sets ignition timing. A loose fitting or distorted key changes ignition timing, which can cause a no spark, weak spark, or a spark that occurs at the wrong time.

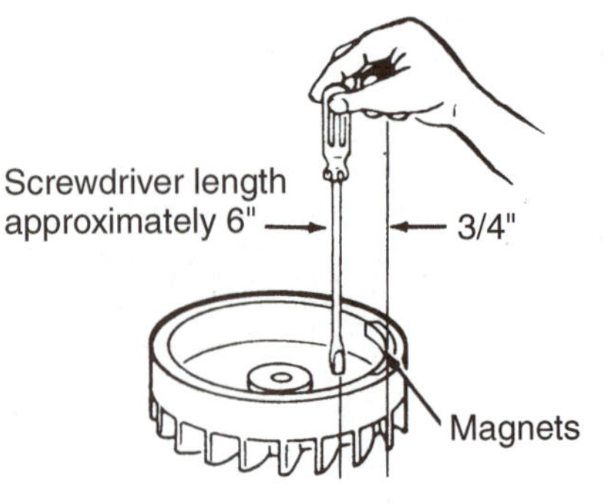

FIGURE 23-21 Checking flywheel magnet magnetism with a screwdriver. (Provided courtesy of Tecumseh Products Company.)

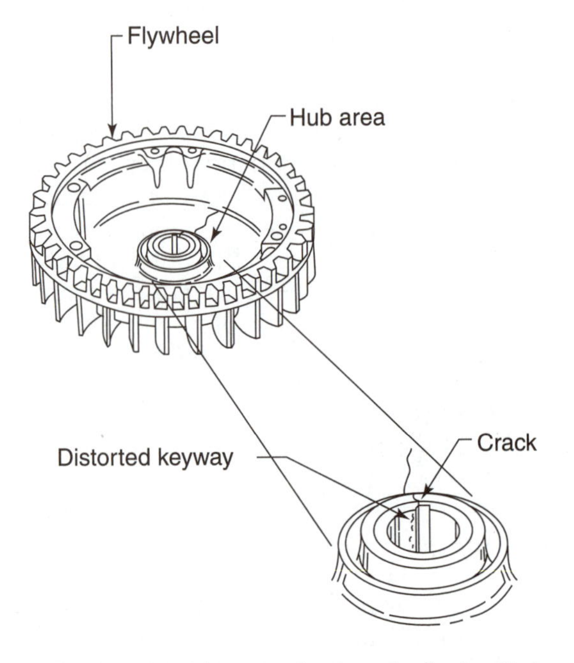

FIGURE 23-22 Areas to check on the flywheel hub.

Some engines used on lawn mowing equipment use a key made from a soft material. If the lawn mower blades strike an object, the key will partially shear (Figure 23-23) and prevent the flywheel momentum from breaking the crankshaft. The partially sheared key will kill the engine. These special soft keys should never be replaced with a hardened steel key.

Inspect the keyway in the flywheel hub for cracks and wear. Put a new key in the keyway and try to move it side to side. Any movement shows ex-

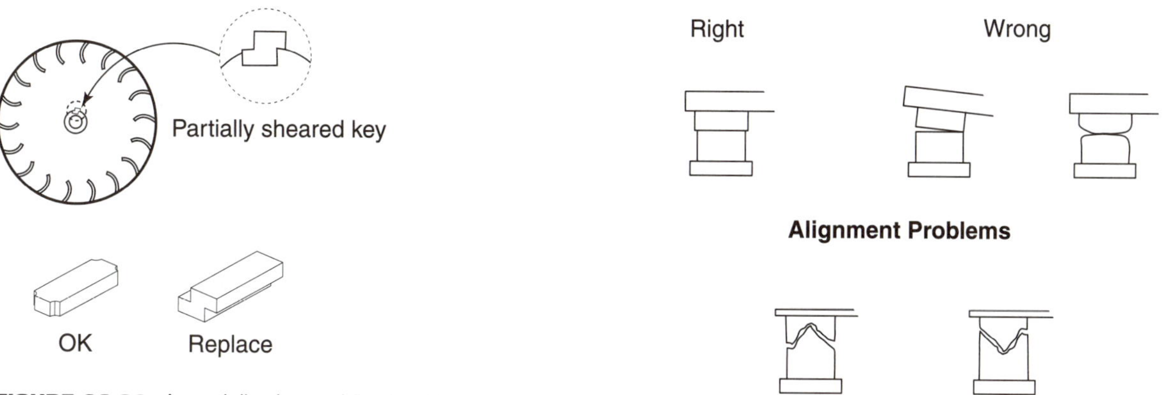

FIGURE 23-23 A partially sheared flywheel key will kill the ignition.

FIGURE 23-24 Common breaker point problems. (Courtesy of American Honda Motor Co., Inc.)

cessive wear. Put the new key in the crankshaft keyseat and make the same check. Wear is much more common in the aluminum flywheel than in the cast-iron crankshaft. A loose key distorts the keyway and causes improper ignition timing. Replace the flywheel if you find any of these problems.

BREAKER POINTS AND CONDENSER SERVICE

> **SERVICE TIP:** When possible, replace worn breaker points and condenser with a solid state conversion kit. This will eliminate the need for breaker point service and make the engine more reliable.

Older magneto systems use breaker points to develop current in the magneto coil. Newer engines do not use breaker points. Breaker points are the least reliable part of any magneto ignition system. The breaker points are always serviced during a tune-up. They are always one of the first places to look when there is a no spark to the spark plug. Breaker point systems use a condenser that requires testing or replacement during a tune-up.

Troubleshooting

Breaker point problems can cause the engine not to start or to misfire when running. Current flow across the breaker points breaks down the contact surfaces. The contact surfaces get burned, oxidized, eroded, and misaligned (Figure 23-24). The contact surfaces can get bad enough to lower the voltage created in the coil. The engine will begin to lose power and waste gasoline. If the breaker points are not replaced, they may get bad enough so the engine stops running.

The condenser's job is to protect the breaker points and improve magneto coil operation. The condenser stores current that tries to jump across open breaker points. If the condenser fails, the engine will not run. A condenser problem is likely if you find eroded breaker point surfaces. This is most likely on breaker points that have not been in service very long. Install a new condenser each time the breaker points are replaced.

Removing Breaker Points

The breaker points on many engines are located behind the flywheel. The common procedure is to remove and inspect the flywheel. When the flywheel is removed, the breaker points can be serviced.

The breaker points must be protected from dirt or water contamination. The breaker points and condenser are often protected by a cover (Figure 23-25). The cover is held in place with screws or a spring clip. Remove the clip or screws get access to the breaker points. The primary wires usually go through the cover. A sealer material is sometimes used where the wires go through the cover to prevent moisture from getting on the breaker points. This area needs to be resealed when the cover is reinstalled.

Plunger-operated breaker points are opened by a flat spot on the crankshaft. The condenser and stationary breaker point are combined into one unit. Remove the holddown screw through the movable breaker point. Disengage the spring from the movable breaker arm. Remove the stationary breaker point and condenser by removing the condenser clamp screw. The primary wire fits through a small hole in the condenser and is held in place

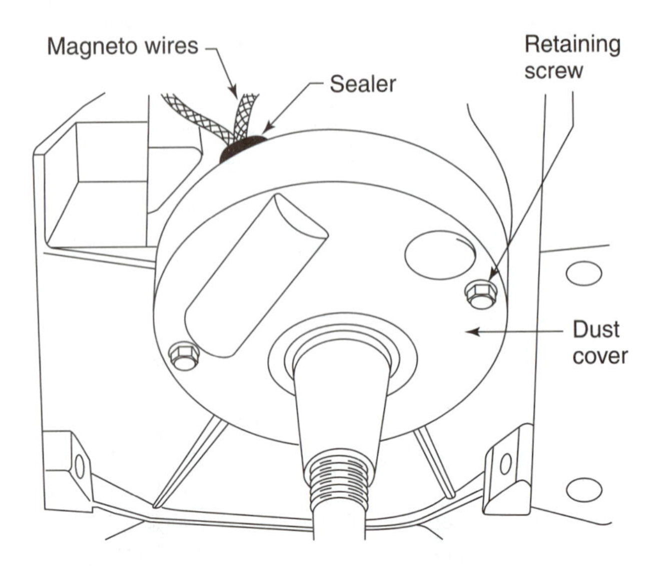

FIGURE 23-25 Remove the dust cover to access the breaker points.

by a small spring. Use the plastic depressor tool supplied with the new condenser to push down on the spring and remove the wire (Figure 23-26).

Cam-operated breaker points are located under a dust cover. The dust cover is held in place by a spring clip. Remove the spring clip and remove the dust cover (Figure 23-27). Before removing the breaker points, carefully observe the wire lead connections. Make a sketch of the wire connections so you can install the wires correctly. Remove the terminal nut that holds the primary magneto lead. Remove any ground leads connected to the breaker point assembly. Remove the stationary breaker point holddown screw. Lift off the stationary breaker point.

The movable point is usually mounted on a stud (Figure 23-28). It is held in place with a lock ring. Remove the lock ring and lift the movable breaker point assembly off the stud. Remove the condenser clamp holddown screw and remove the condenser.

Inspecting the Breaker Points

Clean the breaker point mounting area with a clean shop rag. Use spray electrical parts cleaner to remove any oil from the parts. The plunger used on plunger type breaker points is made from a soft insulator material. It must be carefully inspected and measured. A worn plunger or a worn plunger hole allows engine oil to get past the plunger. The oil can get on the breaker points. Dirt can get between a worn plunger and the hole and seize the plunger and prevent the breaker points from operating. Measure the length of the plunger with an outside

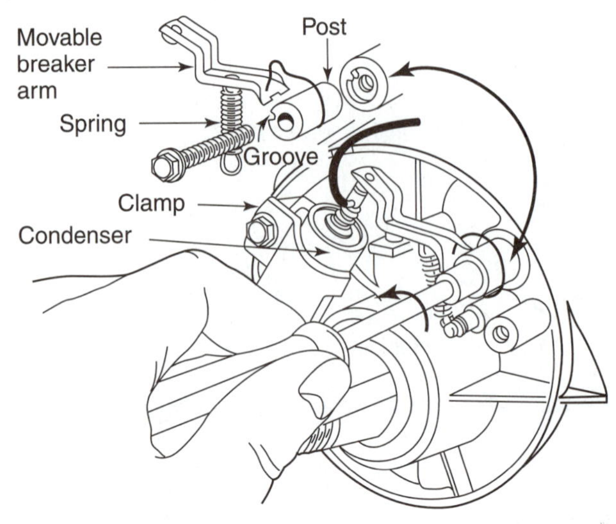

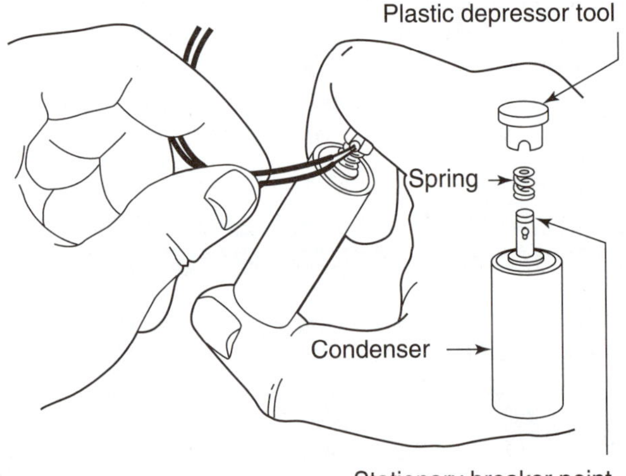

FIGURE 23-26 Disassembling the plunger-operated breaker points.

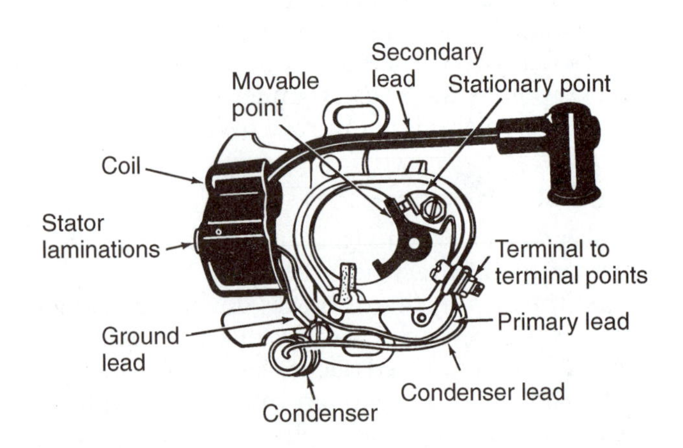

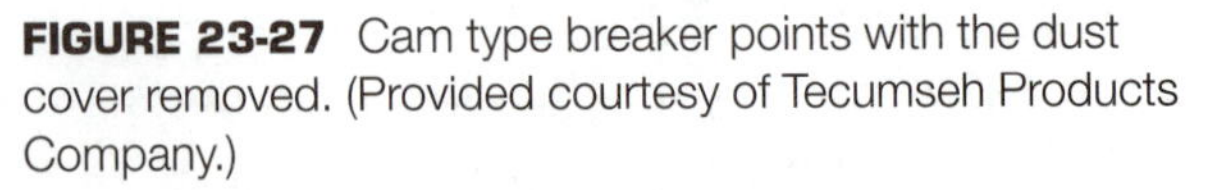

FIGURE 23-27 Cam type breaker points with the dust cover removed. (Provided courtesy of Tecumseh Products Company.)

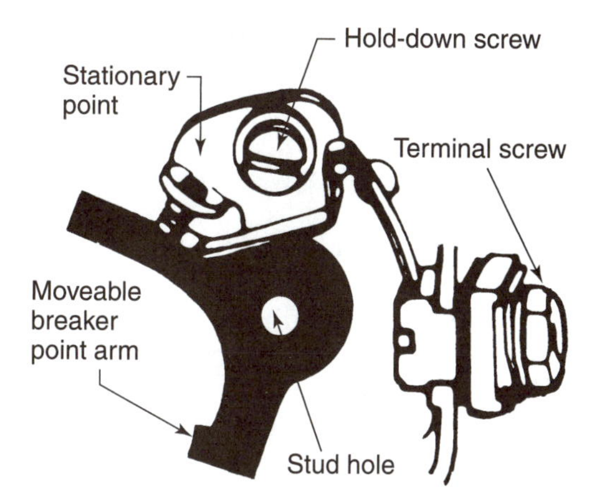

FIGURE 23-28 Parts of the breaker point set. (Provided courtesy of Tecumseh Products Company.)

micrometer. Compare the measurement to specifications. Replace any plunger if it is worn beyond specifications.

CAUTION: Some plungers have a groove at the top. Be sure to note this and install the plunger in the correct direction.

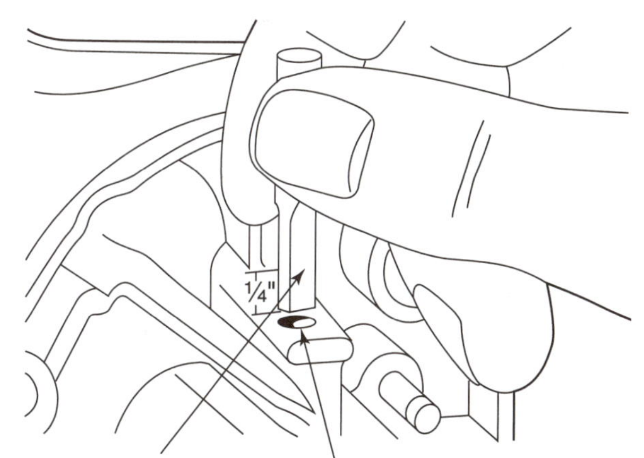

A. Measuring Plunger Hole

Measure length

B. Plunger Measurement

FIGURE 23-29 Measure the plunger for length and the plunger hole for wear with a depth gauge.

A special tapered depth gauge is available to measure the diameter of the plunger hole (Figure 23-29). Push the gauge into the plunger hole. The hole is worn beyond specifications if the gauge fits deeper than ¼ inch (6.35 millimeters). A worn plunger hole can be repaired by installing a bushing.

Breaker points that use a cam to open the points have similar wear areas. The movable breaker point arm is made from a soft insulator material. Wear on the arm causes the breaker points to not open properly. The arm and the movable breaker point are one part. It is always replaced when replacing the breaker points. Wear can also occur to the cam. Carefully inspect the cam lobe area for any signs of wear. The cam has a key that engages the crankshaft. Look for wear on the cam, key, and keyseat (Figure 23-30). Replace any worn parts.

Testing the Condenser

A new condenser is normally included in the breaker points replacement parts kit. Both the new and old condensers should be tested when the ignition system is serviced. Sometimes a new condenser does not test as well as the old one. In this case, the old condenser should be reused. Inspect the condenser for:

- Damaged terminal lead
- Damaged terminal
- Dents or gouges in the housing
- Broken mounting clip

Test the condenser capacitance (capacity) on an electrical parts tester (Figure 23-31) such as the Merc-O-Tronic tester. **Capacitance** is a measure of the condenser's electrical storage ability measured in microfarads. A **microfarad** (mfd) is an electrical

Key
Cam
Crankshaft
Arm
Points
Pivot

FIGURE 23-30 Inspection areas for cam-operated breaker points. (Provided courtesy of Tecumseh Products Company.)

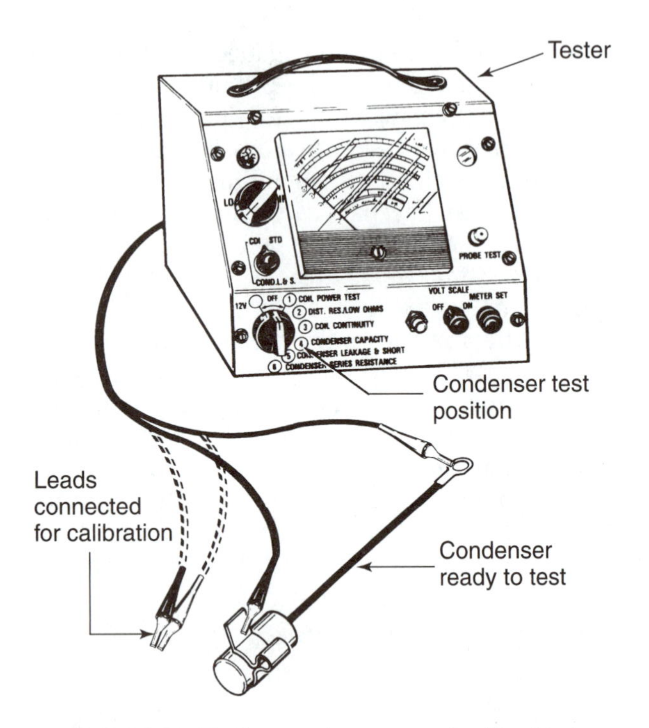

FIGURE 23-31 Testing condenser capacity. (Provided courtesy of Tecumseh Products Company.)

unit of capacitance used to rate the electrical capacity of a condenser.

Follow the test equipment instructions to check for capacitance. The tester is calibrated prior to condenser testing. The red tester lead is connected to the condenser terminal. The black tester lead is connected to the condenser housing. After calibration, the tester selector switch is positioned to test condenser capacity. The condenser capacity will be shown on the tester microfarad scale. Note the reading. The tester manual lists the recommended reading for each condenser by part number. For example, a specified condenser reading may be .16 to .23 mfd. Replace the condenser if you find a capacitance that is too high or too low. The breaker point surfaces will erode rapidly with incorrect capacitance.

Installing and Gapping Breaker Points

SERVICE TIP: Do not touch the surface of new breaker points during installation. Oil from your hands can contaminate the surface and prevent current flow across the breaker point surface. The same goes for the surface of any feeler gauge used to gap the breaker points. Clean any contamination off with electrical parts cleaner and a clean shop rag.

New breaker points are packaged as a tune-up kit. The kit usually has both the breaker points and a replacement condenser. The plunger-operated style breaker points combine the condenser and stationary point. The condenser is always supplied with these breaker points.

When replacing plunger type breaker points, install the movable breaker point in position over the plunger. Install and tighten the holddown fastener. Attach the breaker point spring to hold the points in position. Use the spring depressing tool to depress the spring. Insert the magneto wire into the condenser and stationary point. The spring holds the primary wire in position. Do not tighten the condenser holddown screw. The condenser will have to be moved for breaker point gap adjustment.

The next step is to adjust the gap between the breaker points (Figure 23-32). The **breaker point gap** is the space between the stationary and movable breaker point when the points are in the open position. Find the recommended breaker point gap in the shop service manual. The gap between the points must be set accurately. The gap determines coil buildup time and ignition timing. Rotate the engine crankshaft until the plunger moves up and pushes the breaker points open to their widest gap. Use a clean, flat feeler gauge of the recommended thickness and place it between the breaker points. Adjust the gap by moving the condenser and breaker point closer or further away from the mov-

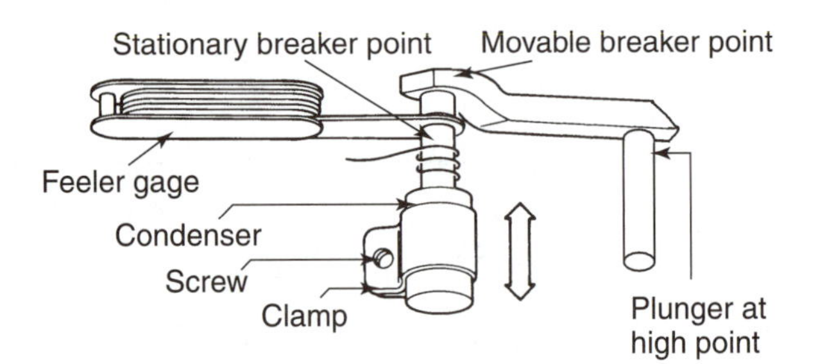

A. Plunger-Operated Breaker Points

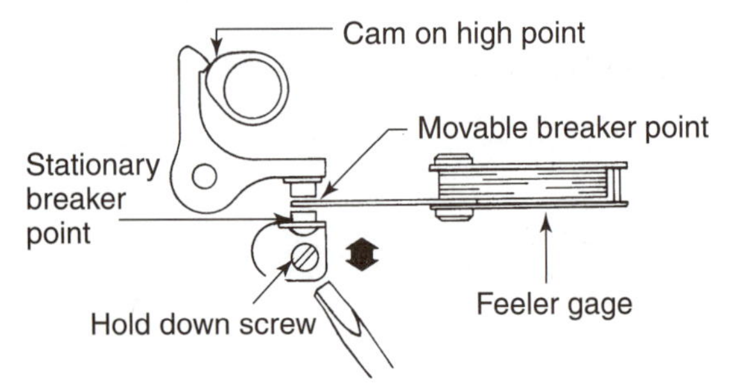

B. Cam-Operated Breaker Points

FIGURE 23-32 Adjusting the point gap on plunger- and cam-operated breaker points.

able point. The gap is adjusted properly when the feeler gauge slides in and out of the gap with a light drag. Tighten the condenser holddown screw after the gap is set.

Install new cam-operated breaker points by placing a new movable point on the mounting stud. Install the lock ring to hold the movable point in position. Position the stationary point and replace the holddown screw. Leave the screw loose for breaker point gap adjustment. Install the new condenser. Connect the primary and condenser wire to the terminal. Be sure to connect any other magneto wires disconnected from this terminal.

Rotate the engine crankshaft until the high part of the breaker point cam lobe contacts the movable breaker point arm. Select and place correct size and cleaned flat feeler gauge between the breaker points. Adjust the gap using a screwdriver to slide the stationary point back and forth. The gap is correct when the feeler gauge fits through the gap with a light drag. Tighten the stationary point screw after setting the gap.

A container of lubricant is often included in the replacement breaker point kit. Cam-operated breaker points often require lubrication between the cam and the movable breaker points. The lubricant prevents wear to the arm on the movable breaker point. Wear here causes the point gap to narrow and eventually close, causing the engine not to run.

CAUTION: Only high-temperature ignition parts lubricant can be used on the breaker point cam. Regular grease would quickly melt off and contaminate the breaker points. Always follow the shop service manual recommendations on cam lubrication.

Install the breaker point dust cover. Place the cover in position and install the screws or spring clip to hold the cover in place. Some covers use a gasket that should be replaced. Some engines require a sealer where the magneto wires pass through the cover. Use the recommended type of sealer on the cover.

TIMING ADJUSTMENT

Ignition timing is the exact time that the spark is introduced in the engine's combustion chamber in relation to piston position. The time that the spark is introduced into the cylinder is measured by the position of the crankshaft (Figure 23-33). When the piston is at top dead center, the crankshaft and connecting rod are lined up vertically. The crankshaft is at an angle to this vertical line when the piston is at a point before or after top dead center. This angle, measured in degrees, is used to describe

CASE STUDY

Sometimes troubleshooting an ignition system can be a challenge. An outdoor power equipment student had a mower in the shop with an ignition problem. The engine had stopped running during a mowing job. The student had used the spark tester and determined that the ignition did not have any spark.

The engine had a CD ignition system. The instructor directed the student to do a dynamic test on the system to find out if the coil or trigger unit was defective. The coil had a strong output. Other static tests failed to uncover the problem. The student tested the kill switch and found it to be working properly.

The student and instructor began to focus on the flywheel. The flywheel magnets had good magnetism. The flywheel used a hard key and it appeared to be in good shape. The instructor had the student put the flywheel key into the flywheel key seat. There was a small amount of play between the two parts. They decided to exchange flywheels with one from a shop engine. The engine passed the spark test with the different flywheel. The engine started and ran fine. The conclusion was that the worn flywheel key seat had changed the relation of the magnets to the coil enough to prevent ignition.

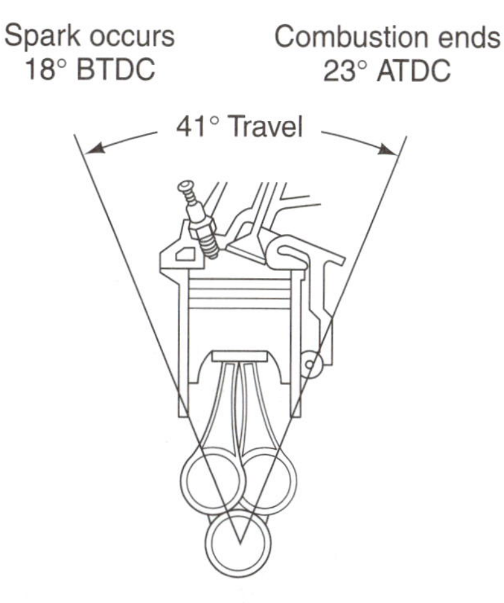

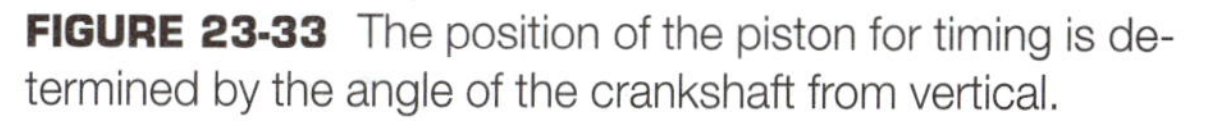

FIGURE 23-33 The position of the piston for timing is determined by the angle of the crankshaft from vertical.

when ignition occurs. Angles before top dead center are called BTDC. Angles after top dead center are called ATDC. **Advanced ignition timing** is an ignition spark that occurs before piston top dead center (BTDC). **Retarded ignition timing** is an ignition spark that occurs after piston top dead center (ATDC).

Ignition timing may be adjustable or fixed. Adjustable timing means that the timing can be changed to advance or retard ignition. Most solid state ignition systems have fixed ignition timing. Timing on these systems is fixed by the gap between the flywheel magnets and the coil armature. If this gap is set correctly, the timing is set correctly.

When the engine uses breaker points, the breaker point gap affects the timing. A wide breaker point gap causes the breaker points to open sooner in relation to crankshaft position. This causes the spark to occur earlier and advances ignition timing. A narrow breaker point gap causes the breaker points to open later in relation to crankshaft position, which causes the spark to occur later and retards ignition timing.

The relationship of the movable breaker point to the cam (or plunger) also affects engine timing. Many breaker point engines have fixed ignition timing. This means that the breaker cam to movable contact point arm relationship is not adjustable. If the breaker point gap is set correctly, the timing should be correct. Engines with plunger operated breaker points have fixed timing.

Some breaker point engines have an ignition timing adjustment. They have a way of changing the relationship between the breaker point cam and movable breaker point arm. Engines with timing adjustments can be identified by slotted holes on the **stator plate,** the ignition system part that holds the breaker points, condenser, armature, and coil. The slotted holes in the stator allow the breaker points to be moved in relation to the cam on the crankshaft (Figure 23-34). Moving the breaker points in a direction opposite to cam rotation causes the timing to be advanced. Moving the breaker points in the same direction as the cam turns causes ignition to be retarded (Figure 23-35).

Ignition timing is measured by measuring the position of the piston in relation to piston TDC (Figure 23-36). The shop service manual will have specifications for this measurement in thousands of an inch or hundredths of a millimeter. The measurement is made with a special tool provided by the engine maker. The tool has a dial indicator with a special sleeve to be mounted in the spark plug hole. The dial indicator plunger fits into the combustion chamber and touches the top of the piston. The dial indicator can then be used to measure movement of the piston up and down in the cylinder.

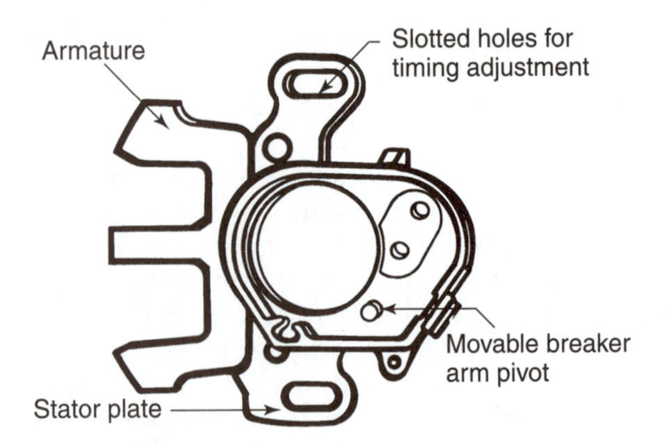

FIGURE 23-34 Slotted holddown screws on the stator are used for a timing adjustment. (Provided courtesy of Tecumseh Products Company.)

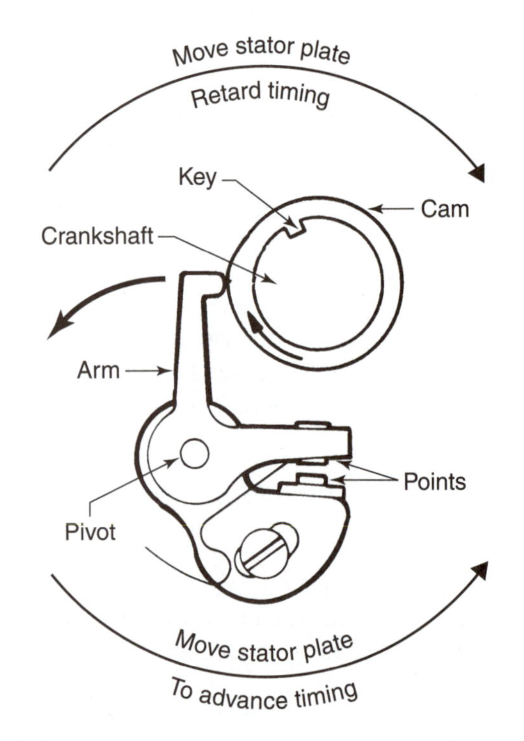

FIGURE 23-35 The stator is moved to move the breaker points in a direction to advance or retard ignition. (Provided courtesy of Tecumseh Products Company.)

Select the correct attachments for the engine being timed. Mount the tool using directions supplied with the tool. Look up the correct timing specifications for the engine in the shop service manual. For example, the specification might be 0.080 inch.

Rotate the crankshaft clockwise when looking at the magneto end of the crankshaft (Figure 23-37). Observe the needle on the dial indicator. It will

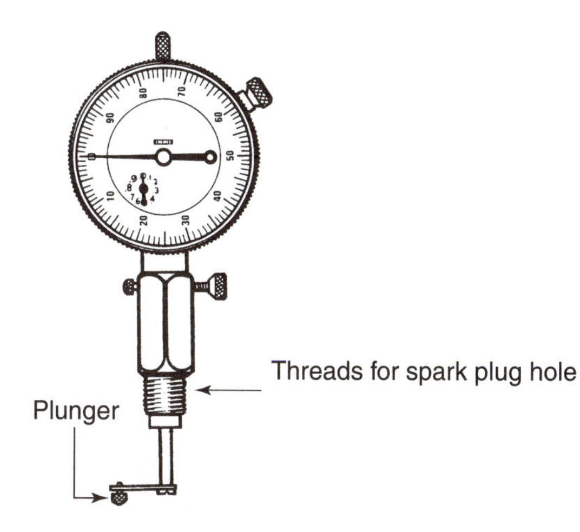

A. Timing Dial Indicator

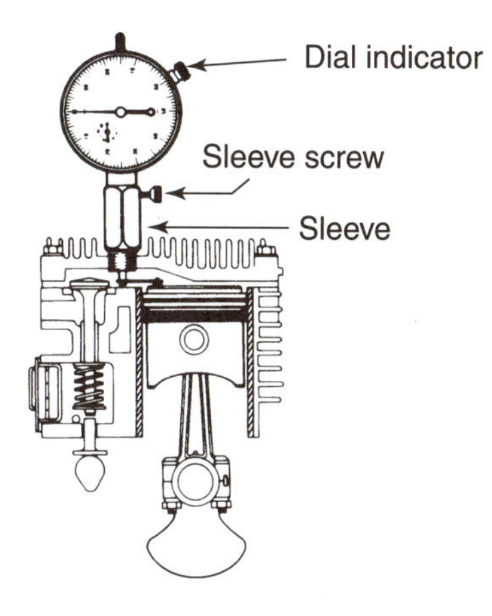

B. Using the Dial Indicator to Set Timing

FIGURE 23-36 A dial indicator inserted into the spark plug hole is used to measure piston position for timing. (Provided courtesy of Tecumseh Products Company.)

Dial screw

Dial at 0

A. Piston at TDC

Dial at .090"

B. Rotate Engine Past Specification

Dial at .080"

C. Rotate Engine Back to Specification

FIGURE 23-37 Position the piston at the specified position to set the timing. (Provided courtesy of Tecumseh Products Company.)

move in one direction then stop and reverse direction at exact TDC. Stop the piston at exact TDC. Turn the dial indicator face to line up the dial zero with the dial indicator needle.

Rotate the crankshaft counterclockwise when looking at the magneto end of the crankshaft. Observe the dial indicator needle and stop when it reads 0.010 inch past the specification (0.090 inch). Then rotate the engine back clockwise to the specified position (0.080 inch). This procedure will correct for any play in the connecting rod and crankshaft.

The piston is in the specified position for adjusting the timing. Loosen the stator holddown screws. Rotate the stator back or forth until the breaker points just start to open. This is an approximate timing adjustment. An exact point opening measurement can be done with continuity tester (Figure 23-38). Disconnect the terminal wires from the breaker points. Connect one lead of the tester to the stationary breaker point. Connect the other tester lead to the moveable breaker point. Turn the tester on and select the continuity test position. Rotate the stator until the points are closed. The continuity tester will sound showing continuity.

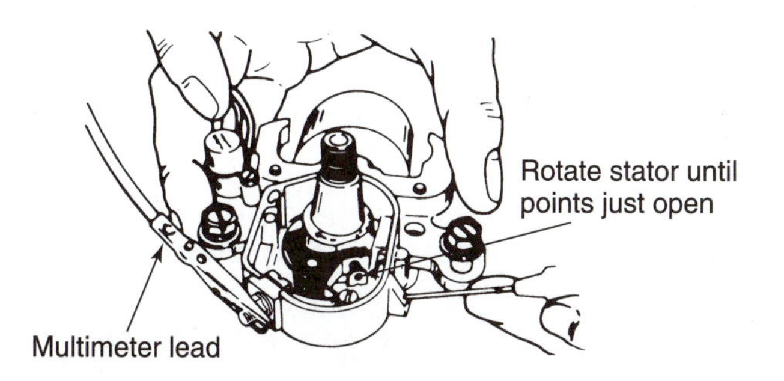

FIGURE 23-38 Rotate the stator until the points just start to open. (Provided courtesy of Tecumseh Products Company.)

Rotate the stator to open the points. The points are open as soon as the tester sound stops. Tighten the holddown screws. Connect any wires that were removed and install the breaker points cover.

SERVICE TIP: If a magneto is removed from an engine, scratch reference marks on the magneto stator and engine to prevent having to time the engine. Anytime the breaker point spacing is changed or adjusted, the timing also must be adjusted

MAGNETO COIL AND SOLID STATE TRIGGER TESTING

A problem with the magneto coil, solid state triggering device, or magneto wiring can result in a no spark to the spark plug condition. A failure in any one of these parts can prevent the system from developing the high voltage necessary for a spark to the spark plug.

Troubleshooting

The first troubleshooting procedure is to do the spark to the spark plug test. A no spark condition can indicate a problem with the magneto coil, solid state triggering device, or magneto wiring. Inspect each of these parts visually. Check the magneto coil, solid state triggering device, and all magneto wiring for signs of damage. Look for cracks or gouges in insulation. Look for signs of overheating. Check the electrical leads to be sure they are not broken and are properly connected. Especially check the wires where they enter the magneto coil. If you do not find a problem, you must test each part electrically.

SERVICE TIP: An improperly connected or improperly positioned (switched to off) engine kill switch can cause a no spark condition. Many technicians have wasted time troubleshooting the ignition to later find this problem.

Magneto Coil and Solid State Trigger Testing

Each engine maker has a specific test procedure for magneto coil and solid state trigger testing. These are found in the shop service manual. Always locate and carefully follow these test procedures.

Some magneto coils and solid state triggers can be tested while installed on the engine. Some engines require that they be removed for testing. Some coils used with breaker points are located under the flywheel. The flywheel has to be removed for inspection and service. Most solid state ignition systems have the coil and trigger device combined in one part. The system is mounted above the flywheel and is accessible without removing the flywheel. Remove the coil and solid state trigger by pulling the spark plug boot off the spark plug. Disconnect any magneto primary and ground wires. Remove the screws that mount the armature.

Static or dynamic electrical testing can be used to test a magneto coil. These tests are done on solid state or breaker point coils. Static tests require you to use a multimeter or ohmmeter to make measurements for specified electrical values. You compare these measurements to specifications to find out if the coil needs replacing. A dynamic tester simulates the off and on current developed by the ignition system primary circuit. This current is then sent into the coil primary. You can see if the coil can develop the high voltage required to jump a spark plug gap. The dynamic tester also includes multimeter functions to do specified static testing. Some engine makers specify static testing and some specify dynamic testing.

Primary and secondary coil circuit resistance is one common static test (Figure 23-39). This test is done with an ohmmeter function of a multimeter. Turn the multimeter on. Select the resistance mode. To test the coil primary resistance, connect one test lead to the magneto coil's primary lead. At the same time, touch the other test lead to the armature iron core. Note the amount resistance. Compare your reading with the coil primary resistance specifications in the shop service manual.

Test the coil secondary circuit resistance by removing the boot from the end of the spark plug wire (Figure 23-40). Connect one tester lead to the end of the spark plug wire. Connect the other tester lead to the armature iron core. Note the amount of

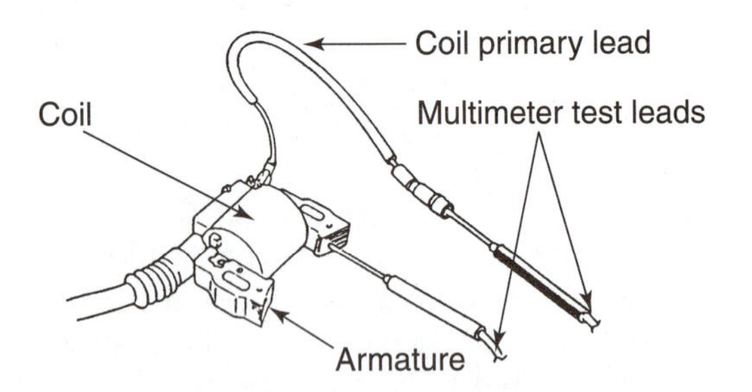

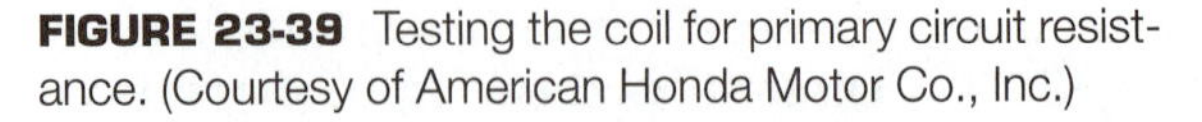
FIGURE 23-39 Testing the coil for primary circuit resistance. (Courtesy of American Honda Motor Co., Inc.)

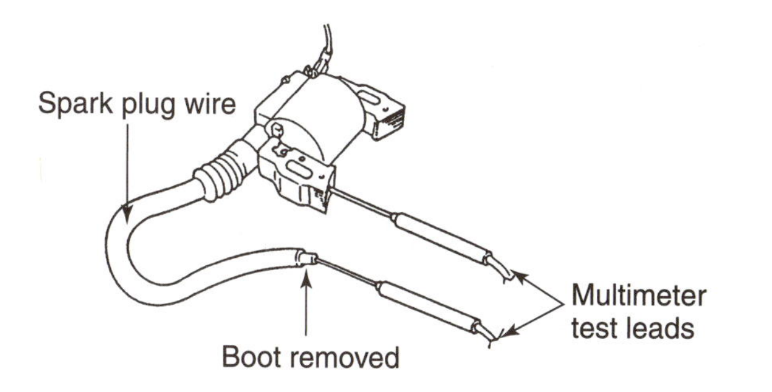

FIGURE 23-40 Testing the coil for secondary circuit resistance. (Courtesy of American Honda Motor Co., Inc.)

resistance. Compare your reading with the coil secondary resistance specifications in the shop service manual.

If you find that either the primary or secondary resistance is higher or lower than specifications, the magneto coil has to be replaced. An infinite reading on either test means that the primary or secondary coil has an open circuit. The coil must be replaced.

The Merc-O-Tronic (Figure 23-41) is a common type of magneto dynamic tester. Both testers are made to test either solid state or breaker points magneto ignitions. Either tester can be used to test several different engine makers' ignition systems. The testers come with an instruction manual. The manual describes the hookup for each test the tester can perform. Typical tests performed by these testers include:

- Condenser capacity (static test)
- Primary coil continuity (static test)
- Coil firing or coil power (dynamic test)
- Insulation leak test (dynamic test)

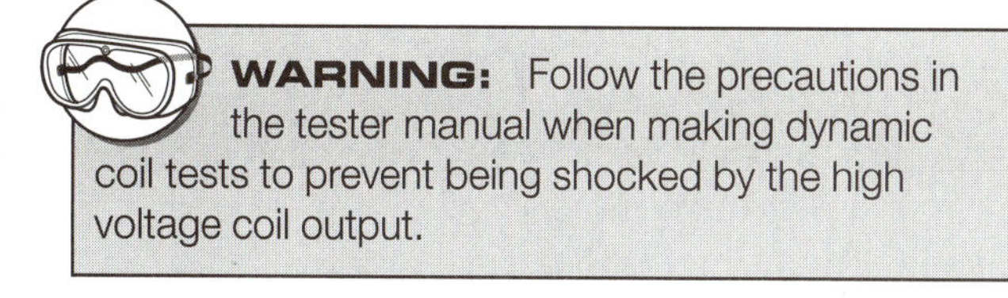

WARNING: Follow the precautions in the tester manual when making dynamic coil tests to prevent being shocked by the high voltage coil output.

The two common dynamic tests are coil power and surface insulation. These tests allow you to see how well the coil works. You can determine if the coil develops enough high-voltage secondary current for a spark across the spark plug gap. The first step in using the tester is to look up the specifications. Coils are listed by coil model (part) number. Each engine maker's coil model number has a specified operating amperage. Operating amperage is the maximum amount of amperage required in the coil primary circuit to develop the required amount of voltage in the secondary circuit. Specifications can be found in the ignition section of the engine shop service manual or in the tester manual (Figure 23-42).

Follow the directions in the tester manual to do a coil power test on a breaker point type coil.

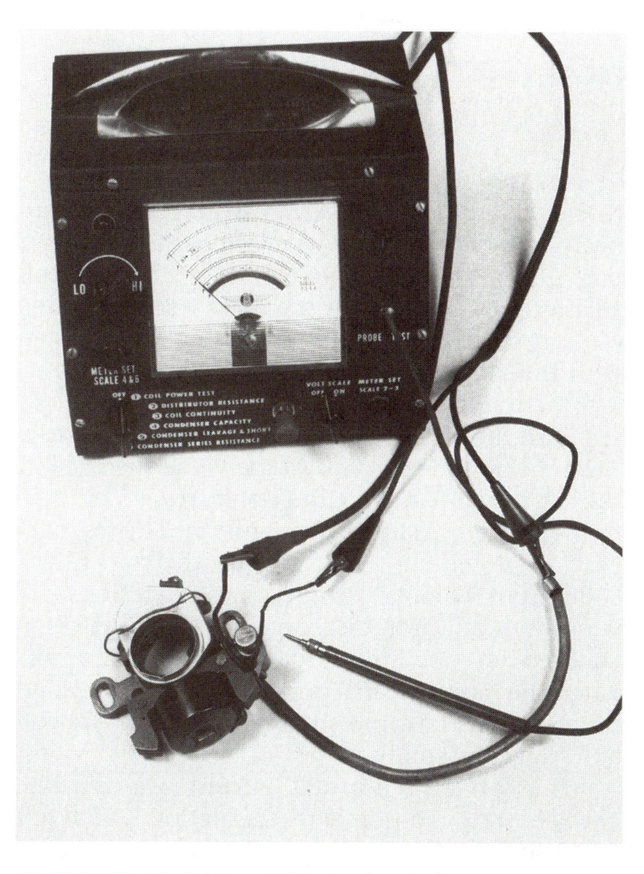

FIGURE 23-41 A Merc-O-Tronic electrical component tester.

COIL	OPERATING AMPERAGE	COIL	OPERATING AMPERAGE
27947	2.8	610371	2.8
29176	2.8	619466	2.8
29632	2.3	610477	2.8
30546	2.7	610586	2.8
30560A	1.6	610633	1.8
30560	1.6	610657	2.25
34431	1.6	610678	2.25

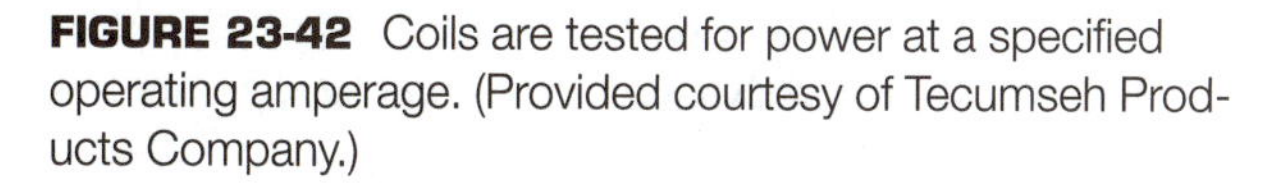

FIGURE 23-42 Coils are tested for power at a specified operating amperage. (Provided courtesy of Tecumseh Products Company.)

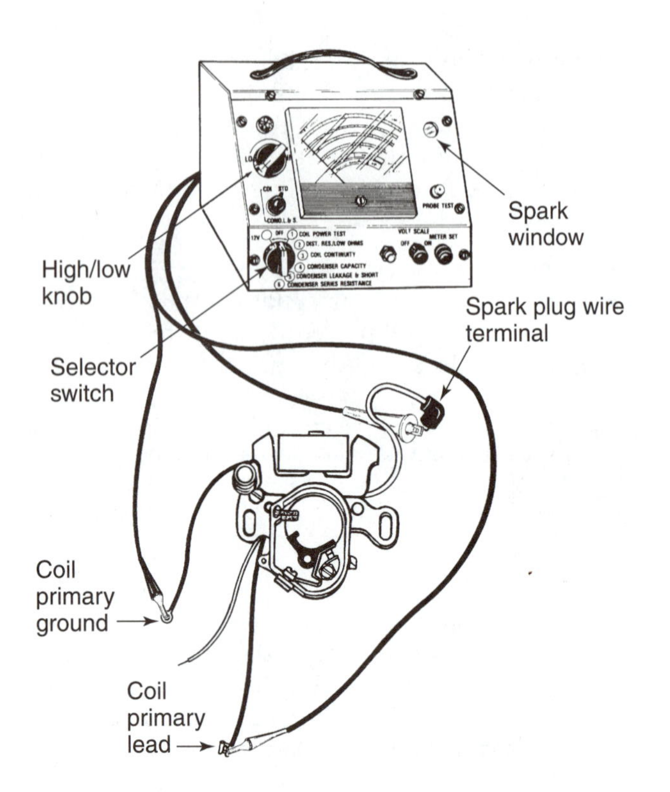

FIGURE 23-43 Testing the coil power on a breaker point coil. (Provided courtesy of Tecumseh Products Company.)

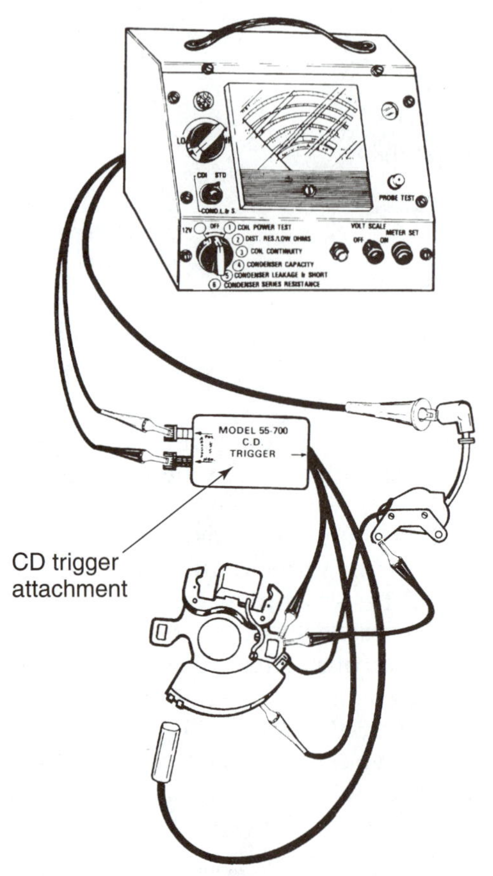

FIGURE 23-44 Testing the coil power on a solid state ignition coil. (Provided courtesy of Tecumseh Products Company.)

Connect a tester lead to the primary ground wire (Figure 23-43). Connect a tester lead to the coil primary coil lead. Connect the large high-voltage tester lead to the coil spark plug terminal wire. Turn the tester selector switch to the Coil Power Test position. Rotate the Low-Hi tester knob clockwise while reading the amperage on the tester scale. During this test, the tester sends its own on and off primary current into the coil. Watch as the scale reading reaches the operating amperage specified for the coil. You should see a steady spark jumping the electrodes in the tester spark window. The spark window has a 5 millimeter gap for the spark to jump. This gap simulates a spark plug electrode gap.

A good coil develops enough voltage for the spark to jump the gap in the window at or below the specified operating amperage. A weak coil requires higher than specified amperage to get the spark to appear in the window. A coil with an open circuit does not develop the spark at any amperage. Replace the coil if it requires higher than specified amperage. Replace the coil if it does not develop any spark.

The solid state ignition coil and trigger are usually one replacement part. They are tested at the same time. Follow the test procedures in the tester manual. One tester lead is connected to the coil primary terminal. Another tester lead is connected to the armature ground. A high-voltage lead is connected to the spark plug wire terminal. The trigger part of the ignition is tested with a trigger coil adapter unit. This part is provided with the tester. Capacitor discharge ignition systems are tested with the same tester using a special CD trigger attachment (Figure 23-44).

Turn the tester switch to the Coil Test position. Set the Coil Index switch to the value specified in the tester manual. Place the tester trigger coil next to the trigger pickup on the ignition. The test trigger coil provides a test pulse into the coil. Check coil output by looking for spark in the tester spark window. Capacitor discharge systems are often checked for one single spark. The spark occurs when the ignition capacitor discharges. If the solid state ignition fails the spark test, replace the coil and trigger assembly.

The coil surface insulation test is a dynamic check of the condition of the insulation on the coil and spark plug cable (Figure 23-45). Follow the directions for this test in the tester manual. The tester connections are similar to the coil power test. The

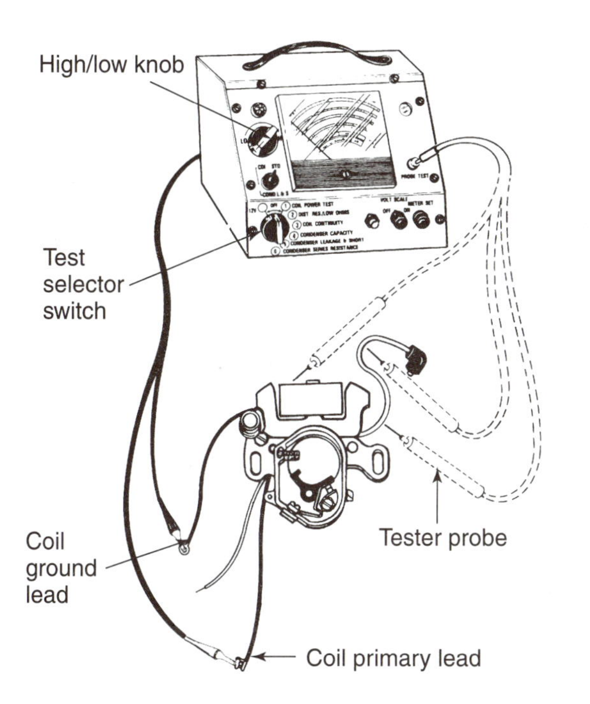

FIGURE 23-45 Testing the ignition surface insulation. (Provided courtesy of Tecumseh Products Company.)

difference is that the high voltage tester lead is not connected to the spark plug terminal. Turn the tester selector switch to the Coil Power Test position. Rotate Low-Hi knob while watching the scale reading. Stop rotating the knob when the coil power amperage specification is reached. Pass the tester probe over the coil insulation and the spark plug wire. Look for a spark discharge between the insulation and the tester probe. Any spark shows cracked or damaged insulation. Replace the coil if you find damaged insulation. The spark plug wire may or may not be replaceable. If you find a damaged spark plug wire, replace the wire or the coil and wire.

FLYWHEEL INSTALLATION AND ARMATURE GAP SETTING

If the flywheel has been removed for inspection or access to ignition parts, replace it using a new key. Use the same holding device used during removal to hold the flywheel. Install the flywheel nut or rewind starter clutch. The shop service manual may specify a torque for the flywheel nut. Use a torque wrench if there is a torque specification.

Install the magneto coil armature and start the screws but do not tighten. The holes in the magneto to armature are slotted. The slots allow the setting a gap between the armature and the flywheel magnets. The **armature air gap** (E gap) is the space between the legs of the magneto armature and the flywheel magnets when the magnets are positioned under the armature (Figure 23-46). This gap cannot be measured accurately with a metal feeler gauge. Magnetism will cause the metal feeler gauge blades to stick to the flywheel magnets. A paper gauge that is the correct thickness is used to set the gap.

An accurate armature air gap setting is very important. The gap has an effect on engine ignition timing. Too wide or narrow of a gap will cause incorrect ignition timing and poor engine performance. If the gap is too narrow, the flywheel may contact the armature causing damage. Too wide a gap can cause poor magnetic contact from the magnets to the armature that results in a no spark condition.

Look up the specification for armature air gap in the shop service manual. Select the correct size paper card. Cards are provided for this purpose by engine makers. Rotate the flywheel until the magnets are lined up directly under the legs of the armature. Place the paper card between the flywheel magnets and armature. Push the armature down against the paper card (Figure 23-47). Tighten the armature screws to secure the armature. Using a paper gauge to set the magneto armature gap is shown in the accompanying sequence of photographs.

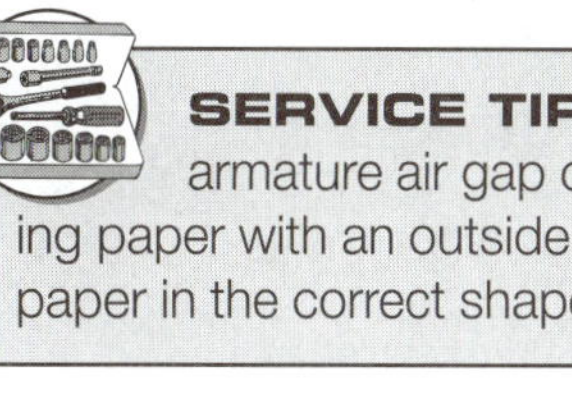

SERVICE TIP: A card for measuring armature air gap can be made by measuring paper with an outside micrometer. Cut the paper in the correct shape to fit on the magnets.

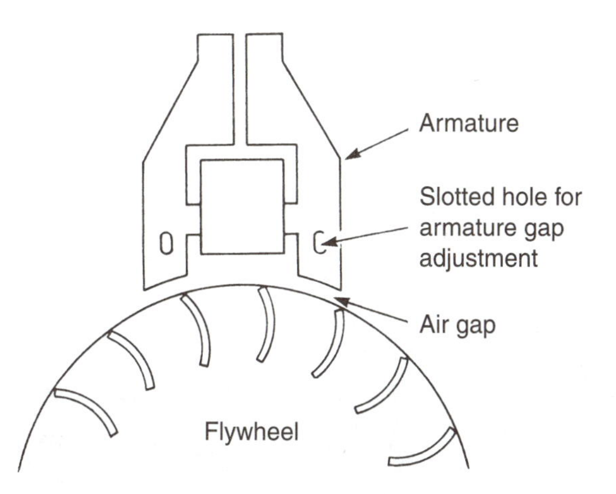

FIGURE 23-46 The armature air gap is adjusted with slotted screws.

PHOTO SEQUENCE 10

Using a Paper Gauge to Set the Magneto Armature Gap

1. Use a paper gauge that is the correct thickness to set the armature gap.

2. Remove the blower (fan) housing to access the magneto armature.

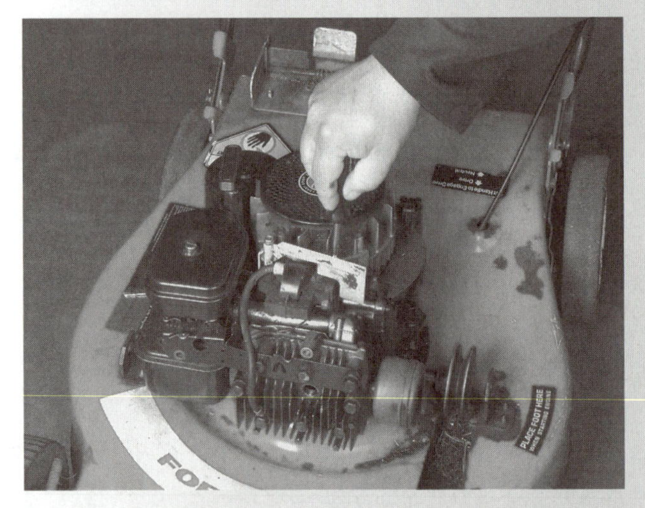

3. Loosen the fasteners that hold the armature.

4. Rotate the flywheel until the magnets are lined up directly under the legs of the armature.

5. Place the paper card between the flywheel magnets and armature.

6. Push the armature down against the paper card and tighten the armature screws to secure the armature.

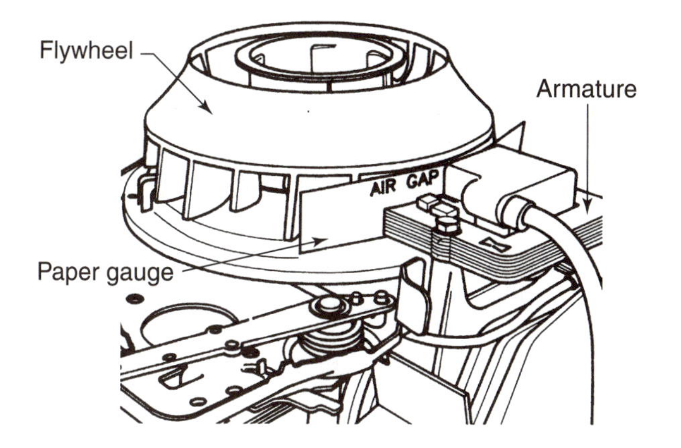

FIGURE 23-47 Measuring and setting air gap with a paper gauge. (Provided courtesy of Tecumseh Products Company.)

REVIEW QUESTIONS

1. Describe how the appearance of the spark plug firing can help determine the source of an engine performance problem.
2. Explain how to use a spark plug gapping tool to measure and adjust a spark plug gap.
3. Describe how to use a thread chaser to repair spark plug threads.
4. Explain how to install a thread insert to repair a spark plug thread.
5. List and describe the parts of the flywheel and key that should be inspected after disassembly.
6. Explain how to use a flywheel puller to remove a flywheel.
7. Explain why eye protection should be worn when using a flywheel puller.
8. Explain how worn breaker points affect ignition operation.
9. Describe the procedure for removing and replacing breaker points.
10. Explain how to test an ignition condenser.
11. Explain how to set the gap on plunger-operated type breaker points.
12. Explain how to set the gap on cam-operated breaker points.
13. Describe how to measure piston position with a dial indicator.
14. Explain how to adjust timing when the piston has been positioned with a dial indicator.
15. Explain how to use a multimeter to test a magneto coil secondary circuit.
16. Describe how to use a multimeter to test a magneto coil primary circuit.
17. Explain how to use a tester to measure magneto coil power on a breaker point type coil.
18. Explain how to use a dynamic tester to test a coil and trigger on a solid state ignition.
19. Describe how use a dynamic tester to check coil surface insulation.
20. Describe how to adjust the armature air gap.

DISCUSSION TOPICS AND ACTIVITIES.

1. Locate a collection of used shop spark plugs. Study the firing end of each of the spark plugs and try to determine engine operating conditions.
2. Locate several used magneto coils. Practice testing the coils for coil power and circuit resistance.
3. A spark occurs in the cylinder at 10 crankshaft degrees BTDC. Combustion ends at 20 crankshaft degrees ATDC. How far has the crankshaft traveled during combustion?

Fuel System Tune-up

OBJECTIVES

Upon completion and review of this chapter, you should be able to:

- Troubleshoot the engine to locate a fuel system problem.
- Service the fuel tank.
- Troubleshoot and service the fuel pump.
- Troubleshoot for carburetor problems.
- Overhaul a vacuum carburetor.
- Overhaul a four-stroke diaphragm suction feed carburetor.
- Overhaul a float carburetor.
- Overhaul a suction feed diaphragm carburetor.
- Adjust a carburetor idle speed, low-speed mixture, and high-speed mixture.
- Adjust the governed speed of an engine.

TERMS TO KNOW

Engine flooding
Engine overspeeding
Engine surging
Float gauge
Float level setting
Governor top speed adjustment
Vapor lock

INTRODUCTION

Servicing the fuel system is an important part of every tune-up. The ignition is usually serviced first, then the fuel system. You will need to determine how much fuel system service is necessary. Often just adjusting the carburetor is all that is necessary. Major fuel system service is necessary if the fuel system parts have not been serviced in a long time. Major service must be done if the fuel system has been used with dirty fuel. What kind of fuel system service is required can be determined by step-by-step troubleshooting.

TROUBLESHOOTING THE FUEL SYSTEM

The first step in fuel system troubleshooting is to find out if the air and fuel mixture is too rich or too lean. A mixture that is too rich can have either too much fuel or not enough air. A mixture that is too lean can have either too much air or not enough fuel. The fastest way to determine if there is a fuel problem is to remove the spark plug. Remove both spark plugs from a multiple cylinder engine. The firing end of the spark plug should have a light brown or tan color. This color shows the mixture has the correct amount of air and fuel. Both spark plugs in a multiple cylinder engine should have the same color.

The air-fuel mixture is too rich if the spark plug firing end is covered with black soot. The problem is too much fuel or not enough air. The black soot from a rich burning cylinder will be relatively dry. An oil fouled spark plug will appear wet. Oil can usually be wiped off the firing end by hand. A too lean fuel mixture is shown by a spark plug firing end that is dry and white in color. The engine is not getting enough fuel or it is getting too much air.

Check all the parts that can affect the air-fuel mixture. Do not immediately blame the carburetor. Inspect the complete fuel system:

- Fuel tank
- Tank strainer
- Tank shutoff valve
- In-line fuel filter
- Fuel lines
- Carburetor
- Air cleaner
- Throttle linkage

If the mixture appears too rich, check the air cleaner and the carburetor choke. Air into the engine will be restricted if the air cleaner is dirty. A stuck choke valve (or linkage) allows extra fuel into the carburetor venturi. The easiest way to find an air cleaner problem is to clean or replace the filter element. Make sure the choke valve stays in the open position during testing to eliminate overchoking as the problem.

WARNING: Do not test an engine by removing the air cleaner element. A back fire through the carburetor throat can cause burn injuries.

If the mixture is too lean, the problem might be:

- Empty fuel tank
- Clogged fuel tank outlet
- Clogged fuel tank strainer
- Clogged shutoff valve
- Clogged in-line fuel filter

Check these problems by disconnecting the fuel line at the carburetor (Figure 24-1). Place the end of the line in a metal container. You should see a strong flow coming out of the fuel line. If there is not a strong flow, remove each part in turn until the flow increases. For example, remove the in-line fuel filter and check the flow again. If the flow is normal with it removed, the filter is the problem.

WARNING: When testing fuel flow, always catch the fuel in a metal container and dispose of it properly. Be sure to service the fuel system in a well-ventilated area away from any source of ignition. Always have a fire extinguisher available when servicing any part of the fuel system. Electric starting systems should have the battery properly disconnected while servicing the fuel system to prevent any possible arcing that could ignite a fire.

An air leak where the carburetor is mounted to the engine allows too much air into the engine (Figure 24-2). This will cause a lean a mixture from too much air. The carburetor is often mounted to the engine with a gasket or O ring seal. Multiple cylinder engines have an intake manifold. The manifold routes the air-fuel mixture to each cylinder from a single carburetor. The intake manifold is attached to each cylinder by a gasket. The carburetor is attached to the intake manifold with a gasket. The air intake system is at less than atmospheric pressure. Any leak at these connections pulls in outside air. The extra air causes a lean mixture. Inspect each of these areas carefully. Tighten or replace any parts found to be defective.

FIGURE 24-1 Remove the fuel line from the carburetor and check for fuel flow.

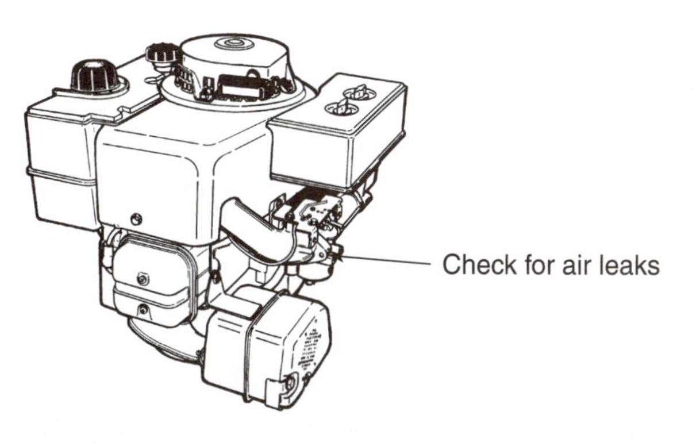

FIGURE 24-2 Check the intake system for air leaks. (Provided courtesy of Tecumseh Products Company.)

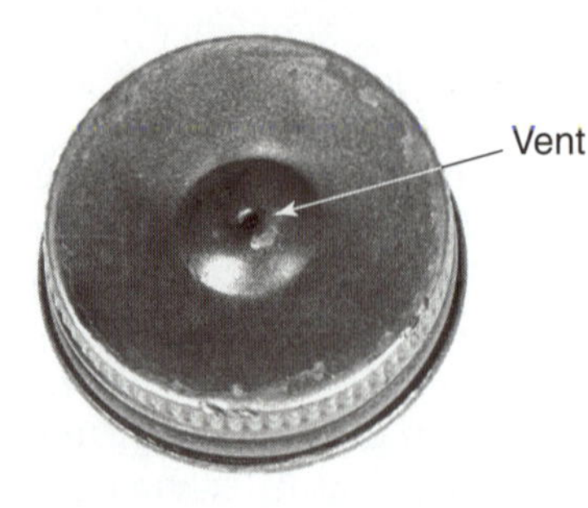

FIGURE 24-3 Check the fuel tank cap to be sure the vent is not plugged.

SERVICE TIP: A good troubleshooting technique is to use duct tape to temporarily seal a suspected air leak. Then see if engine performance is affected. The engine usually slows down when a leak is sealed off because the engine gets less air.

A too lean mixture problem can sometimes be traced to a plugged fuel tank vent. The vent is located in some fuel tank caps (Figure 24-3). It can get plugged from grass or dirt in the air. A plugged vent does not allow air to enter the tank. Air is not able to take the space left by the fuel as it is used. A low pressure is created in the tank above the fuel. This problem can slow the flow of fuel. The air-fuel mixture will be too lean. Correct this problem by trying a new cap and checking engine operation.

The carburetor is the likely problem when none of these troubleshooting steps locates the problem. Before you service the carburetor, check the shop service manual for a carburetor troubleshooting guide. The manual will often have a step-by-step service and adjustment procedure to cure a specific carburetor problem.

FUEL TANK SERVICE

If the fuel is allowed to stay in the tank for a long time, it can become stale. Stale fuel smells like varnish or paint. Stale fuel causes the engine to run poorly or not run at all. Stale fuel must be drained from the fuel tank. Remove the drain plug at the bottom of the fuel tank. Drain the fuel into a metal container. If there is no drain, turn the equipment on its side to drain the tank.

If the equipment cannot be tipped over, the tank must be removed (Figure 24-4). If the fuel tank has a shutoff valve, close it before disconnecting any fuel lines. If there is no shutoff valve, squeeze the fuel line shut with locking pliers. Disconnect the fuel line from the tank to the carburetor. Plug the end of the fuel line to prevent fuel spillage. Remove the fuel tank to engine fasteners to remove the tank. Drain and clean the tank and install it back on the engine.

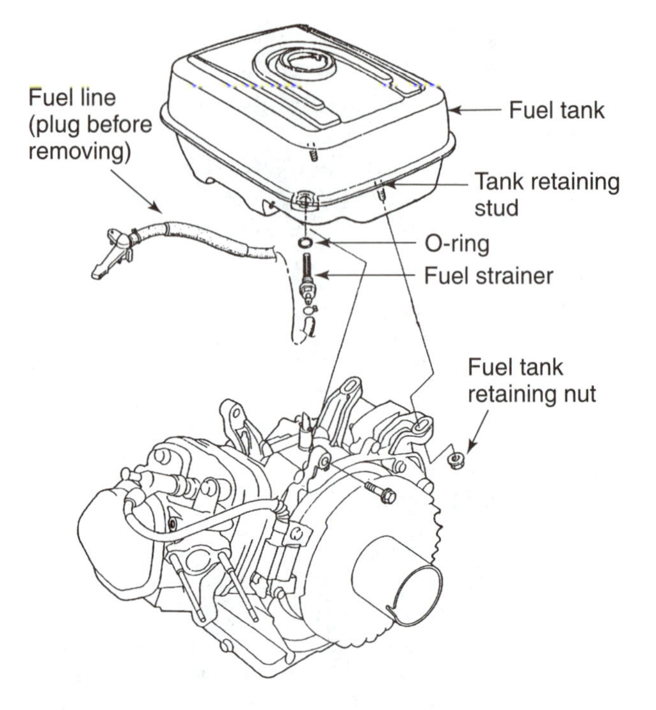

FIGURE 24-4 Remove the retaining fasteners and remove the fuel tank. (Courtesy of American Honda Motor Co., Inc.)

After draining, flush the tank with fresh cleaning solvent to wash out any stale fuel. The inside of the tank should be checked thoroughly for signs of rust. A tank on an engine left outdoors for a long time may get rusty. Unless the rust is completely removed, it can flake off. Rust can pass through the fuel outlet into the carburetor where it can plug up small fuel passages.

The best fix for a rusty fuel tank is to replace the tank. If a tank is not available, fill the tank with small ball bearings (Figure 24-5). Shake the tank up and down. The ball bearings will knock off the rust. Flush the tank with clean solvent and then clean fuel.

Most fuel tanks have a fuel strainer located in the bottom of the fuel tank. The strainer is removed by unscrewing it from the tank. Hold the strainer up to the light. Check to see if it is plugged with dirt, rust, or sediment. Replace or clean the strainer if it is plugged. Install the strainer using new O rings or gaskets.

Carefully inspect all the fuel lines that connect the fuel tank to the carburetor. Flexible fuel hoses can turn hard and brittle with age. Squeeze each hose to see if has become brittle. Check the end of

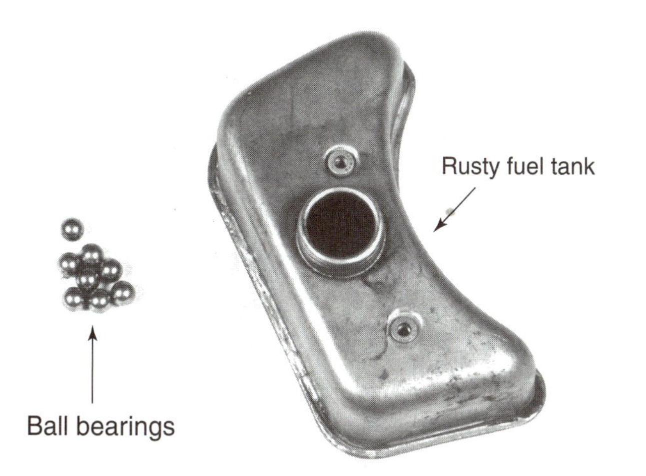

FIGURE 24-5 Shaking ball bearings inside the fuel tank will remove rust.

each hose for cracks. Replace any hose that has cracks or brittleness.

FUEL PUMP TROUBLESHOOTING AND SERVICE

A problem with a mechanical or impulse fuel pump can cause not enough fuel to be pumped from the fuel tank to the carburetor. Troubleshooting procedures are the same for both fuel pumps. A bad fuel pump often results in a no start condition. A bad pump can also cause an engine performance problem. The engine may run but will not operate at high load.

When you have a no start problem, check the spark plug for a lean condition. If the spark plug shows a lean mixture, check the fuel delivery system to the pump. Check the fuel tank cap vent for plugging. Check fuel flow through the fuel tank filter, in-line filter, and fuel lines. Be sure that fuel lines are routed properly and not pinched or twisted. Fuel lines that are incorrectly routed close to hot engine parts can result in vapor lock. **Vapor lock** is a condition in which heated fuel in a fuel line turns from liquid to a vapor and cannot be pumped by a fuel pump. Suspect vapor lock if the engine has performance problems in hot weather.

If the fuel delivery to the pump is satisfactory, you need to do a fuel pump test (Figure 24-6). Remove the fuel line from the carburetor. Connect a length (approximately 18 inches long) of test fuel line to the end of the engine fuel line. Put the end of the test fuel line into a metal container to collect gasoline. Crank the engine by hand or electric starter for approximately 20 seconds. There should be a strong fuel flow into the container. If there is no fuel flow or low fuel flow, the pump is defective and must be replaced.

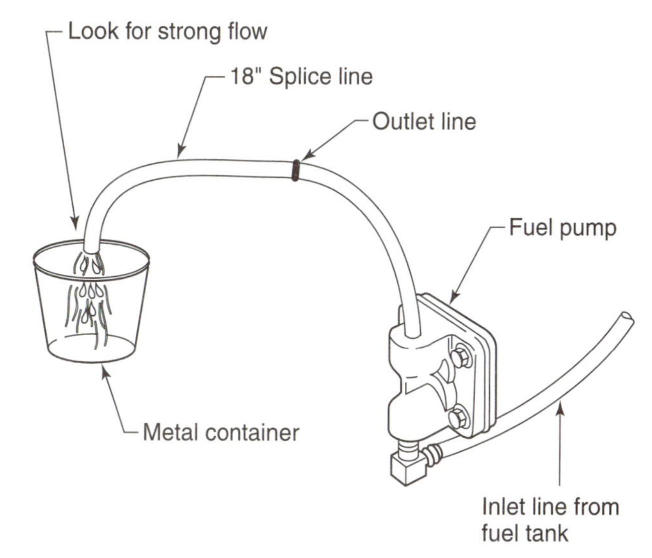

FIGURE 24-6 Checking the fuel flow from the fuel pump.

If the engine will start but not run at high load, make the same checks of the delivery system. Disconnect and plug the fuel line from the pump. Attach a temporary gravity feed gasoline tank and fuel to the engine. Connect a fuel line from the test fuel tank to the carburetor. Operate the engine at high load. If the temporary fuel system solves the problem, replace the fuel pump.

WARNING: Always plug fuel lines from the tank and turn off any tank fuel shutoff valve when servicing the fuel pump to prevent fuel spillage and a possible fire hazard. Use a metal drip pan under the fuel pump during disassembly to collect any fuel.

To remove and replace a mechanical fuel pump, disconnect the fuel lines from the fuel pump inlet and outlet (Figure 24-7). Make careful note of which line is the inlet and which line is the outlet. Remove the pump mounting fasteners. Lift the pump and pump gasket off the engine. Note the pump lever position as you pull the pump away. The replacement pump lever must go in the same position. Compare the new and replacement pump to be sure they are the same. Install a new fuel pump gasket on the engine. Place the pump in position on the engine. Be sure the pump lever is engaged against the camshaft in the correct position. Replace and tighten the fasteners. Connect the fuel

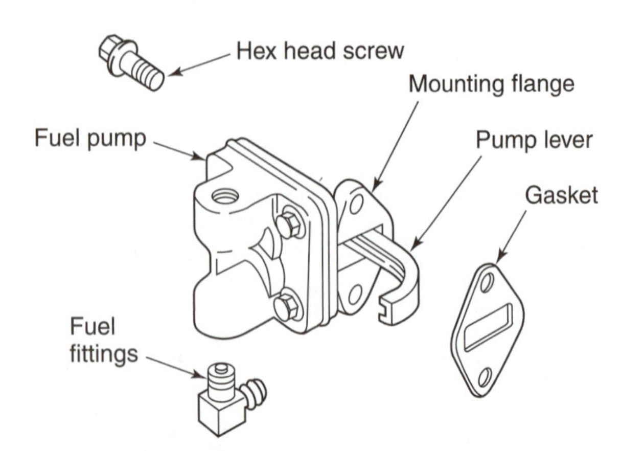

FIGURE 24-7 Replacing the mechanical fuel pump. (Courtesy of Kohler Co.)

lines to the fuel pump inlet and outlet fittings. Start the engine and check all the fuel lines carefully for fuel leaks.

CASE STUDY

A student brought her parents' garden tractor into the shop for help. The tractor was being used to mow a meadow that weekend. The student reported that tractor ran fine all morning but began to misfire and loose power during the afternoon. It had acted as though it was running out of fuel. The student started the tractor and it appeared to be running fine at the moment.

The instructor thought the problem might be temperature related. The weekend had been hot, approaching the 100° mark. He had the student remove the engine cover for an inspection. The tractor fuel tank had been replaced by the student's father for a larger capacity unit. The new tank had required the routing of a new fuel line from the tank to the carburetor. This was done with flexible fuel line which was routed next to the engine cylinder head. The instructor explained to the student that the fuel line may be overheating and creating the opportunity for vapor lock.

The instructor had the student fabricate a new fuel line out of steel. The new fuel line was routed away from the hot cylinder head. The tractor was used in hot conditions the next weekend and ran fine. The problem was vapor lock created by fuel in the heated fuel line turning from a liquid to a vapor. The tractor fuel pump could not pump a vapor so the engine appeared to be running out of fuel.

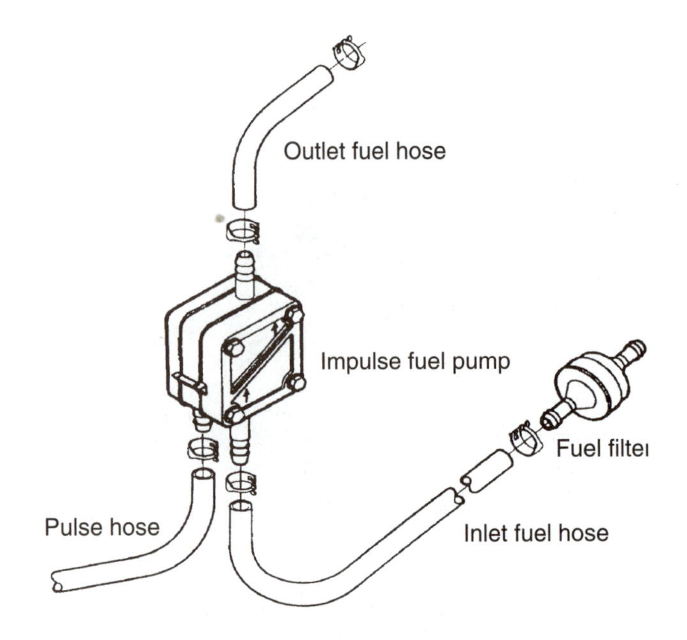

FIGURE 24-8 Replacing the impulse fuel pump. (Courtesy of Onan Corp.)

To replace the impulse type pump (Figure 24-8), disconnect the pulse hose from the fuel pump. Disconnect the inlet and outlet fuel hoses from the pump. Remove the fuel pump to engine fasteners. Install the new pump in the same position as the original. Install the pulse, inlet, and outlet fuel hoses. Start the engine and check for any fuel leaks.

CARBURETOR TROUBLESHOOTING AND SERVICE

A carburetor problem can cause either a too lean or too rich condition. Troubleshoot the suspect carburetor to find the exact cause of the problem. Many times, just disassembling and inspecting the parts fails to find the problem. Always identify the exact carburetor you are working on. Then locate and use the exact service procedures in the shop service manual.

Troubleshooting

Every carburetor has a system or circuit for each engine-operating mode. These are starting, idle, acceleration, and high speed. Try to isolate the carburetor problem to one of these individual circuits. Troubleshooting can be simplified if the problem can be isolated to one circuit. For example, a carburetor has a problem that occurs only during idle. You should spend your time on the carburetor parts that provide air and fuel to the idle circuit.

You should also try to isolate the cause of an air or a fuel problem. A rich mixture can be caused by too much fuel or not enough air. A lean mixture can be caused by too little fuel or too much air.

SERVICE TIP: If you suspect that an engine will not start because of a fuel problem, remove the spark plug. Dip the spark plug firing end into a container of gasoline. Install the spark plug. Try to start the engine. If it runs momentarily on the fuel on the spark plug, the engine probably has a fuel problem.

Carburetor problems can make the engine hard to start. Hard starting may be caused by a rich or lean mixture. Always look for both fuel and air problems. Check out the carburetor parts that provide air and fuel during starting. Common carburetor fuel problems that can cause hard starting are:

- Restrictions in a fuel pickup (lean mixture)
- Damaged adjustment needles (rich or lean mixture)
- Restrictions in the main nozzle (lean mixture)
- Plugged fuel inlet (lean mixture)
- Dirty or stuck inlet needle and seat (rich or lean mixture)

A dirty or stuck inlet needle valve is a common cause of **engine flooding,** a problem in which too much fuel enters the intake system and covers the spark plug electrode. The wet spark plug has a problem igniting the mixture.

Common carburetor air system problems that could cause hard starting are:

- Worn throttle shaft (lean mixture)
- Worn choke shaft (lean mixture)
- Improperly operating choke system (rich or lean mixture)
- Restricted air bleed (rich mixture)

Removing the Carburetor

When troubleshooting shows a problem inside the carburetor, it is removed from the engine for service (Figure 24-9). Shut off the fuel at the fuel tank. If there is a shutoff valve, close the valve. If there is no shutoff valve, clamp the fuel line from the fuel tank. Disconnect and plug the fuel line from the carburetor. Place a metal container under the fuel inlet fitting to catch any fuel in the fuel line. Plug the carburetor fuel line to avoid fuel spillage. Remove the air cleaner and gasket to access the carburetor fasteners. Before disconnecting the governor and throttle linkage, note the position of all

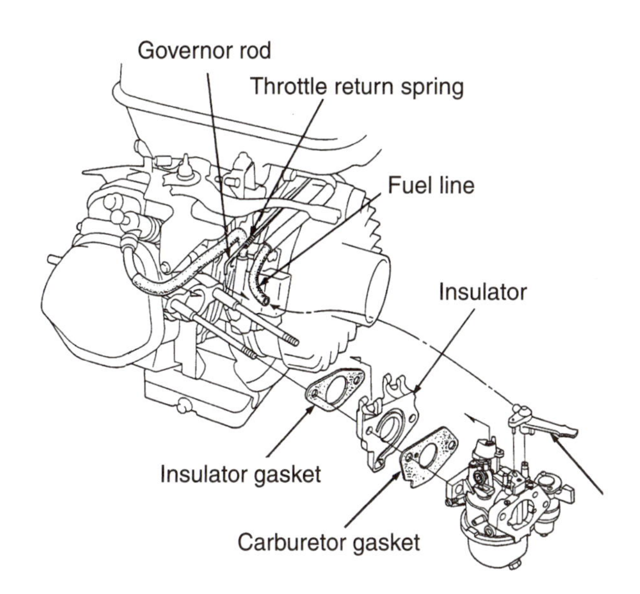

FIGURE 24-9 Removing the carburetor. (Courtesy of American Honda Motor Co., Inc.)

linkages and springs. Make a sketch if necessary to help you install these parts. Remove the carburetor mounting fasteners. Some carburetors have an insulator that fits between the carburetor mount and the engine. This lowers the heat that gets to the carburetor. The insulator has its own gasket. Carefully lift the carburetor off the engine. Remove the carburetor gasket or O ring seal. Hold the float carburetor level to avoid spilling fuel from the float bowl.

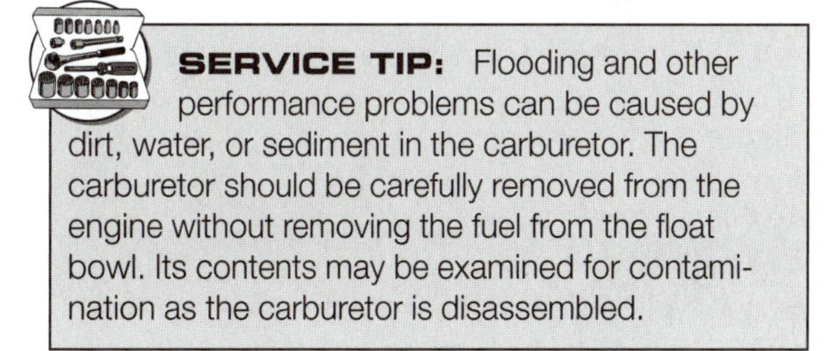

SERVICE TIP: Flooding and other performance problems can be caused by dirt, water, or sediment in the carburetor. The carburetor should be carefully removed from the engine without removing the fuel from the float bowl. Its contents may be examined for contamination as the carburetor is disassembled.

Identifying the Carburetor

The carburetor must be identified by name and model so that the correct service procedures can be located and followed. The identification can be difficult because some engine makers make their own carburetors. Others buy carburetors from carburetor makers. Sometimes one model of engine can have any one of many different carburetor models.

There are several ways to identify a carburetor. First, use the engine model number to locate the correct shop service manual for that engine. The carburetor section of the shop service manual often shows how to identify the carburetor model.

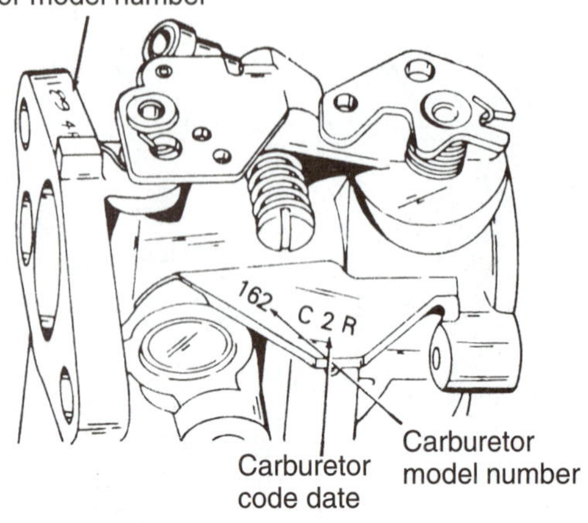

FIGURE 24-10 Use the carburetor model number to locate service procedures. (Provided courtesy of Tecumseh Products Company.)

Sometimes you will have to match up the carburetor with illustrations in the shop service manual. Many carburetors have a model number stamped on the carburetor housing. Use the model number to locate the correct service procedures (Figure 24-10).

Disassembling and Cleaning the Carburetor

Carburetor overhaul is the complete disassembly, cleaning, inspection, and replacement of worn or damaged parts. After disassembly, carburetor parts should be sorted into metal and nonmetal parts. Metal carburetor parts are cleaned with carburetor cleaner. Most shops use spray can type cleaners (Figure 24-11). Nonmetal parts are plastic floats,

FIGURE 24-11 Metal carburetor parts can be cleaned with spray carburetor cleaner. (Courtesy of CRC Industries Inc.)

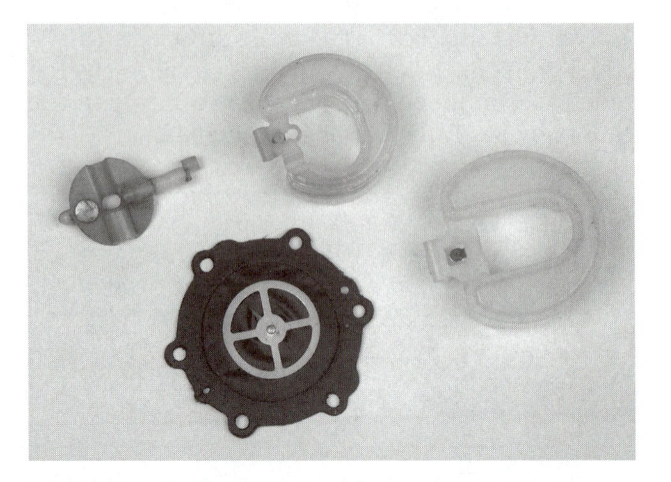

FIGURE 24-12 Nonmetal parts can be damaged by carburetor cleaner.

rubber diaphragms, and nylon carburetor bodies (Figure 24-12). Do not clean nonmetal parts with carburetor cleaner as it may damage these parts. Wash these parts with detergent and water.

Rinse the parts with hot water after spraying to remove all the cleaning chemicals. Place the cleaned parts on a clean shop rag to air dry. All the water used in rinsing must be allowed to dry off the parts. Water in the carburetor causes engine performance problems. Small inside passages may need to be blown clear with compressed air. Do not use drills to clean jets or passages. These tools remove metal and can change the size of the passages, which changes air-fuel mixtures.

WARNING: Always follow the precautions on the carburetor cleaner container. Wear eye and hand protection when using these materials. Wear eye protection and use proper safety precautions when using compressed air.

SERVICE TIP: Monofilament fishing line may be passed through carburetor passages to clean and check passages. This material will not damage the holes and is safer to use than compressed air.

Inspecting the Carburetor

Inspect the disassembled carburetor components to determine if the carburetor can be overhauled or should be replaced. Carburetor parts are often damaged from water contamination in the fuel sys-

FIGURE 24-13 Water contamination damages carburetor parts.

tem (Figure 24-13). Water gets into the system from rain, contaminated fuel, or tank condensation. Water causes carburetor parts to deteriorate. Replace the carburetor if you see this type of damage.

A crushed float is another condition caused by water in the fuel system (Figure 24-14). Sometimes water in the float bowl is subjected to very cold temperatures. The water freezes and the ice expands and crushes the float. A crushed float sinks to the bottom of the carburetor float bowl and the inlet valve is held open, causing the engine to flood.

Fuel that is not completely drained from the fuel system during storage oxidizes. The oxidized fuel eventually forms gum and varnish that clog the carburetor passages (Figure 24-15). Heavy varnish buildup cannot be properly removed. If you find this condition, replace the carburetor.

Engines that are operated in dirty or sandy air can have rapid carburetor wear, which causes poor carburetor performance. Airborne dirt can get between the carburetor body and the choke or throttle shaft. The abrasive dirt causes rapid wear on the shafts (Figure 24-16). Air can leak around worn shafts and cause poor carburetor performance. Dirt that gets into the fuel can cause rapid wear to the float hinge pin. These worn areas cause a sticky or

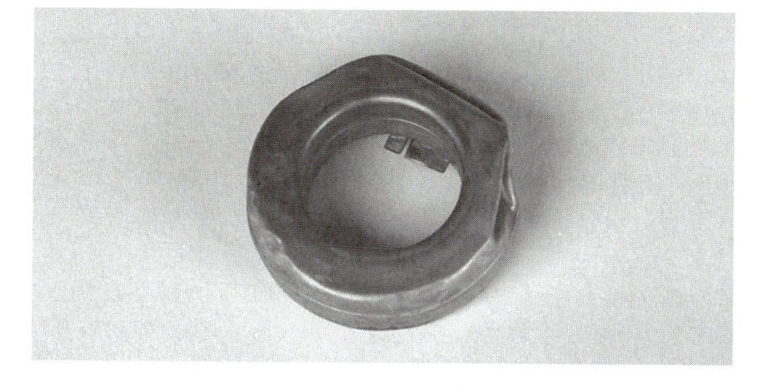

FIGURE 24-14 A crushed float is caused by water freezing in the float bowl.

FIGURE 24-15 Gum and varnish from stale fuel clog carburetor passages.

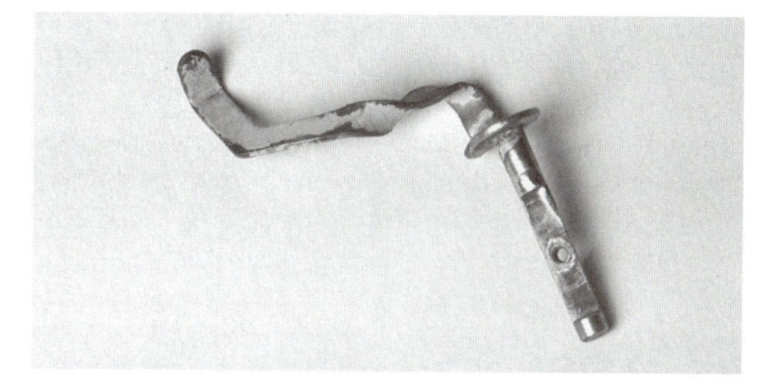

FIGURE 24-16 Worn throttle and choke shafts are caused by dirty operating conditions.

binding float that results in either a flooded or fuel-starved engine. Carburetors with abrasive dirt wear are replaced.

Inspect the high-speed and idle mixture adjusting screws. The screws have a taper on the end. The taper is easily damaged from overtightening or abuse. Broken or worn out tapered ends (Figure 24-17) indicate a carburetor that has been abused. Abused carburetors can be difficult to repair. Replace carburetors with this kind of damage.

Carburetor overhaul kits are available for some carburetors. The kits have the parts normally

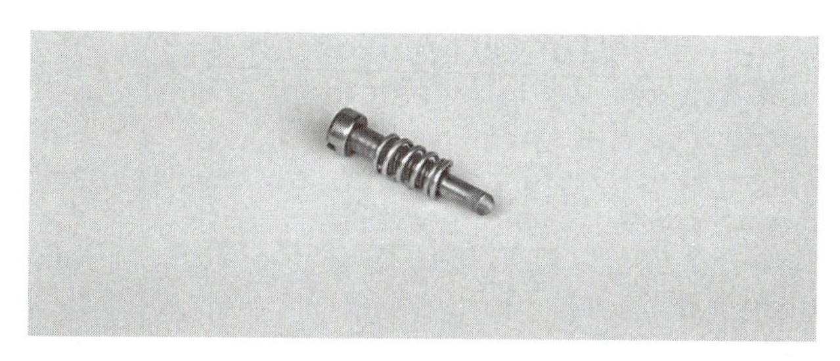

FIGURE 24-17 Worn or broken adjustment screw tapers indicate carburetor abuse.

replaced during an overhaul. Some carburetors do not have kits. Individual parts have to be replaced as necessary. An exploded view and a parts list are helpful for getting the right parts.

Installing the Carburetor

Clean and inspect the carburetor mounting surfaces on the intake manifold or engine. Always use a new gasket or O ring seal when installing the carburetor. Air leaks are common when these sealing parts are reused. An air leak causes a lean air mixture and poor engine performance.

Install the carburetor and other parts that were removed. Tighten the fasteners properly to avoid air leaks. Install the throttle and governor linkage using the sketch made during disassembly or instructions in the shop service manual. Connect the inlet fuel line to the carburetor and open the fuel shutoff valve. Install the air cleaner. Carefully check the fuel line and carburetor for signs of fuel leakage. Any leaks must be repaired before the engine is started. If there are no fuel leaks, start the engine. Allow it to warm up to operating temperature. The carburetor can then be adjusted.

WARNING: Do not prime an engine by pouring gasoline into the carburetor throat. A backfire can cause a fire and burn injuries. Cranking the engine should be all that is necessary to establish fuel flow.

FIGURE 24-18 Remove the carburetor-to-engine screws to remove the carburetor.

FIGURE 24-19 Separate the carburetor from the fuel tank by removing the tank-to-carburetor screws.

VACUUM CARBURETOR OVERHAUL

The vacuum type carburetor is the simplest carburetor design. It has a one-piece housing and few moving parts. Remove the vacuum carburetor and fuel tank at the same time (Figure 24-18). Remove the air cleaner to get access to the carburetor. Carefully note and sketch the position of governor and throttle linkage. Disconnect the linkage. Remove the ground wire if the engine has a throttle linkage stop switch. Remove the two carburetor screws. Lift the carburetor and fuel tank off the engine. Remove and save the carburetor gasket. It will be used to compare with the new one.

Separate the fuel tank from the carburetor by removing the two mounting screws (Figure 24-19). Save the tank-to-carburetor gasket for comparison. Inspect inside the fuel tank for rust, water contamination, dirt or grass, and stale fuel. If the tank does not have rust damage, clean it in solvent and allow it to air dry.

Check the carburetor fuel pickup screen for clogging. There are two styles of fuel pipes used on these carburetors. Older types are made from brass. Newer ones are made from nylon. Both have a check ball and a fine mesh screen. To work properly, the screen must be clean and the check ball must move freely. The pipe requires replacement if the screen and ball cannot be cleaned satisfactorily. Do not clean the nylon type with carburetor cleaner. Nylon fuel pipes are threaded into the carburetor body. Use a socket wrench to remove and replace the nylon fuel pipe (Figure 24-20).

The adjustment screw (needle valve), spring, and adjustment screw seat (jet) are normally replaced each time the carburetor is serviced. Remove the threaded screw with a screw driver (Figure 24-21). Carefully inspect the tapered end of the adjustment screw (Figure 24-22). Look for any nicks, scores, or ridge lines on the tapered surface. Any of these problems will prevent the carburetor from adjusting properly.

FIGURE 24-20 Removing a clogged fuel pipe.

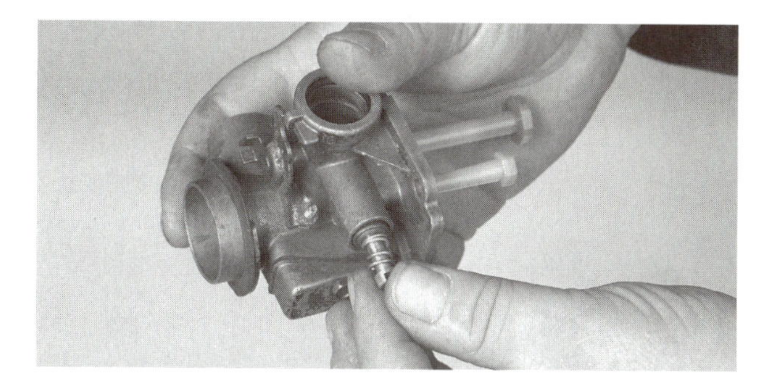

FIGURE 24-21 Removing a mixture adjustment screw.

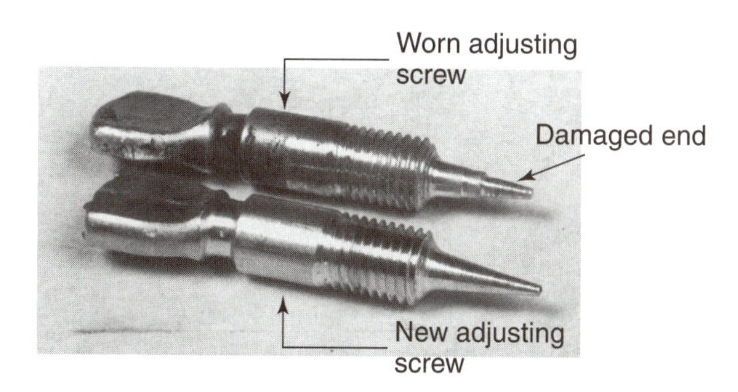

FIGURE 24-22 Replace an adjustment screw if the end is worn or damaged.

Use a wrench to remove the adjustment screw seat (Figure 24-23) from the side of the carburetor. Clean the seat with carburetor cleaner. If the adjustment screw is replaced, the seat must be changed at the same time. A repair kit (Figure 24-24) is available that has an adjustment screw, spring, adjustment screw seat, and washer. Replace the seat with a new washer (gasket). Install the adjustment screw in the seat. Thread it gently down until it bottoms in the seat. Look in the shop service manual for the recommended starting adjustment. Back the adjustment screw off the seat the recommended number of turns. This is the initial setting. You will have to make a final adjustment when the engine is running.

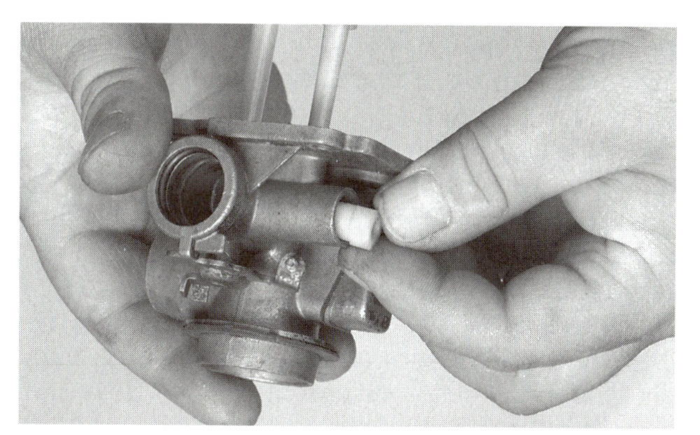

FIGURE 24-23 Removing the adjustment screw seat.

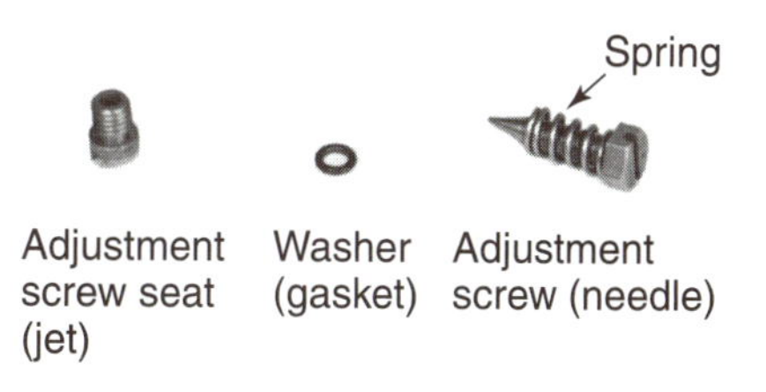

FIGURE 24-24 Parts of a mixture screw repair kit.

Compare the fuel tank and carburetor gaskets with the new ones. They must be exactly the same. Install the carburetor on the fuel tank using the new carburetor-to-tank gasket. Install the fuel tank and carburetor assembly on the engine using the new engine-to-carburetor gasket. Tighten the mounting fasteners to ensure an airtight seal. Connect the throttle and governor linkage. Follow the disassembly sketch or directions in the shop service manual. Replace any disconnected magneto stop wires. Install the air cleaner and fill the fuel tank with fresh fuel. Start the engine and check the carburetor for any fuel leaks. Make the mixture and choke adjustments following the recommendations in the shop service manual.

DIAPHRAGM SUCTION FEED CARBURETOR OVERHAUL

The suction feed carburetor has a diaphragm to pump fuel from the fuel tank up to the carburetor. The diaphragm may be mounted in a chamber on the side of the carburetor. Some carburetors have the diaphragm between the carburetor and the fuel tank. The diaphragm pumps fuel through a fuel pickup tube in the bottom of the fuel tank. The fuel is pumped into a fuel well in the fuel tank. Fuel then enters the carburetor from the fuel well through a second, shorter fuel pickup tube.

In some models, vacuum feed carburetors are mounted directly to the engine intake port with screws and a gasket. Others are mounted to an intake tube fastened to the intake port. The carburetor slips over the intake tube and is sealed to the tube with an O ring. A similar tube is used to connect the crankcase breather to a hose on the carburetor.

Both models of carburetors are removed by removing the carburetor and fuel tank as an assembly. Carefully observe all throttle and governor linkage before disassembly. Make a sketch if necessary for proper installation. Disconnect the governor linkage at the throttle. Leave the governor link and governor spring hooked to the governor blade and control lever. Remove any magneto stop wires connected to the throttle linkage. Remove the fuel tank fasteners. Carburetors with an O ring are pulled off the fuel intake tube. Remove the screws from carburetors with screw mounts and separate the carburetor from the engine.

Remove the tank-to-carburetor screws (Figure 24-25). Lift the carburetor off the fuel tank (Figure 24-26). The pumping diaphragm on some models is located between the fuel tank and the carburetor. Remove the diaphragm being careful not to tear it. You will need the old diaphragm so that it may be compared to the new one. Note the position of the diaphragm spring and spring seat before it is removed. These parts are often installed incorrectly on the wrong side of the diaphragm. Incorrect installation prevents the carburetor from working correctly. Side-mounted diaphragms are removed by removing the screws that retain the side cover.

Use a screwdriver to remove the mounting O ring from the carburetor and save it to compare with the replacement (Figure 24-27). Remove and inspect adjustment screw. Remove the adjustment screw (needle) with a screwdriver. Remove the adjustment screw fitting and seat (jet) from the side of the carburetor with a wrench.

FIGURE 24-25 Removing the screws to separate the carburetor from the fuel tank.

FIGURE 24-26 Lifting the carburetor off the fuel tank.

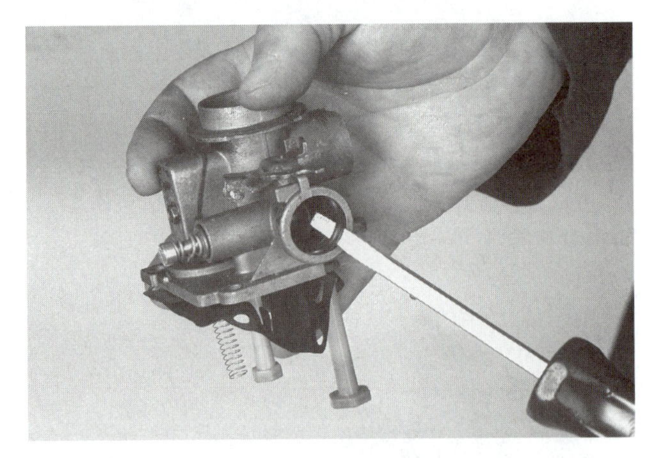

FIGURE 24-27 Removing the carburetor O ring.

Nylon fuel pickup tubes are removed and replaced with a socket wrench. The nylon fuel pickup tube should not be cleaned in carburetor cleaner. Always check the new or cleaned fuel pickup tubes after assembly to make sure they are not plugged. Clean the carburetor body with spray carburetor cleaner. The diaphragm should not be cleaned with carburetor cleaner. Check all the carburetor metering holes to be sure they are clear.

Clean the fuel tank and check it for rust or water contamination. Set a straight edge (scale) across the fuel tank mounting surface (Figure 24-28). Use the recommended thickness gauge and try to slip it under the straight edge. The mounting surface is warped if the gauge slips in. Replace the tank if you find this problem.

Check the throttle shaft for wear by trying to rock it side to side (Figure 24-29). If you find side-to-side movement, the shaft is worn. A worn throt-

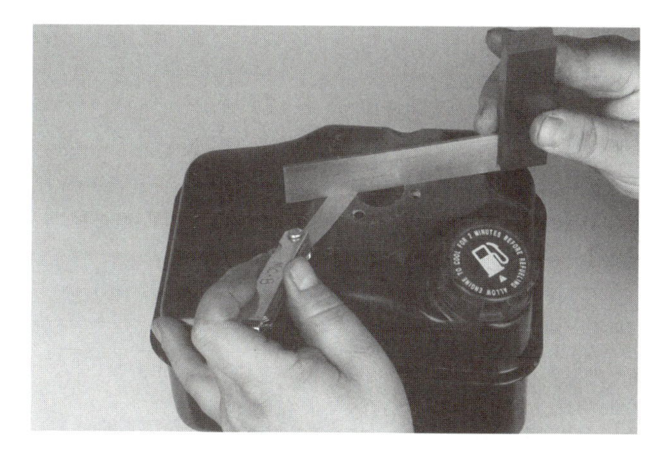

FIGURE 24-28 Checking the fuel tank mounting surface for warping.

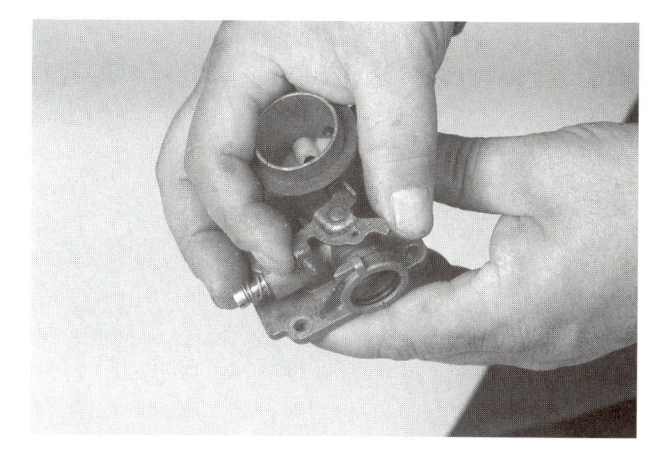

FIGURE 24-29 Rocking the throttle shaft side to side to check for shaft wear.

tle shaft causes an air leak. Replace the carburetor if you find throttle shaft wear.

A new diaphragm is used each time a carburetor is disassembled. Carefully inspect the diaphragm by holding it up to a strong light source. The smallest hole, wrinkle or tear will create a vacuum leak. A leak will prevent the diaphragm from pumping fuel to the carburetor. Compare the new diaphragm to the old one to be sure they are exactly the same. Install the new diaphragm making sure the spring and seat are in proper position (Figure 24-30).

The tank-mounted diaphragm also serves as a gasket between the carburetor and tank. To assemble the carburetor to the fuel tank, first position the diaphragm on the tank. Then place spring seat (cup) and spring on the diaphragm. Install the carburetor and mounting screws. Tighten the screws evenly to avoid distortion. The side-mounted diaphragm is assembled in the same way.

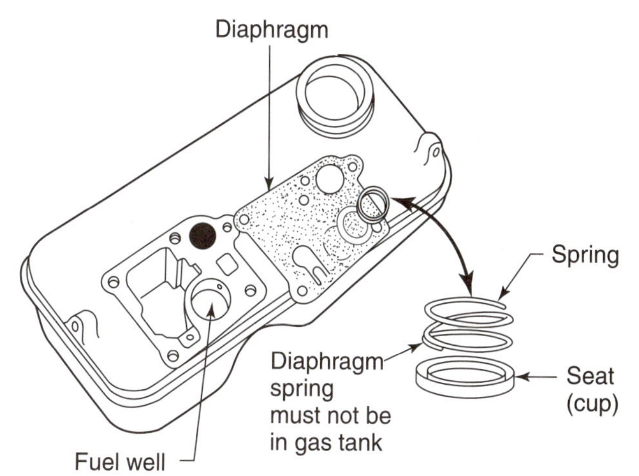

FIGURE 24-30 Install the diaphragm, spring, and seat in the correct position on the fuel tank.

A repair kit is available that has an adjustment screw (needle), spring, screw seat (jet), and washer (gasket). Replace the seat with a new gasket. Gently screw the adjustment screw into the seat and then back it off the recommended number of turns. The shop service manual recommends the number of turns for this initial setting.

When the fuel tank and carburetor are cleaned and reassembled, they are ready for installation. Install a new carburetor to intake port gasket or carburetor to intake tube O ring. Line up the carburetor with the intake tube and breather tube. Be sure the O ring in the carburetor does not distort during installation. Install the carburetor or fuel tank mounting fasteners. Tighten the fasteners properly to ensure an airtight seal.

Connect the throttle and governor linkage following your disassembly sketch or the shop service manual instructions. Replace any disconnected magneto stop wires. Install the air cleaner and fill the fuel tank with fresh fuel. Start the engine and check the carburetor for any fuel leaks. Make the mixture and choke adjustments following the recommendations in the shop service manual.

CAUTION: The fuel tank must be full for initial startup and proper adjustment of the diaphragm type carburetor.

FLOAT CARBURETOR OVERHAUL

Remove the air cleaner assembly to get access to the carburetor. Disconnect and plug the fuel line. Remove throttle and governor linkage and any

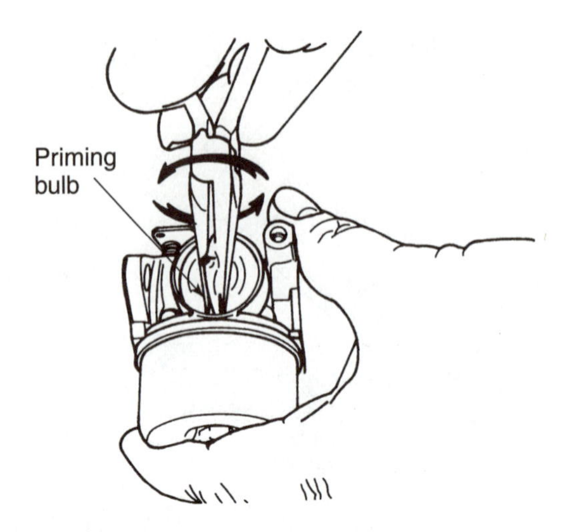

FIGURE 24-31 Removing the priming bulb from the side of a float carburetor. (Provided courtesy of Tecumseh Products Company.)

magneto stop wires. Remove the mounting fasteners and remove the carburetor. The carburetor gasket should be saved for comparison with the new one.

Remove the primer bulb (Figure 24-31) if the carburetor uses one. Use a screwdriver to carefully pry the primer bulb retainer out of the carburetor body. Grip the primer bulb with needle nose pliers and pull the primer bulb out.

WARNING: Wear eye protection to avoid eye injury when removing or replacing retaining rings that hold the primer bulb in place.

A bowl nut is used on many small carburetors to hold the float bowl on the carburetor body (Figure 24-32). Fixed jet emission carburetors commonly use the bowl nut as a main metering jet. Remove the bowl nut and separate the bowl from the body. The main jet and idle transfer passages are usually machined into the bowl nut. Be sure to clean the bowl nut with the carburetor body when cleaning the carburetor.

The float assembly is accessible with the float bowl removed (Figure 24-33). Tip the carburetor upside down and pull out the float hinge pin. The float, needle, and spring clip can then be removed together. Inspect the float carefully. Look for any holes or cracks in hollow floats. These allow gasoline to enter the float and cause it to sink. If the float contains gasoline or is crushed, it must be replaced.

Remove the inlet needle seat by unscrewing it from the body. Use the correct size wrench or

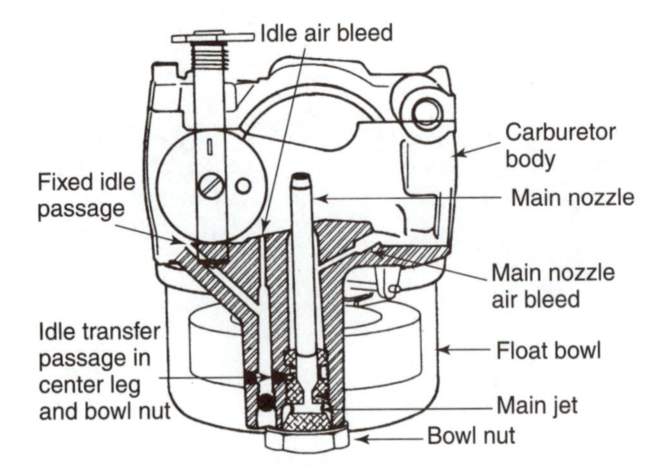

A. Bowl Nut Location

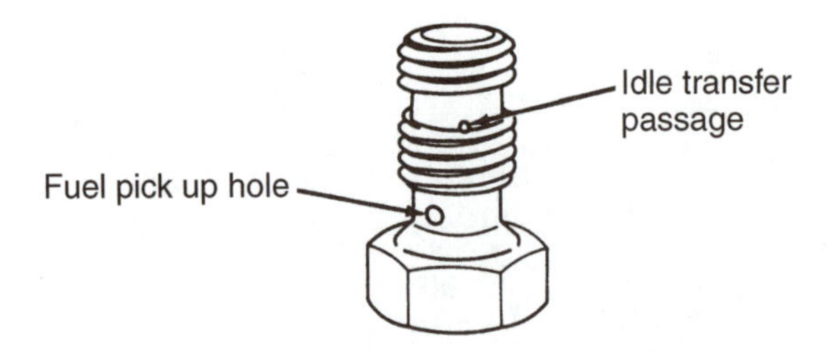

B. Bowl Nut

FIGURE 24-32 The bowl nut holds the float bowl to the carburetor. (Provided courtesy of Tecumseh Products Company.)

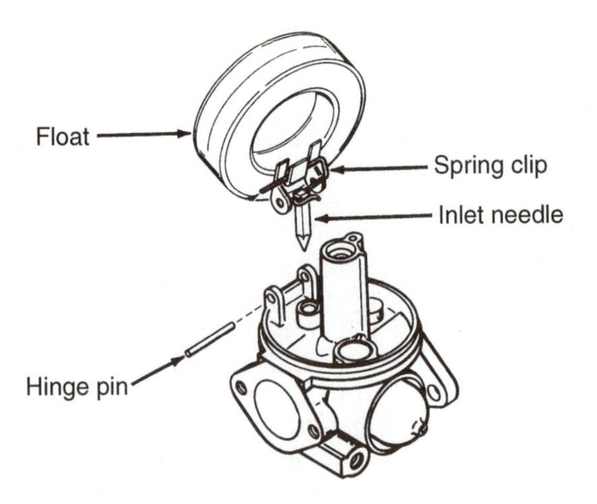

FIGURE 24-33 Removing the float hinge pin to remove the float assembly. (Provided courtesy of Tecumseh Products Company.)

screwdriver to remove the threaded float inlet seat. The wrong tool can damage the soft brass seat. Some carburetors use a pressed in seat that is hooked out with a piece of wire or a crochet hook (Figure

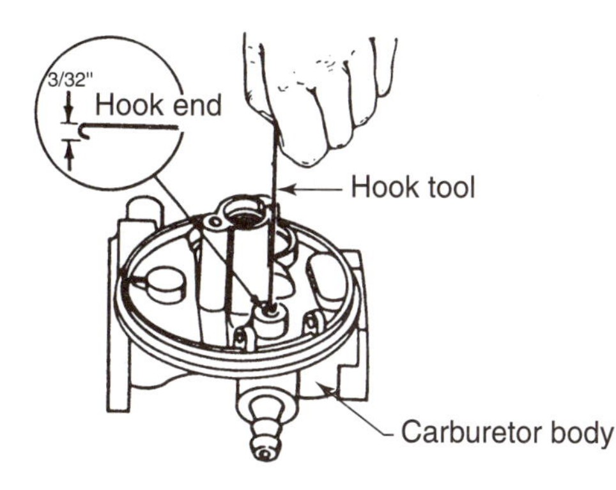

FIGURE 24-34 Removing the inlet valve seat with a hook. (Provided courtesy of Tecumseh Products Company.)

24-34). Be sure not to scratch the polished bore where the inlet seat is located. Unscrew the idle mixture adjusting screw (needle) and seat. Loosen the high-speed adjusting screw (needle valve) packing nut. Remove the packing nut and high-speed adjusting screw together.

The empty carburetor body and bowl nut may be cleaned with carburetor cleaner. Be sure to follow the instructions on the container. Some carburetors have a nonremovable plastic fuel fitting or nylon choke valve. These may be harmed by prolonged (more than 15 minute) contact with the cleaner. Open all plugged or restricted passages with monofilament fishing line or compressed air.

Carburetor repair kits are available for many float carburetors. Use new parts where necessary. Always use new gaskets and seals because old gaskets and seals harden and may leak. Tighten the inlet seat with the gasket (if used) securely in place. Some float valves have a spring clip to connect the float valve to the float tang. Other valves are made from nylon with a clip that fits over the float. Be sure the float clip is properly installed. If not, the float can separate from the inlet needle valve.

The float is checked to see if it is parallel with the carburetor body. Place a scale on the float (Figure 24-35). The scale should be parallel to the carburetor body to ensure that the inlet needle moves up and down in a straight line. If the float is not parallel, check that the clip and needle are installed properly. If the parts are assembled properly, the float tang will require adjustment.

The **float level setting** is a carburetor float adjustment that determines when it closes the inlet needle to establish the level of fuel in the float bowl. The float adjustment is measured on some carburetors by placing a specified size drill between the float and carburetor body. With the drill in place the inlet needle valve should be closed (Figure 24-36).

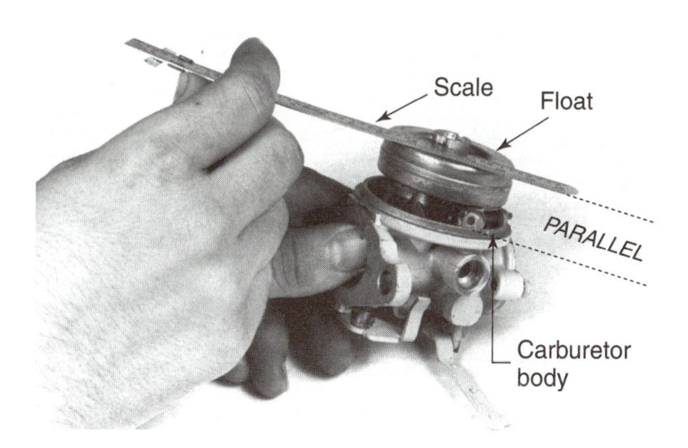

FIGURE 24-35 Measuring for float parallel with a scale.

Sometimes a special float level gauge is supplied with the carburetor repair kit or as a special engine service tool. A **float gauge** is used to measure the position of the float on the carburetor body and determine float level setting. Set the gauge in place on the float at a 90 degree angle to the hinge pin (Figure 24-37). The toe of the float (end opposite the hinge) must be under the first gauge step. The toe should touch the second step without a gap. You may find that the float is too high or too

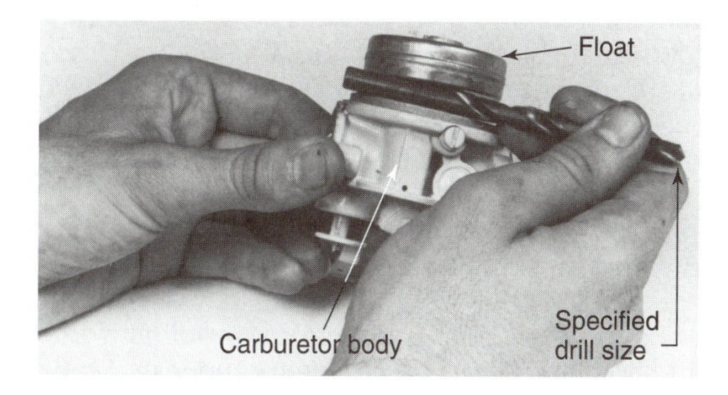

FIGURE 24-36 Measuring the float setting with a drill.

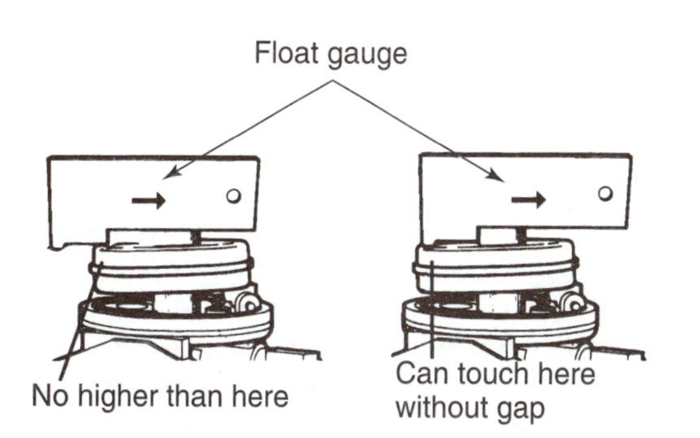

FIGURE 24-37 Measuring the float setting with a float gauge.

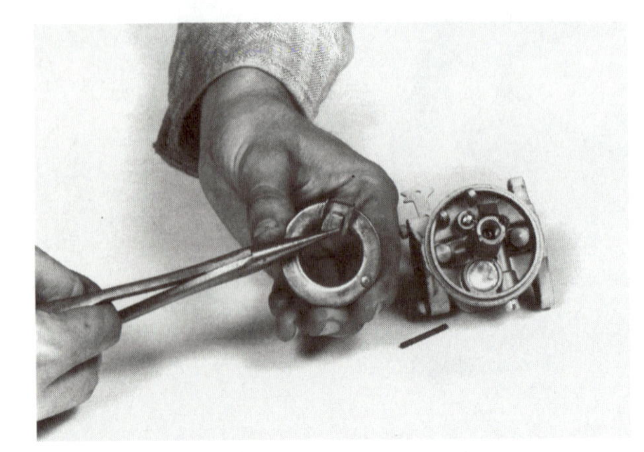

A. Adjusting the Float

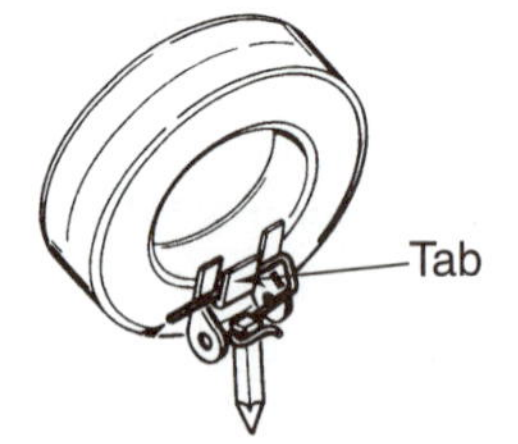

B. Bend Tab to Adjust Level

FIGURE 24-38 The float tab is adjusted to set the float level. (*B:* Provided courtesy of Tecumseh Products Company.)

low. Carefully bend the tab holding the inlet needle to get the proper height (Figure 24-38). Recheck the setting after each adjustment. Some floats are not adjustable. If the gauge shows it is out of adjustment, you must replace the inlet valve.

Install a new float bowl O ring or gasket. Place the float bowl on the carburetor in the correct position. Install the bowl nut with a new fiber washer between the bowl and the nut. Install new idle and high-speed adjustment screws and seats. Screw in the adjustment screws until they just seat. Back off the adjustment screws the specified number of turns listed in the shop service manual. Be careful not to overtighten the high-speed adjusting screw packing nut or it will leak. Install the primer assembly back into the carburetor body.

Install the carburetor on the engine using a new gasket. Connect the throttle and governor linkage following a disassembly sketch or the shop service manual instructions. Replace any disconnected magneto stop wires. Install the air cleaner and fill the fuel tank with fresh fuel. Start the engine and check the carburetor for any fuel leaks. Final adjustment is made when the engine is running and warmed up.

DIAPHRAGM CARBURETOR OVERHAUL

Many two-stroke engines use a small diaphragm carburetor. There are many different makers of these carburetors, but most have the same basic parts and are serviced the same way. Always identify the carburetor and locate specific service information in the appropriate shop service manual.

To remove the carburetor, first remove the air cleaner. Shut off the fuel supply at the fuel tank. Disconnect and plug the fuel line to the carburetor. Disconnect the primer tube if the carburetor has one. Before removing the carburetor, clean the area around the mount to prevent dirt from entering the crankcase. Remove the carburetor mounting fasteners. Note and sketch the governor and throttle linkage connections before they are disconnected. Carefully unhook governor link and spring from throttle lever. Be careful not to bend the governor link. Lift the carburetor off the engine.

Remove the fasteners that hold the diaphragm cover and remove the cover (Figure 24-39). Be careful not to lose the pump valve spring when lifting off the cover. Remove the diaphragm, diaphragm spring, and diaphragm spring cup (seat). Do not discard the diaphragm. It will be needed to compare with the new one. Remove the retaining fastener holding the hinge pin and lift out the inlet lever hinge, hinge pin, spring, and inlet needle assembly (Figure 24-40).

The inlet needle seat is often held in place by a small retaining C ring. A small wire hook can be used to pull the ring out (Figure 24-41). A smaller hook is used to lift out the seat. A small Allen wrench can be used to pry out check valve seat (Fig-

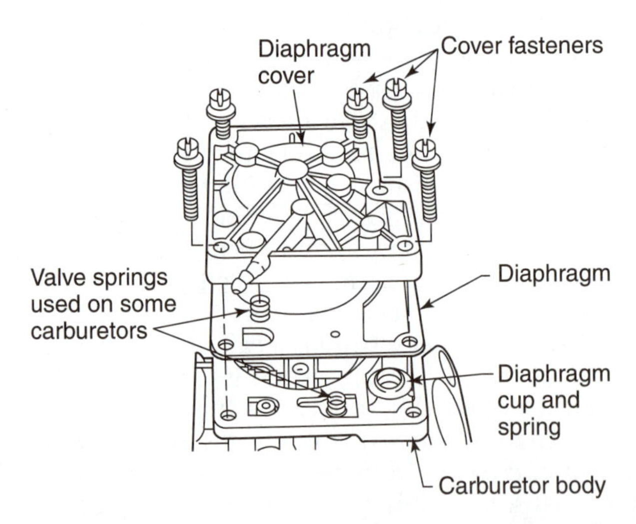

FIGURE 34-39 Removing the diaphragm cover to access the diaphragm.

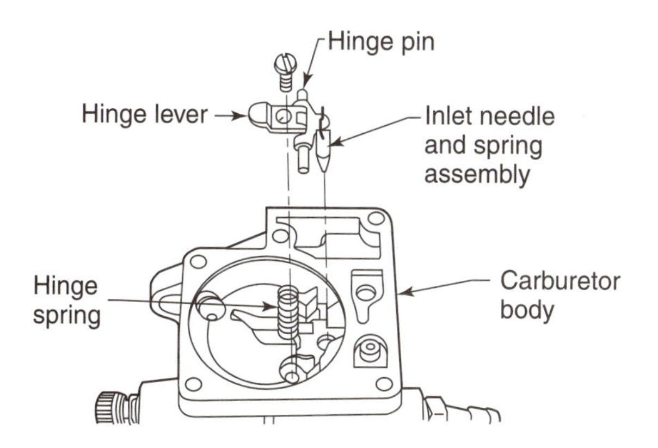

FIGURE 24-40 Removing the inlet needle valve assembly.

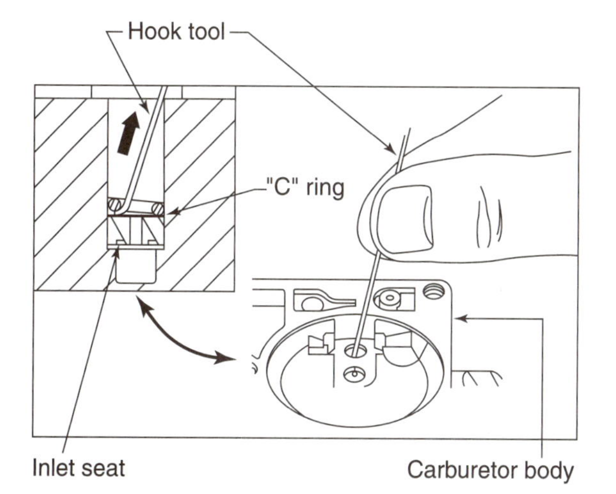

FIGURE 24-41 Removing the inlet seat with a hook tool.

ure 24-42). With check valve retainer removed, the rubber check valve can be removed. Unscrew and remove the mixture adjusting screw (needle valve) assembly. A small O ring is often used to seal around the needle. The O ring may be removed using a small hooked tool. Save the O ring for comparison with the replacement.

When the carburetor is disassembled this far, it may be cleaned with carburetor cleaner. The cleaner can be used on all metal parts. Do not use the cleaner on diaphragms, gaskets, rubber, or nylon parts. Before reassembling, rinse all parts in clean water and allow them to air dry. Do not use cloth rags to wipe the parts. Tiny particles of lint can get into carburetor parts and cause problems. Small passages in the carburetor body can be cleaned with monofilament fishing line or compressed air. Do not use wires or drills to clean orifices.

Check the diaphragm for holes, cracks, or wrinkles by holding it up to a strong light. Inspect the inlet needle and needle valve assembly for wear. All these parts are normally replaced during an over-

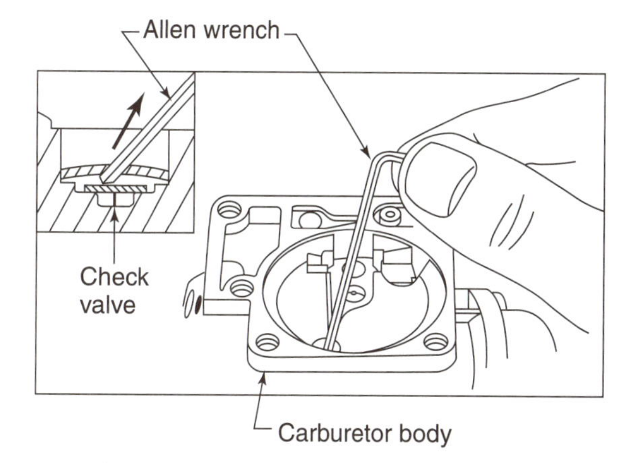

FIGURE 24-42 Using a small Allen wrench to pry out the check valve.

haul. Check the carburetor body for warping by placing a straightedge (scale) across the body (Figure 24-43). Hold the scale down with firm pressure. Try to insert the recommended thickness feeler gauge blade between the body and straightedge. Try the gauge with the straightedge placed in at least two different directions. If the gauge fits, the body is warped and must be replaced.

Insert the new inlet seat and replace the C ring. Place the hinge lever spring in the spring pocket. Assemble the hinge pin, inlet needle lever and inlet needle, and spring. Install these parts in the carburetor body. Install the retaining screw when the parts are in position.

The carburetor inlet valve regulates the amount of fuel into the carburetor venturi. The valve is operated by the diaphragm, diaphragm spring, and hinge lever. Each of these parts can affect the air and fuel mixture. The position of the hinge lever is very important to the carburetor operation. Measure the position of the hinge lever by placing a

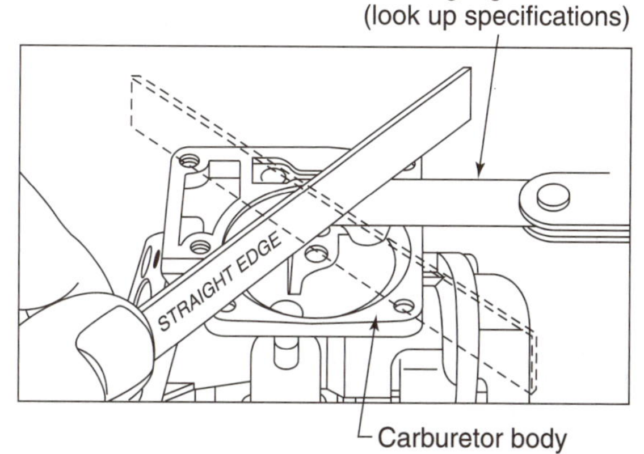

FIGURE 24-43 Using a straightedge and specified feeler gauge to check carburetor body for warping.

straightedge across the hinge lever (Figure 24-44). Measure between the straightedge and carburetor body surface. Compare this measurement to specifications.

The diaphragm hinge lever is adjusted by bending (Figure 24-45). If bending is required, support the needle end of the lever with a screwdriver. Use needle nose pliers to bend the lever. Bend the lever up to raise the lever. Bend down to lower the lever. Recheck the measurement after each adjustment.

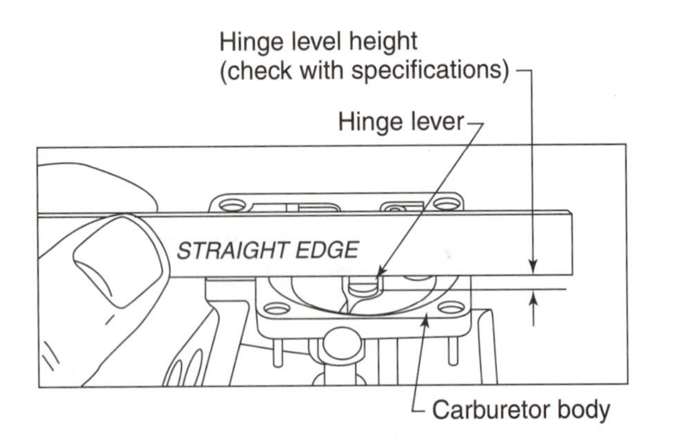

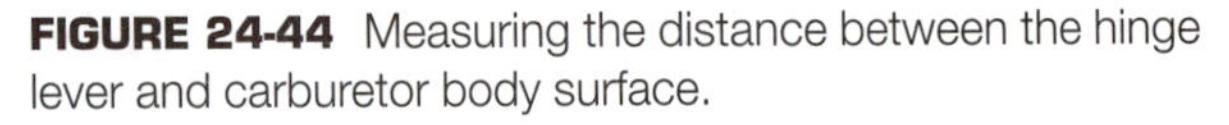
FIGURE 24-44 Measuring the distance between the hinge lever and carburetor body surface.

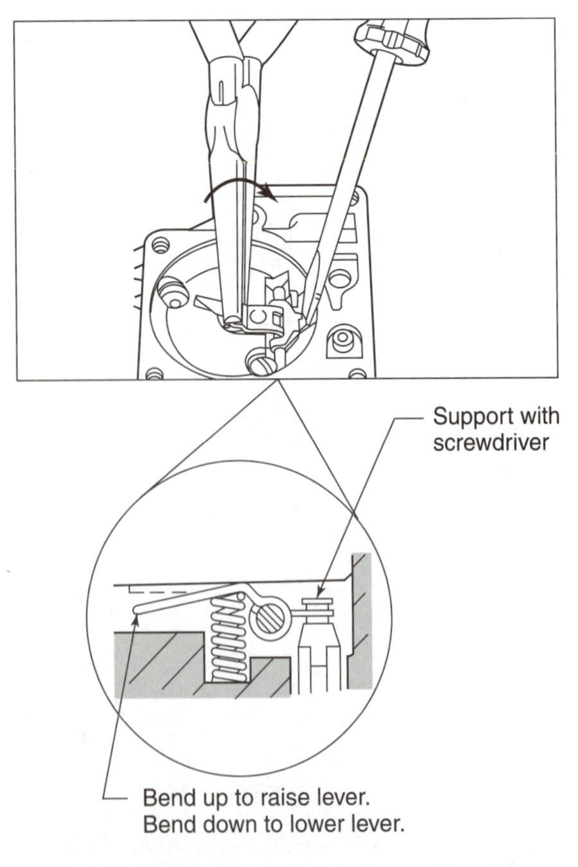

FIGURE 24-45 Adjusting the diaphragm hinge lever.

Install the check valve in position and start the check valve retainer. Use a punch and a light tap with a hammer to seat the retainer. Check the performance of the valve operation by blocking metering holes with your fingers. Use your mouth to blow and suck through the needle valve opening. If the check valve is working correctly, blowing will seat the check valve (stop flow). Sucking gently should allow air to flow.

Compare the new and old diaphragm to make sure they are the same. Place fuel pump spring and cup in position on the body. Install the diaphragm on the carburetor body. Make sure it is in the correct position. Install the cover on the carburetor and start cover fasteners. Tighten the diaphragm cover fasteners firmly. Follow the tightening sequence recommended in the shop service manual. Install the new adjustment screw using a new O ring. Follow the recommendations in the shop service manual for a preliminary adjustment.

Install the reassembled carburetor on the engine. Connect the throttle and governor linkage. Install the fuel line and the air cleaner. Start the engine and check the carburetor for any fuel leaks. Final adjustment is made when the engine is running at operating temperature.

CARBURETOR ADJUSTMENT

> **CAUTION:** Carburetors used on low-exhaust emission type engines often have fixed jets or no adjustment screws so that they cannot be adjusted. This prevents an overly rich high emission adjustment. Always review the correct service procedures in the shop service manual before performing a carburetor adjustment.

Carburetor adjustment is an important part of the tune-up procedure. Adjustment is required after carburetor overhaul. Different carburetors require different types of adjustments. Some carburetors have no adjustments. These often use a primer bulb instead if a choke. Many carburetors have adjustment screws for high-speed mixture (main mixture), low-speed mixture (idle mixture), and idle speed (Figure 24-46). The order in which these adjustments are made is different from one model to another. These adjustments are done on a running engine. The engine has to be warmed up to regular operating temperature. The engine must have the air cleaner installed and in good working order. The intake system must not have any vacuum leaks.

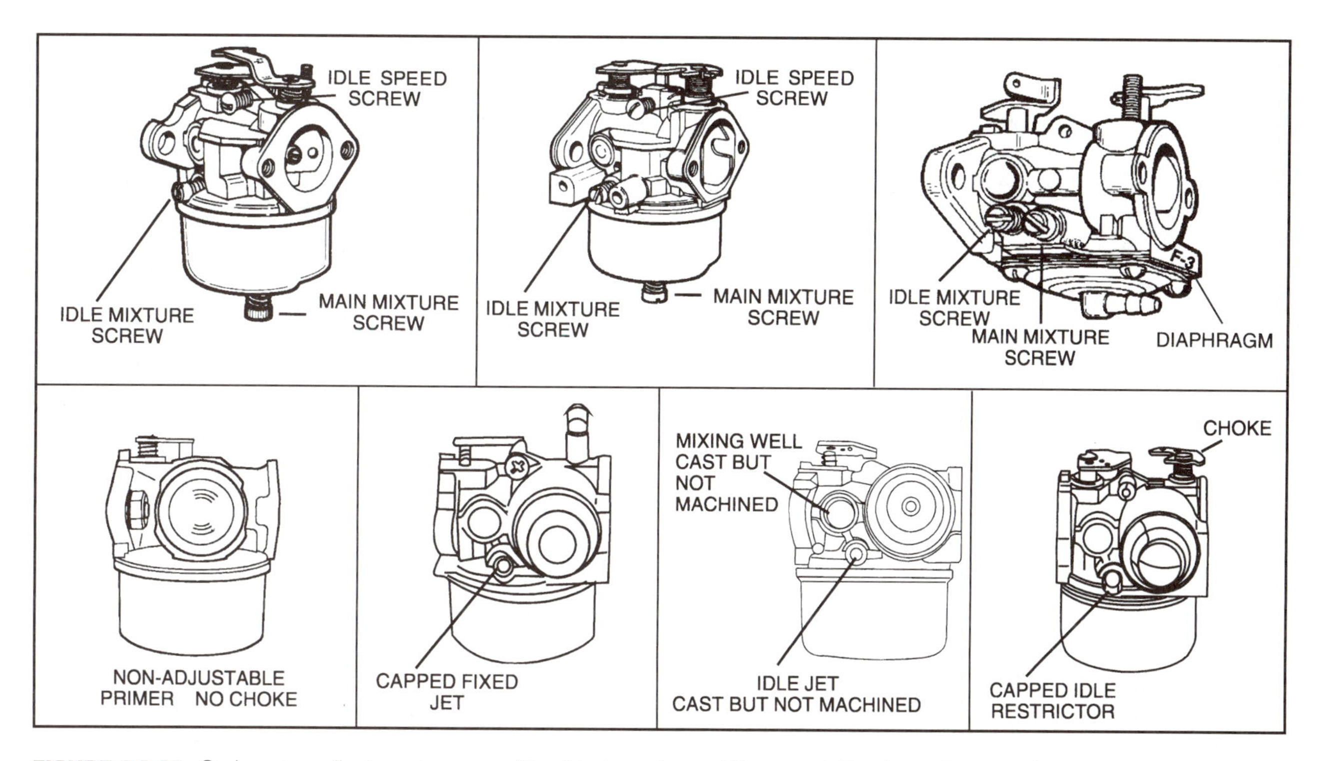

FIGURE 24-46 Carburetor adjustment screws. (Provided courtesy of Tecumseh Products Company.)

Adjusting High-Speed Mixture

The high-speed air and fuel mixture is adjusted by the position of a high-speed mixture adjustment screw. The high-speed adjustment screw has a tapered end that regulates the amount of fuel that can be delivered by the carburetor high-speed circuit. On most carburetors, turning the screw inward reduces the amount of fuel, causing a lean high-speed air-fuel mixture. Turning the screw outward increases the amount of fuel, causing a rich high-speed air-fuel mixture.

An engine with a high-speed mixture that is too lean:

- Runs smoothly at idle
- Misfires at high speed
- Has low power at high speed

An engine with a high-speed fuel mixture that is too rich:

- Runs rough at all speeds
- Has black fuel smoke coming out the exhaust during high speed

An engine mixture needs change when an engine is under load. For example, a lawn mower engine may run fine at high speed when it is not cutting grass, but it may lose power during grass cutting. The engine requires a richer mixture under load. Sometimes the adjustment procedures in the shop service manual recommend adjusting the mixture slightly rich without any load. This is done so that the mixture will be correct during high loads.

Prior to making the adjustment, start up the engine and allow it to run for several minutes to warm up. Locate the carburetor high-speed mixture screw. Some screws have a slot for a screwdriver. Others have a handle so they can be adjusted by hand. Open the engine's throttle so that it runs at the speed specified in the shop service manual. Hold the throttle steady. Turn the screw in or clockwise to lean the high-speed mixture until the engine runs roughly. Turn the screw out slowly to enrich the mixture. The engine will begin to speed up as the mixture gets rich. At some point, the mixture will get too rich, and the engine will slow down. Stop turning when the engine runs as fast as it will run.

Adjusting Low-Speed Mixture

The low-speed mixture adjusting screw is used to adjust the carburetor's mixture at low speed. The low-speed adjustment screw has a tapered end. The tapered end regulates the amount of fuel that is delivered by the carburetor idle and low-speed circuit. On most carburetors, turning the screw inward reduces the amount of fuel, causing a lean, low-speed air-fuel mixture. Turning the screw outward

PHOTO SEQUENCE 11

Typical Procedure for Adjusting a Carburetor

1. Inspect engine for air cleaner condition and any vacuum leaks before doing any adjustment.

2. Start the engine and allow it to warm up. Make sure choke valve is fully open.

3. Open the throttle to operate the engine at the speed specified in the shop service manual.

4. Insert a screwdriver into the high-speed mixture screw. Turn the screw in to lean the mixture until the engine runs rough. Turn the mixture screw out to enrich the mixture until the engine runs as fast as it will run.

5. Allow the engine to return to idle. Insert a screwdriver into the low-speed mixture screw. Turn the screw in to lean the mixture until the engine runs rough. Turn the mixture screw out to enrich the mixture until the engine runs as fast as it will run.

6. Insert a screwdriver into the idle speed adjustment screw. Rotate the screw in a direction to increase or reduce idle speed until the desired speed is set.

increases the amount of fuel into the low-speed circuit, causing a rich air-fuel mixture.

A too-lean, low-speed adjustment results in a no idle condition. Black fuel smoke during idle can be caused by a too rich, low-speed mixture adjustment. An engine that idles roughly may be either too rich or too lean.

Before making the adjustment, start up the engine. Allow it to run for several minutes to warm up. Locate the carburetor low-speed mixture screw. Be careful not to confuse it with the idle speed screw. Close the throttle and allow the engine to idle.

Use a screwdriver in the idle mixture adjusting screw slot. Turn the screw in or clockwise. Turn the screw slowly to lean out the idle mixture. Stop turning when the engine begins to run roughly. Turn the screw slowly out or counterclockwise to enrich the mixture. The engine will speed up as the mixture gets richer. At some point the mixture will be too rich and the engine will slow down. Stop turning the screw when the engine idle reaches its highest speed.

Adjusting Idle Speed

Most carburetors have an idle-speed screw. The idle-speed screw is used to set engine speed when the throttle valve is closed. Idle speed is adjusted after the mixture adjustments are complete. The shop service manual will specify an idle speed in engine revolutions per minute (rpm). Idle speed is most accurately measured with an electronic tachometer. The tachometer is connected to the magneto ignition and counts the ignition impulses. The scale on the tachometer is divided into revolutions per minute. Connect the tachometer to the engine following the instructions provided with the tachometer.

Start the engine and allow it to idle for several minutes and warm up. Observe the idle speed on the tachometer. If the speed is incorrect, locate the idle-speed screw. Use a screwdriver to turn the screw clockwise to open the throttle valve. This will increase idle speed. Turn the screw counterclockwise to close the throttle valve. This will decrease idle speed. Set the idle speed to specifications.

If a tachometer is not available, the speed may be adjusted by trial and error. Open the throttle and increase engine speed momentarily and then close the throttle rapidly. The engine should return to an idle speed without stalling. If it stalls, increase the idle speed. If the engine appears to be idling too fast, lower the idle speed until the engine stalls. Increase idle speed until it can recover from the high-speed throttle closing. A typical procedure for adjusting the high-speed mixture, low-speed mixture, and idle speed is shown in the accompanying sequence of photographs.

GOVERNOR TROUBLESHOOTING AND ADJUSTMENT

The governor is often blamed for engine performance problems that are actually caused by the fuel or ignition system. Governor service should not be attempted unless both of these systems are inspected and received a fuel and ignition tune-up. Problems caused by the governor are engine overspeeding or engine surging.

Engine overspeeding is a governor problem when the engine runs at too high rpm at wide-open throttle (WOT). Overspeeding is complete governor failure that causes the throttle to stick in one position. To correct an overspeeding condition, check:

- External governor linkage
- Governor spring
- Equipment speed control assemblies and cables for breakage or binding

Disconnect the linkage one piece at a time and check for proper movement. Replace any binding or damaged parts.

If linkage inspection fails to uncover the problem, follow shop service manual procedures for adjusting engine governed speed. The engine may require disassembly for inspection or replacement of the air vane or centrifugal governor if the overspeeding problem cannot be fixed by linkage repair. The most common air vane governor problem is limited air vane movement. This problem is usually caused by grass buildup around the vane. Centrifugal governors can have governor rod seizure in the crankcase cover. Another common problem is flyweights that seize on the governor spool.

CAUTION: An overspeeding engine should be shut down or slowed down immediately, before it sustains internal damage

CAUTION: Do not stretch or shorten a governor spring to correct governor speed problems. A governor spring with the incorrect tension will not properly regulate engine speed. When replacing a governor spring, be sure to get the exact part number spring for replacement.

Engine surging is a governor or fuel system problem that causes the engine to constantly speed up rapidly and then slow down rapidly. This problem is also called *governor hunting*. Surging is commonly caused by a too rich or too lean air-fuel mixture. Make sure the fuel system is operating properly before considering this a governor problem. Check all the governor and speed control linkage for signs of binding, wear, or improper hookup. To determine if the problem is related to the governor, use pliers to hold the governor linkage rod in one position. If the speed stabilizes, follow the shop service manual procedure for proper governor adjustment. If the problem continues after adjustment, disassemble the engine and inspect and repair or replace the air vane or mechanical governor.

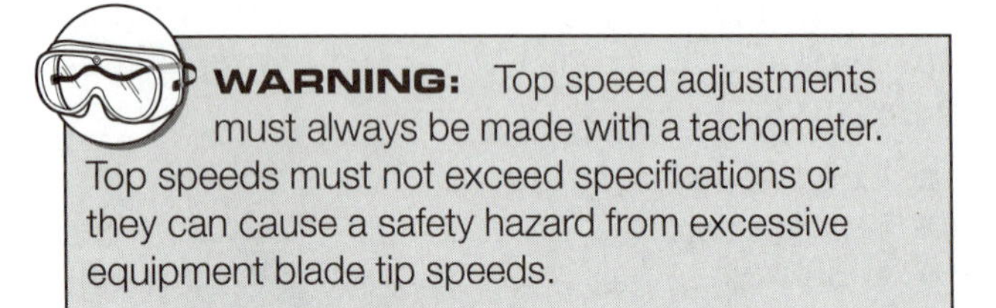

WARNING: Top speed adjustments must always be made with a tachometer. Top speeds must not exceed specifications or they can cause a safety hazard from excessive equipment blade tip speeds.

Governor systems commonly have several adjustments that affect the engine's governed speed. Each engine manufacturer has a specific adjustment procedure in the shop service manual. The **governor top speed adjustment** is a governor linkage adjustment used to set the top governed speed of an engine. This test is done with a tachometer. Some linkages have adjusting holes for different governor spring mounting positions (Figure 24-47). Selecting a different hole will increase or decrease the governor spring tension, thus increasing or decreasing governed speed.

Stop screws that limit speed control linkage travel are often used for governor adjustment (Figure 24-48). Friction clamps are often used to connect the governor shaft to the governor lever (Figure 24-49). These clamps can be loosened and rotated to change the relative position of the governor lever to the governor rod. This adjustment can be used to remove play from the linkage. It is also used to change the tension on the governor spring.

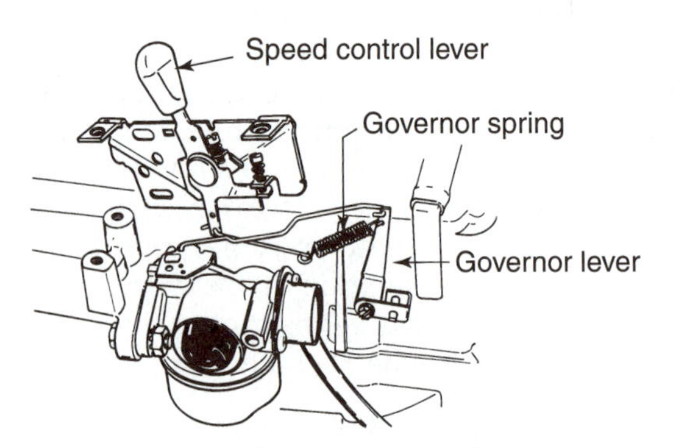

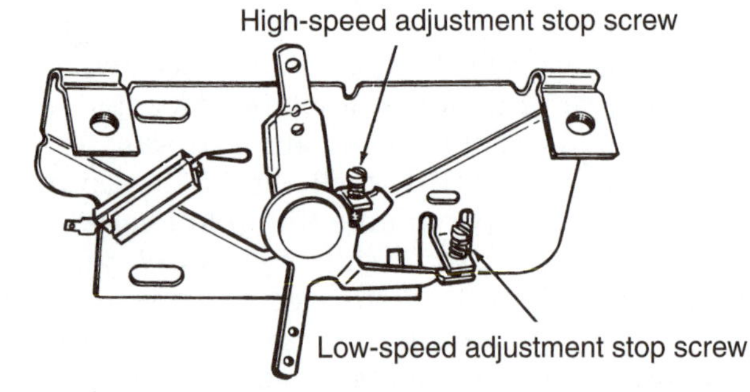

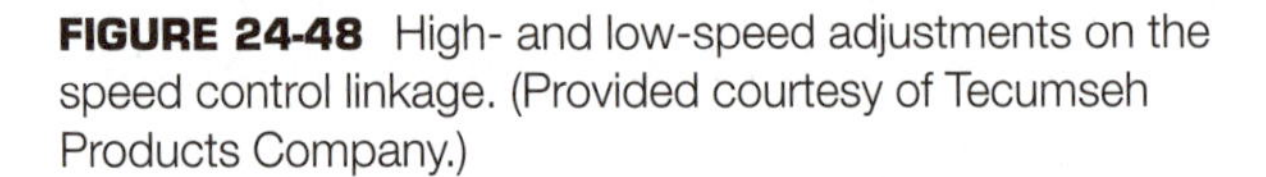

FIGURE 24-48 High- and low-speed adjustments on the speed control linkage. (Provided courtesy of Tecumseh Products Company.)

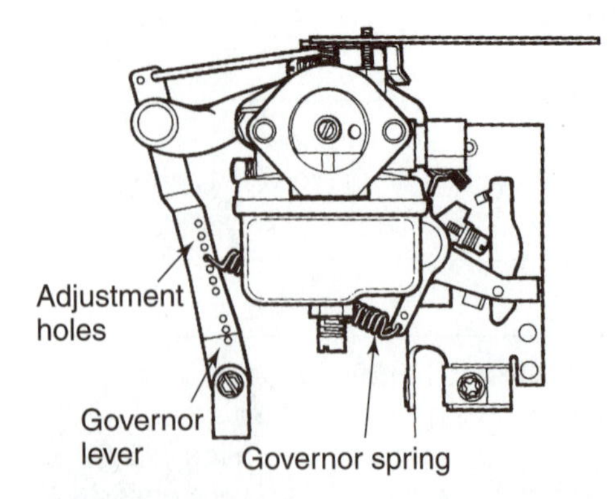

FIGURE 24-47 The governor sensitivity can be adjusted by selecting different holes in the governor lever. (Provided courtesy of Tecumseh Products Company.)

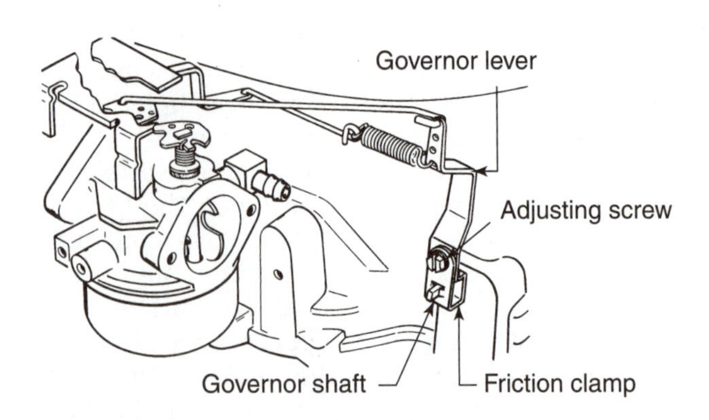

FIGURE 24-49 Friction clamp adjustment on a governor lever. (Provided courtesy of Tecumseh Products Company.)

REVIEW QUESTIONS

1. Describe the troubleshooting steps to determine if an engine is running too rich or too lean.
2. List and describe the common problems for a fuel system that runs too rich.
3. List and describe the common problems for a fuel system that runs too lean.
4. Describe the procedures to follow if stale fuel or rust is found in the fuel tank.
5. Describe the troubleshooting procedures to follow to determine if a fuel pump is operating correctly.
6. List and describe the steps to follow in removing and replacing a mechanical and impulse fuel pump.
7. Describe the steps used to isolate a carburetor problem to an individual carburetor circuit.
8. Explain how to identify the model of a carburetor in order to locate appropriate service information.
9. Explain why drills or wires should not be used to clean out carburetor passages.
10. Explain why nonmetallic parts should not be soaked in carburetor cleaner for long periods.
11. List and explain the inspection procedures on a disassembled carburetor to determine if it should be replaced.
12. Explain why a new gasket or O ring seal is used when the carburetor is installed on the engine.
13. Describe the procedures to follow in overhauling a vacuum type carburetor.
14. Describe the procedures to follow in overhauling a suction feed type carburetor.
15. Describe the procedures to follow in overhauling a float type carburetor.
16. Describe the procedures to follow in overhauling a diaphragm type carburetor.
17. Explain how to adjust the high-speed mixture on a carburetor.
18. Explain how to adjust the low-speed mixture on a carburetor.
19. Explain how to adjust the idle speed on a carburetor.
20. Describe how to troubleshoot and adjust a governor.

DISCUSSION TOPICS AND ACTIVITIES

1. Locate several float type shop carburetors. Disassemble each carburetor and determine if they are suitable for overhaul.
2. Locate several diaphragm type shop carburetors. Disassemble and inspect the parts to determine if they can be overhauled.
3. A carburetor float adjustment specifies that the space between the float and carburetor should be 1/4 inch. Convert this dimension to decimal inch and millimeters to determine what drill size to use.

Engine Service

Engine Disassembly and Failure Analysis

OBJECTIVES

Upon completion and review of this chapter, you should be able to:

- Use the history of a problem to determine the cause of engine failure.
- Examine the outside of an engine to determine the cause of engine failure.
- Disassemble an engine for inspection.
- Inspect internal engine parts for wear and damage to determine the cause of failure.
- Clean each of the engine parts for measuring and service.

TERMS TO KNOW

Blown head gasket	Short block
Crankshaft end play	Valve erosion
Piston ring expander	Valve spring compressor
Ring groove cleaner	Valve sticking

INTRODUCTION

An engine with internal failure usually has one or more of the following symptoms:

- A no start, hard start, or low power condition from low compression
- Excessive blue or white exhaust smoke from oil burning
- Excessive internal engine noise from failed parts
- A locked up or frozen crankshaft from major parts failure

An engine with any of these problems is evaluated to find the cause of the engine failure. Finding the cause of an engine failure is very important. If you do not find the cause, a new or rebuilt engine may soon end up in the same condition. Engine failure analysis has four main steps:

1. Collect all the information possible about the customer's use that led to the failure.
2. Do an external examination of the engine on the power equipment to find out how well the owner maintained the equipment.
3. Disassemble the engine.
4. Examine each part to determine the cause of the failure.

PROBLEM HISTORY

Your first job in analyzing a failed engine is to gather as much information as possible. This information may come from service records on the engine or from interviewing the customer who used the engine.

Engine failure is often caused by parts failure. Part failure is often caused by:

- Many hours of service
- Low or contaminated engine oil
- Overheating
- Abrasives that get into the intake system

Service records and customer questions should be used to get information in these areas. Sometimes a complete service history is available for the engine. Analyze the service history to find out if services were done at the specified intervals. For example, oil changes or air cleaner maintenance may be way overdue. This could be the cause of the problem.

The customer may be able to provide the information. Questions like these will help you determine if parts failure is due to normal wear:

- How many hours did the equipment run before failure?
- How many times was the equipment used?

Questions like these will help you isolate oil-related causes:

- What type of oil was used in the engine?
- How often was oil added?
- How often was oil changed?

Questions like these can help you focus on abrasives in the engine:

- How often was the air cleaner checked or serviced?
- Was the air cleaner ever changed?

FIGURE 25-1 Inspect the outside of the engine for oil leakage and condition.

EXTERNAL EXAMINATION

> **CAUTION:** Do not steam clean or pressure wash the engine or equipment before you do an exterior examination. Cleaning can destroy the evidence you need to find the cause of a problem.

The external examination is an inspection of engine parts that are easily checked without total engine disassembly. Most of these inspection areas take only a few minutes to check. They will provide valuable information on the cause of engine failure. The most important inspection areas are:

- Overall condition of outside of the engine
- Air cleaner condition
- Amount and condition of engine oil
- Governor linkage condition
- Carburetor condition
- Cooling system condition
- Combustion chamber condition

Checking Engine Outside Condition

Inspect the overall condition of the engine outside (Figure 25-1). Check to see if it is relatively clean. Look for an accumulation of oil, dirt, or debris. Look for any oil leaks, especially around the main bearing seals. Leakage here can mean the engine needs major service. Check all oil leakage areas for signs of cracks. Look for cracks in the sides of the crankcase. Cracks can occur from vibration. Look for loose engine mounting bolts that cause this type of vibration. Look for cracks in areas that bulge out. These cracks are caused by broken internal parts coming through the crankcase. Also check for any signs that the engine may have been disassembled or repaired previously. Look for missing or incorrect fasteners. These show the work of an untrained technician. Cleaned parts or parts with mismatched paint are other signs of recent service.

Checking Air Cleaner Condition

The air cleaner is one of the most important parts related to engine service life. Air cleaners are often improperly maintained or have a missing filter element. When this happens, dirt can bypass the filter and enter the engine. Dirt in the engine acts as an abrasive. It rapidly wears out engine parts.

Dirt passing through the carburetor causes rapid wear on the choke and throttle valve shafts, which results in rough idle and poor performance. Dirt drawn through the intake valve passes between the intake valve stem and guide and into the combustion chamber. Wear between the guide and stem causes hard starting, loss of power, and increased oil consumption. Dirt that enters the top of the cylinder bore causes rapid wear to the piston rings and bore, which results in loss of engine performance and increased oil consumption. Dirt that gets by the piston rings enters the oil. The abrasive action of the dirt causes excessive wear on the crankshaft journal and connecting rod, causing early failure.

Inspect the air cleaner assembly thoroughly. Disassemble the air cleaner. Make sure there is a filter element in the air cleaner. Remove the filter

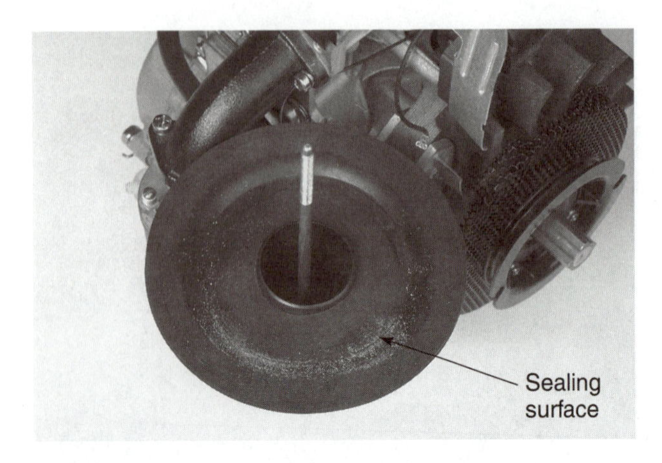

FIGURE 25-2 Checking the air cleaner sealing surface for dirt tracks.

element and check it for obvious tears, deterioration, or plugging. Also note if it is the correct part and that it fits properly in the housing. Check the filter element sealing surfaces. Dirt tracks across the sealing surface show dirt leakage into the engine (Figure 25-2). Place a strong light source inside the paper filter element. Look for punctures in the paper filter material. If light is not visible at the base of the creases, the filter is clogged (Figure 25-3).

Check that all seals, gaskets, and fasteners are properly installed and in good shape. Check the inside surfaces of the air cleaner element for dirt. Check the carburetor throat for signs of dust or dirt (Figure 25-4). Any dirt in these areas shows an air cleaner problem. Check the cover and housing for distortion or damage. This damage would prevent proper sealing and dirt leakage. Check the center hole to see if it is worn elongated from the cover being loose. This is a common source of dirty air entering the engine. Check the wing nut sealing surface for wear.

FIGURE 25-3 Using a light to check the filter element condition.

FIGURE 25-4 Checking the air cleaner housing and carburetor throat for dirt.

Checking Oil Level and Condition

Dirty or low oil is a leading cause of engine failure. Remove the fill plug or pull the dipstick and check the level of oil (Figure 25-5). Low or no oil is a common cause for loss of lubrication and parts failure. Also check the color and thickness of the oil. Oil that is black, thick, and dirty indicates improper changing intervals. Oil that is in the engine 200 hours or more may get so thick that it will not drip off the dipstick. Contaminated oil allows rapid parts wear.

Drain the oil into a clean container. Measure the amount drained out. Compare it to the recommended capacity to determine how much oil was in the engine. Examine the oil closely for color, thickness, and any abnormal smell. Clean oil is brown and dirty oil is black. Thickening or abnormal smell mean too much time between change intervals. Look for any metal chips or wear particles suspended in the oil. If the engine has an oil filter, remove it for inspection. If you find it covered in dirt and sludge, it has not been changed regularly.

Oil-related failures may be caused by the wrong oil viscosity. Either too thin or too thick an oil viscosity can lead to engine lubrication problems. Too thin oil will not protect bearings during high-temperature and high-load conditions. Too thick oil will not carry away heat properly, leading to overheating and parts damage. Talk to the customer about what type of oil is being used. Check this information against the engine maintenance recommendations in the shop service manual.

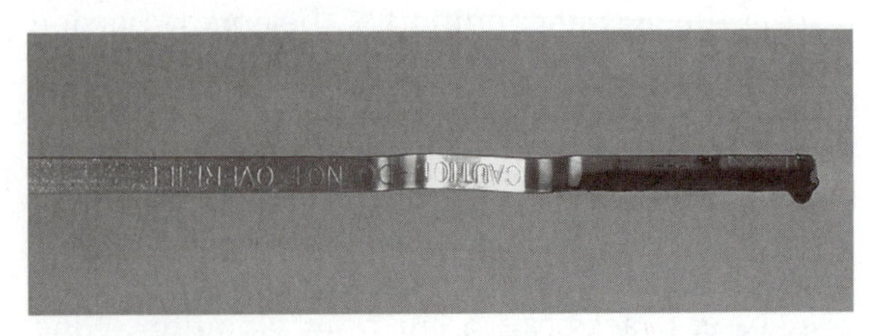

FIGURE 25-5 Checking engine oil for level and condition.

FIGURE 25-6 Checking the condition and operation of the throttle and governor linkages.

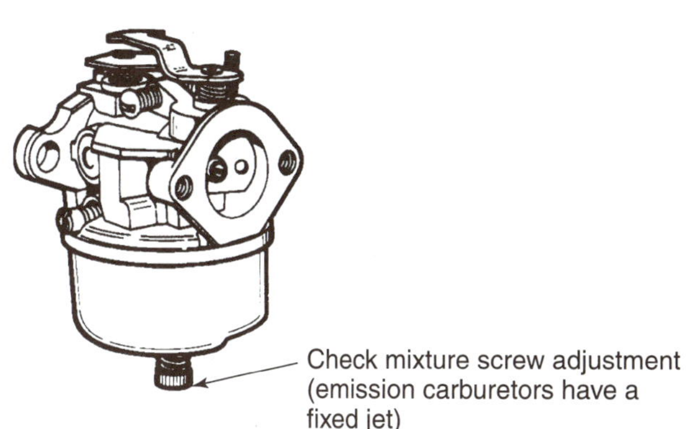

FIGURE 25-7 Check the position of the high-speed mixture screw to troubleshoot lean mixture overheating. (Provided courtesy of Tecumseh Products Company.)

Checking Governor Linkage Condition

A malfunctioning governor can allow excessive engine speed. The excessive speed can result in a broken connecting rod. Some owners remove these parts because they incorrectly think they lower engine power. Check the external governor parts and linkages to see if they are bent, broken, or missing (Figure 25-6). Look for any owner modifications to these parts. Operate the equipment throttle control. Check that the linkage can move freely through its normal range. Be sure the throttle closes properly under spring or manual linkage control.

Checking Carburetor Condition

Carburetor adjustments allow mixture changes for different altitudes and running conditions. An incorrect adjustment may cause the mixture to be adjusted too lean. When the engine runs lean, it runs hotter than normal. Excessive engine heat can cause distortion of the cylinder head and cylinder. Distortion can cause the loss of head bolt torque. The result can be a blown head gasket. A **blown head gasket** is a condition in which a cylinder head gasket ruptures and allows the escape of combustion pressure from the combustion chamber.

The exhaust valve and seat may expand and distort from overheating. Expansion can cause a loss of valve clearance. Loss of valve clearance can cause the exhaust valve to stay open or burn away. Both these problems result in compression loss. Check for these problems by measuring the engine valve clearance. Then check the valve condition during disassembly.

Lean mixture engine operation causes higher internal engine temperatures. Higher internal temperature increases oil temperature. Oil that gets too hot can lose its viscosity. The result may be higher oil consumption.

Check the carburetor to find out if a lean mixture is the cause of a failure. Check the position of the high-speed (main) adjustment screw (Figure 25-7). Turn the mixture screw inward until it seats while counting the number of turns. Compare the adjustment setting to specifications in the shop service manual. A mixture screw that is adjusted too far in is an indication of too lean a mixture. Inspecting the condition of the spark plug firing end for lean operation helps verify this problem.

Checking Cooling System Condition

Cooling system problems cause the same overheating as too lean a carburetor mixture. Cooling system failure can lead to a blown head gasket, loss of valve clearance, or high oil consumption. Inspect each of the cooling system (Figure 25-8)

FIGURE 25-8 A plugged cooling system causes overheating.

components. Check the air intake screen, blower housing, and cooling fins. Be sure they are not blocked by grass clippings. Any blocking that limits air flow will cause overheating and lead to parts failure.

Checking Combustion Chamber Condition

Removing the cylinder head to inspect the combustion chamber parts is a good troubleshooting step. Checking the combustion chamber can help you estimate engine run time. Remove the cylinder head fasteners then remove the cylinder head. Inspect the cylinder head gasket closely. The head gasket will usually not stick to the cylinder and head gasket surfaces if the engine has been run for a short time. A nonsticking head gasket indicates a lack of heat in the head area. Longer engine running times develops more heat, which causes the head gasket to stick to its mating surfaces. A head gasket will usually stick to both the head and cylinder surfaces after approximately 5 hours of running

CASE STUDY

A customer pushed a garden tiller into the shop. The customer had bought the tiller from the shop 2 years ago. The tiller owner was unhappy with the performance of the tiller. He wanted the shop owner to tune it up under the equipment warranty. The tiller owner claimed that the tiller had less than 5 hours of use in the 2 years.

The shop owner looked the tiller over and it not only looked very used but it looked abused. The engine was covered with dirt. The shop owner checked the oil level and condition with the dipstick and the oil was very thick and black. Clearly, the tiller had operated many more hours than the tiller owner was representing.

The shop owner made the following proposal to the tiller owner. The shop would remove the cylinder head and determine the amount of hours the tiller had been run. If the inspection showed the few hours the tiller owner said it had, the shop would do any necessary work to the tiller under warranty. If the inspection showed more hours, the customer would pay for the tear down and the work necessary on the tiller.

The customer thought for a while. He then allowed that his brother in law had borrowed the tiller from time to time. Maybe the tiller had a few more hours on it than he thought. The customer decided to have the work done without any more discussion about the warranty.

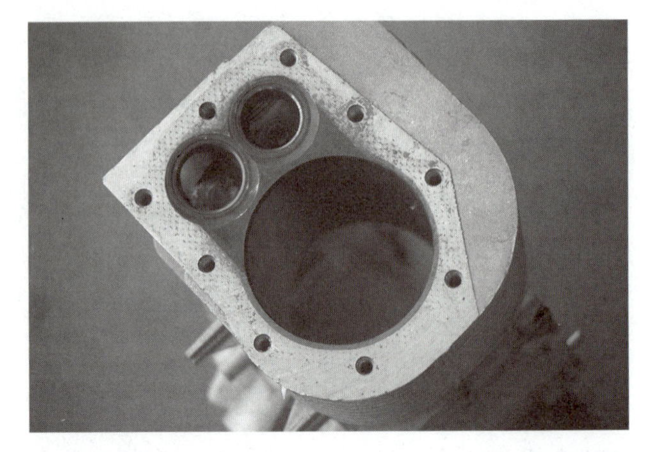

FIGURE 25-9 Combustion chamber deposits on an engine with short running time.

FIGURE 25-10 Combustion chamber deposits on an engine with long running time.

Check the amount of carbon deposits in the combustion chamber. The amount of deposits can help you estimate run time. The longer the engine has run, the more the deposits. Look for deposits that are still very soft and oily and able to be easily wiped off with a cloth (Figure 25-9). These show the engine has a low run time. Deposits may be crusted on and not easily wiped off (Figure 25-10). These show that the engine has probably been run in excess of 5 hours.

ENGINE DISASSEMBLY

Often the engine has to be disassembled to determine the exact cause of failure. Engine disassembly and internal parts inspection is often done to see if the engine can be rebuilt. Most shops charge the customer a labor charge to determine if the engine can be rebuilt. The internal wear and damage is used to estimate the cost and feasibility of engine overhaul. The customer can then decide whether to rebuild or buy a new engine or short block. A **short block** is a new or rebuilt engine sold without any external components. Often the cost of the equipment figures into the decision. Sometimes it is more cost-effective to buy new equipment than to repair the engine on the old one. Engines are more frequently rebuilt on large, costly equipment (tractor) than on small, inexpensive equipment (string trimmer).

Removing External Parts

SERVICE TIP: Engine fasteners and small parts should be separated and stored in plastic bags or baby food containers (Figure 25-11) so that time is not lost looking for the correct fastener during assembly. Exhaust system parts and fasteners are often rusted and difficult to remove. Use a spray penetrating fluid to make these parts easier to remove.

CAUTION: Evaluate the fuel in the gas tank for water contamination and freshness. Then remove all gasoline from the engine tank before starting engine disassembly. Also drain and save the crankcase oil for inspection.

The engine is removed from the power equipment for disassembly. The power takeoff end of the

FIGURE 25-11 Sort and store fasteners and small parts in plastic bags.

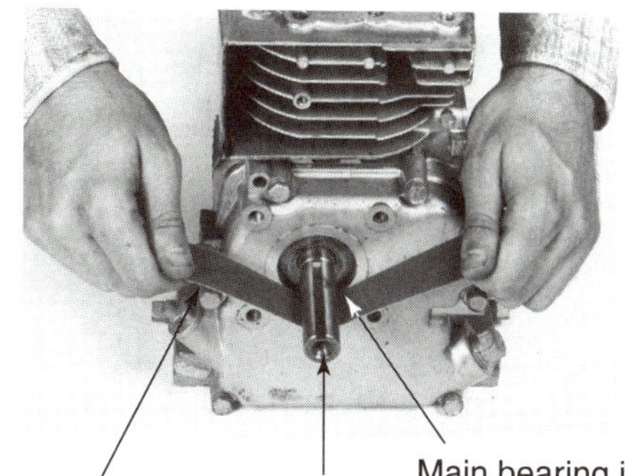

FIGURE 25-12 Using abrasive paper to clean the end of the crankshaft.

crankshaft must be disconnected from the equipment. This means disconnecting any chain, belt, or drive between the equipment and the engine. Many engines have a pulley or blade adapter attached to the crankshaft used to drive the mower blades. A pulley is removed by loosening an Allen set screw that holds it in place. Remove the fasteners that hold the engine to the equipment. As you loosen the fasteners, check them for tightness. Loose mounting bolts can cause vibration that can eventually lead to cracks on the engine mounts. These cracks can spread to the engine crankcase.

Vertical crankshaft engines often have a blade adapter attached to the end of the crankshaft. To remove the adapter, thread a bolt into the end of the crankshaft. Attach a puller to the adapter and adjust the puller forcing screw to push on the bolt. When the drive pulley has been removed, clean the output end of the crankshaft. Burrs or rust on the crankshaft may damage the main bearing when the side cover is removed. A narrow piece of abrasive paper is used to polish the crankshaft (Figure 25-12).

Remove the muffler and any exhaust routing pipes from the exhaust port. Remove the carburetor and intake manifold assembly. Make sure that the fuel is shut off at the tank if the carburetor and fuel tank are separate. Inspect the intake port for any traces of dirt, rust, or any other contamination. Remove the fuel line and fuel tank. When removing the carburetor, make a sketch of the way the throttle and governor linkage and springs are connected. This will save a lot of time when reassembling. Be careful not to bend or stretch any of the linkage.

Removing the Cylinder Head

Remove the fasteners that hold the cooling ducts to the engine. Remove the cooling ducts. Remove the blower housing by removing the fasteners that hold it to the engine. First loosen and then remove the cylinder head bolts (screws), working from the center of the head outward (Figure 25-13). This prevents the head from being distorted. Do not loosen or tighten an aluminum cylinder head when it is hot. Doing so can cause the head to be distorted. Some of the head bolts may be longer than others. Make a sketch of the long and short head bolt location. This will help get them in the correct place during assembly. Carefully note the number, type, and direction of washers used on the head bolts (Figure 25-14). Save the head gasket for failure analysis and to match up with the new one.

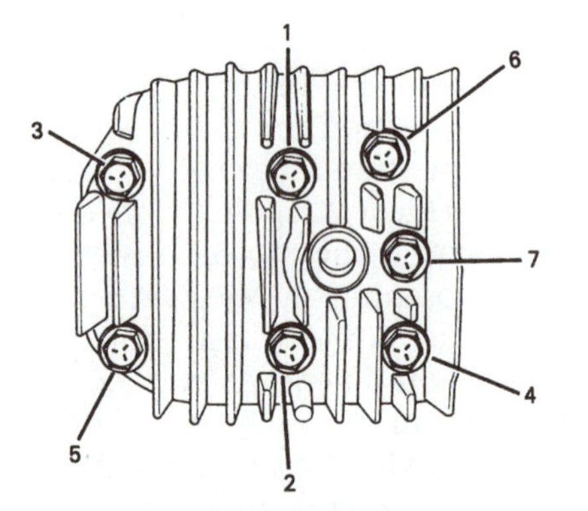

FIGURE 25-13 Loosen the cylinder head bolts from the center outward to avoid distortion. (Provided courtesy of Tecumseh Products Company.)

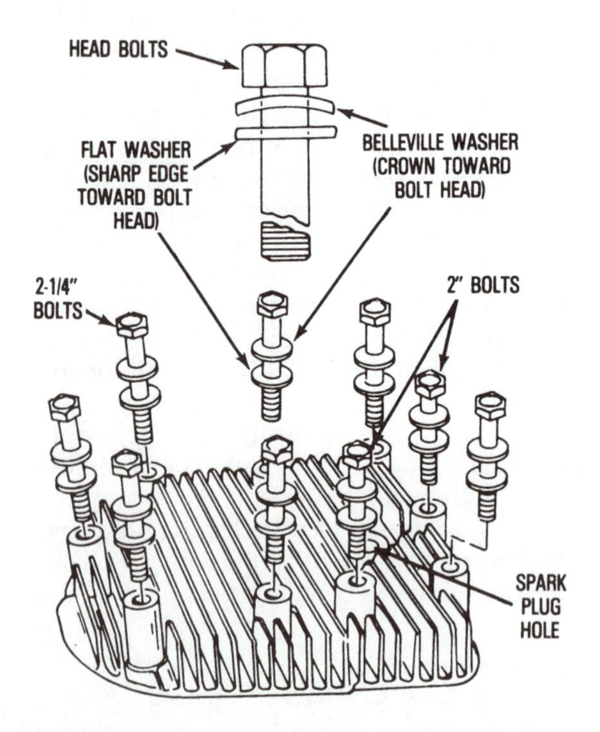

FIGURE 25-14 Note the location of different length head bolts and different types of washers. (Provided courtesy of Tecumseh Products Company.)

Disassembling the Piston, Connecting Rod, and Crankshaft

Remove the flywheel and magneto assembly. If the magneto timing is adjustable, be sure to make marks on the magneto to show where it was timed. The crankshaft end play must be measured before the side cover is removed. **Crankshaft end play** is the end-to-end movement of the crankshaft in the crankcase. There must be enough end play so the crankshaft can turn freely without binding. Too much end play may cause excessive crankshaft and main bearing wear. End play is often adjusted to specifications with gaskets of different thickness on the side cover. End play may be adjusted with different thickness thrust washers between the crankshaft and side cover. The end play is always measured on disassembly. If the end play is not correct, it may be changed during assembly.

Mount a dial indicator (Figure 25-15) to the crankshaft and position the dial indicator plunger on the side cover. Preload the plunger against the side cover. Set the dial indicator to zero. Move the crankshaft back and forth while observing the dial

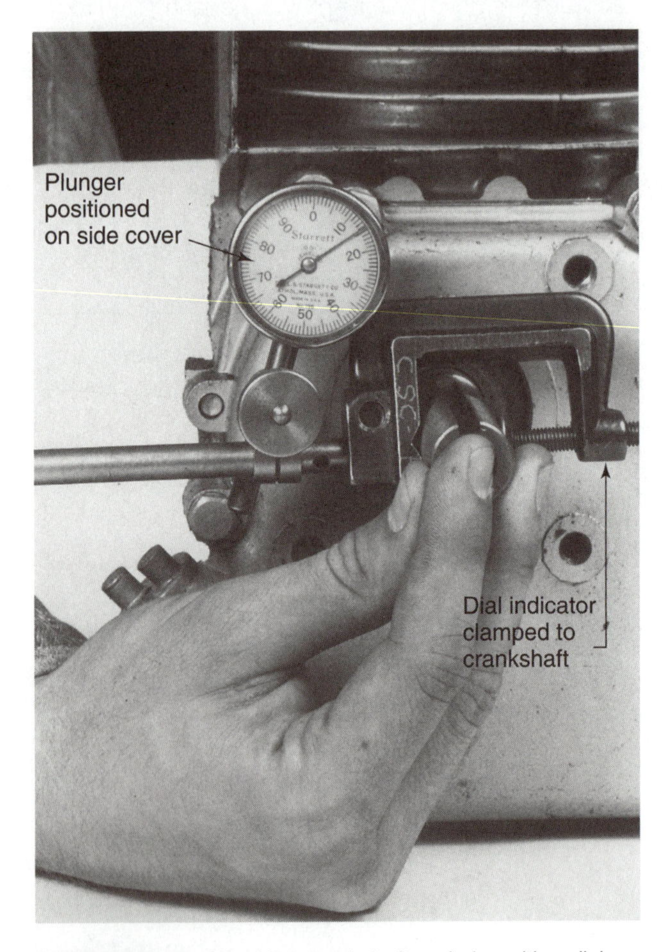

FIGURE 25-15 Checking crankshaft end play with a dial indicator.

PHOTO SEQUENCE 12

Measuring Crankshaft End Play

1. Mount a dial indicator to the side cover and position the dial indicator plunger on the crankshaft.

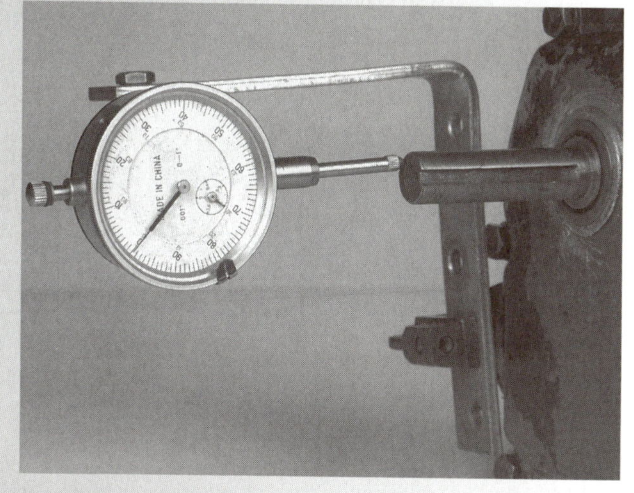

2. Preload the plunger against the crankshaft and set the dial indicator to zero.

3. Move the crankshaft back and forth while observing the dial indicator. The end play will be shown as the total movement (+ and –) of the dial indicator needle.

indicator. The end play will be shown as the total movement (+ and –) of the dial indicator needle. For example, if the needle swings to negative 0.010 then to positive 0.005 the end play is 0.015 inch. Be sure to record the measurement to check against the specifications.

Remove the side cover by removing the fasteners that hold it to the crankcase. Pull it off the crankshaft. If it does not pull off easily, recheck the crankshaft for burrs. With the side cover removed, the crankshaft and camshaft timing gears are visible. The camshaft and crankshaft gears must be in exactly the correct relationship (timed). This timing allows the valves to open or close at the correct time in relation to piston position. If the camshaft to crankshaft timing is off by even one tooth, the engine will not run.

Engines have timing marks (Figure 25-16) on the two gears to show how they should be meshed. There are many types of marking systems. Some engines use holes, punch marks, etched lines, or keyway locations as timing marks. Engines with balance shafts must have these aligned in time. Always check the appropriate shop service manual to determine the types of marks to be located.

Wipe off the gears and carefully look for these marks. Sometimes the marks are hard to see because of heat patterns or burned on oil. If no marks

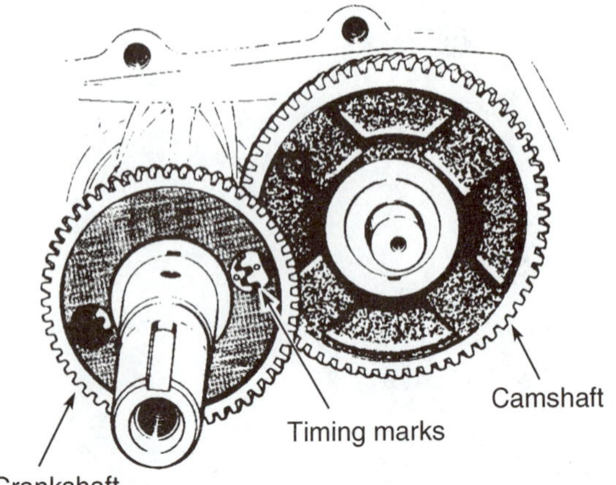

FIGURE 25-16 Crankshaft to camshaft timing marks. (Courtesy of Onan Corp.)

are visible, use a sharp punch to make match marks on meshing camshaft and crankshaft teeth. After marking, remove the camshaft by pulling it out of its bearing. Remove both valve lifters. Use a marking pen to identify the lifter as intake or exhaust. Valve lifters must be assembled back into their original position.

Wipe and inspect the connecting rod bearing cap(s) for a factory marking. These markings can be punched-on dots or etched lines. Stamped numbers 1 and 2 are commonly used on the two connecting rods of multiple cylinder engines. Not all engine makers mark the connecting rods. The caps on some engines fit only one way but many can be reversed. The rod cap big end bore is only perfectly round when the cap is positioned one way. Stamp unmarked connecting rods with punch marks on both cap and rod before removal.

The rod marking is usually done on the camshaft side (Figure 25-17) to help get the connecting rod direction correct during assembly. Many connecting rods have oil holes that must aligned properly. Others are made with an offset in one direction. The connecting rod must reassembled on the piston correctly. It must be installed in the engine in the correct direction.

Many connecting rods have lock tabs that must be bent away from the connecting rod bolts or nuts. Remove the connecting rod cap bolts and rod cap from the rod. Push the connecting rod and piston assembly up and out of the top of the cylinder. Be careful not to scratch the crankshaft journal. If necessary, use a block of wood to tap the piston assembly up through the cylinder.

Remove the crankshaft by pulling it out of the main bearing. The crankshaft may not come out of the main bearing easily. Do not force it out. Recheck the end of the crankshaft for burrs. Any burrs or rust areas that pass through the main bearing can damage the bearing surface. Main bearing seals are removed from the side cover with a large screwdriver (Figure 25-18).

Pistons usually have a direction arrow on the top or side that helps get them in the correct direction (Figure 25-19). If no direction arrow is located, make a new one on the underside of the piston. Pistons and connecting rods must be installed in the same direction in which they were removed. They must be installed in the bore they were removed from on multiple cylinder engines.

Most engines use full-floating piston pins. There are internal expanding snap (retaining) rings on both ends of the piston pin. The rings hold the pin in the piston. Remove these snap rings with needle nose pliers (Figure 25-20). Push the pin out of the piston and connecting rod. If the pin is tight, the piston may be heated in hot water to loosen it.

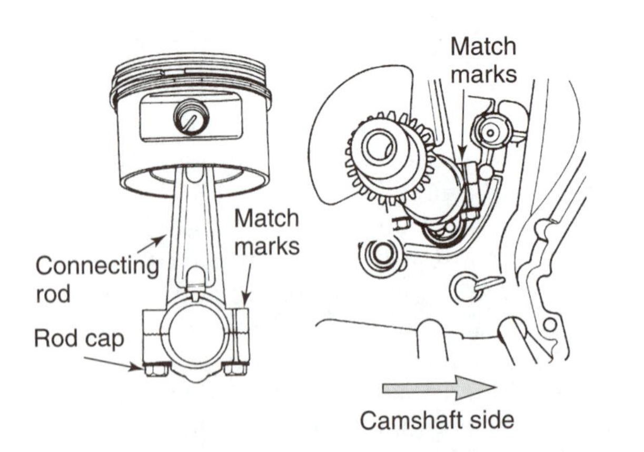

FIGURE 25-17 Match marks allow correct installation of the rod and rod cap. (Provided courtesy of Tecumseh Products Company.)

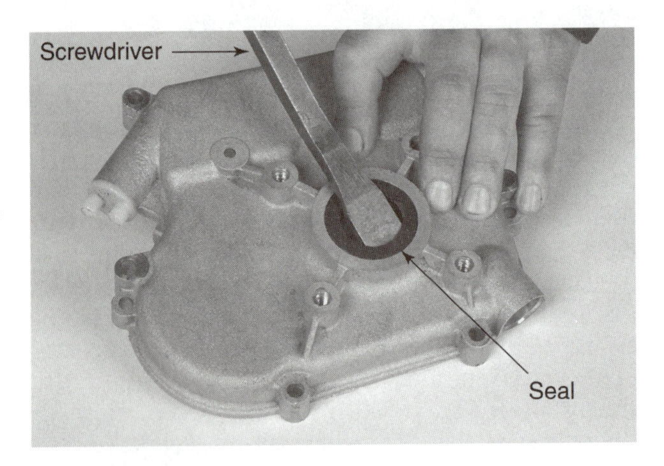

FIGURE 25-18 Main bearing seals are pried out with a screwdriver.

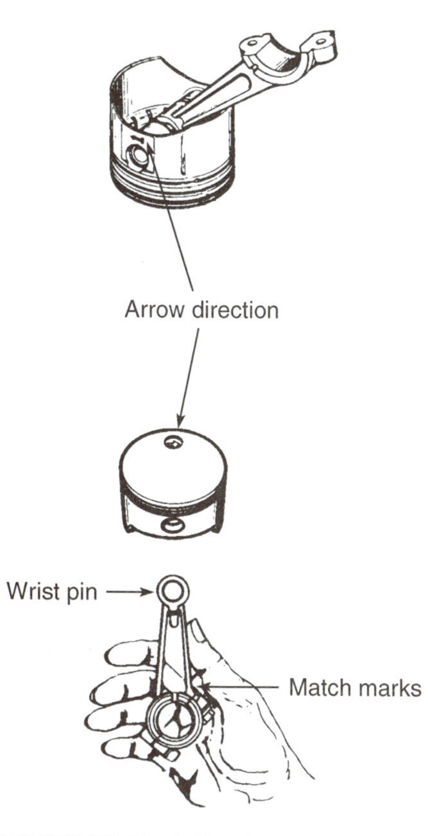

FIGURE 25-19 A direction arrow on the piston shows piston direction in the engine. (Provided courtesy of Tecumseh Products Company.)

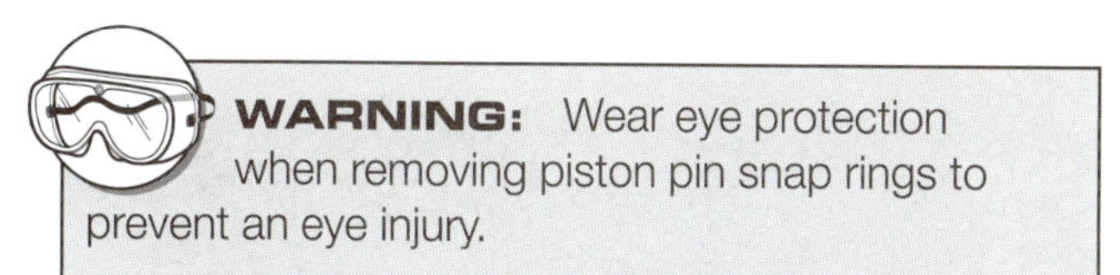

WARNING: Wear eye protection when removing piston pin snap rings to prevent an eye injury.

Piston rings are removed from the piston with a piston ring expander tool. A **piston ring expander** (Figure 25-21) engages the end of piston

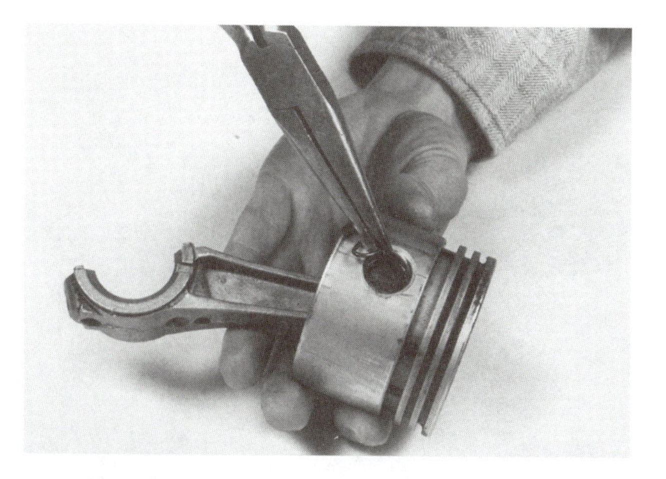

FIGURE 25-20 Removing a piston pin snap ring with needle nose pliers. (Courtesy of Poulan/Weed Eater.)

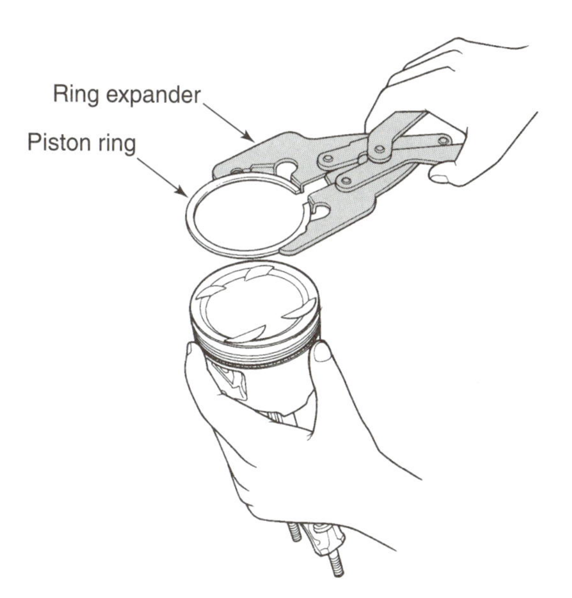

FIGURE 25-21 Using a ring expander tool to remove piston rings. (Courtesy of American Honda Motor Co., Inc.)

rings and expands them for removal from the piston. The tool fits around the piston and engages the two ends of the piston ring. When the handle is squeezed, the piston ring is expanded enough to be removed from the piston. Removing a piston ring without this type of tool can allow ends of the piston ring to scratch and damage the piston. Use masking tape labels on the rings. Identify the rings to the piston groove in which they fit. Also identify each ring by which side of the ring faces up. This will help in failure analysis to determine if the rings have been assembled properly.

Disassembling the Valves

The valves are removed by compressing the valve springs and then removing the valve spring retainers. Three common types of valve spring retainers are pin, collar, and washer (Figure 25-22). The pin

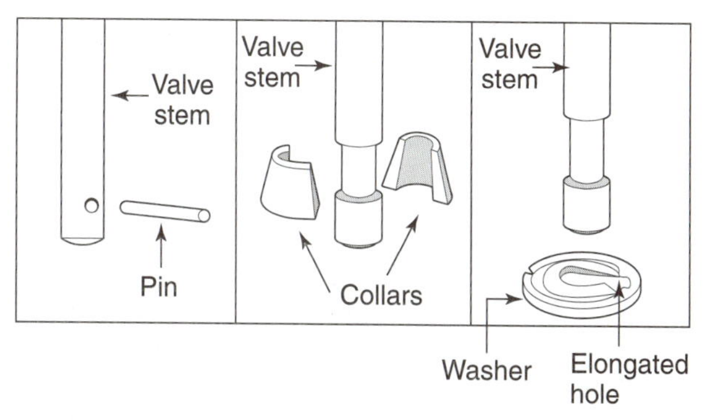

FIGURE 25-22 Three types of valve spring retainers.

type is the oldest type. It is used on older cast-iron engines. The collar retainer has two cone-shaped collars. The washer retainer combines the retainer and washer in one part. The washer has an elongated, figure eight-shaped hole in the center.

SERVICE TIP: Pin and collar type retainers often stick to the valve spring retaining washers and make compressing the valve springs difficult. You can get these apart by pushing the valve upward off its seat with a screwdriver against the valve spring retainer. Gently tapping valve head downward with a soft face hammer will separate the washer from the retainer.

The valve springs must be compressed to remove the valves. Then the retainer, washer, and valve spring are disassembled from the valve stems. The valve springs are compressed with a **valve spring compressor** (Figure 25-23), a tool used to compress a valve spring for valve spring retainer removal. The compressor tool removes the spring pressure from the retainer washer and allows removal.

WARNING: Valve springs and retainers are under high spring pressure. Always wear eye protection to avoid injury from flying parts.

FIGURE 25-23 A valve spring compressor is used to compress valve springs for removal.

The compressor used for L-head valves fits either end of the valve spring. The jaws on the end of the tool are adjusted so that the jaws fit the diameter of the valve spring. The tool is positioned differently for different types of retainers. The tool fits on the top of the spring and on the retaining washer to remove either the collar or pin type retainer. The tool fits on the top of the spring and directly on the bottom of the spring to remove the washer type retainer. The hand crank on the tool is rotated until the spring is fully compressed and the tension is off the valve retainer.

With the spring compressed, the retainer may be removed (Figure 25-24). Pin retainers are pulled

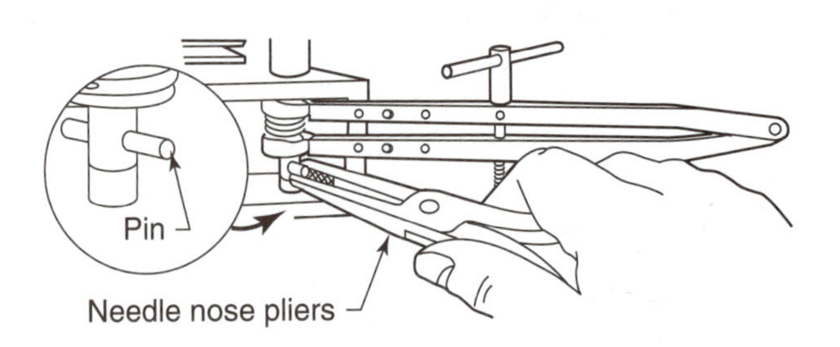

A. Pin

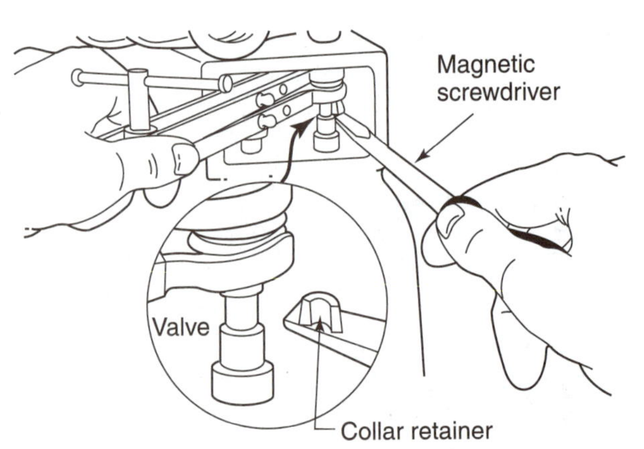

B. Collar

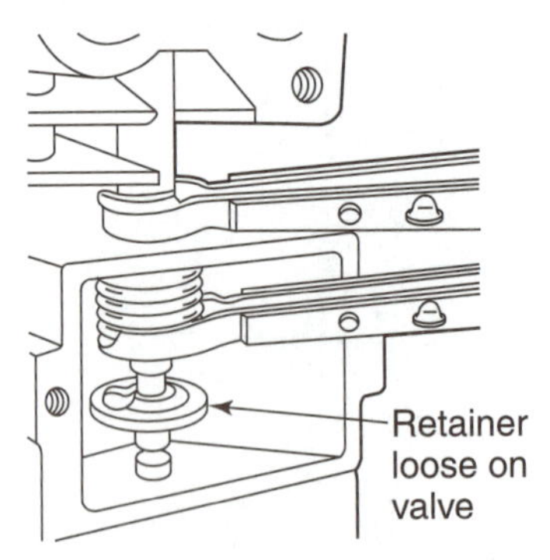

C. Washer

FIGURE 25-24 Removing the three types of valve spring retainers.

out with needle nose pliers. Collar type retainers are removed by picking them off with a magnetic screwdriver. Washer retainers are removed by compressing the spring and leaving the retainer loose on the valve stem. The washer has a large hole in the side that allows it to be pushed to the side and removed. Disassembling an L-head valve assembly with a valve spring compressor is shown in the accompanying sequence of photographs.

Overhead valve engines have valve spring assemblies that are more accessible at the top of the cylinder head. Overhead valve springs often use the collar type valve spring retainer. Engine makers supply a special tool to fit their valve springs. The typical tool has a handle with an anchor at one end. It has a hole to access the valve spring retainers. The anchor on the tool fits over a rocker arm stud to provide a leverage point. Pushing down on the handle compresses the valve spring and retaining washer. The hole in the tool allows access for a magnetic screwdriver to remove the collar retainers. Slowly lifting up on the tool releases the valve spring from the valve stem (Figure 25-25).

The end of the valve gets a constant pounding from the valve lifter or rocker arm. This often forms a burr on the end of the stem. The burr must be filed off. If not, the valve will be difficult to remove. Carefully file the valve stem tip and remove any filings (Figure 25-26). Remove the valves by pushing them out of their guides.

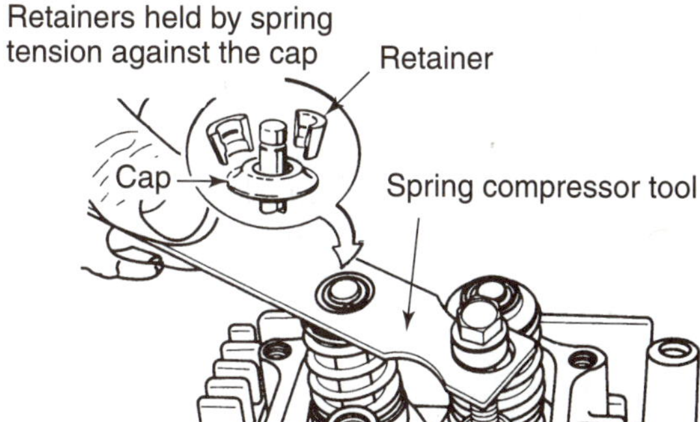

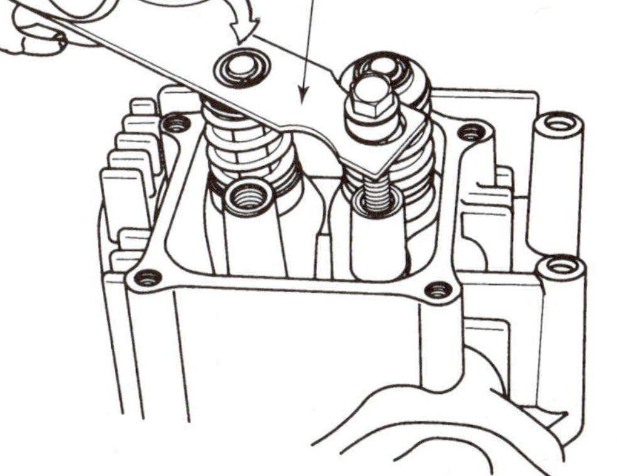

FIGURE 25-25 Compressing an overhead valve spring for removal. (Provided courtesy of Tecumseh Products Company.)

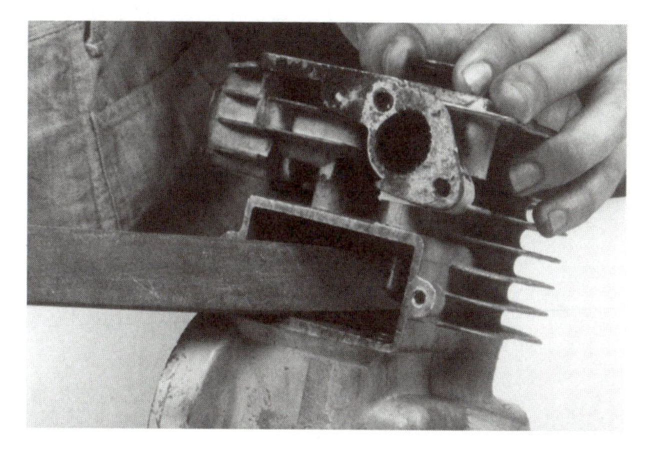

FIGURE 25-26 Removing the burr from the end of the valve stem.

INTERNAL COMPONENT FAILURE ANALYSIS

Spread out the internal parts on a bench. Inspect each of the parts under a strong light. The most important parts to check are the parts that handle the forces of combustion. Focus your attention on the cylinder head, cylinder, valves, crankshaft, main bearings, connecting rod, and piston. These parts have the most critical running tolerances. They are most likely to fail because of the high forces and stresses.

The purpose of your inspection is to determine what part failed and why. The purpose of a failure analysis is to prevent the same type of failure from occurring to the rebuilt (or new) engine. Carefully disassemble the parts so any critical evidence is not disturbed or destroyed. Keep the parts in their original condition until the failure analysis procedure has been completed. Parts should not be cleaned unless it is necessary to make an accurate inspection.

Checking the Combustion Chamber

Turn the cylinder head upside down and inspect the combustion chamber. Look at the color and condition of combustion chamber deposits (Figure 25-27). The deposits will tell a lot about operating conditions. A properly running engine will usually have light brown or gray combustion chamber deposits. Black, oily or gummy deposits mean heavy oil consumption. This is most likely from piston ring or valve guide wear. A crankcase breather problem can also cause oil burning in the combustion chamber. Soft, black, sooty deposits show incomplete combustion. These may be caused by a

PHOTO SEQUENCE 13

Disassembling an L-Head Valve Assembly with a Valve Spring Compressor

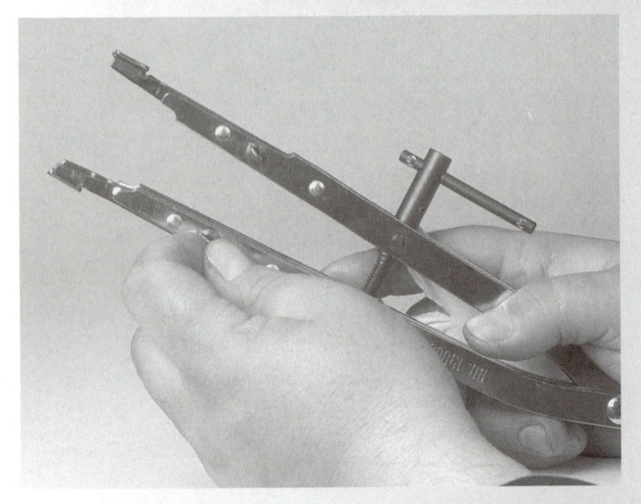

1. Adjust the valve spring compressor jaws on the end of the tool so that they fit the diameter of the valve spring.

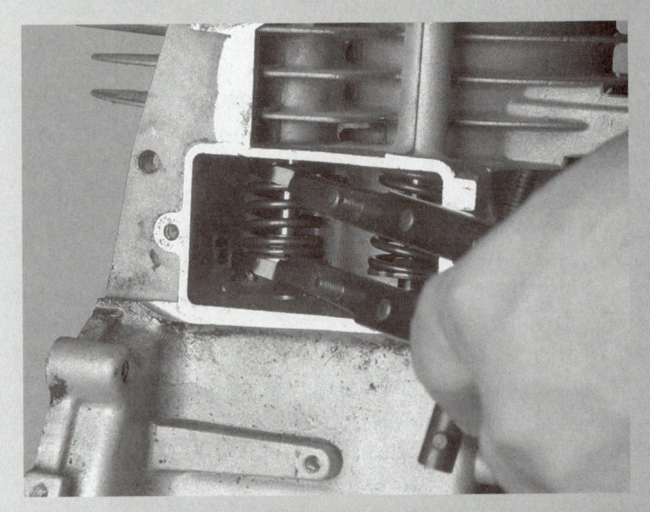

2. To remove the collar type retainer, install the tool so that it fits on the top of the spring and on the retaining washer

3. Rotate the hand crank on the tool until the spring is fully compressed and the tension is off the collars.

4. Collar type retainers are removed by picking them off with a magnetic screwdriver.

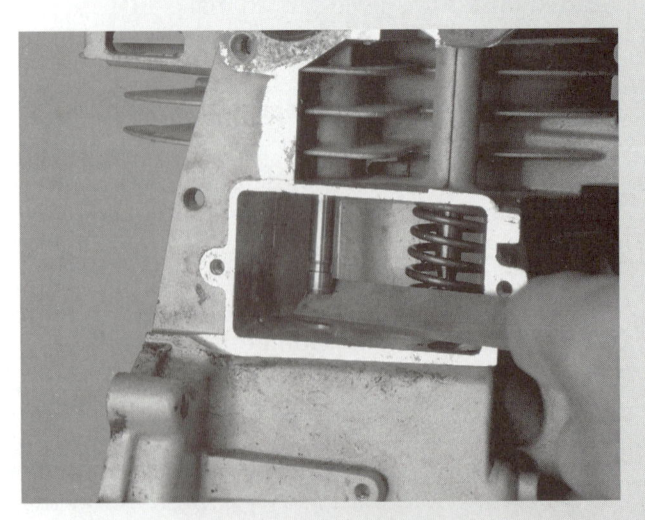

5. Carefully file the valve stem tip to remove any burrs.

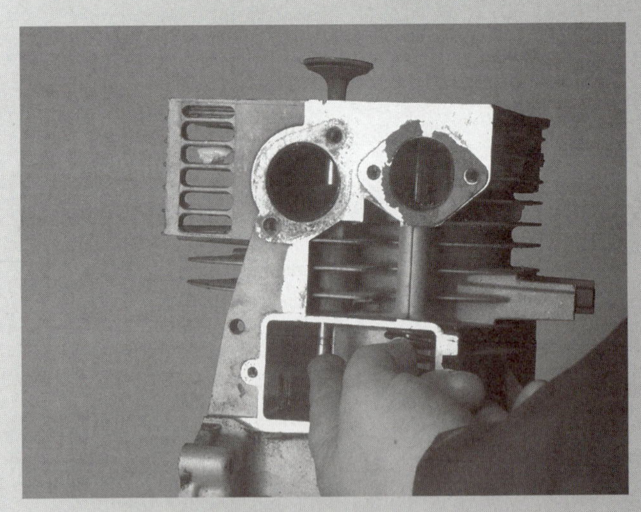

6. Remove the valves by pushing them out of their guides.

A. Light Brown or Grey Deposits—Good Operating Conditions

B. Black, Oily Deposits—Heavy Oil Consumption

C. Soft, Black Deposits—Incomplete Combustion

D. Hard, Crusty, White Deposits—High Combustion Temperatures

FIGURE 25-27 Combustion chamber deposits help determine engine operating conditions.

rich carburetor mixture, dirty air filter, or retarded timing. Hard, crusty, white deposits show high combustion chamber temperatures. These may be caused by a lean carburetor mixture, intake air leak, overadvanced timing, or poor quality gasoline.

The high temperatures that cause the white deposits can also cause cylinder head distortion. A distorted cylinder head can cause the hot exhaust gases to leak past the head gasket. The leaking gas will burn away the gasket. This condition is called a blown head gasket. Look for a burned through section on the head gasket.

Checking the Valves

Check the intake and exhaust valves closely. Valve condition is a good indicator of operating conditions. Leaking valves can cause hard starting, stalling, high fuel consumption, poor compression, loss of power, and exhaust "pop" noises. Common valves problems are burning, sticking, and erosion. Exhaust valves normally show more wear and carbon buildup than intake valves. They handle the high temperature exhaust gas. Intake valves are cooled by the intake of air and fuel mixture.

Look for a bright, uniform sealing ring around the valve face. Look for a small buildup of carbon on the underside of the valve head and upper stem. The deposits should be brownish in color. All these show good operating conditions. Lean mixtures cause high combustion temperatures and leave white deposits on the valve head. High temperatures can lead to valve face-to-seat sealing problems. Valve seating problems can be caused by not enough valve clearance. Carbon deposits lodged between the valve and seat can also prevent valve seating. Under these conditions, the valve is unable to dissipate heat into the valve seat.

Long periods of high temperature operation can cause erosion of the valve face. **Valve erosion** is a condition in which material is eroded or burned away from the valve head area. When the exhaust valve is not sealing, combustion pressure leaks out of the combustion chamber past the open valve. The face continues to burn and erode until a large section disappears (Figure 25-28).

The valve stems can also show poor operating conditions (Figure 25-29). The valve stems should appear shiny. The stems may have a dull wear pattern where they travel in the guides, which shows that abrasive dirt has entered the engine through the air intake. Look for burned oil deposits on the valve stem and in the guide. Deposits can cause the valve stem to stick in its guide. **Valve sticking** is a condition in which the valve stem sticks in the valve guide and prevents proper valve movement.

An engine with valve sticking loses power or "pops" out the exhaust then stalls after a period of

FIGURE 25-28 Exhaust valve head with severe erosion (burning).

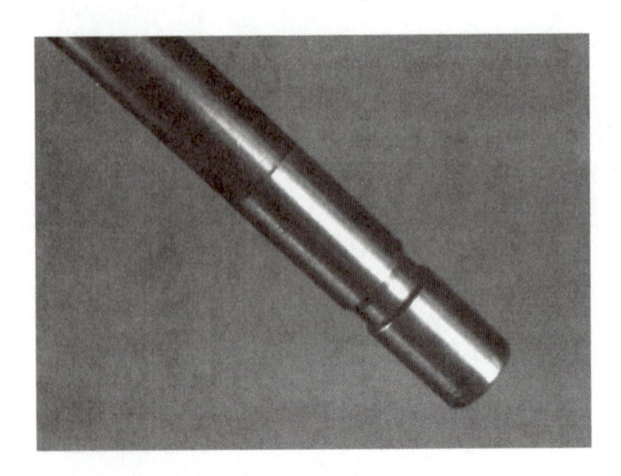

FIGURE 25-29 Valve stem with scoring.

running. Often the engine will not restart until it cools down. Sometimes you can hear a metallic snap when it starts. High valve guide temperatures cause the burned oil deposits that cause valve sticking. The problem usually shows up during hot weather. Engines with a lean air-fuel mixture or an ineffective cooling system usually get this problem.

Checking the Crankshaft and Main Bearings

Carefully inspect the crankshaft main and connecting rod journals for signs of adequate lubrication. A loss of lubrication is the most common cause of bearing failure and crankshaft damage. Each crankshaft journal should be bright and shiny (Figure 25-30). Poor lubrication conditions cause bearing material to transfer to the crankshaft journals (Figure 25-31). If the transfer continues long enough, the bearing and crankshaft will seize together. A low oil or no oil running condition will cause aluminum bearing to seize very quickly.

The connecting rod bearing is under the highest loads and is subject to the worst seizures. During failure, a lack of lubrication causes aluminum to

FIGURE 25-30 A bright, shiny journal shows good lubrication.

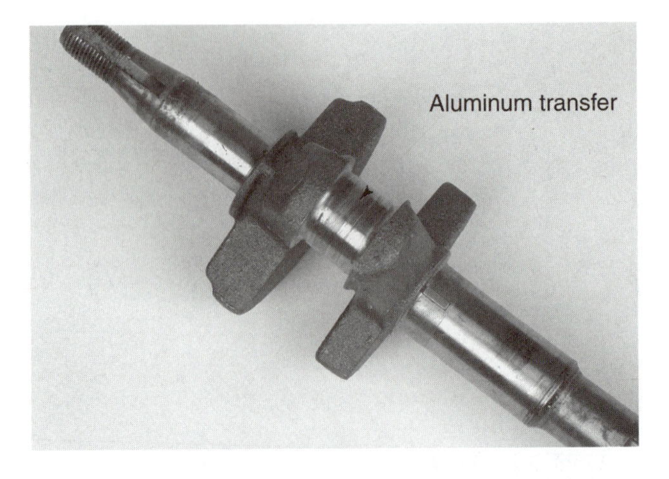

FIGURE 25-31 Poor lubrication causes aluminum transfer to the bearing journal.

transfer to the crankshaft journal from the connecting rod. Temperatures can rise to the melting point of aluminum. This results in a weakening of the connecting rod. When enough metal transfers, the result is a broken connecting rod (Figure 25-32).

Poor lubrication causes metal to transfer from the main bearing to the crankshaft main bearing journal (Figure 25-33). Vertical and horizontal crankshaft engines have different kinds of bearing

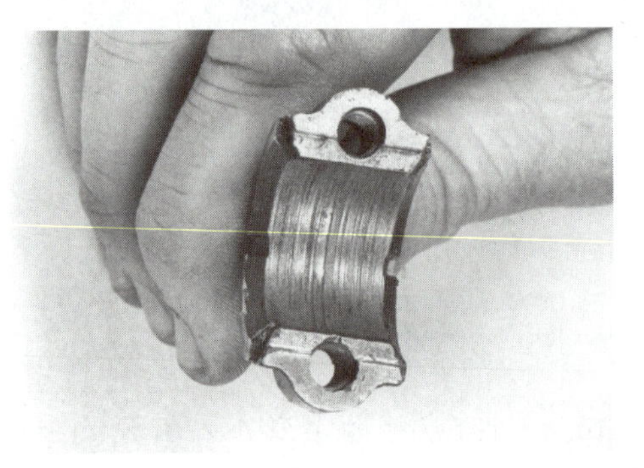

FIGURE 25-32 A seized connecting rod bearing may result in a broken connecting rod.

FIGURE 25-33 Main bearing seizing causes metal transfer.

failures. The vertical crankshaft FWE bearing does not fail as often as the horizontal shaft FWE bearing. Low oil is not as big a problem for vertical crankshafts because these engines use an oil pump. Both types of engines experience more failure at the PTO than at the FWE. This is caused by higher stresses in the PTO area. Seizures at the main bearings will result in a crankshaft that is frozen (welded) to the main bearing. The crankshaft will not rotate.

Checking Connecting Rods

The connecting rod can fail in the crankshaft big end, the piston pin small end, or the beam connecting the two ends. Piston pin end failures are usually caused by a defect during manufacture. Beam failure when there is no damage to the big or small end bearing is usually caused by a manufacturing defect.

The most common failure area is the connecting rod bearing. Poor lubrication or loose rod cap bolts cause most of these failures. Poor lubrication can cause connecting rod bearing seizure. The first seizure often stops the engine without breaking the rod. The engine can often be restarted if oil is added and it is allowed to cool down. The seized connecting rod may continue to work but will overheat. The temperature will be hot enough to burn any oil that comes into contact with the outside surface of the rod, causing a blackening of the connecting rod. Eventually, the rod will seize completely and break. Examine the connecting rod bearing area for any blackening. If you do not see any, the engine may not have had any oil in it when the rod broke. The failure may have come all at once from operating the engine at too high a speed.

Loose connecting rod cap fasteners are another common cause of failure. Be careful not to confuse this damage with bearing seizing. Bearing seizing can loosen the connecting rod fasteners. A failure caused by loose connecting rod fasteners will not usually have aluminum transfer to the crankshaft journal. The connecting rod bearing surface loses its shiny appearance and the surface turns gray.

Checking Pistons and Cylinder Walls

Piston, rings, and cylinder bore problems are usually caused by excessive wear or piston seizure. Excessive wear can often be seen even before any measurements. Inspect the wear pattern on the piston skirt. Normal operation causes a wear pattern that covers about 20–40 percent of the skirt thrust face. The thrust faces are the two sides of the skirt that push (thrust) against the cylinder bore. The thrust faces are the skirt areas 90 degrees to the piston pin hole. Wear is excessive if it covers 50 percent or more of the skirt thrust face. Look for vertical scratches on the skirt (Figure 25-34). This is caused by dirt contamination between the piston and cylinder wall. Look for erosion at the very top edge of the piston and on the piston ring faces. This is also caused by dirt contamination. As the rings wear, oil consumption increases. You will see a carbon ridge at the top of the cylinder.

FIGURE 25-34 A piston with abrasive wear on the thrust face.

Excessive wear between the piston skirt and cylinder wall can cause the piston to rock (slap) against the cylinder walls. The rocking piston puts increased stress on the piston skirts. Eventually the skirt will crack (Figure 25-35). The crack may progress across the skirt and up toward the oil ring groove. The lower part of the skirt eventually breaks off. The piston may completely break up.

Piston seizure occurs when the piston temperature reaches the melting point of aluminum. The skirt welds to the cylinder wall. Piston overheating may be caused by poor cooling, poor lubrication, or dirt contamination. A recently rebuilt engine may not have enough piston clearance. Your external

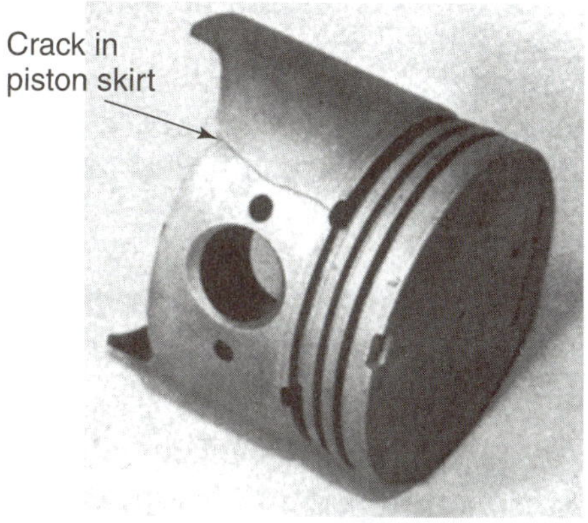

FIGURE 25-35 A piston with a crack in the skirt.

FIGURE 25-36 A piston with a deep groove from seizure.

inspection will probably discover the cause of this problem.

A seized piston has deep grooves (scoring) on the skirt (Figure 25-36). There is metal transfer on the piston and cylinder bore. The scoring is usually found only on the skirt that is closer to the valves. This side has stronger forces on it during the power stroke. It is called the *primary thrust face.* The primary thrust face will be discolored if the failure was due to overheating. Scorched oil deposits cause an overheated piston to turn black.

Seizures caused by abrasive contamination usually result in scoring without any other signs. The same goes for a piston with too tight piston-to-cylinder wall clearance. Scoring often occurs to both skirts. There will probably be deeper scoring on the piston primary thrust face.

Inspect the cylinder walls for the same types of damage as the pistons. Look for scratches and score marks in the bore. Look for metal transfer from the piston skirt. Check for blackened areas that indicate a loss of lubrication.

ENGINE PART CLEANING

The results of the failure analysis should help determine if the engine should be rebuilt. If the engine is to be rebuilt, clean each of the parts. Part cleaning is important. Parts have to be cleaned for accurate wear measurements. Any abrasive dirt not removed from engine parts can damage the new engine and shorten the new engine's service life. How the parts are cleaned depends upon the type of cleaning equipment available in the shop.

Cleaning Equipment

Many shops clean engine parts with spray can type cleaners. Solvent-based spray cleaners thin and wash oil and dirt off of parts. Parts cleaned with these products are sprayed, wiped, and allowed to air dry. Other spray cleaners are detergent based. These are sprayed on parts to get the oil and dirt in suspension. The parts are then flushed with water.

Some shops use a solvent parts washer for engine parts cleaning (Figure 25-37). The cleaning solvent in the washer thins and washes away grease, oil, and sludge. An electric motor in the cleaner pumps solvent through a hose to flush off parts. The solvent is also circulated through a filter to keep it as clean as possible.

Some shops use a hot high-pressure washer (Figure 25-38). The hot high-pressure water washer uses high-temperature water (190–210 degrees) mixed with soap to flush dirt and sludge off the outside of engines and equipment. The water can be heated with kerosene or natural gas. The advantage of this equipment is that it can be used to

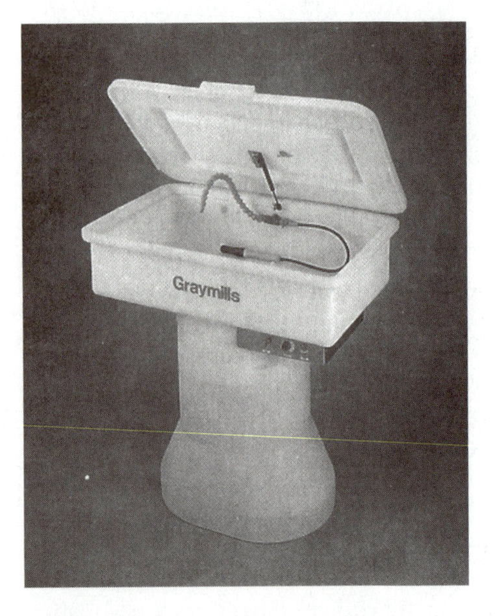

FIGURE 25-37 A solvent parts washer is used to clean steel parts. (Courtesy of Graymills.)

FIGURE 25-38 A hot pressure washer. (Courtesy of Landa Water Cleaning Systems.)

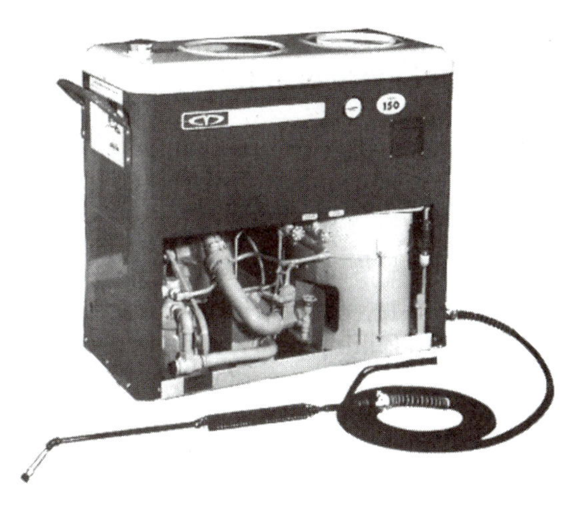

FIGURE 25-39 A steam cleaner. (Courtesy of Malsbary Manufacturing Company.)

clean mowers and other equipment without removing any paint.

A steam cleaner (Figure 25-39) is another common cleaning tool. The steam cleaner develops high-pressure steam that is mixed with soap to clean the outside of engines and equipment. The steam can be generated by a kerosene or natural gas heating source. The high-temperature steam does an effective job of removing dirt and grease from parts.

WARNING: Wear personal protective equipment when using any cleaning equipment. Follow the safety precautions on the label of any cleaning material. High-pressure washers and steam cleaners can cause severe burns and chemical (soap) burns to the skin and eyes.

Cleaning Engine Parts

The cylinder block assembly is the largest engine part to be cleaned. It has many cooling fins and oil passages. Clean each of these passages thoroughly. All abrasive particles have to be removed from the oil passages. If not, they can get into the new bearings.

The cylinder block may be cleaned in a solvent tank, cold tank, or with a spray can cleaner. After the block is clean, flush all deposits from the block with water. Clean oil passageways and holes with a brush and blow them out with air pressure. When the block is dry, oil all the parts that can rust, including the cylinder liner and valve seats. Store the cylinder in a plastic bag to prevent dirt contamination.

Clean the cylinder head the same way as the cylinder block. Aluminum cylinder heads should not be cleaned with a wire brush. A wire brush will scratch the combustion chamber. This can cause hot spots and abnormal combustion. Clean the carbon from the combustion chamber by soaking the cylinder head.

Remove carbon deposits from valves by spraying with cleaner, soaking in a cold tank, or hand wire brushing. Remove all the carbon from the head, face, and stem. Carbon can cause the valve to overheat. It restricts the heat flow from the valve head into the valve seat. Oil the valves with clean engine oil to prevent rust. Store the valves in a plastic bag.

Clean the piston carefully to prevent scratching the aluminum surface. Remove carbon on the head by soaking. Do not wire brush the top of a piston because this will scratch the surface. Carbon in the piston ring grooves can be removed with a **ring groove cleaner**, a scraper tool that fits into the piston ring groove to remove carbon (Figure 25-40). There are different sizes of scrapers on the tool. These fit different sizes of ring grooves. Be careful when using this tool not to cut into the aluminum on the ring groove. Store the cleaned piston in a plastic bag to keep it clean.

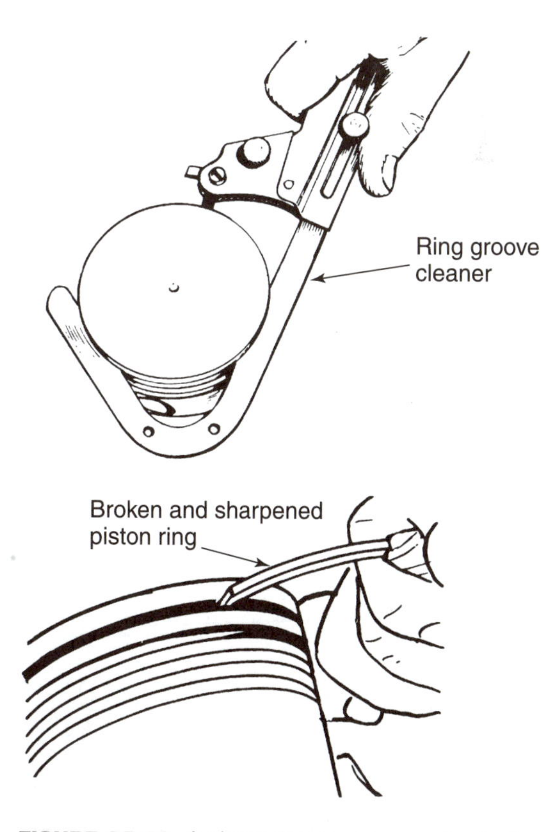

FIGURE 25-40 A piston ring groove scraping tool or a piston ring is used to remove carbon from piston ring grooves. (Courtesy of Onan Corp.)

SERVICE TIP: You can remove carbon from a ring groove with the old piston ring. Break the ring in half. Sharpen one end of the ring with a file. Put masking tape on the other end for a handle. Use the sharpened end to scrape carbon out of the groove.

Wash the crankshaft with solvent in a parts washer. Use a parts-washing brush to loosen deposits. After it is cleaned, allow the crankshaft to air dry. Blow out any crankshaft oil passages with compressed air. Any abrasive dirt in these passages can damage bearings. When the crankshaft is dry, coat the bearing journals with clean engine oil to prevent rusting. Seal the crankshaft in a plastic bag to protect it from dirt.

Clean aluminum covers and connecting rods in a cold tank or with spray cleaner. Store these parts in plastic bags. Wash steel piston pins, valve lifters, valve springs, and washers, as well as the fasteners in the solvent parts washer. Oil the steel parts then to prevent rusting. Store all these parts in plastic bags.

REVIEW QUESTIONS

1. Describe the type of information that should be gathered from a customer when analyzing an engine for failure.
2. List and describe the areas on the outside of the engine to check when analyzing an engine for failure.
3. List and describe the parts of the air cleaner assembly to check when analyzing an engine for failure.
4. Explain how to check oil level and condition when analyzing an engine for failure.
5. Explain how to check the governor linkage for proper operation when analyzing an engine for failure.
6. Explain how to determine if a carburetor is adjusted too lean when analyzing an engine for failure.
7. List and describe the parts of the cooling system that should be examined when analyzing an engine for failure.
8. Explain the areas of the combustion chamber to check when analyzing an engine for failure.
9. Explain the correct procedure to follow when removing a cylinder head.
10. Explain how and why a ring ridge is removed on some cast-iron cylinders.
11. List and describe where match marks must be observed or made when disassembling a piston and connecting rod.
12. List and describe the disassembly procedure for three types of valve spring retainers.
13. List and describe the common types of deposits found in the combustion chamber.
14. List and describe the areas to examine when observing valves for failure analysis.
15. List and describe the areas to examine when observing the crankshaft and main bearings for failure analysis.
16. List and describe the areas to examine when-observing the connecting rod for failure analysis.
17. List and describe the areas to examine when observing the piston for failure analysis.
18. List and describe three types of engine-cleaning equipment.
19. Explain how to remove carbon from a piston ring groove.
20. Explain why and how oil passages in the crankshaft and cylinder block must be cleaned prior to assembly.

DISCUSSION TOPICS AND ACTIVITIES

1. Locate several complete shop engines. Do an outside inspection of each engine and try to determine the cause of failure.
2. Collect a set of shop engine internal parts. Study the wear on the parts and try to determine the cause of failure.
3. A dial indicator is preloaded and set to zero to check crankshaft end play. The end play is checked and the dial moves from +0.006 inch to –0.012 inch. What is the end play measurement?

CHAPTER 26

Valve Service

OBJECTIVES

Upon completion and review of this chapter, you should be able to:

- Inspect the valve, valve seat, and guide for wear.
- Measure a valve guide for wear.
- Install replaceable and insert type valve guides.
- Recondition valve seats.
- Inspect a cylinder head for wear and damage.
- Grind and lap valves.
- Service valve springs.
- Service the camshaft and valve lifters.
- Set nonadjustable valve clearance.
- Assemble and test the valves for sealing.

TERMS TO KNOW

Bushing driver
Peening
Stemming
Valve grinder
Valve guide bushing
Valve guide plug gauge
Valve lapping
Valve lapping compound
Valve lapping tool
Valve seat cutter
Valve seat narrowing
Valve seat puller
Valve spring free length

INTRODUCTION

The valves are the most highly stressed parts in the engine. Valves seal very high pressures. The exhaust valve has high-temperature exhaust gases passing over it on their way out of the cylinder. Under full load, the exhaust valve is often red hot. One valve may fail in a multiple cylinder engine and only a part of the power is affected. The bad cylinder is helped out by the other good cylinders. In a single cylinder engine, one bad valve can cause a great drop in power. The engine may stop running entirely. Good valve service is very important in a single cylinder engine.

VALVE, VALVE SEAT, AND GUIDE INSPECTION

The valve, valve seat, and guide are inspected for wear and damage. Even in normal use, these parts get a great deal of punishment. High pressures and powerful spring tension pound the red hot valve head and seat. Hot gases under high pressure swirl past it. Carbon deposits can form on the valve face. They can prevent the valve from seating and cooling properly. The exhaust valve head may get pitted, burned, warped, or grooved. The damaged valve may begin to leak compression and fail to dissipate heat. Valves also wear at the stem from friction with the guide (Figure 26-1). They get end wear from contact with the valve lifter.

The valve seat is another common wear area. Hot gases can attack and burn the valve seat metal. Carbon particles can build up on the seat. Carbon

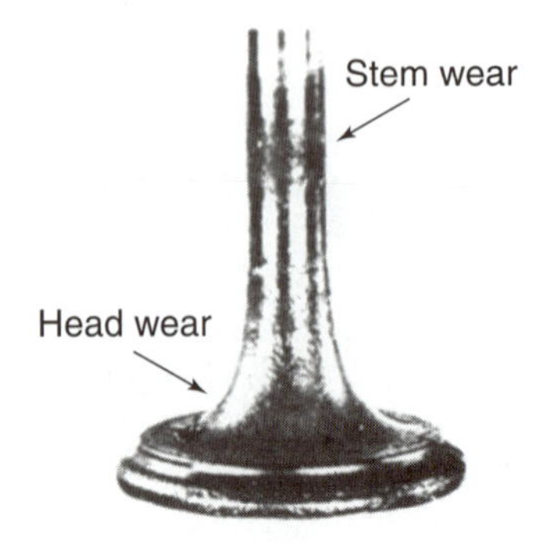

FIGURE 26-1 Valve wear on the valve head and stem. (Courtesy of Federal-Mogul Corporation.)

retains heat, which can cause erosion of the sealing surface. Erosion prevents the valve from sealing properly.

The valve guide has the same friction and erosion forces working on it as the valve stem. The valve guide material is often softer. Wear on the guide is even more rapid than the valve stem. A worn valve guide will not properly support the valve (Figure 26-2). It can quickly cause sealing problems. The clearance between the valve stem and guide must allow free valve movement. It must also allow a small amount of oil to work its way between the stem and guide for lubrication. If there is too much clearance, crankcase oil can work its way up the stem. This oil can get into the combustion chamber. This is a big problem on the intake valve stem. When the intake valve is open, there is a low pressure in the cylinder.

The valve guide also gets rid of valve heat. Cooling fins in the cylinder block are located near the valve guide area. Most valve head heat is dissipated into the valve seat. Heat that is left over travels down the valve stem. This heat goes into the valve guide and cylinder block, and into the cooling fins. Excessive clearance prevents good heat flow. A valve that is not seating properly passes more heat into the valve guide.

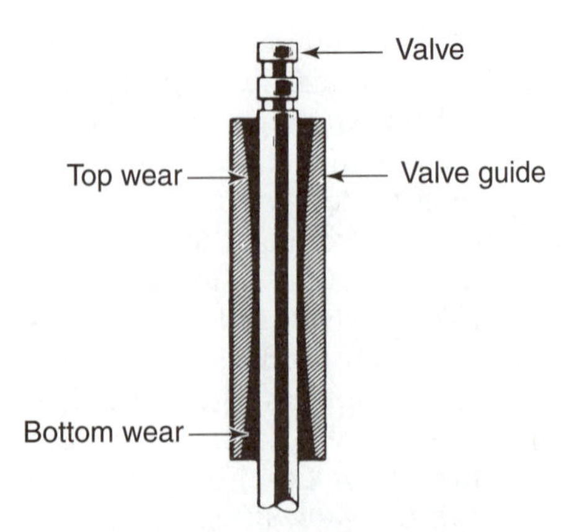

FIGURE 26-2 Valve guides wear at the top and bottom and do not support the valve properly. (Courtesy of Kwik-Way Products.)

MEASURING VALVE GUIDES

The amount of wear between the valve stem and guide may be measured with a plug gauge or a small hole gauge. A **valve guide plug gauge** (Figure 26-3) is a measuring tool that is inserted into a valve guide to determine valve guide wear. The plug gauge, sometimes called a *reject gauge,* is a special tool supplied by the engine maker. Valve guide plug gauges are available for many engines. You must use the correct gauge for the engine being serviced.

Locate the correct plug gauge for the engine. There is usually a number on the gauge. Check this number with the application chart in the shop service manual. Insert the precision ground end of the plug gauge into the valve guide. Measure and note the distance it fits into the guide. Compare this distance to specifications in the shop service manual. The more wear the guide has, the deeper the gauge will fit into the guide. For example, the specifications for the depth of the plug gauge may be 5⁄16 of an inch. If the gauge fits in deeper than 5⁄16th inch the guide requires service. If the gauge goes in less than 5⁄16 of an inch the guide is not worn enough to be serviced.

A micrometer and small hole gauge can be used to determine valve guide clearance. Use an outside micrometer to measure the valve stem (Figure 26-4). Make your measurements in the area where the stem rides in the valve guide. You will be able to see this wear area on the stem. Make measurements at the top, middle, and bottom of the stem wear. Record your measurements.

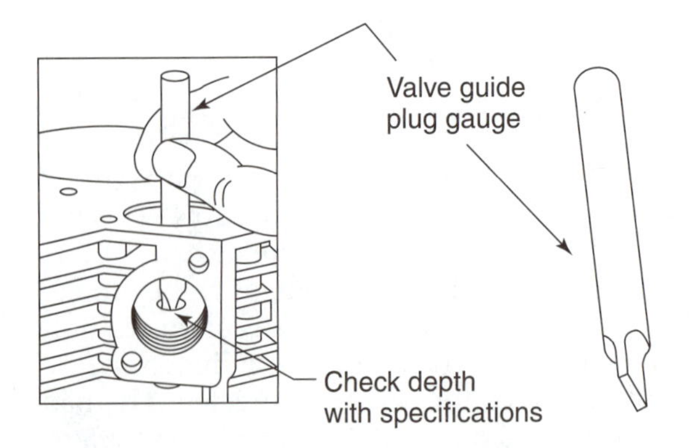

FIGURE 26-3 Measuring valve guide wear with a plug gauge.

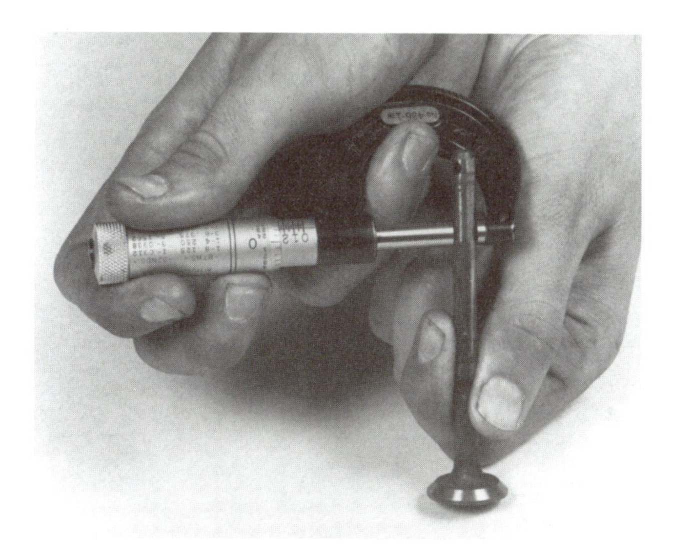

A. Measuring the Valve Stem

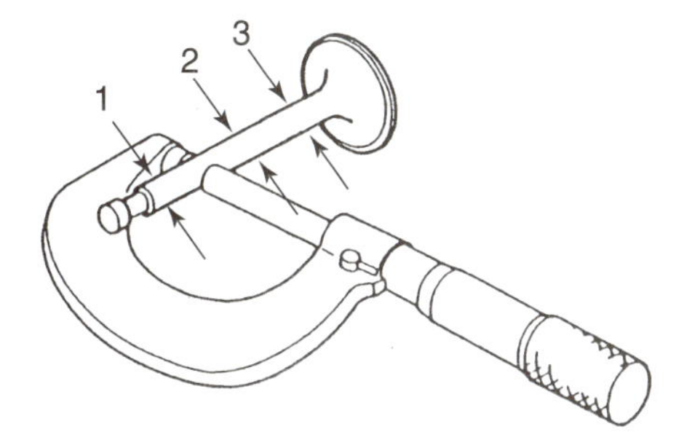

B. Make Three Measurements

FIGURE 26-4 Measure the valve stem in three places in the guide wear area. (*B:* Courtesy of American Honda Motor Co., Inc.)

Insert the correct size small hole gauge into the valve guide (Figure 26-5). Expand the gauge out into contact with the sides of the guide. Pull the small hole gauge out and measure it with the outside micrometer. Measure at the top, middle, and bottom. Valve guides usually wear more at the top and bottom than in the middle. The difference between the smallest valve stem measurement and the largest valve guide measurement is the amount of clearance. For example, the smallest valve stem measurement is 0.312 inch (7.8 millimeter). The largest guide measurement is 0.316 inch (7.9 millimeter). The valve guide clearance is 0.004 inch (0.10 millimeter). Compare your measurement to specifications in the shop service manual to find out if the guides must be serviced. Measuring a valve guide for wear with a micrometer and small hole gauge is shown in the accompanying sequence of photographs.

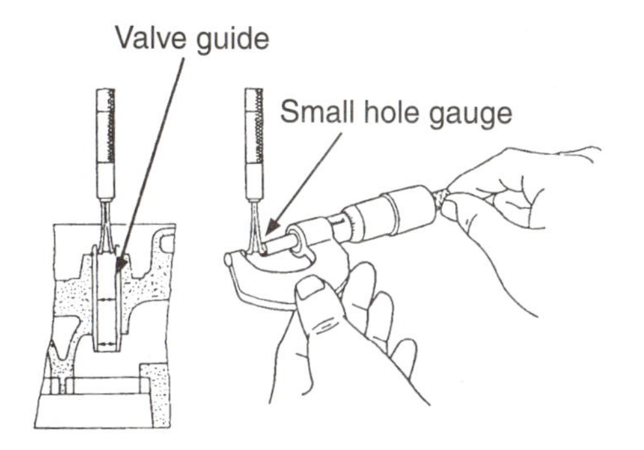

FIGURE 26-5 Measuring the valve guide with a small hole gauge. (Courtesy of Onan Corp.)

VALVE GUIDE SERVICE

Excessive valve guide clearance is repaired in several ways. Most engines have valve guides that can be replaced. Some engines do not have replaceable guides. A thin wall bushing is sometimes used to repair the worn guide. Replacement valves with oversize valve stems are used on some engines.

CAUTION: The valve guide is always serviced before the valve seats because valve seat service equipment is centered in the valve guide. Failure to follow this practice could result in a valve seat that is not concentric with the valve guide.

Removing and Replacing Valve Guides

Most engines have replaceable valve guides so that worn guides can be replaced with new ones. A worn guide may be removed by driving or pulling. Valve guides are often driven out with a **bushing driver,** a tool that fits into a bushing or valve guide and is used with a hammer to drive out a bushing or valve guide. Bushing drivers are available in sets. They may be supplied by the engine maker as a single special tool.

The valve guide depth must be measured before removing the old guide. Use a scale to measure and record the distance the old guide fits out of the cylinder block or cylinder head. Select a bushing driver that fits snugly into the valve guide. Support the cylinder block or cylinder on a block of wood. Use a hammer to drive the guide out through the bottom of the cylinder block (L-head) or cylinder head (overhead) (Figure 26-6).

Some engine makers specify that the valve guide should be pulled rather than driven. This is often required when the guide is made from brass

PHOTO SEQUENCE 14

Measuring a Valve and Valve Guide for Wear with a Micrometer and Small Hole Gauge

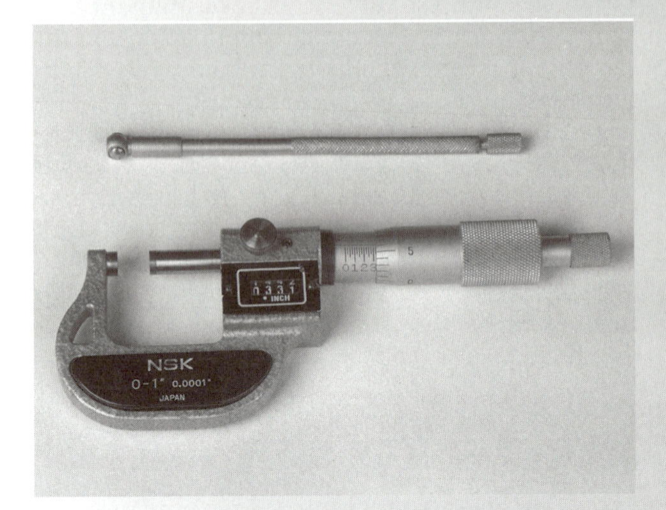

1. Select the correct size small hole gauge and outside micrometer.

2. Clean the valve guide so you can get an accurate measurement.

3. Insert the correct size small hole gauge into the valve guide.

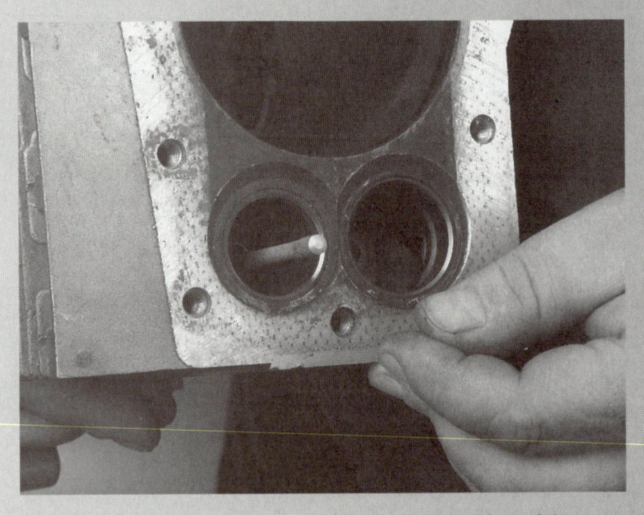

4. Expand the gauge out in contact with the sides of the middle area of the valve guide.

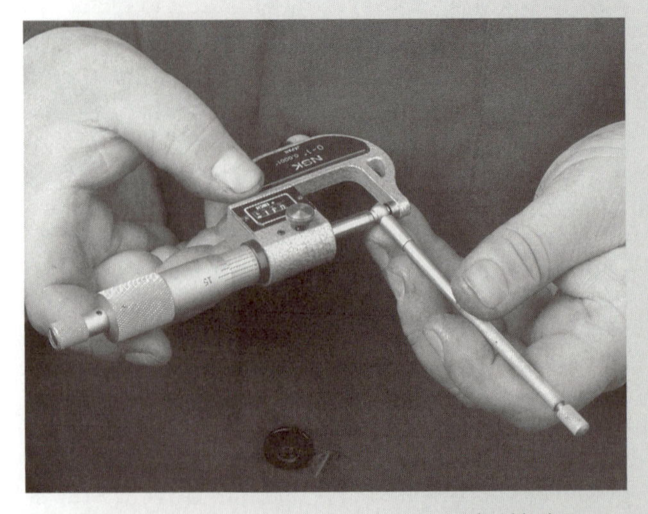

5. Pull the small hole gauge out and measure it with the outside micrometer. Write down your measurement.

6. Repeat this procedure at the top and bottom of the guide. Write this measurement down.

7. Use an outside micrometer to measure the valve stem in the area where the stem rides in the valve guide. You will be able to see this wear area on the stem.

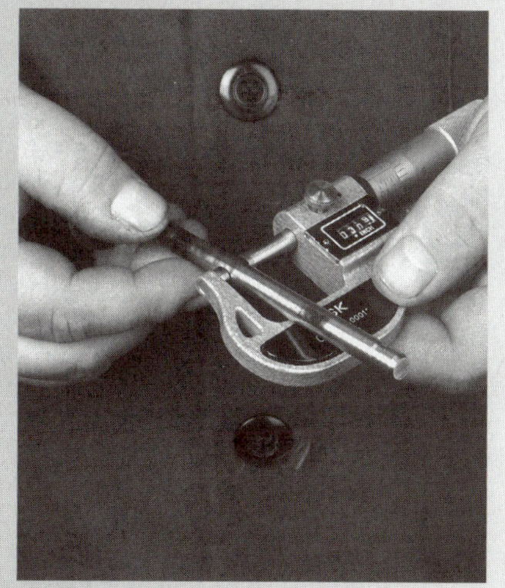

8. Make measurements at the top, middle, and bottom of the stem wear area. Record your measurements.

9. The difference between the smallest valve stem measurement and the largest valve guide measurement is the amount of clearance.

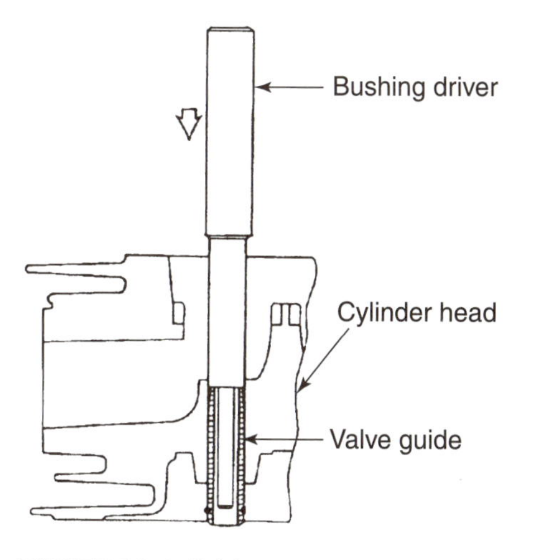

FIGURE 26-6 Driving out a replaceable valve guide. (Courtesy of Onan Corp.)

or other soft material that can break (fracture) when driving. Pulling a valve guide is done with an internal tap, bolt, washer, and jam nut (Figure 26-7). These are often supplied in a set of tools by the engine maker. The tap is used to cut internal threads in the valve guide. The jam nut and washer are assembled on the bolt. The bolt is then threaded into the valve guide. Pulling is done with two wrenches. One wrench is used to prevent the bolt from rotating. The other wrench is used to rotate the jam nut down on the washer. This provides a pulling force to remove the valve guide.

To install the new guide, support the cylinder block or cylinder head on a wooden block. Select the correct size bushing driver. It must fit properly into the guide. If the driver is too tight, it will damage the inside of the new guide. Make a pencil line on the guide showing how deep it goes into the hole. Start the guide squarely into its hole. Drive the guide into the cylinder block or cylinder head to the correct depth (Figure 26-8).

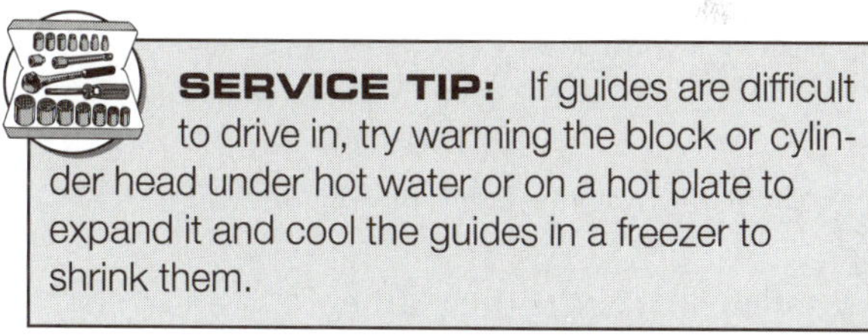

SERVICE TIP: If guides are difficult to drive in, try warming the block or cylinder head under hot water or on a hot plate to expand it and cool the guides in a freezer to shrink them.

Some replacement valve guides must be reamed to the correct size. This procedure is called *finish reaming*. Clean and lightly oil the valve stem of the valve to be used in the guide. Try to insert the stem into the newly installed guide. Sometimes the wear on a used valve stem provides enough clearance and no finish reaming is necessary. The stem should fit smoothly into the guide without any sticking. Finish reaming is necessary if the stem sticks or will not go into the guide.

The size of the finish reamer is determined by valve stem size and the recommended valve guide clearance. Follow the engine maker's finish reamer size recommendations in the shop service manual. If the valve stem is not worn, a standard size reamer may be used to get the correct clearance. If the valve stem is worn, the valve guide must be reamed

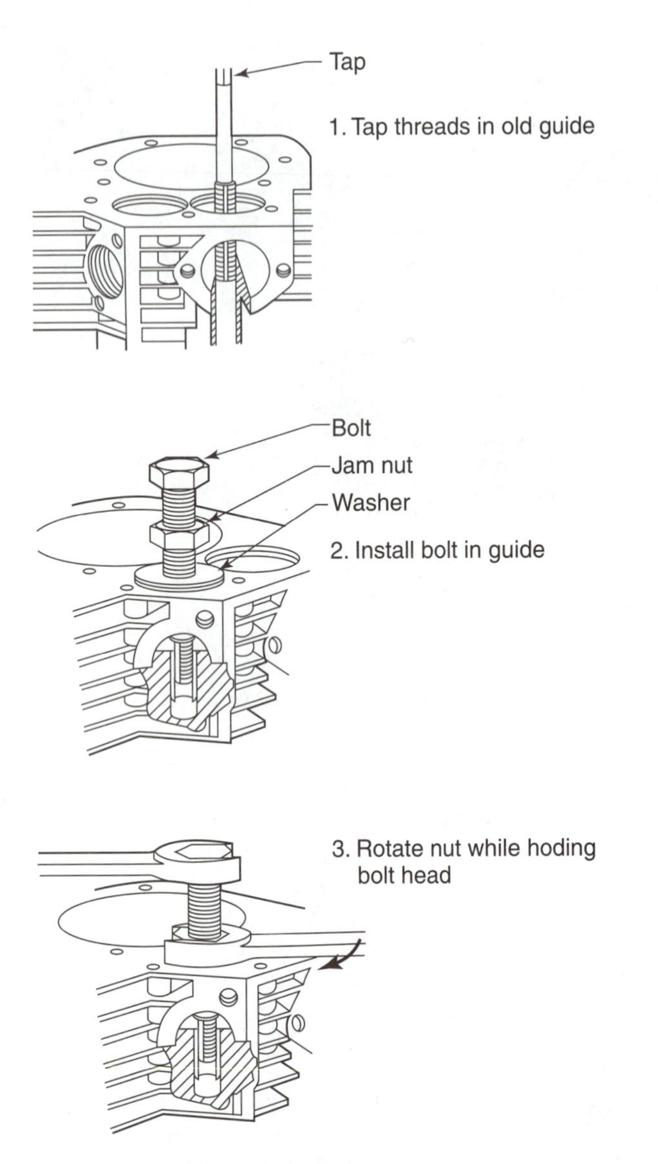

FIGURE 26-7 Pulling out a soft metal guide.

to a smaller size. Insert the reamer in a tap wrench. Start the reamer squarely in the valve guide hole. Rotate the reamer clockwise all the way through the guide (Figure 26-9). Carefully wipe all cuttings off the parts and try the valve stem for proper fit.

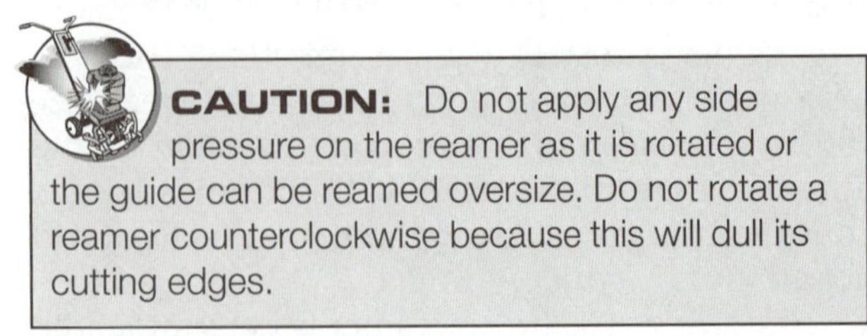

CAUTION: Do not apply any side pressure on the reamer as it is rotated or the guide can be reamed oversize. Do not rotate a reamer counterclockwise because this will dull its cutting edges.

Fitting Oversize Valve Stems

Some engines do not have replaceable valve guides. The guides are part of the aluminum block casting. The valves operate directly in the aluminum. Some

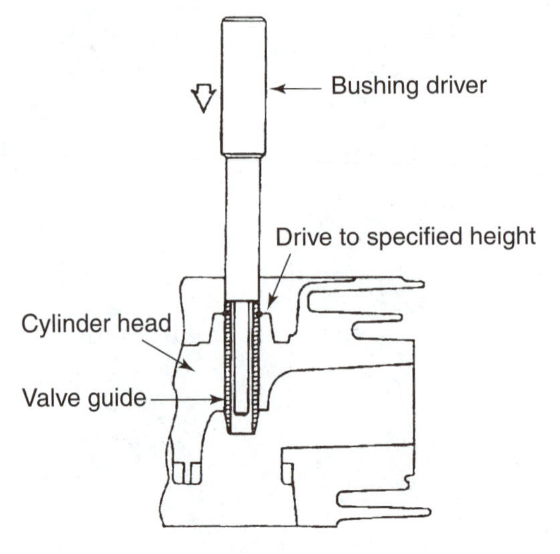

FIGURE 26-8 Installing a replaceable valve guide. (Courtesy of Onan Corp.)

engine makers supply new valves that have stem diameters larger than the original valves. The shop service manual specifies what oversized stems are available. The shop service manual also specifies what size reamer is used with each size valve stem.

Try to insert the new oversize valve stem into the worn guide. Sometimes the oversize stem will fit without any reaming. The oiled stem should fit smoothly with no sticking. You will have to ream if the stem is too tight or does not fit into the guide. Measure the oversize stem with a micrometer. Select the correct size reamer. Install the reamer in a tap wrench. Start the reamer squarely in the guide hole. Rotate the reamer clockwise completely

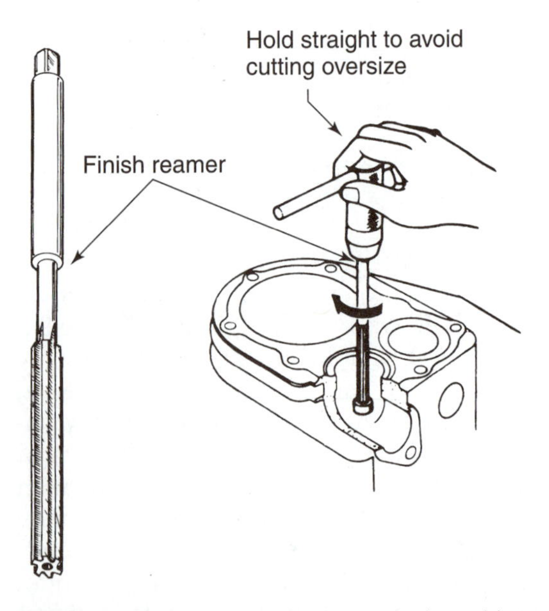

FIGURE 26-9 Finish reaming the new valve guide to size. (Courtesy of American Honda Motor Co., Inc.)

through the hole. Remove the reamer from the hole. Completely clean off all the reamer cuttings. Install the valve in the guide and check the fit.

Installing a Thin Wall Insert

A **valve guide bushing** is a thin wall (thickness) bushing used to repair nonreplaceable valve guides. Thin wall bushings are installed using a set of tools supplied by the engine maker. The tools are a counterbore reamer, centering pilot, and finish reamer (Figure 26-10). The counterbore reamer is used to machine the old valve guide oversize. The counterbore reamer centering end fits in the old valve guide and helps center the reamer. A centering pilot fits in the valve seat to make sure the reamer machines the hole concentric with the valve seat.

The reaming is done to a specific depth determined by the length of the new insert. Make a mark on the reamer at the same height as the insert to get the correct depth. Install the pilot on the valve seat. Insert the counterbore reamer through the pilot and into the old valve guide. Rotate the reamer into the valve guide, turning it clockwise. Stop when the reamer has reached the required depth. Remove the counterbore reamer and pilot. Clean away all the cuttings.

The thin wall insert is made slightly larger in diameter than the hole machined by the reamer to provide a press fit to hold the insert in place. A driver is supplied with the thin wall insert tool set. Place the thin wall insert on the driver. Start the insert squarely into the machined hole. Gently drive the thin wall insert into the machined hole with a hammer. The installed insert is machined to size with a finish reamer (Figure 26-11).

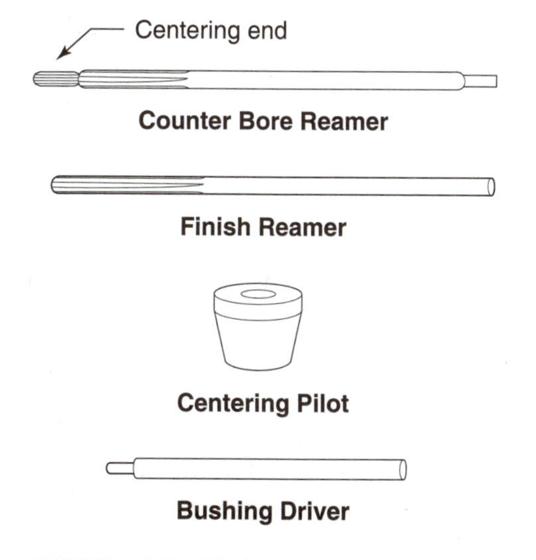

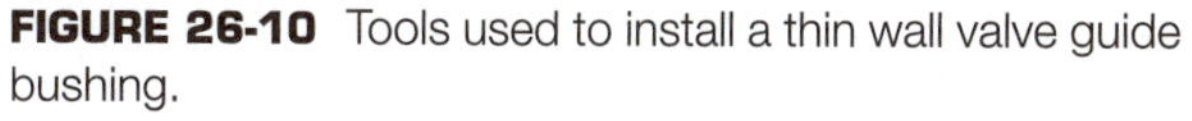

FIGURE 26-10 Tools used to install a thin wall valve guide bushing.

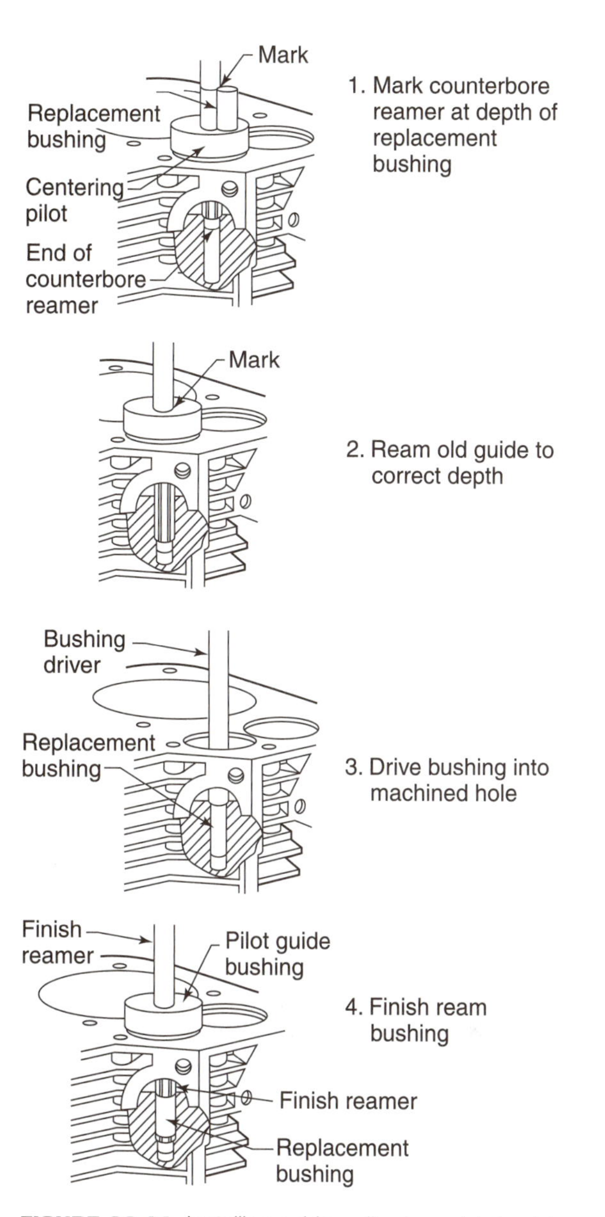

FIGURE 26-11 Installing a thin wall valve guide bushing.

VALVE SEAT SERVICE

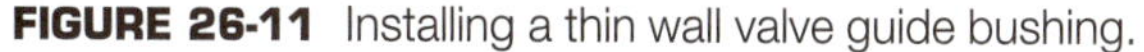

WARNING: Wear eye protection when machining valve seats to prevent eye injuries.

The valve seat is the cast-iron or steel part at the entrance of the valve port. The seat is precision ground to match the valve face. Older cast-iron cylinder blocks may have the seats machined directly in the casting. Aluminum cannot be used as a valve seat because it is too soft. Overhead valve aluminum cylinder heads and aluminum L-head

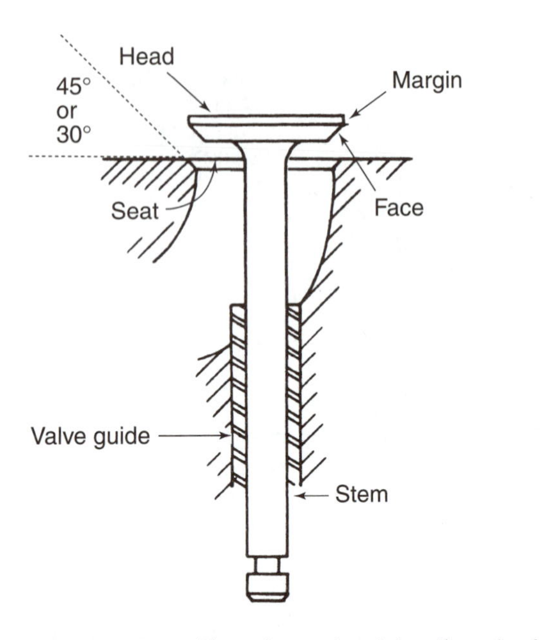

FIGURE 26-12 The valve seat matches the valve face. (Courtesy of Clinton Engine Corporation.)

cylinder blocks have cast-iron or steel valve seats. The seat is removable on some engines. It is installed in the block or head with a press fit. On other engines it is not removable.

The angle ground on the valve seat matches the angle ground on the valve face (Figure 26-12). This angle is 45 degrees on many engines. Some engines use a 30 degree angle. Some engines have a valve interference angle (Figure 26-13). An interference angle is a one degree (or less) difference between the seat and face angles. The seat may be ground to 46 degrees and the valve to 45 degrees. Or the seat may be ground to 45 degrees and the valve to 46 degrees. This provides a thin line contact between the valve and seat. The interference angle helps reduce the buildup of carbon on seating surfaces.

The width of the valve seat is very important for

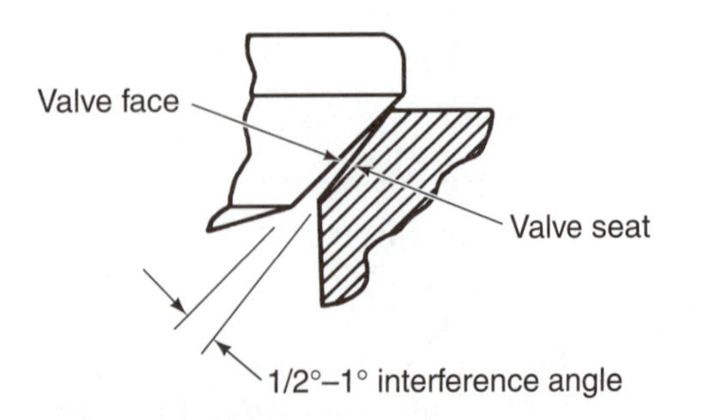

FIGURE 26-13 Valve seat and valve face angles may have an interference angle.

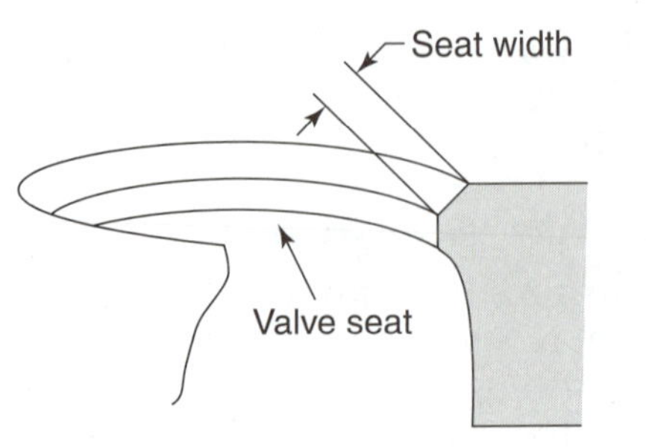

FIGURE 26-14 Valve seat width is important to good valve sealing.

good valve sealing (Figure 26-14). If the seat is too wide, there is a greater chance of a carbon buildup preventing good seating. A wide seat also spreads the valve spring tension over a larger area. Lower tension reduces heat movement away from the valve head and into the valve seat.

Valve seats are always reconditioned during engine service. Reconditioning makes sure there is a good seal between the valve face and valve seat. Valve seats are reconditioned by cutting and or lapping. Damaged replaceable valve seats can be removed and replaced.

Cutting Valve Seats

A **valve seat cutter** (Figure 26-15) is a hardened steel cutting tool used to recondition a valve seat. This method of seat reconditioning provides accurate results and can be done with relatively inexpensive equipment. The cutter has several sharpened steel blades that remove a small amount of metal from the seat. The cutter usually has two sets of blades. There is a set for 45 degree seats on one side. There is a set for 30 degree seats on the other side. Some cutters are actually 46 and 31 degrees to provide an interference angle. The cutters are supplied in a kit. The kit has a variety of cutters necessary for seats of different seat diameters. There are also cutters with larger and smaller angles for valve seat narrowing. The cutter angle is marked on the cutter. The cutters are installed on a pilot. The pilot fits into the valve guide and centers the cutter. The kit has different sizes of pilots for different sizes of valve guides. There is a T handle in the kit used to drive the cutter.

Before beginning the cutting, you must determine if the 30 or 45 degree side of the cutter is to be used. Install one of the valves in the valve guide. Set the cutter next to the guide and match the valve face angle with the cutter angle. If they do not match up, turn the cutter over. Use a black marking pen and paint the valve seat area to allow you to see clearly where you are cutting.

PHOTO SEQUENCE 15

Reconditioning Valve Seats with a Valve Seat Cutter

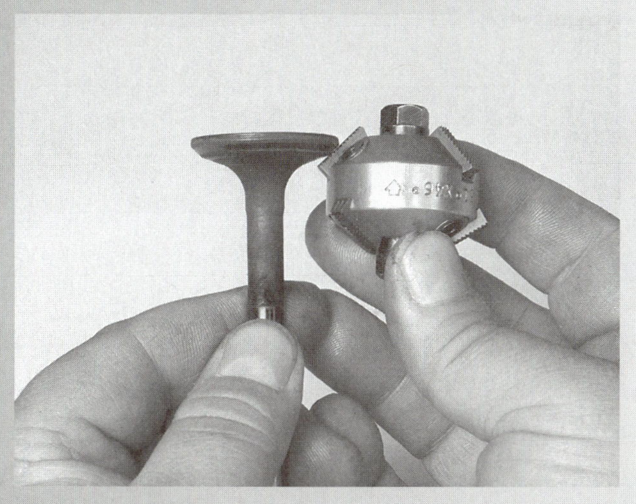

1. Match up the cutter to the valve face to find the correct cutter angle.

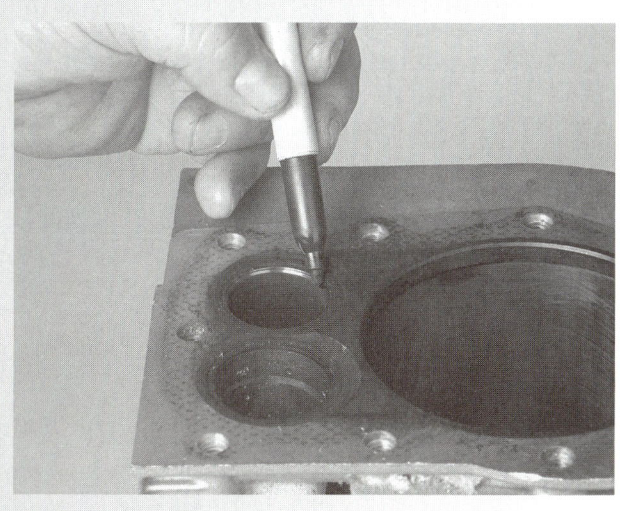

2. Paint the seat area with a felt pen to make the seat visible.

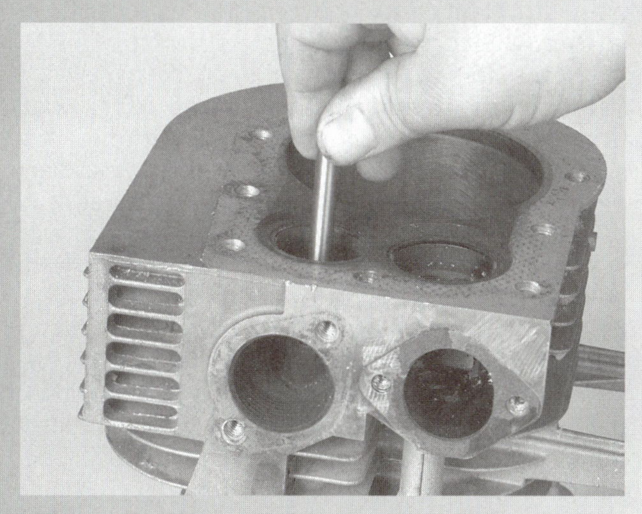

3. Insert the correct size pilot in the valve guide.

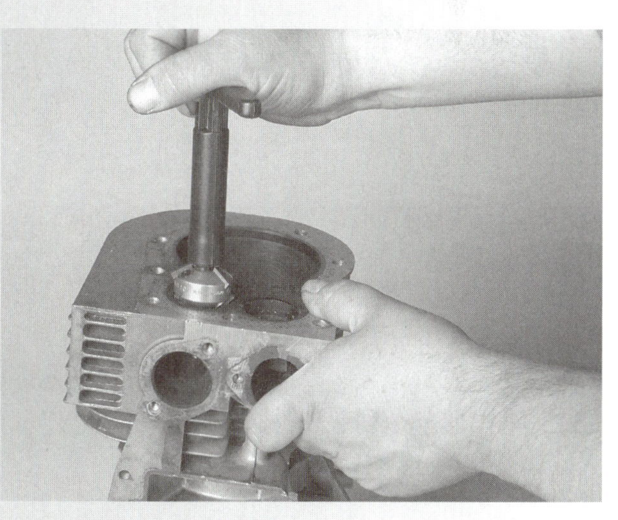

4. Install the T handle on the cutter and rotate the cutter clockwise to recondition the valve seat.

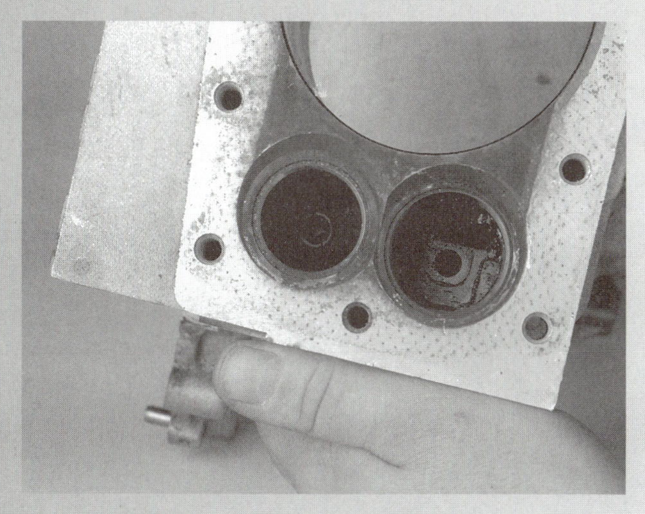

5. Cut the seat until it is shiny and even all the way around.

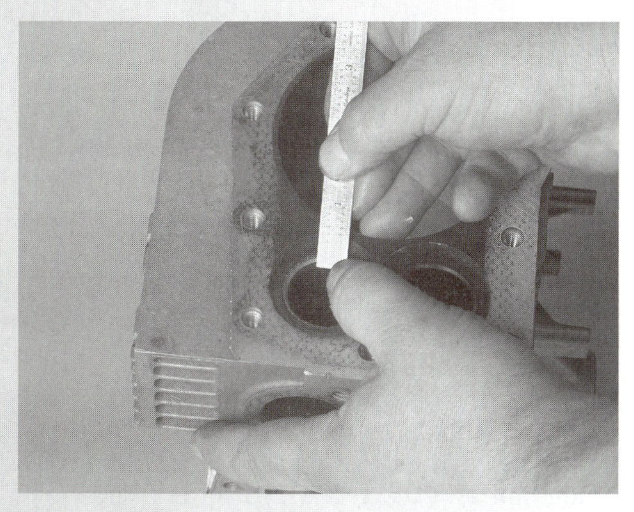

6. Inspect and measure the width of the valve seat.

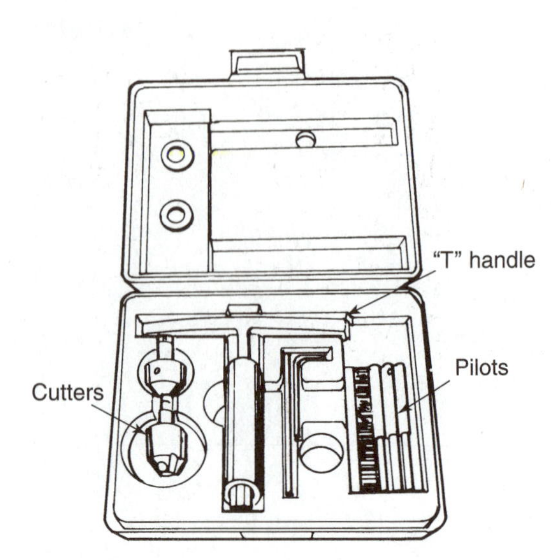

A. Cutter Kit

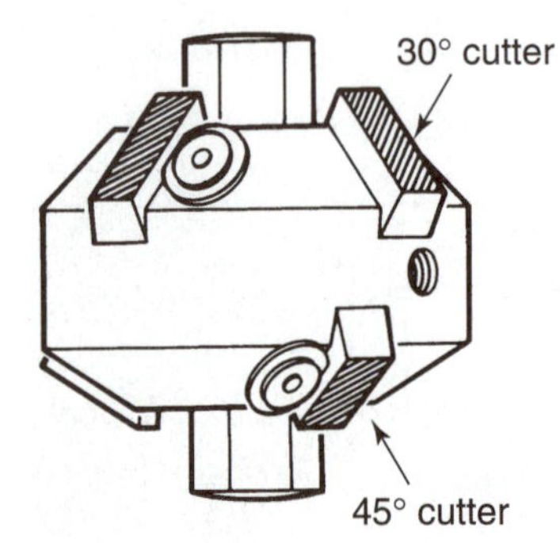

B. Cutter Parts

FIGURE 26-15 A valve seat cutter is used to recondition a valve seat. (Courtesy of American Honda Motor Co., Inc.)

Install the correct diameter pilot in the valve guide. Make sure the pilot fits snugly in the guide to ensure that the valve seat is machined concentric with the valve guide. Install the correct size cutter over the pilot with the correct angle side down. Install the T handle. Rotate the cutter clockwise with the T-handle wrench. Rotate the cutter two or three times. Apply steady pressure directly downward on the cutter. Side pressure can cause the seat to be not concentric to the guide. Too much pressure can cause the cutter to chatter and cause a poor finish.

Remove the cutter and inspect the valve seat. The valve seat should be bright and shiny all the way around. If not, install the cutter and make one or two more cuts. You must cut away all of the oxidized metal until new and solid metal is exposed. The seat width should be even all the way around. Stop cutting as soon as the seat is cut all the way around. Reconditioning a valve seat with a cutter is shown in the accompanying sequence of photographs.

Adjusting Seat Width and Contact Area

SERVICE TIP: Valve seats are easier to see and measure if the valve seat area is painted with a felt marker.

The more material that is cut from the seat, the wider the seat gets. Seat width is very important to good valve sealing. The reconditioned valve seat must provide the proper valve seat to valve face contact (Figure 26-16). The valve face should always be larger than the valve seat. The seat should be wide enough to assist the valve in dissipating heat. It should not be wide enough to collect carbon deposits. Use a scale to measure the width of the seat. The shop service manual has a specification for valve seat width. Most engine makers recommend a seat width between 3/64 inch to 1/16 inch (1.1 millimeters to 1.6 millimeters).

If the seat is too wide it must be narrowed. **Valve seat narrowing** is the removal of material from the top or bottom of the valve seat to reduce seat width. Narrowing is done by cutting material off the top or bottom of the seat. A 45 degree seat can be narrowed by removing material from the top of the seat with a 30 degree cutter. Material can be removed from the bottom of the seat with a 60 degree cutter. A 30 degree seat can be narrowed at the top with a 15 degree cutter. Material from the bottom of the seat can be removed with a 60 degree cutter. When material is removed from both the top and bottom, it is described as a three-angle valve seat.

Before narrowing you must determine valve face-to-seat contact. The valve face must contact the center of the valve seat for good sealing. Paint the valve face with an erasable felt-tipped marker. Place the valve into its guide and snap it down on its seat several times. Be sure the valve does not ro-

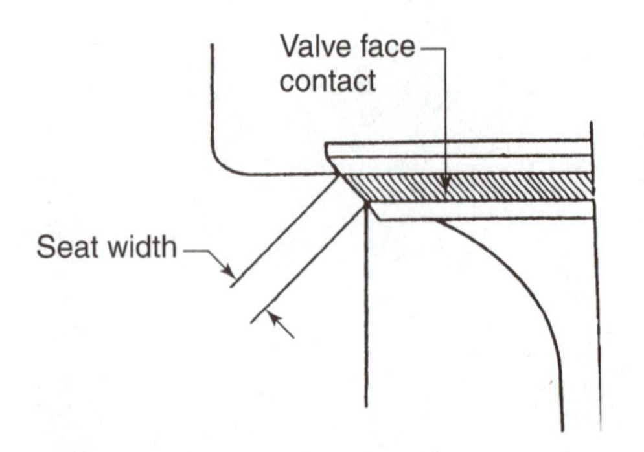

FIGURE 26-16 Checking the valve seat to valve face contact area. (Courtesy of American Honda Motor Co., Inc.)

tate on its seat. The felt marker paint will transfer and leave a visible contact area on the valve face.

The contact area should be centered on the face (Figure 26-17). If the contact is high, select the correct angle cutter (30 or 15 degrees). Remove material from the top of the seat. If the contact pattern is too low, select the correct angle cutter (60 degrees). Remove material from the bottom of the seat. Recheck the contact pattern after narrowing. The job is complete when the contact pattern is centered and the seat width is within specifications.

Lapping Valve Seats

> **CAUTION:** Lapping is a procedure that matches the valve face and valve seat. If the valve face is to be reconditioned, it must be done before lapping.

Valve lapping is a seat-reconditioning method in which an abrasive lapping compound is used to remove metal from the valve seat and valve face. Lapping matches the contour of the valve face with that of the valve seat. When the valve heats up in operation, it can change its shape. The contours no longer line up. For this reason, lapping is not the most precise way to recondition a seat. Valve lapping may be done to recondition the valve and seat when valve seat cutters or valve face grinders are not available. Valve lapping is also done when testing shows that a resurfaced seat is leaking.

> **CAUTION:** Make sure no valve lapping compound gets into the valve guide during lapping. The abrasive compound will wear the valve guide oversize. Be sure to clean up the lapping compound so that it does not get into any engine parts and cause damage.

FIGURE 26-18 Valve lapping compound used to lap valves.

Lapping is done with **valve lapping compound** (Figure 26-18), a mixture of grease and silicon carbide abrasives used to remove metal from the valve seat and valve face. Paint the valve seat with a black felt tipped marker for visibility. Spread a small amount of lapping compound on the valve face and valve seat. Push the valve down on the seat. Rotate the valve back and forth with a **valve lapping tool**, a tool with a small suction cup and handle used to lap valves. The suction cup will stick to the valve head.

Rotate the valve rapidly back and forth with the lapping tool (Figure 26-19). Lift the valve off the seat every few turns. Check the progress of the lapping and add more lapping compound if necessary. You should see a dull shine appear on the valve seat as a ring around the valve face. Stop lapping when there is a continuous dull ring on both the seat and valve face (Figure 26-20).

Replacing Valve Seats

If a valve seat is badly pitted, the cutting may not be enough for reconditioning. The valve seat may have to be replaced. If the cylinder block or

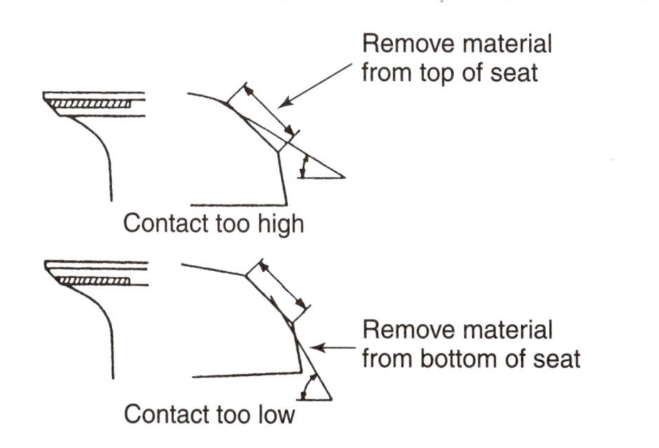

FIGURE 26-17 Narrowing to adjust seat contact and width. (Courtesy of American Honda Motor Co., Inc.)

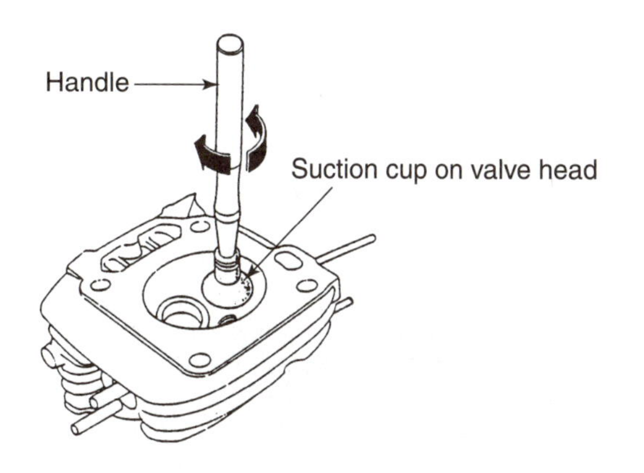

FIGURE 26-19 Lapping a valve seat with a valve lapping tool. (Courtesy of American Honda Motor Co., Inc.)

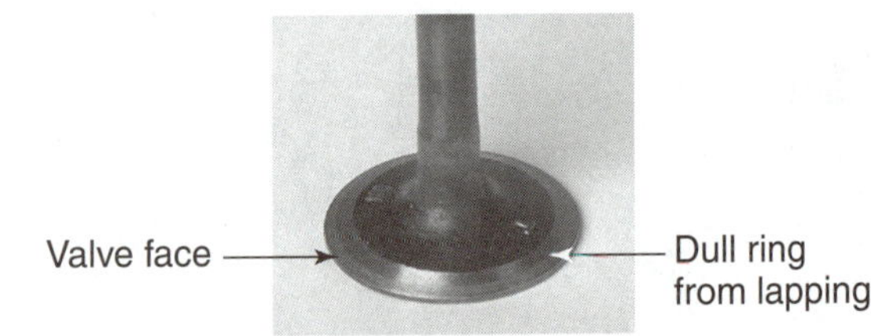

FIGURE 26-20 A lapped valve should have a continuous dull ring around the valve face.

cylinder head has a replaceable valve seat (Figure 26-21), a new one can be installed. If the valve seat is not replaceable, the block or cylinder head may be machined to accept one.

A **valve seat puller** is a pulling tool used to pull a valve seat out of a block or cylinder head (Figure 26-22). Valve seat pullers are special tools supplied by an engine maker. The puller rests on top of the cylinder block or cylinder head. The puller forcing screw fits under the valve seat insert. The puller is connected by an adapter to a slide hammer. Use the slide hammer to force the valve seat out of the block or cylinder head.

If a puller is not available, a valve seat insert can often be pried out with a valve seat pry bar. Insert the pry bar into the port and under the valve seat insert. Push down on the bar to remove the seat (Figure 26-23).

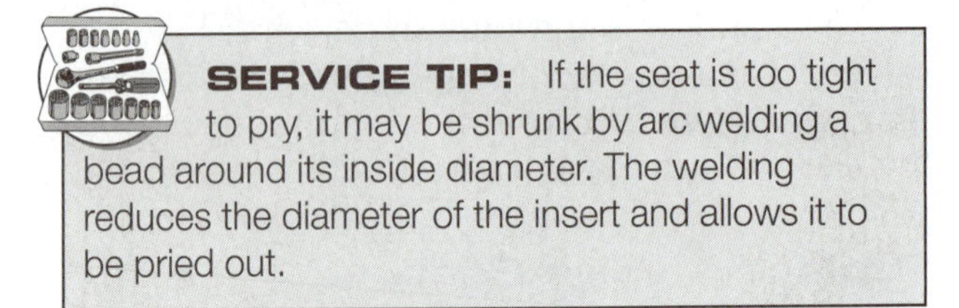

SERVICE TIP: If the seat is too tight to pry, it may be shrunk by arc welding a bead around its inside diameter. The welding reduces the diameter of the insert and allows it to be pried out.

A damaged nonremovable valve seat may be machined with a cutter. Similar to that used for valve seats, the cutter is a special tool supplied by the engine maker. A pilot on the cutter is installed in the valve guide. A T handle is used to rotate the cutter and machine the area for a new valve seat. A stop on the tool prevents cutting the hole too deep.

A new valve seat insert is installed with a hammer and a special driver tool (Figure 26-24). The driver has a pilot on the end that fits in the valve guide. The pilot ensures that the seat is centered in relation to the valve guide. Drive the new valve seat insert into the machined hole until it bottoms.

The new seat must fit tight in its machined hole or it can come loose during operation. A loose

FIGURE 26-21 A replaceable valve seat insert.

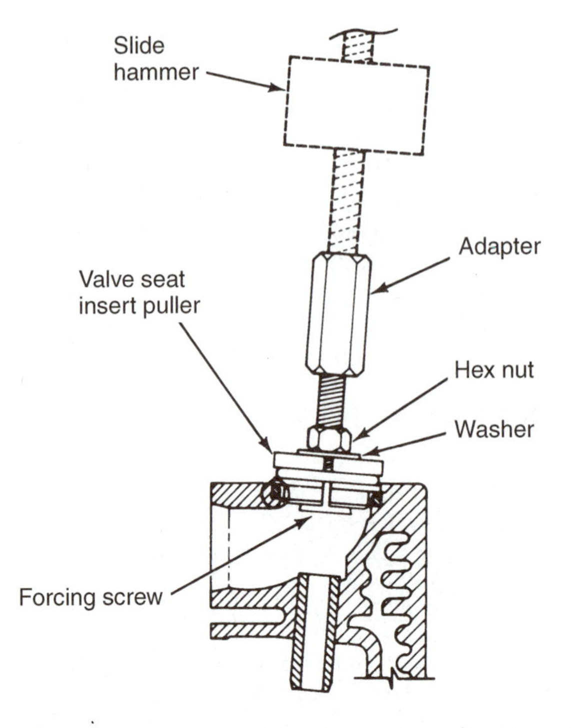

FIGURE 26-22 Replaceable valve seats can be removed with a puller. (Courtesy of Kohler Co.)

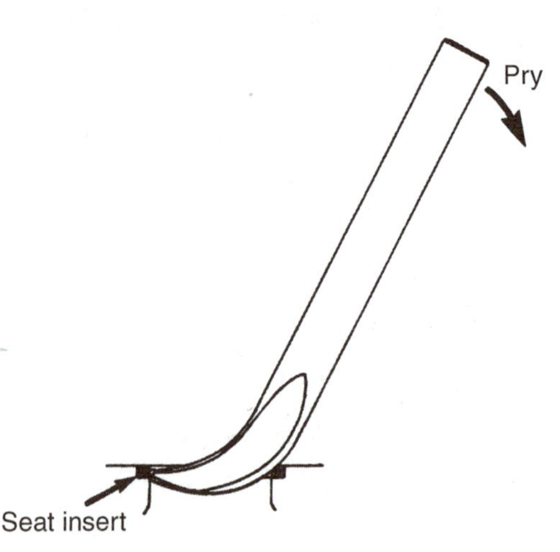

FIGURE 26-23 Removing a valve seat with a pry bar. (Courtesy of K. O. Lee Co.)

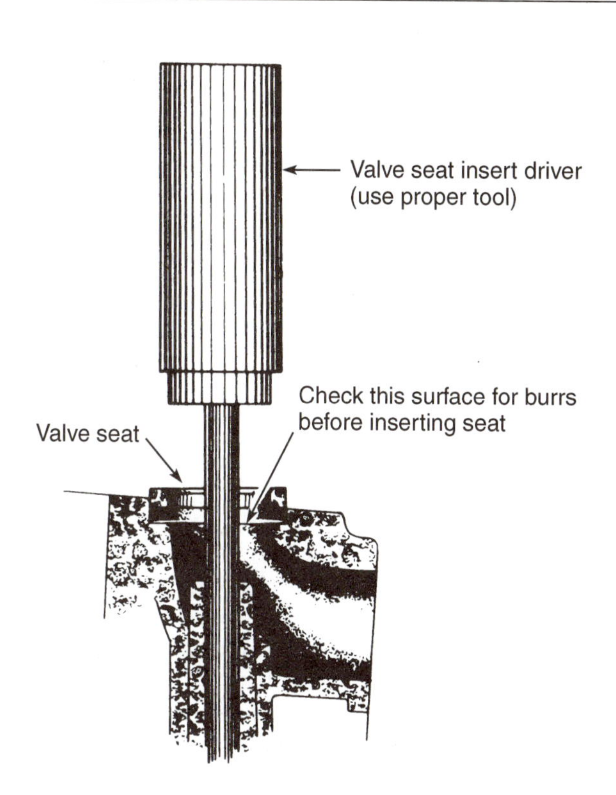

FIGURE 26-24 Driving in a new valve seat. (Courtesy of Onan Corp.)

insert or a newly installed valve seat insert can be tightened by peening. **Peening** is a method used to expand a metal part by making indentions in it with a punch. Use a center punch to peen the insert in the area next to the machined hole (Figure 26-25). Do not peen the valve seat surface. The new seat insert requires a slight cutting to make sure it is in alignment with the valve guide.

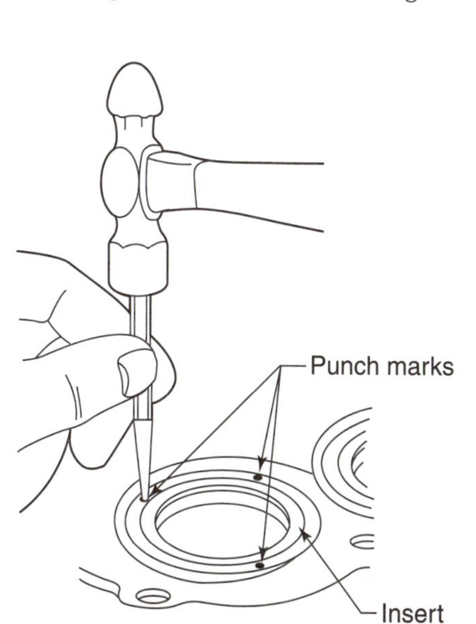

FIGURE 26-25 Peening the metal around a valve seat insert.

CAUTION: Oil the valve seats to prevent rust if the valves are not going to be assembled immediately. Rust can damage freshly machined valve seats.

CYLINDER HEAD INSPECTION

Warping is a common problem with both L-head and overhead cylinder heads. Overheating or uneven head bolt tightening can cause warping. A warped cylinder head will not properly support the head gasket. A head gasket used with a warped head will eventually leak compression pressures.

Place a straightedge on edge in several different directions across the cylinder head sealing surface (Figure 26-26). Look up the recommended thickness feeler gauge to use in the shop service manual. Push down firmly on the straightedge. Try to slip the feeler gauge between the straightedge and cylinder head surface. Try this test in several different directions. If the feeler gauge passes under the straightedge, the cylinder head is warped and should be replaced.

A common problem with overhead valve cylinder heads is cracks that begin near the valve seat area. Carefully clean this area for inspection. Inspect the area of the seats with a strong light for cracks. A crack detection system that uses a dye is available (Figure 26-27). A special spray cleaner is

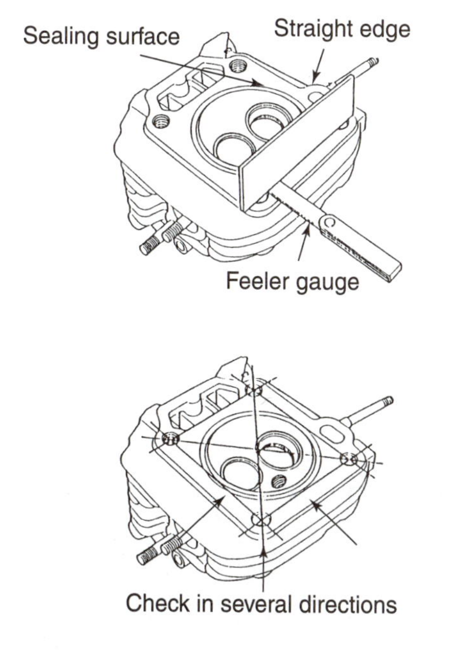

FIGURE 26-26 Checking the cylinder sealing surface for warping. (Courtesy of American Honda Motor Co., Inc.)

A. Spray Dye over Inspection Area

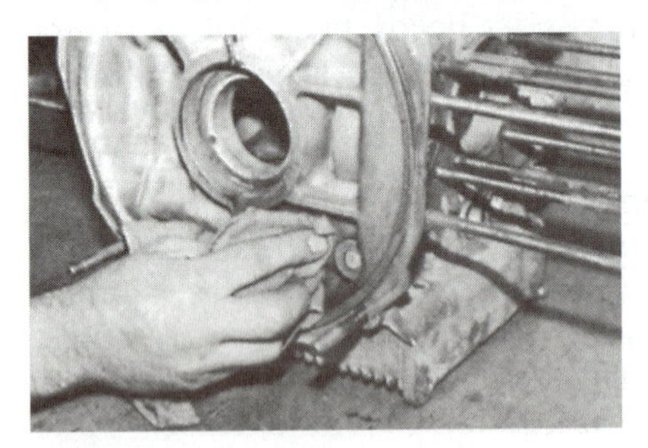

B. Wipe Area to Be Inspected

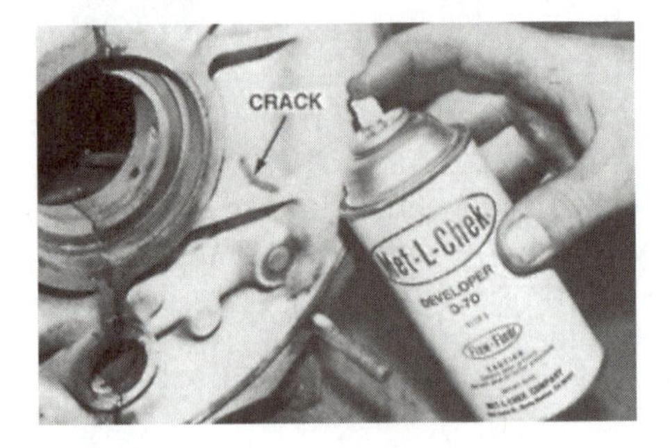

C. Spray Developer over Inspection Area

FIGURE 26-27 A dye system used to check for cracks.

used to prepare the area to be inspected. A dye is sprayed over the inspection area. The dye is drawn by capillary action into any cracks. A developer is then sprayed over the area and makes any dye in a crack very visible. A cracked overhead valve cylinder head is replaced.

VALVE SERVICE

> **WARNING:** Always wear eye protection when using valve-grinding equipment. Grinding stones can fracture and send fragments at the operator.

Valves with burned or eroded areas on the head, neck, or stem are not reconditioned; they are always replaced. Valve reconditioning equipment is expensive. Many shops choose to purchase new valves for the engines they rebuild. This procedure is more cost-effective for many shops than purchasing valve reconditioning equipment.

A valve can be reconditioned by cutting or grinding. Valve cutting is done with a cutter kit similar to that used for valve seats. The valve head is inserted into a cutting unit. The cutters are adjusted to cut 45 or 30 degrees. The end of the valve stem is chucked in a vise. The cutting unit is turned by hand to resurface the valve face.

Valve grinding is done on a machine called a valve grinder. A **valve grinder** (Figure 26-28) is used to grind a new surface on the valve face and tip. Valve grinders rotate the valve in a chuck. The face is ground by a rotating grinding wheel. Each type of valve grinder has specific instructions for use. Always follow the equipment instructions. The valve grinder usually has two grinding wheels: One is used for grinding the valve face and the other is used for valve stems, rocker arms, and other valve train parts. Coolant is pumped over the wheel and valve during grinding to improve the finish of the ground surface.

Stemming is a valve service operation in which a valve grinder is used to regrind and chamfer the valve stem tip (Figure 26-29). This operation is necessary to properly center the valve in the valve grinding chuck. Clamp the valve on its stem in the V-bracket on the valve grinder. Move the valve stem toward the side of the grinding wheel with the grinder feed wheel. When the stem end contacts the wheel, move the valve back and forth across the wheel side. Remove just enough material to resurface the tip. After grinding, the tip is

FIGURE 26-28 A valve-grinding machine. (Courtesy of Sioux Tools Inc.)

FIGURE 26-29 Stemming the valve on a valve grinder.

chamfered. Install the valve in the chamfer fixture. Move the stem toward the wheel. Rotate the stem by hand to grind a slight chamfer into the tip (Figure 26-30).

Before valve face grinding, the grinding wheel is dressed. Dressing or truing is done to clean and condition the grinding wheel surface. A diamond-tipped cutting tool (nib) mounted to a tool post is used to do the dressing. The wheel should be dressed periodically to ensure proper grinding finish on valve faces. To dress the wheel, start the coolant flow over the grinding wheel. Move the dressing tool into the rotating grinding wheel. Move the

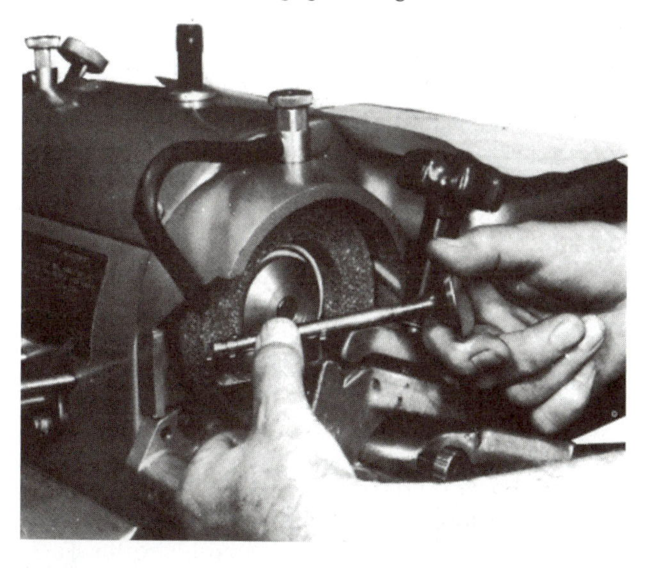

FIGURE 26-30 Chamfering the valve stem. (Courtesy of Sioux Tools Inc.)

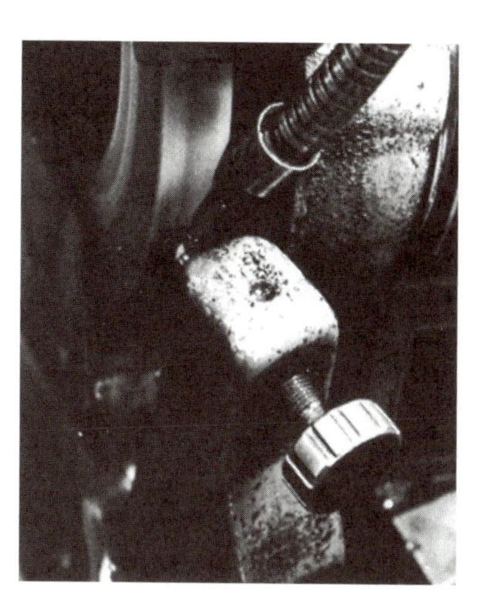

FIGURE 26-31 Dressing the grinding stone with a diamond nib.

wheel back and forth until the entire surface of the grinding wheel is dressed (Figure 26-31).

After dressing, the valve may be installed in the chuck. The chuck is designed to grip the valve in the unworn part of the stem. The chuck has an adjustable sleeve for different size valve stems. It has an adjustable stop to set the gripping position. The valve is centered in the chuck off the valve tip. Open the chuck sleeve and insert the valve. Adjust the chuck so that the valve is held tightly just above the worn stem area. Adjust the carriage stop so that the valve will traverse the entire grinding wheel face. The carriage stop must prevent the valve neck from coming into contact with the grinding wheel (Figure 26-32).

CAUTION: If the valve grinding wheel hits the valve neck, the valve must be discarded. It will break if installed in the engine.

The valve chuck must be adjusted to the correct angle (Figure 26-33) to grind the correct valve face angle. Specifications may call for the valves to be ground to 45 or 30 degrees. An interference angle may be required. Graduations on the chuck carriage allow the chuck to be indexed to the correct angle. Loosen the holddown nut and move the chuck to the correct angle position. Tighten the holddown nut.

Begin grinding by advancing the grinding wheel to grind the valve face. Begin at the left side of the wheel. Move the valve slowly and steadily, right and left across the wheel. Do not allow the valve to pass beyond either edge of the grinding wheel while grinding. Take light cuts by feeding the grinding wheel up to the valve about .001 inch to

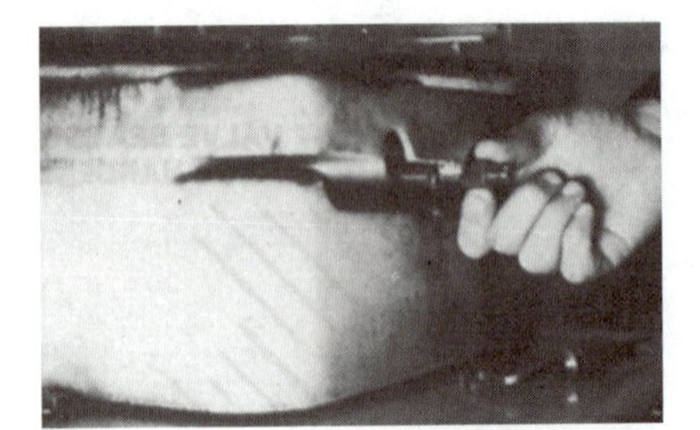

A.

B.

FIGURE 26-32 (*A*) Adjusting the valve grinder carriage stop. (*B*) A valve neck that contacted the valve-grinding wheel because of incorrect adjustment. (Courtesy of Sioux Tools Inc.)

.002 inch (0.03 millimeter to 0.05 millimeter) at a time. Remove just enough material to make a clean, smooth face. When the valve face is ground, you are ready to move the grinding wheel away from the valve. Move the valve to the right until it

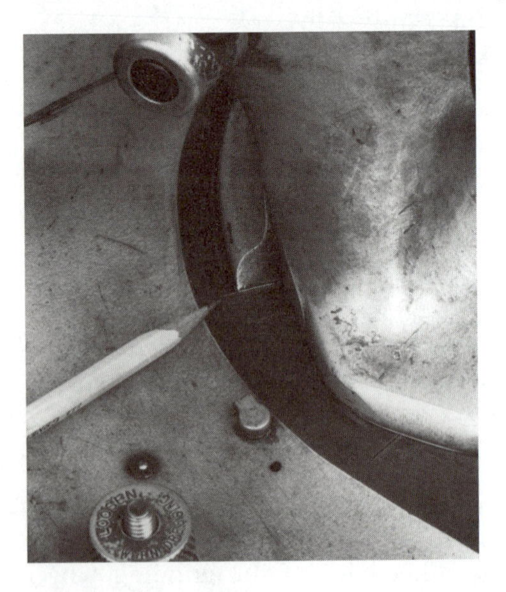

FIGURE 26-33 Adjusting the valve-grinding chuck to the correct grinding angle.

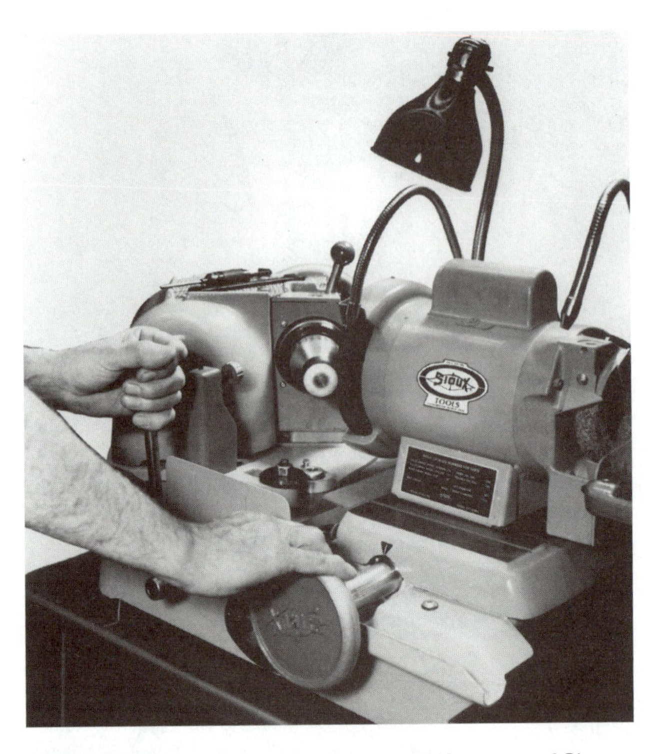

FIGURE 26-34 Grinding the valve face. (Courtesy of Sioux Tools Inc.)

is flush with the right edge of the grinding wheel. Pause a second. Back the grinding wheel away from the valve. Never back the valve away from the wheel. Doing this improperly will result in a rough part of the wheel contacting the valve face which will cause a poor surface finish. See Figure 26-34.

Inspect the valve after grinding. The margin on the valve gets thinner when the valves are ground or cut (Figure 26-35). Badly pitted, burned, or warped valves cannot be refaced without removing too much of the margin. A thin margin makes a sharp edge on the head of the valve. The thin edge will not be able to take high temperatures. This area will burn away quickly and cause a loss of compression. Margin specifications are given in the shop service manual. Most engine makers specify a minimum thickness of 1/32 inch (0.8 millimeters). If margins are not to specified thickness after grinding, the valve must be replaced.

VALVE SPRING SERVICE

Constant compressing and expanding eventually causes the valve springs to lose their tension. Weak valve springs do not properly close the valve. They eventually cause engine performance problems and valve damage. Weak springs become shorter. Each valve spring is checked for free length. **Valve spring free length** is a measurement of valve spring length

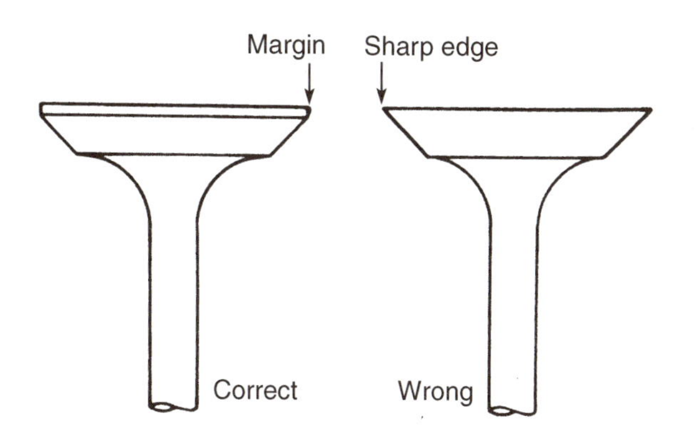

FIGURE 26-35 Grinding reduces the valve margin thickness. (Courtesy of Federal-Mogul Corporation.)

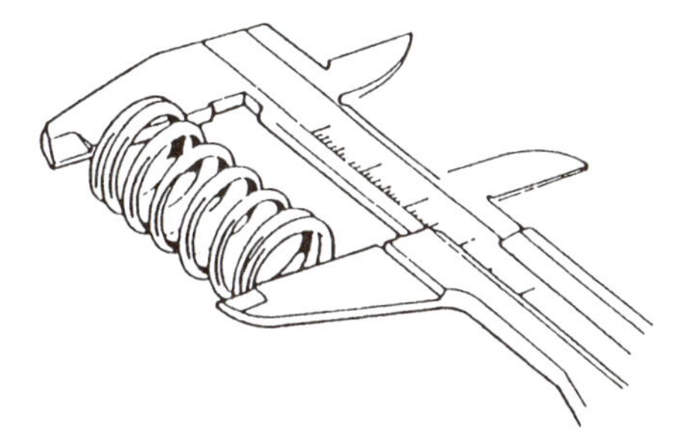

FIGURE 26-36 Measuring valve spring free length. (Courtesy of American Honda Motor Co., Inc.)

when it is not under tension (Figure 26-36). This measurement is an indication of valve spring condition. An outside caliper or scale is used to measure the length of the spring. This measurement is compared with specifications in the shop service manual. A spring that is shorter than specifications is too weak and should be replaced.

CAMSHAFT AND VALVE LIFTER SERVICE

The camshaft lobes and valve lifters are hardened to withstand the forces of a running engine. They must push the valves up against strong spring pressure. They absorb a constant pounding as the valve clearance changes. These forces cause wear on the cam lobe and lifter contact surfaces.

Inspect each camshaft lobe for evidence of any pits or scuffing that would require that the camshaft be replaced. Wear on the lobes can be determined with an outside micrometer. Place an outside micrometer in position to measure from the heel to the top of the lobe (Figure 26-37). Make a similar measurement on all the intake and exhaust lobes. Compare the measurements to specifications in the shop service manual to determine the amount of wear.

The camshaft bearing journals should also be inspected for scuffing or scratching. If you see any of these problems, replace the camshaft. Use an outside micrometer to measure the diameter of each of the journals (Figure 26-38). Compare these measurements with a measurement of the camshaft bearings in the block and side cover. Measure the bearings with a small telescoping gauge. The difference between the camshaft journal and bearing diameters is the amount of clearance. Compare this clearance to specifications in the shop service manual.

Carefully inspect the valve lifter where it contacts the camshaft. Cupping, pitting, and crazing are common types of wear to the bottom of the lifter. Crazing is wear that occurs when low oil conditions cause the hardened surface to break down. If there is any evidence of surface deterioration, the lifter should be replaced.

FIGURE 26-37 Measuring the cam lobe with an outside micrometer.

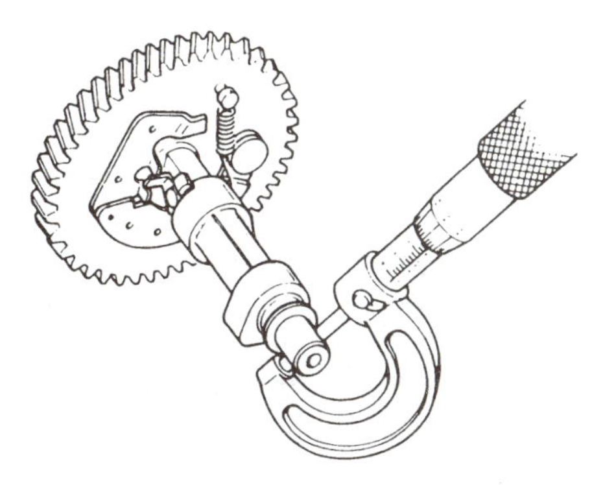

FIGURE 26-38 Measuring the camshaft bearing journal. (Courtesy of American Honda Motor Co., Inc.)

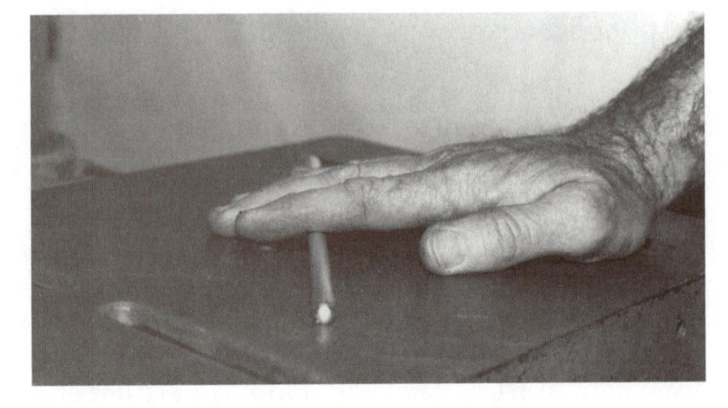

FIGURE 26-39 Checking for a bent push rod.

FIGURE 26-40 Checking nonadjustable valve clearance with a feeler gauge.

Check the push rods on overhead valve engines. Roll each push rod over a flat surface like a drill press table and check for bending (Figure 26-39). A bent push rod will flip up as it rolls. Replace any bent push rods. Do not try to straighten a bent push rod. A straightened push rod will break easily when in the engine.

SETTING NONADJUSTABLE VALVE CLEARANCE

Reconditioning removes metal from the valve seat and valve face, which causes the valves to change position in relation to the camshaft and valve lifter. This change in position reduces or eliminates the valve clearance. This is not a problem for engines with adjustable lifters or rocker arms. The valve clearance is adjusted after the engine is assembled and before it is started.

Many small inexpensive engines have nonadjustable valve clearance. Valve clearance is set by the length of the valve stem. Valve clearance on these engines has to be reset by grinding the end of the valve stem. Resetting the valve clearance requires that the camshaft and valve lifter be reassembled in the cylinder block. Both ends of the camshaft must be supported in its bearing journals. This job is usually done on assembly. Another method is to install the parts temporarily to make the measurement and adjustment.

Look up the specifications for valve clearance in the shop service manual. Rotate the crankshaft until the heel of the camshaft lobe is under the lifter of the valve to be checked. This makes certain that the valve is in the closed position. Insert the valve into its guide. Push down firmly on the valve head. Use the specified feeler gauge to measure the space between the lifter and valve stem (Figure 26-40).

If the clearance is less than specifications, remove the valve from the engine. Install the valve in the stemming V-bracket of the valve grinder. Grind a small amount of material from the end of the stem. Check the clearance again and repeat the steps if more material needs to be removed. Make very small cuts to be sure the clearance is not adjusted oversize.

CASE STUDY

A student in an outdoor power equipment class had just finished an engine overhaul and was having a problem getting the engine to start. The instructor came over to help him diagnose the problem. They checked for spark and the magneto was working fine. They checked for fuel by removing and inspecting a spark plug. The spark plug insulator tip was wet with fuel, so gasoline was getting into the combustion chamber.

The instructor did a rebound test on the flywheel and noticed low compression. The instructor had observed the new piston ring installation and was sure this area was not the problem. The instructor questioned the student about the valve testing. The student had not tested the valves after lapping because he was sure they were going to seal.

The instructor had the student rotate the engine crankshaft until the piston was up at TDC with both valves closed. He directed the student to lock the flywheel in this position with a flywheel holder. The instructor connected a blow gun with a large nozzle to the shop air supply. He directed air into the cylinder and had the student listen. There was a strong hiss of air leaking out of the muffler. The problem was an exhaust valve that was not seating properly. The student had to disassemble the engine and lap the exhaust valve and seat again.

VALVE ASSEMBLY TESTING

The seal between the valve face and seat is checked after reconditioning the valve seat and valve face (Figure 26-41). Testing is done before assembly so a leak can be corrected. Place the cylinder block or cylinder head on its side with the ports pointing upward. Push a valve into its guide and hold it firmly on its seat with your finger. Fill the port with solvent from a spray can. The solvent should not leak out between the valve and seat. A leaking valve and seat will have a steady leak around the head of the valve. If you find a leak, lap the valve and seat. Check the valve again for leakage. Repeat this procedure for the other valve or valves.

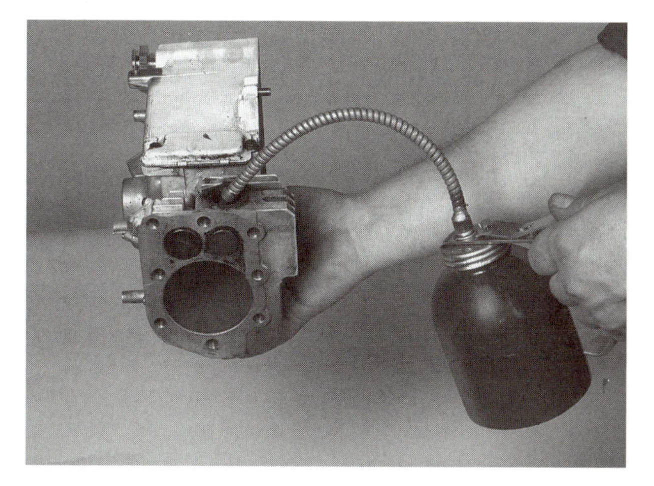

FIGURE 26-41 Checking a valve face and seat for leakage.

REVIEW QUESTIONS

1. Describe the types of wear common to a valve and seat.
2. Explain how to use a small hole gauge and outside micrometer to measure a valve guide.
3. Explain how to use an outside micrometer to measure a valve stem and calculate valve guide clearance.
4. Explain how to remove and replace a replaceable valve guide.
5. Describe how to install a thin wall insert in a nonreplaceable valve guide.
6. Explain how to use a valve cutter to recondition valve seats.
7. Describe how to use a valve grinder to recondition valve seats.
8. Explain how to adjust the width and contact area on a valve face.
9. Describe how to lap a valve seat.
10. Explain how to remove and replace a valve seat.
11. Describe how to inspect a cylinder head to determine warpage.
12. Explain how to recondition a valve face with a valve face cutter.
13. Explain how to stem a valve in a valve-grinding machine.
14. Describe how to recondition the valve face in a valve-grinding machine.
15. Explain how to determine if a valve is usable after grinding in a valve grinder.
16. Explain how to check a valve spring free length.
17. Explain how to measure a valve spring for square.
18. Explain how to measure cam lobes to determine wear.
19. Describe how to use a valve grinder to set the valve clearance on a nonadjustable valve mechanism.
20. Explain how to use solvent to test valve seat to face seal.

DISCUSSION TOPICS AND ACTIVITIES

1. Locate several scrap cylinder blocks. Practice cutting valve seats in each of the blocks.
2. Find several shop engines with assembled valves. Test each valve with solvent to see if any of them leak.
3. An intake valve stem measures 0.141 inch (3.58 millimeters). The valve guide measures 0.161 inch (4.09 millimeter). What is the stem-to-guide clearance in thousandths of an inch and hundredths of a millimeter?

CHAPTER 27

Crankshaft, Connecting Rod, and Bearing Service

OBJECTIVES

Upon completion and review of this chapter, you should be able to:

- Inspect and measure a crankshaft for wear.
- Inspect and measure a connecting rod for wear.
- Inspect a connecting rod bearing for wear.
- Measure and inspect a plain main bearing.
- Recondition a plain main bearing.
- Inspect and service a caged needle, tapered roller, and ball main bearing.

TERMS TO KNOW

Brinelling	Pin fit
Etching	Reject size
Galling	Undersize bearings

INTRODUCTION

The crankshaft, connecting rod, and bearings are some of the most highly stressed internal engine components. The full force of combustion is transferred to these parts through the piston on each power stroke. Each of these parts must be carefully measured and inspected to determine wear.

CRANKSHAFT INSPECTION AND MEASUREMENT

Both the four- and two-stroke crankshafts are inspected and measured in the same way. Carefully inspect the crankshaft for wear or damage to the FWE flywheel threads, FWE keyway, and the PTO keyway (Figure 27-1). Damage in either of these areas usually requires replacement of the crankshaft. Inspect the FWE journal, connecting rod bearing journal (crank pin), and PTO journal for damage. Place the crankshaft on a bench under a strong light. Look over each journal for scoring. Scoring occurs when abrasives get caught between the bearing and the journal. The slightest scoring on a crankshaft journal can shorten the life of a new main or connecting rod bearing. Run your fingernail over any scoring you find. Scoring deep enough to catch a fingernail requires the crankshaft be replaced or reconditioned.

Look for signs of bearing material transfer from a bearing to a journal. Sometimes bearing material transferred to a journal can be cleaned off with abrasive cloth. Use a strip of abrasive paper to gently polish the metal transfer areas. If the metal polishes off, the crankshaft can be measured to find out if it can be used again.

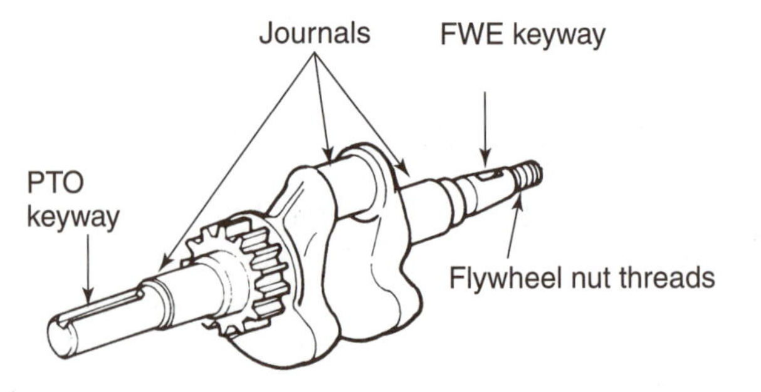

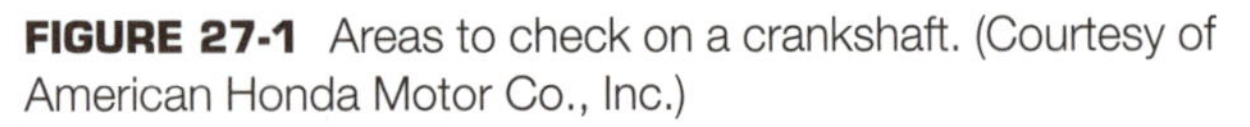
FIGURE 27-1 Areas to check on a crankshaft. (Courtesy of American Honda Motor Co., Inc.)

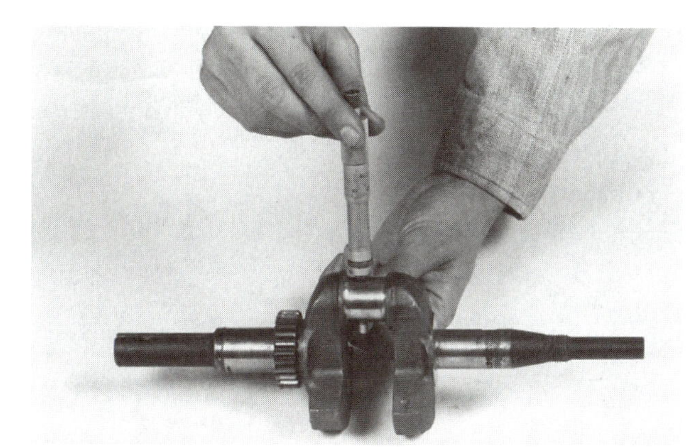

A. Measuring the Crankshaft Journal

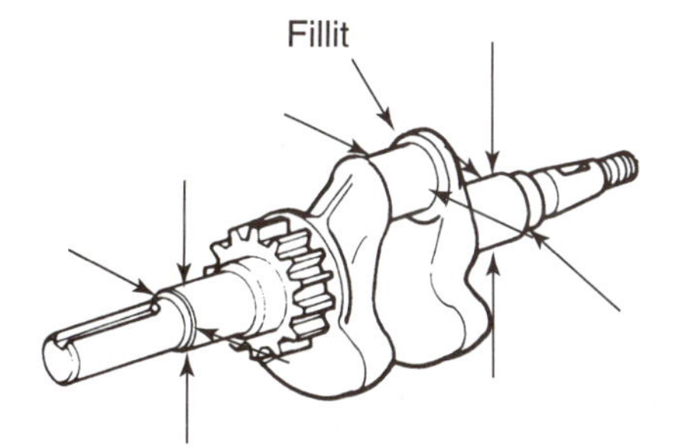

B. Measuring Directions

FIGURE 27-2 Measuring directions to determine crankshaft diameter and out-of-round. (*B:* Courtesy of American Honda Motor Co., Inc.)

Journals that pass a visual inspection are measured with an outside micrometer. Select a micrometer with the correct size range for the journals to be measured. Look up the specifications for main bearing and connecting rod journal diameters in the shop service manual, and write them down for reference.

Wipe the journals with a clean shop rag to make sure that they are free of oil and dirt. Also, wipe off the measuring surfaces of the micrometer. Dirt on the micrometer measuring surfaces will cause a measurement error. Measure the connecting rod journal first (Figure 27-2). This journal gets the most abuse and is likely to be in the worst condition. Make the measurement at least 0.25 inch (6 millimeters) away from the end of the journal to ensure that the micrometer does not measure the small radius (fillet) between the journal and the throw. Write down your measurement. Make another measurement at 90 degrees from the first measurement. Write down this measurement. If the journal is long enough, make one or two more measurements along the journal in both directions.

A difference between two readings taken at 90 degrees from each other means the journal is out-of-round. Out-of-round connecting rod journal wear is common. It is caused by the way the connecting rod applies force to the crankshaft. Multiple cylinder engines may have two connecting rod journals. Repeat the measurement procedure for each of the other connecting rod journals. Record all of your measurements. Repeat this same measurement procedure for each of the main bearings. Record these measurements.

CAUTION: Do not confuse crankshaft diameter specification with crank shaft reject specifications. Some engine makers have both these specifications in the shop service manual. The diameter specification shows the diameter of a new unworn crankshaft. A reject size is the diameter of a worn crankshaft that cannot be reused.

After the measurements have been completed, determine the maximum out-of-round for the worst journal. Check the readings with specifications in the shop service manual to determine if the crankshaft out-of-round is larger than specifications. Specifications for excessive out-of-round are often in a range of 0.001 to 0.003 inch (0.03 to 0.05 millimeter). Excessive out-of-round journals require crankshaft replacement or reconditioning.

The crankshaft may also be worn undersize. Journal undersize specifications are usually in the range of 0.001 to 0.003 inch (0.03 to 0.05 millimeters). Compare your smallest journal measurement with the journal diameter listed in the shop service manual. The specifications in the shop service manual often list a reject size. **Reject size** is a specification dimension that indicates that a part should be rejected, replaced, or reconditioned. A journal that is worn down to reject size requires crankshaft replacement or reconditioning. For example, the reject size for a PTO journal is 1.376 inch. The measurement taken is 1.378 inch. The journal is larger than reject size so it is an acceptable size. However, if the measurement taken is 1.375 inch, the crankshaft requires replacement. Measuring a crankshaft for wear is shown in the accompanying sequence of photographs.

CRANKSHAFT RECONDITIONING

A perfectly round, smooth, and straight crankshaft is required for proper bearing performance. Most small-engine crankshafts are inexpensive enough

PHOTO SEQUENCE 16

Measuring a Crankshaft for Wear

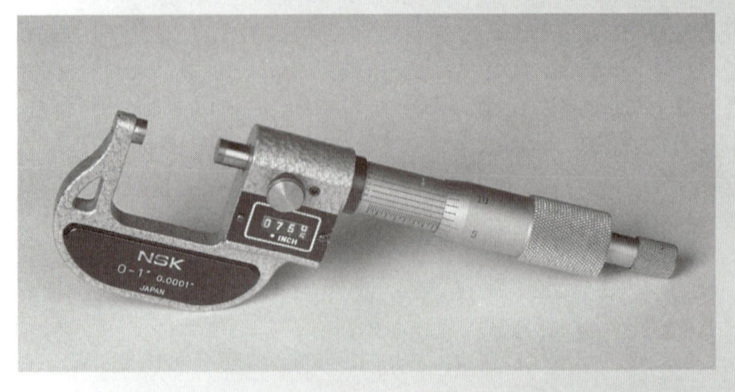

1. Select a micrometer of the correct range to measure the crankshaft journals. Wipe off the measuring surfaces of the micrometer.

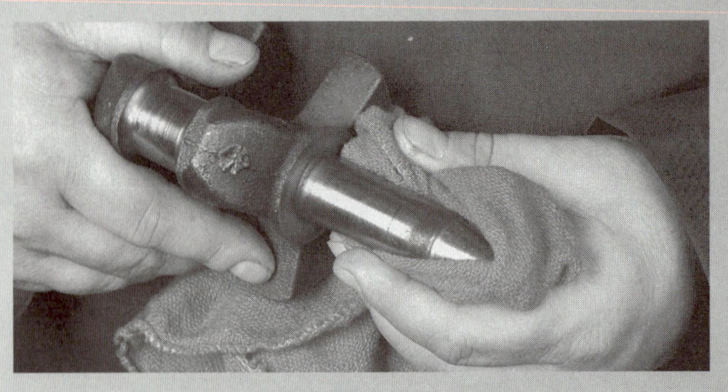

2. Wipe the journals with a clean shop rag to make sure that they are free of oil and dirt.

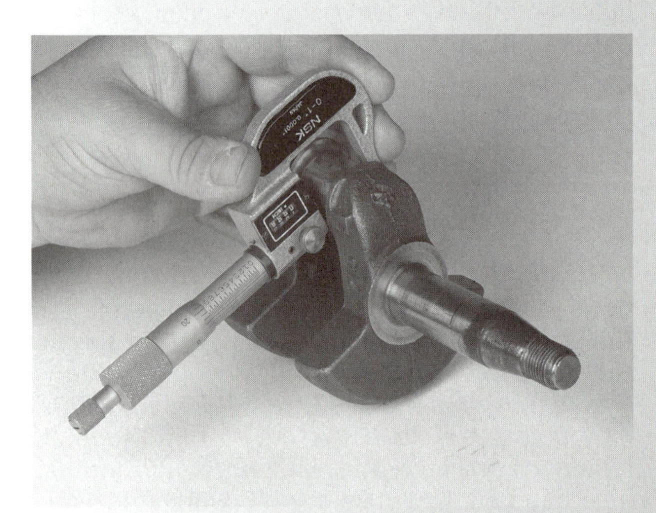

3. Measure the connecting rod journal first because it gets the most abuse and is likely to be in the worst condition. Make the measurement at least 0.25 inch (6 millimeters) away from the end of the journal. Write down your measurement.

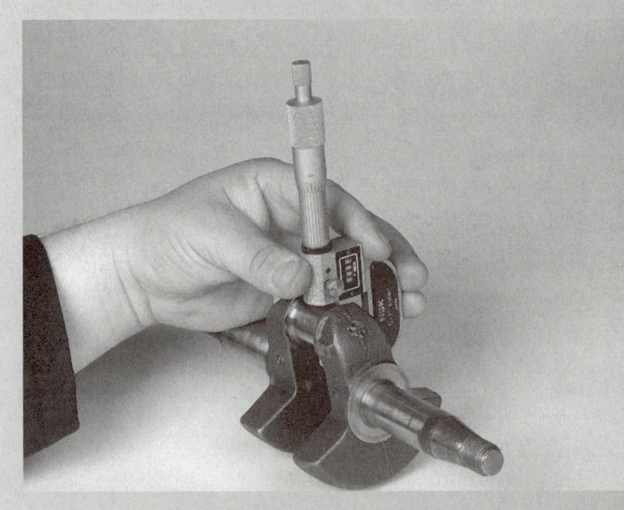

4. Make another measurement at 90 degrees from the first measurement. Write down this measurement. A difference between two readings taken at 90 degrees from each other means the journal is out-of-round.

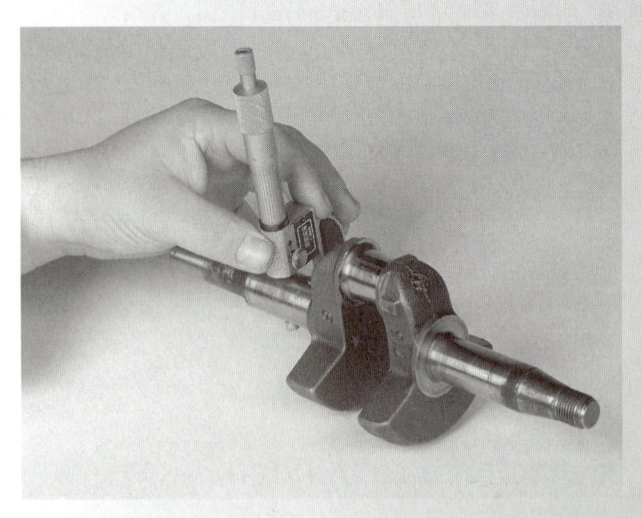

5. Repeat this same measurement procedure for the PTO main bearing journal. Record these measurements.

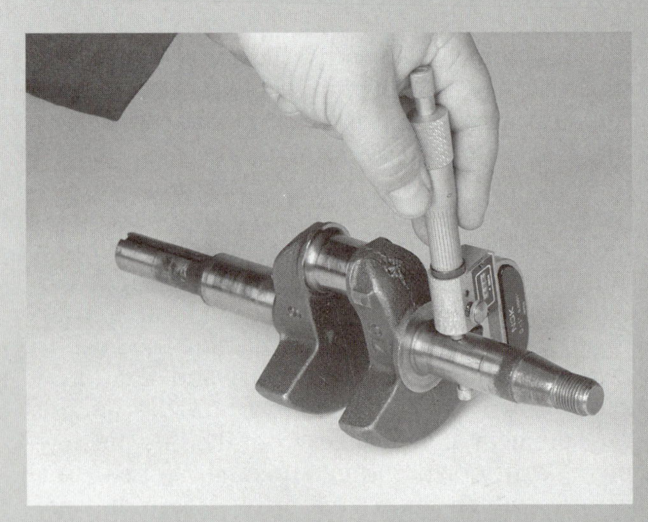

6. Repeat this same procedure for the FWE main bearing journal. Record these measurements.

to be replaced instead of reconditioned. When a new crankshaft is not available or is very expensive, it may be reconditioned. Crankshafts are reconditioned in a machine shop with a crankshaft grinder. The crankshaft grinder removes material from the journal and provides a new bearing surface.

There are two methods for crankshaft reconditioning. One is to weld new metal on the bearing journals. The journals are then ground down to their original size. This method is expensive. It may be the only option if undersize (thicker) bearings are not available. The other method is to grind each of the journals to a standard undersize. **Undersize bearings** are thicker bearings made for journals that have been machined to a standard undersize so that the correct oil clearance is maintained.

CONNECTING ROD INSPECTION AND MEASUREMENT

Inspect the connecting rod for wear or damage. The big end aluminum saddle bore is often used as the connecting rod bearing surface. These connecting rods do not have an insert bearing. Inspect this area for signs of scoring or seizing. Look for metal transfer from seizing. Look for scoring from abrasives in the oil. Replace the connecting rod if you find a saddle bore with either of these problems (Figure 27-3).

The rod cap must be installed on the rod to measure the connecting rod saddle bore. Install the cap on the rod in the correct direction, observing the match marks. Install the connecting rod fasteners. Look up the tightening torque in the shop service manual. Gently clamp the rod and cap in a soft face bench vise or a connecting rod vise (Figure 27-4). This will keep the cap lined up with the rod during tightening. Tighten the fasteners to specifications with a torque wrench.

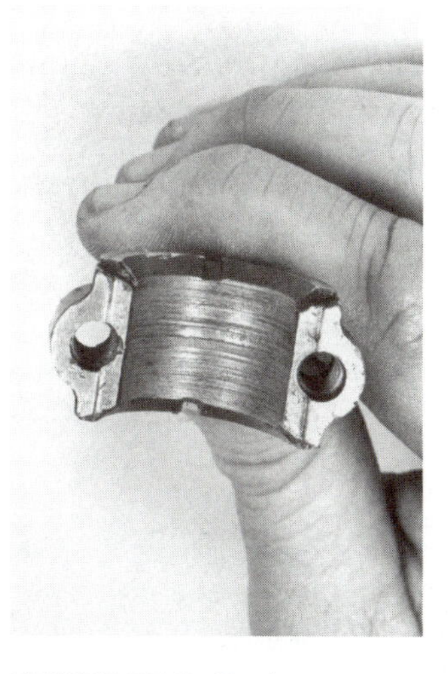

FIGURE 27-3 Replace a connecting rod that shows seizing or scoring in the saddle bore.

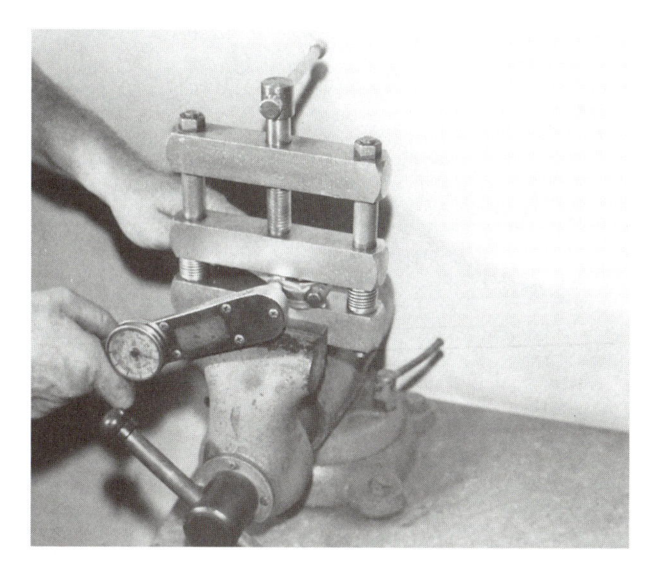

FIGURE 27-4 Tightening the rod cap in a connecting rod vise.

During the power stroke, the top of the saddle bore is placed under a high load. The bottom of the saddle bore is under a low load. This uneven loading causes the saddle bore to eventually stretch out-of-round. A connecting rod with an out-of-round saddle bore will break or seize in a few hours of operation. A rod with this problem should be replaced.

Choose a telescoping gauge that will fit into the saddle bore and expand out against the cap and rod. Place the telescoping gauge into the saddle bore in a direction 90 degrees from the cap mating surfaces (Figure 27-5). Expand the gauge until there is a light drag against the saddle bore. Tighten the handle and remove the gauge. Measure across the gauge with an outside micrometer. Write down this measurement. Place the gauge back in the saddle bore in a direction 90 degrees from the first measurement. Make sure the gauge does not contact the mating surface between the rod and cap. Make sure it does not contact any insert-bearing lock grooves. Write down this second measurement. If there is any difference in your two measurements, the saddle bore is out-of-round.

Saddle bores can also wear oversize. The bore can be round but larger in diameter than specifications. Compare your saddle bore measurements to specifications in the shop service manual. Engine makers often give a saddle bore reject diameter.

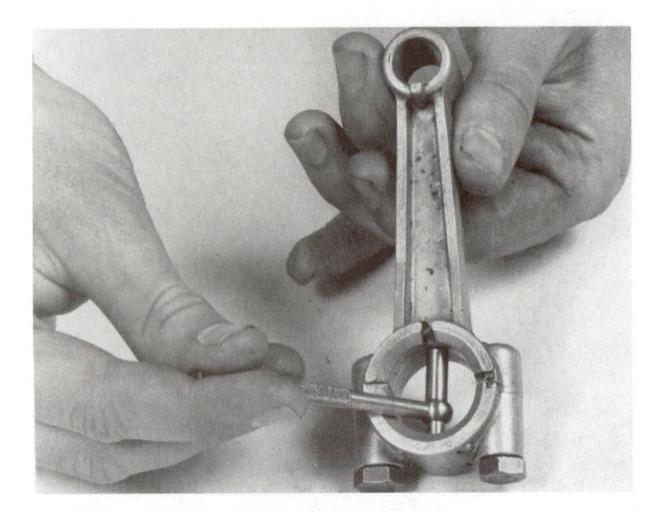

A. Measure with a Telescoping Gauge

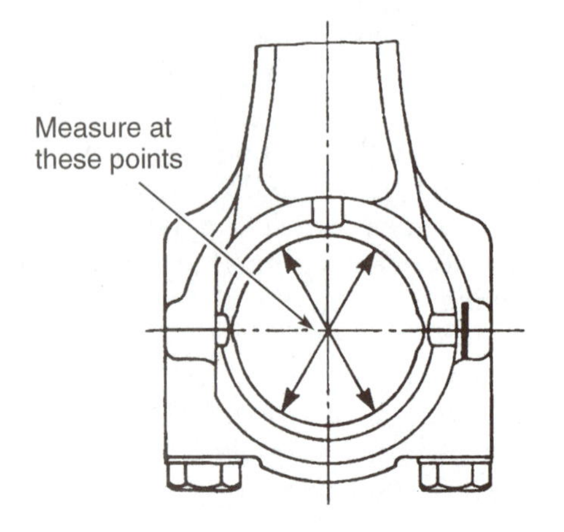

B. Mesasuring Points

FIGURE 27-5 Measuring the connecting rod saddle bore to determine diameter and out-of-round. (*B:* Courtesy of Onan Corp.)

Replace the connecting rod if the saddle bore is worn oversize.

The condition of the connecting rod small end must also be checked. Combustion forces are transferred from the piston to the connecting rod through the small piston pin. The pin is slightly smaller than the bore in the piston and the connecting rod. It is free to rotate (float) in both parts. The pin is often held with two internal snap rings, one on each side of the piston. The floating pin could seize in either the piston or rod without the parts locking up. **Pin fit** is the clearance between the piston pin and the bore in the connecting rod and piston. Pin fit is one of the closest and most precise tolerances in the engine.

Check full floating pins for clearance by inserting the pin into the connecting rod. Try to rock the pin up and down. If you get any movement, there is wear in the small end bore. Wear usually occurs in the rod and not the pin. There is less bearing surface area in the aluminum rod and the steel pin is much harder. Some connecting rods have a replaceable bushing at the small end bore. Most engines use the aluminum of the connecting rod as the bearing. This wear usually is corrected by replacing the connecting rod.

FIGURE 27-6 Checking piston pin small end bore with a special dial gauge. (Courtesy of American Honda Motor Co., Inc.)

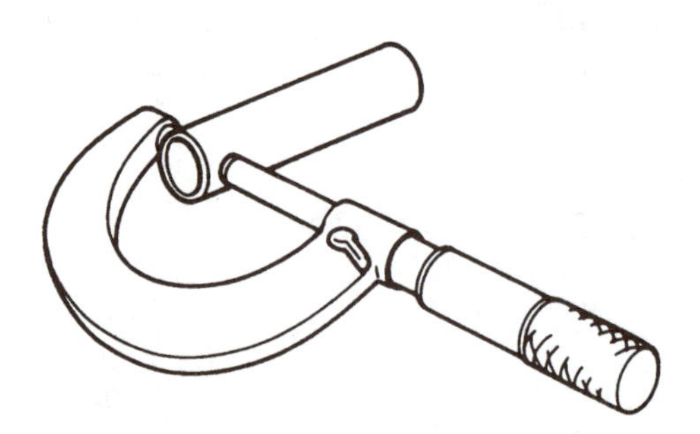

FIGURE 27-7 Measuring the diameter of a piston pin with an outside micrometer. (Courtesy of American Honda Motor Co., Inc.)

Some engines have size and wear specifications for the small end bore. Engine makers may supply a special pin hole dial gauge (Figure 27-6). Use this tool to make several measurements around the small end bore. The piston pin outside diameter is measured with an outside micrometer (Figure 27-7). Compare the measurements to specifications to see if the connecting rod or pin should be replaced. Measuring a connecting rod for wear is shown in the accompanying sequence of photographs.

CONNECTING ROD BEARING INSPECTION

A separate connecting rod bearing is used in some engines. Rod bearings are always replaced during an engine overhaul. Most two-stroke engines use needle bearings in the saddle bore. The needle bearings

PHOTO SEQUENCE 17

Measuring a Connecting Rod for Wear

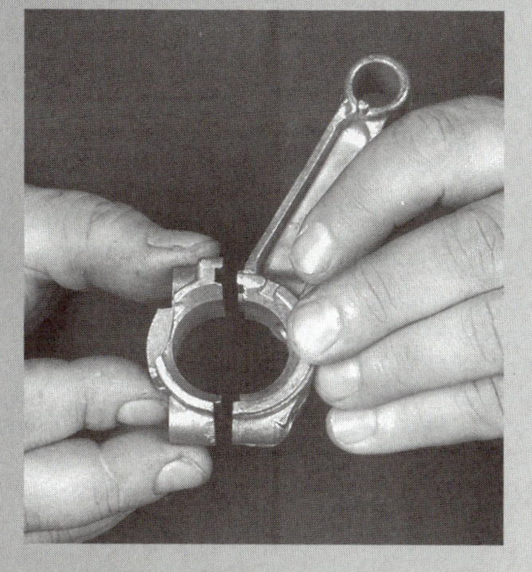

1. Install the rod cap on the rod in the correct direction, observing the match marks.

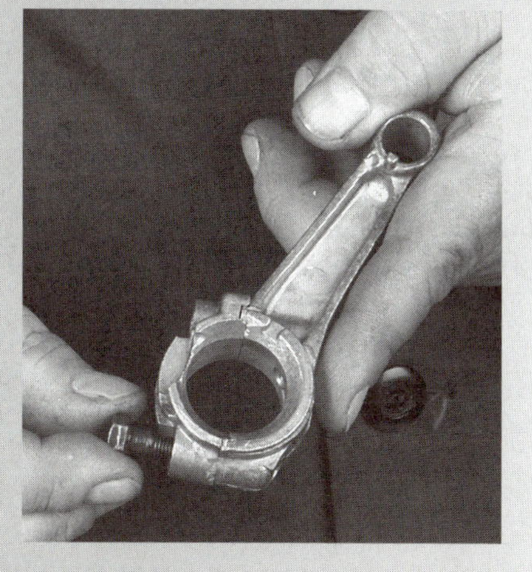

2. Install the connecting rod fasteners. Look up the tightening torque in the shop service manual.

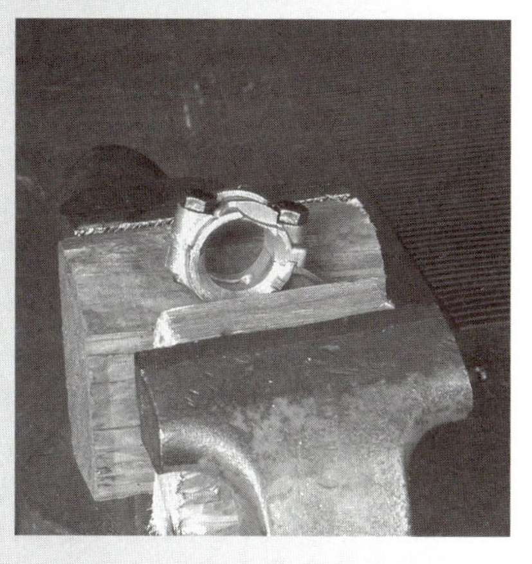

3. Gently clamp the rod and cap in a soft face bench vise or a connecting rod vise to keep the cap lined up with the rod during tightening.

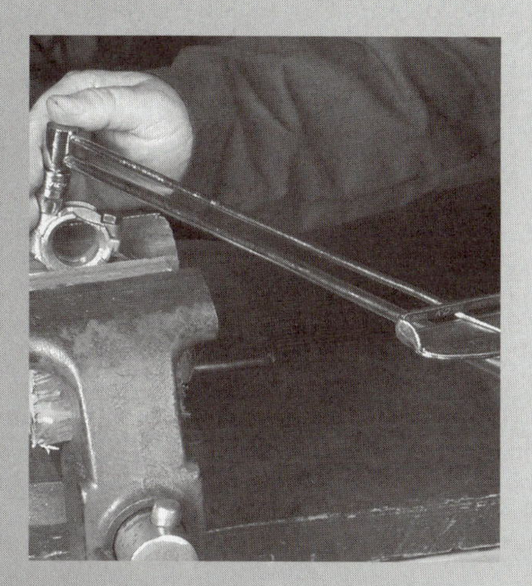

4. Tighten the fasteners to specifications with a torque wrench. Remove the rod from the vise.

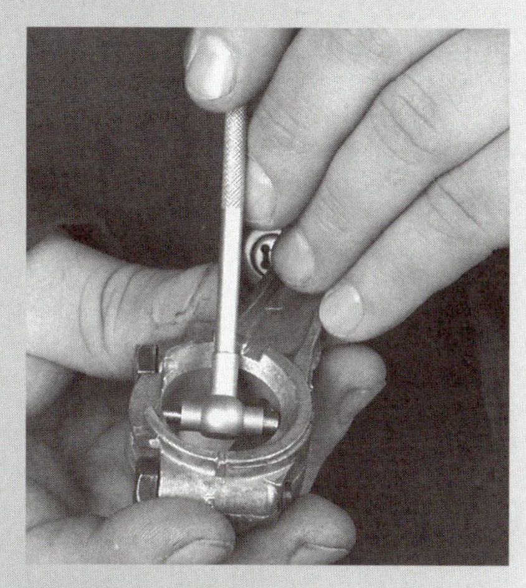

5. Choose a telescoping gauge that will fit into the saddle bore and expand out against the cap and rod. Place the telescoping gauge into the saddle bore in a direction 90 degrees from the cap mating surfaces. Expand the gauge until there is a light drag against the saddle bore. Tighten the handle, and remove the gauge

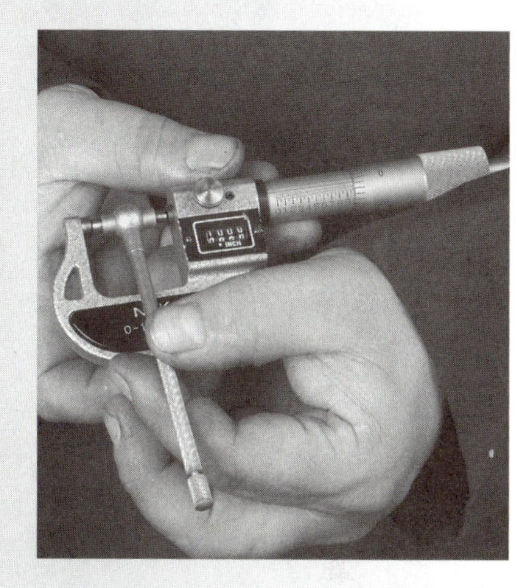

6. Measure across the gauge with an outside micrometer. Write down this measurement.

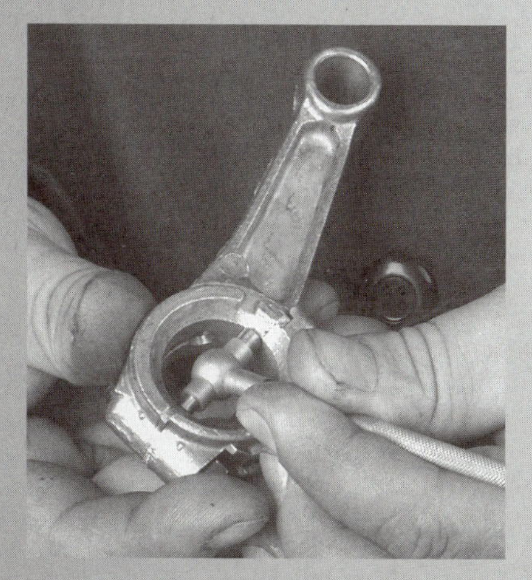

7. Place the gauge back in the saddle bore in a direction 90 degrees from the first measurement and measure in this direction. Write down this second measurement. If there is any difference in your two measurements, the saddle bore is out-of-round.

8. Check full floating pin for clearance by inserting the pin into the connecting rod.

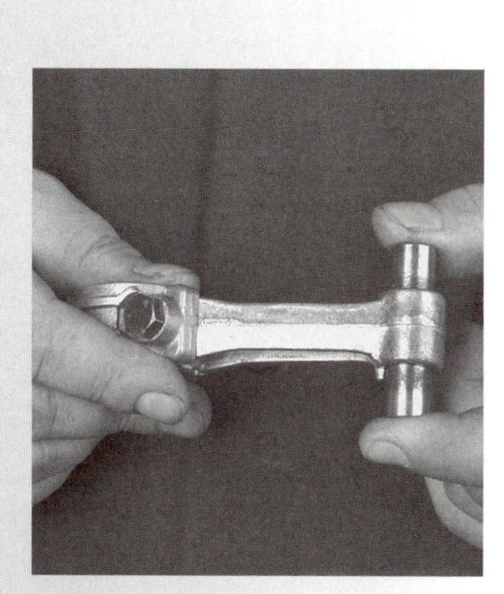

9. Try to rock the pin up and down. If you get any movement, there is wear in the small end bore.

may be free to touch each other or they may be held apart with a cage. Some engines use single row and others use a split row of needle bearings. Some two-stroke rods have a removable cap and others are one piece. The needle bearings for the one piece are placed on the crankshaft journal. The rod slips on or off the bearings (Figure 27-8). Needle bearings are always replaced during a two-stroke engine overhaul. Bearing clearance is set by the correct size crankshaft journal, saddle bore, and needle bearing. Measure the old needle bearing diameter and length with a micrometer (Figure 27-9). Use these measurements as a reference when ordering new needle bearings.

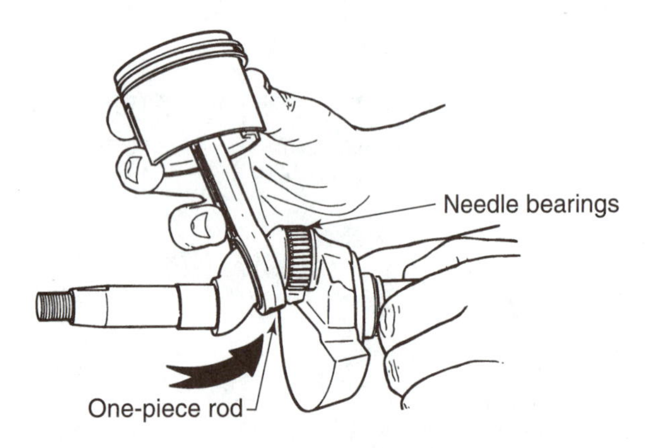

FIGURE 27-8 A one-piece, two-stroke connecting rod. (Provided courtesy of Tecumseh Products Company.)

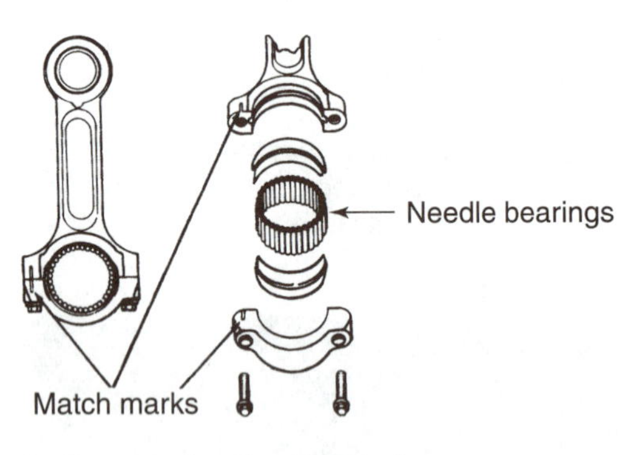

A. Two-Stroke Needle Bearings

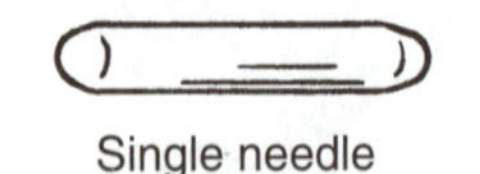

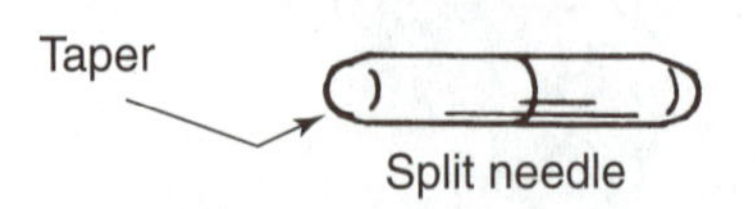

B. Measure Length and Diameter

FIGURE 27-9 A two-piece, two-stroke connecting rod. (Provided courtesy of Tecumseh Products Company.)

CAUTION: The number of needle bearings required on a two-stroke rod is sometimes difficult to determine. Always save and count the needle bearings on disassembly. Use this number as a reference during assembly.

Some four-stroke engines use a precision insert bearing between the crankshaft journal and connecting rod saddle bore (Figure 27-10). The insert bearings are made in two pieces so they can fit in the rod and cap and be assembled around the crankshaft. Each bearing insert has a locking tab. The tab fits into a locking groove in the bore. Insert bearings also use spread to hold them in the bore. Bearing spread is a method used to hold an insert bearing in place by making it slightly larger than the housing it fits. The spread allows the insert to snap into place (Figure 27-11).

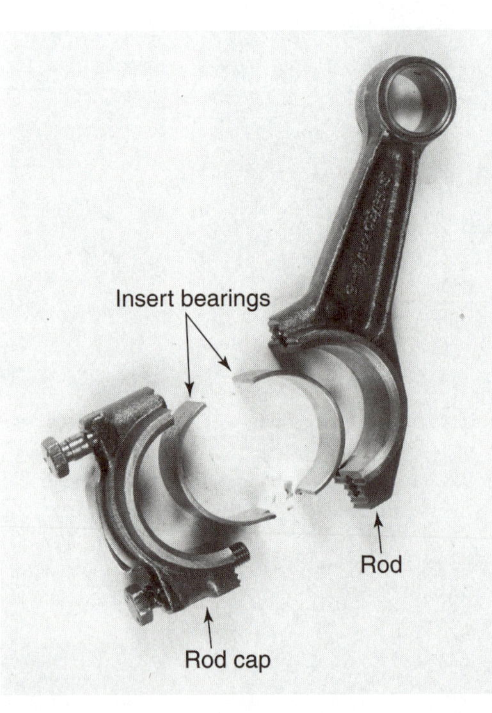

FIGURE 27-10 A connecting rod with bearing inserts.

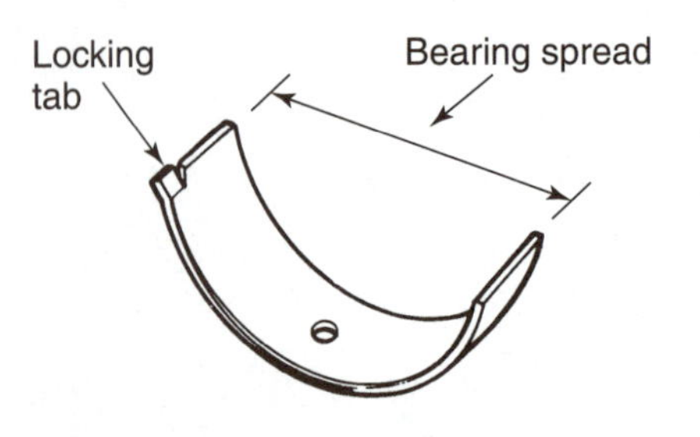

FIGURE 27-11 Inserts are held in place by locking tabs and bearing spread. (Courtesy of American Honda Motor Co., Inc.)

CASE STUDY

A technician was working on the overhaul of an engine with connecting rod inserts. He had already installed the piston and connecting rod assembly into the cylinder block. The upper connecting rod insert bearing was installed on the connecting rod. He installed the bottom insert bearing into the connecting rod cap. He installed the cap on the rod and tightened the rod cap screws with a torque wrench. He knew he should measure the bearing clearance but he was in a hurry to finish the engine. He was confident that the clearance would be fine. He had assembled many engines like this and had never had a problem.

He installed the crankcase cover and cylinder head. He was ready to install the flywheel. The crankshaft needed to be rotated to install the flywheel key. The crankshaft would not rotate; it was locked up solid. He started disassembling the engine to find the problem. The crankshaft would not rotate with the crankcase cover removed. The problem had to be the connecting rod bearing. The technician removed the connecting rod cap and removed the insert bearing. He looked at the back of the insert and immediately knew where he had made his mistake.

The parts supplier had given him an insert bearing for a reconditioned crankshaft with a 0.020 inch undersize. The insert bearing had 0.020 inch stamped on the back. It was 0.020 thicker than the standard size bearing. The crankshaft in this engine was a standard size. The thicker bearing did not have any oil clearance space and locked up against the crankshaft. The technician had lost a significant amount of time locating and correcting this problem. He resolved to check the oil clearance on the next engine.

Insert bearings prevent wear to the bearing surface of the connecting rod. They also allow the use of a reconditioned (undersize) crankshaft. Insert bearings are available in standard sizes to be used with a standard size crankshaft. Thicker inserts are used when the crankshaft journals are ground undersize. These bearings are often made in common undersizes such as 0.010, 0.020, and 0.030 inch (0.025, 0.050, and 0.076 millimeter). The undersize is usually marked on the bearing packaging and on the back of the insert.

FIGURE 27-12 Measuring a plain main bearing with a telescoping gauge.

PLAIN MAIN BEARING INSPECTION AND MEASUREMENT

Many four-stroke engines use a plain main bearing. The crankshaft rotates directly in the aluminum block and side cover. A few engines have a bushing installed in one or both of the main bearings. The PTO side main bearing is under the highest loads and wears the most. Inspect each bearing for scoring and metal transfer. A bearing with a scored or damaged surface must be reconditioned.

Look up the specifications for main bearing diameter or reject size in the shop service manual. Measure the plain main bearing with a telescoping gauge and outside micrometer (Figure 27-12). Measure the bearing in several directions and record your measurements. Check your measurements against specifications to determine if the bearings are out-of-round or worn undersize. An out-of-round or undersize bearing must be reconditioned. Some engine makers supply a plug gauge to measure main bearings (Figure 27-13). The plug gauge measures main bearing reject size. If the gauge fits into the main bearing, the bearing is worn oversize.

PLAIN MAIN BEARING RECONDITIONING

A plain main bearing is reconditioned by reaming and installing a bushing. The reconditioning must be done so that the new main bearings are in

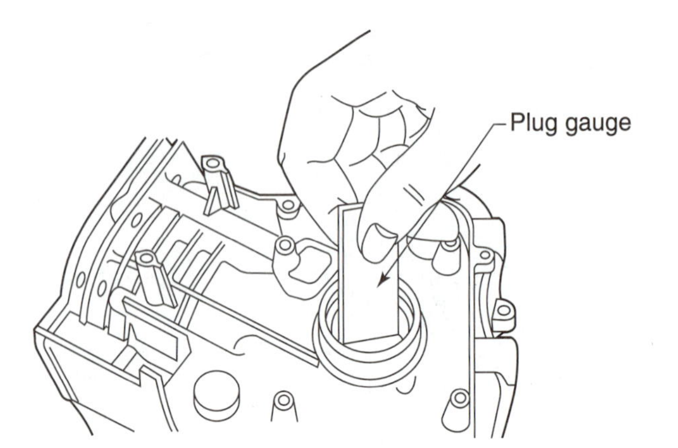

FIGURE 27-13 Measuring a plain main bearing with a plug gauge.

perfect alignment. If they are not, the crankshaft will not rotate freely. Engine makers supply a tool set for plain main bearing reconditioning (Figure 27-14). The tool set has the correct size reamers and pilots for correct size and alignment. The kit also has the tools needed to drive in a replacement bushing.

The cover has to be installed correctly during reconditioning to keep the main bearings in alignment. Install the side or bottom crankcase cover. Look up the specifications for cover screw torque. Use a torque wrench to tighten the screws. To recondition the FWE main bearing, install a pilot guide bushing in the cover main bearing (Figure 27-15). This bushing will center and guide the pilot. Install the correct size counterbore reamer on the pilot. Insert the pilot and counterbore reamer into the FWE main bearing. The end of the pilot goes into the pilot guide bushing in the PTO main bearing. Insert the reamer guide bushing around the reamer. The reamer guide bushing helps center the reamer in the FWE main bearing housing. Turn the reamer clockwise by hand to machine the bearing housing oversize (Figure 27-16). The pilot guide bushing and reamer bushing make sure that

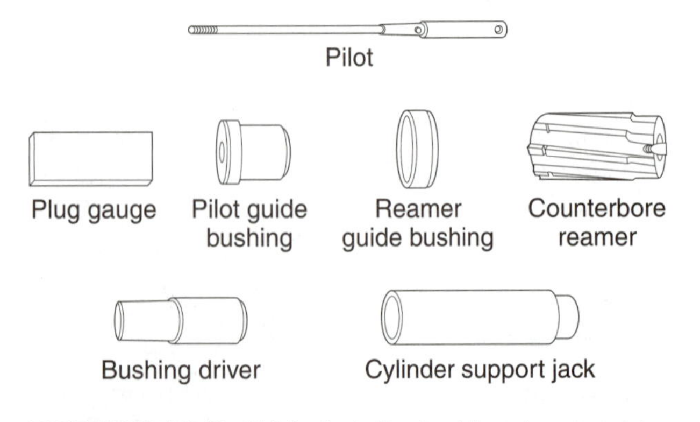

FIGURE 27-14 Tool kit for installing bushings in worn plain main bearings.

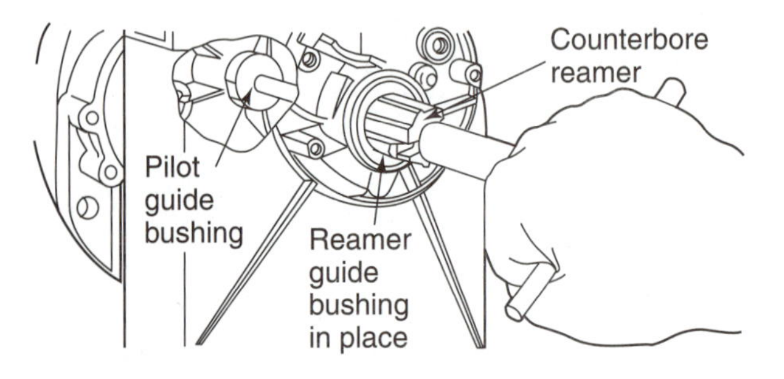

FIGURE 27-15 The counterbore reamer pilot is centered by two guide bushings.

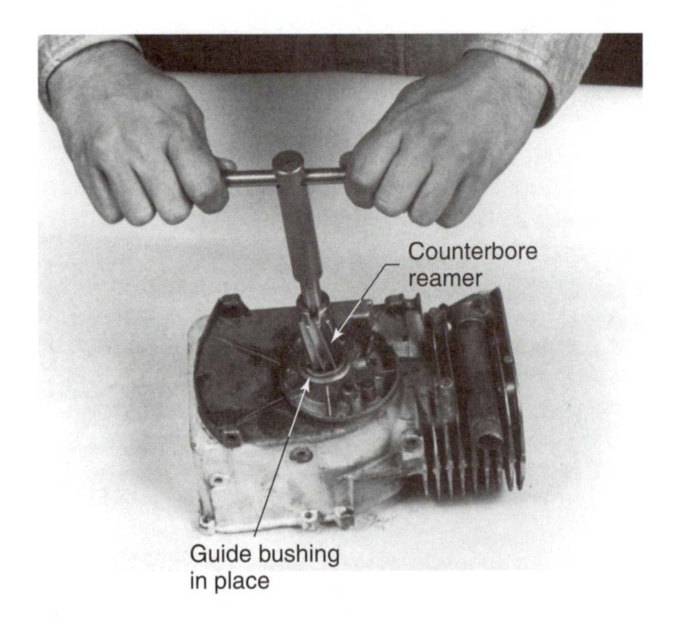

FIGURE 27-16 Reaming the FWE main bearing.

the reaming is done in correct alignment. Reverse this procedure to ream the PTO cover main bearing.

A bushing is installed in the reamed main bearing housing (Figure 27-17). The cylinder block has to be supported when driving the bushing. A special tool called a *cylinder jack* is supplied for this purpose. Position the cylinder jack behind the main bearing. Use a chisel to make a small notch on the edge of the bearing housing. Align the bushing with any oil feed notches or holes. Insert the correct size bushing driver into the bushing and drive it into position. Use a chisel to drive an edge of the bushing into the notch (Figure 27-18). This procedure is called *staking*. Staking is done to prevent the bushing from turning in operation.

Some bushings are made to be the correct size when they are installed. Other bushings require finish reaming after installation. If finish reaming is required, reinstall the guide bushings and attach a finish reamer to the pilot. Ream each main bearing

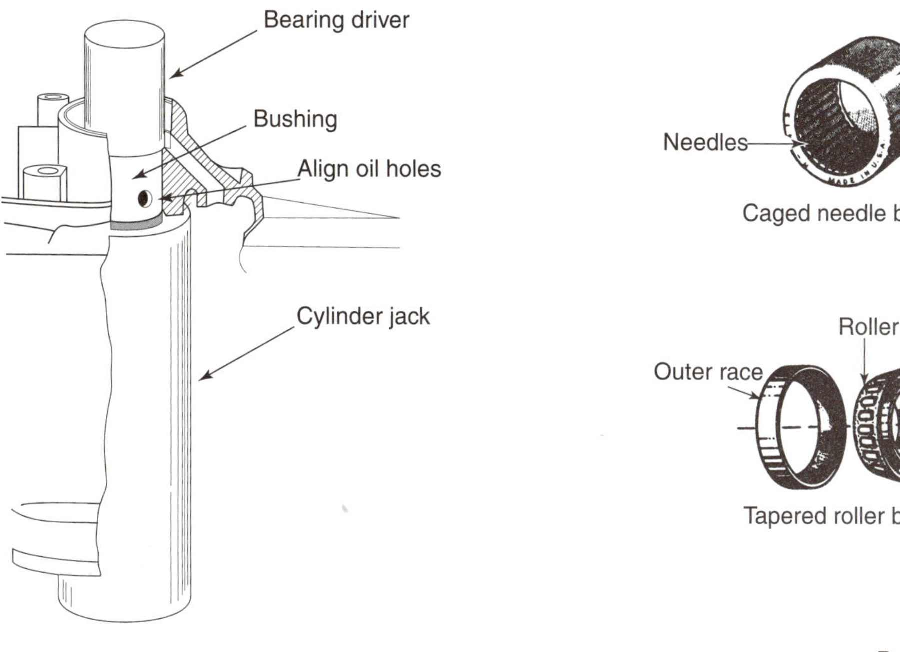

FIGURE 27-17 Installing a bushing in a reamed plain main bearing.

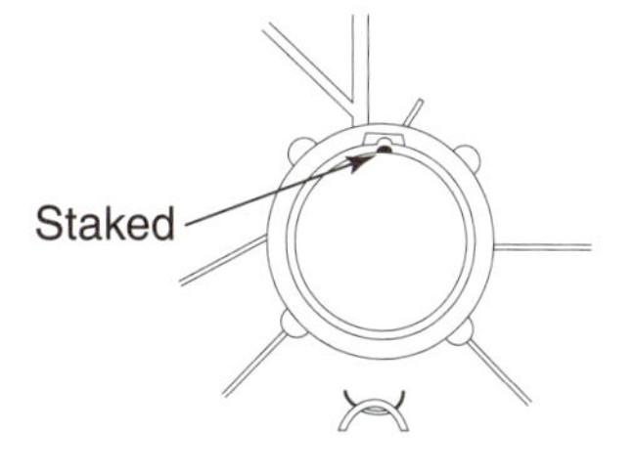

FIGURE 27-18 Staking a bushing to prevent it from rotating.

bushing so that the inside diameter is the correct size according to specifications. Finish reaming often is done with a lubricant such as kerosene on the cutter. The lubricant prevents a rough finish.

Clean away all the cuttings from counterbore and finish reaming. Remove the cover fasteners and remove the cover. Oil the crankshaft main bearings and install the crankshaft into the cylinder block. Install the cover and tighten the fasteners to the correct torque. Rotate the crankshaft by hand. The crankshaft should rotate freely in the new bushings.

CAGED NEEDLE, TAPERED ROLLER, AND BALL BEARING SERVICE

Caged needle, tapered roller, and ball bearings are used as main bearings (Figure 27-19) on many engines. Some smaller two-stroke engines use caged

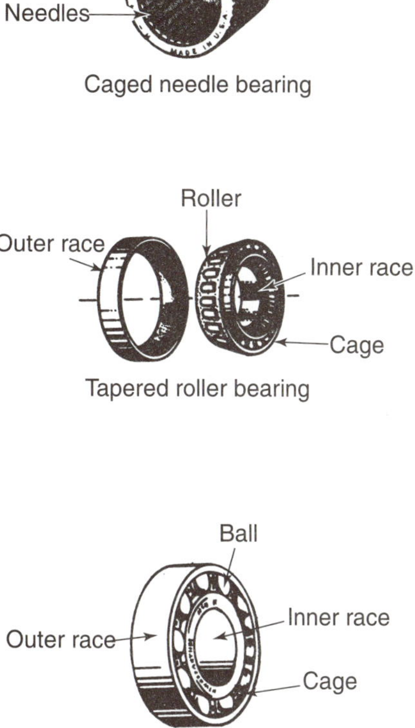

FIGURE 27-19 Common types of main bearings. (Courtesy of Clinton Engines Corporation.)

needle bearings because they are easily lubricated with a gasoline and oil mixture. This bearing is a set of needle bearings contained in a housing or cage. The outside housing is held in the cylinder block or side cover with a press fit or retaining rings. The needle bearings rotate between the bearing housing and the crankshaft. This means the bearings can cause wear to the crankshaft journal but not to the block or side cover. The caged needle bearings are always replaced during an engine overhaul.

Some four-stroke engines use tapered roller bearings as one or both main bearings. These bearings work for high loads as long at the rotational speeds are relatively low. The tapered roller bearing has a set of tapered roller bearings held together with a cage. An inner race is permanently attached to the cage and rollers. An outer race is pressed in to the cylinder block or side cover. The rollers rotate between the two races so there is no relative movement or wear to the block, side cover, or crankshaft journal.

Many two-stroke engines have ball bearings on both ends of the crankshaft. Some four-stroke

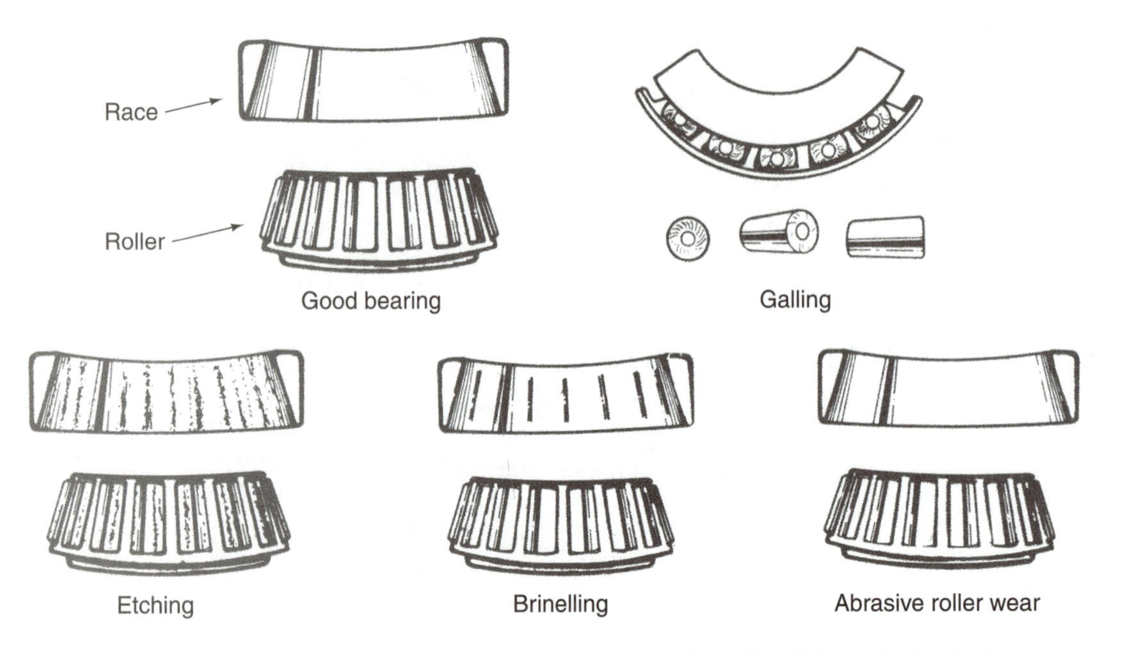

FIGURE 27-20 Different types of roller bearing wear. (Courtesy of Ford Global Technologies, Inc.)

engines use a ball bearing on the crankshaft PTO and a plain bearing on FWE. The ball PTO bearing can handle higher loads and lasts longer than the plain bearing. The ball bearing has a set of ball bearings spaced apart with a cage. The balls fit between an inner and outer race. These parts are permanently assembled and neither race is removable. The ball bearing usually fits on the crankshaft with a press fit between the crankshaft main bearing journal and the ball bearing inner race. Sometimes the ball bearing is a press fit in the cylinder block or cover. The balls rotate between the two races. There is no relative movement or wear to the cylinder block, cover, or crankshaft journal.

Inspecting Bearings

Caged needle bearings are always replaced when rebuilding a two-stroke engine. Keep the old bearing to compare with the new ones. Measure the diameter and length of the old bearings to compare with the new ones.

Tapered roller and ball bearings should be left in place when cleaning the crankshaft (or housing, if so mounted). After cleaning, inspect the bearings and race for wear or damage. Check each individual tapered roller for galling, etching, brinelling, and abrasive wear (Figure 27-20). **Galling** is a bearing problem in which metal transfers from the rollers to the race due to lubricant failure. **Etching** is a bearing problem in which the bearing surface turns black from lack of lubricant. **Brinelling** is a bearing problem in which the surface breaks down from overloading. Abrasive wear is a bearing condition in which the surface is scratched from abrasives in the oil.

Ball bearings are more difficult to inspect. The individual balls are covered by the cage and races. Ball bearings should always be replaced during an engine rebuild. They can be lubricated and rotated to check for damage. A damaged ball bearing feels rough when it is rotated.

Replacing Bearings

Always compare the old and new bearings before the old one is replaced. Place the two bearings side by side and make a comparison. Bearings have an identification number etched on the races (Figure 27-21). Compare these numbers to those on the old bearing to make sure they are the same. This is the best way to determine if you have the correct replacement bearing.

CAUTION: Tapered roller bearings and outside races are always replaced as a set. Installing a new bearing on an old race will result in rapid bearing failure. Always make match marks on a tapered roller bearing and the bearing race during disassembly. If bearings are to be reused, they must go back in their same races. Failure to do this will result in rapid bearing failure.

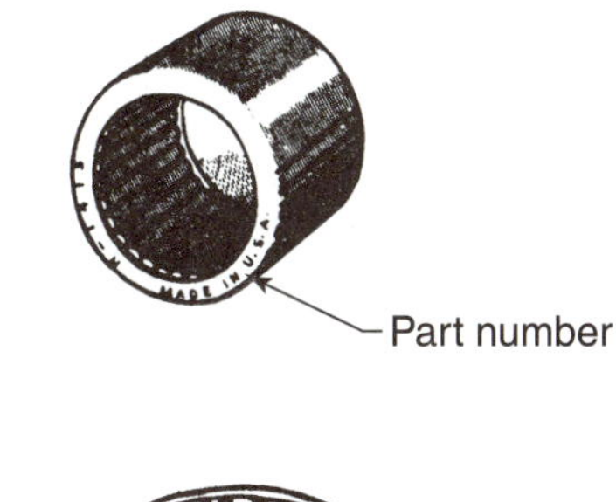

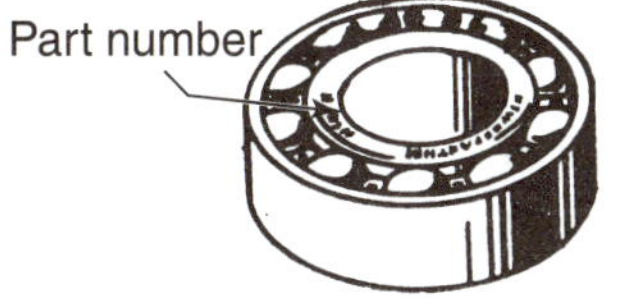

FIGURE 27-21 Bearing identification numbers are used to get the correct replacement bearing. (Courtesy of Clinton Engines Corporation.)

Caged needle bearings are removed or replaced using a bushing driver that fits into the bearing. Place a support tool under the cylinder block or cover for support. Use a hammer to drive the old bearing assembly out and the new one in (Figure 27-22).

WARNING: Wear eye protection when pressing or pulling on bearings because they can explode and cause injury.

A tapered roller bearing is replaced using a bearing driver to drive the old race out of the cylinder block or cover (Figure 27-23). Install the new race in the correct direction using a bearing driver. Be sure to drive the new race all the way into the housing until it seats against the shoulder. The roller bearing and inner race are removed like the ball bearing type.

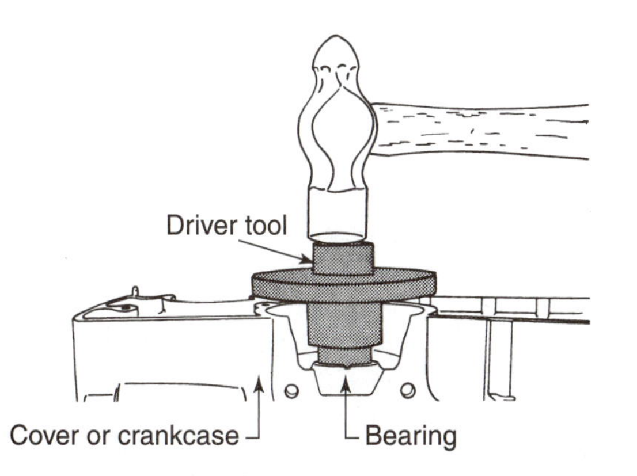

FIGURE 27-22 Driving a caged needle bearing in or out of the crankcase or cover. (Courtesy of Poulan/Weed Eater.)

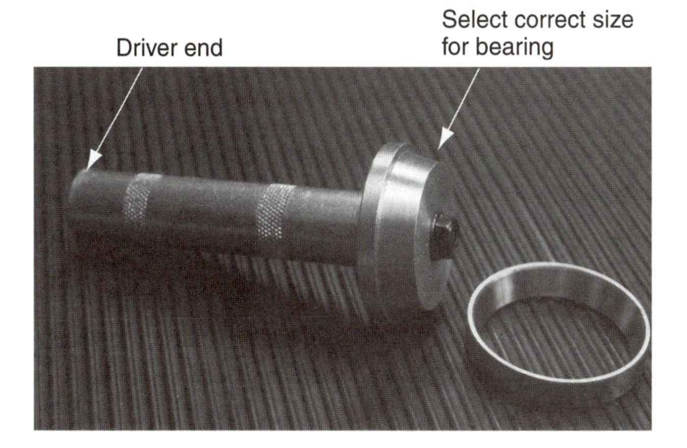

FIGURE 27-23 A bearing driver is used to remove and replace a tapered roller bearing outside race.

Remove a ball bearing held with a press fit in a cylinder block or cover with a press or special pulling tool. To use the press, support the cover on the press table. Select the correct size bearing driver and position it in the bearing. Use the press arbor to push on the tool and force the bearing out of the housing. To install a bearing, turn the cover over and support it on the press table (Figure 27-24). Use the correct size driver to center the bearing over the

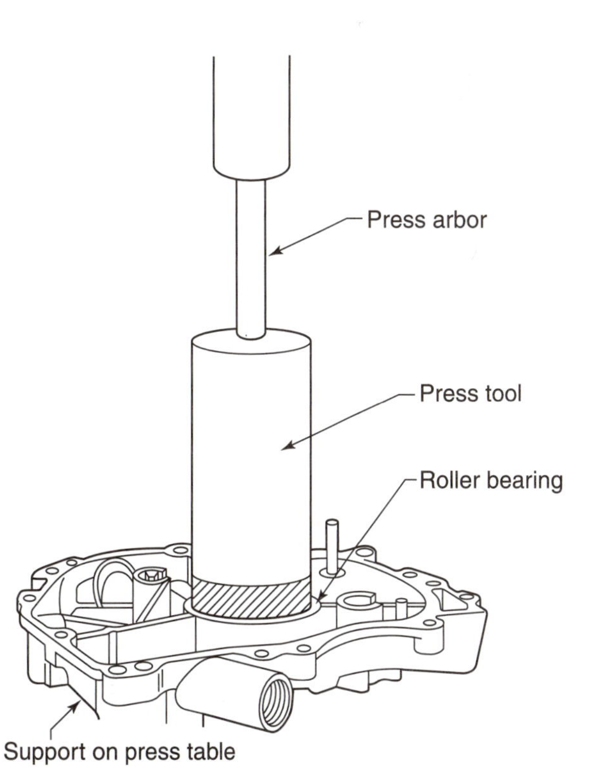

FIGURE 27-24 A press can be used to remove or replace a roller bearing from the cylinder block or cover.

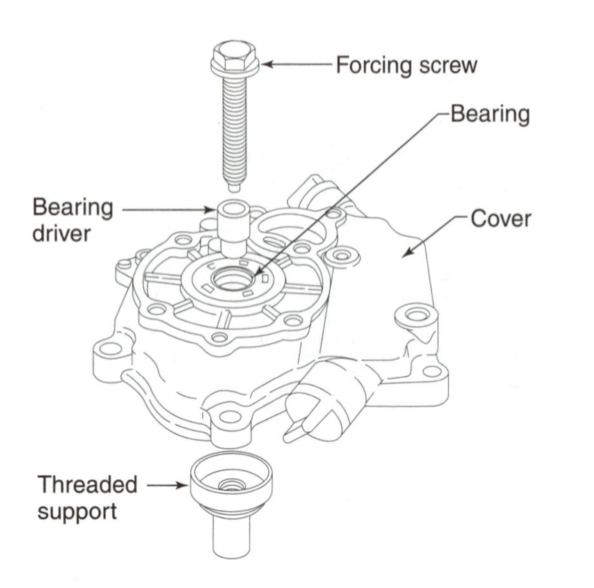

FIGURE 27-25 A puller can be used to remove and replace a roller bearing from the cylinder block or cover.

housing. Press on the driver to push the bearing into place. Be sure it goes in to the correct depth.

Some engine makers supply a special puller installer set to remove and replace cover mounted bearings (Figure 27-25). The tool set has a threaded forcing screw that is used with a bearing driver. There is a selection of thread supports and adapters. Bearings are forced in or out of the housing using a wrench on the puller installer forcing screw.

Crankshaft mounted ball or tapered roller bearings may be removed with a puller or a press. To use a puller, install a bearing spreader (separator) behind the bearing. The bearing spreader is made in two pieces so it can be assembled and installed behind the bearing. Adjust the correct size two-jaw puller to fit on the bearing spreader. Position the center forcing screw on the crankshaft. Use a wrench on the forcing screw. Rotate the forcing screw against the end of the crankshaft until the bearing comes free (Figure 27-26).

To press off a roller bearing, place the crankshaft under the press arbor. Install support blocks under the bearing assembly. Press on the center of the crankshaft to push it though the center of the bearing (Figure 27-27). Be ready to catch the crankshaft so that it does not fall after it goes through the bearing.

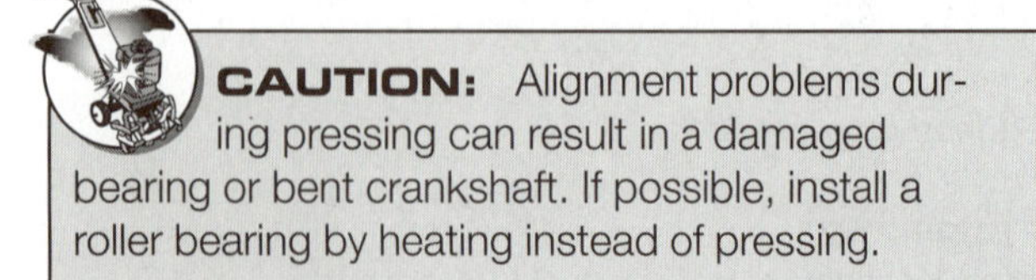

CAUTION: Alignment problems during pressing can result in a damaged bearing or bent crankshaft. If possible, install a roller bearing by heating instead of pressing.

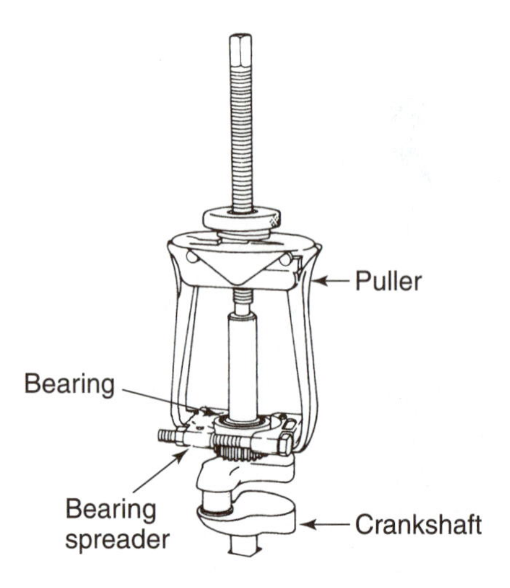

FIGURE 27-26 Removing a roller bearing from a crankshaft with a puller and bearing spreader. (Provided courtesy of Tecumseh Products Company.)

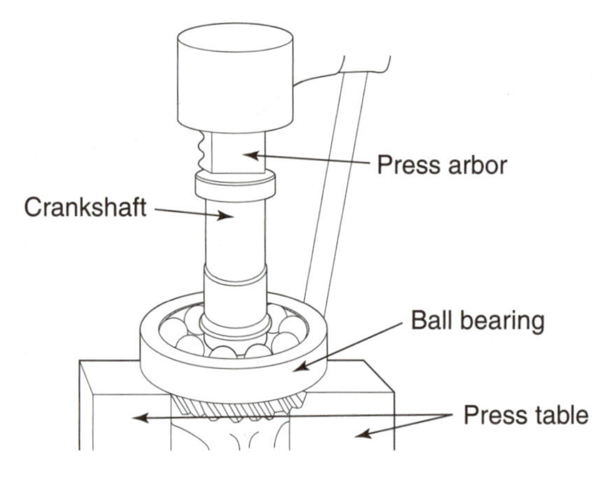

FIGURE 27-27 Removing a roller bearing from a crankshaft with a press.

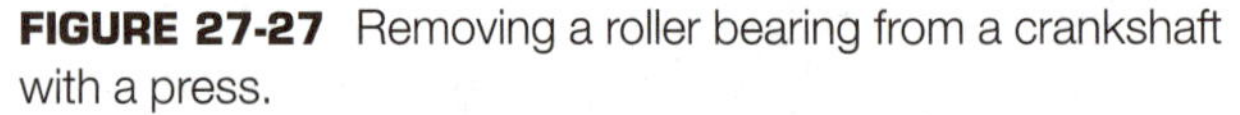

A bearing can be installed back on the crankshaft with a press. Place the crankshaft on the press table. Place the roller bearing on the crankshaft journal. Set the correct size bearing driver on the bearing. Use the press arbor to push against the driver and push the bearing into place on the crankshaft (Figure 27-28).

CAUTION: When heating parts on a hot plate, be sure not to exceed 250°. Oil heated to too high a temperature can ignite and start an oil fire.

The best way to install a roller bearing on a crankshaft is by heating (Figure 27-29). Heat the bearing on an electric hot plate in engine oil to a

FIGURE 27-28 Installing a roller bearing on a crankshaft with a press.

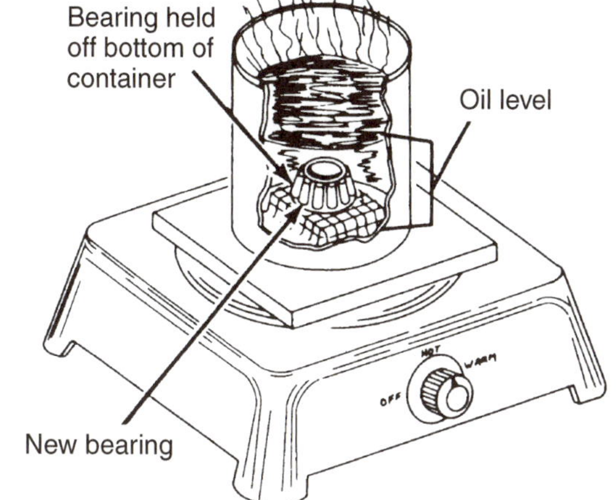

FIGURE 27-29 Heating a bearing in oil for installation. (Provided courtesy of Tecumseh Products Company.)

temperature of 250°F maximum. Heating will expand the bearing large enough to slip over the crankshaft. Use welding gloves to place the bearing on the end of the crankshaft. The bearing should slip down the shaft into position. If the bearing hangs up, tap it gently into place with a soft face hammer.

REVIEW QUESTIONS

1. Describe the types of wear and damage to inspect for on a crankshaft.
2. Explain how to use an outside micrometer to determine crankshaft bearing journal out-of-round.
3. Describe how to use an outside micrometer and the shop service manual to determine if a crankshaft bearing journal is undersize.
4. Explain how a damaged crankshaft journal is reconditioned in a crankshaft grinder.
5. Describe how to check the fit of the piston pin in the small end of the connecting rod.
6. Describe how to visually inspect a connecting rod for wear.
7. Explain why the connecting rod big end saddle bore wears out-of-round.
8. Explain how to use a telescoping gauge to measure the connecting rod big end bore for stretch.
9. Describe how to inspect a plain main bearing for wear.
10. Explain how to use an outside micrometer and telescoping gauge to determine plain main bearing wear and oil clearance.
11. Describe how to ream a plain bearing oversize for a replacement bushing.
12. Describe how to install and stake a main bearing bushing.
13. Explain the advantages of roller over plain main bearings.
14. Explain how to use a puller to remove a ball main bearing from a crankshaft.
15. Explain how to use a press to remove a ball main bearing from a crankshaft.
16. Explain why a new roller bearing race must be used when replacing a new roller bearing.
17. Explain how to remove and install a roller bearing race.

18. Explain how to remove a ball bearing installed in a side cover.
19. Explain why eye protection must be worn when pressing or driving bearings.
20. Explain how to determine the size of a needle bearing to determine the correct replacement part.

DISCUSSION TOPICS AND ACTIVITIES

1. Locate several shop crankshafts. Practice measuring each of the crankshafts to see if any journals are out-of-round.
2. Locate a scrap cylinder block with matching cover. Practice reaming the main bearings for a new bushing.
3. A crankshaft connecting rod journal is measured and found to be 0.748 inch (19 millimeters). The connecting rod saddle bore is measured and found to have a diameter of 0.753 inch (19.13 millimeters). What is the connecting rod bearing oil clearance?

CHAPTER 28

Cylinder and Piston Service

OBJECTIVES

Upon completion and review of this chapter, you should be able to:

- Inspect and measure a four-stroke cylinder.
- Inspect and measure a two-stroke cylinder.
- Remove a ridge from the top of a cylinder.
- Glaze break a cylinder.
- Resize a cylinder with a hone.
- Resize a cylinder with a boring bar.
- Inspect and measure a piston.
- Fit a piston to a cylinder.

TERMS TO KNOW

Boring
Boring bar
Cylinder gauge
Cylinder out-of-round
Cylinder resizing
Cylinder taper
Fitting pistons
Glaze breaker
Glaze breaking
Grit
Hone
Honing
Microfinish number
Piston clearance
Ridge remover
Ring ridge
Ring seating

INTRODUCTION

The cylinder walls are subjected to very high forces. Piston ring pressure continually pushes against them. Combustion forces continually thrust the piston skirt against them. The cylinder walls depend upon splash lubrication in areas that are the hottest in the engine. The piston and piston rings get the full force of the burning gases during the power stroke. Any abrasive dirt that enters the intake system ends up on the cylinder walls. These forces make the cylinder walls, piston, and rings the fastest wearing parts in the engine.

FOUR-STROKE CYLINDER INSPECTION AND MEASUREMENT

The cylinder is inspected and measured to determine what reconditioning is necessary. Inspect the cylinder for scoring (Figure 28-1) caused by abrasives in the incoming air-fuel mixture. If the scoring is deeper at the top of the cylinder, the mixture is probably the source. Abrasives in the oil cause scoring that is deeper at the cylinder bottom. Any score mark deep enough to catch your fingernail requires major cylinder service. Look for a ridge at the top of the cylinder. Move your fingernail up the cylinder. If it catches in the ridge, the cylinder needs major work. If you do not find deep scoring, measure the cylinder to determine the amount of wear.

The area of greatest cylinder wear is where the piston rings reverse direction as the piston goes up and down (Figure 28-2). This area is called the *pocket*. The pocket has the highest temperatures. It gets the least lubrication and the forces from the piston rings. The next highest wear is from the pocket to the end piston ring travel. This area wears from the friction and forces of the piston rings. The ring

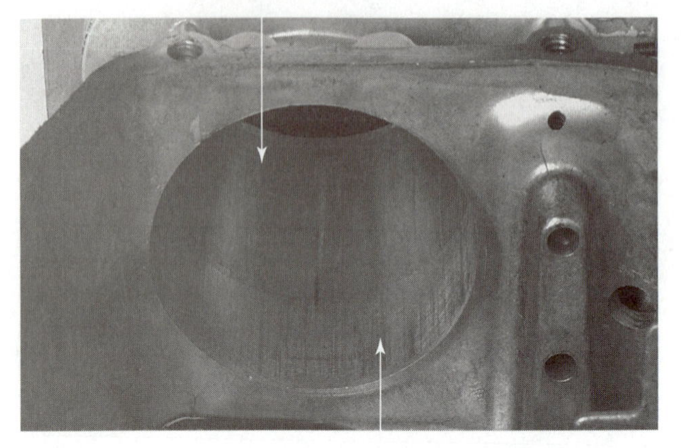

FIGURE 28-1 Inspecting the cylinder for scoring.

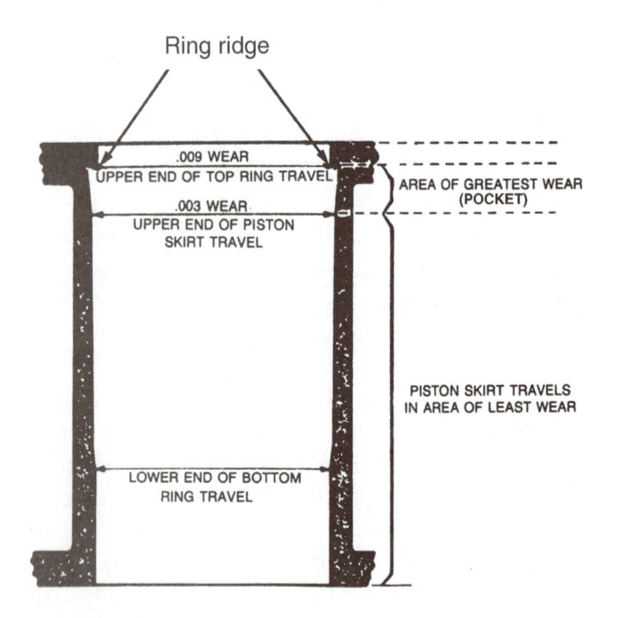

FIGURE 28-2 Common areas of cylinder wear. (Courtesy of Dana Corp.)

travel polishes the cast-iron or aluminum bore and makes it shiny. The next highest wear is just below ring travel. This is the area where just the piston skirt contacts the cylinder walls. This area is cooler and better lubricated. The friction from the piston skirt is less than that of the piston rings. The area of the cylinder below piston skirt travel is the unworn part of the cylinder. It is easy to see because it will look unpolished.

Measuring with a Telescoping Gauge

Cylinders can be measured with a telescoping gauge, cylinder gauge, or a feeler gauge. To use a telescoping gauge, select a gauge that will expand to the size of the cylinder. Insert the gauge into the bottom unworn part of the cylinder and expand it out to contact the cylinder walls. Gently rock the gauge back and forth (Figure 28-3) and feel for a light drag. If the gauge sticks against the walls, it is adjusted too large. If it does not have any drag, it is set too small. When properly adjusted, tighten the lock on the handle. Remove the gauge from the cylinder. Use the correct size outside micrometer to measure the gauge. Write down your measurement.

The unworn cylinder measurement shows the unworn cylinder size. This measurement is used as a reference for looking up specifications and wear limits in the shop service manual. For example, the bore measures 3.350 inch in the unworn area. Look for engine bore specifications with this size in the shop service manual. You may find the size of your measurement is larger than specifications. This can mean the cylinder has been machined oversize.

Cylinder wear is determined by making measurements in the three different cylinder wear areas (Figure 28-4). Make two measurements in each of the wear areas. Measure at the bottom of piston skirt travel, center of ring travel, and top of ring travel. Be sure to make your top measurement below the cylinder ridge. Make one measurement in the crankshaft direction. Make the second meas-

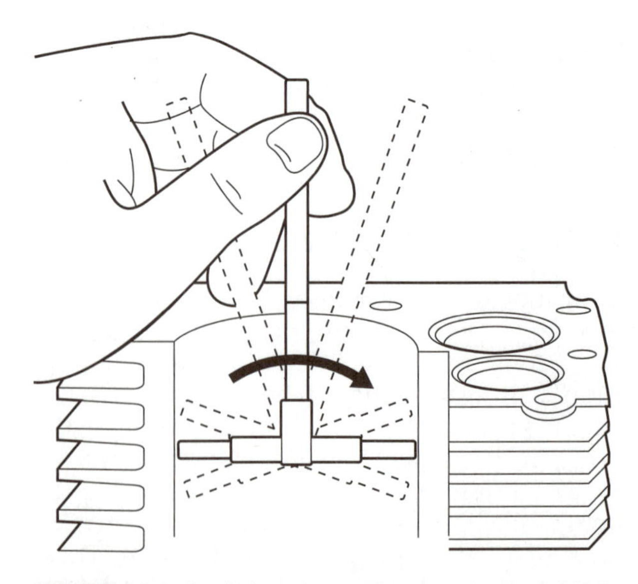

FIGURE 28-3 Rock the telescoping gauge back and forth for an accurate measurement.

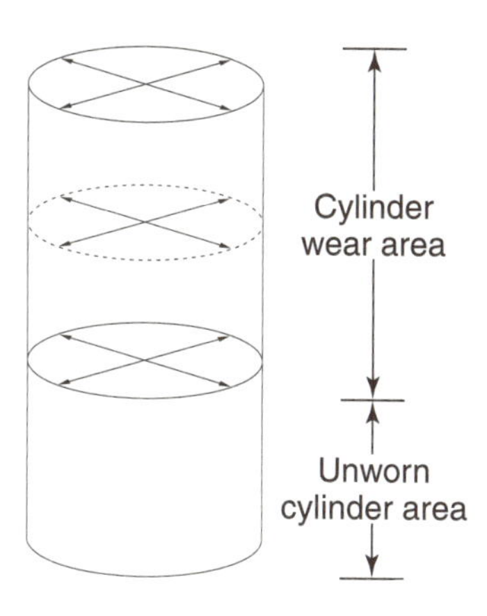

A. Measure at Three Levels

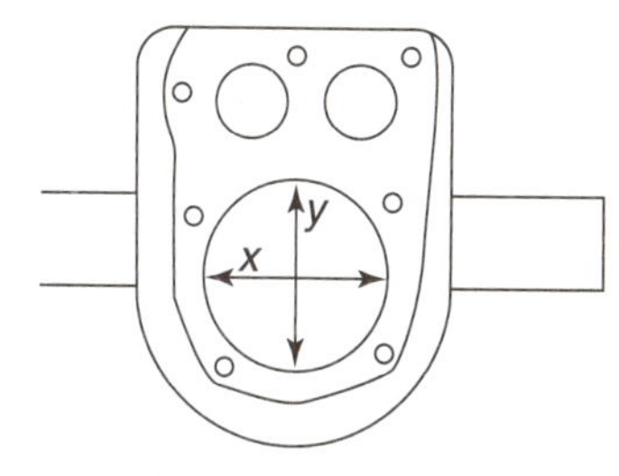

B. Measure in Two Directions

FIGURE 28-4 Measure the cylinder at three levels and in two directions.

urement at 90 degrees to crankshaft direction. Write down each of your measurements.

Cylinder taper is the difference between the size of the cylinder at the top of ring travel and the size at the bottom of piston skirt travel. For example:

Largest measurement at top of ring travel = 3.355 inches
Smallest measurement at bottom of ring travel = 3.350 inches
Cylinder taper = 0.005 inch

The largest difference between any two measurements taken in crankshaft direction and 90 degrees to crankshaft direction is the amount of out-of-round. **Cylinder out-of-round** is egg-shaped wear that occurs as the crankshaft rotates and thrusts the pistons at the sides of the cylinder. For example:

Largest measurement opposite to crankshaft direction = 3.355 inches
Smallest measurement in crankshaft direction = 3.353 inches
Cylinder out-of-round = 0.002 inch

Using a telescoping gauge to measure a four-stroke cylinder is shown in the accompanying sequence of photographs.

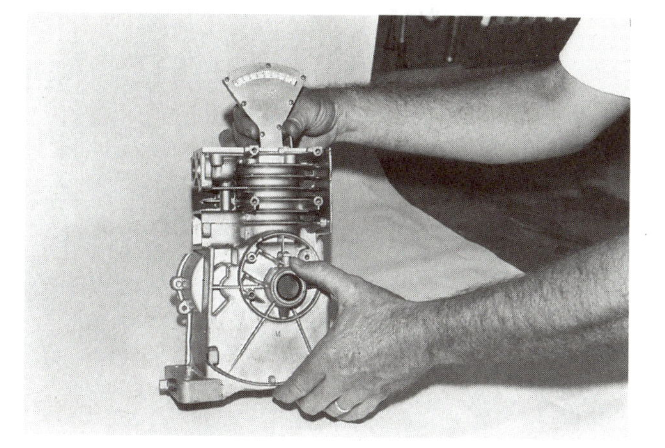

FIGURE 28-5 Measuring the cylinder with a cylinder gauge.

Measuring with a Cylinder Gauge

A **cylinder gauge** (Figure 28-5) is a special measuring tool that shows cylinder taper and out-of-round on the face of a dial indicator in thousandths of an inch or hundredths of a millimeter. The cylinder gauge has a set of precision-machined rails that fit against the cylinder wall. The dial indicator body is made to move in and out in relation to the rails. An adjustable plunger is attached to the dial indicator. It pushes out against the cylinder.

Position the cylinder gauge in the unworn bottom of the cylinder. Make sure the rails are in firm contact with the cylinder wall. Make the plunger longer or shorter to contact the cylinder wall. Preload the dial indicator. Rotate the dial face on the dial indicator to zero. Move the gauge up, down, and around the cylinder. The dial indicator body and plunger will move in and out relative to the rails and register variations in the cylinder size. Move the gauge up the cylinder and note the largest taper reading. Move the gauge from crankshaft direction to 90 degrees from crankshaft direction and note out-of-round.

Measuring with a Feeler Gauge

Cylinder taper may be found with a new piston ring and a feeler gauge (Figure 28-6). This method is not as accurate as precision measuring. It will give a general indication of cylinder wear. Use a new piston ring from the engine to be measured. Place the piston ring in the cylinder. Use the head of the piston to push the ring down into the cylinder about an inch. Make sure the ring is square in the cylinder. Measure the space between the two ends of the ring with a feeler gauge. Push the piston ring down to several other locations and measure the gap at each location. Make your final

PHOTO SEQUENCE 18

Using a Telescoping Gauge to Measure a Four-Stroke Cylinder for Wear

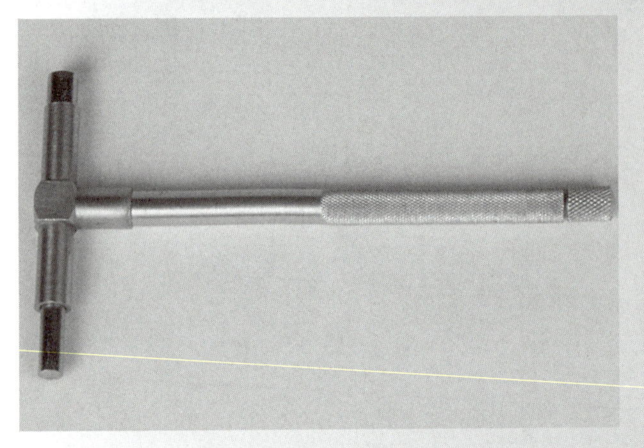

1. Select a telescoping gauge that will expand to the size of the cylinder.

2. Insert the gauge into the bottom unworn part of the cylinder and expand it out to contact the cylinder walls. The unworn cylinder measurement shows the unworn cylinder size.

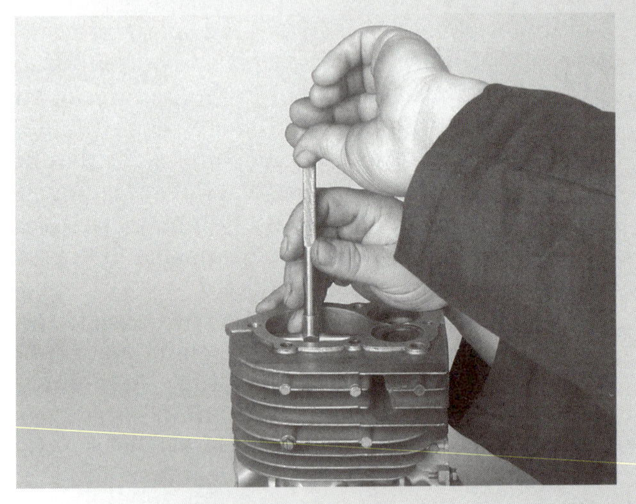

3. Gently rock the gauge back and forth and feel for a light drag. When properly adjusted, tighten the lock on the handle.

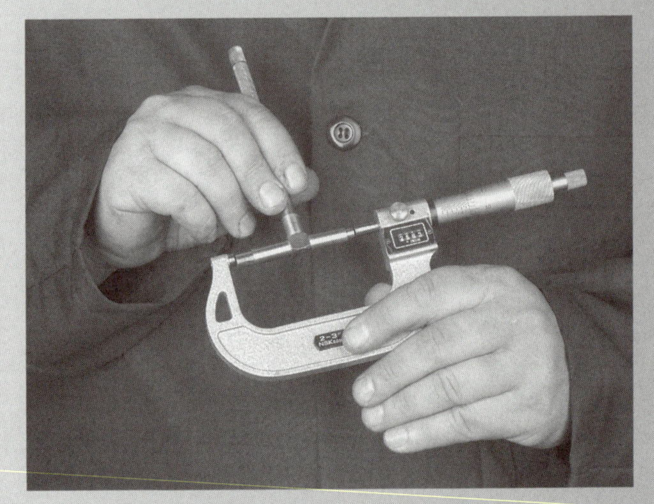

4. Remove the gauge from the cylinder. Use the correct size outside micrometer to measure the gauge. Write down your measurement. This is the unworn cylinder size.

5. Make measurements in the crankshaft direction at the bottom of piston skirt travel, center of ring travel, and top of ring travel. Write down each measurement.

6. Make the same three measurements again, this time at 90 degrees to crankshaft direction. Write down each of your measurements. Differences between top and bottom measurements is taper. Differences between crankshaft direction and 90 degrees to crankshaft direction is out-of-round.

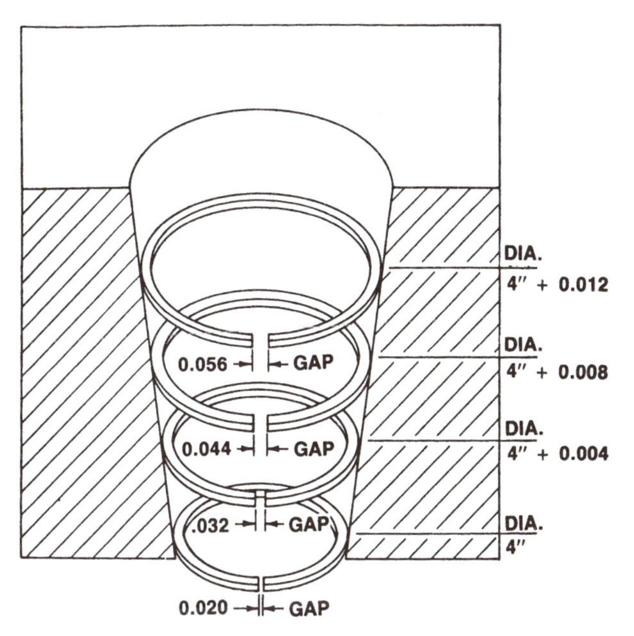

FIGURE 28-6 Measuring a cylinder taper by measuring piston ring gap in different parts of the cylinder. (Courtesy of Hastings Mfg. Co.)

measurement at the bottom of piston ring travel. The difference between your largest and smallest gap is taper.

Interpreting Measurements

Cylinder taper and out-of-round measurements are used to determine the next step in cylinder service. Check your measurements with service limits and reject specifications in the shop service manual. The cylinder can usually be prepared for new piston rings if taper is less than 0.003 inch (0.08 millimeter) and out-of-round is less than 0.002 inch (0.05 millimeter). Sometimes larger wear specifications are allowed for aluminum than for cast-iron cylinders.

New piston rings cannot be installed in a cylinder with taper and out-of-round beyond specifications. New piston rings will not seal the worn oversize cylinder. The engine will begin to burn oil and lose power after very few hours. There are two choices for these cylinders. The engine can be scrapped or the cylinder can be resized. **Cylinder resizing** is the machining of a worn cylinder oversize for larger pistons. The decision to scrap or resize will be determined by factors such as:

- Engine maker recommendations
- Replacement engine cost
- Replacement equipment cost
- Oversize piston availability
- Resizing equipment availability

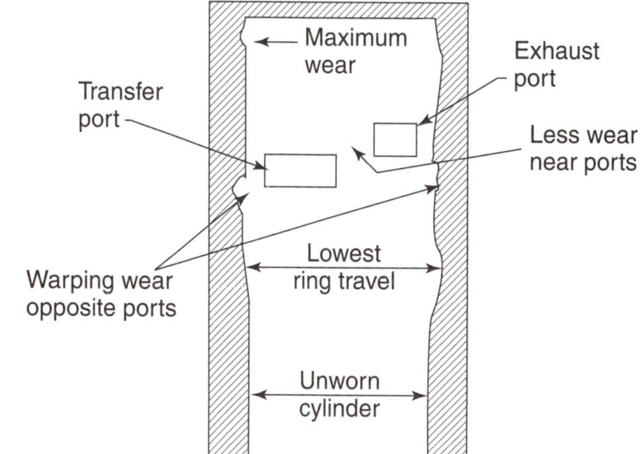

FIGURE 28-7 Common wear areas in a two-stroke cylinder.

TWO-STROKE CYLINDER INSPECTION AND MEASUREMENT

Two-stroke cylinders are inspected and measured in the same way as the four-stroke cylinder. Taper and out-of-round are measured and determined just as in the four-stroke cylinder. The two-stroke cylinder has transfer and exhaust ports in the cylinder that may cause different wear patterns (Figure 28-7). In most two-stroke engines, maximum wear usually occurs in the upper area of the cylinder wall. The wear is caused by heat and lubrication problems in this area. There is normally less wear in the areas near the transfer and exhaust ports. The cylinder is often more rigid in these areas.

Warping is a common two-stroke cylinder problem. Warping causes more wear in one part of the cylinder. This is often from the heat at the exhaust ports or differences in cylinder rigidity. Measure the cylinder at the top of ring travel, just below the port area, and below ring travel. Measure in crankshaft direction and 90 degrees to crankshaft direction. If the measurement just below the ports is larger than either the top or bottom measurement, there is warping. Check the amount of warping against specifications in the shop service manual. If warping is excessive, the cylinder cannot be reused.

RIDGE REMOVING

If the cylinder measurements show the cylinder can be reused, check the cylinder ring ridge. The **ring ridge** is an unworn ridge at the top of the cylinder above piston ring travel. The ring ridge is caused by unworn metal and also by carbon buildup at the top of the cylinder. The old top compression ring does not come in contact with the ridge. The ring wears at the same rate as the ridge area. The corner of a new top compression ring will strike the rounded lower surface of a cylinder ridge. This causes a clicking noise when the engine is running. Eventually the contact will damage the top compression ring and bend the second piston ring land.

Most cylinder ridges can be removed with abrasive paper. Sand the ridge gently with 300 grit abrasive paper. Try to blend the ridge area into the rest of the cylinder. If the ridge is primarily made up of carbon, this procedure is all that will be required.

CAUTION: Using a ridge-removing tool incorrectly can cause damage to the upper end of the cylinder. Practice using the tool on scrap cylinders before attempting it on a good engine. Aluminum cylinders can be damaged rapidly with a ridge-removing tool. It is possible to cut so deeply into the cylinder wall or so far down into ring travel that the engine will have to be scrapped.

The cylinder ridges can be removed with a ridge-removing tool. A **ridge remover** (Figure 28-8) is a cutting tool rotated in the top of the cylinder to remove the cylinder ring ridge. The tool is sometimes called a *ridge reamer* or *ridge cutter*. Follow the instructions supplied with the ridge remover. The ridge remover has a set of expandable supports. Use the adjuster at the top of the tool to adjust the supports out in contact with the cylinder. This centers the tool in the cylinder. The tool has flats on the top for driving with a wrench. Install the correct size wrench on the tool-driving flats. Rotate the tool slowly clockwise. The tool cutter blade will cut and remove the ring ridge. Be careful not to cut down into the ring travel area of the cylinder. Stop cutting when the ridge area bends with the lower part of the cylinder. Stop and check your progress every one or two revolutions of the tool (Figure 28-9).

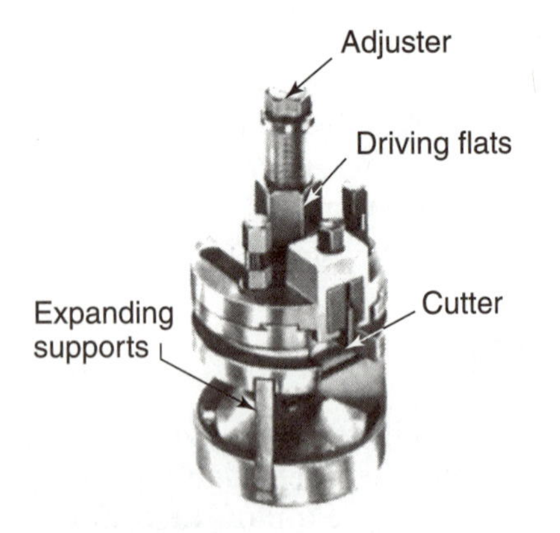

FIGURE 28-8 A ring ridge remover tool. (Courtesy of K-line Industries, Inc.)

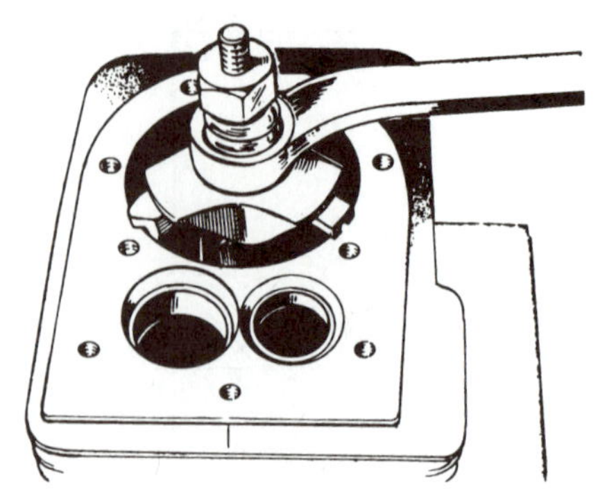

FIGURE 28-9 Removing a ring ridge. (Courtesy of Onan Corp.)

GLAZE BREAKING

The next step in preparing the cylinder is to remove (break) the glaze on the cylinder walls. The movement of the piston rings up and down in the cylinder polishes the cylinder surface. **Glaze breaking** is a procedure used to remove the polish on cylinder walls to improve ring seating. New piston rings have to wear in to the irregularities of the cylinder wall. Until this happens they do not seal properly. **Ring seating** is initial wear that occurs to the face of new piston rings that allows them to conform to irregularities in the cylinder wall surface. A polished cylinder wall can prevent proper ring seating.

Glaze breaking restores the 45 degree crosshatch pattern that was present on the cylinder when the engine was new. The crosshatch pattern is very important (Figure 28-10). Engine lubricat-

FIGURE 28-10 A cylinder requires a crosshatch pattern for proper piston ring seating. (Courtesy of Sealed Power.)

ing oil clings to the small grooves created by the crosshatching. This oil lubricates the piston rings and prevents them from seizing in the cylinder. If the crosshatch pattern is too smooth, the rings will have a hard time seating. The rings need a little cylinder wall roughness to seat properly. If the crosshatch is too rough and too deep, the rings will wear out too fast.

CAUTION: Some piston rings have a coating of molybdenum sprayed on the ring surface. The coating helps prevent scuffing and wear. These rings are called moly rings. In some cases the piston ring maker may specify that the cylinder not be deglazed for "moly" rings. The rough cylinder finish may destroy the coating. Always check the piston ring installation instructions and shop service manual before glaze breaking.

CAUTION: Glaze breaking is not recommended with some aluminum cylinder walls. Always check with the shop service manual for glaze-breaking instructions.

WARNING: Wear face and eye protection when glaze breaking to protect against flying broken stones. Always secure the cylinder assembly firmly during glaze breaking to prevent injury from flying parts.

A **glaze breaker** (Figure 28-11) is a tool with abrasive stones used to remove cylinder wall glaze and restore a crosshatch pattern. A glaze breaker uses spring-loaded 220 grit abrasive balls. **Grit** is the common name for the grain size of an abrasive that may vary from 10 (very coarse) to 600 (very fine). A driver on the end of the glaze breaker is chucked in a slow-moving ½ inch portable drill. The rotational speed should be 300 to 500 rpm.

Clamp or bolt the cylinder to an engine base or bench. Chuck the glaze breaker in the drill chuck. Start the drill and stroke the glaze breaker up and down slowly as it rotates (Figure 28-12). The rotating and stroking will put the proper crosshatch pattern on the cylinder wall. Stop as soon as you see the crosshatch pattern throughout the length of the cylinder. Excessive glaze breaking can make the cylinder oversize. It may not be possible to get a crosshatch pattern in the top cylinder wear pocket.

FIGURE 28-11 A glaze breaker with abrasive balls.

FIGURE 28-12 Breaking the glaze on a cylinder with a glaze breaker.

After glaze breaking, use soap, warm water, and a stiff nonmetallic bristle brush to scrub the cylinders (Figure 28-13). This is the only way to get stone abrasive grit out of the cylinder wall pores after glaze breaking. Soap surrounds the grit and allows it to float out with water. Do not use cleaning solvent for this job. Solvents force the grit back into the pores. Grit works like a grinding compound and wears out the new rings. Wipe the cylinder with clean white paper towels until you can no longer see dirt on the towel. After cleaning cast-iron cylinders, wipe them with an oiled rag to prevent rusting. Using a glaze breaker to remove the cylinder glaze is shown in the accompanying sequence of photographs.

SERVICE TIP: The spring-loaded glaze breaker stones follow the taper or out-of-round contour of the cylinder. This tool cannot be used to correct out-of-round or taper problems.

FIGURE 28-13 Washing the cylinder with brush and soapy water.

CYLINDER RESIZING

A cylinder that has taper or out-of-round measurements that are beyond specifications can be resized. Honing and boring are two common resizing procedures that remove metal from the cylinder to make it oversize. Cylinders are resized for larger (oversize) pistons. Oversize pistons should be selected and be on hand before resizing the cylinder. Some engine makers do not recommend resizing and do not supply oversize pistons.

Some engine makers supply oversize pistons that are larger than standard in steps of 0.010 inch (0.25 millimeter), 0.020 inch (0.59 millimeter), 0.030 inch (0.76 millimeter) and even larger. Piston oversize is determined by cylinder taper. When the cylinder is resized, more metal is removed from the top of a tapered cylinder than the bottom. There must be enough metal in the tapered bottom of the cylinder to allow the resizing. There must be at least a 0.005 inch (0.13 millimeter) difference between the taper size and the oversize piston. For example, the cylinder taper is under 0.005 inch (0.13 millimeter). There is enough material in the tapered part of the cylinder to resize for a 0.010 inch (0.25 millimeter) oversize piston. If the taper were 0.008 inch (0.20 millimeter), there would not be enough material in the cylinder to machine it to 0.010 inch (0.25 millimeter). This cylinder would have to be resized to 0.020 inch (0.51 millimeter).

Honing

WARNING: Be sure that the stones do not come out of the top of the cylinder under power. The stones are not locked into the holder and will fly out, possibly causing injury. Always wear eye and face protection while honing.

CAUTION: Some manufacturers do not recommend honing aluminum cylinders. Check the shop service manual.

Honing is a procedure used to machine worn cylinders oversize with abrasive stones. Honing removes metal from the cylinder to correct taper and out-of-round. A **hone** (Figure 28-14) is a tool that holds and drives abrasive stones used to machine a cylinder oversize. The hone has a rigid holder for the abrasive stones and aluminum guide shoes. It is different from the glaze breaker that has spring-loaded stones. A control knob on the hone expands the stones out against the cylinder wall. This controls the depth of cut. A driver on the hone is chucked in a drill press to rotate the holder.

Engine makers sometimes specify the honed cylinder wall be finished to a specific microfinish number. A **microfinish number** is a measure of the roughness or smoothness of cylinder wall finish. The lower the finish number, the smoother the finish. The microfinish of a honed cylinder is determined by stone grit (Figure 28-15). There are several different grits of honing stones available, ranging from very rough, 70 grit, to very smooth, 600 grit. The piston ring or engine maker usually specifies what finish is best for a piston ring set. A 70-grit stone provides a microfinish between 85 and 105. A 600-grit stone provides a 3 to 5 microfinish. A great deal of honing is necessary to resize a cylinder from standard to 0.020 inch (0.59 millimeter) or 0.030 (0.76 millimeter). Use a rough stone first and finish the cylinder with a smooth stone.

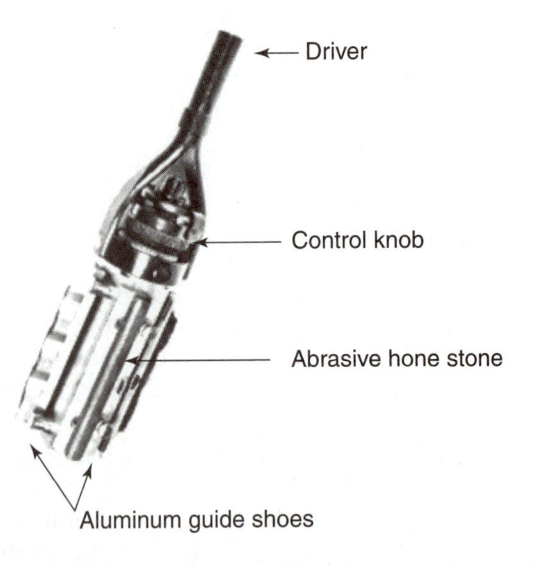

FIGURE 28-14 A hone with abrasive stones is used to remove metal to resize a cylinder. (Courtesy of Sunnen Products Company, St. Louis, Missouri.)

Using a Glaze Breaker to Remove Cylinder Glaze

1. Clamp or bolt the cylinder to an engine base or bench.

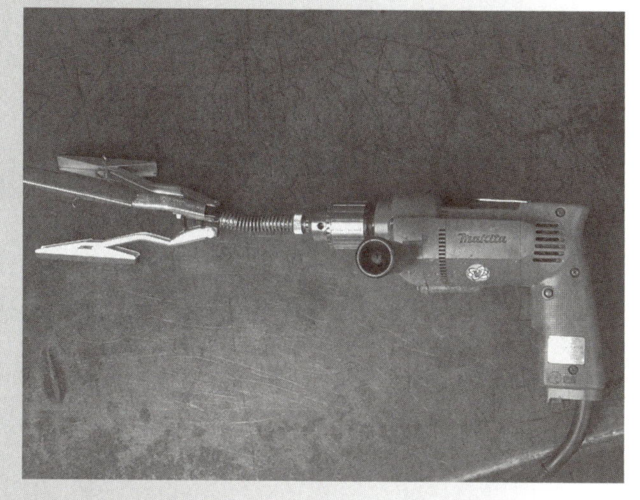

2. Chuck the glaze breaker in the drill chuck.

3. Insert the glaze breaker into the cylinder. Start the drill and stroke the glaze breaker up and down slowly as it rotates.

4. Stop as soon as you see the crosshatch pattern throughout the length of the cylinder. It may not be possible to get a crosshatch pattern in the top cylinder wear pocket.

5. Use soap, warm water, and a stiff nonmetallic bristle brush to scrub the cylinder and remove the metal and abrasive particles.

6. Pour a small amount of clean oil on a clean shop rag and wipe the cylinder to prevent rusting.

	Grit	Approximate Micro inch finish on cast iron
Roughing	70	100
Coarse Finishing	150	32
Medium Finishing	220	20
Polishing	280	12
Fine	400	6
Extra Fine	600	3

FIGURE 28-15 Honing stones are available in different grits for different microfinishes. (Courtesy of Sunnen Products Company, St. Louis, Missouri.)

Select the recommended grit-honing stones and install them into the holder. Some hones use four stones. Some use two stones and two aluminum or felt wipers between the honing stones for support. Others use four honing stones. Mount the cylinder block assembly to a large steel plate with a clamp (Figure 28-16). Be sure the top of the cylinder is absolutely level. Position the plate on the drill press table (Figure 28-17). Install the driver in the drill press chuck. Insert the hone in the cylinder. Move the hone down to the unworn part of the cylinder. Adjust the control knob on the hone out against the cylinder to center the hone in the cylinder. At the same time, center the plate in relation to the drill press chuck. Use a clamp to secure the steel plate to the drill press table. Set the up-and-down stops on the drill press chuck. Set the stones to come up past the cylinder top approximately 1 inch. Set them to come down below the cylinder 1 inch.

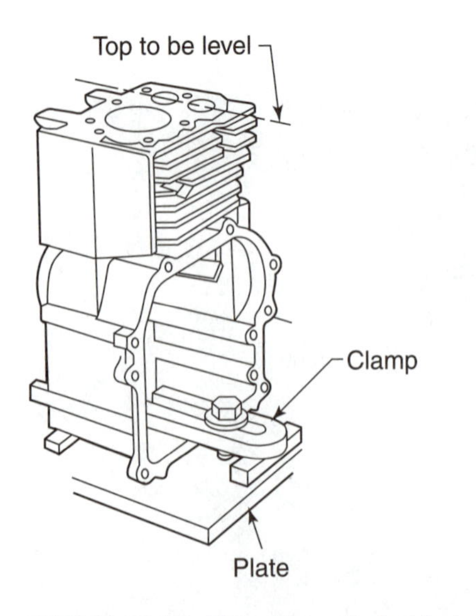

FIGURE 28-16 Mount the cylinder block to a plate for honing.

FIGURE 28-17 Honing the cylinder on a drill press.

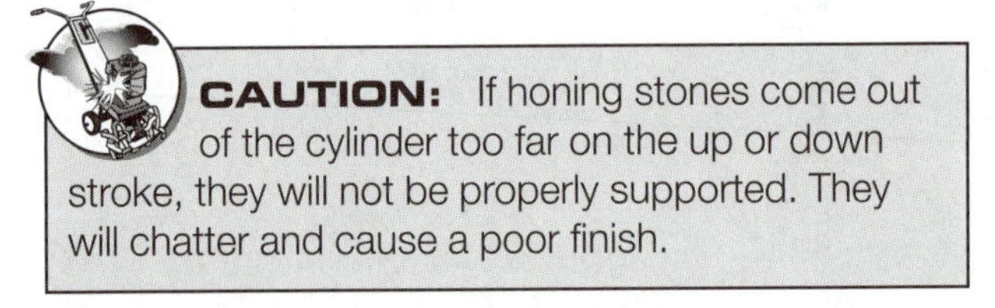

CAUTION: If honing stones come out of the cylinder too far on the up or down stroke, they will not be properly supported. They will chatter and cause a poor finish.

The speed at which the stones are rotated and stroked up and down affects the finish. Adjust the drill press to its slowest speed (approximately 300–700 rpm). Fill a squirt can with honing oil. Honing oil is squirted on the rotating stones during honing. Honing oil flushes the loose abrasive and metal particles from the stones and cylinder wall. This prevents stones from becoming loaded with particles. It also cools the cylinder walls and stones. Overheating causes a cylinder wall finish to have deep scratches or glazed areas.

Begin the honing in the bottom of the cylinder (Figure 28-18). Adjust a firm cutting pressure with the control knob on the honing fixture. Move the hone slowly up and down as it rotates for the prop-

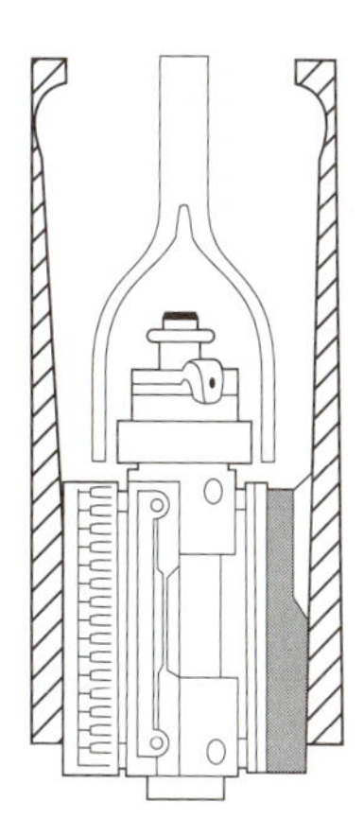

FIGURE 28-18 Start the honing in the unworn part of the cylinder.

er crosshatch pattern. Keep the stones in the tapered bottom of the cylinder in the beginning. As the cylinder is machined and the stones wear, expand the stones out gradually. Wipe and take measurements often to make sure the smallest amount of material is removed to straighten the cylinder. As soon as the bottom measurement is close to the top measurement, stroke the stones through the entire length of the bore. Stroke the hone up and down to produce a crosshatch pattern just like that for glaze breaking.

Have the new piston on hand before honing. It can be used as a go–no-go gauge. Turn the piston upside down and try to insert it in the top and bottom of the cylinder. When the bore is straight (no taper), the piston should fit the same in both ends of the bore. Hone the cylinder to the new piston size plus an additional amount for piston clearance.

Cleaning is very important after honing to remove abrasives and loose metal particles. Scrub the cylinder with a nonmetallic bristle brush and warm, soapy water. Flush with water to float out all the abrasives and metal from the metal pores. Wipe out the bore with paper towels until clean towels show no dirt. Coat cast-iron cylinder surfaces with engine oil to prevent rust.

Boring

Boring is a cylinder resizing operation in which the cylinder is machined oversize with a motor-driven cutter bit. Boring machines a perfectly straight new cylinder bore for oversize pistons. A **boring bar** (Figure 28-19) is a tool that drives a cutter bit through a cylinder to machine it oversize. A large boring bar is used to machine the cylinder when the engine is manufactured. Portable boring bars are used to resize worn cylinders. There are different types of portable boring bars each of which is mounted and operated differently. Each type has step-by-step instructions that must be followed.

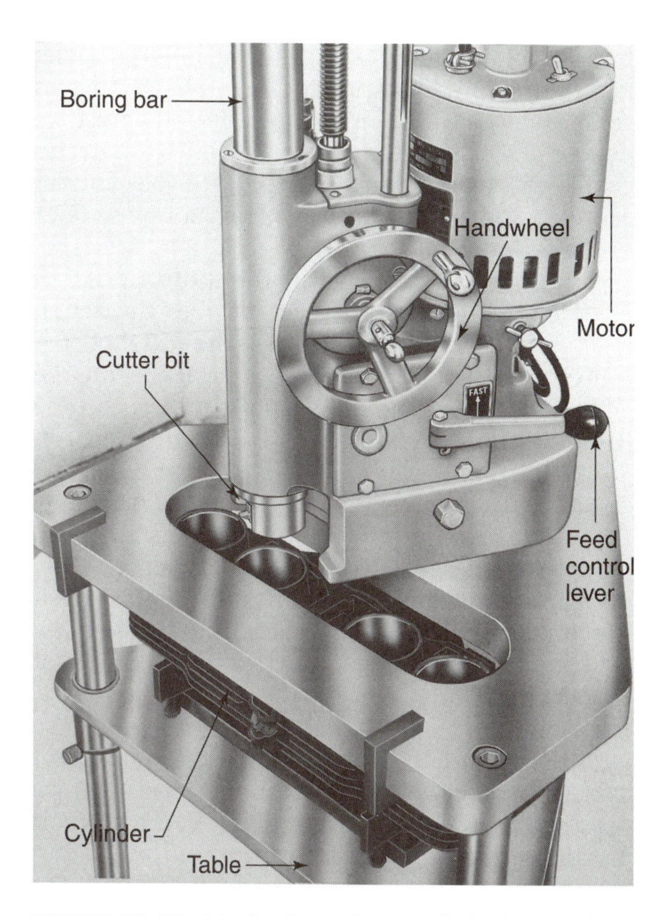

FIGURE 28-19 A boring bar resizes a cylinder by machining it oversize. (Courtesy of Kwik-Way Products.)

The block must be prepared properly for boring. Boring bars use the top of the cylinder to align the cylinder with the cutter bit. This area is called the *deck surface*. The deck surface must be clean and free of any nicks or burrs. If necessary, the surface may be cleaned by carefully filing with a smooth double cut file.

Position the block solidly on the boring bar table so it will not rock or vibrate during boring. Center the boring bar over the cylinder to be bored. Centering the bar is very important to ensure that the newly machined cylinder is in the same position as the old one. The boring bar usually has a handwheel. The handwheel allows the operator to feed the bar down into the cylinder with the power off. Most boring bars have three centering fingers or plungers on the end of the bar. Turning a control knob pushes these three fingers out of the bar. The fingers contact the cylinder wall at three places. The fingers are brought out tightly against the cylinder. In this position, they center the cylinder in the exact center of the boring bar. For normal boring, the centering is done in the unworn lower part of the cylinder to provide the most accuracy in centering. The area used for centering must be perfectly clean. After centering, the cylinder is clamped (anchored) in position. With the cylinder centered and anchored, the boring bar is raised out of the cylinder.

The next step is to prepare the cutter bit or tool bit (Figure 28-20). The cutter bit does the actual machining. Made from tool steel, the cutter bit has a tungsten carbide cutting edge attached on the end. The cutting tip requires sharpening each time it is used. The tool bit is sharpened on a rotating iron disc. The disc surface is covered with diamond dust. The disc usually is mounted on and driven by the boring bar motor. The cutter bit is installed in a fixture that attaches to a pilot shaft over the disc. The fixture holds the cutter bit in the correct position to sharpen the correct angles. The angles on the tip have to be sharpened to the correct shape for proper cutting.

After the tool bit is sharpened, it is ready to be installed in the boring bar. The tool bit has to be set to machine the cylinder to the desired size. The size to which the cylinder will be bored depends upon piston oversize. Choose an oversize piston that allows the cylinder to be completely machined. For example, if cylinder wear is in excess of 0.010 inch (0.25 millimeter), a 0.020 inch (0.59 millimeter) oversize piston should be used. Boring the cylinder 0.030 inch (0.76 millimeter) oversize when 0.010 (0.25 millimeter) would be enough to prevent ever boring again. Piston oversize is marked on the top of the piston. A 10 or 0 .010 on the top of the piston means it is 0.010 inch over standard (Figure 28-21).

FIGURE 28-20 The cutter bit is sharpened in a fixture to get the correct cutting angles.

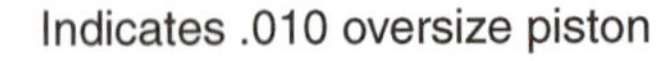

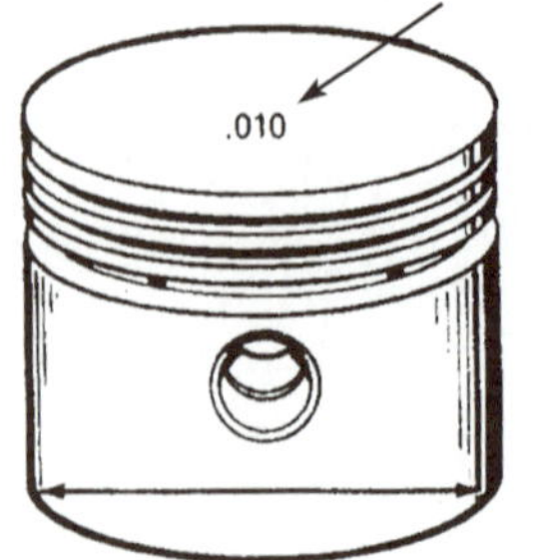

FIGURE 28-21 Piston oversize is marked on the top of the piston. (Provided courtesy of Tecumseh Products Company.)

Have the new piston on hand before boring. Measure the diameter of the new piston and set the boring bar to machine the cylinders at exactly this measurement. The size to which the cylinder is machined is determined by the position of the cutter bit in the boring bar. The position of the tool bit is set accurately with a special micrometer that is part of the boring bar tool kit. In many types of boring bars, the cutter bit fits in a holder that in turn fits into the boring bar. Some boring bars use a standard outside micrometer to set the cutter and holder to the desired size. The cutter bit is installed into the holder and the micrometer adjusted to the desired dimension of the cylinder. Adjust the tool bit and holder to this dimension. The holder is then mounted into the boring bar. Other types of boring bars use a special micrometer (Figure 28-22). It is part of the boring bar and is used to adjust the cutter bit position when it is in the boring bar.

Turn on the motor to rotate the bar and tool bit. Engage the feed control lever to cause the bar to move slowly down the cylinder. The rotating and advancing cutter bit machines the cylinder

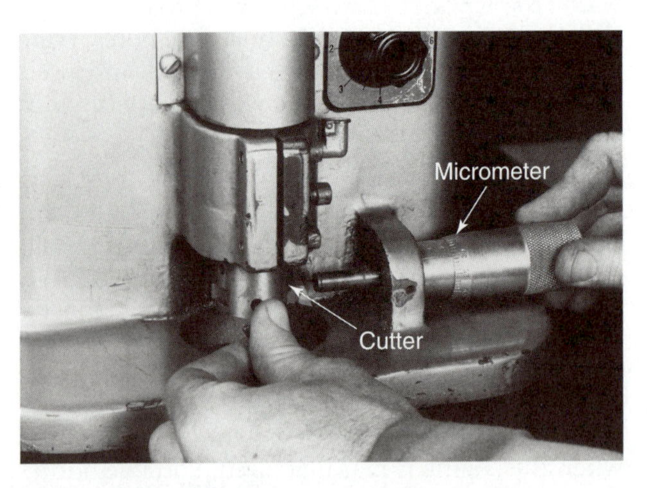

FIGURE 28-22 The size the cutter bit cuts is set by a micrometer on the boring bar.

(Figure 28-23). Motor speed and feed speed controls may be used to vary the quality of the finish. When the tool bit has advanced through the bottom of the cylinder, switch off the bar. Remove the tool bit from the bottom to prevent scratching the new cylinder as the bar is raised up the cylinder. A special long tool bit is then used to put a slight chamfer in the top of the cylinder. This helps get the piston and piston rings into the cylinder.

There are machining marks left by the tool bit going around and down the cylinder. If viewed with a microscope, these marks would look like a fastener thread. This is not a good finish for new piston rings. The bored cylinder is finished with a hone. Honing is the only way to develop a crosshatch pattern. The cylinder is honed with fine grit stones. Just enough material is removed to get the specified amount of clearance between the piston and cylinder wall.

A. Cylinder Being Bored

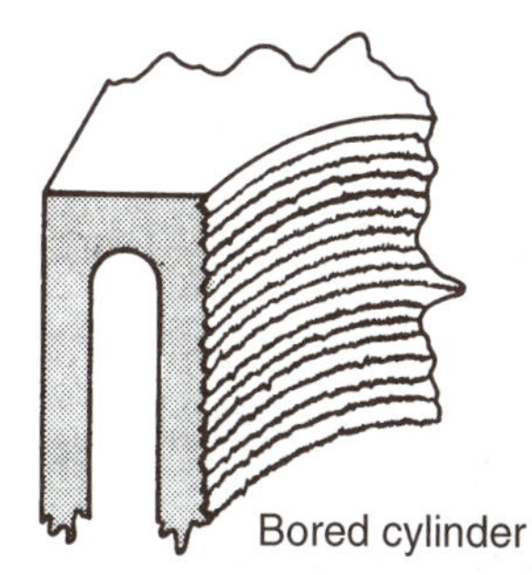

B. The Bored Cylinder Is Honed to Remove Boring Grooves (Threads)

FIGURE 28-23 Boring the cylinder leaves a threaded bore that must be honed. (*A:* Courtesy of Kwik-Way Products; *B:* Courtesy of CHAMPION-Federal Mogul.)

PISTON INSPECTION AND MEASUREMENT

A new piston is measured and fitted to the resized cylinder. If the cylinder is not resized, the original piston may be reused. Inspect and measure the piston to determine if it can be reused. Check the piston for scored or rough surfaces. Abrasives in intake air will cause scoring on the piston head and skirt area (Figure 28-24). Look for metal transfer and deep scoring on the skirts. These show the piston has seized against the cylinder wall. Carefully check for fractures on the skirt thrust surfaces. Cracks in this area show the piston has operated with too much piston clearance. Replace any piston with signs of scoring, seizing, or cracks.

Some two-stroke pistons have a knock pin in the ring grooves that prevents the piston rings from rotating around in the groove. This prevents the end of the ring from getting caught and broken on the edge of the exhaust or transfer port. Carefully inspect the knock pin. Replace any piston that has a loose or damaged knock pin.

Measuring the Piston Skirt

Pistons are measured to determine size for piston-to-cylinder clearance and wear. Select the correct size outside micrometer with the correct range. The piston looks perfectly round but it has several different sizes (Figure 28-25). The ring land area is smaller in diameter than the skirt area. The skirt area of some pistons is oval in shape. These pistons are larger across the skirt at 90 degrees to the piston pinhole. They are smaller across the skirt in the direction of the piston pin. This is done to control clearance. When the skirt is cold, it will contact the cylinder walls and prevent excessive piston clearance. As the piston warms up in the engine, it expands to a

FIGURE 28-24 A piston with scoring on the skirt.

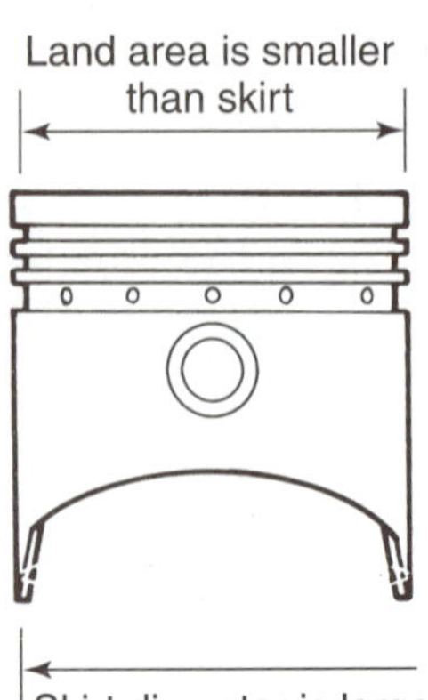

FIGURE 28-25 The piston skirt is larger in diameter than the ring land area. (Courtesy of Onan Corp.)

round shape, which allows it to maintain the proper piston to cylinder wall clearance.

Measure the piston across the skirt at 90 degrees from the piston pin (Figure 28-26). Place the micrometer over the piston. Make the measurement approximately ½ inch (12.70 millimeter) from the bottom of the skirt. Write down your measurement. Compare the piston size to the piston diameter specifications or reject size in the shop service manual. A piston that is more than 0.002 inch (0.05 millimeter) under specifications may be worn or damaged. Undersize pistons should be replaced.

Measuring the Top Piston Ring Groove

The top piston ring groove on a four-stroke engine has severe operating conditions. This ring operates in the hottest part of the cylinder. It gets very little lubrication. The top compression ring is the most important ring on the piston. It seals combustion pressure and also works like a final oil control ring. The ring face and sides have to form a good seal with the cylinder wall and with the sides of the ring groove. The seal depends on the sides of the piston ring grooves being flat, parallel, and smooth.

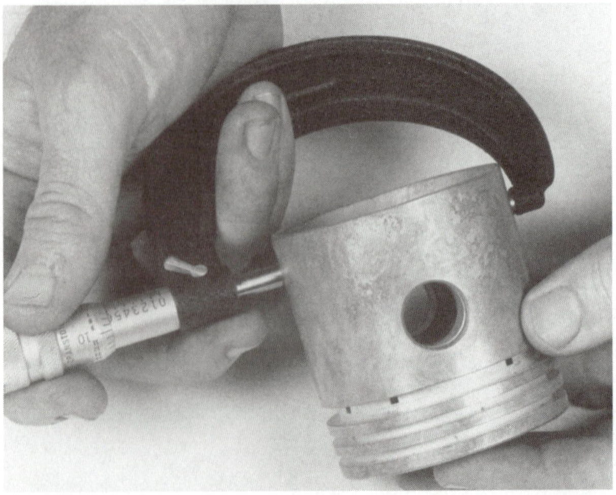

FIGURE 28-26 Measuring the piston across the skirt area.

The top piston ring and top groove become worn due to abrasives and high temperature in the top ring land area (Figure 28-27). Constant rocking of the ring back and forth in the groove also causes wear. A new piston ring should not be installed in a worn piston ring groove. The worn groove forces the upper outside edge of the ring face to contact the cylinder wall, which causes oil to be wiped up into the combustion chamber instead of down into the crankcase. In addition, the continued twisting of the new top ring in the worn groove can cause the ring to break.

Check for wear in the top ring groove. Select a new top piston ring from the ring set for the engine. The ring does not have to be installed on the piston to make this check. Place the ring into the ring groove. Check the shop service manual for the correct feeler gauge to use to check the ring groove. Specifications often recommend a feeler gauge size that should just fit in the groove. Some specifications give a feeler gauge that should not fit in the groove. Try to insert the feeler gauge between the ring and ring groove (Figure 28-28). The piston must be replaced if the specified feeler gauge shows excessive wear. Measuring pistons to determine wear and piston to cylinder clearance is shown in the accompanying sequence of photographs.

Two-stroke engines usually use two compression rings and no oil control ring. Both ring grooves are subject to wear. Measure the ring groove wear in each ring groove with the specified feeler gauge. Replace any piston with excessive ring groove wear.

FIGURE 28-27 A worn top piston ring groove will not support the new piston ring properly. (Courtesy of Dana Corp.)

Measuring Pistons to Determine Wear and Piston-to-Cylinder Clearance

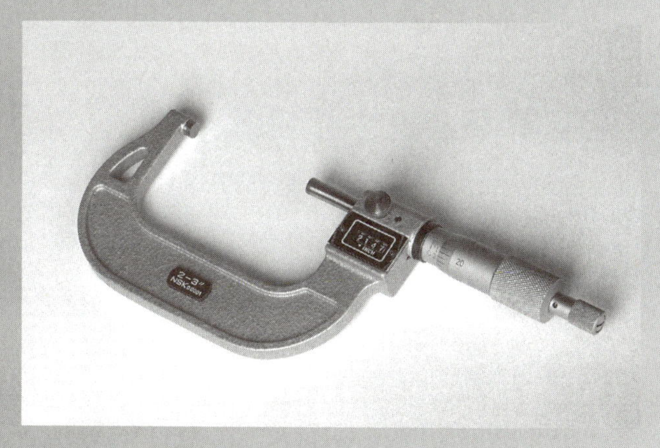

1. Select the correct size outside micrometer to measure the piston skirt.

2. Place the micrometer over the piston. Measure the piston across the skirt at 90 degrees from the piston pin. Make the measurement approximately ½ inch (12.70 millimeters) from the bottom of the skirt. Write down your measurement.

3. Compare the piston size to the piston diameter specifications or reject size in the shop service manual.

4. To check for wear in the top ring groove, select a new top piston ring from the ring set for the engine.

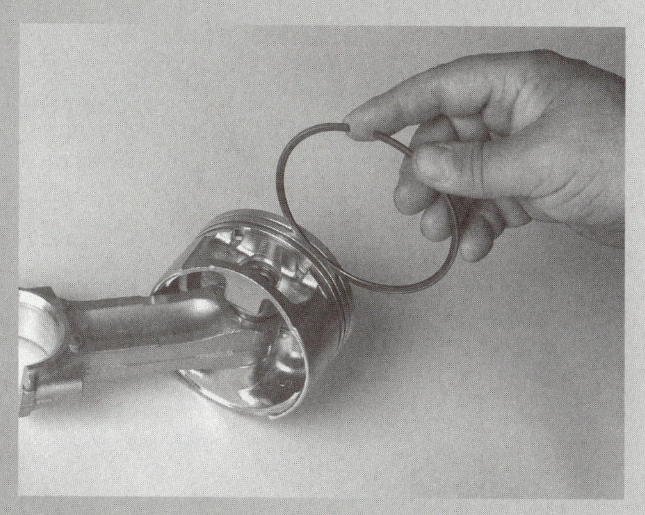

5. Place the ring into the top piston ring groove.

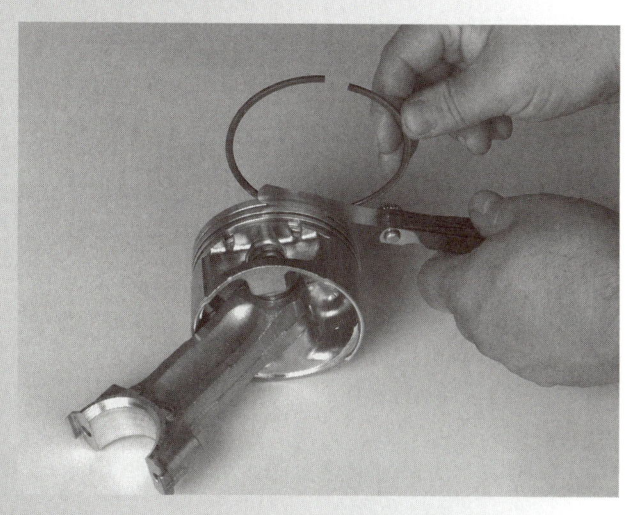

6. Use the feeler gauge specified in the shop service manual. Try to insert the feeler gauge between the ring and ring groove. The piston must be replaced if the specified feeler gauge shows that there is excessive wear.

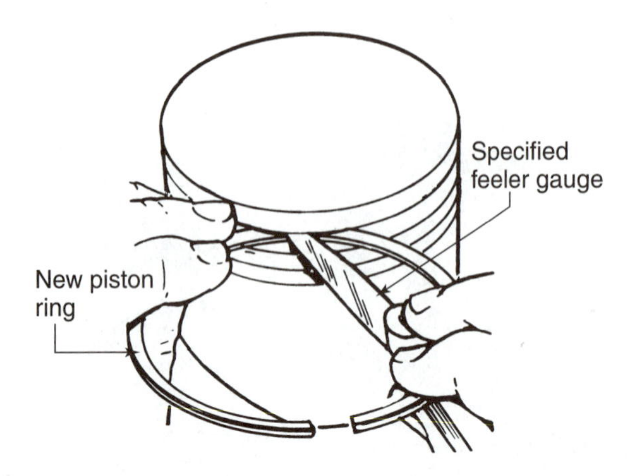

FIGURE 28-28 Measuring piston top ring groove wear. (Courtesy of Onan Corp.)

Checking Piston Pin Fit

The piston pin bosses in the piston form the bearing surfaces for the piston pin. The hardened steel piston pin can eventually cause the aluminum bearing surfaces in the bosses to wear oversize. Piston pin fit is the clearance between the piston pin and the pin-bearing surface in the piston. The piston pin clearance is one of the smallest in the engine. These clearances are difficult to measure accurately with a small hole gauge or telescoping gauge.

Some engine makers have a special dial indicator tool that is used to measure the piston pin bore (Figure 28-29). The piston pin is measured with an outside micrometer. The two measurements are compared to determine the clearance. If these tools are not available, you can check the clearance with the piston pin. Clean and lubricate the piston pin with engine oil. Insert the pin into one piston pin boss. Try to rock the pin back and forth as it is inserted. Any looseness indicates the hole is worn oversize. The pin should slide through the piston boss with a gentle push. If it falls through without any resistance, the piston and pin should be replaced.

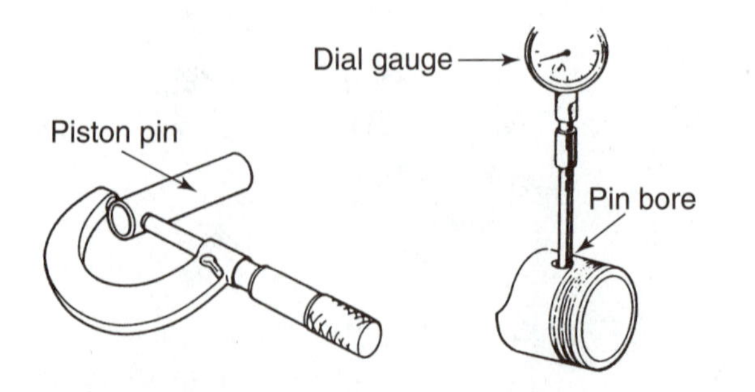

FIGURE 28-29 Measuring piston pin fit with a dial gauge and outside micrometer. (Courtesy of American Honda Motor Co., Inc.)

FITTING A PISTON

The piston-to-cylinder wall clearance is one of the most important measurements during an overhaul. **Piston clearance** is the clearance between the piston skirt thrust faces and the cylinder wall. Exactly the right amount of piston clearance is necessary. If the clearance is too small, there will be no space for a film of oil between the piston skirt and cylinder wall. Without an oil film, the piston will overheat, expand, and seize in the cylinder. Too much piston clearance can allow the piston to rock (slap) in the cylinder. The skirts of the piston may break if they slam against the cylinder walls.

Fitting pistons is the process of honing a resized (honed or bored) cylinder to establish the correct amount of piston clearance. The clearance between the cylinder and piston skirt is determined by comparing a measurement across the piston skirt with the measurement of the bore. The clearance is found by subtracting the piston measurement from the cylinder measurement. For example:

A piston measures 2.873 inches (72.98 millimeter) across the skirt thrust faces.
The unworn cylinder measurement is 2.881 inches (73.17 millimeter).
Piston clearance is 0.008 inch (0.19 millimeter).

Look up the manufacturer's specifications for piston clearance in the shop service manual. If the original piston is being reused, the piston clearance should be within specifications. If not, replace the piston or resize the cylinder. If the clearance is being adjusted at the end of honing or honing after boring, clean and check the cylinder measurements frequently. Compare the cylinder measurements with piston measurements. A clearance that is too small requires more honing to increase clearance. If you make an error and hone the clearance oversize, the cylinder will have to be resized to the next larger oversize.

SERVICE TIP: A piston that has the correct clearance in a cylinder will slowly slide down a clean, lubricated cylinder under its own weight. If it has to be pushed through, it is too tight. If it falls rapidly, it is too loose. Be sure to catch the piston at the end of its travel so that it is not damaged.

CASE STUDY

A student in an outdoor power equipment class had just finished assembling an engine he had overhauled. The engine had started up easily and sounded good. The student was allowing the engine to warm up to make final carburetor adjustments. All of a sudden the engine slowed down and then stopped. As the student pulled on the rope to restart, he felt a great deal of resistance.

Class was about over for the day so the student did not have a chance to investigate further. The next day he pulled the engine out of the storage locker and pulled on the rope. The engine rotated easy and started right away. The same thing happened as the previous day. The engine ran fine for awhile then stopped suddenly.

The student and the instructor began to investigate the problem. The instructor asked for the student's notes and measurements used when fitting the pistons. The engine had a great deal of cylinder wear. The cylinder was honed oversize for 0.020 inch oversize pistons. The student was not sure exactly how much material he had honed from the cylinder to set piston clearance. He did remember that the piston fit into the cylinder with a pretty tight push fit. After hearing this, the instructor was sure there was not enough piston clearance. When the piston got hot and expanded it was seizing up in the cylinder. The engine had to be disassembled and the cylinder honed for the correct clearance.

REVIEW QUESTIONS

1. Describe the areas of the four-stroke cylinder that experience the most wear.
2. Describe the areas of the two-stroke cylinder that experience the most wear.
3. Explain what creates the pocket of wear in the top inch of a cylinder.
4. Explain why the lowest part of a cylinder experiences the least wear.
5. Explain why warping is common in a two-stroke cylinder.
6. Describe how to determine the amount of taper in a cylinder using a telescoping gauge.
7. Describe how to determine the amount of taper and warpage in a cylinder using a cylinder gauge.
8. Describe how to determine the unworn bore size of a cylinder with a telescoping gauge.
9. Explain why a crosshatch pattern is desirable on a cylinder.
10. List and describe the steps to follow in glaze breaking a cylinder.
11. List and describe the steps to follow to resize a cylinder with a hone.
12. Explain why initial honing is done in the unworn part of the cylinder bore.
13. Explain how to use a piston as a go–no-go gauge when honing a cylinder oversize.
14. List and describe the steps to follow to resize a cylinder with a boring bar.
15. Describe how and where to measure a piston to determine its size.
16. Explain how to determine if the top ring groove of a piston is worn excessively.
17. Describe how to check a piston pin fit in the piston.
18. Describe how to use an outside micrometer and telescoping gauge to determine piston-to-cylinder clearance.
19. Describe how to determine piston fit in a newly resized cylinder.
20. Explain how to use a feeler gauge to measure piston ring end gap.

DISCUSSION TOPICS AND ACTIVITIES

1. Locate a scrap cylinder block. Practice removing the cylinder ridge with a ridge removing tool.
2. Locate several empty cylinder blocks. Use a telescoping gauge and outside micrometer to practice measuring cylinder taper and out-of-round.
3. An oversize piston is being fit to a cylinder with a hone. The piston measures 3.010 inch. The recommended piston clearance is 0.004 inch. What will be the finished size of the cylinder?

Engine Assembly

OBJECTIVES

Upon completion and review of this chapter, you should be able to:

- Explain the importance of a clean work area for engine assembly.
- Describe the purpose of and use of gaskets and seals.
- Explain the purpose and use of assembly lubricants and gasket sealants.
- Install the piston on the connecting rod.
- Install four-stroke piston rings.
- Install two-stroke piston rings.
- Install the camshaft and crankshaft.
- Install the four-stroke piston assembly.
- Install the four-stroke connecting rod cap.
- Install the two-stroke piston, connecting rod, and crankshaft.
- Install the crankcase cover.
- Install the cylinder head.
- Start, adjust, and break in an overhauled engine.

TERMS TO KNOW

Assembly lubricant
Break-in period
Connecting rod side clearance
Crankshaft oil clearance
Gasket
Gasket sealant
Lip seal
O ring seal
Oversize piston rings
Plastigage
Piston ring compressor
Piston ring end gap
Ring seating
Seal
Standard piston rings

INTRODUCTION

When all the engine machine work is completed and all the parts have been evaluated, you are ready to begin assembly. You need to make a list of the parts that require replacement. Use the engine model number to order the new replacement parts. When the parts arrive, you are ready to begin the step-by-step engine assembly.

ORDER NEW PARTS

New parts are usually purchased from an authorized dealer or parts distributor. When you make the order, you will need the engine identification information stamped on the blower housing or on the engine decal. The parts dealer needs the information on any oversized parts such as oversize valve stems or resized cylinders. The parts dealer uses this information to determine the part numbers for the parts you need.

Piston Ring Selection

You have several choices on the type of piston rings you get for the engine. All engine makers supply standard ring sets. **Standard piston rings** are piston rings made from the same material and the same size as the rings used in new engine. Standard piston rings are used in cylinders that have less than 0.003 inch (0.078 millimeter) taper and less than 0.002 inch (0.05 millimeter) out-of-round. Some engine makers supply a chrome-plated piston ring for cylinders that exceed these specifications. This ring set has a hard chrome plating on the top compression ring. Some ring sets also have a spring expander that fits behind the oil control ring. This type of ring set seals and controls oil better in a worn cylinder.

Resized cylinders require oversize piston rings. **Oversize piston rings** are piston rings made for common oversized pistons used in resized cylinders. These rings are usually available in 0.010 inch, (0.25 millimeter) 0.020 inch (0.59 millimeter), and 0.030 inch (0.76 millimeter) oversizes.

Gaskets

> **CAUTION:** Gasket sets commonly include gaskets for several different engine models. Always save old gaskets for comparison.

All the gaskets are replaced when an engine is overhauled. Gaskets for an engine may be purchased individually or in a set (Figure 29-1). A complete engine overhaul requires an overhaul gasket set, which includes all of the engine gaskets. A **gasket** is a soft material that, when squeezed between two parts, fills up small irregularities to make a pressure tight seal (Figure 29-2). Gaskets are used because there are microscopic irregularities on the part surfaces. The high and low spots on the parts would allow leakage without a gasket.

Gaskets used in areas of low temperature and pressure are usually made from soft materials such as paper and cork. Cork is easily formed into almost any shape and makes a good gasket. Some cork gaskets are coated with aluminum or rubber compounds to improve their heat resistance. Paper gaskets are commonly used on crankcase covers and cork is used on overhead valve covers.

The high temperatures and pressures in the combustion chamber make it the most difficult area to seal. Sandwich type cylinder head gaskets are usually used. These have two layers of thin soft metal such as aluminum, copper, or steel. They have a soft middle layer of heat resistant core material.

Gaskets should be handled with care to avoid distorting or bending. Store gaskets flat and keep them in cartons until ready for use. A distorted or bent gasket can fail. Never reuse an old gasket. The gasket material between two parts is permanently compressed. If it is reused, it will not be in its exact original position. The compressible material will not readjust itself to surface irregularities and the result will be a leak.

Cleaning the cylinder head, cylinder block, bolts, and bolt holes is very important to long head gasket life. Dirt particles on the sealing surfaces

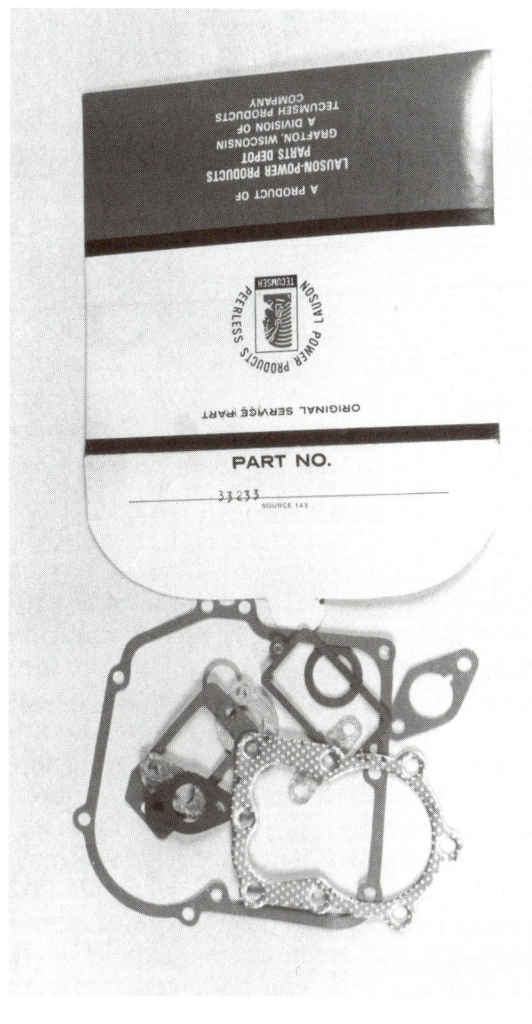

FIGURE 29-1 An engine overhaul gasket set contains all the gaskets needed to assemble the engine.

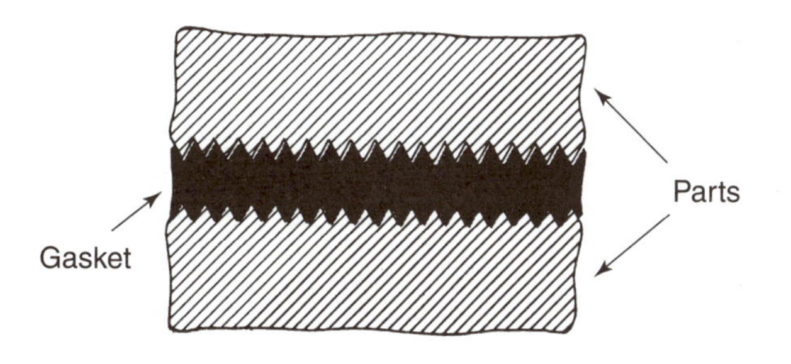

FIGURE 29-2 The gasket compresses into irregularities in the parts surfaces to form a seal.

form pockets in the gasket, which can allow leakage and gasket failure. A dirty bolt or bolt hole thread can cause a large torque-reading error. Bolt threads in the block frequently are overlooked. Inspect bolt holes carefully. Clean the threads with a tap if necessary.

CAUTION: Choose the correct gasket for the job you are doing and then check to see that it fits. Some gaskets may be labeled "This Side Up" or "Front." If so, follow the instructions on the gasket.

Seals

When parts are moving in relation to each other, they cannot be sealed with a gasket. A **seal** is a part used to prevent pressure or fluid loss around a rotating or sliding shaft. A seal is used around the crankshaft PTO and FWE journals. Seals are also used in oil pumps and on oil filters. They prevent the loss of oil from the crankcase. Engines used to power riding equipment may have other shafts that require seals. The O ring and lip are common types of seals.

An **O ring seal** is a solid rubber or synthetic ring used to seal areas of high pressure liquids. The O ring seal is very flexible. It can be stretched into position inside a groove in an engine part. These types of seals are often used on larger engine oil pumps.

A **lip seal** uses a wiper lip to prevent pressure oil leakage around a rotating or sliding shaft (Figure 29-3). Lip seals are often used to seal a lubricant inside a bearing area. They also work to keep dirt out of a bearing area. A seal often is used around the crankshaft PTO and FWE journal. The lip seal has an outer housing that supports a neoprene sealing lip backed up by a coil spring. The sealing is done by the sharp edge of the lip (Figure 29-4). The lip rides on the rotating shaft. The lip works like a squeegee. It wipes the oil from the shaft and prevents it from getting out of the crankcase.

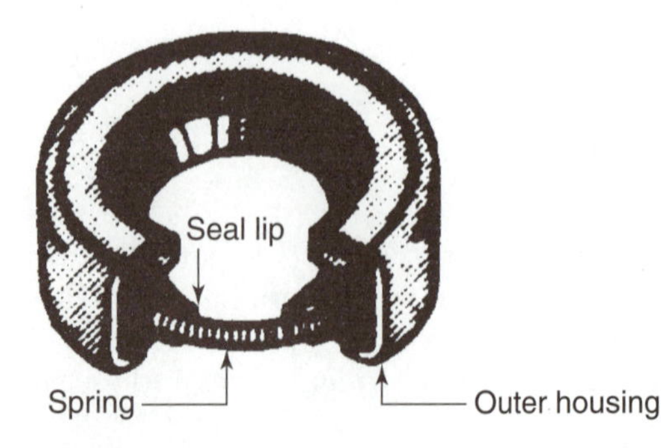

FIGURE 29-3 Lip type seals are used to seal around the rotating crankshaft. (Courtesy of Clinton Engines Corporation.)

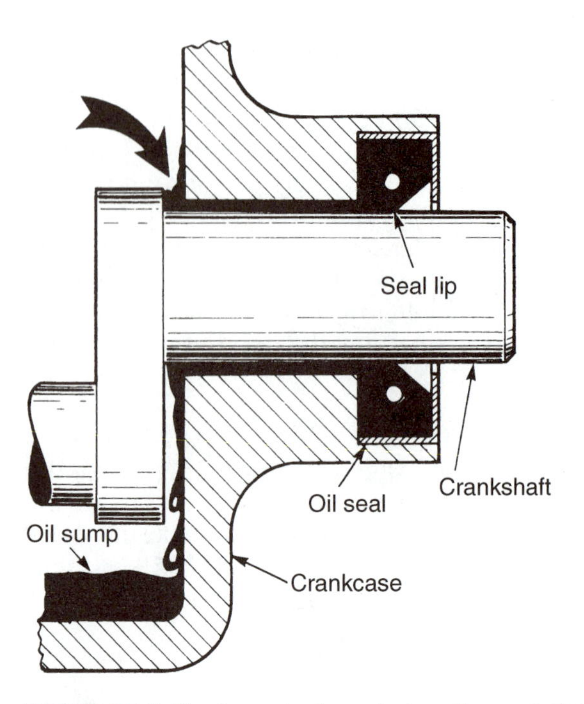

FIGURE 29-4 The lip on the lip seal wipes the crankshaft to prevent oil leakage.

Old seals should not be reused in an overhauled engine. The sealing material used for O rings and lip seals gets hard and brittle after many hours of operation. The lip on a lip type seal wears out from contact with the rotating crankshaft.

Assembly Lubricants and Gasket Sealants

Engine parts are lubricated during assembly to prevent wear during initial engine start up. Engine oil is used for lubricating parts as they are assembled. Sometimes assembly lubricant is used to lubricate parts as they are assembled. **Assembly lubricant** is a thick lubricant grease used to coat bearings on assembly. The assembly lubricant provides lubrication until the engine lubrication system is functioning. This lubricant is often used to hold needle bearings in place on two strokes. It is also used on four-stroke engine bearings if the rebuilt engine is not going to be started right away. Thin engine oil may run out of the bearing area during engine storage. Assembly lubricant is not used on piston rings.

A **gasket sealant** (Figure 29-5) is a liquid compound used to improve gasket sealing, hold a gasket in place, repair a damaged gasket, or form a new gasket. Gasket sealant flows into parts surface irregularities to provide a seal. Gasket sealants work

FIGURE 29-5 Gasket sealant is used to hold a gasket in place or to make a gasket. (Courtesy of Fel-Pro, Inc.)

at different temperature ranges. The sealant must have a high enough temperature range for the area in which it is to be used. Sealants differ in ability to resist different chemicals. For example, some sealants may decompose when used with gasoline, water, antifreeze, or oil. Always follow engine makers' recommendations for the type of sealant. Two-stroke crankcases are often sealed with silicone sealant. A silicone sealant requires that sealing surfaces be clean and dry.

Other Parts for the Parts List

You will want to get other parts for the engine. You should to do a complete ignition and fuel tune-up as a part of your engine overhaul. The customer expects the engine to run well from the beginning. The best way to make sure this happens is to replace any parts that may cause problems. At a minimum your parts list should include:

- New spark plug and any defective ignition parts
- Carburetor rebuild kit
- New fuel filter
- New air cleaner element
- Replacement starter parts
- Engine paint and replacement factory decals

CLEAN ASSEMBLY AREA

A clean work area is very important to good engine overhaul (Figure 29-6). You must be very careful to prevent dirt from entering the engine during assembly. Work carefully to ensure a clean engine. Do the assembly on a clean work surface. Clean your hands and tools before you begin. Make sure the assembly is done as far as possible away from any dirt or dust-producing shop work. Do not take parts out of their plastic bags until they are to be installed. Inspect each part for dirt before it is installed. As parts are installed, wipe them carefully with clean shop rags. Any time the engine is not being worked on, it should be covered. A plastic trash bag sealed closed with masking tape works well for this purpose.

FIGURE 29-6 A clean assembly area prevents dirt from getting into the engine.

INSTALLING VALVES

> **WARNING:** Always wear eye protection when installing valves to protect against eye injury from flying parts.

The valve and spring assembly is installed with the same valve spring compressor used for disassembly. Always reinstall the valves in the same positions from which they were removed. Lubricate the valve stem with engine oil and insert the valve into the valve guide. Install the valve spring over the valve stem and compress it with the compressing tool. Some valve springs have tighter coils at one end. The tight coils and these should be oriented toward the head of the valve. Reinstall the retainers and retaining washer in position on the valve stem. Slowly release and withdraw the compressing tool. Repeat this procedure for each valve.

INSTALLING PISTON ON CONNECTING ROD

Engine assembly begins with the installation of the piston on the connecting rod. The piston pin bosses on some four-stroke piston are offset. The offset is used to control the thrust forces on the piston

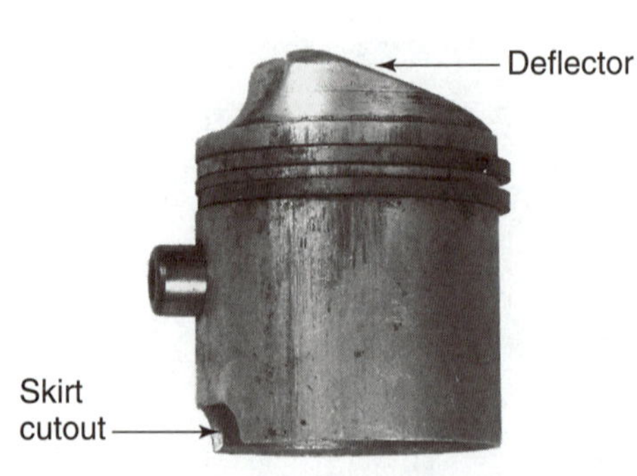

FIGURE 29-7 Two-stroke pistons with deflectors and skirt cutouts must be installed in the correct direction.

skirt. This requires that the piston be installed in the engine in the correct direction. Two-stroke pistons often have deflector tops or transfer passages machined in the skirt. These pistons must be installed in the correct direction or the engine will not run (Figure 29-7).

Connecting rods are made with oiling notches and/or oil dippers. These parts will lubricate properly in only one direction. The connecting rod must be installed on the piston and in the engine in the correct direction. Use the marks you made on the piston and rod during disassembly. Recheck these marks with orientation instructions in the shop service manual.

Different engines use different kinds of match marks (Figure 29-8). Match marks are used to get the piston and rod assembled in the correct position. They are also used to get the parts installed with the proper orientation in the engine. Pistons often have a direction arrow or notch. The arrow or notch must be oriented a certain way in the engine. On some engines, the piston arrows point toward the valves or the carburetor. On other engines, the notch or arrow points toward the flywheel. Some rods have the word "MAG" cast in the side of the rod that should face the magneto side of the engine. Connecting rod oil hole position and piston casting numbers are other ways used to show assembly direction.

CAUTION: There is no universal marking system for piston or rod orientation. Always check the appropriate shop service manual for this information.

WARNING: Wear eye protection when installing piston pin snap rings to prevent eye injury.

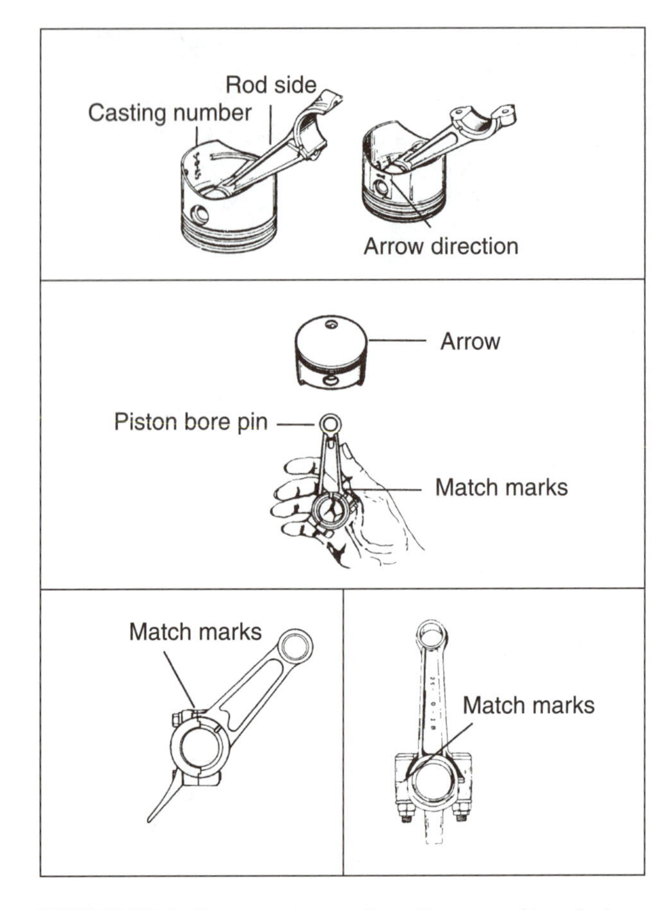

FIGURE 29-8 Common types of markings used to orient the connecting rod and piston in the correct direction. (Provided courtesy of Tecumseh Products Company.)

Be careful when installing the piston pin not to damage the piston or rod. Never lay the piston on a solid object when installing the piston pin. This can cause scratches in the aluminum piston skirt. To prevent damage, support the piston in the palm of your hand. In most engines, the pin may be installed in either direction. Some two-stroke engines have a hollow wrist pin that is closed on one end. Make sure the closed end is toward the exhaust side. Lubricate the pin with engine oil. Push the pin through the piston and rod by hand. If it sticks, heat the piston under hot water to make the hole larger and allow the pin to go through.

Use new piston pin retaining snap rings. The old ones often are stretched out of shape during disassembly. Distorted snap rings may work out of their grooves. If this happens, the piston pin can work out of the piston and contact the cylinder wall. The cylinder will be damaged. Set one end of the snap ring in the piston groove while holding the other end with long nose pliers (Figure 29-9). Rotate the lock into place.

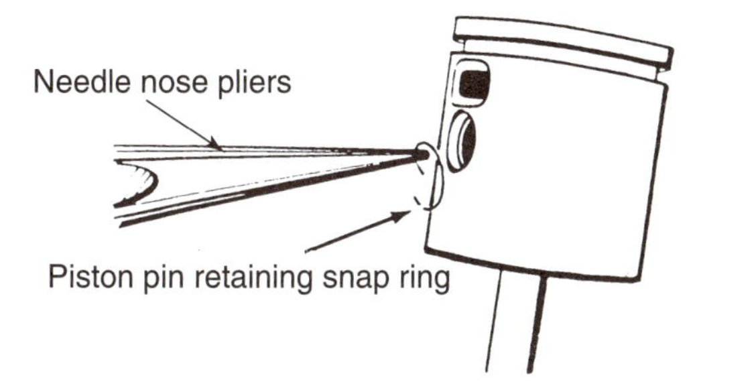

FIGURE 29-9 Installing the piston pins retaining snap rings. (Courtesy of Poulan/Weed Eater.)

INSTALLING FOUR-STROKE PISTON RINGS

Piston rings are always replaced during an engine overhaul. The piston rings contact surface wear from sliding against the cylinder wall. Piston rings lose their ability to seal compression and control oil. Scoring, seizing, and breaking are common piston ring problems in engines with many hours of service.

A new set of piston rings has each ring individually boxed. The rings for each piston ring groove are identified and instructions are provided for correct installation (Figure 29-10). Each new piston ring is measured for end gap before it is installed on a piston. **Piston ring end gap** is the space between the two ends of a piston ring when it is in the cylinder. Remove one of the compression rings from the package. Insert the piston ring into the cylinder. Use the head of the piston to push the piston ring to the cylinder area at the

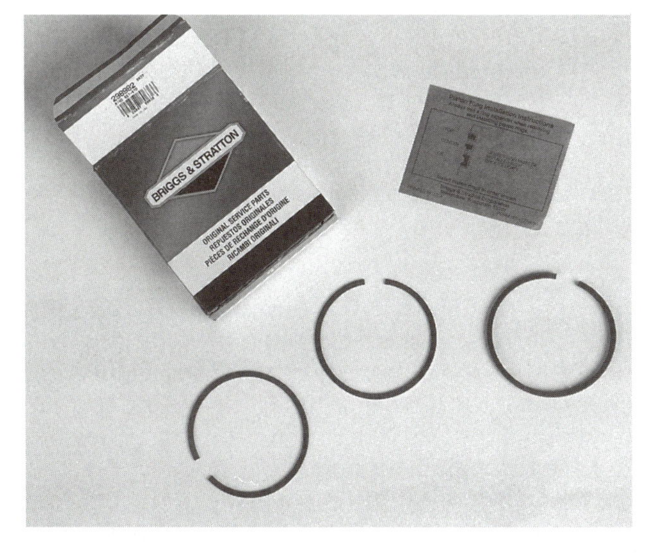

FIGURE 29-10 Piston rings usually come with installation instructions.

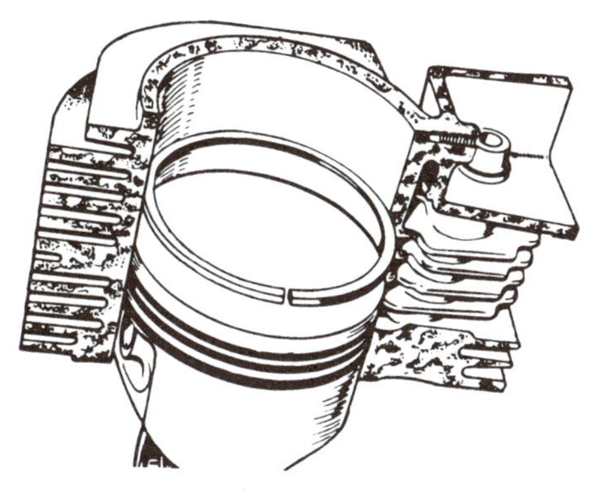

FIGURE 29-11 Use the piston head to position the piston ring at the bottom of ring travel. (Courtesy of Onan Corp.)

bottom of piston ring travel (Figure 29-11). Make sure it is square with the cylinder walls. Use a feeler gauge to measure and record the end gap (Figure 29-12). Repeat this procedure with the other compression ring. If the oil control ring is one piece, measure its end gap. If the oil control ring is not one piece, it cannot be measured for end gap. Record each of your measurements.

CAUTION: After measuring end gap, return each ring to its package. Do not mix up the two compression rings. You may get the rings installed in the wrong piston ring groove.

Look up the specifications for ring end gap in the shop service manual. There usually is a minimum and maximum ring end gap specification. Piston rings must have at least the minimum gap. The end gap provides for expansion between the piston ring and the cylinder. Without enough space, the ring ends may butt together and cause ring breakage and engine seizure. Too much end

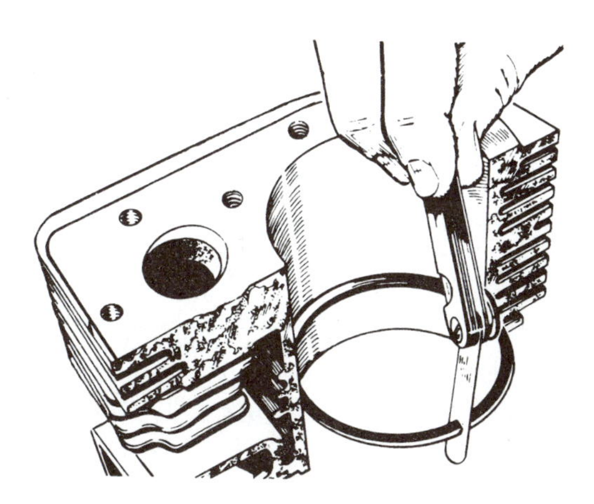

FIGURE 29-12 Using a feeler gauge to measure piston ring end gap. (Courtesy of Onan Corp.)

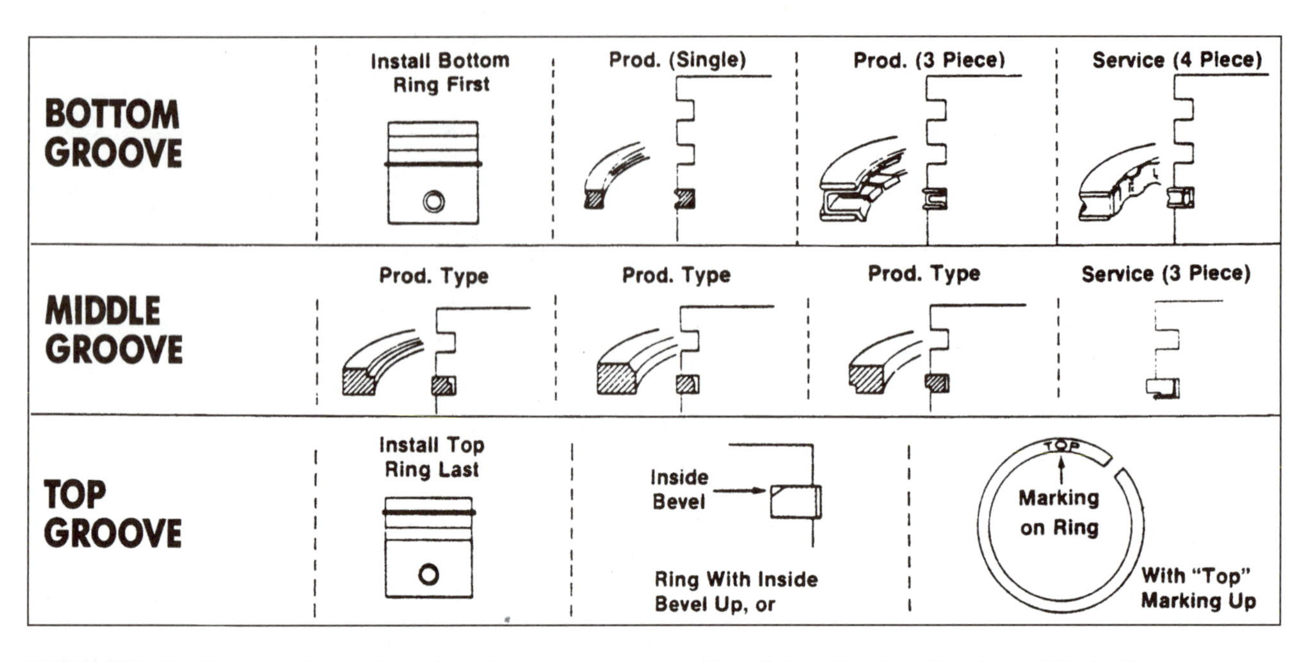

FIGURE 29-13 Piston ring instructions show the correct groove and installation direction. (Courtesy of Kohler Co.)

gap can cause compression and oil leakage through the gap. You may find an end gap that is either too large or too small. This usually means this set of rings is wrong for this engine. Check with your parts supplier to see if it is the wrong ring set.

If the ring end gap is correct, the rings may be installed on the piston. Piston rings must be installed with their top side toward the top of the piston to work correctly. Instructions are supplied with the piston ring package (Figure 29-13). There are also instructions in the shop service manual. Instructions show which ring goes in which groove and in what direction each must be installed. Many rings are not marked. Sometimes a dot or the letter T (for top) is often used to indicate the ring direction.

CAUTION: Be careful not to install a piston ring upside down or in the wrong ring groove. An incorrectly installed ring will not seal compression pressure or control oil. Use a ring expander to install the piston rings on the piston. Do not expand and install piston rings with your hand. This can cause them to be overexpanded and break.

Use a ring expander tool to engage the ends of the piston ring (Figure 29-14). Squeeze the handles to expand the piston ring. Rings are installed over the top of the piston. Do not install rings from the bottom of the piston. The ends of the piston rings may scratch the piston skirts. Install the lowest piston ring first. Install the lowest oil control ring first. Then install the second compression ring and finally the top compression ring. Space the ring ends equally at 120 degree intervals around the piston (Figure 29-15). This will prevent oil or pressure leakage through lined up gaps.

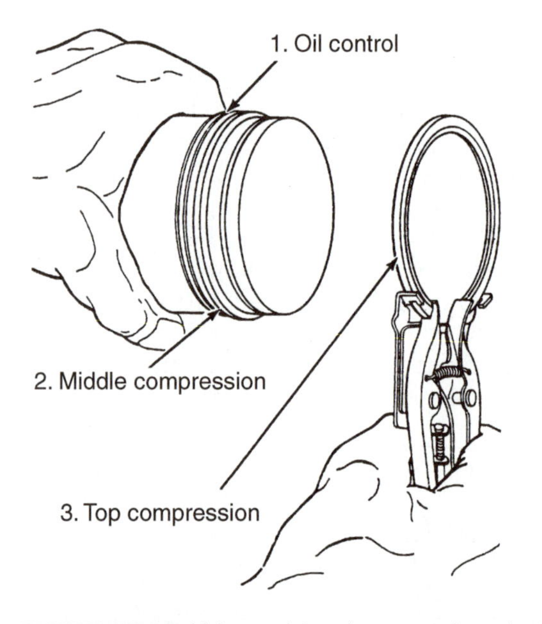

FIGURE 29-14 Using a piston ring expander to install the piston rings in the ring grooves. (Courtesy of Kohler Co.)

Some oil control rings are made in three pieces with two side rails and a center spacer. These parts are installed one at a time and the ends of each are spaced at least 1 inch (25.40 millimeter) apart.

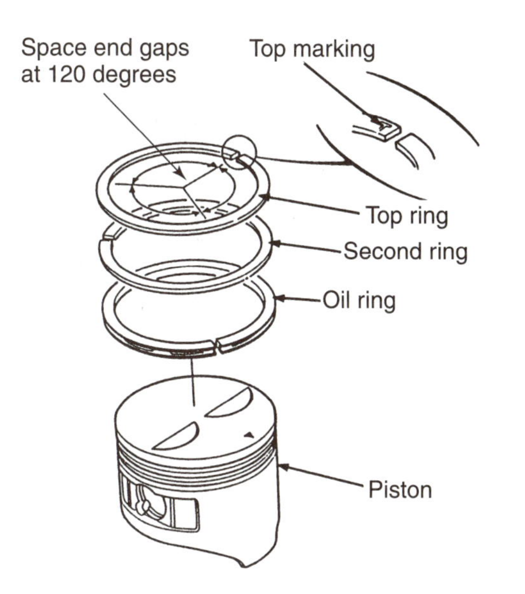

FIGURE 29-15 Rotate the piston end gaps 120 degrees apart. (Courtesy of American Honda Motor Co., Inc.)

INSTALLING TWO-STROKE PISTON RINGS

Two-stroke pistons usually use two piston rings. Check the instructions with the rings or shop service manual for the correct piston ring position and orientation. Some engines have piston rings that can be installed in either ring groove (Figure 29-16). Sometimes the rings may be installed with either side up. Rings that must be installed in one direction are often marked with a T or a dot.

Many two-stroke engines use a half keystone ring in the top piston groove (Figure 29-17). The half keystone has a slanted top. The slanted top must face upward for the ring to operate properly. Two-stroke engines often use a knock pin in the ring groove to prevent the rings from rotating. The ends of the ring must match up with the pin (Figure 28-18).

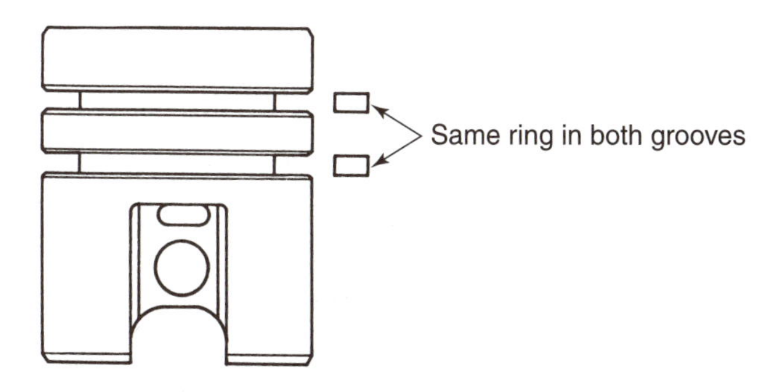

FIGURE 29-16 Two-stroke piston with interchangeable piston rings. (Provided courtesy of Tecumseh Products Company.)

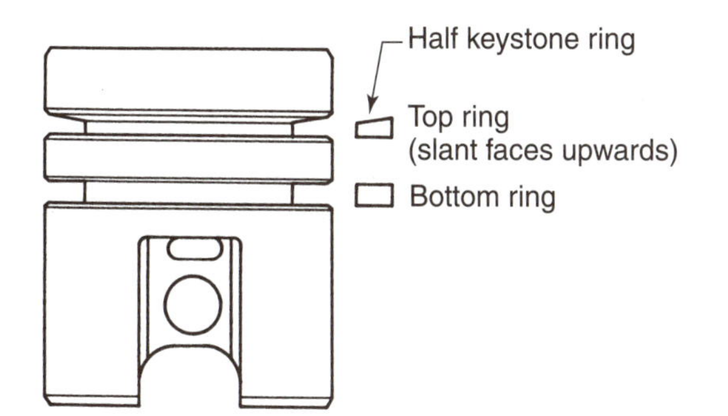

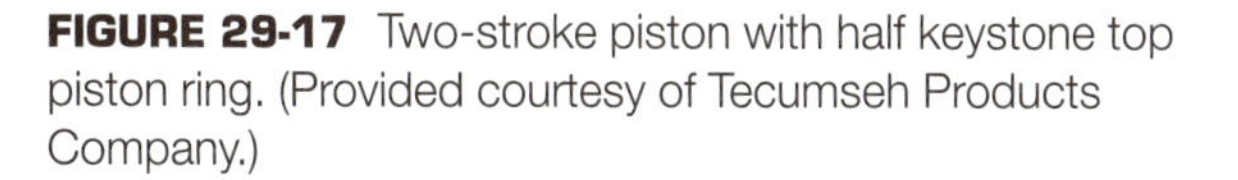

FIGURE 29-17 Two-stroke piston with half keystone top piston ring. (Provided courtesy of Tecumseh Products Company.)

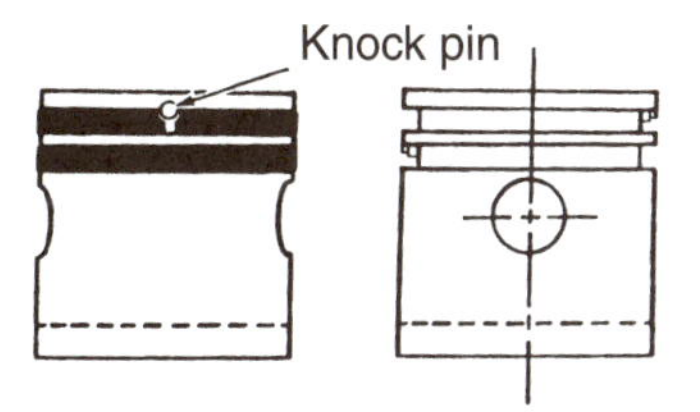

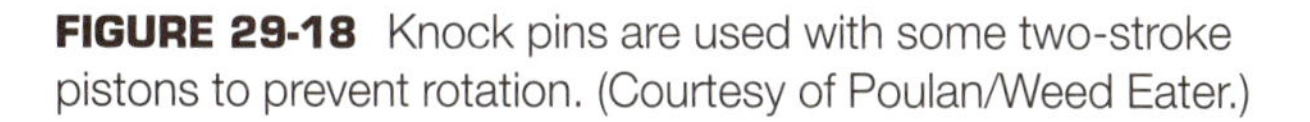

FIGURE 29-18 Knock pins are used with some two-stroke pistons to prevent rotation. (Courtesy of Poulan/Weed Eater.)

CAUTION: Four-stroke and two-stroke piston rings are designed with different internal pressures. A four-stroke ring has the same outward pressure around the complete ring circumference. Two-stroke rings are designed with lower pressures at the ring ends. Do not mix up rings from different types of engines.

INSTALLING FOUR-STROKE CRANKSHAFT AND PISTON ASSEMBLY

Wipe the cylinder block main bearing and crankshaft main bearing journals with a clean rag. Be sure they are absolutely clean. Coat the crankshaft main bearing journal and the main bearing (plain, roller, or ball) with new engine oil. Place the crankshaft into the main bearing in the cylinder block.

The piston and connecting rod assembly is installed in the cylinder block after crankshaft installation. If the connecting rod uses bearing inserts, the rod side insert is installed before the rod goes into the engine. Wipe the rod saddle bore clean with a clean rag. Push the insert bearing into position. Make sure the bearing locking tab seats in the connecting rod notch.

FIGURE 29-19 Submerge the piston and rings in a can of oil to lubricate the piston rings.

CAUTION: The back side seating surface of an insert bearing should not be oiled or greased. Doing so results in poor heat transfer and early bearing failure.

Wipe the cylinder walls clean with a clean rag, then coat them with oil. Fill a clean can with new engine oil and submerge the piston and rings (Figure 29-19). This provides lubrication for the piston rings and cylinders during the first engine startup.

A piston ring compressing tool is used to install the piston and rings into the cylinder bore. A **piston ring compressor** is a tool that is tightened around the piston head to squeeze the rings tightly into their grooves for piston installation. The tool has a thin metal band that is expanded and contracted with an Allen wrench, pliers, or a hose clamp type band (Figure 29-20). Some ring compressors have small steps on the bottom of the tool that prevent it from entering the cylinder.

Expand the compressing tool by depressing its release lever. Place the compressor around the piston rings. Make sure the steps on the tool point downward. Tighten the tool with the wrench to compress the piston rings. Tap the compressor band area lightly with a soft face hammer to seat the rings and allow them to be fully compressed. When the rings are fully compressed, the tool will not compress any further.

Look up the piston direction marking in the shop service manual. A direction arrow on the top of a two-stroke piston points toward the exhaust side of the cylinder. The arrow on a four-stroke piston usually points toward the valves. Insert the piston and connecting rod in the correct direction into the cylinder bore. Lower the piston until the bottom of the compressing tool contacts the cylinder block deck (Figure 29-21). Use your hand to push the piston into the cylinder. If the rings are properly seated in the ring compressor, the piston rings will enter the bore. When all the rings enter the cylinder, release the compressor by pushing on the release lever. Push the piston down the bore until the rod bearing seats on the crankshaft connecting rod journal. Watch the connecting rod to make sure it seats on the crankshaft journal. You may need to rotate the crankshaft slightly to line them up.

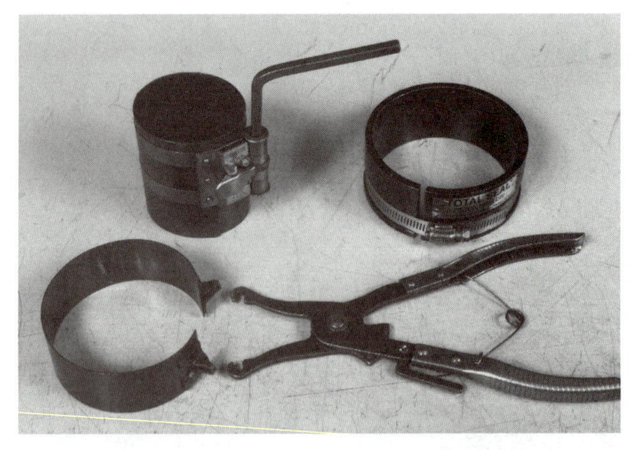

FIGURE 29-20 Types of piston ring compressors used to install the rings into the cylinder.

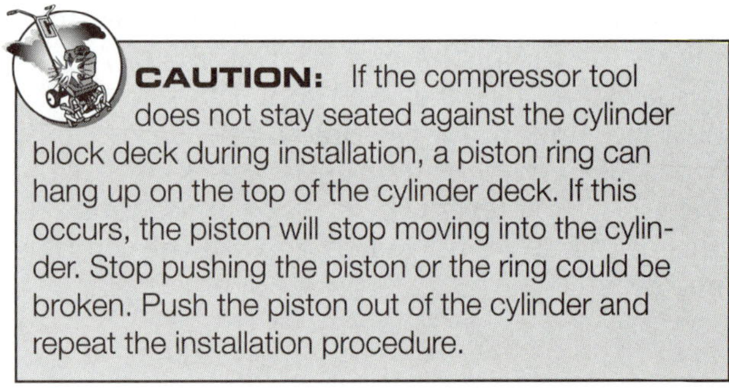

CAUTION: If the compressor tool does not stay seated against the cylinder block deck during installation, a piston ring can hang up on the top of the cylinder deck. If this occurs, the piston will stop moving into the cylinder. Stop pushing the piston or the ring could be broken. Push the piston out of the cylinder and repeat the installation procedure.

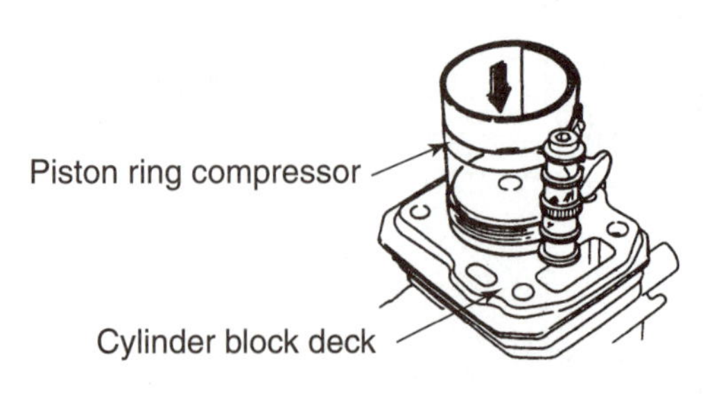

FIGURE 29-21 Compressing the piston rings and installing the piston in the cylinder. (Courtesy of American Honda Motor Co., Inc.)

CASE STUDY

The students in the outdoor power equipment class had just finished getting the demonstration on installing the piston and rings into the cylinder. The instructor usually gave this demonstration on the first engine that was ready for this procedure. His demonstration had gone well and he expected the students would be able to do the job without any difficulty.

The instructor was working with a student who was having a problem determining the direction and position of some piston rings. He heard a crack sound and looked across the bench to where a student was installing a piston in the cylinder. He went over to investigate.

The student had installed the ring compressor around the rings and piston but had not gotten the rings seated properly in the ring compressor. As he pushed the piston down the cylinder, one of the rings hung up on the cylinder deck. During his demonstration the instructor warned the student that this could happen. He told the students if the piston does not push into the cylinder with finger pressure, there could be a problem. The student had felt this resistance and went to the tool panel for a large hammer. He used the hammer handle to pound on the top of the piston. The crack noise was the sound of breaking piston ring.

CAUTION: Watch the position of the connecting rod in relation to the crank shaft as the rod and piston enter the cylinder. Be careful that the rod does not contact the crankshaft and damage either part.

INSTALLING A FOUR-STROKE CONNECTING ROD CAP

With the crankshaft installed, the rod cap is ready for installation. If the cap has an insert bearing, install it in the cap. Make sure the insert locking tab fits tightly in the rod cap notch. The rod cap must be installed in the correct direction. Use the engine maker's match marks or the marks you made during disassembly (Figure 29-22).

The crankshaft to connecting rod bearing oil clearance is measured before the rod cap is permanently installed. **Crankshaft oil clearance** is the space between the connecting rod bearing and the crankshaft. Oil clearance is necessary for proper bearing lubrication and cooling. If the oil clearance is too small, the bearing will overheat. Overheating causes rapid bearing and crankshaft journal wear. If the oil clearance is too large, the oil film between the parts will break down, resulting in early parts failure.

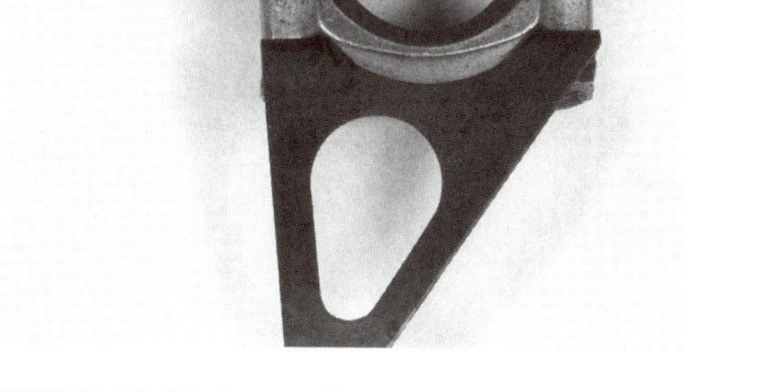

FIGURE 29-22 Locate the engine maker or your match marks on the connecting rod before assembly.

Oil clearance is measured with **Plastigage**, a small diameter soft plastic string used to determine the oil clearance between a bearing and its shaft. Plastigage is available to measure different clearance ranges. Normally only the smallest clearance range (0 to 0.003 inch or 0.08 millimeter) is used for engine assembly work. Plastigage for wider clearances is used for troubleshooting worn engines.

Cut a piece of the Plastigage approximately 1/8 inch (6.35 millimeter) shorter than the connecting rod bearing. Wipe the rod cap bearing surface clean of oil. Place the Plastigage string on the rod bearing. Assemble the rod cap on the connecting rod in the correct direction. Look up the connecting rod fastener torque specification in the shop service manual. Use an inch-pound torque wrench to tighten the connecting rod fasteners (Figure 29-23).

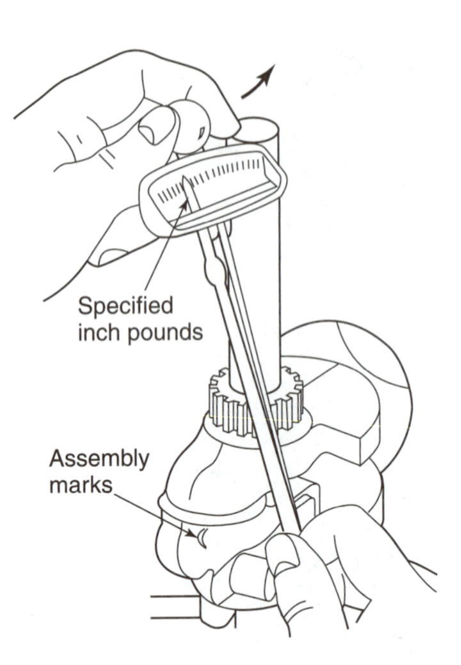

FIGURE 29-23 Using an inch-pound torque wrench to tighten the connecting rod fasteners.

> **CAUTION:** Connecting rod tightening torque specifications are usually specified in inch-pounds. Do not confuse these specifications with foot-pounds. For example, the specification may be 100 inch-pounds. If you tighten to 100 foot-pounds the fastener will break from excessive torque.

Be careful not to rotate the crankshaft because this will ruin the Plastigage. Remove the fasteners and remove the rod cap. The Plastigage will be squeezed flat (Figure 29-24). If the clearance is small, the Plastigage will be very flat. If the clearance is large, the Plastigage will be less flat. Compare the flattened Plastigage with the stripes on the Plastigage package until you find a match. Each stripe on the package has an oil clearance measurement. One side of the package has clearance stripes in thousandths of an inch. The other side has stripes in hundredths of a millimeter.

Look up the specifications for connecting rod oil clearance in the shop service manual. Compare your Plastigage measurement with this specification. If the clearance is not to specifications, the engine cannot be assembled further. You need to determine the cause of the clearance problem. A new connecting rod may be the wrong size. An insert bearing may be undersize. An error may have been made in crankshaft grinding.

If the clearance is within specified limits, lubricate the connecting rod bearing. Install the rod cap. Recheck the cap to be sure the match marks are correct. If the connecting rod has an oil dipper, it must be installed at this time. Make sure the dipper is installed in the correct direction or the engine will not get proper lubrication. Install a new fastener lock plate. Torque the connecting rod fasteners to specifications. Bend the lock plate tabs up around the connecting rod nuts or bolts (Figure 29-25). These tabs are very important in preventing the nuts or bolts from vibrating loose. The procedure for measuring connecting rod oil clearance with Plastigage is shown in the accompanying sequence of photographs.

> **CAUTION:** An incorrectly installed dipper can prevent proper lubrication. On some engines it can also contact internal engine parts and cause the crankshaft not to rotate.

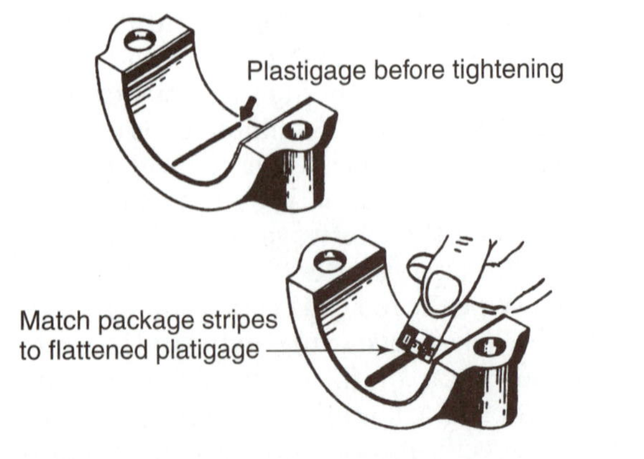

FIGURE 29-24 Measuring the flattened Plastigage to determine oil clearance. (Courtesy of Kohler Co.)

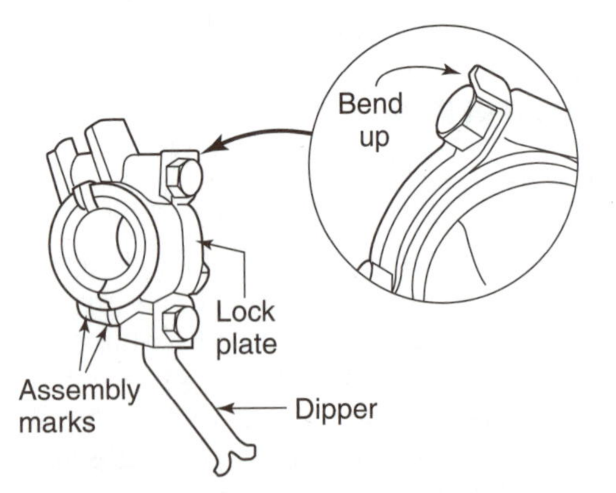

FIGURE 29-25 The lock plate tabs are bent over the connecting rod fasteners to prevent loosening.

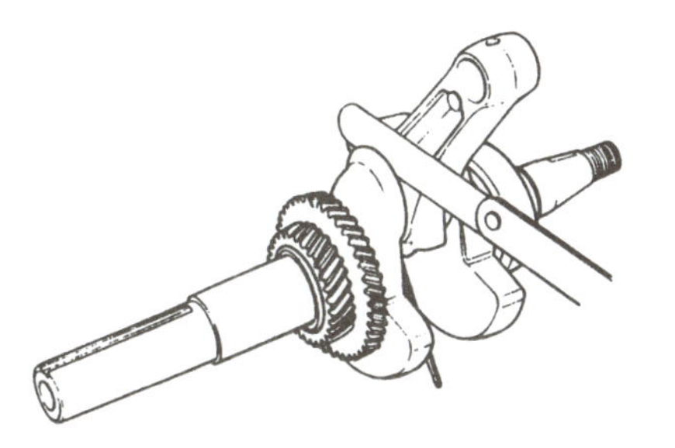

FIGURE 29-26 Checking the connecting rod side clearance. (Courtesy of American Honda Motor Co., Inc.)

SERVICE TIP: Rod cap bolt loosening is a common cause of engine failure. New rod cap fasteners and lock plates should be used when possible.

After final tightening, measure the connecting rod side clearance. **Connecting rod side clearance** is a measurement of the space between the crankshaft and the side of the connecting rod (Figure 29-26). Look up the specification for connecting rod side clearance in the shop service manual. This clearance is measured by inserting the specified thickness feeler gauge between the connecting rod and crankshaft. A clearance less than specifications can cause the rod to expand and lock to the crankshaft. Too large a clearance is a problem on an engine with a full pressure lubrication system. The clearance causes excessive oil from the connecting rod journal to be thrown on the cylinder walls. The connecting rod must be replaced if this measurement is not to specifications.

INSTALLING TWO-STROKE PISTON, CONNECTING ROD, AND CRANKSHAFT

Two-stroke engines often use needle bearings between the connecting rod and crankshaft journal. Some two-stroke engines have a connecting rod with a removable rod cap (Figure 29-27). The piston and rod assembly is installed on the crankshaft before the crankshaft is installed in the engine. Apply assembly lubricant on the crankshaft journal then position a strip of needle bearings on the journal. A split needle bearing housing is often used on the rod and cap. Install the housings in the cap and rod so that the notch in the cap will lock into the V in the rod. Install the rod and cap in position over the needle bearings. Make sure the match marks are aligned. Tighten the cap screws to the specified torque with an inch-pound torque wrench. The crankshaft assembly is now ready for installation into the engine.

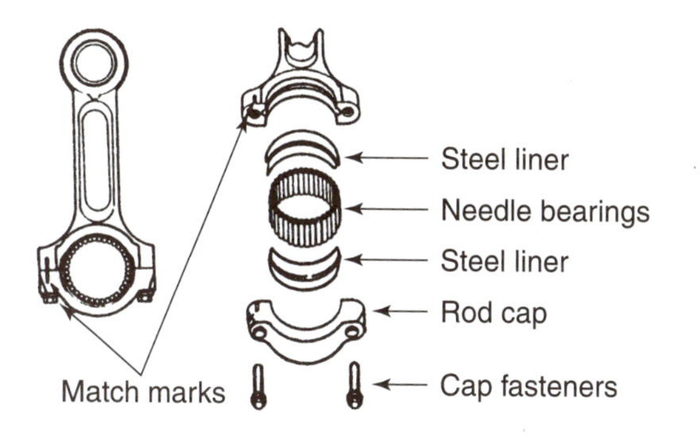

FIGURE 29-27 Installing needle bearings in a two-stroke connecting rod with removable rod cap. (Provided courtesy of Tecumseh Products Company.)

CAUTION: Always count the number of needle bearings installed on the crank shaft journal and compare it with the number specified in the shop service manual. Too many bearings will not allow assembly and too few will result in early engine failure.

Many two-stroke engines use a one-piece connecting rod. The big end of the connecting rod often has a steel liner to support the needle bearings. Be sure the connecting rod and piston are both oriented in the correct direction. Place them into the crankcase. Install a set (called a *strip*) of needle bearings on the crankshaft journal with assembly lubricant. Slide the crankshaft into the crankcase and through the connecting rod until the journal is positioned in the connecting rod (Figure 29-28).

The two-stroke cylinder is often installed over the piston and connecting rod assembly after they are in the engine (Figure 29-29). Lubricate the piston, piston rings, and cylinder walls with engine

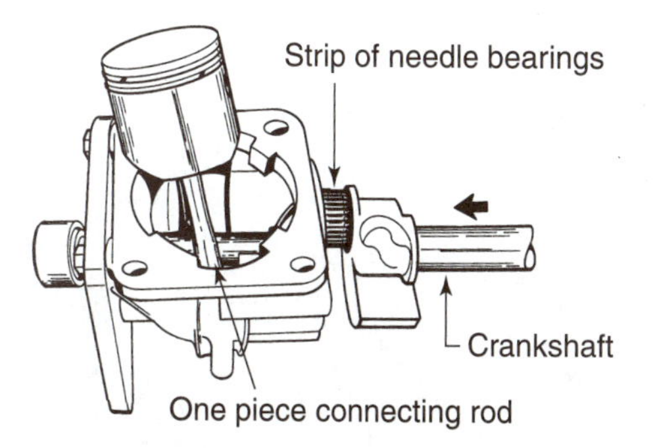

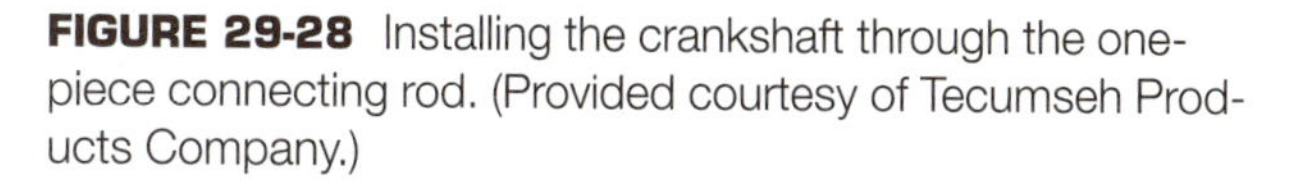

FIGURE 29-28 Installing the crankshaft through the one-piece connecting rod. (Provided courtesy of Tecumseh Products Company.)

PHOTO SEQUENCE 21

Measuring Connecting Rod Oil Clearance with Plastigage

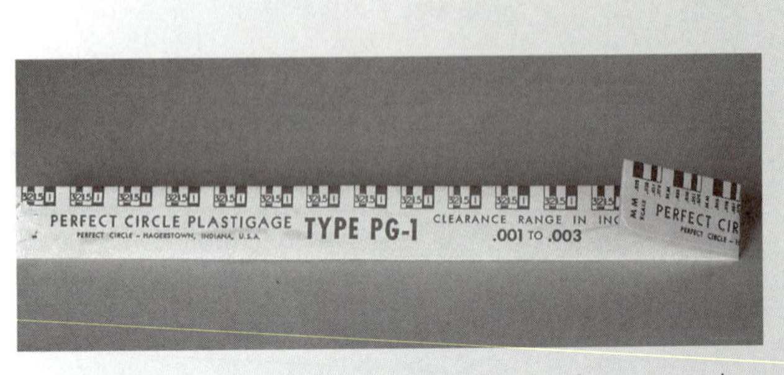

1. Connecting rod to crankshaft oil clearance is measured with small diameter range (0 to 0.003 inch or 0.08 millimeter) soft plastic string called Plastigage.

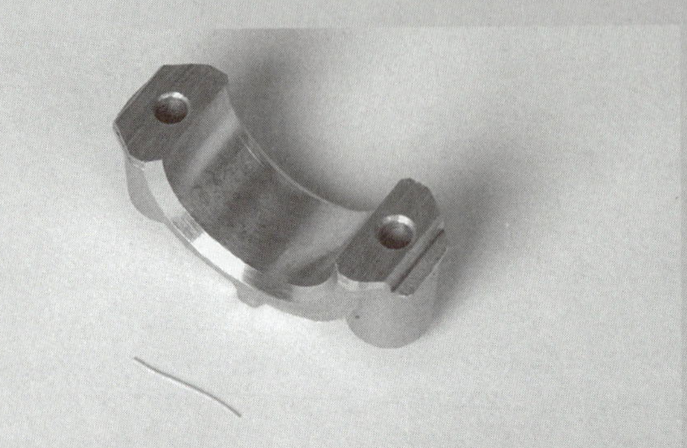

2. Cut a piece of the Plastigage approximately 1/8 inch (6.35 millimeter) shorter than the connecting rod bearing.

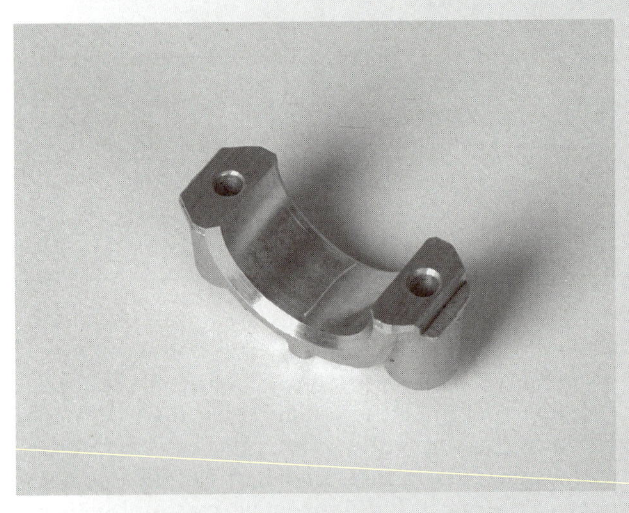

3. Wipe the rod cap bearing surface clean of oil and place the Plastigage string on the rod bearing.

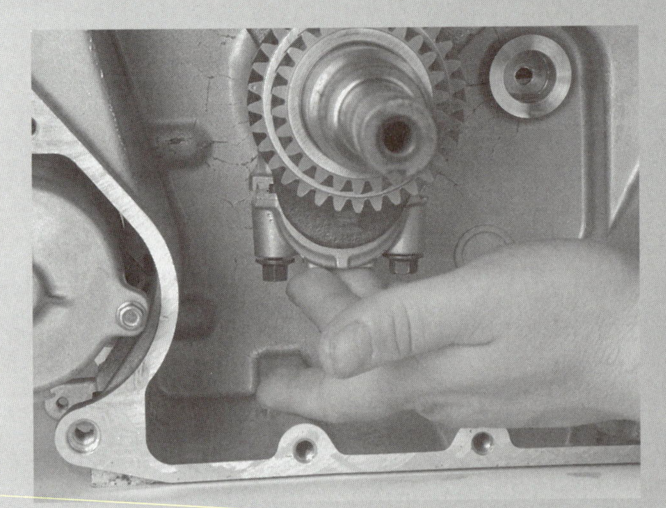

4. Assemble the rod cap on the crankshaft and connecting rod in the correct direction.

5. Look up the connecting rod fastener torque specification in the shop service manual. Use an *inch-pound* torque wrench to tighten the connecting rod fasteners.

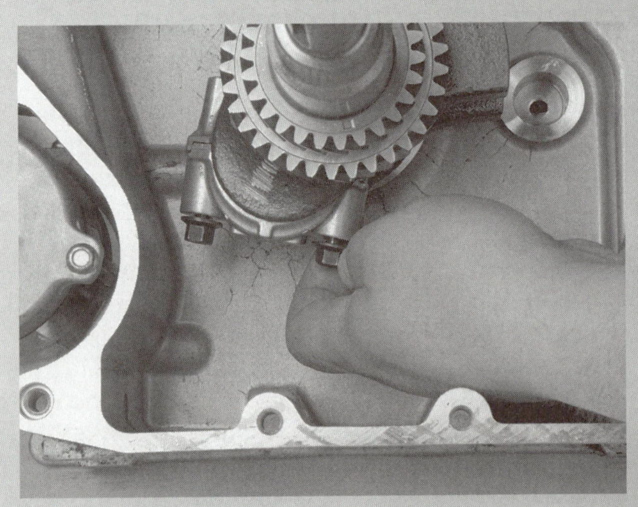

6. Remove the connecting rod fasteners and remove the rod cap.

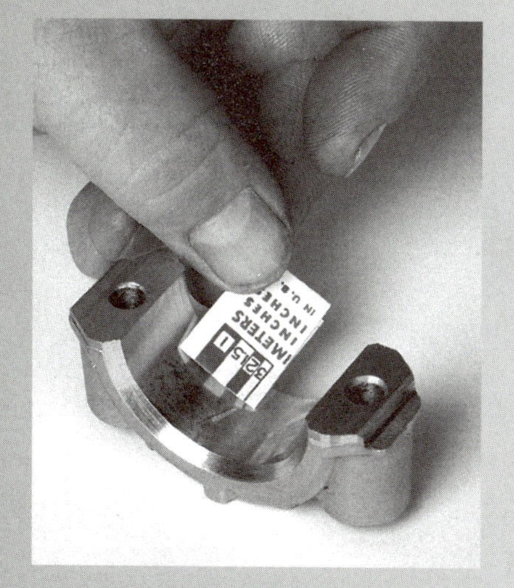

7. The Plastigage will be squeezed flat. Compare the flattened Plastigage with the stripes on the Plastigage package until you find a match.

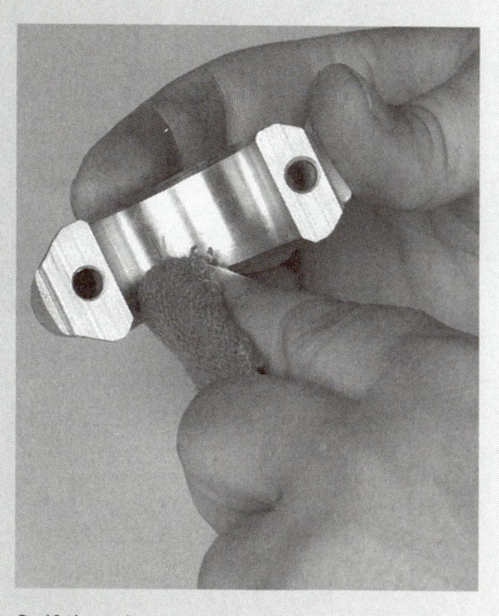

8. If the clearance is correct, wipe the Plastigage string off the connecting rod bearing with a clean shop rag.

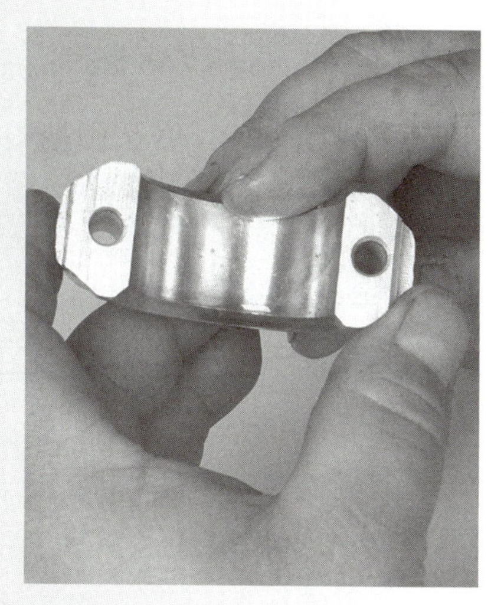

9. If the clearance is within specified limits, lubricate the connecting rod bearing and install the rod cap.

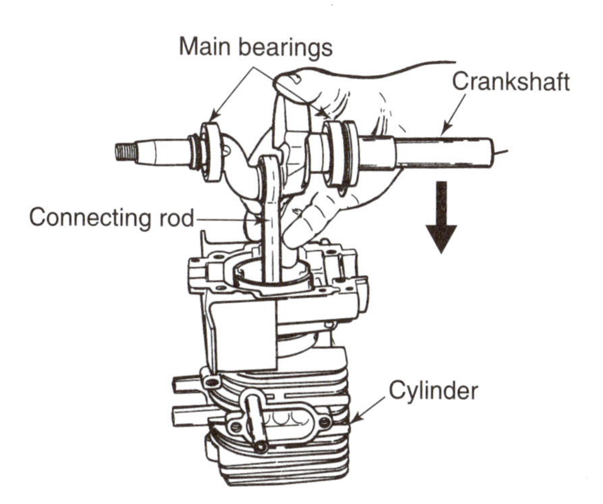

FIGURE 29-29 Installing the piston assembly into the bottom of the two-stroke cylinder. (Provided courtesy of Tecumseh Products Company.)

oil. If the rings do not have knock pins, make sure the end gaps are properly spaced apart. Install any required gaskets, seals, or sealer on the crankcase and cylinder mating surface. Position the cylinder over the piston with one hand and compress the rings with the other. Push the cylinder over the rings and down on the crankcase mating surface. Install and torque the cylinder to crankcase fasteners to specifications. The crankshaft, main bearings, piston, and connecting rod are inserted into the engine as an assembly on some engines. The same method is used to install the piston into the cylinder.

INSTALLING CAMSHAFT AND VALVE LIFTERS

Turn the engine upside down so the valve lifters can be installed without falling out. Clean and lubricate the valve lifters. Follow the marks you made on disassembly. Insert the lifters in the correct position in the cylinder block.

Look up the location and type of the camshaft to crankshaft timing (Figure 29-30) marks in the shop service manual. Rotate the crankshaft until the piston is up at the TDC. Lubricate the camshaft lobes and journals and place the camshaft in its bearing. As you insert the camshaft, rotate the timing gear into the correct timing position.

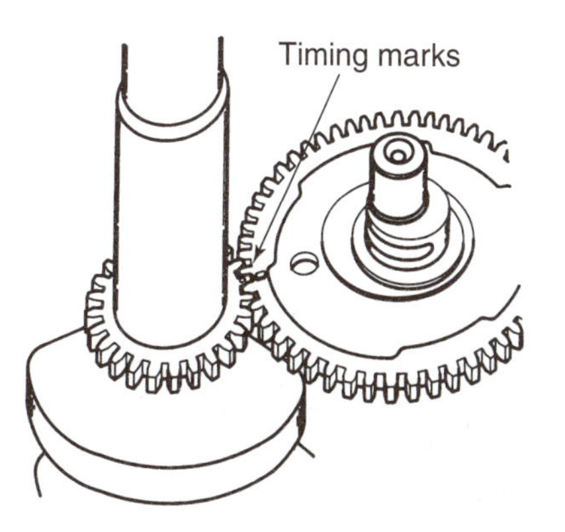

FIGURE 29-30 Align the camshaft to crankshaft timing marks as the camshaft is installed. (Provided courtesy of Tecumseh Products Company.)

CAUTION: If the crankshaft and camshaft are not installed correctly in relation to each other, the valves will not open or close at the correct time in relation to the piston position. If, for example, the piston moves down on an intake stroke, but the intake valve is not open, air and fuel cannot enter the cylinder. Similarly, the other strokes, compression, power, and exhaust, will not operate correctly and the engine will not run.

INSTALLING CRANKCASE COVER

Before installing the side or bottom crankcase cover, make a final check inside the engine. Check that the timing marks are correctly aligned. Make sure the connecting rod cap is installed in the correct position. Make sure any balance shafts oil dipper, oil slinger, or governor parts are installed in the correct position.

Crankshaft end play must be measured and, if necessary, adjusted. Crankshaft end play is set by the thickness of the crankcase cover gasket or a thrust washer. Check the end play specifications in the shop service manual. Compare the specification to the end play measurement you made during disassembly. If the same crankshaft is being used, the engine should have the same end play as it did on disassembly.

The end play may be increased or decreased by selecting thicker or thinner gaskets (or different numbers of gaskets) or thrust washers (Figure 29-31). Most gasket sets contain several cover gaskets of different thickness. Use an outside micrometer to measure the thickness of the old side cover gasket or thrust washer. If the end play was correct on disassembly, choose a gasket of the same thickness as the original. A thicker gasket (or more gaskets) will increase end play. The thicker gasket spaces the two main bearings further apart which allows the crank shaft more end-to-end movement. A thinner gasket (or fewer gaskets) reduces end play. Reuse the same thrust washer if end play was correct or choose a different thickness to correct end play.

Place the crankcase cover gasket on the cylinder block. Slide the crankcase cover over the crankshaft. Tighten the side cover bolts to specifications with a torque wrench. If all the parts are installed correctly, the engine should turn freely by hand. Mount a dial indicator to the crankshaft and check the end play. If the end play is not correct you will have to remove the crankcase cover and correct gasket or thrust washer thickness. The procedure for adjusting crankshaft end play is shown in the accompanying sequence of photographs.

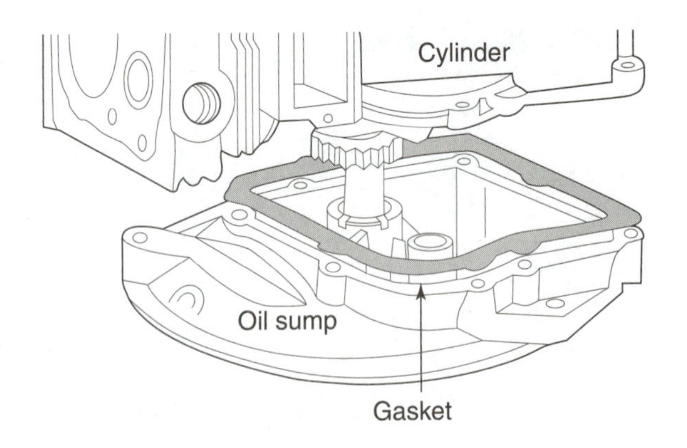

A. Gasket End Play Adjustment

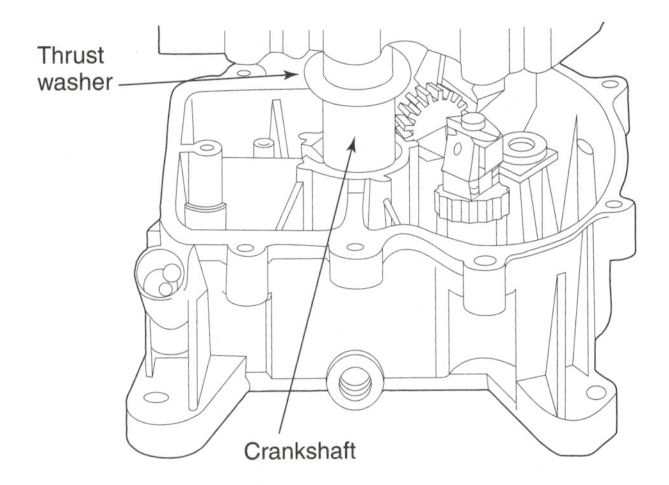

B. Thrust Washer End Play Adjustment

FIGURE 29-31 End play is adjusted by crankcase cover gasket or thrust washer thickness.

CAUTION: The crankshaft should rotate smoothly by hand. If the crankshaft does not rotate or is difficult to turn, disassemble and locate the source of the problem.

Lightly coat the lip of the new FWE and PTO seal with lubricant to prevent them from being damaged during engine startup. Try the new seal lip fit before installation by inserting it over the crankshaft to make sure it fits snugly. Make sure the lip of the seal points toward the inside of the engine. Drive the new seal into the block or crankcase cover using a block of wood or the correct size driver.

INSTALLING CYLINDER HEAD

Before installing the cylinder head, check the head bolts to be sure they are clean and that the threads are in good condition. Replace any damaged bolts. Inspect the cylinder deck area for damage to head bolt hole threads. Repair any damaged threads with a tap. Look up the cylinder head tightening sequence, tightening torque, and bolt length diagram in the shop service manual.

Lay the new and old head gasket side by side and compare them for any differences. Place the head gasket on the block in the correct position and set the cylinder head on top of the gasket. Start each cylinder head bolt by hand. Many cylinder heads use both long and short bolts. Different numbers and types of washers may be used on the bolts. Check the shop service manual or disassembly notes for the positions of the long and short bolts and washers. Many engine makers recommend coating the threads and heads of the bolts with antiseize compound to prevent the bolts from rusting and seizing in their threads. Tighten the cylinder head bolts in the correct order and to the correct torque. The shop service manual often has a tightening sequence diagram to follow (Figure 29-32).

SERVICE TIP: Sometimes a tightening sequence cannot be found for a cylinder head. A general procedure to follow is to tighten from the center outward going in a circle.

CAUTION: Tighten cylinder head bolts in steps not all at once. For example, if the specifications recommend 150 inch-pounds, tighten each bolt to 50, then 100, and finally to 150 inch-pounds. Tightening a head bolt all at once can cause cylinder head to warp.

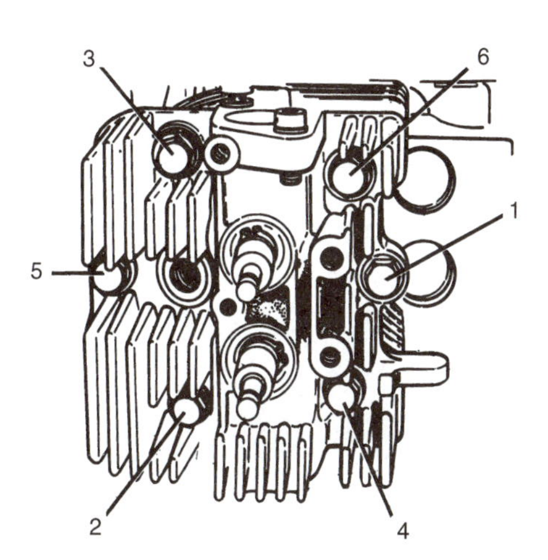

A. Overhead Tightening Sequence

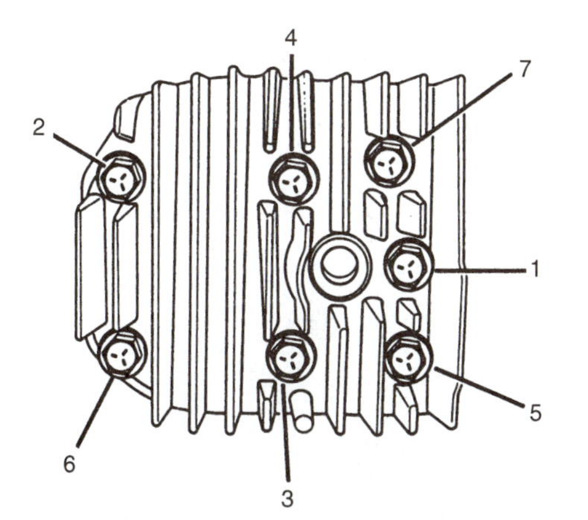

B. L-Head Tightening Sequence

FIGURE 29-32 Cylinder head bolts must be tightened in the correct sequence. (Provided courtesy of Tecumseh Products Company.)

When the overhead valve cylinder head is installed, the pushrods can be installed. Insert each pushrod down its hole in the cylinder head until it seats in the valve lifter (Figure 29-33). Install each rocker arm in position with one end contacting the valve stem and the other engaging the pushrod. Install but do not tighten the rocker arm nuts. Follow the instructions in the shop service manual to adjust each valve (Figure 29-34). Install a new valve cover gasket and replace the valve cover.

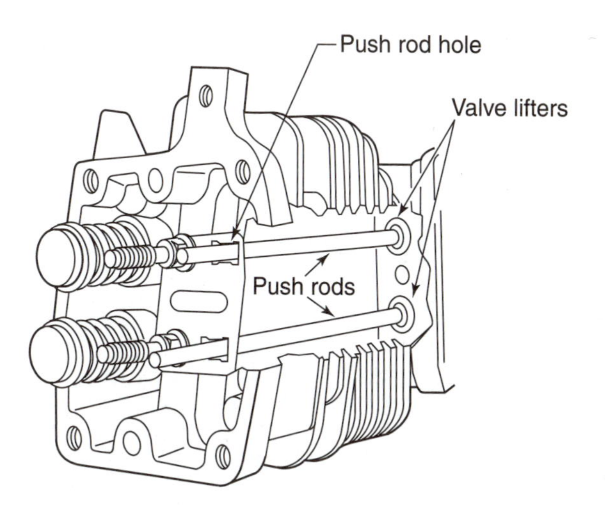

FIGURE 29-33 Installing overhead valve push rods.

PHOTO SEQUENCE 22

Adjusting Crankshaft End Play

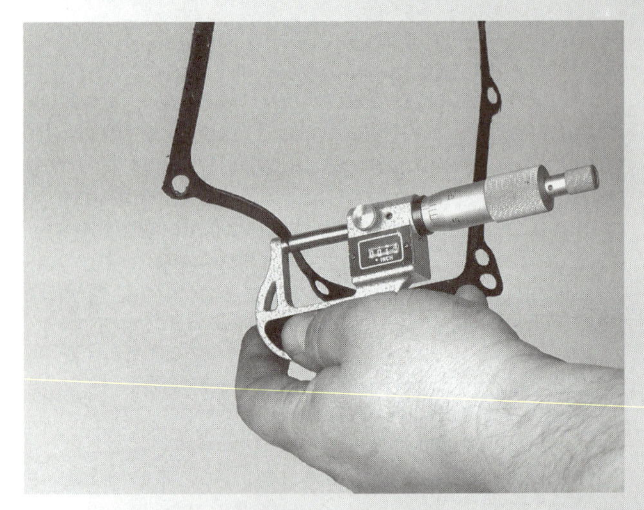

1. Use an outside micrometer to measure the thickness of the old side cover gasket.

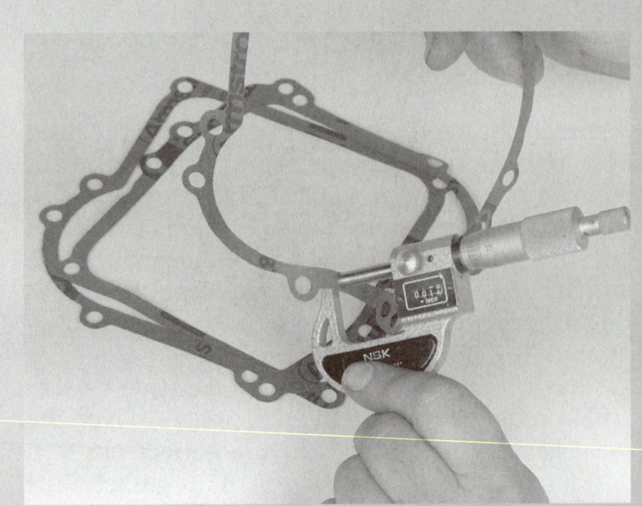

2. Use an outside micrometer to measure and choose a gasket from the engine gasket set. If end play was correct on disassembly, find the same thickness as the original. A thicker gasket (or more gaskets) will increase end play. A thinner gasket (or fewer gaskets) will reduce end play.

3. Install the gasket and install the crankcase cover.

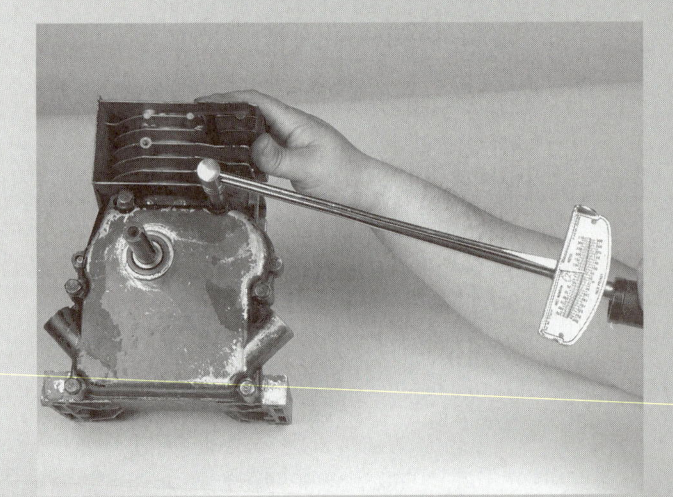

4. Install the cover screws and torque the cover screws to correct torque specifications and in the correct order.

5. Mount a dial indicator to the side cover and position the dial indicator plunger on the crankshaft.

6. Preload the plunger against the crankshaft.

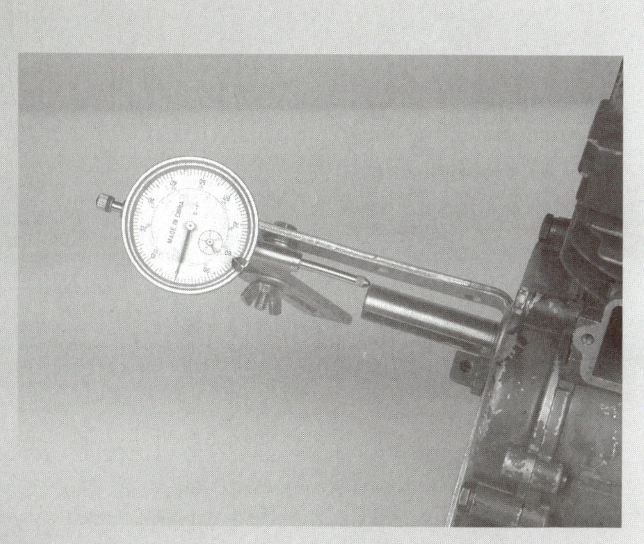

7. Rotate the dial indicator face to set the dial indicator to zero.

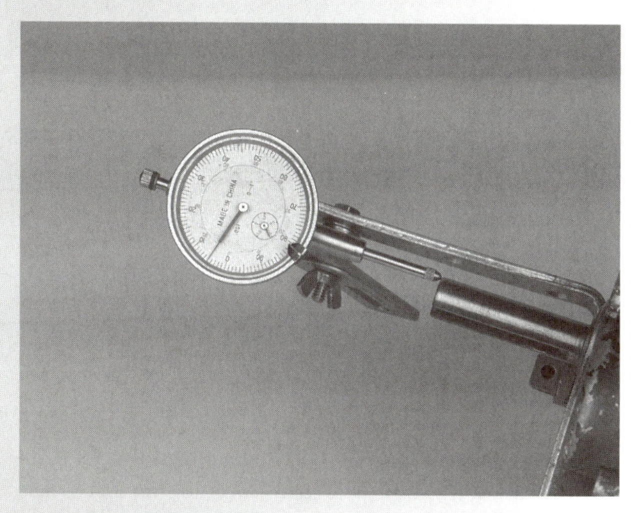

8. Move the crankshaft back and forth while observing the dial indicator.

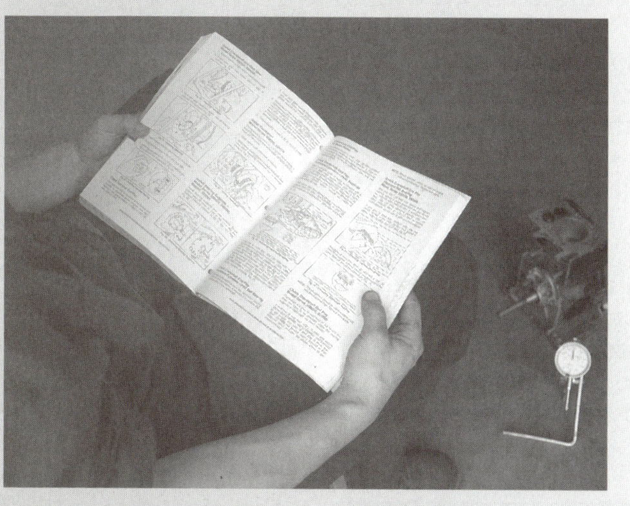

9. Read the end play as total movement of the dial indicator needle and check it against specifications.

SERVICE TIP: Grinding valve seats, faces, or stem tips changes the valve clearance. Clearance must always be checked and adjusted when an engine is assembled.

CAUTION: As soon as the cylinder head is installed, gap and install the new spark plug in the spark plug hole to prevent dirt or other objects from entering the cylinder.

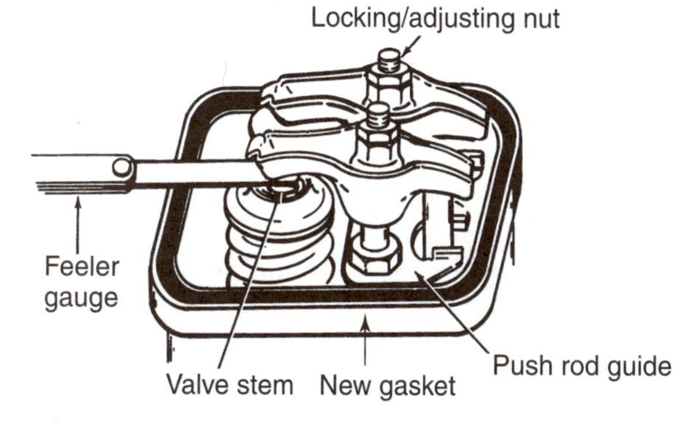

FIGURE 29-34 Adjusting valve clearance. (Provided courtesy of Tecumseh Products Company.)

FINAL ASSEMBLY AND START-UP

SERVICE TIP: After final assembly, the engine should be painted to factory original color and new engine decals installed.

A rebuilt engine should always have new fuel and ignition system components for ease of starting. Install the ignition system and flywheel. Use the same flywheel tool used during disassembly to hold the flywheel. Torque the flywheel nut to specifications. Install the coil armature and set the armature air gap. Use a spark tester to be sure the ignition system is working properly.

Install the carburetor and fuel tank. Use fuel lines to connect the carburetor to the fuel tank. Install a new fuel filter. Follow the recommendations in the shop service manual to adjust the low- and high-speed carburetor adjustment screws. Install and check the governor controls.

Paint and install all the cooling ducting. Rebuild the rewind starter system if necessary. Install the blower housing and starter assembly. Coat the muffler threads with antiseize compound and install the muffler.

Install the engine on the equipment or vehicle or clamp securely to a shop bench. If the engine is to be started in the shop, connect the exhaust to a ventilation system. Fill the crankcase with the correct amount of recommended oil. Fill the fuel tank with fresh fuel.

SERVICE TIP: Engines used on rotary lawnmowers usually use the cutting blade for added flywheel weight. These engines must be mounted on the lawnmower and have the blade properly mounted before they can be started.

Usually several rope pulls or electrical cranks will be necessary to get fuel into the cylinder and allow the engine to start. Start the engine and allow it to run at fast idle for a few minutes. The engine will smoke for a short time until oil on the cylinders and rings burns off. Make the final carburetor mixture and idle speed adjustments. Check and correct any fuel or oil leaks.

WARNING: Make sure the engine is solidly mounted and any blades or adapters connected to the engine crankshaft are secure. Have a fire extinguisher positioned next to the engine.

Check the shop service manual for any cylinder head retightening recommendations. Some engine makers recommend retightening the cylinder heads after the engine has run a short time and cooled down. Some engine makers do not recommend this procedure.

SERVICE TIP: Drain and refill the crankcase oil after 1 hour of engine operation to remove any foreign material left in the engine during assembly.

The engine may smoke for a period of time until the piston ring seats. **Ring seating** is the mating of the ring face with the cylinder wall throughout the complete stroke of the ring. Seating occurs as the very slight irregularities of the ring face and the cylinder wall are worn off. This is called the break-in period. The smoke comes from oil that gets past the rings before they are seated. Oil consumption is often higher than normal for a short period of time until the rings become seated. **Break-in period** is the engine operation time required before oil consumption drops off to an acceptable level. The break-in period is different for different types of piston rings. Check for break-in information on the piston ring instruction sheet. Break in can last as long as 10 hours. The engine should be operated normally during this period. Caution the customer not to operate the engine at extreme loads or speeds until the end of the break-in period. Check the oil frequently during the break-in period.

REVIEW QUESTIONS

1. List and describe the common materials used in gasket construction
2. Explain why a gasket should not be reused.
3. Describe the parts and explain the operation of a lip type seal.
4. Explain the uses for assembly lubricant in engine assembly.
5. Describe the types and uses for gasket sealant.
6. Explain why a piston must be assembled on a connecting rod with the correct orientation.
7. Describe how to install a piston on the connecting rod.
8. List and describe the steps to follow to install piston rings on a four-stroke piston.
9. List and describe the steps to follow to install piston rings on a two-stroke piston.
10. Explain why a camshaft and camshaft gear must be properly meshed when installed in the engine.
11. List and describe the steps to follow to install a four-stroke piston assembly.
12. Explain why the connecting rod must be installed in the engine in the correct direction.
13. Explain why the rod cap must be installed following match marks.
14. List and describe the steps to follow using Plastigage to determine connecting rod journal oil clearance.
15. List and describe the steps to follow to install a two-stroke piston assembly.
16. Explain why two-stroke piston skirt cutouts or head deflectors must be oriented in the correct direction.
17. Explain how to measure and adjust thrust washer thickness to set crankshaft end play.
18. Explain how to measure and adjust cover gasket thickness to set crankshaft end play.
19. Explain why cylinder head bolts are tightened to a specific torque and in a specific sequence.
20. Explain why piston rings require a break-in period for effective oil control.

DISCUSSION TOPICS AND ACTIVITIES

1. Locate a collection of shop connecting rods and pistons. Try to find the different match mark orientation marks on the parts.
2. Locate a scrap piston with rings. Practice getting the rings on and off with a piston ring expander tool.
3. The end play on a crankcase cover was measured during disassembly and found to be 0.008 inch (0.20 millimeters). The specifications recommend an end-play measurement of 0.005 inch (0.13 millimeters). If the old crankcase cover gasket is 0.020 inch (0.51 millimeter) thick, how thick should the new gasket be?

Equipment Operation and Service

Rotary and Reel Walk Behind Mowers

OBJECTIVES

Upon completion and review of this chapter, you should be able to:

- Describe the safety precautions to take when operating or servicing a walk behind lawn mower.
- List and describe the main parts of a rotary lawn mower.
- List and describe the purpose and function of the parts of a rotary lawn mower blade.
- Explain the purpose and describe the operation of a rotary blade slip clutch.
- Describe the parts and operation of a rotary lawn mower blade brake system.
- Troubleshoot a rotary lawn mower that vibrates or cuts unevenly.
- Inspect a rotary lawn mower blade for wear or damage.
- Sharpen and balance a rotary lawn mower blade.
- Describe the operation of a reel lawn mower.
- Adjust the reel and cutter bar on a reel lawn mower.
- Describe the operation of a belt-driven self-propelled mower.
- Describe the parts and operation of a self-propelled mower gear box.
- Inspect and replace a self-propelled mower drive belt.
- Describe the purpose and operation of a one-way clutch used on the self-propelled mower drive wheels.

TERMS TO KNOW

Blade balancer
Control bar assembly
Drive belt
Drive clutch lever
Flywheel brake lever
Lawn mower blade brake
Mower deck
One-way clutch
Reel lawn mower
Rotary lawn mower
Self-propelled lawn mower
Shift lever
Slip clutch
Throttle lever
Transmission
Walk behind lawn mower

INTRODUCTION

One of the most common types of power equipment, a **walk behind lawn mower** is a lawn mower that is controlled by an operator walking behind the equipment. Walk behind lawn mowers are generally powered by a two- or four-stroke engine with the four-stroke models being more common. Some models must be pushed and others are self-propelled; some types use a rotary blade and others use a reel to cut the grass.

MOWER SAFETY

WARNING: Always read and follow the specific safety instructions in the operator's manual before you start or use any power equipment.

The rotary lawn mower uses a large, sharp steel blade that rotates at high speed. Mower cutting blades are made to cut grass, but the cutting blade will cut fingers, hands, and feet just as easily as it cuts grass. Improper use of these mowers has caused many injuries. Follow the specific instructions in the operator or owner's manual and the warning labels on the mower. You must know how to use and control the mower before starting the engine. Use all lawn and garden equipment with caution. Safe procedures can minimize the risk of injury. Always pay attention to how well the equipment is working. Do not try to operate damaged equipment. Follow these general safety rules to prevent injury:

- Read and follow the safety precautions in the mower operator or owner's manual.
- Read and follow all warning labels on the mower.
- Be sure you know how to correctly operate all controls before operating the mower.
- Wear personal protective clothing, shoes, and eye protection when mowing.
- Wear a breathing filter or respirator when mowing in dusty conditions.
- Be sure that all guards are installed and all safety devices operate properly before starting the mower.
- Do not modify or disable any safety device on the mower.
- Do not operate equipment if you are under the influence of alcohol or drugs.
- Be sure you know how to immediately stop the engine and disengage the power before you start an engine.
- Keep children and animals away from your mowing area.
- Do not smoke while you are working on gasoline-powered equipment.
- Do not operate a mower in a fire hazard area.
- Do not operate gasoline equipment in an enclosed area.
- Do not store extra fuel near an open flame or spark.
- Keep your feet and hands in safe locations while starting the engine.
- Do not allow any part of your body or clothing to get near a moving blade.
- Do not allow any part of your body or clothing to get near any moving parts.
- Do not put your feet under the mower deck to lift a mower while it is running.
- Do not reach into the discharge chute to dislodge a clog when the engine is running.
- Stop the engine if you leave the mower unattended.
- Do not use a mower up or down a steep grade, terrace, or incline.
- Do not operate a mower over loose rock or sand.
- Do not pull instead of push a mower.
- Do not operate a mower on wet grass or where there may be bad traction for you or the mower.
- Stop the engine, remove and ground the spark plug wire, and kill the ignition before doing any mower maintenance or repair.
- Allow the engine to cool completely before covering or storing the mower.
- Do not mow when there is insufficient light to see properly.
- Do not run the mower over any loose objects.
- Stop and inspect the mower carefully if you strike any object.
- Stop and repair the problem if the mower begins to vibrate.
- Check parts regularly for looseness, wear, cracks, or other damage and make sure mower is properly adjusted before starting engine.
- Keep the mower blade properly adjusted and sharp.

ROTARY MOWER PARTS AND OPERATION

The most common type of power lawn mower, the **rotary lawn mower**, uses a single horizontally rotating cutting blade to cut grass. Vertical crankshaft engines are used to power the rotary lawnmower. Generally, the vertical crankshaft connects directly to the mower blade.

Mower Parts

The main parts of the rotary lawn mower (Figure 30-1) are the engine, deck, and control bar assembly. The engine is commonly a two- or four-stroke vertical crankshaft model. The **mower deck** is the housing that supports the engine and covers the rotating blade. The deck housing is shaped to house the blade and contain grass cuttings. Grass

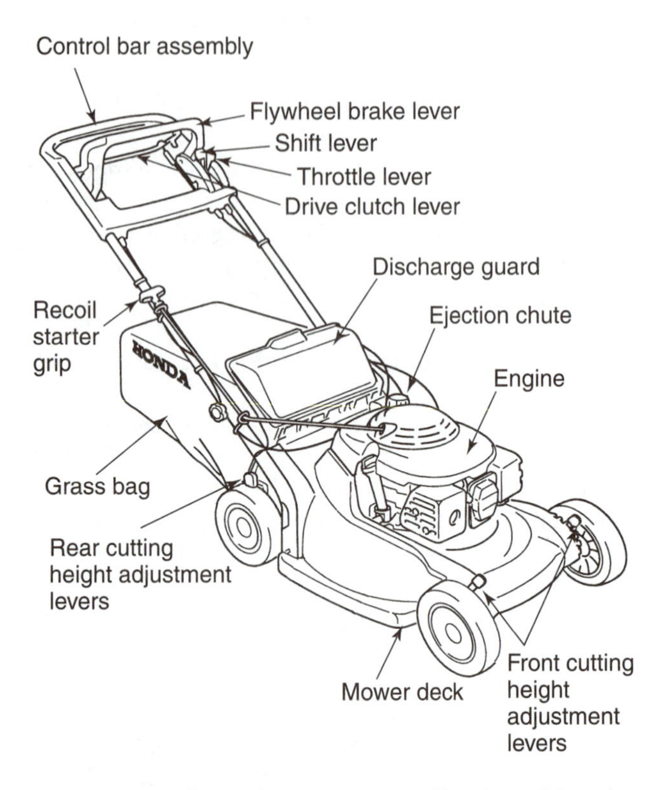

FIGURE 30-1 Parts of a rotary mower. (Courtesy of American Honda Motor Co., Inc.)

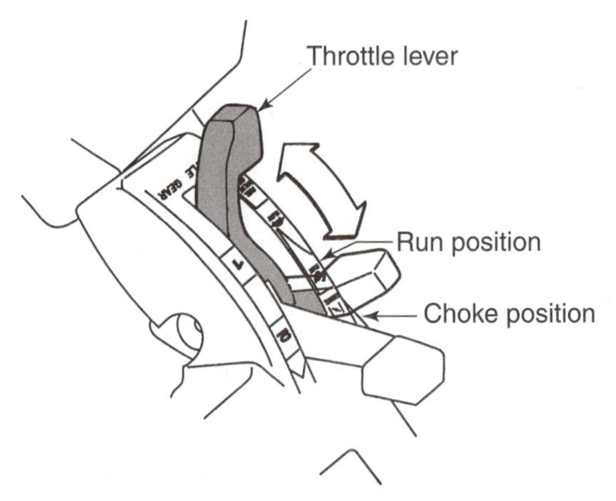

FIGURE 30-2 The throttle lever controls engine speed and choke. (Courtesy of American Honda Motor Co., Inc.)

cuttings are ejected out the side or back of the deck through an ejection chute. A grass bag is mounted on the ejection chute to catch the cut grass. A hinged discharge guard is used on many mowers to prevent grass from being thrown at the operator. A wheel is attached to each corner of the deck. The wheels are usually connected to a height adjustment lever. The height adjustment levers are used to raise or lower the deck and blade to change grass cutting height. The center of the deck has mounting holes for the engine. The base of the vertical engine crankcase bottom cover (sump) mounts directly on the deck. The vertical engine crankshaft passes through the deck to drive the blade.

The **control bar assembly** is the mower handle bar used to control mower direction. Operator controls are mounted on the control bar. These may include throttle control, flywheel brake lever, drive clutch lever, and shift lever. The grip for the recoil starter rope is also located on the control bar assembly. The **throttle lever** is an operator control lever connected to the engine to increase or decrease engine speed at the carburetor. Many throttle levers include a choke position The operator uses the choke position to start the engine. The throttle lever (Figure 30-2) is used to control engine speed. Some mowers have just slow and fast engine speed settings. Some control levers have a fuel shutoff position that is used when the mower is turned off. Older mowers may have an ignition kill switch connected to this lever.

The **flywheel brake lever** (Figure 30-3) is a control lever connected to the engine flywheel brake. The flywheel brake, which is required to prevent injuries to hands, feet, or other body parts, stops the rotating flywheel when the operator's hand is removed from the flywheel brake lever. It prevents an operator from coming in contact with the rotating mower blade. During starting and operation, the operator has to activate the flywheel brake lever. If the lever is not activated the engine will not start or run.

WARNING: Never modify or disable the flywheel brake lever or flywheel brake. This equipment is very important to operator safety.

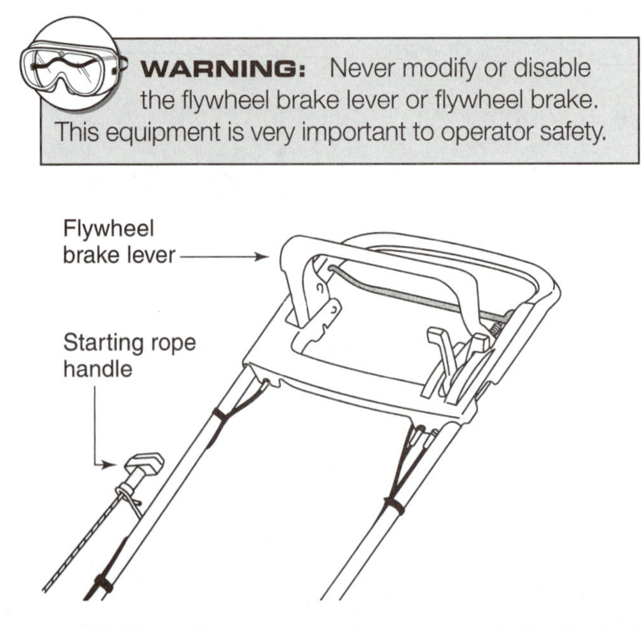

FIGURE 30-3 The flywheel brake lever controls the flywheel brake. (Courtesy of American Honda Motor Co., Inc.)

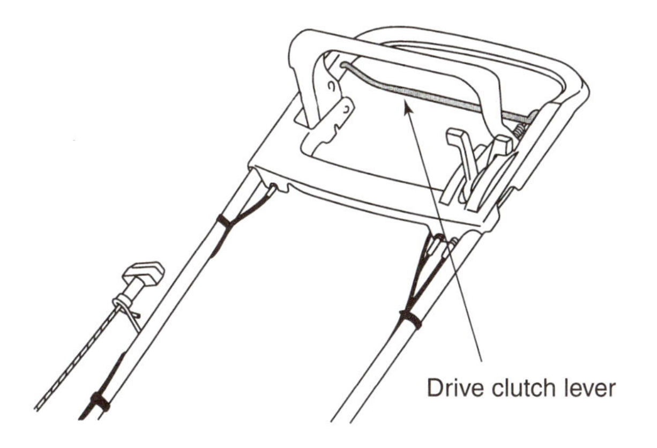

FIGURE 30-4 The drive clutch lever engages and disengages the self-propelled drive. (Courtesy of American Honda Motor Co., Inc.)

Mowers that have a flywheel brake must have the recoil starter grip located on the control bar assembly. The rope has to be pulled at the same time the flywheel brake lever is activated. Older mowers without a flywheel brake lever often have the starter rope and grip mounted to the engine blower housing.

The **drive clutch lever** (Figure 30-4) is an operator control lever on self-propelled mowers that engages and disengages the gear box that drives the rear wheels. The **shift lever** (Figure 30-5) is an operator lever on self-propelled mowers that controls the mower's drive speed. These controls are only used on self-propelled mowers.

Cutting Blade

Many rotary mowers use a single steel cutting blade rotated by the engine crankshaft (Figure 30-6). The blade has a cutting edge on each end. The cutting edges are located on opposite sides of the blade so they can both cut as the blade rotates. The cutting

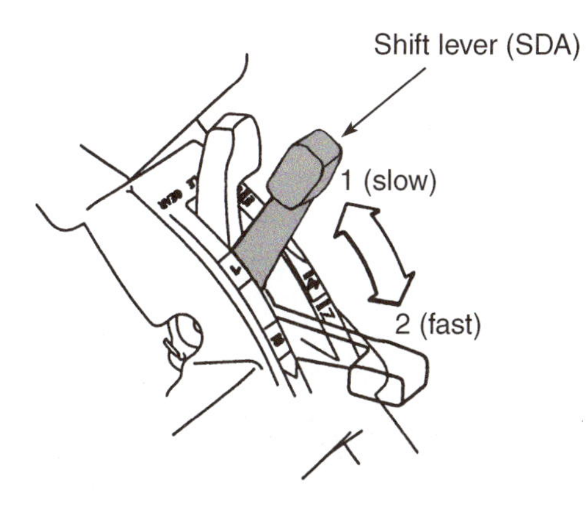

FIGURE 30-5 The shift lever controls the self-propelled mower drive speed. (Courtesy of American Honda Motor Co., Inc.)

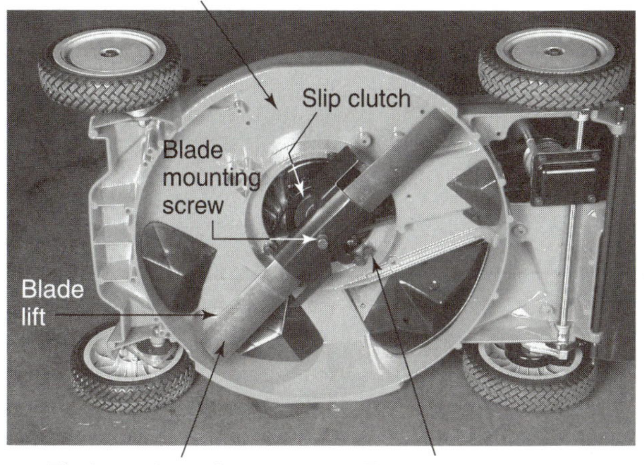

FIGURE 30-6 The blade has a cutting edge to cut grass and a raised sail to vacuum grass.

edge is usually sharpened to a 30-degree angle. The blade is not flat. The area directly behind the cutting edges, called the *sail* or *lift*, is angled upward. The purpose of the lift is to create a vacuum in the deck blade housing, which causes the grass cut by the blade to be pulled into the housing and directed into the ejector chute. From the ejector chute the grass goes into the grass bag.

Sometimes a mower is equipped with two blades. The second blade is used for mulching. Instead of bagging the cut grass, the mulching blade chops the grass and directs it back on the lawn. The mulch helps provide organic material to feed the lawn. The mulching blade is usually mounted above the cutting blade.

CAUTION: When installing a mower blade, the lift section of the blade and the sharpened edge of the blade must point upward. A blade that is installed upside down will not cut properly and can cause injury or damage.

Slip Clutch

A rotating mower blade creates a great deal of inertia. Inexpensive mowers generally have the blade solidly attached to a blade adapter. The blade adapter is solidly attached to the engine crankshaft. If the blade were to strike a solid object, the side forces created could easily bend the crankshaft (Figure 30-7). To avoid this problem, some mowers use a slip clutch that fits between the blade and the end of the crankshaft. The **slip clutch** disconnects the rotary mower blade from the crankshaft if the

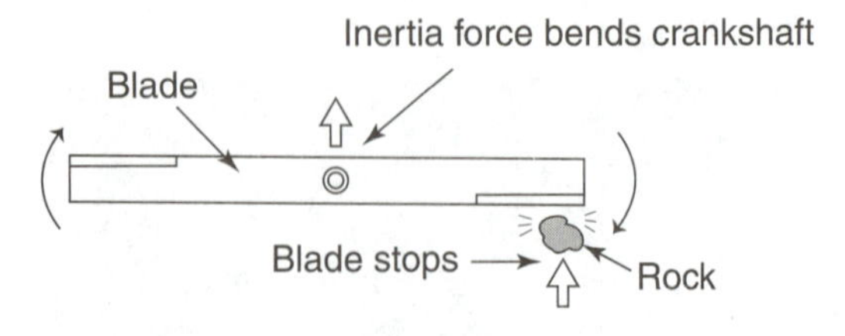

FIGURE 30-7 If the blade tips hit a solid object, the inertia can bend the crankshaft.

blade hits a solid object. Mowers commonly use a lug or friction slip clutch.

The lug slip clutch has housing that fits on the end of the crankshaft (Figure 30-8). The housing is driven by the crankshaft through a key. The housing has two raised tabs called *lugs*. The lugs fit in two holes in the blade. The blade and clutch are held in place by a large washer and hex head screw that threads into the end of the crankshaft. If the blade hits a solid object, the two lugs will shear off, disengaging the blade from the crankshaft and protecting the crankshaft. The engine will no longer rotate the blade. The clutch will have to be replaced to use the mower again.

The friction slip clutch has a housing attached to the crankshaft with a key (Figure 30-9). The bottom of the clutch housing has a flat surface. A large diameter steel, fiber, or rubber friction washer fits on either side of the blade. A bolt and washer are used to hold these parts on the end of the crankshaft. When the bolt is tightened, the blade is held between the two friction washers. The friction of the washers on the blade locks the parts together. The engine drives the blade through the friction lock. If the blade hits a solid object, the friction

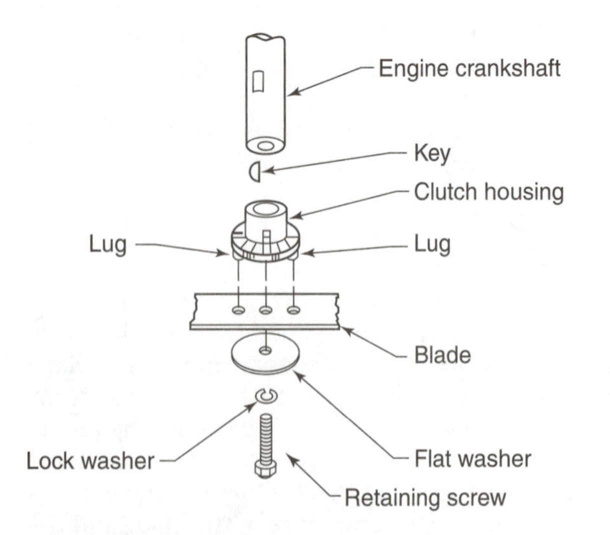

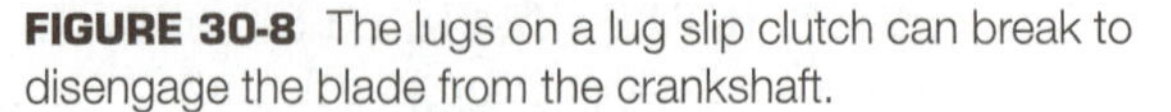

FIGURE 30-8 The lugs on a lug slip clutch can break to disengage the blade from the crankshaft.

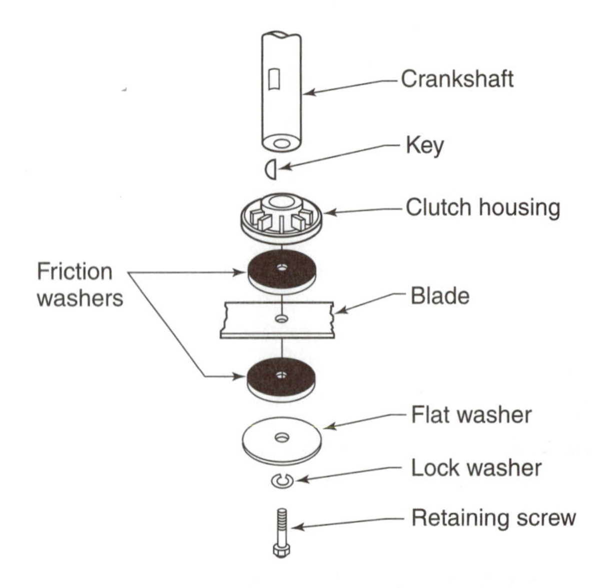

FIGURE 30-9 The friction washers can slip on the blade to disengage it from the crankshaft.

lock will be broken. The blade will be disengaged from the crankshaft. After disengagement, the clutch is usually not damaged. It can be reused after an inspection.

Blade Brake

WARNING: Do not inspect or work on a blade with the engine running. The blade brake clutch could cause the blade to rotate at any time.

All current mowers are required to have a system of stopping the blade from spinning on its own after the engine is shut down. Many mowers use an engine flywheel brake for this purpose. Some mowers have a blade brake. A **lawn mower blade brake** is a brake mechanism attached to the blade assembly that stops the blade. There is a difference between the use of a blade brake and a flywheel brake. The blade break can be combined with a clutch. This allows the blade to stop when the engine is running. The blade brake is usually activated by the same kind of handle bar lever as the flywheel brake.

The handle bar blade brake lever is connected to a cable that enters the mower deck. The cable inner wire is used to release the blade brake and allow the blade to turn. The blade brake and clutch assembly is located between the crankshaft and mower blade. There are several types of brake mechanisms. The disc type brake mechanism is a common type.

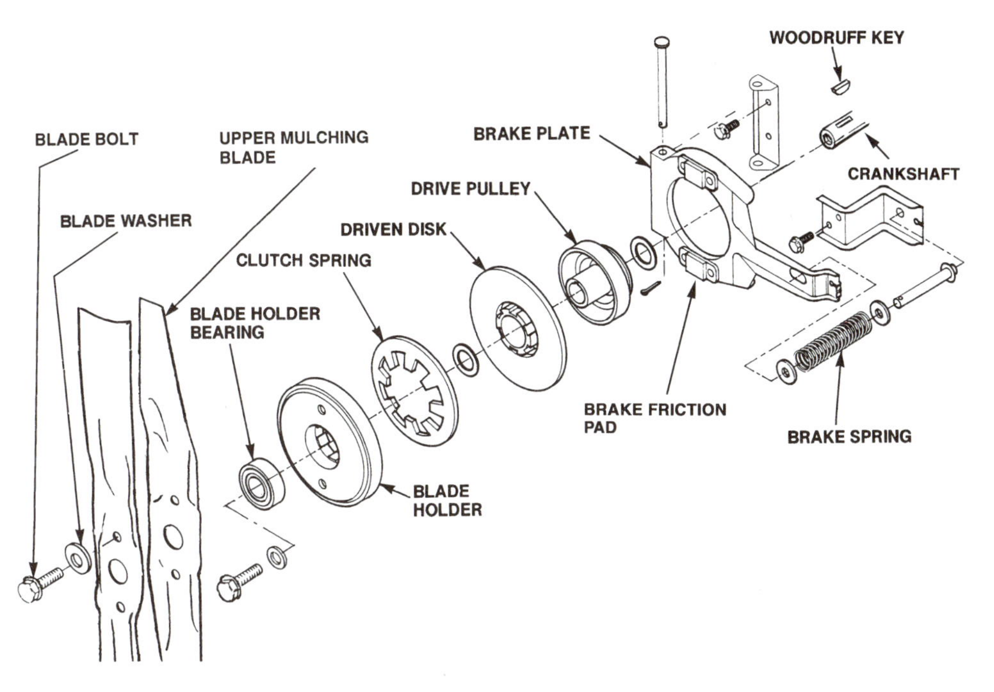

FIGURE 30-10 A blade break clutch can stop the blade and allow the engine to run. (Courtesy of American Honda Motor Co., Inc.)

The disc brake assembly has a drive pulley (sometimes called a *drive disc*) (Figure 30-10). The center of the drive pulley is connected to the engine crankshaft through a key. The drive pulley rotates when the engine is running. A driven disc is connected to the blade through a blade holder. The control cable is connected to a movable brake plate. The plate has two friction brake pads. A large brake return spring holds the brake plate in the "brake on position." In this position, the friction brake pads push the driven disc and stop the driven disc, blade holder, and blade from rotating.

When the operator activates the blade brake lever, the cable moves the brake plate against the brake return spring pressure. The friction brake pads stop pushing on the driven disc. The drive pulley connected to the crankshaft can now rotate the driven disc. The blade can now rotate. There is a clutch spring between the driven disc and the blade holder. Engine speed and centrifugal force must overcome this spring pressure before the driven disc will lock up with the blade holder plate.

ROTARY MOWER TROUBLESHOOTING

Lawn mower problems can be related to engine or equipment problems. Engine troubleshooting is presented in Chapter 21. The most common rotary lawn mower problems are:

- Lawn mower vibrates or rattles excessively
- Lawn mower cuts unevenly

WARNING: Stop the engine, remove and ground the spark plug wire, and kill the ignition before doing any mower inspection, maintenance, or repair.

CAUTION: Check the operator or shop service manual before tipping a rotary lawn mower on its side. Tipping the mower on the wrong side can cause oil to get in the crankcase breather. Unless otherwise specified, tip the mower over with the crankcase breather side up. Mower left and right side is determined by standing behind the mower.

A vibration problem can be caused by loose engine mounting fasteners. Use the correct size wrench to check each fastener. Vibration is often caused by grass and debris lodged under the mower deck. Stop the engine and leave the kill switch in the off position. Remove the spark plug cable from the spark plug and attach it to a good ground. Make sure the fuel tank cap is tight so fuel will not spill when tipping the mower. Tip the mower on its side. Use a 3-foot length of broom handle to scrape and clean the grass and debris out of the deck housing.

CASE STUDY

The Consumer Product Safety Commission studies lawn mower injuries and recommends equipment design changes. It issued the following press release regarding the effectiveness of the blade brake system:

> Americans are using walk-behind rotary lawn mowers with fewer injuries today than at anytime before 1982. According to the U.S. Consumer Product Safety Commission (CPSC), walk-behind mower injuries have declined from about 41,900 in 1983 to about 25,800 emergency room treated injuries in 1989, a reduction of 38 percent. Among the injuries prevented were amputations, fractures, and severe lacerations of fingers, hands, and feet. Blade contact injuries account for 64 percent of all emergency room treatments. CPSC said the decline in injuries was largely due to industrywide compliance with the 1982 federal standard which established blade control requirements for walk-behind mowers. To combat injuries when users accidentally inserted their hands or feet into the path of the moving blade of these mowers, the standard calls for the rotary blade to stop within 3 seconds after the consumer leaves the operator position at the rear of the mower. CPSC estimated that some 80,000 walk-behind mower injuries have been prevented during the 7-year tenure of the standard, with medical and other cost savings of some $680 million. Because of the standard, the agency forecasts that consumers will avoid some 23,000 injuries annually at cost savings of $195 million each year . . .

Source: Consumer Product Safety Commission, press release # 90-131, 1990. Power Lawn Mower Injuries Decline, Washington, DC.

A damaged or out of balance blade also causes a vibration. Inspect the blade for damage. The most common type of damage is a bent blade. If the blade is damaged, it will need to be removed for replacement or sharpening and balancing.

Grass cutting problems may be caused by a dull blade or incorrect blade-cutting height. Mow a test strip of grass and inspect the cut. A dull blade will bend the grass blades but not actually cut them. Stop the engine and leave the kill switch in the off position. Remove the spark plug cable from the spark plug and attach it to a good ground. Make sure the fuel tank cap is tight so fuel will not spill when tipping the mower. Tip the mower on its side. Inspect the cutting edges on both sides of the blade. The outer ½ inch (13 millimeter) of the cutting edge does most of the cutting. Feel this edge to make sure it is sharp. If not, the blade needs to be removed for sharpening or replacement.

A blade that is incorrectly adjusted too high will not cut all of the grass blades. An uneven cut means that the cutting height is adjusted improperly. The blade should be level with the ground. The blade height is adjusted by adjusting the height of the deck. Deck height is adjusted by adjusting the mower wheels up or down in relation to the deck. Some mowers have adjustments on all four wheels. Other mowers have adjustments on only the front wheels. Push the mower on to a flat level surface. Check the front, back, and side-to-side height of the deck off the ground. If you find a ⅛ inch (3 millimeter) difference, adjust the deck height.

ROTARY MOWER SERVICE

WARNING: Stop the engine, remove and ground the spark plug wire, and kill the ignition before doing any mower maintenance or repair.

Mower maintenance and repair requires the correct identification of the mower and engine. There are many different mower manufacturers. Most mower manufacturers do not manufacture engines. They buy engines for their mowers from engine manufacturers. For example a Toro walk behind lawn mower may be equipped with a Briggs & Stratton, Suzuki, or Tecumseh engine. You need to identify the engine to get engine service information and replacement engine parts. You can locate the engine identification information from the blower housing decal or stamping. You must identify the mower to get mower service information and replacement parts. Mowers are identified by a model and serial number plate located on the mower deck (Figure 30-11). Use this information to get service information and replacement parts from an authorized dealer.

Blade Inspection

Blades used on rotary mowers often get damaged from contacting rocks, cement edges, and sprinkler

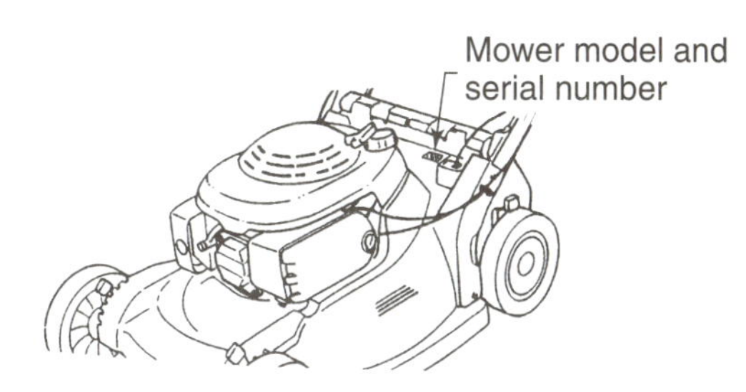

FIGURE 30-11 The mower model and serial number are located on a plate on the mower deck. (Courtesy of American Honda Motor Co., Inc.)

pipes. Damage to the blade dulls the cutting edges and can make the blade out-of-balance. An out-of-balance condition causes the blade to vibrate as it rotates. The vibration is transferred to the engine crankshaft. Eventually, the vibration can result in crankshaft PTO main bearing damage, a bent crankshaft, or a sheared (broken) crankshaft key.

Stop the engine and leave the kill switch in the off position. Remove the spark plug cable from the spark plug and attach it to a good ground. Make sure the fuel tank cap is tight so fuel will not spill when tipping the mower. Tip the mower on its side. Remove and service any blades with visible damage to the cutting edges. Remove the blade by removing the retaining fasteners (Figure 30-12) that secure the blade to the blade adapter, slip clutch, or blade brake.

SERVICE TIP: A blade balancer that includes a bent blade gauge is the best way to accurately check for bent blades.

Use a wire brush and scraper to clean the blade for inspection. Lay the blade on a flat surface and make sure it is not bent (Figure 30-13). A bent blade cannot be repaired; it must be replaced. Check the condition of the sails and cutting edges (Figure 30-14). Damaged sails prevent grass from being

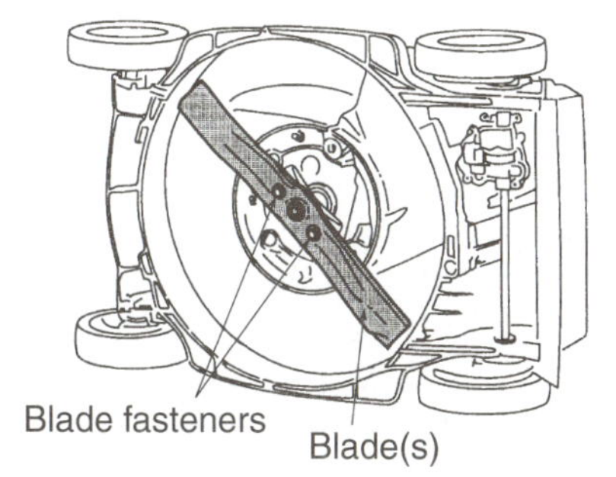

FIGURE 30-12 Remove the retaining fasteners to remove the blade. (Courtesy of American Honda Motor Co., Inc.)

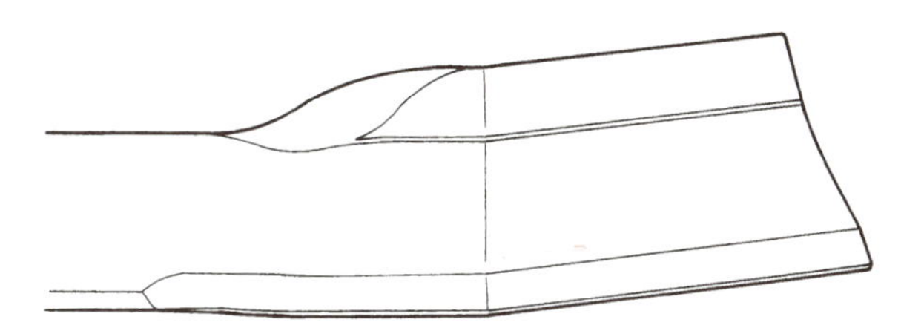

FIGURE 30-13 Place the blade on a flat surface and look for a bent blade. (Courtesy of American Honda Motor Co., Inc.)

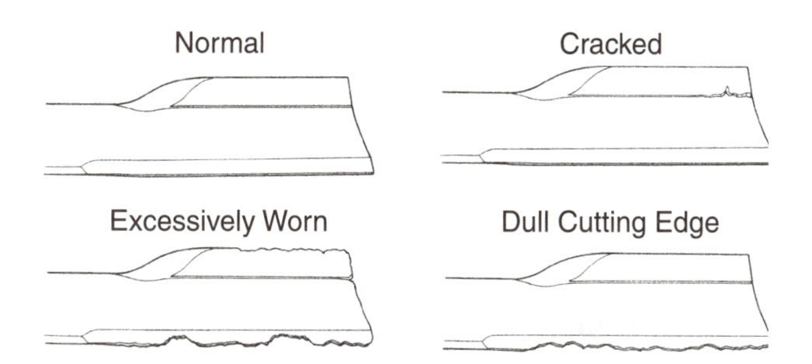

FIGURE 30-14 Common types of blade damage. (Courtesy of American Honda Motor Co., Inc.)

picked up and ejected into the grass bag. Look for cracks near the cutting edges. A cracked blade must be replaced. Inspect the mounting holes or hole in the blade. An out-of-round hole prevents the blade from centering properly and causes vibration. Replace a blade with worn mounting holes. Check the condition of the blade adapter or slip clutch parts. Look for worn slip clutch lugs or friction washers. Replace these parts when you mount the blade.

SERVICE TIP: You can check a blade for cracks by ringing it like a bell. Hang the blade from its mounting hole. Strike the blade gently with a ball peen hammer. A good blade will ring like a bell. A cracked blade will sound like a dull thud.

Blade Sharpening and Balancing

WARNING Wear eye protection to protect from flying metal when sharpening a blade on a bench grinder.

Sharpen each blade cutting edge on a bench grinder or special blade grinder. Remove the minimum amount of metal to sharpen the blade and remove nicks in the cutting surface. Small nicks can be

FIGURE 30-15 Sharpen the blade cutting edge to the original angle. (Courtesy of American Honda Motor Co., Inc.)

removed by sharpening. Large nicks on the cutting surface require the removal of too much material. Blades with large nicks require blade replacement. Try to follow the original 30–40 degree contour of the cutting edge (Figure 30-15). Try to remove the same amount of metal from both cutting edges to keep the blade in balance.

SERVICE TIP: After sharpening the blade, dress the cutting edges with a file. The cutting edges should not be razor sharp. Sharp edges chip easily and get dull.

The blade should be checked for balance after sharpening. A **blade balancer** is a cone-shaped tool used to balance a rotary lawn mower blade. The pointed end of the cone fits in the blade center hole. Place the balancer on a level bench or floor surface. Set the blade center hole over the balancer and observe the blade. A blade that is in balance will stay parallel to the surface supporting the balancer. The heavy end of an out-of-balance blade will dip down. Remove the blade from the balancer. Sharpen the blade, removing additional metal from the heavy side of the blade. Do not grind metal off the blade sail area. Recheck the blade for balance and repeat the procedure if necessary.

If a cone balancer is not available, you can make a rough check for balance with a screwdriver. Hang the blade on the screwdriver through the center blade hole (Figure 30-16). A balanced blade

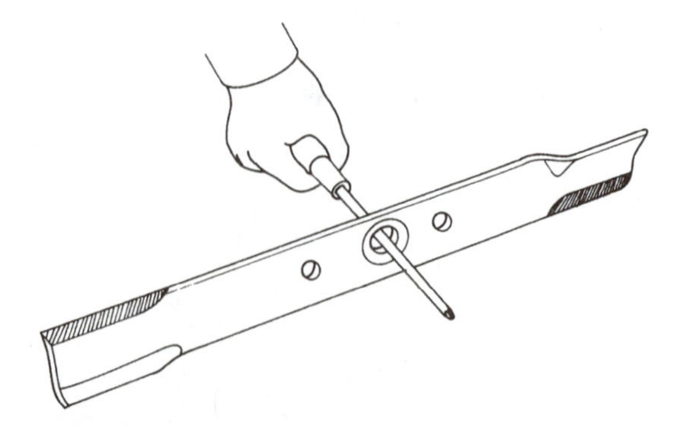

FIGURE 30-16 Checking blade balance on a screwdriver. (Courtesy of American Honda Motor Co., Inc.)

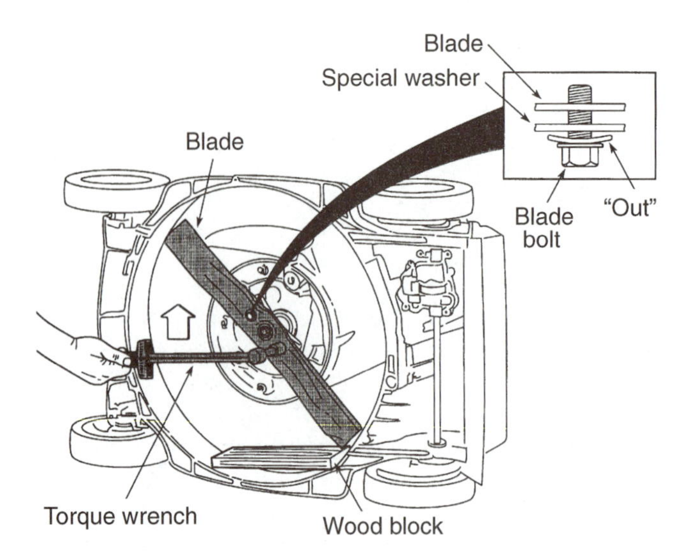

FIGURE 30-17 Tighten the blade retaining fasteners with a torque wrench. (Courtesy of American Honda Motor Co., Inc.)

will stay horizontal. If one end dips down, the blade is out of balance. Sharpen the blade, removing additional metal from the heavy end to get it in balance.

Reinstall the sharpened and balanced blade on the mower blade adapter, slip clutch, or blade brake. Install correct fasteners in the correct order. Look up the recommended torque in the mower owner's or shop service manual. Tighten the retaining fasteners to the correct torque specifications using a torque wrench (Figure 30-17). The procedure for removing, sharpening, balancing, and installing a rotary lawn mower blade is shown in the accompanying sequence of photographs.

WARNING: High quality-fasteners must be used to retain the blade. Low tensile strength fasteners could break and cause damage to the equipment or injury to the operator. Check the shop service manual recommendations.

Adjusting Cutting Height

The rotary blade must be adjusted to the correct height for the type of grass being cut. The cutting height is adjusted by the position of the wheels. Most mowers have adjustment levers on the wheels. To adjust the cutting height, pull the adjustment toward the wheel and move it into another notch (Figure 30-18). Raising the wheel lowers the mower deck and blade, which causes the mower to cut the grass shorter. Lowering the wheel raises the mower deck and blade. The grass will be

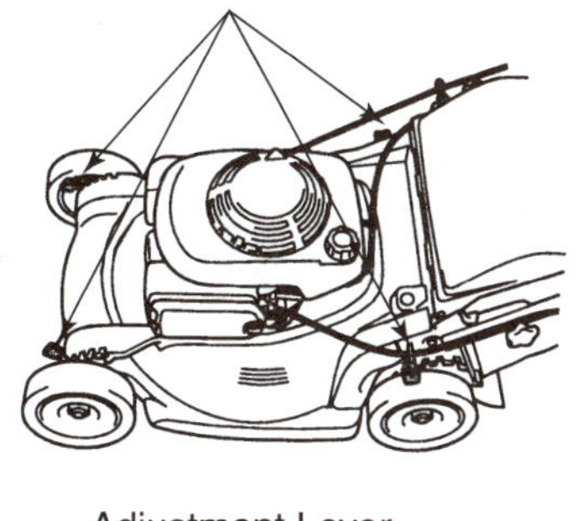

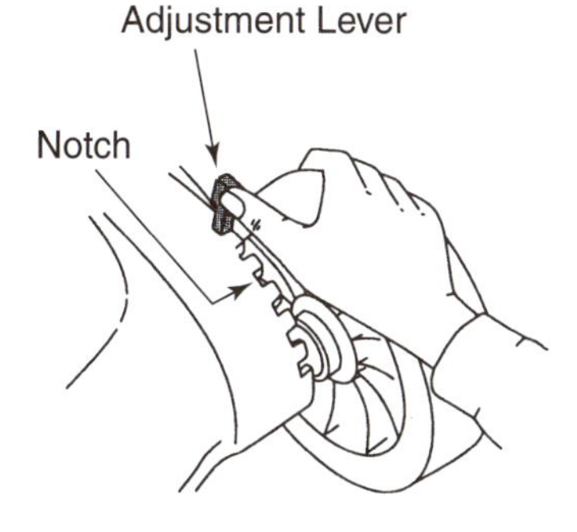

FIGURE 30-18 Adjusting the cutting height by adjusting wheel position. (Courtesy of American Honda Motor Co., Inc.)

cut at a higher level. Make sure that all four adjustment levers are set to the same cutting height adjustment. Check that the deck is the same height off the grass all the way around. Some mowers only have adjustments on two wheels. After the adjustment, cut a test strip of grass. The grass should be cut level and to the correct height.

> **WARNING:** Stop the engine, remove and ground the spark plug wire, and kill the ignition before doing any mower maintenance or repair.

If the height adjustment fails to get the proper result, you should check the blade to make sure it is parallel. The rotary lawn mower blade must rotate in a plane that is parallel to the ground. Stop the engine and leave the kill switch in the off position. Remove the spark plug cable from the spark plug and attach it to a good ground. Make sure the fuel tank cap is tight so fuel will not spill when tipping the mower. Tip the mower on its side. Use a measuring tape to measure the distance between the blade and the top of the mower deck (Figure 30-19). Rotate the blade 180 degrees and take another measurement. Be sure to measure between the same points for both sides. The measurements should be the same. If they are not, the problem could be a bent blade or bent crankshaft. A damaged blade

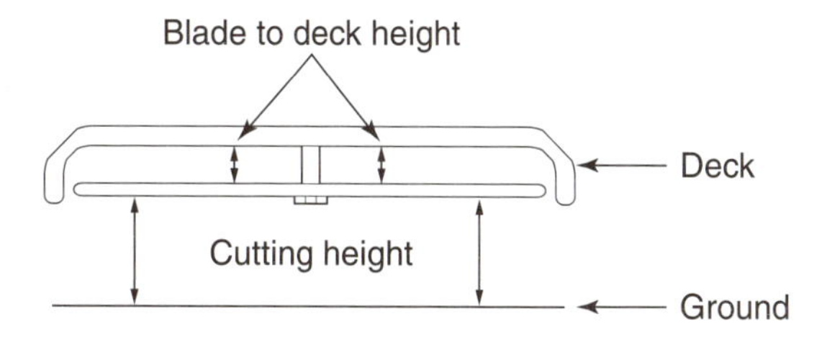

FIGURE 30-19 Checking the blade for parallel.

adapter or worn mounting holes can also cause this problem. Replace the blade or worn mounting parts. Make another measurement. If the blade is still not parallel the problem is most likely a bent crankshaft.

REEL MOWER PARTS AND SERVICE

A **reel lawn mower** is a mower that cuts grass with multiple cutting blades mounted on a rotating reel. Reel lawn mowers were first developed as hand-powered models. Pushing on the mower caused the reel to rotate and cut the grass. Most reel mowers are powered today. The reel mechanism is much heavier than the single rotary blade so reel mowers are larger and heavier than rotary mowers. Most reel mowers are self-propelled so the operator does not have to push the heavy mower.

Mower Parts

The reel lawn mower has the same control bar assembly as the rotary lawn mower. The control bar assembly is connected to the cutting reel assembly. The cutting reel assembly is the reel lawn mower parts assembly that supports the cutting reel, grass bag, and engine (Figure 30-20). Two large rear wheels and two small front wheels support the cutting reel assembly. The small front wheels position the cutting reel close to the ground. The reel mower engine is usually a four-stroke horizontal crankshaft model. The engine drives the reel with a belt or chain drive system. The drive system also drives the rear wheels to self-propel the mower.

A reel mower blade uses a scissors action to cut the grass (Figure 30-21). The reel assembly has a set of curved blades. Each of the blades has a sharpened edge. The action of the blades and cutter bar is similar to the opening and closing of scissors. Blades of grass are caught between the cutter bar and blades and cut. The smoothness of the cut is determined by how often (frequently) the scissors action takes place. The frequency of the cut is

PHOTO SEQUENCE 23

Procedure for Removing, Sharpening, Balancing, and Installing a Rotary Lawn Mower Blade

1. Turn the ignition kill switch to off. Remove and ground the spark plug wire for safety.

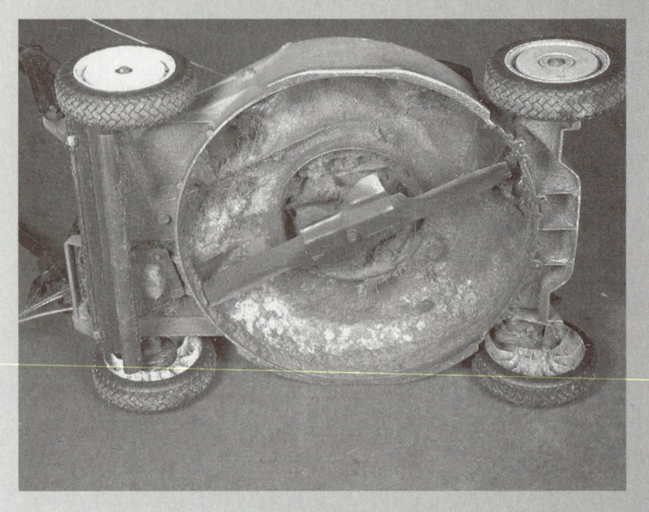

2. Check the manual for the correct direction and tip the mower on the side for blade access.

3. Remove the blade retaining fasteners and remove the blade.

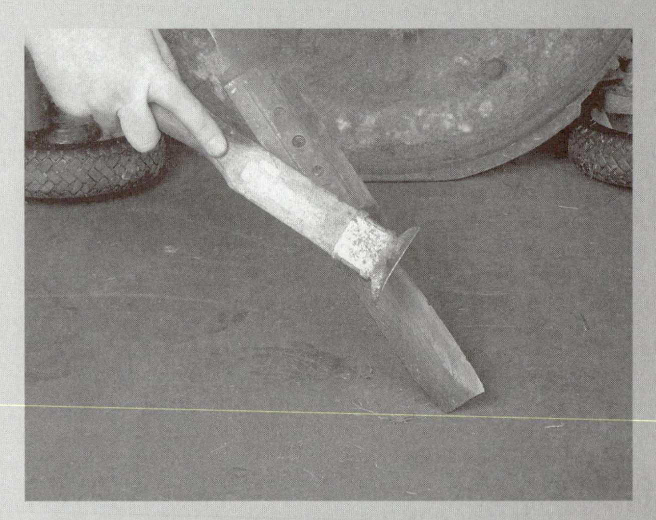

4. Clean the blade with a wire brush.

5. Inspect the blade for bending and damage to the cutting edges.

6. Sharpen the cutting edges on a grinder following the original blade contour.

7. Check the blade balance on a balance cone. Remove metal from the heavy side if necessary.

8. Install the blade and blade fasteners. Torque the blade fasteners to the specified torque with a torque wrench.

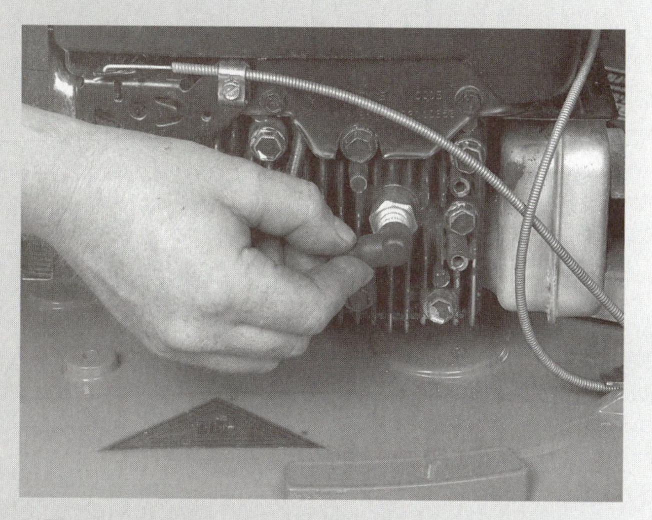

9. Tip mower back on the wheels. Install the spark plug wire back on the spark plug.

determined by reel diameter, number of blades, reel speed, and mower drive speed.

Reel lawn mowers cut grass much more smoothly than rotary mowers. They can cut grass much shorter without scalping. Scalping occurs when the blade cuts through the lawn into the topsoil below. Reel lawn mowers are often used by professional gardeners and on formal turf areas such as golf courses. The reel cutting blades are easily damaged. These mowers must be operated on smooth ground that is free of debris. They cannot be used in deep grass because the blades will bend and run over the grass.

Reel Mower Service

First, inspect each cutting blade along its entire length. Look for nicks or damage to the cutting edges. Small nicks can be removed by careful filing. Dull blades require special sharpening equipment. Coarse valve-lapping compound can be used to sharpen a blade that is slightly dull. Put a small amount of compound along the entire length of the cutter bar. Rotate the blade backwards. The compound will sharpen the blade. Be sure to clean off all of the compound after sharpening.

The cutting bar must be adjusted properly in relation to reel cutting blades for good grass

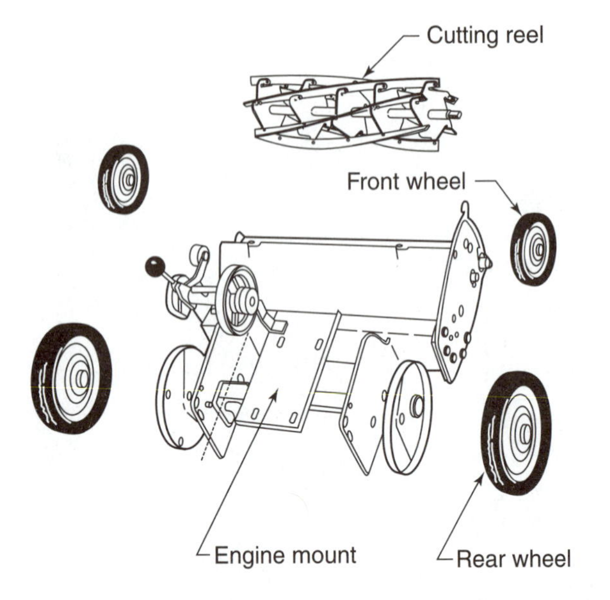

FIGURE 30-20 Basic parts of the cutting reel assembly.

cutting. The cutter bar must be the correct distance away from the reel blades along its entire length. The adjustment is correct when you can rotate the reel with light finger pressure. The blades should just touch the cutting bar along its entire length. Some mowers have an adjustable cutting bar. Other mowers have an adjustable reel.

Before you make the adjustment, clean the cutting bar and blades of all grass and debris so you can see clearly as you make the adjustment. Locate the adjustment screws (Figure 30-22). They are usually located on the side of the mower near the reel. Loosen the adjustment screws. Move the reel or cutting bar until they are just touching each other. Tighten the adjustment screws. Rotate the reel and observe the blade-to-cutting bar contact. You should see the same contact along the entire length of the cutter bar. If the reel is hard to rotate, the cutting bar is too close to the blades. Loosen the adjustment screws and make another adjustment.

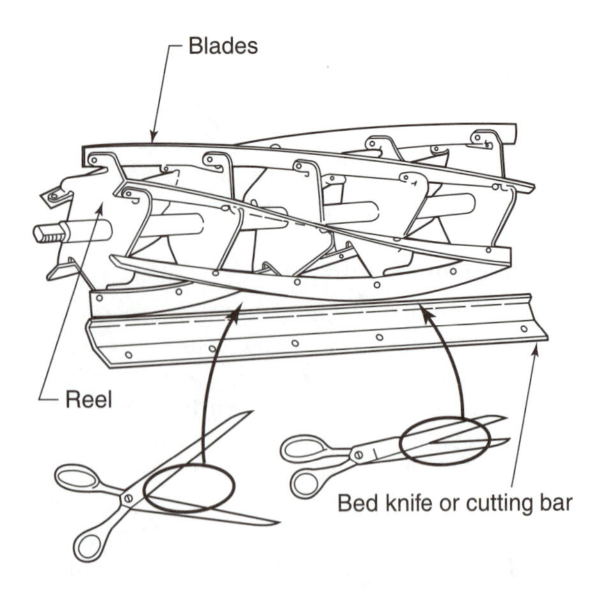

FIGURE 30-21 The blades and cutting bar work like a pair of scissors to cut the grass blades.

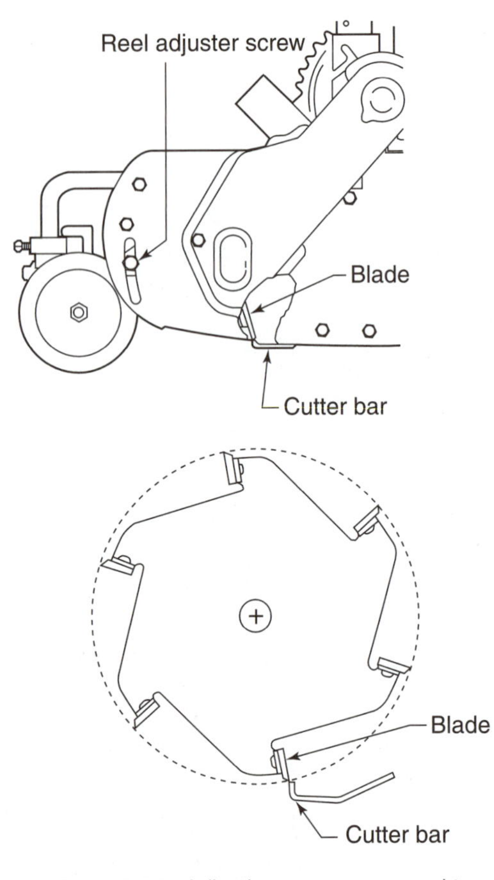

FIGURE 30-22 Adjusting screws are used to position the blades in relation to the cutter bar.

CAUTION: If you do not clean a grease fitting before using a grease gun, dirt will be forced into the bearing and cause early bearing failure.

The reel is usually supported on each end by a tapered roller bearing (Figure 30-23). The bearing must be lubricated with grease at regular intervals. There is usually a grease fitting located in each bearing housing. Wipe off the grease fitting with a clean shop rag. A hand-operated grease gun (Figure 30-24) with the recommended type of grease is used to lubricate the bearings. Push the nozzle of the grease gun over the fitting. Pump the handle on the grease gun several times to force lubricant into the bearing.

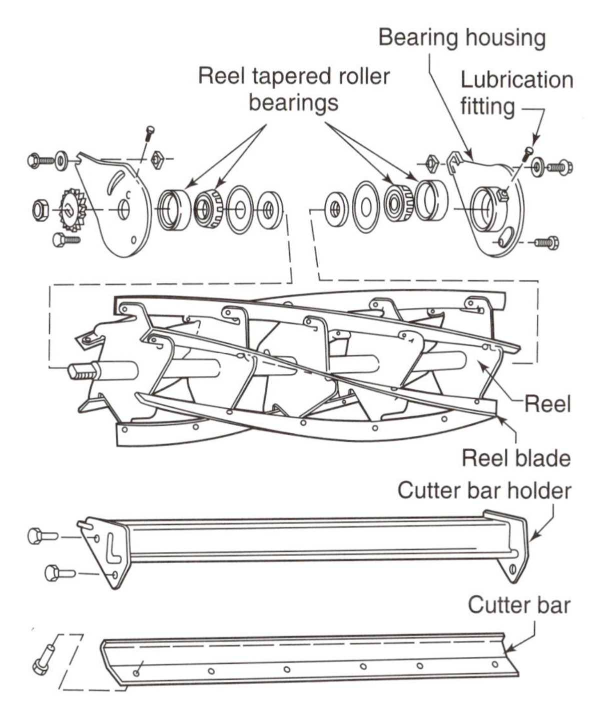

FIGURE 30-23 The reel is supported on two tapered roller bearings.

Remove the nozzle from the grease fitting. Clean up any grease left on the outside of the fitting.

CAUTION: Always wipe up any grease or spilled oil off mower parts. Grease and oil attract dirt that causes parts wear.

SELF-PROPELLED MOWERS

Many rotary and all powered reel lawn mowers are self-propelled. A **self-propelled lawn mower** has a drive system to rotate two of the wheels. These mowers do not require pushing by the operator. When the drive system is engaged, the mower propels itself. There are many different types of drive systems. Rotary mowers usually self-propel the front or rear wheels. Reel type mowers usually propel the rear wheels. The self-propelled system uses a belt drive, gear box, and wheel drive (Figure 30-25).

Belt Drive

A belt drive system commonly uses a belt to transfer power from the PTO end of the crankshaft to a gear box. The belt drive system has a drive pulley that fits on the PTO end of engine crankshaft. A belt drive is usually used to transfer engine torque from the engine PTO to the gear box. A **drive belt** transfers torque from one pulley to another. The drive belt is often called a *V-belt*. It gets its name from its V shape. The sides of the belt form a V shape to fit into the V-shaped groove in a pulley. The friction between the belt and pulley lock the two parts together. The belt must be strong yet flexible. They are constructed from neoprene or oil-resistant artificial rubber. They are usually strengthened with tensile cords (belts) of rayon, polyester, or Kevlar.

The belt transfers torque to a drive pulley (Figure 30-26). The drive pulley is connected to the input shaft on the gear box. Rotation of the engine pulley causes the belt to rotate the drive pulley. An idler pulley is often used between the engine pulley and the drive pulley. The idler pulley maintains a tension on the belt and keeps it on track. Idler pulleys are often used when the belt has to span a long distance.

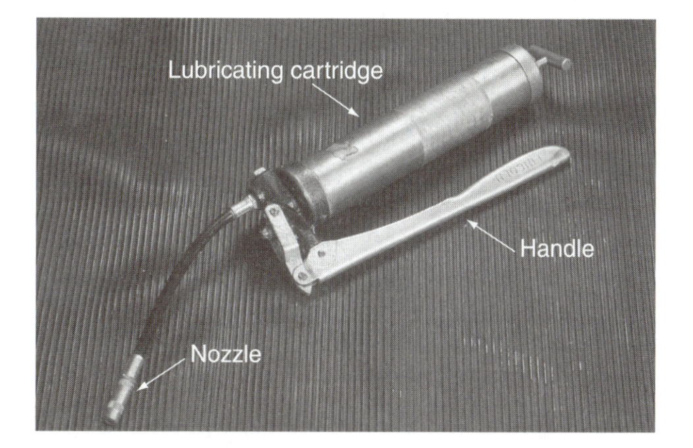

A. Hand Operated Grease Gun

B. Grease Used in Grease Gun

FIGURE 30-24 Reel bearings are lubricated with a hand-operated grease gun.

Transmission (Gearbox)

The **transmission** (gearbox) is a gear system that transfers engine torque to the lawn mower driving

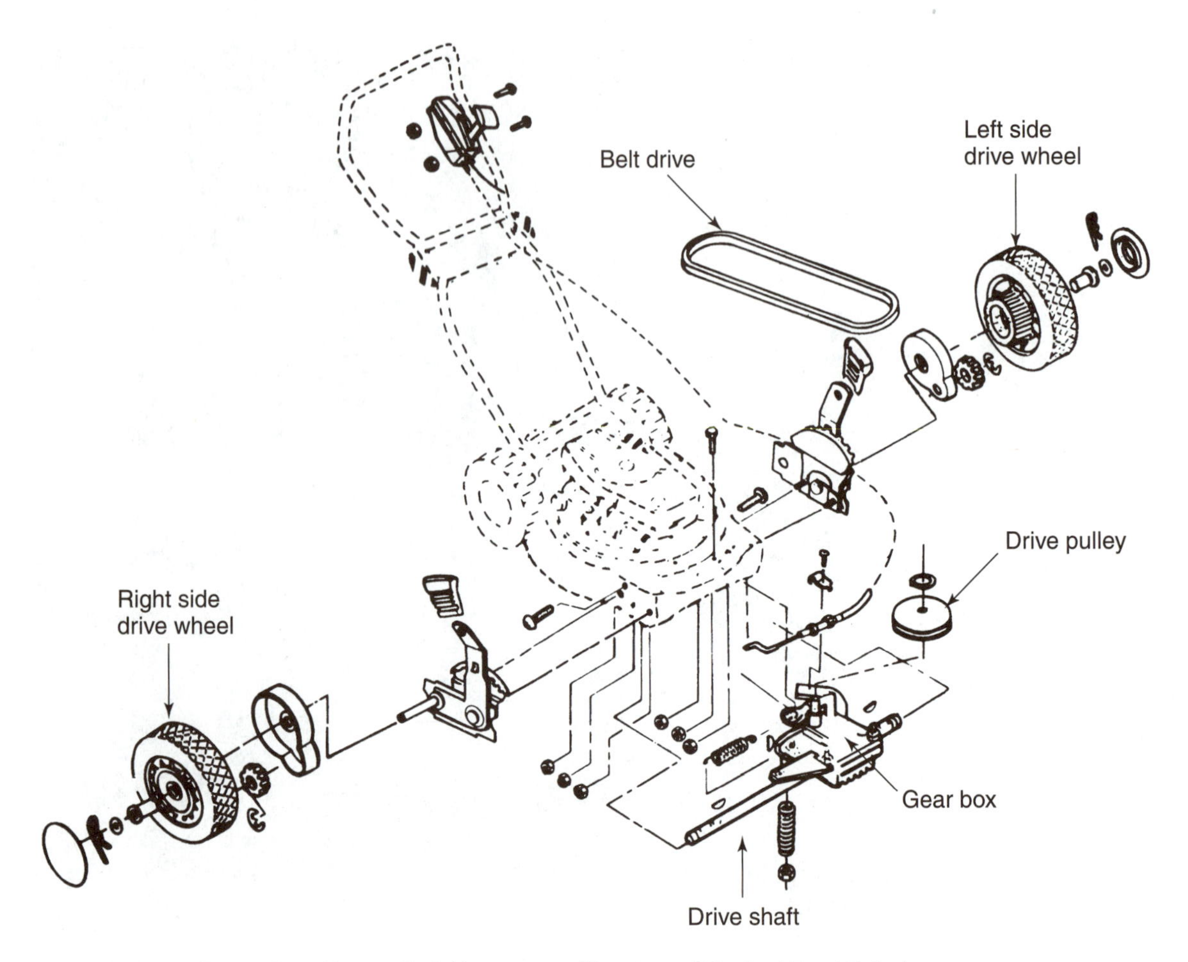

FIGURE 30-25 Parts of a self-propelled drive system. (Courtesy of Poulan/Weed Eater.)

wheels. There are one-, two-, three-speed and higher transmissions. A single speed transmission is the most common. The gears inside the transmission have two basic purposes: They change the direction of the power flow and provide a gear reduction.

The transmission has two main gears (Figure 30-27). One of the gears is called a *worm gear*. The worm gear has a spiral-shaped gear formed on the end of a shaft. The driven pulley is attached to the end of the worm gear shaft. When the engine is running the worm gear rotates. The teeth of a round helical gear are in mesh with the worm gear teeth. The rotating worm gear causes the helical gear to rotate. The rotating helical gear causes a long drive shaft to rotate. Each end of the drive shaft drives one of the mower wheels. The worm gear and helical gear transfer the flow of power through a right angle. The power flow direction is changed from engine rotation direction to wheel rotation direction.

The two gears also provide a torque increase and a speed reduction (Figure 30-28). It is possible to increase torque by connecting the engine to a small gear and the drive wheels to a large gear. When a small gear turns a larger one, the effect is the same as using a larger wrench to tighten a nut or bolt. More turning effort or torque is possible. The larger gear provides more leverage to turn the driving wheels with more torque.

When a small gear turns a large one, the small gear turns around several times before the big one has made a complete revolution. This means that the engine, connected to a small gear, turns at a higher rpm than the drive wheels. The speed difference, as well as the torque increase, depends on the number of teeth on the gears. If two gears in mesh have the same number of teeth, they will turn at the same speed and will not multiply torque. A multiple speed transmission has more than one helical gear. Each gear has a different number of teeth and provides a different driving speed.

The transmission uses a clutch so the operator can engage or disengage the mower drive wheels (Figure 30-29). The helical gear is not connected directly to the drive shaft. It is free to rotate independent of the drive shaft. A jaw clutch next to the

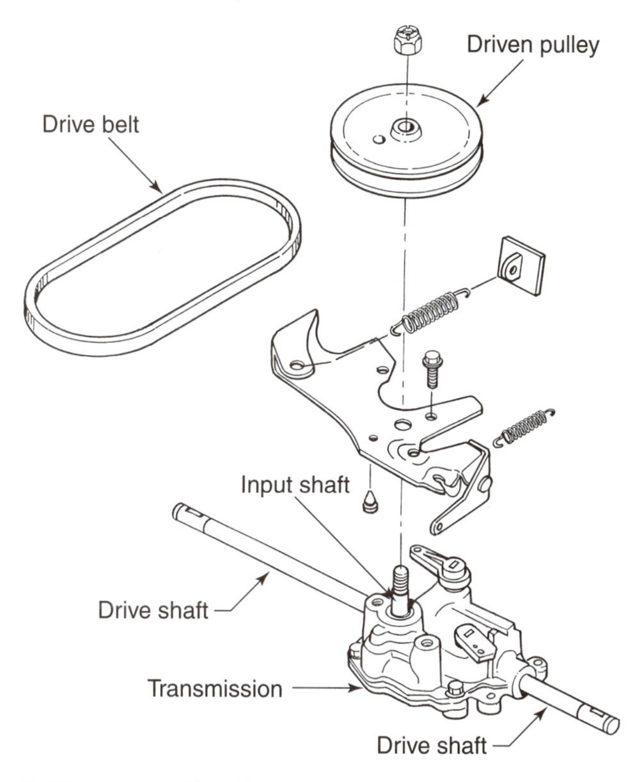

FIGURE 30-26 The drive belt assembly transfers torque from the engine to the gear box. (Courtesy of American Honda Motor Co., Inc.)

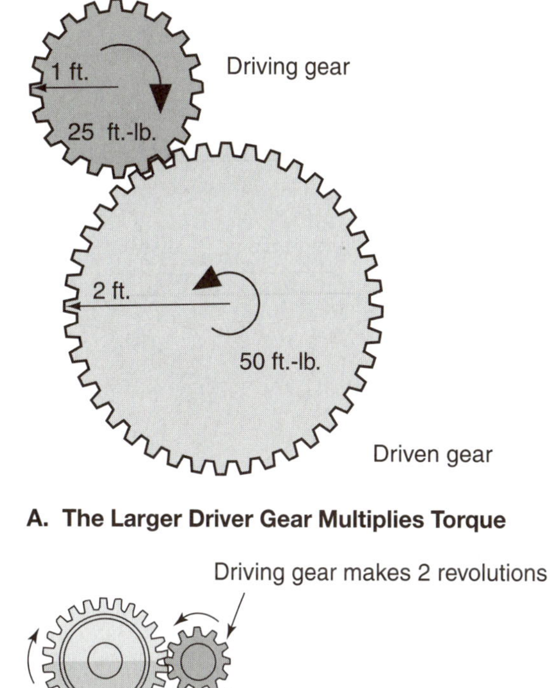

FIGURE 30-28 The larger driven gear multiplies torque and turns more slowly than the drive gear. (*B*: Reproduced by permission of Deere & Company, © 1991. Deere & Company. All rights reserved.)

helical gear is attached to the drive shaft by a key. The jaw clutch has three raised jaws that can be engaged to the side of the helical gear. A clutch yoke is used to lock the jaw clutch to the helical gear. The yoke is connected to a shift lever on the mower handle. When the operator shifts the mower to drive, the yoke moves the jaw clutch onto the helical gear and locks the two parts together. The helical gear is connected to the drive shaft through the jaw clutch and the mower wheels are rotated. When the operator shifts out of drive, the yoke disconnects the jaw clutch from the helical gear. The helical gear is disconnected from drive shaft and does not drive the mower wheels.

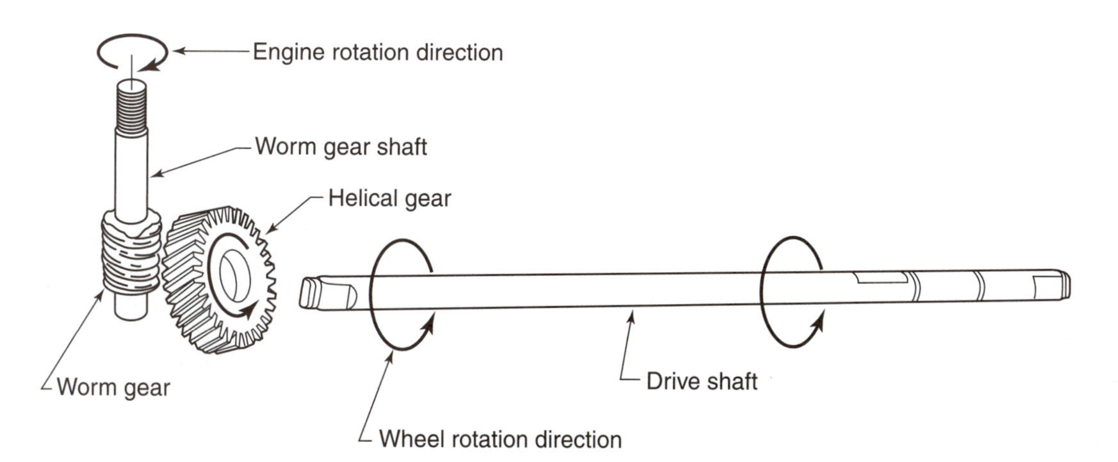

FIGURE 30-27 Power flow through a worm drive gear box.

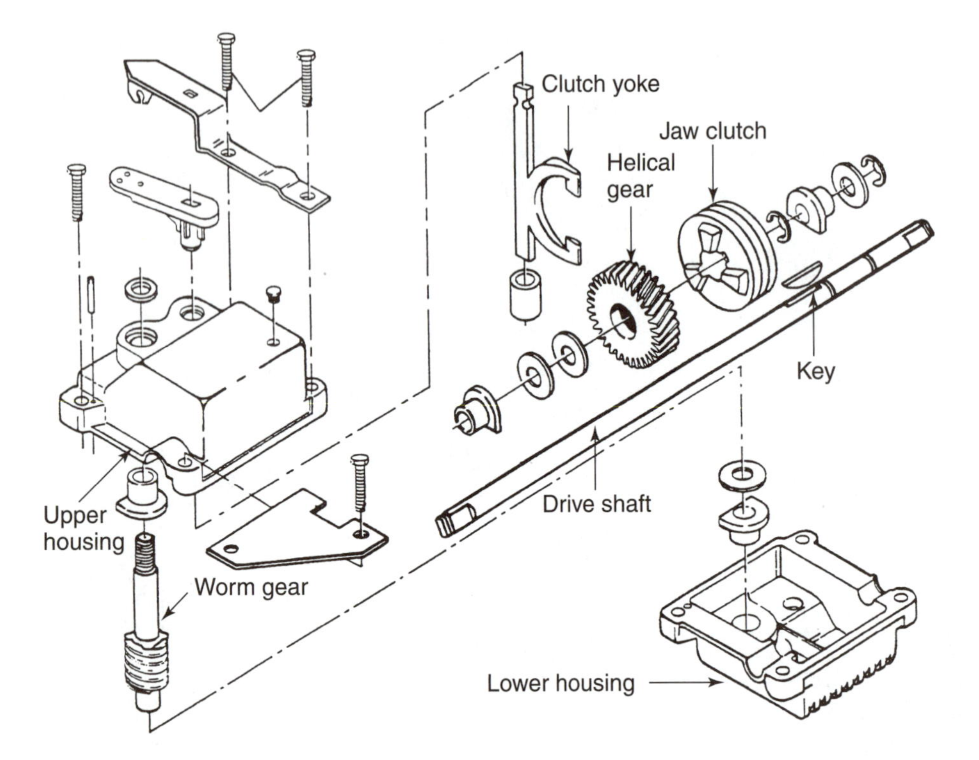

FIGURE 30-29 A jaw clutch locks and unlocks the helical gear to the drive shaft. (Courtesy of Poulan/Weed Eater.)

Wheel One-Way Clutch

Each of the self-propelled mower drive wheels are connected to the drive shaft with a one-way clutch. When a mower is turned during mowing, the inside drive wheel has a smaller turning radius than the outside wheel. If the wheels were connected directly to the drive shaft, the inside wheel would try to spin or skid on the turn. A **one-way clutch** is used on a self-propelled mower drive wheel to allow the inside wheel on a turn to free wheel or rotate slower than the outside wheel.

The one-way clutch uses a spring-loaded ratchet key that fits in the drive shaft and drives the pinion gear (Figure 30-30). The pinion gear drives the wheel. The drive shaft rotates the spring-loaded ratchet key. The ratchet key fits inside the irregularly shaped hole in the middle of the pinion gear. When the mower is going straight ahead, the key wedges in the pinion hole. The clutch is locked up and the drive shaft drives the wheel. When the mower is turned, the inside wheel is slowed by the tight turn. This causes the ratchet key to unlock from the pinion hole. The wheel is unlocked from the drive shaft and freewheels (turns free). The outside wheel stays locked to the drive shaft and drives the mower.

BELT INSPECTION AND REPLACEMENT

WARNING: Stop the engine, remove and ground the spark plug wire, and kill the ignition before doing any mower maintenance or repair. Never try to adjust or inspect a drive belt with the engine running.

The most common problem with a self-propelled mower is a broken or slipping drive belt. A broken or slipping belt prevents the engine pulley from driving the drive pulley. Power will not get to the drive wheels. Drive belts should be inspected for a problem anytime you service the lawn mower. A quick inspection can locate a problem and save your customer a major problem.

To inspect the belts, grab the belt in your hand and twist it so you can see the underside of the V shape (Figure 30-31). Use a flashlight so you can make a close inspection. Check the belt for cracks. Cracks indicate the belt is getting ready to fail. Oil soaked belts can slip and not rotate the drive pulley. Glazed belts have a shiny appearance. This occurs when a belt is not tight enough and the

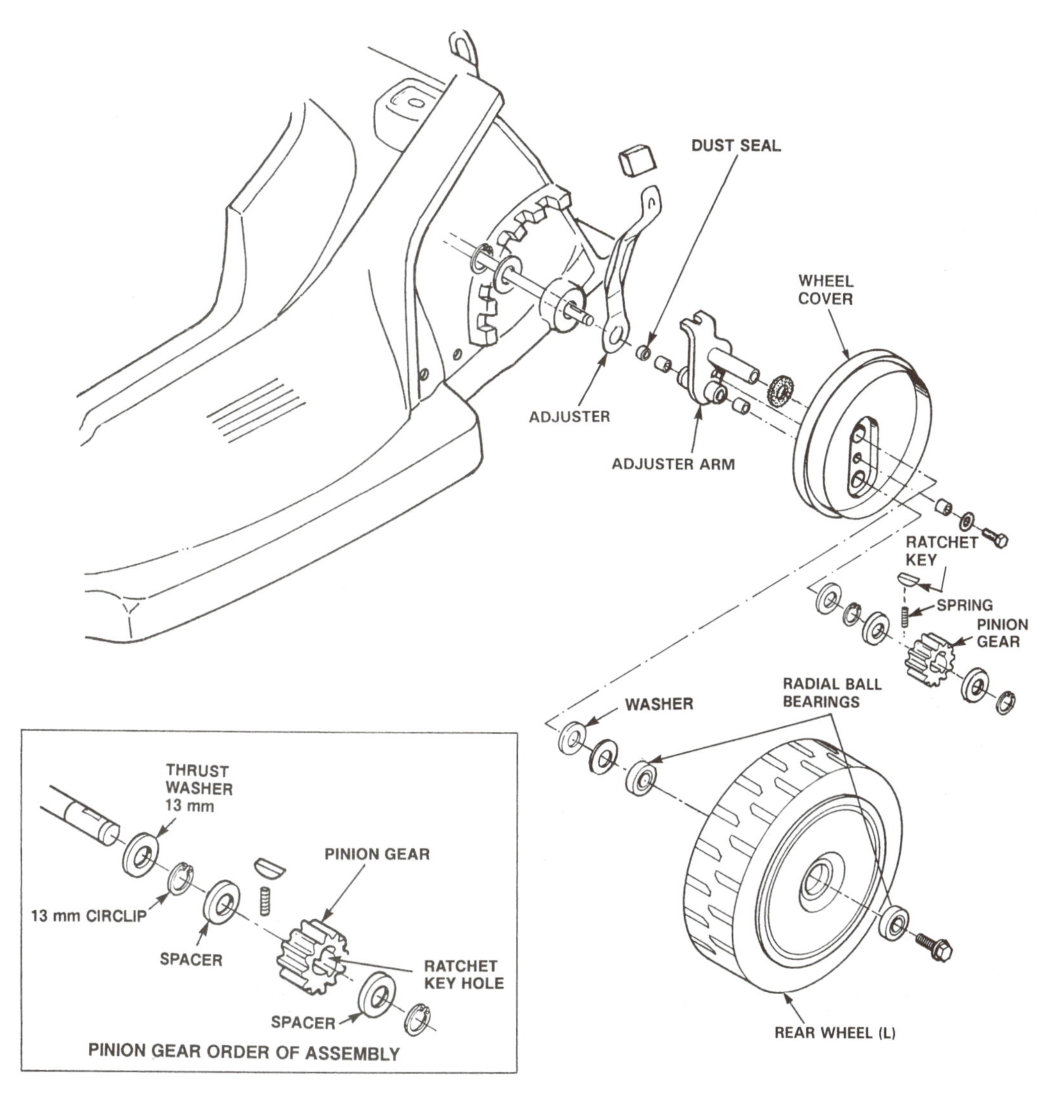

FIGURE 30-30 A one-way clutch allows the inner and outer drive wheels to turn at different speeds. (Courtesy of American Honda Motor Co., Inc.)

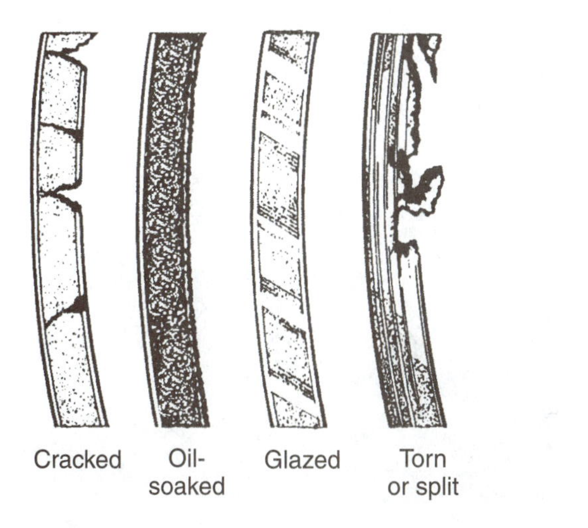

FIGURE 30-31 Replace belts that have any of these problems. (Courtesy of DaimlerChrysler Corporation.)

slipping polishes its surface. Overloading the mower can also cause the belt to slip and glaze. Belts that come off the pulley under power (rollover) are often caused by worn pulley grooves. Torn or split belts have major damage and must be replaced immediately.

Belts that have frayed edges can mean that the pulleys are not properly aligned. When pulley shafts are not parallel, one side of the belt is pulled tighter and is stressed much more than the other side. These belts wear out much faster. Also, the belt enters and leaves the pulley groove at an angle, which causes rapid belt and pulley wear (Figure 30-32).

Belts are usually removed and adjusted by loosening or removing the spring pressure from the idler pulley bracket. After loosening the bracket or removing the spring pressure, slip the belt off each

PHOTO SEQUENCE 24

Replacing a Self-Propelled Mower Drive Belt

1. Kill the ignition. Remove the spark plug wire and attach it to a ground.

2. Tip mower on correct side for under mower service.

3. Remove fasteners and remove the belt guard (baffle).

4. Inspect the belt by twisting it so you can see the underside of the V shape. Replace any belt that is broken, cracked, or glazed.

5. Remove the blade center bolt that secures the blade to the crankshaft and remove the blade, blade adapter, and pulley.

6. Roll the belt off the idler and transmission pulleys and remove the belt.

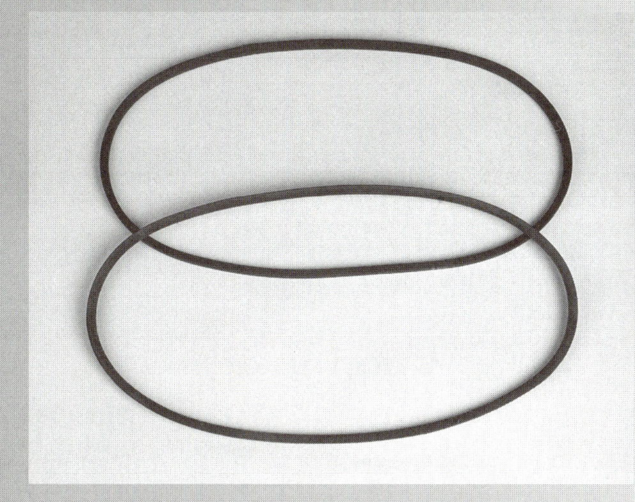

7. Compare the old belt with the new belt to be sure they are the same.

8. Install the new belt by slipping it over each pulley and install and torque the crankshaft center bolt.

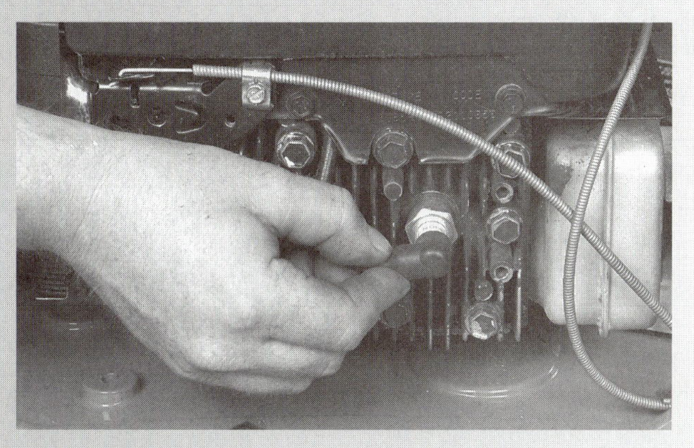

9. Install the belt guard, upright the mower; and install the spark plug wire.

FIGURE 30-32 Misaligned pulleys cause rapid belt and pulley wear. (Courtesy of MTD Products Incorporated.)

pulley. Compare the old belt with the new belt to be sure they are the same (Figure 30-33). Install the new belt by slipping it over each pulley. Tension the belt by moving the idler pulley or releasing the idler bracket spring tension. Make sure the belt fits properly into each pulley. The sides of the belt must grip both sides of the pulley (Figure 30-34). The procedure for replacing a self-propelled mower drive belt is shown in the preceding sequence of photographs.

WARNING: Using the wrong size or type of belt can cause unintentional engagement of a transmission. The mower could drive when it is not in gear. Always use the belt specified by the lawn mower maker.

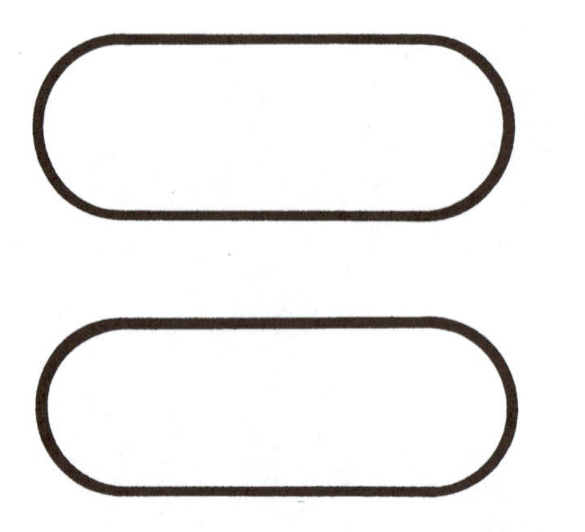

FIGURE 30-33 The new belt must be the correct length. (Courtesy of Simplicity Manufacturing.)

FIGURE 30-33 The belt must grip both sides of the pulley. (Courtesy of Simplicity Manufacturing.)

REVIEW QUESTIONS

1. Explain the safety precautions to take before servicing a lawn mower.
2. Explain why you should always study the operator's manual before using a lawn mower.
3. Explain why you should always observe and follow the instructions on mower warning labels.
4. Explain why safety devices should not be deactivated on a lawn mower.
5. Explain why hands and feet must be in safe locations when starting and using a lawn mower.
6. List and describe the main parts of a rotary lawn mower.
7. List and describe the parts of the blade used on a rotary lawn mower.
8. Explain the operation of a lug type rotary blade slip clutch.
9. Explain the operation of a friction type rotary blade slip clutch.
10. List the parts and explain the operation of a rotary lawn mower blade brake.
11. Describe the most common causes for rotary lawn mower vibration.
12. Describe the common causes for a rotary lawn mower that cuts unevenly.
13. List the common types of rotary lawn mower blade wear and damage.
14. Describe how to sharpen a rotary lawn mower blade.
15. Explain how to balance a rotary lawn mower blade.
16. Explain how a reel lawn mower blade and cutter bar cuts grass.
17. List the parts and explain the operation of a self-propelled mower belt drive.
18. Explain the purpose of the worm and helical gear in a self-propelled mower transmission.
19. Explain the purpose of the jaw clutch in a self-propelled mower transmission
20. Explain why a one-way clutch is used to drive the wheels on a self-propelled lawn mower.

DISCUSSION TOPICS AND ACTIVITIES

1. Locate a scrap rotary lawn mower blade. Practice sharpening and balancing the blade.
2. Locate a scrap self-propelled lawn mower transmission. Disassemble the transmission and trace the power flow in neutral and drive.
3. A mower transmission has a worm gear with 10 teeth and a helical gear with 30 teeth. What is the gear ratio?

CHAPTER 31

Lawn and Garden Tractors

OBJECTIVES

Upon completion and review of this chapter, you should be able to:

- List and describe the basic safety procedures to follow when operating a lawn and garden tractor.
- List and describe the safety precautions to follow to prevent injury to children.
- List and describe the safety precautions to take when operating a tractor on a slope.
- Describe the parts and operation of a two- and four-wheel steering system.
- Describe the parts and explain the operation of a transmission in a transaxle.
- Describe the parts and explain the operation of a differential in a transaxle.
- Explain the parts and operation of a tractor brake system.
- Explain the parts and operation of a tractor electrical system.
- Lubricate the parts of a tractor.
- Service the front and rear wheels of a tractor.
- Adjust a tractor brake.
- Explain the parts and operation of a tractor variable speed belt drive.
- Describe the parts and operation of a tractor hydrostatic transmission.

TERMS TO KNOW

Differential
Hydrostatic transmission
Lawn and garden tractor
Steering system
Tractor chassis
Tractor transmission
Transaxle
Transmission
Variable speed pulley system
Wiring harness

INTRODUCTION

A smaller version of a farm tractor, a **lawn and garden tractor** is used for various residential and light commercial lawn and garden jobs and can be adapted to many different jobs. Lawn and garden tractors use four-stroke engines in the range of 10 to 25 horsepower. Larger tractors use two-cylinder engines. Engines may be air or liquid cooled. Tractors are equipped with an electrical starting and charging system. Most tractors have transmissions with multiple speeds to multiply engine torque. Depending on the type of equipment used, these tractors are classified as garden tractors, lawn tractors, or riding mowers. Different attachments can be added to use the tractor to cut grass, sweep roadways and walk ways, collect leaves, mulch a lawn, till a garden, doze dirt or snow, snow throw, pull trailers, and many other jobs.

LAWN AND GARDEN TRACTOR SAFETY

Lawn and garden tractor accidents have resulted in deaths and serious injuries. Many of these accidents could have been prevented if the operator had followed safe operating procedures. Follow the specific instructions in the operator or owner's manual and the warning labels on the tractor. You must know how to use and control the tractor before using it.

Follow these general safety rules to prevent injury:

- Read and follow the instructions in the operator's manual before you start the tractor.
- Read and follow all the warning messages on tractor decals.
- Make sure that only responsible adults operate the tractor.
- Never operate a tractor when you are under the influence of drugs or alcohol.
- Wear personal protective equipment such as safety shoes and eye protection when operating the tractor.
- Clear the work area of any objects such as rocks, toys, or wire that can be thrown by the tractor blades.
- Be sure the work area is clear of people before using the tractor.
- Stop operating the tractor if people come into the work area.
- Do not allow passengers to ride on a tractor.
- Be aware what is behind you when going in reverse.
- Do not mow in reverse.
- Watch the mower discharge and make sure it is not pointed at anyone.
- Do not operate the mower without a grass catcher or deflector in place.
- Always slow down before you start a turn.
- Do not leave a running tractor unattended.
- Engage the parking brake, disengage the PTO, set the parking brake, stop the engine, and remove the ignition key before you get off the tractor.
- Disengage the PTO to stop the blades when not mowing.
- Stop the engine before you remove any grass catcher or unclog any ejection chute.
- Use the tractor only in daylight or good artificial light.
- Watch for traffic if you mow near roadways.

Child Safety

Children are often attracted to and excited about the mowing activity. You must always be aware of where children are. Children should be under the supervision of an adult. The following are safety rules that will help prevent accidents to children:

- Keep children out of the tractor work area.
- Turn tractor off if a child enters the work area.
- Always look behind and down when backing a tractor.
- Never carry children on the tractor.
- Do not allow children to operate a tractor.
- Be alert for children when approaching blind corners, shrubs, trees, fences or anywhere your vision is obscured.

Slope Operation

Operating a tractor on a slope is a major cause of loss of control accidents. The tractor can tip over on a slope and cause severe injury and death. Use extra caution on a slope. If the slope cannot be mowed safely, do not use the tractor on the slope. The general rule is do not operate a tractor on a slope that is greater than 30 percent. A 30 percent slope is a slope that rises 3 feet vertically in 10 horizontal feet. Always mow up and down a slope. Do not mow across the face of a slope. Be careful when changing direction on a slope. Never start or stop on a slope.

Follow these safety rules when operating a tractor on a slope:

- Follow the tractor operating instructions on wheel weights and counterweights to improve stability.
- Watch for obstructions, ruts, or bumps hidden by tall grass that can tip the tractor over.
- Always use a slow speed so you do not have to stop or shift on a slope.
- Do not make sudden changes in speed or direction while on the slope.
- Plan your work so you do not have to turn while on the slope.
- Do not use the tractor near drop-offs, ditches, or embankments where an edge could cave in and tip the tractor over.
- Avoid mowing on wet grass because reduced traction can cause the tractor to slide.
- Never try to stabilize a tractor by putting your foot on the ground.

Transporting and Storage Safety

Always follow the operation manual instructions when transporting and storing a tractor. Never store a tractor inside where there is an open flame or pilot light from a water heater, gas furnace, or stove. Be sure the tractor is cool to the touch before it is placed in storage. Follow safe procedures for handling fuel and battery.

CASE STUDY

The Consumer Product Safety Commission (CPSC) published the following safety alert press release concerning garden tractors.

> In the wake of a riding lawnmower accident last week that amputated the feet of a five-year old Ohio boy, government safety experts are again warning parents not to allow youngsters to ride on riding mowers or garden tractors. According to the U.S. Consumer Product Safety Commission (CPSC), a Geauga County, Ohio, boy was sitting on the lap of his father who was operating a riding lawnmower. The boy subsequently fell and was trapped under the riding mower. CPSC said injuries and deaths to young children under the age of ten could be reduced or eliminated if children were prohibited from riding on garden tractors and riding mowers, or playing in the area when the machines are being used. One of every five deaths associated with riding mowers and garden tractors involves children under age ten. Deaths most often occur when youngsters fall off the mower and are run over by the machine, or when they run or fall in the mower's path and are run over. A Government survey shows that 54 percent of households with children under ten years of age do allow youngsters to ride on the lawn-mowing equipment. This extremely unsafe practice continues despite labels and warnings to the contrary by outdoor power equipment manufacturers. . . .

Source: Consumer Product Safety Commission press release #90-102, May 25, 1990.

TRACTOR PARTS

A lawn and garden tractor has a number of main parts or systems. The tractor frame (Figure 31-1) is the main structural element that supports the engine and all the operating systems. The frame structure is usually made from sheet steel. It is sometimes called a *weldment* because it is often welded together from several smaller parts. Numerous sheet metal parts such as the fenders, hood, grill, and foot rests are attached to the chassis The **tractor chassis** (Figure 31-2) is the frame and systems used to get engine power to the tractor drive wheels and to brake and steer the tractor.

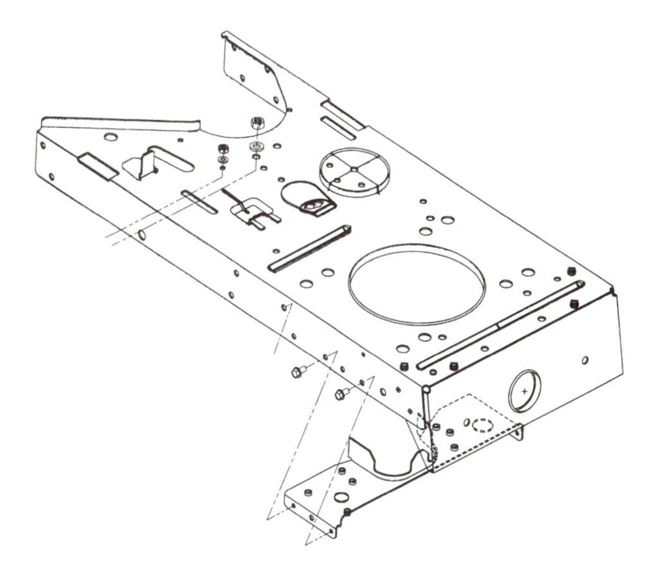

FIGURE 31-1 The tractor frame or weldment supports the engine and all the operating systems. (Courtesy of MTD Products, Incorporated.)

Steering System

The **steering system** is a set of linkages connected to the steering wheel that allows the operator to change the direction of the tractor. The steering system may be connected to turn the front wheels or all four wheels. The tractor steering wheel is attached to a long steering shaft (Figure 31-3 on page 459) that has a steering gear attached to the end. The steering gear meshes with another gear called the *sector gear*. The steering and sector gear changes rotary motion of the steering wheel into straight line motion to move the steering linkage. The small steering gear and large sector gear also provide a gear reduction, which reduces the effort required by the operator and makes the tractor easier to turn.

The sector gear is connected to a linkage rod called a drag link (Figure 31-4 on page 459). The drag link transfers steering motion to one of the two steering spindles. The spindles are connected to each end of the front axle by a shaft that fits into the front axle. When the shaft is installed in the axle, the spindles can move back and forth. The front tractor wheels fit on the spindles. A tie rod is connected from one spindle to another.

When the operator turns the steering wheel, the steering gear turns the sector gear. The sector gear moves the drag link. The drag link moves the left side spindle. The tie rod moves the other spindle. The tractor wheels connected to the spindles change direction and the tractor changes direction.

Some tractors have four-wheel steering (Figure 31-5 on page 459). Four-wheel steering makes the

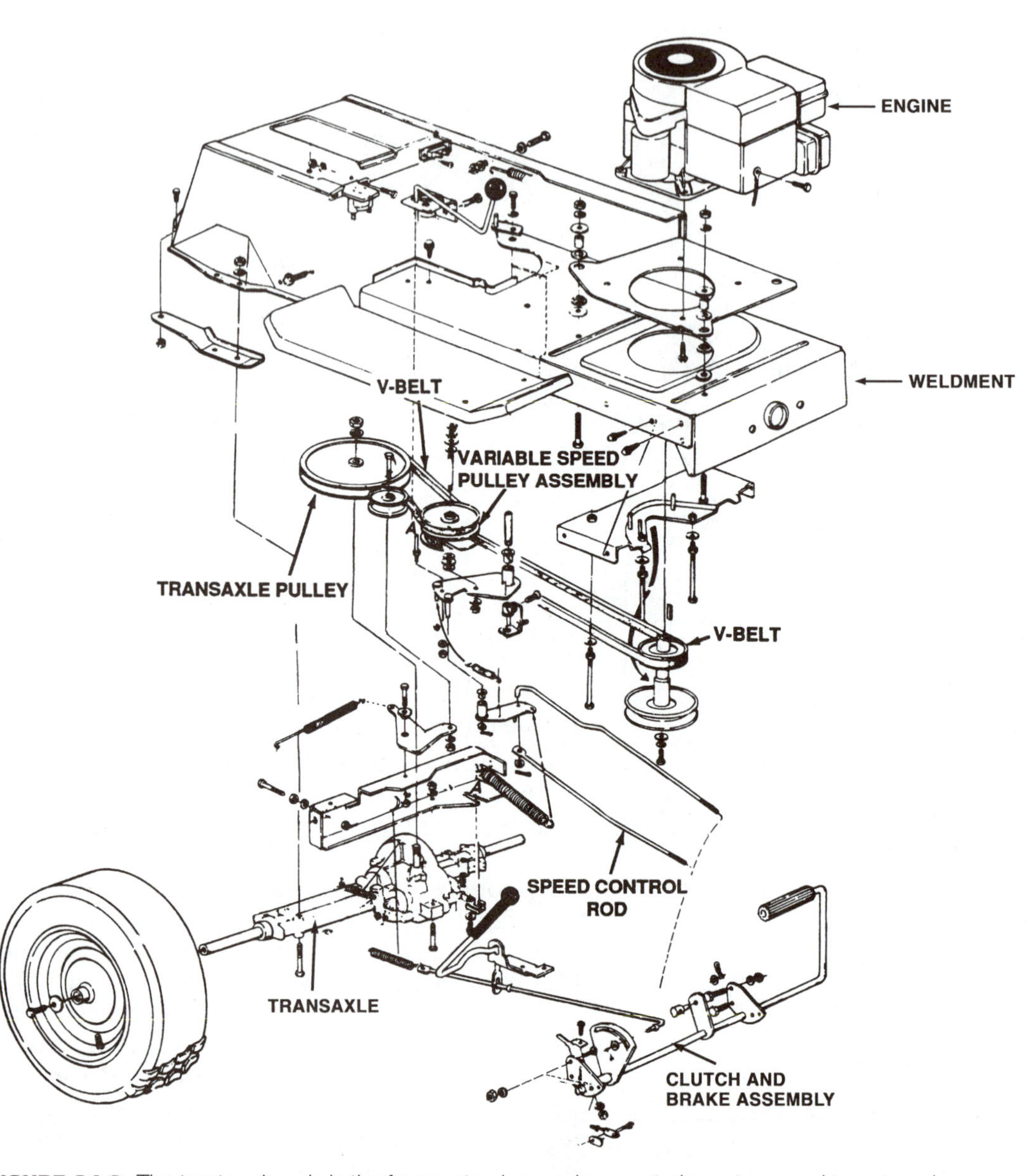

FIGURE 31-2 The tractor chassis is the frame, steering, and power train system used to get engine power to the tractor drive wheels. (Courtesy of MTD Products, Incorporated.)

tractor easier to steer around obstacles. Four-wheel steering has the same basic components as front-wheel steering. In addition there are linkages connected to the rear wheels. The rear driving wheels are supported on universal joints (Figure 31-6 on page 461). A universal joint has two Y-shaped yokes connected to a cross (spider). Needle bearings support the cross in the yokes. The bearings allow the yokes to move back and forth relative to each other. The universal joints make it possible to turn the rear wheels at the same time they are being rotated by the rear axles.

Four-wheel steering linkages are designed so that the front wheels do the steering most of the time. The rear wheels only turn when the operator moves the steering wheel beyond 45 degrees for a tight turn (Figure 31-7 on page 461). At this point, the rear steering linkage changes the direction of the rear wheels. The rear wheels are turned in the opposite direction to the front wheels. The four-wheel steering tractor is able to turn very sharply.

Transaxle

The tractor rear wheels are supported and driven by a transaxle. The **transaxle** is a combination of a transmission and differential used to drive the rear wheels of a tractor. A **transmission** is a gear, belt, or hydraulic system used to change the amount of engine torque delivered to the tractor driving wheels. A **differential** is a gear assembly that allows the

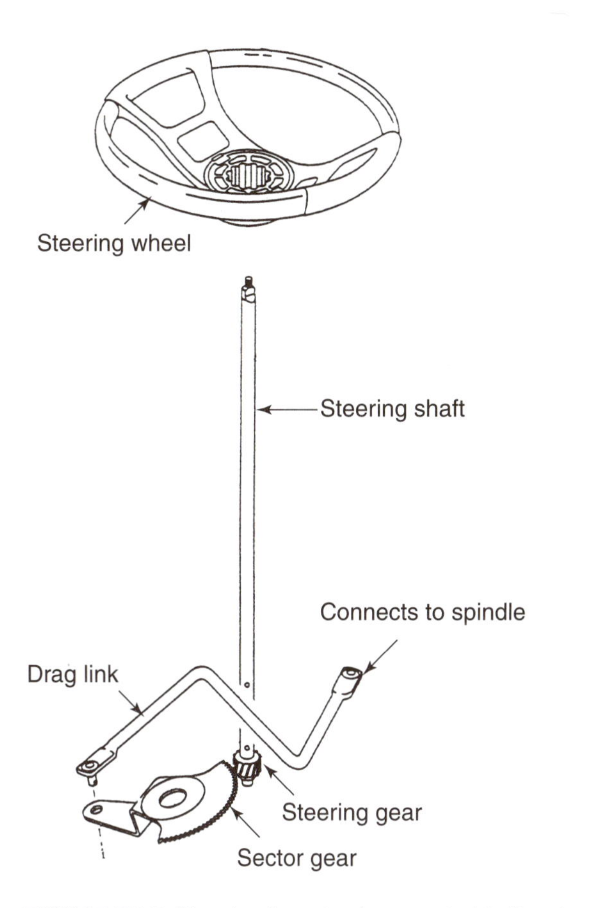

FIGURE 31-3 The steering wheel connected to the steering gear turns the sector gear. (Courtesy of Poulan/Weed Eater.)

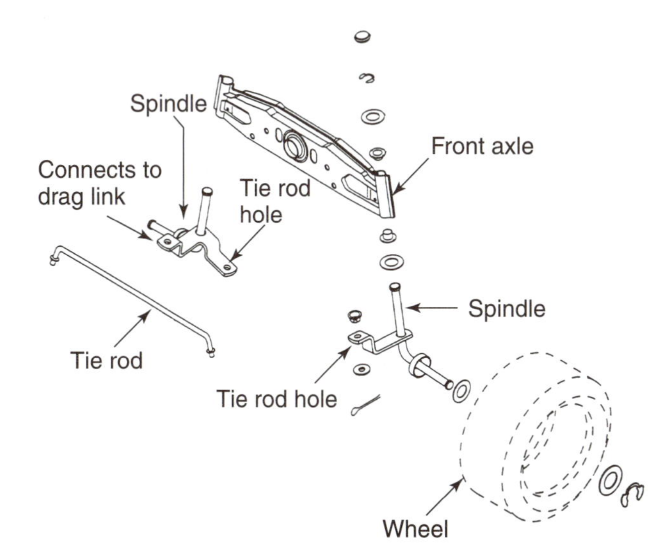

FIGURE 31-4 The steering linkage turns the wheels mounted on spindles. (Courtesy of Poulan/Weed Eater.)

two drive wheels to turn at different speeds when the tractor goes around a corner.

The transaxle is mounted at the rear of the tractor (Figure 31-8 on page 461). Engine torque is delivered into the transaxle through a system of belts to a drive pulley. The rear wheels are driven by axle shafts on each side of the transaxle. Control rods connected to a shift lever and pedal allow the operator to control shifting, clutching, and stopping.

Transmission

The transmission section of the transaxle allows the operator to select different speed reductions and torque increases for different operating conditions. The tractor engine may have enough torque to keep the tractor moving on level ground, but it may not have enough torque to start the tractor rolling from a standstill, to pull the tractor up a slope, or to pull a trailer. The transmission acts like a torque multiplier. The engine is connected to a small gear and the rear wheels are connected to a large gear. When a small gear drives a large gear, there is a speed reduction and a torque increase.

Many tractors use a transmission with a system of gears to multiply torque. Any number of forward speeds are possible. A two-speed transmission has a large driven gear for low (first) gear (Figure 31-9 on page 461). The driven gear is smaller for high (second) gear. The transmission provides a method of reversing the power flow to allow the tractor to back up. It also provides a neutral position so that the engine can run when the tractor is stopped.

The gear transmission uses several gear assemblies. The power flows into the transaxle and transmission from the drive pulley, which rotates an input shaft that enters the top of the transaxle (Figure 31-10 on page 462). There is a bevel gear on the end of the input shaft. The bevel gear meshes with a large bevel gear inside the transmission. The bevel gears change the direction of the power flow from engine rotation direction to axle rotation direction. The size difference in the gears also gives speed reduction and torque multiplication.

The transmission has two rows of gears (Figure 31-11 on page 462). One row of gears is called the *counter shaft assembly*. The other row of gears is called the *shifter shaft assembly*. The bevel gear fits on the counter shaft. Any time the engine is running, the input shaft bevel gears rotate the counter shaft. There is a set of different sized gears on the counter shaft. There is one gear for each forward speed. A three-speed transmission has three gears on the counter shaft. All the counter shaft gears rotate when the counter shaft rotates. The counter shaft and all its gears turn as a unit.

The gears on the counter shaft are in mesh with gears on the shifter shaft. A three-speed transmission has three gears on the shifter shaft in mesh with the three gears on the counter shaft. The shifter shaft gears are called *shift gears*. The shift

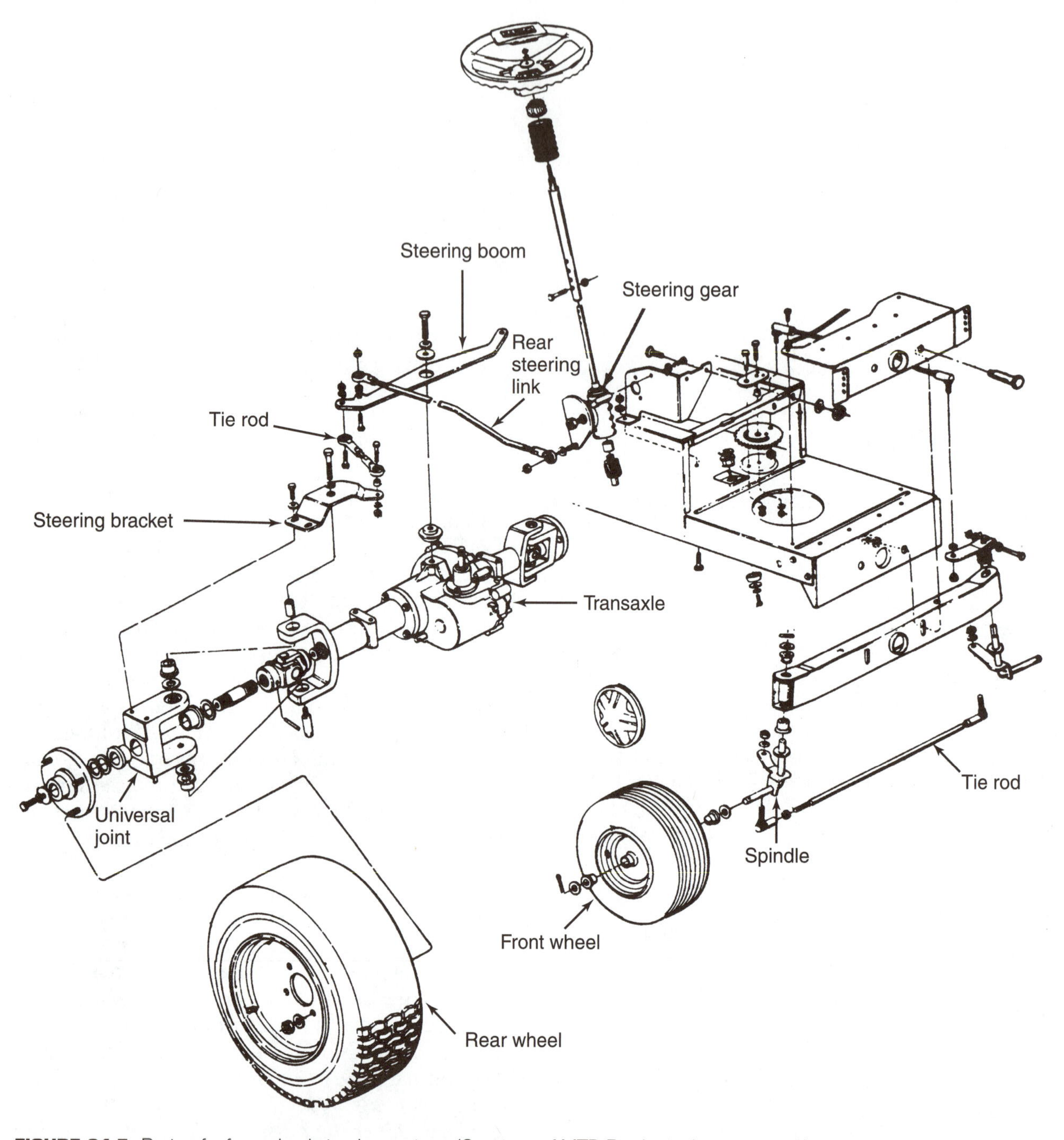

FIGURE 31-5 Parts of a four-wheel steering system. (Courtesy of MTD Products, Incorporated.)

gears are not permanently attached to the shifter shaft. They are free to rotate on it until they are needed. When a shift gear is used, it is locked to the shifter shaft.

A shift collar is used to lock and unlock the shift gears to the shifter shaft (Figure 31-12 on page 462). The tractor operator moves the shift collar with linkage connected to the shift selector lever. When a particular gear is selected, the shift collar locks that shifter gear to the shifter shaft. Power can then be transferred through them. The shift collar moves a long key into the required position to lock the selected gear to the shifter shaft. The gears in the transmission remain in constant mesh with each other. The collar and long key work to synchronize the speed of the shaft and the gear so they may be locked up without any clashing.

Power flows from the selected shifter gear across the shifter shaft to an output shaft (Figure 31-13 on page 462). There is a small output gear on the end of the shifter shaft. The output gear drives a large output pinion gear. The output pinion gear

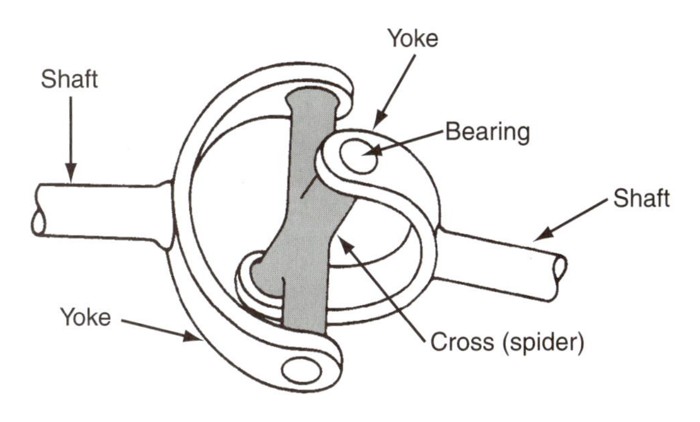

FIGURE 31-6 A universal joint connects two shafts so they can operate at an angle to each other. (Courtesy of General Motors Corporation, Service Technology Group.)

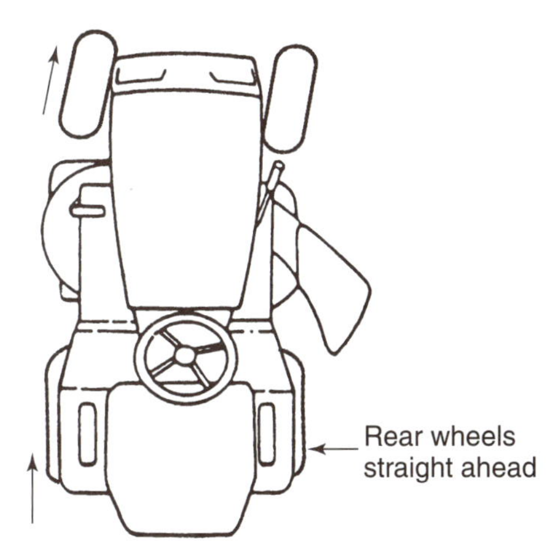

A. Less than 45° Turn of the Steering Wheel

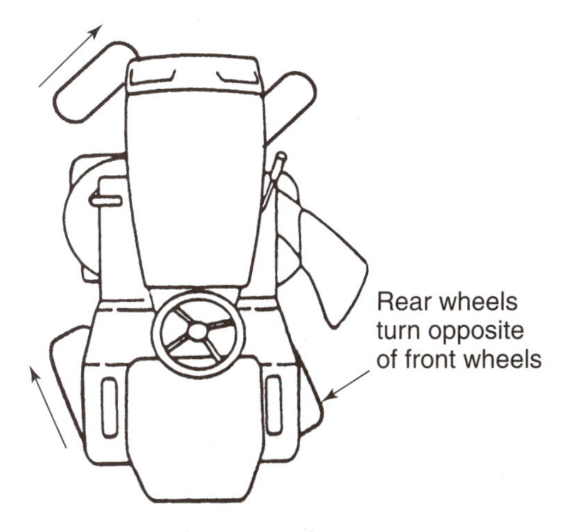

B. More than 45° Turn of the Steering Wheel

FIGURE 31-7 The rear wheels turn in the opposite direction to the front wheels when the steering wheel is turned beyond 45 degrees. (Courtesy of MTD Products, Incorporated.)

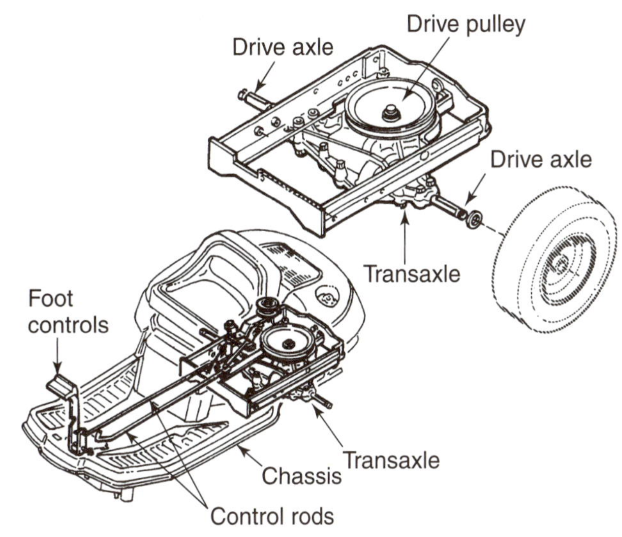

FIGURE 31-8 The transaxle is a combination of a transmission and differential used to drive the rear wheels of a tractor. (Courtesy of Simplicity Manufacturing.)

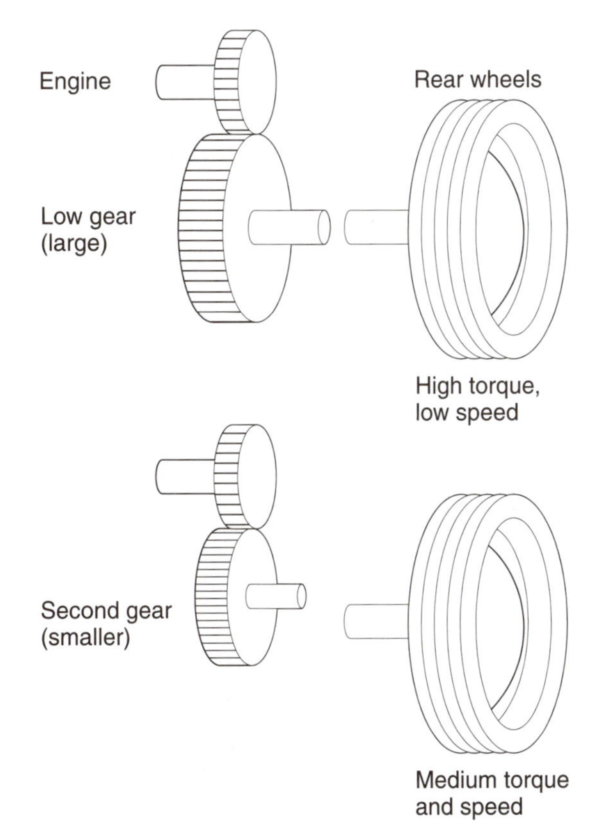

FIGURE 31-9 The engine drives the rear wheels through different sized gears for different amounts of torque multiplication.

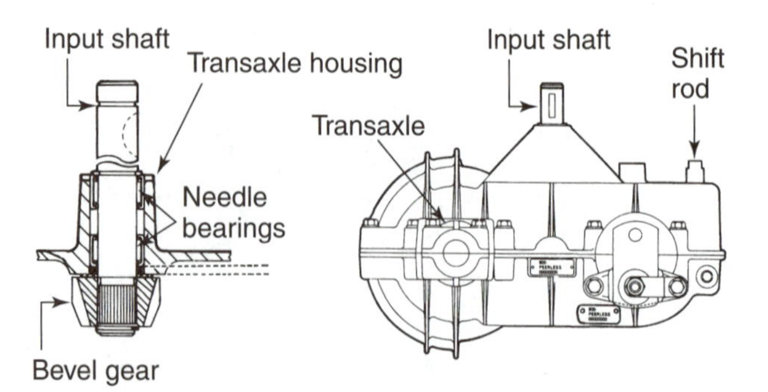

FIGURE 31-10 The input shaft drives a pinion gear to deliver engine power into the transmission. (Provided courtesy of Tecumseh Products Company.)

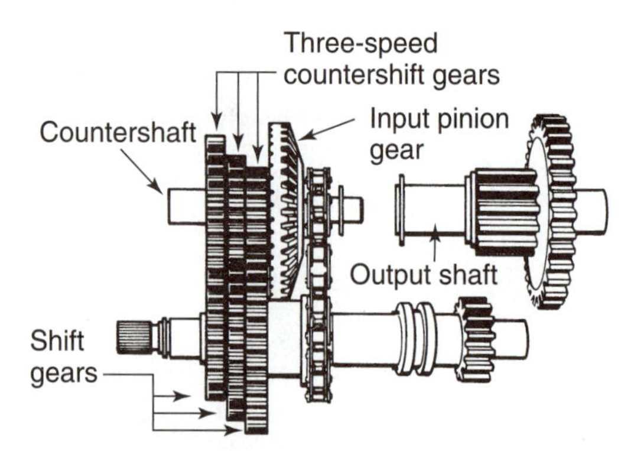

FIGURE 31-11 Power comes into the countershaft gears from the input pinion and flows across to the shifter gears. (Provided courtesy of Tecumseh Products Company.)

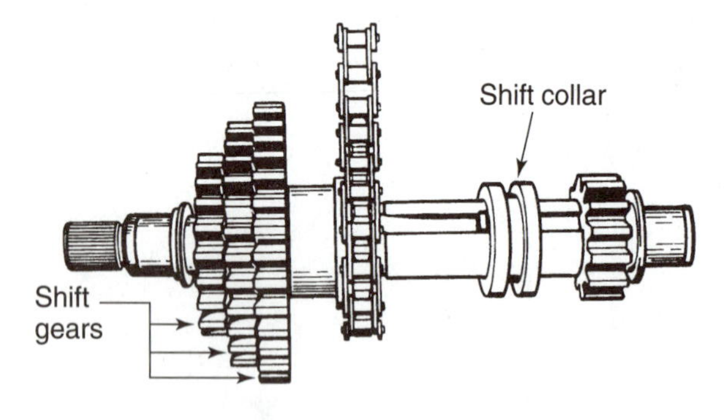

A. Shifter Gear Assembly

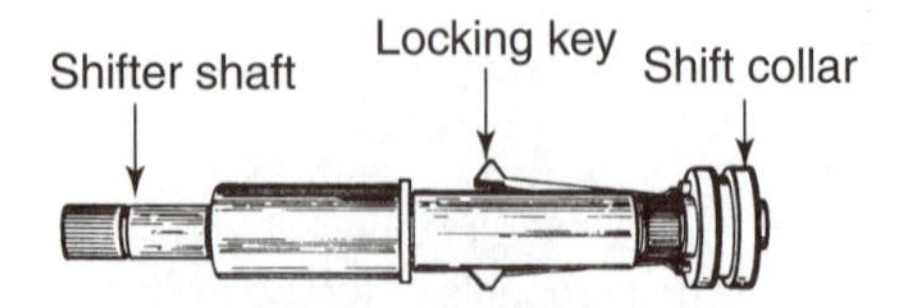

B. Shifter Shaft

FIGURE 31-12 The shift collar moves locking keys in position to lock shift gears to the shifter shaft. (Provided courtesy of Tecumseh Products Company.)

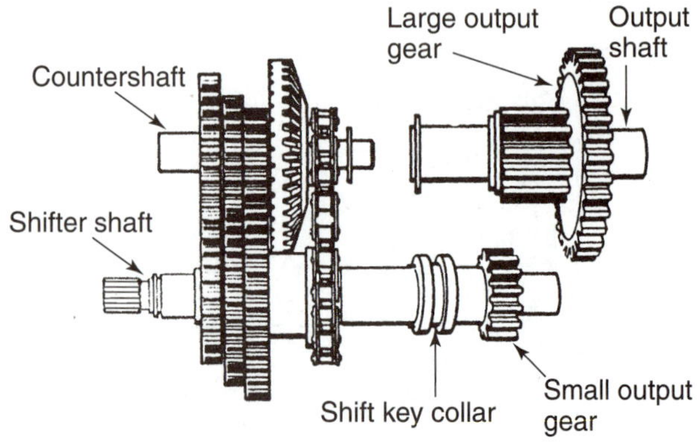

FIGURE 31-13 Power flows from the shifter shaft output gear to an output gear on an output shaft. (Provided courtesy of Tecumseh Products Company.)

is located on a shaft next to the countershaft. The countershaft and output shaft are not connected. The output pinion gear delivers the power into the differential.

When the operator chooses the neutral position, the shifting collar unlocks all the shifter gears from the shifter shaft. Power flows into the countershaft but does not go any farther.

When the operator chooses the reverse position, the output shaft must be turned in the opposite direction. This is done with a chain and sprocket assembly. There is a chain sprocket on the countershaft and one on the shifter shaft. A large chain connects the two sprockets. When reverse is selected, the shift collar locks the shifter shaft sprocket to the shifter shaft. When two gears are in mesh, one gear turns clockwise and the other turns counterclockwise. When the chain drives two sprockets, they both turn in the same direction. The sprocket on the shifter shaft turns the output gear in a direction to reverse the direction of the output pinion (Figure 31-14).

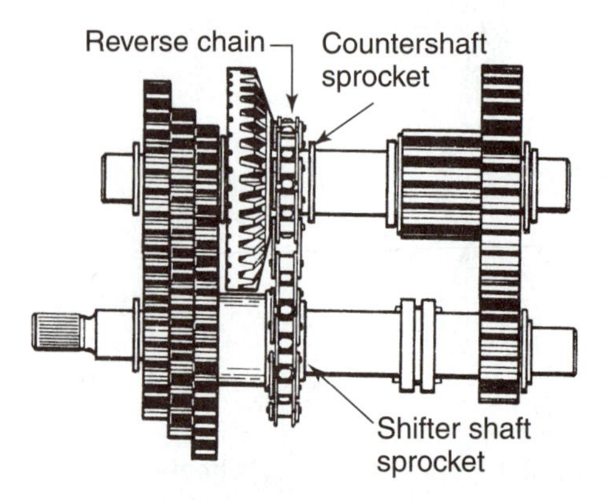

FIGURE 31-14 A chain and sprocket system is used to reverse power flow. (Provided courtesy of Tecumseh Products Company.)

Differential

The purpose of the differential assembly is to allow the two rear drive wheels to turn at different speeds when the tractor goes around a corner. This is necessary because when cornering, the wheel on the inside of the turn goes through a smaller arc or corner than the wheel on the outside. If the wheels were not allowed to turn at different speeds, they would tend to skip around the corner, and steering would be very difficult.

The differential parts are located next to the transmission parts in the transaxle (Figure 31-15). Power enters the differential from the transmission output gear. The output gear is meshed with a large gear in the differential called the *ring gear*. The ring gear is attached to a housing called the *case*. As the ring gear turns, the case attached to it also turns. A shaft through the case also goes through the middle of two small pinion gears (Figure 31-16). As the case turns, this shaft turns the small pinion gears, each of which meshes with a side gear. Each side gear is attached to an axle shaft. Each axle shaft runs through the transaxle housing to one of the tractor rear wheels.

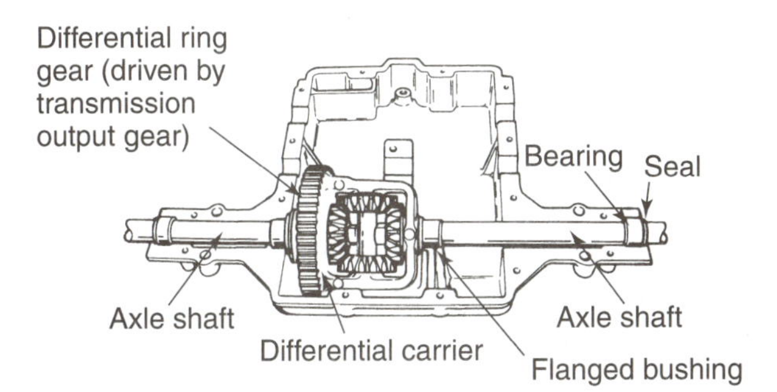

FIGURE 31-15 The differential fits next to the transmission in the transaxle case. (Provided courtesy of Tecumseh Products Company.)

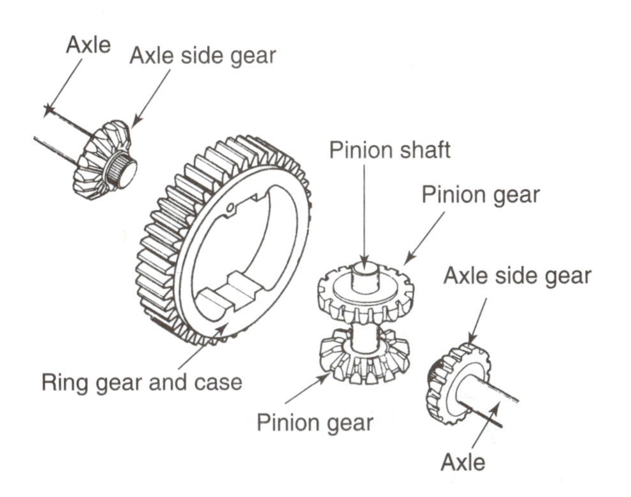

FIGURE 31-16 The axles are driven by the ring gear through axle side gears and pinions. (Provided courtesy of Tecumseh Products Company.)

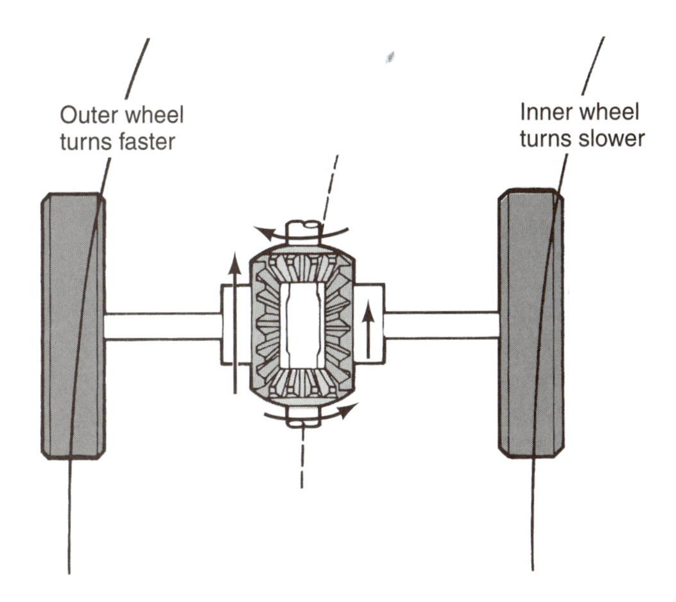

FIGURE 31-17 Power flow through a differential during a right turn. (Courtesy of DaimlerChrysler Corporation.)

When the tractor is traveling in a straight line, the power flow through the system is fairly simple. The ring gear turns the case. The case, through its shaft and pinion gears, turns each of the side gears at the same speed. The axles or drive shafts turn the tractor rear drive wheels.

When the tractor makes a turn, the power flow becomes more complicated. If the tractor is making a right turn, the right drive wheel goes through a sharper corner (Figure 31-17). It travels through a shorter distance than the left drive wheel. The ring gear turns the case. Since the right wheel is going through a sharp corner, the right axle is slowed or stopped momentarily. The pinion gears in the case still turn with the case but they also rotate on the case shaft. They "walk" around the slowed or stopped right side gear and provide all the power to the left side gear. The left wheel turns faster than the right wheel. During a left turn there is more resistance on the left axle, because the left wheel must turn through a sharper corner than the right. The pinions in the case walk around the left side gear and drive the right axle gear.

Brake

The brake system for most tractors is built into the transaxle. The transmission output shaft extends outside of the transaxle case (Figure 31-18). The part of the shaft that comes out of the case is called the brake shaft. A brake disc fits over splines on the brake shaft (Figure 31-19). A brake bracket fits over

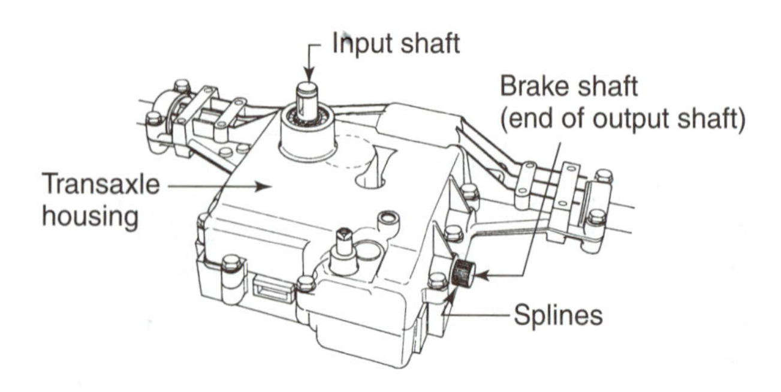

FIGURE 31-18 The brake shaft is part of the transmission output shaft that comes out of the transaxle housing. (Provided courtesy of Tecumseh Products Company.)

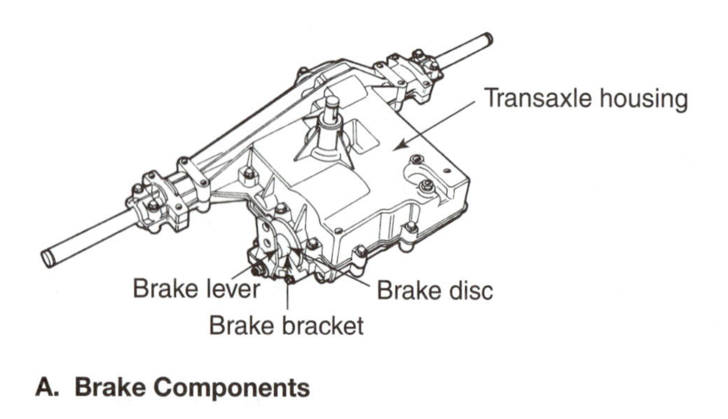

A. Brake Components

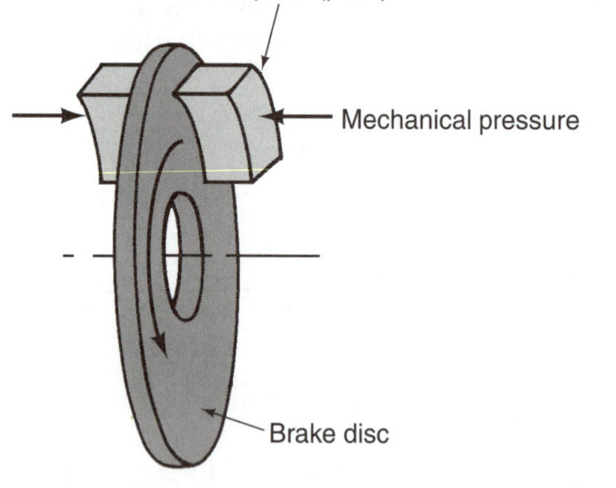

B. Friction Pads Stop the Brake Disc

FIGURE 31-19 Friction pads inside the brake bracket stop the brake disc splined to the brake shaft. (*A:* Provided courtesy of Tecumseh Products Company.)

the bottom half of the brake disc. Two friction brake pads are located inside the brake bracket. One of the brake pads fits on each side of the brake disc. The brake pads are connected to a brake lever. The brake lever is connected by a rod to the brake pedal in the tractor.

When the operator wants to slow or stop the tractor, the brake pedal is depressed. The rod connected to the pedal moves the brake lever on the transaxle. The brake lever brings the two friction pads against the brake disc. The disc is connected through the output shaft to the drive axles and to the rear wheels. Slowing and stopping the disc causes the rear wheel to slow and stop.

Many tractors use the same brake as a parking brake. Hand-operated linkage is connected to a parking brake rod. The parking brake rod is connected into a second hole in the transaxle brake lever. Pulling on the parking brake lever causes the brake pads to contact the brake disc.

Electrical System

Tractors have an electrical system to operate the starter, charge the battery, and activate safety interlocks. The electrical system has a system of wires and connectors that connect the electrical components. The wires are grouped together in a **wiring harness**, which is a system of electrical wires wrapped together and routed to tractor electrical components. The wires connect the battery to the solenoid and then to the starting and charging system on the engine.

The wiring harness also connects the ignition system to the safety interlock switches on the tractor. Since 1987 ANSI has required that tractors must have a safety interlock on the tractor seat. The seat interlock senses the presence (weight) of the operator on the seat (Figure 31-20). When the operator is on the seat, the interlock switch is opened and the engine can run. When the operator comes off the seat, the interlock switch closes and the engine shuts down. This interlock switch prevents injuries to an operator from coming in contact with rotating tractor blades.

A clutch safety start switch is another common safety interlock (Figure 31-21). This switch is connected to the clutch linkage. The switch prevents ignition unless the brake/clutch is depressed. This feature prevents the tractor from moving until the operator is ready to put it in gear. Some tractors have reverse gear interlocks. These prevent the blades from rotating when the tractor is in reverse gear. A power takeoff or blade safety switch prevents starting when the PTO (blades) are engaged.

TRACTOR TROUBLESHOOTING

There is a wide variety of sizes and types of tractors, each with different equipment and parts. Much of this equipment is complex. Each lawn and garden tractor has a service and repair manual. A section of

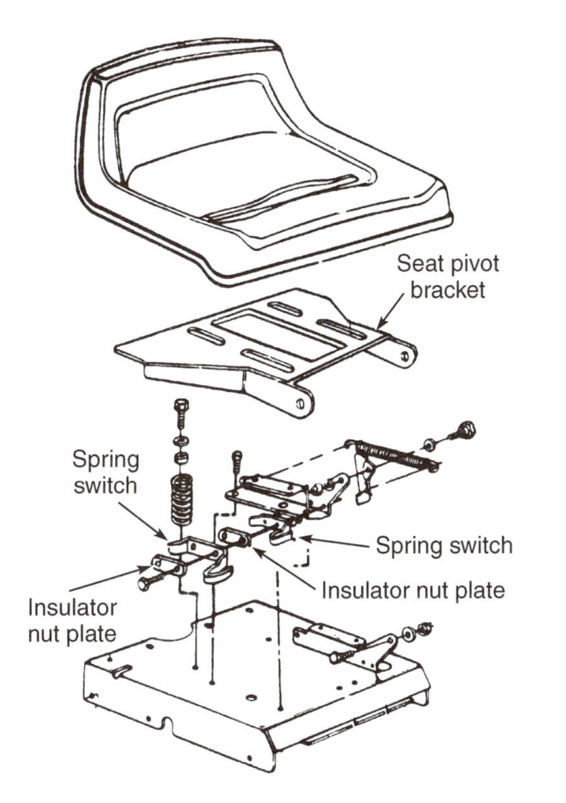

FIGURE 31-20 Operator present safety seat prevents engine starting if the operator is not in the seat. (Courtesy of MTD Products, Incorporated.)

the manual covers the troubleshooting procedures for each of the tractor's systems. You should always locate and follow the specific troubleshooting procedure for the tractor you are working on.

TRACTOR SERVICE

> **WARNING:** Read and follow the instructions in the tractor service and repair manual before you start any mower maintenance or repair.

The tractor service and repair manual provides detailed instructions for maintenance and service. Common service procedures are lubrication, wheel repair, and brake adjustment.

Tractor Lubrication

> **WARNING:** Engage the parking brake, remove the ignition key, and dis connect the spark plug wire(s) to prevent accidental starting before elevating or working on a tractor.

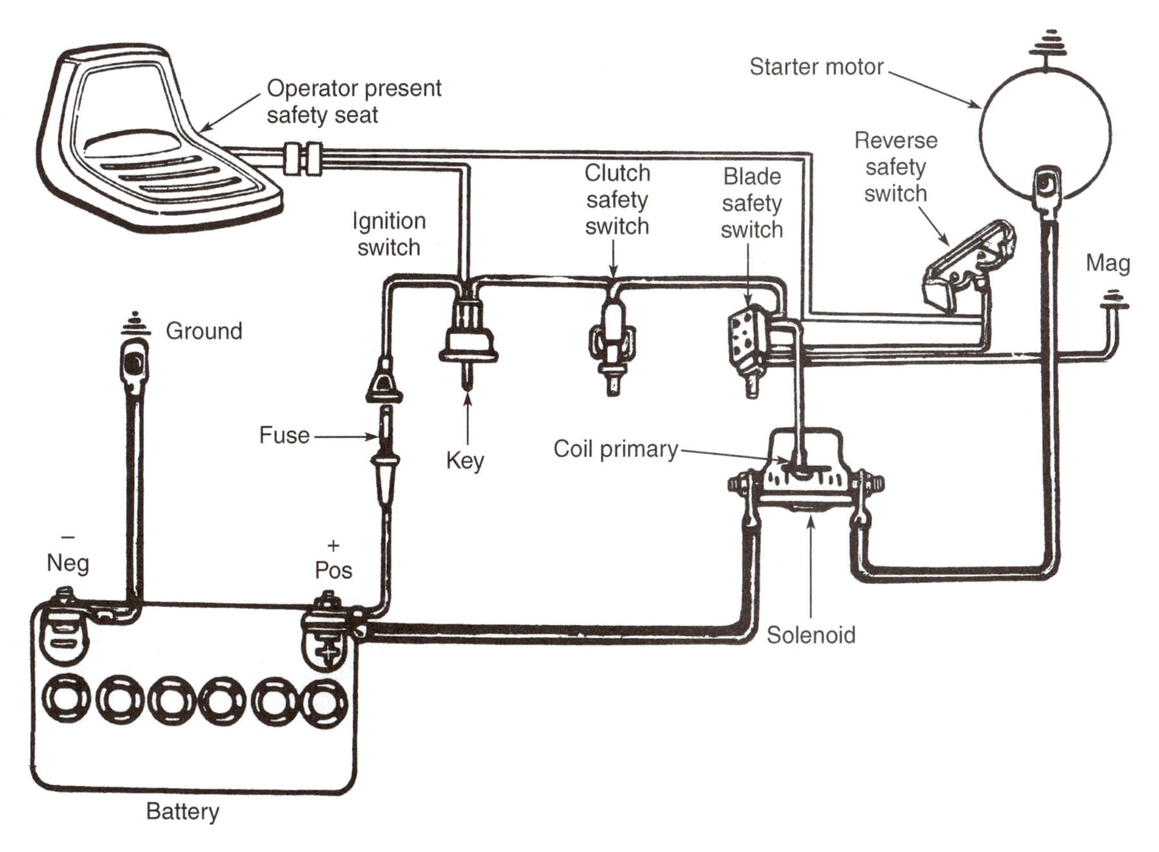

FIGURE 31-21 Typical interlocks in a tractor starting system. (Courtesy of MTD Products, Incorporated.)

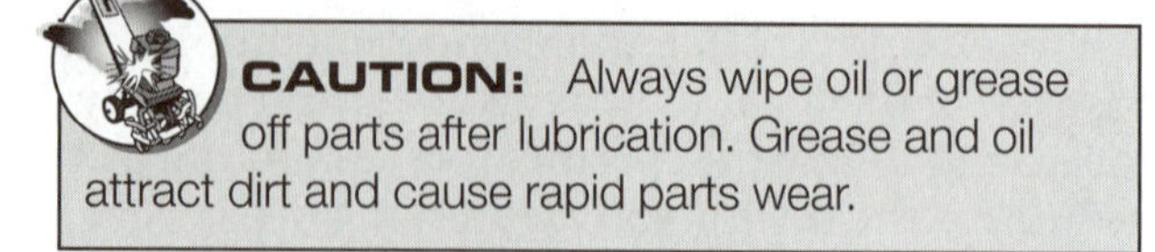

CAUTION: Always wipe oil or grease off parts after lubrication. Grease and oil attract dirt and cause rapid parts wear.

Tractor linkage connections and other parts that rotate in relation to each other require periodic lubrication. The tractor service and repair manual specifies what to lubricate and when. Some parts are lubricated with a squirt can of SAE 30 engine oil. Other parts are lubricated with lithium-based automotive grease in a grease gun. Grease gun lubrication points have a lubrication (zerk) fitting for the grease gun nozzle.

The front axle and steering assembly usually has both grease fittings and oil lubrication points (Figure 31-22). Locate the lubrication fittings. Wipe the lubrication fittings off with a clean shop rag. Insert the grease gun nozzle over the grease fitting and inject grease into the fitting. Wipe excess grease off the fitting. Use the oil squirt can to oil the steering linkage (Figure 31-23).

The mower deck has numerous moving parts, many of which require lubrication with oil or grease (Figure 31-24). Locate the lubrication points. Wipe the points off with a clean shop rag. Squirt oil on the parts. Wipe the excess oil off with a clean shop rag.

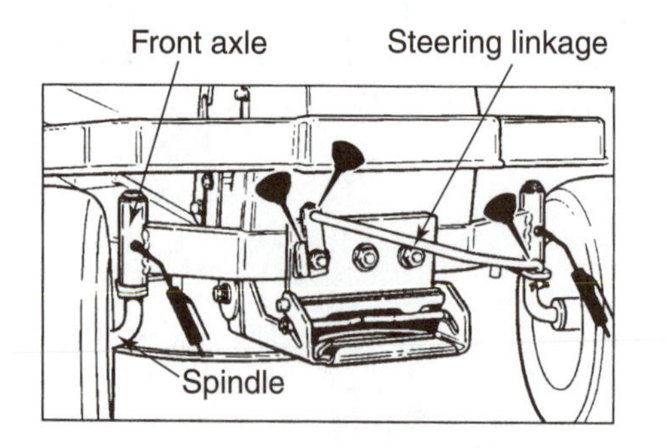

FIGURE 31-22 Grease and oil lubrication points on the front axle. (Courtesy of Simplicity Manufacturing.)

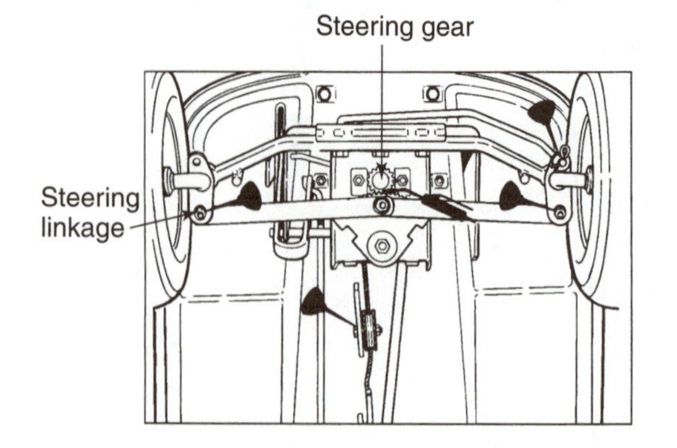

FIGURE 31-23 Grease and oil lubrication points on the steering gear and linkage. (Courtesy of Simplicity Manufacturing.)

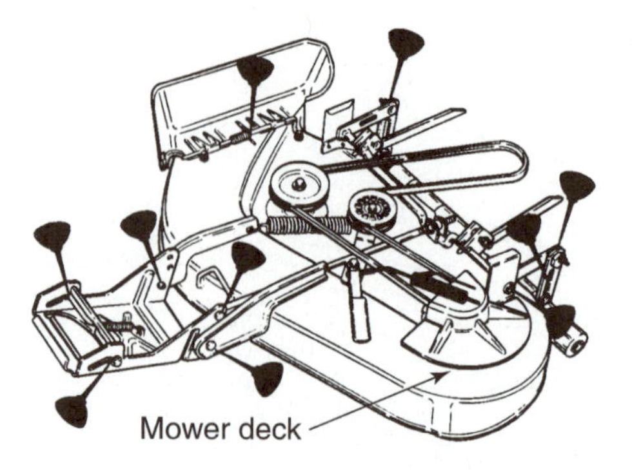

FIGURE 31-24 Lubrication points on a mower deck. (Courtesy of Simplicity Manufacturing.)

Riding mowers may have one or two cutting blades. The blades are supported on bearings located inside housings called *arbors* (Figure 31-25). The arbors usually have lubrication fittings. Locate the arbor lubrication fittings. Wipe the lubrication fittings off with a clean shop rag. Insert the grease gun nozzle over the grease fitting and inject grease into the arbor fitting. Wipe the excess grease off the fitting.

Wheel Repair

The tractor rear wheels are keyed to the transaxle shafts. The front wheels rotate on bearings on the front spindles. These bearings sometimes require replacement. The wheels have to be removed periodically for tire replacement. Tractors use tubeless tires. The tires are removed and replaced on tire mounting equipment at a tire shop.

The tractor must be raised with a hydraulic

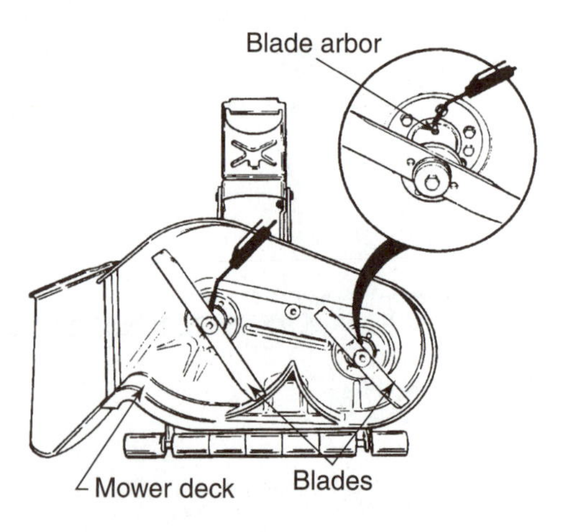

FIGURE 31-25 Lubrication points on the mower blade arbors. (Courtesy of Simplicity Manufacturing.)

floor jack to remove a wheel for repair (Figure 31-26). The handle on the hydraulic floor jack pumps hydraulic fluid to lift the jack pad and raise the tractor. The control knob on the end of the jack handle releases the fluid to lower the tractor. Always use jack safety stands under the tractor in case the jack should fail. Wheel blocks must be used before you raise the tractor. A wheel block is placed in front and in the rear of one of the wheels. Block a wheel that will not be lifted off the floor by the jack. The blocks prevent the tractor from rolling forward or rearward off the jack. Place the jack under a solid part of the tractor chassis. Most tractor service manuals show you where to place the jack (Figure 31-27).

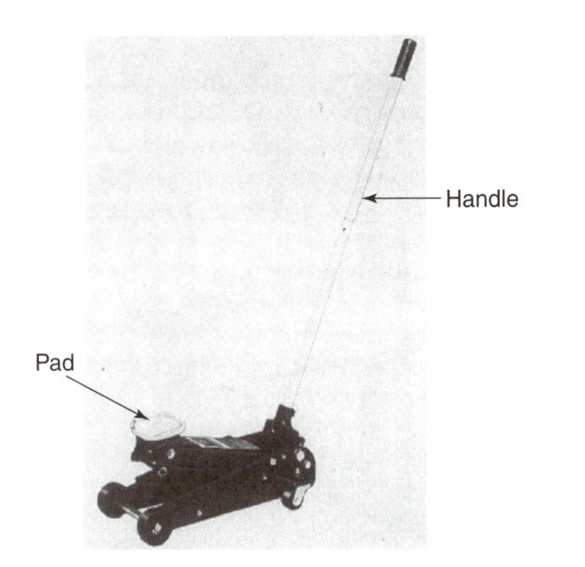

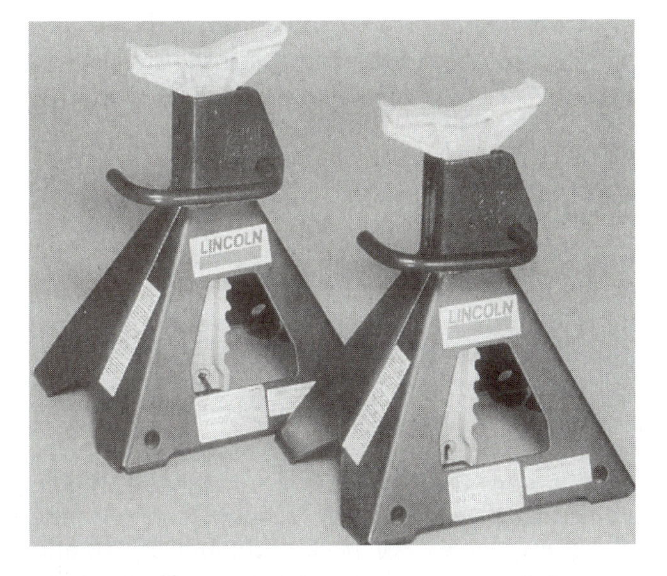

FIGURE 31-26 The tractor is lifted with a hydraulic floor jack and supported with jack safety stands. (Courtesy of Lincoln Automotive, One Lincoln Way, St. Louis, MO 63120-1578.)

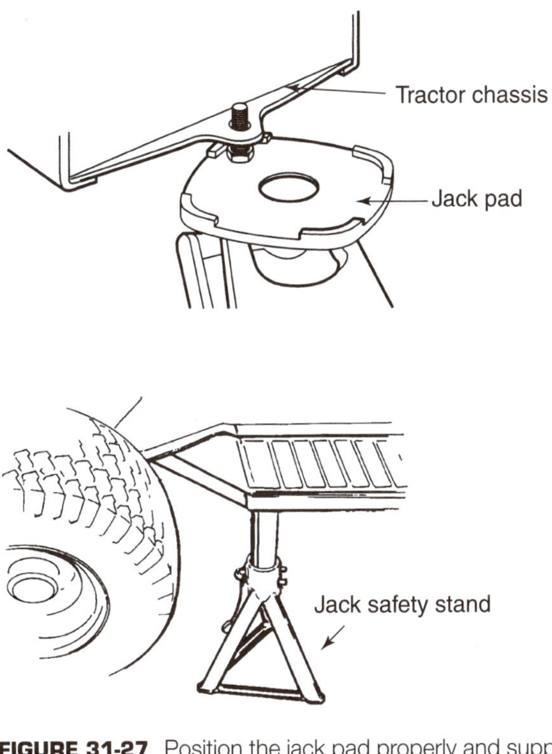

FIGURE 31-27 Position the jack pad properly and support the tractor on jack safety stands. (Courtesy of Simplicity Manufacturing.)

WARNING: Engage the parking brake, remove the ignition key, and disconnect the spark plug wire(s) to prevent accidental starting before elevating or working on a tractor.

Remove a rear wheel by elevating the rear of the tractor. Remove the plastic hub cap. Remove the retaining ring with a screwdriver (Figure 31-28). Remove the washers, spacers, and hubcap retainer. Pull the wheel assembly off the axle. Lubricate the rear axle shafts with grease. Slide the wheel back into position making sure the wheel goes over the

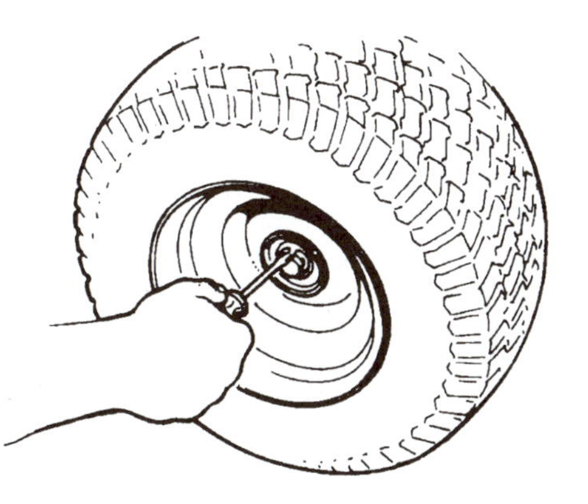

FIGURE 31-28 Removing a retaining ring to remove the rear wheel. (Courtesy of Simplicity Manufacturing.)

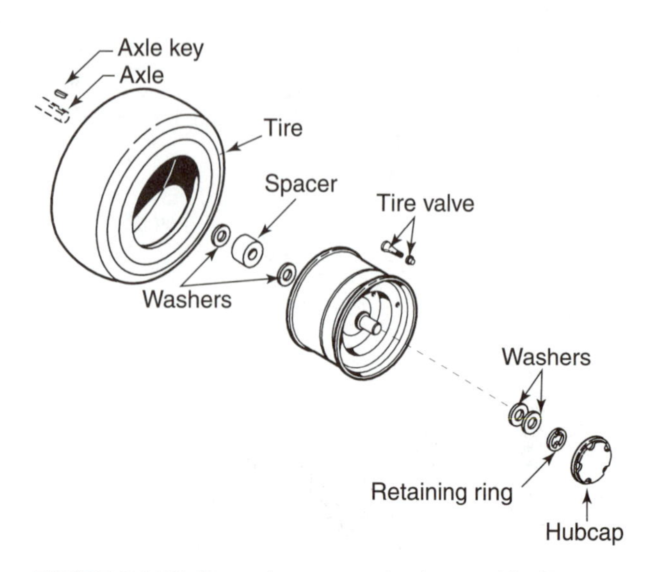

FIGURE 31-29 Removing a rear-wheel assembly. (Courtesy of Simplicity Manufacturing.)

axle key. Install the washers, spacers, and hubcap retainer in the correct order (Figure 31-29). Install the retaining ring around the axle with a screwdriver. Install the hubcap and lower the tractor.

Remove a front wheel by elevating the front end with a jack. Before removing the wheel, check for excessive play in the wheel bearing. Grip the wheel at the top and bottom. Try to move the wheel up and down. Excessive movement indicates a worn bearing. Spin the wheel and listen. A bad bearing often sounds rough as it rotates. Either of these problems requires bearing replacement. Remove the plastic hubcap. Remove the retaining ring with a screwdriver. Remove the washer and then pull the wheel off the spindle (Figure 31-30).

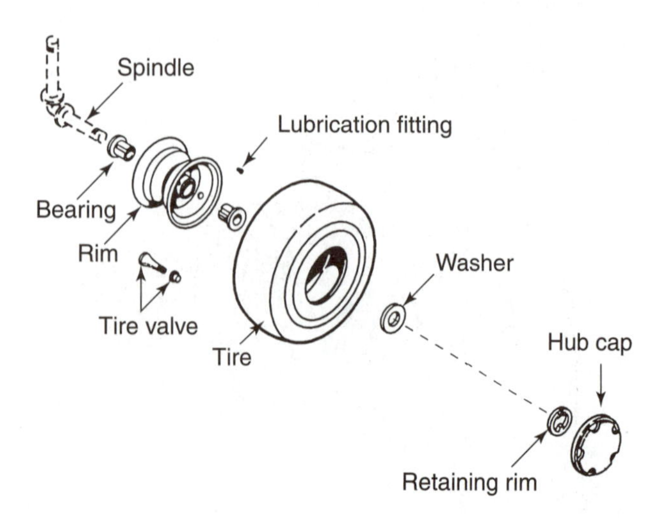

FIGURE 31-30 Removing a front-wheel assembly. (Courtesy of Simplicity Manufacturing.)

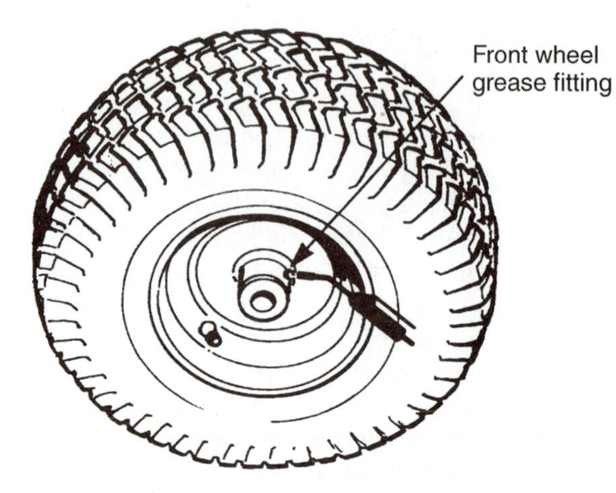

FIGURE 31-31 Lubricating the front-wheel bearing with a grease gun. (Courtesy of Simplicity Manufacturing.)

If the bearing requires replacement, locate the correct-sized bearing driver. Use a hammer or press to drive the old bearing out of the wheel rim. Compare the old and new bearing to be sure you have the correct size. Use a hammer and driver or press to install the new bearing. Make sure it seats in the correct position in the wheel rim.

Install the wheel back on the spindle. Install the washer and retaining ring. Grease the bearing by inserting a grease gun nozzle on the wheel grease fitting (Figure 31-31). Pump the grease gun handle several times to fill the bearing. Install the hubcap and lower the tractor.

Brake Adjustment

WARNING: Do not adjust the brake with the engine running. Turn the ignition to off, remove the ignition key, and block the wheels before adjusting the brakes.

As the brake is used, the friction pads (pucks) wear down and the brake lever tension decreases on the brake disc. Periodically the brake linkage requires adjustment. Different brake systems are adjusted differently. Always locate and follow the specific procedure in the appropriate shop service manual.

To adjust the brake, remove the brake lever cotter pin from the brake castle nut (Figure 31-32). Threading the castle nut in (clockwise) presses the brake pins against the back of the brake pads and brings the brake pad closer to the brake disc. Slide a clean (oil free) flat feeler gauge of the specified thickness between the brake pad and disc. Adjust the castle nut until the feeler gauge is snug. When

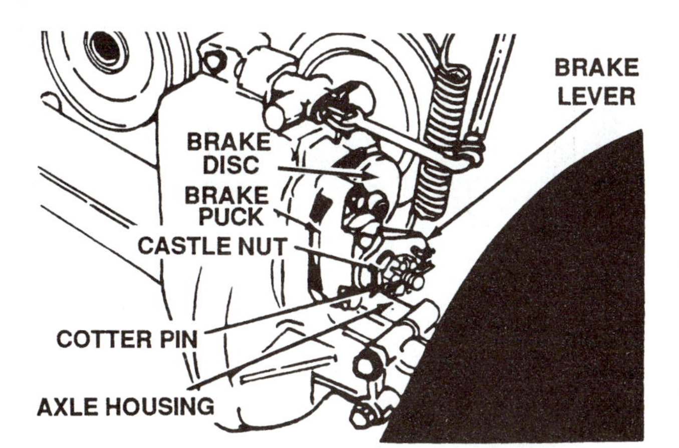

FIGURE 31-32 Adjusting the brake. (Courtesy of MTD Products, Incorporated.)

the adjustment is correct, replace the cotter pin with a new pin. The procedure for adjusting a lawn tractor brake is shown in the accompanying sequence of photographs.

> **CAUTION:** Do not reuse old cotter pins. A used pin could fail and allow the brake castle nut to back off and affect braking.

VARIABLE-SPEED PULLEY TRACTORS

Some tractors are equipped with a variable-speed pulley instead of a gear transmission (Figure 31-33). The **variable-speed pulley system** is a single-speed transmission that uses a pulley that changes diameter to provide a variable ratio between input and output pulleys. The variable-speed pulley system allows the operator to vary the tractor speed while maintaining a constant engine speed. The system is operated by the speed control/clutch/brake pedal. The engine pulley drives a long front belt that goes to a variable-speed pivot assembly. A rear belt goes from variable-speed pivot assembly to the transaxle input pulley. Movement of the pedal linkage changes the spacing between the two sides of the variable-speed pulley. The belt fits in a different part of the pulley as the two sides change position. This changes the effective diameter of the pulley. As the diameter changes, the relative speeds of the front and rear belts change. At slow speed position, the rear belt travels half the speed of the front belt. In the high speed position, the rear belt travels twice the speed of the front belt.

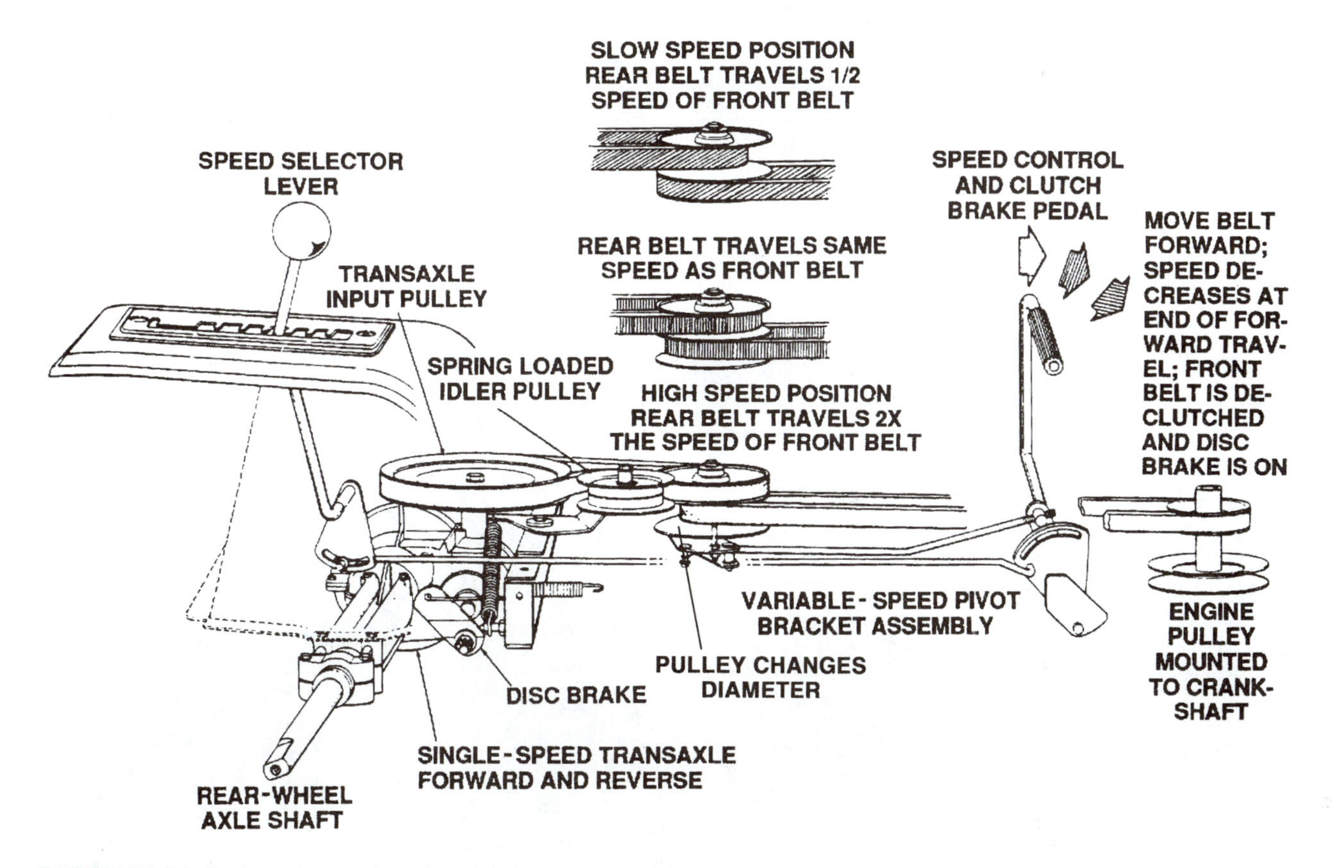

FIGURE 31-33 Parts and operation of a variable speed belt drive. (Courtesy of MTD Products, Incorporated.)

PHOTO SEQUENCE 25

Adjusting a Lawn Tractor Brake

1. Turn the ignition to off, remove the ignition key, and block the wheels before adjusting the brakes.

2. Remove the brake lever cotter pin from the brake castle nut.

3. Use the correct size wrench to turn the castle nut in (clockwise) to move the brake pads and bring the brake pad closer to the brake disc.

4. On some models, use a clean (oil free) flat feeler gauge of the specified thickness between the brake pad and disc to set clearance.

5. Adjust the castle nut until the feeler gauge is snug or the brake pedal has the desired feel.

6. When the adjustment is correct, replace the cotter pin with a new pin.

HYDROSTATIC TRANSMISSION TRACTORS

Some tractors use a fluid drive transmission called a **hydrostatic transmission** (Figure 31-34), a transmission that uses a variable displacement hydraulic pump and hydraulic motor to provide a variable output speed between the input and output shafts. The hydrostatic transmission has a single housing that contains a variable displacement piston hydraulic pump and a fixed displacement piston hydraulic motor. The hydrostatic transmission is controlled by a system of valves that are also contained in the same housing.

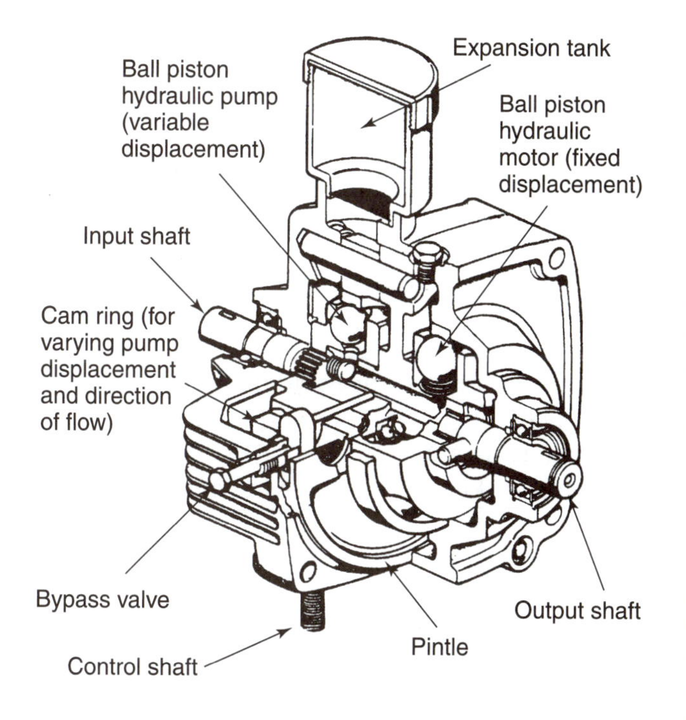

FIGURE 31-34 Parts of a hydrostatic transmission. (Courtesy of MTD Products, Incorporated.)

A single control lever is connected to the hydrostatic transmission pump section. The lever controls the speed and the direction of the output shaft. Varying the displacement ratios between the pump and motor provide infinite speed control. When the operator moves the control lever from neutral to forward, the output shaft turns in one direction. When the operator moves the lever from neutral to reverse, the output shaft turns in the opposite direction. The output shaft increases speed as the operator moves the lever from neutral.

Fluid flow in the hydrostatic transmission is through an internal closed loop between the pump and motor (Figure 31-35). Fluid flow is directed to the pump to the motor and then back to the pump. There is some controlled leakage in the system. This means that the amount of fluid flowing back to the pump is less than that required by the pump. Check valves on the inlet side of the pump open to supply fluid to the pump from the reservoir. Hydrostatic transmission speed control is accomplished by changing the amount of fluid delivered by the pump to the motor. The fluid delivery is controlled by moving the speed control lever.

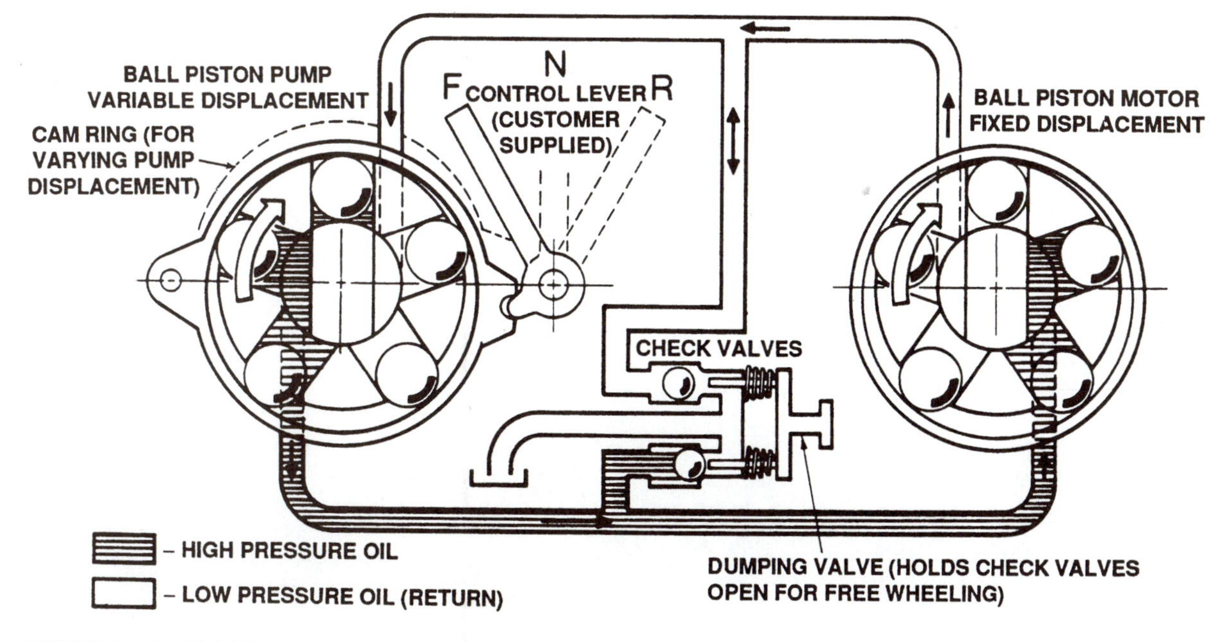

FIGURE 31-35 Fluid flow through a hydrostatic transmission. (Courtesy of MTD Products, Incorporated.)

REVIEW QUESTIONS

1. Explain why you should read and follow the instructions in the operator's manual before starting a tractor.
2. Explain why you should read and follow all the safety messages on the tractor decals.
3. Explain the safety precautions to take when operating a tractor to prevent injury to children.
4. List the safety precautions to take when operating a tractor on a slope.
5. Describe the parts and operation of a two-wheel tractor steering system.
6. Explain the advantage of a four-wheel steering system.
7. Describe the parts and operation of a four-wheel steering system.
8. Explain the purpose of the transmission in a tractor transaxle.
9. Explain the power flow from an engine drive pulley on a transaxle into the transmission counter gear shaft.
10. Explain how transmission shifter gears are locked to their shafts when different gears are selected by the operator.
11. Explain why a chain is used to reverse the transmission power flow in reverse gear.
12. Explain the purpose of a differential in a transaxle.
13. Describe the parts and power flow through a differential when the tractor is turning left.
14. Describe the parts and power flow through a differential when the tractor is turning right.
15. Describe the parts and operation of the transaxle mounted disc brake.
16. Explain the purpose of the tractor wiring harness.
17. Explain why a service manual should be used to troubleshoot tractor equipment problems.
18. Explain the safety precautions to take when jacking up a tractor.
19. Explain the steps in removing and installing a rear tractor wheel.
20. Describe the steps in removing and installing a front wheel bearing.

DISCUSSION TOPICS AND ACTIVITIES

1. Locate and disassemble a scrap tractor transaxle. Trace the power flow through each forward gear and reverse.
2. Trace the power flow through the differential during a left and right turn.
3. An input helical gear is turning clockwise. The input gear is in mesh with a helical output gear. The output gear shaft has a sprocket. The sprocket is driving another sprocket (output) with a chain. In what direction does the output sprocket turn?

CHAPTER
32
Edgers, Tillers, and Snow Throwers

OBJECTIVES

Upon completion and review of this chapter, you should be able to:

- Explain why the owner's manual must be read and followed before starting equipment.
- Describe the safety procedures to follow when using an edger, tiller, or snow thrower.
- Describe the parts and operation of a lawn edger.
- Troubleshoot an edger for clutch and blade problems.
- Change the blade on an edger.
- Describe the parts and operation of a garden tiller.
- Troubleshoot a tiller for drive and tine problems.
- Remove and replace tiller tines.
- Describe the parts and operation of a snow thrower.
- Troubleshoot a snow thrower for snow clearing problems.
- Adjust skid shoes and lubricate a snow thrower.

TERMS TO KNOW

Garden tiller
Lawn edger
Snow thrower
Snow thrower auger
Tiller tine
Tine depth stake

INTRODUCTION

Edgers, tillers, and snow throwers have similar parts and operate in the same basic way. They are walk behind machines that use a handle bar control assembly. They use an engine to drive a rotating blade, tine, or auger. Some type of clutch or transmission is used to engage or disengage the blade, tine, or auger. Four-stroke engines are usually used on larger equipment. Some smaller edgers, snow throwers, and tillers use a two-stroke engine.

EDGER, TILLER, AND SNOW THROWER SAFETY

Edgers, tillers, and snow throwers have a high-speed rotating blade, tine, or auger. If the rotating tool comes in contact with your body, it will cause severe injury. If the tool comes in contact with gravel or other hard objects, it may throw them and cause injury. Always read and understand the equipment owner's manual before operating the equipment. Make sure you understand the controls and the proper use of the equipment. Know how to stop the equipment quickly in an emergency. Follow these general safety rules to prevent injury:

- Understand and follow each danger, warning, caution, and instruction decal installed on equipment (Figure 32-1).
- Be alert for the possibility of the tool striking a solid object and causing a sudden jerking force

FIGURE 32-1 Always follow the safety instructions on equipment decals. (Courtesy of Tanaka/ISM.)

that could cause you to lose control of the equipment.

- Do not allow children to operate or play with equipment.
- Do not allow adults to operate equipment without proper instructions.
- Safe operation of equipment requires complete and unimpaired attention at all times.
- Do not operate equipment if fatigued or during or after consumption of medication, drugs, or alcohol.
- Thoroughly inspect the work area for gravel, sticks, or other foreign objects that could be picked up and thrown by the rotating tool.
- Check for uneven curb, sidewalk, driveway, or other hard surface before edging.
- Protect eyes with nonfogging goggles, ears with ear plugs or protectors, and head with approved safety hard hat.
- A dust filter mask should be worn while tilling or edging a dusty surface.
- Wear sturdy footwear with nonslip soles (steel-toed safety boots are recommended) for good footing.
- Do not operate equipment barefoot or when wearing open sandals or canvas shoes.
- Use sturdy gloves when handling equipment blades to improve your grip and protect your hands.
- Wear sturdy, snug-fitting long pants and clothing that provide complete freedom of movement.
- Do not go barelegged or wear loose clothing or jewelry that may get caught in rotating parts of the equipment.
- Do not attempt to make any adjustments to equipment while the engine is running.
- Do not fill the fuel tank indoors, when engine is running, or while engine is still hot.
- Wipe off any spilled gasoline, allow vapors to dissipate and move unit 10 feet (3 meters) away from the fuel container before starting engine.
- Keep equipment in good condition, all guards and shields in place, and safety devices operating properly.
- Be sure that Start/Stop switch stops engine and the throttle lever moves freely.
- Do not run engine in enclosed or poorly ventilated areas.
- When starting equipment, be sure your feet are not near the rotating tool.
- Do not put hands or feet near rotating parts.
- Maintain a firm grip on both handles at all times to help keep control should sudden jerking of the unit occur.
- While using the equipment, walk with your body well balanced and do not overreach or operate unit with one hand.
- Never operate equipment without someone within calling distance but outside your work area, in case help is needed.
- Operate equipment only when there is good visibility and light.
- Never direct discharge of material toward bystanders.
- Do not allow anyone near equipment while it is in operation.
- The operator is responsible for the safety of bystanders.
- Do not use your hands to clear snow from a snow thrower discharge chute or debris from the auger collector.
- If you hit a solid object, stop the equipment and inspect the tool for damage.
- Stop and inspect the tool if it begins to vibrate.
- Do not use equipment across a sloped surface. Do not use equipment near banks and ditches.
- Use three-wire grounded receptacle when plugging in a snow thrower electric starter.
- Do not overload equipment with too fast a ground speed.
- Disengage the blade, auger, or tines when transporting equipment.

LAWN EDGERS

A **lawn edger** is an engine-powered walk behind cutting tool used to cut the edges of grass and turf. The edger engine may be two or four stroke (Figure 32-2). Edgers usually use engines with a horizontal crankshaft.

Edger Parts and Operation

The edger has a handlebar assembly similar to a walk behind lawnmower (Figure 32-3). The handle usually has a clutch lever to engage and disengage

CASE STUDY

Consumer injuries when using snow throwers caused the Consumer Product Safety Commission to issue the following consumer product safety alert:

> The U. S. Consumer Product Safety Commission (CPSC) wants you and your family to be safe when using snow throwers. Snow throwers are the fourth leading cause of finger amputations among consumer products. In a recent year, there were approximately 1,030 amputations involving snow throwers. CPSC estimates that each year on average there are approximately 5,320 hospital emergency room-treated injuries associated with snow throwers. CPSC has received reports of nine deaths since 1992 involving snow throwers. Two people died after becoming caught in the machine. More commonly, deaths are caused by carbon monoxide poisoning resulting from leaving the engine running in an enclosed area. Injuries most frequently occurred when consumers tried to clear the auger/collector or discharge chute with their hands. CPSC offers the following safety tips for using snow throwers:
>
> - Stop the engine (unplug the spark plug) and use a long stick to unclog wet snow and debris from the machine. Do not use your hands to unclog a snow thrower.
> - Always keep hands and feet away from all moving parts.
> - Never leave the machine unattended if the engine is operating. Shut down the engine if you must leave the machine for any length of time.
> - Never leave the machine running in an enclosed area.
> - Add fuel to the tank outdoors before starting the machine; don't add gasoline to a running or hot engine. Always keep the gasoline can capped, and store gasoline out of the house and away from ignition sources. . . .

Source: Snow Thrower Safety Alert, Consumer Product Safety Commission, CPSC Document #5117, Washington, DC.

FIGURE 32-2 An edger with a two-stroke engine. (Courtesy of Tanaka/ISM.)

the cutting blade from the engine. The engine PTO shaft has a pulley for a V belt. The V belt is connected to a driven pulley on the edger blade assembly.

The edger blade assembly is mounted on a pivot (Figure 32-4). The pivot allows the operator to raise or lower the edger blade assembly with the

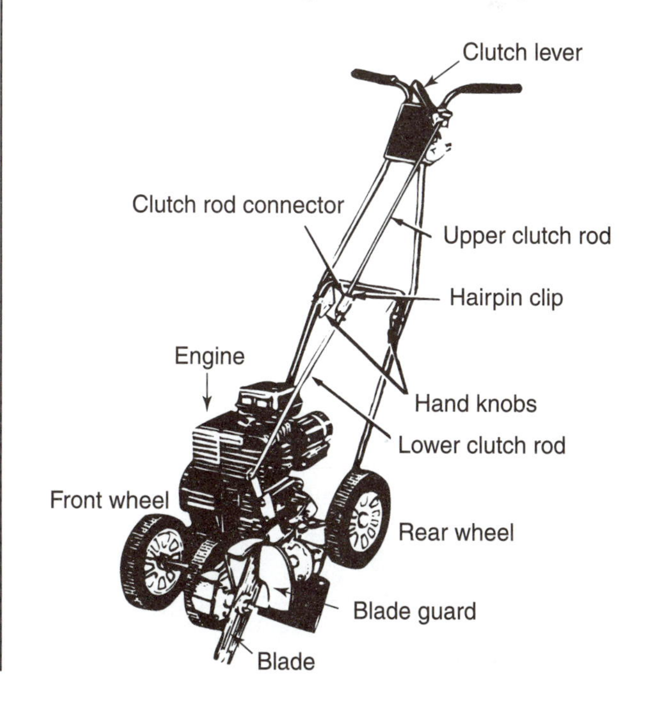

FIGURE 32-3 Parts of a lawn edger. (Courtesy of MTD Products, Incorporated.)

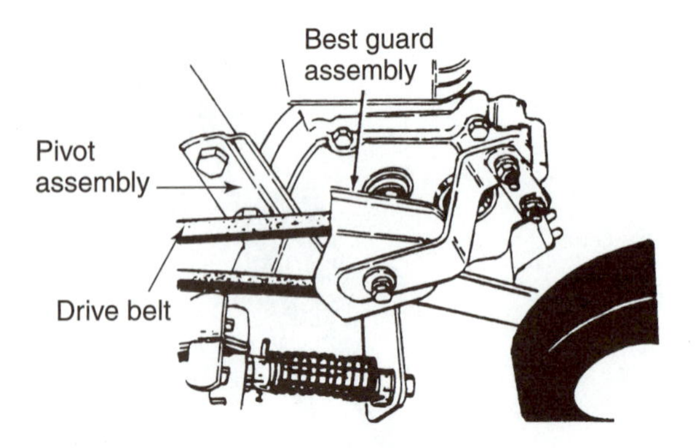

FIGURE 32-4 The edger clutch handle raises and lowers the blade assembly. (Courtesy of MTD Products, Incorporated.)

clutch lever. When the clutch lever is raised, it pulls the edger driven pulley closer to the engine PTO pulley. The belt between the pulleys gets slack and does not drive the edger blade. When the operator lowers the clutch handle, the blade assembly lowers. The driven pulley gets further away from the engine PTO pulley. The belt gets tight and the engine rotates the edger blade. There are notches or holes in the mount for the clutch lever so it can be positioned enabling the operator's hand to be removed. The clutch should always be disengaged when the operator is not edging.

The edger uses a small, thin cutting blade that fits on a shaft rotated by the driven pulley. The blade is often retained on the shaft with a nut that locks the blade between two friction washers (Figure 32-5). The blade is made from thin steel and has two cutting edges. The blade is so thin that it does not require sharpening. The constant friction of the thin blade against concrete wears down the blade length. The blade is covered with a blade guard to protect the operator.

Edger Troubleshooting

Lawn edger problems can be related to engine or equipment problems. Engine troubleshooting is presented in Chapter 21. There are two common problems with edgers:

- Blade fails to cut or does not cut deep enough
- Blade fails to rotate when clutch is engaged

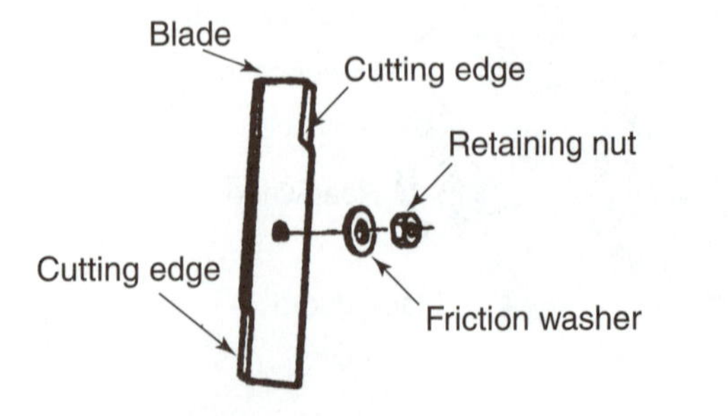

FIGURE 32-5 The edger blade is retained with a nut and friction washers. (Courtesy of Poulan/Weed Eater.)

When the blade fails to cut or does not cut deep enough, the blade may be the problem. Start the engine and lower the clutch handle. Observe the blade to make sure it is rotating.

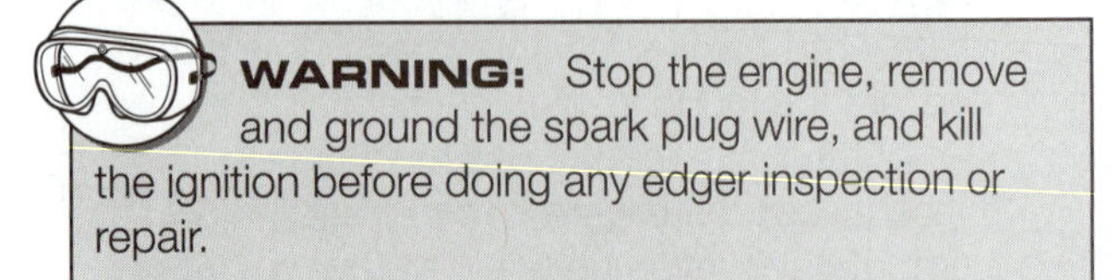

WARNING: Stop the engine, remove and ground the spark plug wire, and kill the ignition before doing any edger inspection or repair.

If the blade is rotating, the blade drive is working. Stop the engine, remove and ground the spark plug wire so the engine cannot start. Check the length of the blade. Blades commonly wear shorter and shorter as they scrape cement until they no longer reach the grass. Most edger blades are 10 inches long. When the blade wears below 8 inches it should be replaced. If this is the problem, replace the blade.

WARNING: Never try to adjust or inspect a drive belt with the engine running.

If the problem is a blade that is not rotating when the clutch is engaged, the problem is most likely in the belt drive. Make sure the kill switch is off and remove and ground the spark plug wire. Disengage the clutch and inspect the V belt. Look for a broken belt. A broken belt prevents the engine pulley from driving the drive pulley. Power will not get to the edger blade.

To inspect for a slipping drive belt, grab the belt in your hand and twist it so you can see the underside of the V shape. Look at the sides of the V. Glazed belts have a shiny appearance because when a belt is not tight enough, the slipping polishes its surface. Check the belts for oil because oil-soaked belts can slip and not rotate the drive pulley. Also check the belt for cracks. Cracks indicate the belt is getting ready to fail. Torn or split belts have major damage and must be replaced immediately.

If the belt is not the problem check the engine PTO drive and blade-driven pulley. A broken key or loose Allen set screw can cause either pulley to spin on its shaft. Try to move the pulley in relation to its shaft. If you see movement, tighten the Allen set screw or replace the key (Figure 32-6).

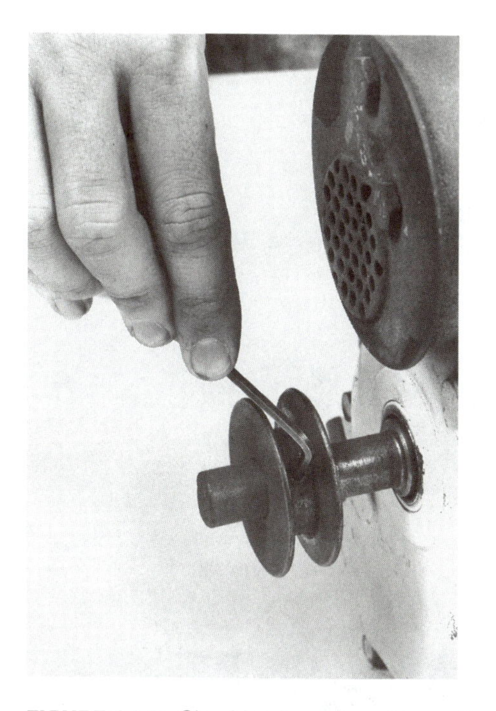

FIGURE 32-6 Checking the edger drive pulley Allen screw.

Replacing the Blade

WARNING: Stop the engine, remove and ground the spark plug wire, and kill the ignition before doing any edger maintenance or repair.

Raise the blade by disengaging the edger clutch. Use a glove on your hand and hold the edger blade to keep it from turning. Use the correct sized wrench to remove the blade retaining nut. The retaining nut on most edgers is standard right-hand thread. Remove the friction washers and the old blade. Compare the old and new blade to make sure the center hole is the same diameter. Install the blade and replace the washers. Tighten the retaining nut securely to prevent it from coming loose under power. The procedure for installing an edger blade is shown in the accompanying sequence of photographs.

Installing an Edger Belt

To install a new edger belt, remove the belt guard assembly by removing the hex bolts, lock washers, and hex nuts (Figure 32-7). Remove the old belt and compare it with the replacement. Make sure the new belt is the correct length. Install the new belt on the front and rear pulleys. Install the belt guard assembly and install the belt guard hex bolts.

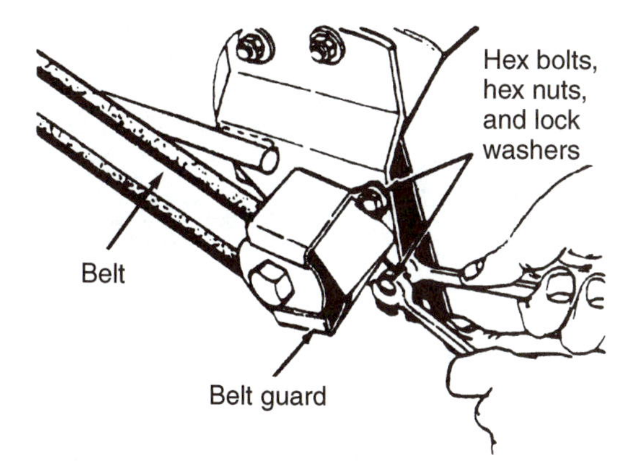

FIGURE 32-7 Removing the belt guard to remove the drive belt. (Courtesy of MTD Products, Incorporated.)

WALK BEHIND GARDEN TILLERS

A **garden tiller** is an engine-powered walk behind cutting tool used to break ground for planting. Tillers require four-stroke, horizontal crankshaft engines with high torque to cut deeply into topsoil. Engines are commonly in the 3–10 horsepower range.

Tiller Parts and Operation

The tiller is operated on rough ground and while making deep cuts in the topsoil. This makes the tilling operation difficult for the operator to control. The tiller has a handlebar control assembly similar to other walk behind equipment (Figure 32-8). The tiller handlebar is usually wider than other walk behind equipment to provide more operator leverage. Tillers commonly have two control levers, one on each handlebar. The tines can only be engaged when the operator has both hands on

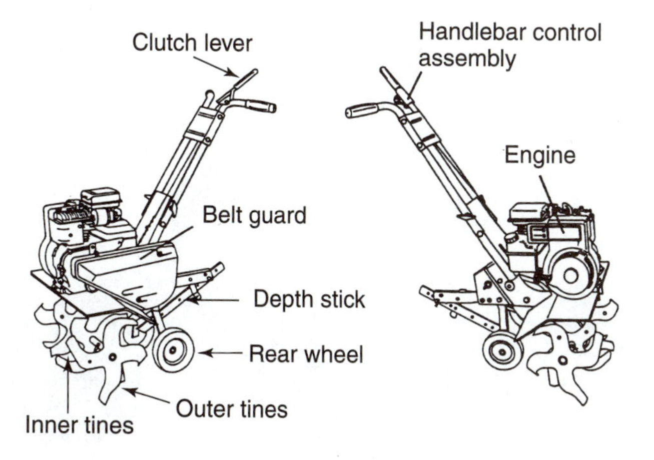

FIGURE 32-8 Parts of a garden tiller. (Courtesy of Poulan/Weed Eater.)

PHOTO SEQUENCE 26

Removing and Replacing an Edger Blade

1. Kill the ignition and remove and ground the spark plug wire for safety.

2. Keep the blade from turning by holding it with a gloved hand. Use a wrench to loosen the blade-retaining nut.

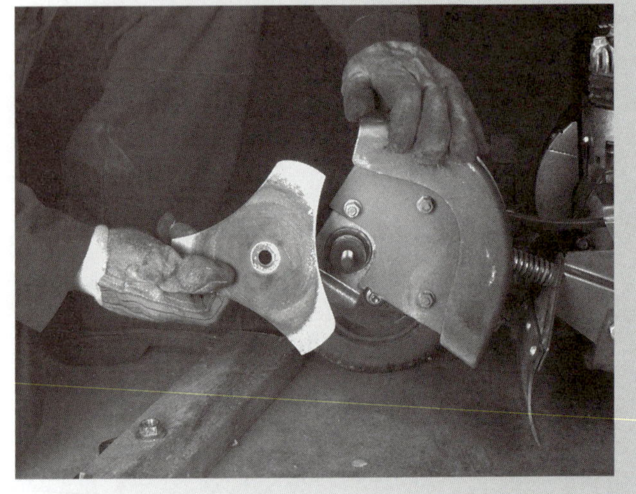

3. Remove the blade-retaining nut, friction washers, and blade.

4. Install the new blade, friction washers, and start the blade-retaining nut.

5. Keep the blade from rotating by holding it with a gloved hand. Tighten the blade-retaining nut.

6. Install the spark plug wire on the spark plug.

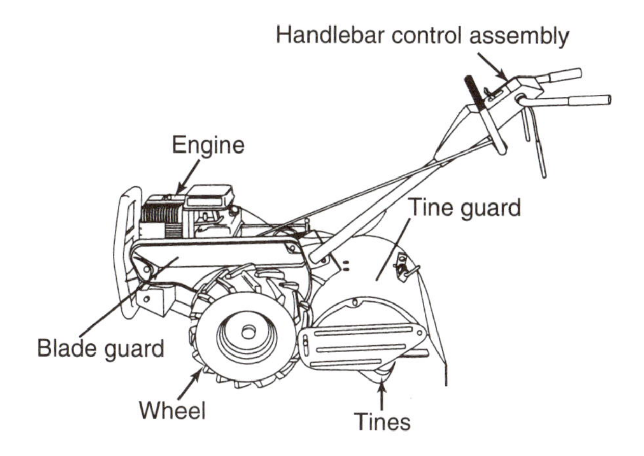

FIGURE 32-9 Tiller with rear-mounted tines. (Courtesy of Poulan/Weed Eater.)

the handlebar. Both levers have to be depressed to engage the tines. The tines will stop rotating if either hand comes off the handlebar. The purpose of these systems is to prevent the operator from coming into contact with the rotating tines.

The tiller engine drives a set of tines. A **tiller tine** is a four-bladed cutting tool shaped to dig into topsoil. Garden tillers commonly use two outside and two inside tines. Small rear wheels support part of the tiller weight. The rotating tines support the tiller weight and also self-propel the tiller. Some tillers use large wheels in the center of the tiller. The tines on these tillers are located under a tine guard at the rear (Figure 32-9).

The tines are driven by the engine through a belt and transmission system. There is a drive pulley on the engine PTO shaft. There is a larger transmission pulley on the transmission input shaft. A V belt transfers engine torque from the drive pulley to the transmission pulley. The transmission pulley is larger than the drive pulley. This feature reduces speed and multiplies engine torque. An idler pulley is used to adjust belt tension. The belt assembly is covered by a belt guard (Figure 32-10).

The belt assembly transfers engine torque into the transmission or chain case (Figure 32-11). The transmission (chain case) is located between the tines. The transmission pulley is mounted on the transmission input shaft. Power flows into the transmission input shaft and through a set of chain driven sprockets. The small chain sprockets drive large chain sprockets to provide forward gear reductions and torque multiplication.

Power flows through the transmission and out a long drive shaft. The drive shaft has a set of holes for mounting the tines. Two tines fit on each side of

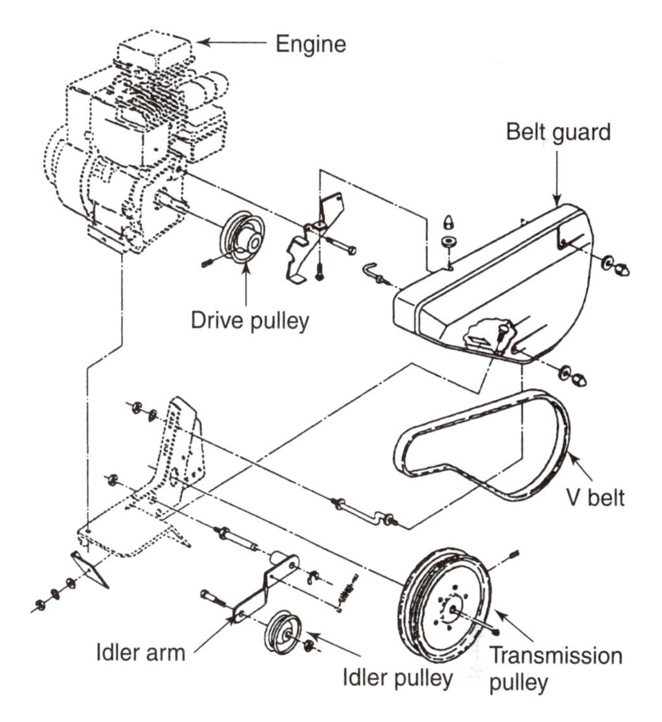

FIGURE 32-10 Tiller drive belt assembly. (Courtesy of Poulan/Weed Eater.)

the drive shaft. The inner left-hand and inner right-hand tines slip over the drive shaft. A clevis pin goes through the tine and drive shaft. The pin is retained by a hair pin clip that fits through a hole in the pin. The outer left-hand and outer right-hand tines slip over the ends of the drive shaft. They are also retained with clevis pins and hairpin clips.

When the engine is running and the operator engages the transmission, the drive shaft rotates all four tines. The tines are shaped to dig into and turn over the topsoil. As they do this, the tiller is pulled forward.

Tiller Troubleshooting

Tiller problems can be related to engine or equipment problems. Engine troubleshooting is presented in Chapter 21. There are two common tiller problems:

- Tines do not rotate
- Tines do not penetrate the ground

WARNING: Stop the engine, remove and ground the spark plug wire to kill the ignition, and remove the ignition key (if used) before doing any tiller inspection or repair.

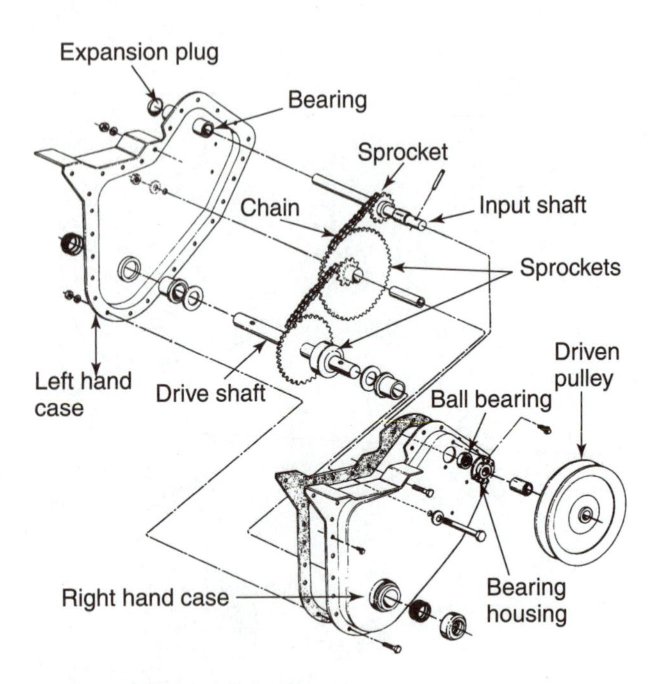

A. Transmission Components

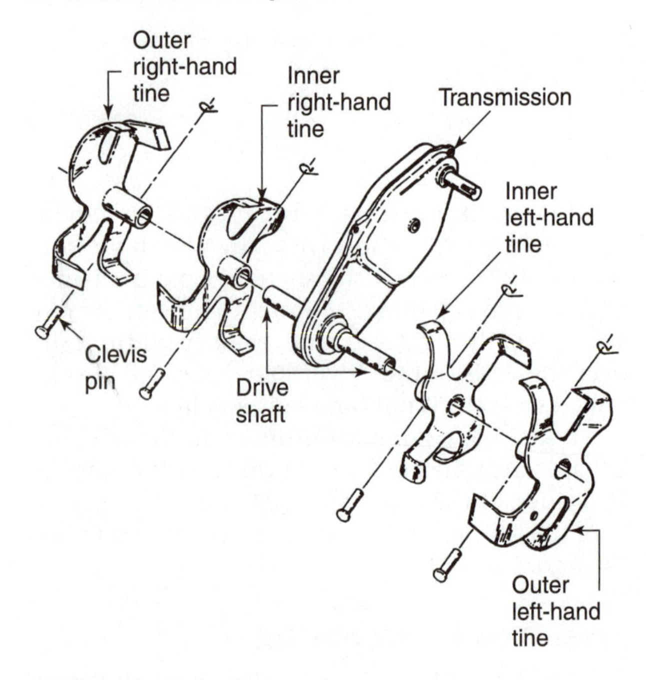

B. Tine Components

FIGURE 32-11 Tiller transmission and tine assembly. (*A:* Courtesy of MTD Products, Incorporated; *B:* Courtesy of Poulan/Weed Eater.)

WARNING: Never try to adjust or inspect a drive belt with the engine running.

When tiller tines do not rotate, first make sure the kill switch is off and remove and ground the spark plug wire. Tines can get jammed with roots and lock up. Use a stick to clear the tines. If the tines still fail to rotate, there is a problem with the drive system. The most likely source of the problem is the drive belt assembly. Remove the belt guard and inspect the V belt (Figure 32-12). Look for a broken belt. A broken belt prevents the engine pulley from driving the transmission (chain case) pulley. If you find a broken belt, replace the belt.

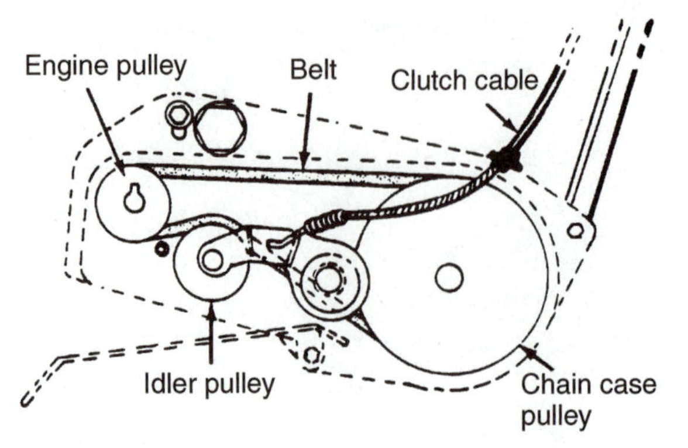

FIGURE 32-12 Inspecting the tiller belt assembly. (Courtesy of MTD Products, Incorporated.)

If the belt is not broken, it may be slipping. To inspect for a slipping drive belt, grab the belt in your hand and twist it so you can see the underside of the V shape. Look for a shiny appearance on the V shape. Glazed belts have a shiny appearance because when a belt is not tight enough, the slipping polishes its surface. Check the belts for oil. Oil-soaked belts can also slip and not rotate the transmission (chain case) pulley.

If the belt is not the problem, check the engine PTO drive and transmission (chain case) pulley. A broken key or loose Allen set screw can cause either pulley to spin on its shaft. Try to move the pulley in relation to its shaft. If you see movement, tighten the Allen set screw or replace the key. If the problem is not in the belt system, the problem is most likely inside the transmission (chain case). The transmission (chain case) will require replacement or overhaul.

When the tines rotate but do not penetrate the ground, check the operator's manual for control settings. The controls may be set incorrectly. Tillers commonly have a depth stake. The **tine depth stake** is an adjustable metal rod on the tiller used to penetrate the ground and set tilling depth. Check the operator's manual for the procedure to set the depth stake. The depth stake is usually adjusted by releasing it from a retaining latch. A series of holes in the depth stake are used to set its depth (Figure 32-13). Remove the depth stake hairpin clip

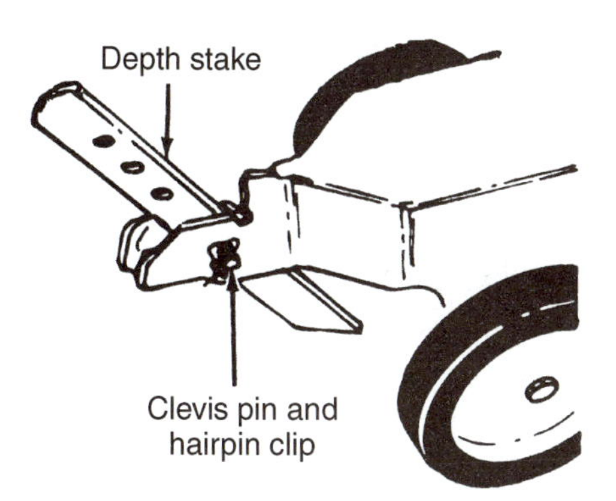

FIGURE 32-13 Setting the tiller depth stake. (Courtesy of MTD Products, Incorporated.)

and clevis pin. Set the depth stake into the ground at the desired depth. Install the pin at the required position on the depth stake hole to set the proper depth. Install the hairpin clip.

If the depth is set correctly, check the condition of the tines. Tines can get bent or jammed with roots. Make sure there are no broken or missing tine clevis pins. Also check for damage to the ends of the drive shafts. Replace or clean the tines as required.

Removing and Replacing Tines

WARNING: Stop the engine, remove and ground the spark plug wire, kill the ignition, and remove the ignition key (if used) before removing the tines.

Make sure the tiller kill switch is in the off position. Disconnect and ground the spark plug wire. Remove the fuel tank drain plug or fuel line and drain the fuel tank into a metal container. Dispose of the engine fuel properly.

SERVICE TIP: You can remove a fuel tank cap and place a plastic sandwich bag over the filler neck. Then replace the cap. This will allow tipping the equipment without spilling fuel. Remove the plastic bag after turning the equipment upright. On some equipment, you can do the same thing with the oil filler cap to prevent oil loss.

Tilt the tiller back on its handles so that the tines are elevated. If necessary, remove the tine shield (Figure 32-14). Use needle nose pliers to remove the hairpin clips from each of the tine clevis pins. Use gloves on your hands and pull each of the

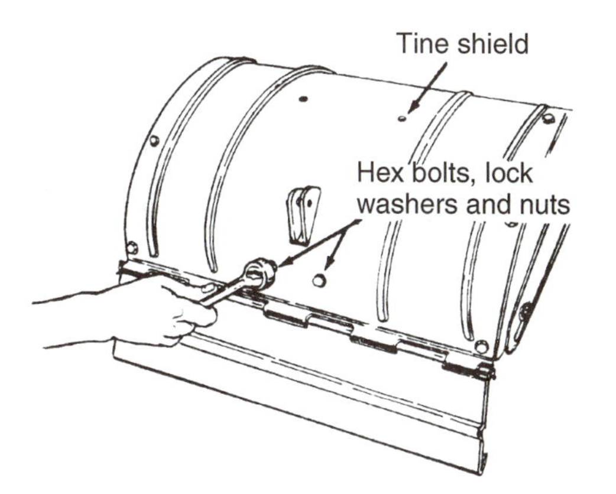

FIGURE 32-14 Removing the tine shield to access the tines. (Courtesy of MTD Products, Incorporated.)

tines off the drive shaft. The tines will probably have to be rocked back and forth to get them to come off. There is often a dust cap between the inner tines and the transmission (chain case) to keep dirt out of the transmission (chain case). It may come off as you pull off the inner tine (Figure 32-15).

WARNING: Wear gloves when handling the tines to protect your hands from cuts.

Clean up the drive shafts by sanding them lightly with fine emery cloth. Wipe the drive shafts clean with a clean shop rag. Compare the new tines with the old ones to make sure they are the same. Lubricate the tin shaft with a light grease. Replace the dust cap if it came off with the tines. Install the inner tines and install new clevis pins and hairpin clips. Then repeat this procedure for the outer tines

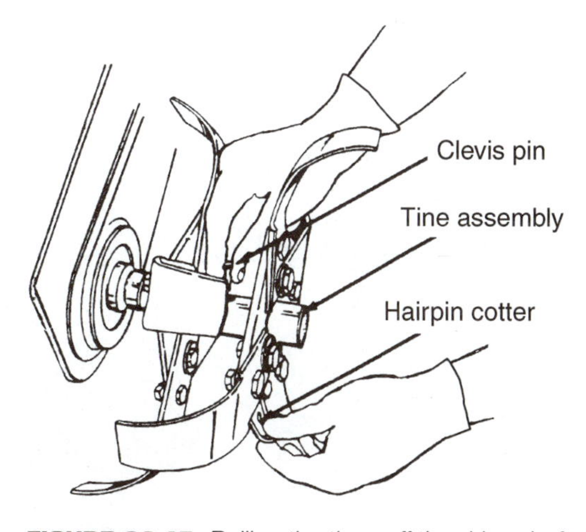

FIGURE 32-15 Pulling the tines off the drive shaft. (Courtesy of MTD Products, Incorporated.)

PHOTO SEQUENCE 27

Removing and Replacing Tiller Tines

1. Remove and ground the spark plug wire, kill the ignition, and remove the ignition key (if used) before removing the tines.

2. Tilt the tiller back on its handles so that the tines are elevated. If necessary, remove the tine shield.

3. Remove the hairpin clips from the tine clevis pins. Use gloved hands and pull each of the tines off the drive shaft.

4. Clean up the drive shafts by sanding them lightly with fine emery cloth. Wipe the drive shafts clean with a clean shop rag.

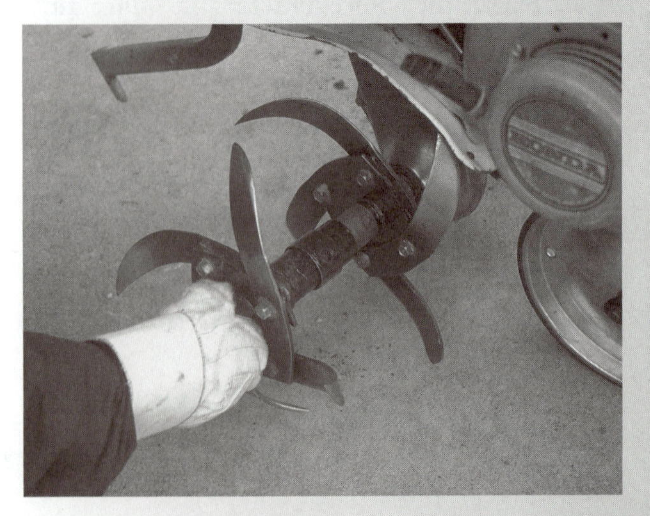

5. Lubricate the tin shaft with a light grease. Install the tines using gloved hands and install new clevis pins and hairpin clips.

6. Tilt the tiller back up in operating condition, install the tine shield, and connect the spark plug wire.

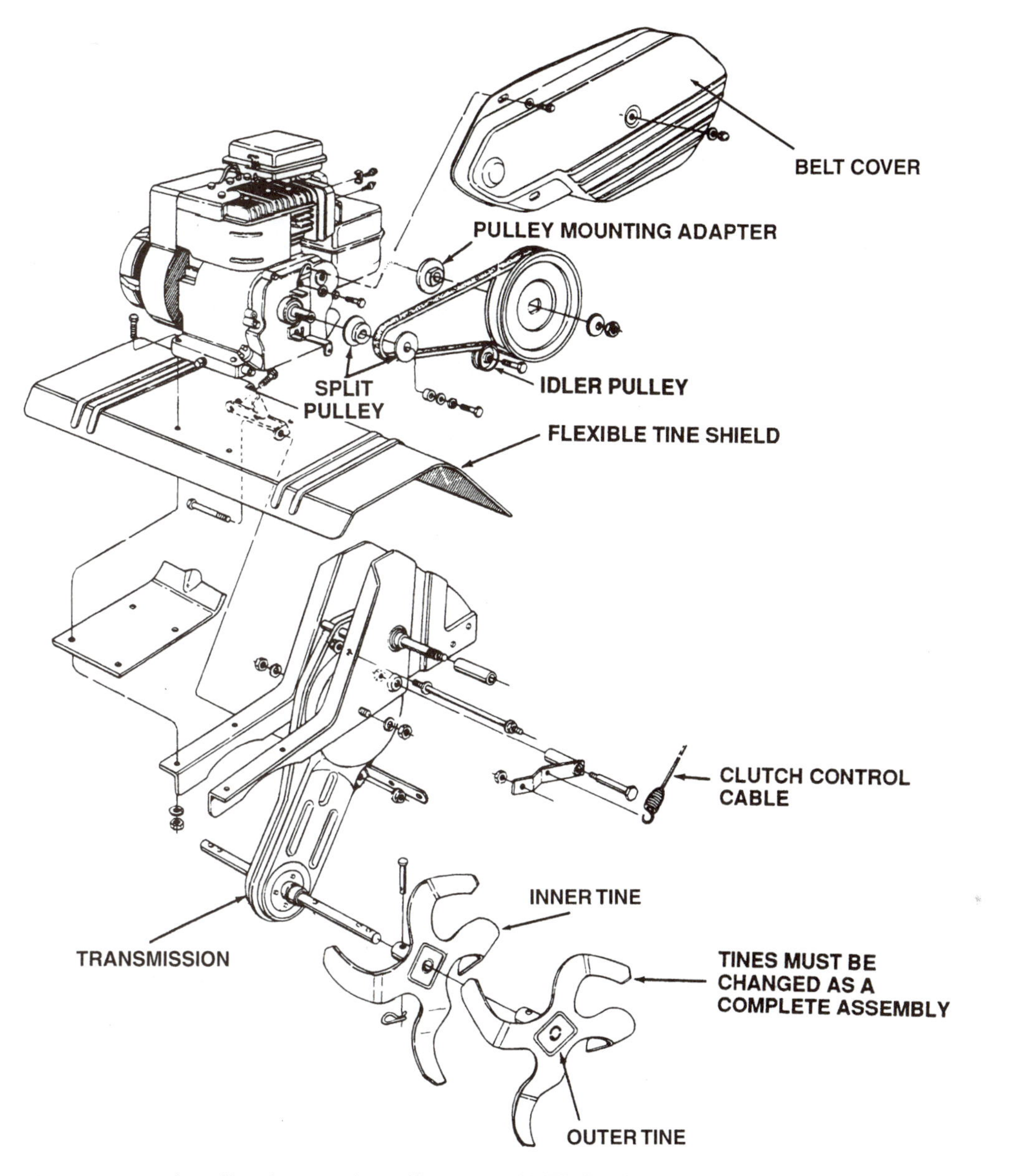

FIGURE 32-16 Installing the new tines. (Courtesy of MTD Products, Incorporated.)

(Figure 32-16). Tilt the tiller back up in operating condition. Install the fuel tank drain plug or fuel line. Refill the fuel tank with fresh fuel. The procedure for removing and replacing tiller tines is shown in the preceding sequence of photographs.

SNOW THROWERS

A **snow thrower** is an engine-powered walk behind tool used to clear snow from driveways and walkways after a snowfall. Snow throwers are rated by clearing width and amount of snow they can remove. The clearing width is the width in inches that can be cleared in one pass. The amount of snow that can be cleared is measured in pounds per minute under perfect conditions. Small snow throwers use two-stroke engines with approximately 3 horsepower. They usually clear widths of 16–20 inches. Larger snow throwers commonly use four-stroke horizontal crankshaft engines. These units may use engines with up to 20 horsepower. Large snow throwers can clear 32 inch widths or greater and move 2,500 pounds of snow per minute. The engines are often equipped with special starting systems and insulation around fuel system parts because they are used in very cold weather. Some snow throwers are self-propelled and others are pulled by the rotating auger. The

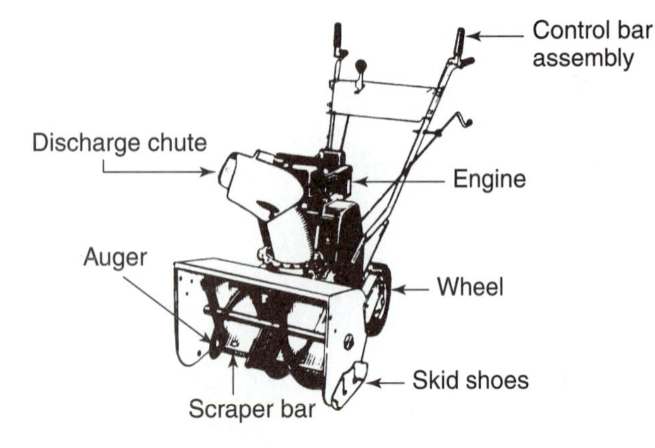

FIGURE 32-17 Parts of a snow thrower. (Courtesy of MTD Products, Incorporated.)

self-propelled system is similar to that used for lawn mowers.

Snow Thrower Parts and Operation

The snow thrower has a handlebar control assembly similar to the tiller (Figure 32-17). The handlebar is usually wide to provide more operator leverage. Controls are often positioned on the ends of the bar to allow the operator to operate them without taking hands off the control bar.

The engine rotates a large auger on the front of the snow thrower. The **snow thrower auger** is a helical-shaped blade used to pick up snow and discharge it through a discharge chute. The auger is contained inside the auger housing. The housing is shaped to collect snow coming off the auger and direct it up a discharge chute. The discharge chute is shaped to direct the snow away from the area being cleared.

The front of the snow thrower rests on skid shoes at the bottom of the auger housing. The rear of the snow thrower rests on a scraper bar. Many snow throwers are propelled by the auger rotating into the snow and pulling the snow thrower. Self-propelled snow throwers use a drive system to rotate the wheels. Wheels for self-propelled snow throwers that provide traction in snow are required. Large wheels with studs are commonly used.

Some snow throwers use a cleated rubber tread driven by a rotating track (Figure 32-18). The flexible track is driven at each end by one of two wheels. The drive wheel is driven by the engine. The other end of the track is supported by an idler wheel. The rubber tread fits over the track. The track and tread provide the best traction in the snow and ice.

A. Snowthrower with Track and Tread

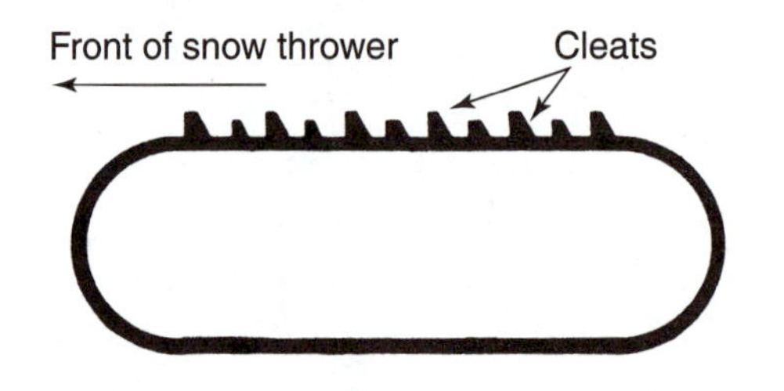

B. Snowthrower Tread

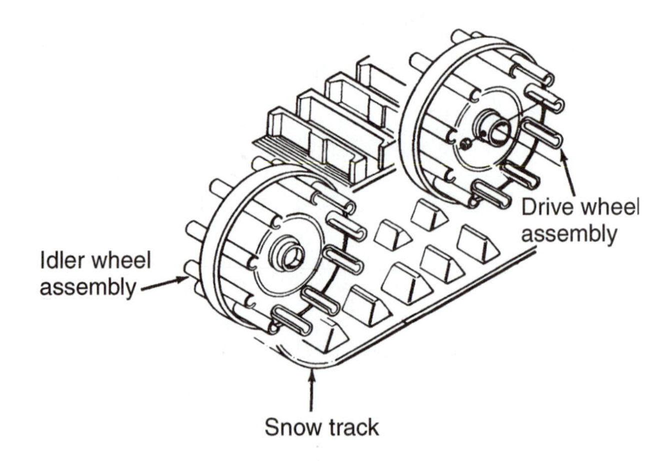

C. Snowthrower Track Assembly

FIGURE 32-18 Some snow throwers use a rubber tread driven by a track for traction. (Courtesy of MTD Products, Incorporated.)

Snow Thrower Drive Systems

The auger on many snow throwers is driven by a belt system (Figure 32-19). The belt system on non-self-propelled snow throwers is covered by a belt cover on the side of the snow thrower. A small engine pulley is used to drive a larger pulley connected to the auger shaft (axle). The small-to-large

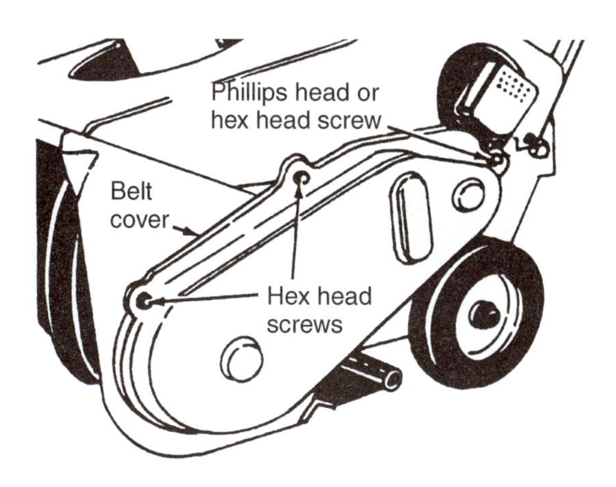

A. Belt Cover

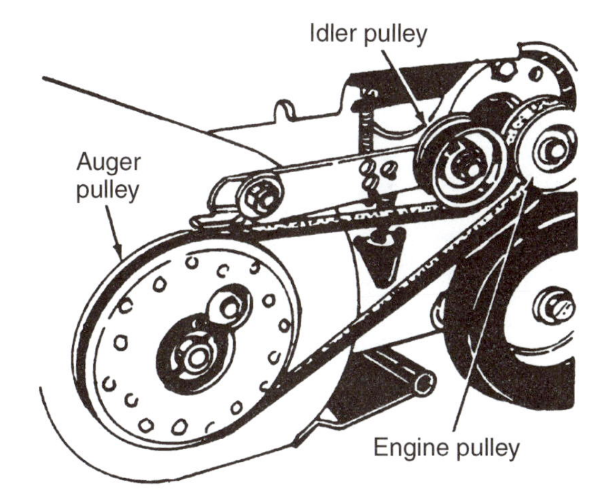

B. Belt System

FIGURE 32-19 A belt drive system used to drive the snow thrower auger. (Courtesy of MTD Products, Incorporated.)

pulley provides speed reduction and torque increase from the engine to the auger. A spring-loaded idler pulley is used to maintain the proper tension on the belt.

The operator engages the belt drive with a clutch lever or control bar on the handlebar. Movement of the clutch lever or bar moves the clutch rod. The clutch rod moves the belt idler pulley in a position to tighten the belt. The belt can then drive the auger drive pulley and the auger rotates.

The auger clutch is commonly connected to a dead man bar similar to that used in mowers. The clutch can only be engaged when the operator has both hands on the handlebar. Some snow throwers have two control levers, one on each handlebar. Both levers must be depressed to engage the auger. The auger stops rotating if either hand comes off the handlebar. The purpose of these systems is to prevent the operator from coming into contact with the rotating auger.

The auger on some snow throwers is driven by a traction drive system. The traction drive system uses a friction drive and a chain system controlled by a traction control lever (Figure 32-20). Engine power is transferred by belt to an engine drive shaft. The engine drive shaft drives a belt that rotates a traction input pulley. The traction input pulley rotates a drive disc. The drive disc contacts and drives the friction disc through frictional contact. Moving the traction lever changes the position of the drive disc in relation to the friction disc. When the center of the drive disc is driving the friction disc, the unit is in a low speed. As the friction disc is moved away from the center of the drive disc (to the outer perimeter), the speed is increased.

Power flows from a hex shaft in the center of the friction disc out to the chain assembly. A small chain sprocket on the hex shaft drives a chain that drives a large sprocket. A small gear in the center of the large sprocket drives a gear that drives the wheel.

Larger snow throwers use a gear box to drive the auger. The gear box is located in the auger housing (Figure 32-21). The transmission has a long input shaft that is rotated by the engine through a belt or chain drive. There is a worm gear on the end of the input shaft (Figure 32-22). The worm gear rotates a helical gear though a right angle at a speed reduction and a torque increase. The helical gear rotates the left and right auger shaft that rotates the augers. A brake pad operated by the control lever is used to stop the auger pulley. The brake pad performs the same function as an idler pulley clutch in other belt systems.

Two-Stage and Dual Auger Snow Throwers

A single stage snow thrower uses the auger to pick up snow and eject it out the chute. Larger snow throwers sometimes use a two-stage system. A two-stage snow thrower uses the auger to pick up the snow and feed it to a high-speed blower impeller that forces it out the ejection chute. The blower impeller is located in the entrance to the ejection chute (Figure 32-23). The blower forces snow out the chute. The blower impeller increases the amount of snow the thrower can discharge and increases its snow clearing ability. The impeller is usually driven by the same input shaft that drives the auger gear box (Figure 32-24).

Some snow throwers increase the snow removal rate with a double auger system With two augers digging into the snow at the same time, a larger area of snow can be cleared (Figure 32-25).

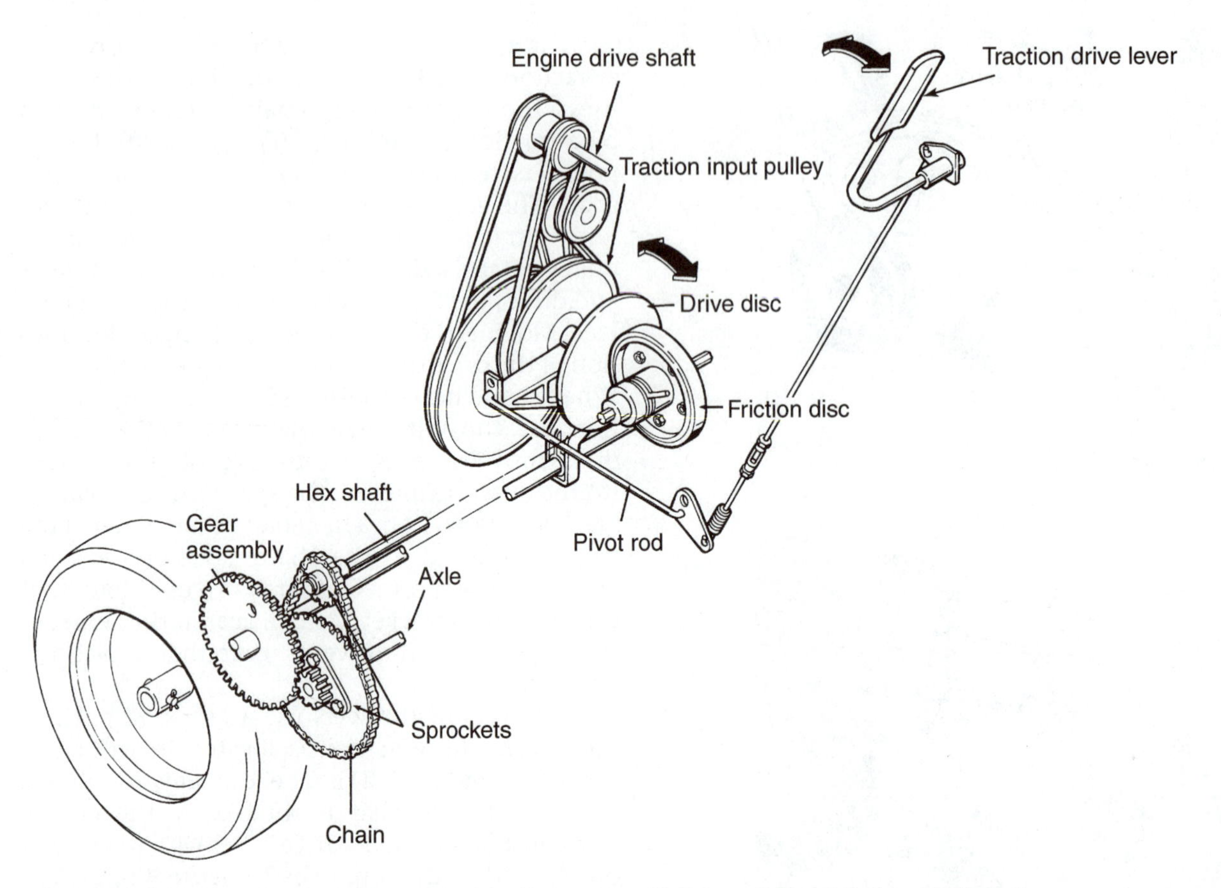

FIGURE 32-20 Parts of a snow thrower traction drive system. (Courtesy of Simplicity Manufacturing)

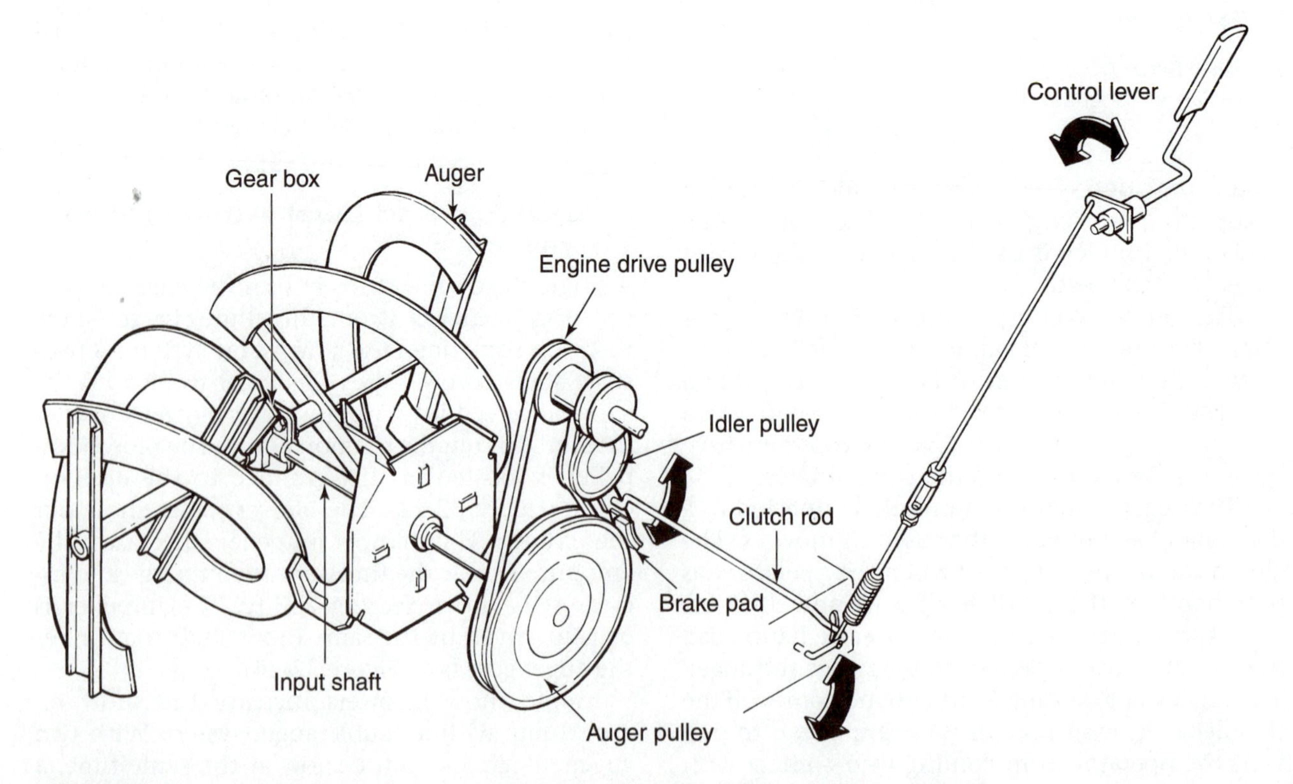

FIGURE 32-21 A gear box drive used to drive the auger. (Courtesy of Simplicity Manufacturing)

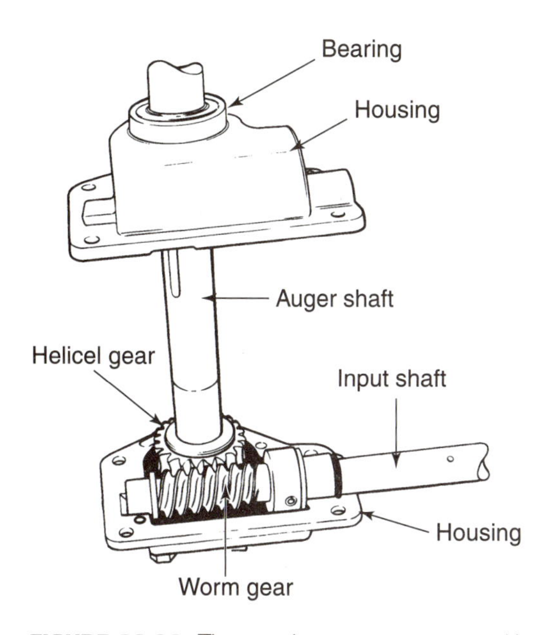

FIGURE 32-22 The gear box uses a worm and helical gear to provide a speed reduction and torque increase from the engine to the auger. (Courtesy of Simplicity Manufacturing.)

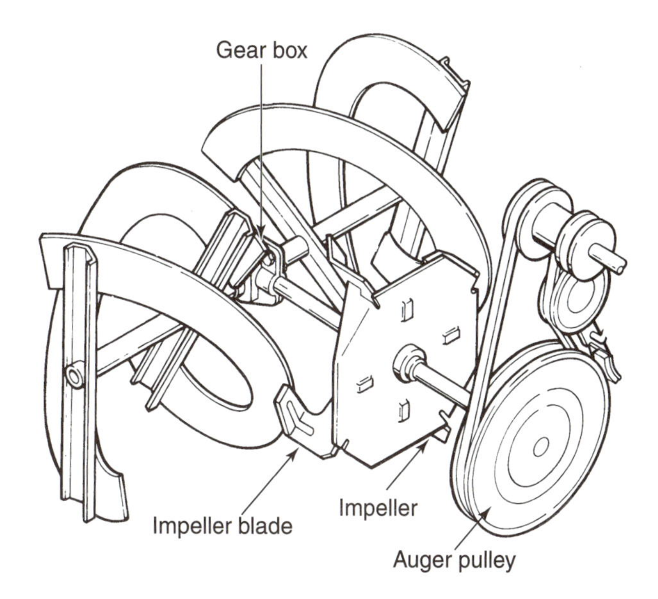

FIGURE 32-24 The two-stage blower impeller is driven by the gear box input shaft. (Courtesy of Simplicity Manufacturing.)

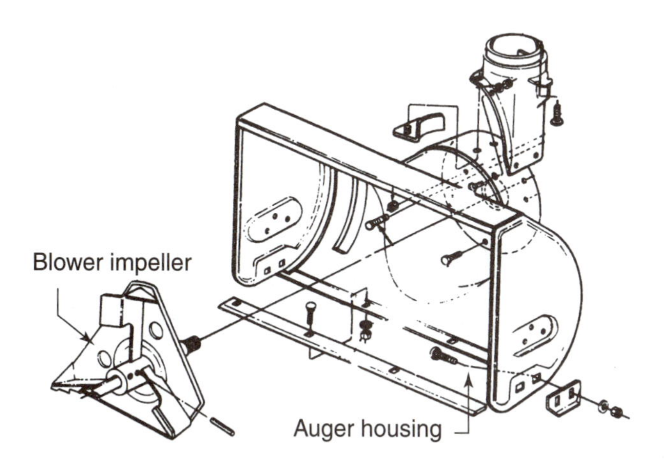

FIGURE 32-23 The two-stage snow thrower has a blower impeller located in the ejection chute. (Courtesy of MTD Products, Incorporated.)

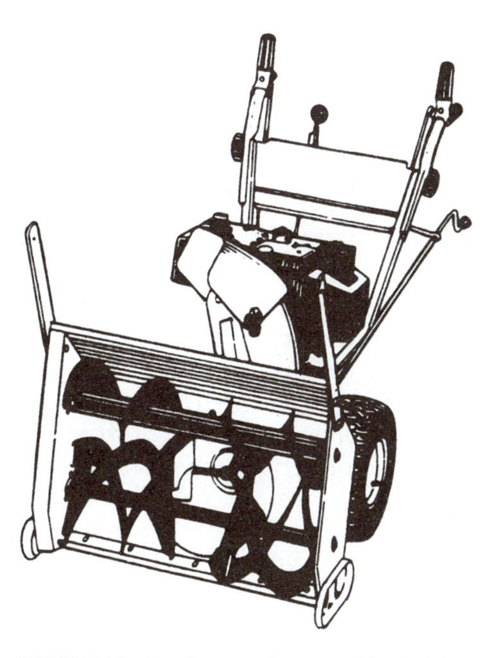

FIGURE 32-25 A snow thrower with double augers. (Courtesy of MTD Products, Incorporated.)

110-Volt Electric Starting Systems

Snow throwers are used in very cold weather. Cold weather causes engine oil to get very thick. It also makes gasoline hard to vaporize. Battery starting systems are not effective in cold weather because the cold temperatures reduce a battery's cranking power. These problems can make an engine difficult to start. Some snow throwers are equipped with a 110-volt alternating current starting system (Figure 32-26). The starting system has a 110-volt alternating current starter motor. Instead of a battery, a power cord is plugged into a wall outlet and then into the snow thrower connector. The operator pushes a starter button to start the engine. After starting the engine, the electrical extension cord is disconnected. Most systems also have a hand-operated recoil starter. This allows the snow thrower to be started by hand when it is warmed up.

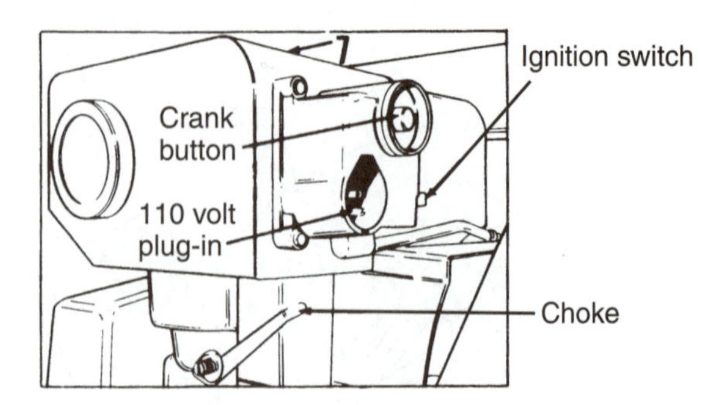

FIGURE 32-26 A 110-volt electric starting system helps start a snow thrower in cold weather. (Courtesy of Simplicity Manufacturing.)

Snow Thrower Troubleshooting

Snow thrower problems can be related to engine or equipment problems. Engine troubleshooting is presented in Chapter 21. The most common snow thrower problems are:

- Snow thrower rattles or vibrates excessively
- Snow thrower does not clear snow

WARNING: Stop the engine, remove and ground the spark plug wire, kill the ignition, and remove the ignition key (if used) before doing any snow thrower inspection or repair.

When a snow thrower rattles or vibrates, check the auger assembly housing and ejection chute for loose fasteners. Equipment vibration can cause these fasteners to work loose. Another common problem is a damaged auger or sheared auger retaining bolt. These problems often occur when the auger runs into a rock or tree branch when clearing snow.

If the auger is covered with snow, brush it off or wait until it melts to make your inspection. Engage the auger control bar (clutch) and rotate the auger by pulling on the starter rope. Check the entire auger surface for damage. Replace the auger if you find damage. Bolts are often used to attach the auger to the drive shaft (Figure 32-27). These can shear off when the auger hits an object. Inspect the bolt and replace any you find broken. Remove the broken fastener with a punch. Be sure to use the recommended tensile strength (grade) auger sheer bolt.

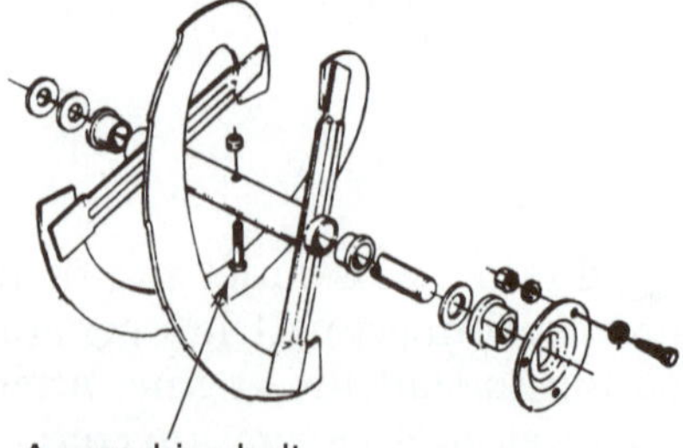

FIGURE 32-27 Remove a sheared auger bolt. (Courtesy of MTD Products, Incorporated.)

WARNING: Do not try to unclog an ejection chute when the snow thrower engine is running. Never put your hands in an ejection chute even if the engine is stopped. Severe injury can occur from contact with the rotating blower fan or auger.

When a snow thrower fails to clear snow, the most common cause is a clogged ejection chute. If the chute gets clogged, snow cannot get out and the auger area gets packed with excessive snow. Stop the engine and remove and ground the spark plug wire. Use a long stick like a broom handle to clear the chute.

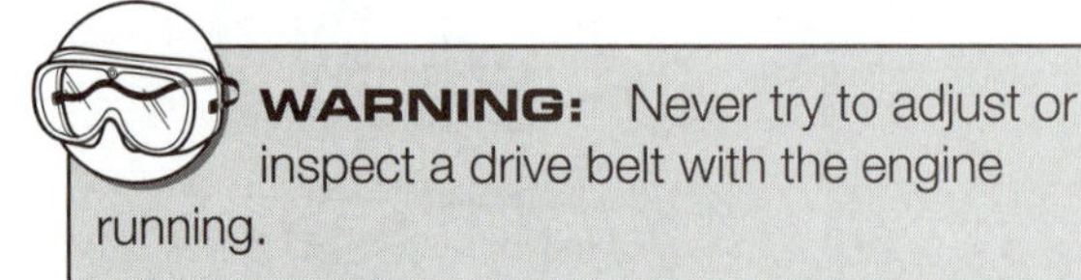

WARNING: Never try to adjust or inspect a drive belt with the engine running.

If a clogged ejection chute is not the problem, inspect the auger for damage. If the auger is not damaged, the problem may be a broken or loose drive belt. Make sure the kill switch is off and remove and ground the spark plug wire. Remove the belt guard and inspect the V belt. Look for a broken belt. A broken belt prevents the engine pulley from driving the auger drive pulley. If you find a broken belt, replace the belt.

If the belt is not broken it may be slipping. Inspect the belt for glazing or oil soaking. If you find glazing, engage the idler belt clutch and check belt tension. Tension specifications can usually be found in the snow thrower operator's manual. If necessary, adjust belt tension.

If the belt is not the problem, check the engine crankshaft and auger drive pulley. A broken key or loose Allen set screw can cause either pulley to spin on its shaft. Try to move the pulley in relation to its shaft. If you see movement, tighten the Allen set screw or replace the key. If the problem is not in the belt system, the problem is most likely inside the gear box. The gear box requires replacement or overhaul.

Skid Shoe Adjustment

> **WARNING:** Stop the engine, remove and ground the spark plug wire to kill the ignition, and remove the ignition key (if used) before doing any snow thrower maintenance or repair.

The snow thrower has adjustable skid shoes located on each side of the auger housing (Figure 32-28). The skid shoes set the height of the scraper bar located on the bottom of the auger housing. The skid shoes and scraper bar wear from friction with the ground. The skid shoe fasteners fit through elongated slots in the skid shoes, allowing the height of the scraper bar and auger to be adjusted.

When the snow thrower is operated on smooth concrete or asphalt, the scraper bar should just clear the surface. It is commonly adjusted to about 1/8 inch (3 millimeters). When the snow thrower is operated over a gravel surface, the scraper bar has to be set higher. Too low a setting will cause the auger to dig into and throw gravel. The adjustment for operating on gravel is commonly 1¼ inches (30 millimeters).

To adjust the scraper shoes, park the snow thrower on a clean concrete or asphalt surface. Stop the engine and allow all the rotating parts time to stop. Kill the ignition and remove and ground the spark plug wire. Check the recommended tire pressure in the snow thrower owner's manual. Use a tire pressure gauge to measure the tire pressure. If necessary, inflate the tires to the correct air pressure. Low or uneven tire pressure affects scraper bar height.

Inspect the condition of the scraper bar. If the scraper bar is worn down to less than 1/16 inch (6.35 mm), it should be replaced. Measure the space between the ground and the scraper bar. If the height is not correct, the shoes need to be adjusted. Find a wooden block that is the thickness of the desired height. Set the block under the scraper bar. Use the correct size wrench to loosen the skid shoe fasteners. Lower or raise the shoes to touch the surface. Tighten the fasteners and remove the wooden block. The procedure for adjusting snow thrower skid shoes is shown in the accompanying sequence of photographs.

Snow Thrower Lubrication

> **WARNING:** Stop the engine, remove and ground the spark plug wire, kill the ignition, and remove the ignition key (if used) before doing any snow thrower maintenance or repair.

Snow thrower linkage connections and other parts that rotate in relation to each other require periodic lubrication. The snow thrower service and repair manual specifies what to lubricate and when. Some parts are lubricated with a squirt can of SAE 30 engine oil. Other parts are lubricated with lithium-based automotive grease in a grease gun. Grease gun lubrication points have a lubrication (zerk) fitting for the grease gun nozzle.

The auger assembly and wheel bearings commonly have grease fittings. Locate the lubrication fittings. Wipe the lubrication fittings off with a clean shop rag. Insert the grease gun nozzle over the grease fitting and inject grease into the fitting. Wipe excess grease off the fitting. Use the oil squirt can to oil moving parts of the control rods and levers on the handlebar assembly.

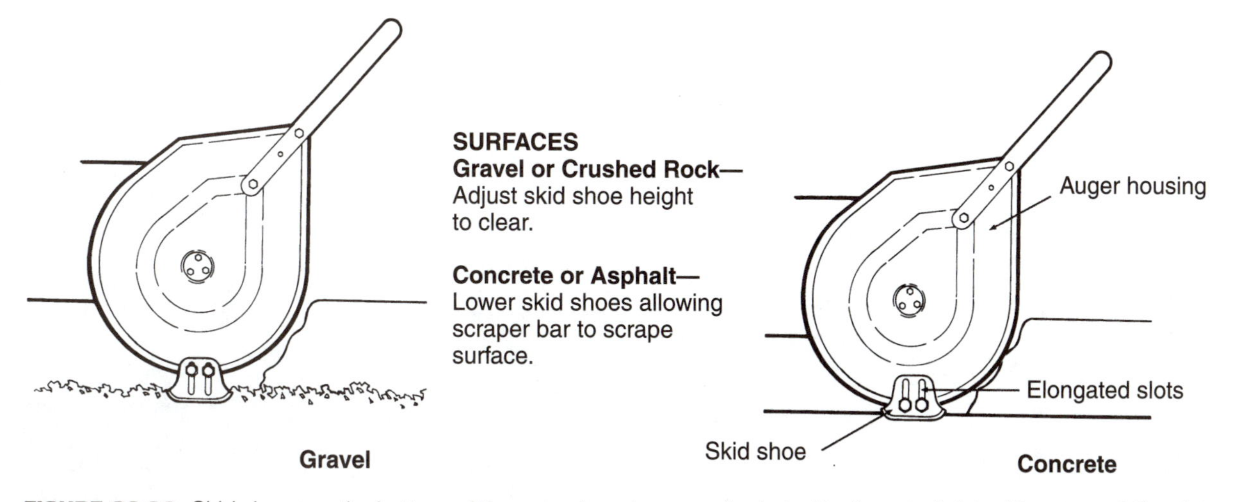

FIGURE 32-28 Skid shoes on the bottom of the auger housing are adjusted with elongated slots. (Courtesy of Simplicity Manufacturing.)

PHOTO SEQUENCE 28

Adjusting Snow Thrower Skid Shoes

1. Park the snow thrower on a clean concrete or asphalt surface. Kill the ignition and remove and ground the spark plug wire.

2. Use a tire pressure gauge to measure the tire pressure. If necessary, inflate the tires to the correct air pressure.

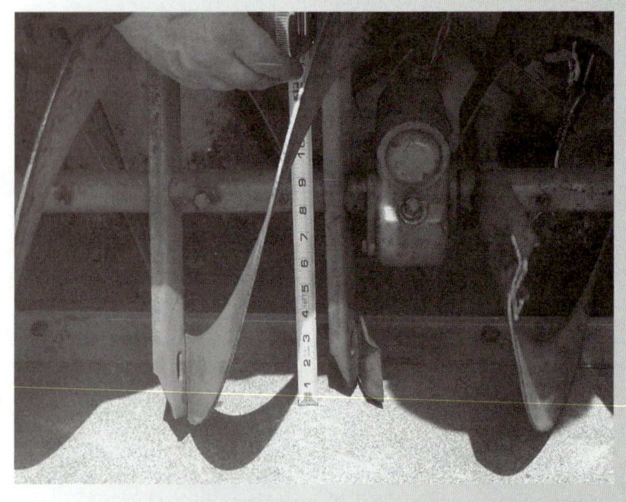

3. Measure the space between the ground and the scraper bar.

4. Find a wooden block that is the thickness of the desired height. Set the block under the scraper bar.

5. Use the correct size wrench to loosen the skid shoe fasteners. Lower or raise the shoes to touch the surface.

6. Tighten the fasteners and remove the wooden block. Install the spark plug wire.

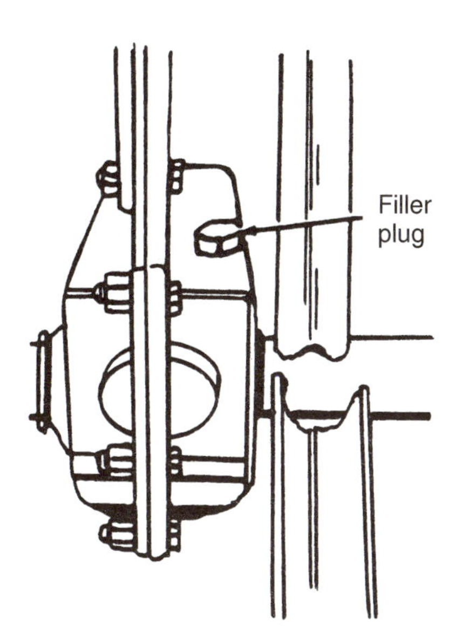

FIGURE 32-29 The gear box lubricating oil is checked and filled through a fill plug hole. (Courtesy of MTD Products, Incorporated.)

Some snow throwers use a gear box to drive the auger. The gear box oil should be checked periodically. There is usually a fill plug on the side of the auger gear box (Figure 32-29). Make sure the auger is on a level surface. Wipe the area around the fill plug with a rag so dirt will not get into the gear box. Use the correct size wrench to remove the fill plug. The gear box lubricant should be full up to the level of the fill plug. If necessary, use the recommended type of gear oil to fill the gear box to the proper level. Install the fill plug. Wipe off any spilled oil.

CAUTION: The gear box is often filled with heavy viscosity gear oil, not engine oil. Check the owner's manual for the correct type of oil to use.

REVIEW QUESTIONS

1. Explain why the operator's manual must be read and understood before starting an edger, tiller, or snow thrower.
2. Explain why it is important to follow safety procedures and warning label messages when using an edger, tiller, or snow thrower.
3. Describe the parts and operation of an edger clutch.
4. Explain how to troubleshoot an edger with a blade that does not cut.
5. Explain how to troubleshoot an edger with a blade that does not rotate.
6. Describe how to change an edger blade.
7. Describe the parts and operation of a tiller drive belt and transmission assembly.
8. Describe the parts and operation of a tiller tine assembly.
9. Explain how to troubleshoot a tiller when the tines do not rotate.
10. Explain how to troubleshoot a tiller when the tines do not penetrate the topsoil.
11. List the steps to remove and replace tiller tines.
12. Describe the parts and operation of a snow thrower.
13. List and explain the troubleshooting steps when a snow thrower will not clear snow.
14. List and explain the steps to adjust snow thrower skid shoes.
15. Describe how to lubricate a snow thrower.

DISCUSSION TOPICS AND ACTIVITIES

1. Trace the power flow on a tiller from the engine PTO shaft down to the tines.
2. Trace the power flow on a snow thrower from the engine PTO shaft down to the auger.
3. An auger transmission worm gear has 7 teeth. The helical gear in mesh with the worm gear has 28 teeth. If the worm gear is rotating at 1,000 rpm, how fast does the helical gear rotate?

CHAPTER 33

Chain Saws, String Trimmers, and Blowers

OBJECTIVES

Upon completion and review of this chapter, you should be able to:

- Explain the general safety precautions to take when operating a chain saw, string trimmer, or blower.
- Describe the specific hazards and safety precautions to take when operating a chain saw.
- Explain the specific hazards and safety precautions to take when operating a string trimmer.
- Describe the specific hazards and safety precautions to take when operating a blower.
- Describe the parts and operation of a chain saw.
- Troubleshoot a chain saw for vibration and cutting problems.
- Remove and replace a chain saw chain and guide bar.
- Tension a chain saw chain.
- Sharpen a chain saw chain.
- Service a chain saw oiling system.
- Remove and replace a chain saw centrifugal clutch.
- Describe the parts and operation of a string trimmer.
- Troubleshoot a string trimmer that will not advance cutting line.
- Remove and replace a string trimmer cutting line.
- Describe the parts and operation of a blower.
- Troubleshoot a blower that has excessive vibration.
- Remove and replace a blower impeller.

TERMS TO KNOW

Blower
Brush cutter
Centrifugal clutch
Chain saw
Chain saw oil pump
Saw chain brake
Saw chain gauge
Saw chain pitch
String trimmer

INTRODUCTION

Chain saws, string trimmers, and blowers are power equipment that is held in the operator's hand. This equipment must be made light so that the operator can hold it for hours at a time. Many of the parts are made from plastic and aluminum to keep the weight as low as possible. Small, light two-stroke engines are used because of their high power-to-weight capabilities.

CHAIN SAW, STRING TRIMMER, AND BLOWER SAFETY

Chain saws have a high-speed rotating chain with sharp edges. A string trimmer has a high-speed rotating plastic string. If the rotating tool comes in contact with your body, it will cause severe injury. Always read and understand the equipment owner's manual before operating a chain saw, string trimmer, or blower. Make sure you understand the controls and proper use of equipment. Know how to stop the equipment quickly in an emergency. Follow these general safety rules to prevent injury:

- Understand and follow each danger, warning, caution, and instruction decal installed on equipment.
- Be alert for the possibility of the tool striking a solid object and causing a sudden jerking force that could cause you to lose control of the equipment.
- Do not allow children to operate or play with equipment.
- Do not allow adults to operate equipment without proper instructions.
- Safe operation of equipment requires complete and unimpaired attention at all times.
- Do not operate equipment if fatigued or during or after consumption of medication, drugs, or alcohol.
- Protect eyes with nonfogging goggles, ears with ear plugs or protectors, and head with approved safety hard hat.
- Wear sturdy footwear with nonslip soles (steel-toed safety boots are recommended) for good footing.
- Do not operate equipment barefoot or when wearing open sandals or canvas shoes.
- Wear sturdy, snug-fitting long pants and clothing that provide complete freedom of movement.
- Do not go barelegged or wear loose clothing or jewelry that may get caught in rotating parts of the equipment.
- Do not attempt to make any adjustments to equipment while the engine is running.
- Do not fill the fuel tank indoors, when engine is running, or while engine is still hot.
- Wipe off any spilled gasoline, allow vapors to dissipate and move unit 10 feet (3 meters) away from the fuel container before starting engine.
- Keep equipment in good condition and all guards and shields in place and safety devices operating properly.
- Be sure that Start/Stop switch stops engine and the throttle lever moves freely.
- Do not run engine in enclosed or poorly ventilated areas.
- Maintain a firm grip on both handles at all times to help keep control should sudden jerking of the unit occur.
- Never operate equipment without someone within calling distance but outside your work area, in case help is needed.
- Operate equipment only when there is good visibility and light.

Chain Saw Safety

The chain saw uses a rotating saw chain to cut wood. The powerful engine creates a strong force on the chain. Wood can provide a strong reactive force on the saw. The reactive force can cause kickback, pushback, and pull-in. These sudden forces can cause the operator to loose control of the saw. Loss of control can result in a severe injury.

Kickback occurs when the tip of the guide bar strikes an object or when wood closes in and pinches the saw chain in the cut. This tip contact can cause a very severe reverse action on the saw. The spinning chain can be forced up and back toward the operator. Pushback occurs when the saw chain on top of the guide bar is stopped suddenly, usually due to pinching or hitting a foreign object in the wood. The reactive force pushes the chain saw back toward the operator. Also caused by pinching or hitting a foreign object in the wood, Pull-in occurs when the saw chain at the bottom of the guide bar is stopped suddenly. The reactive forces pull the chain saw away from the operator (Figure 33-1).

Cutting a tree down is called felling. Never fell a tree that is larger in diameter than the length of the chain saw guide bar. Before felling a tree, determine tree condition, wind direction, surrounding obstacles, and your intended direction of fall. Always plan a retreat path from a falling tree. Cut a felling notch perpendicular to the intended direction of tree fall. The notch should be one-fifth to one quarter of the diameter of the tree and as low to the ground as possible. It should not be higher than it is deep. Begin the felling cut slightly higher than the felling notch on the opposite side of the tree. Do not completely cut through. Leave a space between the felling notch and felling cut for a felling hinge (Figure 33-2).

Follow these specific safety procedures to avoid injury when using a chain saw:

- Wear safety footwear, eye, hearing, and head protection.
- Wear gloves that improve your grip on the saw.
- Wear snug fitting clothing and do not wear scarfs, jewelry, or neck ties that could catch on the chain.
- Always hold the chain saw with both hands using a grip with your thumb and fingers around the chain saw handles.

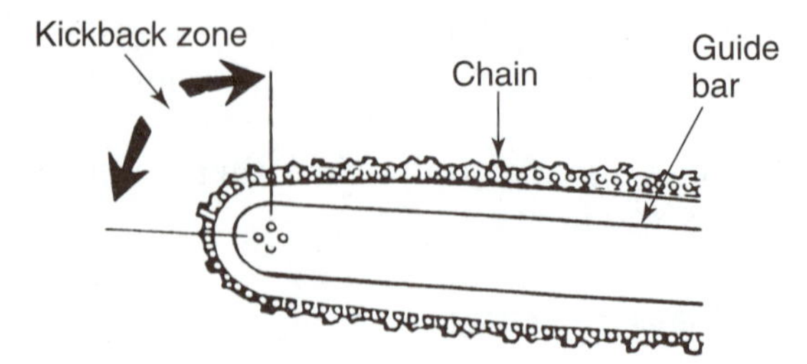

A. Kickback

B. Pushback

C. Pull-in

FIGURE 33-1 Chain saw reactive forces can cause an injury. (Courtesy of Tanaka/ISM.)

- Keep all parts of your body away from the saw chain when the engine is running.
- Always carry the chain saw with the engine stopped, guide bar and saw chain to the rear, and the muffler away from your body.
- Transport the chain saw with a scabbard around the chain and guide bar.

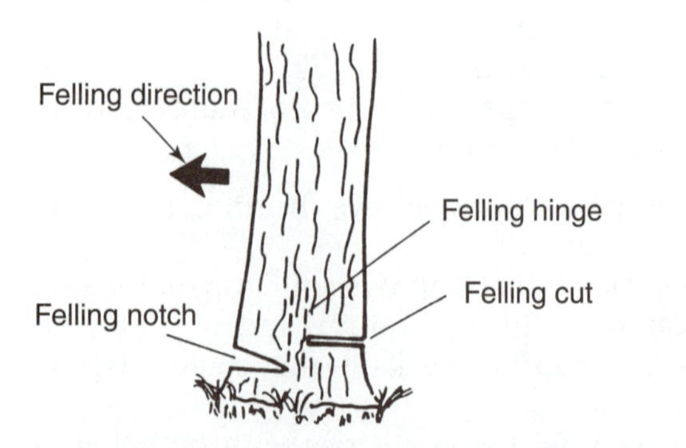

FIGURE 33-2 A felling notch is cut in a tree to fell it in the desired direction. (Courtesy of Tanaka/ISM.)

- Move the chain saw at least 10 feet (3 meters) away from the fueling point before starting the engine.
- Be sure the saw chain is not contacting anything before you start the engine.
- Do not operate a chain saw that is damaged or improperly adjusted.
- Make sure your work area is free of bystanders and animals.
- Be ready to support the saw when the cut is finished so it will not drop on your leg.
- Never make one-hand cuts because the saw can get out of control.
- Keep saw handles dry and free of oil or fuel mixture.
- Shut off the engine before putting the saw down.
- Do not leave the engine running unattended.
- Operate the chain saw only in well-ventilated areas.
- Make sure the chain stops moving when the throttle is released.
- Never start cutting unless your footing is secure.
- When cutting on a slope, stand on the uphill side of the tree.
- Do not put pressure on the saw when nearing the end of a cut because it may cause a kickback.
- Be careful when cutting small brush or saplings that can catch in the saw and pull you off balance.
- Do not stand on a log when cutting off branches because you could fall and loose control of the saw.
- Cut only one limb at a time to help reduce the possibility of a kickback.
- Be alert for springback when cutting a limb that is under tension.
- Use a saw horse when cutting short logs.
- Never allow anyone to hold a log you are sawing.

String Trimmer Safety

When using a string trimmer or brush cutter, follow the same general safety precautions as for other hand-held equipment. Follow all the safety procedures in the equipment owner's manual and warnings on the trimmer. In addition, follow these specific safety procedures:

- Wear eye protection, tight-fitting long pants, and safety shoes when trimming or brush cutting.
- Do not attempt to touch or stop the string or brush cutter blade when it is rotating.

- Always hold the trimmer with both hands when operating the equipment.
- If the trimmer strikes or gets entangled with a foreign object, stop the engine immediately.
- Do not operate the engine faster than the speed necessary to cut, trim, or edge.
- Keep the string guard in place at all times when trimming.
- Always stop the trimmer when walking from one cutting location to another.
- Stop the engine immediately if you feel excessive vibration.

Blower Safety

The engine-powered blower delivers a high volume of air at a very high velocity. Air can be dangerous if it is directed at anyone. The air can also blow objects at bystanders. When using a blower, follow the same general safety precautions as for other hand-held equipment. Follow all the safety procedures in the equipment owner's manual and warnings on the blower. In addition follow, these specific safety procedures:

- Wear hearing protection to avoid hearing loss from exposure to blower noise.
- Wear eye protection for protection against objects thrown by the blower air stream.
- Do not put your hands or other body parts in the air stream from the blower.
- Keep all bystanders and pets away from the area being blown.
- Operate the blower only in a well-ventilated area.

CHAIN SAWS

A **chain saw** (Figure 33-3) is an engine-powered saw that uses a rotating chain with cutting edges to saw wood. Chain saws are very popular because they can save many hours of labor in felling trees and cutting firewood. Chain saws can be operated

CASE STUDY

The Consumer Product Safety Commission is concerned with the dangers in improper use of brush cutters:

Brush Cutters Require Cautious Use by Consumers.

> Consumers are finding a relatively new type of power tool on the market for yard and garden work—brush cutters and combination trimmer/brush cutters. While the product is somewhat similar to the flexible string weed trimmer, it has a much greater potential for serious and disabling injury. The brush cutter uses a rigid cutting blade in place of the flexible plastic string line. The blade, made of steel or rigid plastic materials, permits the cutting of much heavier stands of brush and small diameter saplings, according to manufacturers. However, it can also cut through a hand, arm, or leg, something the flexible line trimmer will not do. Several injury reports received by the Consumer Product Safety Commission (CPSC) indicate the blade has caused severe lacerations and near amputations. These accidents have happened even to professionals using the product. In one case, a man was cutting brush along the bottom of a board fence. His wife, standing nearby trimming a hedge, was severely injured when he lost control and the brush cutter swung in an arc toward his wife. The blade cut into the wife's thigh and then her left hand, nearly amputating it below the wrist. In another incident, two workers were cutting brush. One held down a sapling with an axe, while the other attempted to cut through it with a brush cutter. The blade ricocheted off the sapling and into the assistant, severely cutting his right arm. In another case, a man cutting brush near a chain link fence reported that the blade threw a small piece of wire from the fence into his eye. In another case, the blade was reported to have come off the end of the shaft while it was being operated, causing a foot injury. While only a few reports of injury have been received to date, the Commission's staff believes there may be many more injuries if extra caution is not taken in using the product. The greatest danger appears to be to bystanders or helpers. . . .

Source: Consumer Product Safety Commission Document #5005, Washington, DC.

FIGURE 33-3 A chain saw is used to fell trees and cut firewood. (Courtesy of Tanaka/ISM.)

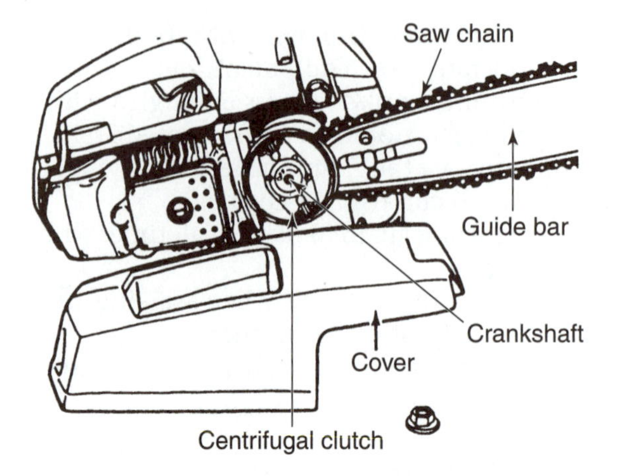

FIGURE 33-5 The centrifugal clutch engages and disengages the crankshaft from the chain saw chain. (Courtesy of Poulan/Weed Eater.)

at any angle when cutting because they use diaphragm carburetors. Large chain saws are used by logging professionals. These have large two-stroke engines and long saw guide bars for felling large diameter trees. The smaller chain saws used by consumers have smaller 3 to 5 horsepower engines and shorter saw bars. They both have the same parts and operation.

Chain Saw Parts and Operation

The chain saw has a two-stroke engine that is completely covered by a housing. A fuel and oil tank is built into the housing. The housing supports a front and rear saw handle. The engine crankshaft is positioned at a right angle to the handles. A recoil starter is mounted on one side of the crankshaft. The PTO shaft on the opposite side drives the saw chain (Figure 33-4).

The engine-driven sprocket rotates cutting saw teeth on a chain saw chain. A sprocket is a ring with teeth that engages and drives the chain. The engine PTO shaft drives the sprocket through a centrifugal clutch (Figure 33-5). A **centrifugal clutch** uses centrifugal force to engage and disengage the engine crankshaft from driven equipment.

The centrifugal clutch engages (connects) or disengages (disconnects) the sprocket to the engine. The centrifugal clutch has a sprocket on one side. The saw chain fits around the sprocket, which is attached to a drum that is free to rotate on the crankshaft. The engine crankshaft drives a clutch assembly that fits inside the drum. The clutch assembly has friction pads that are each attached to a flyweight. A spring holds the flyweights and friction pads away from the drum surface (Figure 33-6).

During engine idle the clutch is disengaged. The engine drives the clutch but the friction pads do not contact the drum. The saw chain is not being driven. When the operator increases the

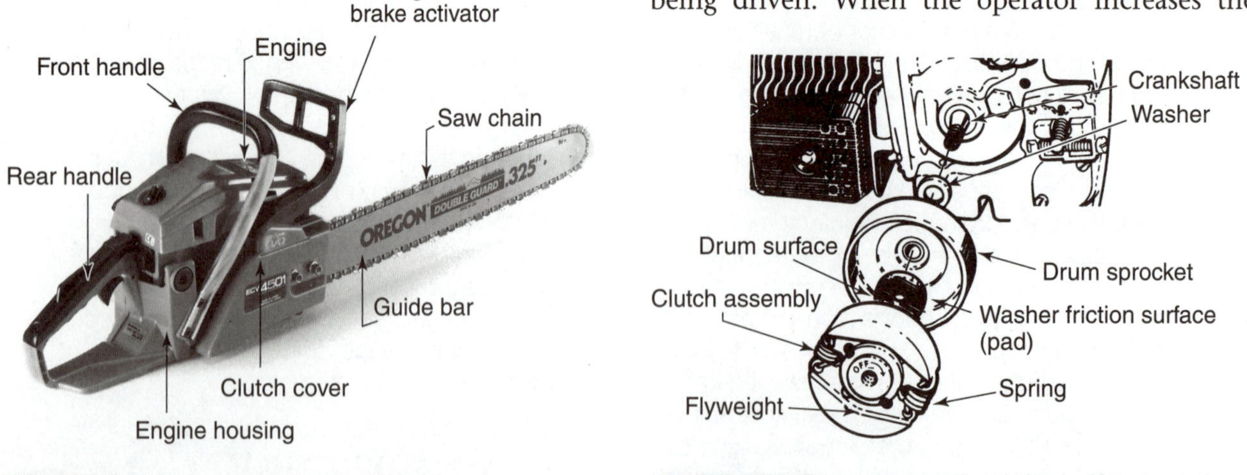

FIGURE 33-4 Parts of a chain saw. (Courtesy of Tanaka/ISM.)

FIGURE 33-6 Parts of a centrifugal clutch. (Courtesy of Poulan/Weed Eater.)

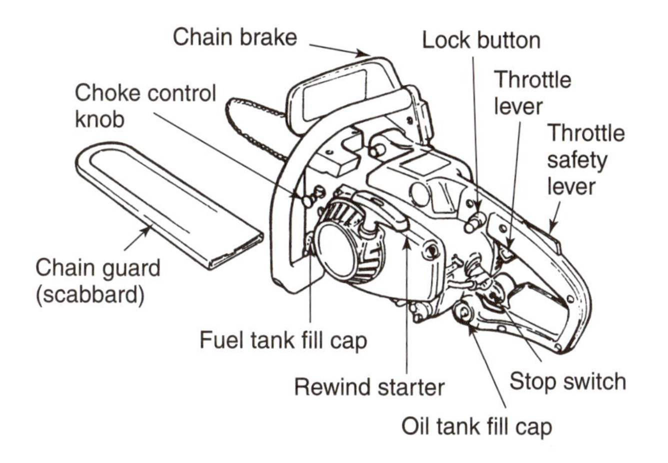

FIGURE 33-7 Chain saw controls. (Courtesy of Tanaka/ISM.)

throttle setting, the crankshaft rotates at a higher rpm. Increased centrifugal force causes the flyweights to overcome spring tension and move outward. The friction pads come in contact with the drum friction surface. The drum begins to rotate and drives the saw. When the operator's hand is removed from the throttle, there is a lower centrifugal force. The clutch spring tension pulls the pads out of contact with the drum. The clutch is disengaged and the chain stops.

The chain saw has several operator controls (Figure 33-7). The carburetor throttle is connected to a throttle lever on the rear handle. The engine is normally operated at the fastest speed when performing cutting operations. A safety lever on top of the handle has to be depressed by the operator's hand for the throttle to open. A throttle lock button is used to lock the throttle in the correct position to start the engine. The choke knob is used to close the choke valve when starting a cold engine. The stop switch is used to kill the ignition.

Saw Chain and Guide Bar

The guide bar supports the saw chain. The guide bar is mounted to the saw at one end. The mounting is adjustable so that the bar can be made longer or shorter to tension the chain. There is a groove that runs along the sides and end (nose) of the bar. The groove and the rail on each side of the groove provide a track for the chain to slide on. Some larger guide bars have a sprocket built in to the nose. The nose sprocket guides the chain around the end of the guide bar. Guide bars come in many different lengths. Guide bars for lawn and garden work are generally 10, 12, 14, and 16 inches long.

The chain is a series of chain plates (links) and rivet pins. The chain is flexible because rivet pins are used to connect the plates together (Figure 33-8). A rivet pin is pressed through each end of a tie strap. The opposite side of each tie strap fits in a cutter plate. The cutter plate has a raised up section with a cutting edge. This part does the wood cutting. There are alternating cutter plates on the left and right side of the chain. There is a hook-shaped part on one end of the cutter plate called the *depth gauge*. The hook on the depth gauge contacts the wood being cut. The hook digs into the wood and helps prevent saw kickback.

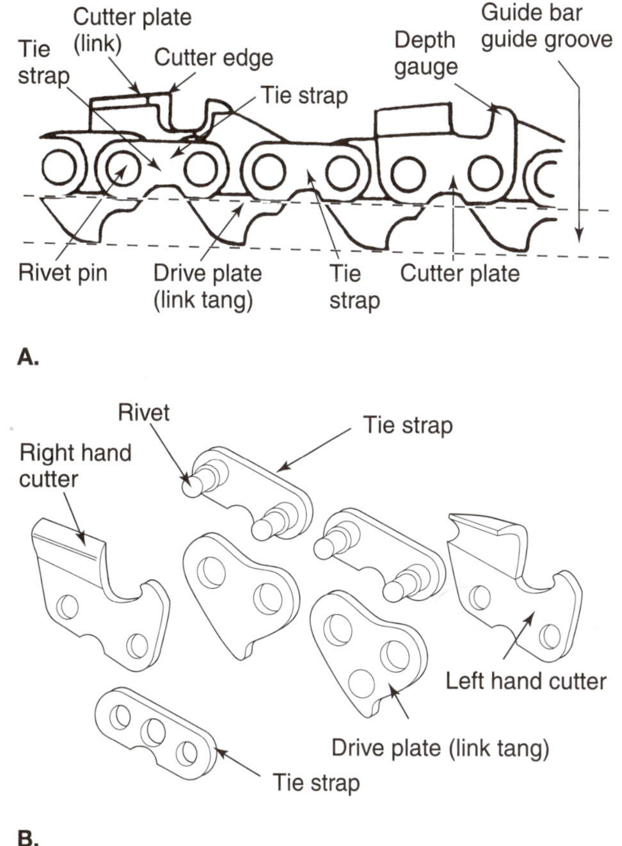

FIGURE 33-8 Parts of a chain saw. (*A:* Courtesy of Tanaka/ISM.)

The rivet pins also go through inner plates. The rivet pin holes in the inner plates are larger in diameter than the pins. This allows the inner plates to swivel in relation to the side plates. The inner plates are called *drive links*. The drive links have a tang that extends below the chain and fits into a guide groove in the chain saw guide bar. The tangs have a hook shape that cleans sawdust out of the guide bar groove as they rotate.

Chains are sized by chain pitch and chain gauge (Figure 33-9). **Saw chain pitch** is the distance between alternate chain rivets divided in

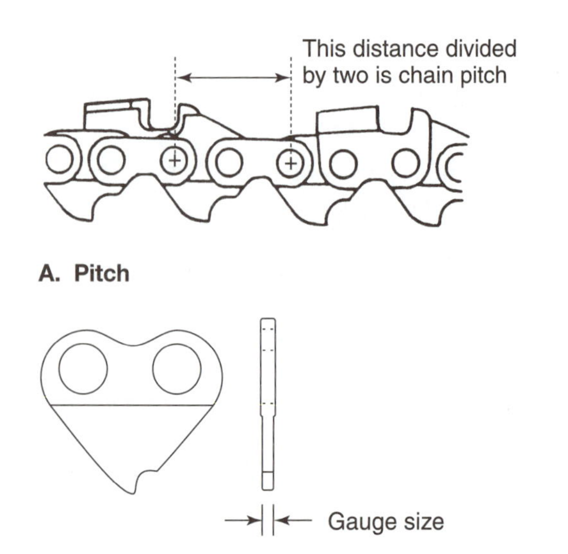

FIGURE 33-9 Chain size is determined by pitch and gauge. (*A:* Courtesy of Tanaka/ISM.)

half. The most common pitch sizes are ¼, ⅜, 0.404, 7⁄16, and ½ inch. **Saw chain gauge** is the thickness of the chain drive tang where it fits in the guide bar groove. The most common gauge sizes are 0.050, 0.058, and 0.063 inch. Chain length is determined by the number of chain links. Longer chains are required for saws with longer guide bars. A replacement chain must have the correct pitch, gauge, and length.

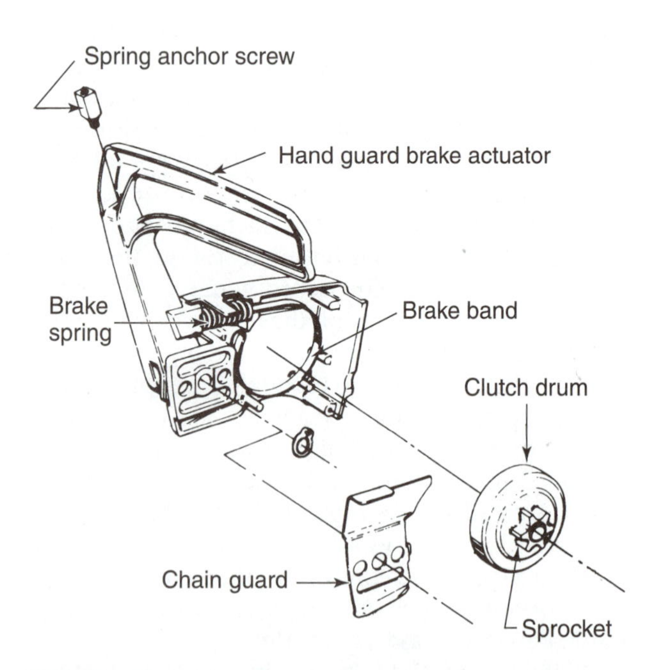

FIGURE 33-10 A chain brake stops the chain from rotating when the operator's hand is off the saw handle. (Courtesy of Poulan/Weed Eater.)

CAUTION: The pitch and gauge size of a replacement chain must be the same as the original chain. If the size is incorrect, the chain can damage the drive sprocket and the guide bar. The chain can bind up or break.

SERVICE TIP: Newer saw chains have been developed that have improved kickback-reducing ability. You can improve the safety of an old saw by installing the newer type of chain.

Chain Brake

WARNING: An incorrectly assembled saw chain brake can cause the chain to rotate when the operator's hand is off the handle, causing operator injury. Always follow service manual instructions when assembling or adjusting a saw brake.

Chain saws have a **saw chain brake,** a band brake that stops the rotation of the saw chain when the operator's hand is removed from the front handle. The chain brake is activated by the hand guard (Figure 33-10). A friction faced brake band fits around the clutch drum. The band is anchored against a spring at one end and the hand guard at the other end. The spring causes the band to fit tightly around the clutch drum. In this position, the clutch drum is prevented from rotating. The chain sprocket attached to the clutch drum is also prevented from rotating. The saw chain, in mesh with the sprocket, is stopped from rotating.

When the operator's hand activates the brake handle, the handle moves the end of the brake spring. The spring is moved in a direction to release the spring pressure on the brake band. The brake band comes away from the clutch drum. The drum is free to rotate and drive the saw chain.

Chain Oil Pump

CAUTION: The oil tank on a chain saw provides oil for the chain. This oil should not be confused with the oil that is mixed with the fuel. Two-stroke oil must be mixed with the gasoline.

The saw chain rivets and side plates require lubricating oil to move freely. The groove in the chain guide bar requires lubrication to keep the chain moving freely. The sawing action and sawdust constantly wipes the lubricating oil off the chain and guide bar. The saw chain and guide bar requires a constant source of oil for lubrication. Chain saws have an oil tank and oil pump to oil the chain. The oil tank is located on the side of the chain saw. There is a filler cap used to add oil periodically.

Oil from the oil tank is routed to the oil pump. The **chain saw oil pump** is a small pump that delivers oil under pressure to lubricate the saw chain. Most chain saws use an automatic oil pump, which provides the oil constantly without the operator having to operate any control. The oil pump fits around and is driven by the engine PTO shaft (Figure 33-11). The oil pump housing has a pumping chamber. The rotating PTO shaft rotates a nylon worm gear in the pump housing. The worm gear meshes with the teeth on a plunger. The plunger rotation causes a low pressure in the pumping chamber. Oil is pulled into the pumping chamber from the oil tank through an oil pick up line. The rotating plunger pressurizes the oil and sends it through a passage or discharge line. The oil is directed on the guide bar and chain to provide lubrication. The oil flow outlet is often called an *oiler*. An adjustment screw in the pump outlet controls the amount of oil delivered to the chain (Figure 33-12).

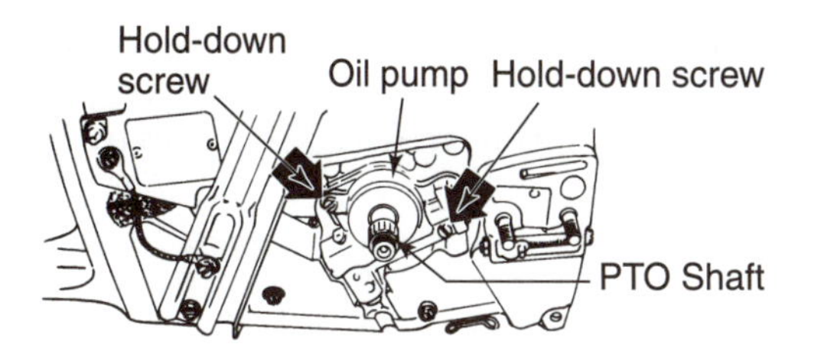

A. Oil Pump Installed

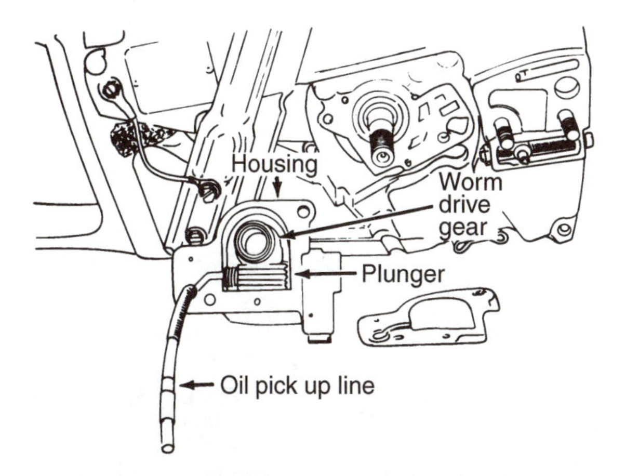

B. Oil Pump Removed

FIGURE 33-11 The oil pump worm drive gear is driven by a crankshaft. (Courtesy of Poulan/Weed Eater.)

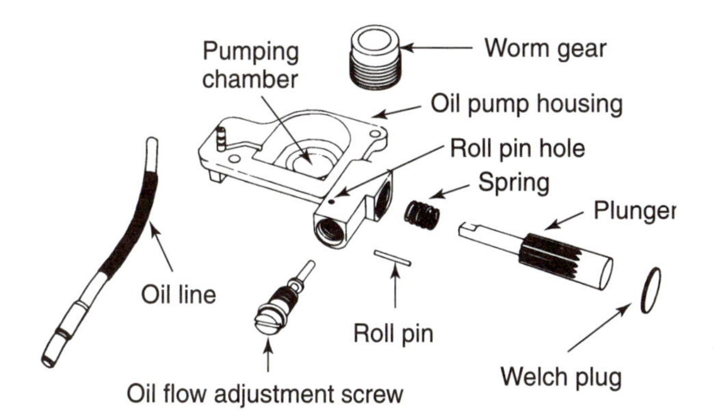

FIGURE 33-12 Parts of a chain saw oil pump. (Courtesy of Poulan/Weed Eater.)

Chain Saw Troubleshooting

Chain saw problems can be related to engine or equipment problems. Engine troubleshooting is presented in Chapter 21. The most common chain saw problems are:

- Chain saw vibrates excessively
- Chain rotates when the engine is at idle
- Chain does not move with acceleration
- Chain clatters or cuts rough
- Chain stops during the cut.
- Chain oil pump fails to lubricate the bar and chain.

WARNING: Stop the engine, remove and ground the spark plug wire, and kill the ignition before doing any chain saw inspection or repair. Wear work gloves to protect hands when handling the saw chain.

When a chain saw vibrates excessively, stop the engine and allow it to cool off. Remove and ground the spark plug wire so the engine will not start during your inspection. Vibration is often caused by a chain that does not have enough tension. The chain requires a tension adjustment. A dull or damaged chain causes a vibration. Sharpening or replacing the chain often solves a vibration problem. A worn or damaged clutch causes a vibration. Replace the clutch to solve this problem.

A saw that rotates when the engine is at idle can be dangerous to the operator. The engine may be idling at too high a speed. Check and adjust the engine idle speed. If the chain still rotates, the problem is most likely a worn or damaged centrifugal clutch. Inspect and replace the clutch if necessary.

A chain that does not move when the engine is accelerated may have a chain that is adjusted too tightly. The groove or rails on the guide bar may be damaged. When this happens, the chain hangs up in the groove and will not rotate. Sometimes a loose chain jumps off the engine drive sprocket and prevents the chain from rotating. A worn or damaged centrifugal clutch prevents the chain from rotating.

When a chain makes a clattering noise or cuts rough, the chain tension may be incorrect. A dull, worn, or improperly sharpened chain also causes this problem. Inspect, replace, or sharpen the chain to solve this problem. A worn centrifugal clutch that slips or grabs can cause rough cutting.

When the chain stops during the cut, the problem may be too much pressure applied to the saw during the cut. If the cutting technique is not the problem, inspect the chain. Improper chain sharpening can cause this problem. Bent or uneven guide bar rails are a common source of this problem. A worn centrifugal clutch also causes this problem.

When there is no oil getting from the oil pump to the chain and guide bar, check the level in the oil tank. If the tank is low, fill it with the recommended type of oil. Some oil systems have an oil filter. Check, and if necessary, replace or clean the filter. The oil hole in the guide bar can get blocked with sawdust. Remove the guide bar and clean the oil hole and passage.

Removing and Replacing a Saw Chain

> **WARNING:** Stop the engine, remove and ground the spark plug wire, and kill the ignition before doing any chain saw maintenance or repair. Wear work gloves to protect hands when handling the saw chain.

Place the chain saw on a flat work surface. The guide bar is commonly mounted on one or two mounting studs. On some saws, a cover plate has to be removed to remove the nuts. Other saws have a cover that is held in place by the studs and nuts. The guide bar engages a chain tension boss. The boss is connected to a tensioning screw. The tensioning screw is used to move the guide bar back and forth to tension the chain. Carefully observe how the bar fits against the boss so you can install it correctly.

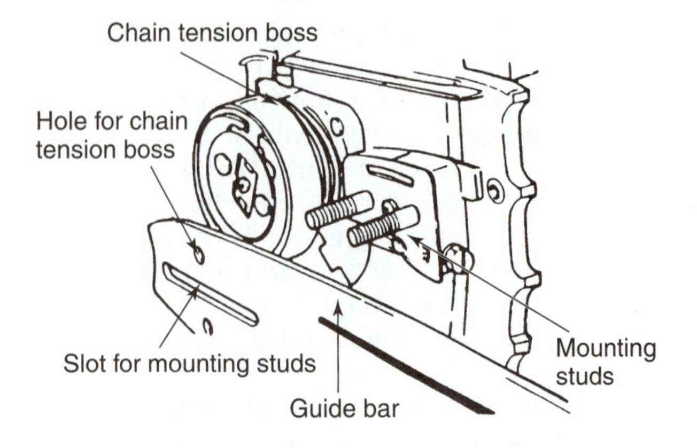

FIGURE 33-13 Removing the chain and guide bar. (Courtesy of Tanaka/ISM.)

Remove the nuts, washers, and guide bar mounting plate. Unhook the chain from the drive sprocket. Remove the chain and guide bar from the saw (Figure 33-13).

Clean and inspect the chain. Look for damage to any tangs, cutting edges or antikickback hooks. If you find any damage, the chain must be replaced. Check the cutting edge on each cutter plate for sharpness. If the edges are dull or if the saw has not been cutting well, replace or sharpen the chain. Flex each link of the chain to locate any tight links. If you find a tight link, replace the chain.

Use a small knife or screwdriver to clean the sawdust out the guide bar groove (Figure 33-14A). Clean out the oil outlets in the guide bar and make

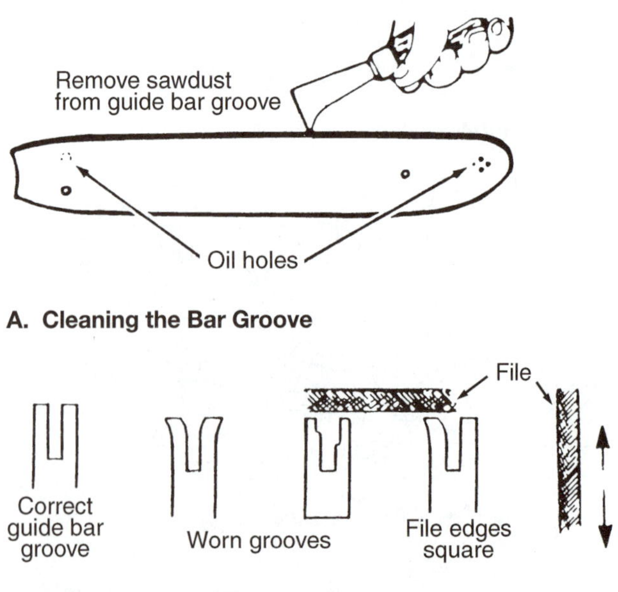

FIGURE 33-14 Clean and inspect the bar groove for wear and repair minor damage with a file. (Courtesy of Tanaka/ISM.)

sure they are clear. Carefully inspect the groove and rails for wear. Worn groove and rails will not properly support the chain. The chain will tilt over during cutting and not cut well. Look for wear inside the groove. Look for any cracked or bent rails. Wear in these areas cannot be repaired. Replace the guide bar if it is worn excessively. Minor groove damage can be repaired with a file (Figure 33-14B). File the top edge and sides of the rails square.

Lay the saw chain out on a bench in a loop. Straighten any kinks. The cutters must face in the direction of chain rotation (Figure 33-15). If they face backward, turn the chain loop over. Feed the chain tangs into the bar groove. Pull the chain so there is a loop at the back of the bar. Hold the chain in place on the bar and hook the loop over the sprocket. Install the bar on its mount, making sure it engages the tensioning boss. Install the mounting plate, stud washers, and nuts. Tighten the stud nuts so they are snug but not tight. The bar must be free to move during the chain tension adjustment.

CAUTION: Different chain tensions are used on bars with and without nose sprockets. Always refer to the owner's manual for the correct chain-tensioning procedure.

Turn the saw upright. Turn the chain-tensioning screw in a direction to move the bar away from the sprocket (usually clockwise) (Figure 33-16). Take up enough slack in the chain so that the chain tangs enter the groove at the bottom of the bar. Do not cinch the chain tightly. Slowly turn the adjuster until the tie straps come close to the bottom bar rails. Check the fit at the middle of the bottom groove. The tie straps should be about the thickness of a dime away from the bar. Use your gloved hand to pull and let go (snap) of the chain to remove any stiffness in the chain. If you notice any droop in the bottom of the chain, adjust it again.

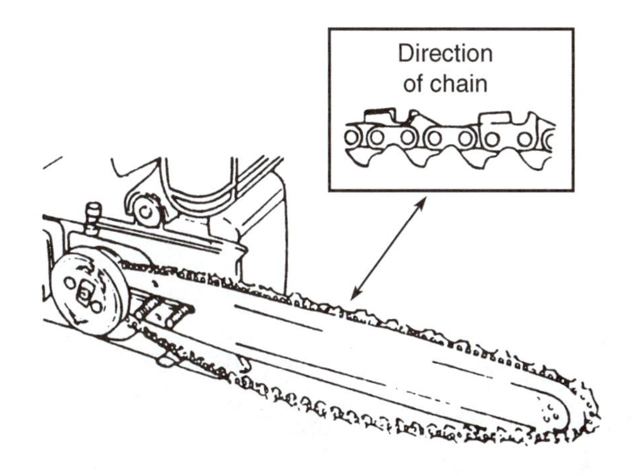

FIGURE 33-15 Install the chain on the guide bar in the correct direction. (Courtesy of Tanaka/ISM.)

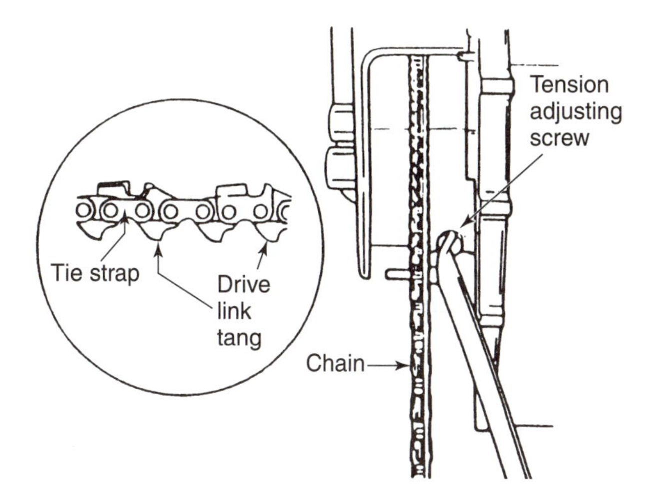

FIGURE 33-16 Adjusting chain tension with the tension-adjusting screw. (Courtesy of Tanaka/ISM.)

WARNING: Chain adjustment changes when the chain warms up during cutting. The chain expands as it warms up. Watch the chain and stop the saw if the tangs begin to almost hang out of the bar rails. Make only light cuts for the first few fuel tank fill ups to make sure the tension is correct.

CAUTION: A hot chain cannot be accurately adjusted. Allow it to cool before you adjust it. If you adjust it when hot, it may be too tight when it cools. A new chain can stretch after a short time. Recheck a new chain after the first use.

Pull the chain around the bar by hand (Figure 33-17). If assembled and tensioned properly, it should move smoothly around in the groove. Tighten the mounting nuts snugly with a wrench. Install the spark plug and start the chain saw. Allow the chain to rotate at a slow speed for a few seconds. Kill the engine. Reset the tension if there is any droop in the chain. The procedure for removing, replacing, and tensioning a chain saw chain is shown in the accompanying sequence of photographs.

PHOTO SEQUENCE 29

Removing, Replacing, and Tensioning a Chain Saw Chain

1. Place saw on flat work bench. Remove and ground the spark plug wire

2. Remove chain cover (if used) and remove the two nuts from the guide bar studs.

3. Unhook the chain from the sprocket and remove the chain and guide bar.

4. Clean and inspect chain cutting edges.

5. Clean the guide bar rails, oil holes, and the guide bar groove with a knife.

6. Install the guide bar on the saw by slipping it over the mounting studs, making sure it engages the chain tension adjusting screw.

7. Install the new chain over the sprocket and into the guide bar groove. Install but do not fully tighten guide bar mounting nuts.

8. Make sure the chain cutting edges are facing in the correct direction.

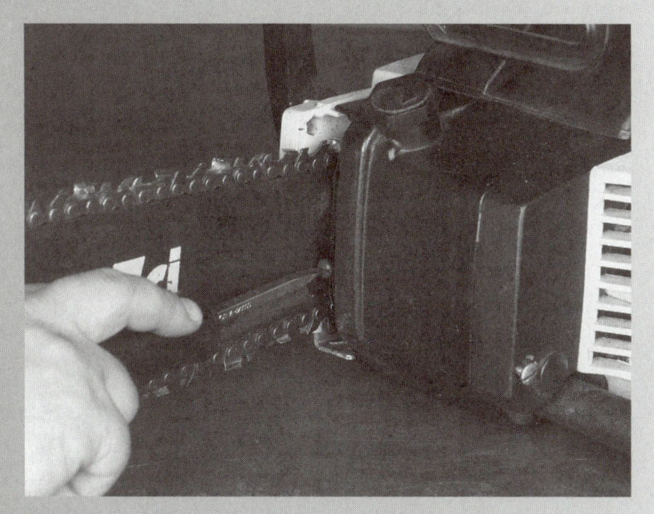

9. Hold nose of guide bar up and adjust the chain tension screw until the chain is snug against the guide bar rails.

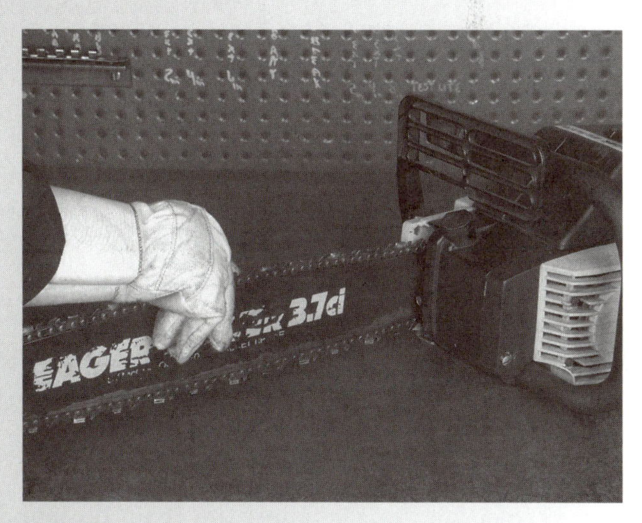

10. Pull chain along the guide bar rail to check for free movement.

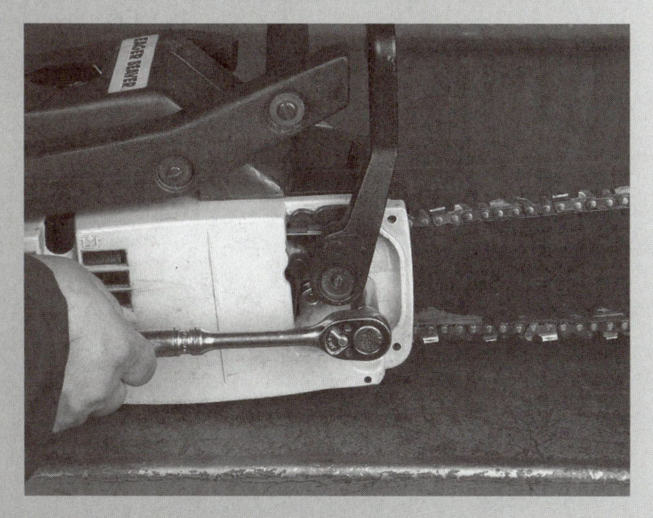

11. Tighten the bar mounting stud nuts. Replace the chain cover if used. Recheck and adjust chain tension if necessary.

12. Install spark plug wire and start chain saw. Allow it to warm up. Recheck and adjust chain tension if necessary.

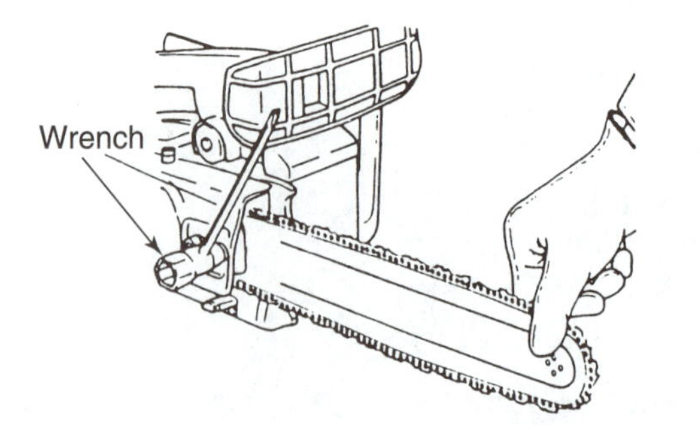

FIGURE 33-17 Pulling the chain around the bar by hand to check the fit. (Courtesy of Tanaka/ISM.)

Sharpening a Saw Chain

> **WARNING:** Sharpening a saw removes metal from the cutters. This eventually reduces the kickback-reducing ability of the chain. Incorrect sharpening angles also reduce the kickback-reducing ability of the chain. A new chain is safer than a sharpened chain.

Cutting into wood eventually dulls the chain cutting edges. If the saw chain contacts sand, dirt, gravel, or roots, the cutting edges will dull very rapidly. Saw chains must be sharpened accurately. An incorrectly sharpened saw causes excessive vibration and will not cut well. If the correct sharpening equipment is not available, have the saw sharpened by a professional or replace the chain. Do not try to sharpen a damaged or worn chain.

A special vise is available for holding a saw chain off the bar for sharpening. If this vise is not available, the chain can be sharpened on the bar. If you sharpen on the bar, adjust the chain tension enough that it will not wobble during sharpening. Do not forget to readjust the tension after sharpening. A depth gauge tool and a sharp file are required to sharpen the chain.

The leading edge of each cutter link has a raised up hook section called the *depth gauge.* A depth gauge measuring tool is used to measure the height of the depth gauges. Place the tool over the chain (Figure 33-18). File off any part of the depth gauge that projects above the tool. Round off the front corner of the depth gauge to match the original shape.

The cutting edge of each cutter is sharpened with a file. Check the chain saw owner's manual for the required angles to file the cutters. Always file cutters from the inside to the outside (Figure 33-19). File away from your body. Make sure each cutter is filed to the same length (Figure 33-20). The cutters are tapered. If some cutters are longer than the others, only the longer cutters will get a chance to cut.

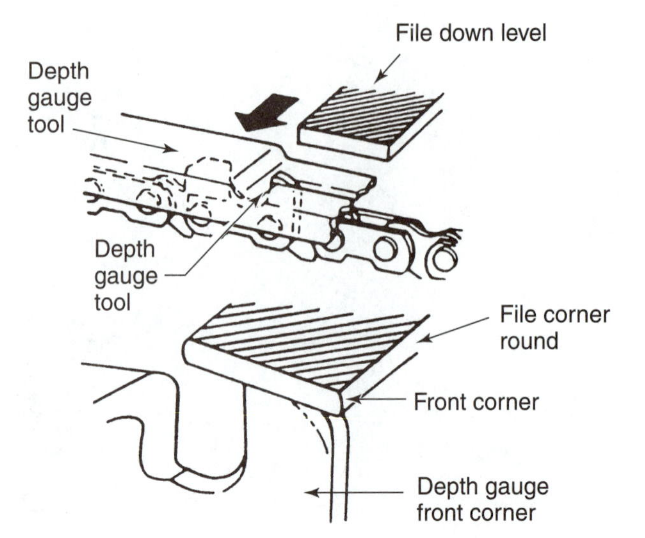

FIGURE 33-18 Using a depth gauge tool to file the chain depth gauges. (Courtesy of Tanaka/ISM.)

FIGURE 33-19 Direction in which to file the cutters. (Courtesy of Tanaka/ISM.)

File the top plate cutting edge to 30 degrees (Figure 33-21). Keep the file level with the top plate cutting edge. File all the left-hand cutters on one side. Then move to the other side and file all the right-hand cutters. File only on the forward stroke, using light but firm pressure. File the side plate to

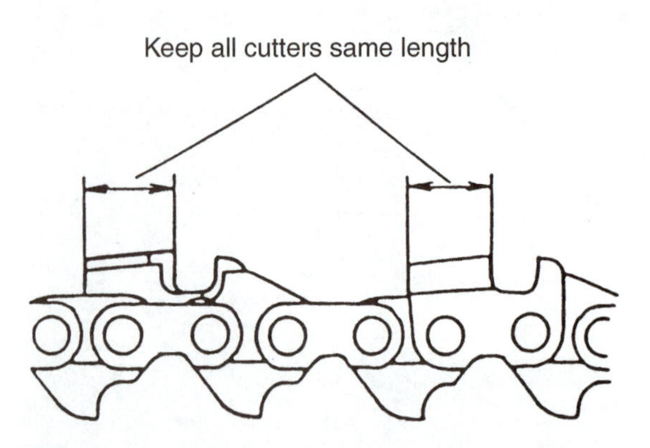

FIGURE 33-20 Make sure all the cutters are the same length. (Courtesy of Tanaka/ISM.)

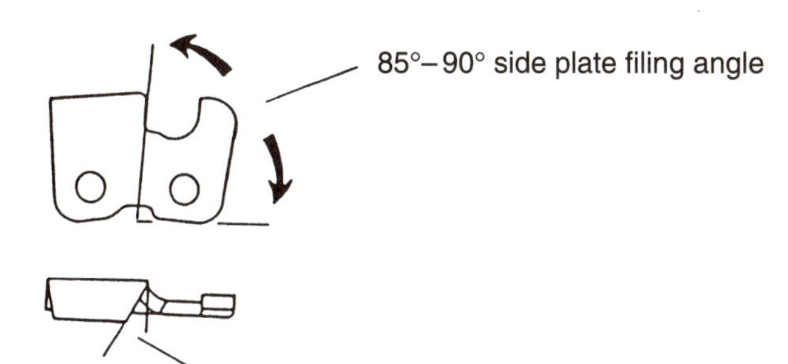

FIGURE 33-21 Filing angles for the top and side of the cutting edge. (Courtesy of Tanaka/ISM.)

85–90 degrees. Recheck the depth gauges after filling the cutters. Check the sharpening job by cutting a test piece of wood.

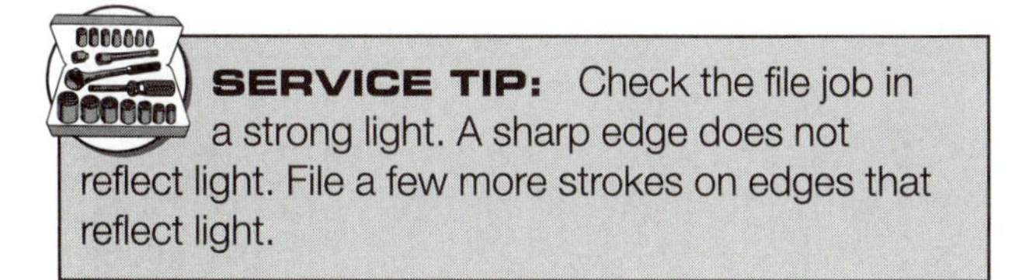

SERVICE TIP: Check the file job in a strong light. A sharp edge does not reflect light. File a few more strokes on edges that reflect light.

Oiler Maintenance and Adjustment

The level of oil in the chain oil tank should be checked and refilled each time the saw is used. Always use clean oil of the viscosity recommended in the saw owner's manual. Dirty oil plugs up oiling passages. Oiler problems are usually caused by a plugged oil passage or a plugged oil filter. Oiler problems show up as dry chain. If not corrected, the chain cutting edges get dull rapidly.

Check and clean the oil passages and holes in the guide bar each time it is removed for chain service. When the bar is off, check the oil filler port in the engine housing (Figure 33-22). If the passages are all clear and the system is not oiling, check the chain oil filter. Drain the oil from the oil

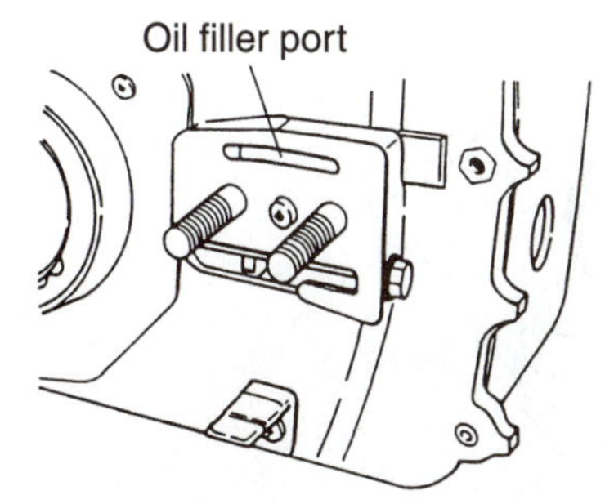

FIGURE 33-22 Checking and cleaning the oil filler port. (Courtesy of Tanaka/ISM.)

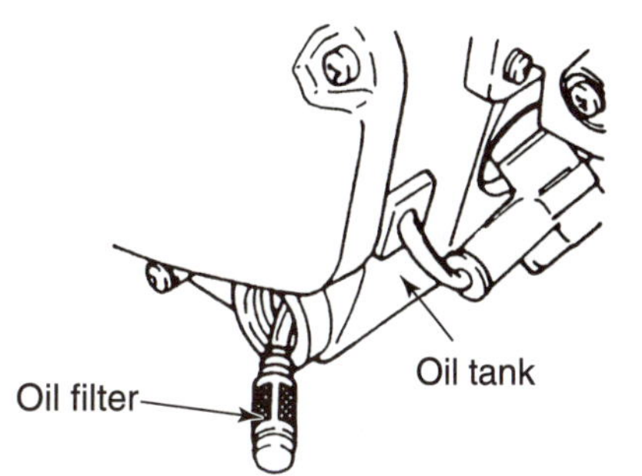

FIGURE 33-23 Checking and cleaning the chain oil filter. (Courtesy of Tanaka/ISM.)

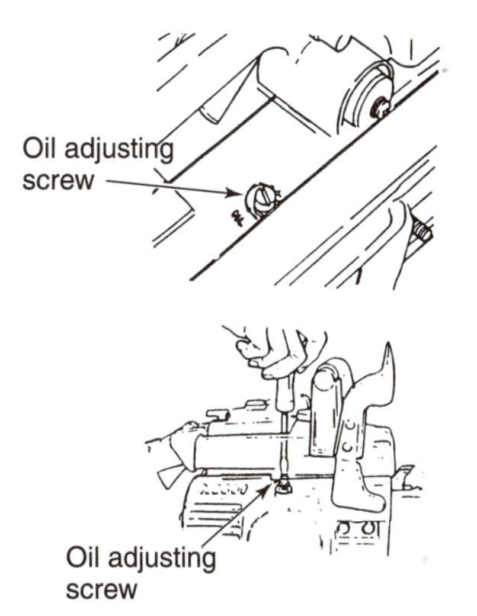

FIGURE 33-24 Adjusting oil flow with the oil-adjusting screw. (Courtesy of Tanaka/ISM.)

tank. Pull the filter line out of the tank (Figure 33-23). Rinse the filter in cleaning solvent and allow it to air dry. Place the filter back into the oil tank. Refill the tank with oil and replace the oil tank cap.

Oiling systems usually have an adjusting screw to control oil output (Figure 33-24). Check the owner's manual for the adjustment procedure. Most systems are adjusted by the saw maker for full output. Oil flow can be reduced for light cutting by adjusting the screw from this setting.

Removing and Replacing a Centrifugal Clutch

WARNING: Wear eye protection when servicing the clutch to protect your eyes in case the clutch spring gets loose. The dust from the clutch friction material may contain asbestos. Wear breathing protection. Do not blow the dust off parts and into the air.

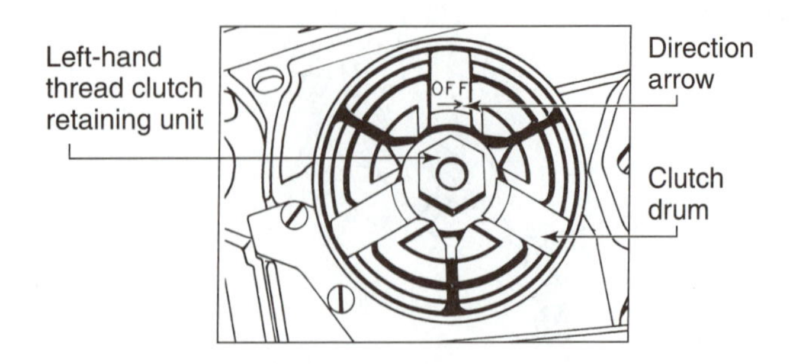

FIGURE 33-25 The clutch-retaining nut often has left-hand threads. (Courtesy of Poulan/Weed Eater.)

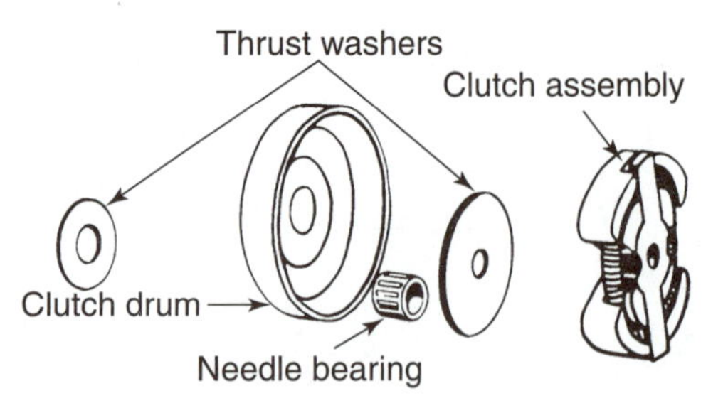

FIGURE 33-27 The clutch drum rotates on a needle bearing. (Courtesy of Poulan/Weed Eater.)

To remove the clutch, first remove the guide bar and saw chain. On some engines, a shroud or cover has to be removed to get access to the clutch assembly. The fan housing and rewind starter have to be removed from the opposite side of the engine. This allows you to hold the flywheel when loosening or tightening the clutch retaining nut. The clutch is removed by removing the clutch-retaining nut on the end of the PTO shaft. This nut is often left-hand thread. The left-hand threads prevent it from loosening during operation. A direction arrow is often etched on the clutch that shows which way to turn the nut (Figure 33-25). Remove the clutch-retaining nut. Slide the clutch off the crankshaft as an assembly. Note the position of any spacing or thrust washers so they can be assembled correctly. Do not separate the drum from the rest of the clutch until it is removed. This will prevent the spring from separating from the assembly.

Remove the clutch drum from the clutch assembly. Remove the clutch spring from around the friction pads (shoes) (Figure 33-26). Remove the friction pads from the center spider. Look at the condition of the friction pads. The pads wear down and get glazed if the clutch has been slipping. Remove the needle bearing from the center of the clutch drum (Figure 33-27). Rotate each needle bearing and inspect for damage. Look for broken needles. Inspect the drum and sprocket for damage or wear. The friction surface in the drum can get glazed from slipping friction pads. Wear can cause grooves in the surface. Inspect the clutch drum sprocket for wear or broken teeth.

CAUTION: A chain and sprocket will stay in the same pitch with each other as they wear. Putting a new chain on a worn sprocket can cause a mismatch. The same problem can occur if you use a new sprocket with an old chain. The result can be a noisy and rough chain operation.

Replace any worn parts you find in the clutch assembly. If the drum and sprocket are in satisfactory condition, you can install new friction pads. If the drum and sprocket are worn, replace the complete clutch assembly. To replace the friction shoes, install them on the center spider. A special tool (Figure 33-28) is available to hold the pads in position while installing the clutch spring. If necessary, install a new needle bearing in the clutch drum. The bearing can be driven in and out with a bearing driver. Check the owner's manual for the correct lubricant to use on the bearing.

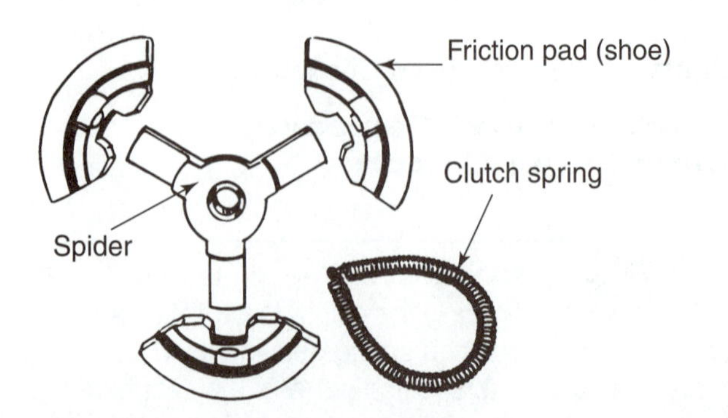

FIGURE 33-26 The spring retains the friction pads on the center spider. (Courtesy of Poulan/Weed Eater.)

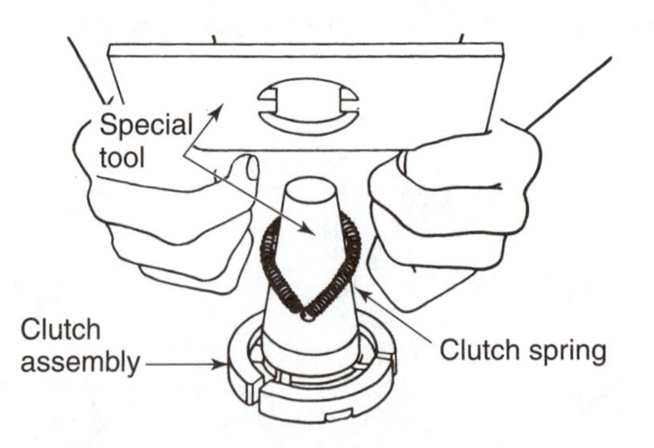

FIGURE 33-28 A special tool is used to assemble the spider, friction pads, and spring. (Courtesy of Poulan/Weed Eater.)

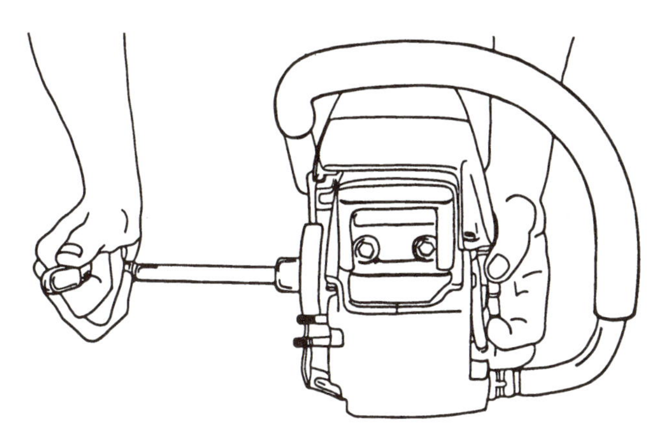

FIGURE 33-29 Holding the flywheel while tightening the clutch-retaining nut. (Courtesy of Poulan/Weed Eater.)

Install the drum over the clutch assembly and install the assembly on the PTO shaft. Install any spacing washers you removed during disassembly. Install the clutch-retaining nut. Hold the flywheel by hand while you tighten the left-hand thread nut (Figure 33-29). Install the fan housing and rewind starter assembly. Install the guide bar and chain. When assembly is complete, start the saw and check the clutch for proper operation.

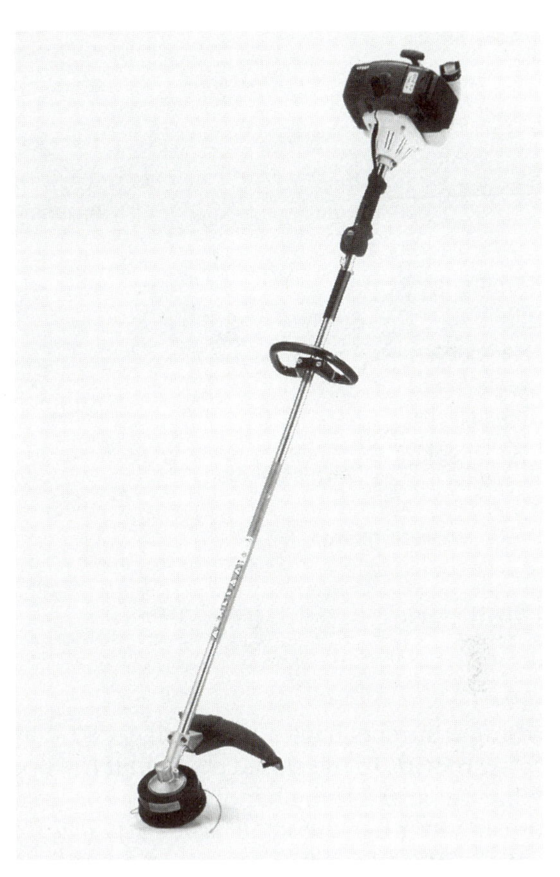

FIGURE 33-30 A string trimmer uses a rotating plastic string to cut, edge, and trim. (Courtesy of Stihl Incorporated.)

STRING TRIMMERS

A **string trimmer** (Figure 33-30) is an engine-powered, hand-held tool that uses a rotating plastic string (line) to cut, edge, or trim. Many string trimmers can be equipped with a blade and used as a brush cutter. A **brush cutter** is an engine-powered, hand-held tool that uses a rotating steel or plastic blade to cut brush. String trimmers use a small, 1–2 horsepower, two-stroke engine with a vertical crankshaft.

String Trimmer Parts and Operation

The main parts of a string trimmer are the engine, control handle, drive shaft, angle transmission, and cutting head (Figure 33-31). The two-stroke engine is mounted behind the operating handles to help balance the weight. A shoulder harness is attached in front of the engine so the operator can balance and support the trimmer weight. Engine power is transferred down a long drive shaft contained in a drive shaft tube. The power enters a small angle transmission at the end of the drive shaft. The angle transmission increases engine torque and changes the power flow direction so that it is parallel to the ground. The output from the transmission rotates the cutting head. A blade guard prevents cuttings from being thrown up at the operator.

The string trimmer controls are located on the handle. There is an ignition kill switch located next to the operator hand. The throttle is operated by a trigger on the handle. Some trimmers have a trigger lock to hold the throttle in an open position.

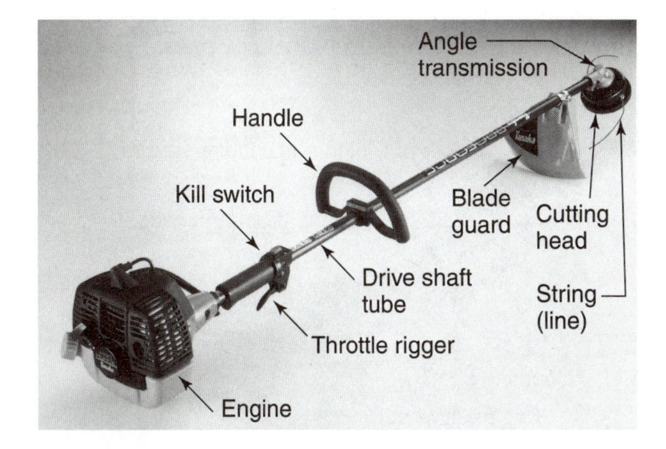

FIGURE 33-31 Parts of a string trimmer. (Courtesy of Tanaka/ISM.)

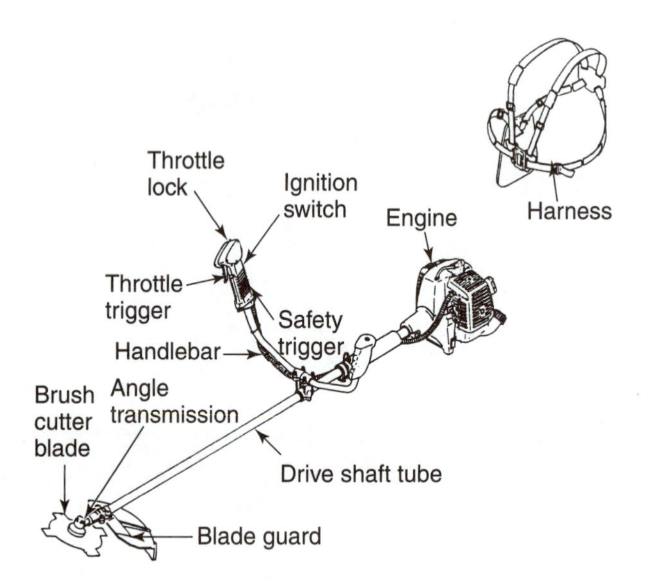

FIGURE 33-32 Parts of a brush cutter. (Courtesy of Tanaka/ISM.)

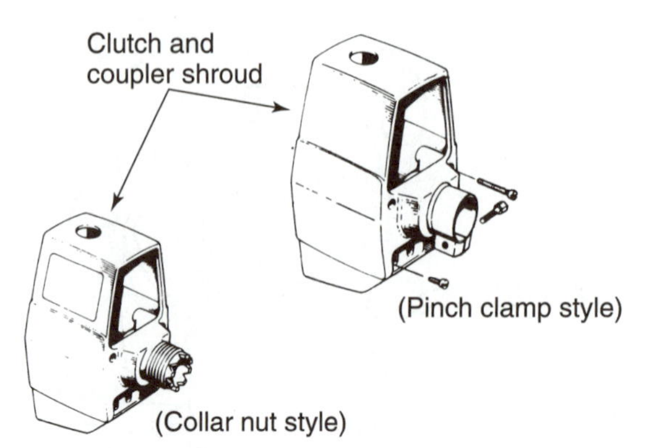

FIGURE 33-34 The drive shaft tube is attached to the end of the clutch and coupler shroud. (Courtesy of Poulan/Weed Eater.)

Many string trimmers have a safety trigger so that the throttle returns to the closed position if the operator's hand comes off the handle.

A brush cutter has the same basic parts (Figure 33-32) as a string trimmer. The main difference is that a metal blade is attached to the angle transmission. The metal blade is able to cut larger diameter brush. Single-use brush cutters may have more powerful engines than string trimmer and brush cutter combination equipment. They also use wider two-hand control handlebars and two shoulder type harnesses.

The string trimmer engine PTO shaft drives the drive shaft through a centrifugal clutch (Figure 33-33). The centrifugal clutch has the same parts and operation as the clutch used on chain saws. The clutch drum has a shaft connected to its end called a *coupling*. The coupling is connected (coupled) to the drive shaft. The clutch and coupler are housed in a coupling shroud. The shroud has a bearing that supports the end of the coupler. The top end of the drive shaft tube is attached to the coupling shroud. The drive shaft tube may be attached to the shroud with a pinch clamp or collar nut (Figure 33-34).

The drive shaft used on a string trimmer may be straight or curved. The straight drive shaft is solid and contained in a straight tube (Figure 33-35). The top of the drive shaft is supported by the clutch drum and coupler bearing. The center of the drive shaft is supported by a bearing located in the middle of the tube. The bottom of the drive shaft is supported by a bearing at the lower end of the tube.

The straight drive shaft is connected to a pinion gear located in the angle transmission housing (Figure 33-36). The pinion gear meshes with and drives the bevel gear. The pinion and bevel gear change the angle of the drive so that it is parallel to the ground. The bevel gear is splined to and drives the bevel gear shaft. The bevel gear shaft drives the brush cutter (or string).

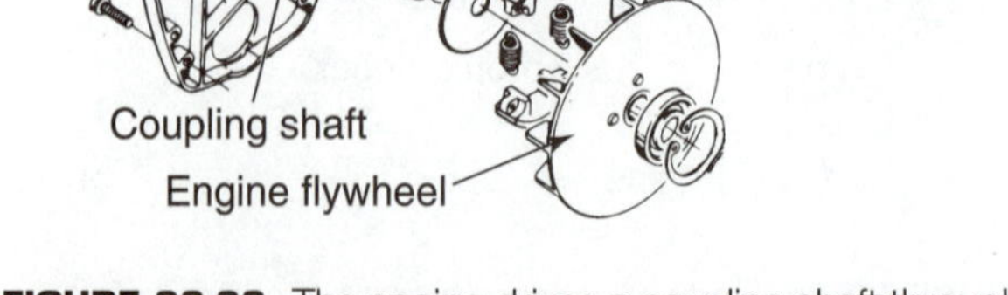

FIGURE 33-33 The engine drives a coupling shaft through a centrifugal clutch. (Courtesy of Poulan/Weed Eater.)

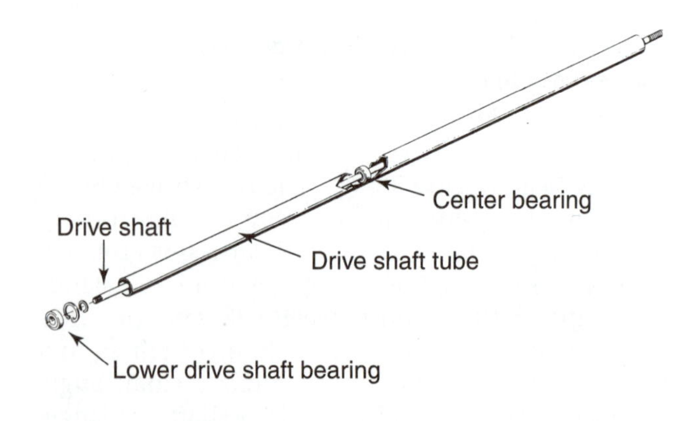

FIGURE 33-35 Straight drive shaft tubes have a solid drive shaft supported by bearings in the tube. (Courtesy of Poulan/Weed Eater.)

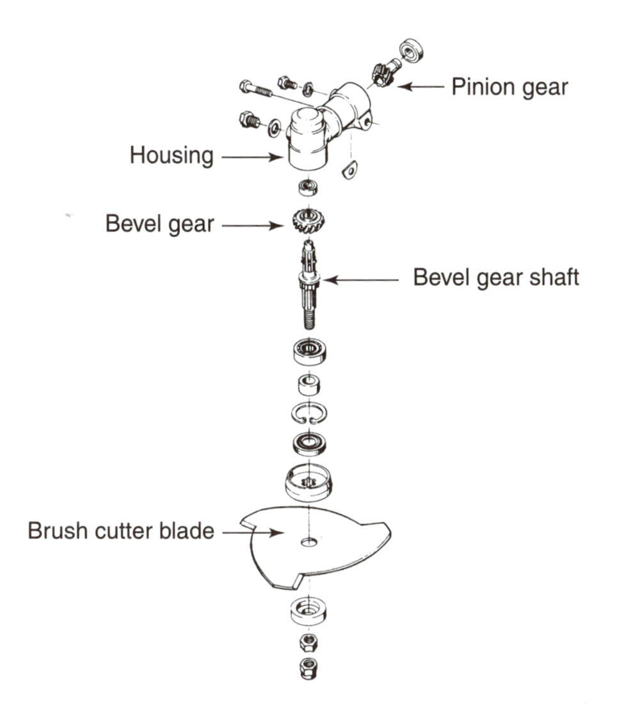

FIGURE 33-36 Parts of an angle transmission. (Courtesy of Poulan/Weed Eater.)

Many string trimmers use a curved drive shaft tube (Figure 33-37). To operate in a curved tube, the drive shaft has to be flexible. The flexible drive shaft is made like a large diameter braided cable. The curved drive shaft eliminates the need for an angle transmission, which lowers the weight and expense of the trimmer. The top end of the drive shaft is connected to and driven by the clutch and coupler assembly. The bottom end of the curved drive shaft is supported by a lower drive shaft bearing.

A coupler bolt usually connects the drive shaft to the outer spool in the cutter head assembly (Figure 33-38). The outer spool rotates with the drive shaft. An inner spool (reel) of plastic string (line) is contained inside the outer spool. The two ends of the plastic string come out of the outer spool and are used to cut grass and weeds. The inner spool is retained in the bottom of the outer spool by a lower housing called a *bump knob*. The ends of the plastic string wear away as they cut grass and weeds. Striking the rotating bump knob against a solid surface causes more line to feed out of the inner spool.

Troubleshooting a String Trimmer

> **WARNING:** Stop the engine, remove and ground the spark plug wire, and kill the ignition before doing any string trimmer inspection or repair.

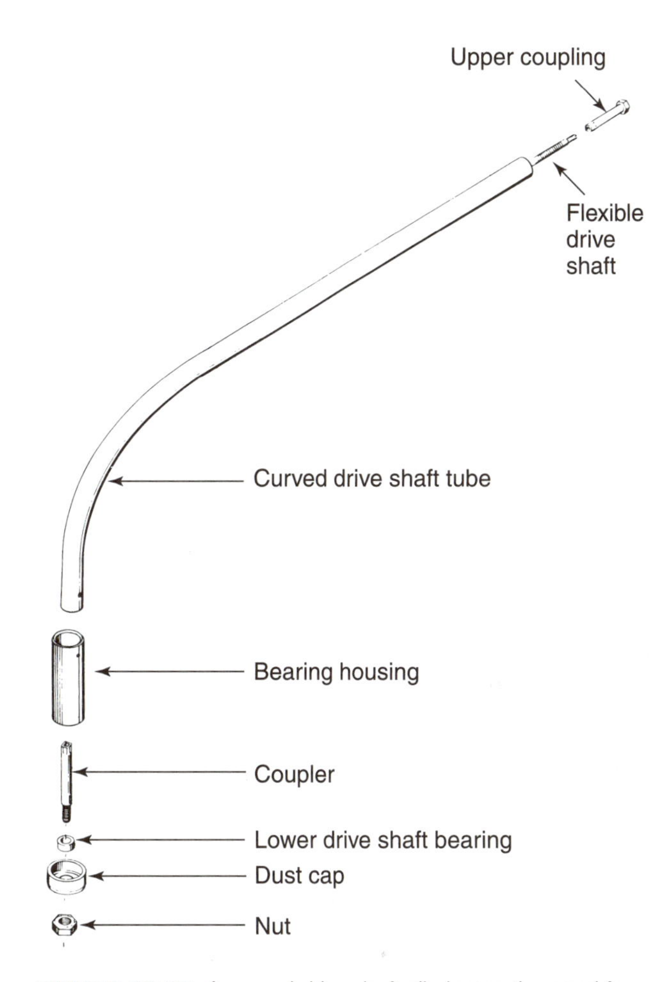

FIGURE 33-37 A curved drive shaft eliminates the need for an angle transmission. (Courtesy of Poulan/Weed Eater.)

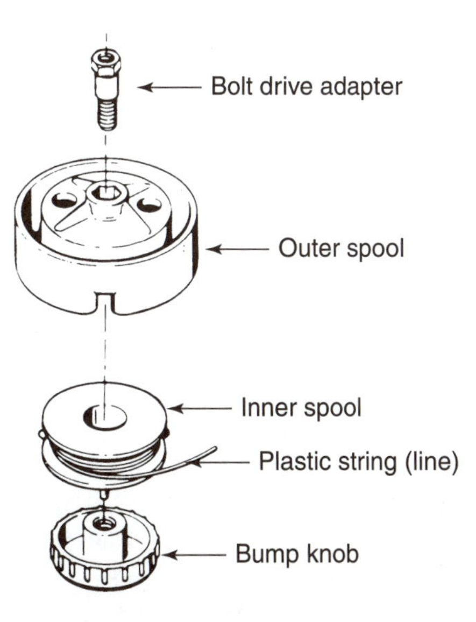

FIGURE 33-38 The drive shaft rotates the cutter head assembly that contains the plastic cutting string. (Courtesy of Poulan/Weed Eater.)

String trimmer problems can be related to engine or equipment problems. Engine troubleshooting is presented in Chapter 21. The most common problem with a string trimmer is that the cutting head will not advance line. A common cause of this problem is that the spool is out of line and the spool must be refilled with new line. If there is line on the spool, the problem may be that the inner reel is bound up and will not rotate. This problem can be caused by dirt in the cutting head or worn indexing teeth on the reel and spool. Line problems can also cause this problem. Sometimes the line gets hot enough to weld (melt) together. Another common line problem is a line that is twisted when it is installed.

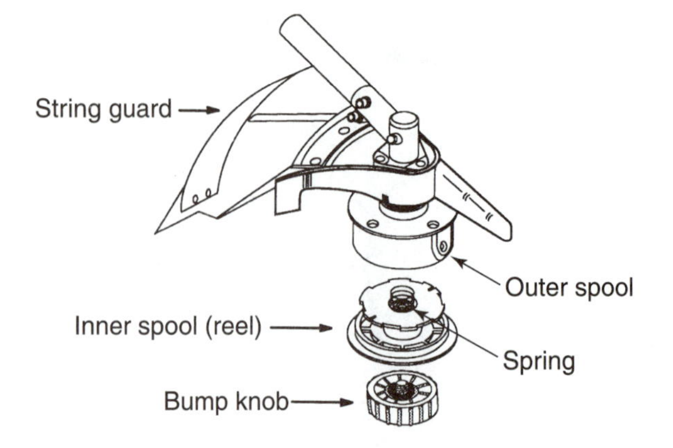

FIGURE 33-39 Unscrew the bump knob to remove the inner spool. (Courtesy of The Toro Company.)

Removing and Replacing Cutting Line

WARNING: Stop the engine, remove and ground the spark plug wire, and kill the ignition before doing any string trimmer maintenance or repair.

WARNING: Be sure to use the line size and type recommended in the string trimmer owner's manual. Using the wrong type of line could cause the line to break and be thrown in the operator's or bystander's direction.

CAUTION: Look up the recommended line type and diameter in the string trimmer owner's manual before replacing the cutting line. The engine may overheat if too large a diameter line is used.

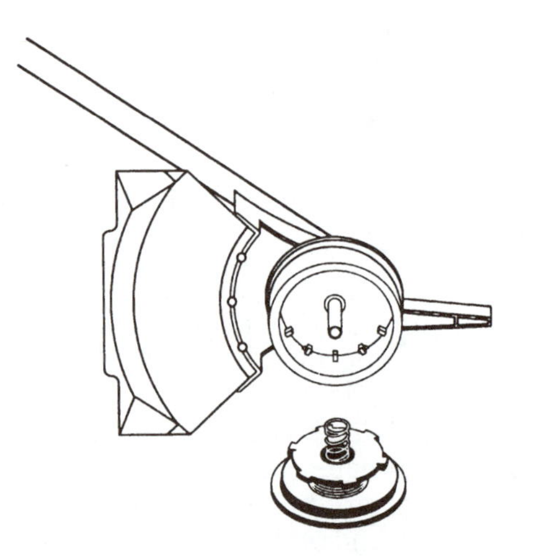

FIGURE 33-40 Removing the inner spool. (Courtesy of The Toro Company.)

When the string trimmer requires new cutting line, you can install a new line reel or you can rewind the existing reel. Stop the engine, kill the ignition, and remove and ground the spark plug wire. Wait for the engine to cool down. Hold the outer spool with one hand and unscrew the bump knob clockwise (Figure 33-39). Inspect the bolt inside the bump knob to make sure it moves freely. Inspect the bump knob for wear and damage. Replace the bump knob if you find any of these problems.

Remove the inner spool (reel) (Figure 33-40). Use a clean shop rag to clean the inner surface of the outer spool. Clean the inner spool and drive shaft. Dirt and grass in this area prevents the line from advancing. Check the indexing teeth on the inner spool and outer spool for wear or damage. Minor burrs on the parts can be sanded with abrasive paper. If these parts are worn, the line will not advance properly. Replace any worn parts.

If you are replacing the existing inner spool with a new one, it can be replaced at this time. If you are rewinding the existing spool, measure approximately 50 feet (15 meters) of new trimming line. Loop the line into two equal lengths. Insert each end of the line through one of the two holes in the inner spool (Figure 33-41). Pull the line so that the loop is small as possible. Wind the lines in even, tight layers on the inner spool (Figure 33-42). Be sure to wind in the correct direction. There is usually a direction arrow on the inner spool. Do not overlap the two ends of the line. Insert the ends of the line into the two holding slots (Figure 33-43).

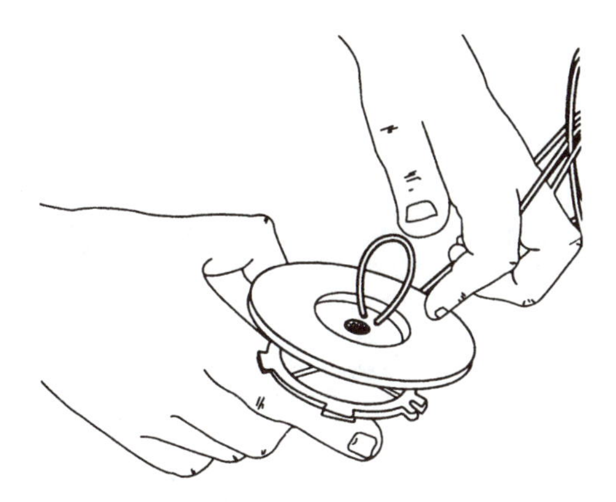

FIGURE 33-41 Inserting the ends of the line into the inner spool. (Courtesy of The Toro Company.)

FIGURE 33-42 Winding the line around the inner spool. (Courtesy of The Toro Company.)

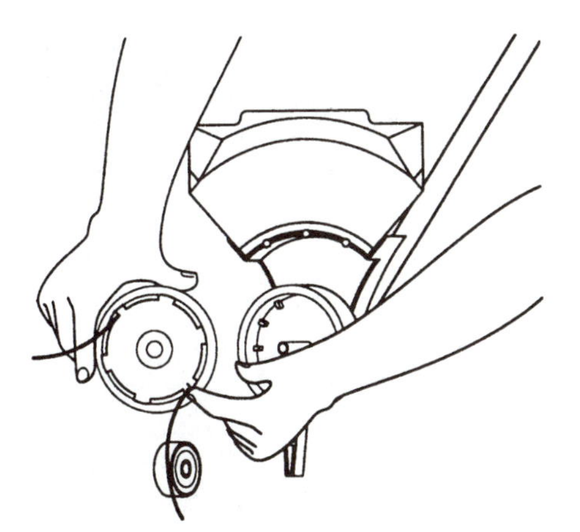

FIGURE 33-43 Inserting the ends of the line into the two holding slots. (Courtesy of The Toro Company.)

To reinstall the new or rewound spool, insert the ends of the line through the eyelets in the outer spool (Figure 33-44). Grip the ends of the line firmly and pull to release the line from the holding slots. Hold the inner reel in position and install the bump knob. Tighten the bump knob in a counterclockwise direction. The procedure for removing and replacing a string trimmer cutting line is shown in the accompanying sequence of photographs.

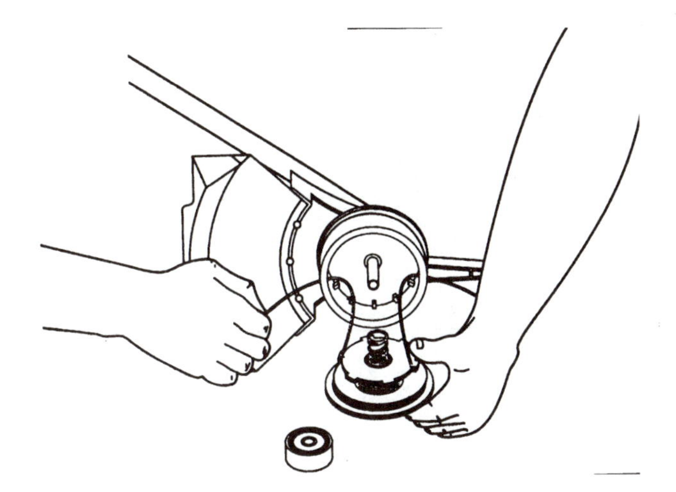

FIGURE 33-44 Inserting the ends of the line through the outer spool eyelets. (Courtesy of The Toro Company.)

Checking Angle Transmission Lubricant

> **WARNING:** The angle transmission gets hot after long periods of use. To avoid burns, allow it to cool down before checking lubricant level.

The gears in the angle transmission require lubrication. Some transmissions use a thick gear oil for lubrication; others use a lithium-based grease. Check the owner's manual for the recommended type of lubricant to use. To check the lubricant level, remove the fill plug (Figure 33-45). The

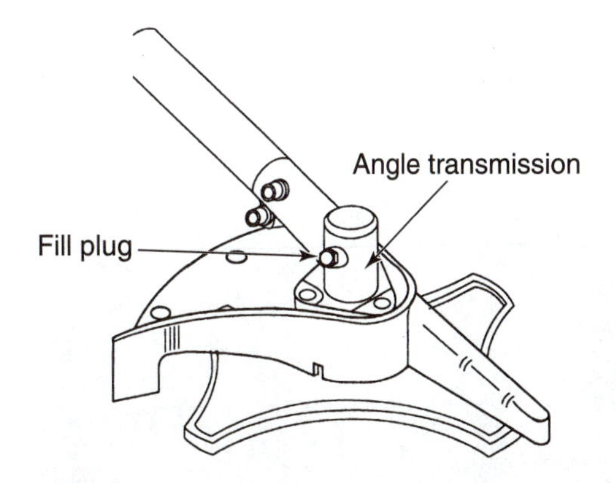

FIGURE 33-45 A fill plug is used to determine lubricant level in an angle transmission. (Courtesy of The Toro Company.)

PHOTO SEQUENCE 30

Removing and Replacing a String Trimmer Cutting Line

1. Remove and ground the spark plug wire and kill the ignition.

2. Hold the outer spool with one hand and unscrew the bump knob clockwise.

3. Remove the inner spool (reel). Use a clean shop rag to clean the inner surface of the outer spool.

4. Reinstall the new or rewound spool by inserting the ends of the line through the eyelets in the outer spool. Grip the ends of the line firmly and pull to release the line from the holding slots.

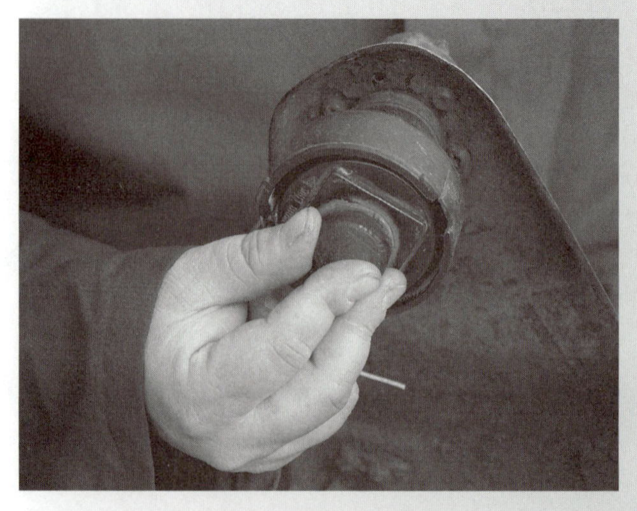

5. Hold the inner reel in position and install the bump knob. Tighten the bump knob in a counterclockwise direction.

6. Install the spark plug wire on the spark plug

lubricant level should be up to the bottom edge of the fill hole. If necessary, add lubricant to the transmission. Install the fill plug.

BLOWERS

A **blower** is an engine-driven air pump used to blow yards and walkways free of leaves and other debris. The blower uses a two-stroke engine to drive the air pump (Figure 33-46). Smaller blowers are held by a handle. The engines used in these blowers are approximately 1 horsepower. The air volume delivered by the blower is approximately 300 cubic feet per minute (cfm). Larger blowers are held on the operator's back like a backpack. The engines used on these blowers have 2 or more horsepower. The air volume delivered by these units is in excess of 500 cubic feet per minute.

Blower Parts and Operation

The shoulder harness and handle carry blowers both have the same main parts (Figure 33-47). A two-stroke engine drives an air pump that fits inside a housing. The air from the pump is directed out a large blow pipe that is directed by the operator toward the area to be blown clean. The common engine controls such as kill switch, throttle lever, choke valve, and rewind starter are located directly on the engine or on the blower handle.

The air pump is an impeller driven by the engine PTO shaft directly at crankshaft speed. The impeller is housed inside a two-piece housing (Figure 33-48). One-half of the housing, called the *main support housing,* is mounted to the side of the engine. The other half, called an *inlet cover housing,* is attached to the main support housing. Like other centrifugal pumps, the impeller rotates inside the housing. Centrifugal force causes air to be thrown off the impeller vanes (blades), causing a low pressure in the center of the impeller. Air is pulled into the housing through the inlet cover. The air flows to the center of the impeller and then off the tips of its vanes. The shape of the housing directs air under pressure out through the housing outlet. From the outlet, air flows out through the blow pipe.

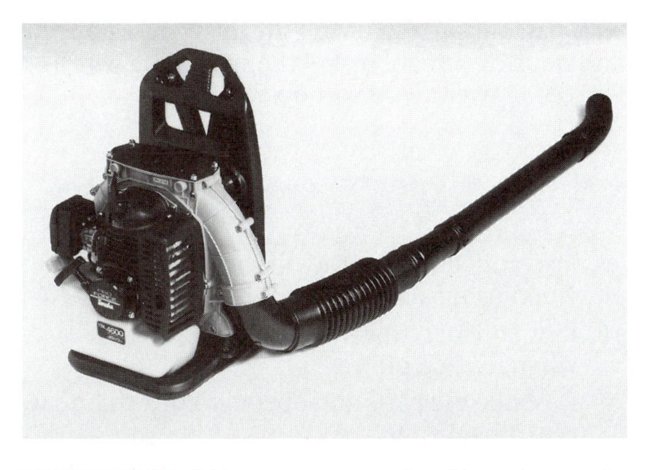

FIGURE 33-46 A blower uses an engine-driven air pump to blow away leaves and debris. (Courtesy of Tanaka/ISM.)

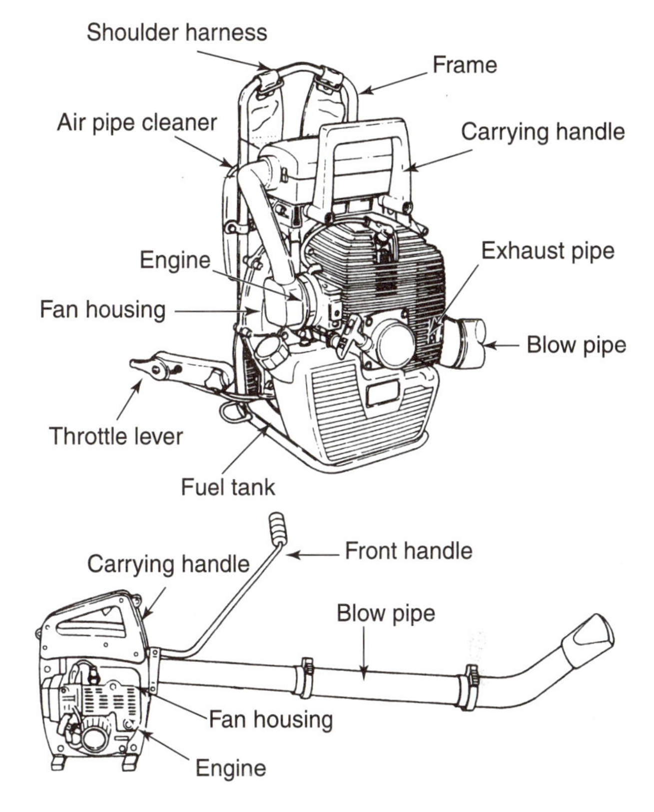

FIGURE 33-47 Parts of a shoulder carry and handle carry blower. (Courtesy of Tanaka/ISM.)

Blower Troubleshooting

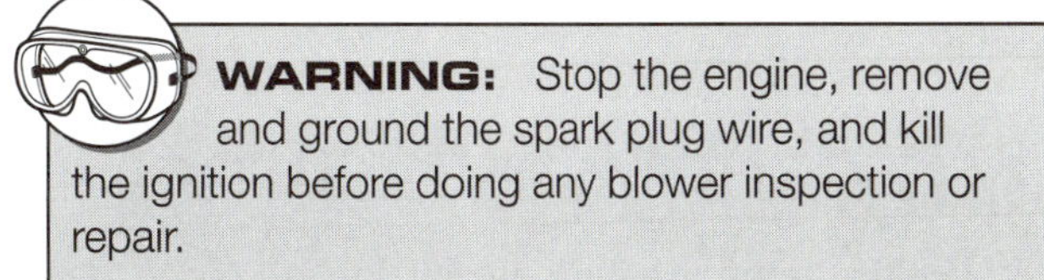

WARNING: Stop the engine, remove and ground the spark plug wire, and kill the ignition before doing any blower inspection or repair.

Blower problems can be related to engine or equipment problems. Engine troubleshooting is presented in Chapter 21. The most common problem with a blower is excessive vibration caused by a loose or damaged impeller. Damage is often caused by dirt or foreign objects getting pulled into the pump.

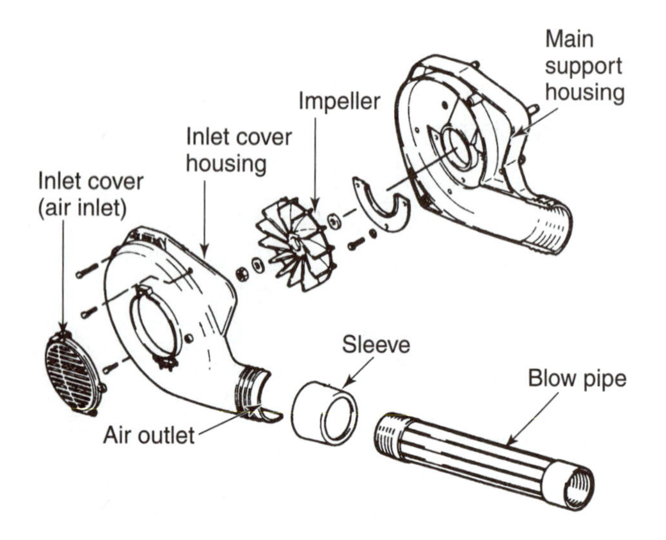

FIGURE 33-48 The engine-driven impeller is located inside a two-piece housing. (Courtesy of Poulan/Weed Eater.)

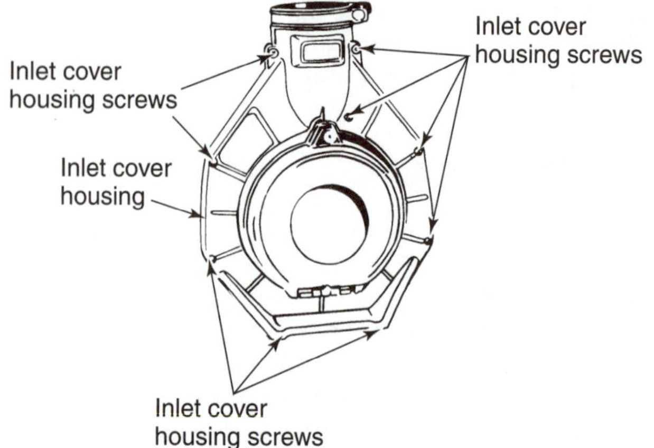

FIGURE 33-49 Remove the inlet cover screws to remove the inlet cover. (Courtesy of Poulan/Weed Eater.)

Removing and Replacing the Impeller

> **WARNING:** Stop the engine, remove and ground the spark plug wire, and kill the ignition before doing any blower maintenance or repair.

If troubleshooting shows a possible problem with the impeller, the inlet cover housing must be removed. Remove the inlet cover housing screws (Figure 33-49). Lift the inlet cover off the main support housing. With the housing off, the impeller can be inspected (Figure 33-50). Look for any cracked or damaged vanes. If the vanes are not damaged, use a wrench to check the tightness of the impeller retaining nut. Tighten the nut if you find it loose. Nuts are commonly left-hand thread.

To replace the impeller, remove the retaining nut. Lift the impeller off the shaft. A puller is usually not necessary to remove the impeller. Observe and remove any washers. Compare the new with the old impeller to be sure it is the correct part. Install the impeller on the shaft. Replace any washers you removed during disassembly. Install and tighten the retaining nut. Install the inlet cover housing and housing screws. Install the spark plug wire and test the blower for correct operation.

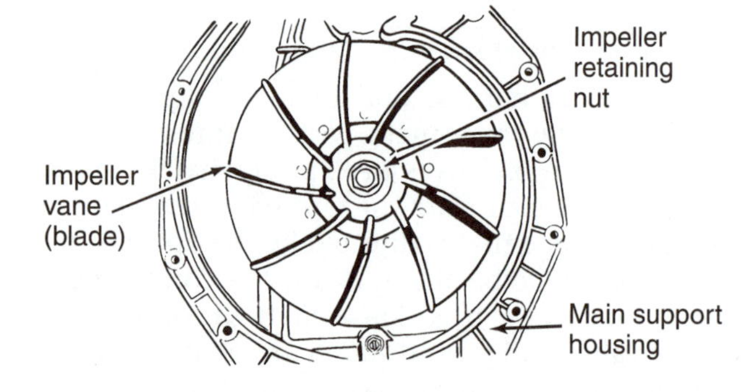

FIGURE 33-50 Inspect the impeller vanes for damage and the retaining nut for tightness. (Courtesy of Poulan/Weed Eater.)

REVIEW QUESTIONS

1. List five general safety precautions to take when operating a chain saw, string trimmer, or blower.
2. List and describe the opposing forces on a saw chain during sawing.
3. Explain why saw kickback is dangerous when operating a chain saw.
4. Explain the hazards and safety precautions to take when felling a tree with a chain saw.
5. List and describe the specific safety hazards when using a string trimmer.
6. List and describe the specific safety hazards when using a blower.
7. Describe the parts and operation of a chain saw centrifugal clutch.
8. Describe the parts of a chain saw chain.
9. Describe the parts of a chain saw guide bar.

10. Describe the parts and operation of a chain saw chain brake.
11. Explain the operation and use of a chain saw oil pump.
12. List and explain the steps to follow in removing and replacing a chain saw chain.
13. Explain the steps to follow when tensioning a chain saw chain.
14. Describe how to sharpen the cutting edges on a chain saw chain.
15. Describe the basic parts and operation of a string trimmer.
16. Describe the purpose and parts of a string trimmer angle transmission.
17. Explain the difference between the drive shaft used in a straight and in a curved drive shaft tube.
18. Explain the steps to follow when removing and replacing a string trimmer cutting line spool.
19. Explain how to install a new line on a string trimmer cutting line spool.
20. Explain how to remove and replace a blower impeller.

DISCUSSION TOPICS AND ACTIVITIES

1. Locate a scrap chain saw chain. Use a depth gauge tool and a file to practice sharpening the chain.
2. Using a scale and micrometer, determine the length, pitch, and gauge of the chain saw chain.
3. One important physical science principle is that for every action there is an equal and opposite reaction. Explain how this principle applies to chain saw kickback, pushback, and pull-in.

Portable Pumps and Generators

OBJECTIVES

After completion and review of this chapter, you should be able to:

- List and describe the specific safety procedures to use when using a portable water pump.
- Describe the parts and operation of a portable water pump.
- Prime a pump.
- Troubleshoot a pump for vibration, poor or no pump output, pump lock-up, and water leaks.
- List and describe the parts of a portable generator.
- Describe the parts and operation of an alternating current generator.
- Use an appliance wattage list to determine required generator output.
- Troubleshoot a generator for abnormal output at the receptacles.
- Inspect and replace generator field current brushes.
- Test generator receptacles and circuit breaker.

TERMS TO KNOW

Circuit breaker
Generator watts rating
Portable generator
Portable pump
Pump output volume
Pump priming

INTRODUCTION

Portable pumps and generators use the power of small engines to pump water or generate electricity. Engine-driven water pumps are often used to remove water in emergency situations, to divert water during construction projects, and to pressurize agriculture chemicals for spraying. Portable generators are used to power electrical devices where a power line is not available or has failed.

PUMP AND GENERATOR SAFETY

Always read and understand the owner's manual before operating a pump or generator. Make sure you understand the controls and proper use of the equipment. You should know how to stop the engine quickly in case of an emergency. Read and follow all the safety procedures in the owner's manual. Be sure you understand each danger, warning, caution, and instruction decal on the equipment. Follow these general safety procedures to avoid injury when using a pump or generator:

- Always operate a pump or generator in a well-ventilated area.
- Do not carry a pump or generator when the engine is running.
- When carrying a pump or generator in a vehicle, secure it from tipping over and spilling fuel.
- Wear sturdy, tight-fitting pants and clothing that will not get caught in rotating parts when operating a pump or generator.
- Wear sturdy gloves when handling a pump or generator.
- Do not operate a pump where it can contact house wiring, conduit, or power lines as electrocution could result.

- Make sure the pump or generator is on a stable platform or ground before starting the engine.
- Do not touch engine parts that might be hot from operation.
- Do not allow anyone, especially children, around the equipment when it is in operation.
- If a pump ingests a foreign object, stop the engine, wait for moving parts to stop, and remove and ground the spark plug wire before working on the problem.
- Do not insert your fingers or any solid object into a pump.
- Never use a portable water pump to pump flammable liquids.
- Never operate a generator indoors because of carbon monoxide poisoning.
- Follow recommended procedures for preventing electrocution when plugging and unplugging the generator high-voltage output connection.
- Make sure generator power cords are in good condition to prevent dangerous electrical shock.

PORTABLE PUMPS

A **portable pump** is an engine-driven centrifugal water pump used to pump water from one location to another. Portable water pumps may use a two- or four-stroke engine. Two-stroke engine pumps (Figure 34-1) are very small and light in weight. Engines for these units develop between 1 and 3 horsepower. The pump and engine weigh less than 15 pounds and can be handled by one person. **Pump output volume** is a rating system used to rate the output of a pump in gallons per hour (GPH) or gallons per minute (GPM). Two-stroke engine pump output volume is usually approximately 1,900 gallons per hour.

FIGURE 34-1 A portable pump with a two-stroke engine. (Courtesy of Tanaka/ISM.)

FIGURE 34-2 A portable pump with a four-stroke engine. (Courtesy of Ace Pump Corporation.)

Four-stroke engine pumps (Figure 34-2) are larger and heavier than the two-stroke units. These units are heavy and often must be handled by two people. Four-stroke pump engines are commonly in the 5 to 10 horsepower range. Many of the engines have overhead valves. The output from the four-stroke engine pumps varies with engine speed. Pump output volume is usually in the 8,000 gallons per hour range at 3,400 engine rpm.

Pump Parts and Operation

The portable pump has an engine and a pump mounted on a base (Figure 34-3). A carry handle is attached to the assembly so the pump can be carried to the desired location. The pump is driven directly by the engine PTO shaft. There is commonly no clutch or belt between the pump and engine. The engine controls such as throttle lever, choke, and kill switch are usually mounted directly on the engine.

The pump is mounted to the base on the PTO side of the engine. The pump has two openings or ports. Pumped water enters the pump suction port (inlet). Water leaves the pump through the pump delivery port (outlet). There are fittings on both ports so that a hose can be connected. The pump

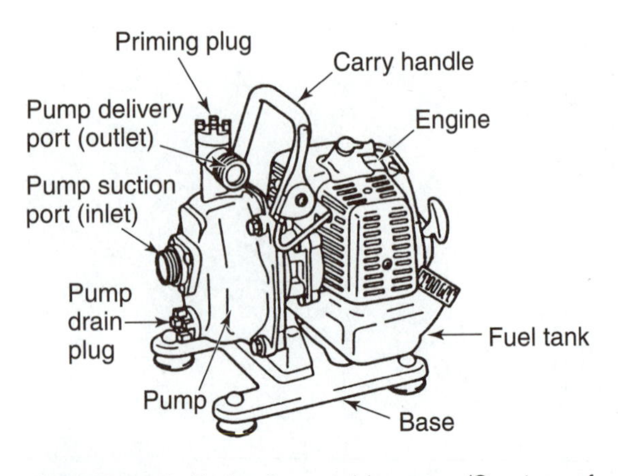

FIGURE 34-3 Parts of a portable pump. (Courtesy of Tanaka/ISM.)

also has a drain valve that can be removed to drain water out of the pump.

A suction hose is connected to the suction port fitting (Figure 34-4). A fitting on the end of the suction hose is connected to a fitting on the suction port. The suction hose has a strainer attached to its end. The strainer is a metal mesh filter. It is used to prevent sand, grit, and other debris from getting into the pump and damaging the pump impeller. The strainer is placed in the water to be pumped. All water entering the pump has to flow through the strainer to get into the suction hose and into the pump.

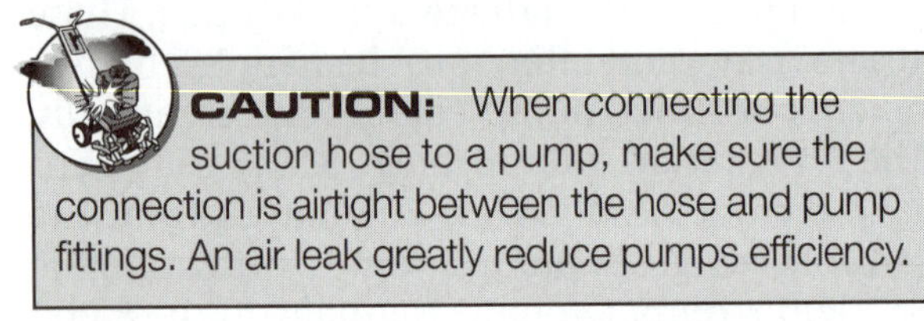

CAUTION: When connecting the suction hose to a pump, make sure the connection is airtight between the hose and pump fittings. An air leak greatly reduce pumps efficiency.

The water pump is a centrifugal type pump. The water pump has an impeller that is driven by

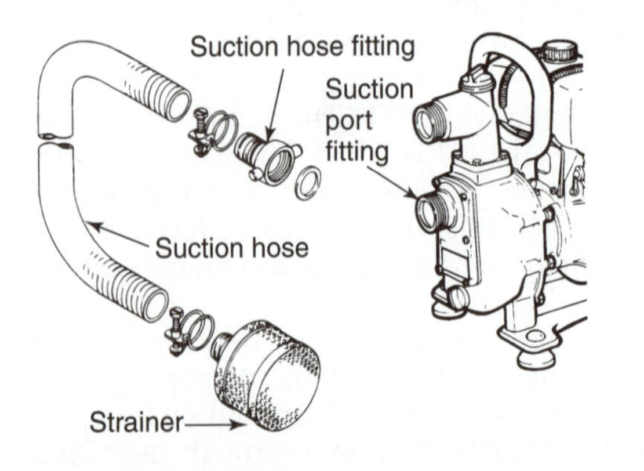

FIGURE 34-4 A strainer is used on the inlet hose to prevent pump damage. (Courtesy of Tanaka/ISM.)

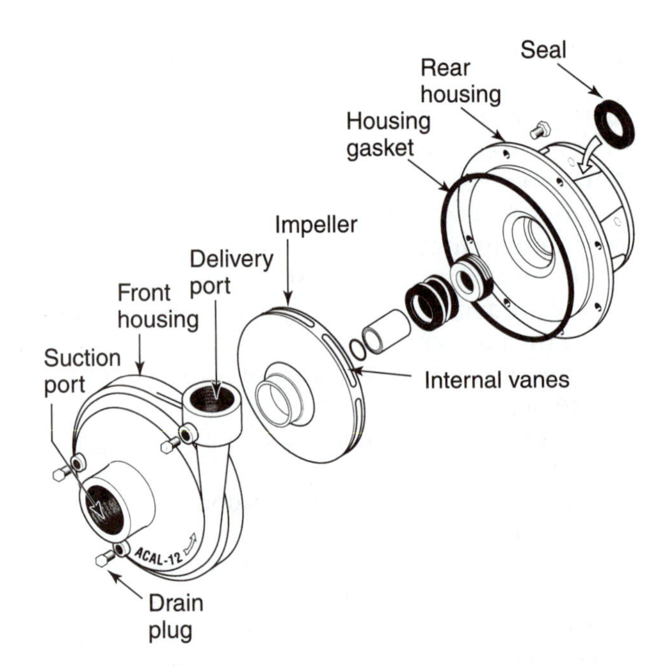

FIGURE 34-5 A water pump has an impeller rotating inside a two-piece housing. (Courtesy of Ace Pump Corporation.)

the engine PTO shaft directly at crankshaft speed. The impeller is housed inside a two-piece housing, (Figure 34-5). One-half of the housing, called the *front housing*, has the suction and delivery ports. The other half is called the *rear housing*. The two housings are bolted together with a gasket in between. Lip type seals are used to seal around the rotating impeller. There is high water pressure inside the pump housing when it is pumping. The gasket and seals prevent water leakage out of the housing.

The impeller has internal vanes (blades). The impeller rotates inside the housing. Centrifugal force causes water to be thrown off the impeller vanes. This causes a low pressure in the center of the impeller. Water is pulled into the housing through the suction port. The water flows to the center of the impeller and then off the tips of its vanes. The shape of the housing directs water under pressure out through the delivery port.

Pump Troubleshooting

WARNING: Stop the engine, wait for all moving parts to stop, remove and ground the spark plug wire, before making any pump inspection.

Pump problems can be related to engine or equipment problems. Engine troubleshooting is presented in Chapter 21. The most common pump problems are vibration, poor or no pump output, pump lock-up, and water leaks.

If the pump vibrates, stop the engine immediately. Vibration is usually caused by foreign material entering the pump housing. Damage is often caused by dirt or foreign objects getting pulled into the pump. The pump housing must be disassembled and the impeller inspected for damage.

When the problem is poor pump output, check the strainer. A plugged strainer prevents water from entering the pump. Clean the strainer. If the pump has run dry, it will require priming. Check the hose connection to the delivery port. An air leak at the hose connection causes poor pump output.

Pump lock-up occurs when the pump comes to a sudden stop. Lock-up occurs when a large foreign object enters the pump and lodges between the impeller and the housing. This problem can occur if the strainer is broken or comes off the suction hose. A locked up pump will usually be severely damaged. A new pump assembly must be installed.

Pump water leaks can be caused by a damaged housing gasket or worn housing seals. Water leaks reduce pump efficiency. The gasket and all the seals should be replaced or a new pump assembly installed.

Pump Priming

WARNING: Stop the engine, wait for all moving parts to stop, and remove and ground the spark plug wire before making any pump inspection, repairs, or adjustments.

The pumping process requires that the pump housing be full of water. If the housing is dry, a low pressure cannot be developed inside the housing. In order to get a dry pump to work, it has to be primed. **Pump priming** is a procedure used to fill the pump housing with water to start the pumping process.

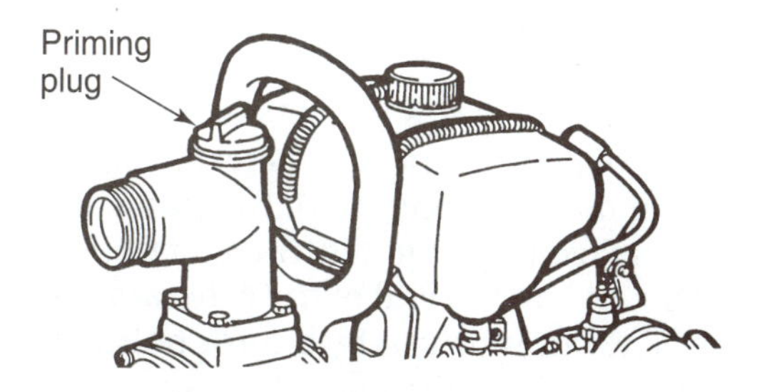

FIGURE 34-6 A pump priming plug is used to prime the pump with water before pumping. (Courtesy of Tanaka/ISM.)

Always look up and follow the priming procedure in the pump owner's manual. Many pumps have a priming plug on top of the pump housing (Figure 34-6). To prime the pump, remove the priming plug. Fill the pump housing with clean water. Place the suction hose and strainer in the water to be pumped. Replace the cap. Replace the spark plug wire and start the engine. If priming was successful, water will be discharged out through the delivery hose.

CAUTION: Never start an engine without priming water in the pump. A dry pump will overheat and parts can be damaged.

CAUTION: Drain all the water out of the pump housing after each use. Water left in the housing causes rust and corrosion to the housing and impeller. If water inside the housing freezes, the housing can be damaged.

PORTABLE GENERATORS

A **portable generator** is an engine-driven alternating current generator that supplies 120-volt power to operate standard electrical devices. Portable generators are commonly used to supply power in emergency situations when there is a power line failure. They are also used to bring power to remote areas for farmers, ranchers, and campers.

Portable generators vary widely in size and output. The power output of a generator is rated in watts. A **generator watts rating** is the maximum power output of a generator determined by multiplying generator voltage (120 volts) by generator current output. For example, a small portable generator can develop 2.9 amps at 120 volts. Its watts rating is 2.9 amps multiplied by 120 volts or 348 watts.

There is also a relationship between watts and engine horsepower. The power output of the generator rates its ability to do work. The engine that powers the generator must meet or exceed this ability to do work. The relationship between engine power and generator power is that 1 horsepower is equal to 746 watts. If a generator has a rating of 650 watts, the engine driving it must have at least 1.14 horsepower (746 watts divided by 650 watts equals 1.14 horsepower). Engines used on generators always have more horsepower than the minimum required. There are always power losses in converting mechanical energy (horsepower) into electrical energy (watts).

Four-stroke engines are usually used to power even the smallest generators. The smallest generators weigh around 20 pounds and can be handled by one person. A generator of this size has a power rating of 350 watts. Large industrial size generators have wheels so they can be moved into position. They often have two-cylinder, liquid-cooled engines. These large generators can develop in excess of 6,000 to 10,000 watts.

Generator Parts and Operation

The main parts of a portable generator (Figure 34-7) are the engine and alternating current (AC) generator. The engine and generator are mounted on a base. The engine PTO shaft drives the generator directly. Usually there is no belt or centrifugal clutch between the engine and generator.

There is an electrical service plug in the panel (control box) on one side of the generator (Figure 34-8). The electrical power developed by the generator is routed up to this panel. There are commonly several three-wire electrical receptacles (plugs) on the panel. These are used to connect an extension cord from the generator to the electrical device being powered by the generator. The engine ignition kill switch is usually located on the panel. An oil level warning light (oil alert) is located on the panel to warn the operator of low oil level. Some models have a voltmeter on the panel that shows generator power output.

A circuit breaker switch is also located on the panel. A **circuit breaker** is an electrical circuit protection device that senses excessive current flow and opens the circuit between the generator and the receptacles. If the circuit opens (trips), the switch has to be reset (moved back to on) to connect the generator to the circuit again. The circuit breaker opens the circuit if the power demands exceed generator output.

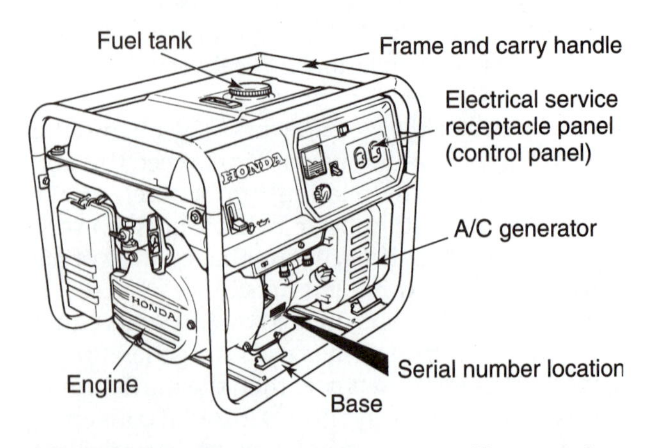

FIGURE 34-7 Parts of a portable generator. (Courtesy of American Honda Motor Co., Inc.)

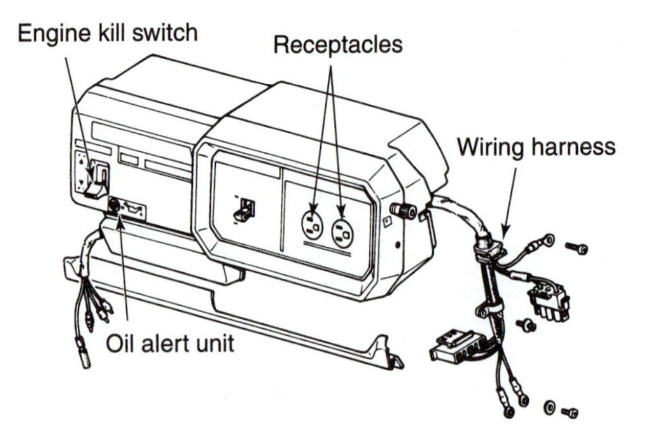

FIGURE 34-8 Parts of a generator control panel. (Courtesy of American Honda Motor Co., Inc.)

The generator section of the portable generator contains a rotor with two rotating magnets and a stationary coil called a *stator* (Figure 34-9). The stator is made up of coils of wire (stator main output winding) wound around a metal lamination (core). The coils of wire provide the conductor for current generation. Metal laminations in the core concentrate the magnetic lines of force. The higher the number of stator coils, the higher alternating current generator's output.

The magnetic field for electrical generation is provided by two large semicircular magnets. There are coils of wire around each magnet called *field windings*. Current is directed into the windings from the stator to strengthen the magnetic field. The field current enters the rotating rotor through sliding carbon contacts called *brushes*. The brushes ride on copper slip rings on the rotor. The rotor is driven by the rotor shaft connected to the engine PTO shaft. The magnetic lines of force (flux) flow from one pole of the rotor through the stator main output winding and back to the other pole of the rotor. The direction of the magnetic flux through the stator main output wiring alternates as the rotor turns. This creates (induces) an alternating current in the stator output winding. The current flowing out of the stator is 120 volt alternating current. It alternates at 60 cycles (hertz) per second.

The process of power generation creates a great deal of heat that must be removed or the stator and rotor can be damaged. A large centrifugal fan (Figure 34-10) is housed on the end of the generator for this purpose. The fan is driven by a rotor bolt that fits in the center of the rotor shaft. The fan is covered by a generator end cover that directs the cooling air over the generator parts.

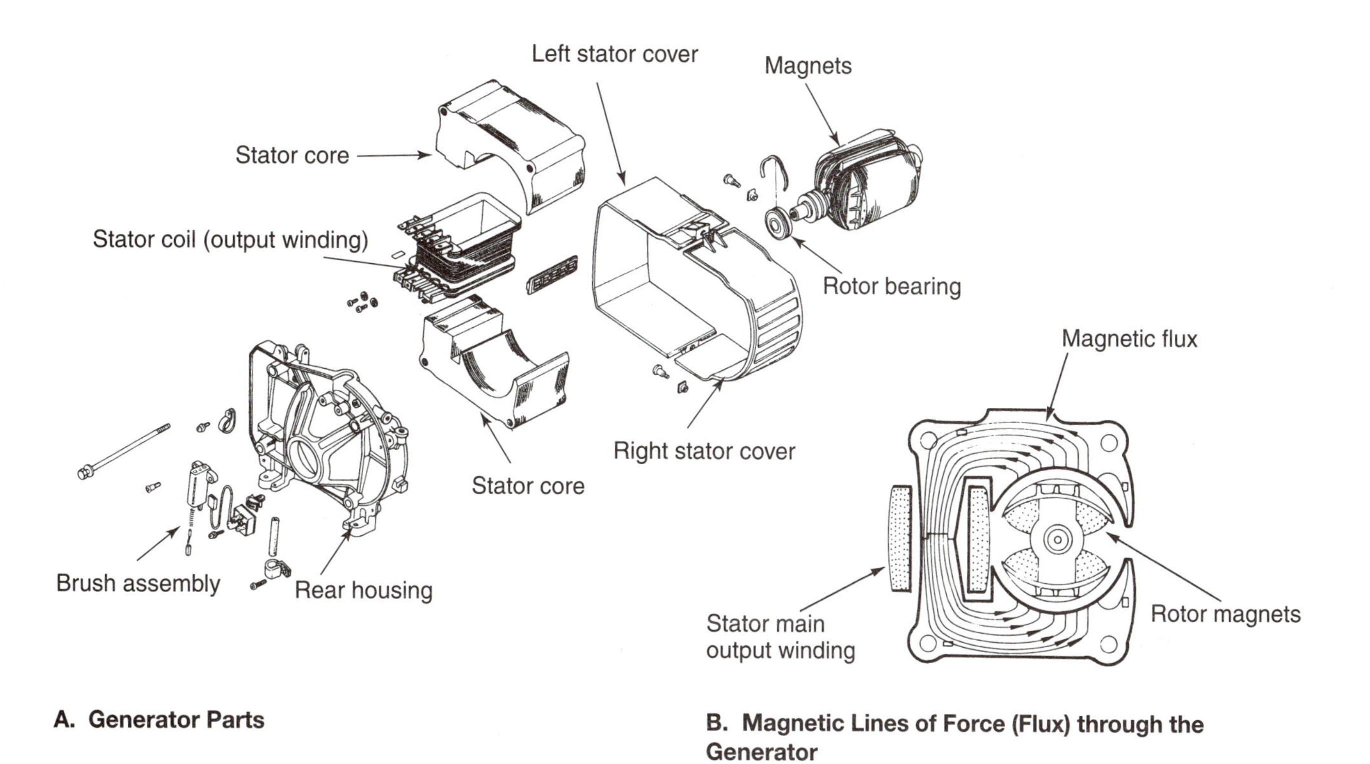

FIGURE 34-9 Rotating magnets develop an alternating current in the wire coils on a stationary stator. (Courtesy of American Honda Motor Co., Inc.)

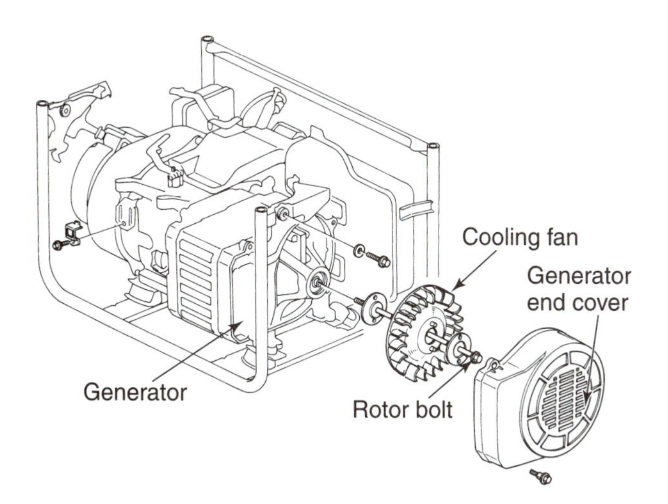

FIGURE 34-10 A fan is used to cool the generator components. (Courtesy of American Honda Motor Co., Inc.)

The alternator output is directed from the stator to the electrical components on the control panel through a wiring harness (Figure 34-11). Generator voltage is regulated by a solid state automatic voltage regulator (AVR) before going to the receptacles.

Generator Throttle Control

The throttle control system on an engine powering a generator must be automatic. The engine must automatically go to idle when the generator power demands are low. The throttle must go to maximum power when the electrical demands of the generator are high. Automatic throttle is achieved with a solenoid mounted on the carburetor (Figure 34-12). The solenoid plunger is connected to a plunger rod, which is connected to the throttle linkage through a throttle spring. Movement of the solenoid plunger causes the throttle spring to open or close the carburetor throttle.

The solenoid is controlled by an electronic part called the *idle control unit*. When the idle control unit detects that the generator is low, current is directed to the solenoid. The solenoid plunger and rod move in a position to increase throttle spring pressure and close the throttle. The engine goes to idle speed. If the generator output increases, the idle control unit stops current flow to the solenoid. A spring inside the solenoid retracts the plunger and plunger rod, lowering the tension on the throttle spring and allowing the engine to go to maximum governed engine speed.

Determining Required Generator Output

The rated generator output wattage has to be matched to the anticipated electrical load. Before

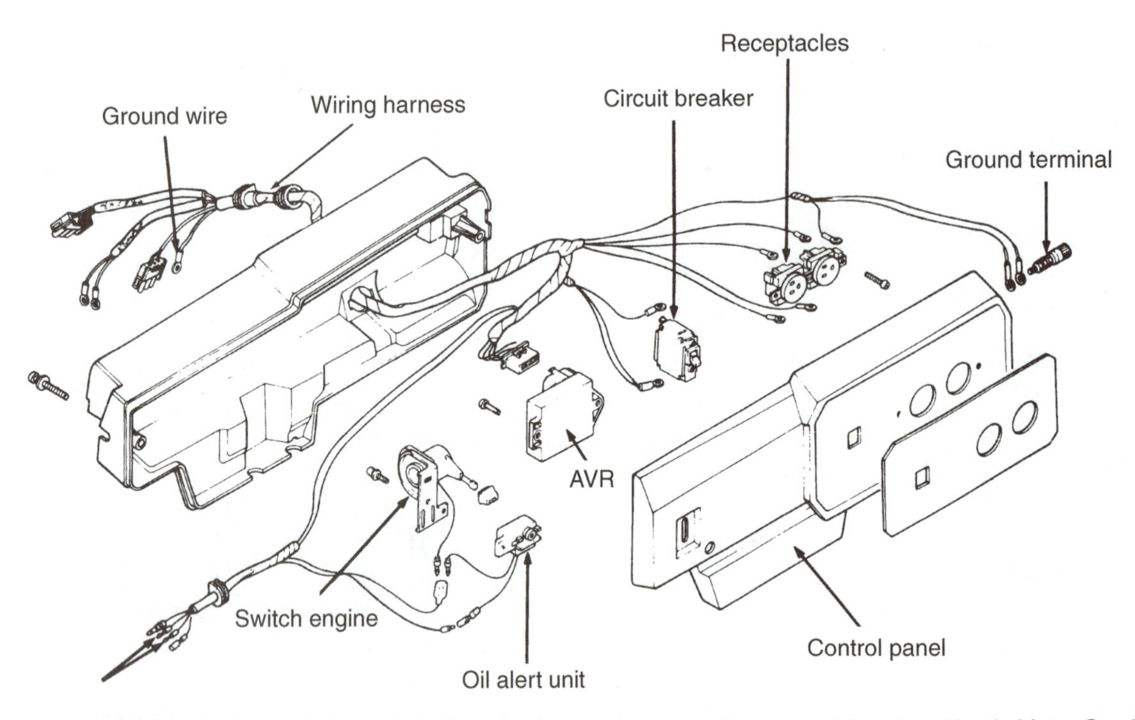

FIGURE 34-11 Electrical components in the control panel assembly. (Courtesy of American Honda Motor Co., Inc.)

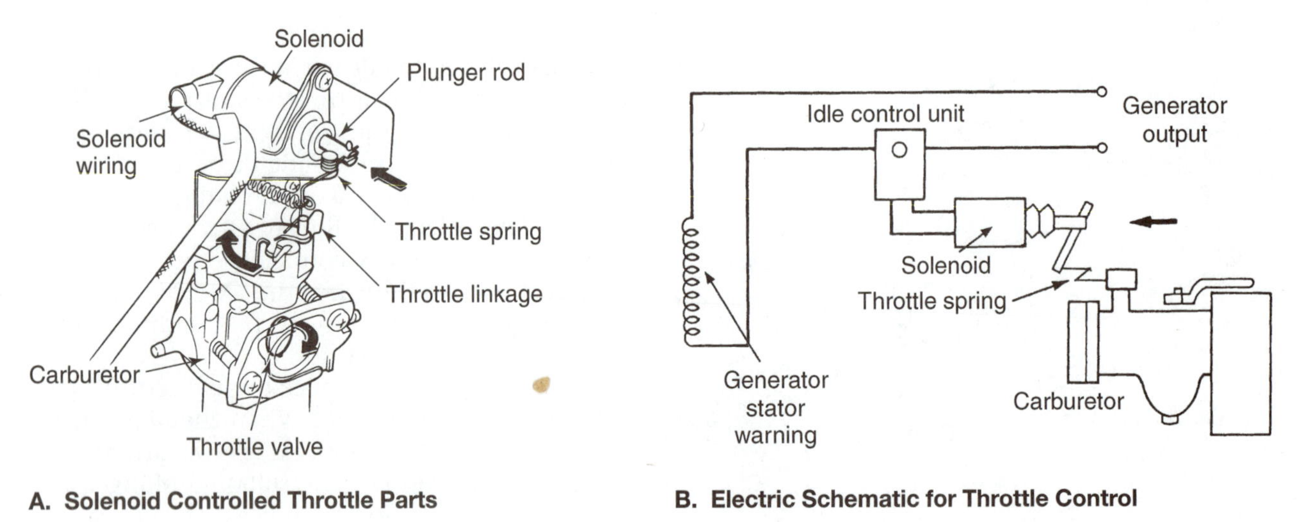

A. Solenoid Controlled Throttle Parts **B. Electric Schematic for Throttle Control**

FIGURE 34-12 Parts and operation of a solenoid-controlled throttle system. (Courtesy of American Honda Motor Co., Inc.)

purchasing a new generator or connecting electrical appliances to a present generator, the wattage requirements must be considered. The generator output must always exceed the power requirements of the electrical equipment you plan to use.

You can determine the power output using an appliance wattage worksheet (Figure 34-13). The worksheet lists all the common household electrical appliances. There is a listing of the running wattage and additional starting wattage requirements for each appliance. If you plan to use more than one of these appliances at one time, you must add up the wattage. When you plan to run the appliances at the same time, use the starting wattage.

Appliance	Running Wattage Requirements	Additional Starting Wattage Requirements	Select
Coffee Maker	1750	0	☐
Dish Washer			☐
Cool Dry	700	1400	☐
Hot Dry	1450	1400	
Electric Fry Pan	1300	0	☐
Electric Range			
6-inch element	1500	0	☐
8-inch element	2100	0	☐
Microwave Oven, 625w	625	800	☐
Refrigerator or Freezer	700	2200	☑
Toaster			
2-slice	1050	0	☐
4-slice	1650	0	☐
Automatic Washer	1150	2300	☐
Clothes Dryer			
Gas	700	1800	☐
Electric	5750	1800	☐
Dehumidifier	650	800	☐
Electric Blanket	400	0	☐
Garage Door Opener			
1/4 Horsepower	550	1100	☐
1/3 Horsepower	725	1400	☐
Furnance Fan, gas or fuel oil			
1/8 Horsepower	300	500	☐
1/6 Horsepower	500	750	☐
1/4 Horsepower	600	1000	☐
2/5 Horsepower	700	1400	☐
3/5 Horsepower	875	2350	☐
Hair Dryers	300-1200	0	☐
Iron	1200	0	☐
Radio	50-200	0	☑
Well Pump			
1/3 Horsepower	750	1400	☐
1/2 Horsepower	1000	2100	☐
Sump Pump			
1/3 Horsepower	800	1300	☐
1/2 Horsepower	1050	2150	☑
Television			
Color	300	0	☐
Black and White	100	0	☐
Vacuum Cleaner			
Standard	800	0	☐
Deluxe	1100	0	☐
Central Air Conditioner			
10,000 BTU	1500	2200	☐
20,000 BTU	2500	3300	☐
24,000 BTU	3800	4950	☐
32,000 BTU	5000	6500	☐
40,000 BTU	6000	7800	☐

FIGURE 34-13 Appliance wattage list. (Source: Generator Selection Worksheet, American Honda Motor Co., Inc. http://www.honda-generators.com)

If you plan to run them individually, use the running wattage.

For example, you need a generator to provide backup power for a ½ horsepower well pump. In addition, you want to operate your refrigerator and a radio if the power in your area fails. You need to be able to operate the pump, refrigerator, and radio at the same time. The list shows the following wattage requirements:

- Refrigerator 2,200 watts (starting wattage)
- Well Pump 2,150 watts (starting wattage)
- Radio 200 watts (running wattage)
- Total required wattage = 4,550 watts

The generator for this application should have a minimum output of 4,550 watts. You should add a safety factor of approximately 20 percent (90 watts) to your calculations to give the system plenty of reserve power. Since generator output is commonly rounded off to the nearest even number, the recommended generator output is 4700 watts.

Generator Troubleshooting

CAUTION: Always locate and follow the troubleshooting guide for the specific model of generator you are troubleshooting. Different generators have different circuit designs. Incorrect testing techniques can damage electrical components or cause injury from shocks.

Generator problems can be related to engine or power generation problems. Engine troubleshooting is presented in Chapter 21. The most common generator problem is abnormal output at one or more of the receptacles. The abnormal power may be none, too low, or too high.

The first check to make when there is a low power problem is the circuit breaker. Make sure the circuit breaker is in the on position. Too high a generator load can trip the breaker into the off position. Also check engine speed. If the engine speed is not to specifications, the generator cannot develop enough power. Check and correct any engine performance problems before troubleshooting any further.

The generator output voltage can be tested with a multimeter. The generator shop manual provides test points to measure the generator output either on the generator or the control panel. The manual also provides voltage specifications. If voltage is below or above specifications, the electrical components on the generator have to be checked to determine the source of the problem. Common electrical problems are worn field coil brushes, open circuit breaker, and open receptacle.

Brush Inspection and Replacement

WARNING: Make sure the ignition kill switch is in the off position. Remove and ground the spark plug wire before doing any static electrical tests or maintenance to the generator.

Worn brushes and dirty rotor slip rings are a common source of low generator output. Locate the brush holders on the rear generator housing. The brushes are removed by disconnecting the brush wire and pulling the brush out of the brush holder. Measure the length of the brush to determine if it is worn excessively (Figure 34-14). Install new brushes if they are worn beyond specifications.

Check the slip rings for dust, rust, or other damage. If necessary, wipe the slip ring surfaces with a clean, lint-free shop rag. If they are rusted, use fine abrasive paper or emery cloth to polish them.

Use a multimeter to measure the resistance of the field windings at the slip rings (Figure 34-15). Compare the resistance to specifications. If the resistance is not to specifications, the field windings are damaged. The rotor requires replacement.

Testing Receptacles and Circuit Breaker

WARNING: Receptacles and circuit breaker tests must be made with the engine off. Voltage in these circuits can cause injury from electrical shocks.

A power problem at any of the AC receptacles can be located with a continuity checker (Figure 34-16). Strip the ends off a piece of wire to use as a

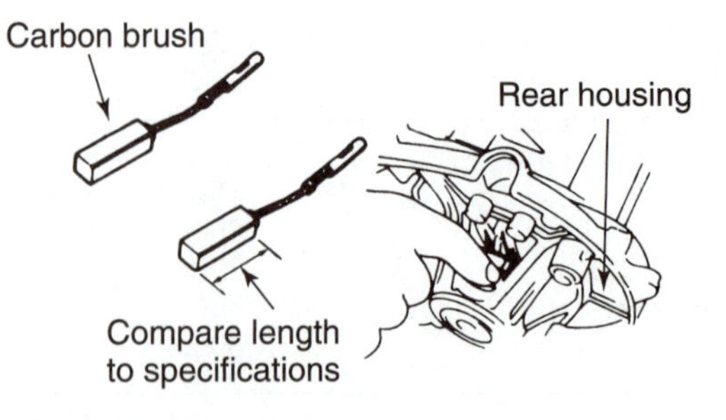

FIGURE 34-14 Measuring brush length to determine brush wear. (Courtesy of American Honda Motor Co., Inc.)

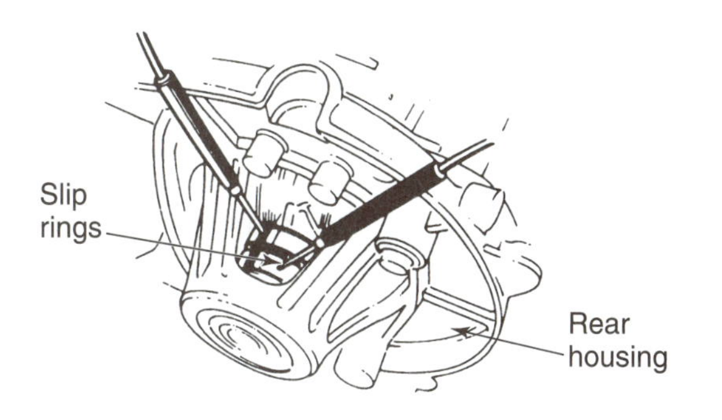

FIGURE 34-15 Checking the resistance of the field windings at the slip rings. (Courtesy of American Honda Motor Co., Inc.)

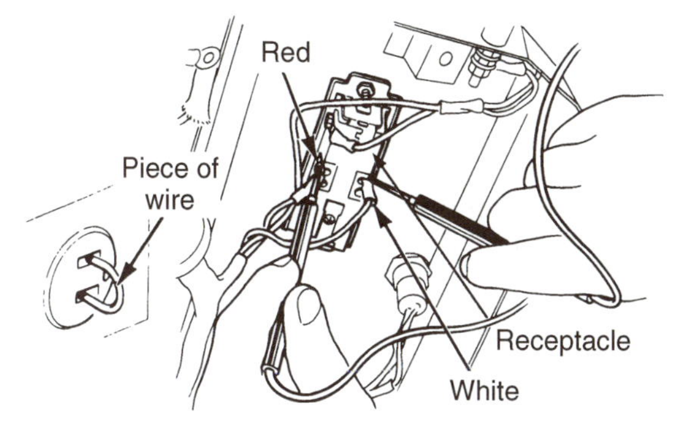

FIGURE 34-16 Testing the continuity of a receptacle. (Courtesy of American Honda Motor Co., Inc.)

CASE STUDY

A customer brought a portable generator into the shop. The owner greeted the customer and began filling out the invoice. The customer explained that he had been using the generator to run his refrigerator and water well while the power company completed a repair to his electrical power service. The engine on the generator had been running fine but stopped suddenly, leaving him without power.

Before he completed the invoice write up, the shop owner pulled on the starter rope to feel for engine compression. He immediately noticed the problem. This model generator had a low oil level alert system. The system shuts down ignition and turns on a warning light if the oil level in the crankcase is too low. The oil alert warning light flashed as he pulled on the starter rope.

The shop owner went into the back room and came out with a container of oil. He poured the oil into the crankcase and checked the oil level on the dip stick. He pulled on the starter rope and the engine started on the first pull. The amazed customer was very happy.

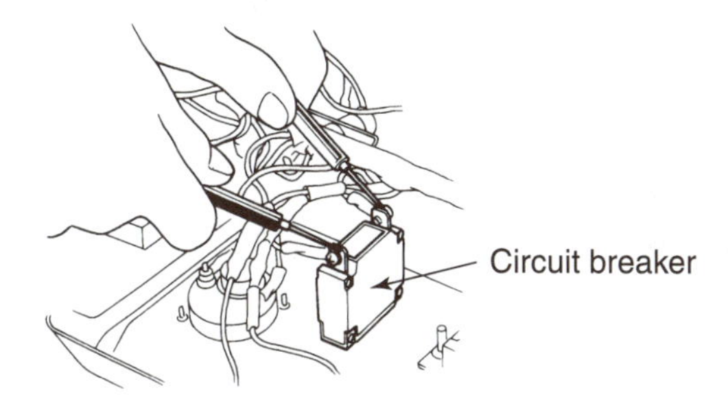

FIGURE 34-17 Testing the continuity of a circuit breaker. (Courtesy of American Honda Motor Co., Inc.

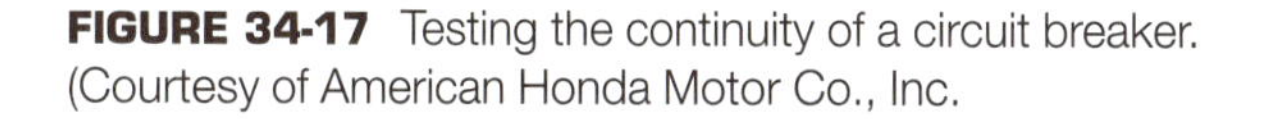

jumper. Connect each end of the jumper wire to the receptacle terminals. Test the receptacle for continuity with a continuity checker on the two receptacle wires behind the control panel. They are usually red and white in color. If there is not continuity, the receptacle is defective and should be replaced.

CAUTION: Make sure you remove the jumper after the test or the receptacle will be short-circuited when the generator starts operating.

Test the circuit breaker by connecting a continuity checker to the two circuit breaker wire terminals behind the control panel (Figure 34-17). There should be continuity between the two wires when the breaker is switched to on. There should not be continuity when the breaker is switched to off. Replace the circuit breaker if it fails either of these tests.

REVIEW QUESTIONS

1. List the safety procedures to follow before inspecting or repairing a pump.
2. Explain how pump output volume is rated.
3. Describe the parts and operation of a portable pump.
4. Explain why a pump has to be primed before the engine is started.
5. Describe the steps to follow in priming a pump.
6. Explain how to troubleshoot a pump for low pump output.
7. Explain the purpose of a portable generator.
8. Explain how watts are used to rate the output of a generator.
9. Describe the relationship between watts and horsepower.
10. Explain the purpose of the magnets in an alternating current generator.
11. Explain the purpose of the stator in an alternating current generator.
12. Describe flow of magnetic lines of force in a generator.
13. Describe the parts and operation of a solenoid operated throttle system.
14. Explain how to remove and measure a brush for wear.
15. Explain how to check a receptacle and circuit breaker for continuity.

DISCUSSION TOPICS AND ACTIVITIES

1. Disassemble a portable pump. Trace the flow of water through the pump from the inlet to the outlet.
2. Make a list of all the appliances in your home. Use a wattage use chart to determine the wattage a generator would have to develop to supply this demand.
3. If 1 horsepower equals 746 watts, what horsepower should an engine have to drive a 5,000 watt generator.

Glossary

Adjustable wrench: wrench with one stationary and one adjustable jaw that may be adjusted using a threaded adjuster for different-sized fasteners.

Advanced ignition timing: ignition spark that occurs before piston top dead center (BTDC).

Aerosol spray cleaner: spray can that has chemicals that break down dirt and grease, allowing them to be removed.

Air bleed: small air passage that allows atmospheric pressure to push on the fuel in the carburetor.

Air blowgun: gun-shaped attachment connected to the end of a compressed air line used to direct a stream of air.

Air cleaner: filter element that traps dirt before it can enter the engine air intake system.

Air cooling system: cooling system in which air is circulated over hot engine parts to remove the heat.

Air ratchet wrench: ratchet handle with a square socket drive lug powered by compressed air for fast removal or installation of fasteners.

Air vane governor: governor that uses air flow coming off the flywheel to regulate throttle opening to engine load.

Allen head screw: screw with a head with an inside, hex-shaped hole.

Allen wrench: hex-shaped wrench that fits into the hollow head of Allen type screws.

Alternating current: electrical current that changes from positive (+) to negative (–) at a regular cycle.

Alternator: generator that develops alternating current that is changed to direct current to charge the battery.

American National Standards Institute (ANSI): national organization that helps identify industrial and public needs for national standards.

Amperage: rate of current flowing in a wire; abbreviated A.

Analog meter: older style test instrument that is read by observing a pointer in relation to a scale.

Antiseize compound: material used on the fastener threads during assembly that prevents seizing.

API service rating: oil rating system that specifies how well an oil performs under severe engine service.

Armature: set of thin, soft iron strips used to make a path for magnetism.

Armature air gap (E gap): space between the legs of the magneto armature and the flywheel magnets when the magnets are positioned under the armature.

Assembly lubricant: thick lubricant grease used to coat bearings on assembly.

Atomization: breaking up of liquid fuel into fine particles that can be mixed with air to form a combustible mixture.

Automatic choke: choke valve connected to a diaphragm or bimetal spring that automatically opens or closes the choke valve.

Back brace: protective device that fits around the back and provides support, especially during lifting.

Back pressure: positive pressure in the exhaust port that opposes the flow of gases out of the port.

Ball bearing: bearing that uses a set of caged balls between an inner and outer bearing race.

Ball peen hammer: type of hammer with a steel head that has one round face and one square face.

Ball rewind starter clutch: clutch that uses a ball and pawl assembly to disconnect the starter pulley from the crankshaft when the engine starts.

Barrel oil pump: oil pump commonly used on vertical crankshaft engines to deliver oil under pressure to engine parts.

Battery carry tool: tool that clamps around a battery and allows carrying the battery by the tool handle.

Battery cell: group of positive and negative lead plates placed into a container with electrolyte.

Battery charging: using direct current from a battery charger to restore battery chemical potential.

Battery terminal brush: wire brush used to clean corrosion from the battery post and battery cable terminal.

Battery terminal pliers: pliers that have jaws specially designed to grip terminal nuts on battery connections.

Battery terminal puller: special puller made to pull battery terminals off battery posts.

Battery vent cap: cap that fits in a battery cell for inspection and for adding water.

Beam and pointer torque wrench: type of torque wrench that has a shaft, called a beam, that bends to show torque with a pointer and scale near the handle.

Bearing clearance: small difference in size between a shaft and a bearing.

Belleville lock washer: lock washer that uses a bent or crowned surface to lock a fastener to a part surface.

Bench grinder: electrically driven power tool that has an abrasive or wire wheel.

Binding head screw: screw with an undercut under the head. Binding head screws are used to connect wires to electrical parts.

Blade balancer: cone shaped tool used to balance a rotary lawn mower blade.

Blower: engine-driven air pump used to blow yards and walkways free of leaves and other debris.

Blower housing: cooling system part that directs the cooling air flow into and out of the flywheel.

Blown head gasket: condition in which a cylinder head gasket ruptures and allows the escape of combustion pressure from the combustion chamber.

Bolt: threaded fastener that fits through a hole in two parts and into a threaded nut to hold the parts together.

Bore: measurement of the engine's cylinder diameter.

Boring: cylinder resizing operation in which the cylinder is machined oversize with a motor driven cutter bit.

Boring bar: tool that drives a cutter bit through a cylinder to machine it oversize.

Bottom dead center (BDC): lowest position a piston can go in a cylinder.

Box-end wrench: wrench with jaws designed to fit on the corners of a hex-shaped fastener.

Brake horsepower (BHP): horsepower measured at the flywheel or PTO end of the crankshaft.

Brass hammer: type of hammer with a head made from soft, heavy brass. It is used to hammer on parts without damaging them.

Breaker bar: socket driver with a long handle, hinged at the drive lug end, used to break loose tight fasteners.

Breaker point: mechanical trigger switch that opens and closes the primary winding circuit.

Breaker point gap: space between the stationary and movable breaker point when the points are in the open position.

Break-in period: engine operation time required before oil consumption drops off to an acceptable level.

Breakover hinge torque wrench: type of torque wrench that has a hinge system near the head and that swivels at a selected torque.

Brienelling: bearing problem in which the surface breaks down from overloading.

Brush cutter: engine-powered hand-held tool that uses a rotating blade to cut brush.

Bushing: thin sleeve made from a soft bearing material pressed into a hole to provide a bearing surface.

Bushing driver: tool that fits into a bushing or valve guide used with a hammer to drive out a bushing or valve guide.

Business owner: person who owns a shop that sells and services outdoor power equipment.

By-pass passage: passageway that connects the intake port to the crankcase. Also called transfer passage.

Cam gear: gear on the end of the camshaft driven by a gear on the crankshaft.

Camshaft: shaft with cam lobes used to open the intake and exhaust valves at the correct time for the four-stroke cycle.

Capacitance: measure of a condenser's electrical storage ability measured in microfarads.

Capacitor discharge ignition (CD): ignition system that uses energy stored and discharged from a capacitor to create a high voltage.

Cape chisel: type of chisel that has a narrow cutting edge for cutting narrow grooves such as keyways.

Carbon: solid particles formed when oil is burned in the combustion chamber.

Carbon monoxide (CO): colorless, odorless, poisonous gas created by the combustion of gasoline that comes out of the engine exhaust when the engine is running.

Carburetor: fuel system part that mixes fuel with air in the proper proportions to burn inside the engine.

Carburetor throat: part of the carburetor that directs air flow in toward the venturi.

Castellated nut: nut that has a hole drilled through its side that, when tightened, matches up with a hole in a bolt or stud. Also called slotted nut.

Center punch: short, steel punch with a hardened conical point ground to a 90-degree angle used to mark the centers of holes to be drilled.

Centrifugal clutch: clutch that uses centrifugal force to engage and disengage the engine crankshaft from driven equipment.

Centrifugal governor: governor that uses centrifugal force to regulate throttle opening.

Chain saw: engine-powered saw that uses a rotating chain with cutting edges to saw wood.

Chain saw oil pump: small pump that delivers oil under pressure to lubricate the saw chain.

Channel lock pliers: pliers with channels cut where the two jaws are attached so that the jaws may be adjusted to wide or narrow openings. Sometimes called water pump pliers.

Charging system: system that develops and regulates a voltage used to recharge the storage battery.

Charging system circuit: wiring system that connects the charging system parts.

Chisel: bar of hardened steel with a cutting edge ground on one end, driven with a hammer to cut metal.

Choke: carburetor valve that restricts air flow and creates a rich mixture for cold starting.

Choke lever: linkage used to manually open and close the choke valve.

Circuit breaker: electrical circuit protection device that senses excessive current flow and opens the circuit between the generator and the receptacles.

Clean air standards: federal Environmental Protection Agency (EPA) and state regulations that limit the pollutants emitted by small air-cooled engines.

Clevis pin: pin that has a flange at one end and a small hole for a cotter pin or safety wire at the other end.

Clutch head screw: screw with a figure eight slot used with a clutch head screwdriver.

Clutch screwdriver: screwdriver with a figure-eight shaped blade made for turning recessed clutch head screws.

Combination pliers: general purpose piers with a slip joint for two different-sized jaw openings.

Combination wrench: wrench with a box-end jaw at one end of the handle and an open end at the other.

Combustion: burning of a mixture of air and fuel.

Combustion chamber: small space between the piston and the top of the cylinder where the burning of the air-fuel mixture takes place.

Compressed air: air that is pressurized by an air compressor and available in the small-engine shop to operate air tools.

Compression: squeezing or compressing of the air-fuel mixture in the combustion chamber.

Compression gauge: tester with a pressure gauge used to measure the amount of compression pressure in an engine cylinder.

Compression ratio: ratio between the cylinder volume with the piston at bottom dead center compared to the volume when the piston is at top dead center.

Compression release: system that lifts the exhaust valve to release compression pressure during cranking.

Compression ring: piston ring located in the groove or grooves near the piston head used to seal in the compression pressure.

Compression stroke: stroke that occurs when the piston moves to the top of the combustion chamber with both ports closed.

Compression test: test of the pressure in an engine cylinder to determine engine condition.

Condenser: electrical part that stores electrical current to prevent it from jumping across the open breaker points.

Conductor: material that allows a good electron flow and will conduct electricity.

Conical flat washer: washer with a conical shape used to take up space and remove play from an assembly.

Connecting rod: rod connected to the center of the piston that transfers movement to the crankshaft.

Connecting rod bearing: bearing used to allow the connecting rod to rotate freely on the crankshaft journal.

Connecting rod big end: end of the connecting rod attached to the crankshaft.

Connecting rod journal: offset part of the crankshaft where the connecting rod is attached.

Connecting rod side clearance: measurement of the space between the crankshaft and the side of the connecting rod.

Connecting rod small end: end of the connecting rod attached to the piston.

Consumer Product Safety Commission (CPSC): federal commission that evaluates the safety of all types of consumer products including outdoor power products.

Continuity: condition in an electrical circuit in which there is a complete path for current flow.

Control bar assembly: mower handle bar used to control mower direction.

Coolant pump: centrifugal pump used to circulate the coolant through the coolant passages and into the radiator.

Coolant reservoir: container connected to the radiator and used to recover coolant that moves out of the radiator as it is heated.

Coolant: liquid circulated around hot engine parts in a liquid cooling system.

Cooling fan: bladed air pump used to pull air through the radiator core to cool the coolant in the radiator.

Cooling fins: thin fins formed on the outside of an engine part used to get the greatest amount of hot metal into contact with the greatest amount of air.

Cooling system: engine system that dissipates heat from engine parts created by the combustion of the air-fuel mixture.

Cotter pin: pin with a split shaft that is installed through holes in fasteners in which the split ends are bent over to prevent part loosening.

Counterbalance system: weighted shaft driven off the crankshaft used to reduce engine vibration.

Crankcase: main housing that supports each end of the crankshaft and allows it to rotate.

Crankcase breather: valve assembly that vents the pressure buildup in a four-stroke engine crankcase.

Crankcase cover: removable cover on the crankcase that allows access to the crankshaft.

Crankshaft: shaft that changes the up-and-down movement of the piston to rotary motion.

Crankshaft end play: end-to-end movement of the crankshaft in the crankcase.

Crankshaft oil clearance: space between the connecting rod bearing and the crankshaft.

Cross-scavenged two-stroke: engine that uses a deflector on the head of the piston for scavenging.

Cycle: sequence of events that is repeated over and over.

Cylinder: hollow metal tube, closed at one end, used for burning the air and fuel.

Cylinder balance test: test performed on a multiple cylinder engine to find out if each cylinder is creating the same amount of power.

Cylinder block: crankcase and cylinder assembly combined into one casting.

Cylinder gauge: special measuring tool that shows cylinder taper and out-of-round on the face of a dial indicator in thousandths of an inch or hundredths of a millimeter.

Cylinder head: top of the cylinder used to form the combustion chamber.

Cylinder head bolts: hex head fasteners used to attach the cylinder head to the cylinder block.

Cylinder head gasket: gasket used between the cylinder head and cylinder block to prevent the loss of pressure from the combustion chamber.

Cylinder leakage tester: tester that fills the cylinder with air and checks for leakage around piston rings and valve seats.

Cylinder liner: cast-iron sleeve installed in an aluminum cylinder block when it is made.

Cylinder out-of-round: egg-shaped wear that occurs as the crankshaft rotates and thrusts the pistons at the sides of the cylinder.

Cylinder resizing: machining of a worn cylinder oversize for larger pistons.

Cylinder taper: difference between the size of the cylinder at the top of ring travel and the size at the bottom of piston skirt travel.

Cylinder wall: inside surface of the cylinder where the piston slides up and down.

Dead man lever: engine speed control that automatically returns engine speed to idle or stops the engine or blade when the operator's hand is removed.

Debris screen: screen in the center of the blower housing that filters incoming cooling air on the way to the flywheel.

Decibel (dB): unit of noise intensity named after Alexander Graham Bell.

Depth micrometer: tool used to measure the distance between two parallel surfaces.

Detonation: situation in which a part of the air-fuel mixture in the combustion chamber explodes rather than burns. Detonation is also called spark knock or ping.

Diagonal cutting pliers: pliers with hardened cutting edges on the two jaws used for cutting cotter pins or wire. Also called diagonals.

Dial indicator: tool that measures very small movements called play, end play, free play, or run out on the face of a dial.

Dial-reading torque wrench: torque wrench that has a dial on the handle with a needle that points to the amount of torque.

Diamond point chisel: type of chisel with a diamond-shaped cutting edge.

Diaphragm carburetor: carburetor that has a flexible diaphragm to regulate the amount of fuel available inside the carburetor.

Die: tool used to repair or make outside threads.

Differential: gear assembly that allows the two drive wheels to turn at different speeds when a tractor goes around a corner.

Digital meter: test instrument that shows test results as numbers on a small screen.

Diode: solid state electrical device that allows current flow in one direction and stops current flow in the opposite direction.

Dipstick: long, flat metal part with a handle and gauge marks used to measure engine oil level.

Direct current: electrical current that does not change from positive (+) to negative (–).

Displacement: the size or volume of the cylinder measured when the piston is at the bottom of the cylinder.

Dog rewind starter clutch: clutch that uses of a set of starter dogs to connect and disconnect the starter pulley from the engine when the engine starts.

Dowel pin: pin that is straight for most of its length with slightly tapered ends.

Downdraft carburetor: carburetor in which the air flows into the venturi in a downward direction.

Draft air cooling system: cooling system in which air is circulated over hot engine parts by the movement of the engine through the air.

Drill gauge: metal plate with holes identified by size, used to determine the size of a drill.

Drill index: set of drills in either fractional, metric, letter, or number sizes.

Drill motor: electrically powered, hand-held motor that drives a drill bit used to drill holes.

Drive belt: belt used to transfer torque from one pulley to another.

Drive clutch lever: operator control lever on self-propelled mowers that engages and disengages the gear box that drives the rear wheel.

Dry weight: engine weight without oil or gasoline.

Ducting: sheet metal or plastic covers that guide the flow of cooling air over the cooling fins.

Efficiency: measure of how well an engine converts energy into work.

Electrical circuit: complete path for current flow in an electrical system.

Electrical connector: fastener that allows two or more wires to be connected together.

Electrical starting system: starter system that uses electrical power to crank the engine for starting.

Electrical terminal: metal fastener attached to the end of a wire so that it can be connected into an electrical system.

Electricity: movement or flow of electrons from one atom to another.

Electromagnetic induction: generating electricity by moving a conductor through a magnetic field.

Emulsion tube: long tube installed in a carburetor passageway used to mix air with fuel.

Engine dynamometer: device used to measure engine torque and calculate engine horsepower.

Engine evaluation record: form with questions used to guide the technician through engine inspection, disassembly, and failure analysis.

Engine flooding: problem in which too much fuel enters the intake system and covers the spark plug electrode.

Engine misfire: engine performance problem in which the air-fuel mixture is not ignited on each engine power stroke.

Engine model number: set of numbers and letters used to identify the size and type of engine.

Engine oil: petroleum or synthetic based lubricant circulated between engine parts to prevent friction and wear.

Engine oil level: level of the oil in the engine crankcase available for the lubrication system.

Engine overspeeding: governor problem in which the engine runs at too high revolutions per minute at wide-open throttle (WOT).

Engine specification sheet: list showing the performance data, internal specifications, and external dimensions of a replacement engine.

Engine surging: governor or fuel system problem that causes the engine to constantly speed up rapidly and then slow down rapidly.

Equipment owner's manual: booklet given with new outdoor power equipment that explains the engine and equipment operation.

E-ring: very small external snap ring. Also called E-clip.

Estimate: form that details the customer's complaint and lists the probable labor, part, and material costs required for a repair.

Etching: bearing problem in which the bearing surface turns black from lack of lubricant.

Exhaust manifold: pipe assembly that routes exhaust gas from the exhaust port to the muffler.

Exhaust port: passage in the combustion chamber used to get the burned exhaust gases out of the cylinder.

Exhaust stroke: stroke that occurs when the exhaust port is opened and the piston moves up to push exhaust gases out of the cylinder.

Exhaust valve: valve that is opened on the exhaust stroke to allow gases to escape from the cylinder.

External snap ring: retaining ring that is expanded to fit over and into a groove machined on a shaft.

Face shield: eye protection device that covers the entire face with a plastic shield.

Feeler gauge: measuring tool using precise thickness blades or round wires to measure the space between two surfaces.

File: hardened steel tool with rows of cutting edges used to remove metal for polishing, smoothing, or shaping.

Fitting pistons: process of honing a resized (honed or bored) cylinder to establish the correct amount of piston clearance.

Flanged cap screw: cap screw that has a flange formed under the head.

Flat chisel: type of chisel used for general cutting of steel. It is driven with a ball peen hammer. Also called a cold chisel.

Flat head screw: screw with a head made so it can fit flush or flat with the surface.

Flat rate schedule: guide that lists the labor time to be billed for equipment repairs.

Flat washer: washer with a flat surface used to spread tightening forces over a wider area.

Float: carburetor part that floats on top of fuel and controls the amount of fuel in the float bowl.

Float bowl: carburetor part that provides a storage area for fuel in the carburetor.

Float carburetor: carburetor that has an internal fuel storage supply controlled by a float assembly.

Float gauge: gauge used to measure the position of the float on the carburetor body to determine float level setting.

Float level setting: carburetor float adjustment that determines when it closes the inlet needle to establish the level of fuel in the float bowl.

Flywheel: heavy wheel on the end of the crankshaft that uses its rotating weight (inertia) to return the piston to the top of the cylinder.

Flywheel brake: brake that stops the rotating flywheel when the operator's hand is removed from the equipment handle dead man switch.

Flywheel brake lever: control lever connected to the engine flywheel brake.

Flywheel holder: tool used to prevent the flywheel from turning when the flywheel nut or other retainer is removed.

Flywheel puller: puller tool used to separate the flywheel from the crankshaft taper.

Flywheel ring gear: ring of gear teeth on the flywheel that are engaged by the starter pinion gear to crank the engine.

Foam air filter element: flexible foam air cleaner filter material that filters intake air.

Fogging: method of coating engine with oil through the carburetor to prevent rust during storage.

Forced air cooling system: cooling system in which air is circulated over hot engine parts by an air pump to remove the heat.

Fouling: rapid buildup of oil or gasoline on the spark plug electrode.

Four-stroke cycle engine: an engine that develops power in a sequence of events using four strokes of the piston.

Friction: resistance to motion created when two surfaces rub against each other.

Fuel filter: filter element connected into the fuel line used to clean fuel before it enters the carburetor.

Fuel fittings: small connecting parts used to connect fuel lines to fuel system parts.

Fuel line: metal, flexible nylon, or synthetic rubber line used to route fuel to fuel system parts.

Fuel pump: pump that creates a low pressure to pull fuel out of the fuel tank then creates a pressure to move the fuel to the carburetor.

Fuel shutoff valve: valve in the fuel tank outlet that allows the fuel flow from the tank to be shut off.

Fuel strainer: wire mesh screen in the fuel tank opening that prevents fuel contamination during filling.

Fuel supply test: troubleshooting procedure used to determine if enough fuel is getting from the fuel tank to the carburetor.

Fuel system: engine system that stores fuel, mixes the fuel with the correct amount of air, and delivers the air-fuel mixture inside the engine.

Fuel tank: storage container attached to the engine or equipment that provides the source of fuel for engine operation.

Fuel tank filler cap: cap on top of the fuel tank that can be removed to refuel the engine.

Full pressure lubrication system: lubrication system in which an oil pump provides oil under pressure to all the engine bearings.

Galling: bearing problem in which metal transfers from the rollers to the race due to lubricant failure.

Garden tiller: engine-powered, walk behind cutting tool used to break ground for planting.

Gasket: soft material which, when squeezed between two parts, fills up small irregularities to make a pressure tight seal.

Gasket sealant: liquid compound used to improve gasket sealing, hold a gasket in place, repair a damaged gasket, or form a new gasket.

Gasohol: engine fuel made by blending gasoline with alcohol distilled from farm grains.

Gasoline: hydrocarbon-based engine fuel refined from crude petroleum oil.

Gear starter: starter that has a gear that engages another gear on the flywheel to crank the engine.

Generator watts rating: maximum power output of a generator determined by multiplying generator voltage (120 volts) by generator current output.

Glaze breaker: tool with abrasive stones used to remove cylinder wall glaze and restore a crosshatch pattern.

Glaze breaking: procedure used to remove the polish on cylinder walls to improve ring seating.

Goggles: eye protection device secured on the face with a headband.

Governor: throttle control system that senses engine load and automatically adjusts engine speed up or down.

Governor system: engine system that regulates the amount of air and fuel mixture to control engine speed.

Governor top speed adjustment: governor linkage adjustment used to set the top governed speed of an engine.

Grade markings: marks on the fastener used to show the grade, material, and tensile strength.

Graphic symbols: small road sign type symbols used to show steps on troubleshooting guides or repair procedures.

Gravity feed fuel system: fuel tank mounted above the carburetor so that gravity will force fuel from the tank to the carburetor.

Grit: common name for the grain size of an abrasive that may vary from 10 (very coarse) to 600 (very fine).

Ground circuit: electrical circuit connected to metal on the engine.

Hacksaw: type of saw that uses a blade made to cut metal.

Half keystone piston ring: piston ring with a top side that tapers downward.

Hammer: tool with a handle and a head used to drive parts or tools.

Helical spring lock washer: washer with two offset ends that dig into the fastener head and the part to lock them together.

Hex flange nut: nut with a flanged surface below the hex used to take the place of a washer.

Hex-head cap screw: screw with a hexagonal head. Also called cap screw.

Hex nut: nut that has six sides or flats so that it can be driven with box-end, open-end, or socket wrench.

High-speed adjustment screw: carburetor screw used to regulate the amount of fuel that goes up the main pickup tube during high-speed operation.

Hone: tool that holds and drives abrasive stones used to machine a cylinder oversize.

Honing: procedure used to machine worn cylinders oversize with abrasive stones.

Horizontal crankshaft: crankshaft that fits in an engine in a horizontal direction.

Horsepower: unit of engine power measurement determined by engine speed and torque.

Horsepower curve: graph showing horsepower at different rpm.

Hydraulic press: hydraulic power tool that uses a hydraulic ram to exert high forces required to remove bearings or straighten parts.

Hydrocarbon (HC): engine emission caused by incomplete combustion in the combustion chamber or evaporation from the fuel system and crankcase.

Hydrostatic transmission: transmission that uses a variable displacement hydraulic pump and hydraulic motor to provide a variable output speed between the input and output shafts.

Ignition module: electronic part that contains the transistor and circuits used to trigger the primary winding.

Ignition system: engine system that produces a high-voltage spark inside the engine at the required time to ignite an air-fuel mixture.

Ignition timing: exact time that the spark is introduced in the engine's combustion chamber in relation to piston position.

Impact nut: hardened nut that is threaded onto the crankshaft and used as a surface for hammering on the flywheel for removal.

Impact wrench: air-operated wrench that provides an impact force while it drives a socket wrench to remove a fastener.

Impulse fuel pump: pump operated by pressure impulses in the intake manifold or the crankcase.

Inch-measuring system: length- or distance-measuring system based on the inch.

Induction: transfer of electrical energy from one conductor coil to another without the coils touching each other.

Inertia starter drive: starter drive that uses inertia to engage and disengage the starter motor pinion from the flywheel ring gear.

Inlet needle valve: carburetor part that fits on the float and controls fuel flow through the fuel inlet.

Inlet seat: carburetor part that houses and provides a matching seat for the tapered end of the float inlet needle valve.

Inline cylinder engine: multiple cylinder engine that has cylinders arranged in a row.

Inside micrometer: tool used to measure the inside of holes or bores.

Insulator: material with atoms that will not part with any of their free electrons and will not conduct current.

Intake manifold: part that delivers the air-fuel mixture from the carburetor into the intake port.

Intake port: passage in the combustion chamber used to let an air-fuel mixture into the cylinder.

Intake stroke: stroke that occurs when the piston moves down, pulling the air-fuel mixture into the combustion chamber.

Intake valve: valve that is opened on the intake stroke to pull the air-fuel mixture into the cylinder.

Internal combustion engine: type of engine in which combustion takes place inside the engine.

Internal snap ring: retaining ring that is compressed to fit into a groove machined inside a hole.

Invoice: form that lists the labor, parts and material, and other costs owed by the customer for an outdoor power equipment repair.

Jet: carburetor part with a small, calibrated hole for regulating the flow of fuel, air, or a mixture of air and fuel.

Key: small metal fastener used with a gear or pulley to lock it to a shaft.

Kill switch: switch used to stop current flow in the ignition system and stop the engine.

Knock pin: stop in the piston ring groove that prevents the piston ring from rotating around the ring groove.

Lawn and garden tractor: smaller version of a farm tractor used for various residential and light commercial lawn and garden jobs.

Lawn edger: engine-powered walk behind cutting tool used to cut the edges of grass and turf.

Lawn mower blade brake: brake mechanism attached to the blade assembly that stops the blade.

L-head valve system: valve system in which the valves are located in the cylinder block.

Lip seal: seal that uses a wiper lip to prevent pressure oil leakage around a rotating or sliding shaft.

Liquefied petroleum gas (LPG or LP): butane- or propane-based fuel stored under pressure as a liquid but released as a gas.

Liquid cooling system: cooling system in which a liquid is circulated around hot engine parts to carry off the heat.

Lock washer: washer designed to prevent fasteners from vibrating loose. There are many different types of lock washers.

Low oil level alert system: system that warns the operator and prevents engine operation if engine oil level is too low.

Low oil pressure switch: electrical switch in the lubricating system that senses and triggers a warning for low engine oil pressure.

Low-speed adjustment screw: carburetor screw used to regulate the amount of fuel that gets behind the closed throttle plate during low-speed or idle operation.

Lubricating system: engine system that provides oil between moving parts to reduce friction, cool engine parts, and flush dirt away from engine parts.

Magneto: ignition part that uses magnetism to create a low voltage that is changed to a high voltage by induction.

Magneto coil: set of two separate, fine copper wire coils in an insulated housing.

Main bearing: bearing used to support each end of the crankshaft allowing it to rotate.

Main bearing journals: round surfaces on either end of a crankshaft that fit in the main bearings.

Maintenance schedule: chart listing the maintenance jobs and the recommended times for doing the jobs.

Manual choke: choke valve that must be operated manually by movement of a choke lever or through linkage from an engine control lever or switch.

Manual starter: engine starter that requires the operator to rotate the crankshaft manually for starting.

Manual throttle control system: engine speed control in which the operator sets the engine speed by positioning the carburetor throttle valve.

Mechanical fuel pump: pump that uses a diaphragm operated by an eccentric on the engine camshaft to pump fuel from the tank to the carburetor.

Metering: mixing of fuel and air needed for efficient burning inside the engine's combustion chamber.

Metric measuring system: length or distance measuring system based on the meter.

Microfarad (mfd): electrical unit of capacitance used to rate the electrical capacity of a condenser.

Microfinish number: measure of the roughness or smoothness of cylinder wall finish.

Motor feed circuit: heavy gauge cable wiring circuit that connects the battery to the starter motor.

Mounting base: part of the engine crankcase or cover that has holes for mounting to the equipment.

Mower deck: housing that supports the engine and covers the rotating blade.

Muffler: part that cools and quiets the exhaust gases coming from an engine.

Multimeter: electrical test instrument used to measure voltage, amperage, resistance, and continuity. Also called a digital volt-ohmmeter (DVOM).

Multiple cylinder engine: engine with more than one cylinder.

Multiple piece crankshaft: crankshaft made in separate parts that is assembled by pressing together.

Needle bearing: small diameter roller bearing often held in a case or cage that works as an outside race.

Needle nose pliers: pliers with long, slim jaws for gripping small objects or gripping objects in small or restricted spaces. Also called long-nose pliers.

Normal combustion: when the spark plug ignites the air-fuel mixture and a wall of flame spreads smoothly across the combustion chamber.

Nut: type of threaded fastener with inside thread used with bolts and studs.

Nut driver: tool that combines a socket, extension, screwdriver type handle.

Octane number: rating of how well a gasoline prevents detonation in the combustion chamber.

Offset screwdriver: screwdriver with blades made at an offset angle to allow driving screws in tight spaces.

Oil bath air cleaner: air cleaner that directs incoming engine air over an oil sump that catches dirt particles.

Oil clearance: space between two parts where oil can flow to create an oil film between the parts.

Oil consumption: lubricating oil loss caused by oil leakage or oil burning inside the combustion chamber.

Oil control ring: piston ring located in the bottom ring groove used to prevent excessive oil consumption.

Oil dipper: dipper attached to a connecting rod used to splash oil in the sump on engine parts.

Oil drain plug: removable plug located in the bottom of the oil sump used to drain engine oil.

Oil fill plug: removable plug located in the oil sump used to check or change oil.

Oil filter: paper filter element used to clean engine oil before it enters the engine parts.

Oil filter tool: tool made to grip and turn the housing of an oil filter for removal.

Oil pressure: pressure created by an oil pump throughout the full pressure oil circulating system.

Oil slinger: gear with paddles used to splash oil on engine parts.

Oil sump: lubricating oil storage area at the bottom of the engine crankcase.

One-way clutch: clutch used on a self-propelled mower drive wheel that allows the inside wheel on a turn to free wheel or rotate slower than the outside wheel.

Open cell voltage check: battery test used to determine if a battery is fully charged.

Open-end wrench: type of wrench with an opening at the end that can slip onto the flats of a screw, bolt, or nut.

Opposed cylinder engine: multiple cylinder engine with cylinders opposite each other.

Original equipment manufacturer (OEM): description used with parts and equipment supplied by the original engine or equipment manufacturer.

O ring seal: solid rubber or synthetic ring used to seal areas of high-pressure liquids.

Outdoor power equipment: equipment powered by small gasoline engines used to perform tasks such as lawn maintenance, garden preparation, snow removal, electrical generation, water removal, and tree service.

Outdoor Power Equipment Institute (OPEI): trade group that represents the manufacturers of outdoor power equipment.

Outlet filter screen: screen in the fuel tank outlet that stains the fuel as it leaves the tank.

Outside micrometer: measuring tool designed to make measurements of the outside of a part. Also called mike.

Overhead valve (OHV) system: valve system that has the valves located in the cylinder head above the piston.

Oversize piston rings: piston rings made for common oversized pistons used in resized cylinders.

Oxides of nitrogen (NOx): engine emission formed when combustion temperatures reach high levels.

Paper air filter element: paper air cleaner filter that allows air to pass through while trapping dirt particles.

Part number: number given to each of the engine or equipment parts by each engine or equipment maker.

Parts manual: book that shows illustrations and lists the identification number of each engine or equipment part.

Parts person: person responsible for ordering, cataloging, storing, and selling small engine and equipment parts.

Peening: method used to expand a metal part by making indentions in it with a punch.

Penetrating fluid: thin oil used to remove rust and corrosion between two threaded parts.

Performance curve: graph that shows the horsepower and torque of an engine compared to engine speed.

Personal protective equipment (PPE): safety equipment worn by a small engine technician for protection against hazards in the work area.

Phillips head screw: screw with a cross-shaped slot in the head for use with a Phillips screwdriver.

Phillips screwdriver: screwdriver with a blade designed to fit into the slots of Phillips head screws.

Pin: small, round metal fastener that fits into a drilled hole to hold two parts together.

Pin fit: clearance between the piston pin and the bore in the connecting rod and piston.

Pin punch: punch used with a hammer to drive out pins from engine parts.

Piston: round metal part that fits inside the cylinder that obtains the force from the burning air-fuel mixture.

Piston clearance: small space between the piston skirt and the cylinder wall that allows the piston to move freely in the cylinder.

Piston head: top of the piston where it gets its push combustion pressures.

Piston pin: pin that goes through the top of the connecting rod and through the piston. Also called wrist pin.

Piston pin bore: hole in the piston for the pin used to connect the piston to the connecting rod.

Piston ring: expandable metal ring in the ring groove of a piston that makes a sliding seal between the combustion chamber and crankcase.

Piston ring compressor: tool that is tightened around the piston head to squeeze the rings tightly into their grooves for piston installation.

Piston ring end gap: space between the two ends of a piston ring when it is in the cylinder.

Piston ring expander: tool that engages the end of piston rings and expands them for removal from the piston.

Piston skirt: piston area below the ring belt.

Pitch gauge: tool with multiple toothed blades that are used to match up with threads in order to identify their pitch.

Plain main bearing: type of main bearing that is machined directly in the aluminum of the crankcase or cover.

Plastic tip hammer: type of hammer with plastic tips on the head for protecting the surface of parts being hammered.

Plastigage: small-diameter soft plastic string used to determine the oil clearance between a bearing and its shaft.

Pliers: tool with two jaws operated with handles used to grip irregularly shaped parts and fasteners.

Portable generator: engine-driven alternating current generator that supplies 120-volt power to operate standard electrical devices.

Portable pump: engine-driven centrifugal water pump used to pump water from one location to another.

Power stroke: stroke that occurs when the compressed air-fuel mixture in the combustion chamber is ignited to force the piston down the cylinder.

Precision insert bearing: thin sleeve bearing made in two halves and inserted in the connecting rod big end and connecting rod cap. Also called insert.

Preignition: combustion started by a red hot part or deposit in the combustion chamber.

Premix: commercially available mixture of fuel and oil used in two-cycle engines.

Premix lubrication system: two-stroke engine lubrication system in which oil and gasoline are mixed together in the fuel tank and enter the engine on the intake stroke.

Pressure gauge: tool used to measure pressure, such as tire air pressure or cylinder compression pressure.

Pressure lubrication: lubrication system in which a pump pressurizes the oil and forces it through passageways into engine parts.

Preventive maintenance: maintenance service done at regular times to increase the life of an engine.

Primary winding: magneto coil influenced by the magneto magnets to develop low voltage.

Primer: rubber squeeze bulb used to force fuel into the air entering the combustion chamber to start the engine.

Puller: tool used to remove gears, bearings, shafts, and other parts off shafts or out of holes.

Pull rope: rope used on the rewind starter pulley to rotate (crank) the crankshaft for starting.

Pump output volume: rating system used to rate the output of a pump in gallons per hour (GPH) or gallons per minute (GPM).

Pump priming: procedure used to fill the pump housing with water to start the pumping process.

Punch: tool used with ball peen hammers to drive pins or to mark the center of a part to be drilled.

Push rod: part that transfers valve lifter motion from the cylinder block up to the rocker arm in the cylinder head.

Radiator: heat exchanger used to remove heat from the coolant in the liquid cooling system.

Radiator cap: cap used to access the coolant in the radiator and to regulate pressure and vacuum in the cooling system.

Radiator hose: hose connecting the radiator to the engine cooling system.

Ratchet handle: socket wrench drive handle that has a freewheeling or ratchet mechanism that allows it to drive a fastener in one direction and to move freely in the other direction. Also called ratchet.

Ratchet recoil starter: type of starter that uses a ratchet assembly to connect and disconnect the starter rope from the engine.

Reamer: tool with cutting edges used to remove a small amount of metal from a drilled hole.

Recommended maximum operating horsepower: 85% to 90% of the maximum horsepower developed by an engine.

Recommended speed range: engine maker's recommended range of engine operating rpm.

Rectifier: charging system part that changes the alternating current output of the alternator to direct current.

Reed valve: spring-loaded flapper type valve that controls the flow of air and fuel into the crankcase.

Reed valve loop-scavenged engine: engine that uses intake ports and a reed valve connected to the crankcase for scavenging.

Reel lawn mower: mower that cuts grass with multiple cutting blades mounted on a rotating reel.

Reject size: specification dimension that indicates that a part should be rejected, replaced, or reconditioned.

Relief valve: valve assembly that controls the oil pressure in a full pressure lubrication system.

Repowering: selection and installation of a new replacement engine for an outdoor power product.

Resistance: electrical resistance to current flow in a circuit.

Respirator: safety device worn over the nose and mouth to protect against chemical and dust breathing hazards.

Retarded ignition timing: ignition spark that occurs after piston top dead center (ATDC).

Retrofit part: part manufactured by an engine or equipment maker used to update a system manufactured earlier.

Rewind starter: manual starter with a spring mechanism that automatically rewinds a rope back into the starting position.

Rewind starter clutch: part of the manual starting system that disconnects the starter pulley from the crankshaft when the engine starts.

Ridge remover: cutting tool rotated in the top of the cylinder to remove the cylinder ring ridge.

Ring belt: ring and land area of the piston.

Ring expander: expandable spring used behind oil control rings to increase ring tension.

Ring groove: groove in the piston to hold the piston rings.

Ring groove cleaner: scraper tool that fits into the piston ring groove to remove carbon.

Ring land: raised space between the ring grooves.

Ring ridge: unworn ridge at the top of the cylinder above piston ring travel.

Ring seating: 1.) initial wear that occurs to the face of new piston rings that allows them to conform to irregularities in the cylinder wall surface. 2.) mating of the ring face with the cylinder wall throughout the complete stroke of the ring.

Rocker arm assembly: assembly on the cylinder head that changes the upward push from a pushrod to a downward motion to open the valve.

Rod cap: removable rod big end part that allows the connecting rod to be assembled around the crankshaft journal.

Roll pin: hollow pin with a split down its length.

Rotary crankcase valve: disc on the engine's crankshaft used to control the flow of air and fuel into the crankcase.

Rotary lawn mower: lawnmower that uses a single horizontally rotating cutting blade to cut grass.

Rotor oil pump: pump that uses the movement of two rotors to move oil under pressure into the parts that require lubrication.

Round nose chisel: type of chisel with a single bevel cutting edge used mostly for cutting semicircular grooves and inside rounded off corners.

rpm: rotational speed of an engine crankshaft in revolutions per minute.

Rule: measuring tool made from a flat length of wood, plastic, or metal graduated in inch or metric units.

Saddle bore: large diameter hole in the big end of the connecting rod that fits around the crankshaft journal.

SAE viscosity rating: rating system developed by the Society of Automotive Engineers (SAE) that rates the thickness or thinness of oil.

Safety container: metal container approved by the Underwriters' Laboratory (UL) for the storage of flammable liquid.

Safety glasses: glasses with impact-resistant lenses, special frames, and side shields.

Safety interlock: switch or circuit that prevents ignition if there is an unsafe condition on the power equipment

Saw chain brake: band brake that stops the rotation of the chain saw chain when the operator's hand is removed from the front handle.

Saw chain gauge: thickness of the chain drive tang where it fits in the guide bar groove.

Saw chain pitch: distance between alternate chain rivets divided in half.

Scavenge phase: replacement of exhaust gas in the combustion chamber in a two-stroke engine with a new mixture of air and fuel.

Screw: fastener that fits through a hole in one part and into a threaded hole in a second part to hold the two parts together.

Screwdriver: tool with a handle at one end and a blade at the other used to turn or drive a screw.

Screw extractor: tool used to remove a bolt or screw that has broken off in a threaded hole.

Seal: part used to prevent pressure or fluid loss around a rotating or sliding shaft.

Secondary winding: magneto coil that develops high voltage through induction.

Self-propelled lawn mower: mower that has a drive system to rotate two of the wheels.

Self-tapping screw: screw with a thread that makes an internal thread as it is screwed into a nonthreaded part.

Semiconductor: material that allows current flow like a conductor under certain conditions but acts like an insulator and stops current flow under other conditions.

Serial number: set of numbers and letters used to identify engine production information.

Service bulletin: revised service procedure or specification given by engine or equipment makers to technicians in the field.

Service tachometer: tester that is connected to the engine ignition system to display engine rpm.

Service technician: person who performs repairs on small engines and outdoor power equipment.

Set screw: screw with a sharp point at the end used to lock a part to a rotating shaft.

Shift lever: operator lever on self-propelled mowers that controls the mower's drive speed.

Shop manager: person who manages the sales and service at an outdoor power equipment shop.

Shop manual: booklet with detailed service information supplied to technicians by the equipment or engine maker.

Short block: new or rebuilt engine sold without any external components.

Sidedraft carburetor: carburetor in which the air flows into the venturi from the side.

Single thread lock nut: nut with arched prongs that grip the bolt or stud threads to prevent loosening.

Single thread nut: nut that has one single thread for fast installation or removal.

Sliding T-handle: socket wrench driver that has a handle shaped like a T.

Slip clutch: part that disconnects the rotary mower blade from the crankshaft if the blade hits a solid object.

Slotted screw: screw with a head made to fit the common screwdriver. Also called machine screw.

Slow charger: battery charger used to charge a battery at a low-amperage rate.

Sludge: mud-like mixture of oil and water that can plug oil passages inside the engine.

Small engine: machine that uses the combustion of a fuel to develop power to operate equipment such as lawnmowers and chainsaws.

Small-hole gauge: measuring tool with a round expandable head used along with an outside micrometer to measure the inside of small holes.

Snap ring: internal or external expanding ring that fits in a groove. Also called retaining ring.

Snap ring pliers: pliers that are made to fit internal or external type snap rings and allow them to be removed or installed safely.

Snow thrower: engine-powered walk behind tool used to clear snow from driveways and walkways after a snowfall.

Snow thrower auger: helical-shaped blade used to pick up snow and discharge it through a discharge chute.

Socket driver: handle with a square drive lug that fits into the square hole in the socket wrench to drive it.

Socket extension: shaft that can be connected from a socket driver to a socket wrench.

Socket wrench: wrench that fits completely around a hex head screw, bolt or nut and can be detached from a handle. Also called a socket.

Solenoid: magnetic switch used to control the electrical circuit between the battery and starter motor.

Solvent tank: metal tank containing cleaning solvent to flush grease and dirt off parts.

Spark arrester: exhaust system part that traps sparks coming out an engine exhaust to prevent a fire.

Spark plug: ignition system part that obtains the high-voltage electricity from the magneto and creates a spark in the combustion chamber.

Spark plug firing end: part of the center electrode and insulator that fits in the combustion chamber.

Spark plug gap: space between the spark plug center and side (ground) electrode.

Spark plug gapping tool: tool with wire feeler gauges for gap measuring and a ground electrode bending tool for gap adjusting.

Spark plug heat range: specification for the transfer of heat from the spark plug firing end.

Spark plug reach: length of the threaded section of a spark plug shell.

Spark plug thread diameter: diameter of the threaded shell measured in millimeters.

Spark plug wire: highly insulated cable used to route the high voltage from the coil secondary winding to the spark plug.

Spark plug wrench: deep socket wrench lined with a rubber insert to protect the porcelain spark plug insulator.

Spark tester: tester used to test for spark by viewing the spark across an air gap.

Specification: information and measurements used to troubleshoot, adjust, and repair an engine.

Speed handle: socket driver shaped like a crank with a swivel handle at one end and a square drive lug at the other end.

Splash lubrication system: lubrication system that depends on the splashing of oil on engine parts for lubrication.

Square nut: nut that has four sides or flats for a wrench.

Stale fuel: gasoline that has oxidized or has picked up water.

Standard piston rings: piston rings made from the same material and the same size as the rings used in the new engine.

Standard screwdriver: screwdriver with a blade and tip designed to drive standard slotted screws.

Starter clutch wrench: tool with lugs that fit into the starter clutch housing used to remove the starter clutch.

Starter control circuit: starter system wiring circuit that connects the starter switch to the solenoid and the magneto.

Starter drive: part used to connect and disconnect the rotating starter motor armature shaft to the engine for cranking.

Starter housing assembly: part of the manual starter that has the pull rope, starter pulley, and rewind spring.

Starter motor: electric motor used to crank the engine for starting.

Starter overrunning clutch: starter drive part that uses rollers in notches to lock and unlock the pinion to the armature to protect the starter motor from overspeeding.

Starter pinion gear: small gear driven by the starter motor that rotates the engine crankshaft.

Starter punch: punch with a taper on the end that allows the starting of pin removal.

Starter rewind spring: spring that rewinds the pull rope on the starter pulley after each engine starting.

Starter switch: multipurpose switch that controls the starter motor through the solenoid and may also be used as the ignition on/off switch.

Starting system: engine system that rotates the engine crankshaft to start the engine.

Stator plate: ignition system part that holds the breaker points, condenser, armature, and coil.

Steering system: set of linkages connected to the steering wheel that allows the operator to change the direction of the tractor.

Stemming: valve service operation in which a valve grinder is used to regrind and chamfer the valve stem tip.

Stepped feeler gauge: type of feeler gauge that has blades with two thicknesses.

Storage battery: rechargeable battery that uses lead plates in electrolyte to develop electrical power.

String trimmer: engine-powered hand-held tool that uses a rotating plastic string to cut, edge, or trim.

Stripping and crimping pliers: special purpose pliers used to strip electrical wire insulation and crimp on solderless connectors.

Stroke: piston movement from the top of the cylinder (TDC) to the bottom (BDC) or from the bottom of the cylinder to the top.

Stud: threaded fastener with threads on one end for a threaded hole and on the opposite end for a nut.

Stud remover: tool that grips a stud and is used with a wrench to remove a stud.

Suction feed diaphragm carburetor: carburetor that combines the features of a vacuum carburetor and the impulse fuel pump.

Surging: engine performance problem in which the engine runs fast and slow at the same throttle setting.

Tachometer: electronic meter used to measure engine rpm.

Tap: tool used to make or repair inside threads.

Tapered pin: pin that has a large end that tapers down to a small end.

Tapered roller bearing: bearing that uses a set of caged tapered roller bearings between an outer and inner race.

Tape rule: metal tape divided into metric or inch units that can be pulled out of a housing.

Telescoping gauge: measuring tool with spring loaded plungers used with a micrometer to measure the inside of holes or bores.

Tensile strength: maximum load in tension (pulling force) that a fastener can hold before it breaks.

Thermal efficiency: percentage of heat energy available in the fuel that is changed into power at the engine's crankshaft.

Thermostat: valve that controls the flow of coolant into the radiator from the engine to regulate engine-operating temperature.

Third port: port connecting the air-fuel mixing part (carburetor) to the crankcase. Also called intake port.

Third port loop-scavenged engine: engine that uses a third port controlled by the piston skirt for scavenging.

Thread chaser: cutting tool rotated through a thread to clean and straighten damaged threads.

Thread diameter: largest diameter on an internal or external thread.

Threaded fastener: type of fastener that uses the wedging action of threads to hold two parts together.

Thread insert: new steel thread inserted in a specially tapped hole in a damaged aluminum thread.

Thread-locking compound: material used on studs and other fasteners that locks the threads together to prevent them from vibrating loose.

Thread pitch: distance between the peaks or crests of the internal or external threads.

Thread series: number of threads found in a specified length of threads.

Throttle lever: operator control lever connected to the engine to increase or decrease engine speed at the carburetor.

Throttle valve: valve that is opened or closed by the operator to regulate the amount of air and fuel mixture that enters the engine.

Tiller tine: four-bladed cutting tool shaped to dig into topsoil.

Timing gear: gear on one end of the crankshaft used to drive the engine's camshaft.

Timing marks: marks on the camshaft and crankshaft gear that are aligned so the camshaft opens the valves at the correct time in relation to crankshaft position.

Tine depth stake: adjustable metal rod on the tiller used to penetrate the ground and set tilling depth.

Toothed washer: lock washer with small internal or external teeth that are made to lock against the fastener and part.

Toothed washer lock nut: nut that has a permanently installed toothed lock washer.

Top dead center (TDC): highest position a piston can go in a cylinder.

Torque: rotary unit of force that causes or tries to cause rotation.

Torque curve: graph showing torque at different engine rpm.

Torque indicating wrench: socket driver made to tighten a fastener to a specified amount of torque. Also called torque wrench.

Torx screwdriver: screwdriver with a blade made to fit the deep, six-sided recessed slot in a Torx type screw head.

Tractor chassis: frame and systems used to get engine power to the tractor drive wheels as well as to brake and steer the tractor.

Tractor transmission gear, belt, or hydraulic system used to change the amount of engine torque delivered to the tractor driving wheels.

Transaxle: combination of a transmission and differential used to drive the rear wheels of a tractor.

Transistor: solid state electrical device used to open and close an electrical circuit.

Transmission: 1.) gear system that transfers engine torque to the lawn mower driving wheels.

Transmission: 2.) gear, belt, or hydraulic system used to change the amount of engine torque delivered to the power equipment driving wheels.

Troubleshooting: step-by-step procedure followed to locate and correct an engine problem.

Troubleshooting guide: chart or diagram showing the steps to follow to locate and correct a service problem.

Twist drill: cutting tool mounted or chucked in an electric drill motor to cut holes. Also called drill.

Two-stroke cycle engine: engine that uses two piston strokes and one revolution of the crankshaft to develop power. Also called two stroke.

Two-stroke exhaust port: passage in the cylinder that allows exhaust gases to flow out of the cylinder.

Two-stroke intake port: passage in the cylinder that allows the air-fuel mixture to enter the cylinder.

Two-stroke lower end: parts located at the bottom of the engine such as crankcase, main bearings, crankshaft, and connecting rod.

Two-stroke upper end: parts located at the top of the engine such as cylinder head, cylinder, piston, piston rings, and piston pin.

Undersize bearings: thicker bearings made for journals that have been machined to a standard undersize so that the correct oil clearance is maintained.

Updraft carburetor: carburetor in which the air flows into the venturi in an upward direction.

Vacuum carburetor: carburetor that uses vacuum to pull fuel out of the fuel tank and mixes it with the air entering the engine.

Vacuum gauge: tool used to measure low pressure such as intake manifold vacuum.

Valve: part that is opened or closed to control the flow of air and fuel or burned exhaust gas through the engine's ports.

Valve clearance: small space in the valve parts that allows for heat expansion. Also called valve lash.

Valve cover: removable part used to get access to the rocker arm area.

Valve erosion: condition in which material is eroded or burned away from the valve head area.

Valve face: precision ground area of the valve head that seals against the valve seat.

Valve grinder: grinding machine used to grind a new surface on the valve face and tip.

Valve guide: part that supports and guides the valve stem in the cylinder block or cylinder head.

Valve guide bushing: thin wall (thickness) bushing used to repair nonreplaceable valve guides.

Valve guide plug gauge: measuring tool inserted into a valve guide to determine valve guide wear.

Valve head: part of the valve that when closed, seals the intake or exhaust port.

Valve lapping: seat reconditioning method in which an abrasive lapping compound is used to remove metal from the valve seat and valve face.

Valve lapping compound: mixture of grease and silicon carbide abrasives used to remove metal from the valve seat and valve face.

Valve lapping tool: tool with a small suction cup and handle used to lap valves.

Valve lifter: valve mechanism part that transfers cam lobe up and down movement to the valve stem. Also called tappet.

Valve-locking groove: groove on the end of the valve stem used to retain a valve spring.

Valve margin: round area of the valve head above the valve face.

Valve rotator: part connected between the valve stem and valve spring retainer that causes valve rotation.

Valve seat: precision ground opening in the port that makes a pressure tight seal with the valve face.

Valve seat cutter: hardened steel cutting tool used to recondition a valve seat.

Valve seat narrowing: removal of material from the top or bottom of the valve seat to reduce seat width.

Valve seat puller: pulling tool used to pull a valve seat out of a block or cylinder head.

Valve spring: coil spring used to hold the valve in a closed position.

Valve spring compressor: tool used to compress a valve spring for valve spring retainer removal.

Valve spring free length: measurement of valve spring length when it is not under tension.

Valve spring retainer: part that holds the compressed valve spring in position on the valve stem.

Valve stem: shaft connected to the valve head.

Valve sticking: condition in which the valve stem sticks in the valve guide and prevents proper valve movement.

Valve timing: opening and closing of the intake and exhaust valve in relation to crankshaft position.

Valve tip: end of the valve stem.

Vaporization: changing of a liquid such as gasoline into a vapor or gas.

Vapor lock: condition in which heated fuel in a fuel line turns from liquid to a vapor and cannot be pumped by a fuel pump.

Variable speed pulley system: single-speed transmission that uses a pulley that changes diameter to provide a variable ratio between input and output pulleys.

V-cylinder engine: multiple cylinder engine with cylinders arranged in two rows in the shape of a V.

Vent: air passage that allows atmospheric pressure to act on the fuel in a float bowl.

Venturi: restricted area in a carburetor air passage used to create a low pressure for fuel delivery.

Vertical crankshaft: crankshaft that fits in an engine in a vertical direction.

Viscosity: thickness or thinness of a fluid.

Vise grip pliers: pliers with a lever-operated lock on the lower jaw that allow the jaws to be locked tightly on a part.

Volatility: ease with which a fuel changes from a liquid to a vapor state.

Voltage regulator: charging system part that senses battery charge and limits alternator output to prevent battery overcharging.

Voltage: amount of pressure pushing the current through an electrical circuit; abbreviated V.

Volumetric efficiency: actual volume of a cylinder compared to the volume that is filled during engine operation.

Walk behind lawn mower: lawn mower that is controlled by operator walking behind the equipment.

Warranty: legal document provided by the engine or equipment maker that says that certain parts of the engine or equipment will be repaired at no cost to the owner.

Warranty claim: request by an engine or equipment owner for a repair to be covered (paid for) by the engine or equipment maker.

Washer: fastener used with nuts or screw heads to protect surfaces and prevent loosening.

Windup spring starter: starter that uses a handle to wind up a large spring that is then released to crank the engine.

Wing nut: nut with small gripping handles called *wings* that allow the nut to be loosened or tightened by hand.

Wiper ring: piston ring located in the center ring groove used to control oil and compression pressure.

Wiring harness: system of electrical wires wrapped together and routed to tractor electrical components.

Wrench: tool used to tighten or loosen hex head screws, bolts, and nuts.

Index